Fifth Edition

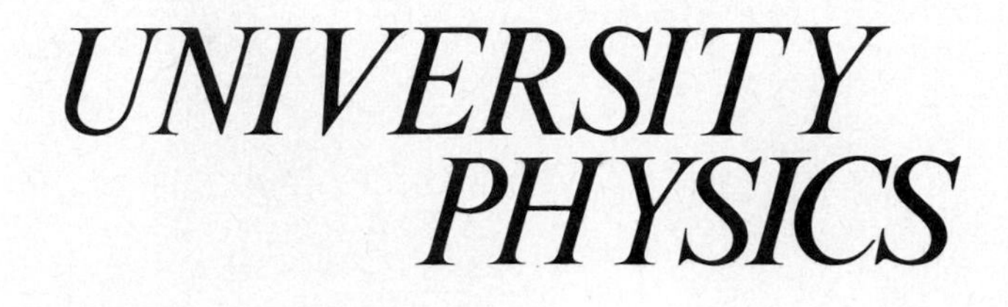

Fifth Edition

UNIVERSITY PHYSICS

Francis W. Sears
Professor Emeritus
Dartmouth College

Mark W. Zemansky
Professor Emeritus
The City College of
The City University of New York

Hugh D. Young
Associate Professor of Physics
Carnegie-Mellon University

ADDISON-WESLEY PUBLISHING COMPANY
Reading, Massachusetts · Menlo Park, California
London · Amsterdam · Don Mills, Ontario · Sydney

This book is in the
ADDISON-WESLEY SERIES IN PHYSICS

Fifth printing, July 1979

 Printed in the United States of America. Published simultaneously in Canada. Library of Congress Catalog Card No. 75-20989.

ISBN 0-201-06936-9
BCDEFGHIJK-DO-79

Preface

University Physics is intended for students of science and engineering who are taking an introductory calculus course concurrently. The complete text may be taught in an intensive two- or three-semester course and is also adaptable to a variety of shorter courses. Primary emphasis is on physical principles and problem solving; historical background and specialized practical applications have been given a place of secondary importance. Many worked-out examples and an extensive collection of problems are included with each chapter. *University Physics* is available as a single volume or as two separate parts. Part I includes mechanics, heat, and sound, and Part II includes electricity and magnetism, optics, and atomic and nuclear physics.

In this new edition, the basic philosophy and outline and the balance between depth of treatment and breadth of subject-matter coverage are unchanged from previous editions. We have tried to preserve those features that users of previous editions have found desirable, while incorporating a number of changes that should enhance the book's usefulness. Here are the most important changes.

1. The mks system of units, with the conventions and nomenclature of the Système Internationale, has become the principal unit system in the book. In this system the joule is the fundamental unit of energy of all forms, including heat. In the first half of the book, however, some examples and problems using English units have been retained.

2. The material on atomic and nuclear physics (Chapters 44 through 46) has been completely rewritten and expanded into three chapters, now including an elementary discussion of physics of solids, high-energy physics, and elementary particles.

3. A new chapter on relativistic mechanics has been added. Positioned somewhat arbitrarily at the end of the mechanics material, this chapter can be taken up earlier or later, or may be omitted completely if desired.

4. The chapter on electromagnetic waves has been completely rewritten, to exhibit more clearly and in simpler context the relation of wave propagation to the basic principles of electromagnetism.

5. Several sections have been added to broaden subject coverage. Among these are: 16–7 Examples (of calorimetric calculations), 19–25 (practical aspects of) Energy Conversion, 23–5 Musical Intervals and Scales, 23–7 Applications of Acoustic Phenomena, 28–10 Physiological Effects of Electric Currents, 38–7 Absorption (of light), 38–8 Illumination, 41–10 Defects of Vision, 42–1 Coherent Sources (of light), 42–13 Holography, 45–7 Semiconductors, 45–8 Semiconductor Devices, and 46–11 Radiation and the Life Sciences.

6. Some material has been reorganized. The material on surface tension has been shortened and incorporated into the hydrostatics chapter; the treatment of thermoelectricity has been shortened to a single section. The chapter on electromagnetic in-

duction has been rearranged to exhibit more clearly the various applications of Faraday's law to moving conductors and to stationary conductors in varying fields. The material on inductance and associated problems has been removed from this chapter and placed in a separate chapter.

7. Many sections have been completely rewritten for improved clarity and pedagogical effectiveness. These include the beginning of Chapter 7 (Work and Energy), Chapter 8 (Impulse and Momentum), Chapter 11 (Harmonic Motion); Section 25–4 (Gauss's Law); Chapter 26 (Potential); Section 27–7 (Polarization and Electric Displacement); Section 29–7 (*R–C* Circuits); Section 32–6 (Ampère's Law); and Chapter 35 (Magnetic Properties of Matter), to cite only a few examples. In a few cases, such as the opening sections of Chapters 7 and 8, the rewriting may create the illusion of de-emphasizing the use of calculus. Not so; the treatment is just as rigorous as in previous editions, but has been rearranged to follow the pedagogical principle of moving from the simple to the complex.

8. About 300 new problems have been added, bringing the total to over 1400. The added problems provide greater variety and also broader subject coverage than in previous editions. The authors have resisted the temptation to key problems to specific sections of text. Learning to select the principles appropriate for a specific problem is, after all, part of learning to solve problems. In addition, many problems require material from more than one section.

9. In every case where material has been rewritten, sound pedagogical principles and the authors' own teaching experience have guided the revision. In some instances we have shifted from a sequence in which a principle or concept is presented initially in its full generality to one that begins with special cases and then progresses to the more general statement. We hope thus to help the student attain the same final level of sophistication as previously by climbing a less steep slope.

The text is adaptable to a wide variety of course outlines. The entire text can be used for an intensive course two or three semesters in length. For a less intensive course, many instructors will want to omit certain chapters or sections to tailor the book to their individual needs. The format of this edition facilitates this kind of flexibility. For example, any or all of the chapters on relativity, hydrostatics, hydrodynamics, acoustics, magnetic properties of matter, electromagnetic waves, optical instruments, and several others can be omitted without loss of continuity.

Conversely, however, many topics that were regarded a few years ago as of peripheral importance and were purged from introductory courses have now come to the fore again in the life sciences, earth and space sciences, and environmental problems. An instructor who wishes to stress these kinds of applications will find this text a useful source for discussion of the appropriate principles.

In any case, it should be emphasized that instructors should not feel constrained to work straight through the book from cover to cover. Many chapters are, of course, inherently sequential in nature, but within this general limitation instructors should be encouraged to select among the contents those chapters that fit their needs, omitting material that is not relevant for the objectives of a particular course.

Again, we wish to thank our many colleagues who have contributed suggestions for this new edition. In particular, Prof. Robert Folk, Prof. Shelden H. Radin, and Prof. Charles W. Smith have read the entire manuscript, and their critical and constructive comments are greatly appreciated.

Hanover, New York, F.W.S.
and Pittsburgh M.W.Z.
November 1975 H.D.Y.

Contents

PART I

Mechanics, Heat, and Sound

Chapter 1

Composition and Resolution of Vectors

1-1 THE FUNDAMENTAL INDEFINABLES OF MECHANICS

Physics has been called the science of measurement. To quote from Lord Kelvin (1824–1907), "I often say that when you can measure what you are speaking about, and express it in numbers, you know something about it; but when you cannot express it in numbers, your knowledge is of a meagre and unsatisfactory kind; it may be the beginning of knowledge, but you have scarcely, in your thoughts, advanced to the state of *Science*, whatever the matter may be."

A definition of a quantity in physics must provide a set of rules for calculating it in terms of other quantities that can be measured. Thus, when momentum is defined as the product of "mass" and "velocity," the rule for calculating momentum is contained within the definition, and all that is necessary is to know how to measure mass and velocity. The definition of velocity is given in terms of length and time, but there are no simpler or more fundamental quantities in terms of which length and time may be expressed. *Length and time are two of the indefinables of mechanics.* It has been found possible to express all the quantities of mechanics in terms of only three indefinables. The third may, with equal justification, be taken to be "mass" or "force." *We shall choose mass as the third indefinable of mechanics.*

In geometry, the fundamental indefinable is the "point." The geometer asks his disciple to build any picture of a point in his mind, provided the picture is consistent with what the geometer says about the point. In physics, the situation is not so subtle. Physicists from all over the world staff international committees at whose meetings the rules of measurement of the indefinables are adopted. The rule for *measuring* an indefinable takes the place of a definition, and such a rule is sometimes called an *operational definition.*

1-2 STANDARDS AND UNITS

The set of rules for measuring the indefinables of mechanics is determined by an international committee called the *General Conference on Weights and Measures,* to which all the major countries send delegates. One of the chief functions of the Conference is to decide on a standard for each indefinable. A standard may be an actual object, in which case its main characteristic must be *durability.* Thus, in 1889 a meter bar of platinum-iridium alloy was chosen as the *standard of length,* because this alloy is particularly stable in its chemical structure. However, the preservation of a bar of this material as a world standard entails a number of cumbersome provisions, such as making a large number of replicas for

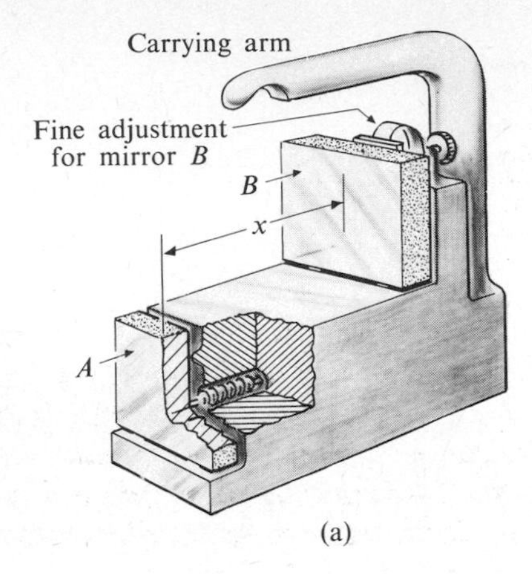

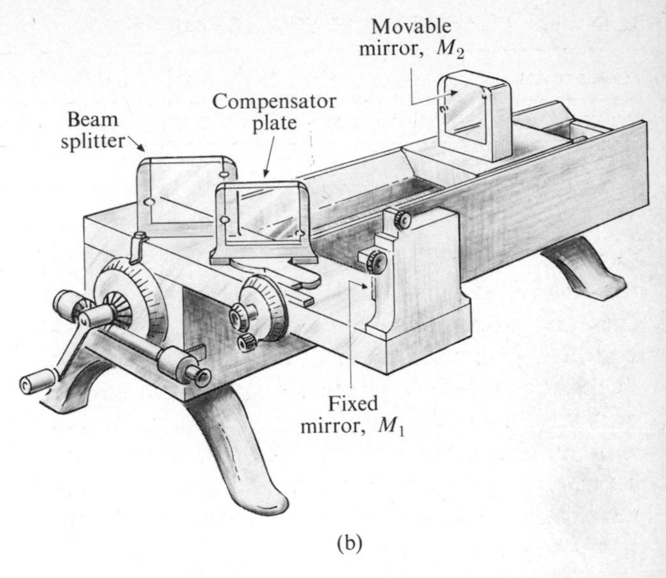

Fig. 1–1 *(a) Etalon and (b) Michelson interferometer for use in measuring the distance x in terms of the wavelength of light.*

all the major countries and comparing these replicas with the world standard at periodic intervals. On October 14, 1960, the General Conference changed the standard of length to an *atomic constant, namely, the wavelength of the orange-red light emitted by the individual atoms of krypton-86* in a tube filled with krypton gas in which an electrical discharge is maintained. Such a standard is much more readily reproducible than one based on a specific material object.

Table 1–1 STANDARDS AND UNITS AS OF 1969

	Standard	Measuring device	Unit
Length	Wavelength of orange-red light from krypton-86	Optical interferometer	1 meter = 1,650,763.73 wavelengths
Mass	Platinum-iridium cylinder, 1 kilogram	Equal-arm balance	1 kilogram
Time	Periodic time associated with a transition between two energy levels of cesium-133 atom	Atomic clock	1 second = 9,192,631,770 cesium periods

The *standard of mass* is the mass of a cylinder of platinum-iridium, designated as *one kilogram* and kept at the International Bureau of Weights and Measures at Sèvres, near Paris.

Before 1960, the *standard of time* was the interval of time between successive appearances of the sun overhead, averaged over a year, and called the *mean solar day*. Between 1960 and 1967 it was changed to the *tropical year 1900*, that is, the time it took the sun to move from a certain point in the heavens, known as the *vernal equinox*, back to the same point in 1900. In October 1967, the standard was changed again to the *periodic time of the radiation corresponding to the transition between the two hyperfine energy levels of the fundamental state of the atom of cesium-133.*

The three standards are listed in Table 1–1.

After the choice of a standard, the next step is to decide upon an instrument and a technique for comparing the standard with an unknown. Consider, for example, the distance x between two mirrors, A and B, of the device called an *etalon*, shown in Fig. 1–1(a). To find the number of wavelengths of orange-red light of krypton-86 in the distance x requires the use of an *optical interferometer*, one type of which (due to Michelson) is shown in Fig. 1–1(b). A mova-

Table 1–2 PREFIXES FOR POWERS OF TEN

Power of ten	10^{-12}	10^{-9}	10^{-6}	10^{-3}	10^{-2}	10^{3}	10^{6}	10^{9}	10^{12}
Prefix	pico-	nano-	micro-	milli-	centi-	kilo-	mega-	giga-	tera-
Abbreviation	p	n	μ	m	c	k	M	G	T

ble mirror M_2 on the Michelson interferometer is first made to coincide in position with A on the etalon. Then the mirror is moved slowly until it coincides with B, during which time gradations of orange and black, known as *interference fringes*, move past the cross hair in the field of view of a telescope and are counted. The motion of one complete fringe corresponds to a motion of mirror M_2 of exactly one-half wavelength. A length known as *one meter* is defined in this way as:

1 meter = 1,650,763.73 wavelengths of orange-red light of krypton-86.

The metric system of units is used exclusively in defining the standards of mass, length, and time. Most nations other than the United States and Great Britain also use the metric system exclusively for commerce and industry as well; Britain is making long-range plans to convert to the metric system, and many well-informed people believe it would be very advantageous for the United States to convert. One advantage of the metric system is that the various units for a quantity are always related by factors of ten. Thus, some units of length in common use in science and technology are:

1 angstrom unit = 1 Å = 10^{-10}m (used by spectroscopists),

1 nanometer = 1 nm = 10^{-9}m (used by optical designers),

1 micrometer = 1 μm = 10^{-6}m (used commonly in biology),

1 millimeter = 1 mm = 10^{-3}m and
1 centimeter = 1 cm = 10^{-2}m (used most often),

1 kilometer = 1 km = 10^{3}m (a common European unit of distance).

The words "nanometer," "micrometer," and "kilometer" are all accented on the *first* syllable, *not* the second, just like the words "millimeter" and "centimeter." The prefix "nano" is pronounced "nanno." A common set of prefixes is used with all units. These and their standard abbreviations are shown in Table 1–2. Thus,

1 kilometer = 1 km = 10^{3} meter = 10^{3}m,

1 kilogram = 1 kg = 10^{3} grams = 10^{3}g,

1 kilowatt = 1 kW = 10^{3} watts = 10^{3}W.

It is convenient to memorize Table 1–2, to have the information available when needed.

Units of length used in everyday life and in engineering in both the United States and the United Kingdom are defined as follows:

1 inch = 1 in. = { 41,929.399 wavelengths of Kr light, or *exactly* 2.54 cm,

1 foot = 1 ft = 12 in.,

1 yard = 1 yd = 3 ft,

1 mile = 1 mi = 5280 ft.

The device used to subdivide the standard of mass, the kilogram, into equal submasses is the *equal-arm balance*, which will be discussed in Chapter 5. Frequently used units of mass are:

1 microgram = 1 μg = 10^{-9}kg,

1 milligram = 1 mg = 10^{-6}kg,

1 gram = 1 g = 10^{-3}kg,

1 pound mass = 1 lbm = 0.45359237 kg.

The clock used to define the standard time interval is the *cesium clock*, a large, complex, and expensive laboratory instrument. It is extraordinarily precise and maintains its frequency constant to one part in one hundred billion (10^{11}) or better. Furthermore, it may be compared with other high-precision clocks in an hour or so, instead of the years required for comparison with the old astronomical standard. In the atomic clock, a beam of cesium-133 atoms passes through a long metal cylinder and interacts with microwaves brought in by a wave guide from a generator controlled by a quartz oscillator. The *unit* of time used throughout the world is called the *second* and is defined to be

$$1 \text{ second} = 1 \text{ s} = 9{,}192{,}631{,}770 \text{ Cs periods.}$$

Other common units of time are:

1 nanosecond	= 1 ns	= 10^{-9} s,
1 microsecond	= 1 μs	= 10^{-6} s,
1 millisecond	= 1 ms	= 10^{-3} s,
1 minute	= 1 min	= 60 s,
1 hour	= 1 hr	= 3600 s,
1 day	= 1 day	= 86,400 s.

1–3 SYMBOLS FOR PHYSICAL QUANTITIES

We shall adopt the convention that an algebraic symbol representing a physical quantity, such as F, p, or v, stands for both a *number* and a *unit.* For example, F might represent a force of 10 N (where N stands for newton), p a pressure of 15 N m^{-2}, and v a velocity of 15 m s^{-1}.

When we write

$$x = v_0 t + \tfrac{1}{2}at^2,$$

if x is in meters, then the terms $v_0 t$ and $\frac{1}{2}at^2$ must be in meters also. Suppose t is in seconds. Then the units of v_0 must be m s^{-1} and those of a must be m s^{-2}. (The factor $\frac{1}{2}$ is a *pure number,* without units.) The units of v_0 could be written as m/s rather than m s^{-1}, but the negative-exponent form is usually more convenient and will be used in all such expressions in this book.

As a numerical example, let $v_0 = 10$ m s^{-1}, $a = 4$ m s^{-2}, $t = 10$ s. Then the preceding equation would be written

$$x = (10 \text{ m } \not{s}^{-1}) \cdot (10 \not{s}) + \tfrac{1}{2} \cdot (4 \text{ m } \not{s}^{-2}) \cdot (10 \not{s})^2.$$

The units are treated like algebraic symbols. The s's cancel in the first term and the s^2's in the second, and

$$x = 100 \text{ m} + 200 \text{ m} = 300 \text{ m}.$$

The beginning student will do well to include the units of all physical quantities, as well as their magnitudes, in all his calculations. This will be done consistently in the numerical examples throughout this book.

1–4 FORCE

Mechanics is the branch of physics which deals with the motion of material bodies and with the forces that bring about the motion. We shall postpone a discussion of motion until Chapter 4, and start with a study of forces.

When we push or pull on a body, we are said to exert a *force* on it. Forces can also be exerted by inanimate objects: a stretched spring exerts forces on the bodies to which its ends are attached; compressed air exerts a force on the walls of its container; a locomotive exerts a force on the train it is pulling or pushing. The force of which we are most aware in our daily lives is the force of gravitational attraction exerted on every physical body by the earth, called the *weight* of the body. Gravitational forces (and electrical and magnetic forces also) can act through empty space without contact. A force on an object resulting from direct contact with another object is called a *contact force*; viewed on an atomic scale, contact forces arise chiefly from electrical attraction and repulsion of the electrons and nuclei making up the atoms of material.

To describe a force, we need to describe the *direction* in which it acts, as well as its *magnitude,* which is a quantitative description of "how much" or "how hard" the force pushes or pulls, in terms of a standard unit of force. In Chapter 5 we shall see how a unit of force can be defined in terms of the units of mass, length, and time. In the meter-kilogram-second (mks) system, this unit is the *newton*, abbreviated N. A more familiar unit is the *pound*, which can be defined as the force with which the earth attracts a standard body (i.e., its weight) with a mass of 1 pound-mass as defined in Section 1–2. A particular location on the earth's surface must be specified, since the attraction of the earth for a given body varies by as much as 0.5% from one point to another. If great precision is not required, it suffices to take any point at sea level and 45° latitude.

In order for an unknown force to be compared with the force unit and thereby *measured*, some observable effect produced by a force must be used. One such effect is to alter the dimensions or shape of a body on which the force is exerted; another is to alter the state of motion of the body. Both of these effects can be used in the measurement of forces. In this chapter we shall consider only the former; the latter will be discussed in Chapter 5.

An instrument commonly used to measure forces is the spring balance, which consists of a coil spring enclosed in a case for protection and carrying at one end a pointer that moves over a scale. A force exerted on the balance changes the length of the spring, and the change can be read on the scale. The balance can be calibrated as follows. The standard pound is first suspended from the balance at sea level and 45° latitude and the position of the pointer is marked 1 lb. Any number of duplicates of the standard can then be prepared by suspending a body from the balance and adding or removing material until the index again stands at 1 lb. Then when two, three, or more of these are suspended simultaneously from the balance, the force stretching it is 2 lb, 3 lb, etc., and the corresponding positions of the pointer can be labeled 2 lb, 3 lb, etc. This procedure makes no assumption about the elastic properties of the spring except that the force exerted on it is always the same when the pointer stands at the same position. The calibrated balance can then be used to measure the magnitude of an unknown force. An analogous procedure can be used to calibrate a spring balance in newtons

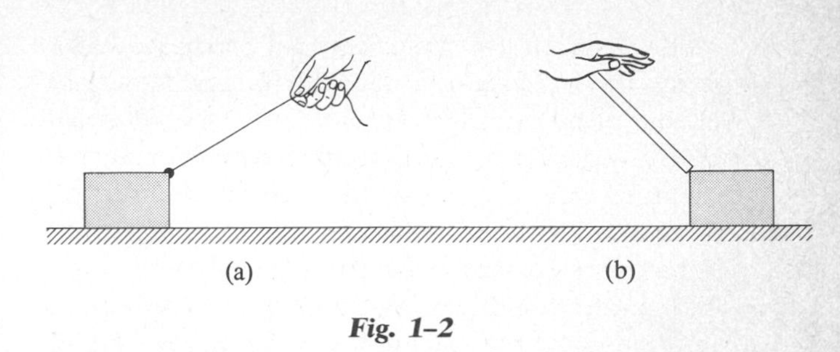

Fig. 1–2

1–5 GRAPHICAL REPRESENTATION OF FORCES. VECTORS

Suppose we are to slide a box along the floor by pulling it with a string or pushing it with a stick, as in Fig. 1–2. That is, we are to slide it by exerting a force on it. The point of view which we now adopt is that the motion of the box is caused not by the *objects* which push or pull on it, but by the *forces* which these exert. For concreteness, assume the magnitude of the push or pull to be 10 N. To write "10 N" on the diagram would not completely describe the force, since it would not indicate the *direction* in which the force acts. One might write "10 N, 30° above horizontal to the right," or "10 N, 45° below horizontal to the right," but all the above information may be conveyed more briefly if we adopt the convention of representing a force by an arrow. The length of the arrow, to some chosen scale, indicates the size or *magnitude* of the force, and the direction in which the arrow points indicates the *direction* of the force. Thus, Fig. 1–3 is the force diagram corresponding to Fig. 1–2. (There are other forces acting on the box, but these are not shown in the figure.)

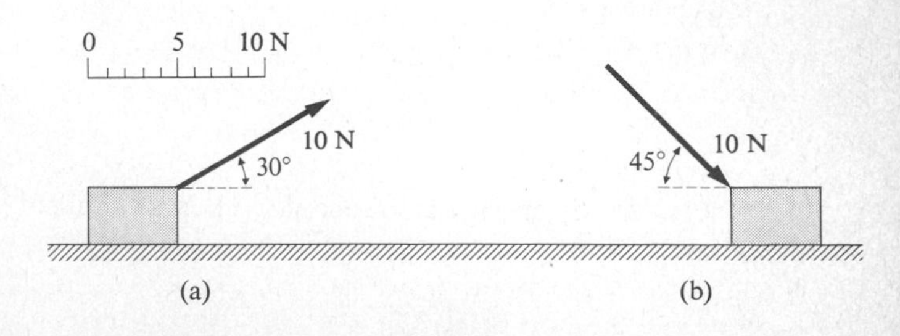

Fig. 1–3

Force is not the only physical quantity which requires the specification of a direction in space as well as a magnitude. For example, the velocity of an aircraft is not completely specified by stating that it is 300 miles per hour; we need to know the direction also. The concept of volume, on the other hand, has no direction associated with it.

Quantities such as volume, which involve a magnitude only, are called *scalar quantities.* Those such as force and velocity, which involve both a magnitude and a direction in space, are called *vector quantities.* Any vector quantity can be represented by an arrow, and this arrow is called a vector (or, if a more specific statement is needed, a force vector or a velocity vector).

Some vector quantities, of which force is one, are not *completely* specified by their magnitude and direction alone. Thus, the effect of a force depends also on its *line of action* and its *point of application.* (The line of action is a line of indefinite length, of which the force vector is a segment.) For example, if one is pushing horizontally against a door, the effectiveness of a force of given magnitude and direction depends on the perpendicular distance of its line of action from the hinges. If a body is deformable, as all are to some extent, the deformation depends upon the point of application of the force. However, since many actual objects are deformed only very slightly by the forces acting on them, we shall assume for the present that all objects considered are perfectly rigid. The point of application of a given force acting on a rigid body may be transferred to any other point on the line of action without altering the effect of the force. Thus a *force applied to a rigid body may be regarded as acting anywhere along its line of action.*

A vector quantity is represented by a letter in boldface type. The same letter in ordinary type represents the magnitude of the quantity. Thus the magnitude of a force $\boldsymbol{F}$ is represented by F.

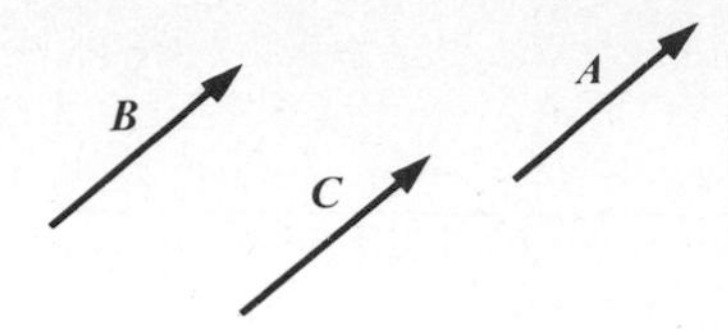

Fig. 1-4 *The vectors **A, B,** and **C** are mathematically equal.*

1–6 VECTOR ADDITION. RESULTANT OF A SET OF FORCES

Two vector quantities are said to be equal if they have the same magnitude and direction. In Fig. 1–4 the vectors **A, B,** and **C,** which may represent physical quantities, are all equal, and we may write symbolically

$$\mathbf{A} = \mathbf{B} = \mathbf{C}.$$

Two vector quantities which are equal need not have the same physical effect. For example, as already pointed out, two forces with the same magnitude and direction may have different points of application, and a complete description must include the point of application in addition to the magnitude and direction. Thus vector equality has a rather specialized meaning, and in this text boldface "equals" signs will be used as a reminder of this meaning.

The *vector sum* of two vector quantities is defined as follows. Let **A** and **B** in Fig. 1–5(a) be two given vectors. Draw the vectors as in (b) at any convenient point, with the initial point of **B** at the endpoint of **A.** The vector sum **C** is then defined as the vector from the initial point of **A** to the endpoint of **B.** This relationship may be expressed symbolically as:

$$\mathbf{C} = \mathbf{A} + \mathbf{B}.$$

Clearly, vector addition is not the same operation as addition of ordinary numbers; in this book a

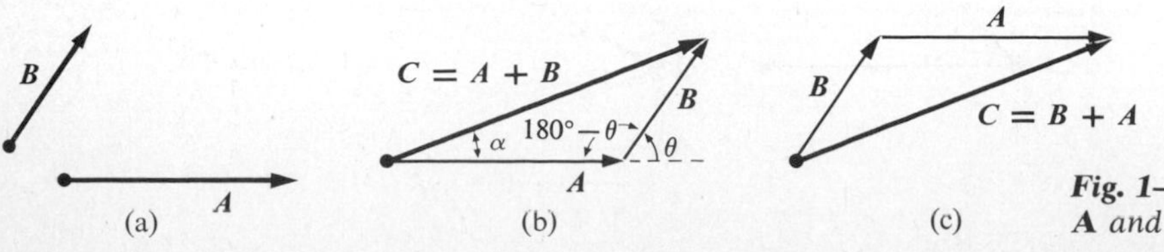

Fig. 1–5 *Vector **C** is the vector sum of vectors **A** and **B. C = A + B = B + A.***

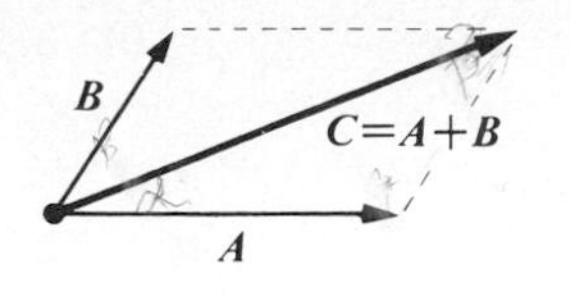

Fig. 1–6 *Parallelogram method for obtaining the vector sum of two vectors.*

boldface "plus" sign will be used to denote vector addition.

An alternative procedure is to draw the vectors as in Fig. 1–5(c), with the initial point of **A** at the endpoint of ***B***. The vector ***C*** has the same magnitude and direction as in (b), and hence the two vector sums are mathematically equal. The order in which the vectors are added is therefore immaterial, and vector addition obeys the same *commutative law* as algebraic addition:

$$\mathbf{A} + \mathbf{B} = \mathbf{B} + \mathbf{A}.$$

The magnitude and direction of the vector sum ***C*** can be found from measurements on a carefully drawn diagram. They can also be computed by the methods of trigonometry. Thus, if θ represents the angle between vectors **A** and ***B***, as in Fig. 1–5(b), the magnitude of ***C*** is given by the law of cosines:

$$C^2 = A^2 + B^2 - 2AB\cos(180° - \theta).$$

Since, for any angle θ, $\cos(180° - \theta) = -\cos\theta$, this may also be written

$$C^2 = A^2 + B^2 + 2AB\cos\theta.$$

The angle α between ***C*** and **A** can be found from the law of sines:

$$\frac{\sin\alpha}{B} = \frac{\sin(180° - \theta)}{C} = \frac{\sin\theta}{C}.$$

Another useful method of finding the sum of two vectors is shown in Fig. 1–6, where vectors **A** and ***B*** are both drawn from a common point. The vector sum ***C*** is the diagonal of a parallelogram of which the given vectors form two sides.

Figure 1–7 illustrates a special case in which two vectors are parallel, as in (a), or antiparallel, as in (b). If they are parallel, the magnitude of the vector sum ***C*** equals the sum of the magnitudes of **A** and ***B***. If they are antiparallel, the magnitude of the vector sum equals the *difference* of the magnitudes of **A** and ***B***. The vectors in Fig. 1–7 have been displaced slightly sidewise to show them more clearly, but they actually lie along the same geometrical line.

When more than two vectors are to be added, we may first find the vector sum of any two, add this vectorially to the third, and so on. This process is illustrated in Fig. 1–8, which shows in part (a) four vectors **A, *B*, *C*,** and ***D***. In Fig. 1–8(b), vectors **A** and ***B*** are first added by the triangle method, giving a

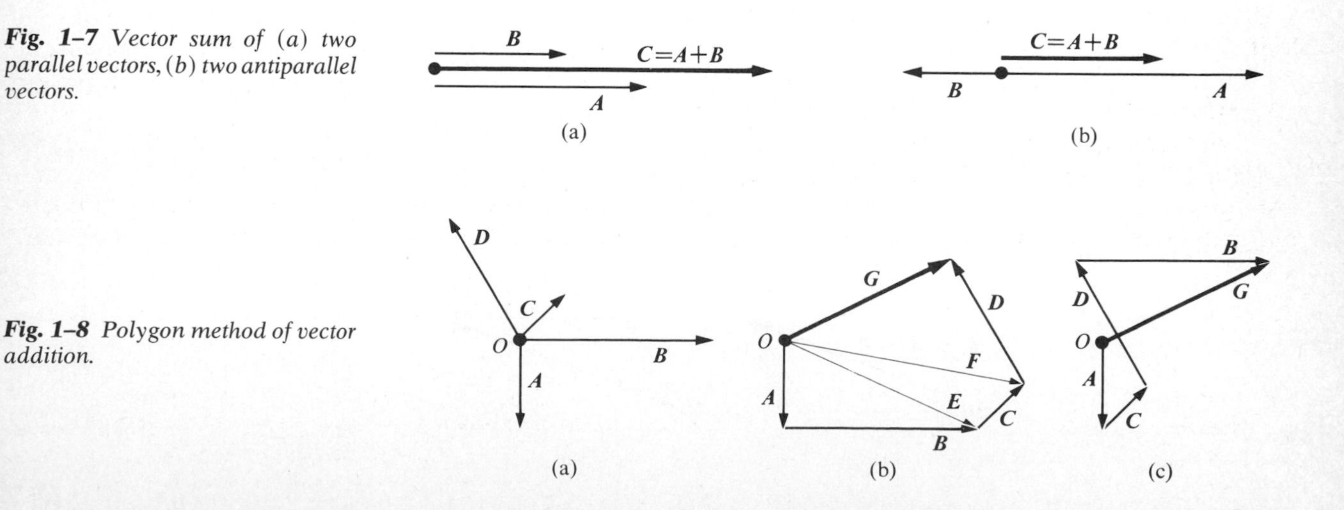

Fig. 1–7 *Vector sum of (a) two parallel vectors, (b) two antiparallel vectors.*

Fig. 1–8 *Polygon method of vector addition.*

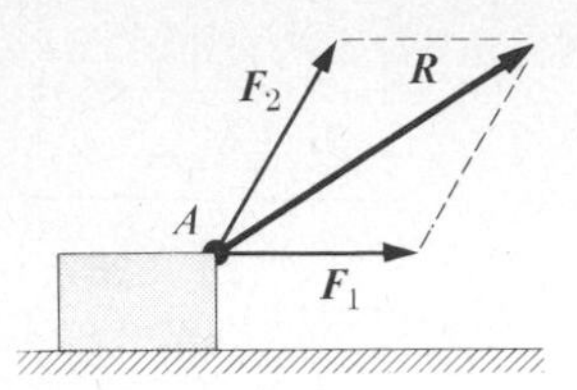

*Fig. 1–9 A force represented by the vector **R**, equal to the vector sum of $\boldsymbol{F}_1$ and $\boldsymbol{F}_2$, produces the same effect as the forces $\boldsymbol{F}_1$ and $\boldsymbol{F}_2$ acting simultaneously.*

vector sum ***E***; vectors ***E*** and ***C*** are then added by the same process, to obtain the vector sum ***F***; finally, ***F*** and ***D*** are added, to obtain the vector sum

$$\boldsymbol{G} = \mathbf{A} + \boldsymbol{B} + \boldsymbol{C} + \boldsymbol{D}.$$

Evidently the vectors ***E*** and ***F*** need not have been drawn; we need only draw the given vectors in succession, with the tail of each at the head of the one preceding it, and complete the polygon by a vector ***G*** from the tail of the first to the head of the last vector. The order in which the vectors are drawn makes no difference, as shown in Fig. 1–8(c); the reader may wish to try other possibilities.

Now consider the following physical problem. Two forces, represented by the vectors $\boldsymbol{F}_1$ and $\boldsymbol{F}_2$ in Fig. 1–9, are applied simultaneously at the same point *A* of a body. Is it possible to produce the same effect by applying a *single* force at *A*, and if so, what should be its magnitude and direction? The question can be answered only by experiment; investigation shows that a single force, represented in magnitude, direction, and line of action by the vector sum ***R*** of the original forces, is in all respects equivalent to them. This single force is called the *resultant* of the original forces. Hence the mathematical process of *vector addition* of two force vectors corresponds to the physical operation of finding the *resultant of two forces,* simultaneously applied at a given point.

The vector sum of $\boldsymbol{F}_1$ and $\boldsymbol{F}_2$ can be obtained by the construction of Fig. 1–5, where the vectors are drawn tail-to-head at any convenient point. The vector sum then has the same magnitude and direction as the resultant ***R***, but not necessarily the same line of action. That is, the line of action of ***R*** passes through point *A*. This again illustrates that although a mathematical vector may be displaced in any way (retaining its original magnitude and direction), a force acting on a rigid body can be displaced only along its action line.

1–7 COMPONENTS OF A VECTOR

Any two vectors whose vector sum equals a given vector are called the *components* of that vector. In Fig. 1–6, for example, vectors **A** and ***B*** are components of the vector ***C***. Evidently, a given vector has an infinite number of pairs of possible components. If the *directions* of the components are specified, however, the problem of finding the components, or of *resolving* the vector into components, has a unique solution. Thus suppose we are given the vector **A** in Fig. 1–10(a), and we wish to represent it in terms of components in the direction of the lines *Op* and *Oq*, that is, to *resolve* it into components in these directions. From the tip of vector *A* draw the dotted construction lines parallel to *Op* and *Oq*, forming a parallelogram. Vectors $\mathbf{A}_p$ and $\mathbf{A}_q$, from *O* to the points of intersection of the construction lines with *Op* and *Oq*, are then the desired components, since they are in the specified directions and the given vector is their vector sum.

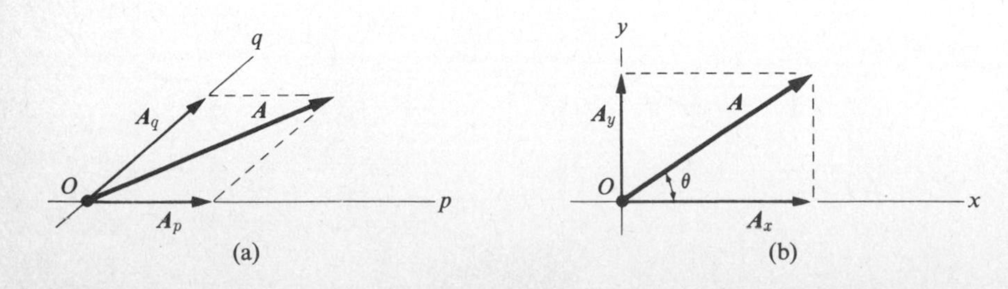

*Fig. 1–10 (a) Vectors $\mathbf{A}_p$ and $\mathbf{A}_q$ are the components of **A** in the directions Op and Oq. (b) Vectors $\mathbf{A}_x$ and $\mathbf{A}_y$ are the rectangular components of **A** in the directions of the x- and y-axes.*

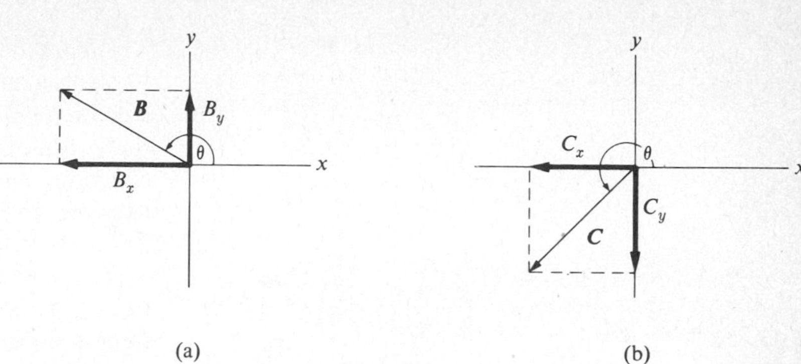

Fig. 1–11 *Components of a vector may be positive or negative numbers.*

The special case in which the specified directions are at right angles to each other is of particular importance. In Fig. 1–10(b), lines Ox and Oy are the axes of a rectangular coordinate system. The parallelogram obtained by drawing the dotted construction lines from the tip of vector **A** then becomes a rectangle, and the components $\mathbf{A}_x$ and $\mathbf{A}_y$ are called the *rectangular components* of **A**.

The magnitudes of the rectangular components of a vector are easily computed. If θ is the angle which the vector **A** makes with the x-axis, then

$$A_x = A \cos \theta, \qquad A_y = A \sin \theta,$$

where A, A_x, and A_y are the magnitudes of the corresponding vectors.

The components of a vector are themselves *scalar* quantities, and they may be positive or negative. In Fig. 1–11, the component B_x is negative, since its direction is opposite to that of the positive x-axis, and also since the cosine of an angle in the second quadrant is negative; B_y is positive, but both C_x and C_y are negative.

The application of the above concepts to a physical problem is illustrated in Fig. 1–12, where a force **F** is exerted on a body at point O. The rectangular components of **F** in the directions Ox and Oy are $\mathbf{F}_x$ and $\mathbf{F}_y$, and it is found that simultaneous application of the forces $\mathbf{F}_x$ and $\mathbf{F}_y$, as in Fig. 1–12(b), is equivalent in all respects to the effect of the original force. *Any force can be replaced by its rectangular components, acting at the same point.*

As a numerical example, let

$$F = 10 \text{ N}, \qquad \theta = 30°.$$

Then,

$$F_x = F \cos \theta = (10 \text{ N})(0.866) = 8.66 \text{ N},$$

$$F_y = F \sin \theta = (10 \text{ N})(0.500) = 5.00 \text{ N},$$

and the effect of the original 10-N force is equivalent to the simultaneous application of a horizontal force, to the right, of 8.66 N, and a lifting force of 5.00 N.

The axes used to obtain rectangular components of a vector need not be vertical and horizontal. For

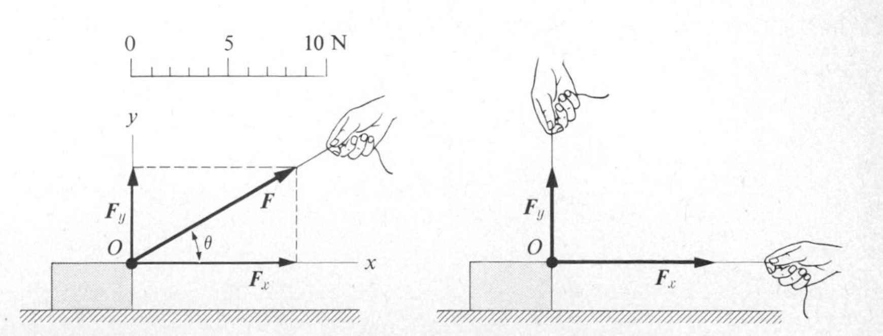

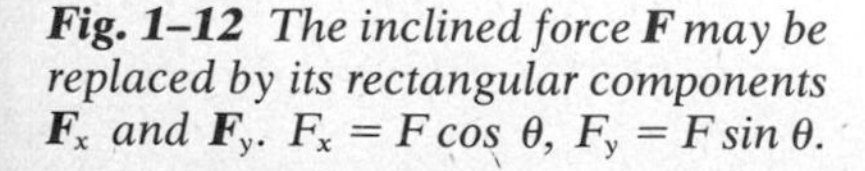

Fig. 1–12 *The inclined force* **F** *may be replaced by its rectangular components* $\mathbf{F}_x$ *and* $\mathbf{F}_y$. $F_x = F \cos \theta$, $F_y = F \sin \theta$.

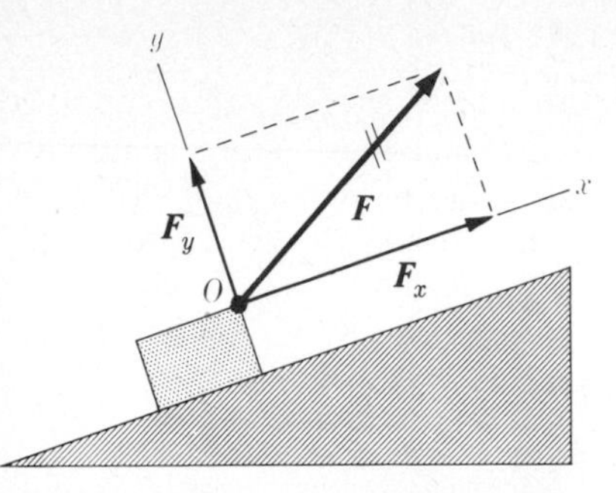

Fig. 1–13 *F_x and F_y are the rectangular components of **F**, parallel and perpendicular to the sloping surface of the inclined plane.*

example, Fig. 1–13 shows a block being pulled up an inclined plane by a force ***F***, represented by its components F_x and F_y, parallel and perpendicular to the sloping surface of the plane. This choice of axes simplifies the analysis of this particular problem.

1-8 RESULTANT BY RECTANGULAR RESOLUTION

Although in principle the polygon method is a satisfactory way to find the resultant of a number of forces, it is awkward for computation because one must, in general, solve a number of oblique triangles. Therefore the usual analytical method of finding the resultant is first to resolve all forces into rectangular components along any convenient pair of axes and then combine these into a single resultant. This makes it possible to work with *right* triangles only.

Figure 1–14(a) shows three concurrent forces $\boldsymbol{F}_1$, $\boldsymbol{F}_2$, and $\boldsymbol{F}_3$, whose resultant we wish to find. Let a pair of rectangular axes be constructed in any arbitrary direction. Simplification results if one axis coincides with one of the forces, which is always possible. In Fig. 1–14(b), the x-axis coincides with $\boldsymbol{F}_1$. Let us first resolve each of the given forces into x- and y-components. According to the usual conventions of analytic geometry, x-components toward the right are considered positive and those toward the left, negative. Upward y-components are positive and downward y-components are negative.

Force $\boldsymbol{F}_1$ lies along the x-axis and need not be resolved. The components of $\boldsymbol{F}_2$ are

$$F_{2x} = F_2 \cos \theta, \qquad F_{2y} = F_2 \sin \theta.$$

Both of these are positive, and $\boldsymbol{F}_{2x}$ has been slightly displaced upward to show it more clearly. The numerical values of the components of $\boldsymbol{F}_3$ are

$$F_{3x} = F_3 \cos \phi, \qquad F_{3y} = F_3 \sin \phi.$$

Both of these are negative.

We now imagine $\boldsymbol{F}_2$ and $\boldsymbol{F}_3$ to be removed and replaced by their rectangular components. To indicate this, the vectors $\boldsymbol{F}_2$ and $\boldsymbol{F}_3$ are crossed out lightly. All of the x-components can now be combined into a single force $\boldsymbol{R}_x$ whose magnitude equals the algebraic sum of the magnitudes of the x-components, and all of the y-components can be combined into a single force $\boldsymbol{R}_y$. The Greek letter $\sum$ is often used to

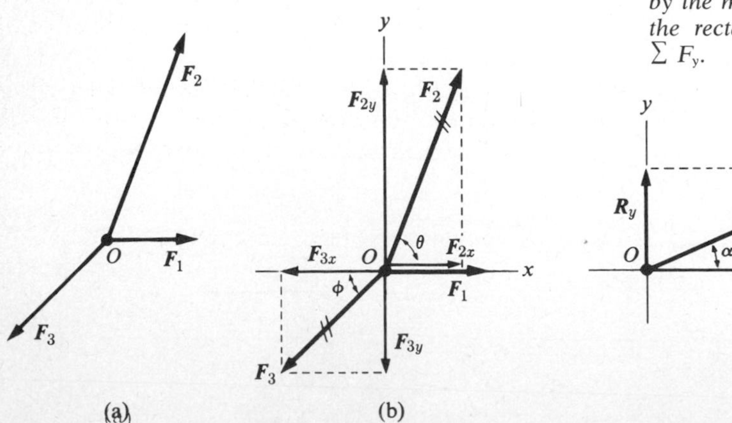

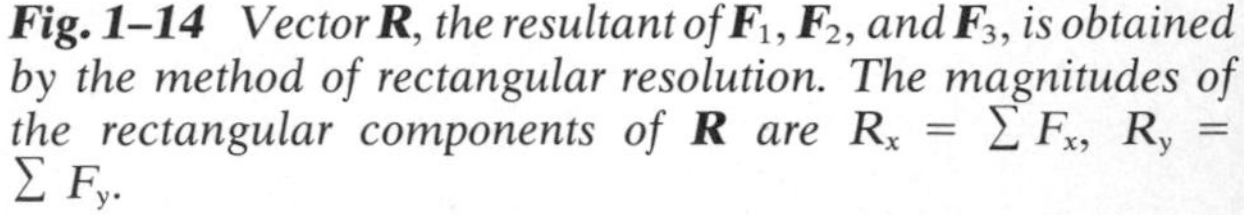

Fig. 1–14 *Vector **R**, the resultant of $\boldsymbol{F}_1$, $\boldsymbol{F}_2$, and $\boldsymbol{F}_3$, is obtained by the method of rectangular resolution. The magnitudes of the rectangular components of **R** are $R_x = \sum F_x$, $R_y = \sum F_y$.*

denote a sum, and $\sum F_x$ is read as *sum of the x-components of the forces.* Using the notation, we write

$$R_x = \sum F_x, \qquad R_y = \sum F_y.$$

Finally, these can be combined, as in part (c) of the figure, to form the resultant R whose magnitude is:

$$R = \sqrt{R_x^2 + R_y^2},$$

since R_x and R_y are perpendicular to each other.

The angle α between R and the x-axis can now be found from any one of its trigonometric functions. For example, $\tan \alpha = R_y/R_x$. As with the individual vectors, the components R_x and R_y may be positive or negative, and the angle α in Fig. 1–14 may be in any of the four quadrants.

Example In Fig. 1–14, let $F_1 = 120$ N, $F_2 = 200$ N, $F_3 = 150$ N, $\theta = 60°$, and $\phi = 45°$. The computations can be arranged systematically as in the following table:

Force	Angle	x-component	y-component
$F_1 = 120$ N	0	+120 N	0
$F_2 = 200$ N	60°	+100 N	+173 N
$F_3 = 150$ N	45°	−106 N	−106 N
		$\sum F_x = +114$ N	$\sum F_y = +67$ N

$$R = \sqrt{(114\text{ N})^2 + (67\text{ N})^2} = 132\text{ N},$$

$$\alpha = \tan^{-1}\frac{67\text{ N}}{114\text{ N}} = \tan^{-1} 0.588 = 30.4°.$$

1–9 VECTOR DIFFERENCE

It is sometimes necessary to subtract one vector from another. The process of subtracting one *algebraic* quantity from another is equivalent to adding the negative of the quantity to be subtracted, that is,

$$a - b = a + (-b).$$

Similarly, the negative $-\boldsymbol{B}$ of a *vector* quantity $\boldsymbol{B}$ is defined as a vector of the same length (or magnitude) but opposite in direction. Then the vector operation $\mathbf{A} - \boldsymbol{B}$ is defined to be the vector sum of $\mathbf{A}$ and $(-\boldsymbol{B})$. That is,

$$\mathbf{A} - \boldsymbol{B} = \mathbf{A} + (-\boldsymbol{B}).$$

The boldface **+**, **−**, and **=** signs remind us of the vector nature of these operations.

Vector subtraction is illustrated in Fig. 1–15. The given vectors are shown in (a). In (b), the vector sum of $\mathbf{A}$ and $-\boldsymbol{B}$, or the vector difference $\mathbf{A} - \boldsymbol{B}$, is found by the parallelogram method. Part (c) shows a second method; vectors $\mathbf{A}$ and $\boldsymbol{B}$ are drawn from a common origin, and the vector difference $\mathbf{A} - \boldsymbol{B}$ is the vector from the tip of $\boldsymbol{B}$ to the tip of $\mathbf{A}$. The vector difference $\mathbf{A} - \boldsymbol{B}$ is therefore seen to be the vector that must be added to $\boldsymbol{B}$ in order to give $\mathbf{A}$, since

$$\boldsymbol{B} + (\mathbf{A} - \boldsymbol{B}) = \mathbf{A}.$$

Vector differences may also be found by the method of rectangular resolution. Both vectors are resolved into x- and y-components. The difference between the x-components is the x-component of the desired vector difference, and the difference between the y-components is the y-component of the vector difference.

Vector subtraction is not often used when dealing with forces, but we shall use it frequently in connection with velocities and accelerations.

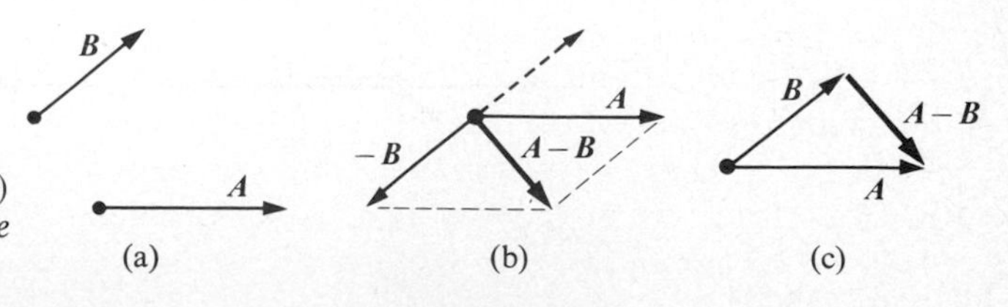

Fig. 1–15 *The vector difference* $\mathbf{A} - \boldsymbol{B}$ *is found in part (b) by the parallelogram method and in part (c) by the triangle method.*

1–10 A WORD ABOUT PROBLEMS

To obtain a benefit commensurate with the time expended, problem-solving should be considered as much more than merely substituting numbers for the symbols in a formula, or fitting together the pieces of a jigsaw puzzle. To merely thumb through the book until you find a formula that seems to fit, or a worked-out example that resembles the problem, is a waste of time and effort. One purpose which problems can serve is to enable you to find out for yourself whether or not you understand the assigned material, after having listened to a lecture and studied the text. Do your studying *before* you tackle the problems, instead of beginning with the problems and not referring to your book or lecture notes until you find yourself stymied. No problem assignment of reasonable length can hope to cover *every* important point, and you will miss a great deal if you read only enough to enable you to "do" your problems.

Every problem involves one or more general physical laws or definitions. After reading the statement of a problem, ask yourself what these laws or definitions are and be sure that you know them. This means that you should be able to state them to yourself clearly and explicitly, and not be satisfied with the comforting feeling that you "understand" Newton's second law although you can't quite put it in so many words.

Nearly every quantity of physical interest is expressed in terms of a unit of that quantity (some quantities are pure numbers). No answer is complete unless the proper units are given.

Unless otherwise stated, the numerical data given in the problems are to be assumed correct to three significant figures (for example, 2 m implies 2.00 m). Numerical answers should therefore be rounded off to three significant figures. That is, if long division leads to the result 6.575938 · · · m s^{-1}, give the answer as 6.58 m s^{-1}. (This is automatically taken care of if you use a slide rule.) Answers are provided to about half the problems, and are rounded off to two or three significant figures.

One of the best ways of making sure you understand the principles covered by a particular problem is to work it backwards. That is, if the problem gives x and asks you to compute y, then make up another problem in which y is given and x is to be found. Another helpful procedure is to ask yourself how the result would be altered if the given conditions had been somewhat different. Suppose the friction force had been 10 N instead of 5 N? Suppose the slope of the plane had been 60° instead of 30°?

Helpful advice on the techniques of studying and of solving problems will be found in *How to Study Physics* by Chapman and *How to Study, How to Solve* by Dadourian (published by Addison-Wesley).

PROBLEMS

1–1 Find graphically the magnitude and direction of the resultant of the three forces in Fig. 1–16. Use the polygon method.

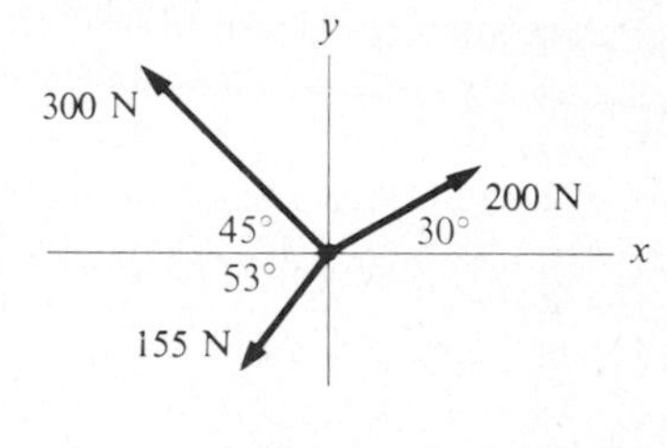

Fig. 1–16

1–2 Two men and a boy want to push a crate in the direction marked x in Fig. 1–17. The two men push with forces $\mathbf{F}_1$ and $\mathbf{F}_2$, whose magnitudes and directions are indicated in the figure. Find the magnitude and direction of the smallest force which the boy should exert.

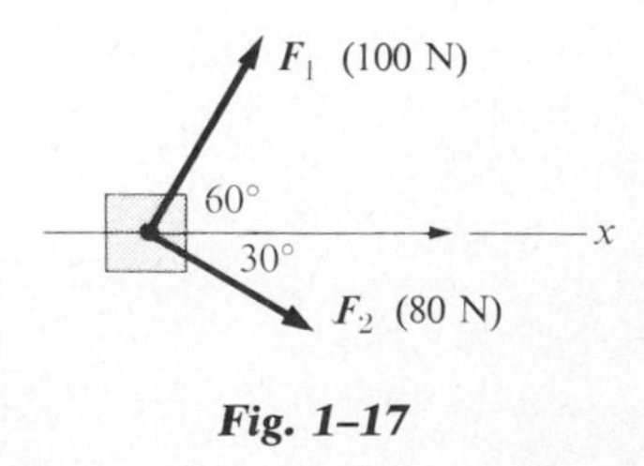

Fig. 1–17

1–3 Find graphically the resultant of two 10-N forces applied at the same point:

a) when the angle between the forces is 30°;
b) when the angle between them is 130°. Use any convenient scale.

1–4 Two men pull horizontally on ropes attached to a post, the angle between the ropes being 45°. If man A exerts a force of 300 N and man B a force of 200 N, find the magnitude of the resultant force and the angle it makes with A's pull. Solve:

a) graphically, by the parallelogram method, and
b) graphically, by the triangle method. Let 1 in. = 100 N in (a) and (b).

1–5 Find graphically the vector sum $\boldsymbol{A} + \boldsymbol{B}$ and the vector difference $\boldsymbol{A} - \boldsymbol{B}$ in Fig. 1–18.

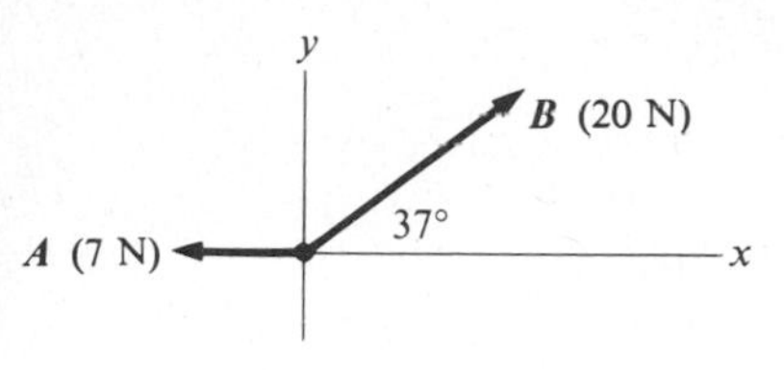

Fig. 1–18

1–6 Vector $\boldsymbol{A}$ is 2 in. long and is 60° above the x-axis in the first quadrant. Vector $\boldsymbol{B}$ is 2 in. long and is 60° below the x-axis in the fourth quadrant. Find graphically (a) the vector sum $\boldsymbol{A} + \boldsymbol{B}$, and (b) the vector differences $\boldsymbol{A} - \boldsymbol{B}$ and $\boldsymbol{B} - \boldsymbol{A}$.

1–7 a) Find graphically the horizontal and vertical components of a 40-N force the direction of which is 50° above the horizontal to the right. Let $\frac{1}{16}$ in. = 1 N. b) Check your results by calculating the components.

1–8 A box is pushed along the floor as in Fig. 1–2 by a force of 40 N making an angle of 30° with the horizontal. Using a scale of 1 in. = 10 N, find the horizontal and vertical components of the force by the graphical method. Check your results by calculating the components.

1–9 A block is dragged up an inclined plane of slope angle 20° by a force $\boldsymbol{F}$ making an angle of 30° with the plane. (a) How large a force $\boldsymbol{F}$ is necessary in order that the component F_x parallel to the plane shall be 16 N? (b) How large will the component F_y then be? Solve graphically, letting 1 in. = 8 N.

1–10 The three forces shown in Fig. 1–16 act on a body located at the origin. (a) Find the x- and y-components of each of the three forces. (b) Use the method of rectangular resolution to find the components, magnitude, and direction of the resultant of the forces. (c) Find the magnitude and direction of a fourth force which must be added to make the resultant force zero. Indicate the fourth force by a diagram.

1–11 Use the method of rectangular resolution to find the resultant of the following set of forces and the angle it makes with the horizontal: 200 N, along the x-axis toward the right; 300 N, 60° above the x-axis to the right; 100 N, 45° above the x-axis to the left; 200 N, vertically down.

1–12 A vector $\boldsymbol{A}$ of length 10 units makes an angle of 30° with a vector $\boldsymbol{B}$ of length 6 units. Find the magnitude of the vector difference $\boldsymbol{A} - \boldsymbol{B}$ and the angle it makes with vector $\boldsymbol{A}$: (a) by the parallelogram method; (b) by the triangle method; (c) by the method of rectangular resolution.

1–13 Two forces, $\boldsymbol{F}_1$ and $\boldsymbol{F}_2$, act at a point. The magnitude of $\boldsymbol{F}_1$ is 8 N and its direction is 60° above the x-axis in the first quadrant. The magnitude of $\boldsymbol{F}_2$ is 5 N and its direction is 53° below the x-axis in the fourth quadrant. (a) What are the horizontal and vertical components of the resultant force? (b) What is the magnitude of the resultant? (c) What is the magnitude of the vector difference $\boldsymbol{F}_1 - \boldsymbol{F}_2$?

1–14 Two forces, $\boldsymbol{F}_1$ and $\boldsymbol{F}_2$, act upon a body in such a manner that the resultant force $\boldsymbol{R}$ has a magnitude equal to that of $\boldsymbol{F}_1$ and makes an angle of 90° with $\boldsymbol{F}_1$. Let $F_1 = R = 10$ N. Find the magnitude of the second force, and its direction (relative to $\boldsymbol{F}_1$).

1–15 The resultant of four forces is 1000 N in the direction 30° west of north. Three of the forces are 400 N, 60° north of east; 200 N, south; and 400 N, 53° west of south. Find the rectangular components of the fourth force.

1–16 Vector $\boldsymbol{A}$ has components $A_x = 2$ cm, $A_y = 3$ cm, and vector $\boldsymbol{B}$ has components $B_x = 4$ cm, $B_y = -2$ cm. Find (a) the components of the vector sum $\boldsymbol{A} + \boldsymbol{B}$; (b) the magnitude and direction of $\boldsymbol{A} + \boldsymbol{B}$; (c) the components of the vector difference $\boldsymbol{A} - \boldsymbol{B}$; (d) the magnitude and direction of $\boldsymbol{A} - \boldsymbol{B}$.

1–17 A car drives 5 mi east, then 4 mi south, then 2 mi west. Find the magnitude and direction of the resultant displacement.

1–18 A sailboat sails 2 km east, then 4 km southeast, then an additional distance in an unknown direction. Its final position is 5 km directly east of the starting point. Find the magnitude and direction of the third leg of the journey.

1–19 Vector $\boldsymbol{M}$, of magnitude 5 cm, is at 36.9° counterclockwise from the $+x$-axis. It is added to vector $\boldsymbol{N}$, and the resultant is a vector of magnitude 5 cm, at 53.1° counterclockwise from the $+x$-axis. Find (a) the components of $\boldsymbol{N}$; (b) the magnitude and direction of $\boldsymbol{N}$.

1–20 The velocity of an airplane relative to the surface of the earth, $\boldsymbol{v}_{\mathrm{PE}}$, equals the vector sum of its velocity relative to the air, $\boldsymbol{v}_{\mathrm{PA}}$, and the velocity of the air relative to the earth, $\boldsymbol{v}_{\mathrm{AE}}$. Thus

$$\boldsymbol{v}_{\mathrm{PE}} = \boldsymbol{v}_{\mathrm{PA}} + \boldsymbol{v}_{\mathrm{AE}}.$$

a) Find graphically the magnitude and direction of the velocity $\boldsymbol{v}_{\mathrm{PE}}$ if the velocity $\boldsymbol{v}_{\mathrm{PA}}$ is 100 mi hr^{-1} due north, and if the wind velocity $\boldsymbol{v}_{\mathrm{AE}}$ is 40 mi hr^{-1} from east to west. Let 1 cm = 20 mi hr^{-1}.

b) Find the direction of the velocity $\boldsymbol{v}_{\mathrm{PA}}$ if its magnitude is 100 mi hr^{-1}, if the direction of $\boldsymbol{v}_{\mathrm{PE}}$ is due north, and if $\boldsymbol{v}_{\mathrm{AE}}$ is 40 mi hr^{-1} from east to west. What is then the magnitude of $\boldsymbol{v}_{\mathrm{PE}}$?

Chapter 2

Equilibrium of a Particle

2–1 INTRODUCTION

The science of mechanics is based on three natural laws which were clearly stated for the first time by Sir Isaac Newton (1642–1727) and were published in 1686 in his *Philosophiae Naturalis Principia Mathematica* (The Mathematical Principles of Natural Science). It should not be inferred, however, that the science of mechanics began with Newton. Many men had preceded him in his field, the most outstanding being Galileo Galilei (1564–1642), who in his studies of accelerated motion had laid much of the groundwork for Newton's three laws.

In this chapter, we shall make use of only two of Newton's laws, the first and the third. Newton's second law will be discussed in Chapter 5.

2–2 EQUILIBRIUM. NEWTON'S FIRST LAW

One effect of a force is to alter the dimensions or shape of a body on which the force acts; another is to alter the state of motion of the body.

The motion of a rigid body can be considered as made up of its motion as a whole, or its *translational* motion, together with any *rotational* motion the body may have. In the most general case, a single force acting on a body produces a change in both its translational and rotational motion. However, when several forces act on a body simultaneously, their effects can compensate one another, with the result that there is no change in either the translational or rotational motion. When this is the case, the body is said to be in *equilibrium*. This means (1) that the body as a whole either remains at rest or moves in a straight line with constant speed, and (2) that the body is either not rotating at all or is rotating at a constant rate.

Let us consider some (idealized) experiments from which the laws of equilibrium can be deduced. Figure 2–1 represents a flat, rigid object of arbitrary shape on a level surface having negligible friction. If a single force $\boldsymbol{F}_1$ acts on the body, as in Fig. 2–1(a), and if the body is originally at rest, it at once starts to move and to rotate clockwise. If originally in motion, the effect of the force is to change the translational motion of the body in magnitude or direction (or both) and to increase or decrease its rate of rotation. In either case, the body does not remain in equilibrium.

Equilibrium can be maintained, however, by the application of a second force $\boldsymbol{F}_2$ as in Fig. 2–1(b), provided that $\boldsymbol{F}_2$ is *equal in magnitude* to $\boldsymbol{F}_1$, is *opposite in direction* to $\boldsymbol{F}_1$, and has the same *line of action* as $\boldsymbol{F}_1$. The resultant of $\boldsymbol{F}_1$ and $\boldsymbol{F}_2$ is then zero. If the lines of action of the two forces are not the same, as in Fig. 2–1(c), the body will be in transla-

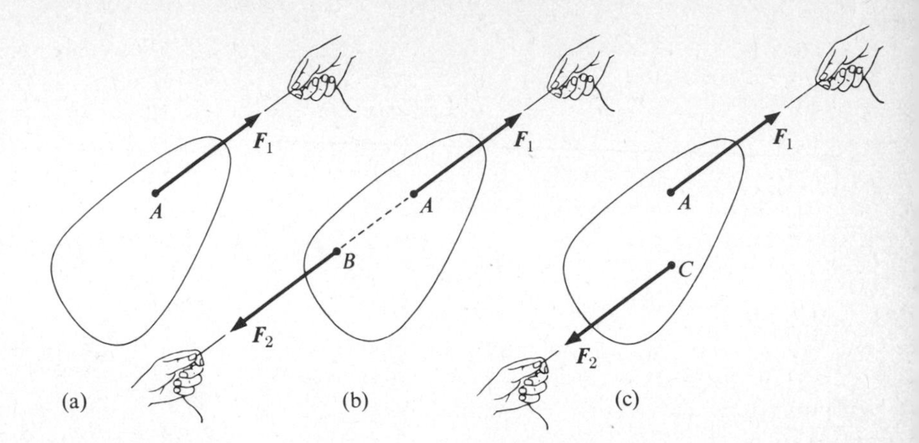

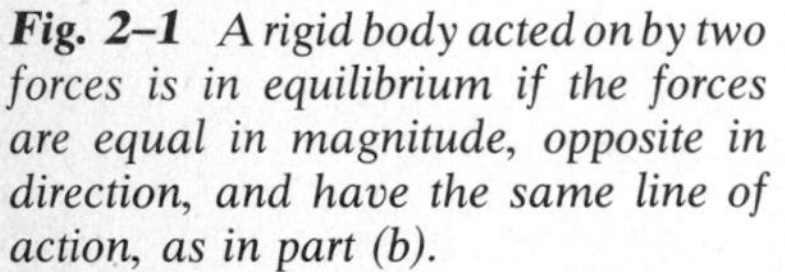

Fig. 2–1 *A rigid body acted on by two forces is in equilibrium if the forces are equal in magnitude, opposite in direction, and have the same line of action, as in part (b).*

tional but not in rotational equilibrium. (Such a combination of two forces is called a *couple*; see Section 3–4.)

If the forces $\boldsymbol{F}_1$ and $\boldsymbol{F}_2$ are equal in magnitude and opposite in direction, then one equals the negative of the other. That is,

$$\boldsymbol{F}_2 = -\boldsymbol{F}_1.$$

Then if $\boldsymbol{R}$ represents the resultant of $\boldsymbol{F}_1$ and $\boldsymbol{F}_2$,

$$\boldsymbol{R} = \boldsymbol{F}_1 + \boldsymbol{F}_2 = \boldsymbol{F}_1 - \boldsymbol{F}_1 = \boldsymbol{0}.$$

For brevity, we shall often speak of two forces as simply being "equal and opposite," meaning that their magnitudes are equal and that one is the negative of the other.

In Fig. 2–2(a), a body in equilibrium is acted on by three nonparallel coplanar forces, $\boldsymbol{F}_1$, $\boldsymbol{F}_2$, and $\boldsymbol{F}_3$. Any force applied to a rigid body may be regarded as acting anywhere along its line of action. Therefore, let any two of the force vectors, say $\boldsymbol{F}_1$ and $\boldsymbol{F}_2$, be transferred to the point of intersection of their lines of action and their resultant $\boldsymbol{R}$ obtained, as in Fig. 2–2(b). The forces are now reduced to two, $\boldsymbol{R}$ and $\boldsymbol{F}_3$; for equilibrium these must (1) be equal in magnitude, (2) be opposite in direction, and (3) have the same line of action. It follows from the first two conditions that the resultant of the three forces is zero. The third condition can be fulfilled only if the line of action of $\boldsymbol{F}_3$ passes through the point of intersection of the lines of action of $\boldsymbol{F}_1$ and $\boldsymbol{F}_2$. When the lines of action of several forces pass through a common point, the forces are said to be *concurrent*; the body in Fig. 2–2 can be in equilibrium only when the three forces are concurrent.

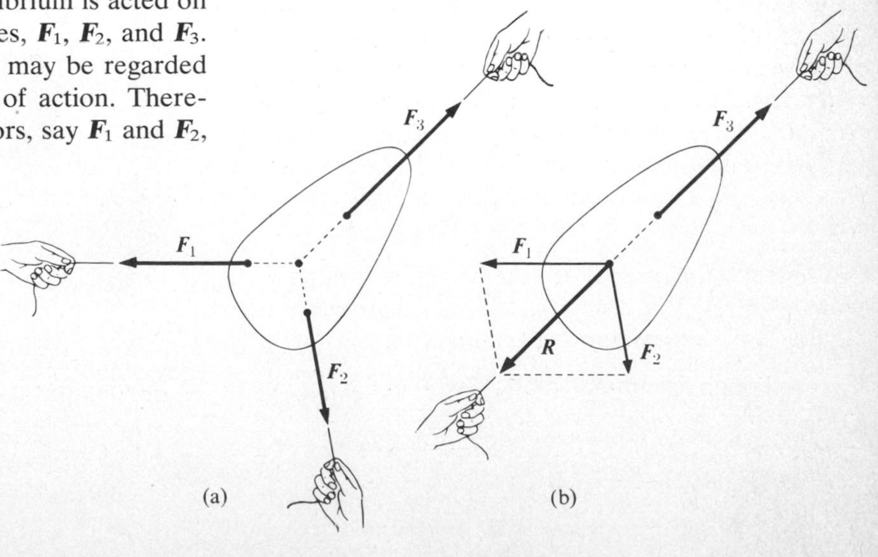

Fig. 2–2 *When a body acted on by three nonparallel coplanar forces is in equilibrium, the forces are concurrent and the resultant of any two is equal in absolute magnitude and opposite in direction (or sign) to the third.*

The construction in Fig. 2–2 provides a satisfactory graphical method for the solution of problems in equilibrium. For an analytical solution, it is usually simpler to deal with the rectangular components of the forces. We have shown that the magnitudes of the rectangular components of the resultant $\boldsymbol{R}$ of any set of coplanar forces are

$$R_x = \sum F_x, \qquad R_y = \sum F_y.$$

When a body is in equilibrium, the resultant of all the forces acting on it is zero. Both rectangular components are then zero and hence, for a body in equilibrium,

$$\boldsymbol{R} = \boldsymbol{0}, \qquad \text{or} \qquad \sum F_x = 0, \qquad \sum F_y = 0.$$

These equations are called the first condition of equilibrium.

The *second condition for equilibrium,* to be developed more fully in Chapter 3, is a mathematical statement of the fact that the forces must have no tendency to rotate the body. In particular, when only two forces act they must have the same line of action, and when three forces act they must be concurrent, in order for the body to be in equilibrium. There is an exceptional case, three forces with parallel lines of action, which will be discussed in Chapter 3.

The first condition of equilibrium ensures that a body shall be in *translational* equilibrium; the second, that it be in *rotational* equilibrium. The statement that a body is in complete equilibrium when both conditions are satisfied is the essence of *Newton's first law of motion.* Newton did not state his first law in exactly these words. His original statement (translated from the Latin in which the *Principia* was written) reads:

"*Every body continues in its state of rest, or of uniform motion in a straight line, unless it is compelled to change that state by forces impressed on it.*"

Although rotational motion was not explicitly mentioned by Newton, it is clear from his work that he fully understood the conditions that the forces must satisfy when the rotation is zero or is constant.

2–3 DISCUSSION OF NEWTON'S FIRST LAW OF MOTION

Newton's first law of motion is not as self-evident as it may seem. In the first place, this law asserts that in the absence of any applied force a body either remains at rest or moves uniformly in a straight line. It follows that *once a body has been set in motion it is no longer necessary to exert a force on it to maintain it in motion.*

This assertion appears to be contradicted by everyday experience. Suppose we exert a force with the hand to move a book along a level table top. After the book has left our hand, and we are no longer exerting a force on it, it does *not* continue to move indefinitely, but slows down and eventually comes to rest. If we wish to keep it moving uniformly we must continue to exert some forward force on it. But this force is required only because a frictional force is exerted on the sliding book by the table top, in a direction opposite to the motion of the book. The smoother the surfaces of the book and table, the smaller the frictional force and the smaller the force we must exert to keep the book moving. The first law asserts that if the frictional force could be eliminated completely, no forward force at all would be required to keep the book moving, once it had been set in motion. More than that, however, the law implies that if the *resultant* force on the book is zero, as it is when the frictional force is balanced by an equal forward force, the book also continues to move uniformly. In other words, *zero resultant force is equivalent to no force at all.*

Furthermore, the first law defines by implication what is known as an *inertial reference system.* To understand what is meant by this term we must recognize that the motion of a given body can be specified only with respect to, or relative to, some other body. Its motion relative to one body may be very different from that relative to another. Thus a passenger in an aircraft which is making its take-off run may continue at rest relative to the aircraft but be moving faster and faster relative to the earth.

A *reference system* means a set of coordinate axes attached to (or moving with) some specified body or bodies. Suppose we consider a reference system attached to the aircraft referred to above. Everyone knows that during the take-off run, while the aircraft is going faster and faster, a passenger feels the back of his seat pushing him forward, although he remains at rest relative to a reference system attached to the aircraft. Newton's first law does not therefore correctly describe the situation; a forward force *does* act on the passenger, but nevertheless (relative to the aircraft) he remains at rest.

Suppose, on the other hand, that the passenger is standing in the aisle on roller skates. He then starts to move *backward*, relative to the aircraft, when the take-off run begins, even though no backward force acts on him. Newton's first law again does not correctly describe the facts.

We can now define an inertial reference system as one relative to which a body *does* remain at rest or move uniformly in a straight line when no force (or no resultant force) acts on it. That is, *an inertial reference system is one in which Newton's first law correctly describes the motion of a body not acted on by any force.*

An aircraft gaining speed during take-off is evidently *not* an inertial system. For many purposes a reference system attached to the earth can be considered an inertial system, although it is not precisely so, due to effects related to the earth's rotation and other motions. But if a frame of reference *is* inertial, then a second frame moving uniformly (i.e., with constant speed in a straight line, without rotation) relative to it is *also* inertial, since any body in uniform motion as seen by an observer in the first frame will also appear to an observer in the second frame to be in uniform motion. The speed and direction appear different to the two observers, but both observe that Newton's first law is obeyed.

Thus there is no single, unique inertial frame of reference, there are infinitely many; but the motion of any one relative to any other is uniform in the above sense. It follows that the concepts of "absolute rest" and "absolute motion" have no physical meaning. Experiments show that a frame of reference at rest relative to the so-called *fixed stars* is inertial, within the limits of experimental error, and so every frame in uniform motion relative to this one is also inertial. An aircraft in steady flight over the earth is just as suitable an inertial frame of reference as the earth itself.

Finally, Newton's first law contains a qualitative definition of the concept of *force*, or at least of one aspect of the force concept, as "that which changes the state of motion of a body." (Of course, forces also produce other effects, such as changing the length of a coil spring.) When a body at rest relative to the earth is observed to start moving, or when a moving body speeds up, slows down, or changes its direction, we can conclude that force is acting on it. This effect of a force can be used to define the ratio of two forces and to define a *unit* of force, and we shall show in Chapter 5 how this is done.

2–4 NEWTON'S THIRD LAW OF MOTION

Any given force is but one aspect of a mutual interaction between *two* bodies. It is found that *whenever one body exerts a force on another, the second always exerts on the first a force which is equal in magnitude, is opposite in direction, and has the same line of action.** A single isolated force is therefore an impossibility.

The two forces involved in every interaction between two bodies are often called an "action" and a "reaction," but this does not imply any difference in their nature, or that one force is the "cause" and the other its "effect." *Either* force may be considered the "action," and the other the "reaction" to it.

This property of forces was stated by Newton in his *third law of motion.* In his words,

"*To every action there is always opposed an equal reaction: or, the mutual actions of two bodies upon each other are always equal, and directed to contrary parts.*"

As an example, suppose that a man pulls on one end of a rope attached to a block, as in Fig. 2–3. The weight of the block and the force exerted on it by the surface are not shown. The block may or may not be in equilibrium. The resulting action–reaction pairs of

*There are instances, such as the motion of electrically charged particles, where this law is not obeyed, but it is correct for all the macroscopic forces we shall encounter in mechanics.

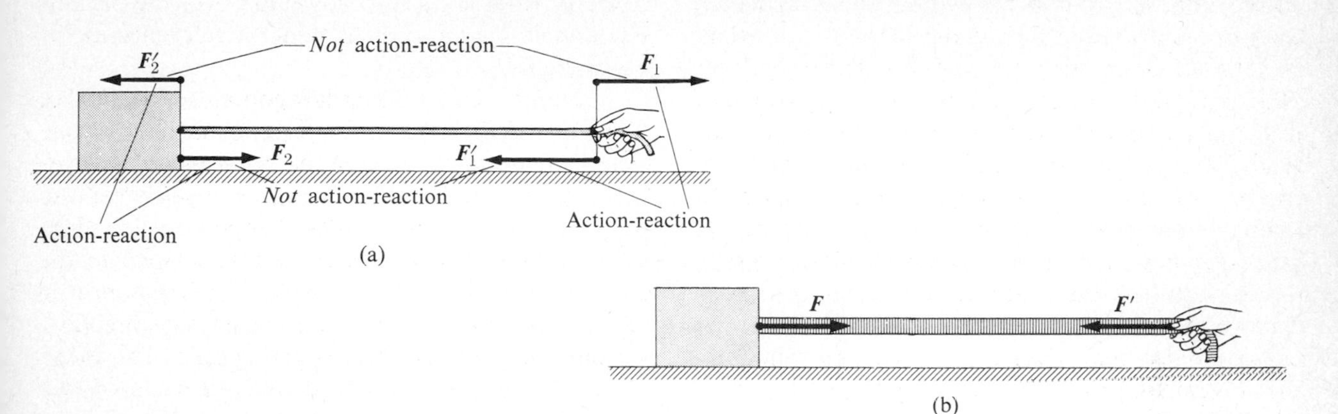

Fig. 2–3 *(a) Forces* $\boldsymbol{F}_1$ *and* $\boldsymbol{F}_1'$ *form one action–reaction pair, and forces* $\boldsymbol{F}_2$ *and* $\boldsymbol{F}_2'$ *another.* $\boldsymbol{F}_1$ *is always equal to* $\boldsymbol{F}_1'$, $\boldsymbol{F}_2$ *is always equal to* $\boldsymbol{F}_2'$. $\boldsymbol{F}_1$ *and* $\boldsymbol{F}_2'$ *are equal only if the rope is in equilibrium, and force* $\boldsymbol{F}_2'$ *is not the reaction to* $\boldsymbol{F}_1$. *(Actually all forces lie along the rope.) (b) If the rope is in equilibrium, it can be considered to transmit a force from the man to the block, and vice versa.*

forces are indicated in the figure. (The lines of action of all the forces lie along the rope; the force vectors have been offset from this line to show them more clearly.) Vector $\boldsymbol{F}_1$ represents the force exerted on the rope by the man. Its reaction is the equal and opposite force $\boldsymbol{F}_1'$ exerted on the man by the rope. Vector $\boldsymbol{F}_2$ represents the force exerted on the block by the rope. The reaction to it is the equal and opposite force $\boldsymbol{F}_2'$, exerted on the rope by the block:

$$\boldsymbol{F}_1' = -\boldsymbol{F}_1, \qquad \boldsymbol{F}_2' = -\boldsymbol{F}_2.$$

It is very important to realize that the forces $\boldsymbol{F}_1$ and $\boldsymbol{F}_2'$, although they are opposite in direction and have the same line of action, do *not* constitute an action–reaction pair. For one thing, both of these forces act on the *same* body (the rope) while an action and its reaction necessarily act on *different* bodies. Furthermore, the forces $\boldsymbol{F}_1$ and $\boldsymbol{F}_2'$ are not necessarily equal in magnitude. If the block and rope are moving to the right with increasing speed, the rope is not in equilibrium and $\boldsymbol{F}_1$ is greater than $\boldsymbol{F}_2'$. Only in the special case when the rope remains at rest or moves with constant speed are the forces $\boldsymbol{F}_1$ and $\boldsymbol{F}_2'$ equal in magnitude, but this is an example of Newton's *first* law, not his *third*. Even when the speed of the rope is changing, however, the action–reaction forces $\boldsymbol{F}_1$ and $\boldsymbol{F}_1'$ are equal in magnitude to *each other*, and the action–reaction forces $\boldsymbol{F}_2$ and $\boldsymbol{F}_2'$ are equal in magnitude to *each other*, although then $\boldsymbol{F}_1$ is not equal in magnitude to $\boldsymbol{F}_2'$.

In the special case when the rope is in equilibrium, and when no forces act on it except those at its ends, F_2' equals F_1 by Newton's *first* law. Since F_2 *always* equals F_2', by Newton's *third* law, then in this special case F_2 also equals F_1 and the force exerted on the block by the rope is equal to the force exerted on the rope by the man. The rope can therefore be considered to "transmit" to the block, without change, the force exerted on it by the man. This point of view is often useful, but it is important to remember that it applies only under the restricted conditions as stated.

If we adopt this point of view, the rope itself need not be considered and we have the simpler force diagram of Fig. 2–3(b), where the man is considered to exert a force $\boldsymbol{F}$ directly on the block.

The reaction is the force $\boldsymbol{F}'$ exerted by the block on the man. The only effect of the rope is to transmit these forces from one body to the other.

A body, like the rope in Fig. 2–3, which is subjected to pulls at its ends is said to be in *tension.* The tension at any point equals the force exerted at that point. Thus in Fig. 2–3(a) the tension at the righthand end of the rope equals the magnitude of $\boldsymbol{F}_1$ (or of $\boldsymbol{F}_1'$) and the tension at the lefthand end equals the magnitude of $\boldsymbol{F}_2$ (or of $\boldsymbol{F}_2'$). If the rope is in equilibrium and if no forces act except at its ends, as in Fig. 2–3(b), the tension is the same at both ends. If, for example, in Fig. 2–3(b) the magnitudes of $\boldsymbol{F}$ and $\boldsymbol{F}'$ are each 50 N, the tension in the rope is 50 N (*not* 100 N).

2–5 EQUILIBRIUM OF A PARTICLE

The materials found in nature and the processes which occur are rarely simple. As a first step in dealing with a problem of nature it is often necessary to *idealize* the situation and to make *simplifying assumptions* concerning the processes. Suppose, for example, a baseball is thrown into the air and it is desired to calculate where it lands and with what velocity. The first step in treating this problem consists in idealizing the baseball by ignoring the details of its surface and all departures from sphericity during its motion. In other words, we replace the baseball by an ideal object, i.e., a rigid smooth sphere. The next step is to ignore the rotation of the baseball and the forces brought into play by the air dragged around by the rotating ball. A further step is to regard as negligible the buoyancy and the resistance of the air. We are then left with a problem quite different from the original one; in fact we may be accused of having robbed the problem of almost all of its reality. This is true, but what remains approximates the original problem at low velocities, and has the virtue that it is amenable to simple mathematical treatment, whereas the original problem at high velocities would require the most advanced methods.

In general, the forces acting on a rigid body do not all pass through one point (they are *nonconcurrent*) and, as a result, the rigid body undergoes rotational as well as translational motion. There are, however, many situations of great interest where the rotation of a body is of only minor consequence and is not pertinent in the solution of the problem. The planetary motion, for example, of the earth about the sun taking place under the action of the gravitational force between the two bodies may be studied alone without regard to the earth's rotation. *A body whose rotation is ignored as irrelevant is called a particle. A particle may be so small that it is an approximation to a point, or it may be of any size, provided that the action lines of all the forces acting on it intersect in one point.* Thus a particle serves as an *idealized model* for a body whose rotational motion is of no consequence in a given situation.

In the remainder of this chapter, we shall limit ourselves to examples and problems involving the equilibrium of particles only. It is surprising how many situations of interest and importance in engineering, in the life and earth sciences, and in everyday life involve the equilibrium of particles. In many of these, it is important to know how to calculate one or two of the forces acting on a particle when the others are given. In order to do this, it is best to adhere scrupulously to the following rules.

1. Make a simple line sketch of the apparatus or structure, showing dimensions and angles.

2. Choose some object (a knot in a rope, for example) as the particle in equilibrium. Draw a separate diagram of this object and show by arrows (use a colored pencil) *all* of the forces exerted *on* it by other bodies. This is called the *force diagram* or *free-body diagram.* When a system is composed of several particles, it may be necessary to construct a separate free-body diagram for each one. Do *not* show, in the free-body diagram of a chosen particle, any of the forces exerted *by* it. These forces (which are the reactions to the forces acting *on* the chosen particle) all act *on other bodies* and appear in the free-body diagrams of those bodies.

3. Construct a set of rectangular axes and resolve any inclined forces into rectangular components. Cross out lightly those forces which have been resolved.

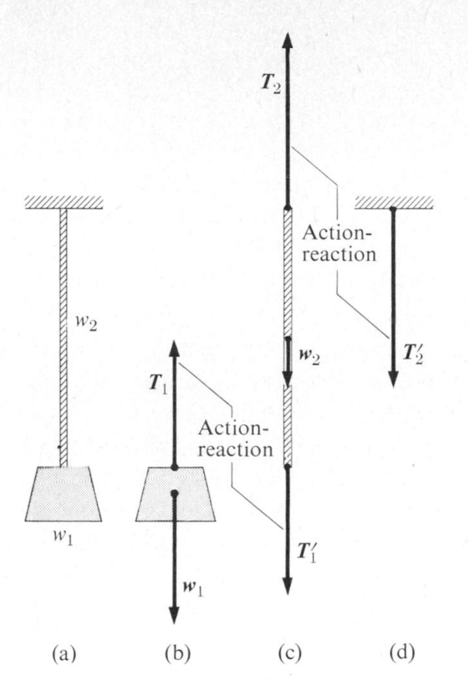

Fig. 2–4 (a) Block hanging at rest from vertical cord. (b) The block is isolated and all forces acting on it are shown. (c) Forces on the cord. (d) Downward force on the ceiling. Lines connect action–reaction pairs.

4. Set the algebraic sum of all x-forces (or force components) equal to zero, and the algebraic sum of all y-forces (or components) equal to zero. This provides two independent equations which can be solved simultaneously for two unknown quantities (which may be forces, angles, distances, etc.)

A force which will be encountered in many problems is the *weight* of a body, that is, the force of gravitational attraction exerted on the body by the earth. We shall show in the next chapter that the line of action of this force always passes through a point called the *center of gravity* of the body.

The force of gravitational attraction exerted on a body by the earth is but one aspect of a mutual interaction between the earth and the body. At the same time, the body attracts the earth with a force equal in magnitude and opposite in direction to the force the earth exerts on the body. Thus if a body weighs 10 N, the earth pulls down on it with a force of 10 N, and the body pulls *up* on the earth with a force of 10 N, another example of an action–reaction pair.

Example 1 To begin with a simple example, consider the body in Fig. 2–4(a), hanging at rest from the ceiling by a vertical cord. Part (b) of the figure is the free-body diagram for the body. The forces on it are its weight $\boldsymbol{w}_1$ and the upward force $\boldsymbol{T}_1$ exerted on it by the cord. If we take the x-axis horizontal and the y-axis vertical, there are no x-components of force, and the y-components are the forces $\boldsymbol{w}_1$ and $\boldsymbol{T}_1$. Then from the condition that $\sum F_y = 0$, we have

$$\sum F_y = T_1 - w_1 = 0,$$

$$T_1 = w_1. \qquad \text{(First Law)}$$

In order that both forces have the same line of action, the center of gravity of the body must lie vertically below the point of attachment of the cord.

Let us emphasize again that the forces $\boldsymbol{w}_1$ and $\boldsymbol{T}_1$ are *not* an action–reaction pair, although they are equal in magnitude, opposite in direction, and have the same line of action. The weight $\boldsymbol{w}_1$ is a force of attraction exerted on the body by the earth. Its

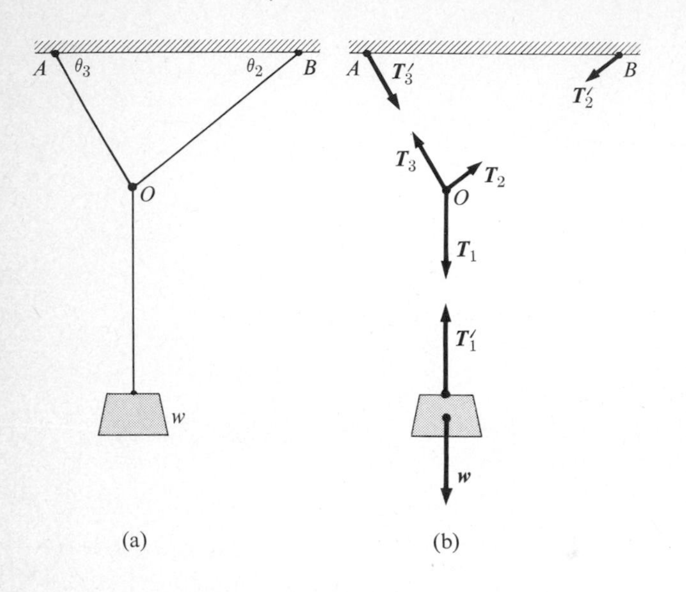

Fig. 2–5 *(a) Block hanging in equilibrium. (b) Forces acting on the block, on the knot, and on the ceiling.*

reaction is an equal and opposite force of attraction exerted on the earth by the body. This reaction is one of the set of forces acting *on the earth*, and therefore it does not appear in the free-body diagram of the suspended block.

The reaction to the force $\boldsymbol{T}_1$ is an equal downward force, $\boldsymbol{T}_1'$,

$$T_1 = T_1'. \qquad \text{(Third Law)}$$

The force $\boldsymbol{T}_1'$ is shown in part (c), which is the free-body diagram of the *cord.* The other forces on the cord are its own weight $\boldsymbol{w}_2$ and the upward force $\boldsymbol{T}_2$ exerted on its upper end by the ceiling. Since the cord is also in equilibrium,

$$\sum F_y = T_2 - w_2 - T_1' = 0$$

$$T_2 = w_2 + T_1'. \qquad \text{(First Law)}$$

The reaction to $\boldsymbol{T}_2$ is the downward force $\boldsymbol{T}_2'$ in part (d), exerted on the ceiling by the cord:

$$T_2 = T_2'. \qquad \text{(Third Law)}$$

As a numerical example, let the body weigh 20 N and the cord weigh 1 N. Then

$$T_1 = w_1 = 20 \text{ N},$$

$$T_1' = T_1 = 20 \text{ N},$$

$$T_2 = w_2 + T_1' = 1 \text{ N} + 20 \text{ N} = 21 \text{ N},$$

$$T_2' = T_2 = 21 \text{ N}.$$

If the weight of the cord were so small as to be negligible, then in effect no forces would act on it except at its ends. The forces $\boldsymbol{T}_2$ and $\boldsymbol{T}_2'$ would then each equal 20 N and, as explained earlier, the cord could be considered to transmit a 20-N force from one end to the other without change. We could then consider the upward pull of the cord on the block as an "action" and the downward pull on the ceiling as its "reaction." The tension in the cord would then be 20 N.

Example 2 In Fig. 2–5(a), a block of weight w hangs from a cord which is knotted at O to two other cords fastened to the ceiling. We wish to find the tensions in these three cords. The weights of the cords are negligible.

In order to use the conditions of equilibrium to compute an unknown force, we must consider some body which is in equilibrium and on which the desired forces act. The hanging block is one such body and, as shown in the preceding example, the tension in the vertical cord supporting the block is equal to the weight of the block. The inclined cords do not exert forces on the block, but they do act on the knot at O. Hence we consider the *knot* as a particle in equilibrium whose own weight is negligible.

The free-body diagrams for the block and the knot are shown in Fig. 2–5(b), where $\boldsymbol{T}_1$, $\boldsymbol{T}_2$, and $\boldsymbol{T}_3$ represent the forces exerted *on the knot* by the three cords and $\boldsymbol{T}_1'$, $\boldsymbol{T}_2'$, and $\boldsymbol{T}_3'$ are the reactions to these forces.

Consider first the hanging block. Since it is in equilibrium,

$$T_1' = w. \qquad \text{(First Law)}$$

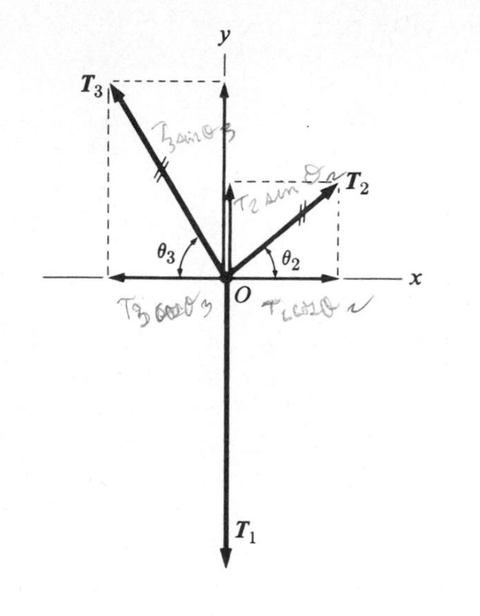

Fig. 2–6 Forces on the knot in Fig. 2–5, resolved in x- and y-components.

Since $\boldsymbol{T}_1$ and $\boldsymbol{T}_1'$ form an action–reaction pair,

$$T_1' = T_1. \quad \text{(Third Law)}$$

Hence,

$$T_1 = w.$$

To find the forces $\boldsymbol{T}_2$ and $\boldsymbol{T}_3$ we resolve these forces (see Fig. 2–6) into rectangular components, noting that components upward or to the right (the positive axis directions) are positive, while those downward or to the left (negative axis directions) are negative.

$$\sum F_x = T_2 \cos\theta_2 - T_3 \cos\theta_3 = 0,$$

$$\sum F_y = T_2 \sin\theta_2 + T_3 \sin\theta_3 - T_1 = 0.$$

Since w is known, $\boldsymbol{T}_1$ is also known, and these are two simultaneous equations which may be solved for $\boldsymbol{T}_2$ and $\boldsymbol{T}_3$.

As a numerical example, let

$$w = 50\text{ N}, \qquad \theta_2 = 30°, \qquad \theta_3 = 60°.$$

Then $T_1 = 50$ N, and the two preceding equations become

$$T_2\left(\frac{\sqrt{3}}{2}\right) - T_3\left(\frac{1}{2}\right) = 0,$$

$$T_2\left(\frac{1}{2}\right) + T_3\left(\frac{\sqrt{3}}{2}\right) = 50\text{ N}.$$

Simultaneous solution yields the results

$$T_2 = 25\text{ N}, \qquad T_3 = 43.3\text{ N}.$$

The reader may wish to show that the general solutions are given by

$$T_2 = \frac{T_1 \cos\theta_3}{\sin(\theta_2 + \theta_3)}, \qquad T_3 = \frac{T_1 \cos\theta_2}{\sin(\theta_2 + \theta_3)}.$$

These forms have the advantage that it is easy to see how the results are affected by changing the angles. When either θ_2 or θ_3 decreases, both tensions increase, as might be expected intuitively. Working out the general results has the added advantage of decreasing the amount of numerical computation needed to obtain values for T_2 and T_3.

Finally, we know from Newton's *third law* that the inclined cords exert on the ceiling the forces T_2' and T_3', respectively.

Example 3 Figure 2–7(a) shows a strut AB, pivoted at end A, attached to a wall by a cable, and carrying a load w at end B. The weights of the strut and of the cable are negligible.

Figure 2–7(b) shows the forces acting on the strut: $\boldsymbol{T}_1$ is the force exerted by the vertical cable, $\boldsymbol{T}_2$ is the force exerted by the inclined cable, and $\boldsymbol{C}$ is the force exerted by the pivot. Force $\boldsymbol{T}_1$ is known both in magnitude and in direction, force $\boldsymbol{T}_2$ is known in direction only, and neither the magnitude nor the direction of $\boldsymbol{C}$ is known. However, the forces $\boldsymbol{T}_1$ and $\boldsymbol{T}_2$ intersect at the outer end of the strut, and since the strut is in equilibrium under three forces, the line of action of force $\boldsymbol{C}$ must also pass through the outer end of the strut. In other words, the direction of force $\boldsymbol{C}$ is along the line of the strut. (This is not true in general. If the weight of the strut is included, or if additional forces are present, the forces are not all

Fig. 2–7

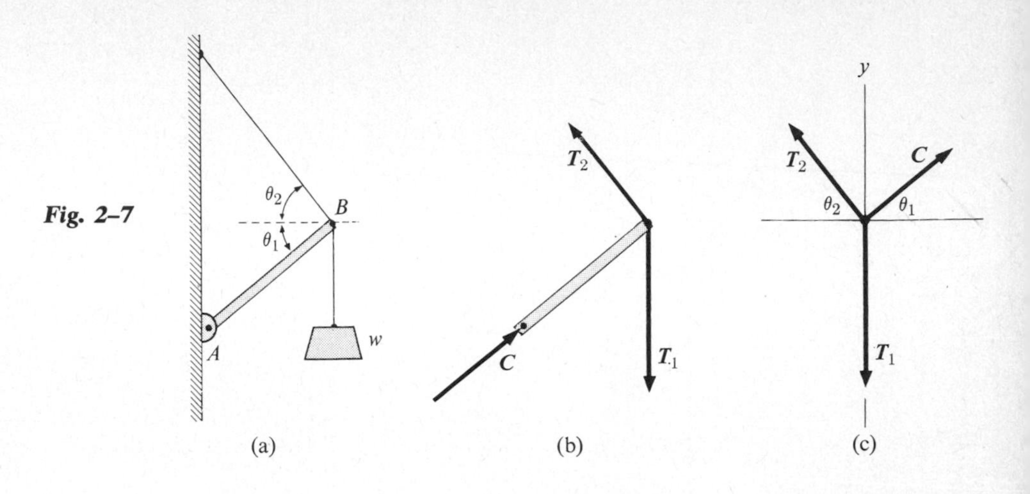

concurrent, and the force on the strut at point A is in general *not* along the line of the strut.)

Hence the resultant of $\boldsymbol{T}_1$ and $\boldsymbol{T}_2$ is also along this line and the strut, in effect, is acted on by forces at its ends, directed toward each other along the line of the strut. The effect of these forces is to compress the strut, and it is said to be in *compression.*

If the forces acting on a strut are not all applied at its ends, the direction of the resultant force at the ends is not along the line of the strut. This is illustrated in the next example.

In Fig. 2–7(c), the force $\boldsymbol{C}$ has been transferred along its line of action to the point of intersection B of the three forces. The force diagram for a particle at B is exactly like that of Fig. 2–6, and the problem is solved in the same way. The conditions of equilibrium provide two independent equations among the five quantities T_1, T_2, C, θ_1, and θ_2. Hence if any three are known, the other two may be calculated.

Example 4 In Fig. 2–8 a ladder in equilibrium leans against a vertical frictionless wall. The forces on the ladder are (1) its weight $\boldsymbol{w}$, (2) the force $\boldsymbol{F}_1$ exerted on the ladder by the vertical wall, which is perpendicular to the wall if there is no friction, and (3) the force

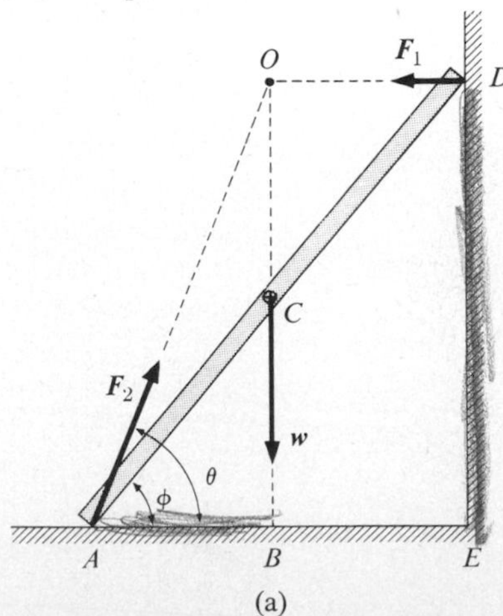

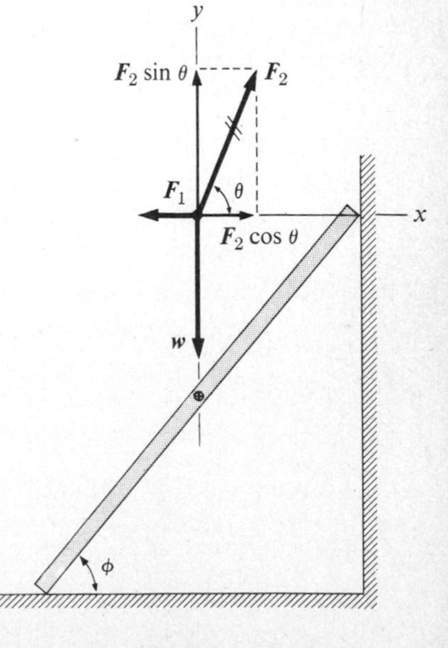

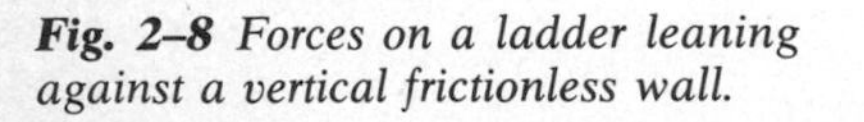
Fig. 2–8 *Forces on a ladder leaning against a vertical frictionless wall.*

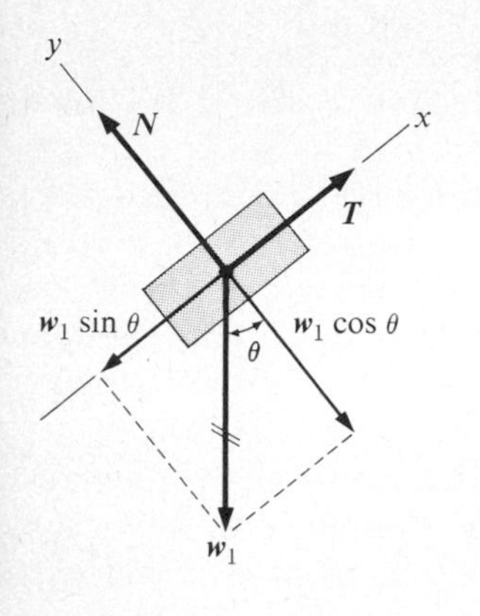

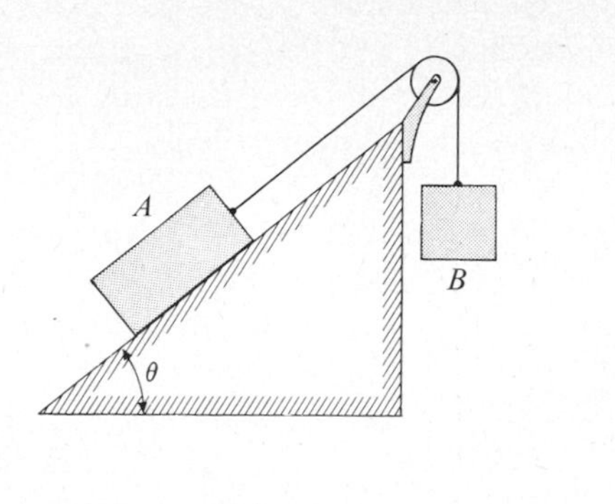

Fig. 2–9 *Forces on a block in equilibrium on a frictionless inclined plane.*

$\boldsymbol{F}_2$ exerted by the ground on the base of the ladder. The force $\boldsymbol{w}$ is known in magnitude and in direction, the force $\boldsymbol{F}_1$ is known in direction only, and the force $\boldsymbol{F}_2$ is unknown in both magnitude and direction. As in the preceding example, the ladder is in equilibrium under three forces, *which must be concurrent.* Since the lines of action of $\boldsymbol{F}_1$ and $\boldsymbol{w}$ are known, their point of intersection (point O) can be located. The line of action of $\boldsymbol{F}_2$ must then pass through this point also. Note that neither the direction of $\boldsymbol{F}_1$ nor that of $\boldsymbol{F}_2$ lies along the line of the ladder. In part (b) the forces have been transferred to the point of intersection of their lines of action and, applying the equations for the equilibrium of a particle at this point, we get

$$\sum F_x = F_2 \cos\theta - F_1 = 0, \qquad (2\text{–}1)$$

$$\sum F_y = F_2 \sin\theta - w = 0. \qquad (2\text{–}2)$$

As a numerical example, suppose the ladder weighs 400 N, is 10 m long, has its center of gravity at its center, and makes an angle $\phi = 53°$ with the ground. We wish to find the angle θ and the forces $\boldsymbol{F}_1$ and $\boldsymbol{F}_2$. To calculate θ, we first find the lengths of AB and BO. From the right triangle ABC, we have

$$\overline{AB} = \overline{AC}\cos\phi = (5\text{ m})(0.60) = 3.0\text{ m},$$

and from the right triangle AED,

$$\overline{DE} = \overline{AD}\sin\phi = (10\text{ m})(0.80) = 8\text{ m}.$$

Then, from the right triangle AOB, since $\overline{OB} = \overline{DE}$, we have

$$\tan\theta = \frac{\overline{OB}}{\overline{AB}} = \frac{8\text{ m}}{3\text{ m}} = 2.67.$$

Then

$$\theta = 69.5°, \qquad \sin\theta = 0.937, \qquad \cos\theta = 0.350.$$

$$F_2 = \frac{w}{\sin\theta} = \frac{400\text{ N}}{0.937} = 427\text{ N},$$

and from Eq. (2–1),

$$F_1 = F_2\cos\theta = (427\text{ N})(0.350) = 150\text{ N}.$$

The ladder presses against the wall and the ground with forces which are equal and opposite to F_1 and F_2, respectively.

In the next chapter we shall explain another method of solving this problem, using the concept of the *moment* of a force.

Example 5 In Fig. 2–9, block A of weight w_1 rests on a frictionless inclined plane of slope angle θ. The center of gravity of the block is at its center. A flexible cord is attached to the center of the right face of the block, passes over a frictionless pulley, and is attached to a second block B of weight $\boldsymbol{w}_2$. The weight of the cord and friction in the pulley are negligible. If w_1 and θ are given, find the weight w_2 for which the system is in equilibrium, that is, for which it remains at rest or moves in either direction at constant speed.

The free-body diagrams for the two blocks are shown at the left and right. The forces on block B are its weight $\boldsymbol{w}_2$ and the force $\boldsymbol{T}$ exerted on it by the cord. Since it is in equilibrium,

$$T = w_2. \qquad \text{(First Law)} \qquad (2\text{–}3)$$

Block A is acted on by its weight $\boldsymbol{w}_1$, the force $\boldsymbol{T}$ exerted on it by the cord, and the force $\boldsymbol{N}$ exerted on it by the plane. We can use the same symbol $\boldsymbol{T}$ for the force exerted on each block by the cord because, as explained in Section 2–4, these forces are equivalent to an action–reaction pair and have the same magnitude. The force $\boldsymbol{N}$, if there is no friction, is perpendicular or *normal* to the surface of the plane. Since the lines of action of $\boldsymbol{w}_1$ and $\boldsymbol{T}$ intersect at the center of gravity of the block, the line of action of $\boldsymbol{N}$ passes through this point also. It is simplest to choose x- and y-axes parallel and perpendicular to the surface of the plane, because then only the weight $\boldsymbol{w}_1$ needs to be resolved into components. The conditions of equilibrium give

$$\left.\begin{aligned} \sum F_x &= T - w_1 \sin\theta = 0, \qquad (2\text{–}4)\\ \sum F_y &= N - w_1 \cos\theta = 0. \qquad (2\text{–}5) \end{aligned}\right\} \text{(First Law)}$$

Thus if $w_1 = 100$ N and $\theta = 30°$, we have, from Eqs. (2–3) and (2–4),

$$w_2 = T = w_1 \sin\theta = (100\text{ N})(0.500) = 50\text{ N},$$

and from Eq. (2–5),

$$N = w_1 \cos\theta = (100\text{ N})(0.866) = 86.6\text{ N}.$$

Note carefully that, in the *absence of friction*, the same weight w_2 of 50 N is required whether the system remains at rest or moves with constant speed in *either* direction. This is not the case when friction is present.

2–6 FRICTION

Whenever the surface of one body slides over that of another, *each* body exerts a frictional force on the other, parallel to the surfaces. The force *on* each body is opposite to the direction of its motion relative to the other. Thus when a block slides from left to right along the surface of a table, a frictional force to the left acts on the block and an equal force toward the right acts on the table. Frictional forces may also act when there is no relative motion. A horizontal force on a heavy packing case resting on the floor may not be enough to set the case in motion, because of an equal and opposite frictional force exerted on the case by the floor.

The origin of these frictional forces is not fully understood, and the study of them is an important field of research. When one unlubricated metal slides over another, there appears to be an actual momentary welding of the metals together at the "high spots" where they make contact. The observed friction force is the force required to break these tiny welds. The mechanism of the friction force between two blocks of wood, or between two bricks, must be quite different.

In Fig. 2–10(a) a block is at rest on a horizontal surface, in equilibrium under the action of its weight $\boldsymbol{w}$ and the upward force $\boldsymbol{P}$ exerted on it by the surface. The lines of action of $\boldsymbol{w}$ and $\boldsymbol{P}$ have been displaced slightly to show these forces more clearly.

Suppose now that a cord is attached to the block as in Fig. 2–10(b) and the tension $\boldsymbol{T}$ in the cord is gradually increased. Provided the tension is not too great, the block remains at rest. The force $\boldsymbol{P}$ exerted on the block by the surface is inclined toward the left as shown, since the sum of the three forces $\boldsymbol{P}$, $\boldsymbol{w}$, and $\boldsymbol{T}$ must be zero. The component of $\boldsymbol{P}$ parallel to the surface is called the *force of static friction*, $\boldsymbol{f}_s$. The other component, $\boldsymbol{N}$, is the *normal* force exerted on the block by the surface. From the conditions of equilibrium, the force of static friction $\boldsymbol{f}_s$ is equal in magnitude and opposite in direction to the force $\boldsymbol{T}$, and the normal force $\boldsymbol{N}$ is equal and opposite to the weight $\boldsymbol{w}$.

As the force $\boldsymbol{T}$ is increased further, a limiting value is reached at which the block breaks away from the surface and starts to move. In other words, there is a certain *maximum* value which the force of static friction f_s can have. Figure 2–10(c) is the force diagram when $\boldsymbol{T}$ is just below its critical value. If the force $\boldsymbol{T}$ exceeds this value, the block is no longer in equilibrium.

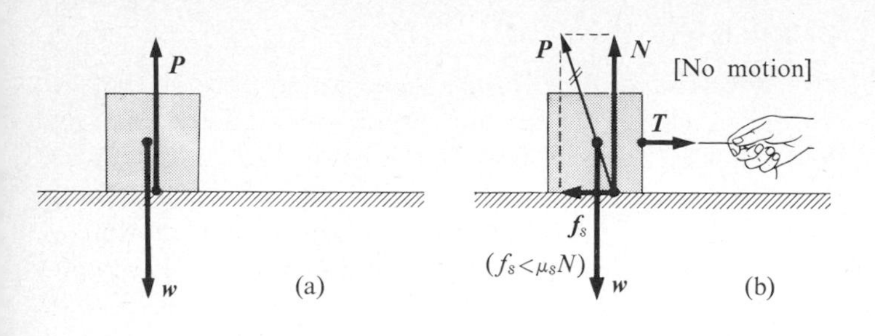

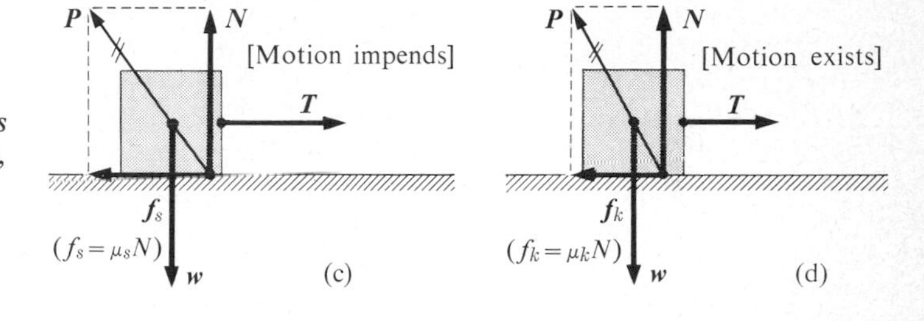

Fig. 2–10 *The magnitude of the friction force f is less than or equal to $\mu_s N$ when there is no relative motion, and is equal to $\mu_k N$ when motion exists.*

Experiment shows that for a given pair of surfaces, the magnitude of the maximum value of f_s depends on that of the normal force $\mathbf{N}$; when the normal force increases, a greater force is required to make the block slide. In general the details of this relationship are quite complex, but we may often find it useful to assume simply that the maximum value of f_s is *directly proportional* to N, with proportionality factor μ_s, called the *coefficient of static friction.* The actual force of static friction can have any magnitude between zero (when there is no applied force parallel to the surface) and a maximum value given by $\mu_s N$. Thus,

$$\boxed{f_s \leq \mu_s N.} \tag{2–6}$$

The equality sign holds only when the applied force $\boldsymbol{T}$, parallel to the surface, has such a value that motion is about to start (Fig. 2–10(c)). When $\boldsymbol{T}$ is less than this value (Fig. 2–10(b)), the inequality sign holds and the magnitude of the friction force must be computed from the conditions of equilibrium.

As soon as sliding begins, it is found that the friction force *decreases.* This new friction force, for a given pair of surfaces, also depends on the magnitude of the normal force, and again it is convenient, though often rather imprecise, to represent the relation as a proportionality. The proportionality factor, μ_k is called the *coefficient of sliding friction* or *kinetic* friction. Thus, when the block is in motion, the force of sliding or kinetic friction is given by

$$\boxed{f_k = \mu_k N.} \tag{2–7}$$

This is illustrated in Fig. 2–10(d).

The coefficients of static and sliding friction depend primarily on the nature of the surfaces in contact, being relatively large if the surfaces are rough, and small if they are smooth. The coefficient of sliding friction varies somewhat with the relative velocity, but for simplicity we shall assume it to be independent of velocity. It is also nearly independent of the contact area. However, since two real surfaces actually touch each other only at a relatively small number of high spots, the true contact area is very different from the overall area. Equations (2–6) and (2–7) are useful empirical relations, but do not represent fundamental physical laws like Newton's laws. Typical numerical values are given in Table 2–1.

Table 2–1 lists the coefficients of friction of solids only. Liquids and gases show frictional effects also, but the simple equation $f = \mu N$ does not hold. It will be shown in Chapter 13 that there exists a property of liquids and gases called *viscosity* which determines the friction force between two surfaces

Table 2–1 COEFFICIENTS OF FRICTION

Materials	Static, μ_s	Kinetic, μ_k
Steel on steel	0.74	0.57
Aluminum on steel	0.61	0.47
Copper on steel	0.53	0.36
Brass on steel	0.51	0.44
Zinc on cast iron	0.85	0.21
Copper on cast iron	1.05	0.29
Glass on glass	0.94	0.4
Copper on glass	0.68	0.53
Teflon on Teflon	0.04	0.04
Teflon on steel	0.04	0.04

sliding over each other with a layer of liquid gas between them. Gases have the lowest viscosities of all materials at normal temperatures, and therefore, to reduce friction to a value close to zero, it is convenient to have an object slide on a layer of gas.

This is the principle of the *hovercraft*, a type of vehicle containing powerful blowers that maintain a cushion of air between the vehicle and the ground so that it literally floats on a layer of air. A familiar laboratory example of the principle is the linear air track, conceived originally by H. V. Neher and R. B. Leighton at the California Institute of Technology, and developed further by John Stull of Alfred University. Elastic bumpers are provided for the inverted V-shaped riders. (See Fig. 2–11(a).) When one rider only is placed on the air track and is given a push, it will hit the stationary bumper and then proceed to move back and forth many times before coming to rest. The frictional force is velocity-dependent, but at typical speeds the effective coefficient of friction is of the order of 0.001.

A similar device is the frictionless air table shown in Fig. 2–11(b), developed by Harold A. Daw of New Mexico State University. The pucks, made of plastic, are supported by over 1600 tiny air jets about an inch apart. Two-dimensional collisions, both elastic and inelastic, may be demonstrated on this table.

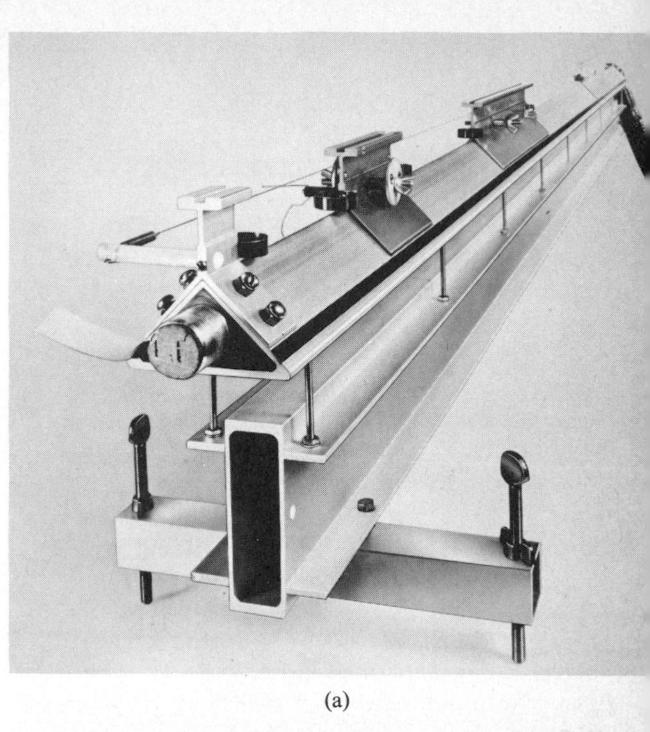

(a)

(b)

Fig. 2–11 *(a) The Ealing–Stull linear air track. Inverted Y-shaped sliders ride on a layer of air streaming through many fine holes in the inverted V-shaped surface. (b) The Ealing–Daw two-dimensional air table. Plastic pucks slide on a cushion of air issuing from more than a thousand minute holes in the tabletop.* [*Courtesy of the Ealing Corporation*]

Example 1 In Fig. 2–10, suppose that the block weighs 20 N, that the tension T can be increased to 8 N before the block starts to slide, and that a force of 4 N will keep the block moving at constant speed once it has been set in motion. Find the coefficients of static and kinetic friction.

From Fig. 2–10(c) and the data above, we have

$$\left.\begin{aligned}\sum F_y &= N - w = N - 20\text{ N} = 0,\\ \sum F_x &= T - f_s = 8\text{ N} - f_s = 0,\end{aligned}\right\}\text{ (First Law)}$$

$$f_s = \mu_s N \quad \text{(motion impends).}$$

Hence we have

$$\mu_s = \frac{f_s}{N} = \frac{8\text{ N}}{20\text{ N}} = 0.40.$$

From Fig. 2–10(d), we have

$$\left.\begin{aligned}\sum F_y &= N - w = N - 20\text{ N} = 0,\\ \sum F_x &= T - f_k = 4\text{ N} - f_k = 0,\end{aligned}\right\}\text{ (First Law)}$$

$$f_k = \mu_k N \quad \text{(motion exists).}$$

Hence

$$\mu_k = \frac{f_k}{N} = \frac{4\text{ N}}{20\text{ N}} = 0.20.$$

Example 2 What is the friction force if the block in Fig. 2–10(b) is at rest on the surface and a horizontal force of 5 N is exerted on it?

We have

$$\sum F_x = T - f_s = 5\text{ N} - f_s = 0, \qquad \text{(First Law)}$$

$$f_s = 5\text{ N}.$$

Note that in this case $f_s < \mu_s N$.

Example 3 What force $\boldsymbol{T}$, at an angle of 30° above the horizontal, is required to drag a block weighing 20 N to the right at constant speed, as in Fig. 2–12, if the coefficient of kinetic friction between block and surface is 0.20?

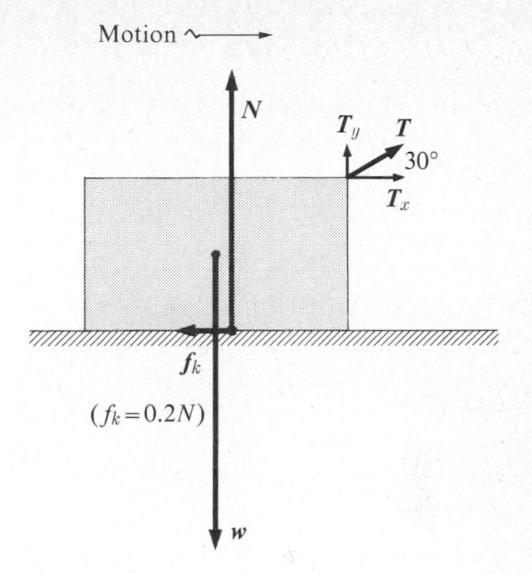

Fig. 2–12 Forces on a block being dragged to the right on a level surface at constant speed.

The forces on the block are shown in the diagram. From the first condition of equilibrium,

$$\sum F_x = T\cos 30^\circ - 0.2N = 0,$$

$$\sum F_y = T\sin 30^\circ + N - 20\text{ N} = 0.$$

Simultaneous solution gives

$$T = 4.15\text{ N},$$

$$N = 17.9\text{ N}.$$

Note that in this example the normal force N is not equal to the weight of the block, but is less than the weight by the vertical component of the force $\boldsymbol{T}$.

Example 4 In Fig. 2–13, a block has been placed on an inclined plane and the slope angle θ of the plane has been adjusted until the block slides down the plane at constant speed, once it has been set in motion. Find the angle θ.

The forces on the block are its weight $\boldsymbol{w}$ and the normal and frictional components of the force exerted by the plane. Take axes perpendicular and parallel

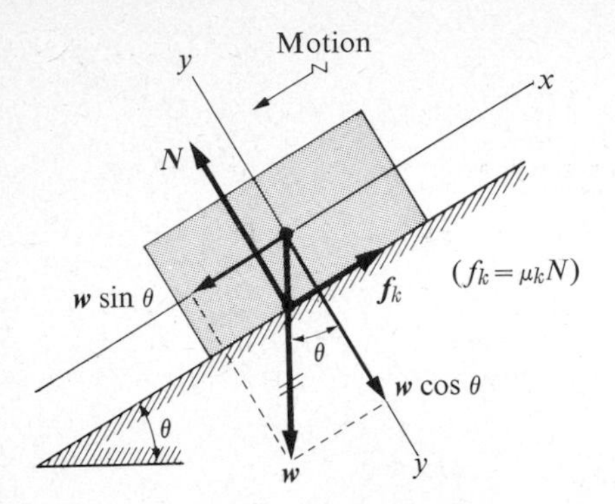

Fig. 2-13 *Forces on a block sliding down an inclined plane (with friction) at constant speed.*

to the surface of the plane. Then

$$\left.\begin{aligned}\sum F_x &= \mu_k N - w \sin\theta = 0,\\ \sum F_y &= N - w\cos\theta = 0.\end{aligned}\right\}\quad \text{(First Law)}$$

Hence

$$\mu_k N = w \sin\theta, \qquad N = w\cos\theta.$$

Dividing the former by the latter, we get

$$\mu_k = \tan\theta.$$

It follows that a block, regardless of its weight, slides down an inclined plane with constant speed if the tangent of the slope angle of the plane equals the coefficient of kinetic friction. Measurement of this angle then provides a simple experimental method for determining the coefficient of kinetic friction.

PROBLEMS

Section 2-5 should be read carefully before you begin the solution of problems in this chapter. If difficulties arise, they will not result from any complicated mathematics or from a failure to remember the "formula"—after all, the only "formulas" are

$$\sum F_x = 0, \qquad \sum F_y = 0,$$

—but rather from (1) a failure to *select some one body to talk about*, and (2) *a failure to recognize precisely what forces are exerted* on *the selected body*. Once the forces acting on the body have been clearly and correctly shown in a force diagram, the physics of the problem is finished; the rest of the computations might be (and in practice often are) turned over to a machine.

2-1 Imagine that you are holding a book weighing 4 N at rest on the palm of your hand. Complete the following sentences.

a) A downward force of magnitude 4 N is exerted on the book by __________.
b) An upward force of magnitude __________ is exerted on __________ by the hand.
c) Is the upward force (b) the reaction to the downward force (a)?
d) The reaction to force (a) is a force of magnitude __________, exerted on __________ by __________. Its direction is __________.
e) The reaction to force (b) is a force of magnitude __________, exerted on __________ by __________. Its direction is __________.
f) That the forces (a) and (b) are equal and opposite is an example of Newton's __________ law.
g) That forces (b) and (e) are equal and opposite is an example of Newton's __________ law.

Suppose now that you exert an upward force of magnitude 5 N on the book.

h) Does the book remain in equilibrium?
i) Is the force exerted on the book by the hand equal and opposite to the force exerted on the book by the earth?
j) Is the force exerted on the book by the earth equal and opposite to the force exerted on the earth by the book?
k) Is the force exerted on the book by the hand equal and opposite to the force exerted on the hand by the book?

Finally, suppose that you snatch your hand away while the book is moving upward.

l) How many forces then act on the book?
m) Is the book in equilibrium?
n) What balances the downward force exerted on the book by the earth?

2-2 A block is given a push along a table top, and slides off the edge of the table.

a) What forces are exerted on it while it is falling from the table to the floor?

b) What is the reaction to each force, that is, on what body and by what body is the reaction exerted? Neglect air resistance.

2–3 Two 10-N weights are suspended at opposite ends of a rope which passes over a light frictionless pulley. The pulley is attached to a chain which goes to the ceiling.

a) What is the tension in the rope?

b) What is the tension in the chain?

2–4 In Fig. 2–5, let the weight of the hanging block be 50 N. Find the tensions T_2 and T_3 (a) if $\theta_2 = \theta_3 = 60°$, (b) if $\theta_2 = \theta_3 = 10°$, (c) if $\theta_2 = 60°$, $\theta_3 = 0$, and (d) if $AB = 5$ m, $AO = 3$ m, $OB = 4$ m.

2–5 Find the tension in each cord in Fig. 2–14 if the weight of the suspended body is 200 N.

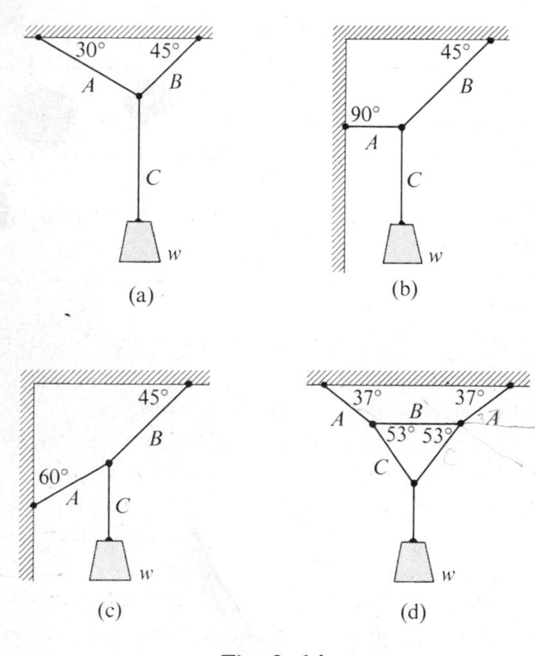

Fig. 2–14

2–6 Find the tension in each cable, and the magnitude and direction of the force ***C*** exerted on the strut by the pivot, in each of the arrangements in Fig. 2–15. Let the weight of the suspended object in each case be 1000 N. Neglect the weight of the strut.

2–7

a) In which of the arrangements in Fig. 2–16 can the tension T be computed, if the only quantities known are those explicitly given?

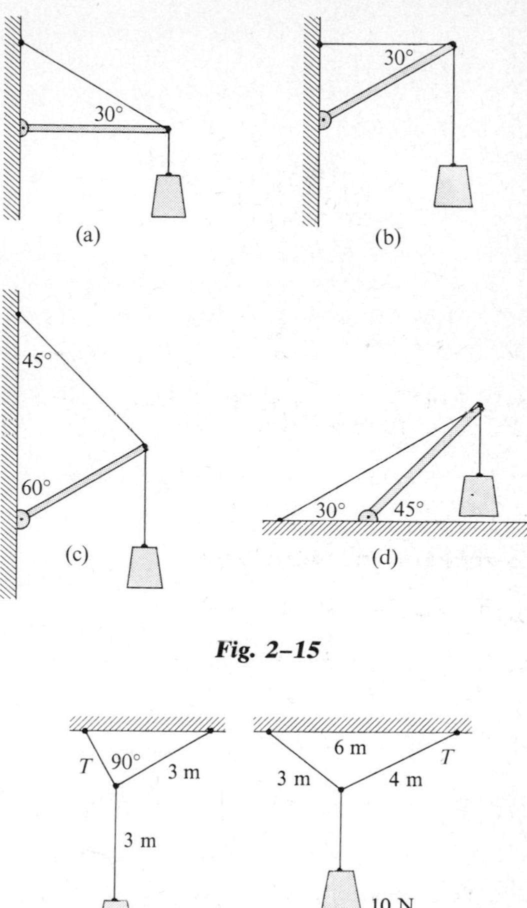

Fig. 2–15

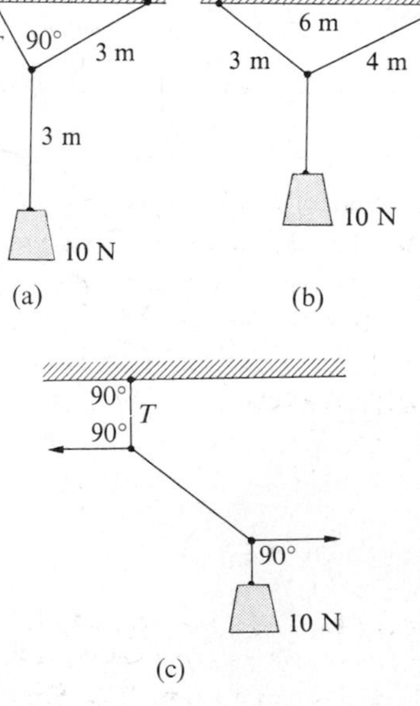

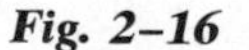
Fig. 2–16

b) For each case in which insufficient information is given, state one additional quantity, a knowledge of which would permit solution.

2–8 A horizontal boom 4 m long is hinged to a vertical wall at one end, and a 500-N body hangs from its outer end. The boom is supported by a guy wire from its outer end to a point on the wall directly above the boom.

a) If the tension in this wire is not to exceed 1000 N, what is the minimum height above the boom at which it may be fastened to the wall?

b) By how many newtons would the tension be increased if the wire were fastened 0.5 m below this point, the boom remaining horizontal? Neglect the weight of the boom.

2–9 One end of a rope 25 m long is attached to an automobile. The other end is fastened to a tree. A man exerts a force of 500 N at the midpoint of the rope, pulling it 1 m to the side. What is the force exerted on the automobile?

2–10 Find the largest weight w which can be supported by the structure in Fig. 2–17 if the maximum tension the upper rope can withstand is 1000 N and the maximum compression the strut can withstand is 2000 N. The vertical rope is strong enough to carry any load required. Neglect the weight of the strut.

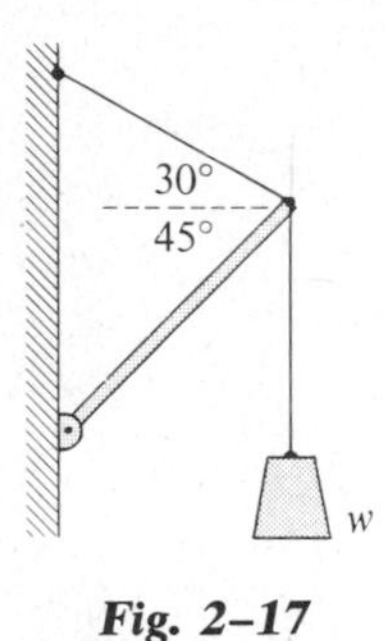

Fig. 2–17

2–11

a) Block A in Fig. 2–18 weighs 100 N. The coefficient of static friction between the block and the surface on which it rests is 0.30. The weight w is 20 N and the system is in equilibrium. Find the friction force exerted on block A.

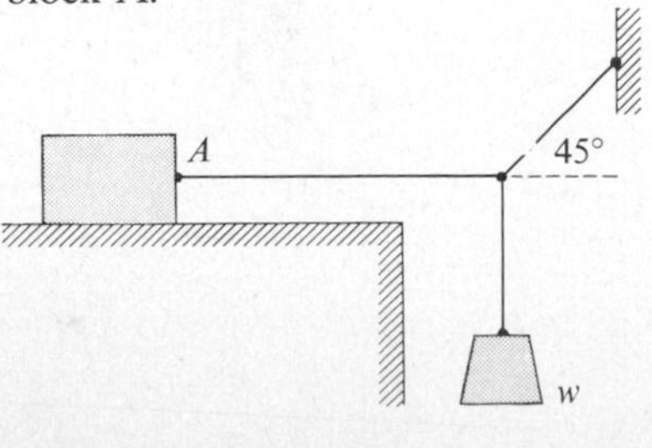

Fig. 2–18

b) Find the maximum weight w for which the system will remain in equilibrium.

2–12 A block hangs from a cord 10 m long. A second cord is tied to the midpoint of the first, and a horizontal pull equal to half the weight of the block is exerted on it, the second cord being always kept horizontal.

a) How far will the block be pulled to one side?

b) How far will it be lifted?

2–13 A flexible chain of weight w hangs between two hooks at the same height, as shown in Fig. 2–19. At each end the chain makes an angle θ with the horizontal.

a) What is the magnitude and direction of the force $\boldsymbol{F}$ exerted by the chain on the hook at the left?

b) What is the tension T in the chain at its lowest point?

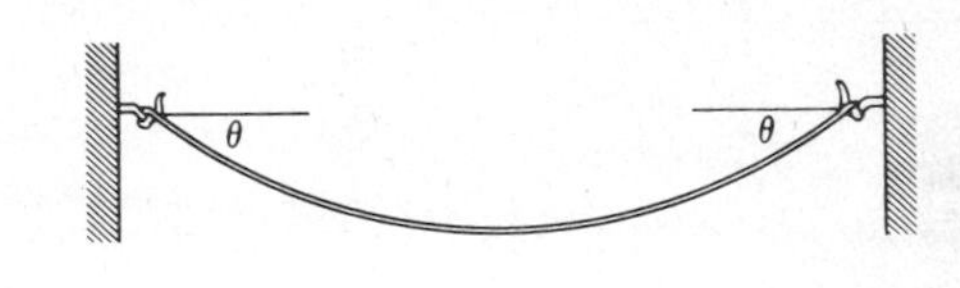

Fig. 2–19

2–14 A 30-N block is pulled at constant speed up a frictionless inclined plane by a weight of 10 N hanging from a cord attached to the block and passing over a frictionless pulley at the top of the plane. (See Fig. 2–9.) Find (a) the slope angle of the plane, (b) the tension in the cord, and (c) the normal force exerted on the block by the plane.

2–15

a) A block rests upon a rough horizontal surface. A horizontal force $\boldsymbol{T}$ is applied to the block and is slowly increased from zero. Draw a graph with T along the x-axis and the friction force f along the y-axis, starting at $T = 0$ and showing the region of no motion, the point where motion impends, and the region where motion exists.

b) A block of weight w rests on a rough horizontal plank. The slope angle of the plank θ is gradually increased until the block starts to slip. Draw two graphs, both with θ along the x-axis. In one graph show the ratio of the normal force to the weight N/w as a function of θ. In the second graph, show the ratio of the friction force to the weight f/w. Indicate the region of no motion, the point where motion impends, and the region where motion exists.

2–16 A block weighing 20 N rests on a horizontal surface. The coefficient of static friction between block and surface is 0.40 and the coefficient of sliding friction is 0.20.

a) How large is the friction force exerted on the block?
b) How great will the friction force be if a horizontal force of 5 N is exerted on the block?
c) What is the minimum force which will start the block in motion?
d) What is the minimum force which will keep the block in motion once it has been started?
e) If the horizontal force is 10 N, how great is the friction force?

2–17 A block is pulled to the right at constant velocity by a 10-N force acting 30° above the horizontal. The coefficient of sliding friction between the block and the surface is 0.5. What is the weight of the block?

2–18 A block weighing 14 N is placed on an inclined plane and connected to a 10-N block by a cord passing over a small frictionless pulley, as in Fig. 2–9. The coefficient of sliding friction between the block and the plane is (1/7). For what two values of θ will the system move with constant velocity? [Hint: $\cos\theta = \sqrt{1 - \sin^2\theta}$.]

2–19 A block weighing 100 N is placed on an inclined plane of slope angle 30° and is connected to a second hanging block of weight w by a cord passing over a small frictionless pulley, as in Fig. 2–9. The coefficient of static friction is 0.40 and the coefficient of sliding friction is 0.30.

a) Find the weight w for which the 100-N block moves up the plane at constant speed.
b) Find the weight w for which it moves down the plane at constant speed.
c) For what range of values of w will the block remain at rest?

2–20 What force P at an angle ϕ above the horizontal is needed to drag a box of weight w at constant speed along a level floor if we are given that the coefficient of sliding friction between box and floor is μ?

2–21 A box of weight W is pushed at constant speed up a plane inclined at angle ϕ by a horizontal force of magnitude P. If the coefficient of kinetic friction is μ, derive an expression for P in terms of the other quantities. Show that, if ϕ exceeds a certain critical value, the box cannot be pushed up, no matter how large P is. Express the critical angle in terms of μ.

2–22 A block of wood of weight W is pushed at constant speed along the horizontal surface of a circular saw table by a stick which exerts a force P at an angle ϕ downward from the horizontal. If the coefficient of kinetic friction is μ, find the force P required, in terms of the other quantities. Show that if ϕ exceeds a certain critical value, the block cannot be pushed, no matter how great P is. Express the critical angle in terms of μ.

2–23 A rope is stretched across a glacial crevasse 20 m wide. At the center of the rope sits a mountaineer weighing 800 N (about 180 lb). The center of the rope is 4 m lower than the ends. What is the tension in the rope?

2–24 A concrete bucket weighing 2000 lb is suspended from a crane by a cable 50 ft long. What horizontal force is required to push the bucket to a position 5 ft to the side of the straight-down position?

2–25 A safe weighing 2000 N is to be lowered at constant speed down skids 4 m long, from a truck 2 m high.

a) If the coefficient of sliding friction between safe and skids is 0.30, will the safe need to be pulled down or held back?
b) How great a force parallel to the skids is needed?

2–26

a) If a force of 86 N parallel to the surface of a 20° inclined plane will push a 120-N block up the plane at constant speed, what force parallel to the plane will push it down at constant speed?
b) What is the coefficient of sliding friction?

2–27 Block A in Fig. 2–20 weighs 4 N and block B weighs 8 N. The coefficient of sliding friction between all surfaces is 0.25. Find the force $\boldsymbol{P}$ necessary to drag block B to the left at constant speed (a) if A rests on B and moves with it, (b) if A is held at rest, and (c) if A and B are connected by a light flexible cord passing around a fixed frictionless pulley.

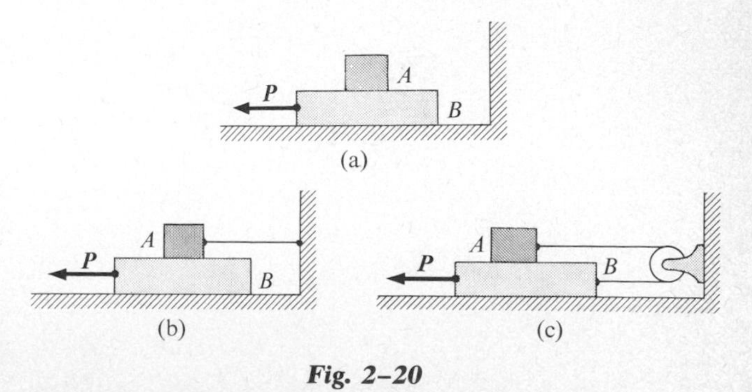

Fig. 2–20

2–28 Block A, of weight w, slides down an inclined plane S of slope angle 37° at constant velocity while the plank B, also of weight w, rests on top of A. The plank is attached by a cord to the top of the plane (Fig. 2–21).

a) Draw a diagram of all the forces acting on block A.

b) If the coefficient of kinetic friction is the same between the surfaces A and B and between S and A, determine its value.

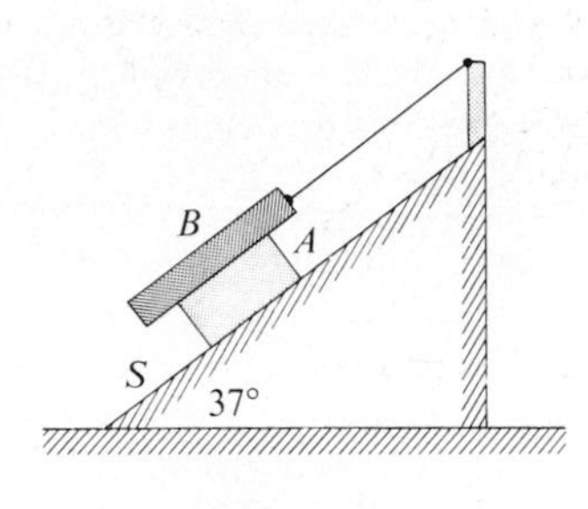

Fig. 2–21

2–29 Two blocks, A and B, are placed as in Fig. 2–22 and connected by ropes to block C. Both A and B weigh 20 N and the coefficient of sliding friction between each block and the surface is 0.5. Block C descends with constant velocity.

a) Draw two separate force diagrams showing the forces acting on A and B.

b) Find the tension in the rope connecting blocks A and B.

c) What is the weight of block C?

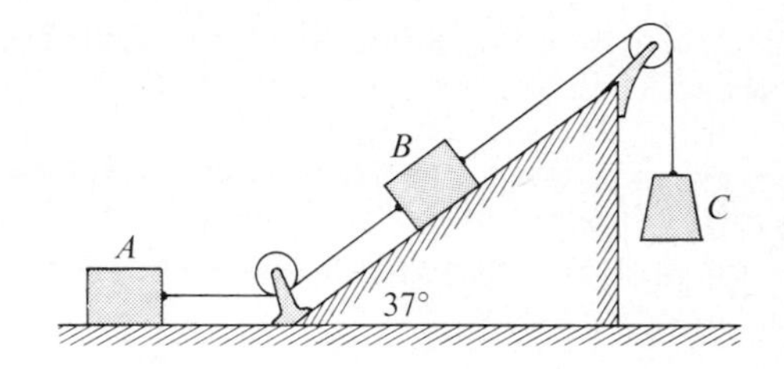

Fig. 2–22

Chapter 3

Equilibrium. Moment of a Force

3–1 MOMENT OF A FORCE

The effect produced on a body by a force of given magnitude and direction depends on the position of the *line of action* of the force. Thus in Fig. 3–1 the force $\boldsymbol{F}_1$ would produce a counterclockwise rotation (together with a translation toward the right), while $\boldsymbol{F}_2$ would produce clockwise rotation.

The line of action of a force can be specified by giving the perpendicular distance from some reference point to the line of action. In many instances, we shall be studying the motion of a body which is free to rotate about some axis, and which is acted on by a number of coplanar forces all lying in a plane perpendicular to the axis. It is then most convenient to select as the reference point the point at which the axis intersects the plane of the forces. The perpendicular distance from this point to the line of action of a force is called the *force arm* or the *moment arm* of the force about the axis. The product of the magnitude of a force and its force arm is called the *moment* of the force about the axis, or the *torque.*

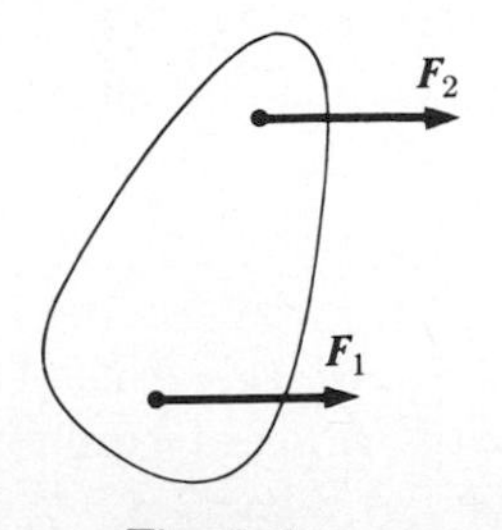

Fig. 3–1

Thus Fig. 3–2 is a top view of a flat object, pivoted about an axis perpendicular to the plane of the diagram and passing through point O. The body is acted on by the forces $\boldsymbol{F}_1$ and $\boldsymbol{F}_2$, lying in the plane of the diagram. The moment arm of $\boldsymbol{F}_1$ is the perpendicular distance OA, of length ℓ_1, and the moment arm of $\boldsymbol{F}_2$ is the perpendicular distance OB of length ℓ_2.

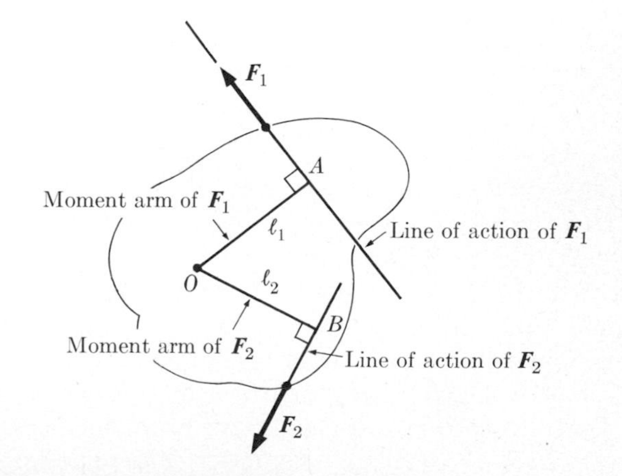

Fig. 3–2 The moment of a force about an axis is the product of the force and its moment arm.

The effect of the force $\boldsymbol{F}_1$ is to produce counterclockwise rotation about the axis, while that of $\boldsymbol{F}_2$ is to produce clockwise rotation. To distinguish between these directions of rotation, we shall adopt the convention that counterclockwise moments are positive, and that clockwise moments are negative. Hence the moment Γ_1 (Greek "gamma") of the force $\boldsymbol{F}_1$ about the axis through O is

$$\Gamma_1 = +F_1\ell_1,$$

and the moment Γ_2 of F_2 is

$$\Gamma_2 = -F_2\ell_2.$$

In the mks system, where the unit of force is the newton and the unit of length the meter, the unit of torque is the newton-meter. If forces are expressed in pounds and lengths in feet, torques are expressed in pound-feet.

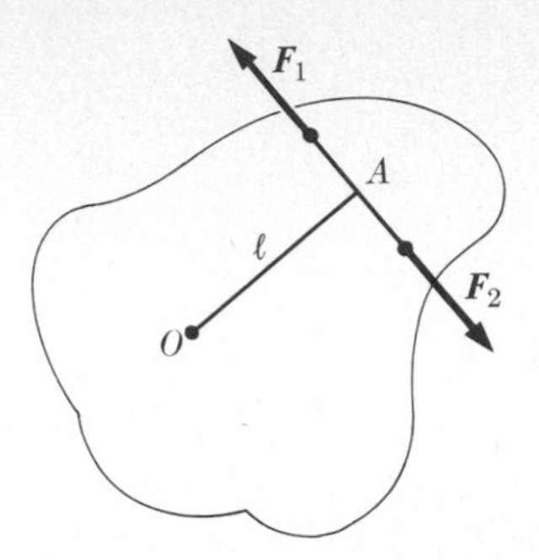

Fig. 3–3 When a body is in equilibrium, the resultant moment of the forces about any axis is zero.

3–2 THE SECOND CONDITION OF EQUILIBRIUM

We saw in Section 2–2 that when a body is in equilibrium under the action of several coplanar forces, the vector sum of the forces must be zero. In addition, if there are only two forces they must have the same line of action, and if there are three they must be *concurrent* (unless their lines of action are all parallel). The first of these requirements is satisfied by the *first condition for equilibrium*,

$$\sum F_x = 0, \qquad \sum F_y = 0.$$

The second requirement can be simply expressed in terms of the moments of the forces. Figure 3–3 again shows a flat object acted on by two forces $\boldsymbol{F}_1$ and $\boldsymbol{F}_2$. If the object is in equilibrium, the magnitudes of $\boldsymbol{F}_1$ and $\boldsymbol{F}_2$ are equal and both forces have the same line of action. Hence they have the same moment arm OA, of length ℓ, about an axis perpendicular to the plane of the body and passing through any arbitrary point O. Their moments about the axis are therefore equal in magnitude and opposite in sign, and the algebraic sum of their moments is zero.

In the case of three forces, we can replace any two by their vector sum, acting at the point of intersection of their lines of action. By the first condition, the third force must be equal and opposite, and we see that the second condition is satisfied only when it has the same line of action as the sum of the first two, in which case its moment must be the negative of that of the sum.

When a body is in equilibrium under the action of four or more forces, the lines of action need not all intersect at the same point, but the second condition of equilibrium can still be stated in terms of moments:

$$\sum \Gamma = 0 \qquad \text{(about any arbitrary axis).}$$

Although the choice of axis is arbitrary, one must, of course, use the *same* axis for *all* the moments.

Example 1 A rigid rod whose own weight is negligible (Fig. 3–4) is pivoted at point O and carries a body of weight w_1 at end A. Find the weight w_2 of a second body which must be attached at end B if the rod is to be in equilibrium, and find the force exerted on the rod by the pivot at O.

Figure 3–4(b) is the free-body diagram of the rod. The forces T_1 and T_2 are equal respectively to w_1 and w_2. The conditions of equilibrium, taking moments about an axis through O, perpendicular to the

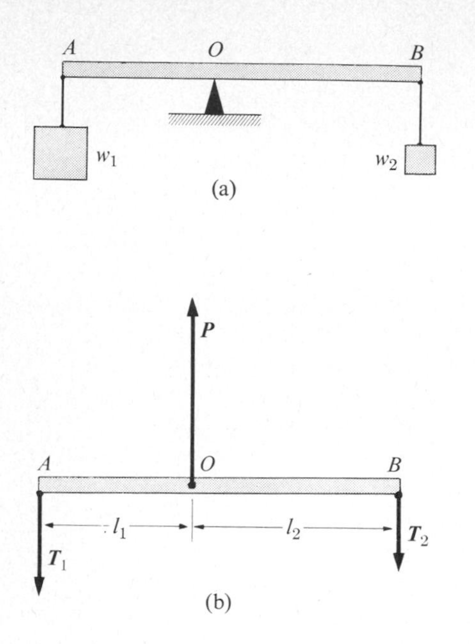

Fig. 3–4 *A rod in equilibrium under three parallel forces.*

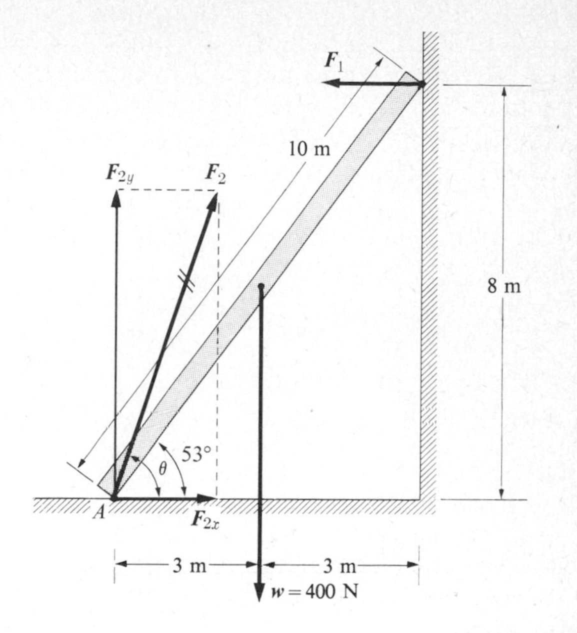

Fig. 3–5 *Forces on a ladder in equilibrium, leaning against a frictionless wall.*

diagram, give

$$\sum F_y = P - T_1 - T_2 = 0, \qquad \text{(First condition)}$$
$$\sum \Gamma_o = T_1\ell_1 - T_2\ell_2 = 0. \qquad \text{(Second condition)}$$

Let $\ell_1 = 1.2$ m, $\ell_2 = 1.6$ m, $w_1 = 20$ N. Then, from the equations above,

$$P = 35\text{ N}, \qquad T_2 = w_2 = 15\text{ N}.$$

To illustrate that the resultant moment about *any* axis is zero, let us compute moments about an axis through point A:

$$\begin{aligned}\sum \Gamma_A &= P\ell_1 - T_2(\ell_1 + \ell_2)\\ &= (35\text{ N})(1.2\text{ m}) - (15\text{ N})(2.8\text{ m}) = 0.\end{aligned}$$

The point about which moments are computed need not lie on the rod. To verify this, let the reader calculate the resultant moment about a point 0.4 m to the left of A and 0.4 m above it.

Example 2 In Fig. 3–5 a ladder 10 m long, of weight 400 N, with its center of gravity at its center, leans in equilibrium against a vertical frictionless wall and makes an angle of 53° with the horizontal (forming a 3–4–5 right triangle). We wish to find the magnitudes and directions of the forces $\boldsymbol{F}_1$ and $\boldsymbol{F}_2$.

If the wall is frictionless, $\boldsymbol{F}_1$ is horizontal. The direction of $\boldsymbol{F}_2$ is unknown (except in special cases, its direction does *not* lie along the ladder). Instead of considering its magnitude and direction as unknowns, it is simpler to resolve the force $\boldsymbol{F}_2$ into (unknown) x- and y-components and solve for these. The magnitude and direction of $\boldsymbol{F}_2$ may then be computed. The first condition of equilibrium therefore provides the equations

$$\left.\begin{aligned}\sum F_x &= F_2\cos\theta - F_1 = 0,\\ \sum F_y &= F_2\sin\theta - 400\text{ N} = 0.\end{aligned}\right\} \qquad \text{(First condition)}$$

In writing the second condition, moments may be computed about an axis through any point. The resulting equation is simplest if one selects a point through which two or more forces pass, since these forces then do not appear in the equation. Let us therefore take moments about an axis through point A.

$$\sum \Gamma_A = F_1(8 \text{ m}) - (400 \text{ N})(3 \text{ m}) = 0. \qquad \text{(Second condition)}$$

From the second equation, $F_2 \sin\theta = 400$ N, and from the third,

$$F_1 = \frac{1200 \text{ N-m}}{8 \text{ m}} = 150 \text{ N}.$$

Then from the first equation,

$$F_2 \cos\theta = 150 \text{ N}.$$

Hence

$$\boldsymbol{F}_2 = \sqrt{(400 \text{ N})^2 + (150 \text{ N})^2} = 427.2 \text{ N}.$$

$$\theta = \tan^{-1}\frac{400 \text{ N}}{150 \text{ N}} = 69.5°.$$

(Note that this problem has already been solved by a different method in Example 4 in Section 2–5.)

Example 3 Figure 3–6(a) shows a human arm lifting a dumbbell, and Fig. 3–6(b) shows a free-body diagram for the forearm, showing the forces involved. The forearm is in equilibrium under the action of the weight w, the tension T in the tendon connected to the biceps muscle, and the forces exerted on it by the upper arm, through the elbow joint. We want to find the tendon tension and the components of force at the elbow, three unknown scalar quantities in all.

First we represent the tendon force in terms of its components T_x and T_y, using the given angle θ and the unknown magnitude T:

$$T_x = T\cos\theta, \qquad T_y = T\sin\theta.$$

Next we note that if moments about the elbow joint are taken, the resulting moment equation does not contain the unknowns E_x and E_y, since their lines of action pass through this point, as also does T_x. The moment equation is then simply

$$\ell w - dT_y = 0.$$

From this we immediately find

$$T_y = \frac{\ell w}{d} \quad \text{and} \quad T = \frac{\ell w}{d\sin\theta}.$$

To find E_x and E_y we could now use the first conditions for equilibrium, $\sum F_x = 0$ and $\sum F_y = 0$. Instead, for added practice in using moments, we take moments about the point A where the tendon is attached:

$$(\ell - d)w + dE_y = 0 \quad \text{and} \quad E_y = -\frac{(\ell - d)w}{d}.$$

The negative sign of this result shows that our initial guess for the direction of E_y was wrong; it is actually vertically *downward.*

Finally, we take moments about point B in the figure:

$$\ell w - hE_x = 0 \quad \text{and} \quad E_x = \frac{\ell w}{h}.$$

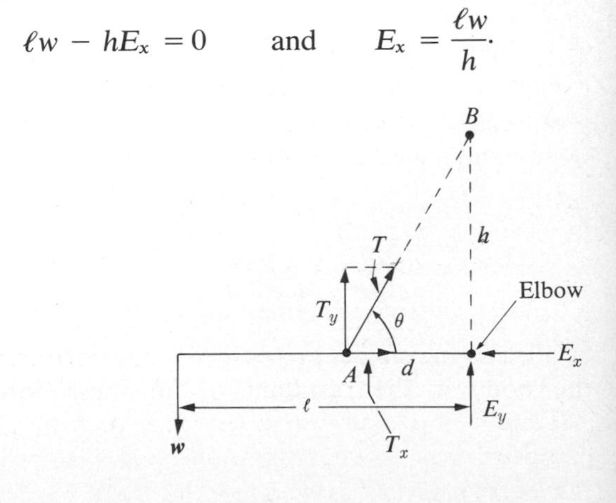

Fig. 3–6

(a)

(b)

We note that each stage of this calculation is simplified by choosing the point for calculating moments so as to eliminate one or more of the unknown quantities. In the last step, the force T has no moment about point B; thus when the moments of T_x and T_y are computed separately, they must add to zero. The reader is invited to verify this statement in detail.

As a specific example, suppose $w = 50\text{ N}$, $d = 0.1\text{ m}$, $\ell = 0.5\text{ m}$, and $\theta = 80°$. Since $\tan\theta = h/d$, we find

$$h = d\tan\theta = (0.1\text{ m})(5.67) = 0.567\text{ m}.$$

From the previous general results, we find

$$T = \frac{(0.5\text{ m})(50\text{ N})}{(0.1\text{ m})(0.985)} = 254\text{ N},$$

$$E_y = \frac{(0.5\text{ m} - 0.1\text{ m})(50\text{ N})}{0.1\text{ m}} = 200\text{ N},$$

$$E_x = \frac{(0.5\text{ m})(50\text{ N})}{0.567\text{ m}} = 44\text{ N}.$$

The magnitude of the force at the elbow is

$$E = \sqrt{E_x^2 + E_y^2} = 204\text{ N}.$$

As mentioned above, we have not explicitly used the first condition for equilibrium, that the vector sum of the forces be zero. As a check, the reader should verify that this condition *is* satisfied by the above results.

Units of meters have been used in the above example; centimeters could have been used. In fact, *any* unit of distance may be used in a moment equation, provided the *same* unit is used throughout a given equation.

3–3 CENTER OF GRAVITY

Every particle of matter in a body is attracted by the earth, and the single force which we call the *weight* of the body is the resultant of all these forces of attraction. The direction of the force on each particle is toward the center of the earth, but the distance to the earth's center is so great that for all practical purposes the forces can be considered parallel to one another. Hence the weight of a body is the resultant of a large number of parallel forces.

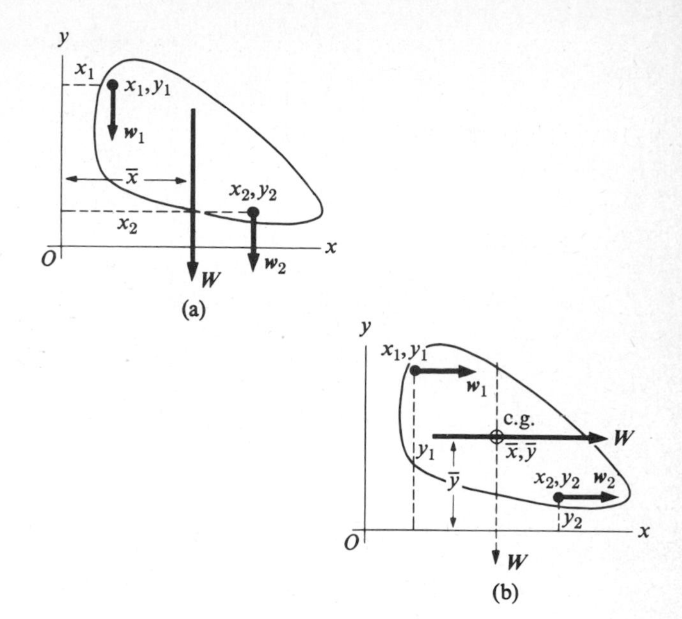

Fig. 3–7 *The body's weight W is the resultant of a large number of parallel forces. The line of action of* **W** *always passes through the center of gravity.*

Figure 3–7(a) shows a flat object of arbitrary shape in the xy-plane, the y-axis being vertical. Let the body be subdivided into a large number of small particles of weights w_1, w_2, etc., and let the coordinates of these particles be x_1 and y_1, x_2 and y_2, etc. The total weight W of the object is

$$W = w_1 + w_2 + \cdots = \sum w. \qquad (3\text{–}1)$$

Each particle's weight also contributes to the total torque acting on the body. Computing moments about point O, we see that the torque associated with particle 1 is w_1x_1, that for particle 2 is w_2x_2, and so on, and that the total torque is

$$w_1x_1 + w_2x_2 + \cdots = \sum wx.$$

For convenience in the present discussion we are taking clockwise torques to be positive, the opposite choice from that used in preceding sections.

Along what line must the total weight W act in order for the total torque to equal the above expres-

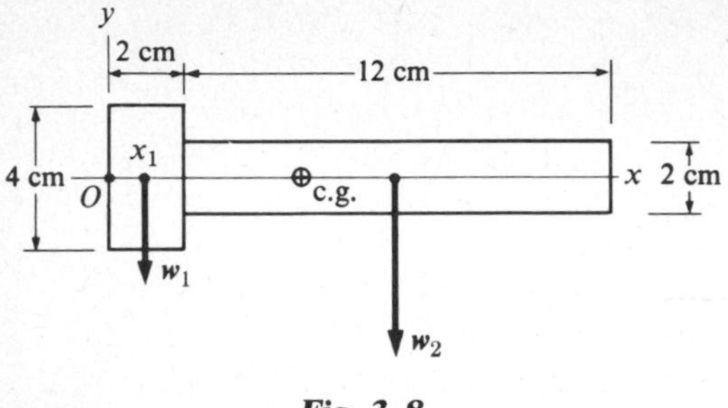

Fig. 3-8

sion? Suppose it acts along a line a distance $\bar{x}$ to the right of the origin. Then it must be true that $w_1x_1 + w_2x_2 + \cdots = W\bar{x}$, or

$$\bar{x} = \frac{w_1x_1 + w_2x_2 + \cdots}{W} = \frac{\sum wx}{W} = \frac{\sum wx}{\sum w}. \tag{3-2}$$

That is, the torque due to the weight of the object can always be obtained by assuming the total weight to be acting along a line determined by Eq. (3-2). Stated another way, in computing the torque due to the weight of an object, we must consider the weight as acting at a point whose x-coordinate $\bar{x}$ is given by Eq. (3-2). (See Fig. 3-7(a).)

Now let the object and the reference axes be rotated 90° clockwise or, which amounts to the same thing, let us consider the gravitational forces to be rotated 90° counterclockwise, as in Fig. 3-7(b). The total weight W is unaltered, but now it must act along a line a distance $\bar{y}$ above the origin, such that $W\bar{y} = \sum wy$, or

$$\bar{y} = \frac{w_1y_1 + w_2y_2 + \cdots}{w_1 + w_2 + \cdots} = \frac{\sum wy}{\sum w} = \frac{\sum wy}{W}. \tag{3-3}$$

The point of intersection of the lines of action of $\boldsymbol{W}$ in the two parts of Fig. 3-7 has the coordinates $\bar{x}$ and $\bar{y}$ and is called the *center of gravity* of the object. By considering some arbitrary orientation of the object, one can show that the line of action of $\boldsymbol{W}$ always passes through the center of gravity, and thus that the torque of $\boldsymbol{W}$ can always be obtained correctly by taking $\boldsymbol{W}$ to act at the center of gravity.

If the centers of gravity of each of a number of bodies have been determined, the coordinates of the center of gravity of the combination can be computed from Eqs. (3-2) and (3-3), letting w_1, w_2, etc., be the weights of the bodies and x_1 and y_1, x_2 and y_2, etc., be the coordinates of the center of gravity of each.

Symmetry considerations are often useful in finding the position of the center of gravity. Thus the center of gravity of a homogeneous sphere, cube, circular disk or rectangular plate is at its center. That of a cylinder or right circular cone is on the axis of symmetry, and so on.

Example 1 Locate the center of gravity of the machine part in Fig. 3-8, consisting of a disk 4 cm in diameter and 2 cm long, and a rod 2 cm in diameter and 12 cm long, constructed of a homogeneous material.

By symmetry, the center of gravity lies on the axis and the center of gravity of each part is midway between its ends. The volume of the disk is 8π cm^3 and that of the rod is 12π m^3. Since the weights of the two parts are proportional to their volumes,

$$\frac{w(\text{disk})}{w(\text{rod})} = \frac{w_1}{w_2} = \frac{8\pi}{12\pi} = \frac{2}{3}.$$

Take the origin O at the left face of the disk, on the axis. Then

$$x_1 = 1 \text{ cm}, \qquad x_2 = 8 \text{ cm},$$

and

$$\bar{x} = \frac{w_1(1 \text{ cm}) + \frac{3}{2}w_1(8 \text{ cm})}{w_1 + \frac{3}{2}w_1} = 5.2 \text{ cm}.$$

The center of gravity is on the axis, 5.2 cm to the right of O.

The center of gravity of a flat object can be located experimentally as shown in Fig. 3-9. In part (a) the body is suspended from some arbitrary point

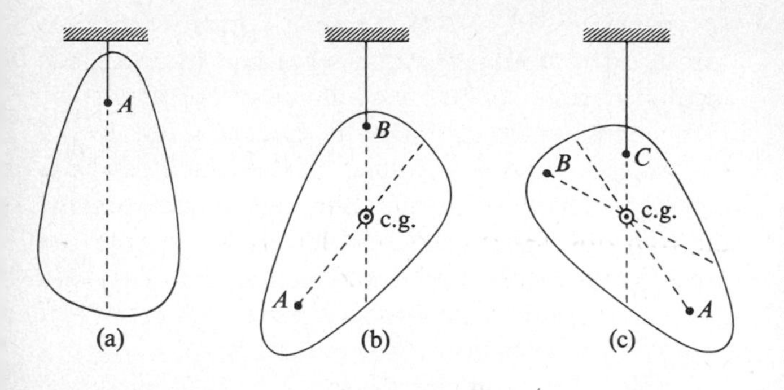

Fig. 3–9 Locating the center of gravity of a flat object.

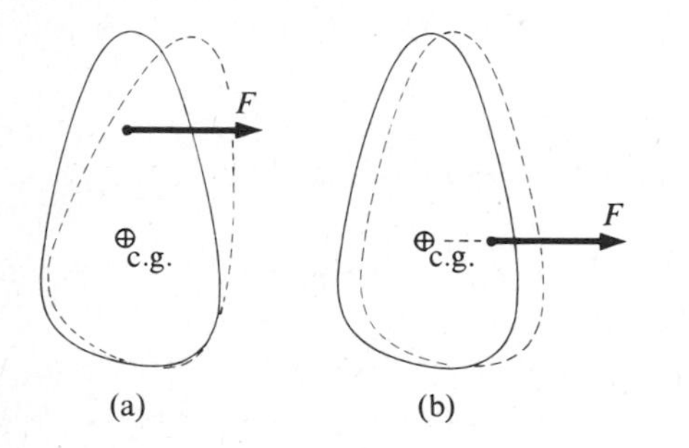

Fig. 3–10 *A body is in rotational but not translational equilibrium when acted on by a force whose line of action passes through the center of gravity, as in (b).*

A. When allowed to come to equilibrium, the center of gravity must lie on a vertical line through A. When the object is suspended from a second point B, as in part (b), the center of gravity lies on a vertical line through B and hence lies at the point of intersection of this line and the first. If the object is now suspended from a third point C, as in part (c), a vertical line through C will be found to pass through the point of intersection of the first two lines. The extent to which the three lines fail to meet at a single point gives an indication of the experimental error in this method of locating the center of gravity.

The center of gravity of a body has another important property. A force $\boldsymbol{F}$ whose line of action lies at one side or the other of the center of gravity, as in Fig. 3–10(a), will change both the translational and rotational motion of the body on which it acts. However, if the line of action passes through the center of gravity, as in part (b), only the translational motion is affected and the body remains in rotational equilibrium. Thus when a body is tossed in the air with a whirling motion, it continues to rotate at a constant rate, since the line of action of its weight passes through the center of gravity.

Example 2 Figure 3–11(a) shows an arrangement to maintain proper position of a factured femur. The thigh muscles tend to contract and pull the broken ends past each other; the tension T_3 prevents this, while T_1 and T_2 support the weight of the lower leg. It is essential that no vertical force be exerted at the

Fig. 3–11 *Traction arrangement for fractured femur. (a) Weights and pulleys support lower leg and provide tension to counteract muscle tension at fracture. There must be no sideways force at fracture. (b) Free-body diagram for lower leg, showing forces exerted by weights and pulleys.*

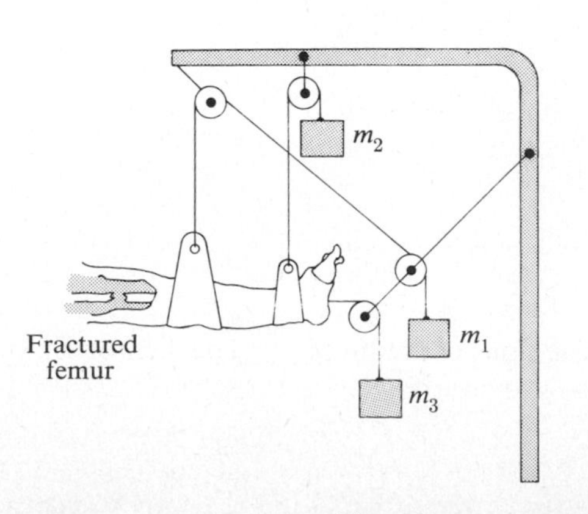

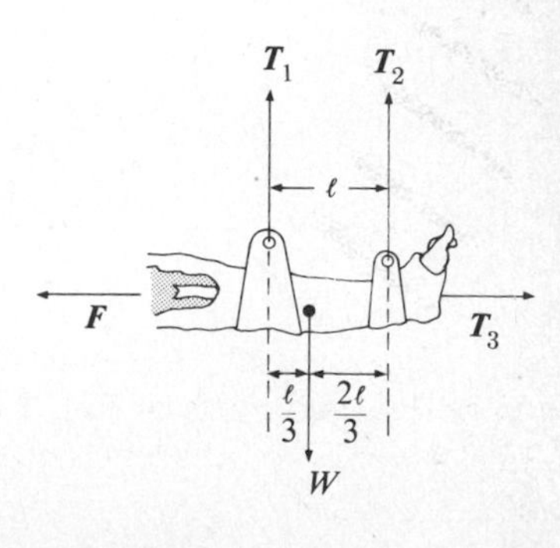

position of the fracture; otherwise the broken ends would be pulled out of position.

Figure 3–11(b) shows a free-body diagram for the lower leg and the approximate location of its center of gravity. The force F is the muscle tension at the fracture; clearly, this must equal T_3, which in practice is determined by experience and trial. The other conditions for equilibrium are

$$T_1 + T_2 = W,$$

$$\frac{T_1\ell}{3} - \frac{2T_2\ell}{3} = 0,$$

where we have taken moments about the center of gravity. Simultaneous solution gives

$$T_1 = \tfrac{2}{3}W, \qquad T_2 = \tfrac{1}{3}W.$$

3–4 COUPLES

It sometimes happens that the forces on a body reduce to two forces of equal magnitude and opposite direction, having lines of action which are parallel but which do not coincide. Such a pair of forces is called a *couple*. A common example is afforded by the forces on a compass needle in the earth's magnetic field, as shown in Fig. 3–12. The north and south poles of the needle are acted on by equal forces, one toward the north and the other toward the south. Except when the needle points in the N–S direction, the two forces do not have the same line of action.

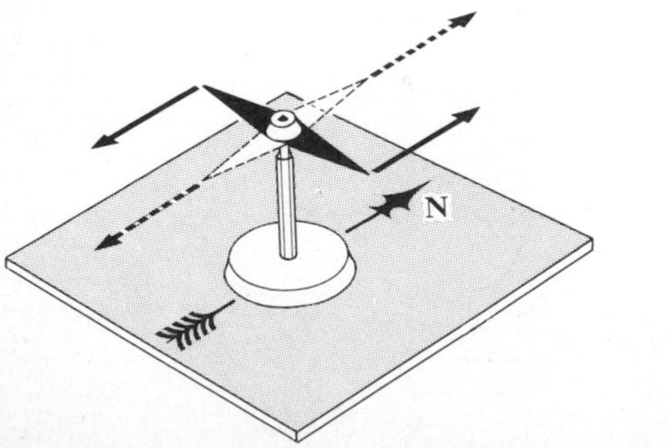

Fig. 3–12 *Forces on the poles of a compass needle.*

Figure 3–13 shows a couple consisting of two forces, each of magnitude F, separated by a perpendicular distance ℓ. The resultant R of the forces is

$$R = F - F = 0.$$

The fact that the resultant force is zero means that a couple has no effect in producing translation (as a whole) of the body on which it acts. The only effect of a couple is to produce rotation.

The resultant torque of the couple in Fig. 3–13 about an arbitrary point O is

$$\begin{aligned}\sum \Gamma_O &= x_1F - x_2F \\ &= x_1F - (x_1 + \ell)F \\ &= -\ell F.\end{aligned}$$

Since the distances x_1 and x_2 do not appear in the result, we conclude that the torque of the couple is the same about all points in the plane of the forces forming the couple and is equal to the product of the magnitude of either force and the perpendicular distance between their lines of action.

A body acted on by a couple can be kept in equilibrium only by another couple of the same moment and in the opposite direction. As an example, the ladder in Fig. 3–5 can be considered as acted on by two couples, one formed by the forces $\boldsymbol{F}_2 \sin\theta$ and $\boldsymbol{w}$, the other by the forces $\boldsymbol{F}_2 \cos\theta$ and $\boldsymbol{F}_1$. The moment of the first is

$$\Gamma_1 = (3\ \text{m})(400\ \text{N}) = 1200\ \text{N-m}.$$

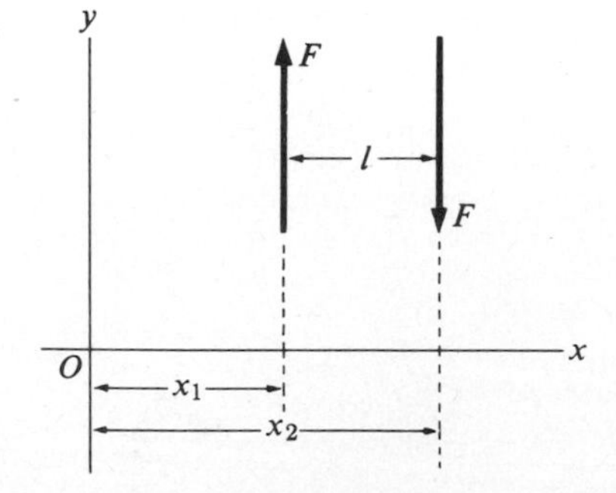

Fig. 3–13 *Two equal and opposite forces having different lines of action are called a couple. The moment of the couple is the same about all points, and is equal to ℓF.*

The moment of the second is

$$\Gamma_2 = (8\text{ m})(150\text{ N}) = 1200\text{ N-m}.$$

The first moment is clockwise and the second is counterclockwise.

PROBLEMS

Section 2–5 and the note at the beginning of the problems in Chapter 2 are equally applicable to these problems. The only difference is that now we have a third "formula," $\sum \Gamma$ (about any axis) = 0.

3–1 A force $\boldsymbol{F}$ in the xy-plane has components of magnitudes F_x and F_y, respectively. The coordinates of its point of application are (x,y). The magnitude of the moment of force about the origin is $\Gamma_O = xF_y - yF_x$. Show that this equation is correct if the point (x,y) lies in any one of the four quadrants, and also that it is correct regardless of the direction of the force $\boldsymbol{F}$.

3–2 You are given (a) a meter stick through which a number of holes have been bored so that its center of gravity is not at its midpoint, (b) a knife point, on which the meter stick can be pivoted, (c) a body whose weight is known to be $\boldsymbol{w}$, and (d) a spool of thread. Using this equipment only, explain, with the aid of a diagram, how you would determine the weight of the meter stick.

3–3 A uniform plank 15 m long, weighing 400 N, rests symmetrically on two supports 8 m apart, as shown in Fig. 3–14. A boy weighing 640 N starts at point A and walks toward the right.

a) Construct in the same diagram two graphs showing the upward forces $\boldsymbol{F}_A$ and $\boldsymbol{F}_B$ exerted on the plank at points A and B, as functions of the coordinate x of the boy. Let 1 cm = 100 N vertically, and 1 cm = 1 m horizontally.

b) Find from your diagram how far beyond point B the boy can walk before the plank tips.

Fig. 3–14

c) How far from the right end of the plank should support B be placed in order that the boy walk just to the end of the plank without causing it to tip?

3–4 The strut in Fig. 3–15 weighs 200 N and its center of gravity is at its center. Find (a) the tension in the cable and (b) the horizontal and vertical components of the force exerted on the strut at the wall.

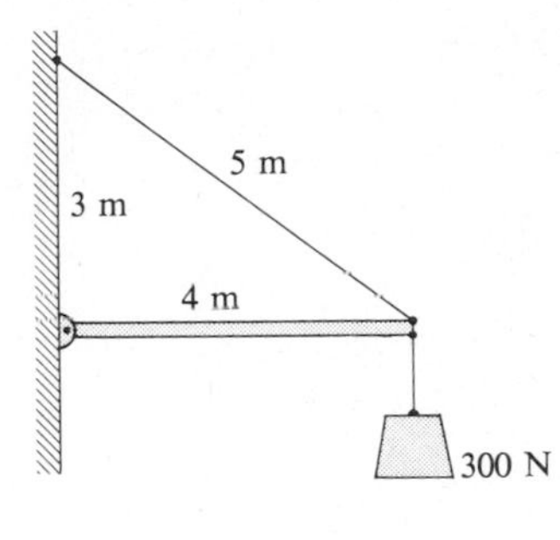

Fig. 3–15

3–5 Find the tension in the cable BD in Fig. 3–16, and the horizontal and vertical components of the force exerted on the strut AB at pin A, using:

a) the first and second conditions of equilibrium ($\sum F_x = 0$, $\sum F_y = 0$, $\sum \Gamma = 0$), taking moments about an axis through point A perpendicular to the plane of the diagram;

b) the second condition of equilibrium only, taking moments first about an axis through A, then about an axis through B, and finally about an axis through D. The weight of the strut can be neglected.

c) Represent the computed forces by vectors in a diagram drawn to scale, and show that the lines of action of the forces exerted on the strut at points A, B, and C intersect at a common point.

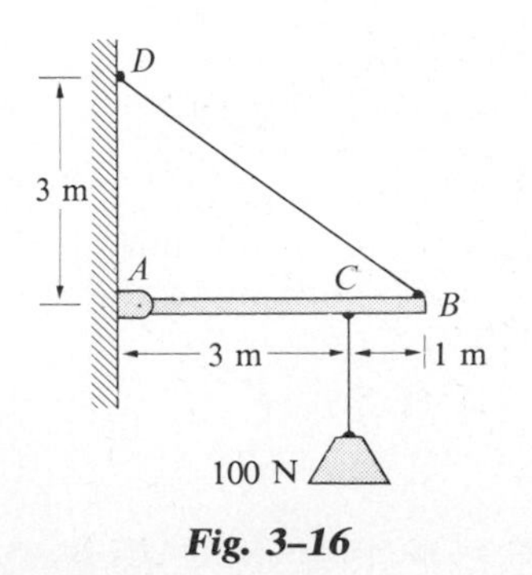

Fig. 3–16

3–6 End A of the bar AB in Fig. 3–17 rests on a frictionless horizontal surface, while end B is hinged. A horizontal force $\boldsymbol{P}$ of 60 N is exerted on end A. Neglect the weight of the bar. What are the horizontal and vertical components of the force exerted by the bar on the hinge at B?

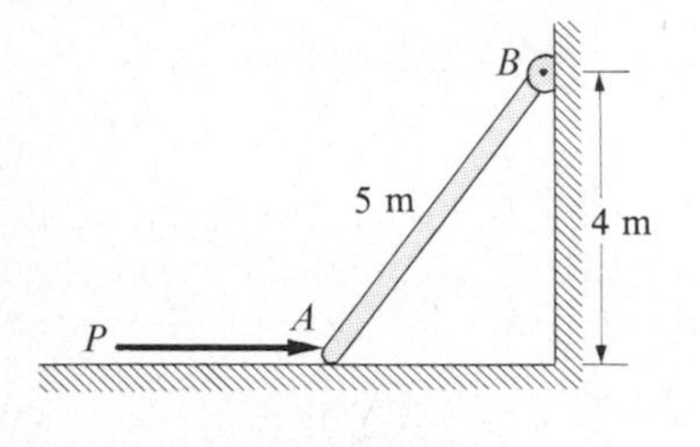

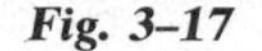
Fig. 3–17

3–7 A single force is to be applied to the bar in Fig. 3–18 to maintain it in equilibrium in the position shown. The weight of the bar can be neglected.

a) What are the x- and y-components of the required force?

b) What is the tangent of the angle which the force must make with the bar?

c) What is the magnitude of the required force?

d) Where should the force be applied?

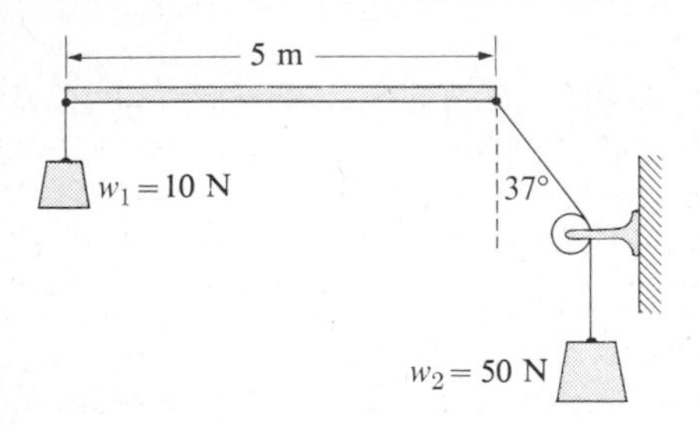

Fig. 3–18

3–8 A circular disk 0.5 m in diameter, pivoted about a horizontal axis through its center, has a cord wrapped around its rim. The cord passes over a frictionless pulley P and is attached to a body of weight 240 N. A uniform rod 2 m long is fastened to the disk, with one end at the center of the disk. The apparatus is in equilibrium, with the rod horizontal, as shown in Fig. 3–19.

a) What is the weight of the rod?

b) What is the new equilibrium direction of the rod when a second body weighing 20 N is suspended from the outer end of the rod, as shown by the dotted line?

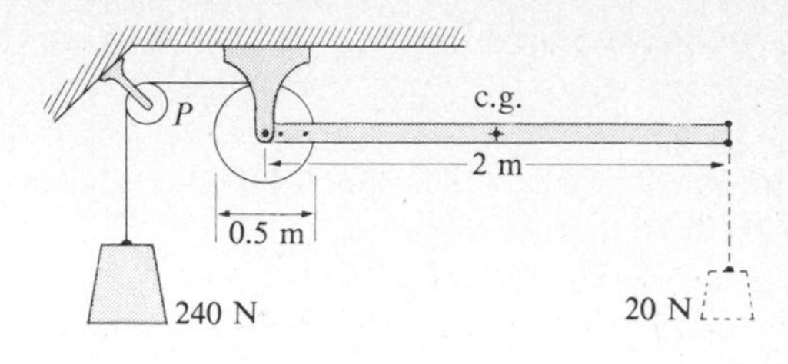

Fig. 3–19

3–9 A roller whose diameter is 1.0 m weighs 360 N. What horizontal force is necessary to pull the roller over a brick 0.1 m high when the force is applied (a) at the center, (b) at the top?

3–10 The boom in Fig. 3–20 is uniform and weighs 2500 N. (a) Find the tension in the guy wire, and the horizontal and vertical components of the force exerted on the boom at its lower end. (b) Does the line of action of this force lie along the boom?

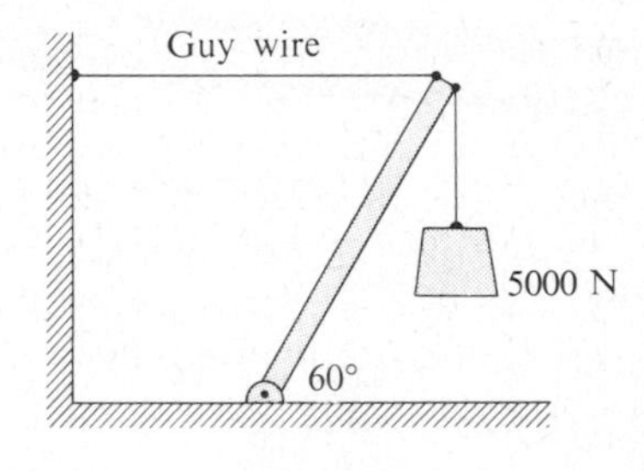

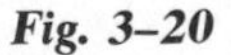
Fig. 3–20

3–11 Two ladders, 4 m and 3 m long, respectively, are hinged at point A and tied together by a horizontal rope 0.6 m above the floor, as in Fig. 3–21. The ladders weigh 400 N and 300 N respectively, and the center of gravity of each is at its center. If the floor is frictionless, find (a) the upward force at the bottom of each ladder, (b) the tension in the rope, and (c) the force which one ladder exerts on the other at point A. (d) If a load of 1000 N is now suspended from point A, find the tension in the rope.

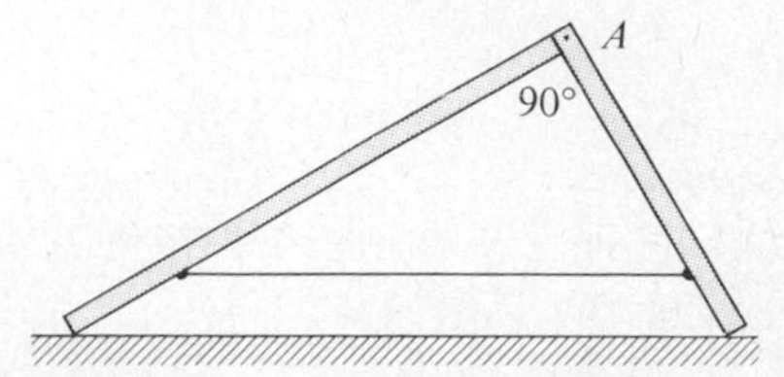

Fig. 3–21

3–12 A uniform ladder 10 m long rests against a vertical frictionless wall with its lower end 6 m from the wall. The ladder weighs 400 N. The coefficient of static friction between the foot of the ladder and the ground is 0.40. A man weighing 800 N climbs slowly up the ladder.

a) What is the maximum frictional force which the ground can exert on the ladder at its lower end?

b) What is the actual frictional force when the man has climbed 5 m along the ladder?

c) How far along the ladder can the man climb before the ladder starts to slip?

3–13 One end of a meter stick is placed against a vertical wall, as in Fig. 3–22. The other end is held by a light cord making an angle θ with the stick. The coefficient of static friction between the end of the meter stick and the wall is 0.30.

a) What is the maximum value the angle θ can have if the stick is to remain in equilibrium?

b) Let the angle θ be 10°. A body of the same weight as the meter stick is suspended from the stick as shown by dotted lines, at a distance x from the wall. What is the minimum value of x for which the stick will remain in equilibrium?

c) When $\theta = 10°$, how large must the coefficient of static friction be so that the body can be attached at the left end of the stick without causing it to slip?

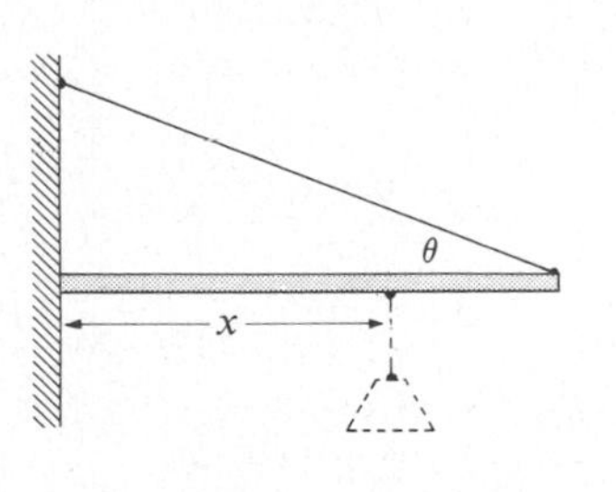

Fig. 3–22

3–14 One end of a post weighing 500 N rests on a rough horizontal surface with $\mu_s = 0.3$. The upper end is held by a rope fastened to the surface and making an angle of 37° with the post, as in Fig. 3–23. A horizontal force $\boldsymbol{F}$ is exerted on the post as shown.

a) If the force $\boldsymbol{F}$ is applied at the midpoint of the post, what is the largest value it can have without causing the post to slip?

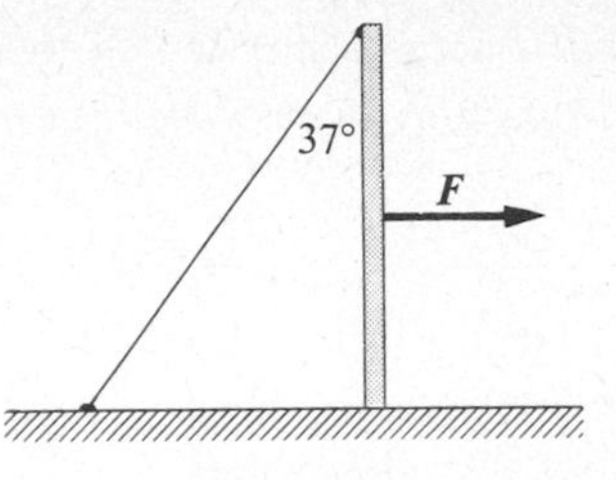

Fig. 3–23

b) How large can the force be, without causing the post to slip, if its height above the surface equals 9/10 of the length of the post?

3–15 A uniform smooth rod of length ℓ and weight w rests at equilibrium in a smooth semispherical bowl of radius R, as shown in Fig. 3–24, where $R < \ell/2 < 2R$. If θ is the angle of equilibrium and P is the force exerted by the edge of the bowl on the rod, prove that (a) $P = (\ell/4R)w$, (b) $(\cos 2\theta)/(\cos\theta) = \ell/4R$.

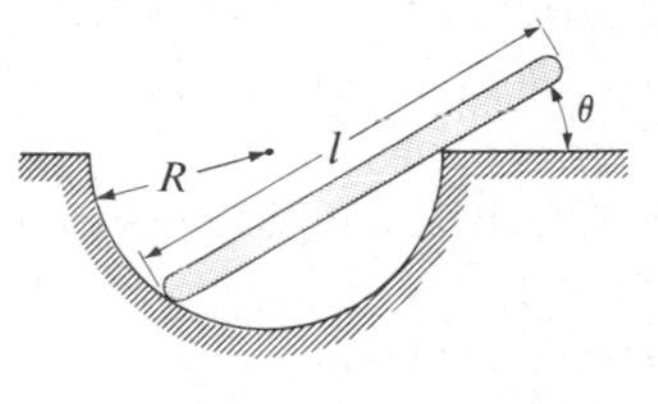

Fig. 3–24

3–16 A door 3.5 m high and 1.5 m wide is hung from hinges 3 m apart and 0.25 m from the top and bottom of the door. The door weighs 300 N, its center of gravity is at its center, and each hinge carries half the weight of the door. Find the horizontal component of the force exerted on the door at each hinge.

3–17 A gate 4 m long and 2 m high weighs 400 N. Its center of gravity is at its center, and it is hinged at A and B. To relieve the strain on the top hinge, a wire CD is connected as shown in Fig. 3–25. The tension in CD is increased until the horizontal force at hinge A is zero.

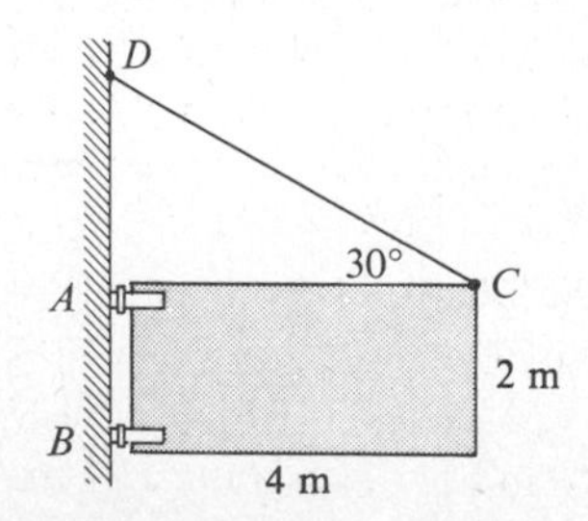

Fig. 3–25

a) What is the tension in the wire CD?

b) What is the magnitude of the horizontal component of force at hinge B?

c) What is the combined vertical force exerted by hinges A and B?

3–18 A uniform rectangular block, 0.5 m high and 0.25 m wide, rests on a plank AB, as in Fig. 3–26. The coefficient of static friction between block and plank is 0.40.

a) In a diagram drawn to scale, show the line of action of the resultant normal force exerted on the block by the plank when the angle $\theta = 15°$.

b) If end B of the plank is slowly raised, will the block start to slide down the plank before it tips over? Find the angle θ at which it starts to slide, or at which it tips.

c) What would be the answer to part (b) if the coefficient of static friction were 0.60? If it were 0.50?

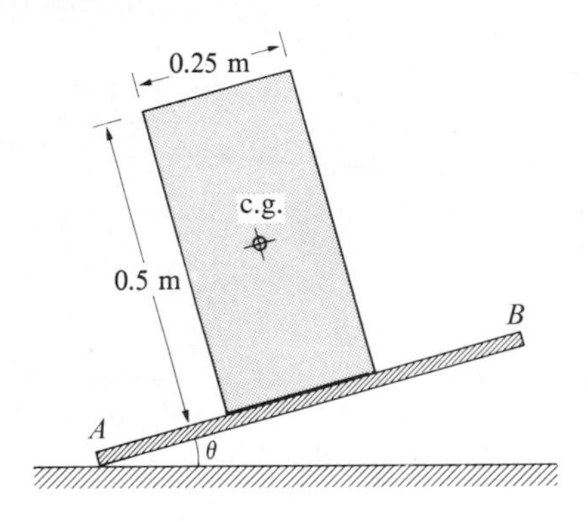

Fig. 3–26

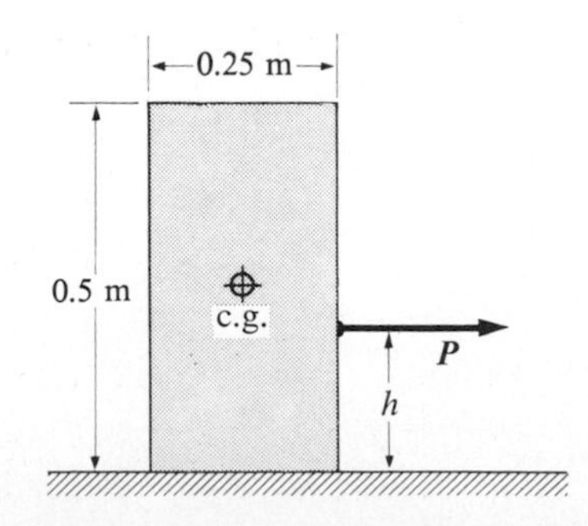

Fig. 3–27

3–19 A rectangular block 0.25 m wide and 0.5 m high is dragged to the right along a level surface at constant speed by a horizontal force $\boldsymbol{P}$, as shown in Fig. 3–27. The coefficient of sliding friction is 0.40, the block weighs 25 N, and its center of gravity is at its center.

a) Find the magnitude of the force $\boldsymbol{P}$.

b) Find the position of the line of action of the normal force $\boldsymbol{N}$ exerted on the block by the surface, if the height $h = 0.125$ m.

c) Find the value of h at which the block just starts to tip.

3–20 A garage door is mounted on an overhead rail, as in Fig. 3–28. The wheels at A and B have rusted so that they do not roll, but slide along the track. The coefficient of sliding friction is 0.5. The distance between the wheels is 2 m, and each is 0.5 m in from the vertical sides of the door. The door is symmetrical and weighs 800 N. It is pushed to the left at constant velocity by a horizontal force $\boldsymbol{P}$.

a) If the distance h is 1.5 m, what is the vertical component of the force exerted on each wheel by the track?

b) Find the maximum value h can have without causing one wheel to leave the track.

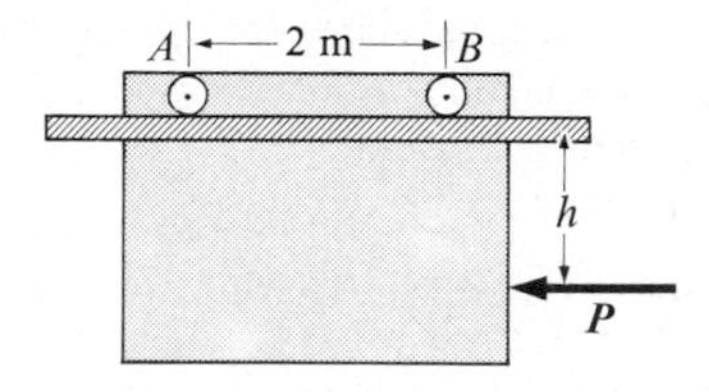

Fig. 3–28

3–21 The objects in Fig. 3–29 are constructed of wire bent into the shapes shown. Find the position of the center of gravity of each.

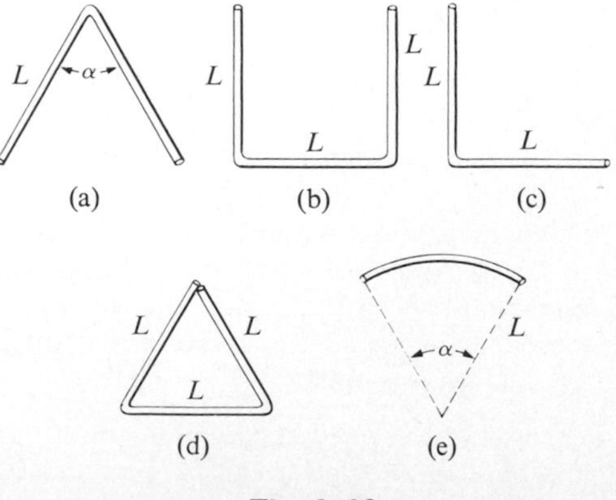

Fig. 3–29

3–22 A certain automobile has a wheelbase (distance between front and rear axles) of 120 inches. If 60 percent of the weight rests on the front wheels, how far behind the front wheels is the center of gravity?

3–23 Two men are carrying a ladder 20 ft long, weighing 200 lb. If one man can lift a maximum of 80 lb and lifts at one end, at what point should the other man lift?

3–24 A station wagon weighing 4000 lb has a wheelbase (distance between front and rear axles) of 120 inches. Ordinarily 2200 lb rests on the front wheels and 1800 lb on the rear wheels. A box weighing 500 lb is now placed on the tailgate, 4 ft behind the rear axle. How much total weight now rests on the front wheels? On the rear wheels?

3–25 A door 1.0 m wide and 2.5 m high is supported by two hinges, one 0.5 m from the top and the other 0.5 m from the bottom. Each hinge supports half the total weight of the door, which is 200 N. Assuming the door's center of gravity is at its center, find (a) the components of force exerted on the door by each hinge; (b) the magnitude and direction of the force exerted by each hinge.

3–26 Figure 3–30 is a top view of a triangular flat plate *ABC* resting on a horizontal frictionless surface. The length of side *AB* is 20 cm and that of *CB* is 15 cm. The plate is acted on by a couple consisting of the horizontal forces $\mathbf{F}_1$ and $\mathbf{F}'_1$, each of magnitude 150 units.

a) In part (a) of the figure, the plate is kept in equilibrium by a second couple consisting of the forces $\mathbf{F}_2$ and $\mathbf{F}'_2$. Find the magnitude of each if the length ab = 10 cm.

b) Show that the plate is still in equilibrium (that is, the first and second conditions of equilibrium are satisfied) if the couple formed by $\mathbf{F}_2$ and $\mathbf{F}'_2$ is displaced to the right, as in part (b).

c) Can equilibrum be maintained by the couple formed by the forces $\mathbf{F}_3$ and $\mathbf{F}'_3$ in part (c) of the diagram? If so, find the magnitude of each.

d) Show in a diagram how the plate could be kept in equilibrium by a couple consisting of two forces perpendicular to side *AC* and applied at corners *A* and *C*. Find the magnitude of the required forces.

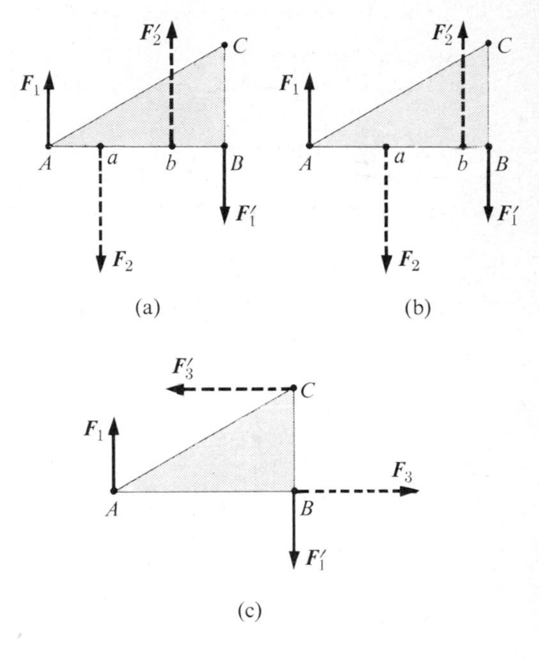

Fig. 3–30

3–27 Where is the center of gravity of the system made up of the earth and the moon?

Chapter 4

Rectilinear Motion

4–1 MOTION

Mechanics deals with the relations of force, matter, and motion. The preceding chapters have been concerned with forces, and we are now ready to discuss the mathematical methods of describing motion. This branch of mechanics is called *kinematics.*

Motion may be defined as a continuous change of position. In most actual motions, different points in a body move along different paths. The complete motion is known if we know how each point in the body moves, so to begin we consider only a moving point, or a very small body called a *particle.*

The position of a particle is conveniently specified by its projections onto the three axes of a rectangular coordinate system. As the particle moves along any path in space, its projections move in straight lines along the three axes. The actual motion can be reconstructed from the motions of these three projections, so we shall begin by discussing the motion of a single particle along a straight line, or *rectilinear motion.*

4–2 AVERAGE VELOCITY

Consider a particle moving along the x-axis, as in Fig. 4–1(a). The curve in Fig. 4–1(b) is a graph of its coordinate x plotted as a function of time t. At a time t_1 the particle is at point P in Fig. 4–1(a), where its coordinate is x_1, and at a later time t_2 it is at point Q, whose coordinate is x_2. The corresponding points on the coordinate-time graph in part (b) are lettered p and q.

The *displacement* of a particle as it moves from one point of its path to another is defined as the vector $\Delta \boldsymbol{x}$ drawn from the first point to the second.* Thus in Fig. 4–1(a) the vector PQ, of magnitude $x_2 - x_1 = \Delta x$, is the displacement. The *average velocity* of the particle is defined as the ratio of the displacement to the time interval $t_2 - t_1 = \Delta t$. We shall represent average velocity by the symbol $\bar{\boldsymbol{v}}$ (the bar signifying an average value):

$$\bar{\boldsymbol{v}} = \frac{\Delta \boldsymbol{x}}{\Delta t}.$$

Average velocity is a vector, since the ratio of a vector to a scalar is itself a vector. Its direction is the same as that of the displacement vector. The magni-

* The Greek letter Δ is often used together with another symbol to represent a *change* in the quantity represented by that symbol. Thus Δt, read as a symbol, means "change in t." The elapsed time interval between time t_1 and t_2 is often written as $t_2 - t_1 = \Delta t$, and similarly $\Delta \boldsymbol{x}$ is the change in the vector $\boldsymbol{x}$ drawn from the origin to the position of the particle.

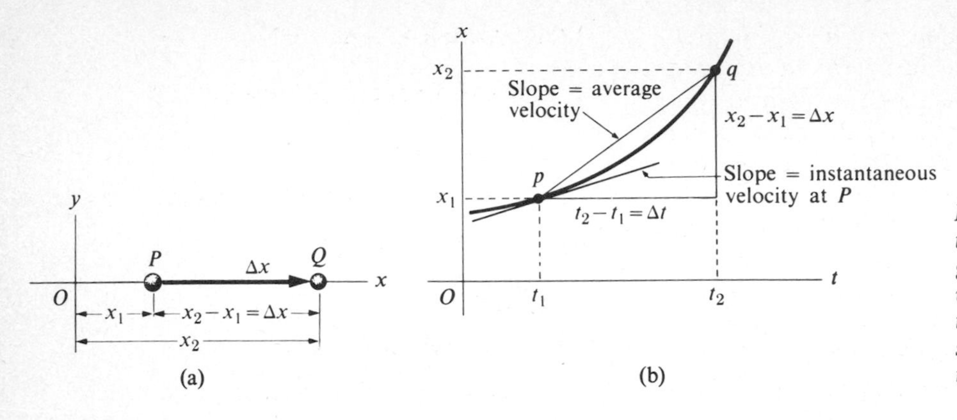

***Fig. 4–1** (a) Particle moving on the x-axis. (b) Coordinate-time graph of the motion. The average velocity between t_1 and t_2 equals the slope of the chord pq. The instantaneous velocity at p equals the slope of the tangent at p.*

tude of the average velocity is therefore

$$\bar{v} = \frac{x_2 - x_1}{t_2 - t_1} = \frac{\Delta x}{\Delta t}. \tag{4–1}$$

In Fig. 4–1(b), the average velocity is represented by the slope of the chord pq; that is, the ratio of vertical to horizontal intervals on the triangle of which pq is the hypotenuse, evaluated using the scales to which x and t are plotted. The slope is the ratio of the "rise," $x_2 - x_1$ or Δx, to the "run," $t_2 - t_1$ or Δt.

Equation (4–1) can be written

$$x_2 - x_1 = \bar{v}(t_2 - t_1). \tag{4–2}$$

Since our time-measuring device can be started at any instant, we can let $t_1 = 0$ and let t_2 be any arbitrary later time t. Then if x_0 is the coordinate when $t = 0$ (called the *initial position*), and x is the coordinate at time t, Eq. (4–2) becomes

$$x - x_0 = \bar{v}t. \tag{4–3}$$

If the particle is at the origin when $t = 0$, then $x_0 = 0$ and Eq. (4–3) simplifies further to

$$x = \bar{v}t. \tag{4–4}$$

4–3 INSTANTANEOUS VELOCITY

The velocity of a particle at some one instant of time, or at some one point of its path, is called its *instantaneous velocity*. This concept requires careful definition.

Suppose we wish to find the instantaneous velocity of the particle in Fig. 4–1 at the point P. The average velocity between points P and Q is associated with the entire displacement $\mathbf{\Delta x}$, and with the entire time interval Δt. Imagine the second point Q to be taken closer and closer to the first point P, and let the average velocity be computed over these shorter and shorter displacements and time intervals. The instantaneous velocity at the first point can then be defined as the *limiting value* of the average velocity when the second point is taken closer and closer to the first. Although the displacement then becomes extremely small, the time interval by which it must be divided becomes small also and the quotient therefore is not necessarily a small quantity.

In the notation of calculus, the limiting value of $\Delta x/\Delta t$, as Δt approaches zero, is written dx/dt and is called the *derivative* of x with respect to t.

Then if $\mathbf{v}$ represents the instantaneous velocity, its magnitude is

$$v = \lim_{\Delta t \to 0} \frac{\Delta x}{\Delta t} = \frac{dx}{dt}. \tag{4–5}$$

Instantaneous velocity is also a vector, whose direction is the limiting direction of the displacement vector $\mathbf{\Delta x}$. Since Δt is necessarily positive, it follows that v has the same algebraic sign as Δx. Hence a positive velocity indicates motion toward the right along the x-axis, if we use the usual convention of signs.

As point Q approaches point P in Fig. 4–1(a), point q approaches point p in Fig. 4–1(b). In the limit, the slope of the chord pq equals the slope of the *tangent* to the curve at point p, due allowance being made for the scales to which x and t are plotted. *The instantaneous velocity at any point of a coordinate-time graph therefore equals the slope of the tangent to the graph at that point.* If the tangent slopes upward to the right, its slope is positive, the velocity is positive, and the motion is toward the right. If the tangent slopes downward to the right, the velocity is negative. At a point where the tangent is horizontal, its slope is zero and the velocity is zero.

If we express distance in meters and time in seconds, velocity is expressed in meters per second (m s^{-1}). Other common units of velocity are feet per second (ft s^{-1}), centimeters per second (cm s^{-1}), miles per hour (mi hr^{-1}), and knots (1 knot = 1 nautical mile per hour).

Example Suppose the motion of the particle in Fig. 4–1 is described by the equation $x = a + bt^2$, where $a = 20$ cm and $b = 4$ cm s^{-2}.

a) Find the displacement of the particle in the time interval between $t_1 = 2$ s and $t_2 = 5$ s.

At time $t_1 = 2$ s, the position is

$$x_1 = 20 \text{ cm} + (4 \text{ cm s}^{-2})(2 \text{ s})^2 = 36 \text{ cm}.$$

At time $t_2 = 5$ s,

$$x_2 = 20 \text{ cm} + (4 \text{ cm s}^{-2})(5 \text{ s})^2 = 120 \text{ cm}.$$

The displacement is therefore

$$x_2 - x_1 = 120 \text{ cm} - 36 \text{ cm} = 84 \text{ cm}.$$

b) Find the average velocity in this time interval:

$$\bar{v} = \frac{x_2 - x_1}{t_2 - t_1} = \frac{84 \text{ cm}}{3 \text{ s}} = 28 \text{ cm s}^{-1}.$$

This corresponds to the slope of the chord pq in Fig. 4–1(b).

c) Find the instantaneous velocity at time $t_1 = 2$ s.

The position at time $t = 2 \text{ s} + \Delta t$ is

$$x = 20 \text{ cm} + (4 \text{ cm s}^{-2})(2 \text{ s} + \Delta t)^2$$
$$= 36 \text{ cm} + (16 \text{ cm s}^{-1})\Delta t + (4 \text{ cm s}^{-2})(\Delta t)^2.$$

The displacement during the interval Δt is

$$\Delta x = 36 \text{ cm} + (16 \text{ cm s}^{-1})\Delta t + (4 \text{ cm s}^{-2})(\Delta t)^2 - 36 \text{ cm}$$
$$= (16 \text{ cm s}^{-1})\Delta t + (4 \text{ cm s}^{-2})(\Delta t)^2.$$

The average velocity during Δt is

$$\bar{v} = \frac{\Delta x}{\Delta t} = 16 \text{ cm s}^{-1} + (4 \text{ cm s}^{-1})\Delta t.$$

For the instantaneous velocity at $t = 2$ s, we let Δt approach zero in the expression for $\bar{v}$: $v = 16$ cm s^{-1}. This corresponds to the slope of the tangent at point p in Fig. 4–1(b). The instantaneous velocity may also be obtained directly by calculating the derivative of x with respect to t. Starting again with $x = a + bt^2$ and using Eq. (4–5), we find

$$v = \frac{dx}{dt} = \frac{d}{dt}(a + bt^2) = 2bt.$$

The instantaneous velocity at time $t_1 = 2$ s is

$$v = (2)(4 \text{ cm s}^{-2})(2 \text{ s}) = 16 \text{ cm s}^{-1}.$$

The term *speed* has two different meanings. It may mean the *magnitude* of the instantaneous velocity; two automobiles traveling at 50 mi hr^{-1}, one north and the other south, both have a speed of 50 mi hr^{-1}. In another sense, the *average speed* of a body means the total length of path covered, divided by the elapsed time. Thus, if an automobile travels 90 mi in 3hr, its average speed is 30 mi hr^{-1}, even if the trip starts and ends at the same point. The average *velocity*, in the latter case, would be zero, since the total displacement is zero.

4–4 AVERAGE AND INSTANTANEOUS ACCELERATION

When the velocity of a moving body changes continuously as the motion proceeds, the body is said to move with *accelerated motion*,

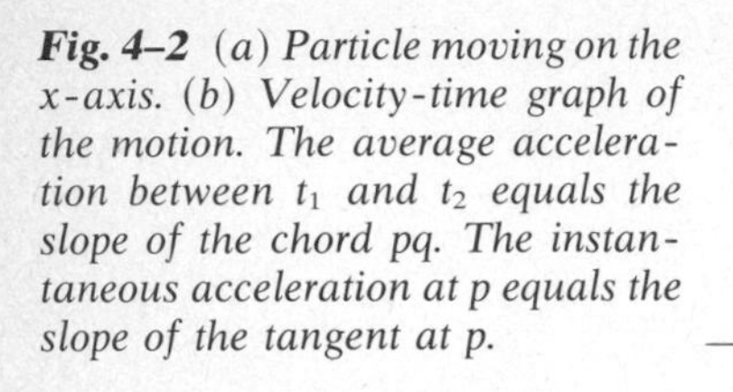

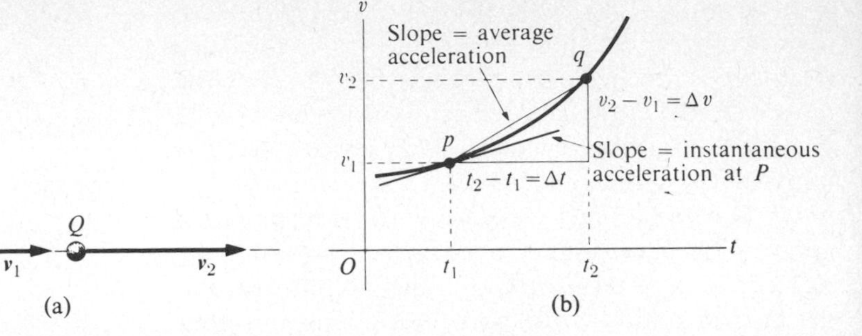

Fig. 4–2 *(a) Particle moving on the x-axis. (b) Velocity-time graph of the motion. The average acceleration between t_1 and t_2 equals the slope of the chord pq. The instantaneous acceleration at p equals the slope of the tangent at p.*

Figure 4–2(a) shows a particle moving along the x-axis. The vector $\mathbf{v}_1$ represents its instantaneous velocity at point P, and the vector $\mathbf{v}_2$ represents its instantaneous velocity at point Q. Figure 4–2(b) is a graph of the magnitude of the instantaneous velocity $\mathbf{v}$ plotted as a function of time, points p and q corresponding to P and Q in part (a).

The *average acceleration* of the particle as it moves from P to Q is defined as the ratio of the change in velocity to the elapsed time.

$$\bar{\mathbf{a}} = \frac{\mathbf{v}_2 - \mathbf{v}_1}{t_2 - t_1} = \frac{\Delta \mathbf{v}}{\Delta t}, \tag{4–6}$$

where t_1 and t_2 are the times corresponding to the velocities $\mathbf{v}_1$ and $\mathbf{v}_2$. Since $\mathbf{v}_1$ and $\mathbf{v}_2$ are vectors, the quantity $\mathbf{v}_2 - \mathbf{v}_1$ is a *vector difference* and must be found by the methods explained in Section 1–9. However, since in rectilinear motion both vectors lie in the same straight line, the magnitude of the vector difference in this special case equals the difference between the magnitudes of the vectors. The more general case, in which $\mathbf{v}_1$ and $\mathbf{v}_2$ are not in the same direction, will be considered in Chapter 6.

In Fig. 4–2(b), the magnitude of the average acceleration is represented by the slope of the chord pq.

The *instantaneous acceleration* of a body, that is, its acceleration at some one instant of time or at some one point of its path, is defined in the same way as instantaneous velocity. Let the second point Q in Fig. 4–2(a) be taken closer and closer to the first point P, and let the average acceleration be computed over shorter and shorter intervals of time. The instantaneous acceleration at the first point is defined as the limiting value of the average acceleration when the second point is taken closer and closer to the first:

$$a = \lim_{\Delta t \to 0} \frac{\Delta v}{\Delta t} = \frac{dv}{dt}. \tag{4–7}$$

The direction of the instantaneous acceleration is the limiting direction of the vector change in velocity, $\Delta \mathbf{v}$.

Instantaneous acceleration plays an important part in the laws of mechanics. Average acceleration is less frequently used. Hence from now on when the term "acceleration" is used we shall understand it to mean instantaneous acceleration.

The definition of acceleration just given applies to motion along any path, straight or curved. When a particle moves in a curved path, the *direction* of its velocity changes and this change in direction also gives rise to an acceleration, as will be explained in Chapter 6.

As point Q approaches point P in Fig. 4–2(a), point q approaches point p in Fig. 4–2(b) and the slope of the chord pq approaches the slope of the tangent to the velocity-time graph at point q. *The instantaneous acceleration at any point of the graph therefore equals the slope of the tangent to the graph at that point.*

The acceleration $a = dv/dt$ can be expressed in various ways. Since $v = dx/dt$, it follows that

$$a = \frac{dv}{dt} = \frac{d}{dt}\left(\frac{dx}{dt}\right) = \frac{d^2x}{dt^2}.$$

The acceleration is therefore the *second* derivative of the coordinate with respect to time.

We can also use the chain rule and write

$$a = \frac{dv}{dt} = \frac{dv}{dx}\frac{dx}{dt} = v\frac{dv}{dx}, \qquad (4\text{–}8)$$

which expresses the acceleration in terms of the *space* rate of change in velocity, dv/dx.

If we express velocity in feet per second and time in seconds, acceleration is expressed in feet per second, per second (ft s^{-1} s^{-1}). This is usually written as ft s^{-2}, and is read "feet per second squared." Other common units of acceleration are meters per second squared (m s^{-2}) and centimeters per second squared (cm s^{-2}).

When the absolute value of the magnitude of the velocity of a body is decreasing (in other words, when the body is slowing down), the body is said to be *decelerated* or to have a *deceleration*.

Example Suppose the velocity of the particle in Fig. 4–2 is given by the equation

$$v = m + nt^2,$$

where $m = 10$ cm s^{-1} and $n = 2$ cm s^{-3}.

a) Find the change in velocity of the particle in the time interval between $t_1 = 2$ s and $t_2 = 5$ s.

At time $t_1 = 2$ s,

$$v_1 = 10 \text{ cm s}^{-1} + (2 \text{ cm s}^{-3})(2 \text{ s})^2 = 18 \text{ cm s}^{-1}.$$

At time $t_2 = 5$ s,

$$v_2 = 10 \text{ cm s}^{-1} + (2 \text{ cm s}^{-3})(5 \text{ s})^2 = 60 \text{ cm s}^{-1}.$$

The change in velocity is therefore

$$v_2 - v_1 = 60 \text{ cm s}^{-1} - 18 \text{ cm s}^{-1} = 42 \text{ cm s}^{-1}.$$

b) Find the average acceleration in this time interval.

$$\bar{a} = \frac{v_2 - v_1}{t_2 - t_1} = \frac{42 \text{ cm s}^{-1}}{3 \text{ s}} = 14 \text{ cm s}^{-2}.$$

This corresponds to the slope of the chord pq in Fig. 4–2(b).

c) Find the instantaneous acceleration at time $t_1 = 2$ s.

At time $t = 2 \text{ s} + \Delta t$,

$$v = 10 \text{ cm s}^{-1} + (2 \text{ cm s}^{-3})(2 \text{ s} + \Delta t)^2$$
$$= 18 \text{ cm s}^{-1} + (8 \text{ cm s}^{-2})\Delta t + (2 \text{ cm s}^{-3})(\Delta t)^2.$$

The change in velocity during Δt is

$$\Delta v = 18 \text{ cm s}^{-1} + (8 \text{ cm s}^{-2})\Delta t + (2 \text{ cm s}^{-3})(\Delta t)^2 - 18 \text{ cm s}^{-1}$$
$$= (8 \text{ cm s}^{-2})\Delta t + (2 \text{ cm s}^{-3})(\Delta t)^2.$$

The average acceleration during Δt is

$$\bar{a} = \frac{\Delta v}{\Delta t} = 8 \text{ cm s}^{-2} + (2 \text{ cm s}^{-3})\Delta t.$$

The instantaneous acceleration at $t = 2$ s, obtained by letting Δt approach zero, is $a = 8$ cm s^{-2}. This corresponds to the slope of the tangent at point p in Fig. 4–2(b).

The instantaneous acceleration can also be obtained directly by taking the time derivative of the instantaneous velocity: Starting with $v = m + nt^2$ and using Eq. (4–7), we find

$$a = \frac{dv}{dt} = \frac{d}{dt}(m + nt^2) = 2nt.$$

The instantaneous acceleration when $t = 2$ s is

$$a = (2)(2 \text{ cm s}^{-3})(2 \text{ s}) = 8 \text{ cm s}^{-2}.$$

4–5 RECTILINEAR MOTION WITH CONSTANT ACCELERATION

The simplest kind of accelerated motion is rectilinear motion in which the acceleration is constant, that is,

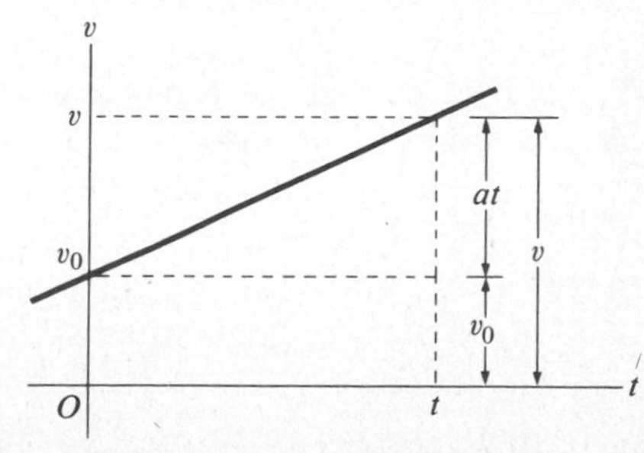

Fig. 4–3 Velocity–time graph for rectilinear motion with constant acceleration.

in which the velocity changes at the same rate throughout the motion. The velocity–time graph is then a straight line, as in Fig. 4–3, the velocity increasing by equal amounts in equal intervals of time. The slope of a chord between any two points on the line is the same as the slope of a tangent at any point, and the average and instantaneous accelerations are equal. Hence in Eq. (4–6) the average acceleration $\bar{a}$ can be replaced by the constant acceleration a and we have

$$a = \frac{v_2 - v_1}{t_2 - t_1}. \tag{4–9}$$

Now let $t_1 = 0$ and let t_2 be any arbitrary time t. Let v_0 represent the velocity when $t = 0$ (called the *initial* velocity), and let v be the velocity at time t. Then the preceding equation becomes

$$a = \frac{v - v_0}{t - 0},$$

or

$$\boxed{v = v_0 + at.} \tag{4–10}$$

The equation can be interpreted as follows: The acceleration a is the constant rate of change of velocity, or the change per unit time. The term at is the product of the change in velocity per unit time, a, and the duration of the time interval, t; therefore it equals the *total* change in velocity. The velocity v at the time t then equals the velocity v_0 at the time $t = 0$, plus the change in velocity at. Graphically, the ordinate v at time t, in Fig. 4–3, can be considered as the sum of two segments: one of length v_0 equal to the initial velocity, the other of length at equal to the change in velocity in time t.

To find the displacement of a particle moving with constant acceleration, we make use of the fact that when the acceleration is constant and the velocity–time graph is a straight line, as in Fig. 4–3, the average velocity in any time interval equals one-half the sum of the velocities at the beginning and the end of the interval. Hence the average velocity between zero and t is

$$\bar{v} = \frac{v_0 + v}{2}. \tag{4–11}$$

(This is *not* true in general, when the acceleration is not constant and the velocity–time graph is curved as in Fig. 4–2.)

By definition, the average velocity is

$$\bar{v} = \frac{x_2 - x_1}{t_2 - t_1}.$$

Now let $t_1 = 0$ and let t_2 be any arbitary time t. Let x_0 represent the position when $t = 0$ (the *initial* position) and let x be the position at time t. Then the preceding equation becomes

$$x - x_0 = \bar{v}t. \tag{4–12}$$

Substituting the expression for $\bar{v}$ in Eq. (4–11), we obtain

$$x - x_0 = \left(\frac{v_0 + v}{2}\right)t. \tag{4–13}$$

Two more very useful equations can be obtained from Eqs. (4–10) and (4–13), first by eliminating v and then by eliminating t. When we substitute in Eq. (4–13) the expression for v in Eq. (4–10), we get:

$$x - x_0 = \left(\frac{v_0 + v_0 + at}{2}\right)t,$$

or

$$\boxed{x - x_0 = v_0 t + \tfrac{1}{2}at^2.} \tag{4–14}$$

When Eq. (4–10) is solved for t and the result substituted in Eq. (4–13), we have

$$x - x_0 = \left(\frac{v_0 + v}{2}\right)\left(\frac{v - v_0}{a}\right) = \frac{v^2 - v_0^2}{2a},$$

or finally,

$$\boxed{v^2 = v_0^2 + 2a(x - x_0).} \tag{4–15}$$

Equations (4–10), (4–13), (4–14), and (4–15) are the *equations of motion with constant acceleration.*

The curve in Fig. 4–4 is the coordinate-time graph for motion with constant acceleration. That is, it is a graph of Eq. (4–14); the curve is a parabola. The slope of the tangent at $t = 0$ equals the initial

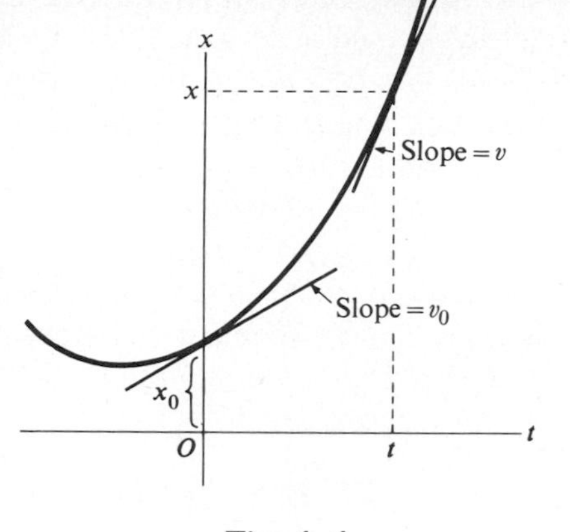

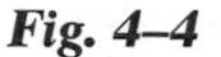
Fig. 4–4

velocity v_0, and the slope of the tangent at time t equals the velocity v at that time. It is evident that the slope continually increases, and measurements would show that the rate of increase is constant, that is, that the acceleration is constant.

A special case of motion with constant acceleration is that in which the acceleration is zero. The *velocity* is then constant and the equations of motion become simply

$$v = \text{constant}, \qquad x = x_0 + vt.$$

4–6 VELOCITY AND COORDINATE BY INTEGRATION

If the coordinate x of a particle moving on the x-axis is given as a function of time, the velocity can be found by differentiation, from the definition $v = dx/dt$. A second differentiation gives the acceleration, since $a = dv/dt$. We now consider the converse process: given the acceleration, to find the velocity and the cordinate. This can be done by the methods of *integral* calculus. We shall discuss first the indefinite and then the definite integral.

Suppose we are given the acceleration $a(t)$ as a function of time. Then, since

$$\frac{dv}{dt} = a(t),$$

we have

$$dv = a(t)dt, \qquad \int dv = \int a(t)dt,$$

and

$$v = \int a(t)dt + C_1, \tag{4–16}$$

were C_1 is an integration constant whose value can be determined if the velocity is known at any time. It is customary to express C_1 in terms of the velocity v_0 when $t = 0$.

When the integral above has been evaluated, we have the velocity $v(t)$ as a function of time. Then, since

$$\frac{dx}{dt} = v(t),$$

we have

$$dx = v(t)dt, \qquad \int dx = \int v(t)dt,$$

$$x = \int v(t)dt + C_2, \tag{4–17}$$

where C_2 is a second integration constant whose value can be determined if the coordinate is known at any time. It is customary to express C_2 in terms of the coordinate x_0 when $t = 0$.

If the acceleration is given as a function of x, we can use Eq. (4–8):

$$v\frac{dv}{dx} = a(x), \qquad \int v\,dv = \int a(x)dx,$$

$$\frac{v^2}{2} = \int a(x)dx + C_3. \tag{4–18}$$

Example Derive the equations of motion with constant acceleration, using the indefinite integral.

If a is constant, then from Eq. (4–16),

$$v = at + C_1.$$

But $v = v_0$ when $t = 0$, so

$$v_0 = 0 + C_1$$

and

$$v = v_0 + at,$$

which is Eq. (4–10).

From Eq. (4–17), when a is constant,

$$x = \int (v_0 + at)\,dt = v_0t + \tfrac{1}{2}at^2 + C_2.$$

If $x = x_0$ when $t = 0$, then $C_2 = x_0$ and

$$x = x_0 + v_0t + \tfrac{1}{2}at^2,$$

which is Eq. (4–14).

From Eq. (4–18), when a is constant,

$$\frac{v^2}{2} = ax + C_3 .$$

If $v = v_0$ when $x = x_0$, then $C_3 = v_0^2/2 - ax_0$ and

$$v^2 = v_0^2 + 2a(x - x_0),$$

which is Eq. (4–15).

Consider next the definite integral. Let the area under the velocity–time graph in Fig. 4–5, between the vertical lines at t_1 and t_2, be subdivided into narrow rectangular strips of width Δt. The ordinate of the graph at any time t equals the instantaneous velocity v at that time. If the velocity remained constant at this value, the displacement Δx in the time interval between t and $t + \Delta t$ would equal $v\Delta t$. But this is the *area* of the shaded strip, since its height is v and its width is Δt. The *sum* of the areas of such rectangles, between t_1 and t_2, is approximately equal to the *total* displacement $x_2 - x_1$ in this time interval:

$$x_2 - x_1 \approx \sum v\Delta t.$$

The smaller the time intervals Δt, the more closely does $v\Delta t$ approach the actual displacement. In the limit, as Δt approaches zero, the sum of the areas becomes exactly equal to the total area under the curve, and to the total displacement $x_2 - x_1$. The limit of the sum of the areas is the definite integral from t_1 to t_2, so

$$x_2 - x_1 = \int_{t_1}^{t_2} v\, dt. \tag{4–19}$$

The displacement in any time interval is therefore equal to the area between a velocity–time graph and the time axis, bounded by vertical lines at the beginning and end of the interval.

In the same way, the area under an acceleration–time graph can be subdivided into vertical strips of height a and width Δt. If the acceleration remained constant, the change in velocity Δv in time Δt would equal $a\Delta t$, the area of a rectangular strip. The total change in velocity, $v_2 - v_1$, in the time interval from t_1 and t_2, is approximately equal to the sum of all such areas.

$$v_2 - v_1 \approx \sum a\Delta t.$$

In the limit, as Δt approaches zero,

$$v_2 - v_1 = \int_{t_1}^{t_2} a\, dt. \tag{4–20}$$

The change in velocity in any time interval is therefore equal to the area between an acceleration–time graph and the time axis, bounded by vertical lines at the beginning and end of the interval.

Example Use the definite integral to find the velocity and coordinate, at any time t, of a body moving on the x-axis with constant acceleration. The initial velocity is v_0 and the initial coordinate is x_0.

As the limits of integration, we take $t_1 = 0$ and $t_2 = t$. Then, from Eq. (4–20),

$$v - v_0 = \int_0^t a\, dt = at,$$

which is Eq. (4–10).

From Eq. (4–19),

$$x - x_0 = \int_0^t (v_0 + at)\, dt = v_0 t + \tfrac{1}{2}at^2,$$

which is Eq. (4–14).

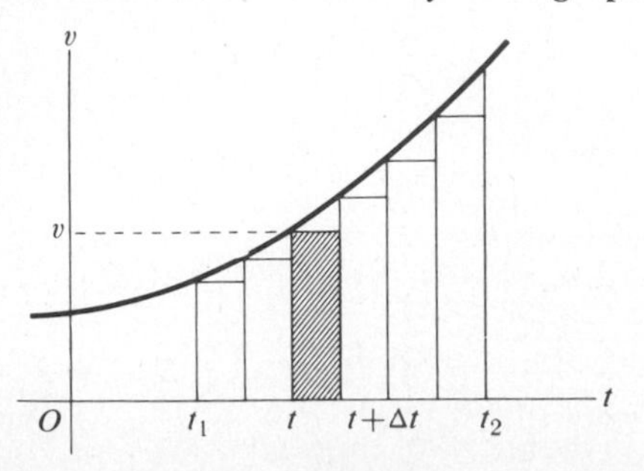

Fig. 4–5 *The area under a velocity–time graph equals the displacement.*

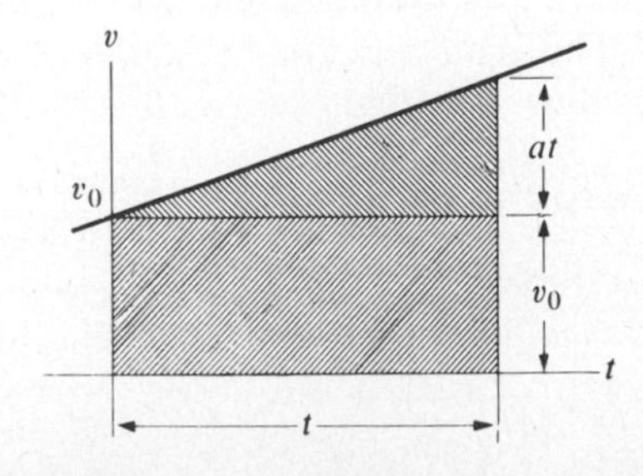

Fig. 4–6 *The area under a velocity–time graph equals the displacement $x - x_0$.*

It is not always necessary to use the methods of integral calculus to find the area under a graph. Figure 4–6 is the velocity–time graph for motion with constant acceleration. The area under the graph, between $t = 0$ and $t = t$, can be subdivided into a rectangle and a triangle. The area of the rectangle is $v_0 t$ and that of the triangle is $\frac{1}{2}(t)(at) = \frac{1}{2}at^2$. Since the displacement equals the total area,

$$x - x_0 = v_0 t + \tfrac{1}{2}at^2,$$

which is Eq. (4–14).

4–7 FREELY FALLING BODIES

The most common example of motion with (nearly) constant acceleration is that of a body falling toward the earth. In the absence of air resistance it is found that all bodies, regardless of their size or weight, fall with the same acceleration at the same point on the earth's surface, and if the distance covered is small compared to the radius of the earth, the acceleration remains constant throughout the fall. The effect of air resistance and the decrease in acceleration with altitude will be neglected. This idealized motion is spoken of as "free fall," although the term includes rising as well as falling motion.

The acceleration of a freely falling body is called acceleration due to gravity, or the acceleration of gravity, and is denoted by the letter g. At or near the earth's surface its magnitude is approximately 32 ft s^{-2}, 9.8 m s^{-2}, or 980 cm s^{-2}. More precise values, and small variations with latitude and elevation, will be considered later. On the surface of the moon, the acceleration of gravity is due to the attractive force exerted on a body by the moon rather than earth. On the moon, $g = 1.67$ m s^{-2} = 5.47 ft s^{-2}. Near the surface of the *sun*, g = 274 m s^{-2}!

Note. The quantity "g" is sometimes referred to simply as "gravity," or as "the force of gravity," both of which are incorrect. "Gravity" is a phenomenon, and the "force of gravity" means the force with which the earth attracts a body, otherwise known as the weight of the body. The letter "g" represents the *acceleration* caused by the force resulting from the phenomenon of gravity.

Example 1 A body is released from rest and falls freely. Compute its position and velocity after 1, 2, 3, and 4 s. Take the origin 0 at the elevation of the starting point, the y-axis vertical, and the upward direction as positive.

The initial coordinate y_0 and the initial velocity v_0 are both zero. The acceleration is downward, in the negative y-direction, so $a = -g = -32$ ft s^{-2}.

From Eqs. (4–14) and (4–10),

$$y = v_0 t + \tfrac{1}{2}at^2 = 0 - \tfrac{1}{2}gt^2 = (-16 \text{ ft s}^{-2})t^2,$$

$$v = v_0 + at = 0 - gt = (-32 \text{ ft s}^{-2})t.$$

When $t = 1$ s, $y = (-16 \text{ ft s}^{-2})(1 \text{ s})^2 = -16$ ft, and $v = (-32 \text{ ft s}^{-2})(1 \text{ s}) = 32 \text{ ft s}^{-1}$. The body is therefore 16 ft below the origin (y is negative) and has a downward velocity (v is negative) of magnitude 32 ft s^{-1}.

The position and velocity at 2, 3, and 4 s are found in the same way. The results are illustrated in Fig. 4–7.

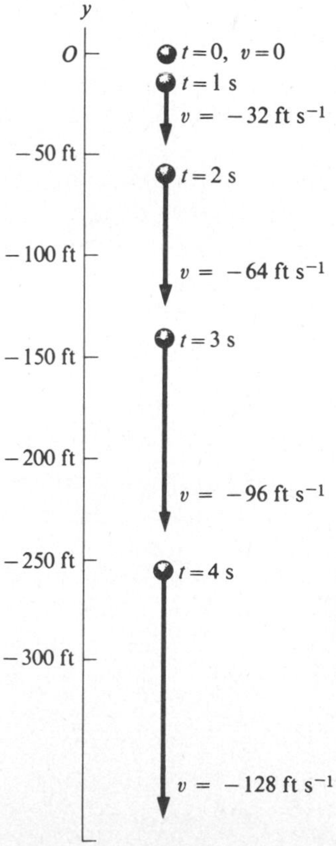

Fig. 4–7 *Position and velocity of a freely falling body.*

Example 2 A ball is thrown (nearly) vertically upward from the cornice of a tall building, leaving the thrower's hand with a speed of 48 ft s^{-1} and just missing the cornice on the way down, as in Fig. 4–8. The downward path is shown displaced to the right for clarity. Find (a) the position and velocity of the ball, 1 s and 4 s after leaving the thrower's hand; (b) the velocity when the ball is 20 ft above its starting point; (c) the maximum height reached and the time at which it is reached. Take the origin at the elevation at which the ball leaves the thrower's hand, the y-axis vertical and positive upward.

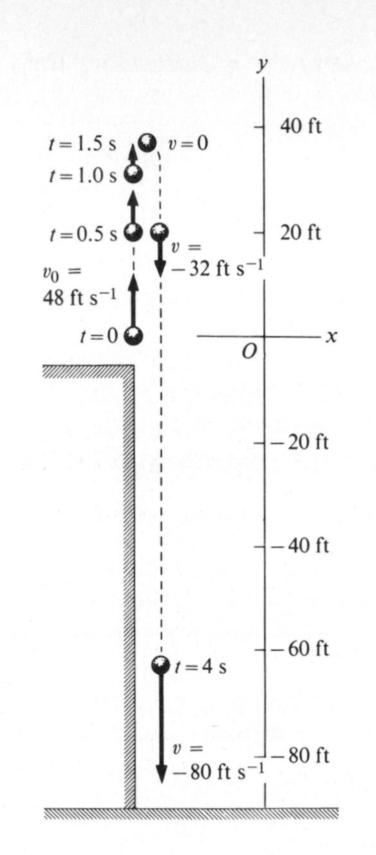

Fig. 4–8 Position and velocity of a body thrown vertically upward.

The initial position y_0 is zero. The initial velocity v_0 is +48 ft s^{-1}, and the acceleration is −32 ft s^{-2}. The velocity at any time is

$$v = v_0 + at = 48 \text{ ft s}^{-1} - (32 \text{ ft s}^{-2})t. \quad (4\text{–}21)$$

The coordinate at any time is

$$y = v_0 t + \tfrac{1}{2}at^2 = (48 \text{ ft s}^{-1})t - (16 \text{ ft s}^{-2})t^2.$$

The velocity at any coordinate is

$$v^2 = v_0^2 + 2ay = (48 \text{ ft s}^{-1})^2 - (64 \text{ ft s}^{-2})y. \quad (4\text{–}22)$$

a) When $t = 1$ s,

$$y = +32 \text{ ft}, \qquad v = +16 \text{ ft s}^{-1}.$$

The ball is 32 ft above the origin (y is positive) and it has an upward velocity (v is positive) of 16 ft s^{-1} (less than the initial velocity, as expected).

When $t = 4$ s,

$$y = -64 \text{ ft}, \qquad v = -80 \text{ ft s}^{-1}.$$

The ball has passed its highest point and is 64 ft *below* the origin (y is negative). It has a *downward* velocity (v is negative) of magnitude 80 ft s^{-1}. Note that it is not necessary to find the highest point reached, or the time at which it was reached. The equations of motion give the position and velocity at *any* time, whether the ball is on the way up or the way down.

b) When the ball is 20 ft above the origin

$$y = +20 \text{ ft}$$

and

$$v^2 = 1024 \text{ ft}^2 \text{ s}^{-2}, \qquad v = \pm 32 \text{ ft s}^{-1}.$$

The ball passes this point *twice*, once on the way up and again on the way down. The velocity on the way up is +32 ft s^{-1}, and on the way down it is −32 ft s^{-1}.

c) At the highest point, $v = 0$. Hence

$$0 = (48 \text{ ft s}^{-1})^2 - (64 \text{ ft s}^{-2})y,$$

and

$$y = +36 \text{ ft}.$$

The time can now be found either from Eq. (4–21), setting $v = 0$, or from the equation following

Eq. (4–21), setting $y = 36$ ft. From either equation, we get

$$t = 1.5 \text{ s}.$$

Although at the highest point the velocity is instantaneously zero, the *acceleration* at this point is still -32 ft s^{-1}. The ball stops for an instant, but its velocity is continuously changing; the acceleration is constant throughout.

Figure 4–9 is a "multiflash" photograph of a freely falling golf ball. This photograph was taken with aid of a stroboscopic light source, developed originally by Dr. Harold E. Edgerton of the Massachusetts Institute of Technology. This source produces a series of intense flashes of light. The interval between successive flashes is controllable at will, and the duration of each flash is so short (a few millionths of a second) that there is no blur in the image of even a rapidly moving body. The camera shutter is left open during the entire motion, and as each flash occurs, the position of the ball at that instant is recorded on the photographic film.

The equally spaced light flashes subdivide the motion into equal time intervals Δt. Since the time intervals are all equal, the velocity of the ball between any two flashes is directly proportional to the separation of its corresponding images in the photograph. If the velocity were constant, the images would be equally spaced. The increasing separation of the images during the fall shows that the velocity is continually increasing or the motion is accelerated. By comparing two successive displacements of the ball, we can find the *change* in velocity in the corresponding time interval. Careful measurement, preferably on an enlarged print, shows that this change in velocity is the same in each time interval. In other words, the motion is one of *constant* acceleration.

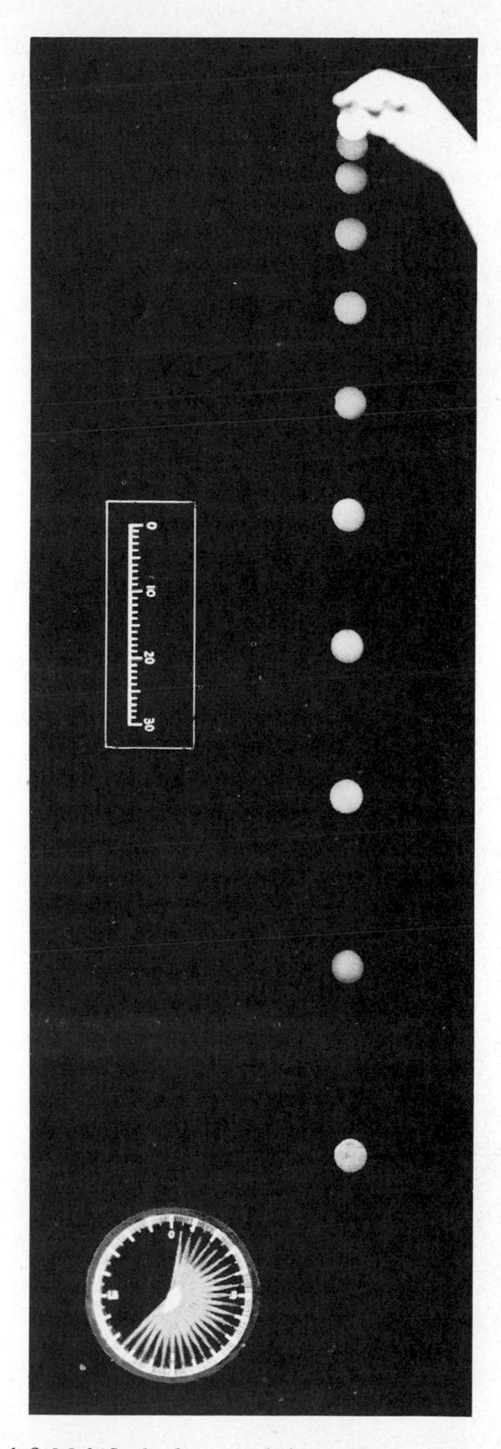

Fig. 4–9 *Multiflash photograph (retouched) of freely falling golf ball.*

4–8 RECTILINEAR MOTION WITH VARIABLE ACCELERATION

Motion with constant acceleration approximates the motion of some falling bodies, and of cars and of airplanes at the start of their journey. There are many important types of motion, however, in which the acceleration is variable, and it is worth while to develop the technique of dealing with such cases. Consider, for example, the motion of a body in the positive x-direction with an acceleration whose direction is opposite that of the velocity and whose magnitude is proportional to the speed. For such a motion

$$a = -kv,$$

where k is a constant. If the initial speed is v_0, how do the speed and the distance vary with the time?

Since $a = dv/dt$, we have

$$\frac{dv}{dt} = -kv, \quad \text{or} \quad \frac{dv}{v} = -k\,dt,$$

and since $v = v_0$ when $t = 0$,

$$\int_{v_0}^{v} \frac{dv}{v} = -k\int_0^t dt.$$

Integration yields the result

$$\ln\frac{v}{v_0} = -kt,$$

which may be written as

$$v = v_0 e^{-kt}, \tag{4–23}$$

showing that the velocity decays exponentially with the time. Exponential curves are found frequently in many diverse fields of physics. The decrease in activity of a radioactive substance, the discharge of a capacitor, and the dying out of mechanical, acoustical, and electrical oscillations are all described by exponential equations. Equation (4–23) shows that an infinite time would be required to reduce the speed to zero.

To find the distance x as a function of the time, we replace v by dx/dt. Thus

$$\frac{dx}{dt} = v_0 e^{-kt}.$$

Suppose that $x = 0$ when $t = 0$. Then

$$\int_0^x dx = v_0\int_0^t e^{-kt}dt,$$

and

$$x = -\frac{v_0}{k}\left[e^{-kt}\right]_0^t,$$

$$x = \frac{v_0}{k}(1 - e^{-kt}). \tag{4–24}$$

It follows from this equation that although an infinite time is necessary for the body to come to rest, in this infinite time the body will have gone only a finite distance v_0/k.

Other examples of variable acceleration will be found in the problems at the end of this chapter and also in Chapter 5.

4–9 VELOCITY COMPONENTS, RELATIVE VELOCITY

Velocity is a vector quantity involving both magnitude and direction. A velocity may therefore be resolved into components, or a number of velocity components combined into a resultant. As an example of the former process, suppose that a ship is steaming 30° E of N at 20 mi hr^{-1} in still water. Its velocity may be represented by the arrow in Fig. 4–10, and one finds by the usual method that its velocity component toward the east is 10 mi hr^{-1}, while toward the north it is 17.3 mi hr^{-1}.

The velocity of a body, like its position, can be specified only relative to some other body. The second body may be in motion relative to a third, and so on. Thus when we speak of "the velocity of an automobile," we usually mean its velocity relative to the earth. But the earth is in motion relative to the sun, the sun is in motion relative to some other star, and so on.

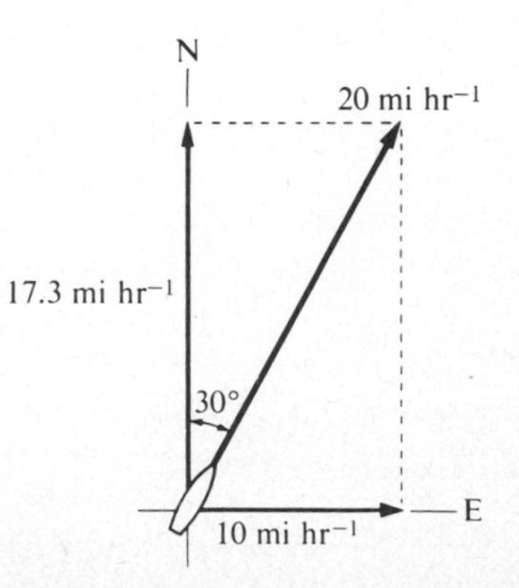

***Fig. 4–10** Resolution of a velocity vector into components.*

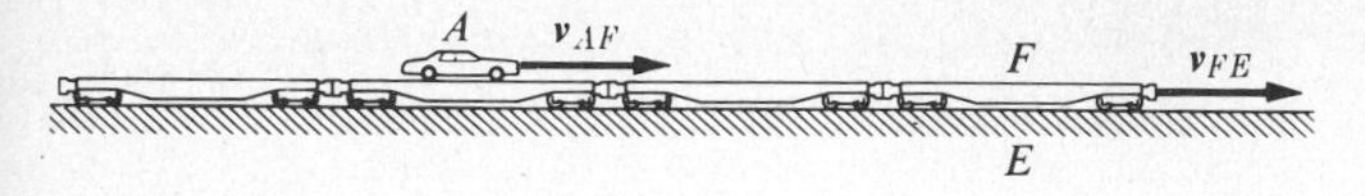

Fig. 4–11 *Vector $\mathbf{v}_{FE}$ is the velocity of the flatcars F relative to the earth E, and $\mathbf{v}_{AF}$ is the velocity of automobile A relative to the flatcars.*

Suppose a long train of flatcars is moving to the right along a straight level track, as in Fig. 4–11, and that a daring automobile driver is driving to the right along the flatcars. The vector $\mathbf{v}_{FE}$ in Fig. 4–11 represents the velocity of the flatcars F relative to the earth E, and the vector $\mathbf{v}_{AF}$ the velocity of the automobile A relative to the flatcars F. The velocity of the automobile relative to the earth, v_{AE}, is evidently equal to the sum of the relative velocities v_{AF} and v_{FE}:

$$v_{AE} = v_{AF} + v_{FE}. \tag{4–25}$$

Thus if the flatcars are traveling relative to the earth at 30 mi hr^{-1} ($= v_{FE}$) and the automobile is traveling relative to the flatcars at 40 mi hr^{-1} ($= v_{AF}$), the velocity of the automobile relative to the earth (v_{AE}) is 70 mi hr^{-1}.

Imagine now that the flatcars are wide enough so that the automobile can drive on them in any direction. The velocity of the automobile relative to the earth is then the vector sum of its velocity relative to the flatcars and the velocity of the flatcars relative to the earth. The preceding equation is therefore a special case of the more general *vector* equation,

$$\boxed{\mathbf{v}_{AE} = \mathbf{v}_{AF} + \mathbf{v}_{FE}} \tag{4–26}$$

Thus if the automobile were driving transversely across the flatcars at 40 mi hr^{-1}, its velocity relative to the earth would be 50 mi hr^{-1} at an angle of 53° with the railway track.

In the special case in which $\mathbf{v}_{AF}$ and $\mathbf{v}_{FE}$ are along the same line, as in Fig. 4–11, the velocity v_{AE} is the *algebraic* sum of v_{AF} and v_{FE}. Thus if the automobile were traveling to the *left* with a velocity of 40 mi hr^{-1} relative to the flatcars, $v_{AF} = -40$ mi hr^{-1} and the velocity of the automobile relative to the earth would be -10 mi hr^{-1}; that is, it would be traveling to the left, relative to the earth.

Equation (4–26) can be extended to include any number of relative velocities. For example, if a bug B crawls along the floor of the automobile with a velocity relative to the automobile of v_{BA}, his velocity relative to the earth is the vector sum of his velocity relative to the automobile and that of the velocity of the automobile relative to the earth:

$$\mathbf{v}_{BE} = \mathbf{v}_{BA} + \mathbf{v}_{AE}.$$

When this is combined with Eq. (4–26), we get

$$\mathbf{v}_{BE} = \mathbf{v}_{BA} + \mathbf{v}_{AF} + \mathbf{v}_{FE}. \tag{4–27}$$

This equation illustrates the general rule for combining relative velocities.

1. Write each velocity with a double subscript in the *proper order*, meaning "velocity of (first subscript) relative to (second subscript)."

2. When adding relative velocities, the first letter of any subscript is to be the same as the last letter of the preceding subscript.

3. The first letter of the subscript of the first velocity in the sum, and the second letter of the subscript of the last velocity, are the subscripts, in that order, of the relative velocity represented by the sum. This somewhat lengthy statement should be clear when it is compared with Eq. (4–27).

Any of the relative velocities in an equation like Eq. (4–26) can be transferred from one side of the equation to the other, with sign reversed. Thus Eq. (4–26) can be written

$$\mathbf{v}_{AF} = \mathbf{v}_{AE} - \mathbf{v}_{FE}.$$

The velocity of the automobile relative to the flatcar equals the *vector difference* between the velocities of automobile and flatcar, both relative to the earth.

One more point should be noted. The velocity of body A relative to body B, $\mathbf{v}_{AB}$, is the negative of the velocity of B relative to A, $\mathbf{v}_{BA}$:

$$\mathbf{v}_{AB} = -\mathbf{v}_{BA}.$$

That is, $\boldsymbol{v}_{AB}$ is equal in magnitude and opposite in direction to $\boldsymbol{v}_{BA}$. If the automobile in Fig. 4–11 is traveling to the right at 40 mi hr^{-1}, relative to the flatcars, the flatcars are traveling to the left at 40 mi hr^{-1}, relative to the automobile.

Example 1 An automobile driver A, traveling relative to the earth at 65 km hr^{-1} on a straight level road, is ahead of motorcycle officer B traveling in the same direction at 80 km hr^{-1}. What is the velocity of B relative to A?

We have given

$$v_{AE} = 65 \text{ km hr}^{-1}, \qquad v_{BE} = 80 \text{ km hr}^{-1},$$

and we wish to find v_{BA}.

From the rule for combining velocities (along the same line), we get:

$$v_{BA} = v_{BE} + v_{EA}.$$

But

$$v_{EA} = -v_{AE},$$

so

$$\begin{aligned} v_{BA} &= v_{BE} - v_{AE} \\ &= 80 \text{ km hr}^{-1} - 65 \text{ km hr}^{-1} \\ &= 15 \text{ km hr}^{-1}, \end{aligned}$$

and the officer is overtaking the driver at 15 km hr^{-1}.

Example 2 How would the relative velocity be altered if B were ahead of A?

Not at all. The relative *positions* of the bodies do not matter. The velocity of B relative to A is still + 15 km hr^{-1}, but he is now pulling ahead of A at this rate.

Example 3 The compass of an aircraft indicates that it is headed due north, and its airspeed indicator shows that it is moving through the air at 120 mi hr^{-1}. If there is a wind of 50 mi hr^{-1} from west to east, what is the velocity of the aircraft relative to the earth?

Let subscript A refer to the aircraft, and subscript F to the moving air (which now corresponds to the flatcar in Fig. 4–11). Subscript E refers to the earth. We have given

$$\boldsymbol{v}_{AF} = 120 \text{ mi hr}^{-1}, \qquad \text{due north}$$

$$\boldsymbol{v}_{FE} = 50 \text{ mi hr}^{-1}, \qquad \text{due east},$$

and we wish to find the magnitude and direction of $\boldsymbol{v}_{AE}$:

$$\boldsymbol{v}_{AE} = \boldsymbol{v}_{AF} + \boldsymbol{v}_{FE}.$$

The three relative velocities are shown in Fig. 4–12. It follows from this diagram that

$$\boldsymbol{v}_{AE} = 130 \text{ mi hr}^{-1}, \qquad 22.5°\text{E of N}.$$

Example 4 In what direction should the pilot head in order to travel due north? What will then be his velocity relative to the earth? (The magnitude of his airspeed, and the wind velocity, are the same as in the preceding example.)

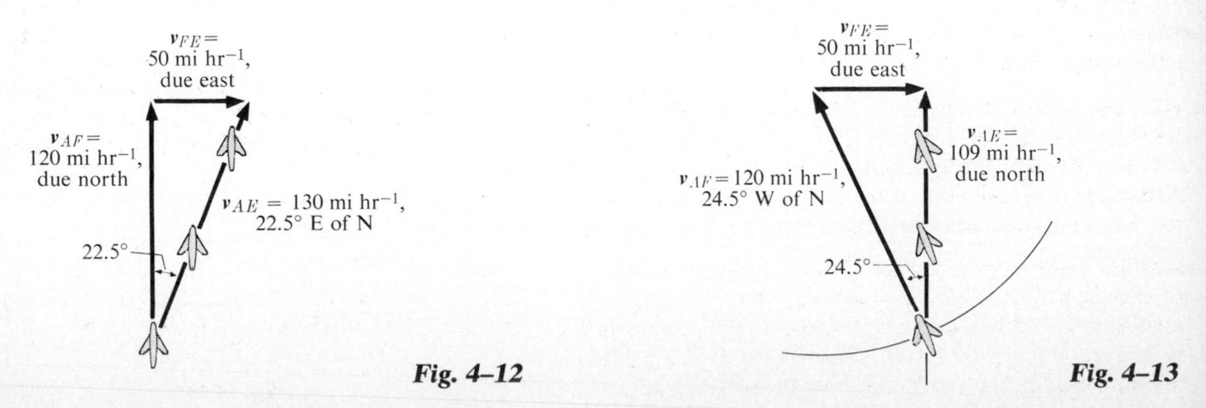

Fig. 4–12 **Fig. 4–13**

We now have given:

$$\boldsymbol{v}_{AF} = 120 \text{ mi hr}^{-1}, \qquad \text{direction unknown,}$$

$$\boldsymbol{v}_{FE} = 50 \text{ mi hr}^{-1}, \qquad \text{due east,}$$

and we wish to find $\boldsymbol{v}_{AE}$, whose magnitude is unknown but whose direction is due north. (Note that both this and the preceding example require us to determine two unknown quantities. In the former example, these were the *magnitude and direction* of $\boldsymbol{v}_{AE}$. In this example, the unknowns are the *direction* of $\boldsymbol{v}_{AF}$ and the *magnitude* of $\boldsymbol{v}_{AE}$.)

The three relative velocities must still satisfy the *vector equation*

$$\boldsymbol{v}_{AE} = \boldsymbol{v}_{AF} + \boldsymbol{v}_{FE}.$$

The problem can be solved graphically as follows. First, construct the vector $\boldsymbol{v}_{FE}$ (see Fig. 4–13), known in magnitude and direction. At the head of this vector draw a construction line of indefinite length due north, in the known direction of $\boldsymbol{v}_{AE}$. With the tail of $\boldsymbol{v}_{FE}$ as a center, construct a circular arc of radius equal to the known magnitude of $\boldsymbol{v}_{AF}$. Vectors $\boldsymbol{v}_{AF}$ and $\boldsymbol{v}_{AE}$ may then be drawn from the point of intersection of this arc and the construction line to the ends of the vector $\boldsymbol{v}_{FE}$. We find on solving the right triangle that the magnitude of $\boldsymbol{v}_{AE}$ is 109 mi hr^{-1} and the direction of $\boldsymbol{v}_{AF}$ is 24.5° W of N. That is, the pilot should head 24.5° W of N, and his ground speed will be 109 mi hr^{-1}.

PROBLEMS

4–1 Suppose that a runner on a straight track covers a distance of 1 mile in exactly 4 minutes. What was his average velocity in (a) mi hr^{-1}? (b) ft s^{-1}? (c) cm s^{-1}?

4–2 A body moves along a straight line, its distance from the origin at any instant being given by the equation $x = 8t - 3t^2$, where x is in centimeters and t is in seconds. Find the average velocity of the body in the interval from $t = 0$ to $t = 1$ s, and in the interval from $t = 0$ to $t = 4$ s.

4–3 The motion of a certain body along the x-axis is described by the equation $x = (10 \text{ cm s}^{-2})t^2$. Compute the instantaneous velocity of the body at time $t = 3$ s, by letting Δt first equal 0.1 s, then 0.01 s, and finally 0.001 s. What limiting value do the results seem to be approaching?

4–4 An automobile is provided with a speedometer calibrated to read m s^{-1} rather than mi hr^{-1}. The following series of speedometer readings was obtained during a start.

Time (s)	0	2	4	6	8	10	12	14	16
Velocity (m s^{-1})	0	0	2	5	10	15	20	22	22

a) Compute the average acceleration during each 2-s interval. Is the acceleration constant? Is it constant during any part of the time?

b) Make a velocity-time graph of the data above, using scales of 1 in. = 2 s horizontally, and 1 in. = 5 m s^{-1} vertically. Draw a smooth curve through the plotted points. What distance is represented by 1 sq. in.? What is the displacement in the first 8 s? What is the acceleration when $t = 8$ s? When $t = 13$ s? When $t = 15$ s?

4–5 The graph in Fig. 4–14 shows the velocity of a body plotted as a function of time.

a) Find the instantaneous acceleration at $t = 3$ s, at $t = 7$ s, and at $t = 11$ s.

b) How far does the body go in the first 5 s? The first 9 s? The first 13 s?

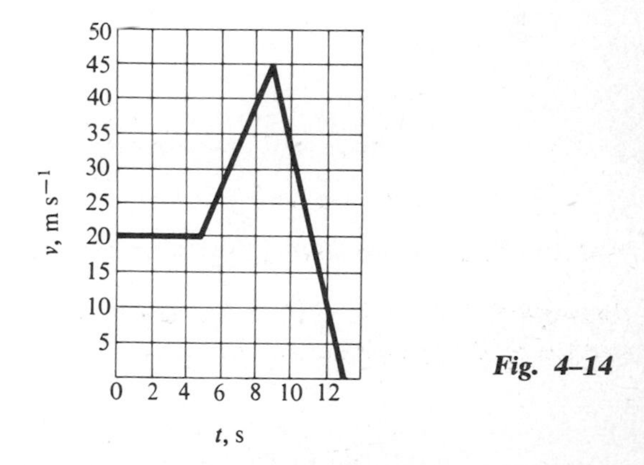

Fig. 4–14

4–6 Figure 4–15 is a graph of the acceleration of a body moving on the x-axis. Sketch the graphs of its velocity and coordinate as functions of time, if $x = v = 0$ when $t = 0$.

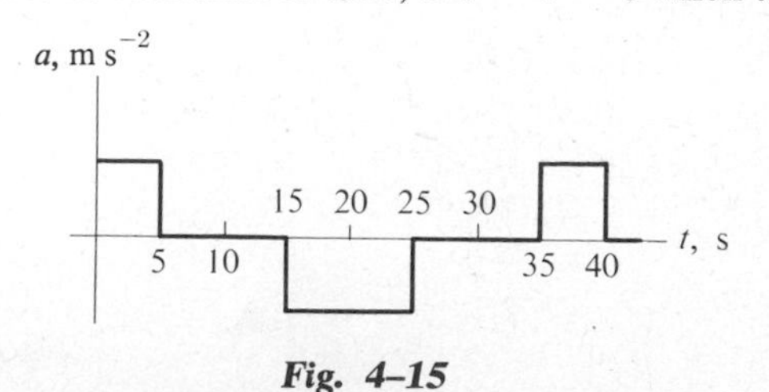

Fig. 4–15

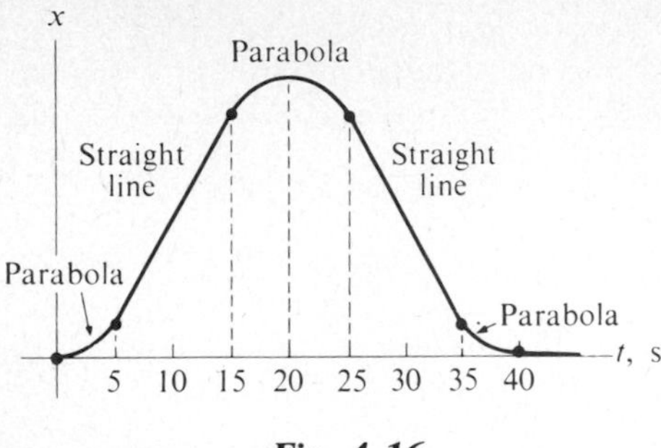

Fig. 4–16

4–7 Figure 4–16 is a graph of the coordinate of a body moving on the x-axis. Sketch the graphs of its velocity and acceleration as functions of the time.

4–8 Each of the following changes in velocity takes place in a 10-s interval. What is the magnitude, the algebraic sign, and the direction of the average acceleration in each interval?

a) At the beginning of the interval a body is moving toward the right along the x-axis at 5 m s^{-1}, and at the end of the interval it is moving toward the right at 20 m s^{-1}.

b) At the beginning it is moving toward the right at 20 m s^{-1}, and at the end it is moving toward the right at 5 m s^{-1}.

c) At the beginning it is moving toward the left at 5 m s^{-1}, and at the end it is moving toward the left at 20 m s^{-1}.

d) At the beginning it is moving toward the left at 20 m s^{-1}, and at the end it is moving toward the left at 5 m s^{-1}.

e) At the beginning it is moving toward the right at 20 m s^{-1}, and at the end it is moving toward the left at 20 m s^{-1}.

f) At the beginning it is moving toward the left at 20 m s^{-1}, and at the end it is moving toward the right at 20 m s^{-1}.

g) In which of the above instances is the body decelerated?

4–9 The makers of a certain automobile advertise that it will accelerate from 15 to 50 mi hr^{-1} in 13 s. Compute (a) the acceleration in ft s^{-2}, and (b) the distance the car travels in this time, assuming the acceleration to be constant.

4–10 An airplane taking off from a landing field has a run of 500 m. If it starts from rest, moves with constant acceleration, and makes the run in 30 s, with what velocity in m s^{-1} did it take off?

4–11 An automobile starts from rest and acquires a velocity of 40 km hr^{-1} in 15 s.

a) Compute the acceleration in kilometers per hour per second, assuming it to be constant.

b) If the automobile continues to gain velocity at the same rate, how many more seconds are needed for it to acquire a velocity of 60 km hr^{-1}?

c) Find the distances covered by the automobile in parts (a) and (b).

4–12 A body moving with constant acceleration covers the distance between two points 60 m apart in 6 s. Its velocity as it passes the second point is 15 m s^{-1}.

a) What is the acceleration?

b) What is its velocity at the first point?

4–13 A ball is released from rest and rolls down an inclined plane, requiring 4 s to cover a distance of 100 cm. (a) What was its acceleration in cm s^{-2}? (b) How many centimeters would it have fallen vertically in the same time?

4–14 The "reaction time" of the average automobile driver is about 0.7 s. (The reaction time is the interval between the perception of a signal to stop and the application of the brakes.) If an automobile can decelerate at 16 ft s^{-2}, compute the total distance covered in coming to a stop after a signal is observed: (a) from an initial velocity of 30 mi hr^{-1}, (b) from an initial velocity of 60 mi hr^{-1}.

4–15 At the instant the traffic lights turn green, an automobile that has been waiting at an intersection starts ahead with a constant acceleration of 2 m s^{-2}. At the same instant a truck, traveling with a constant velocity of 10 m s^{-1}, overtakes and passes the automobile. (a) How far beyond its starting point will the automobile overtake the truck? (b) How fast will it be traveling?

4–16 The engineer of a passenger train traveling at 30 m s^{-1} sights a freight train whose caboose is 200 m ahead on the same track. The freight train is traveling in the same direction as the passenger train with a velocity of 10 m s^{-1}. The engineer of the passenger train immediately applies the brakes, causing a constant deceleration of 1 m s^{-2}, while the freight train continues with constant speed. (a) Will there be a collision? (b) If so, where will it take place?

4–17 A sled starts from rest at the top of a hill and slides down with a constant acceleration. The sled is 140 ft from the top of the hill 2 s after passing a point which is 92 ft from the top. Four seconds after passing the 92-ft point it is 198 ft from the top, and 6 s after passing the point it is 266 ft from the top. (a) What is the average velocity of the sled during each of the 2-s intervals after passing the 92-ft point? (b) What is the acceleration of the sled? (c) What was the velocity of the sled when it passed the 92-ft point? (d) How long did it take to go from the top to the 92-ft point? (e) How far did the sled go during the first second after passing the 92-ft point? (f) How long does it take the sled to go from the 92-ft point to the midpoint between the 92-ft and the 140-ft mark? (g) What is the velocity of the sled as it passes the midpoint in part (f)?

4–18 A subway train starts from rest at a station and accelerates at a rate of 2 m s^{-2} for 10 s. It then runs at constant speed for 30 s, and decelerates at 4 m s^{-2} until it stops at the next station. Find the *total* distance covered.

4–19 A body starts from rest, moves in a straight line with constant acceleration, and covers a distance of 64 ft in 4 s. (a) What was the final velocity? (b) How long a time was required to cover half the total distance? (c) What was the distance covered in one-half the total time? (d) What was the velocity when half the total distance had been covered? (e) What was the velocity after one-half the total time?

4–20 The speed of an automobile going north is reduced from 30 to 20 m s^{-1} in a distance of 125 m. Find (a) the magnitude and direction of the acceleration, assuming it to be constant, (b) the elapsed time, (c) the distance in which the car can be brought to rest from 20 m s^{-1}, assuming the acceleration of part (a).

4–21 An automobile and a truck start from rest at the same instant, with the automobile initially at some distance behind the truck. The truck has a constant acceleration of 2 m s^{-2} and the automobile an acceleration of 3 m s^{-2}. The automobile overtakes the truck after the truck has moved 75 m.

a) How long does it take the auto to overtake the truck?

b) How far was the auto behind the truck initially?

c) What is the velocity of each when they are abreast?

4–22

a) With what velocity must a ball be thrown vertically upward in order to rise to a height of 20 m?

b) How long will it be in the air?

4–23 A ball is thrown vertically downward from the top of a building, leaving the thrower's hand with a velocity of 30 ft s^{-1}.

a) What will be its velocity after falling for 2 s?

b) How far will it fall in 2 s?

c) What will be its velocity after falling 30 ft?

d) If it moved a distance of 3 ft while in the thrower's hand, find its acceleration while in his hand.

e) If the ball was released at a point 120 ft above the ground, in how many seconds will it strike the ground?

f) What will the velocity of the ball be when it strikes the ground?

4–24 A balloon, rising vertically with a velocity of 5 m s^{-1}, releases a sandbag at an instant when the balloon is 20 m above the ground.

a) Compute the position and velocity of the sandbag at the following times after its release: $\frac{1}{4}$ s, $\frac{1}{2}$ s, 1 s, 2 s.

b) How many seconds after its release will the bag strike the ground?

c) With what velocity will it strike?

4–25 A stone is dropped from the top of a tall cliff, and 1 s later a second stone is thrown vertically downward with a velocity of 20 m s^{-1}. How far below the top of the cliff will the second stone overtake the first?

4–26 A ball dropped from the cornice of a building takes 0.25 s to pass a window 3 m high. How far is the top of the window below the cornice?

4–27 A ball is thrown nearly vertically upward from a point near the cornice of a tall building. It just misses the cornice on the way down, and passes a point 160 ft below its starting point 5 s after it leaves the thrower's hand.

a) What was the initial velocity of the ball?

b) How high did it rise above its starting point?

c) What were the magnitude and direction of its velocity at the highest point?

d) What were the magnitude and direction of its acceleration at the highest point?

e) What was the magnitude of its velocity as it passed a point 64 ft below the starting point?

4–28 A juggler performs in a room whose ceiling is 3 m above the level of his hands. He throws a ball vertically upward so that it just reaches the ceiling.

a) With what initial velocity does he throw the ball?

b) What time is required for the ball to reach the ceiling?

He throws a second ball upward with the same initial

velocity, at the instant that the first ball is at the ceiling.

c) How long after the second ball is thrown do the two balls pass each other?

d) When the balls pass each other, how far are they above the juggler's hands?

4–29 An object is thrown vertically upward. It has a speed of 32 ft s^{-1} when it has reached one-half its maximum height.

a) How high does it rise?

b) What are its velocity and acceleration 1 s after it is thrown?

c) 3 s after?

d) What is the average velocity during the first half second?

4–30 A student determined to test the law of gravity for himself walks off a skyscraper 300 m high, stopwatch in hand, and starts his free fall (zero initial velocity). Five seconds later, Superman arrives at the scene and dives off the roof to save the student.

a) What must Superman's initial velocity be in order that he catch the student just before the ground is reached?

b) What must be the maximum height of the skyscraper so that even Superman can't save him? (Assume that Superman's acceleration is that of any freely falling body.)

4–31 A ball is thrown vertically upward from the ground and a student gazing out of the window sees it moving upward past him at 5 m s^{-1}. The window is 10 m above the ground.

a) How high does the ball go above the ground?

b) How long does it take to go from a height of 10 m to its highest point?

c) Find its velocity and acceleration $\frac{1}{2}$ s after it left the ground.

4–32 A ball is thrown vertically upward from the ground with a velocity of 30 m s^{-1}.

a) How long will it take to rise to its highest point?

b) How high does the ball rise?

c) How long after projection will the ball have a velocity of 10 ms^{-1} upward?

d) Of 10 m s^{-1} downward?

e) When is the displacement of the ball zero?

f) When is the magnitude of the ball's velocity equal to half its velocity of projection?

g) When is the magnitude of the ball's displacement equal to half the greatest height to which it rises?

h) What are the magnitude and direction of the acceleration while the ball is moving upward?

i) While moving downward?

j) When at the highest point?

4–33 A ball rolling on an inclined plane moves with a constant acceleration. One ball is released from rest at the top of an inclined plane 18 m long and reaches the bottom 3 s later. At the same instant that the first ball is released, a second ball is projected upward along the plane from its bottom with a certain initial velocity. The second ball is to travel part way up the plane, stop, and return to the bottom so that it arrives simultaneously with the first ball.

a) Find the acceleration.

b) What must be the initial velocity of the second ball?

c) How far up the plane will it travel?

4–34 The rocket-driven sled Sonic Wind No. 2, used for investigating the physiological effects of large accelerations and decelerations, runs on a straight, level track 3500 ft long. Starting from rest, it can reach a speed of 1000 mi hr^{-1} in 1.8 s.

a) Compute the acceleration, assuming it to be constant.

b) What is the ratio of this acceleration to that of a freely-falling body, g?

c) What is the distance covered?

d) A magazine article states that at the end of a certain run the speed of the sled was decreased from 632 mi hr^{-1} to zero in 1.4 s, and that as the sled decelerated its passenger was subjected to more than 40 times the pull of gravity (that is, the deceleration was greater than 40 g). Are these figures consistent?

4–35 The first stage of a rocket to launch an earth satellite will, if fired vertically upward, attain a speed of 4000 mi hr^{-1} at a height of 36 mi above the earth's surface, at which point its fuel supply will be exhausted.

a) Assuming constant acceleration, find the time to reach a height of 36 mi.

b) How much higher would the rocket rise if it continued to "coast" vertically upward?

4–36 Suppose the acceleration of gravity were only 1.0 m s^{-2}, instead of 10 m s^{-2}.

a) Estimate the height to which you could jump vertically from a standing start.

b) How high could you throw a baseball?

c) Estimate the maximum height of a window from which you would care to jump to a concrete sidewalk below. (Each story of an average building is about 4 m high.)

d) With what speed, in meters per second, would you strike the sidewalk?

e) How many seconds would be required?

4–37 After the engine of a moving motorboat is cut off, the boat has an acceleration in the opposite direction to its velocity and directly proportional to the square of its velocity. That is, $dv/dt = -kv^2$, where k is a constant.

a) Show that the magnitude v of the velocity at a time t after the engine is cut off is given by

$$\frac{1}{v} = \frac{1}{v_0} + kt.$$

b) Show that the distance x traveled in a time t is

$$x = \frac{1}{k}\ln(v_0 kt + 1).$$

c) Show that the velocity after traveling a distance x is

$$v = v_0 e^{-kx}.$$

As a numerical example, suppose the engine is cut off when the velocity $v_0 = 20$ ft s^{-1}, and that the velocity decreases to 10 ft s^{-1} in a time of 15 s.

d) Find the numerical value of the constant k, and the unit in which it is expressed.

e) Find the acceleration at the instant the engine is cut off.

f) Construct graphs of x, v, and a for a time of 20 s. Let 1 in. = 5 s horizontally, and 1 in. = 100 ft, 5 ft s^{-1}, and 0.5 ft s^{-2} vertically.

4–38 The equation of motion of a body suspended from a spring and oscillating vertically is $y = A \sin \omega t$, where A and ω are constants.

a) Find the velocity of the body as a function of time.

b) Find its acceleration as a function of time.

c) Find its velocity as a function of its coordinate.

d) Find its acceleration as a function of its coordinate.

e) What is the maximum distance of the body from the origin?

f) What is its maximum velocity?

g) What is its maximum acceleration?

h) Sketch graphs of y, v, and a as functions of time.

4–39 The acceleration of a body suspended from a spring and oscillating vertically is $a = -Ky$, where K is a constant and y is the coordinate measured from the equilibrium position. Suppose that a body moving in this way is given an initial velocity v_0 at the coordinate y_0. Find the expression for the velocity v of the body as a function of its coordinate y. [*Hint:* Use the expression $a = v\,dv/dy$.]

4–40 The motion of a body falling from rest in a resisting medium is described by the equation

$$\frac{dv}{dt} = A - Bv,$$

where A and B are constants. In terms of A and B, find

a) the initial acceleration, and

b) the velocity at which the acceleration becomes zero (the terminal velocity).

c) Show that the velocity at any time t is given by

$$v = \frac{A}{B}(1 - e^{-Bt}).$$

4–41 Two cars, A and B, travel in a straight line. The distance of A from the starting point is given as a function of time by $x_A = 4t + 4t^2$, and the distance of B from the starting point is $x_B = 2t^2 + 2t^3$.

a) Which car is ahead just after they leave the starting point?

b) At what times are the cars at the same point?

c) At what times is the velocity of B relative to A zero?

d) At what times is the distance from A to B neither increasing nor decreasing?

4–42 A hypothetical spaceship takes a straight-line path from the earth to the moon, a distance of about 400,000 km. Suppose it accelerates at 10 m s^{-2} for the first 10 min of the trip, then travels at constant speed until the last 10 min, when it decelerates at 10 m s^{-2}, just coming to rest as it reaches the moon.

a) What is the maximum speed attained?

b) What total time is required for the trip?

c) What fraction of the total distance is traveled at constant speed?

4–43 A "moving sidewalk" in an airport terminal building moves 1 m s^{-1} and is 150 m long. If a man steps on at one end and walks 2 m s^{-1} relative to the moving sidewalk, how much time does he require to reach the opposite end if he walks (a) in the same direction the sidewalk is moving; (b) in the opposite direction?

4–44 The driver of a car wishes to pass a truck which is traveling at a constant speed of 20 m s^{-1} (about 50 mi hr^{-1}). The car's maximum acceleration at this speed is 0.5 m s^{-2}. Initially the vehicles are separated by 25 m, and the car pulls back into the truck's lane after it is 25 m ahead of the truck. The car is 5 m long and the truck 20 m.

a) How much time is required for the car to pass the truck?

b) What distance does the car travel during this time?

c) What is the final speed of the car, assuming its acceleration while passing the truck is constant?

4–45 Two piers A and B are located on a river, one mile apart. Two men must make round trips from pier A to pier B and return. One man is to row a boat at a velocity of 4 mi hr^{-1} relative to the water, and the other man is to walk on the shore at a velocity of 4 mi hr^{-1}. The velocity of the river is 2 mi hr^{-1} in the direction from A to B. How long does it take each man to make the round trip?

4–46 A passenger on a ship traveling due east with a speed of 18 knots observes that the stream of smoke from the ships' funnels makes an angle of 20° with the ship's wake. The wind is blowing from south to north. Assume that the smoke acquires a velocity (with respect to the earth) equal to the velocity of the wind, as soon as it leaves the funnels. Find the velocity of the wind.

4–47 An airplane pilot wishes to fly due north. A wind of 60 mi hr^{-1} is blowing toward the west. If the flying speed of the plane (its speed in still air) is 180 mi hr^{-1}, in what direction should the pilot head? What is the speed of the plane over the ground? Illustrate with a vector diagram.

4–48 An airplane pilot sets a compass course due west and maintains an air speed of 240 km hr^{-1}. After flying for $\frac{1}{2}$ hr, he finds himself over a town which is 150 km west and 40 km south of his starting point.

a) Find the wind velocity, in magnitude and direction.

b) If the wind velocity were 120 km hr^{-1} due south, in what direction should the pilot set his course in order to travel due west? Take the same air speed of 240 km hr^{-1}.

4–49 When a train has a speed of 10 m s^{-1} eastward, raindrops which are falling vertically with respect to the earth make traces which are inclined 30° to the vertical on the windows of the train.

a) What is the horizontal component of a drop's velocity with respect to the earth? With respect to the train?

b) What is the velocity of the raindrop with respect to the earth? With respect to the train?

4–50 A river flows due north with a velocity of 2 m s^{-1}. A man rows a boat across the river, his velocity relative to the water being 3 m s^{-1} due east.

a) What is his velocity relative to the earth?

b) If the river is 1000 m wide, how far north of his starting point will he reach the opposite bank?

c) How long a time is required to cross the river?

4–51

a) In what direction should the rowboat in Problem 4–50 be headed in order to reach a point on the opposite bank directly east from the start?

b) What will be the velocity of the boat relative to the earth?

c) How long a time is required to cross the river?

4–52 A motorboat is observed to travel 10 mi hr^{-1} relative to the earth in the direction 37° north of east. If the velocity of the boat due to the wind is 2 mi hr^{-1} eastward and that due to the current is 4 mi hr^{-1} southward, what is the magnitude and direction of the velocity of the boat due to its own power?

Chapter 5

Newton's Second Law. Gravitation

5-1 INTRODUCTION

In the preceding chapters we have discussed separately the concepts of force and acceleration. We have made use, in problems in equilibrium, of Newton's first law, which states that when the resultant force on a body is zero, the acceleration of the body is also zero. The next logical step is to ask how a body behaves when the resultant force on it is *not* zero. The answer to this question is contained in Newton's second law, which states that when the resultant force is not zero the body moves with accelerated motion, and the acceleration, with a given force, depends on a property of the body known as its *mass.*

This part of mechanics, which includes the study both of motion and of the forces that bring about the motion, is called *dynamics.* In its broadest sense, dynamics includes nearly the whole of mechanics. Statics treats of special cases in which the acceleration is zero, and kinematics deals with motion only.

It will be assumed in this chapter that all velocities are small compared with the velocity of light, so that relativistic considerations do not arise. We shall also assume that unless otherwise stated all velocities and accelerations are measured relative to an inertial reference system (Section 2-3), and shall consider rectilinear motion only. Motion in a curved path will be discussed in the next chapter.

5-2 NEWTON'S SECOND LAW. MASS

We know from experience that an object at rest never starts to move by itself; a push or pull must be exerted on it by some other body. Similarly, a force is required to slow down or stop a body already in motion, and to make a moving body deviate from straight-line motion requires a sideways force. All these processes (speeding up, slowing down, or changing direction) involve a change in either the magnitude or direction of the velocity. Thus in each case the body is *accelerated,* and an external force must act on it to produce the acceleration.

Let us consider some experiments. A small body (a particle) rests on a level frictionless surface, and moves to the right along the x-axis of an inertial reference system, as in Fig. 5-1(a). A horizontal force $\boldsymbol{F}$, measured by a spring balance calibrated as described in Section 1-4, is exerted on the body. We find that the velocity of the body increases as long as the force acts. In other words, the body has an acceleration $\boldsymbol{a}$ toward the right. If the magnitude of the force $\boldsymbol{F}$ is kept constant, the velocity is found to increase at a constant rate. If the force is altered, the rate of change of velocity alters in the same proportion. Doubling the force doubles the rate of change of velocity, halving the force halves the rate of change, etc. If the force is reduced to zero, the rate

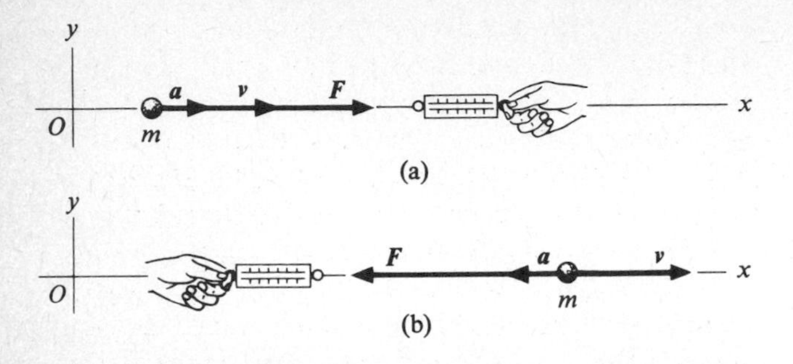

Fig. 5-1 *The acceleration* $\boldsymbol{a}$ *is proportional to the force* $\boldsymbol{F}$ *and is in the same direction as the force.*

of change of velocity is zero and the body continues to move with constant velocity.

Before the time of Galileo and Newton, it was generally believed that a force was necessary just to keep a body moving, even on a level frictionless surface or in "outer space." Galileo and Newton realized that *no* force is necessary to keep a body moving, once it has been set in motion, and that the effect of a force is not to *maintain* the velocity of a body, but to *change* its velocity. The *rate of change* of velocity, for a given body, is directly proportional to the force exerted on it.

In Fig. 5-1(b), the velocity of the body is also toward the right, but the force is toward the left. Under these conditions the body moves more slowly. If the force continues to act, it will ultimately reverse the body's direction of motion. The acceleration will then be toward the *left*, in the same direction as the force $\boldsymbol{F}$. Hence we conclude not only that the *magnitude* of the acceleration is proportional to that of the force, but that the *direction* of the acceleration is the *same* as that of the force, regardless of the direction of the velocity.

To say that the rate of change of the velocity of a body is directly proportional to the force exerted on it is to say that the ratio of the force to the rate of change of velocity is a constant, regardless of the magnitude of the force. This constant ratio of force to rate of change of velocity is called the *mass* m of the body. Thus

$$m = \frac{F}{dv/dt} = \frac{F}{a},$$

or

$$\boldsymbol{F} = m\frac{d\boldsymbol{v}}{dt} = m\boldsymbol{a}. \qquad (5\text{-}1)$$

By writing this as a vector equation, we automatically include the experimental fact that the direction of the acceleration $\boldsymbol{a}$ is the same as that of the force $\boldsymbol{F}$. Mass is, however, a *scalar* quantity.

The mass of a body can be thought of as the *force per unit of acceleration.* For example, if the acceleration of a certain body is found to be 5 ft s^{-2} when the force is 20 lb, the mass of the body is

$$m = \frac{20\ \text{lb}}{5\ \text{ft s}^{-2}} = 4\ \text{lb ft}^{-1}\ \text{s}^2,$$

and a force of 4 lb must be exerted on the body for every ft s^{-2} of acceleration.

When a large force is needed to give a body a certain acceleration (i.e., speed it up, slow it down, or deviate it if it is in motion), the mass of the body is large; if only a small force is needed for the same acceleration, the mass is small. Thus the mass of a body is a quantitative measure of the property described in everyday language as *inertia.*

In *rectilinear* motion, the force $\boldsymbol{F}$ acting on a body and its velocity $\boldsymbol{v}$ always have the same action line, as in Fig. 5-1. If the direction of the force is *not* the same as that of the velocity, the body is deflected sideways and moves in a curved path. We shall see in the next chapter, however, that Eq. (5-1) applies in this case also, except that the change in velocity $\Delta\boldsymbol{v}$, or the acceleration $\boldsymbol{a}$, includes the change in *direction* as well as the change in *magnitude* of the velocity. In every case, then, *the vector force equals the product of the mass and the vector acceleration.*

When two vectors are equal, their rectangular components are equal also. Hence the vector equation (5-1) is equivalent (for forces and accelerations in the xy-plane) to the pair of scalar equations

$$F_x = m\frac{dv_x}{dt} = ma_x, \quad F_y = m\frac{dv_y}{dt} = ma_y. \qquad (5\text{-}2)$$

This means that each component of the force can be considered to produce its own component of acceleration. It follows that if any number of forces act on a body simultaneously, as is often the case in problems of practical interest, the forces may be resolved into x- and y-components, the algebraic sums $\sum F_x$ and $\sum F_y$ may be computed, and the components of acceleration are given by

$$\sum F_x = m\frac{dv_x}{dt} = ma_x, \quad \sum F_y = m\frac{dv_y}{dt} = ma_y.$$

This pair of equations is equivalent to the single vector equation

$$\boxed{\sum \boldsymbol{F} = m\frac{d\boldsymbol{v}}{dt} = m\boldsymbol{a},} \qquad (5\text{–}3)$$

where we write the lefthand side explicitly as $\sum \boldsymbol{F}$ to emphasize that the acceleration is determined by the *resultant* of *all* of the *external* forces acting *on* the body.

Equation (5–3) is the mathematical statement of Newton's *second law of motion.* If we think of the equation as solved for $\boldsymbol{a}$ or $d\boldsymbol{v}/dt$, the law can be stated: *The rate of change of the velocity of a particle, or its acceleration, is equal to the resultant of all external forces exerted on the particle divided by the mass of the particle, and is in the same direction as the resultant force.* The acceleration is to be measured relative to an inertial system.

It is necessary to state the law for a *particle*, because when a resultant force acts on an extended body, the body may be set in rotation and not all particles in it have the same acceleration. We shall discuss this more fully later on, but it may be said at this point that the acceleration of the center of gravity of a body is the same as that of a particle whose mass equals that of the body.

5-3 SYSTEMS OF UNITS

Nothing was said in the preceding discussion regarding the *units* in which force, mass, and acceleration are to be expressed. It is evident, however, from the equation $\boldsymbol{F} = m\boldsymbol{a}$, that these must be such that *unit force imparts unit acceleration to unit mass.*

Several systems of units are in common use in the United States, including the meter-kilogram-second (mks) system, the centimeter-gram-second (cgs) system, and the British system. This book will use principally the mks system, not only because it is convenient and widely used in scientific work, but also because there seems little doubt that the metric system (mks, cgs, or a combination) will eventually be adopted worldwide for commerce and industry as well as scientific work. The United States and Britain are the only two major countries that do not use the metric system, and Britain is beginning to convert.

In the mks system, the mass of the standard kilogram is the unit of mass, and the unit of acceleration is 1 m s^{-2}. The unit force in this system is then *that force which gives a standard kilogram an acceleration of* 1 m s^{-2}. This force is called *one newton* (1 N). One newton is approximately equal to one-quarter of a pound-force (more precisely, 1 N = 0.22481 lb.) Thus in the mks system,

$$F(\text{N}) = m(\text{kg}) \times a(\text{m s}^{-2}).$$

The mass unit in the *centimeter-gram-second* (cgs) system is one gram, equal to 1/1000 kg, and the unit of acceleration is 1 cm s^{-2}. The unit force in this system is then *that force which gives a body whose mass is one gram, an acceleration of* 1 cm s^{-2}. This force is called *one dyne* (1 dyn). Since 1 kg = 10^3 g and 1 m s^{-2} = 10^2 cm s^{-2}, it follows that 1 N = 10^5 dyn. In the cgs system,

$$F(\text{dyn}) = m(\text{g}) \times a(\text{cm s}^{-2}).$$

In the mks and cgs systems, we first selected units of mass and acceleration, and defined the unit of force in terms of these. In the *British engineering system*, we first select a unit of *force* (1 lb) and a unit of acceleration (1 ft s^{-2}), and then define the unit of mass as *the mass of a body whose acceleration is* 1 ft s^{-2} *when the resultant force on the body is* 1 lb. This unit of mass is called *one slug.* (The origin of the name is obscure; it may have arisen from the concept of mass as inertia or *sluggishness.*) Then in the engineering system,

$$F(\text{lb}) = m\,(\text{slugs}) \times a(\text{ft s}^{-2}).$$

The pound is used in everyday life as a unit of quantity of matter (e.g., a pound of butter) but properly speaking it is a unit of *force* or *weight.* Thus a pound of butter is that quantity which has a weight

Table 5–1

System of units	Force	Mass	Acceleration
mks	newton (N)	kilogram (kg)	$m\ s^{-2}$
cgs	dyne (dyn)	gram (g)	$cm\ s^{-2}$
engineering	pound (lb)	slug	$ft\ s^{-2}$

of 1 lb. A useful fact in converting between mks and British units is that an object with a *mass* of 1 kg has a *weight* of about 2.205 lb.

The units of force, mass, and acceleration in the three systems are summarized in Table 5–1.

Example 1 A constant horizontal force of 2 N is applied to a body of mass 4 kg, resting on a level frictionless surface. The acceleration of the body is

$$a = \frac{F}{m} = \frac{2\ \text{N}}{4\ \text{kg}} = 0.5\ \text{m s}^{-2}.$$

Since the force is constant, the acceleration is constant also. Hence if the initial position and velocity of the body are known, the velocity and position at any later time can be found from the equations of motion with constant acceleration.

Example 2 A body of mass 0.2 kg is given an initial velocity of $0.4\ \text{m s}^{-1}$ toward the right along a level laboratory tabletop. The body is observed to slide a distance of 1.0 m along the table before coming to rest. What was the magnitude and direction of the friction force f acting on it?

In the absence of further information, let us assume that the friction force is constant. The acceleration is then constant also; from the equations of motion with constant acceleration, we have

$$v^2 = v_0^2 + 2ax, \qquad 0 = (0.4\ \text{m s}^{-1})^2 + (2a)(1.0\ \text{m}),$$
$$a = -0.08\ \text{m s}^{-2}.$$

The negative sign means that the acceleration is toward the *left* (although the velocity is toward the right). The friction force on the body is

$$f = ma = (0.2\ \text{kg})(-0.08\ \text{m s}^{-2}) = -0.016\ \text{N},$$

and is toward the left also. (A force of equal magnitude, but directed toward the right, is exerted on the table by the sliding body.)

Example 3 The acceleration of a certain body is found to be $5\ \text{ft s}^{-2}$ when the resultant force on the body is 20 lb. The mass of the body is

$$m = \frac{F}{a} = \frac{20\ \text{lb}}{5\ \text{ft s}^{-2}} = 4\ \text{lb ft}^{-1}\ \text{s}^2 = 4\ \text{slugs}.$$

5–4 NEWTON'S LAW OF UNIVERSAL GRAVITATION

Throughout our study of mechanics we have been continually encountering the force of gravitational attraction between a body and the earth. We now wish to study this phenomenon of gravitation in more detail.

The law of universal gravitation was discovered by Newton, and was published by him in 1686. There seems to be some evidence that Newton was led to deduce the law from speculations concerning the fall of an apple toward the earth, but his first published calculations to justify its correctness had to do with the motion of the moon around the earth.

Newton's law of gravitation may be stated: *Every particle of matter in the universe attracts every other particle with a force which is directly proportional to the product of the masses of the particles and inversely proportional to the square of the distance between them.* Thus

$$F_g = G\frac{mm'}{r^2}, \tag{5–4}$$

where F_g is the gravitational force on either particle, m and m' are their masses, r is the distance between them and G is a universal constant called the *gravitational constant,* whose numerical value depends on the units in which force, mass, and length are expressed.

The gravitational forces acting on the particles form an action–reaction pair. Although the masses of the particles may be different, forces of *equal* magni-

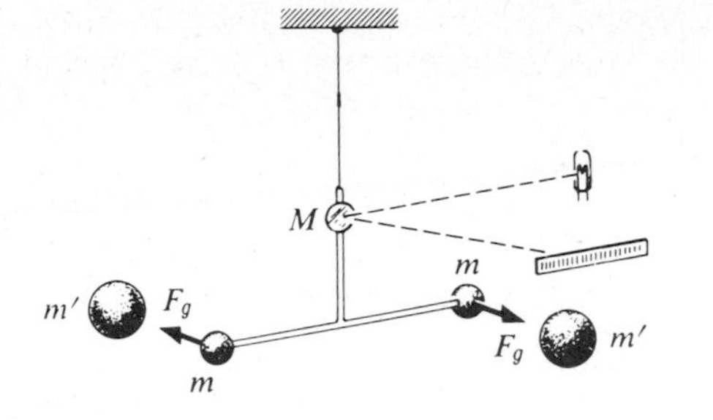

Fig. 5-2 *Principle of the Cavendish balance.*

tude act on each, and the action line of both forces lies along the line joining the particles.

Newton's law of gravitation refers to the force between two *particles*. It can also be shown, however, that the force of attraction exerted on or by a homogeneous sphere is the same as if the mass of the sphere were concentrated at its center. The proof is not difficult but is too long to give here, and we shall simply state as a fact that *the gravitational force exerted on or by a homogeneous sphere is the same as if the entire mass of the sphere were concentrated in a point at its center.* Thus if the earth were a homogeneous sphere, of mass m_E, the force exerted by it on a small body of mass m, at a distance r from its center, would be

$$F_g = G\frac{mm_E}{r^2}.$$

A force of the same magnitude would be exerted *on* the earth by the body.

The magnitude of the gravitational constant G can be found experimentally by measuring the force of gravitational attraction between two bodies of known masses m and m', at a known separation. For bodies of moderate size the force is extremely small, but it can be measured with an instrument invented by the Rev. John Michell and first used for this purpose by Sir Henry Cavendish in 1798. The same type of instrument was also used by Coulomb for studying forces of electrical and magnetic attraction and repulsion.

The Cavendish balance consists of a light, rigid T-shaped member (Fig. 5-2) supported by a fine vertical fiber such as a quartz thread or a thin metallic ribbon. Two small spheres of mass m are mounted at the ends of the horizontal portion of the T, and a small mirror M, fastened to the vertical portion, reflects a beam of light onto a scale. To use the balance, two large spheres of mass m' are brought up to the positions shown. The forces of gravitational attraction between the large and small spheres result in a *couple* which twists the system through a small angle, thereby moving the reflected light beam along the scale.

By using an extremely fine fiber, the deflection of the mirror may be made sufficiently large so that the gravitational force can be measured quite accurately. The gravitational constant, measured in this way, is found to be

$$G = 6.670 \times 10^{-11}\ \mathrm{N\,m^2\,kg^{-2}}.$$

Example 1 The mass m of one of the small spheres of a Cavendish balance is 0.001 kg, the mass m' of one of the large spheres is 0.5 kg, and the center-to-center distance between the spheres is 0.05 m. The gravitational force on each sphere is

$$F_g = 6.67 \times 10^{-11}\ \mathrm{N\,m^2\,kg^{-2}}\frac{(0.001\ \mathrm{kg})(0.5\ \mathrm{kg})}{(0.05\ \mathrm{m})^2}$$
$$= 1.33 \times 10^{-11}\ \mathrm{N},$$

or about one hundred-billionth of a newton!

Example 2 Suppose the spheres in Example 1 are placed 0.05 m from each other at a point in space far removed from all other bodies. What is the acceleration of each, relative to an inertial system?

The acceleration a of the smaller sphere is

$$a = \frac{F_g}{m} = \frac{1.33 \times 10^{-11}\ \mathrm{N}}{1.0 \times 10^{-3}\ \mathrm{kg}} = 1.33 \times 10^{-8}\ \mathrm{m\,s^{-2}}.$$

The acceleration a' of the larger sphere is

$$a' = \frac{F_g}{m'} = \frac{1.33 \times 10^{-11}\ \mathrm{N}}{0.5\ \mathrm{kg}} = 2.67 \times 10^{-11}\ \mathrm{m\,s^{-2}}.$$

In this case, the accelerations are *not* constant because the gravitational force increases as the spheres approach each other.

5–5 MASS AND WEIGHT

The *weight* of a body can now be defined more generally than in the preceding chapters as *the resultant gravitational force exerted on the body by all other bodies in the universe.* At or near the surface of the earth, the force of the earth's attraction is so much greater than that of any other body that for practical purposes all other gravitational forces can be neglected and the weight can be considered as arising solely from the gravitational attraction of the earth. Similarly, at the surface of the moon, or of another planet, the weight of a body results almost entirely from the gravitational attraction of the moon or the planet. Thus if the earth were a homogeneous sphere of radius R and mass m_E, the weight w of a small body of mass m at or near its surface would be

$$w = F_g = \frac{Gmm_E}{R^2}. \quad (5\text{–}5)$$

The weight of a given body varies by a few tenths of a percent from point to point on the earth's surface, partly because of local deposits of ore, oil, or other substances whose density differs from the average, and partly because the earth is not a perfect sphere but is flattened somewhat at the poles. Also, the weight of a given body decreases inversely with the square of its distance from the earth's center, and at a radial distance of two earth radii, for example, it has decreased to one-quarter of its value at the earth's surface.

The *apparent* weight of a body at the surface of the earth differs slightly in magnitude and direction from the earth's force of gravitational attraction because of the rotation of the earth about its axis. For the present, we shall ignore the small difference between the apparent weight of a body and the force of the earth's gravitational attraction, and shall assume that the earth is an inertial reference system. Then when a body is allowed to fall freely, the force accelerating it is its weight $\mathbf{w}$ and the acceleration produced by this force is the acceleration due to gravity, $\mathbf{g}$. The general relation

$$\mathbf{F} = m\mathbf{a}$$

therefore becomes, for the special case of a freely falling body,

$$\boxed{\mathbf{w} = m\mathbf{g}.} \quad (5\text{–}6)$$

Since

$$w = mg = G\frac{mm_E}{R^2},$$

it follows that

$$g = \frac{Gm_E}{R^2}, \quad (5\text{–}7)$$

showing that the acceleration due to gravity is the same for *all* bodies (since m cancelled out) and is very nearly constant (since G and m_E are constants and R varies only slightly from point to point on the earth).

The weight of a body is a force, and must be expressed in terms of the unit of force in the particular system of units one is using. Thus in the mks system the unit of weight is 1 N, in the cgs system it is 1 dyn, and in the engineering system it is 1 lb. Equation (5–6) indicates the relation between the mass and weight of a body in any consistent set of units.

For example, the weight of a standard kilogram, at a point where $g = 9.80 \text{ m s}^{-2}$, is

$$w = mg = 1 \text{ kg} \times 9.80 \text{ m s}^{-2} = 9.80 \text{ N}.$$

At a second point where $g = 9.78 \text{ m s}^{-2}$, the weight is

$$w = 9.78 \text{ N}.$$

Thus, unlike the mass of a body, which is a constant, the weight varies from one point to another. On the moon, where $g = 1.67 \text{ m s}^{-2}$, the weight is 1.67 N, but the mass is still 1 kg.

The weight of a body whose mass is 1 slug, at a point where $g = 32.0 \text{ ft s}^{-2}$, is

$$w = mg = 1 \text{ slug} \times 32.0 \text{ ft s}^{-2} = 32.0 \text{ lb},$$

and the mass of a man who weighs 160 lb at this point is

$$m = \frac{w}{g} = \frac{160 \text{ lb}}{32.0 \text{ ft s}^{-2}} = 5 \text{ slugs}.$$

If we insert for the weight w in the equation $w = mg$, the gravitational force F_g as given by Newton's law of gravitation, we obtain, after cancelling the mass m,

$$m_E = \frac{R^2 g}{G},$$

where R is the earth's radius. All of the quantities on the right are known, so the mass of the earth, m_E, can be calculated. Taking $R = 6370 \text{ km} = 6.37 \times 10^6 \text{ m}$, and $g = 9.80 \text{ m s}^{-2}$, we find

$$m_E = 5.98 \times 10^{24} \text{ kg}.$$

The volume of the earth is

$$V = \frac{4}{3}\pi R^3 = 1.08 \times 10^{21} \text{ m}^3.$$

The mass of a body divided by its volume is known as its average *density*. (The density of water is $1 \text{ g cm}^{-3} = 1000 \text{ kg m}^{-3}$.) The average density of the earth is therefore

$$\frac{m_E}{V} = 5.5 \text{ g cm}^{-3} = 5500 \text{ kg m}^{-3}.$$

The density of most rock near the earth's surface, such as granites and gneisses, is about $3 \text{ g cm}^{-3} = 3000 \text{ kg m}^{-3}$, so the interior of the earth must have much higher density. Some rock found on the surface, notably basaltic rock, has a density of about 5 g cm^{-3}.

As with many other physical quantities, the mass of a body can be measured in several different ways. One is to use the relation by which the quantity is defined, which in this case is the ratio of the force on the body to its acceleration. A measured force is applied to the body, its acceleration is measured, and the unknown mass is obtained by dividing the force by the acceleration. This method is used exclusively to measure masses of atomic particles.

The second method consists of finding by trial some other body whose mass (a) is equal to that of the given body, and (b) is already known. Consider first a method of determining when two masses are equal. At the same point on the earth's surface all bodies fall freely with the same acceleration g. Since the weight w of a body equals the product of its mass m and the acceleration g, it follows that if, at the same point, the weights of the two bodies are equal, their *masses* are equal also. The *equal-arm balance* is an instrument by means of which one can determine very precisely when the weights of two bodies are equal, and hence when their masses are equal.

It will be seen from the preceding discussion that the property of matter called *mass* makes itself evident in two very different ways. The force of gravitational attraction between two particles is said to be proportional to the product of their masses, and in this sense mass can be considered as *that property of matter by virtue of which every particle exerts a force of attraction on every other particle.* We may call this property *gravitational mass.* On the other hand, Newton's second law is concerned with an entirely different property of matter, namely, the fact that a force (not necessarily gravitational) must be exerted on a particle in order to accelerate it, i.e., to change its velocity, either in magnitude or direction. This property can be called *inertial mass.* It is not at all obvious that the gravitational mass of a particle should be the same as its inertial mass, but experiment shows that the two are in fact the same. That is, if we have to push twice as hard on body A as we do on body B to produce a given acceleration, then the force of gravitational attraction between body A and some third body C is twice as great as the gravitational attraction between body B and body C, the distance between them being the same. Thus when an equal-arm balance is used to compare masses, it is actually *gravitational* mass that is being measured. Therefore the property represented by m in Newton's second law can be operationally defined as the result obtained by the prescribed methods of using an equal-arm balance.

5–6 APPLICATIONS OF NEWTON'S SECOND LAW

We now give a number of applications of Newton's second law to specific problems. In all these examples, and in the problems at the end of the chapter, it will be assumed that the acceleration due to gravity is 9.80 m s^{-2} or 32.0 ft s^{-2}, unless otherwise specified. In later chapters the approximate value $g = 10 \text{ m s}^{-2}$, which is in error by only about 2%, will often be used for convenience.

Example 1 A block whose mass is 10 kg rests on a horizontal surface. What constant horizontal force T is required to give it a velocity of 4 m s^{-1} in 2 s,

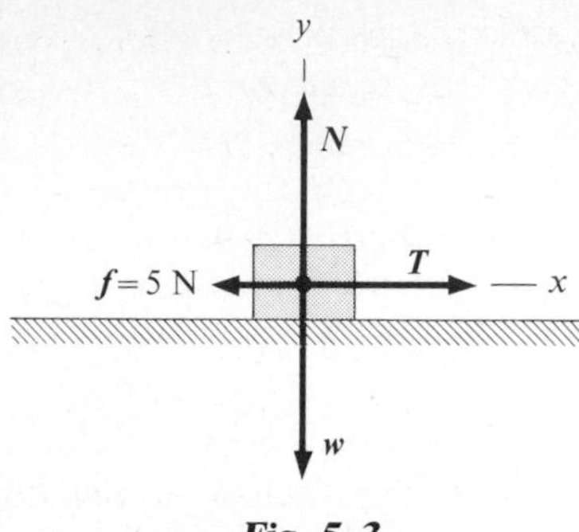

Fig. 5–3

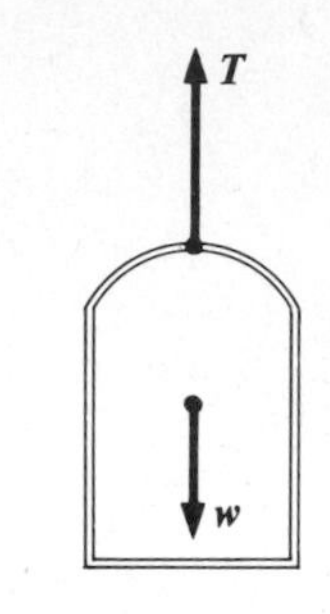

Fig. 5–4 The resultant force is **T** − **w**.

starting from rest, if the friction force between the block and the surface is constant and is equal to 5 N? Assume that all forces act at the center of the block. (See Fig. 5–3.)

The mass of the block is given. The y-component of its acceleration is zero. The x-component of acceleration can be found from the data on the velocity acquired in a given time. Since the forces are constant, the x-acceleration is constant and, from the equations of motion with constant acceleration,

$$a_x = \frac{v - v_0}{t} = \frac{4\ \text{m s}^{-1} - 0}{2\ \text{s}} = 2\ \text{m s}^{-2}.$$

The resultant of the x-forces is

$$\sum F_x = T - f,$$

and that of the y-forces is

$$\sum F_y = N - w.$$

Hence, from Newton's second law in component form,

$$T - f = ma_x, \qquad N - w = ma_y = 0.$$

From the second equation, we find that

$$N = w = mg = (10\ \text{kg})(9.80\ \text{m s}^{-2}) = 98.0\ \text{N},$$

and from the first,

$$T = f + ma_x = 5\ \text{N} + (10\ \text{kg})(2\ \text{m s}^{-2}) = 25\ \text{N}.$$

Example 2 An elevator and its load have a total mass of 800 kg. Find the tension T in the supporting cable when the elevator, originally moving downward at 10 m s^{-1}, is brought to rest with constant acceleration in a distance of 25 m. (See Fig. 5–4).

The weight of the elevator is

$$w = mg = (800\ \text{kg})(9.8\ \text{m s}^{-2}) = 7840\ \text{N}.$$

From the equations of motion with constant acceleration,

$$v^2 = v_0^2 + 2ay, \qquad a = \frac{v^2 - v_0^2}{2y}.$$

The initial velocity v_0 is -10 m s^{-1}; the final velocity v is zero. If we take the origin at the point where the deceleration begins, then $y = -25$ m. Hence

$$a = \frac{0 - (-10\ \text{m s}^{-1})^2}{2(-25\ \text{m})} = 2\ \text{m s}^{-2}.$$

The acceleration is therefore positive (upward). From the free-body diagram (Fig. 5–4) the resultant force is

$$\sum F = T - w = T - 7840\ \text{N}.$$

Since $F = ma$,

$$T - 7840\ \text{N} = (800\ \text{kg})(2\ \text{m s}^{-2}) = 1600\ \text{N},$$

$$T = 9440\ \text{N}.$$

The tension must be *greater* than the weight by 1600 N to cause the upward acceleration when the elevator stops.

Example 3 In the above example, with what force do the feet of a passenger press downward on the floor, if the passenger's mass is 80 kg?

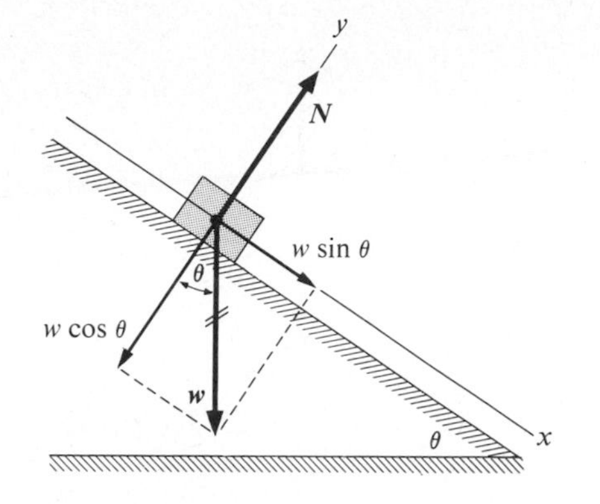

Fig. 5–5 *A block on a frictionless inclined plane.*

We first find the force the floor exerts *on* the passenger's feet, which is the reaction force to the force required. The forces on the passenger are his weight (80 kg)(9.8 m s^{-2}) = 784 N and the upward force F caused by the floor. Thus Newton's second law applied to the man, whose acceleration is the same as that of the elevator, gives

$$F - 784 \text{ N} = (80 \text{ kg})(2 \text{ m s}^{-2}) = 160 \text{ N},$$

and

$$F = 944 \text{ N}.$$

Thus the man pushes down on the floor with a force of 944 N while the elevator is stopping, and he *feels* a greater strain in his legs and feet than while standing at rest. If he stands on a bathroom scale calibrated in newtons, what will the scale read?

Example 4 What is the acceleration of a block on a frictionless plane inclined at an angle θ with the horizontal?

The only forces acting on the block are its weight $\boldsymbol{w}$ and the normal force $\boldsymbol{N}$ exerted by the plane (Fig. 5–5).

Take axes parallel and perpendicular to the surface of the plane and resolve the weight into x- and y- components. Then

$$\sum F_y = N - w \cos \theta,$$

$$\sum F_x = w \sin \theta.$$

But we know that $a_y = 0$, so, from the equation $\sum F_y = ma_y$, we find that $N = w \cos \theta$. From the equation $\sum F_x = ma_x$, we have

$$w \sin \theta = ma_x,$$

and since $w = mg$,

$$a_x = g \sin \theta.$$

The mass does not appear in the final result, which means that any block, regardless of its mass, will slide on a frictionless inclined plane with an acceleration down the plane of $g \sin \theta$. In particular, when $\theta = 0$, $a_x = 0$, and when the plane is vertical, $\theta = 90°$ and $a_x = g$, as we should expect.

Example 5 Refer to Fig. 2–3 (Chapter 2). Let the mass of the block be 4 kg and that of the rope be 0.5 kg. If the force F_1 is 9 N, what are the forces $\boldsymbol{F}_1'$, $\boldsymbol{F}_2$, and $\boldsymbol{F}_2'$? The surface on which the block moves is level and frictionless.

We know from Newton's third law that $F_1 = F_1'$ and that $F_2 = F_2'$. Hence $F_1' = 9$ N. The force F_2 could be computed by applying Newton's second law to the block, if its acceleration were known, or the force F_2' could be computed by applying this law to the rope if its acceleration were known. The acceleration is not given, but it can be found by considering the block and rope together as a single system. The vertical forces on this system need not be considered. Since there is no friction, the resultant *external* force acting *on* the system is the force F_1. (The forces F_2 and F_2' are *internal* forces when we consider block and rope as a single system, and the force F_1' does not act on the system, but *on the man.*) Then, from Newton's second law,

$$\sum F = ma,$$

$$9 \text{ N} = (4 \text{ kg} + 0.5 \text{ kg})a,$$

$$a = 2 \text{ m s}^{-2}.$$

We can now apply Newton's second law to the block:

$$\sum F = ma,$$

$$F_2 = (4 \text{ kg})(2 \text{ m s}^{-2}) = 8 \text{ N}.$$

If we consider the rope alone, the resultant force on it is

$$\sum F = F_1 - F_2' = 9\text{ N} - F_2',$$

and from the second law,

$$9\text{ N} - F_2' = (0.5\text{ kg})(2\text{ m s}^{-2}) = 1\text{ N},$$

$$F_2' = 8\text{ N}.$$

In agreement with Newton's third law, which was tacitly used when the forces F_2 and F_2' were omitted in considering the system as a whole, we find that F_2 and F_2' are equal in magnitude. Note, however, that the forces F_1 and F_2 are *not* equal and opposite (the rope is not in equilibrium) and that these forces are *not* an action–reaction pair.

Example 6 In Fig. 5–6, a block of weight w_1 (mass $= m_1$) moves on a level frictionless surface, connected by a light flexible cord passing over a small frictionless pulley to a second hanging block of weight w_2(mass $= m_2$). What is the acceleration of the system, and what is the tension in the cord connecting the two blocks?

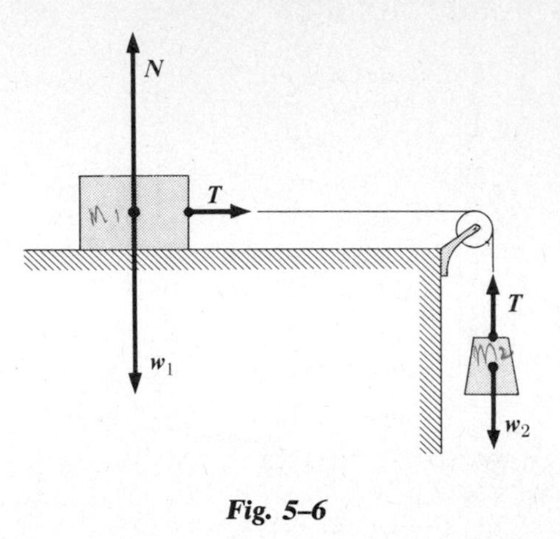

Fig. 5–6

The diagram shows the forces acting on each block. The forces exerted on the blocks by the cord can be considered an action–reaction pair, so we have used the same symbol T for each. For the block on the surface,

$$\sum F_x = T = m_1 a,$$

$$\sum F_y = N - w_1 = 0.$$

Since the cord connecting the two blocks is inextensible, the accelerations are the same. Applying Newton's second law to the hanging block, we obtain

$$\sum F_y = w_2 - T = m_2 a.$$

Addition of the first and third equations gives

$$w_2 = (m_1 + m_2)a,$$

or

$$a = \frac{w_2}{m_1 + m_2},$$

which says that the acceleration of the *entire system* equals the *resultant external force* (w_2) divided by the *total mass* ($m_1 + m_2$). Since $w_2 = m_2 g$,

$$a = g\frac{m_2}{m_1 + m_2}.$$

Eliminating a from the first and third equations, we get

$$T = w_2\frac{m_1}{m_1 + m_2},$$

so that T is *only a fraction of* w_2. Although the earth pulls on the *hanging* block with a force $\mathbf{w}_2$, the force exerted on the *sliding* block is only a fraction of $\mathbf{w}_2$. It is not the earth that pulls on the sliding block, but the cord, whose tension *must be less than* $\mathbf{w}_2$ if the hanging block is to accelerate downward.

Example 7 A body of mass m is suspended from a spring balance attached to the roof of an elevator, as shown in Fig. 5–7. What is the reading of the balance if the elevator has an acceleration a, relative to the earth? Consider the earth's surface to be an inertial reference system.

The forces on the body are its weight $\mathbf{w}$ (the gravitational force $\mathbf{f}_g$ exerted on it by the earth) and the upward force $\mathbf{T}$ exerted on it by the balance. The body is at rest relative to the elevator, and hence has an acceleration $\boldsymbol{a}$ relative to the earth. (We take the downward direction as positive.) The resultant force on the body is $w - T$, so, from Newton's second law,

$$w - T = ma, \qquad T = w - ma.$$

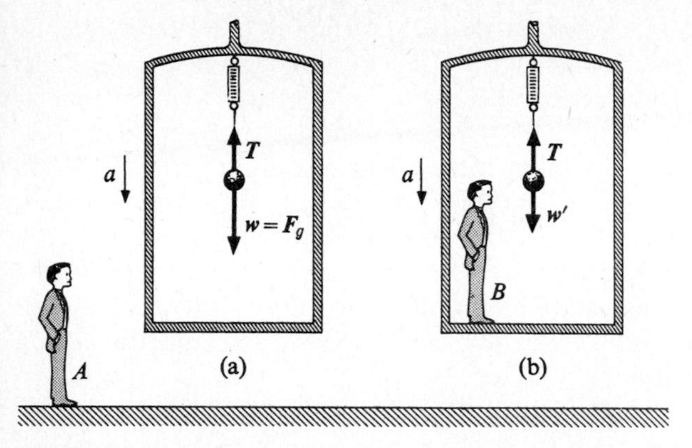

Fig. 5–7 *(a) To observer A, the body has a downward acceleration a, and he writes* $\mathbf{w} - \mathbf{T} = m\mathbf{a}$. *(b) To observer B, the acceleration of the body is zero. He writes* $\mathbf{w}' = \mathbf{T}$.

By Newton's third law, the body pulls down on the balance with a force equal and opposite to $\mathbf{T}$, or equal to $(w - ma)$, so the balance reading equals $(w - ma)$.

If the same body were suspended in equilibrium from a balance attached to the earth, the balance reading would equal the weight w. To an observer riding in the elevator, the body *appears* to be in equilibrium and hence *appears* to be acted on by a downward force $\mathbf{w}'$ equal in magnitude to the balance reading, as in Fig. 5–7(b). This *apparent* force $\mathbf{w}'$ can be called the *apparent weight* of the body. The gravitational force $\mathbf{w}$ will be called the *true weight.* Then

$$w' = w - ma. \tag{5–8}$$

If the elevator is at rest, or moving vertically (either up or down) with constant velocity, $a = 0$ and the apparent weight equals the true weight. If the acceleration is downward, as in Fig. 5–7, so that a is positive, the apparent weight is less than the true weight; the body appears "lighter." If the acceleration is *upward,* a is negative, the apparent weight is greater than the true weight, and the body appears "heavier." If the elevator falls freely, $a - g$, and since the true weight w also equals mg, the apparent weight is zero and the body *appears* "weightless." It is in this sense that an astronaut orbiting the earth in a space capsule is said to be "weightless."

The physiological effect of this condition is exactly the same as though the body were in outer space with no gravitational force at all. The physiological effect of prolonged weightlessness is an interesting medical problem which is just beginning to be explored, with the advent of manned spacecraft. Gravity plays some role in blood distribution in the body, and one reaction to weightlessness is a decrease in volume of blood through increased excretion of water. In some cases astronauts returning to earth have experienced a temporary impairment of sense of balance and a greater tendency toward motion sickness.

Example 8 Figure 5–8(a) represents a simple *accelerometer.* A small body is fastened at one end of a light rod pivoted freely at the point P. When the system has an acceleration a toward the right, the rod makes an angle θ with the vertical. (In a practical

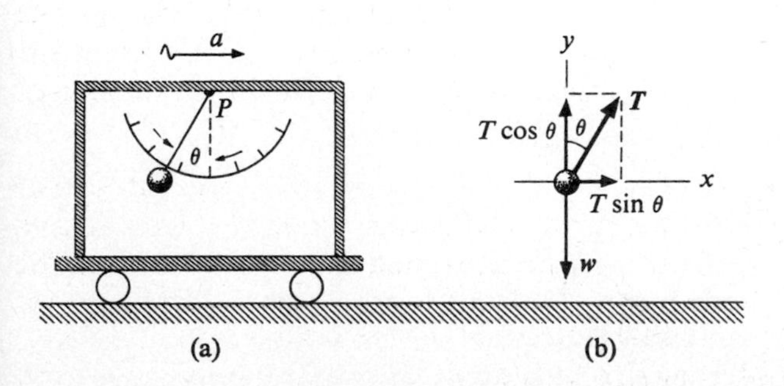

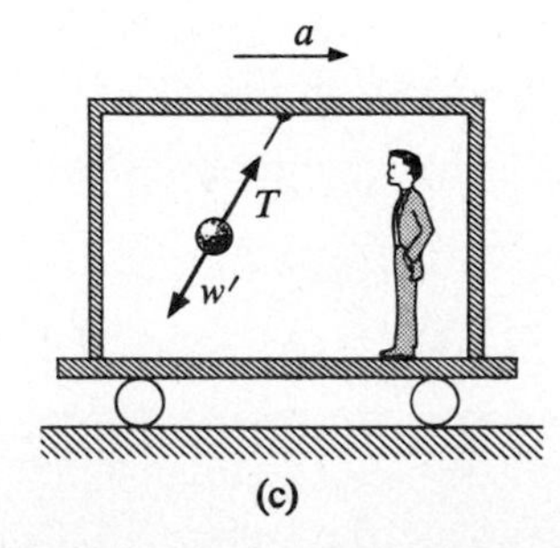

Fig. 5–8 *(a) A simple accelerometer. (b) The forces on the body are* $\mathbf{w}$ *and* $\mathbf{T}$. *(c) The apparent weight* $\mathbf{w}'$ *is equal and opposite to* $\mathbf{T}$.

instrument, some form of damping must be provided to keep the rod from swinging when the acceleration changes. The rod might hang in a tank of mineral oil.)

As shown in the free-body diagram, Fig. 5–8(b), two forces are exerted on the body; its weight **w** and the tension **T** in the rod. (We neglect the weight of the rod.) The resultant horizontal force is

$$\sum F_x = T \sin \theta,$$

and the resultant vertical force is

$$\sum F_y = T \cos \theta - w.$$

The x-acceleration is the acceleration a of the system, and the y-acceleration is zero. Hence

$$T \sin \theta = ma, \qquad T \cos \theta = w.$$

When the first equation is divided by the second, and w replaced by mg, we get

$$a = g \tan \theta,$$

and the acceleration a is proportional to the tangent of the angle θ.

Let us consider this problem from the standpoint of an observer riding with the accelerometer, as in Fig. 5–8(c). To him, the body *appears* in equilibrium and hence appears to be acted on by a force $\mathbf{w}'$, equal and opposite to **T**, and which he calls the *apparent weight* of the body. To obtain the expression for $\mathbf{w}'$, let us write Newton's second law in general vector form. The resultant force on the body is the vector sum of **w** and **T**, and hence

$$\mathbf{w} + \mathbf{T} = m\mathbf{a}, \qquad \mathbf{T} = -\mathbf{w} + m\mathbf{a}.$$

The apparent weight $\mathbf{w}'$ is equal and opposite to **T**, so $\mathbf{w}' = -\mathbf{T}$ and

$$\mathbf{w}' = \mathbf{w} - m\mathbf{a}. \tag{5–9}$$

This is the general form of Eq. (5–8) for the apparent weight $\mathbf{w}'$ of a body, in a system which has an acceleration $\mathbf{a}$ relative to an inertial system. In this example, *the apparent weight differs in both magnitude and direction from the true weight.*

Up to this point, applications of Newton's second law have been limited to cases where the resultant force acting on a body was constant, thereby imparting to the body a constant acceleration. Such cases are very important, but make only slight demands on one's mathematical knowledge. When the resultant force is variable, however, the acceleration is not constant and the simple equations of motion with constant acceleration do not apply. We conclude this section with two examples of motion under the action of a variable force.

Example 9 Assume the earth to be a nonrotating, homogeneous sphere. Discuss the motion of a freely falling body (or a body projected vertically upward), taking into account the variation of the gravitational force on the body with its distance from the earth's center. Neglect air resistance.

The gravitational force on the body at a distance r from the earth's center is Gmm_E/r^2, and from Newton's second law its acceleration is

$$g = \frac{w}{m} = -\frac{Gm_E}{r^2},$$

where the positive direction is upward (or better, radially outward).

It was shown in Eq. (4–8) that the acceleration can be expressed as

$$g = v\frac{dv}{dr}.$$

Then

$$v\frac{dv}{dr} = -\frac{Gm_E}{r^2}, \quad \int_{v_1}^{v_2} v\,dv = -Gm_E\int_{r_1}^{r_2}\frac{dr}{r^2},$$

where v_1 and v_2 are the velocities at the radial distances r_1 and r_2. It follows that

$$v_2^2 - v_1^2 = 2Gm_E\left(\frac{1}{r_2} - \frac{1}{r_1}\right). \tag{5–10}$$

As an illustration, let us find the initial velocity v_1 required to project a body vertically upward so that it rises to a height above the earth's surface equal to the earth's radius R. Then $v_2 = 0$, $r_1 = R$, $r_2 = 2R$, and

$$v_1^2 = \frac{Gm_E}{R}. \tag{5-11}$$

Let g_0 represent the acceleration of gravity at the earth's surface, where $r = R$. Then

$$Gm_E = g_0 R^2,$$

and Eq. (5–11) can be written

$$v_1^2 = g_0 R. \tag{5-12}$$

How does this compare with the velocity that would be required if the acceleration had the *constant* value g_0?

Example 10 A resisting force that varies directly with the speed is found frequently in nature. Any small spherical body of radius r, like a raindrop, an oil droplet, or a steel sphere, moving with a small velocity v through a viscous fluid (liquid or gas) is subjected to a force R, where

$$\boldsymbol{R} = -6\pi\eta r\boldsymbol{v},$$

and η is the viscosity. This relation is known as *Stokes' law*. Letting

$$k = 6\pi\eta r,$$

we may write Stokes' law simply as

$$\boldsymbol{R} = -k\boldsymbol{v}.$$

A small sphere falling through a viscous fluid is subjected to three vertical forces, as shown in Fig. 5–9: the weight $\boldsymbol{w}$, the buoyant force $\boldsymbol{B}$, and the resisting force $\boldsymbol{R}$.

Let us suppose that the sphere starts from rest and that the positive y-direction is downward. Under these conditions

$$\sum F_y = w - B - kv = ma.$$

At first, when $v = 0$, the resisting force is zero and the initial acceleration a_0 is positive:

$$a_0 = \frac{w - B}{m}. \tag{5-13}$$

The sphere speeds up and, after a while, when v becomes large enough, the resisting force equals $w - B$ and there is no resultant force acting on the sphere. At this moment the acceleration is zero and the speed undergoes no further increase. The maximum or *terminal speed* v_T may therefore be calculated by setting $a = 0$; thus

$$w - B - kv_T = 0$$

or

$$v_T = \frac{w - B}{k}. \tag{5-14}$$

Figure 5–10 shows how the acceleration, velocity, and position vary with time.

To find the relation between the speed and the time during the interval before the terminal speed is reached, we go back to Newton's second law,

$$m\frac{dv}{dt} = w - B - kv.$$

After rearranging terms and replacing $(w - B)/k$ by v_T, we get

$$\frac{dv}{v - v_T} = -\frac{k}{m}\,dt.$$

Since $v = 0$ when $t = 0$,

$$\int_0^v \frac{dv}{v - v_T} = -\frac{k}{m}\int_0^t dt,$$

whence

$$\ln\frac{v_T - v}{v_T} = -\frac{k}{m}t,$$

or

$$1 - \frac{v}{v_T} = e^{-(k/m)t},$$

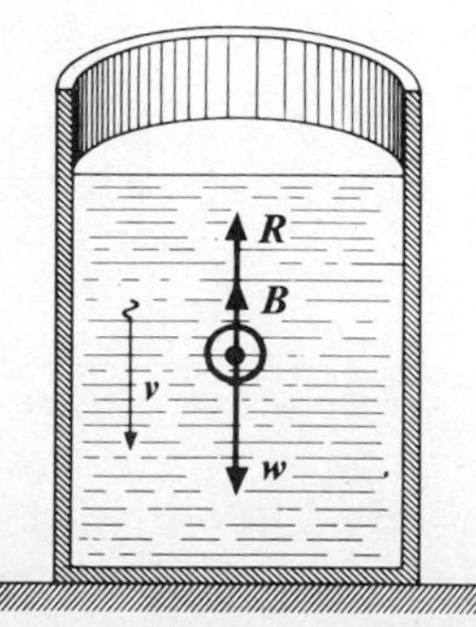

Fig. 5–9 *Forces acting on a small sphere falling through a viscous fluid.*

and finally

$$v = v_T(1 - e^{-(k/m)t}). \tag{5–15}$$

An important concept related to an exponentially varying quantity is the *relaxation time*, t_R, whose meaning can be seen from Fig. 5–10(b). Suppose the acceleration remained constant at the initial value a_0, as indicated by the dotted line. The relaxation time can be defined as the time that *would* be required to reach the terminal velocity with this constant acceleration, and evidently

$$t_R = \frac{v_T}{a_0} = \frac{(w - B)/k}{(w - B)/m} = \frac{m}{k}.$$

Equation (5–15) can now be written more simply as

$$v = v_T(1 - e^{-t/t_R}). \tag{5–16}$$

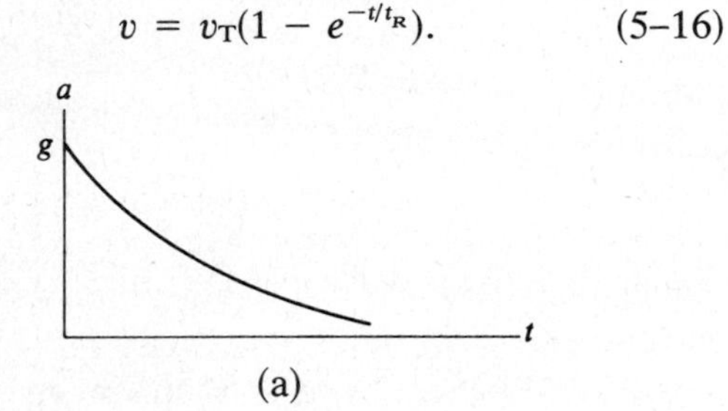

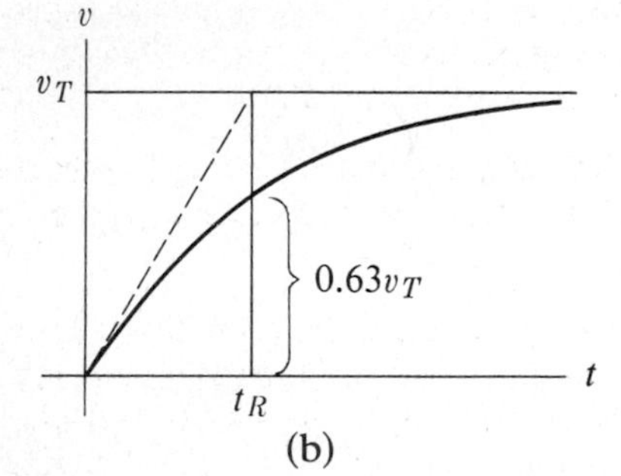

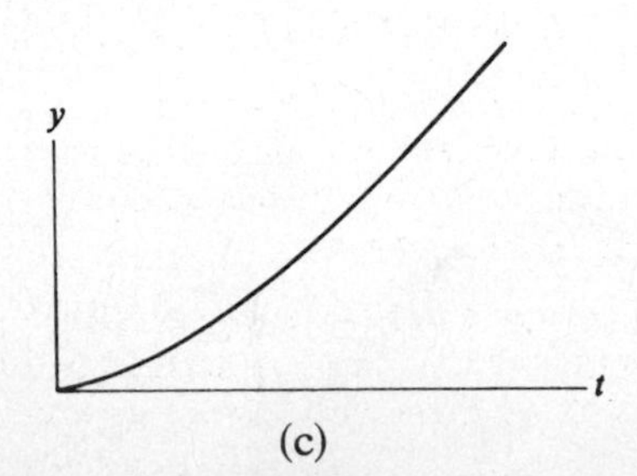

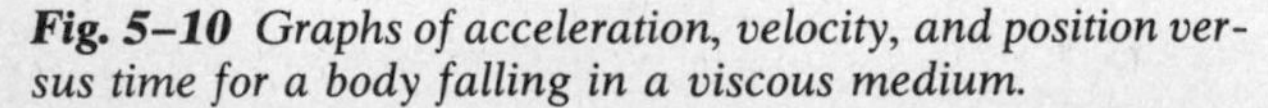
Fig. 5–10 *Graphs of acceleration, velocity, and position versus time for a body falling in a viscous medium.*

At the instant when t equals the relaxation time, $t/t_R = 1$ and

$$v = v_T(1 - e^{-1}) = 0.63v_T.$$

Thus about 63% of the final velocity is actually acquired in a time equal to the relaxation time.

PROBLEMS

For problem work, use the approximate value $g = 9.80 \text{ m s}^{-2}$. A force diagram should be constructed for each problem.

5–1

a) What is the mass of a body which weighs 1 N at a point where $g = 9.80 \text{ m s}^{-2}$?

b) What is the mass of a body which weighs 1 dyn at a point where $g = 980 \text{ cm s}^{-2}$?

c) What is the mass of a standard pound body?

5–2

a) At what distance from the earth's center would a standard kilogram weigh 1 N?

b) At what distance would a 1-g body weigh 1 dyn?

c) At what distance would a 1-slug body weigh 1 lb?

5–3 If action and reaction are always equal in magnitude and opposite in direction, why don't they always cancel each other and leave no net force for acceleration of the body?

5–4 The mass of a certain object is 10 kg.

a) What would its mass be if taken to the planet Mars?

b) Is the expression $F = ma$ valid on Mars?

c) Newton's second law is sometimes written in the form $F = wa/g$ instead of $F = ma$. Would this expression be valid on Mars?

5–5 A constant horizontal force of 10 lb acts on a body on a *smooth horizontal* plane. The body starts from rest and is observed to move 250 ft in 5 s.

a) What is the mass of the body?

b) If the force ceases to act at the end of 5 s, how far will the body move in the next 5 s?

5–6 A .22 rifle bullet, traveling at 360 m s^{-1}, strikes a block of soft wood, which it penetrates to a depth of 0.1 m. The mass of the bullet is 1.8 g. Assume a constant retarding force.

a) How long a time was required for the bullet to stop?

b) What was the decelerating force, in newtons? in pounds?

5–7 A body of mass 15 kg rests on a frictionless horizontal plane and is acted on by a horizontal force of 30 N.

a) What acceleration is produced?

b) How far will the body travel in 10 s?

c) What will be its velocity at the end of 10 s?

5–8 A body of mass 0.5 kg is at rest at the origin, $x = 0$, on a horizontal frictionless surface. At time $t = 0$ a force of 0.001 N is applied to the body parallel to the x-axis, and 5 s later this force is removed.

a) What are the position and velocity of the body at $t = 5$ s?

b) If the same force is again applied at $t = 15$ s, what are the position and velocity of the body at $t = 20$ s?

5–9 A body of mass 10 kg is moving with a constant velocity of 5 m s^{-1} on a horizontal surface. The coefficient of sliding friction between body and surface is 0.20.

a) What horizontal force is required to maintain the motion?

b) If the force is removed, how soon will the body come to rest?

5–10 A hockey puck leaves a player's stick with a velocity of 10 m s^{-1} and slides 40 m before coming to rest. Find the coefficient of friction between the puck and the ice.

5–11 An electron (mass $= 9 \times 10^{-31}$ kg) leaves the cathode of a radio tube with zero initial velocity and travels in a straight line to the anode, which is 1 cm away. It reaches the anode with a velocity of 6×10^{6} m s^{-1}. If the acceleration force was constant, compute (a) the accelerating force, in newtons, (b) the time to reach the anode, (c) the acceleration. (The gravitational force on the electron may be neglected.)

5–12 Give arguments either for or against the statement that "the only reason an apple falls downward to meet the earth instead of the earth falling upward to meet the apple is that the earth, being so much more massive, exerts the greater pull."

5–13 In an experiment using the Cavendish balance to measure the gravitational constant G, it is found that a sphere of mass 0.8 kg attracts another sphere of mass 0.004 kg with a force of 13×10^{-11} N when the distance between the centers of the spheres is 0.04 m. The acceleration of gravity at the earth's surface is 9.80 m s^{-2}, and the radius of the earth is 6400 km. Compute the mass of the earth from these data.

5–14 Two spheres, each of mass 6.4 kg, are fixed at points A and B (Fig. 5–11). Find the magnitude and direction of the initial acceleration of a sphere of mass 0.010 kg if released from rest at point P and acted on only by forces of gravitational attraction of the spheres at A and B.

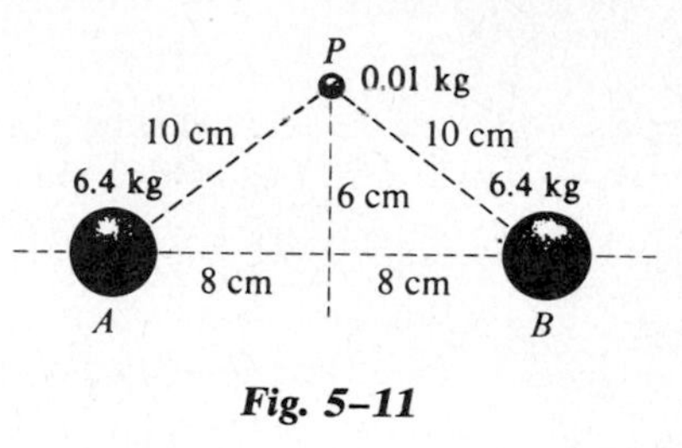

Fig. 5–11

5–15 The mass of the moon is about one eighty-first, and its radius one-fourth, that of the earth. What is the acceleration due to gravity on the surface of the moon?

5–16 In round numbers, the distance from the earth to the moon is 250,000 mi, the distance from the earth to the sun is 93 million mi, the mass of the earth is 6×10^{27} g, and the mass of the sun is 2×10^{33} g. Approximately what is the ratio of the gravitational pull of the sun on the moon to that of the earth on the moon?

5–17 An elevator with mass 2000 kg rises with an acceleration of 1 m s^{-2}. What is the tension in the supporting cable?

5–18 A 4-kg block is accelerated upward by a cord whose breaking strength is 60 N. Find the maximum acceleration which can be given the block without breaking the cord.

5–19 A 5-kg block is supported by a cord and pulled upward with an acceleration of 2 m s^{-2}.

a) What is the tension in the cord?

b) After the block has been set in motion, the tension in the cord is reduced to 49 N. What sort of motion will the block perform?

c) If the cord is now slackened completely, the block is observed to move up 2 m farther before coming to rest. With what velocity was it traveling?

5–20 A block weighing 10 lb is held up by a string which can be moved up or down. What conclusions can you draw regarding magnitude and direction of the acceleration and velocity of the upper end of the string when the tension in the string is (a) 5 lb? (b) 10 lb? (c) 15 lb?

5–21 A body hangs from a spring balance supported from the roof of an elevator.

a) If the elevator has an upward acceleration of 4 ft s^{-2} and the balance reads 45 lb, what is the true weight of the body?

b) Under what circumstances will the balance read 35 lb?

c) What will the balance read if the elevator cable breaks?

5–22 A transport plane is to take off from a level landing field with two gliders in tow, one behind the other. Each glider has a mass of 1200 kg, and the friction force or drag on each may be assumed constant and equal to 2000 N. The tension in the towrope between the transport plane and the first glider is not to exceed 10,000 N.

a) If a velocity of 40 m s^{-1} is required for the take-off, how long a runway is needed?

b) What is the tension in the towrope between the two gliders while the planes are accelerating for the take-off?

5–23 If the coefficient of friction between tires and road is 0.5, what is the shortest distance in which an automobile can be stopped when traveling at 60 mi hr^{-1}?

5–24 A 40-kg packing case is on the floor of a truck. The coefficient of static friction between the case and the truck floor is 0.40, and the coefficient of sliding friction is 0.25. Find the magnitude and direction of the friction force acting on the case (a) when the truck is accelerating at 2 m s^{-2}, (b) when it is decelerating at 3 m s^{-2}.

5–25 A balloon is descending with a constant acceleration a, less than the acceleration due to gravity g. The weight of the balloon, with its basket and contents, is w. What weight, W, of ballast should be released so that the balloon will begin to be accelerated upward with constant acceleration a? Neglect air resistance.

5–26 A 64-lb block is pushed up a 37° inclined plane by a horizontal force of 100 lb. The coefficient of sliding friction is 0.25. Find (a) the acceleration, (b) the velocity of the block after it has moved a distance of 20 ft along the plane, and (c) the normal force exerted by the plane. Assume that all forces act at the center of the block.

5–27 A block rests on an inclined plane which makes an angle θ with the horizontal. The coefficient of sliding friction is 0.50, and the coefficient of static friction is 0.75.

a) As the angle θ is increased, find the minimum angle at which the block starts to slip.

b) At this angle, find the acceleration once the block has begun to move.

c) How long a time is required for the block to slip 20 ft along the inclined plane?

5–28

a) What constant horizontal force is required to drag a 16-lb block along a horizontal surface with an acceleration of 4 ft s^{-2} if the coefficient of sliding friction between block and surface is 0.5?

b) What weight, hanging from a cord attached to the 16-lb block and passing over a small frictionless pulley, will produce this acceleration?

5–29 A 20-kg box rests on the flat floor of a truck. The coefficient of friction between box and floor is 0.1. The truck stops at a stop sign and then starts, with an acceleration of 2 m s^{-2}. If the box is 5 m from the rear of the truck when it starts, how much time elapses before it falls off the rear of the truck? How far does the truck travel in this time?

5–30 A 160-lb man stands on a bathroom scale in an elevator. As the elevator starts moving, the scale reads 200 lb.

a) Find the acceleration of the elevator (magnitude and direction).

b) What is the acceleration if the scale reads 120 lb?

c) If the scale reads zero, should the man worry? Explain.

5–31 A short commuter train consists of a locomotive and two cars. The mass of the locomotive is 6000 kg and that of each car 2000 kg. The train pulls away from a station with an acceleration of 0.5 m s^{-2}.

a) Find the tension in the coupler joining the locomotive to the first car, and in the coupler joining the two cars.

b) What total horizontal force must the locomotive wheels exert on the track?

5–32 A loaded elevator with very worn cables has a total mass of 2000 kg, and the cables can withstand a maximum tension of 24,000 N.

a) What is the maximum acceleration for the elevator if the cables are not to break?

b) What is the answer for part (a) if the elevator is taken to the moon?

5–33 A block of mass 5 kg resting on a horizontal surface is connected by a cord passing over a light frictionless pulley to a hanging block of mass 5 kg. The coefficient of friction between the block and the horizontal surface is 0.5. Find (a) the tension in the cord, and (b) the acceleration of each block.

5–34 A block having a mass of 2 kg is projected up a long 30° incline with an initial velocity of 22 m s^{-1}. The coefficient of friction between the block and the plane is 0.3.

a) Find the friction force acting on the block as it moves up the plane.
b) How long does the block move up the plane?
c) How *far* does the block move up the plane?
d) How long does it take the block to slide from its position in part (c) to its starting point?
e) With what velocity does it arrive at this point?
f) If the mass of the block had been 5 kg instead of 2 kg, would the answers in the preceding parts be changed?

5–35 A 30-lb block on a level frictionless surface is attached by a cord passing over a small frictionless pulley to a hanging block originally at rest 4 ft above the floor. The hanging block strikes the floor in 2 s. Find (a) the weight of the hanging block, (b) the tension in the string while both blocks were in motion.

5–36 Two blocks, each having mass 20 kg, rest on frictionless surfaces, as shown in Fig. 5–12. Assuming the pulleys to be light and frictionless, compute (a) the time required for block A to move 1 m down the plane, starting from rest, (b) the tension in the cord connecting the blocks.

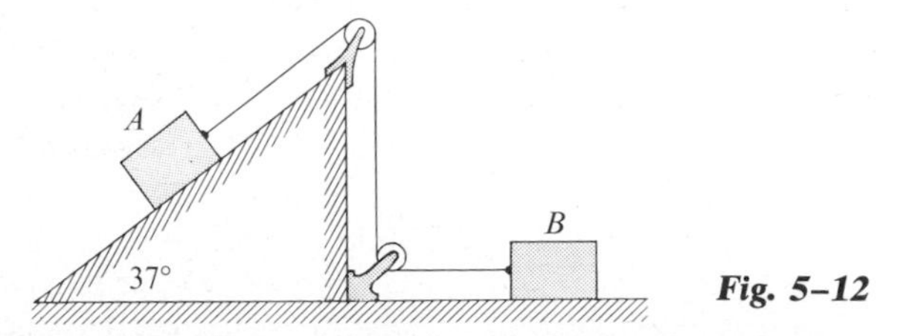

Fig. 5–12

5–37 A block of mass 0.2 kg rests on top of a block of mass 0.8 kg. The combination is dragged along a level surface at constant velocity by a hanging block of mass 0.2 kg as in Fig. 5–13(a).

a) The first 0.2-kg block is removed from the 0.8-kg block and attached to the hanging block, as in Fig. 5–13(b). What is now the acceleration of the system?
b) What is the tension in the cord attached to the 0.8-kg block in part (b)?

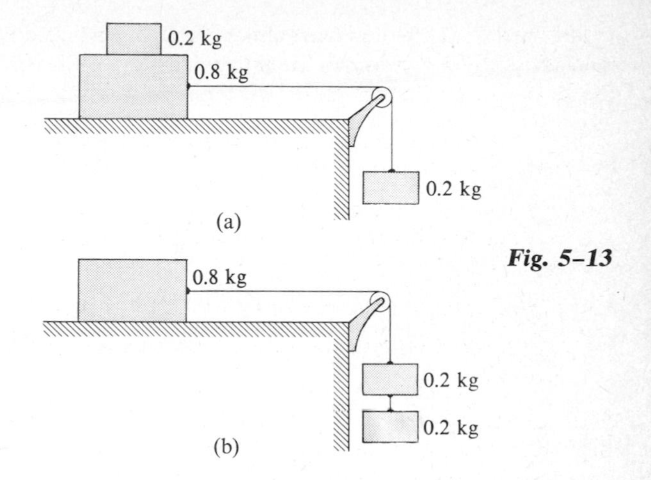

Fig. 5–13

5–38 Block A in Fig. 5–14 has a mass of 2 kg and block B 20 kg. The coefficient of friction between B and the horizontal surface is 0.1.

a) What is the mass of block C if the acceleration of B is 2 m s^{-2} toward the right?
b) What is the tension in each cord when B has the acceleration stated above?

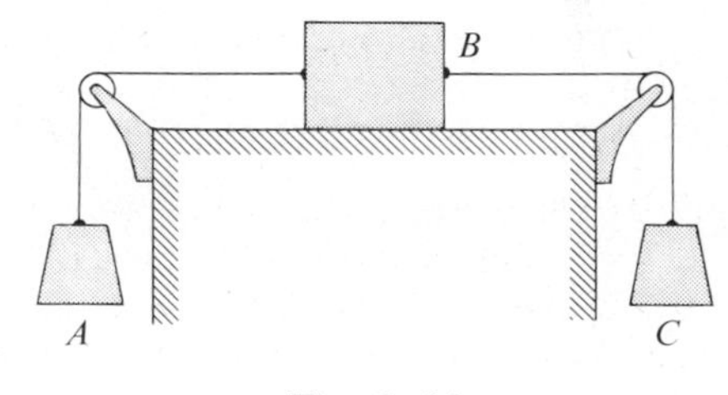

Fig. 5–14

5–39 Two blocks connected by a cord passing over a small frictionless pulley rest on frictionless planes, as shown in Fig. 5–15.

a) Which way will the system move?
b) What is the acceleration of the blocks?
c) What is the tension in the cord?

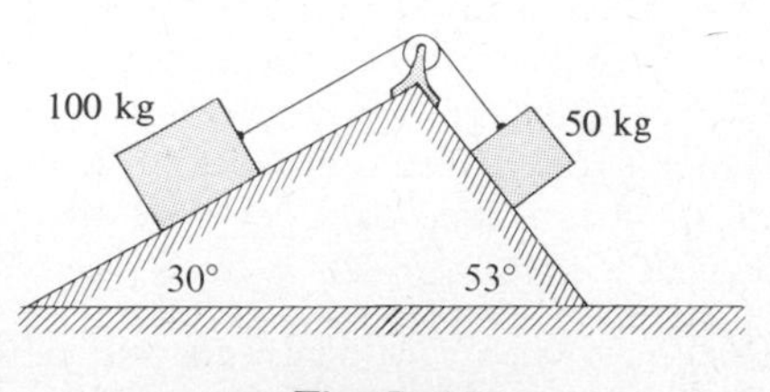

Fig. 5–15

5–40 Two 0.2-kg blocks hang at the ends of a light flexible cord passing over a small frictionless pulley, as in Fig. 5–16. A 0.1-kg block is placed on the right, and removed after 2 s.

a) How far will each block move in the first second after the 0.1-kg block is removed?

b) What was the tension in the cord before the 0.1-kg block was removed? After it was removed?

c) What was the tension in the cord supporting the pulley before the 0.1-kg block was removed? Neglect the weight of the pulley.

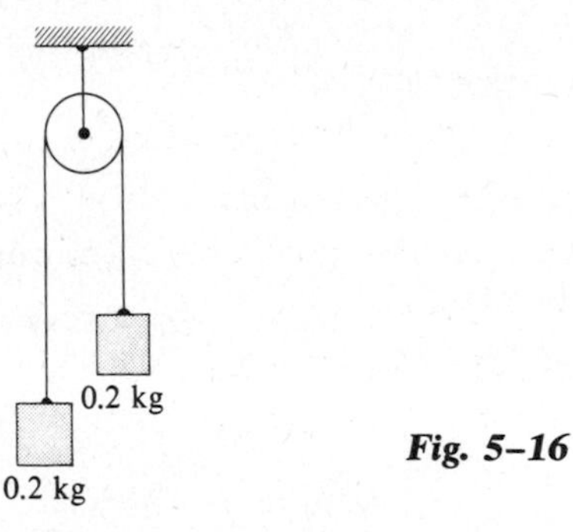

Fig. 5–16

5–41 In terms of m_1, m_2, and g, find the acceleration of both blocks in Fig. 5–17. Neglect all friction and the masses of the pulleys.

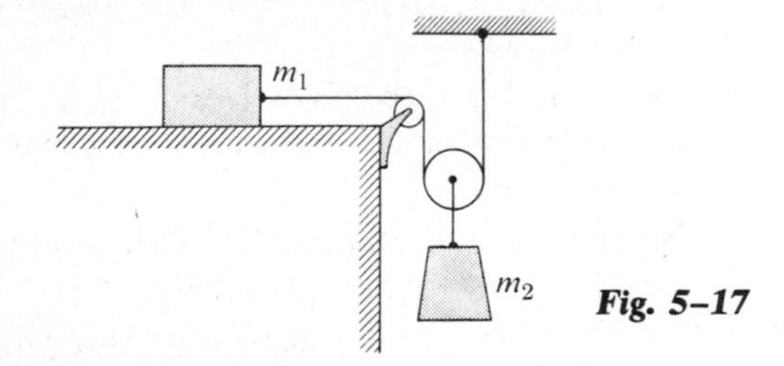

Fig. 5–17

5–42 The masses of bodies A and B in Fig. 5–18 are 20 kg and 10 kg, respectively. They are initially at rest on the floor and are connected by a weightless string passing over a weightless and frictionless pulley. An upward force $\boldsymbol{F}$ is applied to the pulley. Find the accelerations a_1 of body A and a_2 of body B when F is (a) 98 N, (b) 196 N, (c) 394 N, (d) 788 N.

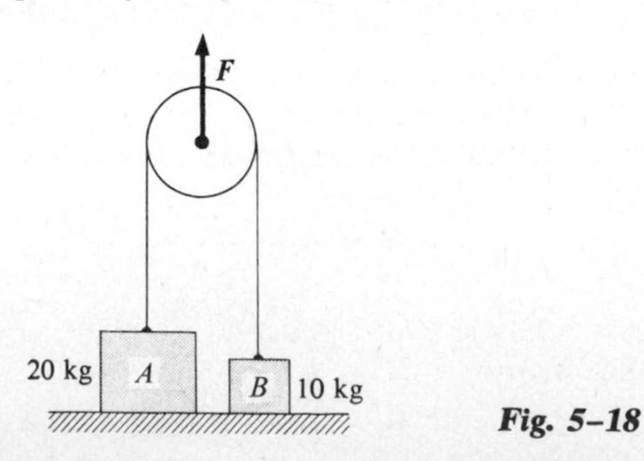

Fig. 5–18

5–43 The two blocks in Fig. 5–19 are connected by a heavy uniform rope of mass 4 kg. An upward force of 200 N is applied as shown.

a) What is the acceleration of the system?

b) What is the tension at the top of the heavy rope?

c) What is the tension at the midpoint of the rope?

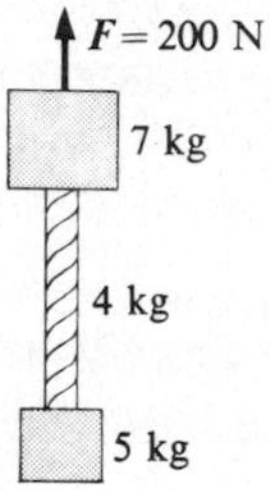

Fig. 5–19

5–44 Two blocks with masses of 4 kg and 8 kg, respectively, are connected by a string and slide down a 30° inclined plane, as in Fig. 5–20. The coefficient of sliding friction between the 4-kg block and the plane is 0.25, and between the 8-kg block and the plane it is 0.50.

a) Calculate the acceleration of each block.

b) Calculate the tension in the string.

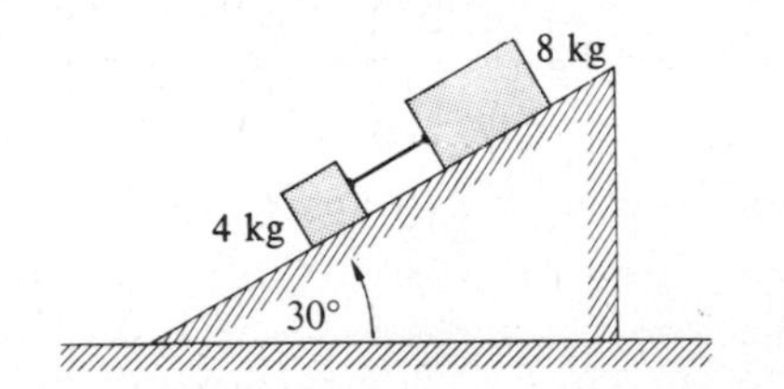

Fig. 5–20

5–45 Two bodies with masses of 5 kg and 2 kg, respectively, hang 1 m above the floor from the ends of a cord 3 m long passing over a frictionless pulley. Both bodies start from rest. Find the maximum height reached by the 2-kg body.

5–46 A man who weighs 160 lb stands on a platform which weighs 80 lb. He pulls a rope which is fastened to the platform and runs over a pulley on the ceiling. With what force does he have to pull in order to give himself and the platform an upward acceleration of 2 ft s^{-2}?

5–47 What acceleration must the cart in Fig. 5–21 have in order that the block A will not fall? The coefficient of friction between the block and the cart is μ. How would the behavior of the block be described by an observer on the cart?

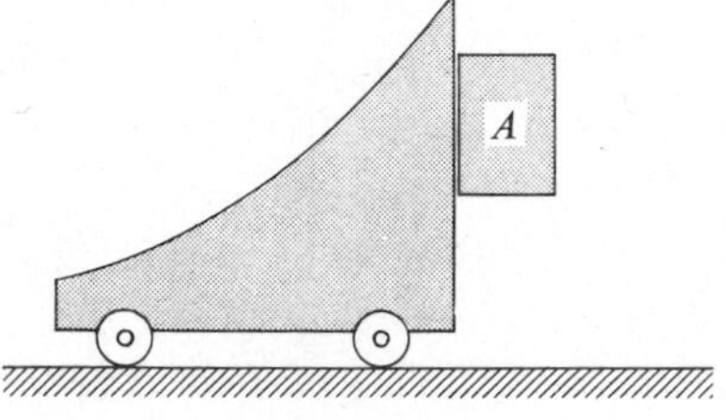

Fig. 5–21

5–48 The left end of the weightless rod shown in Fig. 5–22 is hinged to a cart. A heavy particle is attached to the right end. If the cart has an acceleration a to the right, find the angle θ. How would this situation be described by an observer on the cart?

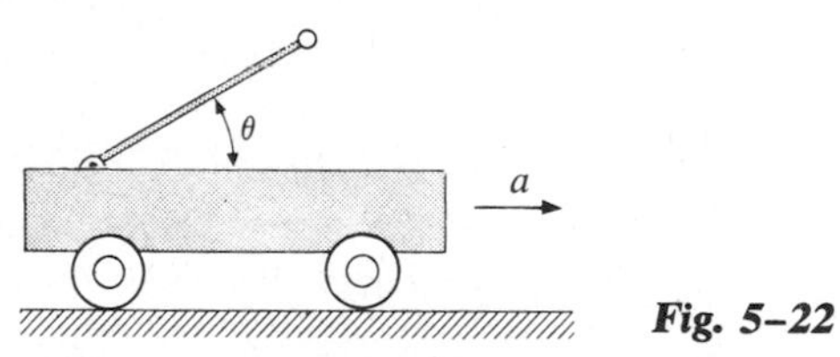

Fig. 5–22

5–49 Which way will the accelerometer in Fig. 5–8 deflect under the following conditions?

a) The cart is moving toward the right and traveling faster.
b) The cart is moving toward the right and traveling slower.
c) The cart is moving toward the left and traveling faster.
d) The cart is moving toward the left and traveling slower.
e) The cart is at rest on a sloping surface.
f) The cart is given an upward velocity on a frictionless inclined plane. It first moves up, then stops, and then moves down. What is the deflection in each stage of the motion?

5–50

a) In terms of the acceleration of gravity at the surface of the earth, g_0, and the earth's radius R, find the velocity with which a body must be projected vertically upward, in the absence of air resistance, to rise to an infinite distance above the earth's surface. This is called the *escape velocity*.
b) In terms of the same quantities, find the velocity with which a body will strike the earth's surface if it falls from rest toward the earth from an infinitely distant point.
c) Compute both of these velocities in miles per hour.
d) Explain why the velocities are not infinitely large.

5–51 The mass of the motorboat in Problem 4–37 is 100 slugs. Find the force decelerating the boat when its speed is (a) 20 ft per sec, and (b) 10 ft per sec. (c) If the boat is being towed at 10 ft per sec, what is the tension in the towline?

5–52 A body of mass $m = 5$ kg falls from rest in a viscous medium. The body is acted on by a net constant downward force of 20 N, and by a viscous retarding force proportional to its speed and equal to $5v$, where v is the speed in meters per second.

a) Find the initial acceleration, a_0.
b) Find the acceleration when the speed is 3 m s^{-1}.
c) Find the speed when the acceleration equals $0.1a_0$.
d) Find the terminal velocity, v_T.
e) Find the relaxation time, t_R.
f) Find the coordinate, velocity, and acceleration 2 s after the start of the motion.
g) Find the time required to reach a speed $0.9v_T$.
h) Construct a graph of v versus t, for a time of 3 s.

5–53 A body falls from rest through a medium which exerts a resisting force that varies directly with the square of the velocity ($R = -kv^2$).

a) Draw a diagram showing the direction of motion and indicate with the aid of vectors all of the forces acting on the body.
b) Apply Newton's second law and infer from the resulting equation the general properties of the motion.
c) Show that the body acquires a terminal velocity and calculate it.
d) Find the relaxation time.
e) Derive the equation for the velocity at any time.

5–54 A particle of mass m, originally at rest, is subjected to a force whose direction is constant but whose magnitude varies with the time according to the relation

$$F = F_0\left[1 - \left(\frac{t - T}{T}\right)^2\right],$$

where F_0 and T are constants. The force acts only for the time interval $2T$.

a) Make a rough graph of F versus t.

b) Prove that the speed v of the particle after a time $2T$ has elapsed is equal to $4F_0T/3m$.

c) Choose numbers for v, T, and m that might be appropriate to a batted baseball, and calculate the force F_0. Judge whether the answer is sensible.

5-55 A fifty-lb monkey with downcast eyes has a firm hold on a light rope that passes over a frictionless pulley and is attached to a fifty-lb. bunch of bananas, as shown in Fig. 5-23. The monkey happens to glance upward, sees the bananas, and starts to climb the rope to get at the bananas.

a) As he climbs, do the bananas move up, down, or remain at rest?

b) As he climbs, does the distance between him and the bananas decrease, increase, or remain constant?

c) The monkey releases his hold on the rope. What about the distance between him and the bananas while he is falling?

d) Before he reaches the ground, he grabs the rope to stay his fall. What do the bananas do?

Fig. 5-23

Chapter 6

Motion in a Plane

6–1 MOTION IN A PLANE

Thus far we have discussed only motion along a straight line, or *rectilinear motion.* In this chapter we shall consider *plane motion,* that is, motion in a curved path which lies in a fixed plane. Examples of such motion are the flight of a thrown or batted baseball, a projectile shot from a gun, a ball whirled at the end of a cord, the motion of the moon or of a satellite around the earth, and the motion of the planets around the sun.

If the motion is referred to a set of rectangular coordinate axes x and y, the equation of the path expresses y as a function of x, $y = f(x)$. Very often one is interested in the position of the moving body as a function of time. If s is the distance along the path from some fixed point to the position of the body, its position at any time is given by an equation of the form $s = f(t)$. It is usually simpler, however, to deal with the x- and y-coordinates separately and to describe the motion by the two equations

$$x = f_1(t), \qquad y = f_2(t).$$

These can be considered as *parametric equations* of the path, expressing the coordinates x and y in terms of the parameter t.

We shall consider several kinds of problems in plane motion. In some the motion of the particle is known and we wish to determine its velocity and acceleration and the resultant force acting on it. An example is that of a ball attached to a cord and whirled in a circle at constant speed. What is the tension in the cord? In others, the force acting on a particle may be known at every point of space and we may wish to find the equation of motion of the particle. Examples are the orbit of a planet around the sun, or the path followed by a rocket.

6–2 AVERAGE AND INSTANTANEOUS VELOCITY

Consider a particle moving along the curved path in Fig. 6–1(a). Points P and Q represent two positions of the particle. Its displacement as it moves from P to Q is the vector $\Delta \mathbf{s}$. Just as in the case of rectilinear motion, the average velocity $\bar{\mathbf{v}}$ of the particle is defined as the vector displacement $\Delta \mathbf{s}$ divided by the elapsed time Δt:

$$\text{Average velocity } \bar{\mathbf{v}} = \Delta \mathbf{s}/\Delta t. \tag{6–1}$$

The average velocity would be the same for any path that would take the particle from P to Q in the time interval Δt.

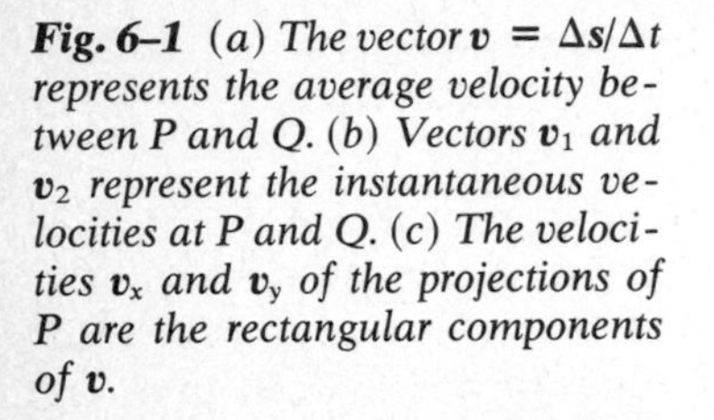

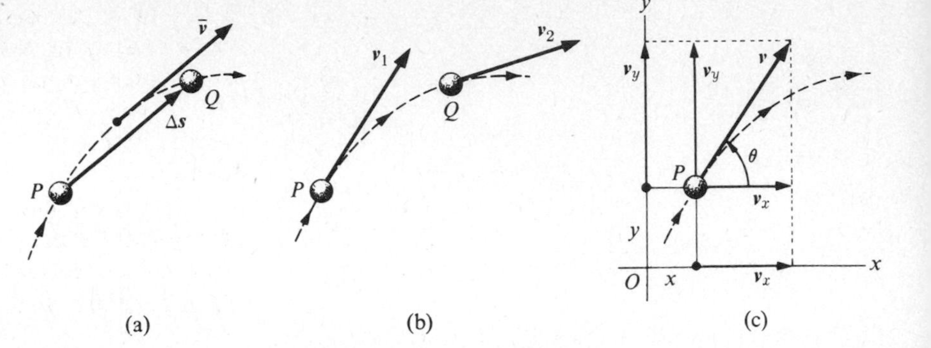

Fig. 6–1 *(a) The vector $\boldsymbol{v} = \Delta \boldsymbol{s}/\Delta t$ represents the average velocity between P and Q. (b) Vectors $\boldsymbol{v}_1$ and $\boldsymbol{v}_2$ represent the instantaneous velocities at P and Q. (c) The velocities $\boldsymbol{v}_x$ and $\boldsymbol{v}_y$ of the projections of P are the rectangular components of $\boldsymbol{v}$.*

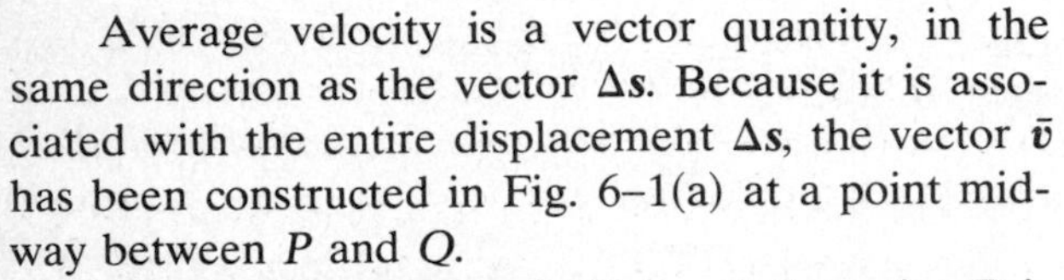

Average velocity is a vector quantity, in the same direction as the vector $\Delta \boldsymbol{s}$. Because it is associated with the entire displacement $\Delta \boldsymbol{s}$, the vector $\bar{\boldsymbol{v}}$ has been constructed in Fig. 6–1(a) at a point midway between P and Q.

The *instantaneous* velocity $\boldsymbol{v}$ at the point P is defined in magnitude and direction as the *limit approached by the average velocity* when point Q is taken closer and closer to point P:

$$\text{Instantaneous velocity } \boldsymbol{v} = \lim_{\Delta t \to 0} \frac{\Delta \boldsymbol{s}}{\Delta t} = \frac{d\boldsymbol{s}}{dt}. \tag{6–2}$$

As point Q approaches point P, the direction of the vector $\Delta \boldsymbol{s}$ approaches that of the tangent to the path at P, so that the instantaneous velocity vector at any point is *tangent* to the path at that point. The instantaneous velocities at point P and Q are shown in Fig. 6–1(b).

In Fig. 6–1(c), the motion of a particle is referred to a rectangular coordinate system. As the particle moves along its path, its projections onto the x- and y-axes move along these axes in rectilinear motion. If in a time interval Δt these projections move by amounts Δx and Δy, respectively, then the average velocities of the projections which we may call $\bar{v}_x$ and $\bar{v}_y$, are given by

$$\bar{v}_x = \frac{\Delta x}{\Delta t}, \qquad \bar{v}_y = \frac{\Delta y}{\Delta t}.$$

We see that these are equal to the x- and y-components of the average velocity vector. Similarly, the instantaneous velocities of these projections, denoted by v_x and v_y, are equal respectively to the x- and y-components of the instantaneous velocity $\boldsymbol{v}$, and we may write

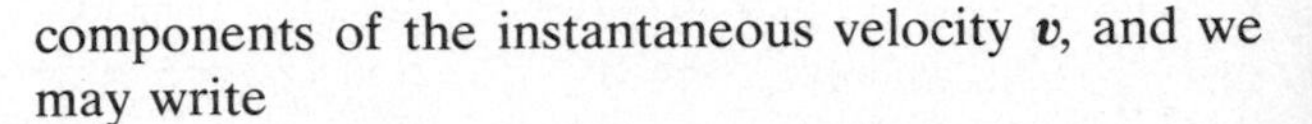

$$v_x = \lim_{\Delta t \to 0} \frac{\Delta x}{\Delta t} = \frac{dx}{dt}, \qquad v_y = \lim_{\Delta t \to 0} \frac{\Delta y}{\Delta t} = \frac{dy}{dt}.$$

The magnitude of the instantaneous velocity is given by

$$|\boldsymbol{v}| = \sqrt{v_x^2 + v_y^2},$$

and the angle θ in Fig. 6–1(c) by

$$\tan \theta = \frac{v_y}{v_x}.$$

Thus we may represent velocity, a vector quantity, in terms of its components or in terms of its magnitude and direction, just as with force, another vector quantity.

It is well to remember that the direction of the instantaneous velocity of a particle at any point in its path is *always* in a direction tangent to the path at that point, no matter how complicated the motion.

6–3 AVERAGE AND INSTANTANEOUS ACCELERATION

In Fig. 6–2(a), the vectors $\boldsymbol{v}_1$ and $\boldsymbol{v}_2$ represent the instantaneous velocities, at points P and Q, of a particle moving in a curved path. The velocity $\boldsymbol{v}_2$ necessarily differs in *direction* from the velocity $\boldsymbol{v}_1$. The diagram has been constructed for a case in which it also differs in *magnitude*, although in special cases the magnitude of the velocity may remain constant.

The *average acceleration* $\bar{\boldsymbol{a}}$ of the particle as it moves from P to Q is defined, just as in the case of

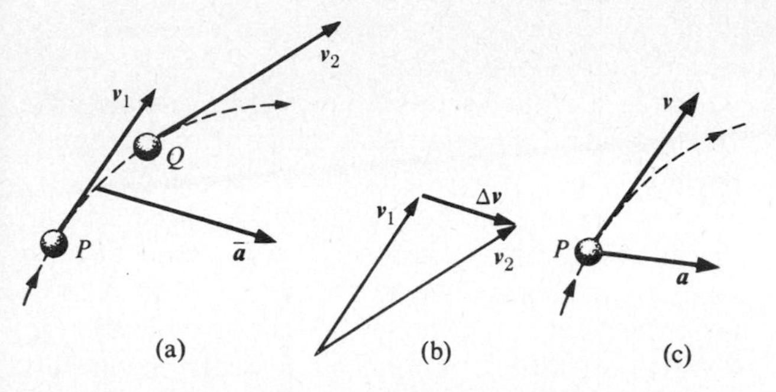

Fig. 6-2 (*a*) *The vector* $\boldsymbol{a} = \Delta\boldsymbol{v}/\Delta t$ *represents the average acceleration between P and Q.* (*b*) *Construction for obtaining* $\Delta\boldsymbol{v} = \boldsymbol{v}_2 - \boldsymbol{v}_1$. (*c*) *Instantaneous velocity* $\boldsymbol{v}$ *and instantaneous acceleration* $\boldsymbol{a}$ *at point P. Vector* $\boldsymbol{v}$ *is tangent to path; vector* $\boldsymbol{a}$ *points toward concave side of path.*

rectilinear motion, as the *vector change in velocity*, $\Delta\boldsymbol{v}$, divided by the time interval Δt:

$$\text{Average acceleration } \bar{\boldsymbol{a}} = \frac{\Delta \boldsymbol{v}}{\Delta t}. \tag{6-3}$$

Average acceleration is a vector quantity, in the same direction as the vector $\Delta\boldsymbol{v}$.

The vector change in velocity, $\Delta\boldsymbol{v}$, means the vector difference $\boldsymbol{v}_2 - \boldsymbol{v}_1$:

$$\Delta\boldsymbol{v} = \boldsymbol{v}_2 - \boldsymbol{v}_1,$$

or

$$\boldsymbol{v}_2 = \boldsymbol{v}_1 + \Delta\boldsymbol{v}.$$

As explained in Section 1-9, the vector difference $\Delta\boldsymbol{v}$ can be found by drawing the vectors $\boldsymbol{v}_1$ and $\boldsymbol{v}_2$ from a common point, as in Fig. 6-2(b), and constructing the vector from the tip of $\boldsymbol{v}_1$ to the tip of $\boldsymbol{v}_2$. Then $\boldsymbol{v}_2$ is the vector sum of $\boldsymbol{v}_1$ and $\Delta\boldsymbol{v}$.

The average acceleration vector, $\bar{\boldsymbol{a}} = \Delta\boldsymbol{v}/\Delta t$, is shown in Fig. 6-2(a) at a point midway between P and Q.

The *instantaneous acceleration* $\boldsymbol{a}$ at point P is defined in magnitude and direction as the limit approached by the average acceleration when point Q approaches point P and $\Delta\boldsymbol{v}$ and Δt both approach zero:

$$\text{Instantaneous acceleration } \boldsymbol{a} = \lim_{\Delta t \to 0} \frac{\Delta \boldsymbol{v}}{\Delta t} = \frac{d\boldsymbol{v}}{dt}. \tag{6-4}$$

The instantaneous acceleration vector at point P is shown in Fig. 6-2(c). Note that it does *not* have the same direction as the velocity vector. Reference to the construction of Fig. 6-2(b) will show that the acceleration vector must always lie on the *concave* side of the curved path.

6-4 COMPONENTS OF ACCELERATION

Figure 6-3(a) again shows the motion of a particle referred to a rectangular coordinate system. The accelerations of the projections of the particle onto the x- and y-axes are

$$a_x = \frac{dv_x}{dt} = \frac{d^2x}{dt^2}, \qquad a_y = \frac{dv_y}{dt} = \frac{d^2y}{dt^2}.$$

These accelerations, however, are also the rectangular components of the acceleration $\boldsymbol{a}$ of the

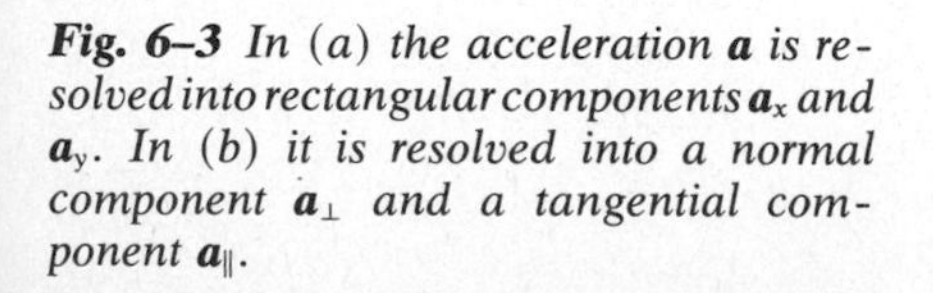

Fig. 6-3 In (*a*) the acceleration $\boldsymbol{a}$ is resolved into rectangular components $\boldsymbol{a}_x$ and $\boldsymbol{a}_y$. In (*b*) it is resolved into a normal component $\boldsymbol{a}_\perp$ and a tangential component $\boldsymbol{a}_\parallel$.

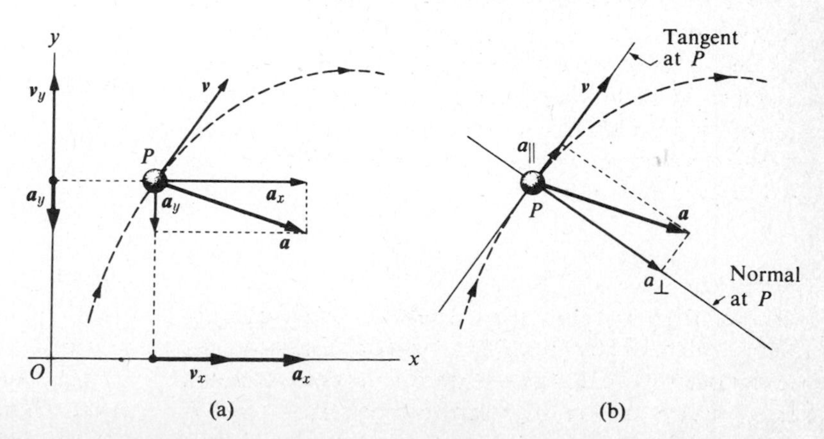

particle. Thus, if the components a_x and a_y are known, the magnitude and direction of the acceleration $\boldsymbol{a}$ can be obtained, just as with velocity. When the acceleration is known, the force on the particle can be found, in magnitude and direction, from Newton's second law,

$$\boldsymbol{F} = m\boldsymbol{a}.$$

Conversely, if the force $\boldsymbol{F}$ is known at every point, the acceleration $\boldsymbol{a}$ and its components a_x and a_y can be found from Newton's second law.

The acceleration of a particle moving in a curved path can also be resolved into rectangular components $a_\perp$ and $a_\parallel$, in directions *normal* and *parallel* (or tangential) to the path, as shown in Fig. 6–3(b). Unlike the rectangular components referred to a set of fixed axes, the normal and tangential components do not have fixed directions in space. They do, however, have a direct physical significance. The parallel component $a_\parallel$ arises from a change in the *magnitude* of the velocity vector $\boldsymbol{v}$, while the normal component $a_\perp$ arises from a change in the *direction* of the velocity.

This is illustrated in Fig. 6–4, which corresponds to Fig. 6–2(b). The vector from O to A represents the velocity $\boldsymbol{v}_1$ of a particle at point P in Fig. 6–2(a), and the vector from O to B represents the velocity $\boldsymbol{v}_2$ at point Q. The change in velocity, $\Delta\boldsymbol{v}$, is given by the vector from A to B.

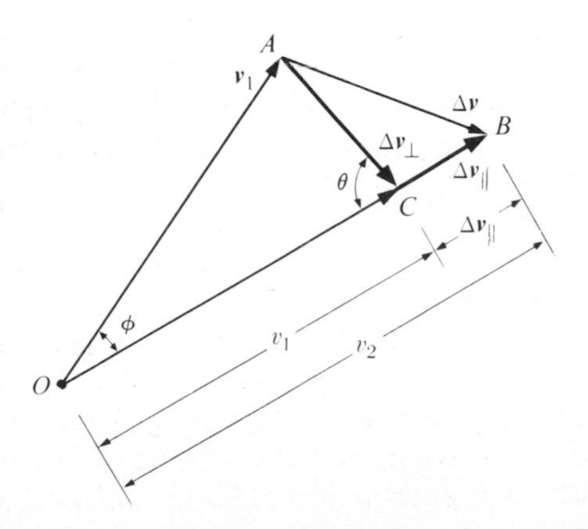

Fig. 6–4 *The vector $\Delta\boldsymbol{v}$ is resolved into normal and tangential components $\Delta\boldsymbol{v}_\perp$ and $\Delta\boldsymbol{v}_\parallel$. The normal component is the change in velocity resulting from the change in direction of $\boldsymbol{v}$; the tangential component is the change resulting from a change in the magnitude of $\boldsymbol{v}$.*

Note that the vectors $\Delta\boldsymbol{v}$ in Figs. 6–4 and 6–2(b) have the same direction and (apart from the difference in scale of the two diagrams) the same length.

The vector from O to C in Fig. 6–4 has the same length as $\boldsymbol{v}_1$. The vector $\Delta\boldsymbol{v}$ can be resolved into components represented by the vectors from A to C, and from C to B. The length of the vector from C to B equals the difference in *length* between the vectors $\boldsymbol{v}_2$ and $\boldsymbol{v}_1$. That is, this vector represents the change in *magnitude* of the velocity, and when divided by Δt, gives the component of average acceleration resulting from this change in magnitude.

If the magnitude of the velocity did *not* change between points P and Q, the velocity $\boldsymbol{v}_2$ at point Q would be represented by the vector from O to C in Fig. 6–4. In this case, however, *there would still be a change in the vector velocity*, represented by the vector from A to C. This change would result from the change in *direction* of the velocity vector, and when divided by Δt would give the component of average acceleration resulting from this change in direction. That is, *motion in a curved path with constant speed is accelerated motion*, because velocity is a vector quantity which can change in magnitude, in direction, or, as in Fig. 6–4, in both.

Now suppose that point Q in Fig. 6–2(a) approaches point P. The vector from O to B in Fig. 6–4 then swings upward toward the vector $\boldsymbol{v}_1$. The angle ϕ becomes smaller and smaller and the angle θ approaches 90°. The vector from A to C becomes more and more nearly perpendicular to $\boldsymbol{v}_1$ and the vector from C to B becomes more and more nearly parallel to $\boldsymbol{v}_1$. In the limit, the vector from A to C becomes normal to $\boldsymbol{v}_1$ (and hence normal to the path) and the vector from C to B becomes parallel to $\boldsymbol{v}_1$ (and hence tangent to the path). Thus although the vectors labeled $\Delta\boldsymbol{v}_\perp$ and $\Delta\boldsymbol{v}_\parallel$ in Fig. 6–4 are not normal and parallel to $\boldsymbol{v}_1$ in this figure, they become so in the limit as point Q approaches point P. The limiting value of $\Delta\boldsymbol{v}_\perp/\Delta t$ equals the normal component of acceleration $\boldsymbol{a}_\perp$, and the limiting value of $\Delta\boldsymbol{v}_\parallel/\Delta t$ equals the parallel component of acceleration $\boldsymbol{a}_\parallel$.

Is the acceleration resulting from a change in the *direction* of a velocity as "real" as that arising from a change in its *magnitude?* From a purely kinematical viewpoint the answer is of course "Yes," since by definition acceleration equals the vector rate of change of velocity. A more satisfying answer is that a force must be exerted on a body to change its direction of motion, as well as to increase or decrease its speed. In the absence of an external force, a body continues to move not only with constant speed but also *in a straight line.* When a body moves in a *curved* path, a transverse force must be exerted on it to deviate it sidewise. The ratio of transverse (or normal) force to normal acceleration is the mass of the body, and is equal to the ratio of tangential force to tangential acceleration. That is, if $\boldsymbol{F}_{\perp}$ and $\boldsymbol{F}_{\parallel}$ are the normal and tangential components of the force on a body moving in a curved path, Newton's second law (with the same value of m) applies to both of these components:

$$F_{\perp} = ma_{\perp}, \qquad F_{\parallel} = ma_{\parallel}.$$

These considerations are useful in atomic physics. The usual experimental method of measuring the mass of an individual ion is to project it into a magnetic field. The magnetic field exerts a transverse force on the ion and the mass of the ion is obtained by dividing this force by the measured transverse acceleration.

It follows from the discussion above that if the force on a particle is always normal to the path, then $F_{\parallel} = 0$, $a_{\parallel} = 0$, and the particle has no tangential component of acceleration. The *magnitude* of the velocity then remains constant and the only effect of the force is to change the direction of motion, that is, to deviate the particle sidewise.

If the force has no normal component, then $F_{\perp} = 0$, $a_{\perp} = 0$, and there is no change in the *direction* of the velocity; that is, the particle moves in a straight line.

6–5 MOTION OF A PROJECTILE

Any object that is given an initial velocity and which subsequently follows a path determined by the gravitational force acting on it and by the frictional resistance of the atmosphere is called a *projectile.* Thus the term applies to a missile shot from a gun, a rocket after its fuel is exhausted, a bomb released from an airplane, or a thrown or batted baseball. The motion of a freely falling body discussed in Chapter 4 is a special case of projectile motion. The path followed by a projectile is called its *trajectory.*

The gravitational force on a projectile is directed toward the center of the earth and is inversely proportional to the square of the distance from the earth's center. Here we shall consider only trajectories which are of sufficiently short range so that the gravitational force can be considered constant in magnitude and direction. The motion will be referred to axes fixed with respect to the earth. Since this is not an inertial system, it is not strictly correct to use Newton's second law to relate the force on the projectile to its acceleration. However, for trajectories of short range, the error is very small. Finally, all effects of air resistance will be ignored, so that our results apply only to motion in a vacuum on a flat, nonrotating earth. These simplifying assumptions form the basis of an idealized *model* of the physical problem, in which we neglect unimportant details and focus attention on the most important aspects of the phenomenon.

Since the only force on a projectile in this idealized case is its weight, considered constant in magnitude and direction, the motion is best referred to a set of rectangular coordinate axes. We shall take the x-axis horizontal, the y-axis vertical, and the origin at the point where the projectile starts its free flight, for example, at the muzzle of a gun or the point where it leaves the thrower's hand. The x-component of the force on the projectile is then zero and the y-component is the weight of the projectile, $-mg$. Then from Newton's second law,

$$a_x = \frac{F_x}{m} = 0, \qquad a_y = \frac{F_y}{m} = \frac{-mg}{m} = -g.$$

That is, the horizontal component of acceleration is zero and the vertical component is downward and equal to that of a freely falling body. Since zero acceleration means constant velocity, the motion can be described as a combination of *horizontal motion with constant velocity and vertical motion with constant acceleration.*

Consider next the velocity of the projectile. In Fig. 6–5, x- and y-axes have been constructed with the origin at the point where the projectile begins its free flight. We shall set $t = 0$ at this point. The velocity at the origin is represented by the vector $\mathbf{v}_0$, called the *initial velocity* or the *muzzle velocity* if the projectile is shot from a gun. The angle θ_0 is the *angle of departure.* The initial velocity has been resolved into a horizontal component v_{0x}, of magnitude $v_0 \cos \theta_0$, and a vertical component v_{0y}, of magnitude $v_0 \sin \theta_0$.

Since the horizontal velocity component v_x is constant, we have at any later time t,

$$v_x = v_{0x} = v_0 \cos \theta_0 .$$

The vertical acceleration is $-g$, so the vertical velocity component at time t is

$$v_y = v_{0y} - gt = v_0 \sin \theta_0 - gt.$$

These components can be added vectorially to find the resultant velocity v. Its magnitude is

$$v = \sqrt{v_x^2 + v_y^2},$$

and the angle θ it makes with the horizontal is given by

$$\tan \theta = \frac{v_y}{v_x}.$$

The velocity vector $\mathbf{v}$ is tangent to the trajectory, so its direction is the same as that of the trajectory.

The coordinates of the projectile at any time can now be found from the equations of motion with constant velocity, and with constant acceleration, developed in Sec. 4–5. The x-coordinate is

$$x = v_{0x}\, t = (v_0 \cos \theta_0)t, \qquad (6\text{–}5)$$

and the y-coordinate is

$$y = v_{0y}\, t - \tfrac{1}{2}gt^2 = (v_0 \sin \theta_0)t - \tfrac{1}{2}gt^2. \qquad (6\text{–}6)$$

The two preceding equations give the equation of the trajectory in terms of the parameter t. The equation in terms of x and y can be obtained by eliminating t. We find $t = x/v_0 \cos \theta_0$ and

$$y = (\tan \theta_0)x - \frac{g}{2v_0^2 \cos^2 \theta_0} x^2. \qquad (6\text{–}7)$$

The quantities v_0, $\tan \theta_0$, $\cos \theta_0$, and g are constants, so the equation has the form

$$y = ax - bx^2,$$

which will be recognized as the equation of a *parabola.*

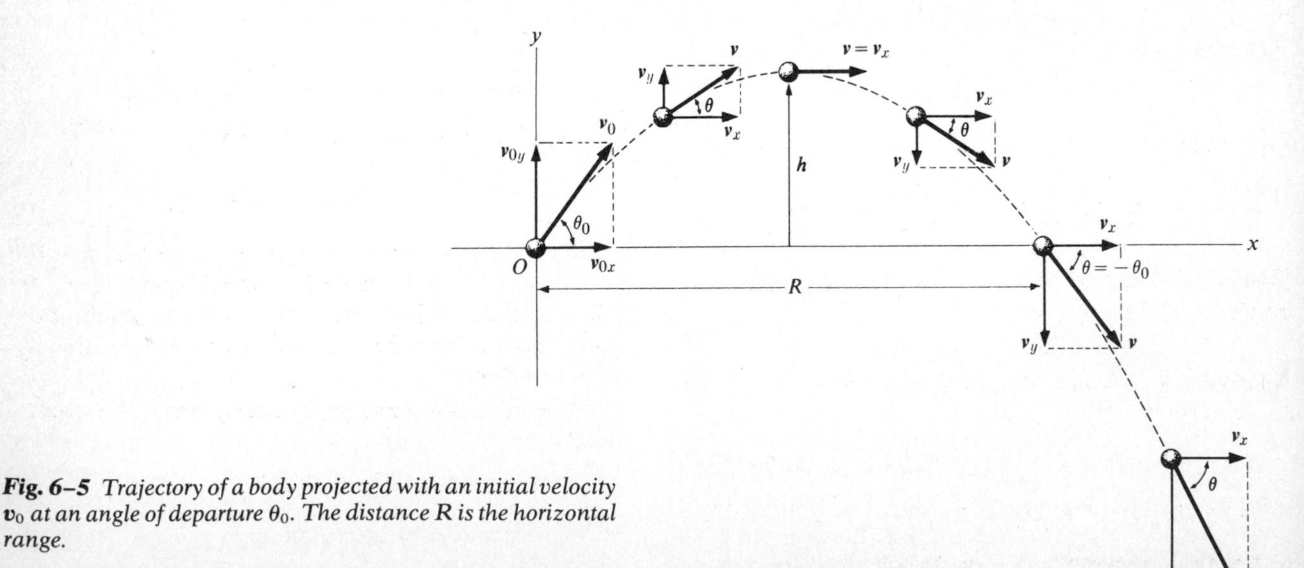

Fig. 6–5 *Trajectory of a body projected with an initial velocity* $\mathbf{v}_0$ *at an angle of departure* θ_0. *The distance R is the horizontal range.*

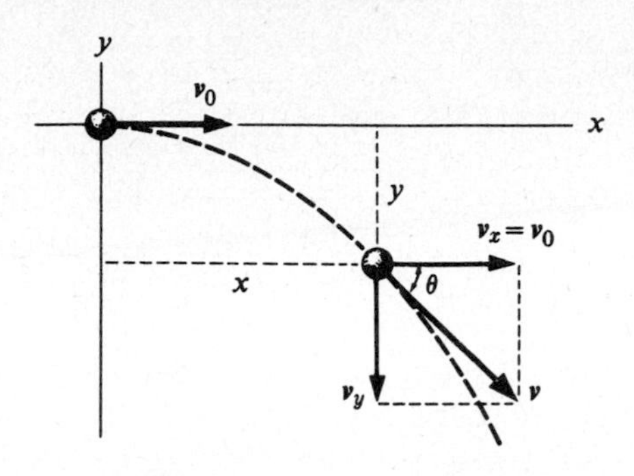

Fig. 6–6 Trajectory of a body projected horizontally.

Example 1 A ball is projected horizontally with a velocity v_0 of magnitude 5 m s^{-1}. Find its position and velocity after $\frac{1}{2}$ s (see Fig. 6–6).

In this case, the departure angle is zero. The initial vertical velocity component is therefore zero. The horizontal velocity component equals the initial velocity and is constant.

The x- and y-coordinates, when $t = \frac{1}{2}$ s, are

$$x = v_x t = (5\text{ m s}^{-1})(\tfrac{1}{2}\text{ s}) = 2.5\text{ m},$$

$$y = -\tfrac{1}{2}gt^2 = -\tfrac{1}{2}(10\text{ m s}^{-2})(\tfrac{1}{2}\text{ s})^2 = -1.25\text{ m}.$$

The components of velocity are

$$v_x = v_0 = 5\text{ m s}^{-1},$$

$$v_y = -gt = (-10\text{ m s}^{-2})(\tfrac{1}{2}\text{ s}) = -5\text{ m s}^{-1}.$$

The resultant velocity is

$$v = \sqrt{v_x^2 + v_y^2} = 5\sqrt{2}\text{ m s}^{-1}.$$

The angle θ is

$$\theta = \tan^{-1}\frac{v_y}{v_x} = \tan^{-1}\frac{5\text{ m s}^{-1}}{-5\text{ m s}^{-1}} = -45°.$$

That is, the velocity at $t = \frac{1}{2}$ s is 45° *below* the horizontal.

Example 2 In Fig. 6–5, let $v_0 = 160$ ft s^{-1}, $\theta_0 = 53°$. Then

$$v_{0x} = v_0 \cos\theta_0 = (160\text{ ft s}^{-1})(0.60) = 96\text{ ft s}^{-1},$$

$$v_{0y} = v_0 \sin\theta_0 = (160\text{ ft s}^{-1})(0.80) = 128\text{ ft s}^{-1}.$$

a) Find the position of the projectile, and the magnitude and direction of its velocity, when $t = 2.0$ s.

$$x = (96\text{ ft s}^{-1})(2.0\text{ s}) = 192\text{ ft},$$

$$y = (128\text{ ft s}^{-1})(2.0\text{ s}) - \tfrac{1}{2}(32\text{ ft s}^{-2})(2.0\text{ s})^2 = 192\text{ ft}.$$

$$v_x = 96\text{ ft s}^{-1},$$

$$v_y = 128\text{ ft s}^{-1} - (32\text{ ft s}^{-2})(2.0\text{ s}) = 64\text{ ft s}^{-1},$$

$$\theta = \tan^{-1}\frac{64\text{ ft s}^{-1}}{96\text{ ft s}^{-1}} = \tan^{-1}0.667 = 33.5°.$$

b) Find the time at which the projectile reaches the highest point of its flight, and find the elevation of this point.

At the highest point, the vertical velocity v_y is zero. If t_1 is the time at which this point is reached,

$$v_y = 0 = 128\text{ ft s}^{-1} - (32\text{ ft s}^{-2})t_1,$$

$$t_1 = 4\text{ s}.$$

The elevation h of the point is the value of y when $t = 4$ s.

$$h = (128\text{ ft s}^{-1})(4\text{ s}) - \tfrac{1}{2}(32\text{ ft s}^{-2})(4\text{ s})^2 = 256\text{ ft}.$$

c) Find the *horizontal range R*, that is, the horizontal distance from the starting point to the point at which the projectile returns to its original elevation and at which, therefore, $y = 0$. Let t_2 be the time at which this point is reached. Then

$$y = 0 = (128\text{ ft s}^{-1})t_2 - \tfrac{1}{2}(32\text{ ft s}^{-2})t_2^2.$$

This quadratic equation has two roots,

$$t_2 = 0 \quad\text{and}\quad t_2 = 8\text{ s},$$

corresponding to the two points at which $y = 0$. Evidently the time desired is the second root, $t_2 = 8$ s, which is just twice the time to reach the highest point. The time of descent therefore equals the time of rise.

The horizontal range R is the value of x when $t = 8$ s:

$$R = v_x t_2 = (96\text{ ft s}^{-1})(8\text{ s}) = 768\text{ ft}.$$

The vertical velocity at this point is

$$v_y = 128\text{ ft s}^{-1} - (32\text{ ft s}^{-2})(8\text{ s}) = -128\text{ ft s}^{-1}.$$

That is, the vertical velocity has the same magnitude as the initial vertical velocity, but the opposite direction. Since v_x is constant, the angle below the horizontal at this point equals the angle of departure.

d) If unimpeded, the projectile continues to travel beyond its horizontal range. For example, the projectile might have been fired from the edge of a high cliff, so that negative values of y are possible. It is left as an exercise to compute the position and velocity at a time 10 s after the start, corresponding to the last position shown in Fig. 6–5. The results are:

$$x = 960 \text{ ft}, \qquad y = -320 \text{ ft},$$

$$v_x = 96 \text{ ft s}^{-1}, \qquad v_y = -192 \text{ ft s}^{-1}.$$

Example 3 Figure 6–7 illustrates an interesting experimental demonstration of the properties of projectile motion. A ball (shown by the open circle) is projected directly toward a second ball (the solid circle). The second ball is released from rest at the instant the first is projected, and it is found that the balls collide as shown regardless of the value of the initial velocity. To show that this must happen, we note that the initial elevation of the second ball is $x \tan \theta_0$, and that in time t it falls a distance $\frac{1}{2}gt^2$. Its elevation at the instant of collision is therefore

$$y = x \tan \theta_0 - \tfrac{1}{2}gt^2.$$

But t is also the time required for the first ball to traverse the horizontal distance x with constant horizontal velocity $v_0 \cos \theta_0$, so

$$x = v_0 \cos \theta_0 t.$$

Solving this for t and substituting in the above equation, we find

$$y = x \tan \theta_0 - \tfrac{1}{2}g\left(\frac{x}{v_0 \cos \theta}\right)^2,$$

which is the same as the elevation of the first ball as given by Eq. (6–7).

For any given initial velocity, there is one particular angle of departure for which the horizontal range is a maximum. To find this angle, we first work out a general algebraic expression for the range. From Eq. (6–6), the projectile returns to earth ($y = 0$) at a time t_2 given by $2v_0 \sin \theta_0/g$. The range R is the value of x at this time, and is

$$R = v_x t_2 = (v_0 \cos \theta_0)\left(\frac{2v_0 \sin \theta_0}{g}\right) = \frac{2v_0^2 \sin \theta_0 \cos \theta_0}{g} = \frac{v_0^2 \sin 2\theta_0}{g}. \tag{6–8}$$

Since the sine of an angle can never be greater than unity, we see that the *maximum* value of R is v_0^2/g, corresponding to the case $2\theta_0 = 90°$ or $\theta_0 = 45°$. Thus the maximum horizontal range is attained when the departure angle is 45°.

From the standpoint of gunnery, what we usually want to know is what the departure angle should be for a given muzzle velocity v_0 in order to hit a target whose position R is known. Let us assume target and gun are at the same elevation. Then, from Eq. (6–8),

$$\theta_0 = \tfrac{1}{2} \sin^{-1} \frac{Rg}{v_0^2}.$$

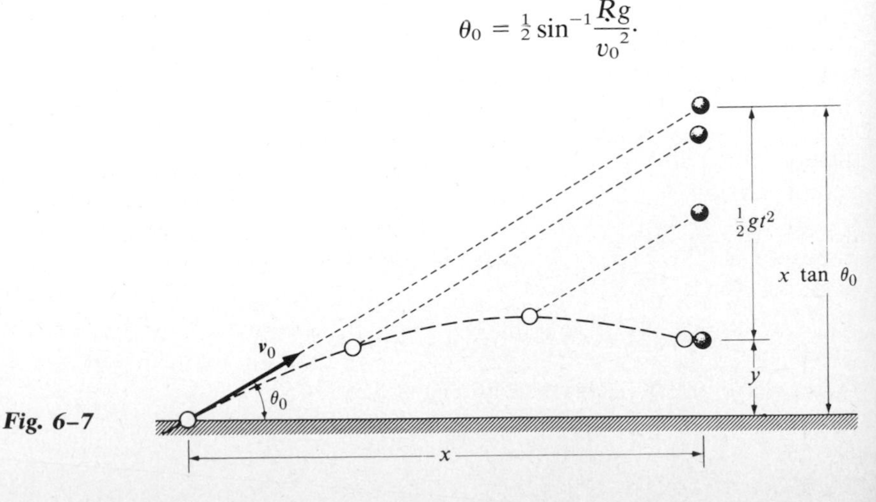

Fig. 6–7

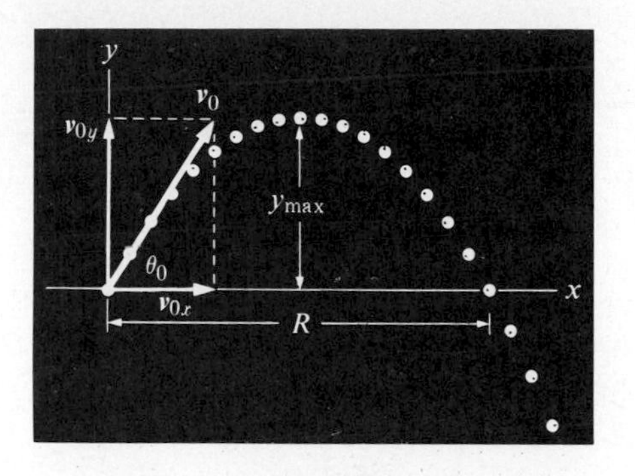

Fig. 6–8 Trajectory of a body projected at an angle with the horizontal.

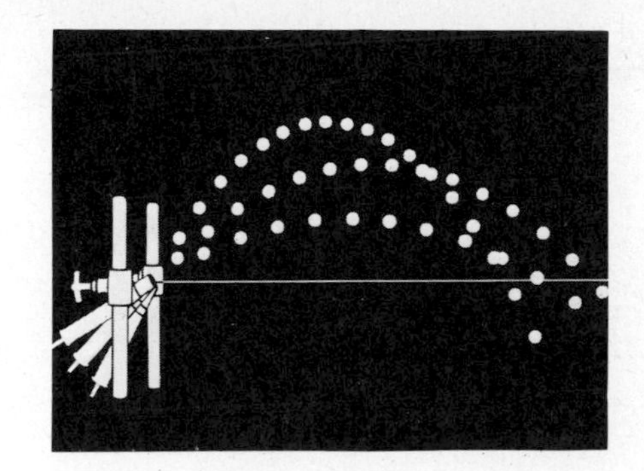

Fig. 6–9 An angle of departure of 45° gives the maximum horizontal range.

Provided R is less than the maximum range, this equation has two solutions for values of θ_0 between 0° and 90°.

Thus if $R = 800$ ft, $g = 32$ ft s^{-2}, and $v_0 = 200$ ft s^{-1},

$$\theta_0 = \tfrac{1}{2}\sin^{-1}\frac{(800\text{ ft})(32\text{ ft s}^{-2})}{(200\text{ ft s}^{-2})^2}$$

$$= \tfrac{1}{2}\sin^{-1}0.64.$$

But $\sin^{-1}0.64 = 40°$, or $180° - 40°$. Therefore

$$\theta_0 = 20° \qquad \text{or} \qquad 70°.$$

Either of these angles gives the same range. Of course both the time of flight and the maximum height reached are greater for the high-angle trajectory.

Figure 6–8 is a multiflash photograph of the trajectory of a ball; x- and y-axes and the initial velocity vector have been added. The horizontal distances between consecutive positions are all equal, showing that the horizontal velocity component is constant. The vertical distances first decrease and then increase, showing that the vertical motion is accelerated.

Figure 6–9 is a composite photograph of the three trajectories of a ball projected from a spring gun with departure angles of 30°, 45°, and 60°. It will be seen that the horizontal ranges are (nearly) the same for the 30° and 60° angles and that both are less than the range when the angle is 45°. As the departure angle is altered, there is a variation in the initial velocity which is imparted to the ball.

If the departure angle is *below* the horizontal, as for instance in the motion of a ball after it rolls off a sloping roof, or the trajectory of a bomb released from a dive bomber, exactly the same principles apply. The horizontal velocity component remains constant and equal to $v_0 \cos \theta_0$. The vertical motion is the same as that of a body projected *downward* with an initial velocity $v_0 \sin \theta_0$, which is now a negative quantity.

6–6 CIRCULAR MOTION

The acceleration of a particle moving in a curved path can be resolved into components normal and tangential to the path. There is a simple relation between the normal component of acceleration, the speed of the particle, and the radius of curvature of the path. We now derive this for the special case of motion in a circle.

Figure 6–10(a) represents a particle moving in a circular path of radius R with center at O. Vectors $\boldsymbol{v}_1$ and $\boldsymbol{v}_2$ represent its velocities at points P and Q. The vector change in velocity, $\Delta\boldsymbol{v}$, is obtained in Fig. 6–10(b), which is the same as Fig. 6–4. Vectors $\Delta v_\perp$ and $\Delta v_\parallel$ are the normal and tangential components of $\Delta\boldsymbol{v}$.

The triangles OPQ and opq in Fig. 6–10(a) and (b) are similar, since both are isosceles triangles and their long sides are mutually perpendicular. Hence

$$\frac{\Delta v_\perp}{v_1} = \frac{\Delta s}{R} \qquad \text{or} \qquad \Delta v_\perp = \frac{v_1}{R}\Delta s.$$

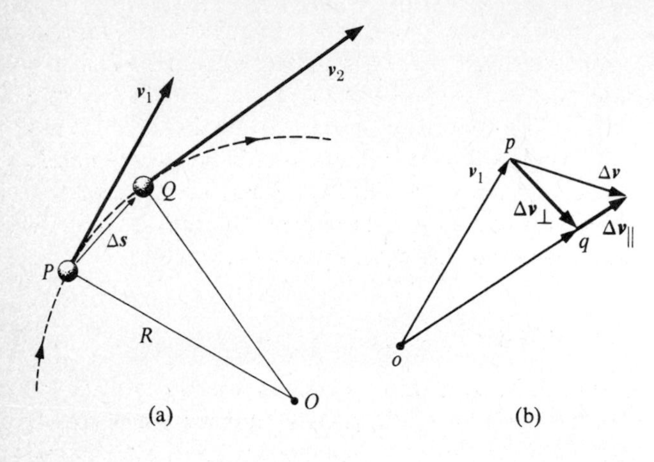

Fig. 6-10 *Construction for finding change in velocity, $\Delta \boldsymbol{v}$, of particle moving in a circle.*

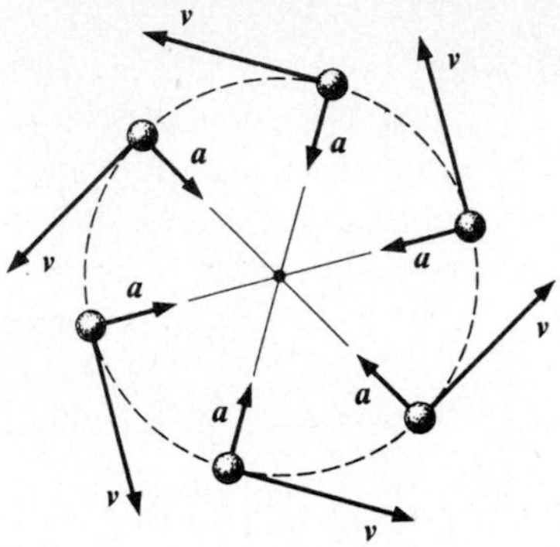

Fig. 6-11 *Velocity and acceleration vectors of a particle in uniform circular motion.*

The magnitude of the average normal acceleration $\bar{a}_\perp$ is therefore

$$\bar{a}_\perp = \frac{\Delta v_\perp}{\Delta t} = \frac{v_1}{R}\frac{\Delta s}{\Delta t}.$$

The *instantaneous* acceleration $a_\perp$ at point P is the limiting value of this expression, as point Q is taken closer and closer to point P:

$$a_\perp = \lim_{\Delta t \to 0} \frac{v_1}{R}\frac{\Delta s}{\Delta t} = \frac{v_1}{R} \lim_{\Delta t \to 0} \frac{\Delta s}{\Delta t}.$$

But the limiting value of $\Delta s/\Delta t$ is the speed v_1 at point P, and since P can be any point of the path, we can drop the subscript from v_1 and let v represent the speed at any point. Then

$$\boxed{a_\perp = \frac{v^2}{R}.} \qquad (6\text{–}9)$$

The magnitude of the instantaneous normal acceleration is therefore equal to the square of the speed divided by the radius. The direction is inward along the radius, toward the center of the circle. Because of this it is called a *central,* a *centripetal,* or a *radial* acceleration. (The term "centripetal" means "seeking a center.")

The unit of radial acceleration is the same as that of an acceleration resulting from a change in the *magnitude* of a velocity. Thus if a particle travels with a speed of 4 m s^{-1} in a circle of radius 2 m, its radial acceleration is

$$a = \frac{(4\ \mathrm{m\,s^{-1}})^2}{2\ \mathrm{m}} = 8\ \mathrm{m\,s^{-2}}.$$

If the *speed* of the particle changes, it will also have a *tangential* component of acceleration, defined as

$$a_\parallel = \lim_{\Delta t \to 0} \frac{\Delta v_\parallel}{\Delta t}.$$

If the speed is constant, there is no tangential component of acceleration and the acceleration is purely normal, resulting from the continuous change in *direction* of the velocity. In general, there will be both tangential and normal components of acceleration.

Figure 6–11 shows the directions of the velocity and acceleration vectors at a number of points, for a particle revolving in a circle with a velocity of constant magnitude.

A *centrifuge* is a device for whirling an object with a high velocity. The consequent large radial acceleration is equivalent to increasing the value of g, and such processes as sedimentation (settling of particles or precipitates out of a solution) which would otherwise take place only slowly, can be greatly accelerated in this way. Very high speed centrifuges, called ultracentrifuges, have been operated at speeds as high as 180,000 rev min^{-1}, and some small experimental units have been driven as fast as 1,300,000 rev min^{-1}.

Centrifuges producing accelerations up to the order of 100,000 g are used routinely in medical laboratories. Applications include increasing of sedimentation rates and actual measurements of molecular weights of large molecules such as proteins and viruses by measurement of sedimentation rate.

Example The moon revolves about the earth in a circle (very nearly) of radius R = 239,000 mi or 12.6×10^8 ft, and requires 27.3 days, or 23.4×10^5 s, to make a complete revolution.

a) What is the acceleration of the moon toward the earth?

The velocity of the moon is

$$v = \frac{2\pi R}{T} = \frac{2\pi(12.6 \times 10^8\ \text{ft})}{23.4 \times 10^5\ \text{s}} = 3383\ \text{ft s}^{-1}.$$

Its radial acceleration is therefore

$$a = \frac{v^2}{R} = \frac{(3383\ \text{ft s}^{-1})^2}{12.6 \times 10^8\ \text{ft}} = 0.00908\ \text{ft s}^{-2}.$$

b) If the gravitational force exerted on a body by the earth is inversely proportional to the square of the distance from the earth's center, the acceleration produced by this force should vary in the same way. Therefore, if the acceleration of the moon is caused by the gravitational attraction of the earth, the ratio of the moon's acceleration to that of a falling body at the earth's surface should equal the ratio of the square of the earth's radius (3950 mi or 2.09×10^7 ft) to the square of the radius of the moon's orbit. Is this true?

The ratio of the two accelerations is

$$\frac{8.96 \times 10^{-3}\ \text{ft s}^{-2}}{32.2\ \text{ft s}^{-2}} = 2.82 \times 10^{-4}.$$

The ratio of the squares of the distances is

$$\frac{(2.09 \times 10^7\ \text{ft})^2}{(12.6 \times 10^8\ \text{ft})^2} = 2.75 \times 10^{-4}.$$

The agreement is very close, although not exact because we have used average values.

Newton used the above calculation to justify his hypothesis that gravitation was truly *universal* and that the earth's pull extended out indefinitely into space. The numerical values available in Newton's time were not highly precise. While he did not obtain as close an agreement as that above, he states that he found his results to "answer pretty nearly," and he concluded that his hypothesis was verified.

6–7 CENTRIPETAL FORCE

Having obtained an expression for the normal or *radial* acceleration of a particle moving in a circle, we can now use Newton's second law to find the radial force on the particle. Since the magnitude of the radial acceleration equals v^2/R, and its direction is toward the center, the magnitude of the radial force on a particle of mass m is

$$F = m\frac{v^2}{R}. \tag{6–10}$$

The direction of this force is toward the center also, and it is called a *centripetal force.* It is unfortunate that it has become common practice to characterize the force by the adjective "centripetal," since this seems to imply that there is some difference in nature between centripetal forces and other forces. This is not the case. Centripetal forces, like other forces, are pushes and pulls exerted by sticks and strings, or arise from the action of gravitational or other causes. The term "centripetal" refers to the *effect* of the force, that is, to the fact that it results in a change in the *direction* of the velocity of the body on which it acts, rather than a change in the *magnitude* of this velocity.

Anyone who has ever tied an object to a cord and whirled it in a circle will realize the necessity of exerting this inward, centripetal force. If the cord breaks, the direction of the velocity ceases to change (unless other forces are acting) and the object flies off along a tangent to the circle.

Example 1 A small body of mass 0.2 kg revolves uniformly in a circle on a horizontal frictionless surface, attached by a cord 0.2 m long to a pin set in the surface. If the body makes two complete revolu-

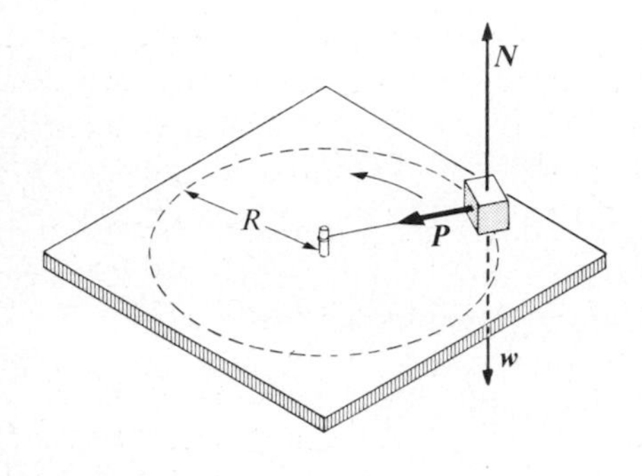

Fig. 6–12

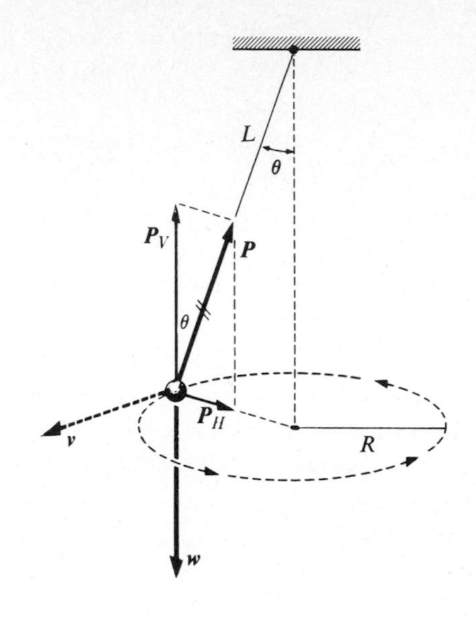

Fig. 6–13 The conical pendulum.

tions per second, find the force P exerted on it by the cord. (See Fig. 6–12.)

The circumference of the circle is

$$2\pi(0.2\text{ m}) = 0.4\pi\text{ m}$$

so the velocity is 0.8π m s^{-1}. The magnitude of the centripetal acceleration is

$$a = \frac{v^2}{R} = \frac{(0.8\pi\text{ m s}^{-1})^2}{0.2\text{ m}} = 31.6\text{ m s}^{-2}.$$

Since the body has no vertical acceleration, the forces $\mathbf{N}$ and $\boldsymbol{w}$ are equal and opposite and the force $\boldsymbol{P}$ is the resultant force. Therefore

$$P = ma = (0.2\text{ kg})(31.6\text{ m s}^{-2})$$
$$= 6.3\text{ N}.$$

Example 2 Figure 6–13 represents a small body of mass m revolving in a horizontal circle with velocity $\boldsymbol{v}$ of constant magnitude at the end of a cord of length L. As the body swings around its path, the cord sweeps over the surface of a cone. The cord makes an angle θ with the vertical, so the radius of the circle in which the body moves is $R = L \sin\theta$ and the magnitude of the velocity $\boldsymbol{v}$ equals $2\pi L \sin\theta/T$, where T is the time for one complete revolution.

The forces exerted on the body when it is in the position shown are its weight $\boldsymbol{w}$ and the tension $\boldsymbol{P}$ in the cord. (Note that the force diagram in Fig. 6–13 is exactly like that in Fig. 5–8(b). The only difference is that in this case the acceleration a is the *radial* acceleration, v^2/R.) Let $\boldsymbol{P}$ be resolved into a horizontal component $\boldsymbol{P}_{\mathrm{H}}$ and a vertical component $\boldsymbol{P}_{\mathrm{V}}$, of magnitudes $P\sin\theta$ and $P\cos\theta$, respectively. The body has no vertical acceleration, so the vertical forces $P\cos\theta$ and w are equal in magnitude, and the resultant inward, radial, or centripetal force is the horizontal component $P\sin\theta$, which is equal to the mass m times the radial (or centripetal) acceleration:

$$P\sin\theta = m\frac{v^2}{R}, \qquad P\cos\theta = w.$$

When the first of these equations is divided by the second, and w is replaced by mg, the result is

$$\tan\theta = \frac{v^2}{Rg}. \tag{6–11}$$

When we make use of the relations $R = L\sin\theta$ and $v = 2\pi L\sin\theta/T$, Eq. (6–11) becomes

$$\cos\theta = \frac{gT^2}{4\pi^2 L}, \tag{6–12}$$

or

$$T = 2\pi\sqrt{L(\cos\theta)/g}. \qquad (6\text{–}13)$$

Equation (6–12) indicates how the angle θ depends on the time of revolution T and the length L of the cord. For a given length L, $\cos\theta$ decreases as the time is made shorter, and the angle θ increases. The angle never becomes 90°, however, since this requires that $T = 0$ or $v = \infty$.

Equation (6–13) is similar in form to the expression for the time of swing of a simple pendulum, which will be derived in Chapter 11. Because of this similarity, the present device is called a *conical pendulum.*

Some readers may wish to add to the forces shown in Fig. 6–13 an outward, "centrifugal" force, to "keep the body out there," or to "keep it in equilibrium." ("Centrifugal" means "fleeing a center.") Let us examine this point of view. In the first place, to look for a force to "*keep* the body out there" is an example of faulty observation, because the body doesn't stay there! A moment later it will be at a different position on its circular path. At the instant shown it is moving in the direction of the velocity vector $\boldsymbol{v}$, and unless a resultant force acts on it, it will, according to Newton's first law, continue to move in this direction. If an outward force *were* acting on it, equal and opposite to the inward component of the force P, there would be no resultant inward force to deviate it sidewise from its present direction of motion.

Those who wish to add a force to "keep the body in equilibrium" forget that the term *equilibrium* refers to a state of rest, or of motion *in a straight line* with constant speed. Here, the body is *not* moving in a straight line, but in a circle. It is *not* in equilibrium, but has an acceleration toward the center of the circle and must be acted on by a resultant or *un*balanced force to produce this acceleration. In this example there is *no outward force* on the body!

Example 3 Figure 6–14(a) represents an automobile or a railway car rounding a flat curve of radius R, on a level road or track. The forces acting on it are its weight $\boldsymbol{w}$, the normal force $\boldsymbol{N}$, and the centripetal force $\boldsymbol{P}$. The force $\boldsymbol{P}$ must be provided by friction, in the case of an automobile, or by a force exerted by the rails against the flanges on the wheels of a railway car.

In order not to have to rely on friction, or to reduce wear on the rails and flanges, the road or the track may be banked, as shown in Fig. 6–14(b). The normal force $\boldsymbol{N}$ then has a vertical component of magnitude $N\cos\theta$, and a horizontal component of magnitude $N\sin\theta$ toward the center, which provides the centripetal force. The banking angle θ can be computed as follows. If v is the velocity and R the radius, then

$$N\sin\theta = \frac{mv^2}{R}.$$

Since there is no vertical acceleration, $N\cos\theta = w$.

Dividing the first equation by the second, and replacing w by mg, we obtain

$$\tan\theta = \frac{v^2}{Rg}.$$

The tangent of the angle of banking is proportional to the square of the speed and inversely

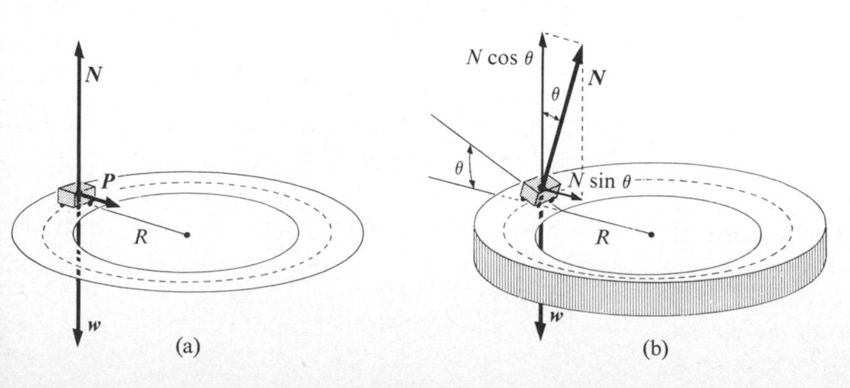

Fig. 6–14 *(a) Forces on a vehicle rounding a curve on a level track. (b) Forces when the track is banked.*

proportional to the radius. For a given radius no one angle is correct for all speeds. Hence in the design of highways and railroads, curves are banked for the *average speed* of the traffic over them. The same considerations apply to the correct banking angle of a plane when it makes a turn in level flight.

Note that the banking angle is given by the same expression as that for the angle with the vertical made by the cord of a conical pendulum. In fact, the force diagrams in Figs. 6–13 and 6–14 are identical.

6–8 MOTION IN A VERTICAL CIRCLE

Figure 6–15 represents a small body attached to a cord of length R and whirling in a *vertical* circle about a fixed point O to which the other end of the cord is attached. The motion, while circular, is not *uniform*, since the speed increases on the way down and decreases on the way up.

The forces on the body at any point are its weight $\mathbf{w} = m\mathbf{g}$ and the tension $\mathbf{T}$ in the cord. Let the weight be resolved into a normal component, of magnitude $w \cos\theta$, and a tangential component of magnitude $w \sin\theta$, as in Fig. 6–15. The resultant tangential and normal forces are then

$$F_{\parallel} = w \sin\theta \qquad \text{and} \qquad F_{\perp} = T - w\cos\theta.$$

The tangential acceleration, from Newton's second law, is

$$a_{\parallel} = \frac{F_{\parallel}}{m} = g\sin\theta,$$

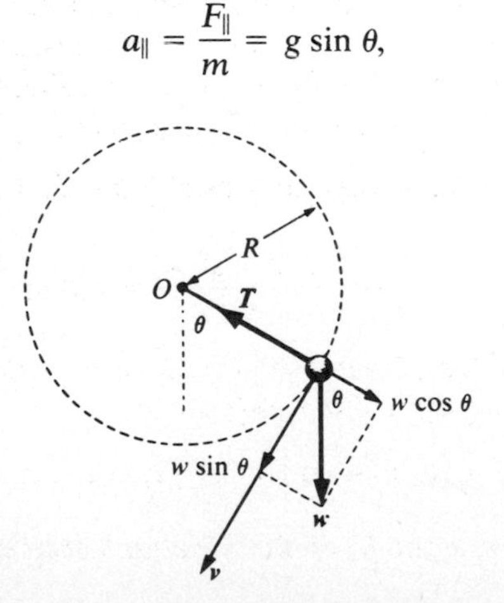

Fig. 6–15 Forces on a body whirling in a vertical circle with center at O.

and is the same as that of a body sliding on a frictionless inclined plane of slope angle θ. The normal (radial) acceleration $a_{\perp} = v^2/R$, is

$$a_{\perp} = \frac{F_{\perp}}{m} = \frac{T - w\cos\theta}{m} = \frac{v^2}{R},$$

and the tension in the cord is therefore

$$T = m\left(\frac{v^2}{R} + g\cos\theta\right). \tag{6–14}$$

At the lowest point of the path, $\theta = 0$, $\sin\theta = 0$, $\cos\theta = 1$. Hence at this point $F_{\parallel} = 0$, $a_{\parallel} = 0$, and the acceleration is purely radial (upward). The magnitude of the tension, from Eq. (6–14), is

$$T = m\left(\frac{v_{\text{bot}}^2}{R} + g\right).$$

At the highest point, $\theta = 180°$, $\sin\theta = 0$, $\cos\theta = -1$, and the acceleration is again purely radial (downward). The tension is

$$T = m\left(\frac{v_{\text{top}}^2}{R} - g\right). \tag{6–15}$$

With motion of this sort, it is a familiar fact that there is a certain critical speed v_c at the highest point, below which the cord becomes slack and the path is no longer circular. To find this speed, set $T = 0$ in Eq. (6–15):

$$0 = m\left(\frac{v_c^2}{R} - g\right), \qquad v_c = \sqrt{Rg}.$$

The multiflash photographs of Fig. 6–16 illustrate another case of motion in a vertical circle, a small ball "looping-the-loop" on the inside of a vertical circular track. The inward normal force exerted on the ball by the track takes the place of the tension T shown in Fig. 6–15.

In Fig. 6–16(a), the ball is released from an elevation such that its speed at the top of the track is greater than the critical speed, $\sqrt{Rg}$. In Fig. 6–16(b), the ball starts from a lower elevation and reaches the top of the circle with a speed such that its own weight is slightly larger than the requisite centripetal force. In other words, the track would have to pull *outward*

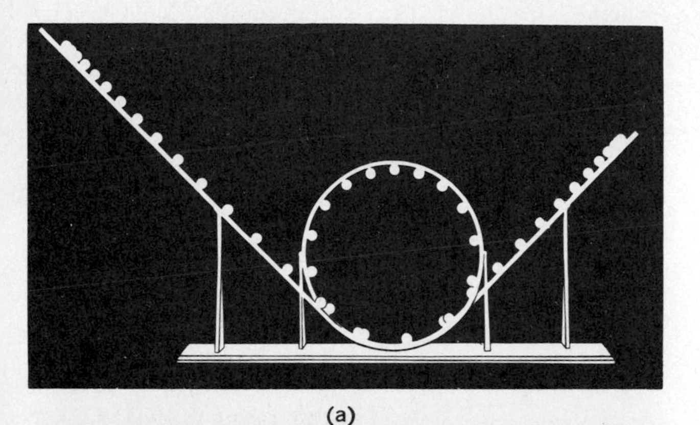

(a)

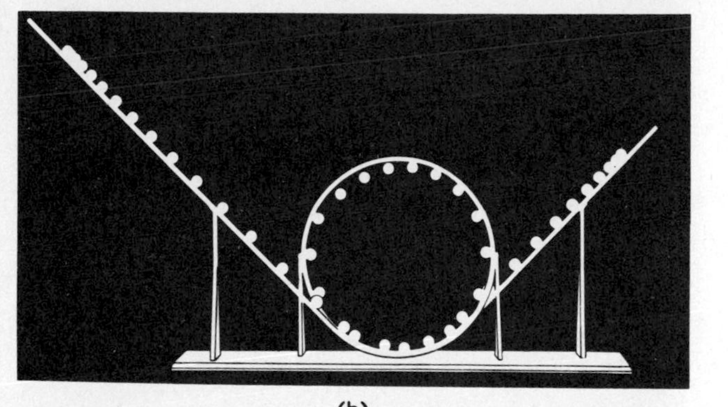

(b)

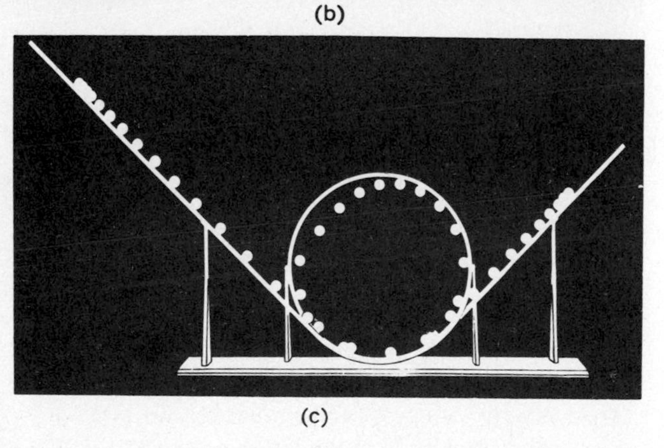

(c)

Fig. 6–16 Multiflash photographs of a ball looping-the-loop in a vertical circle.

to maintain the circular motion. Since this is impossible, the ball leaves the track and moves for a short distance in a parabola. This parabola soon intersects the circle, however, and the remainder of the trip is completed successfully. In Fig. 6–16(c), the start is made from a still lower elevation, the ball leaves the track sooner, and the parabolic path is clearly evident.

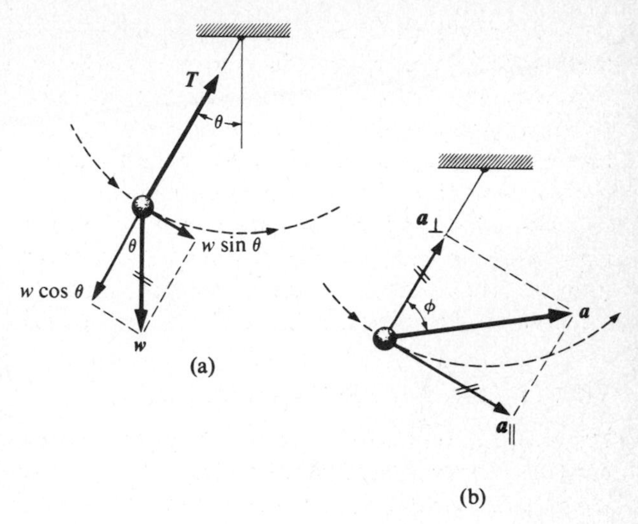

Fig. 6–17 (a) Forces on a body swinging in a vertical circle. (b) The radial and tangential components of acceleration are combined to obtain the resultant acceleration a.

Example In Fig. 6–17, a small body of mass $m = 0.10$ kg swings in a vertical circle at the end of a cord of length $R = 1.0$ m. If its speed $v = 2.0\ \mathrm{m\,s^{-1}}$ when the cord makes an angle $\theta = 30°$ with the vertical, find (a) the radial and tangential components of its acceleration at this instant, (b) the magnitude and direction of the resultant acceleration, and (c) the tension T in the cord.

a) The radial component of acceleration is

$$a_\perp = \frac{v^2}{R} = \frac{(2.0\ \mathrm{m\,s^{-1}})^2}{1.0\ \mathrm{m}} = 4.0\ \mathrm{m\,s^{-2}}.$$

The tangential component of acceleration, due to the tangential force $mg \sin\theta$, is

$$a_\parallel = g\sin\theta = (9.8\ \mathrm{m\,s^{-2}})(0.50) = 4.9\ \mathrm{m\,s^{-2}}.$$

b) The magnitude of the resultant acceleration (see Fig. 6–17b) is

$$a = \sqrt{a_\perp^{\,2} + a_\parallel^{\,2}} = 6.3\ \mathrm{m\,s^{-2}}.$$

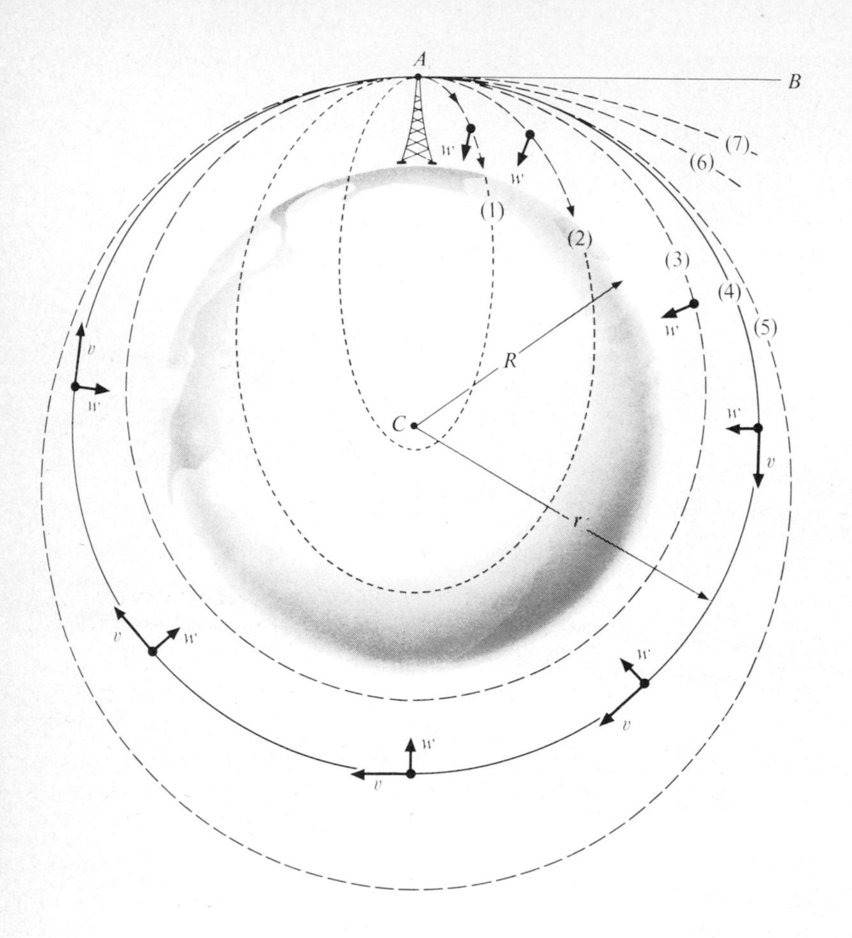

Fig. 6-18 Trajectories of a body projected from point A in the direction AB with different initial velocities.

The angle ϕ is

$$\phi = \tan^{-1} \frac{a_{\parallel}}{a_{\perp}} = 51°.$$

c) The tension in the cord is given by $F_{\perp} = ma_{\perp}$: $T - mg \cos \theta = mv^2/R$, so

$$T = m\left(\frac{v^2}{R} + g \cos \theta\right) = 1.3 \text{ N}.$$

Note that the magnitude of the tangential acceleration is not constant but is proportional to the sine of the angle θ. Hence the equations of motion with constant acceleration *cannot* be used to find the speed at other points of the path. We shall show in the next chapter, however, how the speed at any point can be found from *energy* considerations.

6-9 MOTION OF A SATELLITE

In discussing the trajectory of a projectile in Section 6-5, we assumed that the gravitational force on the projectile (its weight $\boldsymbol{w}$) had the same direction and magnitude at all points of the trajectory. These conditions are approximately satisfied if the projectile remains near the earth's surface and if its trajectory is small compared to the earth's radius. Under these conditions the trajectory is a parabola.

In reality, the gravitational force is directed toward the center of the earth and is inversely proportional to the square of the distance from the earth's center, so that it is not constant in either magnitude or direction. It can be shown that under an inverse square force directed toward a fixed point the trajectory must always be a *conic section* (ellipse, circle, parabola, or hyperbola).

Suppose that a very tall tower could be constructed as in Fig. 6–18, and that a projectile were launched from point A at the top of the tower in the "horizontal" direction AB. If the initial velocity is not too great, the trajectory will be like that numbered (1), which is a portion of an ellipse with the earth's center C at one focus. (If the trajectory is so short that changes in the magnitude and direction of $\boldsymbol{w}$ can be neglected, the ellipse *approximates* a parabola.)

The trajectories numbered from (2) to (7) illustrate the effect of increasing the initial velocity. (Any effect of the earth's atmosphere is neglected.) Trajectory (2) is again a portion of an ellipse. Trajectory (3), which just misses the earth, is a *complete* ellipse and the projectile has become an earth satellite. Its velocity when it returns to point A is the same as its initial velocity, and in the absence of retarding forces it will repeat its motion indefinitely. (The earth's rotation will have moved the tower to a different point by the time the satellite returns to point A, but does not affect the orbit.)

Trajectory (4) is a special case in which the orbit is a circle. Trajectory (5) is again an ellipse, (6) is a parabola, and (7) is a hyperbola. Trajectories (6) and (7) are not closed orbits.

All manmade earth satellites have orbits like (3), (4), or (5). Many are very nearly circles; for simplicity we shall consider circular orbits only. Let us calculate the velocity required for such an orbit, and the time of one revolution. The centripetal acceleration of the satellite in its circular orbit is produced by the gravitational force on the satellite, which is equal to the product of the mass and the centripetal (or radial) acceleration. The acceleration may in turn be obtained from the velocity of the satellite and the radius of the orbit.

Thus

$$w = F_g = G\left(\frac{mm_E}{r^2}\right) = m\left(\frac{v^2}{r}\right),$$

and from the last two terms,

$$v^2 = \frac{Gm_E}{r}, \qquad v = \sqrt{\frac{Gm_E}{r}}. \tag{6–16}$$

The larger the radius r, the smaller the orbital velocity.

The speed of the satellite may also be expressed in terms of the acceleration of gravity g at the earth's surface, which according to Eq. (5–7) is given by $g = Gm_E/R^2$. Combining this relation with Eq. (6–16), we obtain

$$v = R\sqrt{\frac{g}{r}}. \tag{6–17}$$

The acceleration, given by $a_\perp = v^2/r$ (*not* v^2/R!) can also be expressed in terms of g:

$$a_\perp = \frac{R^2}{r^2}\,g. \tag{6–18}$$

This is, of course, the acceleration of gravity at radius r; the satellite, like any projectile, is a freely falling body. As expected, the acceleration is less than g at the earth's surface, in the ratio of the square of the radii.

The period T, or the time required for one complete revolution, equals the circumference of the orbit, $2\pi r$, divided by the velocity v:

$$T = \frac{2\pi r}{v} = \frac{2\pi r}{R\sqrt{g/r}} = \frac{2\pi}{R\sqrt{g}}\,r^{3/2}, \tag{6–19}$$

and the larger the radius, the longer the period.

Example An earth satellite revolves in a circular orbit at a height of 300 km (about 200 miles) above the earth's surface.

a) What is the velocity of the satellite, assuming the earth's radius to be 6400 km and g to be $9.80\ \text{m s}^{-2}$?

From Eq. (6–17),

$$\begin{aligned} v &= R\sqrt{\frac{g}{r}} \\ &= (6.40 \times 10^6\ \text{m})\left(\frac{9.80\ \text{m s}^{-2}}{6.70 \times 10^6\ \text{m}}\right)^{1/2} \\ &= 7740\ \text{m s}^{-1} = 25{,}400\ \text{ft s}^{-1} \\ &= 17{,}300\ \text{mi hr}^{-1}. \end{aligned}$$

b) What is the period T?

$$T = \frac{2\pi r}{v} = 90.6 \text{ min} = 1.51 \text{ hr}.$$

c) What is the radial acceleration of the satellite?

$$a_\perp = \frac{v^2}{r} = 8.94 \text{ m s}^{-2}.$$

This is equal to the free-fall acceleration at a height of 300 km above the earth.

6–10 EFFECT OF THE EARTH'S ROTATION ON g

Because of the earth's rotation it is not precisely an inertial frame of reference, and the acceleration of a falling body is not precisely equal to the ratio of its true weight (the earth's gravitational attraction) to its mass. The small correction associated with earth's rotation will be discussed next.

Figure 6–19 is a cutaway view of our rotating earth, with three observers each holding a body of mass m, hanging from a string. Each body is attracted toward the earth's center with a force $F_g = Gmm_E/R^2$, which we now designate by $\mathbf{w}_0$. Let us consider the earth's center as the origin of an inertial reference system. The string exerts a force T on the body.

Except at the pole, each body is carried along by the earth's rotation and moves in a circle with center on the earth's axis. It therefore has a radial acceleration $a_\perp$ equal to v^2/R, toward the axis. The resultant or vector sum of the forces $\mathbf{T}$ and $\mathbf{w}_0$ must therefore be such as to provide the requisite radial acceleration. That is, in general vector form,

$$\mathbf{T} + \mathbf{w}_0 = m\mathbf{a}_\perp .$$

At some arbitrary latitude θ, $\mathbf{T}$ and $\mathbf{w}_0$ have directions as shown in the diagram, such that their resultant $\mathbf{F}$ points toward the point O' (the center of the circle in which the body moves) and equals $m\mathbf{a}_\perp$. Thus we see that, except at the pole and the equator, a plumb line does not point toward the earth's center.

At the pole, where $a_\perp = 0$, the force $\mathbf{T}$ is equal and opposite to the true weight $\mathbf{w}_0$.

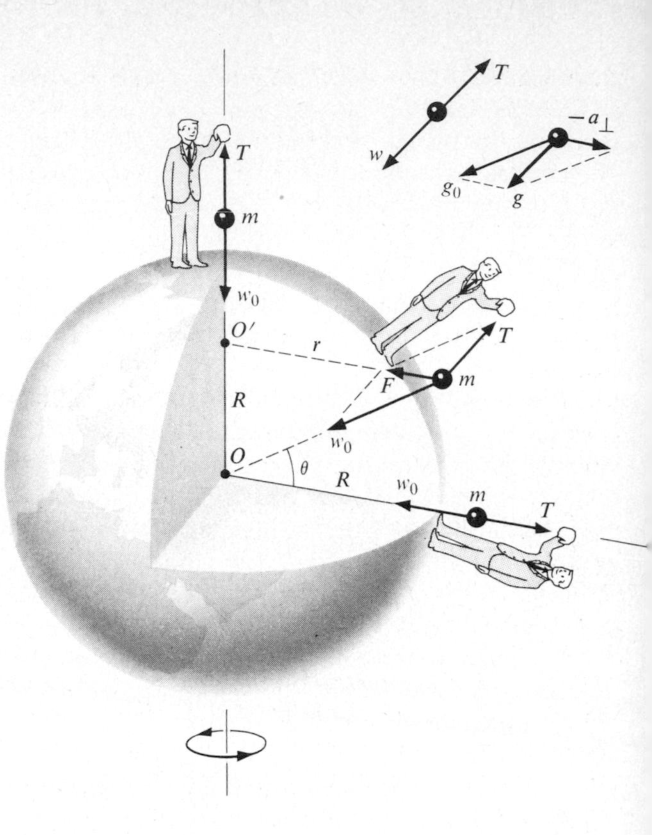

*Fig. 6–19 The resultant of the forces **T** and **w**₀ is the centripetal force, equal to mv^2/r.*

At the equator, where the direction of the radial acceleration is toward the earth's center O, $\mathbf{T}$ and $\mathbf{w}_0$ have the same action line but the magnitude of $\mathbf{w}_0$ is greater than that of $\mathbf{T}$.

Since each body appears in equilibrium to its respective observer, the *apparent* weight of each body, which we now represent by $\mathbf{w}$, is a force equal and opposite to $\mathbf{T}$, as shown in the inset diagram for the body at latitude θ. That is,

$$\mathbf{w} = -\mathbf{T} = \mathbf{w}_0 - m\mathbf{a}_\perp .$$

At an intermediate latitude, the apparent weight $\mathbf{w}$ differs in both magnitude and direction from the true weight $\mathbf{w}_0$.

At the pole, the apparent and true weights are equal. At the equator,

$$w = w_0 - ma_\perp ,$$

Table 6–1 VARIATIONS OF g WITH LATITUDE AND ELEVATION

Station	North latitude	Elevation, m	g, $\mathrm{m\,s^{-2}}$	g, $\mathrm{ft\,s^{-2}}$
Canal Zone	9°	0	9.78243	32.0944
Jamaica	18°	0	9.78591	32.1059
Bermuda	32°	0	9.79806	32.1548
Denver	40°	1638	9.79609	32.1393
Cambridge, Mass.	42°	0	9.80398	32.1652
Standard station			9.80665	32.1740
Greenland	70°	0	9.82534	32.2353

where $a_\perp$ is the radial acceleration at the equator. Dividing by m, we get

$$\frac{w}{m} = \frac{w_0}{m} - a_\perp,$$

where w/m is the observed acceleration g relative to the surface of the earth of a body falling at the equator, and w_0/m is the acceleration g_0 relative to the center of the earth. Thus, at the equator, $g = g_0 - a_\perp$.

Let us calculate the magnitude of the radial acceleration $a_\perp$ at the equator. The equatorial velocity v, equal to the earth's circumference divided by the time of one rotation, is

$$v = \frac{2\pi(6.4 \times 10^6\ \mathrm{m})}{8.64 \times 10^4\ \mathrm{s}} = 465\ \mathrm{m\,s^{-1}},$$

and hence

$$a_R = \frac{v^2}{R} = 0.034\ \mathrm{m\,s^{-2}} = 3.4\ \mathrm{cm\,s^{-2}}.$$

Thus if the free-fall acceleration g_0 at the equator is $9.880\ \mathrm{m\,s^{-2}}$, the observed acceleration g is $9.766\ \mathrm{m\,s^{-2}}$. Table 6–1 lists the measured free-fall acceleration g at a number of points. A part of the variation results from the fact that the earth is not spherical, that the observation points are at different elevations, or that there are local variations in earth density. It will be seen, however, that the change in the radial acceleration between equator and poles has the right magnitude to account for most of the differences.

A similar discussion can be applied to the phenomenon of "weightlessness" in satellites. A space vehicle in orbit is a freely falling body, with an acceleration $\mathbf{a}_\perp$ toward the earth's center equal to the value of the acceleration of gravity $\mathbf{g}$ at its orbit radius. The apparent weight $\mathbf{w}$ is given as before by

$$\mathbf{w} = \mathbf{w}_0 - m\mathbf{a}_\perp = m\mathbf{g} - m\mathbf{a}_\perp.$$

But, in this case,

$$\mathbf{g} = \mathbf{a}_\perp,$$

so

$$\mathbf{w} = 0.$$

It is in this sense that an astronaut in the vehicle is said to be "weightless" or in a state of "zero g." The vehicle is like the freely falling elevator discussed in Section 5–6, Example 7, except that it has a large and constant tangential speed along its orbit.

PROBLEMS

6–1 A ball rolls off the edge of a tabletop 4 ft above the floor, and strikes the floor at a point 6 ft horizontally from the edge of the table.

a) Find the time of flight.
b) Find the initial velocity.
c) Find the magnitude and direction of the velocity of the ball just before it strikes the floor. Draw a diagram to scale.

6–2 A block slides off a horizontal tabletop 1 m high with a velocity of $3\ \mathrm{m\,s^{-1}}$. Find (a) the horizontal distance from the table at which the block strikes the floor, and (b) the horizontal and vertical components of its velocity when it reaches the floor.

6–3 A level-flight bomber, flying at $300\ \mathrm{ft\,s^{-1}}$, releases a bomb at an elevation of 6400 ft.

a) How much time is required for the bomb to reach the earth?
b) How far does it travel horizontally?
c) Find the horizontal and vertical components of its velocity when it strikes.

6–4 A block passes a point 3 m from the edge of a table with a velocity of $4\ \mathrm{m\,s^{-1}}$. It slides off the edge of the table, which is 1 m high, and strikes the floor 1 m from the edge of the table. What was the coefficient of sliding friction between block and table?

6–5 A golf ball is driven horizontally from an elevated tee with a velocity of 25 m s^{-1}. It strikes the fairway 2.5 s later.

a) How far has it fallen vertically?

b) How far has it traveled horizontally?

c) Find the horizontal and vertical components of its velocity, and the magnitude and direction of its resultant velocity, just before it strikes.

6–6 A level-flight bombing plane, flying at an altitude of 1024 ft with a velocity of 240 ft s^{-1}, is overtaking a motor torpedo boat traveling at 80 ft s^{-1} in the same direction as the plane. At what distance astern of the boat should a bomb be released in order to hit the boat?

6–7 A .22 calibre rifle bullet is fired in a horizontal direction with a muzzle velocity of 300 m s^{-1}. In the absence of air resistance, how far will it have dropped in traveling a horizontal distance of (a) 20 m? (b) 40 m? (c) 60 m? (d) How far will it drop in one second?

6–8 A ball is projected with an initial upward velocity component of 20 m s^{-1} and a horizontal velocity component of 25 m s^{-1}.

a) Find the position and velocity of the ball after 2 s, 3 s, 6 s.

b) How long a time is required to reach the highest point of the trajectory?

c) How high is this point?

d) How long a time is required for the ball to return to its original level?

e) How far has it traveled horizontally during this time? Show your results in a neat sketch, large enough to show all features clearly.

6–9 A batted baseball leaves the bat at an angle of 30° above the horizontal, and is caught by an outfielder 400 ft from the plate.

a) What was the initial velocity of the ball?

b) How high did it rise?

c) How long was it in the air?

6–10 A spring gun projects a golf ball at an angle of 45° above the horizontal. The horizontal range is 10 m.

a) What is the maximum height to which the ball rises?

b) For the same initial speed, what are the two angles of departure for which the range is 6 m?

c) Sketch all three trajectories to scale in the same diagram.

6–11 Suppose the departure angle θ_0 in Fig. 6–7 is 15° and the distance x is 5 m. Where will the balls collide if the muzzle velocity of the first is (a) 20 m s^{-1}, (b) 5 m s^{-1}? Sketch both trajectories. (c) Will a collision take place if the departure angle is below horizontal?

6–12 If a baseball player can throw a ball a maximum distance of 60 m over the ground, what is the maximum vertical height to which he can throw it? Assume the ball to have the same initial speed in each case.

6–13 A player kicks a football at an angle of 37° with the horizontal and with an initial velocity of 48 ft s^{-1}. A second player standing at a distance of 100 ft from the first in the direction of the kick starts running to meet the ball at the instant it is kicked. How fast must he run in order to catch the ball before it hits the ground?

6–14 A baseball leaves the bat at a height of 4 ft above the ground, traveling at an angle of 45° with the horizontal, and with a velocity such that the horizontal range would be 400 ft. At a distance of 360 ft from home plate is a fence 30 ft high. Will the ball be a home run?

6–15 A projectile shot at an angle of 60° above the horizontal strikes a building 30 m away from a point 15 m above the point of projection.

a) Find the velocity of projection.

b) Find the magnitude and direction of the velocity of the projectile when it strikes the building.

6–16

a) What must be the velocity of a projectile fired vertically upward to reach an altitude of 60 km?

b) What velocity is required to reach the same height if the gun makes an angle of 45° with the vertical?

c) Compute the time required to reach the highest point in both trajectories.

d) How far would a plane traveling at 500 km hr^{-1} move in this time?

6–17 The angle of elevation of an antiaircraft gun is 70° and the muzzle velocity is 2700 ft s^{-1}. For what time after firing should the fuse be set if the shell is to explode at an altitude of 5000 ft?

6–18 A trench mortar fires a projectile at an angle of 53° above the horizontal with a muzzle velocity of 200 ft s^{-1}. A tank is advancing directly toward the mortar on level ground at a speed of 10 ft s^{-1}. What should be the distance from mortar to tank at the instant the mortar is fired in order to score a hit?

6–19 A projectile is fired with an initial speed v_0 at a departure angle θ_0, from the foot of an inclined plane of slope angle α (Fig. 6–20).

a) What is the range R measured along the plane?

b) Show that the expression for R reduces to that for the horizontal range when $\alpha = 0$.

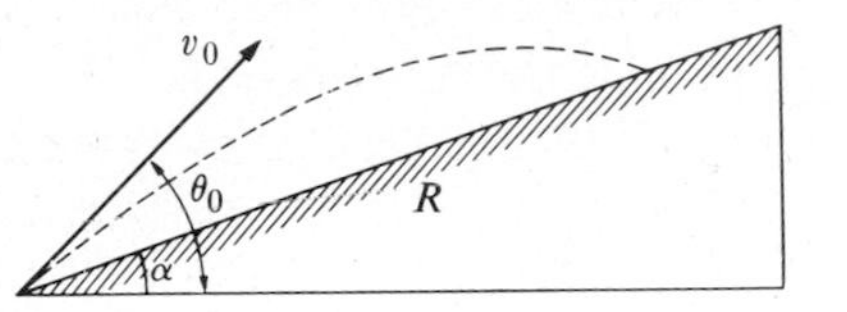

Fig. 6–20

6–20 A bomber, diving at an angle of 53° with the vertical, releases a bomb at an altitude of 800 m. The bomb is observed to strike the ground 5 s after its release.

a) What was the velocity of the bomber?

b) How far did the bomb travel horizontally during its flight?

c) What were the horizontal and vertical components of its velocity just before striking?

6–21 A 10-kg stone is dropped from a cliff in a high wind. The wind exerts a steady horizontal 50-N force on the stone as it falls. Is the path of the stone a straight line, a parabola, or some more complicated path? Explain.

6–22 A lump of ice slides 10 m down a smooth, sloping roof making an angle of 30° with the horizontal. The edge of the roof is 10 m above a sidewalk which extends 5 m out from the side of the building. Will the ice land on the sidewalk or in the street?

6–23 The second stage of a three-stage rocket designed to launch an earth satellite will exhaust its fuel at a height of about 130 mi above the earth's surface, when its speed will be approximately 11,000 mi hr^{-1}. The rocket will then "coast" to a height of about 300 mi, at which point its velocity will be horizontal. Assume a flat earth and a constant value of g.

a) Find the direction of the trajectory at the 130-mi height.

b) Find the horizontal distance traveled while the rocket rises from 130 mi to 300 mi.

c) Find the speed at the 300-mi height.

d) At a height of 300 mi, the third stage will separate from the second and the latter will fall to the earth. How far will it travel horizontally while falling?

e) How long a time will be required?

6–24 A man is riding on a flatcar traveling with a constant velocity of 30 ft s^{-1} (Fig. 6–21). He wishes to throw a ball through a stationary hoop 16 ft above the height of his hands in such a manner that the ball will move horizontally as it passes through the hoop. He throws the ball with a velocity of 40 ft s^{-1} with respect to himself.

a) What must be the vertical component of the initial velocity of the ball?

b) How many seconds after he releases the ball will it pass through the hoop?

c) At what horizontal distance in front of the hoop must he release the ball?

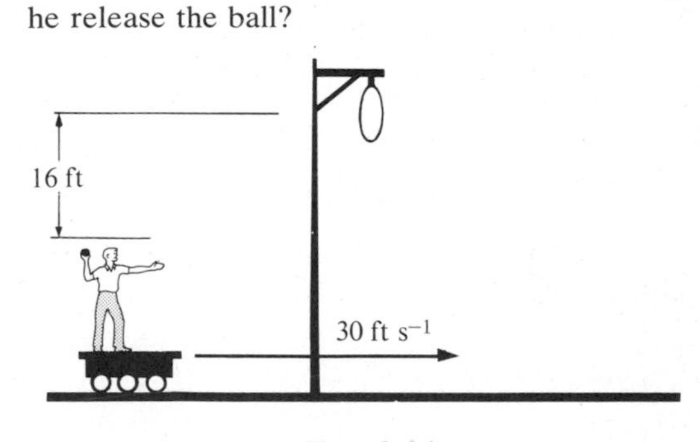

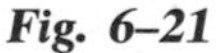
Fig. 6–21

6–25 A particle moves in the xy-plane with coordinates given as functions of time by the equations $x = 2t + t^3$, $y = 6 - t^2$, where x and y are measured in meters and t in seconds.

a) Find the x and y components of velocity as functions of time.

b) Find the magnitude and direction of the velocity at time $t = 2$ s.

c) If the particle has a mass of 2 kg, find the components of force on the particle at time $t = 2$ s.

d) In part (c), find the magnitude and direction of force at time $t = 2$ s.

6–26 A particle moves in the xy-plane with coordinates given as functions of time by the equations $x = R \cos \omega t$, $y = R \sin \omega t$, where R and ω are constants.

a) Show that the particle moves in a circle centered at the origin, with radius R.

b) Determine the components of velocity of the particle. From these find the magnitude of velocity. Show that it is constant and equal to $R\omega$.

c) Determine the components of acceleration of the particle. From these find the magnitude of acceleration. Show that it is constant and equal to $R\omega^2$. Use the result of part (b) to show that this is also equal to v^2/R.

d) Comparing the acceleration components to the expressions for the x and y coordinates, show that the acceleration is always in a direction opposite to that from the origin to the particle.

6–27 At time $t = 0$ a body is moving east at 0.1 m s^{-1}. At time $t = 2$ s it is moving 25° north of east at 0.14 m s^{-1}. Find graphically the magnitude and direction of its change in velocity and the magnitude and direction of its average acceleration.

6–28 An automobile travels around a circular track 2000 m in circumference at a constant speed of 30 m s^{-1}.

a) Show in a scale diagram the velocity vectors of the automobile at the beginning and end of a time interval $\Delta t = 5$ s. Let 1 cm = 5 m s^{-1}.

b) Find graphically the change in velocity, $\Delta \boldsymbol{v}$, in this time interval.

c) Find the magnitude of the average acceleration during this interval, $\Delta v/\Delta t$.

d) What is the radial acceleration, v^2/R?

e) If the track is 10 m wide, what should be the elevation of the outer circumference above the inner circumference, for the speed above?

6–29 The radius of the earth's orbit around the sun (assumed circular) is 93×10^6 mi, and the earth travels around this orbit in 365 days.

a) What is the magnitude of the orbital velocity of the earth, in miles per hour?

b) What is the radial acceleration of the earth toward the sun, in feet per second squared?

6–30 A model of a helicopter rotor has four blades, each 2 m long, and is rotated in a wind tunnel at 1500 rev min^{-1}.

a) What is the linear speed of the blade tip, in meters per second?

b) What is the radial acceleration of the blade tip, in terms of the acceleration of gravity, g?

c) A pressure-measuring device with a mass of 0.1 kg is mounted at the blade tip. Find the centripetal force on it, and compare with its weight.

6–31 A highway curve of radius 1600 ft is to be banked so that a car traveling 50 mi hr^{-1} will have no tendency to skid sideways. At what angle should it be banked?

6–32 A flat (unbanked) curve on a highway has a radius of 800 ft, and a car rounds the curve at a speed of 50 mi hr^{-1}. What must be the minimum coefficient of friction to prevent sliding?

6–33 A stone of mass 1 kg is attached to one end of a string 1 m long, of breaking strength 500 N, and is whirled in a horizontal circle on a frictionless tabletop. The other end of the string is kept fixed. Find the maximum velocity the stone can attain without breaking the string.

6–34 An unbanked circular highway curve on level ground makes a turn of 90°. The highway carries traffic at 60 mi hr^{-1}, and the centripetal force on a vehicle is not to exceed $\frac{1}{10}$ of its weight. What is the minimum length of the curve, in miles?

6–35 A coin placed on a 12-in.record will revolve with the record when it is brought up to a speed of $33\frac{1}{3}$ rev min^{-1}, provided the coin is not more than 4 in. from the axis.

a) What is the coefficient of static friction between coin and record?

b) How far from the axis can the coin be placed, without slipping, if the turntable rotates at 45 rev min^{-1}?

6–36

a) At how many revolutions per second must the apparatus of Fig. 6–22 rotate about the vertical axis in order that the cord shall make an angle of 45° with the vertical?

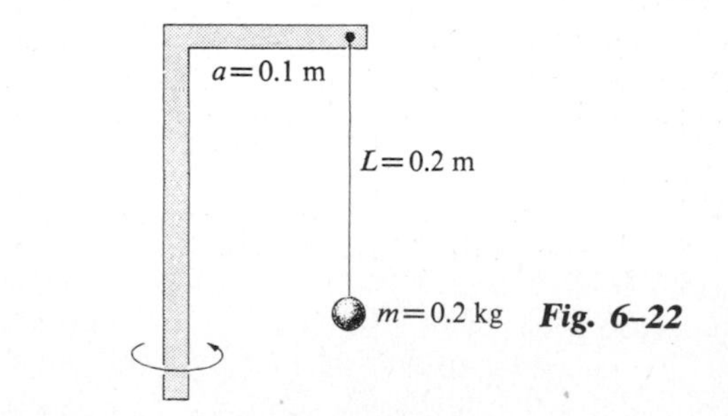

Fig. 6–22

b) What is then the tension in the cord?

c) Find the angle θ which the cord makes with the vertical if the system is rotating at 1.5 rev s^{-1}. (Set up the general equation relating the angle θ to the number of revolutions per second, n, the lengths a and L, and the acceleration of gravity, g. Then find by trial the angle θ which satisfies this equation.)

6–37 The "Giant Swing" at a country fair consists of a vertical central shaft with a number of horizontal arms attached at its upper end. Each arm supports a seat suspended from a cable 5 m long, the upper end of the cable being fastened to the arm at a point 4 m from the central shaft.

a) Find the time of one revolution of the swing if the cable supporting a seat makes an angle of 30° with the vertical.

b) Does the angle depend on the weight of the passenger for a given rate of revolution?

6–38 The 4-kg block in Fig. 6–23 is attached to a vertical rod by means of two strings. When the system rotates about the axis of the rod, the strings are extended as shown in the diagram.

a) How many revolutions per minute must the system make in order that the tension in the upper cord shall be 60 N?

b) What is then the tension in the lower cord?

6–39 A bead can slide without friction on a circular hoop of radius 0.1 m in a vertical plane. The hoop rotates at a constant rate of 2 rev s^{-1} about a vertical diameter, as in Fig. 6–24.

a) Find the angle θ at which the bead is in vertical equilibrium. (Of course it has a radial acceleration toward the axis.)

b) Is it possible for the beads to "ride" at the same elevation as the center of the hoop?

c) What will happen if the hoop rotates at 1 rev s^{-1}?

6–40 A curve of 200-m radius on a level road is banked at the correct angle for a velocity of 15 m s^{-1}. If an automobile rounds this curve at 30 m s^{-1}, what is the minimum coefficient of friction between tires and road so that the automobile will not skid? Assume all forces to act at the center of gravity.

6–41 An airplane in level flight is said to make a *standard turn* when it makes a complete circular turn in 2 min.

a) What is the banking angle of a standard turn if the speed of the airplane is 100 m s^{-1}?

b) What is the radius of the circle in which it turns?

c) What is the centripetal force on the airplane, expressed as a fraction (or multiple) of its weight?

6–42 The radius of the circular track in Fig. 6–16 is 0.4 m and the mass of the ball is 0.1 kg.

a) Find the critical velocity at the highest point of the track.

b) If the actual velocity at the highest point of the track is twice the critical velocity, find the force exerted by the ball against the track.

6–43 The pilot of a divebomber who has been diving at a velocity of 400 mi hr^{-1} pulls out of the dive by changing his course to a circle in a vertical plane.

a) What is the minimum radius of the circle in order that the acceleration at the lowest point shall not exceed "$7g$"?

b) How much does a 180-lb pilot apparently weigh at the lowest point of the pullout?

6–44 A cord is tied to a pail of water and the pail is swung in a vertical circle of radius 1 m. What must be the minimum velocity of the pail at the highest point of the circle if no water is to spill from the pail?

6–45 The radius of a Ferris wheel is 5 m and it makes one revolution in 10 s.

a) Find the difference between the apparent weight of a passenger at the highest and lowest points, expressed

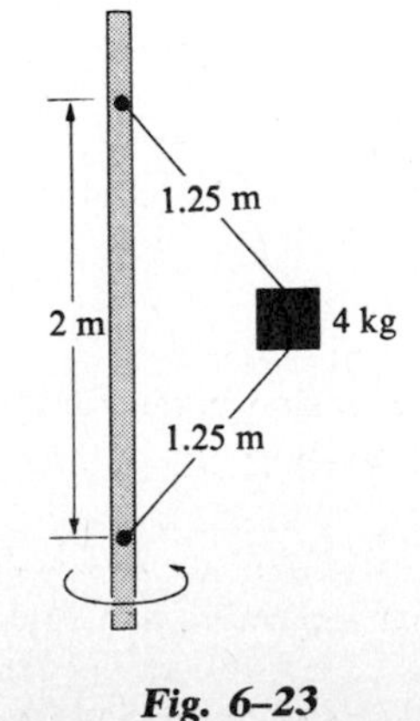

Fig. 6–23

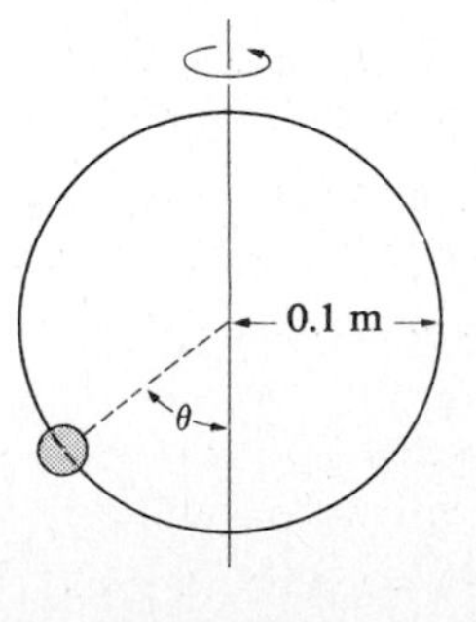

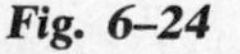

Fig. 6–24

as a fraction of his weight. (That is, find the difference between the upward force exerted on the passenger by the seat at these two points.)

b) What would the time of one revolution be if his apparent weight at the highest point were zero?

c) What would then be his apparent weight at the lowest point?

d) What would happen, at this rate of revolution, if his seat belt broke at the highest point, and if he did not hang onto his seat?

6–46 A ball is held at rest in position A in Fig. 6–25 by two light cords. The horizontal cord is cut and the ball starts swinging as a pendulum. What is the ratio of the tension in the supporting cord, in position B, to that in position A?

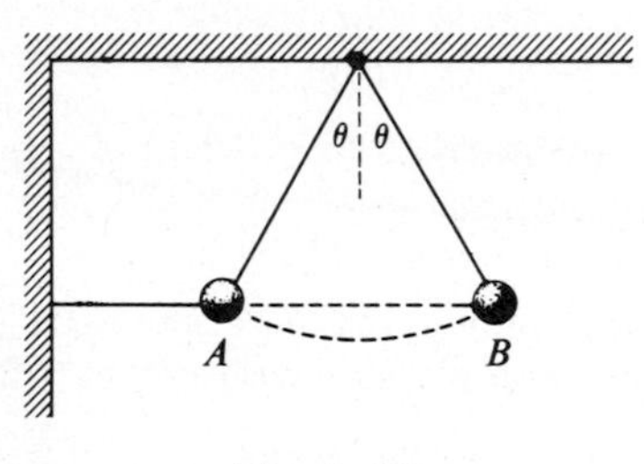

Fig. 6–25

6–47 The questions are often asked: "What keeps an earth satellite moving in its orbit?" and "What keeps the satellite up?"

a) What are your answers to these questions?

b) Are your answers applicable to the moon?

6–48 What is the period of revolution of a manmade satellite of mass m which is orbiting the earth in a circular path of radius 8000 km? (Mass of earth $= 5.98 \times 10^{24}$ kg.)

6–49 There exist several satellites moving in a circle in the earth's equatorial plane and at such a height above the earth's surface that they remain always above the same point. Find the height of such a satellite.

6–50 It is desired to launch a satellite 600 km above the earth in a circular orbit. If suitable rockets are used to reach this elevation, what horizontal orbital velocity must be imparted to the satellite? The radius of the earth is 6380 km.

6–51 An earth satellite rotates in a circular orbit of radius 4400 mi (about 400 mi above the earth's surface) with an orbital speed of 17,000 mi hr^{-1}.

a) Find the time of one revolution.

b) Find the acceleration of gravity at the orbit.

6–52 What would be the length of a day if the rate of revolution of the earth were such that $g = 0$ at the equator?

6–53 The weight of a man as determined by a spring balance at the equator is 180 lb. By how many ounces does this differ from the true force of gravitational attraction at the same point?

Chapter 7

Work and Energy

7-1 WORK

In everyday life, the word *work* is applied to any form of activity that requires the exertion of muscular or mental effort. In physics, however, the term is used in a very specific sense. Figure 7-1 represents a body moving in a horizontal direction which we shall take as the x-axis. A constant force, $\boldsymbol{F}$, at an angle θ with the direction of motion, is exerted on the body by some outside agent. The work W of this force (or the work done by the force), when its point of application undergoes a displacement $\mathbf{s}$, is defined as the product of the magnitudes of the displacement and of the component of the force in the direction of the displacement.

The magnitude of the component of $\boldsymbol{F}$ in the direction of $\mathbf{s}$ is $F \cos \theta$. Then

$$W = F \cos \theta \cdot s. \tag{7-1}$$

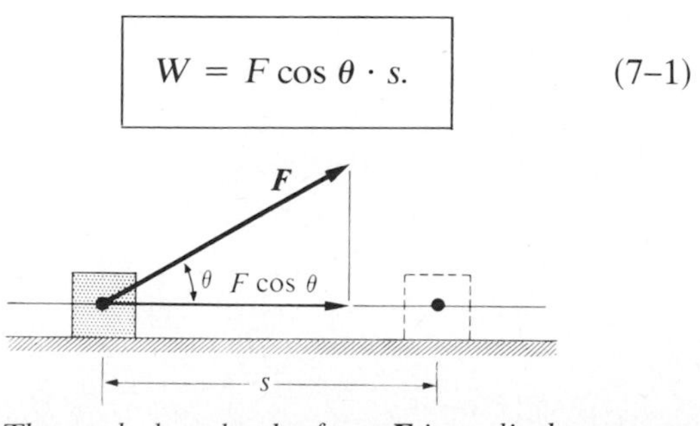

Fig. 7-1 *The work done by the force* $\boldsymbol{F}$ *in a displacement* $\mathbf{s}$ *is* $(F \cos \theta) \cdot s$.

Work is a *scalar* quantity, equal to the product of $F \cos \theta$ and s.

Work is done only when a force is exerted on a body while the body at the same time moves in such a way that the force has a component along the line of motion of its point of application. If the component of the force is in the *same direction* as the displacement, the work W is *positive*. If it is *opposite* to the displacement, the work is *negative*. If the force is at *right angles* to the displacement, it has no component in the direction of the displacement and the work is *zero*.

Thus, when a body is lifted, the work done by the lifting force is positive; when a spring is stretched, the work done by the stretching force is positive; when a gas is compressed in a cylinder, again the work of the compressing force is positive. On the other hand, the work of the gravitational force on a body being lifted is negative, since the (downward) gravitational force is opposite to the (upward) displacement. When a body slides on a fixed surface, the work of the frictional force exerted *on the body* is negative, since this force is always opposite to the displacement of the body. No work is done by the frictional force acting *on the fixed surface* because there is no motion of this surface. Also, although it would be considered "hard work" to hold a heavy object stationary at arm's length, no work

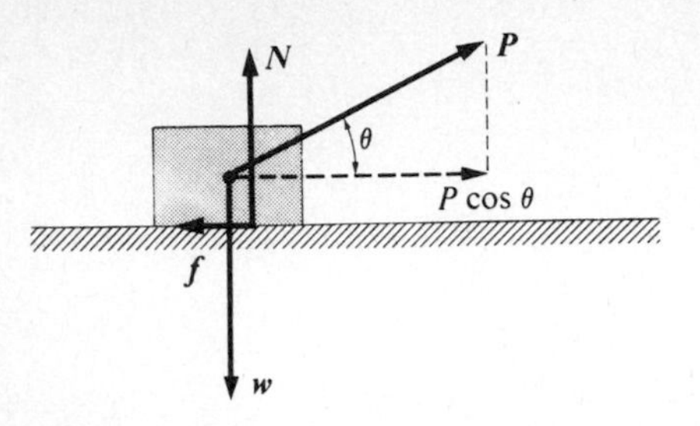

Fig. 7-2 An object on a rough horizontal surface moving to the right under the action of a force **P** *inclined at an angle θ.*

would be done in the technical sense because there is no motion. Even if one were to walk along a level floor while carrying the object, no work would be done, since the (vertical) supporting force has no component in the direction of the (horizontal) motion. Similarly, the work of the normal force exerted on a body by a surface on which it moves is zero, as is the work of the centripetal force on a body moving in a circle.

The unit of work in any system is the unit of force multiplied by the unit of distance. In the mks system the unit of force is the newton and the unit of distance is the meter; thus in this system the unit of work is one *newton meter* (1 N m) which is also called one *joule* (1 J). The unit of work in the cgs system is one *dyne centimeter* (1 dyn cm), called one *erg*. Since 1 m = 100 cm and 1 N = 10^5 dyn, it follows that

$$1\text{ N m} = 10^7\text{ dyn cm}, \qquad \text{or} \qquad 1\text{ J} = 10^7\text{ erg}.$$

In the engineering system, the unit of work is one *foot pound* (1 ft lb):

$$1\text{ J} = 0.7376\text{ ft lb}, \qquad 1\text{ ft lb} = 1.356\text{ J}.$$

When several external forces act on a body, we may wish to consider the works of the separate forces. Each of these may be computed from the definition of work in Eq. (7-1). Then, since work is a scalar quantity, the total work is the algebraic sum of the individual works.

Example Figure 7-2 shows a box being dragged along a horizontal surface by a constant force **P** making a constant angle θ with the direction of motion. The other forces on the box are its weight **w**, the normal upward force **N** exerted by the surface, and the friction force *f*. What is the work of each force when the box moves a distance s along the surface to the right?

The component of **P** in the direction of motion is $P\cos\theta$. The work of the force **P** is therefore

$$W_P = (P\cos\theta)s.$$

The forces **w** and **N** are both at right angles to the displacement. Hence,

$$W_w = 0, \qquad W_N = 0.$$

The friction force *f* is opposite to the displacement, so the work of the friction force is

$$W_f = -fs.$$

Since work is a scalar quantity, the total work W of all forces on the body is the algebraic (not the vector) sum of the individual works:

$$\begin{aligned} W &= W_P + W_w + W_N + W_f \\ &= (P\cos\theta)s + 0 + 0 - fs \\ &= (P\cos\theta - f)s. \end{aligned}$$

But $(P\cos\theta - f)$ is the *resultant* force on the body. Hence *the total work of all forces is equal to the work of the resultant force.*

Suppose that $w = 100$ N, $P = 50$ N, $f = 15$ N, $\theta = 37°$, and $s = 20$ m. Then

$$\begin{aligned} W_P &= (P\cos\theta)s \\ &= (50\text{ N})(0.8)(20\text{ m}) = 800\text{ N m}, \\ W_f &= -fs = (-15\text{ N})(20\text{ m}) = -300\text{ N m}, \\ W &= W_P + W_f = 500\text{ N m}. \end{aligned}$$

As a check, the total work may be expressed as

$$\begin{aligned} W &= (P\cos\theta - f)s \\ &= (40\text{ N} - 15\text{ N})(20\text{ m}) = 500\text{ N m}. \end{aligned}$$

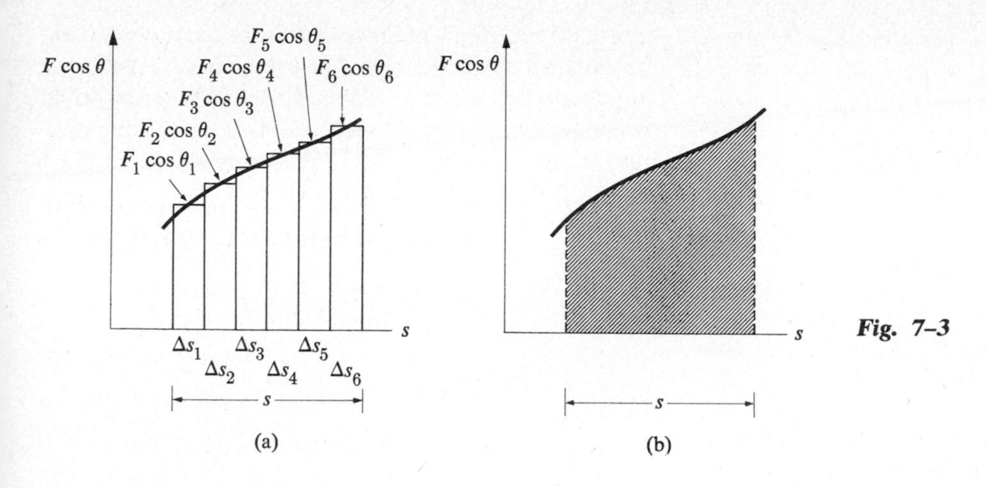

Fig. 7-3

7-2 WORK DONE BY A VARYING FORCE

In the preceding section we defined the work of a *constant* force. In many important cases, however, the work is done by a force which varies in magnitude or direction during the displacement of the body on which it acts. Thus when a spring is stretched slowly, the force required to stretch it increases steadily as the spring elongates; when a body is projected vertically upward, the gravitational force exerted on it by the earth decreases inversely with the square of its distance from the earth's center.

The work of a varying force can be found as follows. In Fig. 7-3(a), the curved line is a graph of the component $F \cos \theta$, for a force which varies in some arbitrary way (in magnitude and direction) with the distance s. Let the distance be subdivided into short segments Δs_1, Δs_2, etc., and approximate the varying force component by one that remains constant at the value $F_1 \cos \theta_1$ over the distance Δs_1, then increases to the constant value $F_2 \cos \theta_2$ over the distance Δs_2, etc., as indicated by the zigzag line. The work that would be done by the constant force $F_1 \cos \theta_1$ in the displacement Δs_1 would be $F_1 \cos \theta_1 \, \Delta s_1$; the work done in the displacement Δs_2 would be $F_2 \cos \theta_2 \, \Delta s_2$, etc. The work total W would be

$$W = F_1 \cos \theta_1 \, \Delta s_1 + F_2 \cos \theta_2 \, \Delta s_2 + \cdots.$$

We may take these intervals smaller and smaller and, in the limit of very small intervals, the sum becomes an integral:

$$W = \int_{s_1}^{s_2} F \cos \theta \, ds. \qquad (7\text{-}2)$$

We also see that the total work during the displacement from s_1 to s_2 is represented graphically by the *area* under the graph of $F \cos \theta$ between these limits, as shown in Fig. 7-3(b).

Example 1 Compute the work required to stretch a spring a distance X from its unstretched length, if the force required increases in direct proportion to the amount of elongation.

To keep a spring stretched an amount x beyond its unstretched length requires a force F at one end and a force of equal magnitude and opposite direction at the other end. If the elongation x is not too great, F is directly proportional to x, and

$$F = kx, \qquad (7\text{-}3)$$

where k is a constant called the *force constant* or the *stiffness* of the spring. This direct proportion between force and elongation, for elongations that are not too great, was discovered by Robert Hooke in 1678 and is known as *Hooke's law.* It will be discussed more fully in a later chapter.

Suppose that forces equal in magnitude and opposite in direction are exerted on the ends of a spring, and that the forces are gradually increased from zero. Let one end of the spring be kept fixed. The work of the force at the fixed end will be zero, but work will be done by the varying force $\boldsymbol{F}$ at the moving end. This force is in the same direction as the displacement. In Fig. 7-4, the force F is plotted vertically and the displacement x of the moving end representing the elongation of the spring, is plotted horizontally.

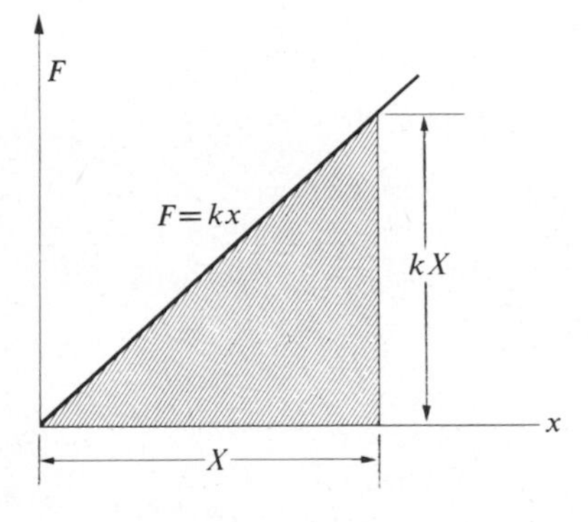

Fig. 7-4 *The work done in stretching a spring is equal to the area of the shaded triangle.*

The work required to stretch the spring from $x = 0$ (no elongation) to $x = X$ is

$$W = \int_0^X F\,dx = \int_0^X kx\,dx = \tfrac{1}{2}kX^2. \qquad (7\text{-}4)$$

This result can also be obtained graphically; the area of the shaded triangle in Fig. 7-4, representing the total work, is equal to half the product of base and altitude, or

$$W = \tfrac{1}{2}(X)(kX) = \tfrac{1}{2}kX^2$$

in agreement with the above result.

The work of a force can be expressed in several ways. If θ is the angle between the force vector $\boldsymbol{F}$ and the infinitesimal displacement vector ds, the component $F_{\parallel}$ in the direction of ds is equal to $F\cos\theta$, so

$$W = \int_{s_1}^{s} F_{\parallel}\,ds = \int_{s_1}^{s_2} F\cos\theta\,ds.$$

In Chapter 1 we discussed the *addition* and *subtraction* of vectors. The definition of work suggests a third process of vector algebra, namely, *scalar multiplication* of two vectors. In computing the work of a force, we multiply the magnitude of the vector ds by the magnitude of the component of another vector $\boldsymbol{F}$ in the direction of ds. The product is called the *scalar product* or the *dot product* of the vectors. Thus if $\mathbf{A}$ and $\boldsymbol{B}$ are any two vectors, their scalar or dot product is defined as a scalar, equal to the product of the magnitudes of the vectors and the cosine of the angle θ between their positive directions:

$$\mathbf{A}\cdot\boldsymbol{B} = AB\cos\theta.$$

Evidently, $\mathbf{A}\cdot\boldsymbol{B} = \boldsymbol{B}\cdot\mathbf{A}$. In this new notation, the work of a force can be written as

$$W = \int_{s_1}^{s_2} \boldsymbol{F}\cdot ds.$$

In the special case in which the force $\boldsymbol{F}$ is constant in magnitude and makes a constant angle with the direction of motion of its point of application, $F\cos\theta$ is constant and may be taken outside the integral sign. If in addition we measure diplacements from the starting point of the motion and let $s_1 = 0$ and $s_2 = s$, the work of the force is

$$W = \int_0^s F\cos\theta\,ds = F\cos\theta\int_0^s ds = (F\cos\theta)s.$$

If the force is constant and in the same direction as the motion or opposite to the motion, the angle θ equals zero or 180°, $\cos\theta = \pm 1$, and

$$W = \pm Fs.$$

That is, in this *very special case only*, we can say, "Work equals force times distance." It is important to remember, however, that the *general* definition of the work of a force is

$$\boxed{W = \int_{s_1}^{s_2} \boldsymbol{F}\cdot \boldsymbol{ds} = \int_{s_1}^{s_2} F_{\parallel}\,ds = \int_{s_1}^{s_2} F\cos\theta\,ds.} \qquad (7\text{-}5)$$

Example 2 A small object of weight $\boldsymbol{w}$ hangs from a string of length ℓ, as shown in Fig. 7-5. A *variable* horizontal force $\boldsymbol{P}$, which starts at zero and gradually increases, is used to pull the object very slowly (so that equilibrium exists at all times) until the string makes an angle θ with the vertical. Calculate the work of the force $\boldsymbol{P}$.

Since the object is in equilibrium, the sum of the horizontal forces equals zero, whence

$$P = T\sin\theta.$$

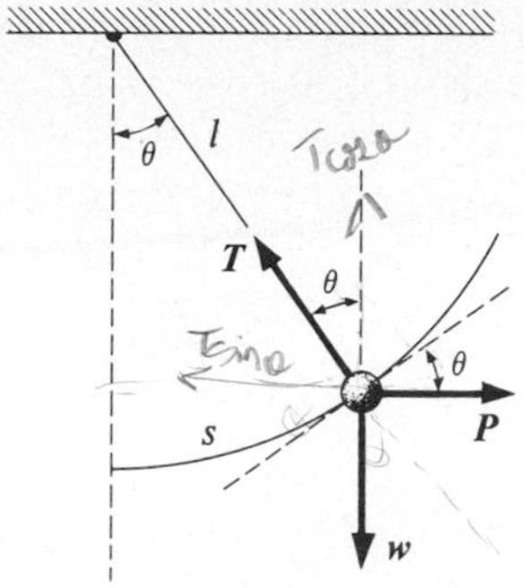

Fig. 7–5 A variable horizontal force **P** acts on a small object while the displacement varies from zero to s.

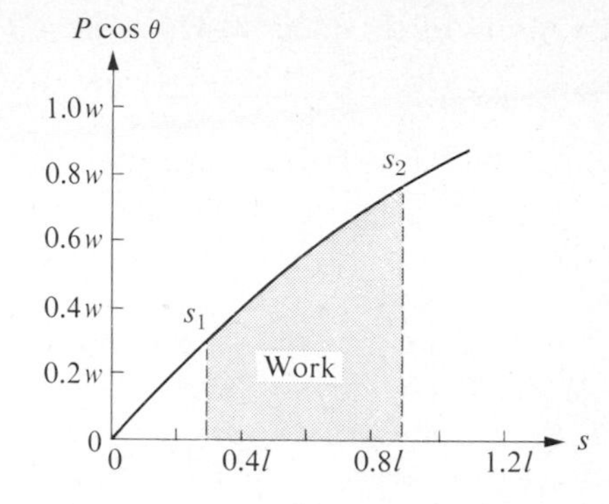

Fig. 7–6 The work of a force **P** when its point of application moves from $s_1 = 0.3\ell$ to $s_2 = 0.9\ell$ is the shaded area under the curve of $P\cos\theta$, plotted against s.

Equating the sum of the vertical forces to zero, we find

$$w = T\cos\theta.$$

Dividing these two equations, we get

$$P = w\tan\theta.$$

The point of application of **P** swings through the arc s. Since $s = \ell\theta$, $ds = \ell\, d\theta$ and

$$W = \int \mathbf{P}\cdot d\mathbf{s} = \int P\cos\theta\, ds$$

$$= \int_0^\theta w\tan\theta\cos\theta\,\ell\, d\theta$$

$$= w\ell\int_0^\theta \sin\theta\, d\theta = w\ell(1-\cos\theta). \qquad (7\text{–}6)$$

A graph of $P\cos\theta$ ($= w\sin\theta$) plotted against s ($= \ell\theta$) is shown in Fig. 7–6. The work during any displacement is represented by the area under this curve. The shaded area is the work when the point of application of **P** moves from a position where $s_1 = 0.3\ell$ to a place where $s_2 = 0.9\ell$.

7–3 WORK AND KINETIC ENERGY

Consider again the body illustrated in Fig. 7–2. The constant resultant force F on the body is $F = P\cos\theta - f$, and from Newton's second law,

$$F = ma.$$

Suppose the speed of the body increases from v_1 to v_2 while the body undergoes a displacement s. Then, since a is constant,

$$v_2^2 = v_1^2 + 2as,$$

$$a = \frac{v_2^2 - v_1^2}{2s}.$$

Hence

$$F = m\frac{v_2^2 - v_1^2}{2s},$$

and

$$Fs = \tfrac{1}{2}mv_2^2 - \tfrac{1}{2}mv_1^2. \qquad (7\text{–}7)$$

The product Fs is the work W of the resultant force F. The quantity $\frac{1}{2}mv^2$, one-half the product of the mass of the body and the square of its speed, is called its *kinetic energy*, E_k.

$$\boxed{E_k = \tfrac{1}{2}mv^2.} \qquad (7\text{–}8)$$

The first term on the right side of Eq. (7–7) containing the final speed v_2, is the final kinetic energy of the body, E_{k2}, and the second term is the initial kinetic energy, E_{k1}. The difference between these terms is the *change* in kinetic energy, and we have the important result that *the work of the resultant external force on a body is equal to the change in kinetic energy of the body.* As we have shown, the work of the resultant force is equal to the algebraic sum of the works of the individual forces

$$W = E_{k2} - E_{k1} = \Delta E_k. \tag{7–9}$$

Kinetic energy, like work, is a *scalar* quantity. The kinetic energy of a moving body depends only on its speed, or the *magnitude* of its velocity, but not on the *direction* in which it is moving. The *change* in kinetic energy depends only on the work $W = Fs$ and not on the individual values of F and s. That is, the force F could have been large and the displacement s small, or the reverse might have been true. If the mass m and the speeds v_1 and v_2 are known, the *work* of the resultant force can be found without any knowledge of the force F and the displacement s.

If the work W is *positive*, the final kinetic energy is greater than the initial kinetic energy and the kinetic energy *increases*. If the work is *negative*, the kinetic energy *decreases*. In the special case in which the work is *zero*, the kinetic energy remains *constant*.

In computing the kinetic energy of a body, consistent units must be used for m and v. In the mks system, m must be in kilograms and v in meters per second. In the cgs system m must be expressed in grams and v in centimeters per second. In the engineering system, m must be in slugs and v in feet per second. The corresponding units of kinetic energy are 1 kg m^2 s^{-2}, 1 g cm^2 s^{-2}, and 1 slug ft^2 s^{-2}. However, the unit of kinetic energy in any system is equal to the unit of work in that system, and kinetic energy is customarily expressed in joules, ergs, or foot pounds. That is,

$$1 \text{ kg m}^2\text{ s}^{-2} = 1 \text{ N m}^{-1}\text{ s}^2 \cdot \text{m}^2\text{ s}^{-2} = 1 \text{ N m} = 1 \text{ J}.$$

In the same way,

$$1 \text{ g cm}^2\text{ s}^{-2} = 1 \text{ erg}, \qquad 1 \text{ slug ft}^2\text{ s}^{-2} = 1 \text{ ft lb}.$$

Example Refer again to the body in Fig. 7–2 and the numerical values given in the example of Section 7–1. The total work of the external forces was shown to be 500 N m. Hence the kinetic energy of the body increases by 500 N m. To verify this, suppose the initial speed v_1 is 4 m s^{-1}. The mass of the body is $m = w/g = (100 \text{ N})/(10 \text{ m s}^{-2}) = 10$ kg. The initial kinetic energy is

$$E_{k1} = \tfrac{1}{2}mv_1^2 = \tfrac{1}{2}(10\text{ kg})(4\text{ m s}^{-1})^2 = 80\text{ J}.$$

To find the final kinetic energy we must first find the acceleration:

$$a = \frac{F}{m} = \frac{40\text{ N} - 15\text{ N}}{10\text{ kg}} = 2.5\text{ m s}^{-2}.$$

Then

$$v_2^2 = v_1^2 + 2as = (4\text{ m s}^{-1})^2 + 2(2.5\text{ m s}^{-2})(20\text{ m})$$
$$= 116\text{ m}^2\text{ s}^{-2},$$

and

$$E_{k2} = \tfrac{1}{2}(10\text{ kg})(116\text{ m}^2\text{ s}^{-2}) = 580\text{ J}.$$

The increase in kinetic energy is therefore 500 J.

Equation (7–9) was derived for the special case of straight-line motion with a constant resultant force, but it is also valid for a curved path and a variable force. To show this, we consider the curve in Fig. 7–7, representing the path of a particle of mass m moving in the xy-plane and acted on by a *resultant* force $\boldsymbol{F}$ which may vary in magnitude and direction from point to point of the path. Let us resolve the force into a component $\boldsymbol{F}_{\parallel}$ along the path and a component $\boldsymbol{F}_{\perp}$ normal to the path.

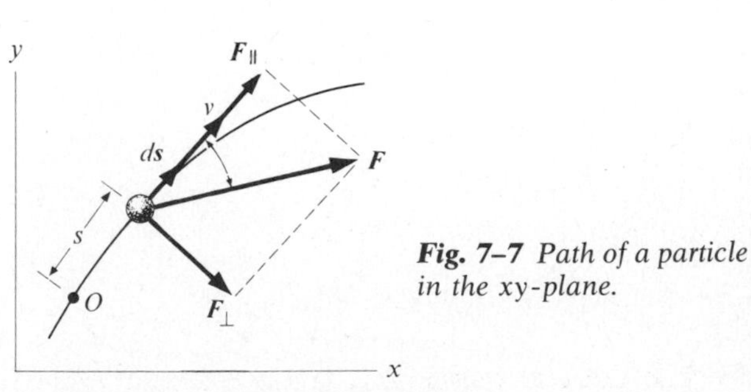

Fig. 7–7 *Path of a particle in the xy-plane.*

The component $\boldsymbol{F}_{\perp}$, at right angles to the velocity $\boldsymbol{v}$, is a *centripetal* force, and its only effect is to change the *direction* of the velocity. The effect of the component $\boldsymbol{F}_{\parallel}$ is to change the *magnitude* of the velocity.

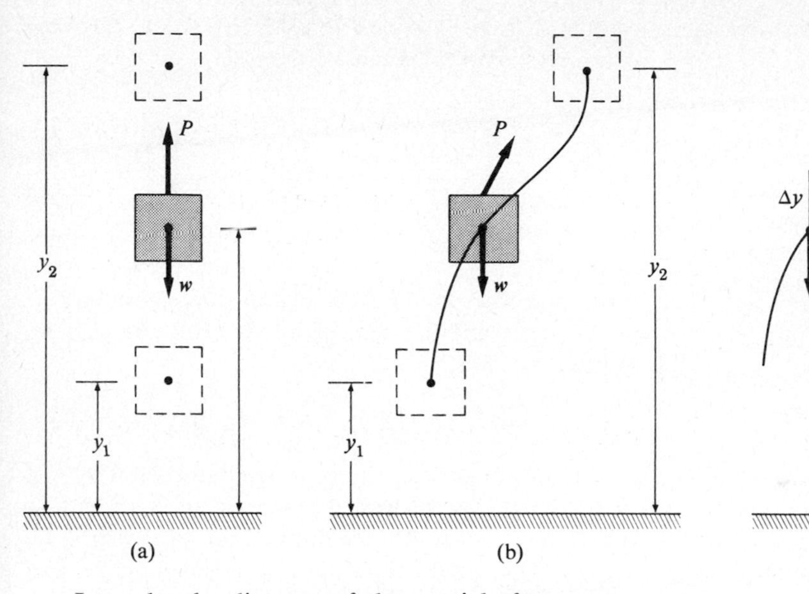

Δy θ Δs $180° - \theta$ w

(c)

Fig. 7–8 *Work of the gravitational force* **w** *during the motion of an object from one point in a gravitational field to another.*

Let s be the distance of the particle from some fixed point O, measured along the path. In general, the magnitude of $F_\parallel$ will be a function of s. From Newton's second law,

$$F_\parallel = m\frac{dv}{dt}.$$

Since $F_\parallel$ is a function of s, we use the chain rule and write

$$\frac{dv}{dt} = \frac{dv}{ds}\frac{ds}{dt} = v\frac{dv}{ds}.$$

Then we have

$$F_\parallel = mv\frac{dv}{ds},$$

$$F_\parallel\, ds = mv\, dv.$$

If v_1 is the velocity when $s = s_1$, and v_2 the velocity when $s = s_2$, it follows that

$$\int_{s_1}^{s_2} F_\parallel\, ds = \int_{v_1}^{v_2} mv\, dv. \qquad (7\text{–}10)$$

The integral on the left is the *work* W of the force $\boldsymbol{F}$, between the points s_1 and s_2. This integral can be evaluated only when the component $F_\parallel$ is known as a function of s, or when both F and s are known as functions of another variable. The integral on the right side of Eq. (7–10), however, can always be evaluated:

$$\int_{v_1}^{v_2} mv\, dv = \tfrac{1}{2}mv_2^{\,2} - \tfrac{1}{2}mv_1^{\,2}$$

$$= E_{k2} - E_{k1}.$$

Equation (7–10) can therefore be written

$$W = E_{k2} - E_{k1},$$

which is Eq. (7–9).

7–4 GRAVITATIONAL POTENTIAL ENERGY

Suppose a body of mass m (and of weight $\boldsymbol{w} = m\boldsymbol{g}$) moves vertically, as in Fig. 7–8(a), from a point where its center of gravity is at a height y_1 above an arbitrarily chosen plane (the reference level) to a point at a height y_2. For the present, we shall consider only displacements near the earth's surface, so that variations of gravitational force with distance from the earth's center can be neglected. The downward gravitational force on the body is then constant and equal to $\boldsymbol{w}$. Let $\boldsymbol{P}$ represent the resultant of all other forces acting on the body and let W' be the work of these forces. The direction of the gravitational force $\boldsymbol{w}$ is opposite to the upward displacement, and the work of this force is

$$W_{\text{grav}} = -w(y_2 - y_1) = -(mgy_2 - mgy_1). \quad (7\text{–}11)$$

The reader should convince himself that the work of the gravitational force is given by $-mg(y_2 - y_1)$ whether the body moves up or down.

Now suppose that the body starts at the same elevation y_1 but is moved up to the elevation y_2 along some arbitrary path, as in Fig. 7–8(b). Part (c) of the figure is an enlarged view of a small portion of the path. The work of the gravitational force is

$$W_{\text{grav}} = \int_{s_1}^{s_2} w \cos\theta \, ds.$$

Let ϕ represent the angle between ds and its vertical component dy. Then $dy = ds \cos\phi$, and since $\phi = 180° - \theta$,

$$\cos\phi = -\cos\theta, \qquad \cos\theta \, ds = -dy,$$

and

$$W_{\text{grav}} = -\int_{y_1}^{y_2} w \, dy = -w(y_2 - y_1)$$
$$= -(mgy_2 - mgy_1). \tag{7–12}$$

The work of the gravitational force, therefore, depends only on the initial and final *elevations* and not on the path. If these points are at the same elevation the work is zero.

Let W' be the work done by forces other than the gravitational force. Since the *total* work equals the change in kinetic energy,

$$W' + W_{\text{grav}} = E_{k2} - E_{k1},$$
$$W' - (mgy_2 - mgy_1) = (\tfrac{1}{2}mv_2^2 - \tfrac{1}{2}mv_1^2). \tag{7–13a}$$

The quantities $\frac{1}{2}mv_2^2$ and $\frac{1}{2}mv_1^2$ depend only on the final and initial *speeds*; the quantities mgy_2 and mgy_1 depend only on the final and initial *elevations.* Let us therefore rearrange this equation, transferring the quantities mgy_2 and mgy_1 from the "work" side of the equation to the "energy" side:

$$W' = (\tfrac{1}{2}mv_2^2 - \tfrac{1}{2}mv_1^2) + (mgy_2 - mgy_1). \tag{7–13b}$$

The left side of Eq. (7–13b) contains only the work of the force $\boldsymbol{P}$. The terms on the right depend only on the final and initial states of the body (its speed and elevation) and not specifically on the way in which it moved. The quantity mgy, the product of the weight mg of the body and the height y of its center of gravity above the reference level, is called its *gravitational potential energy, E_p.*

$$\boxed{E_p(\text{gravitational}) = mgy.} \tag{7–14}$$

The first expression in parentheses on the right of Eq. (7–13b) is the change in kinetic energy of the body, and the second is the change in its gravitational potential energy. Equation (7–13b) can also be written

$$W' = (\tfrac{1}{2}mv_2^2 + mgy_2) - (\tfrac{1}{2}mv_1^2 + mgy_1). \tag{7–13c}$$

The sum of the kinetic and potential energy of the body is called its *total mechanical energy.* The first expression in parentheses on the right of Eq. (7–13c) is the final value of the total mechanical energy, and the second is the initial value. Hence, *the work of all forces acting on the body,* ***with the exception of the gravitational force,*** *equals the change in the total mechanical energy* of the *body.* If the work W' is positive, the mechanical energy increases. If W' is negative, the energy decreases.

In the special case in which the *only* force on the body is the gravitational force, the work W' is zero. Equation (7–13c) can then be written

$$\tfrac{1}{2}mv_2^2 + mgy_2 = \tfrac{1}{2}mv_1^2 + mgy_1 .$$

Under these conditions, then, the *total mechanical energy remains constant,* or is *conserved.* This is a special case of the principle of *the conservation of mechanical energy.*

Example 1 A man holds a ball of mass $m = 0.2$ kg at rest in his hand. He then throws the ball vertically upward. In this process, his hand moves up 0.5 m and the ball leaves his hand with an upward velocity of 20 m s^{-1}. Discuss the motion of the ball from the work–energy standpoint, assuming $g = 10$ m s^{-2}.

First, consider the throwing process. Take the reference level at the initial position of the ball. Then $E_{k1} = 0$, $E_{p1} = 0$. Take point 2 at the point where the ball leaves the thrower's hand. Then

$$E_{p2} = mgy_2 = (0.2\text{ kg})(10\text{ m s}^{-2})(0.5\text{ m}) = 1.0\text{ J},$$
$$E_{k2} = \tfrac{1}{2}mv_2^2 = \tfrac{1}{2}(0.2\text{ kg})(20\text{ m s}^{-1})^2 = 40\text{ J}.$$

Let P represent the upward force exerted on the ball by the man in the throwing process. The work W' is then the work of this force and is equal to the sum of changes in kinetic and potential energy of the ball.

The kinetic energy of the ball increases by 40 J and its potential energy by 1 J. The work W' of the upward force P is therefore 41 J.

If the force P is constant, the work of this force is given by

$$W' = P(y_2 - y_1),$$

and the force P is then

$$P = \frac{W'}{y_2 - y_1} = \frac{41\text{ J}}{0.5\text{ m}} = 82\text{ N}.$$

However, the *work* of the force P is 41 J, whether or not the force is constant.

Now consider the flight of the ball after it leaves the thrower's hand. In the absence of air resistance, the only force on the ball is then its weight $\mathbf{w} = m\mathbf{g}$. Hence the total mechanical energy of the ball remains constant. The calculations will be simplified if we take a new reference level at the point where the ball leaves the thrower's hand. Calling this point 1, we have

$$E_{k1} = 40\text{ J}, \qquad E_{p1} = 0,$$
$$E_k + E_p = 40\text{ J},$$

and the total mechanical energy at any point of the path equals 40 J.

Suppose we wish to find the speed of the ball at a height of 15 m above the reference level. Its potential energy at this elevation is 30 J. (Why?) Its kinetic energy is therefore 10 J. To find its speed, we have

$$\tfrac{1}{2}mv^2 = E_k, \qquad v = \pm\sqrt{2E_k/m} = \pm 10\text{ m s}^{-1}.$$

The significance of the $\pm$ sign is that the ball passes this point *twice*, once on the way up and again on the way down. Its *potential* energy at this point is the same whether it is moving up or down. Hence its kinetic energy is the same and its *speed* is the same. The algebraic sign of the speed is + when the ball is moving up and − when it is moving down.

Next, let us find the height of the highest point reached. At this point $v = 0$ and $E_k = 0$. Therefore $E_p = 40$ J, and the ball rises to a height h above the point where it leaves the thrower's hand, given by $mgh = (2\text{ N})h = 40$ J. Thus $h = 20$ m.

Finally, suppose we were asked to find the speed at a point 30 m above the reference level. The potential energy at this point would be 60 J. But the *total* energy is only 40 J, so the ball never reaches a height of 30 m.

Example 2 A body slides down a curved track which is one quadrant of a circle of radius R, as in Fig. 7–9. If it starts from rest and there is no friction, find its speed at the bottom of the track. The motion of this body is exactly the same as that of a body attached to one end of a string of length R, the other end of which is held at point O.

The equations of motion with constant acceleration cannot be used, since the acceleration decreases during the motion. (The slope angle of the track becomes smaller and smaller as the body descends.) However, if there is no friction, the only force on the body in addition to its weight is the normal force N exerted on it by the track. The work of this force is zero because at each point it is perpendicular to the small element of displacement near that point. Thus

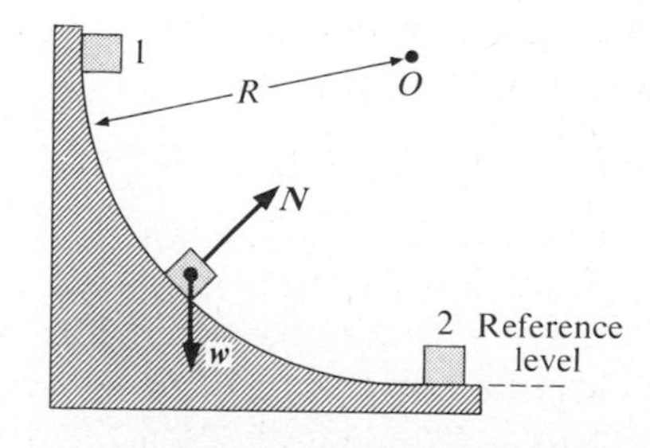

Fig. 7–9 An object sliding down a frictionless curved track.

$W' = 0$ and mechanical energy is conserved. Take point 1 at the starting point and point 2 at the bottom of the track. Take the reference level at point 2. Then $y_1 = R$, $y_2 = 0$, and

$$E_{k2} + E_{p2} = E_{k1} + E_{p1},$$
$$\tfrac{1}{2}mv_2^2 + 0 = 0 + mgR,$$
$$v_2 = \pm\sqrt{2gR}.$$

The speed is therefore the same as if the body had fallen *vertically* through a height R. (What is now the significance of the $\pm$ sign?)

As a numerical example, let $R = 1$ m. Then

$$v = \pm\sqrt{2(9.8\ \text{m s}^{-2})(1\ \text{m})} = \pm 4.43\ \text{m s}^{-1}.$$

Example 3 Suppose a body of mass 0.5 kg slides down a track of radius $R = 1$ m, like that in Fig. 7–9, but its speed at the bottom is only 3 m s^{-1}. What was the work of the frictional force acting on the body?

In this case, $W' = W_f$, and

$$\begin{aligned} W_f &= (\tfrac{1}{2}mv_2^2 - \tfrac{1}{2}mv_1^2) + (mgy_2 - mgy_1) \\ &= \tfrac{1}{2}(0.5\ \text{kg})(9\ \text{m}^2\ \text{s}^{-2}) - 0 \\ &\quad + 0 - (0.5\ \text{kg})(9.8\ \text{m s}^{-2})(1\ \text{m}) \\ &= 2.25\ \text{J} - 4.9\ \text{J} = -2.65\ \text{J}. \end{aligned}$$

The frictional work was therefore -2.65 J, and the total mechanical energy *decreased* by 2.65 J. The mechanical energy of a body is *not* conserved when friction forces act on it.

Example 4 In the absence of air resistance, the only force on a projectile is its weight, and the mechanical energy of the projectile remains constant. Figure 7–10 shows two trajectories of a projectile with the same initial speed (and hence the same total energy) but with different angles of departure. At all points at the same elevation the potential energy is the same; hence the kinetic energy is the same and the speed is the same.

Example 5 A small object of weight $\boldsymbol{w}$ hangs from a string of length ℓ, as shown in Fig. 7–11. A *variable* horizontal force $\boldsymbol{P}$ which starts at zero and gradually increases is used to pull the object very slowly until the string makes an angle θ with the vertical. Calculate the work of the force $\boldsymbol{P}$.

The sum of the works W' of all the forces other than the gravitational force must equal the change of kinetic energy plus the change of gravitational potential energy. Hence

$$W' = W_p + W_T = \Delta E_k + \Delta E_p.$$

Since $\boldsymbol{T}$ is perpendicular to the path of its point of application, $W_T = 0$; and since the body was pulled very slowly at all times, the change of kinetic

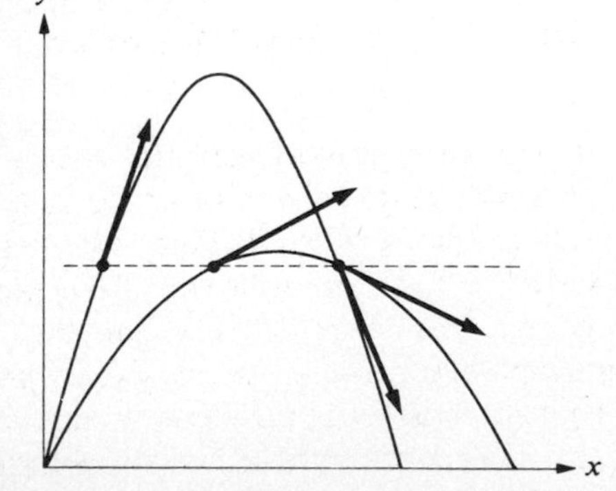

Fig. 7–10 *For the same initial speed, the speed is the same at all points at the same elevation.*

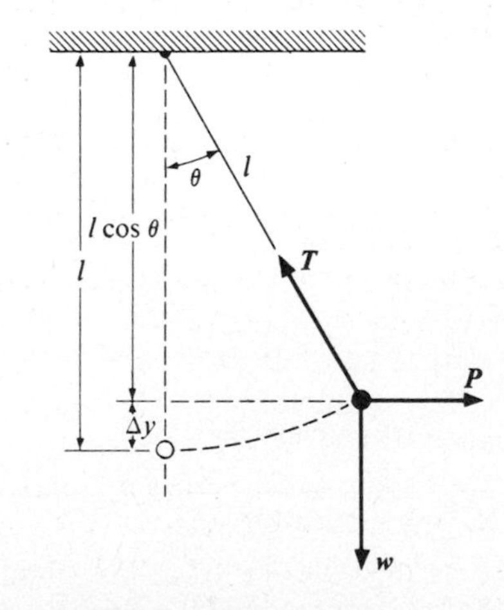

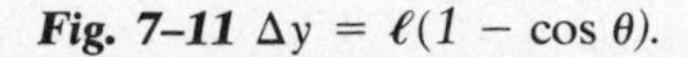

Fig. 7–11 $\Delta y = \ell(1 - \cos\theta)$.

energy is also zero. Hence

$$W_P = \Delta E_p = w\,\Delta y,$$

where Δy is the distance that the object has been raised. From Fig. 7–1, Δy is seen to be $\ell(1 - \cos\theta)$. Therefore

$$W_P = w\ell(1 - \cos\theta).$$

The reader is referred to Example 2 in Section 7–2, where this same problem was solved by performing the integration in the equation $W_p = \int P \cos\theta\, ds$. The two answers, of course, are identical, but how much more simply the energy principle leads to the final result!

Thus far in this section it has been assumed that the changes in elevation are small enough so that the gravitational force on a body can be considered constant. But as discussed in Section 5–5, the gravitational force is given in general by

$$F_g = \frac{Gmm_E}{r^2},$$

where m_E is the mass of the earth and r is the distance from the earth's center. When r increases from r_1 to r_2, the work of the gravitational force is

$$W_{\text{grav}} = -Gmm_E \int_{r_1}^{r_2} \frac{dr}{r^2} = \frac{Gmm_E}{r_2} - \frac{Gmm_E}{r_1}.$$

Setting the total work equal to the change in kinetic energy and rearranging terms, we get, instead of Eq. (7–13),

$$\boxed{W' = \left(\tfrac{1}{2}mv_2^2 - \frac{Gmm_E}{r_2}\right) - \left(\tfrac{1}{2}mv_1^2 - \frac{Gmm_E}{r_1}\right).}$$

The quantity $-G(mm_E/r)$ is therefore the general expression for the gravitational potential energy of a body attracted by the earth:

$$\boxed{E_p(\text{gravitational}) = -G\frac{mm_E}{r}.} \qquad (7\text{–}14)$$

The total mechanical energy of the body, the sum of its kinetic energy and potential energy, is

$$E = E_k + E_p = \tfrac{1}{2}mv^2 - G\frac{mm_E}{r}.$$

If the only force on the body is the gravitational force, then $W' = 0$ and the total mechanical energy remains constant, or is *conserved.*

Example 6 Use the principle of conservation of mechanical energy to find the velocity with which a body must be projected vertically upward, in the absence of air resistance, (a) to rise to a height above the earth's surface equal to the earth's radius, R, and (b) to escape from the earth.

a) Let v_1 be the initial velocity. Then

$$r_1 = R, \qquad r_2 = 2R, \qquad v_2 = 0,$$

and

$$\tfrac{1}{2}mv_1^2 - G\frac{mm_E}{R} = 0 - G\frac{mm_E}{2R},$$

or

$$v_1^2 = \frac{Gm_E}{R},$$

in agreement with Eq. (5–11) obtained in Example 9, Section 5–6.

b) When v_1 is the escape velocity, $r_1 = R$, $r_2 = \infty$, $v_2 = 0$. Then

$$\tfrac{1}{2}mv_1^2 - G\frac{mm_E}{R} = 0 \qquad \text{or} \qquad v_1^2 = \frac{2Gm_E}{R}.$$

It may be puzzling at first sight that the general expression for gravitational potential energy should contain a minus sign. The reason for this lies in the choice of a reference state or reference level in which the potential energy is considered zero. If we set $E_p = 0$ in Eq. (7–14) and solve for r, we get

$$r = \infty.$$

That is, *the gravitational potential energy of a body is now considered to be zero when the body is at an infinite distance from the earth.* Since the potential energy *decreases* as the body approaches the earth, it must be *negative* at any finite distance from the earth. The *change* in potential energy of a body as it moves from one point to another is the same, whatever the choice of reference level, and it is only changes in potential energy that are significant.

Finally, we show that the general expression for the change in potential energy reduces to Eq. (7–11) for small changes in elevation near the earth's surface. The increase in potential energy of a body of mass m, when its distance from the earth's center increases from r_1 to r_2, is

$$E_{p2} - E_{p1} = -G\frac{mm_E}{r_2} - \left(-G\frac{mm_E}{r_1}\right)$$
$$= Gmm_E\left(\frac{1}{r_1} - \frac{1}{r_2}\right)$$
$$= Gmm_E\left(\frac{r_2 - r_1}{r_1 r_2}\right)$$

If points 1 and 2 are at elevations y_1 and y_2 above the earth's surface, then

$$r_2 - r_1 = y_2 - y_1\,.$$

Furthermore, if y_1 and y_2 are both small compared with the earth's radius R, the product $r_1 r_2$ is very nearly equal to R^2. The preceding equation then reduces to

$$E_{p2} - E_{p1} = \frac{Gmm_E}{R^2}(y_2 - y_1).$$

But

$$\frac{Gm_E}{R^2} = g,$$

where g is the acceleration of gravity at the earth's surface, so

$$E_{p2} - E_{p1} = mg(y_2 - y_1),$$

which is the same as the expression derived earlier when variations in g were neglected.

7-5 ELASTIC POTENTIAL ENERGY

Figure 7-12 shows a body of mass m on a level surface. One end of a spring is attached to the body and the other end of the spring is fixed. Take the origin of coordinates at the position of the body when the spring is unstretched (Fig. 7-12a). An outside agent exerts a force $\boldsymbol{P}$ of sufficient magnitude to cause the spring to stretch. As soon as the slightest extension takes place, a force $\boldsymbol{F}$ is created within the spring which acts in a direction opposite that of $\boldsymbol{P}$. The force $\boldsymbol{F}$ is called an *elastic force*. If the force $\boldsymbol{P}$ is reduced or made zero, the elastic force will restore the spring to its original unstretched condition. It may therefore be referred to as a *restoring force*.

As shown in Section 7-2, for a force obeying Hooke's law, stretching the string to a final elongation x requires an amount of work $\frac{1}{2}kx^2$. Thus the work done *by the spring* (the restoring force) during this elongation is $-\frac{1}{2}kx^2$. Similarly, in a displacement from an initial elongation x_1 to a final elongation x_2, the elastic restoring force of the spring does an amount of work W_{el} given by

$$W_{el} = -\tfrac{1}{2}kx_2^2 - (-\tfrac{1}{2}kx_1^2).$$

Let W' stand for the work of the applied force $\boldsymbol{P}$. Setting the total work equal to the change in kinetic energy of the body, we have

$$W' + W_{el} = \Delta E_k\,,$$
$$W' - (\tfrac{1}{2}kx_2^2 - \tfrac{1}{2}kx_1^2) = (\tfrac{1}{2}mv_2^2 - \tfrac{1}{2}mv_1^2).$$

The quantities $\frac{1}{2}kx_2^2$ and $\frac{1}{2}kx_1^2$ depend only on the initial and final positions of the body and not specifically on the way in which it moved. Let us therefore transfer them from the "work" side of the equation to the "energy" side. Then

$$\boxed{W' = (\tfrac{1}{2}mv_2^2 - \tfrac{1}{2}mv_1^2) + (\tfrac{1}{2}kx_2^2 - \tfrac{1}{2}kx_1^2).} \quad (7\text{-}15)$$

The quantity $\frac{1}{2}kx^2$, one-half the product of the force constant and the square of the coordinate of the body, is called the *elastic potential energy* of the body, E_p. (The symbol E_p is used for any form of potential energy.)

$$\boxed{E_p\ (\text{elastic}) = \tfrac{1}{2}kx^2.} \quad (7\text{-}16)$$

Hence the work W' of the force $\boldsymbol{P}$ equals the sum of the change in the kinetic energy of the body and the change in its elastic potential energy.

Equation (7-15) can also be written

$$W' = (\tfrac{1}{2}mv_2^2 + \tfrac{1}{2}kx_2^2) - (\tfrac{1}{2}mv_1^2 + \tfrac{1}{2}kx_1^2).$$

The sum of the kinetic and potential energies of the body is its total mechanical energy; and *the work*

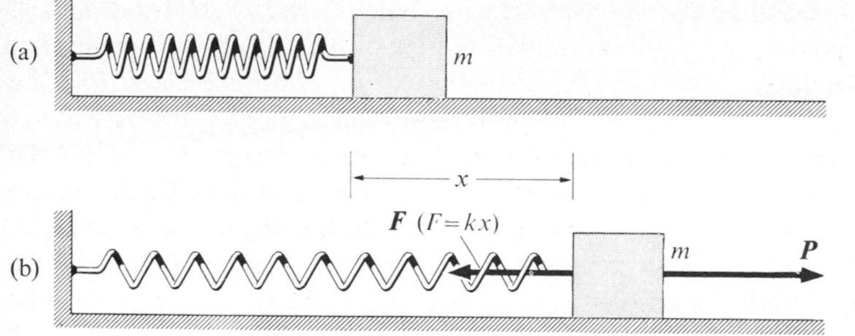

Fig. 7–12 *When an applied force* ***P*** *produces an extension x of a spring, an elastic restoring force* ***F*** *is created within the spring, where F = kx.*

of all forces acting on the body, ***with the exception of the elastic force,*** *equals the change in the total mechanical energy of the body.*

If the work W' is positive, the mechanical energy increases. If W' is negative, it decreases. In the special case in which W' is zero, the mechanical energy remains constant or is *conserved.*

Example 1 Let the force constant k of the spring in Fig. 7–12 be 24 N m^{-1}, and let the mass of the body be 4 kg. The body is initially at rest, and the spring is initially unstretched. Suppose that a constant force ***P*** of 10 N is exerted on the body, and that there is no friction. What will be the speed of the body when it has moved 0.5 m?

The equations of motion with constant acceleration cannot be used, since the resultant force on the body varies as the spring is stretched. However, the speed can be found from energy considerations:

$$W' = \Delta E_k + \Delta E_p,$$

$$(10\text{ N})(0.5\text{ m}) = \tfrac{1}{2}(4\text{ kg})v_2^2 - 0 + \tfrac{1}{2}(24\text{ N m}^{-1})(0.25\text{ m}^2) - 0,$$

$$v_2 = 1\text{ m s}^{-1}.$$

Example 2 Suppose the force ***P*** ceases to act when the body has moved 0.5 m. *How much farther* will the body move before coming to rest?

The elastic force is now the only force, and mechanical energy is conserved. The kinetic energy is $\frac{1}{2}mv^2 = 2$ J, and the potential energy is $\frac{1}{2}kx^2 = 3$ J. The total energy is therefore 5 J (equal to the work of the force ***P***). When the body comes to rest, its kinetic energy is zero and its potential energy is therefore 5 J. Hence

$$\tfrac{1}{2}kx_{\max}^2 = 5\text{ J}, \qquad x_{\max} = 0.645\text{ m}.$$

7-6 CONSERVATIVE AND DISSIPATIVE FORCES

We have seen that when an object acted on by gravity is moved from any position above a zero reference level to any other position, the work of the gravitational force is independent of the path and equal to the difference between the final and the initial values of a function called the *gravitational potential energy.* If the gravitational force alone acts on the object, the total mechanical energy (the sum of the kinetic and gravitational potential energies) is constant or conserved; the gravitational force is therefore called a *conservative force.* Thus, if the object is ascending, the work of the gravitational force is accomplished at the expense of the kinetic energy. When the object descends to its original level, the work of the gravitational force serves to increase the kinetic energy back to its original value. Thus the work done by gravity on ascent is completely recovered on descent. Complete recoverability is an important aspect of the work of a conservative force.

When an object attached to a spring is moved from one value of the spring extension to any other value, the work of the elastic force is also independent of the path and equal to the difference between the final and initial values of a function called the *elastic potential energy*. If the elastic force alone acts on the object, the sum of the kinetic and elastic potential energies is conserved; therefore the elastic force is also a conservative force. If the object moves so as to increase the extension of the spring, the work of the elastic force is accomplished at the expense of the kinetic energy. If, however, the extension is decreasing, then the work of the elastic force serves to increase the kinetic energy so that this work is completely recovered also.

To summarize, we see that the work of a conservative force has the following properties:

1. It is independent of the path.
2. It is equal to the difference between the final and initial values of an energy function.
3. It is completely recoverable.

Contrast a conservative force with a friction force exerted on a moving object by a fixed surface. The work of the friction force *does depend* on the path; the longer the path between two given points, the greater the work. There is *no* function the difference of two values of which equals the work of the friction force. When we slide an object on a rough fixed surface back to its original position, the friction force reverses, and instead of recovering the work done in the first displacement, we must again do work on the return trip. In other words, frictional work is *not* completely recoverable. When the friction force acts alone, the total mechanical energy is *not* conserved. The friction force is therefore called a *nonconservative* or a *dissipative* force. *The mechanical energy of a body is conserved only when no dissipative forces act on it.*

We find that when friction forces act on a moving body, another form of energy is involved. The more general principle of conservation of energy includes this other form of energy, along with kinetic and potential energy; and, when it is included the *total* energy of any system remains constant. We shall study this general conservation principle more fully in a later chapter.

Example Example 3 in Section 7–4 illustrates the motion of a body acted on by a dissipative friction force. The initial mechanical energy of the body is its initial potential energy of 4.9 J. Its final mechanical energy is its final kinetic energy of 2.25 J. The frictional work W_f is −2.65 J. A quantity of thermal energy equivalent to 2.65 J is developed as the body slides down the track. The sum of this energy and the final mechanical energy equals the initial mechanical energy, and the total energy of the system is conserved.

7–7 INTERNAL WORK

Figure 7–13(a) shows a man standing on frictionless roller skates on a level surface, facing a rigid wall. Suppose that he sets himself in motion backward by pushing against the wall. The external forces *on the man* are his weight $\boldsymbol{w}$, the upward forces $\boldsymbol{N}_1$ and $\boldsymbol{N}_2$ exerted by the ground, and the horizontal force $\boldsymbol{P}$ exerted by the wall. (The latter is the reaction to the force with which the man pushes against the wall.) The works of $\boldsymbol{w}$ and of $\boldsymbol{N}$ are zero because they are perpendicular to the motion. The force $\boldsymbol{P}$ is the unbalanced horizontal force which imparts to the system a horizontal acceleration. *The work of* $\boldsymbol{P}$, *however, is zero because there is no motion of its point of application.* We are therefore confronted with a curious situation in which a force is responsible for acceleration, but its work, being zero, is not equal to the increase in kinetic energy of the system!

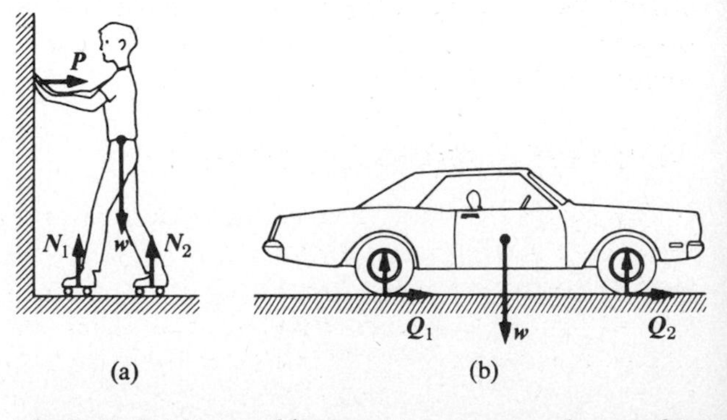

Fig. 7–13 *(a) External forces acting on a man who is pushing against a wall. The work of these forces is zero. (b) External forces on an automobile. The work of these forces is zero. In both cases, the work of the internal force is responsible for the increase in kinetic energy.*

At this point the concept of *internal work* is useful. Although internal forces play no role in accelerating the system, their points of application may move in such a way that work is done. In this case, an *internal* muscular force is exerted within the man away from the wall. (Think of his lengthening arm as an expanding spring.) Since the point of application of this force moves in the same direction, the work W_i of the internal force is not zero. This is the work responsible for the increase in kinetic energy.

When both external and internal forces act on the particles of a system, the total work W of *all* forces, external and internal, is the sum of the works W_e and W_i and is equal to the change in the total kinetic energy of the system:

$$W = W_e + W_i = \Delta E_k .$$

The same principle holds in the case of an accelerated automobile. The portions of the rubber tires that are in contact with the rough roadway push back on the ground; the reactions to these forces, designated by $\boldsymbol{Q}_1$ and $\boldsymbol{Q}_2$ in Fig. 7-13(b), are the external horizontal forces acting *on* the automobile which are responsible for imparting the horizontal acceleration to the system. Since, however, the portions of the tires momentarily in contact with the road are at rest with respect to the road, the works of the forces $\boldsymbol{Q}_1$ and $\boldsymbol{Q}_2$ are zero. As a result of the expanding gases in the cylinders of the automobile engine there are many internal forces, some of which do work; these internal forces are responsible for the increase in kinetic energy.

7-8 INTERNAL POTENTIAL ENERGY

If any of the *external* forces on the particles of a system are conservative (gravitational or elastic), the work of these forces can be transferred to the energy side of the work-energy equation and called the change in *external potential energy of the system*. In many instances, the *internal* forces depend only on the distances between *pairs* of particles. The internal work then depends only on the initial and final distances between the particles, and not on the particular way in which they moved. The work of these internal forces can then also be transferred to the energy side of the work-energy equation and called the change in *internal potential energy of the system*. Internal potential energy is a property of the system as a whole and cannot be assigned to any specific particle.

Let W' represent the works of all external and internal forces that have *not* been transferred to the energy side of the work-energy equation and called changes in external and internal potential energy. Let $E_p{}^e$ and $E_p{}^i$ represent the external and internal potential energies. The work-energy equation then takes the form

$$W' = \Delta E_k + \Delta E_p{}^e + \Delta E_p{}^i, \qquad (7\text{-}17)$$

where the total mechanical energy now includes the kinetic energy of the system and both its external and internal potential energy. If the work W' is zero, the total mechanical energy is conserved.

Example 1 Consider a system consisting of two particles in "outer space," very far from all other matter. No external forces act on the system, no external work is done when the bodies move, and the system has no external potential energy. A gravitational force of attraction acts between the bodies, which depends only on the distance between them. We can therefore say that the system has an internal potential energy. If the particles start from rest and accelerate toward each other, the total kinetic energy of the system increases and its internal potential energy decreases. The sum of its kinetic energy and internal potential energy remains constant.

Energy considerations *alone* do not suffice to tell us how much kinetic energy is gained by each body separately. This can be determined, however, from *momentum* considerations, as will be explained in the next chapter.

Example 2 In Section 7-5 we computed the elastic potential energy of a body acted on by a force exerted by a spring. This force was an *external* force acting on the body, and the elastic potential energy $\frac{1}{2}kx^2$ should properly be called the *external* elastic potential energy *of the body*.

Let us now consider the spring itself. The spring is a *system* composed of an enormous number of molecules which exert internal forces on one another. When the spring is stretched, the distances between its molecules change. The intermolecular forces depend only on the distances between pairs of molecules, so that a stretched spring has internal elastic potential energy. To calculate this, suppose the spring is slowly stretched from its no-load length by equal and opposite forces applied at its ends. The work of these forces was shown to equal $\frac{1}{2}kx^2$, where x is the elongation of the spring above its no-load length. Let us retain this work on the left side of the work–energy equation so that it becomes the work W' in Eq. (7–17). There is no change in the kinetic energy of the spring and no change in its external potential energy. If we call the internal potential energy zero when $x = 0$, the *change* in internal potential energy in the stretching process equals the final potential energy $E_p{}^i$. Then

$$W' = \Delta E_p{}^i = E_p{}^i, \qquad E_p{}^i = \tfrac{1}{2}kx^2.$$

Hence the *internal elastic potential energy of a stretched spring* is equal to $\frac{1}{2}kx^2$. Note that the internal potential energy can be calculated without any detailed information regarding the intermolecular forces.

Example 3 A block of mass m, initially at rest, is dropped from a height h onto a spring whose force constant is k. Find the maximum distance y that the spring will be compressed. (See Fig. 7–14).

This is a process for which the principle of the conservation of mechanical energy holds. At the moment of release, the kinetic energy is zero. At the moment when maximum compression occurs, there is also no kinetic energy. Hence, the loss of gravitational potential energy of the block equals the gain of elastic potential energy of the spring. As shown in Fig. 7–14, the total fall of the block is $h + y$, whence

$$mg(h + y) = \tfrac{1}{2}ky^2,$$

or

$$y^2 - \frac{2mg}{k}y - \frac{2mgh}{k} = 0.$$

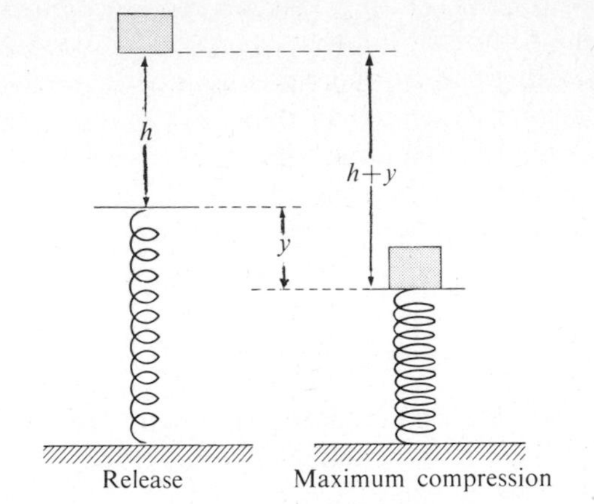

Fig. 7–14 *The total fall of the block is $h + y$.*

Therefore,

$$y = \frac{1}{2}\left[\frac{2mg}{k} \pm \sqrt{\left(\frac{2mg}{k}\right)^2 + \frac{8mgh}{k}}\right].$$

The positive root is the desired result; the negative root corresponds to the height to which the block and spring would rebound if they were fastened together after contact.

7–9 POWER

Time considerations are not involved in the definition of work. The same amount of work is done in raising a given weight through a given height whether the work is done in 1 s, or 1 hr, or 1 yr. In many instances, however, it is necessary to consider the *rate* at which work is done as well as the total amount of work accomplished. The rate at which work is done by a working agent is called the *power* developed by that agent.

If a quantity of work ΔW is done in a time interval Δt, the average power $\bar{P}$ is defined as

$$\text{Average power} = \frac{\text{work done}}{\text{time interval}},$$

$$\bar{P} = \frac{\Delta W}{\Delta t}.$$

If the rate at which work is done is not constant, this ratio may vary; in this case we may define an *instantaneous* power as the limiting value of this quotient as Δt approaches zero:

$$P = \lim_{\Delta t \to 0} \frac{\Delta W}{\Delta t} = \frac{dW}{dt}. \tag{7–18}$$

The mks unit of power is one joule per second (1 J s^{-1}), which is called one *watt* (1 W). Since this is a rather small unit, the kilowatt ($1 \text{ kW} = 10^3 \text{ W}$) and the megawatt ($1 \text{ MW} = 10^6 \text{ W}$) are commonly used. The cgs power unit is one erg per second (1 erg s^{-1}). No single term is assigned to this unit.

In the engineering system, where work is expressed in foot-pounds and time in seconds, the unit of power is one foot-pound per second. Since this unit is inconveniently small, a larger unit called the *horsepower* (hp) is in common use: $1 \text{ hp} = 550 \text{ ft lb s}^{-1} = 33{,}000 \text{ ft lb min}^{-1}$. That is, a 1-hp motor running at full load is doing 33,000 ft-lb of work every minute it runs.

A common misconception is that there is something inherently *electrical* about a watt or a kilowatt. This is not the case. It is true that electrical power is usually expressed in watts or kilowatts, but the power consumption of an incandescent lamp could equally well be expressed in horsepower, or an automobile engine rated in kilowatts.

From the relations between the newton, pound, meter, and foot, we can show that 1 hp = 746 W = 0.746 kW, or about $\frac{3}{4}$ of a kilowatt, a useful figure to remember.

Since power is energy or work per unit time, the units of power may be used to define new units of work or energy. One such unit in common use is the kilowatt-hour, a unit of energy commonly used for electrical energy.

One kilowatt-hour is the work done in one hour by an agent working at the constant rate of one kilowatt.

Since such an agent does 1000 J of work each second, the work done in 1 hr is $3600 \times 1000 =$ 3,600,000 J:

$$1 \text{ kwh} = 3.6 \times 10^6 \text{ J} = 3.6 \text{ MJ}.$$

Note that the horsepower-hour and the kilowatt-hour are units of *work* or *energy*, not power.

Although energy is an abstract physical quantity, it nevertheless has a monetary value. A newton of force or a meter per second of velocity are not things which are bought and sold as such, but a kilowatt-hour of energy is a quantity offered for sale at a definite market rate. In the form of electrical energy, a kilowatt-hour can be purchased at a price varying from a few tenths of a cent to a few cents, depending on the locality and the quantity purchased.

7–10 POWER AND VELOCITY

Suppose that a force $\boldsymbol{F}$ is exerted on a particle while the particle moves a distance Δs along its path. If $F_\parallel$ is the magnitude of the component of $\boldsymbol{F}$ tangent to the path, then the work of $\boldsymbol{F}$ is given by $\Delta W = F_\parallel \, \Delta s$, and the average power is

$$\bar{P} = \frac{\Delta W}{\Delta t} = F_\parallel \frac{\Delta s}{\Delta t} = F_\parallel \bar{v}.$$

The instantaneous power is therefore

$$P = F_\parallel v, \tag{7–19}$$

where v is the instantaneous velocity. Equation (7–19) may also be written in terms of the scalar product:

$$P = \boldsymbol{F} \cdot \boldsymbol{v}. \tag{7–20}$$

Example A jet airplane engine develops a thrust of 15,000 N (roughly 3000 lb). When the plane is flying at 300 m s^{-1} (roughly 600 mph), what horsepower is developed?

$$P = Fv = (1.5 \times 10^4 \text{ N})(300 \text{ m s}^{-1}) = 4.5 \times 10^6 \text{ W}$$

$$= \frac{4.5 \times 10^6}{746} \text{ hp} = 6030 \text{ hp}.$$

7-11 MASS AND ENERGY

At sufficiently high speeds (comparable to the speed of light) the laws of newtonian mechanics are no longer precisely correct but must be replaced by the more general relations predicted by the special relativity, to be studied in Chapter 14. One aspect of this generalization is that kinetic energy is no longer precisely $\frac{1}{2}mv^2$, but is given instead by

$$E_k = \frac{mc^2}{\sqrt{1 - v^2/c^2}} - mc^2, \tag{7-21}$$

where c is the speed of light, $c = 3.00 \times 10^8 \text{ m s}^{-1}$.

This appears quite different from $\frac{1}{2}mv^2$, but when v is small it reduces to $\frac{1}{2}mv^2$. To see this, we apply the binomial theorem to the radical in Eq. (7-21):

$$(1 - a)^{-\frac{1}{2}} = 1 + \tfrac{1}{2}a + \tfrac{3}{8}a^2 + \cdots,$$

where, in this case, $a = v^2/c^2$. Equation (7-21) becomes

$$E_k = mc^2\left(1 + \frac{v^2}{2c^2} + \frac{3v^4}{8c^4} + \cdots\right) - mc^2$$

$$= \tfrac{1}{2}mv^2 + \tfrac{3}{8}m\left(\frac{v^4}{c^2}\right) + \cdots$$

When v is much smaller than c, powers of v^2/c^2 become very small and may be neglected, and what remains is simply $\frac{1}{2}mv^2$.

The relativistic kinetic energy expression was predicted theoretically by Lorentz and Einstein, and has been verified experimentally with a variety of observations on rapidly moving particles such as electrons. One interesting feature is that the expression becomes very large as v becomes nearly equal to c, and predicts an infinitely large energy at $v = c$. This suggests physically that no particle may have a speed as large as c, since this would require an infinite quantity of energy. The relation of Eq. (7-21) to the nonrelativistic expression $\frac{1}{2}mv^2$ is shown in Fig. 7-15.

Equation (7-21) may be reinterpreted in an interesting way. Suppose a particle with mass m has

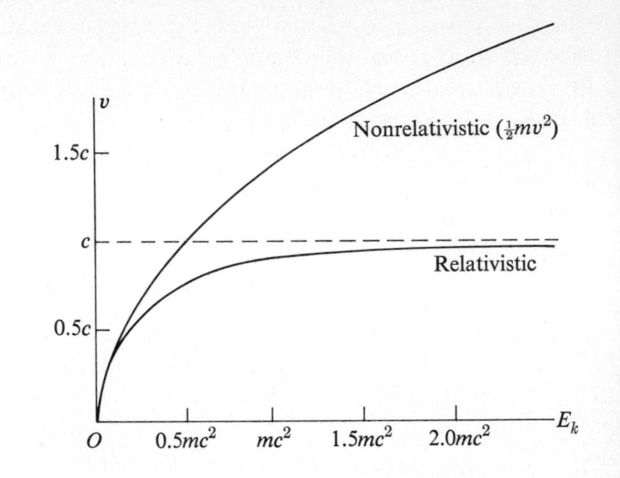

Fig. 7-15 Relation between the speed of a particle and its kinetic energy. The speed never exceeds c, no matter how great the kinetic energy. At speeds small compared to c, the kinetic energy is approximately (1/2)mv², but at larger speeds this expression deviates more and more from the relativistically correct equation.

an intrinsic energy equal to mc^2, not related directly to its speed. Then the total energy of a moving particle is the sum

$$E = E_k + mc^2 = \frac{mc^2}{\sqrt{1 - v^2/c^2}}. \tag{7-22}$$

We may then call mc^2 the "rest energy" of the particle. The validity of this concept may be tested by considering situations in which the mass of a particle changes; we may investigate whether, when mass changes, there is a corresponding change in kinetic energy, so that the *total* energy is conserved.

Here is an example from the field of nuclear physics. When the nucleus of a lithium atom is struck by a rapidly moving proton (the nucleus of a hydrogen atom), a momentary union of the two nuclei takes place, after which the compound nucleus breaks up into two alpha particles. (Alpha particles are the nuclei of helium atoms.) The alpha particles recoil in almost opposite directions and move initially with very high velocities. Their combined kinetic energy is much *greater* than the kinetic energy of the original proton. The source of this kinetic energy is the so-called "binding energy" of the nuclear particles, which is a form of internal potential energy. That is, the internal potential energy of the assemblage of protons and neutrons that makes up the unstable composite nucleus is larger than the potential energy when the same number of particles is combined in the form of two helium nuclei. A crude analogy is that of two masses forced apart by a compressed spring, but tied together by a cord. If the cord is cut, the internal potential energy of the spring is transformed into kinetic energy of the recoiling masses.

The mass of a proton is 1.6715×10^{-27} kg. The mass of a lithium nucleus is 11.6399×10^{-27} kg, and that of an alpha particle is 6.6404×10^{-27} kg. Hence the mass of the original system is

$$(1.6715 + 11.6399) \times 10^{-27}\,\text{kg} = 13.3114 \times 10^{-27}\,\text{kg}.$$

The mass of the two alpha particles is

$$2(6.6404 \times 10^{-27}\,\text{kg}) = 13.2808 \times 10^{-27}\,\text{kg}.$$

Thus, the final mass is less than the initial by 0.0306×10^{-27} kg, corresponding to an energy difference $(0.0306 \times 10^{-27}\,\text{kg})(3.00 \times 10^{8}\,\text{m s}^{-1})^2 = 0.275 \times 10^{-11}$ J. In order for energy to be conserved, the final total kinetic energy must be *greater* than the initial by this amount. Experiments have confirmed this prediction and hence also confirmed the validity of the "rest energy" concept.

Mass changes occur also in ordinary chemical reactions, but they are so small as not to be directly observable. For example, when two molecules of hydrogen H_2, and one of oxygen O_2, combine to form two molecules of water (H_2O), an energy of about 4×10^{-19} J is released for each water molecule produced. This corresponds to a mass change of $4 \times 10^{-19}\,\text{J}/c^2 = 0.44 \times 10^{-37}$ kg. Since the mass of a water molecule is about 3×10^{-26} kg, this is a mass change of less than 1 part in 10^{11}; such small mass changes in chemical reactions have never been observed, but in nuclear reactions the mass changes may be as large as 10^{-4}, roughly ten million times as great, and are readily observable.

Thus the concepts of *mass* and *energy* emerge as two aspects of a single conserved entity that may be called *mass–energy*. The relation between mass changes and energy changes can also be described by reinterpreting Eq. (7–21) to mean that the mass of a body actually depends on its speed. In this case, mass increase with speed is another way of describing kinetic energy. Although the concept of "relativistic mass increase" is widely used in the literature, it can be misleading. In any event, it is not necessary, and will not be used in this book.

PROBLEMS

7–1 The locomotive of a freight train exerts a constant force of 60,000 N on the train while drawing it at 50 km hr^{-1} on a level track. How much work does it do in a distance of 1 km?

7–2 A 40-kg block is pushed a distance of 5 m along a level floor at constant speed by a force at an angle of 30° below the horizontal. The coefficient of friction between block and floor is 0.25. How much work is done by the force?

7–3 A horse is towing a canal boat, the towrope making an angle of 10° with the towpath. If the tension in the rope is 100 lb, how much work is done while moving 100 ft along the towpath?

7–4 A block is pushed 2 m along a fixed horizontal surface by a horizontal force of 2N. The opposing force of friction is 0.4 N.

a) How much work is done by the 2-N force?
b) What is the work of the friction force?

7–5 A body is attracted toward the origin with a force given by $F = -6x^3$ where F is in pounds and x in feet. (a) What force is required to hold the body at point a, 1 ft from the origin? (b) At point b, 2 ft from the origin? (c) How much work must be done to move the body from point a to point b?

7–6 The force exerted by a gas in a cylinder on a piston whose area is A is given by $F = pA$, where p is the force per unit area, or *pressure*. The work W in a displacement of the piston from x_1 to x_2 is

$$W = \int_{x_1}^{x_2} F\,dx = \int_{x_1}^{x_2} pA\,dx = \int_{V_1}^{V_2} p\,dV,$$

where dV is the accompanying infinitesimal change of volume of the gas.

a) During an expansion of a gas at constant temperature (isothermal) the pressure depends on the volume according to the relation

$$p = \frac{nRT}{V},$$

where n and R are constants and T is the constant temperature. Calculate the work in expanding isothermally from volume V_1 to volume V_2.

b) During an expansion of a gas at constant entropy (adiabatic) the pressure depends on the volume according to the relation

$$p = \frac{K}{V^{\gamma}},$$

where K and γ are constants. Calculate the work in expanding adiabatically from V_1 to V_2.

7–7

a) Compute the kinetic energy of an 1800-lb automobile traveling at 30 mi hr^{-1}.

b) How many times as great is the kinetic energy if the velocity is doubled?

7–8 Compute the kinetic energy, in joules, of a 2-g rifle bullet traveling at 500 m s^{-1}.

7–9 An electron strikes the screen of a cathode-ray tube with a velocity of 10^7 m s^{-1}. Compute its kinetic energy in joules. The mass of an electron is 9×10^{-31} kg.

7–10 What is the potential energy of an 800-kg elevator at the top of the Empire State Building 380 m above street level? Assume the potential energy at street level to be zero.

7–11 What is the increase in potential energy of a 1-kg body when lifted from the floor to a table 1 m high?

7–12 A meter stick whose mass is 0.3 kg is pivoted at one end, as in Fig. 7–16, and displaced through an angle of 60°. What is the increase in its potential energy?

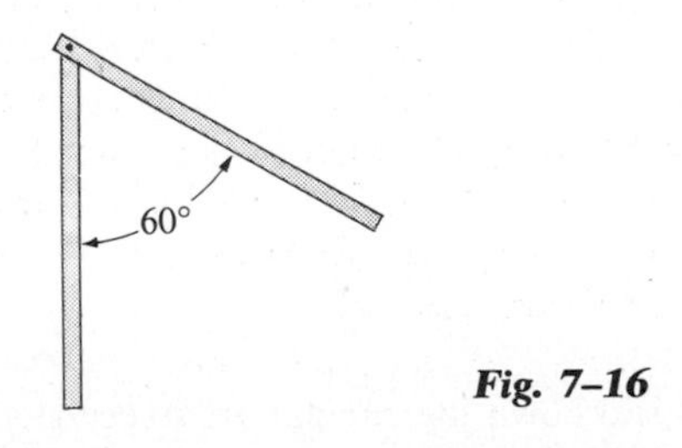

Fig. 7–16

7–13 The force in newtons required to stretch a certain spring a distance of x m beyond its unstretched length is given by $F = 100x$.

a) What force will stretch the spring 0.1 m? 0.2 m? 0.4 m?

b) How much work is required to stretch the spring 0.1 m? 0.2 m? 0.4 m?

7–14 The scale of a certain spring balance reads from zero to 1200 N and is 0.1 m long.

a) What is the potential energy of the spring when it is stretched 0.1 m? 0.05 m?

b) When a 60-kg mass hangs from the spring?

7–15 A body moves a distance of 10 m under the action of a force which has the constant value of 5.5 N for the first 6 m and then decreases to a value of 2 N, as shown by the graph in Fig. 7–17.

a) How much work is done in the first 6 m of the motion?

b) How much work is done in the last 4 m?

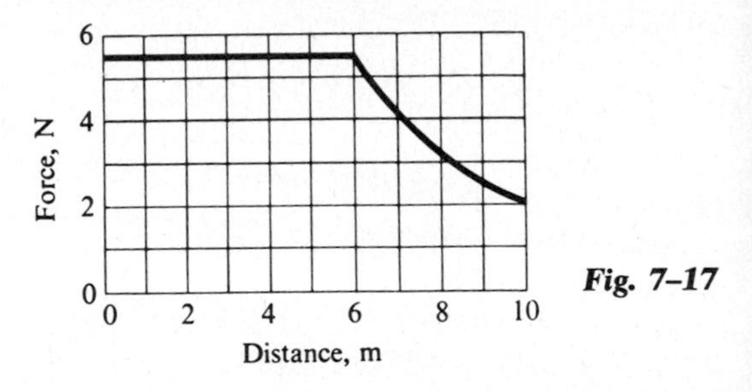

Fig. 7–17

7–16 A block weighing 16 lb is pushed 20 ft along a horizontal frictionless surface by a horizontal force of 8 lb. The block starts from rest.

a) How much work is done? What becomes of this work?

b) Check your answer by computing the acceleration of the block, its final velocity, and its kinetic energy.

7–17 In the preceding problem, suppose the block had an initial velocity of 10 ft s^{-1}, other quantities remaining the same.

a) How much work is done?

b) Check by computing the final velocity and the increase in kinetic energy.

7–18 A 5-kg block is lifted vertically at a constant velocity of 4 m s^{-1} through a height of 12 m.

a) How great a force is required?

b) How much work is done? What becomes of this work?

7–19 A 12-kg block is pushed 20 m up the sloping surface of a plane inclined at an angle of 37° to the horizontal, by a constant force F of 120 N acting parallel to the plane. The coefficient of friction between the block and plane is 0.25.

a) What is the work of the force F?

b) Compute the increase in kinetic energy of the block.

c) Compute the increase in potential energy of the block.

d) Compute the work done against friction. What becomes of this work?

e) What can you say about the sum of (b), (c), and (d)?

7–20 A man of mass 80 kg sits on a platform suspended from a movable pulley and raises himself by a rope passing over a fixed pulley (Fig. 7–18). Assuming no friction losses, find

a) the force he must exert,

b) the increase in his energy when he raises himself 1 m. Answer part (b) by calculating his increase in potential energy, and also by computing the product of the force on the rope and the length of rope passing through his hands.

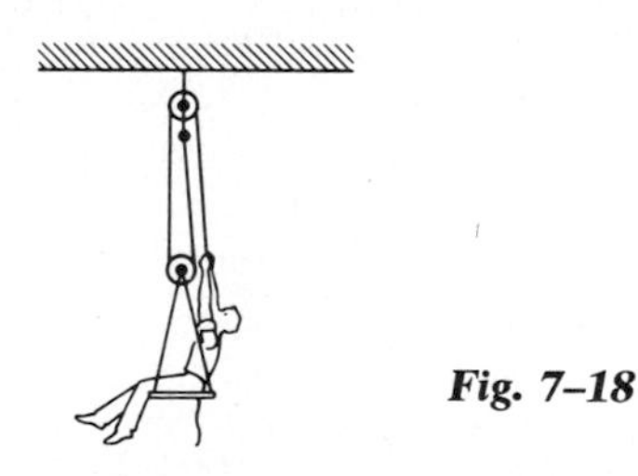

Fig. 7–18

7–21 A barrel of mass 120 kg is suspended by a rope 10 m long.

a) What horizontal force is necessary to hold the barrel sideways 2 m from the vertical?

b) How much work is done in moving it to this position?

7–22 The system in Fig. 7–19 is released from rest with the 12-kg block 3 m above the floor. Use the principle of conservation of energy to find the velocity with which the block strikes the floor. Neglect friction and inertia of the pulley.

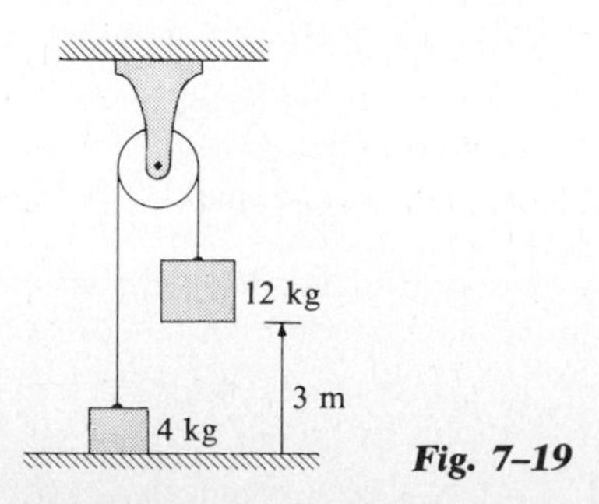

Fig. 7–19

7–23 The spring of a spring gun has a force constant of 500 N m^{-1}. It is compressed 0.05 m and a ball of mass 0.01 kg is placed in the barrel against the compressed spring.

a) Compute the maximum velocity with which the ball leaves the gun when released.

b) Determine the maximum velocity if a constant resisting force of 10 N acts on the ball.

7–24 A block of mass 1 kg is forced against a horizontal spring of negligible mass, compressing the spring an amount $x_1 = 0.2$ m. When released, the block moves on a horizontal tabletop a distance $x_2 = 1.0$ m before coming to rest. The spring constant k is 100 N m^{-1} (Fig. 7–20). What is the coefficient of friction, μ, between the block and the table?

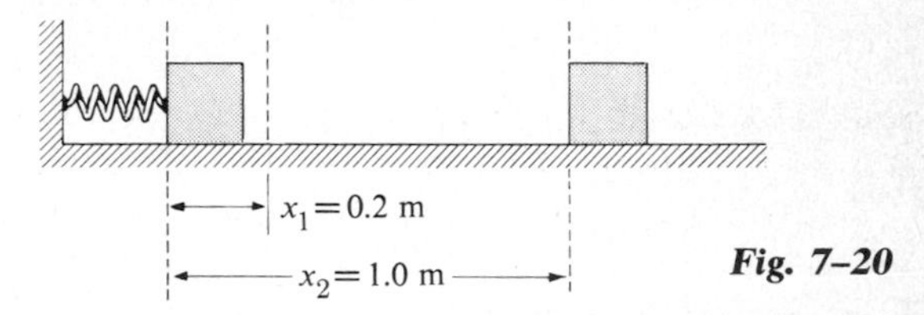

Fig. 7–20

7–25 A 2-kg block is dropped from a height of 0.4 m onto a spring whose force constant k is 1960 N m^{-1}. Find the maximum distance the spring will be compressed.

7–26 A 16-lb projectile is fired from a gun with a muzzle velocity of 800 ft s^{-1} at an angle of departure of 45°. The angle is then increased to 90° and a similar projectile is fired with the same muzzle velocity.

a) Find the maximum height attained by each projectile.

b) Show that the total energy at the top of the trajectory is the same in the two cases.

c) Using the energy principle, find the height attained by a similar projectile if fired at an angle of 30°.

7–27 A block of mass 2 kg is released from rest at point A on a track which is one quadrant of a circle of radius 1 m (Fig. 7–21). It slides down the track and reaches point B with a velocity of 4 m s^{-1}. From point B it slides on a level surface a distance of 3 m to point C, where it comes to rest.

a) What was the coefficient of sliding friction on the horizontal surface?

b) How much work was done against friction as the body slid down the circular arc from A to B?

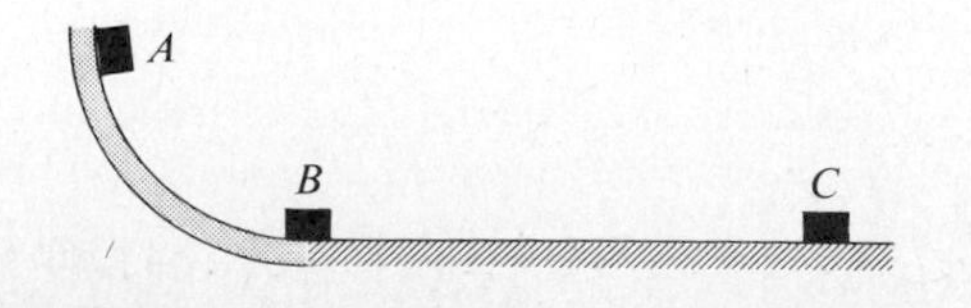

Fig. 7–21

7–28 A small sphere of mass m is fastened to a weightless string of length 0.5 m to form a pendulum. The pendulum is swinging so as to make a maximum angle of 60° with the vertical.

a) What is the velocity of the sphere when it passes through the vertical position?

b) What is the instantaneous acceleration when the pendulum is at its maximum deflection?

7–29 A ball is tied to a cord and set in rotation in a vertical circle. Prove that the tension in the cord at the lowest point exceeds that at the highest point by six times the weight of the ball.

7–30 A small body of mass m slides without friction around the loop-the-loop apparatus shown in Fig. 7–22. It starts from rest at point A at a height $3R$ above the bottom of the loop. When it reaches point B at the end of a horizontal diameter of the loop, compute

a) its radial acceleration,

b) its tangential acceleration, and

c) its resultant acceleration. Show these accelerations in a diagram, approximately to scale.

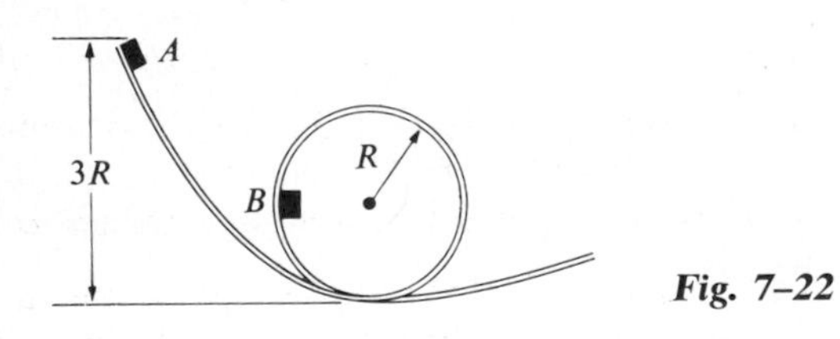

Fig. 7–22

7–31 A meter stick, pivoted about a horizontal axis through its center, has a body of mass 2 kg attached to one end and a body of mass 1 kg attached to the other. The mass of the meter stick can be neglected. The system is released from rest with the stick horizontal. What is the velocity of each body as the stick swings through a vertical position?

7-32 A variable force $\boldsymbol{P}$ is maintained tangent to a frictionless cylindrical surface of radius a, as shown in Fig. 7–23. By slowly varying this force, a block of weight w is moved and the spring to which it is attached is stretched from position 1 to position 2. The spring, of force constant k, is unstretched in position 1. Calculate the work of the force $\boldsymbol{P}$, (a) by integration, (b) by use of the energy principle.

7–33 A 5-kg block is pushed up a frictionless plane inclined at 30° to the horizontal. It is pushed 2.0 m along the plane by a constant force parallel to the plane, of magnitude 100 N. If its speed at the bottom is 2 m s^{-1}, what is its speed at the top?

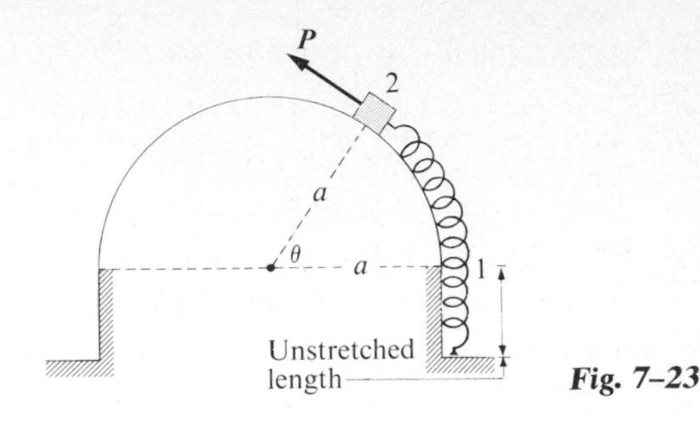

Fig. 7–23

7–34 In Problem 7–33, suppose the plane is not frictionless but has $\mu = 0.1$.

a) What is the speed of the block at the top?

b) What fraction of the work done by the force is dissipated by friction?

7–35 An 80-kg man jumps from a height of 2 m onto a platform mounted on springs. As the springs compress, the platform is pushed down a maximum distance of 0.2 m below its initial position, and it then rebounds.

a) What is the man's speed at the instant the platform is depressed 0.1 m?

b) If the man had just stepped gently onto the platform how much would it have been pushed down?

7–36 In Problem 7–30, if point A is at a height h (not equal to $3R$) above the bottom of the loop, what is the minimum value of h (in terms of R) so that the body moves around the loop without falling off at the top?

7–37 A ball of mass 0.5 kg is tied to a string of length 1.0 m, and the other end of the string is tied to a rigid support. The ball is held straight out from the point of support, with the string pulled taut, and is then released.

a) What is the speed of the ball at the lowest point of its motion?

b) What is the tension in the string at this point?

7–38 A skier starts at the top of a large frictionless spherical snowball, with very small initial velocity, and skis straight down the side. At what point does he lose contact with the snowball and fly off at a tangent? That is, at the instant he loses contact with the snowball, what angle does a radial line from the center to the skier make with the vertical?

7–39 A small 4-lb body is fastened to a weightless rod 5 ft long to form a pendulum, as shown in Fig. 7–24. The body is pulled aside until the rod makes an angle of 53° with the vertical.

a) With what tangential speed v_A must the body be started from point A so that it will reach C, the highest point, with a tangential speed of 10 ft s^{-1}?

b) With what speed does it pass through the lowest point B when started from A with the speed v_A?

c) What is the tension in the rod as the body passes through B?

d) If the body is started from A with the tangential speed v_A of part (a) in the direction opposite to that shown, with what speed will it then arrive at C?

Imagine that the rod is replaced by a weightless string and that the body is given the same speed v_A as in part (a) in the same direction as that shown in Fig. 7–24.

e) At what point D does the string become slack?

f) What is the speed at D?

g) What happens after the body passes D?

h) What is the minimum speed that the body can have at the top point C when tied to a string?

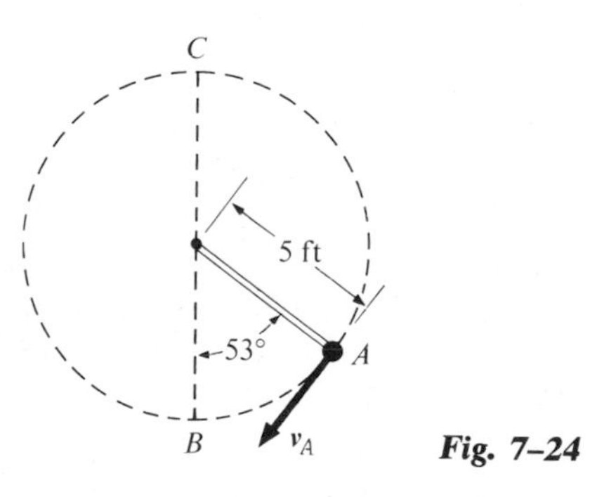

Fig. 7–24

7–40 The potential energy of a diatomic molecule, as given by Lennard-Jones, is the following function of the distance r between the atoms:

$$E_p(r) = \epsilon_0\left[\left(\frac{r_0}{r}\right)^{12} - 2\left(\frac{r_0}{r}\right)^{6}\right].$$

Prove that (a) r_0 is the intermolecular separation when the potential energy is a minimum, (b) the minimum potential energy is $-\epsilon_0$, and (c) the intermolecular separation when $E_p(r) = 0$ is equal to $r_0/\sqrt[6]{2}$. (d) Sketch the graph of $E_p(r)$.

7–41 A particle originally at rest at the origin is constrained to move along the x-axis. Its potential energy is a function of x, $E_p(x)$, and its total energy is a constant E. Prove that the time t to go to a point where the coordinate is x is

$$t = \int_0^x \frac{dx}{\sqrt{(2/m)[E - E_p(x)]}}.$$

7–42 An earth satellite of mass m revolves in a circle at a height above the earth equal to twice the earth's radius, $2R$. In terms of m, R, the gravitational constant G, and the mass of the earth m_E, what are the values of (a) the kinetic energy of the satellite, (b) its gravitational potential energy, and (c) its total mechanical energy?

7–43 What average horsepower is developed by an 80-kg man when climbing in 10 s a flight of stairs which rises 6 m vertically? Express this power in watts and kilowatts.

7–44 The hammer of a pile driver has a mass of 500 kg and must be lifted a vertical distance of 2 m in 3 s. What horsepower engine is required?

7–45 A ski tow is to be operated on a 37° slope 800 ft long. The rope is to move at 8 mi hr^{-1} and power must be provided for 80 riders at one time, with average weight 150 lb. Estimate the horsepower required to operate the tow.

7–46

a) If energy costs 5 cents per kwh, how much is one horsepower-hour worth?

b) How many ft-lb can be purchased for one cent?

7–47 Compute the monetary value of the kinetic energy of the projectile of a 14-in. naval gun, at the rate of 2 cents per kwh. The projectile has a mass of 600 kg and its muzzle velocity is 800 m s^{-1}.

7–48 At 5 cents per kwh, what does it cost to operate a 10-hp motor for 8 hr?

7–49 The engine of an automobile develops 20 hp when the automobile is traveling at 50 km hr^{-1}.

a) What is the resisting force in newtons?

b) If the resisting force is proportional to the velocity, what horsepower will drive the car at 25 km hr^{-1}? At 100 km hr^{-1}?

7–50 The engine of a motorboat delivers 40 hp to the propeller while the boat is moving at 20 mi hr^{-1}. What would be the tension in the towline if the boat were being towed at the same speed?

7–51 A man whose mass is 70 kg walks up to the third floor of a building. This is a vertical height of 12 m above the street level.

a) How many joules of work has he done?

b) By how much has he increased his potential energy?

c) If he climbs the stairs in 20 s, what was his rate of working in horsepower?

7–52 A pump is required to lift 800 kg (about 200 gallons) of water per minute from a well 10 m deep and eject it with a speed of 20 m s^{-1}.

a) How much work is done per minute in lifting the water?

b) How much in giving it kinetic energy?

c) What horsepower engine is needed?

7–53 A 2000-kg elevator starts from rest and is pulled upward with a constant acceleration of 4 m s^{-2}.

a) Find the tension in the supporting cable.

b) What is the velocity of the elevator after it has risen 15 m?

c) Find the kinetic energy of the elevator 3 s after it starts.

d) How much is its potential energy increased in the first 3 s?

e) What horsepower is required when the elevator is traveling 8 m s^{-1}?

7–54 An automobile of mass 1200 kg has a speed of 30 m s^{-1} on a horizontal road when the engine is developing 50 hp. What is its speed, with the same horsepower, if the road rises 1 m in 20 m? Assume all friction forces to be constant.

7–55

a) If 20 hp are required to drive a 1200 kg automobile at 50 km hr^{-1} on a level road, what is the retarding force of friction, windage, etc?

b) What power is necessary to drive the car at 50 km hr^{-1} up a 10 percent grade (ie., one rising 10 m vertically in 100 m horizontally)?

c) What power is necessary to drive the car at 50 km hr^{-1} *down* a 2 percent grade?

d) Down what grade would the car coast at 50 km hr^{-1}?

7–56 The total consumption of electrical energy in the United States is of the order of 10^{19} joules per year.

a) What is the average rate of energy consumption, in watts?

b) If the population of the United States is 200 million, what is the average rate of energy consumption per person?

7–57 The sun transfers energy to the earth by radiation, at a rate of approximately 1.4 kW per square meter of surface. If this energy could be collected and converted to electrical energy with 100 percent efficiency, how great an area would be required to collect the energy cited in the previous problem?

7–58

a) Calculate the distance from the surface of the moon to the point P where the gravitational fields of the moon and of the earth cancel.

b) If the potential energy of a 1-kg object is zero at an infinite distance from the moon and the earth, what is its potential energy at point P?

c) Assume that the Apollo 8 space capsule acquired a speed of 12,000 m s^{-1} while still close to the earth and then coasted on an orbit designed to bring it close to the moon. What would be its speed at point P?

d) What would be its speed almost at the surface of the moon?

e) At what speed would a space vehicle execute a circular orbit a short distance from the surface of the moon?

f) What would be the period of the motion in part (e)? [*Note*: The distance between the centers of the moon and of the earth is 3.84×10^8 m; the radius of the moon is 1.74×10^6 m; the radius of the earth is 6.37×10^6 m; the mass of the moon is 7.34×10^{22} kg; the mass of the earth is 5.98×10^{24} kg; the gravitational constant is 6.67×10^{-11} N m^2 kg^{-2}.]

7–59 A nuclear bomb containing 20 kg of plutonium explodes. The rest mass of the products of the explosion is less than the original rest mass by one ten-thousandth of the original rest mass.

a) How much energy is released in the explosion?

b) If the explosion takes place in 1 μs, what is the average power developed by the bomb?

c) How much water could the released energy lift to a height of 1 km?

7–60 In a hypothetical nuclear-fusion reactor two deuterium nuclei combine or "fuse" to form one helium nucleus. The mass of a deuterium nucleus, expressed in atomic mass units (u) is 2.0147 u; that of a helium nucleus is 4.0039 u

($1\ \text{u} = 1.66 \times 10^{-27}$ kg).

a) How much energy is released when 1 kg of deuterium undergoes fusion?

b) The annual consumption of electrical energy in the United States is of the order of 10^{19} J. How much deuterium must react to produce this much energy?

7–61 Compute the kinetic energy of an electron (mass 9.11×10^{-31} kg) using both the nonrelativistic and relativistic expressions, and compare the two results, for speeds of

a) $1.0 \times 10^{7}\ \text{m s}^{-1}$;

b) $2.0 \times 10^{8}\ \text{m s}^{-1}$.

Chapter 8

Impulse and Momentum

8–1 IMPULSE AND MOMENTUM

In the preceding chapter, the concepts of work and energy were developed from Newton's laws of motion. We shall see next how two similar concepts, those of *impulse* and *momentum*, also arise from these laws.

Let us consider again a particle of mass m moving along a straight line. For the moment we assume that the force $\boldsymbol{F}$ on this particle is constant and directed along the line of motion. If the particle's velocity at some initial time $t = 0$ is v_0, the velocity v at a later time t is given by

$$v = v_0 + at,$$

where the constant acceleration a is given by F/m. Making this substitution in the above equation, multiplying through by m, and rearranging, we obtain

$$Ft = mv - mv_0. \tag{8–1}$$

The lefthand side, the product of the force and the time during which it acts, is called the *impulse* of the force. More generally, if a constant force $\boldsymbol{F}$ acts from time t_1 to time t_2, the impulse of the force is defined to be

$$\text{Impulse} = \boldsymbol{F}(t_2 - t_1).$$

The right side of Eq. (8–1) contains the product of mass and velocity of the particle at two different times. This product is also given a special name, *momentum*. It is also sometimes called *linear momentum* to distinguish it from a similar quantity called *angular* momentum, to be discussed later. The symbol $\boldsymbol{p}$ is often used for momentum.

$$\text{Momentum} = \boldsymbol{p} = m\boldsymbol{v}.$$

In terms of these newly-defined quantities, the content of Eq. (8–1) is that the impulse of the force from time zero to time t is equal to the change of momentum during that interval, that is, the momentum at the end of the interval minus that at the beginning. We see that there is nothing special about the two times zero and t, so we may also say that if the particle's velocity at time t_1 is v_1, and its velocity at time t_2 is v_2, then

$$\boldsymbol{F}(t_2 - t_1) = m\boldsymbol{v}_2 - m\boldsymbol{v}_1. \tag{8–2}$$

The above relation between impulse and momentum change has the same form as the relation between work and kinetic-energy change developed in Chapter 7. There are important differences, however. First, impulse is a product of a force and a *time* interval, while work is a product of a force and a distance, and depends on the angle between force and displacement. Furthermore, both force and velocity are *vector* quantities, so impulse and momentum are also vector quantities, unlike work and kinetic energy, which are scalars. In motion along a straight line, where only one component of a vector is involved, the force and velocity may have components along this line which are either positive or negative.

Example 1 A particle of mass 2 kg moves along the x-axis with an initial velocity of 3 m s^{-1}. A force $F = -6$ N (i.e., in the negative x-direction) is applied for a period of 3 s. Find the final velocity.

From Eq. (8–2),

$$(-6\text{ N})(3\text{ s}) = (2\text{ kg})v_2 - (2\text{ kg})(3\text{ m s}^{-1})$$

or

$$v_2 = -6\text{ m s}^{-1}.$$

The particle's final velocity is in the negative x-direction.

The unit of impulse, in any system, equals the product of the units of force and time in that system. Thus, in the mks system, the unit is one newton second (1 N · s), in the cgs system it is one dyne second (1 dyn · s), and in the British system it is one pound second (1 lb · s).

The unit of momentum in the mks system is one kilogram meter per second (1 kg m s^{-1}), in the cgs system it is one gram centimeter per second (1 g · cm · s^{-1}), and in the British system it is one slug foot per second (1 slug · ft · s^{-1}). Since

$$1\text{ kg}\cdot\text{m}\cdot\text{s}^{-1} = (1\text{ kg}\cdot\text{m}\cdot\text{s}^{-2})(1\text{ s}) = 1\text{ N}\cdot\text{s},$$

it follows that the unit of momentum in any system equals the unit of impulse in that system.

The concept of impulse can be generalized to include forces that vary with time; both the magnitude and the direction of the force may vary. We consider a particle of mass m moving in the xy-plane, as in Fig. 8–1, acted on by a varying resultant force $\mathbf{F}$. Newton's second law states that

$$\mathbf{F} = m\frac{d\mathbf{v}}{dt},$$

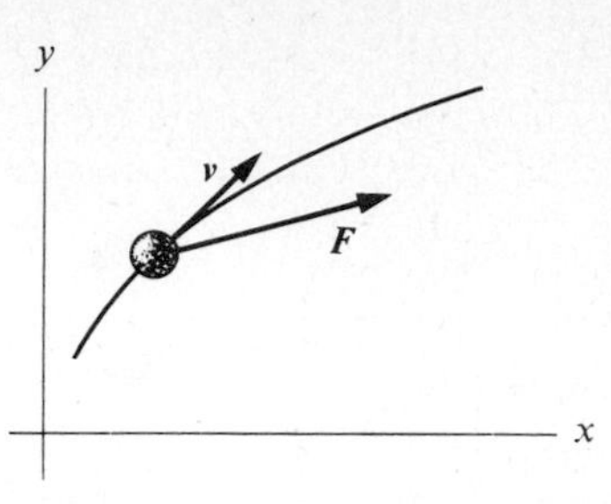

Fig. 8–1 Particle moving in the xy-plane.

or

$$\mathbf{F}\,dt = m\,d\mathbf{v}.$$

If $\mathbf{v}_1$ is the velocity when $t = t_1$, and $\mathbf{v}_2$ the velocity when $t = t_2$, it follows that

$$\int_{t_1}^{t_2} \mathbf{F}\,dt = \int_{\mathbf{v}_1}^{\mathbf{v}_2} m\,d\mathbf{v}. \tag{8–3}$$

The integral on the left is the *impulse of the force* $\mathbf{F}$ in the time interval $t_2 - t_1$, and is a *vector quantity:*

$$\boxed{\text{Impulse} = \int_{t_1}^{t_2} \mathbf{F}\,dt.}$$

This integral can be evaluated, of course, only when the force is known as a function of the time. The integral on the right, however, always yields the result

$$\int_{\mathbf{v}_1}^{\mathbf{v}_2} m\,d\mathbf{v} = m\mathbf{v}_2 - m\mathbf{v}_1.$$

Thus Eq. (8–3) may be written as

$$\int_{t_1}^{t_2} \mathbf{F}\,dt = m\mathbf{v}_2 - m\mathbf{v}_1, \tag{8–4}$$

In the special case when $\mathbf{F}$ is constant, this reduces to Eq. (8–2).

Equation (8–4) is also equivalent (for forces and velocities in the xy-plane) to the two scalar equations

$$\int_{t_1}^{t_2} F_x\,dt = mv_{x_2} - mv_{x1},$$

$$\int_{t_1}^{t_2} F_y\,dt = mv_{y_2} - mv_{y1}, \tag{8–5}$$

The impulse of any force component, or any force whose direction is constant, can be represented graphically by plotting the force vertically and the time horizontally, as in Fig. 8–2. The *area* under the curve, between vertical lines at t_1 and t_2, is equal to the impulse of the force in this time interval.

If the impulse of a force is *positive* the momentum of the body on which it acts *increases* algebraically. If the impulse is *negative*, the momentum *decreases*. If the impulse is *zero*, there is no change in momentum.

Example 2 Consider the changes in momentum produced by the following forces:

a) A body moving on the x-axis is acted on for 2 s by a constant force of 10 N toward the right.

b) The body is acted on for 2 s by a constant force of 10 N toward the right and then for 2 s by a constant force of 20 N toward the left.

c) The body is acted on for 2 s by a constant force of 10 N toward the right, then for 1 s by a constant force of 20 N toward the left. The three forces are shown graphically in Fig. 8–3.

a) The impulse of the force is (+10 N)(2 s) = +20 N s. Hence the momentum of *any* body on which the force acts increases by 20 kg m s^{-1}. This change is the same whatever the mass of the body and whatever the magnitude and direction of its initial velocity.

Suppose the mass of the body is 2 kg and that it is initially at rest. Its *final* momentum then equals its *change* in momentum and its final *velocity* is 10 m s^{-1} toward the right. (The reader should verify by computing the acceleration.)

Had the body been initially moving toward the right at 5 m s^{-1}, its initial momentum would have been 10 kg m s^{-1} toward the right.

Had the body been moving initially toward the *left* at 5 m s^{-1}, its initial momentum would have been −10 kg m s^{-1}, its final momentum +10 kg m s^{-1}, and its final velocity 5 m s^{-1} toward the right. That is, the constant force of 10 N toward the right would first have brought the body to rest and then given it a velocity in the direction opposite to its initial velocity.

b) The impulse of this force is

$$(+10\text{ N})(2\text{ s}) - (20\text{ N})(2\text{ s}) = -20\text{ N s}.$$

The momentum of *any* body on which it acts is decreased by 20 kg m s^{-1}. The reader should examine various possibilities, as in the preceding example.

c) The impulse of this force is

$$(+10\text{ N})(2\text{ s}) - (20\text{ N})(1\text{ s}) = 0.$$

Hence the momentum of any body on which it acts is not changed. Of course, the momentum of the body is increased during the first 2 s but it is *decreased* by an equal amount in the next second. As an exercise, describe the motion of a body of mass 2 kg, moving initially to the left at 5 m s^{-1}, and acted on by this force. It will help to construct a graph of velocity versus time.

Example 3 A ball of mass 0.4 kg is thrown against a brick wall. When it strikes the wall it is moving horizontally to the left at 30 m s^{-1}, and it rebounds horizontally to the right at 20 m s^{-1}. Find the impulse of the force exerted on the ball by the wall.

The initial momentum of the ball is (0.4 kg)(−30 m s^{-1}) = −12 kg m s^{-1}. The final momentum is +8 kg m s^{-1}. The *change* in momentum is

$$mv_2 - mv_1 = 8\text{ kg ms}^{-1} - (-12\text{ kg m s}^{-1})$$
$$= 20\text{ kg ms}^{-1}.$$

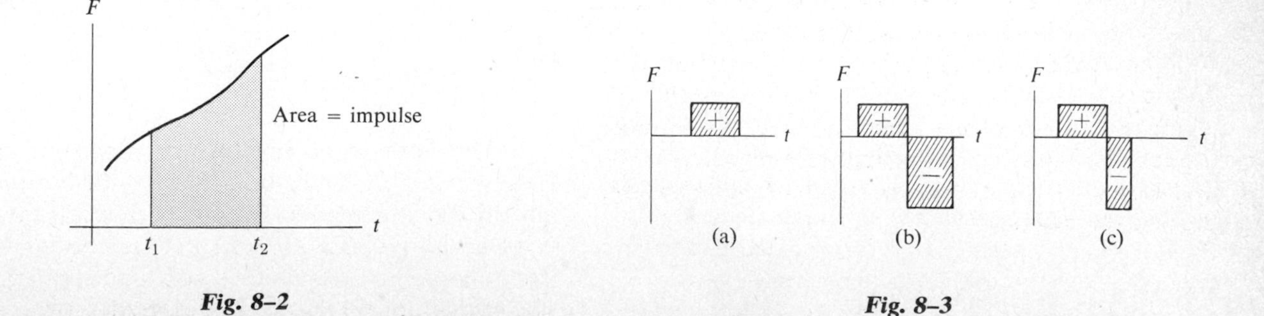

Fig. 8–2

Fig. 8–3

Hence, the impulse of the force exerted on the ball was 20 N s. Since the impulse is *positive*, the force must be toward the right.

Note that the *force* exerted on the ball cannot be found without further information regarding the collision. The general nature of the force–time graph is shown by one of the curves in Fig. 8–4. The force is zero before impact, rises to a maximum, and decreases to zero when the ball leaves the wall. If the ball is relatively rigid, like a baseball, the time of collision is small and the maximum force is large, as in curve (a). If the ball is more yielding, like a tennis ball, the collision time is larger and the maximum force is less, as in curve (b). In any event, the *area* under the force–time graph must equal 20 N s.

For an idealized case in which the force is constant and the collision time is 1 m s (10^{-3} s), as represented by the horizontal straight line, the force is 20,000 N.

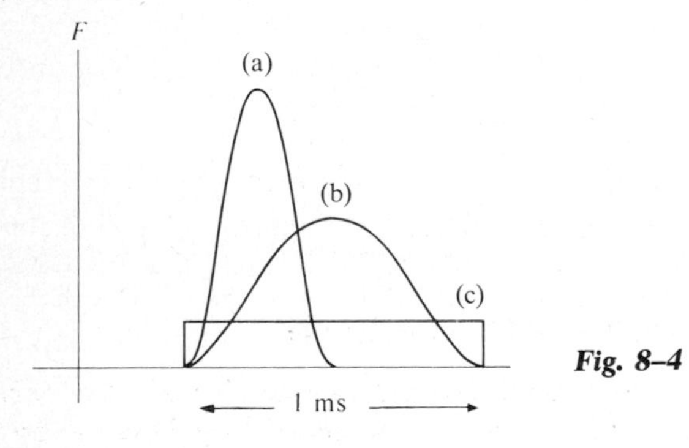

Fig. 8–4

8-2 CONSERVATION OF MOMENTUM

Whenever there is a force of interaction between two particles, the momentum of each particle is changed as a result of the force exerted on it by the other. (The force may be gravitational, electric, magnetic, or of any other origin.) Furthermore, since, by Newton's *third* law, the force on one particle is always equal in magnitude and opposite in direction to that on the other, the impulses of the forces are equal in magnitude and opposite in direction. It follows that *the vector change in momentum of either particle*, in any time interval, *is equal in magnitude and opposite in direction to the vector change in momentum of the other.* The *net change in momentum of the* **system** (the two particles together) *is therefore zero.*

The pair of action–reaction forces are *internal* forces of the system, and we conclude that *the total momentum of a system of bodies cannot be changed by internal forces between the bodies.* Hence, if the *only* forces acting on the particles of a system are *internal* forces (that is, if there are no *external* forces), the total momentum of the system remains constant in magnitude and direction. This is the *principle of conservation of linear momentum*: *When no resultant external force acts on a system, the total momentum of the system remains constant in magnitude and direction.*

The principle of conservation of momentum is one of the most fundamental and important principles of mechanics. Note that it is more general than the principle of conservation of mechanical energy; mechanical energy is conserved *only* when the internal forces are *conservative.* The principle of conservation of momentum holds whatever the nature of the internal forces, conservative or not.

Example 1 Figure 8–5 shows a body A of mass m_A moving toward the right on a level frictionless surface (or a level air track) with a velocity $\mathbf{v}_{A1}$. It collides with a second body B of mass m_B moving toward the left with a velocity $\mathbf{v}_{B1}$. Since there is no friction and the resultant vertical force on the system is zero, the only forces on the bodies are the internal action–reaction forces which they exert on each other in the collision process, and the momentum of the system remains constant in magnitude and direction.

Let $\mathbf{v}_{A2}$ and $\mathbf{v}_{B2}$ represent the velocities of A and B after the collision. Then

$$m_A\mathbf{v}_{A1} + m_B\mathbf{v}_{B1} = m_A\mathbf{v}_{A2} + m_B\mathbf{v}_{B2}. \quad (8\text{–}6)$$

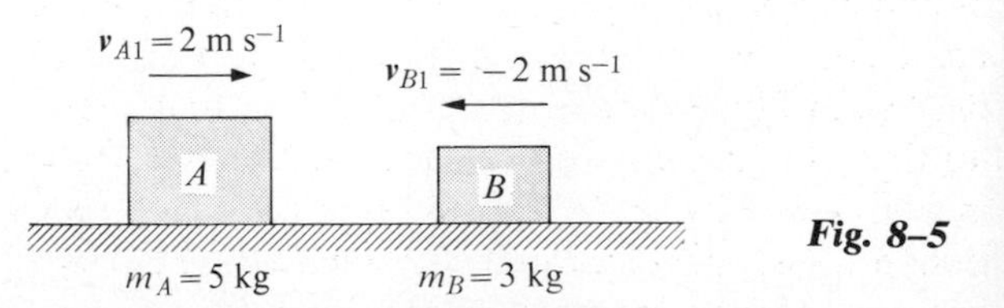

Fig. 8–5

Example 2 In Fig. 8–6, body A of mass m_A is initially moving toward the right with a velocity $\mathbf{v}_{A1}$. It collides with body B, initially at rest, after which the bodies separate and move with velocities $\mathbf{v}_{A2}$ and

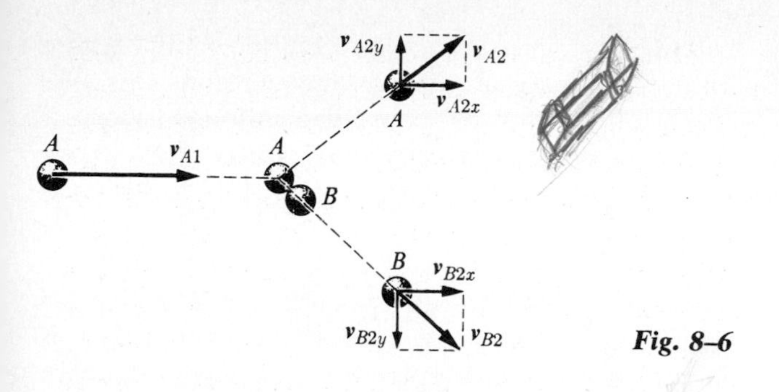

Fig. 8–6

$\boldsymbol{v}_{B2}$. No forces act on the system except those in the collision process.

This example illustrates the *vector* nature of momentum, that is the x- and y-components of momentum are *both* conserved. Let us take the x-axis in the direction of $\boldsymbol{v}_{A1}$. The initial x-momentum is $m_A\boldsymbol{v}_{A1}$ and the initial y-momentum is zero.

The final x-momentum of the system is

$$m_A v_{A2x} + m_B v_{B2x},$$

and the final y-momentum is

$$m_A v_{A2y} - m_B v_{B2y}.$$

Therefore

$$m_A v_{A2x} + m_B v_{B2x} = m_A v_{A1}, \tag{8–7}$$

$$m_A v_{A2y} - m_B v_{B2y} = 0. \tag{8–8}$$

8–3 COLLISIONS

Suppose we are given the masses and initial velocities of two colliding bodies, and we wish to compute their velocities after the collision. If the collision is "head on," as in Fig. 8–5, Eq. (8–6) provides one equation for the velocities $\boldsymbol{v}_{A2}$ and $\boldsymbol{v}_{B2}$. If it is of the type shown in Fig. 8–6, Eqs. (8–7) and (8–8) provide two equations for the four velocity components v_{A2x}, v_{A2y}, v_{B2x}, and v_{B2y}. Hence, momentum considerations *alone* do not suffice to completely determine the final velocities; we must have more information about the collision process.

If the forces of interaction between the bodies are *conservative*, the total kinetic energy is the same before and after the collision and the collision is said to be *completely elastic*. Such a collision is closely approximated if one end of an inverted U-shaped steel spring is attached to one of the bodies, as in Fig. 8–7. When the bodies collide the spring is momentarily compressed and some of the original kinetic energy is momentarily converted to elastic potential energy. The spring then expands, and when the bodies separate, this potential energy is reconverted to kinetic energy.

At the opposite extreme from a completely elastic collision is one in which the colliding bodies stick together and move *as a unit* after the collision. Such a collision is called a *completely inelastic* collision, and will result if the bodies in Fig. 8–5 are provided with a coupling mechanism like that between two freight cars, or if the bodies in Fig. 8–7 are locked together at the instant when their velocities become equal and the spring is compressed.

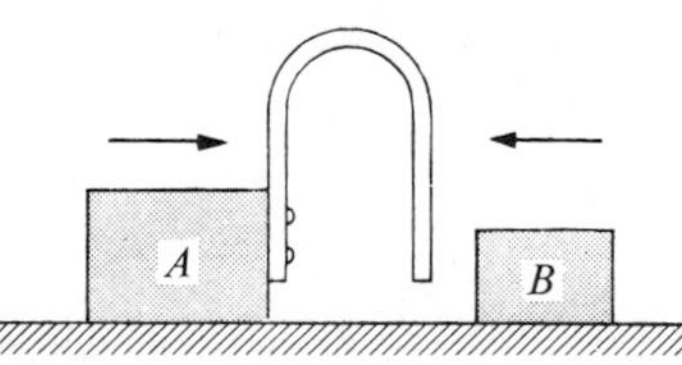

Fig. 8–7

8–4 INELASTIC COLLISIONS

For the special case of a completely inelastic collision between two bodies A and B, we have, from the definition of such a collision,

$$\boldsymbol{v}_{A2} = \boldsymbol{v}_{B2} = \boldsymbol{v}_2.$$

When this is combined with the principle of conservation of momentum, we obtain

$$m_A\boldsymbol{v}_{A1} + m_B\boldsymbol{v}_{B1} = (m_A + m_B)\boldsymbol{v}_2, \tag{8–9}$$

and the final velocity can be computed if the initial velocities and the masses are known.

The kinetic energy of the system before collision is

$$E_{k1} = \tfrac{1}{2}m_A v_{A1}^2 + \tfrac{1}{2}m_B v_{B1}^2.$$

The final kinetic energy is

$$E_{k2} = \tfrac{1}{2}(m_A + m_B)v_2^2.$$

For the special case in which body B is *initially* at rest, $v_{B1} = 0$, and the ratio of the final to the initial kinetic energy is

$$\frac{E_{k2}}{E_{k1}} = \frac{(m_A + m_B)v_2^2}{m_A v_{A1}^2}.$$

Inserting the expression for v_2 from Eq. (8–9), we find that this reduces to

$$\frac{E_{k2}}{E_{k1}} = \frac{m_A}{m_A + m_B}.$$

The right side is necessarily less than unity, so *the total kinetic energy always decreases in an inelastic collision.*

Example 1 Suppose the collision in Fig. 8–5 is completely inelastic and that the masses and velocities have the values shown. The velocity after the collision is then

$$v_2 = \frac{m_A v_{A1} + m_B v_{B1}}{m_A + m_B} = 0.5 \text{ m s}^{-1}.$$

Since v_2 is positive, the system moves to the right after the collision.

The kinetic energy of body A before the collision is

$$\tfrac{1}{2} m_A v_{A1}^2 = 10 \text{ J},$$

and that of body B is

$$\tfrac{1}{2} m_B v_{B1}^2 = 6 \text{ J}.$$

The total kinetic energy before collision is therefore 16 J.

Note that the kinetic energy of body B is positive, although its velocity v_{B1} and its momentum mv_{B1} are both negative. The kinetic energy after the collision is $\frac{1}{2}(m_A + m_B)v_2^2 = 1$ J.

Hence, far from remaining constant, the final kinetic energy is only $\frac{1}{16}$ of the original, and $\frac{15}{16}$ is "lost" in the collision. If the bodies couple together like two freight cars, most of this energy is converted to elastic waves which are eventually absorbed.

If there is a spring between the bodies, as in Fig. 8–7, and the bodies are locked together when their velocities become equal, the energy is trapped as potential energy in the compressed spring. If all these forms of energy are taken into account, the *total* energy of the system is conserved although its *kinetic* energy is not. However, *momentum is always conserved* in a collision, whether or not the collision is elastic.

Example 2 The *ballistic pendulum* is a device for measuring the velocity of a bullet. The bullet is allowed to make a completely inelastic collision with a body of much greater mass. The momentum of the system immediately after the collision equals the original momentum of the bullet, but since the velocity is very much smaller, it can be determined more easily. Although the ballistic pendulum has now been superseded by other devices, it is still an important laboratory experiment for illustrating the concepts of momentum and energy.

In Fig. 8–8, the pendulum, consisting perhaps of a large wooden block of mass m', hangs vertically by two cords. A bullet of mass m, traveling with a velocity v, strikes the pendulum and remains embedded in it. If the collision time is very small compared with the time of swing of the pendulum, the supporting cords remain practically vertical during this time. Hence, no external horizontal forces act on the system during the collision, and the horizontal momentum is conserved. Then if V represents the velocity of bullet and block immediately after the collision,

$$mv = (m + m')V, \qquad v = \frac{m + m'}{m} V.$$

The kinetic energy of the system, immediately after the collision, is $E_k = \frac{1}{2}(m + m')V^2$.

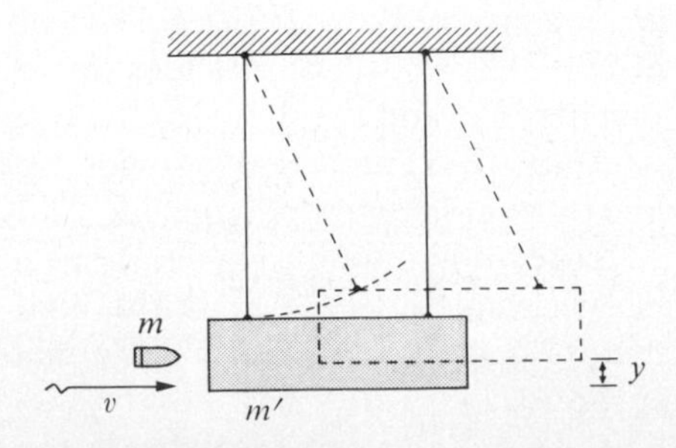

Fig. 8–8 *The ballistic pendulum.*

The pendulum now swings to the right and upward until its kinetic energy is converted to gravitational potential energy. (Small frictional effects can be neglected.) Hence,

$$\tfrac{1}{2}(m + m')V^2 = (m + m')gy,$$

$$V = \sqrt{2gy},$$

and

$$v = \frac{m + m'}{m}(\sqrt{2gy}).$$

By measuring m, m', and y, we can compute the original velocity v of the bullet.

It is important to remember that kinetic energy is not conserved *in the collision.* The ratio of the kinetic energy of bullet and pendulum, after the collision, to the original kinetic energy of the bullet, is

$$\frac{\tfrac{1}{2}(m + m')V^2}{\tfrac{1}{2}mv^2} = \frac{m}{m + m'}.$$

Thus, if $m' = 1000$ g and $m = 1$ g, only about 0.1 percent of the original energy remains as kinetic energy; 99.9 percent is converted to internal energy.

8–5 ELASTIC COLLISIONS

Consider next a perfectly elastic "head-on" or *central* collision between two bodies A and B. The bodies separate after the collision and have different velocities, $\mathbf{v}_{A2}$ and $\mathbf{v}_{B2}$. Since kinetic energy and momentum are *both* conserved, we have:

Conservation of kinetic energy

$$\tfrac{1}{2}m_A v_{A1}{}^2 + \tfrac{1}{2}m_B v_{B1}{}^2 = \tfrac{1}{2}m_A v_{A2}{}^2 + \tfrac{1}{2}m_B v_{B2}{}^2.$$

Conservation of momentum

$$m_A v_{A1} + m_B v_{B1} = m_A v_{A2} + m_B v_{B2}.$$

Hence if the masses and initial velocities are known, we have two independent equations, from which the final velocities can be found. Simultaneous solution of these equations gives

$$(v_{B2} - v_{A2}) = -(v_{B1} - v_{A1}), \qquad (8\text{–}10)$$

$$v_{A2} = \frac{2m_B v_{B1} + v_{A1}(m_A - m_B)}{m_A + m_B}, \qquad (8\text{–}11)$$

$$v_{B2} = \frac{2m_A v_{A1} - v_{B1}(m_A - m_B)}{m_A + m_B}. \qquad (8\text{–}12)$$

The difference $(v_{B2} - v_{A2})$ is the velocity of B relative to A after the collision, while $(v_{B1} - v_{A1})$ is its relative velocity before the collision. Thus, from Eq. (8–10), *the relative velocity of two particles in a central, completely elastic collision is unchanged in magnitude but reversed in direction.*

For the special case in which body B is at rest before the collision, $v_{B1} = 0$ and Eqs. (8–11) and (8–12) simplify to

$$v_{A2} = \frac{m_A - m_B}{m_A + m_B}v_{A1}, \qquad v_{B2} = \frac{2m_A}{m_A + m_B}v_{A1}.$$

If the masses of A and B are equal, $v_{A2} = 0$ and $v_{B2} = v_{A1}$. That is, the first body comes to rest and the second moves off with a velocity equal to the original velocity of the first. Both the momentum and kinetic energy of the first are completely transferred to the second.

When the masses are unequal, the kinetic energies after the collision are

$$(E_{k2})_A = \tfrac{1}{2}m_A v_{A2}{}^2 = \left(\frac{m_A - m_B}{m_A + m_B}\right)^2 (E_{k1})_A,$$

$$(E_{k2})_B = \tfrac{1}{2}m_B v_{B2}{}^2 = \left(\frac{4m_A m_B}{(m_A + m_B)^2}\right)(E_{k1})_A.$$

It is of interest to derive the expression for the *fractional decrease* in kinetic energy of body a, that is, the ratio of its decrease in kinetic energy to its original kinetic energy. Since, in an elastic collision, the energy lost by A equals the energy gained by B, this ratio is

$$\frac{(E_{k2})_B}{(E_{k1})_A} = \frac{4m_A m_B}{(m_A + m_B)^2} = 4\left(\frac{m_A}{m_B}\right)\frac{1}{[1 + (m_A/m_B)]^2}.$$

The fractional decrease in kinetic energy of body A is plotted in Fig. 8–9 as a function of the ratio m_A/m_B (note the logarithmic scale). The energy loss approaches zero as m_A/m_B approaches zero (a very small body colliding with a very large one) and also as m_A/m_B approaches infinity (a very large body colliding with a very small one). In the former case, the first body rebounds with practically its original velocity. In the latter, the first body continues to

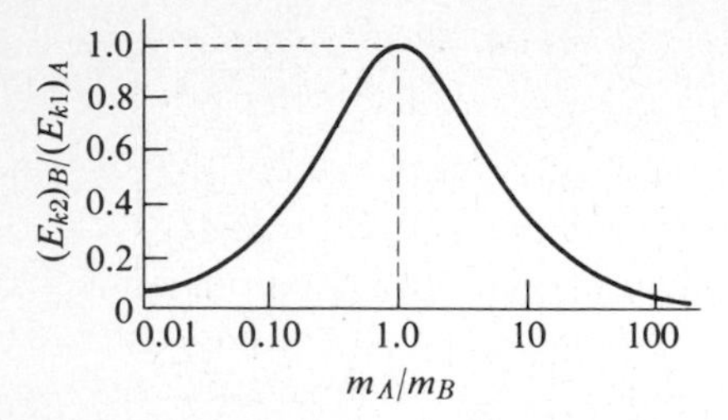

Fig. 8–9 Fractional loss in energy in a head-on elastic collision, plotted as a function of the ratio of the masses of the colliding bodies.

move with very nearly its original velocity. The maximum fractional energy decrease occurs when $m_A = m_B$, and for this ratio of masses the fractional energy loss equals unity, as shown above.

This fact has an important bearing on the problem of slowing down the rapidly moving neutrons in the moderator of a nuclear reactor. As neutrons pass through matter, they make occasional elastic collisions with nuclei. In a head-on collision, the greatest loss of kinetic energy results when the mass of the nucleus equals that of the neutron, that is, when the nucleus is that of ordinary hydrogen. The greater the nuclear mass, the less of the neutron's energy is transferred to the nucleus.

Example Suppose the collision illustrated in Fig. 8–5 is completely elastic. What are the velocities of A and B after the collision?

From the principle of conservation of momentum,

$$(5\text{ kg})(2\text{ m s}^{-1}) + (3\text{ kg})(-2\text{ m s}^{-1}) = (5\text{ kg})v_{A2} + (3\text{ kg})v_{B2},$$

$$5v_{A2} + 3v_{B2} = 4\text{ m s}^{-1}.$$

Since the collision is completely elastic,

$$v_{B2} - v_{A2} = -(v_{B1} - v_{A1}) = 4\text{ m s}^{-1}.$$

Solving these equations simultaneously, we obtain

$$v_{A2} = -1\text{ m s}^{-1}, \qquad v_{B2} = 3\text{ m s}^{-1}.$$

Both bodies therefore reverse their directions of motion, A traveling to the left at 1 m s^{-1} and B to the right at 3 m s^{-1}.

The total kinetic energy after the collision is

$$\tfrac{1}{2}(5\text{ kg})(-1\text{ m s}^{-1})^2 + \tfrac{1}{2}(3\text{ kg})(3\text{ m s}^{-1})^2 = 16\text{ J},$$

which equals the total kinetic energy before the collision.

8–6 RECOIL

Figure 8–10 shows two blocks A and B, between which there is a compressed spring. When the system is released from rest, the spring exerts equal and opposite forces on the blocks until it has expanded to its natural unstressed length. It then drops to the surface, while the blocks continue to move. The original momentum of the system is zero, and if frictional forces can be neglected, the resultant external force on the system is zero. The momentum of the system therefore remains constant and equal to zero. Then if v_A and v_B are the velocities acquired by A and B, we have

$$m_A v_A + m_B v_B = 0, \qquad \frac{v_A}{v_B} = -\frac{m_B}{m_A}.$$

The velocities are of opposite sign and their magnitudes are inversely proportional to the corresponding masses.

The original kinetic energy of the system is also zero. The final kinetic energy is

$$E_k = \tfrac{1}{2}m_A v_A{}^2 + \tfrac{1}{2}m_B v_B{}^2.$$

The source of this energy is the original elastic potential energy of the system. The ratio of kinetic energies is

$$\frac{\tfrac{1}{2}m_A v_A{}^2}{\tfrac{1}{2}m_B v_B{}^2} = \frac{m_A}{m_B}\left(\frac{v_A}{v_B}\right)^2 = \frac{m_B}{m_A}.$$

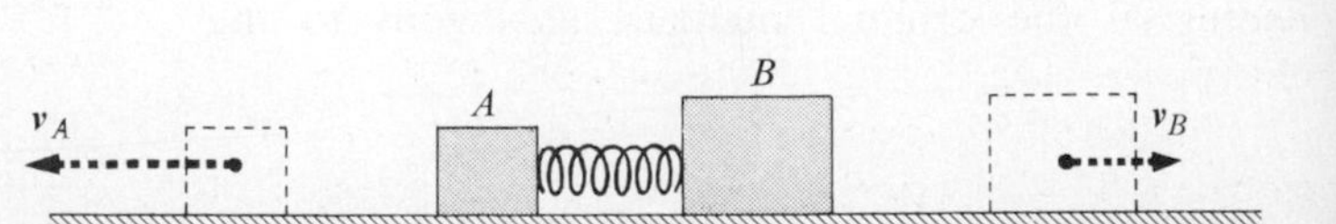

Fig. 8–10 Conservation of momentum in recoil.

Thus, although the momenta are equal in magnitude, the kinetic energies are inversely proportional to the corresponding masses, the body of smaller mass receiving the larger share of the original potential energy. The reason is that the change in *momentum* of a body equals the *impulse* of the force acting on it, while the change in *kinetic energy* equals the *work* of the force. The forces on the two bodies are equal in magnitude and act for equal times, so they produce equal and opposite changes in momentum. The points of application of the forces, however, do not move through equal distances (except when $m_A = m_B$), since the acceleration, velocity, and displacement of the smaller body are greater than those of the larger. Hence more *work* is done on the body of smaller mass.

Considerations like those above also apply to the firing of a rifle. The initial momentum of the system is zero. When the rifle is fired, the bullet and the powder gases acquire a forward momentum, and the rifle (together with any system to which it is attached) acquires a rearward momentum of the same magnitude. The ratio of velocities cannot be expressed as simply as in the case discussed above. The bullet travels with a definite velocity, but different portions of the powder gases have different velocities. Because of the relatively large mass of the rifle compared with that of the bullet and powder charge, the velocity and kinetic energy of the rifle are much smaller than those of the bullet and powder gases.

In the processes of radioactive decay and nuclear fission, a nucleus splits into two or more parts, which fly off in different directions. Although the nature of the forces which act in these processes is not yet completely understood, it has been verified by many experiments that the principle of conservation of momentum applies to them also. The total (vector) momentum of the products equals the original momentum of the system. The source of the kinetic energy of the products is the so-called *binding energy* of the original nucleus, analogous to the chemical energy of the powder charge in a rifle.

Figure 8–11 shows the explosive break-up (fission) of one uranium atom into two fragments of approximately equal mass. The atom was at rest in the thin foil that stretches horizontally in the picture, then it was "triggered off" by a passing neutron, and the two fission products recoiled from each other in opposite directions. While they cannot, of course, be directly observed, the two fragments give themselves away indirectly by the thin streaks of fog or clouds which can be made to condense along their paths.* It is evident that the fragments are slowed down to a stop after a few centimeters in the medium through which they have to travel, but, from such data as the length and density of the tracks, we can deduce the speeds of separation. From the law of conservation of momentum, applied to the fission fragments, we can then obtain directly the ratio of their masses, an important aid in the identification of the nuclear reaction.

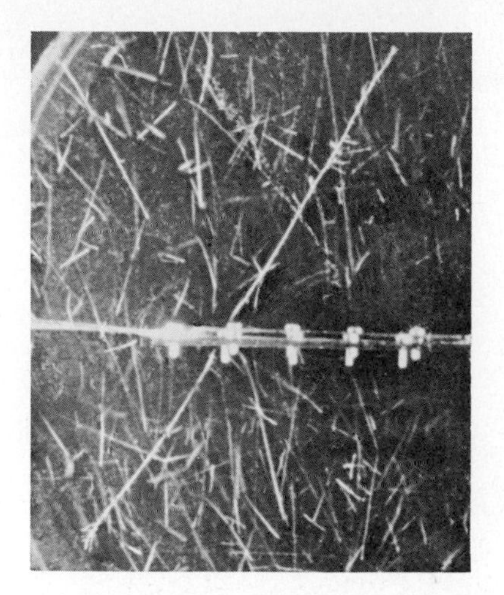

Fig. 8–11 *Cloud-chamber photograph of the fission of an atom of uranium. [From Boggild, Brostøm, and Lauritzen,* Phys. Rev. ***59,*** *275, 1941.]*

In this simplified account we are neglecting the contribution to momentum of minor fission products not apparent in this picture, and of the incident neutron. In the "background" can be seen the result of collisions between other incident neutrons and the atmosphere in the cloud chamber.

---

* The equipment for this purpose, a most important tool in contemporary research, is called a *cloud chamber.* It originated in a design devised in 1895 by the British Nobel Prize physicist C. T. R. Wilson.

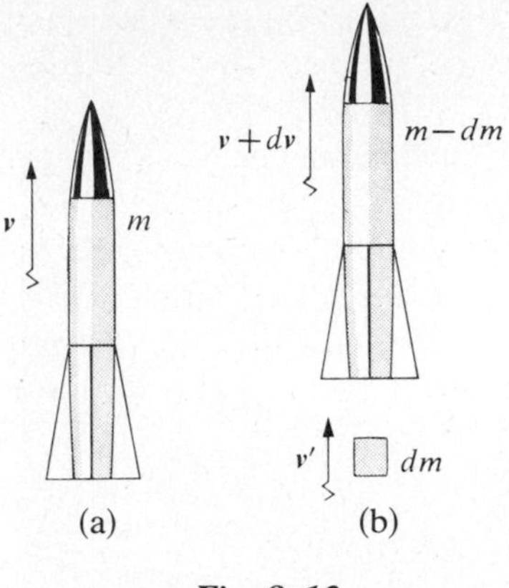

Fig. 8–12

8–7 ROCKET PROPULSION

A rocket is propelled by the ejection of a portion of its mass to the rear. The forward force on the rocket is the reaction to the backward force on the ejected material, and as more material is ejected, the mass of the rocket decreases. The problem is best handled by impulse–momentum considerations. In order not to bring in too many complicating forces, we shall consider a rocket fired vertically upward, and neglect air resistance and variations in g.

Figure 8–12(a) represents the rocket at a time t after take-off, when its mass is m and its upward velocity is v. The total momentum at this instant is thus mv. In a short time interval dt, a mass dm of gas is ejected from the rocket. Let v_r represent the downward velocity of this gas *relative to the rocket.* The velocity v' of the gas relative to the earth is then

$$v' = v - v_r,$$

and its momentum is

$$dm\, v' = dm(v - v_r).$$

At the end of the time interval dt, the mass of rocket and unburned fuel has decreased to $m - dm$, and its velocity has increased to $v + dv$. Its momentum is therefore

$$(m - dm)(v + dv).$$

Thus the *total* momentum at time $t + dt$ is

$$(m - dm)(v + dv) + dm(v - v_r).$$

Figure 8–12(b) represents rocket and ejected gas at this time.

We now make use of the impulse–momentum relation; the product of the resultant external force F on a system, and the time interval dt during which it acts, is equal to the change in momentum of the system. If air resistance is neglected, the external force F on the rocket is its weight, $-mg$. (We take the upward direction as positive.) The change in momentum, in time dt, is the difference between the momentum of the system at the end and at the beginning of the time interval. Hence

$$-mg\, dt = [(m - dm)(v + dv) + dm(v - v_r)] - mv$$
$$= mdv - dm\, v_r - dm\, dv.$$

The term $dm\, dv$ may be dropped because it is a product of two small quantities and thus is much smaller than the other terms. Dropping this term, dividing by dt, and rearranging, we obtain

$$m\left(\frac{dv}{dt}\right) = v_r\left(\frac{dm}{dt}\right) - mg. \qquad (8\text{–}13)$$

The ratio dv/dt is the acceleration of the rocket, so the left side of this equation (mass times acceleration) equals the resultant force on the rocket. The first term on the right equals the upward thrust on the rocket, and the resultant force equals the difference between this thrust and the weight of the rocket, mg. It will be seen that the upward thrust is proportional both to the relative velocity v_r of the ejected gas and to the mass of gas ejected per unit time, dm/dt.

The acceleration is

$$\frac{dv}{dt} = \frac{v_r}{m}\left(\frac{dm}{dt}\right) - g. \qquad (8\text{–}14)$$

As the rocket rises, the value of g decreases, according to Newton's law of gravitation. (In "outer space," far from all other bodies, g becomes negligibly small.) If the values of v_r and dm/dt remain approximately constant while the fuel is being consumed and m continuously decreases, the acceleration *increases* until all the fuel is burned.

Equation (8–13) can be integrated to find a relation between the velocity at any time and the remaining mass. From Eq. (8–13),

$$dv = v_r \frac{dm}{m} - g\,dt.$$

Now dm is a positive quantity, representing the mass ejected in time dt, so the change in mass of the rocket in that time is $-dm$. Thus, in computing the total mass change *in the rocket* we must change the sign of the term containing dm.

Let the mass and velocity at time $t = 0$ be m_0 and v_0, respectively; then

$$\int_{v_0}^{v} dv = -\int_{m_0}^{m} v_r \frac{dm}{m} - \int_0^t g\,dt,$$

and

$$v - v_0 = -v_r \ln \frac{m}{m_0} - gt,$$

$$v = v_0 + v_r \ln \frac{m_0}{m} - gt. \qquad (8\text{–}15)$$

Example 1 In the first second of its flight, a rocket ejects 1/60 of its mass with a relative velocity of $2400\ \mathrm{m\,s^{-1}}$. What is the acceleration of the rocket?

We have: $dm = m/60$, $dt = 1$ s. From Eq. (8–14),

$$\frac{dv}{dt} = \frac{2400\ \mathrm{m\,s^{-1}}}{(60)(1\ \mathrm{s})} - 10\ \mathrm{m\,s^{-2}}$$

$$= 30\ \mathrm{m\,s^{-2}}.$$

Example 2 Suppose the ratio of initial mass to final mass for the rocket above is 4 and that the fuel is consumed in a time $t = 60$ s. The velocity at the end of this time, from Eq. (8–15), is

$$v = (2400\ \mathrm{m\,s^{-1}})(\ln 4) - (10\ \mathrm{m\,s^{-2}})(60\ \mathrm{s})$$

$$= 2727\ \mathrm{m\,s^{-1}}.$$

At the start of the flight, when the velocity of the rocket is zero, the ejected gases are moving downward, relative to the earth, with a velocity equal to the relative velocity v_r. When the velocity of the rocket has increased to v_r, the ejected gases have a velocity zero relative to the earth. When the rocket velocity becomes greater than v_r, the velocity of the ejected gases is in the same direction as that of the rocket. Thus the velocity acquired by the rocket can be greater (and is often much greater) than the relative velocity v_r. In the example above, where the final velocity of the rocket was $2727\ \mathrm{m\,s^{-1}}$ and the relative velocity was $2400\ \mathrm{m\,s^{-1}}$, the last portion of the ejected fuel had an upward velocity of $(2727 - 2400)\mathrm{m\,s^{-1}} = 327\ \mathrm{m\,s^{-1}}$.

8–8 GENERALIZATIONS

The momentum–impulse relations and the principle of conservation of momentum have been developed from Newton's laws of motion, which are strictly valid only for bodies moving with speeds small compared to the speed of light. Thus it should not be surprising that when v is *not* small compared to c, modifications are needed.

The correct relativistic generalization, as predicted theoretically by Lorentz and Einstein and also verified experimentally, is to define momentum not as $m\mathbf{v}$, but as

$$\text{Momentum} = \mathbf{p} = \frac{m\mathbf{v}}{\sqrt{1 - v^2/c^2}}. \qquad (8\text{–}16)$$

When v is much smaller than c, this reduces to $m\mathbf{v}$, but in general the momentum of a particle is larger than $m\mathbf{v}$. Equation (8–16) can be interpreted as saying that the mass of the particle increases by the factor $1/(1 - v^2/c^2)^{1/2}$ when it is moving, but an equally valid and, in some ways, less misleading view is simply that the definition of momentum is no longer $m\mathbf{v}$, but instead is given by Eq. (8–16). In either case, it is found that in collisions where there is no resultant external force, *the principle of conservation of momentum is still obeyed,* provided the generalized definition of momentum is used.

A further generalization of the concept of momentum is needed when we deal with particles that have no mass in the usual sense. One such particle is the *photon,* or *quantum* of electromagnetic wave energy, which will be discussed more fully in later chapters. An electromagnetic wave, such as light or a radio wave, transfers energy in packages of size proportional to the frequency of the radiation. For a wave of frequency f, the energy E in each package or *photon* is given by

$$E = hf, \tag{8–17}$$

where h is a fundamental physical constant called *Planck's constant.* In mks units, $h = 6.62 \times 10^{-34}$ J s.

The theory of relativity predicts that any particle with energy must also have momentum, whether it has mass or not. For a massless particle the momentum turns out to be the energy divided by the speed of light. Thus, from Eq. (8–17) we obtain, for the momentum of a massless particle,

$$\boxed{p = \frac{hf}{c} = \frac{h}{\lambda},} \tag{8–18}$$

where $\lambda = c/f$ is the wavelength of the radiation.

Example A sodium atom initially at rest emits a photon of yellow light, recoiling in the opposite direction to the direction of emission. What is the atom's recoil velocity?

The wavelength of the light is about $\lambda = 589 \times 10^{-9}$ m, and the momentum of a photon is

$$p = \frac{h}{\lambda} = \frac{6.62 \times 10^{-34}\ \text{J s}}{589 \times 10^{-9}\ \text{m}} = 1.12 \times 10^{-27}\ \text{kg m s}^{-1}.$$

For conservation of momentum, the atom must recoil with a momentum of this magnitude in the opposite direction. The mass of a sodium atom is about 3.8×10^{-26} kg, so its speed after emission is

$$v = \frac{1.12 \times 10^{-27}\ \text{kg m s}^{-1}}{3.8 \times 10^{-26}\ \text{kg}} = 0.029\ \text{m s}^{-1}.$$

In conclusion, we emphasize again that the principle of conservation of momentum, like that of conservation of energy, has much greater generality than the newtonian context in which it was originally developed. As indicated briefly above, the definition of momentum can be generalized to relativistic mechanics and to quantum mechanics. With these generalizations, conservation of momentum is believed to be a *universal* conservation law; *no exception has ever been found.*

PROBLEMS

8–1

a) What is the momentum of a 10,000-kg truck whose velocity is 20 m s^{-1}? What velocity must a 5,000-kg truck attain in order to have (b) the same momentum, (c) the same kinetic energy?

8–2 A baseball has a mass of about 0.2 kg.

a) If the velocity of a pitched ball is 30 m s^{-1}, and after being batted it is 50 m s^{-1} in the opposite direction, find the change in momentum of the ball and the impulse of the blow.

b) If the ball remains in contact with the bat for 0.002 s, find the average force of the blow.

8–3 A bullet having a mass of 0.05 kg, moving with a velocity of 400 m s^{-1}, penetrates a distance of 0.1 m into a wooden block firmly attached to the earth. Assume the decelerating force constant. Compute (a) the deceleration of the bullet, (b) the decelerating force, (c) the time of deceleration, (d) the impulse of the collision. Compare the answer to part (d) with the initial momentum of the bullet.

8–4 A bullet emerges from the muzzle of a gun with a velocity of 300 m s^{-1}. The resultant force on the bullet, while it is in the gun barrel, is given by

$$F = 400 - \frac{4 \times 10^5}{3} t,$$

where F is in newtons and t in seconds.

a) Construct a graph of F versus t.

b) Compute the time required for the bullet to travel the length of the barrel, assuming the force becomes zero just at the end of the barrel.

c) Find the impulse of the force.

d) Find the mass of the bullet.

8–5 A box, initially sliding on the floor of a room, is eventually brought to rest by friction. Is the momentum of the box conserved? If not, does this process contradict the principle of conservation of momentum? What becomes of the original momentum of the box?

8–6 Compare the damage to an automobile (and its occupants) in the following circumstances:

a) The automobile makes a completely inelastic head-on collision with an identical automobile traveling with the same speed in the opposite direction, and (b) it makes a completely inelastic head-on collision with a vertical rock cliff.

c) Which would be worse (for the occupants of a light car), to collide head-on with a truck traveling in the opposite direction with a momentum of equal magnitude, or to collide head-on with a truck having the same kinetic energy?

8–7

a) An empty freight car of mass 10,000 kg rolls at 2 m s^{-1} along a level track and collides with a loaded car of mass 20,000 kg, standing at rest with brakes released. If the cars couple together, find their speed after the collision.

b) Find the decrease in kinetic energy as a result of the collision.

c) With what speed should the loaded car be rolling toward the empty car, in order that both shall be brought to rest by the collision?

8–8 When a bullet of mass 20 g strikes a ballistic pendulum of mass 10 kg, the center of gravity of the pendulum is observed to rise a vertical distance of 7 cm. The bullet remains embedded in the pendulum.

a) Calculate the original velocity of the bullet.

b) What fraction of the original kinetic energy of the bullet remains as kinetic energy of the system immediately after the collision?

c) What fraction of the original momentum remains as momentum of the system?

8–9 A bullet weighing 0.01 lb is shot through a 2-lb wood block suspended on a string 5 ft long. The center of gravity of the block is observed to rise a distance of 0.0192 ft. Find the speed of the bullet as it emerges from the block if the initial speed is 1000 ft s^{-1}.

8–10 When a bullet of mass 10 g strikes a ballistic pendulum of mass 2 kg, the center of gravity of the pendulum is observed to rise a vertical distance of 10 cm. The bullet remains embedded in the pendulum. Calculate the velocity of the bullet.

8–11 A frame of mass 200 g, when suspended from a certain coil spring, is found to stretch the spring 10 cm. A lump of putty of mass 200 g is dropped from rest onto the frame from a height of 30 cm (Fig. 8–13). Find the maximum distance the frame moves downward.

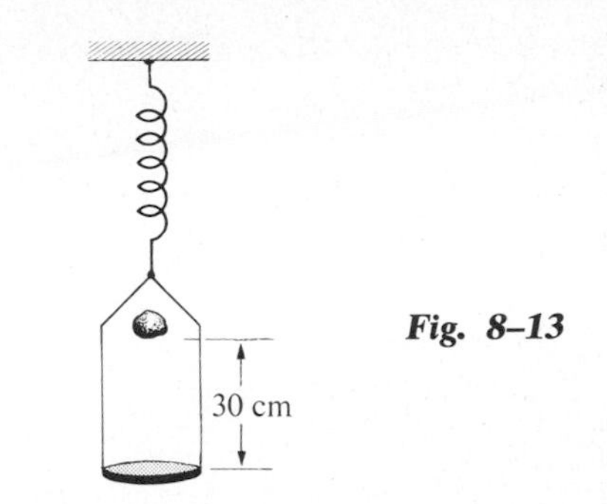

Fig. 8–13

8–12 A bullet of mass 2 g, traveling in a horizontal direction with a velocity of 500 m s^{-1}, is fired into a wooden block of mass 1 kg, initially at rest on a level surface. The bullet passes through the block and emerges with its velocity reduced to 100 m s^{-1}. The block slides a distance of 20 cm along the surface from its initial position.

a) What was the coefficient of sliding friction between block and surface?

b) What was the decrease in kinetic energy of the bullet?

c) What was the kinetic energy of the block at the instant after the bullet passed through it?

8–13 A rifle bullet weighing 0.02 lb is fired with a velocity of 2500 ft s^{-1} into a ballistic pendulum weighing 10 lb, suspended from a cord 3 ft long. Compute (a) the vertical height through which the pendulum rises, (b) the initial kinetic energy of the bullet, (c) the kinetic energy of bullet and pendulum after the bullet is embedded in the pendulum.

8–14 A 5-g bullet is fired horizontally into a 3-kg wooden block resting on a horizontal surface. The coefficient of sliding friction between block and surface is 0.20. The bullet remains embedded in the block, which is observed to slide 25 cm along the surface. What was the velocity of the bullet?

8–15 A bullet of mass 2 g, traveling at 500 m s^{-1}, is fired into a ballistic pendulum of mass 1 kg suspended from a cord 1 m long. The bullet penetrates the pendulum and emerges with a velocity of 100 m s^{-1}. Through what vertical height will the pendulum rise?

8–16 A rifle bullet of mass 0.01 kg strikes and embeds itself in a block of mass 0.99 kg which rests on a horizontal frictionless surface and is attached to a coil spring, as shown in Fig. 8–14. The impact compresses the spring 10 cm. Calibration of the spring shows that a force of 1.0 N is required to compress the spring 1 cm.

a) Find the maximum potential energy of the spring.

b) Find the velocity of the block just after the impact.

c) What was the initial velocity of the bullet?

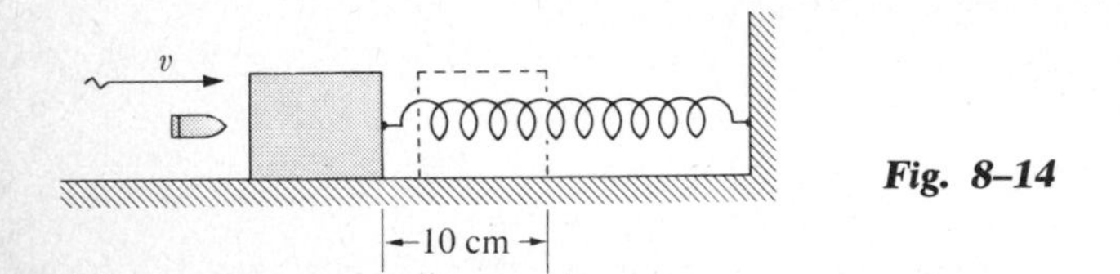

Fig. 8–14

8–17 A 4000-lb automobile going eastward on Chestnut Street at 40 mi hr^{-1} collides with a 4-ton truck which is going southward across Chestnut Street at 15 mi hr^{-1}. If they become coupled on collision, what is the magnitude and direction of their velocity immediately after colliding?

8–18 On a frictionless table, a 3-kg block moving 4 m s^{-1} to the right collides with an 8-kg block moving 1.5 m s^{-1} to the left.

a) If the two blocks stick together, what is the final velocity?

b) If the two blocks make a completely elastic head-on collision, what are their final velocities?

c) How much mechanical energy is dissipated in the collision of part (a)?

8–19 Two blocks of mass 300 g and 200 g are moving toward each other along a horizontal frictionless surface with velocities of 50 cm s^{-1} and 100 cm s^{-1}, respectively.

a) If the blocks collide and stick together, find their final velocity.

b) Find the loss of kinetic energy during the collision.

c) Find the final velocity of each block if the collision is completely elastic.

8–20

a) Prove that when a moving body makes a perfectly inelastic collision with a second body of equal mass, initially at rest, one-half of the original kinetic energy is "lost."

b) Prove that when a very heavy particle makes a perfectly elastic collision with a very light particle that is at rest, the light one goes off with twice the velocity of the heavy one.

8–21 A 10-g block slides with a velocity of 20 cm s^{-1} on a smooth level surface and makes a head-on collision with a 30-g block moving in the opposite direction with a velocity of 10 cm s^{-1}. If the collision is perfectly elastic, find the velocity of each block after the collision.

8–22 A block of mass 200 g, sliding with a velocity of 12 cm s^{-1} on a smooth, level surface, makes a perfectly elastic head-on collision with a block of mass m, initially at rest. After the collision, the velocity of the 200-g block is 4 cm s^{-1} in the same direction as its initial velocity. Find (a) the mass m, and (b) its velocity after the collision.

8–23 A body of mass 600 g is initially at rest. It is struck by a second body of mass 400 g initially moving with a velocity of 125 cm s^{-1} toward the right along the x-axis. After the collision, the 400-g body has a velocity of 100 cm s^{-1} at an angle of 37° above the x-axis in the first quadrant. Both bodies move on a horizontal frictionless plane.

a) What is the magnitude and direction of the velocity of the 600-g body after the collision?

b) What is the loss of kinetic energy during the collision?

8–24 A small steel ball moving with speed v_0 in the positive x-direction makes a perfectly elastic, noncentral collision with an identical ball originally at rest. After impact, the first ball moves with speed v_1 in the first quadrant at an angle θ_1 with the x-axis and the second with speed v_2 in the fourth quadrant at an angle θ_2 with the x-axis.

a) Write the equations expressing conservation of linear momentum in the x-direction, and in the y-direction.

b) Square these equations and add them.

c) At this point, introduce the fact that the collision is perfectly elastic.

d) Prove that $\theta_1 + \theta_2 = \pi/2$.

8–25 A jet of liquid of cross-sectional area A and density ρ moves with speed v_J in the positive x-direction and impinges against a perfectly smooth blade B which deflects the stream at right angles but does not slow it down, as shown in Fig. 8–15.

a) If the blade is *stationary*, prove that the rate of arrival of mass at the blade is $\Delta m/\Delta t = \rho A v_J$.

b) If the impulse–momentum theorem is applied to a small mass Δm, prove that the x-component of the force acting on this mass for the time interval Δt is given by

$$F_x = -\frac{\Delta m}{\Delta t} v_J.$$

c) Prove that the *steady* force exerted *on* the blade in the x-direction is

$$F_x = \rho A v_J^2.$$

If the blade moves to the right with a speed v_B ($v_B < v_J$), derive the equations for:

d) the rate of arrival of mass at the moving blade,

e) the force F_x on the blade, and

f) the power delivered to the blade.

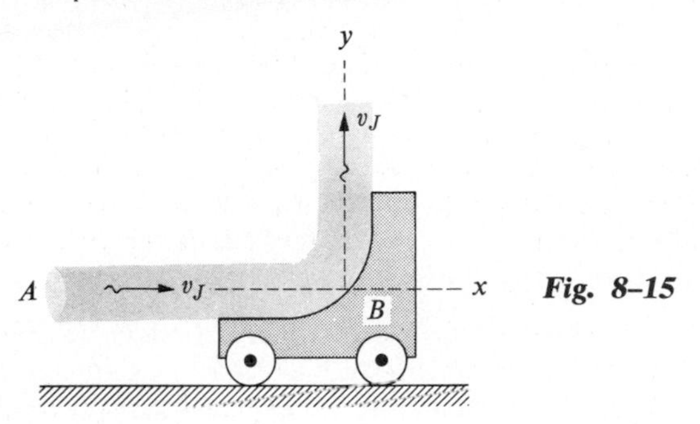

Fig. 8–15

8–26 A stone whose mass is 100 g rests on a horizontal frictionless surface. A bullet of mass 2.5 g, traveling horizontally at 400 m s^{-1}, strikes the stone and rebounds horizontally at right angles to its original direction with a speed of 300 m s^{-1}.

a) Compute the magnitude and direction of the velocity of the stone after it is struck.

b) Is the collision perfectly elastic?

8–27 A hockey puck *B* rests on a smooth ice surface and is struck by a second puck, *A*, which was originally traveling at 30 m s^{-1} and which is deflected 30° from its original direction (Fig. 8–16). Puck *B* acquires a velocity at 45° with the original velocity of *A*.

a) Compute the speed of each puck after the collision.

b) Is the collision perfectly elastic? If not, what fraction of the original kinetic energy of puck *A* is "lost"?

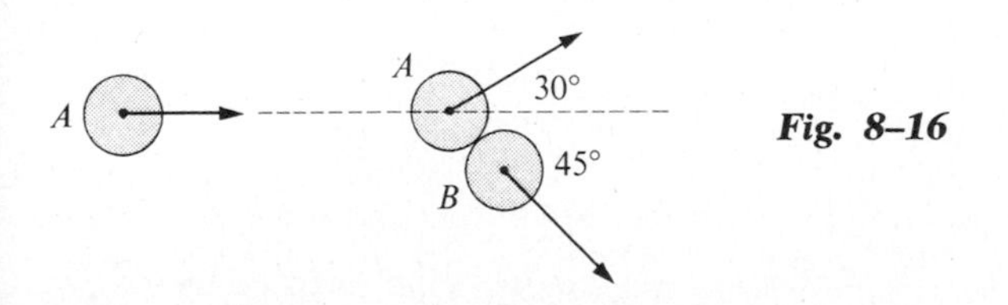

Fig. 8–16

8–28 Imagine that a ball of mass 200 g rolls back and forth between opposite sides of a billiard table 1 m wide, with a velocity that remains constant in magnitude but reverses in direction at each collision with the cushions. The magnitude of the velocity is 4 m s^{-1}.

a) What is the change in momentum of the ball at each collision?

b) How many collisions per unit time are made by the ball with one of the cushions?

c) What is the average time rate of change of momentum of the ball as a result of these collisions?

d) What is the average force exerted by the ball on a cushion?

e) Sketch a graph of the force exerted on the ball as a function of time, for a time interval of 5 s. [*Note*: This problem illustrates how one computes the average force exerted by the molecules of a gas on the walls of the containing vessel.]

8–29 A projectile is fired at an angle of departure of 60° and with a muzzle velocity of 400 m s^{-1}. At the highest point of its trajectory the projectile explodes into two fragments of equal mass, one of which falls vertically.

a) How far from the point of firing does the other fragment strike if the terrain is level?

b) How much energy was released during the explosion?

8–30 A railroad handcar is moving along straight frictionless tracks. In each of the following cases, the car initially has a total mass (car and contents) of 200 kg and is traveling with a velocity of 4 m s^{-1}. Find the *final velocity* of the car in each of the three cases.

a) A 20-kg mass is thrown sideways out of the car with a velocity of 2 m s^{-1} relative to the car.

b) A 20-kg mass is thrown backwards out of the car with a velocity of 4 m s^{-1} relative to the car.

c) A 20-kg mass is thrown into the car with a velocity of 6 m s^{-1} relative to the ground and opposite in direction to the velocity of the car.

8–31 A bullet weighing 0.02 lb is fired with a muzzle velocity of 2700 ft s^{-1} from a rifle weighing 7.5 lb.

a) Compute the recoil velocity of the rifle, assuming it free to recoil.

b) Find the ratio of the kinetic energy of the bullet to that of the rifle.

8–32 Block *A* in Fig. 8–17 has a mass of 1 kg, and block *B* has a mass of 2 kg. The blocks are forced together, compressing a spring *S* between them, and the system is released from rest on a level frictionless surface. The spring is not fastened to either block and drops to the surface after it has expanded. Block *B* acquires a speed of 0.5 m s^{-1}. How much potential energy was stored in the compressed spring?

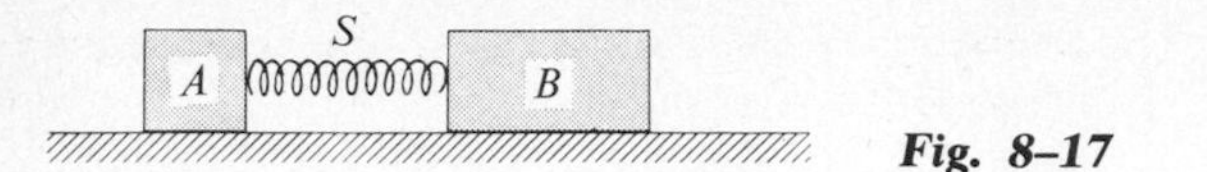

Fig. 8–17

8–33 A steel ball of mass 0.5 kg is dropped from a height of 4 m onto a horizontal steel slab. The collision is elastic, and the ball rebounds to its original height.

a) Calculate the impulse delivered to the ball during impact.

b) If the ball is in contact with the slab for 0.002 s, find the average force on the ball during impact.

8–34 A baseball of mass 0.25 kg is struck by a bat. Just before impact, the ball is traveling horizontally at 40 m s^{-1}, and it leaves the bat at an angle of 30° above horizontal with a speed of 60 m s^{-1}. If the ball and bat were in contact for 0.05 s, find the horizontal and vertical components of the average force on the ball.

8–35 In Problem 8–27, suppose the collision is perfectly elastic, and A is deflected 30° from its initial direction. Find the final speed of each puck and the direction of B's velocity.

8–36 A neutron of mass m collides elastically with a nucleus of mass M. Show that if the neutron's initial kinetic energy is E_0, the maximum kinetic energy it can *lose* during the collision is $4mME_0/(M + m)^2$.

8–37 An open-topped freight car of mass 10,000 kg is coasting without friction along a level track. It is raining very hard, with the rain falling vertically downward. The car is originally empty and moving with a velocity of 1 m s^{-1}. What is the velocity of the car after it has traveled long enough to collect 1000 kg of rain water?

8–38 A neutron of mass 1.67×10^{-27} kg, moving with a velocity of 2×10^4m s^{-1}, makes a head-on collision with a boron nucleus of mass 17.0×10^{-27} kg, originally at rest.

a) If the collision is completely inelastic, what is the final kinetic energy of the system, expressed as a fraction of the original kinetic energy?

b) If the collision is perfectly elastic, what fraction of its original kinetic energy does the neutron transfer to the boron nucleus?

8–39 A nucleus, originally at rest, decays radioactively by emitting an electron of momentum 9.22×10^{-21} kg m s^{-1}, and at right angles to the direction of the electron a neutrino with momentum 5.33×10^{-21} kg m s^{-1}.

a) In what direction does the residual nucleus recoil?

b) What is its momentum?

c) If the mass of the residual nucleus is 3.90×10^{-25} kg, what is its kinetic energy?

8–40 An 80-kg man standing on ice throws a 0.2-kg ball horizontally with a speed of 30 m s^{-1}.

a) With what speed and in what direction will the man begin to move?

b) If the man throws 4 such balls every 3 s, what is the average force acting on him? [*Hint*: Average force equals average rate of change of momentum.]

8–41 Find the average recoil force on a machine gun firing 120 shots per minute. The mass of each bullet is 10 g, and the muzzle velocity is 800 m s^{-1}.

8–42 A rifleman, who together with his rifle has a mass of 100 kg, stands on roller skates and fires 10 shots horizontally from an automatic rifle. Each bullet has a mass of 10 g and a muzzle velocity of 800 m s^{-1}.

a) If the rifleman moves back without friction, what is his velocity at the end of the 10 shots?

b) If the shots were fired in 10 s, what was the average force exerted on him?

c) Compare his kinetic energy with that of the 10 bullets.

8–43 A rocket burns 0.05 kg of fuel per second, ejecting it as a gas with a velocity of 5,000 m s^{-1}.

a) What force does this gas exert on the rocket? Give the result in dynes and newtons.

b) Would the rocket operate in free space?

c) If it would operate in free space, how would you steer it? Could you brake it?

8–44 This problem illustrates the advantage of using a multi-stage rather than a single-stage rocket. Suppose that the first stage of a two-stage rocket has a total mass of 12,000 kg, of which 9000 kg is fuel. The total mass of the second stage is 1000 kg, of which 750 kg is fuel. Assume that the relative velocity v_r of ejected material is constant, and neglect any effect of gravity. (The latter effect is small during the firing period if the rate of fuel consumption is large.)

a) Suppose that the entire fuel supply carried by the two-stage rocket were utilized in a single-stage rocket of the same total mass of 13,000 kg. What would be the velocity of the rocket, starting from rest, when its fuel was exhausted?

b) What is the velocity when the fuel of the first stage is exhausted, if the first stage carries the second stage with it to this point? This velocity then becomes the initial velocity of the second stage.

c) What is the final velocity of the second stage?

d) Sputnik I's velocity was about 8 km s^{-1}. What value of v_r would be required to give the second stage of the above rocket a velocity of this magnitude?

8–45 If a single-stage rocket, fired vertically from rest at the earth's surface, burns its fuel in a time of 30 s, and the relative velocity $v_r = 3000$ m s^{-1}, what must be the mass ratio m_0/m for a final velocity v of 8 km s^{-1} (about equal to the orbital velocity of an earth satellite)?

8–46

a) Show that the acceleration of a rocket fired vertically upward is given by

$$a = -\frac{v_r}{m}\frac{dm}{dt} - g.$$

b) Suppose that the rate of ejection of mass by the rocket is constant, that is, the rate of decrease of mass is $dm/dt = -km_0$, where k is a positive constant and m_0 is the initial mass. What is the numerical value of k, and in what units is it expressed, for the rocket in Prob. 8–45?

c) Show that the mass at any time t is given by the equation

$$m = m_0(1 - kt).$$

d) Show that the acceleration at any time is equal to

$$\frac{v_r k}{1 - kt} - g.$$

e) Find the initial acceleration of the rocket in Prob. 8–45, in terms of the acceleration of gravity, g.

f) Find the acceleration 15 s after the motion starts.

g) Sketch the acceleration–time graph.

8–47 At what speed does the momentum of a particle differ by 1 percent from the value obtained using the nonrelativistic expression mv? Is the correct relativistic value greater or less than that obtained from the nonrelativistic expression?

8–48 By combining the relativistic energy and momentum expressions, Eqs. (7–22) and (8–16), show that these two quantities are related by

$$E^2 = (mc^2)^2 + (pc)^2,$$

where p is the relativistic momentum and E the *total* energy, including rest energy. Hence, show that when the total energy of a particle is much greater than its rest energy, the energy–momentum relation is approximately the same as for a massless particle. Note that for this to occur the speed of the particle must be very close to c.

8–49 Construct a right triangle in which one of the angles is α, where $\sin\alpha = v/c$. (v is the speed of a particle, c the speed of light.) If the base of the triangle (the side adjacent to α) is the rest energy mc^2, show that (a) the hypotenuse is the *total energy* and (b) the side opposite α is c times the relativistic momentum.

c) Describe a simple graphical procedure for finding the kinetic energy E_k.

8–50 The mass of an electron is 9.11×10^{-31} kg. Comparing the classical definition of momentum with its relativistic generalization, by how much is the classical expression in error if (a) $v = 0.01c$; (b) $v = 0.5c$; (c) $v = 0.9c$?

8–51 A radioactive isotope of cobalt, ^{60}Co, emits an electromagnetic photon (γ ray) of wavelength 0.932×10^{-12} m. The cobalt nucleus contains 27 protons and 33 neutrons, each with a mass of about 1.66×10^{-27} kg. If the nucleus is at rest before emission, what is the speed afterward? Is it necessary to use the relativistic generalization of momentum?

8–52 In Problem 8–51, suppose the cobalt atom is in a metallic crystal containing 0.01 moles of cobalt (about 6.02×10^{21} atoms) and that the entire crystal recoils as a unit, rather than just the single nucleus. Find the recoil velocity. (This recoil of the entire crystal rather than a single nucleus is called the *Mössbauer effect,* in honor of its discoverer, who first observed it in 1958.)

Chapter 9

Rotation

9-1 INTRODUCTION

Thus far our discussion of the principles of mechanics has been concerned primarily with particles. In problems involving a body whose size is not negligibly small, we nevertheless have assumed that its motion can be described in terms of the motion of a *point.* That is, the point mass serves as a *model* to represent a body whose size and shape are not relevant in the particular problem under discussion.

There are many problems in which this model is inadequate. One important class of problems involves *rigid bodies,* which can undergo both translational and rotational motion. The rigid body, a body with a perfectly definite and unchanging shape, is itself an idealized model, since real materials always deform somewhat when forces are exerted, but it is a useful model in cases where such deformation may be neglected.

The most general type of motion which a rigid body can undergo is a combination of *translation* and *rotation.* In preceding chapters we have discussed translational motion of a point, along a straight line or along a curve. We now discuss motion of rotation about a fixed axis, that is, motion of *rotation without translation.* We shall see that many of the equations describing rotation about a fixed axis are exactly analogous to those encountered in rectilinear motion. If the axis is *not* fixed, the problem becomes more complicated, and we shall not attempt to give a complete discussion of the general case of translation plus rotation.

9-2 ANGULAR VELOCITY

Figure 9-1 represents a rigid body of arbitrary shape rotating about a fixed axis through point O and perpendicular to the plane of the diagram. Line OP is a line fixed with respect to the body and rotating with it. The position of the entire body is evidently completely specified by the angle θ which the line OP makes with some reference line fixed in space, such as Ox. The motion of the body is therefore analogous to the rectilinear motion of a particle whose position is completely specified by a single coordinate such as

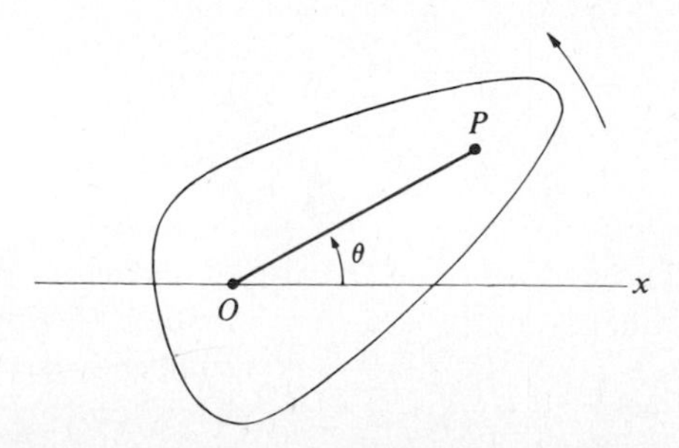

Fig. 9-1 *Body rotating about a fixed axis through point O.*

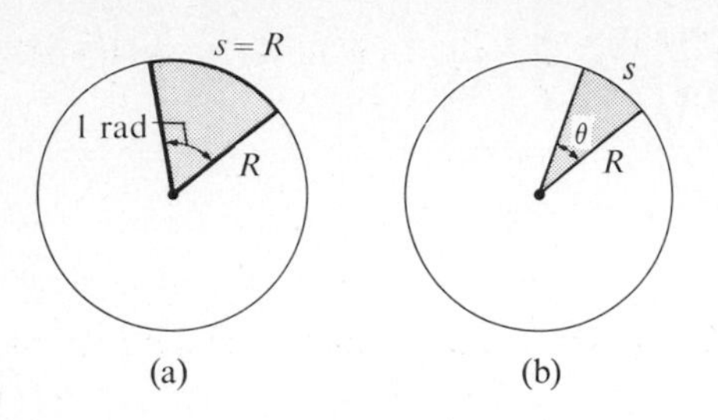

Fig. 9–2 *An angle θ in radians is defined as the ratio of the arc s to the radius R.*

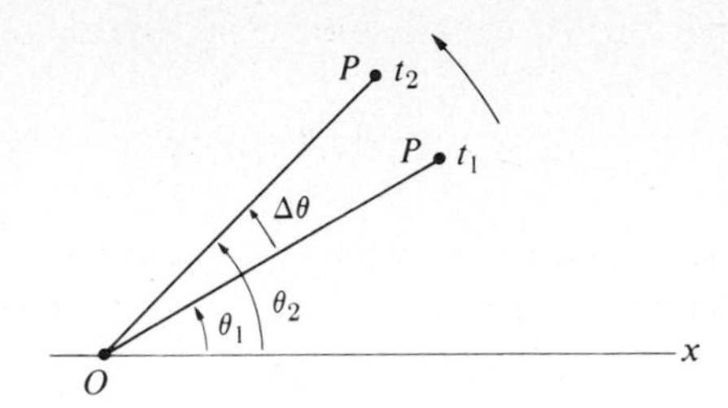

Fig. 9–3 *Angular displacement Δθ of a rotating body.*

x or y. The equations of motion are greatly simplified if the angle θ is expressed in *radians.*

One radian (1 rad) is the angle subtended at the center of a circle by an arc of length equal to the radius of the circle (Fig. 9–2(a)). Since the radius is contained 2π times ($2\pi = 6.28\ldots$) in the circumference, there are 2π or $6.28\ldots$ rad in one complete revolution or 360°. Hence,

$$
\begin{aligned}
1 \text{ rad} &= \frac{360^\circ}{2\pi} = 57.3 \text{ degrees},\\
360^\circ &= 2\pi \text{ rad} = 6.28 \text{ rad},\\
180^\circ &= \pi \text{ rad} = 3.14 \text{ rad},\\
90^\circ &= \pi/2 \text{ rad} = 1.57 \text{ rad},\\
60^\circ &= \pi/3 \text{ rad} = 1.05 \text{ rad},\\
45^\circ &= \pi/4 \text{ rad} = 0.79 \text{ rad},
\end{aligned}
$$

and so on.

In general (Fig. 9–2(b)), if θ represents any arbitrary angle subtended by an arc of length s on the circumference of a circle of radius R, then θ (in radians) is equal to the length of the arc s divided by the radius R:

$$\theta = \frac{s}{R}, \qquad s = R\theta. \tag{9–1}$$

An angle in radians, being defined as the ratio of a length to a length, is a pure number. If $s = 1.5\ m$ and $R = 1$ m, the angle is usually described as $\theta = 1.5$ rad, but it would be equally correct to say simply $\theta = 1.5$.

In Fig. 9–3, a reference line OP in a rotating body makes an angle θ_1 with the reference line Ox, at a time t_1. At a later time t_2 the angle has increased to θ_2. The *average angular velocity* of the body, $\bar{\omega}$, in the time interval between t_1 and t_2, is defined as the ratio of the *angular displacement* $\theta_2 - \theta_1$, or $\Delta\theta$, to the elapsed time $t_2 - t_1$ or Δt:

$$\bar{\omega} = \frac{\Delta\theta}{\Delta t}.$$

The *instantaneous angular velocity* ω is defined as the limit approached by this ratio as Δt approaches zero:

$$\omega = \lim_{\Delta t \to 0} \frac{\Delta\theta}{\Delta t} = \frac{d\theta}{dt}. \tag{9–2}$$

Since the body is rigid, *all* lines in it rotate through the same angle in the same time, and the angular velocity is characteristic of the body as a whole. If the angle θ is in radians, the unit of angular velocity is one radian per second (1 rad s^{-1}). Other units, such as the revolution per minute (rev min^{-1}), are in common use.

9–3 ANGULAR ACCELERATION

If the angular velocity of a body changes, it is said to have an angular acceleration. If ω_1 and ω_2 are the instantaneous angular velocities at times t_1 and t_2, the *average angular acceleration* $\bar{\alpha}$ is defined as

$$\bar{\alpha} = \frac{\omega_2 - \omega_1}{t_2 - t_1} = \frac{\Delta\omega}{\Delta t},$$

and the *instantaneous angular acceleration* α is defined as the limit of this ratio when Δt approaches zero:

$$\boxed{\alpha = \lim_{\Delta t \to 0} \frac{\Delta\omega}{\Delta t} = \frac{d\omega}{dt}.} \qquad (9\text{–}3)$$

The unit of angular acceleration is 1 rad s^{-2} or 1 s^{-2}. Angular velocity and angular acceleration are exactly analogous to linear velocity and acceleration.

Since $\omega = d\theta/dt$, the angular acceleration can be written

$$\alpha = \frac{d}{dt}\frac{d\theta}{dt} = \frac{d^2\theta}{dt^2}.$$

Also, by the chain rule,

$$\alpha = \frac{d\omega}{d\theta}\frac{d\theta}{dt} = \omega\frac{d\omega}{d\theta}. \qquad (9\text{–}4)$$

9–4 ROTATION WITH CONSTANT ANGULAR ACCELERATION

The simplest type of accelerated rotational motion is that in which the angular acceleration is constant. When this is the case, the expressions for the angular velocity and angular coordinate can readily be found by integration. We have

$$d\omega/dt = \alpha = \text{constant},$$

$$\int d\omega = \int \alpha\, dt,$$

$$\omega = \alpha t + C_1.$$

If ω_0 is the angular velocity when $t = 0$, the integration constant $C_1 = \omega_0$ and

$$\boxed{\omega = \omega_0 + \alpha t.} \qquad (9\text{–}5)$$

Then, since $\omega = d\theta/dt$,

$$\int d\theta = \int \omega_0\, dt + \int \alpha t\, dt, \qquad \theta = \omega_0 t + \tfrac{1}{2}\alpha t^2 + C_2.$$

The integration constant C_2 is the value of θ when $t = 0$, say θ_0.

$$\boxed{\theta = \theta_0 + \omega_0 t + \tfrac{1}{2}\alpha t^2.} \qquad (9\text{–}6)$$

If we write the angular acceleration as

$$\alpha = \omega\frac{d\omega}{d\theta},$$

then

$$\int \alpha\, d\theta = \int \omega\, d\omega + C_3, \qquad \alpha\theta = \tfrac{1}{2}\omega^2 + C_3.$$

If the angle θ again has the value θ_0 when $t = 0$, and if the initial angular velocity is ω_0, then $C_3 = \theta_0 - \frac{1}{2}\omega_0{}^2$ and

$$\boxed{\omega^2 = \omega_0{}^2 + 2\alpha(\theta - \theta_0).} \qquad (9\text{–}7)$$

Equations (9–5), (9–6), and (9–7) are exactly analogous to the corresponding equations for linear motion with constant acceleration:

$$v = v_0 + at,$$

$$x = x_0 + v_0 t + \tfrac{1}{2}at^2,$$

$$v^2 = v_0{}^2 + 2a\,(x - x_0)$$

Table 9–1 shows the similarity between the equations for motion with constant linear acceleration and those for motion with constant angular acceleration.

Example The angular velocity of a body is 4 rad s^{-1} at time $t = 0$, and its angular acceleration is constant and equal to 2 s^{-2}. A line *OP* in the body is horizontal at time $t = 0$. a) What angle does this line make with the horizontal at time $t = 3$ s? b) What is the angular velocity at this time?

a) $$\theta = \theta_0 + \omega_0 t + \tfrac{1}{2}\alpha t^2$$
$$= 0 + (4 \text{ rad s}^{-1})(3 \text{ s}) + \tfrac{1}{2}(2 \text{ rad s}^{-2})(3 \text{ s})^2$$
$$= 21 \text{ rad} = \frac{21}{2\pi} \text{ rev} = 3.34 \text{ rev}.$$

b) $$\omega = \omega_0 + \alpha t$$
$$= 4 \text{ rad s}^{-1} + (2 \text{ rad s}^{-2})(3 \text{ s}) = 10 \text{ s}^{-1}.$$

Table 9–1

Motion with constant linear acceleration	Motion with constant angular acceleration
$a = \text{constant}$	$\alpha = \text{constant}$
$v = v_0 + at$	$\omega = \omega_0 + \alpha t$
$x = x_0 + v_0 t + \frac{1}{2}at^2$	$\theta = \theta_0 + \omega_0 t + \frac{1}{2}\alpha t^2$
$x = x_0 + \dfrac{v_0 + v}{2}t$	$\theta = \theta_0 + \dfrac{\omega_0 + \omega}{2}t$
$v^2 = v_0^2 + 2a(x - x_0)$	$\omega^2 = \omega_0^2 + 2\alpha(\theta - \theta_0)$

Alternatively, from Eq. (9–7),

$$\begin{aligned}\omega^2 &= \omega_0^2 + 2\alpha(\theta - \theta_0)\\ &= (4\ \text{rad s}^{-1})^2 + 2(2\ \text{rad s}^{-2})(21\ \text{rad})\\ &= 100\ \text{rad}^2\ \text{s}^{-2},\\ \omega &= 10\ \text{rad s}^{-1}.\end{aligned}$$

9–5 RELATION BETWEEN ANGULAR AND LINEAR VELOCITY AND ACCELERATION

In Section 6–6 we discussed the linear velocity and acceleration of a *particle* revolving in a circle. When a *rigid body* rotates about a fixed axis, every point in the body moves in a circle whose center is on the axis and which lies in a plane perpendicular to the axis. There are some useful and simple relations between the angular velocity and acceleration of the rotating body and the linear velocity and acceleration of points within it.

Let r be the distance from the axis to some point P in the body, so that the point moves in a circle of radius r, as in Fig. 9–4. When the radius makes an angle θ with the reference axis, the distance s to the point P, measured along the circular path, is

$$s = r\theta. \tag{9–8}$$

Differentiating both sides of this equation with respect to t and noting that r is constant, we find

$$\frac{ds}{dt} = r\frac{d\theta}{dt}.$$

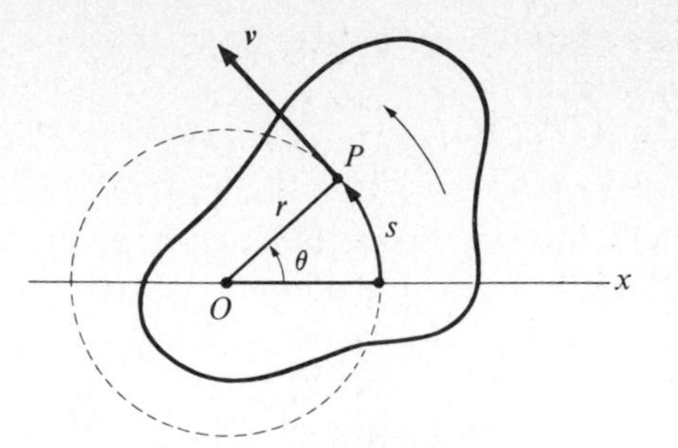

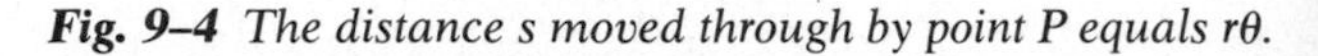
Fig. 9–4 *The distance s moved through by point P equals rθ.*

But ds/dt is the magnitude of the linear velocity v of point P, and $d\theta/dt$ is the angular velocity ω of the rotating body. Hence

$$v = r\omega \tag{9–9}$$

and the magnitude v of the linear velocity equals the product of the angular velocity ω and the distance r of the point from the axis.

Differentiating Eq. (9–9) with respect to t gives

$$\frac{dv}{dt} = r\frac{d\omega}{dt}.$$

But dv/dt is the magnitude of the tangential component of acceleration $a_{\parallel}$ of point P, and $d\omega/dt$ is the angular acceleration α of the rotating body, so

$$a_{\parallel} = r\alpha \tag{9–10}$$

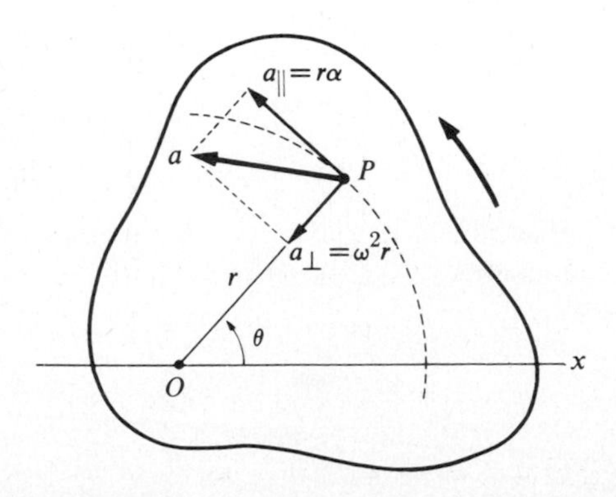

Fig. 9–5 *Nonuniform rotation about a fixed axis through point O. The tangential component of acceleration of point P equals rα; the radial component equals $\omega^2 r$.*

and the tangential component of acceleration equals the product of the angular acceleration and the distance from the axis.

The *radial* component of acceleration $a_\perp = v^2/r$ of the point P can also be expressed in terms of the angular velocity,

$$a_\perp = \frac{v^2}{r} = \omega^2 r. \qquad (9\text{–}11)$$

This is true at each instant *even when* ω *and* v *are not constant.*

The tangential and radial components of acceleration of any arbitrary point P in a rotating body are shown in Fig. 9–5.

9–6 TORQUE AND ANGULAR ACCELERATION. MOMENT OF INERTIA

We are now ready to consider the *dynamics* of rotation about a fixed axis, that is, the relation between the forces on a pivoted body and its angular acceleration.

Figure 9–6 represents a rigid body pivoted about a fixed axis through point O, perpendicular to the plane of the diagram.

The solid circle represents one of the particles of the body, of mass m_i. The particle is acted on by an external force $\mathbf{F}_i$, and by an internal force $\mathbf{f}_i$, the resultant of the forces exerted on it by all of the other particles of the body. It will suffice to consider only the case in which the forces $\mathbf{F}_i$ and $\mathbf{f}_i$ lie in a plane perpendicular to the axis. From Newton's second law,

$$\mathbf{F}_i + \mathbf{f}_i = m_i \mathbf{a}_i.$$

Let us resolve the forces and accelerations into radial and tangential components. Then

$$F_i \cos\theta_i + f_i \cos\phi_i = m_i a_{i\perp} = m_i r_i \omega^2,$$

$$F_i \sin\theta_i + f_i \sin\phi_i = m_i a_{i\parallel} = m_i r_i \alpha.$$

The first of these equations does not concern us further. When both sides of the second are multiplied by the distance r_i of the particle from the axis, we obtain

$$F_i r_i \sin\theta_i + f_i r_i \sin\phi_i = m_i r_i^2 \alpha. \qquad (9\text{–}12)$$

The first term on the left is the moment Γ_i of the external force about the axis, and the second is the moment of the internal force.

Equations corresponding to Eq. (9–12) can be written for all the particles of the body. When these equations are *added,* the moments of the *internal* forces cancel, since the resultant moment of every internal action-reaction force pair is zero. The sum of the left sides of the equations is then simply the resultant moment Γ of the *external* forces about the axis, or

$$\Gamma = \sum \Gamma_i = \sum F_i r_i \sin\theta_i.$$

Since the body is rigid, all particles have the same angular acceleration α and therefore

$$\Gamma = (m_1 r_1^2 + m_2 r_2^2 + \cdots)\alpha = (\sum m_i r_i^2)\alpha. \qquad (9\text{–}13)$$

The sum $\Sigma m_i r_i^2$ is called the *moment of inertia* of the body about the axis through point O, and is represented by I:

$$I = \sum m_i r_i^2. \qquad (9\text{–}14)$$

Equation (9–13) then becomes

$$\Gamma = I\alpha = I\frac{d\omega}{dt}. \qquad (9\text{–}15)$$

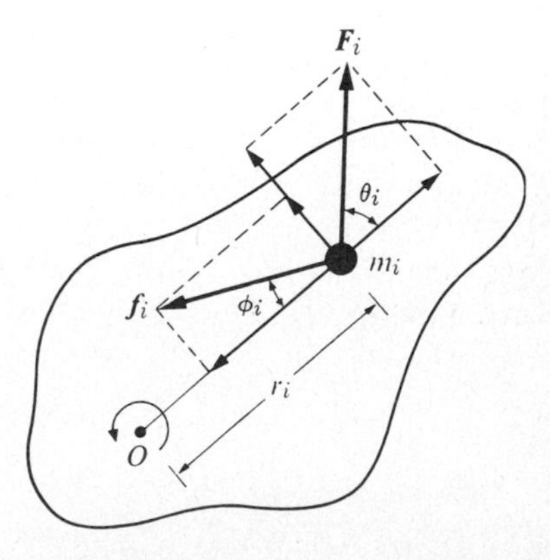

Fig. 9–6 *An external force* $\mathbf{F}_i$ *and an internal force* $\mathbf{f}_i$ *acting on a particle of mass* m_i *in a rigid body rotating about a fixed axis.*

That is, *when a rigid body is pivoted about a fixed axis, the resultant external torque about the axis equals the product of the moment of inertia of the body about the axis, and the angular acceleration.*

Thus the *angular* acceleration of a rigid body about a fixed axis is given by an equation having exactly the same form as that for the *linear* acceleration of a particle:

$$F = ma = m\frac{dv}{dt}.$$

The resultant torque Γ, about the axis corresponds to the resultant force F, the angular acceleration α corresponds to the linear acceleration a, and the moment of inertia I about the axis corresponds to the mass m.

The concept of moment of inertia can be thought of either as the sum of the product of the mass of each particle in a rigid body and the square of its distance from the axis, $I = \Sigma m_i r_i^2$, or as the ratio of the resultant torque to the angular acceleration, $I = \Gamma/\alpha$.

The analogy between translational and rotational quantities is displayed in Table 9–2.

Example A wheel of radius R, mass m_2, and moment of inertia I is mounted on an axle supported in fixed bearings, as in Fig. 9–7. A light flexible cord is wrapped around the rim of the wheel and carries a body of mass m_1. Friction in the bearings can be neglected. Discuss the motion of the system.

We must consider the resultant *force* on the suspended body and the resultant *torque* on the wheel. Let $\boldsymbol{T}$ represent the tension in the cord and $\boldsymbol{P}$ the upward force exerted on the shaft of the wheel by the bearings.

The resultant force on the suspended body is $w_1 - T$, and from Newton's second law for linear motion,

$$w_1 - T = m_1 a.$$

We have taken the downward direction as positive in order that a positive (counterclockwise) angular displacement of the wheel shall correspond to a positive linear displacement of the suspended body.

The forces $\boldsymbol{P}$ and $\boldsymbol{w}_2$ have no moment about the axis of the wheel. The resultant torque on the wheel, about the axis, is TR, and from Newton's second law

TABLE 9–2 ANALOGY BETWEEN TRANSLATIONAL AND ROTATIONAL QUANTITIES

Concept	Translation	Rotation	Comments
Displacement	s	θ	$s = r\theta$
Velocity	$v = ds/dt$	$\omega = d\theta/dt$	$v = r\omega$
Acceleration	$a = dv/dt$	$\alpha = d\omega/dt$	$a_T = r\alpha$
Resultant force, moment	F	Γ	$\Gamma = Fr$
Equilibrium	$F = 0$	$\Gamma = 0$	
Acceleration constant	$v = v_0 + at$ $s = v_0 t + \frac{1}{2}at^2$ $v^2 = v_0^2 + 2as$	$\omega = \omega_0 + \alpha t$ $\theta = \omega_0 t + \frac{1}{2}\alpha t^2$ $\omega^2 = \omega_0^2 + 2\alpha\theta$	
Mass, moment of inertia	m	I	$I = \sum m_i r_i^2$
Newton's second law	$F = ma$	$\Gamma = I\alpha$	
Work	$W = \int F\,ds$	$W = \int \Gamma\,d\theta$	
Power	$P = Fv$	$P = \Gamma\omega$	
Potential energy	$E_p = mgy$		
Kinetic energy	$E_k = \frac{1}{2}mv^2$	$E_k = \frac{1}{2}I\omega^2$	
Impulse	$\int F\,dt$	$\int \Gamma\,dt$	
Momentum	mv	$L = I\omega$	

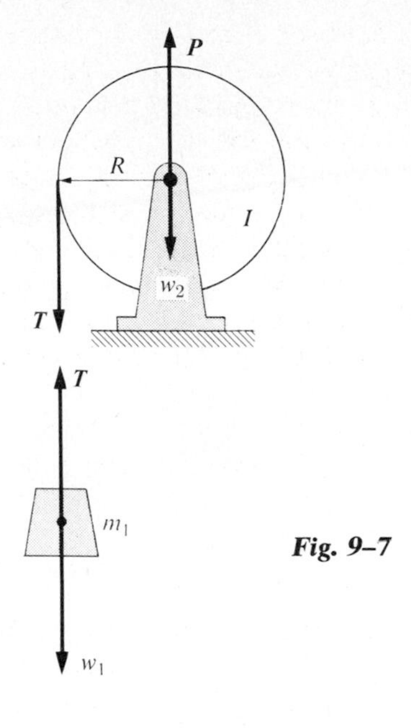

Fig. 9-7

for rotation,

$$TR = I\alpha$$

Since the linear acceleration of the suspended body equals the tangential acceleration of the rim of the wheel, we have

$$a = R\alpha.$$

Simultaneous solution of these equations gives

$$a = g\frac{1}{1 + (I/m_1R^2)}.$$

If the system starts from rest, the linear speed v of the suspended body, after descending a distance y (the acceleration is constant) is given by

$$v^2 = 2ay = 2\left[g\frac{1}{1 + (I/m_1R^2)}\right]y.$$

9-7 CALCULATION OF MOMENTS OF INERTIA

The moment of inertia of a body about an axis can be found *experimentally* by pivoting the body about the axis, applying a measured torque Γ to the body, and measuring the resulting angular acceleration α. The moment of inertia is then given by

$$I = \frac{\Gamma}{\alpha}.$$

The moment of inertia can be *calculated* from the defining equation $I = \Sigma m_i r_i^2$, for any system consisting of discrete point masses.

Example 1 Three small bodies, which can be considered as particles, are connected by light rigid rods, as in Fig. 9-8. What is the moment of inertia of the system (a) about an axis through point A, perpendicular to the plane of the diagram, and (b) about an axis coinciding with the rod BC?

a) The particle at point A lies on the axis. Its distance *from* the axis is zero and it contributes nothing to the moment of inertia. Therefore

$$I = \sum m_i r_i^2 = (10 \text{ g})(5 \text{ cm})^2 + (20 \text{ g})(4 \text{ cm})^2$$
$$= 570 \text{ g cm}^2.$$

b) The particles at B and C both lie on the axis. The moment of inertia is

$$I = \sum m_i r_i^2 = (30 \text{ g})(4 \text{ cm})^2$$
$$= 480 \text{ g cm}^2.$$

This illustrates the important fact that the moment of inertia of a body, unlike its mass, is not a unique property of the body but depends on the axis about which it is computed.

For a body which is not composed of discrete point masses but is a continuous distribution of matter, the summation expressed in the definition of moment of inertia, $I = \Sigma m_i r_i^2$, must be evaluated by the methods of calculus. The body is imagined to be subdivided into volume elements, each of mass Δm.

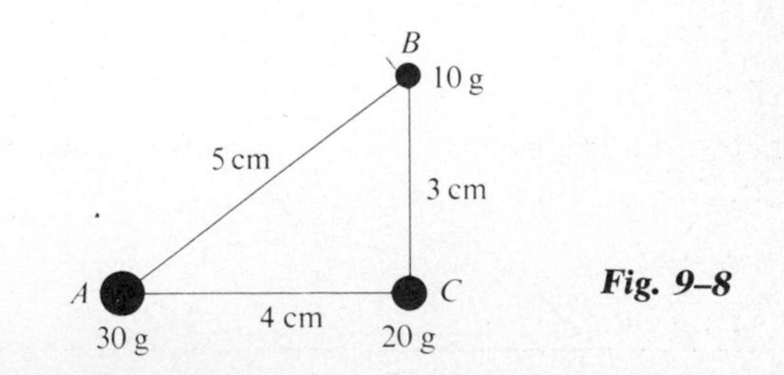

Fig. 9-8

Let r be the distance from any element to the axis of rotation. If each mass Δm is multiplied by the square of its distance from the axis and all the products $r^2\ \Delta m$ summed over the whole body, the moment of inertia is

$$I = \lim_{\Delta m \to 0} \sum r^2 \Delta m = \int r^2\, dm. \qquad (9\text{–}16)$$

If dV and dm represent the volume and mass, respectively, of an element, the density ρ is defined by the relation $dm = \rho\, dV$. Equation (9–16) may therefore be written

$$I = \int r^2\, dV.$$

If the density of a body is the same at all points, the body is said to be *uniform* or *homogeneous*, in which case

$$I = \rho \int r^2\, dV.$$

In using this equation, any convenient volume element may be chosen, *provided that all points within the element are the same distance r from the axis.*

The evaluation of integrals of this type may present considerable difficulty if the body is irregular, but for bodies of simple shape the integration can be carried out relatively easily. Three examples are given below.

Example 2 *Uniform slender rod, axis perpendicular to length.* Figure 9–9 shows a slender uniform rod of mass m and length ℓ. We wish to compute its moment of inertia about an axis through A, at an arbitrary distance h from one end. Select as an element of volume a short section of length dx and cross-sectional area S, at a distance x from point A. Then

$$dm = \rho\, dV = \rho S\, dx = \frac{\rho S \ell}{\ell}\, dx = \frac{m}{\ell}\, dx.$$

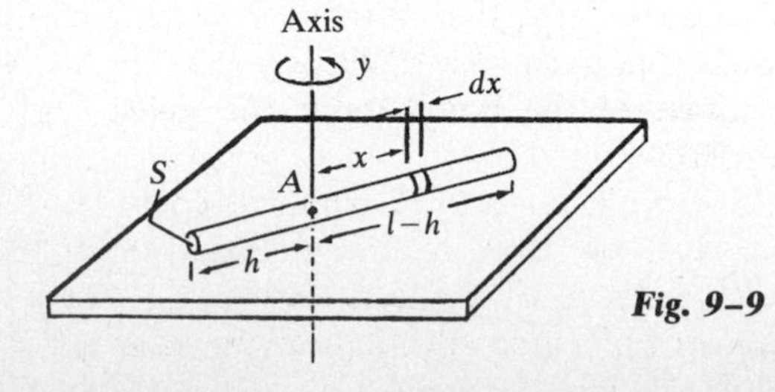

Fig. 9–9

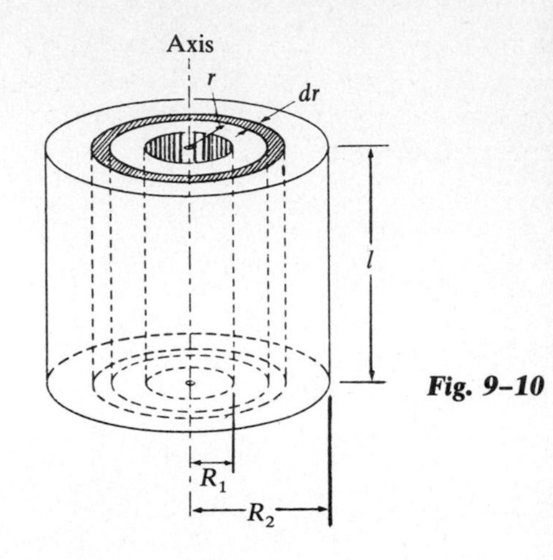

Fig. 9–10

Now

$$I_A = \int x^2\, dm = \frac{m}{\ell} \int_{-h}^{\ell - h} x^2\, dx$$

$$= \frac{m}{\ell} \frac{x^3}{3} \Bigg]_{-h}^{\ell - h} = \tfrac{1}{3} m(\ell^2 - 3\ell h + 3h^2).$$

From this general expression, we can find the moment of inertia about an axis through any point on the rod. For example, if the axis is at the left end, $h = 0$ and

$$I = \tfrac{1}{3} m\ell^2. \qquad (9\text{–}17)$$

If the axis is at the right end, $h = \ell$ and

$$I = \tfrac{1}{3} m\ell^2,$$

as would be expected. If the axis passes through the center,

$$h = \ell/2 \quad \text{and} \quad I = \tfrac{1}{12} m\ell^2. \qquad (9\text{–}18)$$

Example 3 *Hollow or solid cylinder, axis of symmetry* Figure 9–10 shows a hollow cylinder of length ℓ and inner and outer radii R_1 and R_2. We choose as the most convenient volume element the thin cylindrical shell of radius r, thickness dr, and length ℓ. If ρ is the density of the material, that is, the mass per unit volume, then

$$dm = \rho\, dV = \rho(2\pi r\, dr) \times \ell.$$

The moment of inertia is given by

$$I = \int r^2\, dm = 2\pi\ell \int_{R_1}^{R_2} \rho r^3\, dr.$$

If the body were of nonuniform density, one would have to know ρ as a function of r before the integration could be carried out. For a homogeneous solid, however, ρ is constant, and

$$I = 2\pi\ell\rho\int_{R_1}^{R_2} r^3\,dr$$
$$= \frac{\pi\ell\rho}{2}(R_2^4 - R_1^4).$$

The mass m of the entire cylinder is the product of its density and its volume. The volume is given by

$$\pi\ell(R_2^2 - R_1^2).$$

Hence

$$m = \pi\ell\rho(R_2^2 - R_1^2),$$

and the moment of inertia is therefore

$$I = \tfrac{1}{2}m(R_1^2 + R_2^2). \qquad (9\text{–}19)$$

If the cylinder is solid, $R_1 = 0$, and if we let R represent the outer radius,

$$I = \tfrac{1}{2}mR^2. \qquad (9\text{–}20)$$

If the cylinder is very thin-walled (like a stovepipe), $R_1 = R_2$ (very nearly) and if R represents this common radius,

$$I = mR^2.$$

Note that the moment of inertia of a cylinder about an axis coinciding with its axis of symmetry does not depend on the length ℓ. Two hollow cylinders of the same inner and outer radii, one of wood and one of brass, but having the same mass m, have equal moments of inertia even though the length of the former is much greater. Moment of inertia depends only on the *radial* distribution of mass, not on its distribution along the axis. Thus Eq. (9–19) holds for a very short cylinder like a washer, and Eq. (9–20) for a thin disk.

Example 4 *Uniform sphere, axis through center.* Divide the sphere into thin disks, as indicated in Fig. 9–11. The radius r of the disk shown is

$$r = \sqrt{R^2 - x^2}.$$

Its volume is

$$dV = \pi r^2\,dx = \pi(R^2 - x^2)\,dx$$

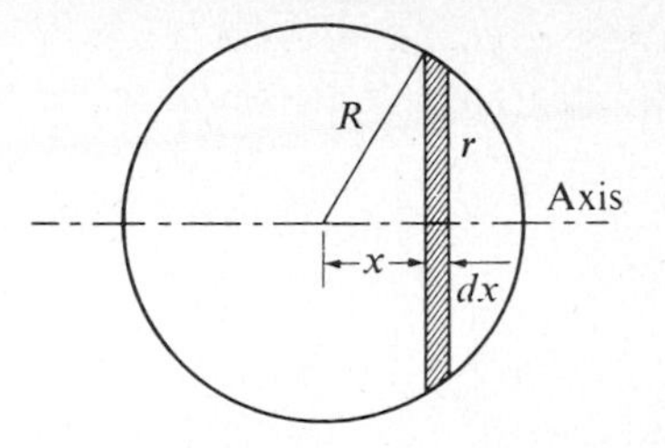

Fig. 9–11

and its mass is

$$dm = \rho\,dV.$$

Hence from Eq. (9–20) its moment of inertia is

$$dI = \frac{\pi\rho}{2}(R^2 - x^2)^2\,dx,$$

and for the whole sphere,

$$I = (2)\frac{\pi\rho}{2}\int_0^R (R^2 - x^2)^2\,dx,$$

since by symmetry the right hemisphere has the same moment of inertia as the left. Carrying out the integration, we obtain

$$I = \frac{8\pi\rho}{15}R^5.$$

The mass m of the sphere is

$$m = \rho V = \frac{4\pi\rho R^3}{3}.$$

Hence

$$I = \tfrac{2}{5}mR^2.$$

The moments of inertia of a few simple bodies are shown in Fig. 9–12.

Whatever the shape of a body, it is always possible to find a radial distance from any given axis at which the mass of the body could be concentrated without altering the moment of inertia of the body about that axis. This distance is called the *radius of gyration* of the body about the given axis, and is represented by k.

If the mass m of the body actually were concentrated at this distance, the moment of inertia would be that of a particle of mass m at a distance k from an axis, or mk^2. Since this equals the moment of inertia

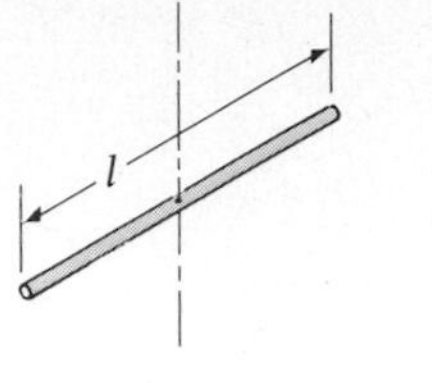

$\frac{1}{12}ml^2$

(a) Slender rod, axis through center

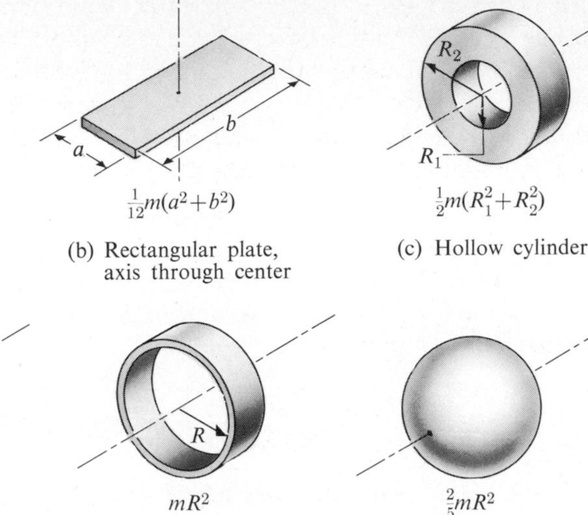

$\frac{1}{12}m(a^2+b^2)$

(b) Rectangular plate, axis through center

$\frac{1}{2}m(R_1^2+R_2^2)$

(c) Hollow cylinder

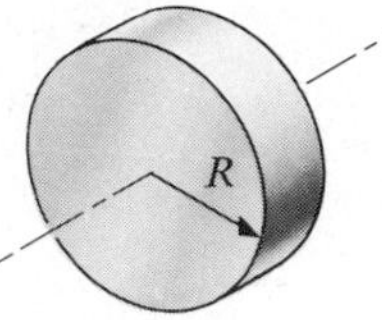

$\frac{1}{2}mR^2$

(d) Solid cylinder

mR^2

(e) Thin-walled hollow cylinder

$\frac{2}{5}mR^2$

(f) Solid sphere

***Fig. 9–12** Moments of inertia.*

I, then

$$mk^2 = I, \qquad k = \sqrt{I/m}. \tag{9–21}$$

Example 5 What is the radius of gyration of a slender rod of mass m and length ℓ about an axis perpendicular to its length and passing through the center?

The moment of inertia about an axis through the center is

$$I_0 = \tfrac{1}{12}m\ell^2.$$

Hence

$$k_0 = \sqrt{\frac{\frac{1}{12}m\ell^2}{m}} = \frac{\ell}{2\sqrt{3}} = 0.289\ell.$$

The radius of gyration, like the moment of inertia, depends on the location of the axis.

Note carefully that, in general, the mass of a body cannot be considered as concentrated at its center of gravity for the purpose of computing its moment of inertia. For example, when a thin rod is pivoted about an axis through one end, perpendicular to one end, the moment of inertia is $I = mL^2/3$ while, if all the mass were concentrated at the center, the moment of inertia about this axis would be $m(L/2)^2$ or $mL^2/4$.

9–8 KINETIC ENERGY, WORK, AND POWER

When a rigid body rotates about a fixed axis, the velocity v_i of a particle at a perpendicular distance r_i from the axis equals $r_i\omega$, where ω is the angular velocity. The kinetic energy of the particle is then

$$\tfrac{1}{2}m_iv_i^2 = \tfrac{1}{2}m_ir_i^2\omega^2,$$

and the total kinetic energy of the body is

$$E_k = \sum\tfrac{1}{2}m_ir_i^2\omega^2 = \tfrac{1}{2}(\sum m_ir_i^2)\omega^2.$$

But $\sum m_ir_i^2$ equals the moment of inertia I about the axis, so

$$\boxed{E_k = \tfrac{1}{2}I\omega^2.} \tag{9–22}$$

Thus the kinetic energy of a rigid body rotating about a fixed axis is given by an expression exactly analogous to that for the kinetic energy of a particle in linear motion, the moment of inertia I corresponding to the mass m and the angular velocity ω corresponding to the linear velocity v. (See Table 9–2.)

Example 1 Let us consider the motion of the system in Fig. 9–7 from the energy standpoint. Looking at the system as a whole, the external forces are the forces $\boldsymbol{P}$ and $\mathbf{w}_2$, which do no work, and the force $\mathbf{w}_1$, which is conservative. We can therefore apply the principle of conservation of energy, setting the decrease in potential energy of the body, as it descends a distance y, equal to the sum of the increase in *translational* kinetic energy of the body and the increase in *rotational* kinetic energy of the wheel:

$$m_1 g y = \tfrac{1}{2} m_1 v^2 + \tfrac{1}{2} I \omega^2 .$$

But

$$v = \omega R,$$

and again we find that

$$v^2 = 2\left[g \frac{1}{1 + (I/m_1 R^2)}\right] y.$$

In Fig. 9–13, an external force $\boldsymbol{F}$ is applied at point P of a rigid body rotating about a fixed axis through O, perpendicular to the plane of the diagram. As the body rotates through a small angle $d\theta$, point P moves a distance $ds = r\, d\theta$ and the work done by the force $\boldsymbol{F}$ is

$$W = \int F_{\parallel}\, ds = \int F_{\parallel} r\, d\theta.$$

But $F_{\parallel} r$ is the moment Γ of the force about the axis, so

$$W = \int_{\theta_1}^{\theta_2} \Gamma\, d\theta. \tag{9–23}$$

If more than one force acts on the body, the total work equals the work of the resultant moment.

From Eq. (9–15),

$$\Gamma = I\alpha = I\omega \frac{d\omega}{d\theta}.$$

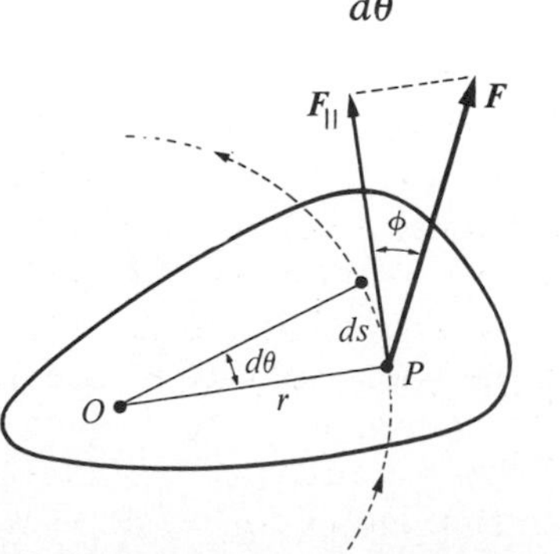

Fig. 9–13 *The work done by the force* $\boldsymbol{F}$ *in an angular displacement* $d\theta$.

Hence

$$\Gamma\, d\theta = I\omega\, d\omega ,$$

and

$$W = \int_{\theta_1}^{\theta_2} \Gamma\, d\theta = \int_{\omega_1}^{\omega_2} I\omega\, d\omega$$

$$= \tfrac{1}{2} I \omega_2^2 - \tfrac{1}{2} I \omega_1^2. \tag{9–24}$$

That is, *the work of the resultant moment equals the increase in kinetic energy*, in analogy with the work–energy equation for linear motion.

The power developed by the force Γ in Fig. 9–13, if v is the velocity of its point of application, is

$$P = F_{\parallel} v = F_{\parallel} r \omega,$$

and since $F_{\parallel} r = \Gamma$,

$$\boxed{P = \Gamma\omega,} \tag{9–25}$$

the rotational analog of $P = F_{\parallel} v$. The power output of an engine therefore equals the product of torque and angular velocity. The transmission of an automobile engine is often called a "torque converter." An automobile engine can deliver full power only when running at its rated angular velocity and delivering the corresponding torque. When accelerating, or climbing a steep hill, the torque converter permits the engine to run at rated speed, torque, and power, while the same power is transmitted to the drive shaft at a larger torque and a smaller angular velocity.

Example 2 The manufacturers of an automobile state that its engine develops 345 hp and a torque of 475 lb ft. What is the corresponding angular velocity?

$$\omega = \frac{P}{\Gamma} = \frac{(345)(550 \text{ ft lb s}^{-1})}{475 \text{ lb ft}}$$

$$= 400 \text{ rad s}^{-1} \approx 3800 \text{ rev min}^{-1}.$$

9–9 ANGULAR MOMENTUM

The equation

$$\Gamma = I\alpha$$

can be written as

$$\Gamma = I\frac{d\omega}{dt} = \frac{d}{dt}(I\omega). \tag{9–26}$$

The product of moment of inertia I and angular velocity ω is the rotational analog of the product of mass m and linear velocity v. The latter product is the linear momentum, and by analogy we call the product $I\omega$ the *angular momentum L:*

$$\boxed{L = I\omega.} \tag{9–27}$$

(Although in this special case of a rigid body rotating about a fixed axis the angular momentum is equal to $I\omega$, this is not the general definition of this quantity.)

Equation (9–26) can now be written

$$\Gamma = \frac{dL}{dt}. \tag{9–28}$$

The resultant external torque is equal to the rate of change of angular momentum, just as the resultant external force equals the rate of change of linear momentum.

Multiplying by dt and integrating, we get

$$J = \int_0^t \Gamma\, dt = L - L_0. \tag{9–29}$$

The integral $J = \int \Gamma\, dt$ is called the *angular impulse* of the torque and is analogous to the impulse of a force, $\int F\, dt$. The preceding equation is therefore the analog of the impulse–momentum principle in linear motion. That is, *the resultant angular impulse of the torque on a body is equal to the change in angular momentum of the body.*

9–10 CONSERVATION OF ANGULAR MOMENTUM

Here is an example of the usefulness of the concepts of angular impulse and angular momentum. Figure 9–14 shows two disks with moments of inertia I and I', rotating with initial constant angular velocities ω_0 and ω_0', respectively. At some time t_0, the disks are pushed together by a force directed along the axis, without any torque on either disk. After a short time interval, say Δt, the disks reach a common final angular velocity ω.

During the interval Δt, the larger disk exerts a torque Γ' on the smaller, and the smaller exerts a torque Γ on the larger. Both Γ and Γ' vary during the contact, becoming zero after the common final angular velocity is reached. At each instant the two torques are equal in magnitude and opposite in direction, because of Newton's third law. That is, at each instant $\Gamma = -\Gamma'$; the total impulses on the two bodies are thus related by $J = -J'$.

Now according to Eq. (9–29), the impulse on each disk equals the change of angular momentum of that disk:

$$J = I\omega - I\omega_0,$$

$$J' = I'\omega - I'\omega_0'.$$

But because $J = -J'$, these are equal and opposite:

$$I\omega - I\omega_0 = -(I'\omega - I'\omega_0')$$

or

$$\boxed{I\omega_0 + I'\omega_0' = (I + I')\omega.} \tag{9–30}$$

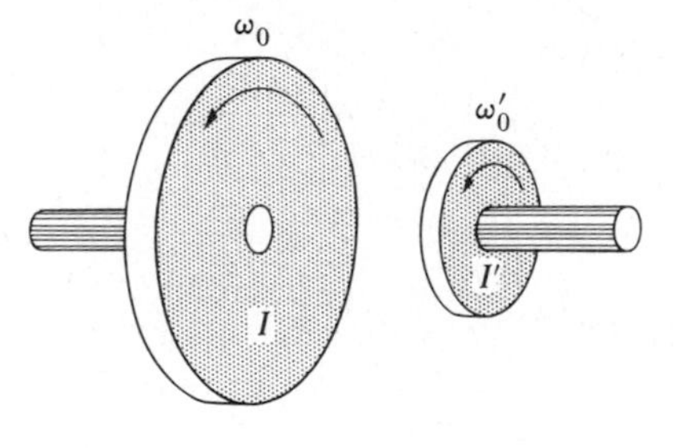

(a)

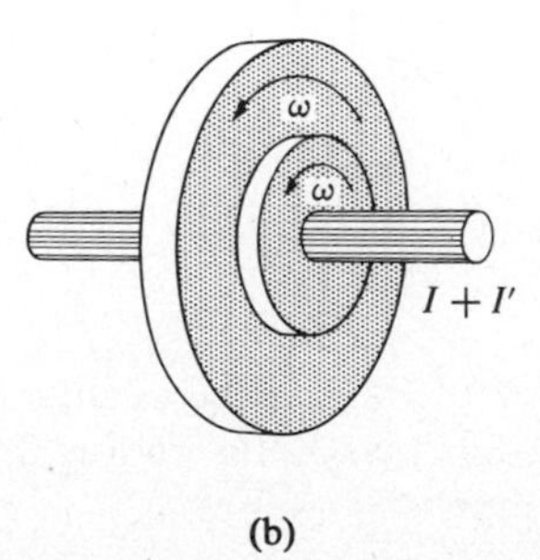

(b)

Fig. 9–14 *An impulsive torque acts when two rotating disks engage.*

The left side of Eq. (9–30) is the total angular momentum of the system (consisting of both disks) before contact, and the right side is the total angular momentum after contact. We have therefore derived the important result that the total angular momentum of the whole system is unaltered. When both disks are regarded as one system, then the torques Γ and Γ' are *internal* torques; during the engaging process no *external* torque acts. *When the resultant external torque on a system is zero, the angular momentum of the system remains constant*; hence any internal interaction between the parts of a system cannot alter its total angular momentum. This is the *principle of conservation of angular momentum*, and it ranks with the principles of conservation of linear momentum and conservation of energy as one of the most fundamental of physical laws.

A circus acrobat, a diver, or a skater performing a pirouette on the toe of one skate, all take advantage of the principle. Suppose an acrobat has just left a swing as in Fig. 9–15, with arms and legs extended and with a counterclockwise angular momentum. When he pulls his arms and legs in, his moment of inertia I becomes much smaller. Since his angular momentum $I\omega$ remains constant and I decreases, his angular velocity ω *increases.*

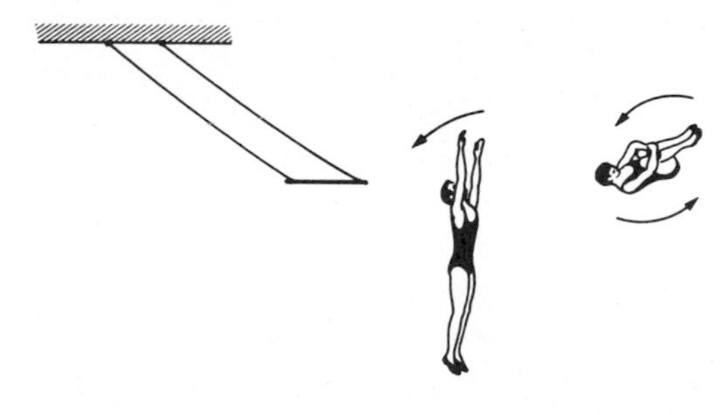

Fig. 9–15 *Conservation of angular momentum.*

Example 1 In the situation of Fig. 9–14, suppose the large disk has mass 2 kg, radius 0.2 m, and initial angular velocity 50 rad s^{-1}, and the small disk mass 4 kg, radius 0.1 m, and initial angular velocity 200 rad s^{-1} (about 1900 rpm). Find the common final angular velocity after the disks are pushed into contact. Is kinetic energy conserved during this process?

The moments of inertia of the two disks are

$$I = \tfrac{1}{2}(2\text{ kg})(0.2\text{ m})^2 = 0.04\text{ kg m}^2,$$

and

$$I' = \tfrac{1}{2}(4\text{ kg})(0.1\text{ m})^2 = 0.02\text{ kg m}^2.$$

From conservation of angular momentum, we have

$$(0.04\text{ kg m}^2)(50\text{ rad s}^{-1}) + (0.02\text{ kg m}^2)(200\text{ rad s}^{-1}) = (0.04\text{ kg m}^2 + 0.02\text{ kg m}^2)\omega,$$

$$\omega = 100\text{ rad s}^{-1}.$$

The initial kinetic energy is

$$E_{k_0} = \tfrac{1}{2}(0.04\text{ kg m}^2)(50\text{ rad s}^{-1})^2 + \tfrac{1}{2}(0.02\text{ kg m}^2)(200\text{ rad s}^{-1})^2 = 450\text{ J}.$$

The final kinetic energy is

$$E_k = \tfrac{1}{2}(0.04\text{ kg m}^2 + 0.02\text{ kg m}^2)(100\text{ rad s}^{-1})^2 = 300\text{ J}.$$

One-third of the kinetic energy was lost during this "angular collision," which is the rotational analog of an inelastic collision. We should not expect kinetic energy to be conserved, even though the resultant external force and torque are zero, since nonconservative (frictional) internal forces act during the contact.

Example 2 A man stands at the center of a turntable, holding his arms extended horizontally, with a 5-kg mass in each hand, as in Fig. 9–16. He is set rotating about a vertical axis with an angular velocity of one revolution in 2 s. Find his new angular velocity if he drops his hands to his sides. The moment of inertia of the man may be assumed constant and equal to 6 kg m^2. The original distance of the weights from the axis is 1 m, and their final distance is 0.2 m.

If friction in the turntable is neglected, no external torques act about a vertical axis and the angular momentum about this axis is constant. That

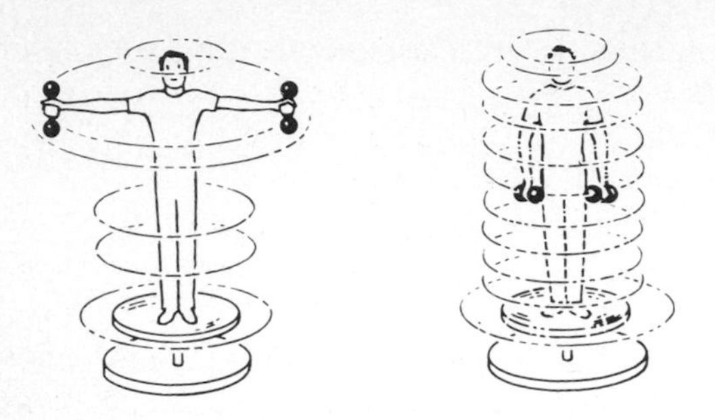

Fig. 9–16 *Conservation of angular momentum about a fixed axis.*

is,

$$I\omega = (I\omega)_0 = I_0\omega_0,$$

where I and ω are the final moment of inertia and angular velocity, and I_0 and ω_0 are the initial values of these quantities:

$$I = I_{\text{man}} + I_{\text{weights}},$$

$$I_0 = 6\text{ kg m}^2 + 2(5\text{ kg})(1.0\text{ m})^2 = 16\text{ kg m}^2,$$

$$I = 6\text{ kg m}^2 + 2(5\text{ kg})(0.2\text{ m})^2 = 6.4\text{ kg m}^2,$$

$$\omega_0 = 2\pi(1/2)\text{ rad s}^{-1},$$

$$\omega = \omega_0\left(\frac{I_0}{I}\right) = \pi\frac{16\text{ kg m}^2}{6.4\text{ kg m}^2} = 2.5\pi\text{ rad s}^{-1}.$$

That is, the angular velocity is more than doubled.

The initial kinetic energy is

$$E_{k0} = \tfrac{1}{2}(16\text{ kg m}^2)(\pi\text{ rad s}^{-1})^2 = 79\text{ J}.$$

The final kinetic energy is

$$E_k = \tfrac{1}{2}(6.4\text{ kg m}^2)(2.5\pi\text{ rad s}^{-1})^2 = 197\text{ J}.$$

Where did the extra energy come from?

9–11 VECTOR REPRESENTATION OF ANGULAR QUANTITIES

A rotational quantity associated with an axis, such as angular velocity, angular momentum, torque, etc., can be represented by a *vector* along the axis. Thus angular velocity is a vector with magnitude equal to the number of radians through which the body turns in unit time (our original definition of angular velocity), with direction along the axis of rotation in the direction a righthand-thread screw would advance if turned with the body, as shown in Fig. 9–17. Thus far we have considered only problems of rotation about a fixed axis, for which the angular-velocity vector always lies along the axis of rotation. For motion where the direction of the axis changes, the vector nature of angular velocity is an essential feature of the description of motion. The concepts of moment and angular momentum must also be generalized to vector quantities.

We consider first the vector generalization of moment. In Fig. 9–18, a force $\boldsymbol{F}$ lying in the xy-plane is applied at point P of a rigid body. Point O is at a perpendicular distance r from point P, and $\boldsymbol{r}$ is the vector from O to P. There may or may not be an axis through point O. If there *were* an axis through O, perpendicular to the plane of $\boldsymbol{r}$ and $\boldsymbol{F}$, the moment Γ of the force $\boldsymbol{F}$ about this axis would be

$$\Gamma = rF.$$

We define the *vector moment* of the force $\boldsymbol{F}$, *about the* **point** O, as a vector $\boldsymbol{\Gamma}$ whose magnitude equals the moment rF and whose direction is perpendicular to the plane of $\boldsymbol{r}$ and $\boldsymbol{F}$, as shown. The sense of $\boldsymbol{\Gamma}$ along the axis is specified by the *right-hand screw rule:* Rotate the first vector ($\boldsymbol{r}$) through the smaller angle (θ) that will bring it into parallelism with the second vector ($\boldsymbol{F}$), as shown by the broken line in the

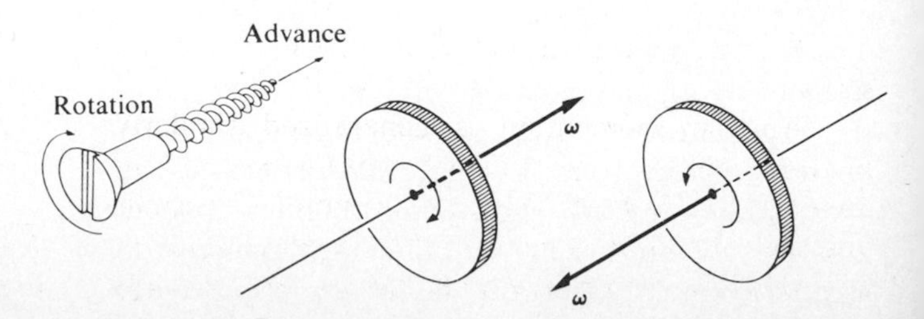

Fig. 9–17 *Vector angular velocity of a rotating body.*

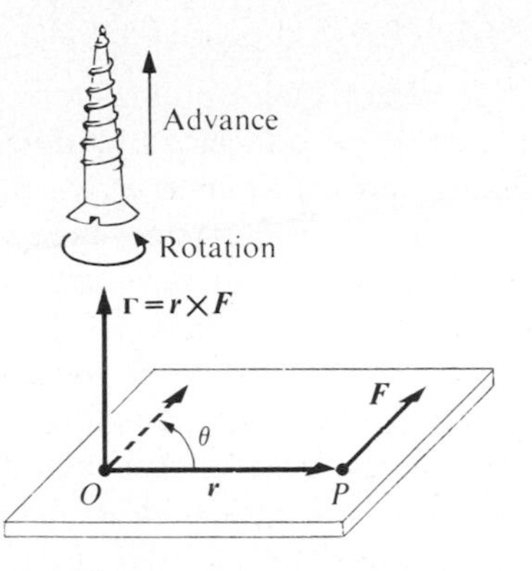

Fig. 9–18 (*a*) *Vector* **Γ** *is the vector moment of the force* **F** *about the point O:* $\mathbf{\Gamma} = \mathbf{r} \times \mathbf{F}$.

diagram. The vector **Γ** then points in the direction of *advance* of a right-hand screw when it is *rotated* through the same angle θ. The vector moment of **F** about point O has the same magnitude and direction whatever the direction of an actual axis through O, *or even if there is no actual axis through this point.*

Since **r** and **F** are vectors, the definition above suggests that we define a new product of two vectors called the *vector product* or *cross product.* The vector product of any two vectors **A** and **B** is written $\mathbf{A} \times \mathbf{B}$ and is defined as a *vector* of magnitude $AB \sin \theta$, where θ is the angle between **A** and **B.** Then if **C** represents this product

$$\mathbf{C} = \mathbf{A} \times \mathbf{B}, \qquad C = AB \sin \theta. \qquad (9\text{–}31)$$

The direction and sense of the vector product is specified by the right-hand screw rule given above. The vector movement of the force **F** about the point O can therefore be written in general as

$$\boxed{\mathbf{\Gamma} = \mathbf{r} \times \mathbf{F},} \qquad (9\text{–}32)$$

where

$$\Gamma = rF \sin \theta$$

For the special case in Fig. 9–18, $\theta = 90°$ and $\sin \theta = 1$.

Angular momentum is generalized similarly. Figure 9–19 shows a horizontal flat plate rotating about a fixed vertical axis through point O with an angular velocity ω. A particle of the plate of mass m, at a distance r from the axis, has a linear velocity of magnitude $v = r\omega$ and a linear momentum of magnitude mv. The angular momentum **L** of the particle is now defined as a vector quantity equal to the vector product of **r** and $m\mathbf{v}$:

$$\boxed{\mathbf{L} = \mathbf{r} \times m\mathbf{v}.} \qquad (9\text{–}33)$$

As an example, if a particle of mass m moves counterclockwise in a circle of radius r, in the plane of this page, with speed v, its angular momentum is a vector pointing out of the page, with magnitude mvr. This can also be expressed in terms of the angular velocity; since $v = r\omega$, we have

$$L = mvr = m(r\omega)r = mr^2\omega = I\omega. \qquad (9\text{–}34)$$

We have used the fact that the moment of inertia I of the particle about an axis through the center of the circle is mr^2.

The total angular momentum of a rigid body is the vector sum of the angular momenta of the particles composing it. If the body lies in a single plane, as in Fig. 9–19, and rotates about an axis perpendicular to this plane, then all these vectors lie along the axis. Then, the *vector* sum becomes an *arithmetic* sum and the magnitude of the angular momentum of the plate is

$$L = \sum rmv.$$

But $v = r\omega$, and ω has the same value for all particles. Hence

$$L = (\sum mr^2)\omega = I\omega,$$

and this definition of angular momentum leads to the same expression for L as that stated in Sec. 9–9.

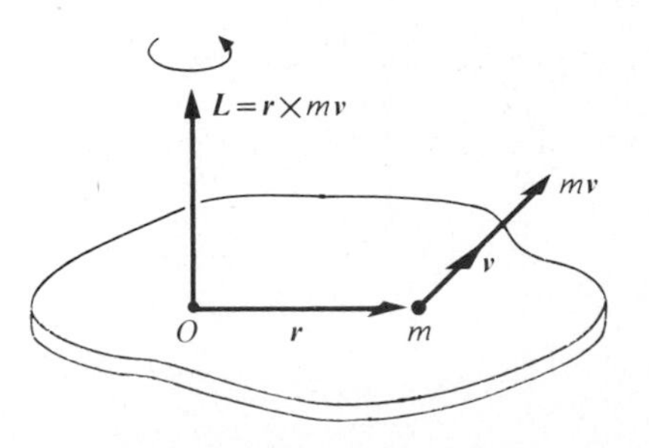

Fig. 9–19 *Vector* **L** *represents the moment of momentum, or angular momentum, of a particle of mass m about point O:* $\mathbf{L} = \mathbf{r} \times m\mathbf{v}$.

Now suppose that an external torque Γ is exerted on the plate. In a small time interval Δt the angular momentum of the plate changes by ΔL, where $\Delta L = \Gamma\,\Delta t$.

Since the *vectors* $\mathbf{\Delta L}$ and $\mathbf{\Gamma}$ are both along the axis of rotation, this can be written as a *vector* equation:

$$\mathbf{\Delta L} = \mathbf{\Gamma}\Delta t. \tag{9–35}$$

An analysis of the general case, where $\mathbf{L}$ and $\mathbf{\Gamma}$ may *not* be in the same direction, shows that the result above is always true. That is, *the* ***vector change*** *in angular momentum of a body, $\mathbf{\Delta L}$, is equal in magnitude and direction to the vector impulse $\mathbf{\Gamma}\,\Delta t$ of the resultant external torque on the body.*

Let us now apply these vector concepts of torque and angular momentum to a specific example. In Fig. 9–20, a disk is mounted on a shaft through its center, the shaft being supported by fixed bearings. If the disk is rotating as indicated, its angular momentum vector $\mathbf{L}$ points toward the right, along the axis.

Suppose a cord is wrapped around the rim of the disk and a force $\mathbf{F}$ is exerted on the cord. The magnitude of the resultant torque on the disk is $\Gamma = FR$, and the torque vector $\mathbf{\Gamma}$ also points along the axis. In a time Δt, the torque produces a vector change $\mathbf{\Delta L}$ in the angular momentum, equal to $\mathbf{\Gamma}\,\Delta t$ and having the same direction as $\mathbf{\Gamma}$. When this change is added vectorially to the original angular momentum $\mathbf{L}$, the resultant is a vector of length $\mathbf{L} + \mathbf{\Delta L}$, in the same direction as $\mathbf{L}$. In other words, the *magnitude* of the angular momentum is increased, its *direction* remaining the same. An increase in the magnitude of the angular momentum simply means that the body rotates more rapidly.

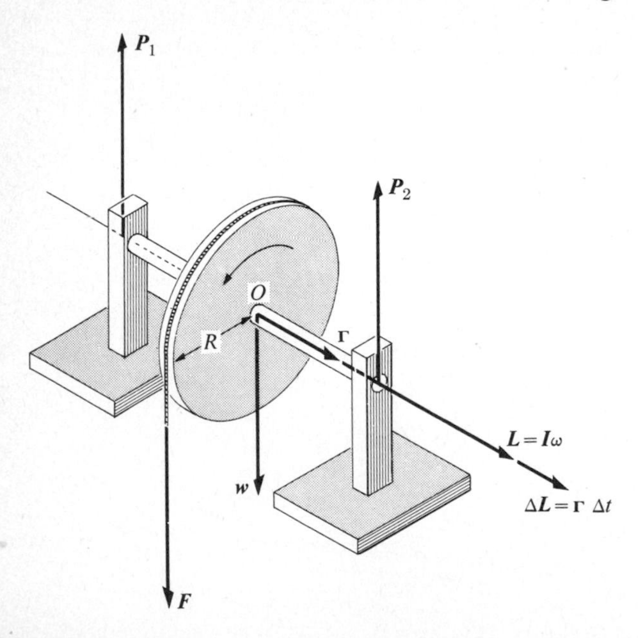

Fig. 9–20 Vector $\mathbf{\Delta L}$ is the change in angular momentum produced in time Δt by the moment $\mathbf{\Gamma}$ of the force $\mathbf{F}$. Vectors $\mathbf{\Delta L}$ and $\mathbf{\Gamma}$ are in the same direction.

The lengthy argument above appears at first to be nothing more than a difficult way of solving an easy problem in rotation about a fixed axis. However, the vector nature of torque and angular momentum are essential to an understanding of the gyroscope, to be discussed in the next section.

Example Two forces $\mathbf{P}$ and $\mathbf{P}'$, each of magnitude 20 N, are applied for 3 s to a disk of radius 0.5 m and moment of inertia 40 kg m^2, pivoted about an axis through its center, as in Fig. 9–21. The initial angular velocity of the disk is 5 rad s^{-1}. Show in a vector diagram the torque, the initial angular momentum, and the final angular momentum.

From the preceding discussion, the vectors representing the torque $\mathbf{\Gamma}$ due to the couple, and the initial angular momentum $\mathbf{L_0}$, are directed as in Fig. 9–21. The magnitude of the initial angular momentum is

$$L_0 = I\omega_0 = (40\text{ kg m}^2)(5\text{ rad s}^{-1})$$
$$= 200\text{ kg m}^2\text{ s}^{-1}.$$

The angular impulse is the product of the constant torque Γ and the time interval Δt. Since this must equal the change in angular momentum, we have

$$\Gamma\Delta t = I\omega - I\omega_0 = \Delta(I\omega) = \Delta L.$$

But $\Gamma = (20\text{ N})(0.5\text{ m}) = 10\text{ N m}$, and $\Delta t = 3$ s. Hence

$$\Delta L = (10\text{ N m})(3\text{ s})$$
$$= 30\text{ N m s} = 30\text{ kg m}^2\text{ s}^{-1}.$$

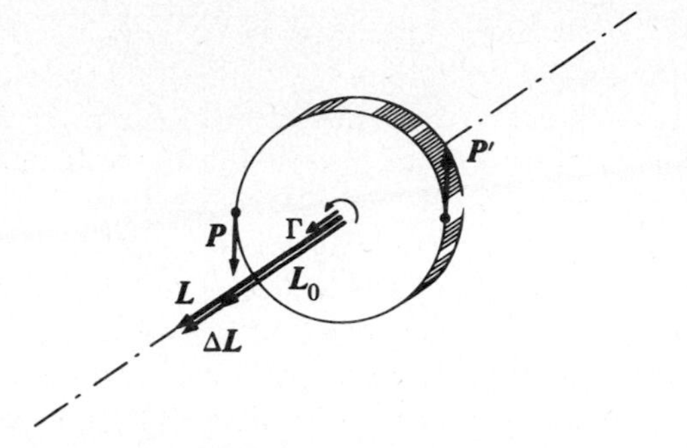

Fig. 9-21 *Vector* $\Delta \boldsymbol{L}$ *is the change in angular momentum produced by the couple* $\boldsymbol{P} - \boldsymbol{P}'$.

This increase has been represented by the vector $\Delta \boldsymbol{L}$ in Fig. 9-21.

The final angular momentum, $\boldsymbol{L}$, is the vector sum of $\boldsymbol{L}_0$ and $\Delta \boldsymbol{L}$. Since both are in the same direction, the vector sum is simply the arithmetic sum. That is

$$L = L_0 + \Delta L = 230 \text{ kg m}^2 \text{ s}^{-1}.$$

9-12 THE TOP AND THE GYROSCOPE

A symmetrical body rotating about an axis, one point of which is fixed, is called a *top*. If the fixed point is at the center of gravity, the body is called a *gyroscope*. The axis of rotation of a top or gyroscope can itself rotate about the fixed point, so the *direction* of the angular momentum vector can change.

Figure 9-22 illustrates the usual mounting of a toy gyroscope (more properly called a top, since the fixed point O is not at the center of mass). The top is spinning about its axis of symmetry, and if the axis is initially set in motion in the direction shown, with the proper angular velocity, the system continues to rotate uniformly about the pivot at O, the spin axis remaining horizontal.

If the axis of the top were fixed in space, its angular momentum would equal the product of its moment of inertia about the axis and its angular velocity about the axis, and would point along the axis. Because the axis itself is rotating, the angular momentum vector no longer lies on the axis. However, if the angular velocity of the axis is small compared with the angular velocity about the axis, the component of angular momentum arising from the former effect is small, and we shall neglect it. The angular momentum vector $\boldsymbol{L}$, about the fixed point O, can then be drawn along the axis as shown and, as the top rotates about O, its angular momentum vector rotates with it.

The upward force $\boldsymbol{P}$ at the pivot has no moment about O. The resultant external moment is that due to the weight $\boldsymbol{w}$; its magnitude is

$$\Gamma = wR.$$

The direction of $\boldsymbol{\Gamma}$ is *perpendicular* to the axis, as shown. In a time Δt (compare with the analysis of Fig. 9-20), this torque produces a change $\Delta \boldsymbol{L}$ in the angular momentum, having the same direction as $\boldsymbol{\Gamma}$ and given by

$$\Delta \boldsymbol{L} = \boldsymbol{\Gamma} \, \Delta t.$$

The angular momentum $\boldsymbol{L} + \Delta \boldsymbol{L}$, after a time Δt, is the vector sum of $\boldsymbol{L}$ and $\Delta \boldsymbol{L}$. Since $\Delta \boldsymbol{L}$ is perpendicular to $\boldsymbol{L}$, the new angular momentum vector has the same *magnitude* as the old but a different *direction*. The tip of the angular momentum

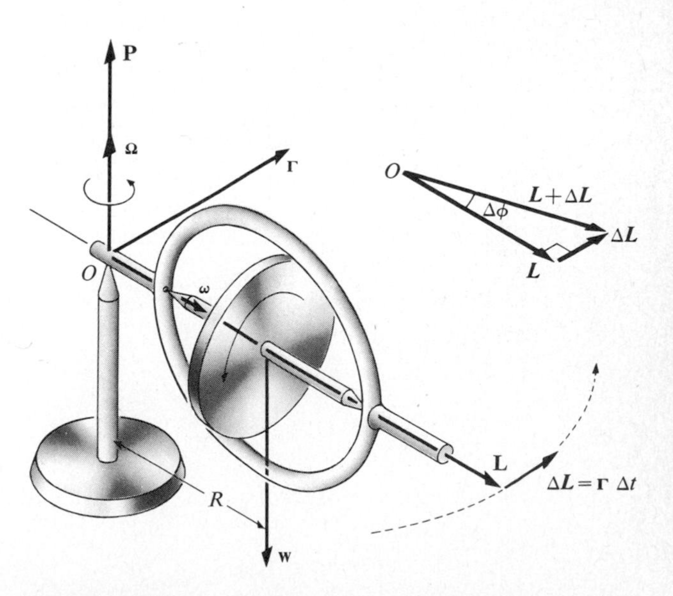

Fig. 9-22 *Vector* $\Delta \boldsymbol{L}$ *is the change in angular momentum produced in time* Δt *by the moment* $\boldsymbol{\Gamma}$ *of the force* $\boldsymbol{w}$. *Vectors* $\Delta \boldsymbol{L}$ *and* $\boldsymbol{\Gamma}$ *are in the same direction. (Compare with Fig. 9-20.)*

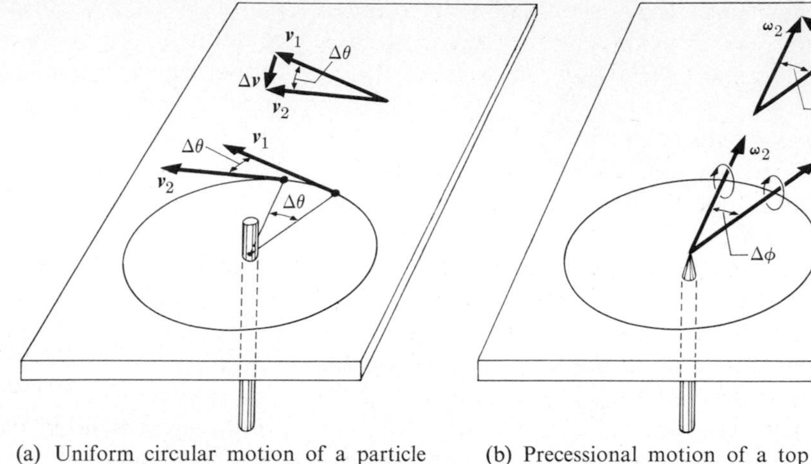

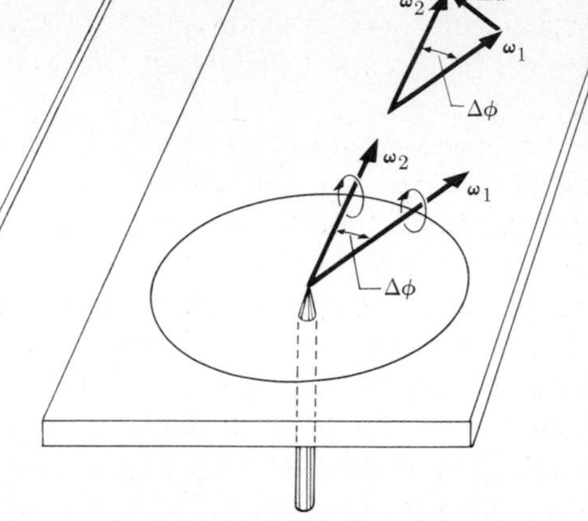

(a) Uniform circular motion of a particle

(b) Precessional motion of a top

(a)	(b)
$v_1 = v_2 = v$	$\omega_1 = \omega_2 = \omega$
$\boldsymbol{v}_2 = \boldsymbol{v}_1 + \Delta\boldsymbol{v}$	$\boldsymbol{\omega}_2 = \boldsymbol{\omega}_1 + \Delta\boldsymbol{\omega}$
$\Delta v \simeq v\,\Delta\theta$	$\Delta\omega \simeq \omega\,\Delta\phi$
$\lim_{\Delta t\to 0}\frac{\Delta v}{\Delta t} = v\lim_{\Delta t\to 0}\frac{\Delta\theta}{\Delta t}$	$\lim_{\Delta t\to 0}\frac{\Delta\omega}{\Delta t} = \omega\lim_{\Delta t\to 0}\frac{\Delta\phi}{\Delta t}$
$a = v\omega\left(= \frac{v^2}{R} = R\omega^2\right)$	$\alpha = \omega\Omega$
$F = ma$	$\Gamma = I\alpha$
$F = mv\omega$	$\Gamma = I\omega\Omega$

Fig. 9–23 *Analogy between uniform circular motion of a particle and precessional motion of a top.*

vector moves as shown, and as time goes on it swings around a horizontal circle. Since the angular momentum vector lies along the gyroscope axis, the axis turns also, rotating in a horizontal plane about the point O. This motion of the axis of rotation is called *precession.*

The angle $\Delta\phi$ turned through by the vector $\boldsymbol{L}$ in time Δt (see the small inset diagram) is

$$\Delta\phi = \frac{\Delta L}{L} = \frac{\Gamma\Delta t}{L}.$$

The *angular velocity of precession* Ω is $\Delta\phi/\Delta t$, and therefore

$$\boxed{\Omega = \frac{\Gamma}{L}.} \tag{9–36}$$

The angular velocity of precession is therefore *inversely* proportional to the angular momentum. If this is large, the precessional angular velocity will be small.

From a purely kinematic point of view, precessional motion of a top (see Fig. 9–22) is the rotational analog of uniform circular motion of a particle. The analogy is illustrated in Fig. 9–23, where the diagram and calculation for uniform circular motion are shown on the left and the corresponding diagram and calculation for precessional motion of a top are displayed on the right.

Why doesn't the top in Fig. 9–22 fall? The answer is that the upward force $\boldsymbol{P}$ exerted on it by the pivot is just equal in magnitude to its weight $\boldsymbol{w}$, so that the resultant vertical *force* is zero and the vertical acceleration of the center of gravity is zero. In other words, the vertical component of its linear momentum remains zero, since there is no resultant vertical force. However, the resultant *torque* of these forces is *not* zero and the angular momentum changes.

If the top were not rotating, it would have no angular momentum $\boldsymbol{L}$ initially. Its angular momentum $\Delta\boldsymbol{L}$ after a time Δt would be that acquired from the torque acting on it and would have the same direction as the torque. In other words, the top would rotate about an axis through O in the direction of the vector $\boldsymbol{\Gamma}$. But if the top is originally rotating, the change in its angular momentum produced by the torque adds vectorially to the large angular momentum it already has, and since $\Delta\boldsymbol{L}$ is horizontal and perpendicular to $\boldsymbol{L}$, the result is a motion of precession, with both the angular momentum vector and the axis remaining horizontal.

To understand why the vertical force P should equal w, we must look further into the way in which the precessional motion in Fig. 9–22 originated. If the frame of the top is initially held at rest, say by supporting the projecting portion of the frame opposite O with one's finger, the upward forces exerted by the finger and by the pivot are equal to $w/2$. If the finger is suddenly removed, the upward force at O, at the first instant, is still $w/2$. The resultant vertical *force* is not zero, and the center of gravity has an initial downward acceleration. At the same time, precessional motion begins, although with a smaller angular velocity than that in the final steady state. The result of this motion is that the end of the frame at O presses down on the pivot with a greater force, so that the upward force O increases and eventually becomes *greater* than w. When this happens, the center of gravity starts to accelerate *upward*. The process repeats itself, and the motion consists of a precession together with an up-and-down oscillation of the axis, called *nutation*.

To start the top off with pure precession (without nutation), it is necessary to give the outer end of the axis a push in the direction in which it would normally precess. This causes the end of the frame at O to bear down on the pivot so that the upward force at O increases. When this force equals w, the vertical forces are in equilibrium, the outer end can be released, and precession goes on as in Fig. 9–22.

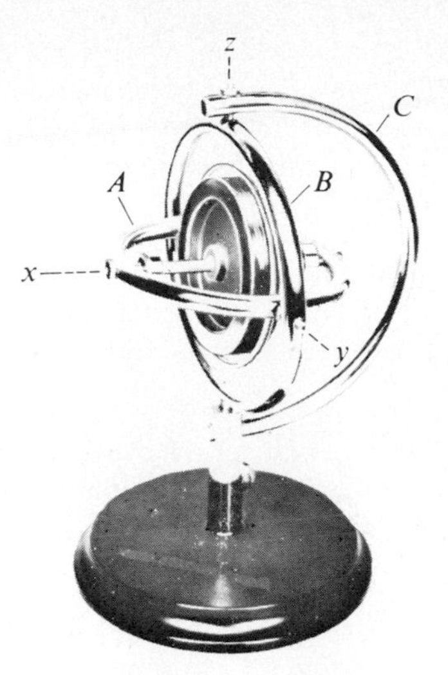

Fig. 9-24 *Photograph of a gimbal-mounted gyro.* [*Courtesy of Sperry Gyroscope Co.*]

Figure 9–24 is a photograph of a gyroscope mounted in gimbals. This mounting scheme permits the gyro axis x to take any possible orientation in space while the outer frame C remains stationary. If the center of mass of the gyro wheel is at the intersection of axes x, y, and z, no torque is exerted on it except for the small frictional torques at the pivot bearings.

PROBLEMS

9–1

a) What angle in radians is subtended by an arc 6 ft in length, on the circumference of a circle whose radius is 4 ft?

b) What angle in radians is subtended by arc of length 78.54 cm on the circumference of a circle of diameter 100 cm? What is this angle in degrees?

c) The angle between two radii of a circle is 0.60 radian. What length of arc is intercepted on the circumference of a circle of radius 200 cm? of radius 200 ft?

9–2 Compute the angular velocity, in rad s^{-1}, of the crankshaft of an automobile engine rotating at 4800 rev min^{-1}.

9–3 An electric motor running at 1800 rev min^{-1} has on its shaft three pulleys, of diameters 5, 10, and 15 cm, respectively. Find the linear velocity of the surface of each pulley in cm s^{-1}. The pulleys may be connected by a belt to a similar set on a countershaft, the 5-cm to the 15-cm, the 10-cm to the 10-cm, or the 15-cm to the 5-cm. Find the three possible angular velocities of the countershaft in rev min^{-1}.

9–4 A wheel 0.6 m in diameter starts from rest and accelerates uniformly to an angular velocity of 100 rad s^{-1} in 20 s. Find the angular acceleration and the angle turned through.

9–5 The angular velocity of a flywheel decreases uniformly from 1000 rev min^{-1} to 400 rev min^{-1} in 5 s. Find the angular acceleration and the number of revolutions made by the wheel in the 5-s interval. How many more seconds are required for the wheel to come to rest?

9–6 A flywheel requires 3 s to rotate through 234 radians. Its angular velocity at the end of this time is 108 rad s^{-1}. Find its constant angular acceleration.

9–7 A flywheel whose angular acceleration is constant and equal to 2 rad s^{-2}, rotates through an angle of 100 radians in 5 s. How long had it been in motion at the beginning of the 5-s interval if it started from rest?

9–8

a) Distinguish clearly between tangential and radial acceleration.

b) A flywheel rotates with constant angular velocity. Does a point on its rim have a tangential acceleration? a radial acceleration?

c) A flywheel is rotating with constant angular acceleration. Does a point on its rim have a tangential acceleration? a radial acceleration? Are these accelerations constant in magnitude?

9–9 A wheel 1.0 m in diameter is rotating about a fixed axis with an initial angular velocity of 2 rev s^{-1}. The acceleration is 3 rev s^{-2}.

a) Compute the angular velocity after 6 s.

b) Through what angle has the wheel turned in this time interval?

c) What is the tangential velocity of a point on the rim of the wheel at $t = 6$ s?

d) What is the resultant acceleration of a point on the rim of the wheel at $t = 6$ s?

9–10 A wheel having a diameter of 1 ft starts from rest and accelerates uniformly to an angular velocity of 900 rev min^{-1} in 5 s.

a) Find the position at the end of 1 s of a point originally at the top of the wheel.

b) Compute and show in diagram the magnitude and direction of the acceleration at the end of 1 s.

9–11 A flywheel of radius 30 cm starts from rest and accelerates with a constant angular acceleration of 0.50 rad s^{-2}. Compute the tangential acceleration, the radial acceleration, and the resultant acceleration, of a point on its rim (a) at the start, (b) after it has turned through 120°, (c) after it has turned through 240°.

9–12 A wheel starts from rest and accelerates uniformly to an angular velocity of 900 rev min^{-1} in 20 s. At the end of 1 s, (a) find the angle through which the wheel has rotated; and (b) compute and show in a diagram the magnitude and direction of the tangential and radial components of acceleration of a point 6 in from the axis.

9–13 Find the required angular velocity of an ultracentrifuge, in rpm, in order that the radial acceleration of a point 1 cm from the axis shall equal 300,000g (i.e., 300,000 times the acceleration of gravity).

9–14 An automobile engine is idling at 500 rev min^{-1}. When the accelerator is depressed, the angular velocity increases to 3000 rev min^{-1} in 5 s. Assume a constant angular acceleration.

a) What are the initial and final angular velocities, expressed in radians per second?

b) What was the angular acceleration, in radians per second squared?

c) How many revolutions did the engine make during the acceleration period?

d) The flywheel of the engine is 0.5 m in diameter. What is the linear speed of a point at its rim when the angular speed is 3000 rev min^{-1}?

e) What was the tangential acceleration of the point during the acceleration period?

f) What is the radial acceleration of the point when the angular speed is 3000 rev min^{-1}?

9–15 (a) Prove that when a body starts from rest and rotates about a fixed axis with constant angular acceleration, the radial acceleration of a point in the body is directly proportional to its angular displacement. (b) Through what angle will the body have turned when the resultant acceleration makes an angle of 60° with the radial acceleration?

9–16 A particle moves in the xy-plane according to the law

$$x = R \cos \omega t, \qquad y = R \sin \omega t,$$

where x and y are the coordinates of the body, t is the time, and R and ω are constants. (a) Eliminate t between these equations to find the equation of the curve in which the body moves. [*Hint:* Square each equation.] What is this curve? (b) Differentiate the original equations to find the x- and y-components of the velocity of the particle. Combine these expressions to obtain the magnitude and direction of the resultant velocity. (c) Differentiate again to obtain the magnitude and direction of the resultant acceleration. (This problem illustrates an alternative method of deriving the expressions for the speed and radial acceleration of a particle moving in a circle.)

9–17 Find the moment of inertia of a rod 4 cm in diameter and 2 m long, of mass 8 kg, (a) about an axis perpendicular to the rod and passing through its center, (b) about an axis perpendicular to the rod and passing through one end, (c) about a longitudinal axis through the center of the rod.

9–18 The four bodies shown in Fig. 9–25 have equal masses m. Body A is a solid cylinder of radius R. Body B is a hollow thin cylinder of radius R. Body C is a solid square with length of side $= 2R$. Body D is the same size as C, but hollow (i.e., made up of four thin sticks). The bodies haves axes of rotation perpendicular to the page and through the center of gravity of each body.

a) Which body has the smallest moment of inertia?
b) Which body has the largest moment of inertia?

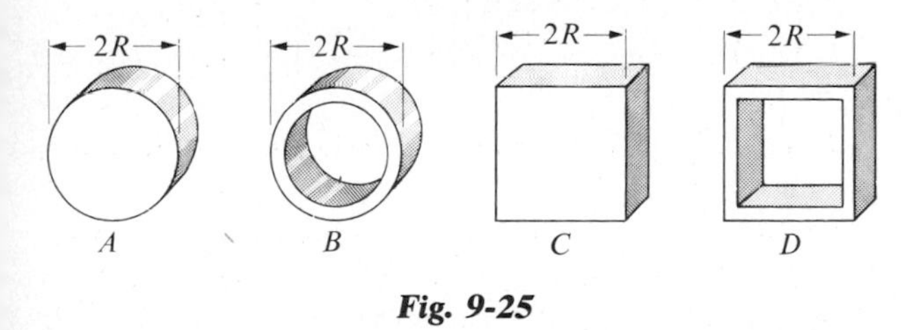

Fig. 9-25

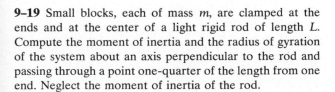

9–19 Small blocks, each of mass m, are clamped at the ends and at the center of a light rigid rod of length L. Compute the moment of inertia and the radius of gyration of the system about an axis perpendicular to the rod and passing through a point one-quarter of the length from one end. Neglect the moment of inertia of the rod.

9–20 The inner radius of a hollow cylinder is 10 cm, the outer radius is 20 cm, and the length is 30 cm. What is the radius of gyration of the cylinder about its axis?

9–21 The uniform thin rectangular plate in Fig. 9–26 has a length a, a width b, and a mass m. Find its moment of inertia (a) about the axis AA through its center O, (b) about the axis BB at one edge.

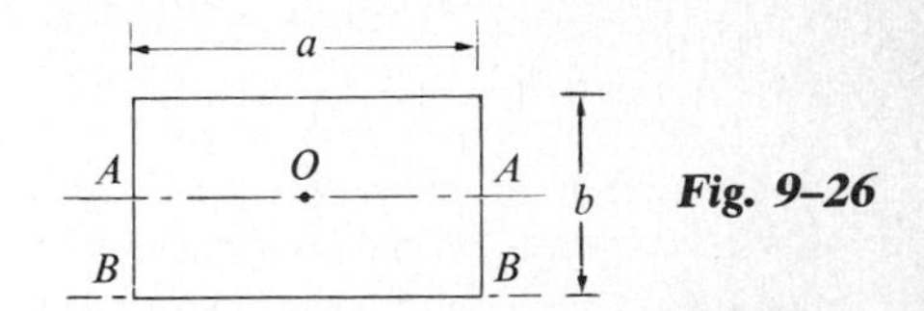

Fig. 9–26

9–22 (a) Prove that the moment of inertia of the thin flat plate in Fig. 9–27, about the z-axis, equals the sum of its moments of inertia about the x- and y-axes. (b) Given that the moment of inertia of a disk about an axis through its center and perpendicular to its plane is $mR^2/2$, use the relation above to find its moment of inertia about a diameter. (c) Derive the result of part (b) by direct integration of the defining equation, $I = \int r^2\, dm$. (d) What is the moment of inertia of a disk about an axis tangent to its edge?

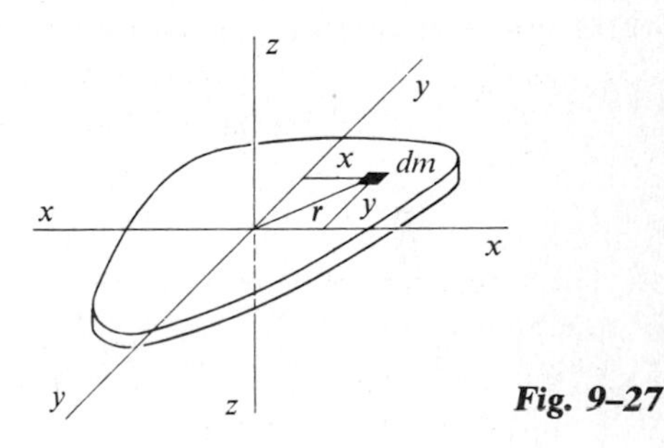

Fig. 9–27

9–23 A 60-kg grindstone is 1 m in diameter and has a radius of gyration of 0.25 m. A tool is pressed down on the rim with a normal force of 50 N. The coefficient of sliding friction between the tool and stone is 0.6, and there is a constant friction torque of 5 N m between the axle of the stone and its bearings.

a) How much force must be applied normally at the end of a crank handle 0.5 m long to bring the stone from rest to 120 rev min^{-1} in 9 s ?
b) After attaining a speed of 120 rev min^{-1}, what must the normal force at the end of the handle become to maintain a constant speed of 120 rev min^{-1} ?
c) How long will it take the grindstone to come from 120 rev min^{-1} to rest if it is acted on by the axle friction alone?

9–24 A flywheel consists of a solid disk 1 ft in diameter and 1 in. thick, and two projecting hubs 4 in. in diameter and 3 in. long. If the material of which it is constructed weighs 480 lb ft^{-3}, find (a) its moment of inertia, and (b) its radius of gyration about the axis of rotation.

9–25 A grindstone 1.0 m in diameter, of mass 50 kg, is rotating at 900 rev min^{-1}. A tool is pressed normally against the rim with a force of 200 N, and the grindstone comes to rest in 10 s. Find the coefficient of friction between the tool and the grindstone. Neglect friction in the bearings.

9–26 A constant torque of 20 N m is exerted on a pivoted wheel for 10 s, during which time the angular velocity of the wheel increases from zero to 100 rev min^{-1}. The external torque is then removed and the wheel is brought to rest by friction in its bearings in 100 s. Compute (a) the moment of inertia of the wheel, (b) the friction torque, (c) the total number of revolutions made by the wheel.

9–27 A cord is wrapped around the rim of a flywheel 0.5 m in radius, and a steady pull of 50 N is exerted on the cord, as in Fig. 9–28(a). The wheel is mounted in frictionless bearings on a horizontal shaft through its center. The moment of inertia of the wheel is 4 kg m^2.

a) Compute the angular acceleration of the wheel.

b) Show that the work done in unwinding 5 m of cord equals the gain in kinetic energy of the wheel.

c) If a mass having a weight of 50 N hangs from the cord, as in Fig. 9–28(b), compute the angular acceleration of the wheel. Why is this not the same as in part (a)?

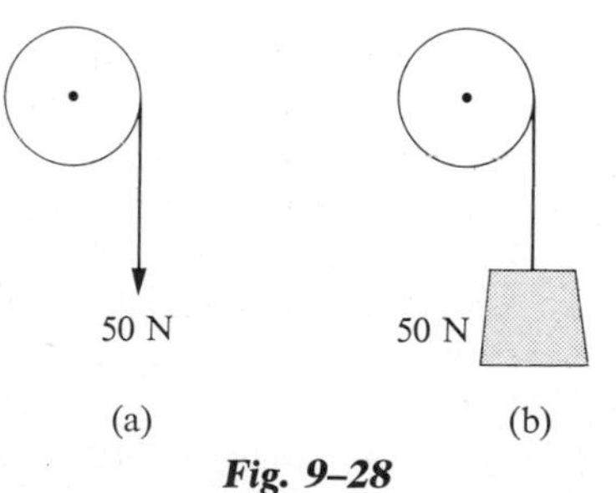

Fig. 9–28

9–28 A solid cylinder of mass 15 kg, 0.3 m in diameter, is pivoted about a horizontal axis through its center, and a rope wrapped around the surface of the cylinder carries at its end a block of mass 8 kg.

a) How far does the block descend in 5 s, starting from rest?

b) What is the tension in the rope?

c) What is the force exerted on the cylinder by its bearings?

9–29 A bucket of water weighing 64 lb is suspended by a rope wrapped around a windlass in the form of a solid cylinder 1 ft in diameter, also weighing 64 lb. The bucket is released from rest at the top of a well and falls 64 ft to the water.

a) What is the tension in the rope while the bucket is falling?

b) With what velocity does the bucket strike the water?

c) What was the time of fall? Neglect the weight of the rope.

9–30 A 16-lb block rests on a horizontal frictionless surface. A cord attached to the block passes over a pulley, whose diameter is 6 in., to a hanging block also weighing 16 lb. The system is released from rest, and the blocks are observed to move 16 ft in 2 s.

a) What was the moment of inertia of the pulley?

b) What was the tension in each part of the cord?

9–31 Figure 9–29 represents an *Atwood's machine.* Find the linear accelerations of blocks A and B, the angular acceleration of the wheel C, and the tension in each side of the cord (a) if the surface of the wheel is frictionless; (b) if there is no slipping between the cord and the surface of the wheel. Let the masses of blocks A and B be 4 kg and 2 kg, respectively, the moment of inertia of the wheel about its axis be 0.2 kg m^2, and the radius of the wheel be 0.1 m.

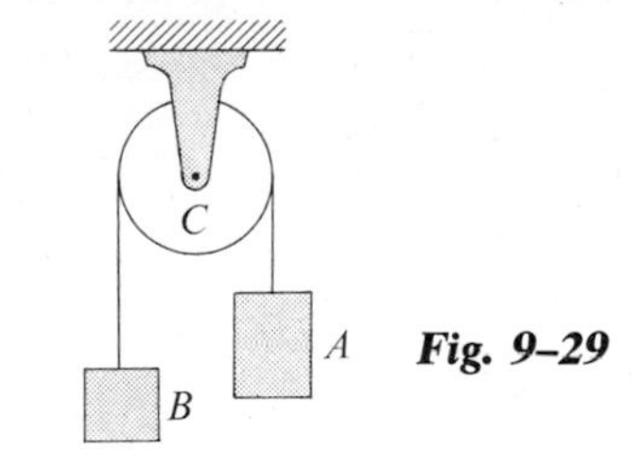

Fig. 9–29

9–32 A flywheel 1.0 m in diameter is pivoted on a horizontal axis. A rope is wrapped around the outside of the flywheel, and a steady pull of 50 N is exerted on the rope. It is found that 10 m of rope are unwound in 4 s.

a) What was the angular acceleration of the flywheel?

b) What is its final angular velocity?

c) What is its final kinetic energy?

d) What is its moment of inertia?

9–33 A light rigid rod 1.0 m long has a small block of mass 0.05 kg attached at one end. The other end is pivoted, and the rod rotates in a vertical circle. At a certain instant, the rod makes an angle of 53° with the vertical, and the tangential speed of the block is 4 m s^{-1}.

a) What are the horizontal and vertical components of the velocity of the block?

b) What is the moment of inertia of the system?

c) What is the radial acceleration of the block?

d) What is the tangential acceleration of the block?

e) What is the tension or compression in the rod?

9–34

a) Compute the torque developed by an airplane engine whose output is 2000 hp at an angular velocity of 2400 rev min^{-1}.

b) If a drum 0.5 m in diameter were attached to the motor shaft, and the power output of the motor were used to raise a weight hanging from a rope wrapped around the shaft, how large a weight could be lifted?

c) With what velocity would it rise?

9–35 A block of mass $m = 5$ kg slides down a surface inclined 37° to the horizontal, as shown in Fig. 9–30. The coefficient of sliding friction is 0.25. A string attached to the block is wrapped around a flywheel on a fixed axis at O. The flywheel has a mass $M = 20$ kg, an outer radius $R = 0.2$ m, and a radius of gyration with respect to the axis of $k_O = 0.1$ m.

a) What is the acceleration of the block down the plane?

b) What is the tension in the string?

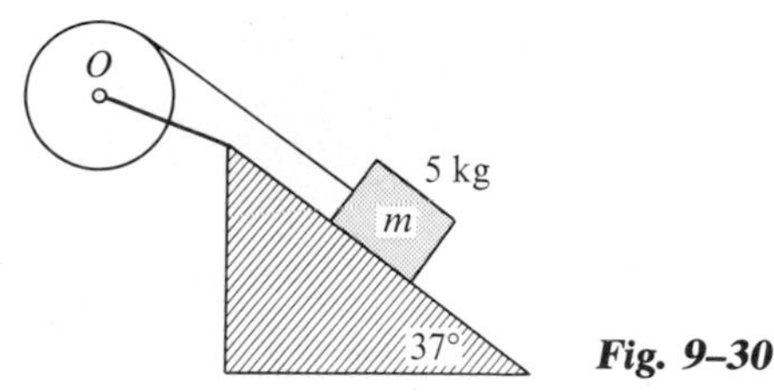

Fig. 9–30

9–36 A body of mass m is attached to a light cord wound around the shaft of a wheel, as in Fig. 9–31. The radius of the shaft is r, and the shaft is supported in fixed, frictionless bearings. When released from rest, the body descends a distance of 1.75 m in 5 s. Find the moment of inertia of the wheel and shaft, in terms of m and r.

9–37 A grinding wheel 0.2 m in diameter, of mass 3 kg, is rotating at 3600 rev min^{-1}.

a) What is its kinetic energy?

b) How far would it have to fall to acquire the same kinetic energy?

9–38 A disk of mass m and radius R is pivoted about a horizontal axis through its center, and a small body of mass m is attached to the rim of the disk. If the disk is released from rest with the small body at the end of a horizontal radius, find the angular velocity when the small body is at the bottom.

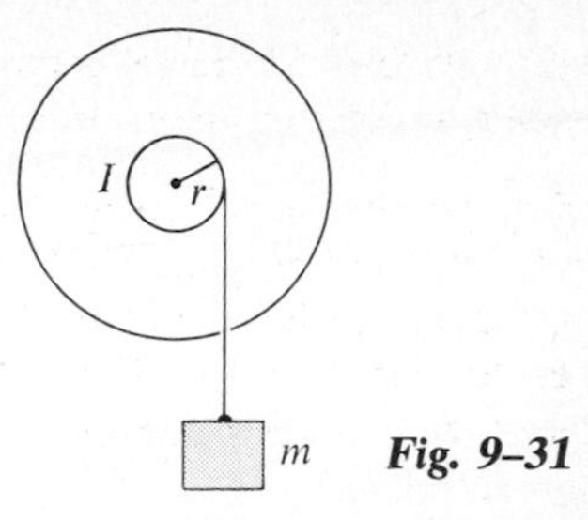

Fig. 9–31

9–39 The flywheel of a gasoline engine is required to give up 380 ft lb of kinetic energy while its angular velocity decreases from 600 rev min^{-1} to 540 rev min^{-1}. What moment of inertia is required?

9–40 The flywheel of a punch press has a moment of inertia of 25 kg m^2 and it runs at 300 rev min^{-1}. The flywheel supplies all the energy needed in a quick punching operation.

a) Find the speed in rev min^{-1} to which the flywheel will be reduced by a sudden punching operation requiring 4000 J of work.

b) What must be the constant power supply to the flywheel in horsepower to bring it back to its initial speed in 5 s?

9–41 A magazine article described a passenger bus in Zurich, Switzerland, which derived its motive power from the energy stored in a large flywheel. The wheel was brought up to speed periodically, when the bus stopped at a station, by an electric motor which could then be attached to the electric power lines. The flywheel was a solid cylinder of mass 1000 kg, diameter 1.8 m, and its top speed was 3000 rev min^{-1}.

a) At this speed, what is the kinetic energy of the flywheel?

b) If the average power required to operate the bus is 25 hp, how long can it operate between stops?

9–42 A grindstone in the form of a solid cylinder has a radius of 0.5 m and a mass of 50 kg.

a) What torque will bring it from rest to an angular velocity of 300 rev min^{-1} in 10 s?

b) What is its kinetic energy when rotating at 300 rev min^{-1} ?

9–43 The flywheel of a motor has a mass of 300 kg and a radius of gyration of 1.5 m. The motor develops a constant torque of 2000 N m and the flywheel starts from rest.

a) What is the angular acceleration of the flywheel?

b) What will be its angular velocity after making 4 revolutions?

c) How much work is done by the motor during the first 4 revolutions?

9–44 The flywheel of a stationary engine has a moment of inertia of 20 slug ft^2.

a) What constant torque is required to bring it up to an angular velocity of 900 rev min^{-1} in 10 s, starting from rest?

b) What is its final kinetic energy?

9–45 In Fig. 9–32, the steel balls A and B have a mass of 500 g each, and are rotating about the vertical axis with an angular velocity of 4 rad s^{-1} at a distance of 15 cm from the axis. Collar C is now forced down until the balls are at a distance of 5 cm from the axis. How much work must be done to move the collar down?

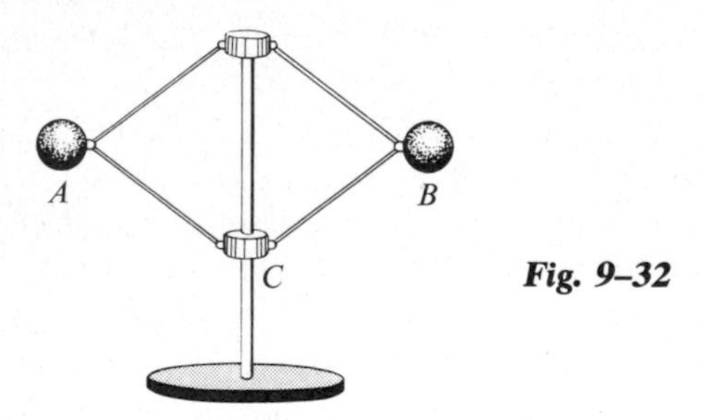

Fig. 9–32

9–46 A man sits on a piano stool holding a pair of dumbbells at a distance of 0.6 m from the axis of rotation of the stool. He is given an angular velocity of 5 rad s^{-1}, after which he pulls the dumbbells in until they are only 0.2 m distant from the axis. The moment of inertia of the man about the axis of rotation is 5 kg m^2 and may be considered constant. Each dumbbell has a mass of 5 kg and may be considered a point mass. Neglect friction.

a) What is the initial angular momentum of the system?

b) What is the angular velocity of the system after the dumbbells are pulled in toward the axis?

c) Compute the kinetic energy of the system before and after the dumbbells are pulled in. Account for the difference, if any.

9–47 A block of mass 0.05 kg is attached to a cord passing through a hole in a horizontal frictionless surface, as in Fig. 9–33. The block is originally revolving at a distance of 0.2 m from the hole, with an angular velocity of 3 rad s^{-1}. The cord is then pulled from below, shortening the radius of the circle in which the block revolves to 0.1 m. The block may be considered a point mass.

a) What is the angular velocity?

b) Find the change in kinetic energy of the block.

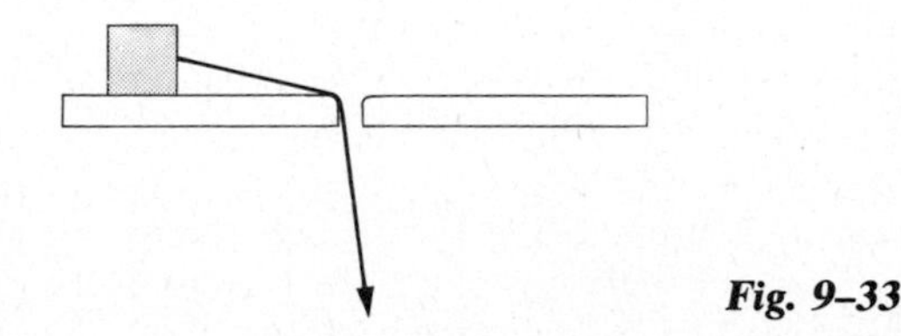

Fig. 9–33

9–48 A small block weighing 8 lb is attached to a cord passing through a hole in a horizontal frictionless surface. The block is originally revolving in a circle of radius 2 ft about the hole, with a tangential velocity of 12 ft s^{-1}. The cord is then pulled slowly from below, shortening the radius of the circle in which the block revolves. The breaking strength of the cord is 144 lb. What will be the radius of the circle when the cord breaks?

9–49 A block of mass M rests on a turntable which is rotating at constant angular velocity ω. A smooth cord runs from the block through a hole in the center of the table down to a hanging block of mass m. The coefficient of friction between the first block and the turntable is μ. (See Fig. 9–34.) Find the largest and smallest values of the radius r for which the first block will remain at rest relative to the turnable.

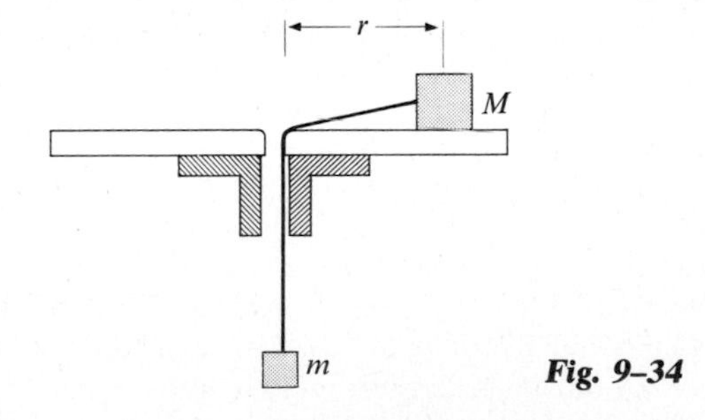

Fig. 9–34

9–50 A uniform rod of mass 0.03 kg and 0.2 m long rotates in a horizontal plane about a fixed vertical axis through its center. Two small bodies, each of mass 0.02 kg, are mounted so that they can slide along the rod. They are initially held by catches at positions 0.05 cm on each side of the center of the rod, and the system is rotating at 15 rev min^{-1}. Without otherwise changing the system, the catches are released and the masses slide outward along the rod and fly off at the ends.

a) What is the angular velocity of the system at the instant when the small masses reach the ends of the rod?

b) What is the angular velocity of the rod after the small masses leave it?

9–51 A turntable rotates about a fixed vertical axis, making one revolution in 10 s. The moment of inertia of the turntable about this axis is 1200 kg m^2. A man of mass 80 kg, initially standing at the center of the turntable, runs out along a radius. What is the angular velocity of the turntable when the man is 2 m from the center?

9–52 Disks A and B are mounted on a shaft SS and may be connected or disconnected by a clutch C, as in Fig. 9–35. The moment of inertia of disk A is one-half that of disk B. With the clutch disconnected, A is brought up to an angular velocity ω_0. The accelerating torque is then removed from A and it is coupled to disk B by the clutch. (Bearing friction may be neglected.) It is found that 2000 J of heat are developed in the clutch when the connection is made. What was the original kinetic energy of disk A?

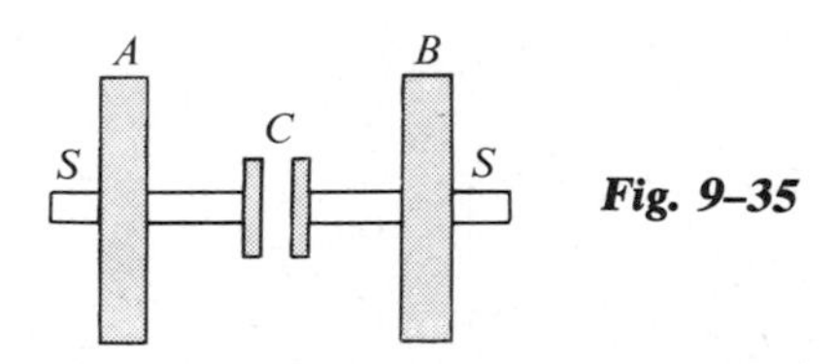

Fig. 9–35

9–53 A man of mass 100 kg stands at the rim of a turntable of radius 2 m and moment of inertia 4000 kg m^2, mounted on a vertical frictionless shaft at its center. The whole system is initially at rest. The man now walks along the outer edge of the turntable with a velocity of 1 m s^{-1}, relative to the earth.

a) With what angular velocity and in what direction does the turntable rotate?

b) Through what angle will it have rotated when the man reaches his initial position on the turntable?

c) Through what angle will it have rotated when he reaches his initial position relative to the earth?

9–54 A man of mass 60 kg runs around the edge of a horizontal turntable mounted on a vertical frictionless axis through its center. The velocity of the man, relative to the earth, is 1 m s^{-1}. The turntable is rotating in the opposite direction with an angular velocity of 0.2 rad s^{-1}. The radius of the turntable is 2 m and its moment of inertia about the axis of rotation is 400 kg m^2. Find the final angular velocity of the system if the man comes to rest, relative to the turntable.

9–55 Two flywheels, A and B, are mounted on shafts which can be connected or disengaged by a friction clutch C (see Fig. 9–35). The moment of inertia of wheel A is 4 slug ft^2. With the clutch disengaged, wheel A is brought up to an angular velocity of 600 rev min^{-1}. Wheel B is initially at rest. The clutch is now engaged, accelerating B and decelerating A, until both wheels have the same angular velocity. The final angular velocity of the system is 400 rev min^{-1}.

a) What was the moment of inertia of wheel B?

b) How much energy was lost in the process? (Neglect all bearing friction.)

9–56 The stabilizing gyroscope of a ship has a mass of 50,000 kg, its radius of gyration is 2 m, and it rotates about a vertical axis with an angular velocity of 900 rev min^{-1}.

a) How long a time is required to bring it up to speed, starting from rest, with a constant power input of 100 hp?

b) Find the torque needed to cause the axis to precess in a vertical fore-and-aft plane at the rate of 1 degree s^{-1}.

9–57 The mass of the rotor of a toy gyroscope is 150 g and its moment of inertia about its axis is 1500 g cm^2. The mass of the frame is 30 g. The gyroscope is supported on a single pivot, as in Fig. 9–36, with its center of gravity distant 4 cm horizontally from the pivot, and is precessing in a horizontal plane at the rate of one revolution in 6 s.

a) Find the upward force exerted by the pivot.

b) Find the angular velocity with which the rotor is spinning about its axis, expressed in rev min^{-1}.

c) Copy the diagram, and show by vectors the angular momentum of the rotor and the torque acting on it.

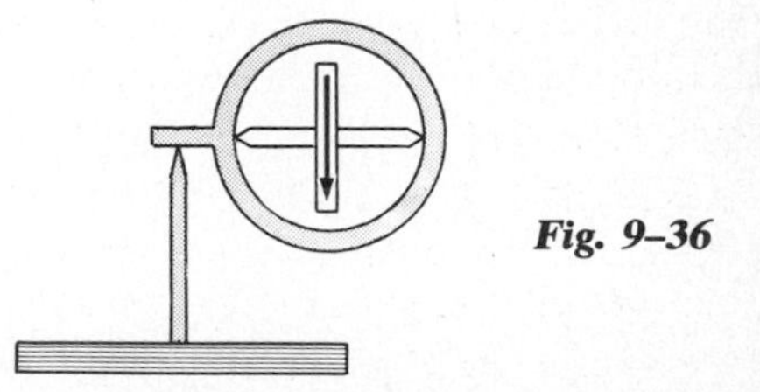

Fig. 9–36

9–58 The moment of inertia of the front wheel of a bicycle is 0.4 kg m^2, its radius is 0.5 m, and the forward speed of the bicycle is 5 m s^{-1}. With what angular velocity must the front wheel be turned about a vertical axis to counteract the capsizing torque due to a mass of 60 kg, 0.02 m horizontally to the right or left of the line of contact of wheels and ground? (Bicycle riders: Compare your own experience and see if your answer seems reasonable.)

9–59 The rotor of a small control gyro can be accelerated from rest to an angular velocity of 50,000 rev min^{-1} in 0.2 s. The moment of inertia of the rotor about its spin axis is 385 g cm^2.

a) What is the angular acceleration (assumed constant) as the rotor is brought up to speed?

b) What torque is needed to produce this angular acceleration?

c) Suppose the data above refer to the gyro in Fig. 9–24, and that the rotor is spinning clockwise when viewed from the left along the x-axis. Which way will the spin axis turn if ring B is forced to rotate clockwise, as seen from above, about the z-axis?

d) Suppose that rings A and B are locked in place. What torque must be exerted about the y-axis to rotate the outer frame about this axis at 1 deg s^{-1}?

9–60 A demonstration gyro wheel is constructed by removing the tire from a bicycle wheel 1.0 m in diameter, wrapping lead wire around the rim, and taping it in place. The shaft projects 0.2 m at each side of the wheel and a man holds the ends of the shaft in his hands. The mass of the system is 5 kg and its entire mass may be assumed to be located at its rim. The shaft is horizontal and the wheel is spinning about the shaft at 5 rev s^{-1}. Find the magnitude and direction of the force each hand exerts on the shaft under the following conditions:

a) The shaft is at rest.

b) The shaft is rotating in a horizontal plane about its center at 0.04 rev s^{-1}.

c) The shaft is rotating in a horizontal plane about its center at 0.20 rev s^{-1}.

d) At what rate must the shaft rotate in order that it may be supported at one end only?

Chapter 10

Elasticity

10–1 STRESS

The preceding chapter dealt with the motion of a rigid body, an idealized model used to represent the motion of a body when rotational motion may not be neglected but when the changes in the *shape* of the body are so small as to be negligible. Real materials always yield to some extent under the influence of applied forces; we shall now develop models to describe the relation between deformation of real materials and the forces responsible for these deformations.

Ultimately, the change in shape or volume of a body when outside forces act on it is determined by the forces between its molecules. Although molecular theory is at present not sufficiently advanced to enable us to calculate the elastic properties of, say, a block of copper starting from the properties of a copper atom, the study of the solid state is an active subject in many research laboratories and our knowledge of it is steadily increasing. In this chapter we shall, however, confine ourselves to quantities that are directly measurable, and not attempt any detailed molecular explanation of the observed behavior.

Figure 10–1(a) shows a bar of uniform cross-sectional area A subjected to equal and opposite pulls $\boldsymbol{F}$ at its ends. The bar is said to be in *tension.* Consider a section through the bar at right angles to its length, as indicated by the dotted line. Since every portion of the bar is in equilibrium, that portion at the right of the section must be pulling on the portion at the left with a force $\boldsymbol{F}$, and vice versa. If the section is not too near the ends of the bar, these pulls are

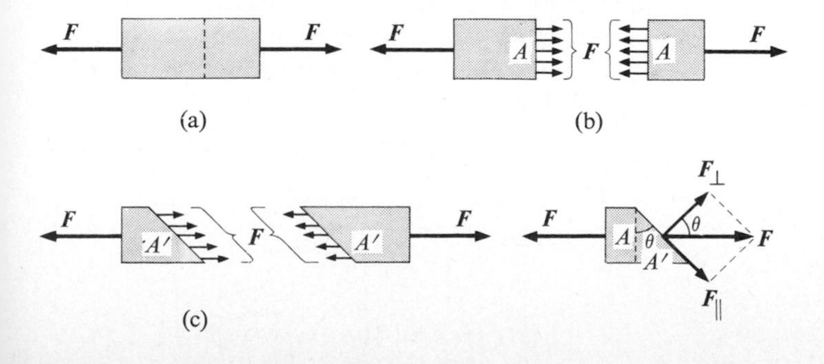

Fig. 10–1 *(a) A bar in tension. (b) The stress at a perpendicular section equals F/A. (c) and (d) The stress at an inclined section can be resolved into a normal stress $F_\perp/A'$, and a tangential or shearing stress $F_\parallel/A'$.*

uniformly distributed over the cross-sectional area A, as indicated by the short arrows in Fig. 10–1(b). We define the *stress* S at the section as the ratio of the force F to the area A:

$$\boxed{\text{Stress} = \frac{F}{A}.} \tag{10–1}$$

The stress is called a *tensile* stress, meaning that each portion *pulls* on the other, and it is also a *normal* stress because the distributed force is perpendicular to the area. Units of stress are 1 newton per square meter (1 N m^{-2}), 1 dyne per square centimeter (1 dyn cm^{-2}), and 1 pound per square foot (1 lb ft^{-2}). The hybrid unit 1 lb in^{-2} is also commonly used.

Example A human biceps (upper arm muscle) may exert a force of the order of 600 N on the bones to which it is attached. If the muscle has a cross-sectional area at its center of 50 cm^2 = 0.005 m^2, and the tendon attaching its lower end to the bones below the elbow joint has a cross section of 0.5 cm^2 = 5×10^{-5} m^2, find the tensile stress in each of these.

In each case the stress is the force per unit area. For the muscle,

$$\text{Tensile stress} = \frac{600\ \text{N}}{0.005\ \text{m}^2} = 120{,}000\ \text{N m}^{-2}$$
$$= 1.2 \times 10^5\ \text{N m}^{-2}.$$

For the tendon,

$$\text{Tensile stress} = \frac{600\ \text{N}}{5 \times 10^{-5}\ \text{m}^2} = 1.2 \times 10^7\ \text{N m}^{-2}.$$

The maximum force a muscle can exert depends on its cross-sectional area, but the maximum *stress* is nearly the same for a wide variety of muscles.

Returning to Fig. 10–1, we consider next a section through the bar in some arbitrary direction, as in Fig. 10–1(c). The resultant force exerted on the portion at either side of this section, by the portion at the other, is equal and opposite to the force $\boldsymbol{F}$ at the end of the section. Now, however, the force is distributed over a larger area A' and is not at right angles to the area. If we represent the resultant of the distributed forces by a single vector of magnitude F, as in Fig. 10–1(d), this vector can be resolved into a component $F_\perp$ normal to the area A', and a component $F_\parallel$ tangent to the area. The *normal* stress is defined, as before, as the ratio of the component $F_\perp$ to the area A'. The ratio of the component $F_\parallel$ to the area A' is called the *tangential* stress or, more commonly, the *shearing* stress at the section:

$$\begin{aligned} \text{Normal stress} &= \frac{F_\perp}{A'} \\ \text{Tangential (shearing) stress} &= \frac{F_\parallel}{A'}. \end{aligned} \tag{10–2}$$

Stress is not a vector quantity since, unlike a force, we cannot assign to it a specific direction. The *force* acting on the portion of the body on a specified side of a section has a definite direction. Stress is one of a class of physical quantities called *tensors.*

A bar subjected to *pushes* at its ends, as in Fig. 10–2, is said to be in *compression.* The stress on the dotted section, illustrated in part (b), is also a normal stress but is now a *compressive* stress, since each portion pushes on the other. It should be evident that if we take a section in some arbitrary direction, it will be subject to both a tangential (shearing) and a normal stress, the latter now being a compression.

As another example of a body under stress, consider the block of square cross section in Fig. 10–3(a), acted on by two equal and opposite couples produced by the pairs of forces $\boldsymbol{F}_x$ and $\boldsymbol{F}_y$ distributed over its surfaces. The block is in equilibrium, and any portion of it is in equilibrium also. Thus the distributed forces over the diagonal face in part (b) must have a resultant $\boldsymbol{F}$ whose components are equal to $\boldsymbol{F}_x$ and $\boldsymbol{F}_y$. The stress at this section is therefore a pure compression, although the stresses at the right face and the bottom are both shearing stresses. Similarly, we see from Fig. 10–3(c) that the other diagonal face is in pure tension.

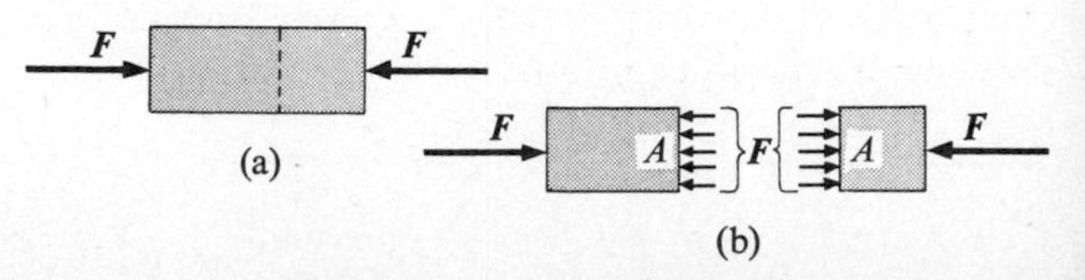

Fig. 10–2 *A bar in compression.*

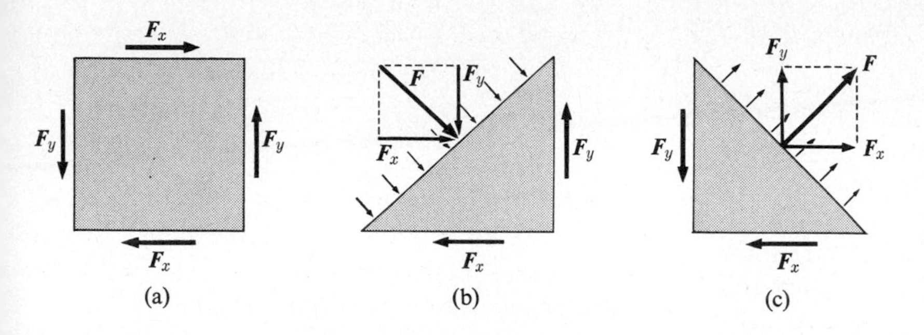

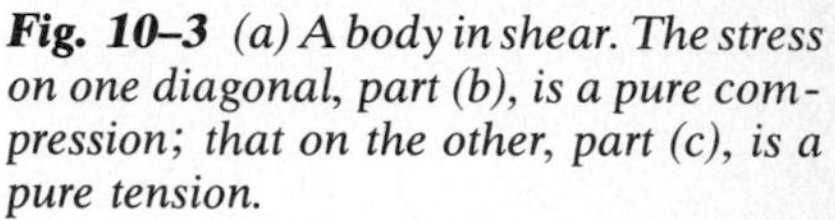
Fig. 10–3 *(a) A body in shear. The stress on one diagonal, part (b), is a pure compression; that on the other, part (c), is a pure tension.*

Consider next a fluid under pressure. The term "fluid" means a substance that can flow; hence the term applies to both liquids and gases. If there is a shearing stress at any point in a fluid, the fluid slips sidewise so long as the stress is maintained. Hence in a fluid at rest, the shearing stress is everywhere zero. Figure 10–4 represents a fluid in a cylinder provided with a piston, on which is exerted a downward force. The triangle is a side view of a wedge-shaped portion of the fluid. If for the moment we neglect the weight of the fluid, the only forces on this portion are those exerted by the rest of the fluid, and since these forces can have no shearing (or tangential) component, they must be normal to the surfaces of the wedge. Let $\boldsymbol{F}_x$, $\boldsymbol{F}_y$, and $\boldsymbol{F}$ represent the forces against the three faces. Since the fluid is in equilibrium, it follows that

$$F \sin \theta = F_x, \qquad F \cos \theta = F_y.$$

Also,

$$A \sin \theta = A_x, \qquad A \cos \theta = A_y.$$

Dividing the upper equations by the lower, we find

$$\frac{F}{A} = \frac{F_x}{A_x} = \frac{F_y}{A_y}.$$

Hence the force per unit area is the *same*, regardless of the direction of the section, and is always a compression. Any one of the preceding ratios defines the *hydrostatic pressure p* in the fluid,

$$\boxed{p = \frac{F}{A}, \qquad F = pA.} \tag{10–3}$$

Units of pressure are 1 N m^{-2}, 1 dyn cm^{-2}, or 1 lb ft^{-2}. Like other types of stress, pressure is not a vector quantity and no direction can be assigned to it. *The force against any area within (or bounding) a fluid at rest and under pressure is normal to the area, regardless of the orientation of the area.* This is what is meant by the common statement that "the pressure in a fluid is the same in all directions."

The stress within a solid can also be a hydrostatic pressure, provided the stress at all points of the surface of the solid is of this nature. That is, the force per unit area must be the same at *all* points of the surface, and the force must be normal to the surface and directed inward. This is *not* the case in Fig. 10–2, where forces are applied at the ends of the bar only, but it is automatically the case if a solid is immersed in a fluid under pressure.

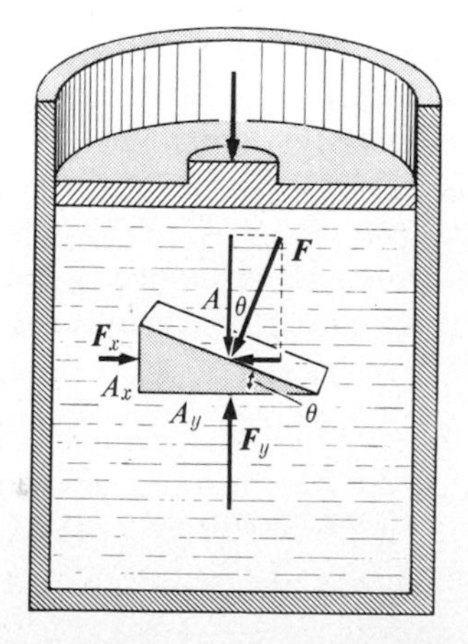

Fig. 10–4 *A fluid under hydrostatic pressure. The force on a surface in any direction is normal to the surface.*

10–2 STRAIN

The term *strain* refers to the relative change in dimensions or shape of a body which is subjected to stress. Associated with each type of stress described in the preceding section is a corresponding type of strain.

Figure 10–5 shows a bar whose natural length is ℓ_0 and which elongates to a length ℓ when equal and opposite pulls are exerted at its ends. The elongation, of course, does not occur at the ends only; every element of the bar stretches in the same proportion as does the bar as a whole. The *tensile strain* in the bar is defined as the ratio of the increase in length to the original length:

$$\text{Tensile strain} = \frac{\ell - \ell_0}{\ell_0} = \frac{\Delta\ell}{\ell_0}. \qquad (10\text{–}4)$$

The *compressive strain* of a bar in compression is defined in the same way, as the ratio of the decrease in length to the original length.

Figure 10–6(a) illustrates the nature of the deformation when shearing stresses act on the faces of a block, as in Fig. 10–3. The dotted outline *abcd* represents the unstressed block, and the full lines $a'b'c'd'$ represent the block under stress. The centers of the stressed and unstressed block coincide in part (a). In part (b), the edges *ad* and $a'd'$ coincide. The lengths of the faces under shear remain very nearly constant, while all dimensions parallel to the diagonal *ac* increase in length, and those parallel to the diagonal *bd* decrease in length. Note that this is to be expected in view of the nature of the corresponding internal stresses (see Fig. 10–3). This type of strain is called a *shear strain*, and is defined as the ratio of the displacement x of corner b to the transverse dimension h:

$$\text{Shearing strain} = \frac{x}{h}. \qquad (10\text{–}5)$$

Like other types of strain, shearing strain is a pure number.

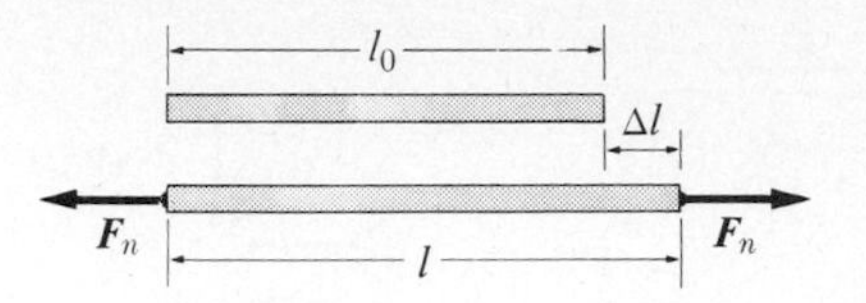

Fig. 10–5 *The longitudinal strain is defined as* $\Delta\ell/\ell_0$.

The strain produced by a hydrostatic pressure, called a *volume strain*, is defined as the ratio of the change in volume, ΔV, to the original volume V. It also is a pure number:

$$\text{Volume strain} = \frac{\Delta V}{V}. \qquad (10\text{–}6)$$

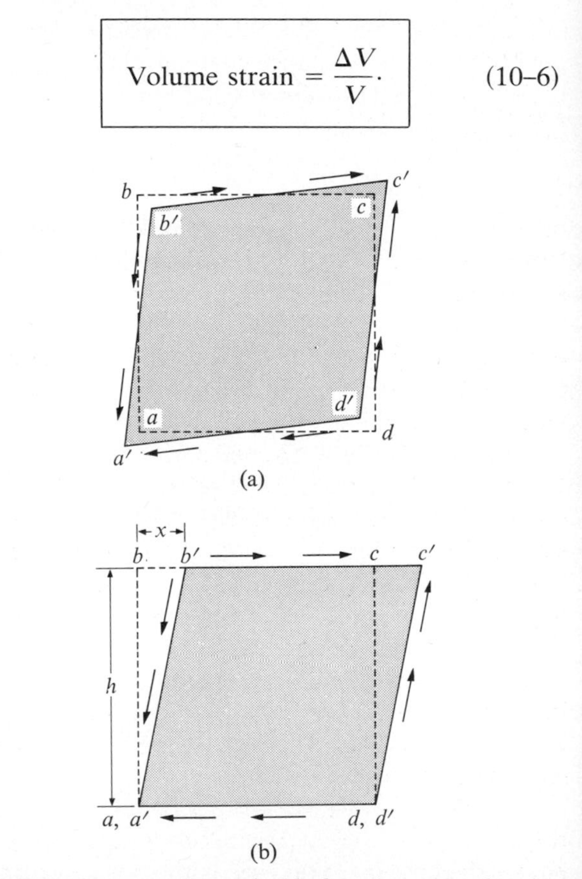

Fig. 10–6 *Change in shape of a block in shear. The shearing strain is defined as* x/h.

10–3 ELASTICITY AND PLASTICITY

The relation between each of the three kinds of stress and its corresponding strain plays an important role in the branch of physics called the *theory of elasticity*,

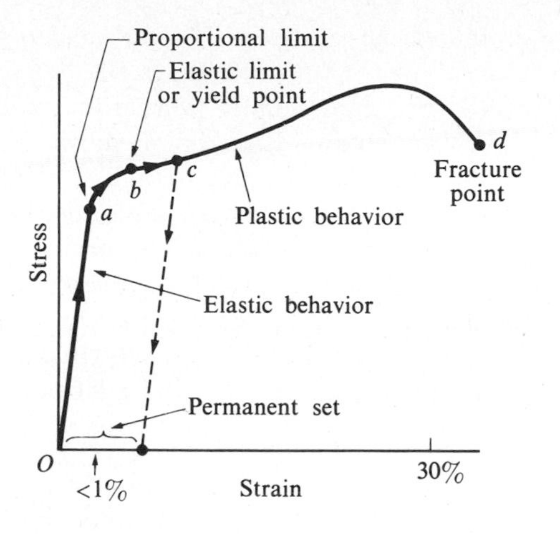

Fig. 10–7 *Typical stress–strain diagram for a ductile metal under tension.*

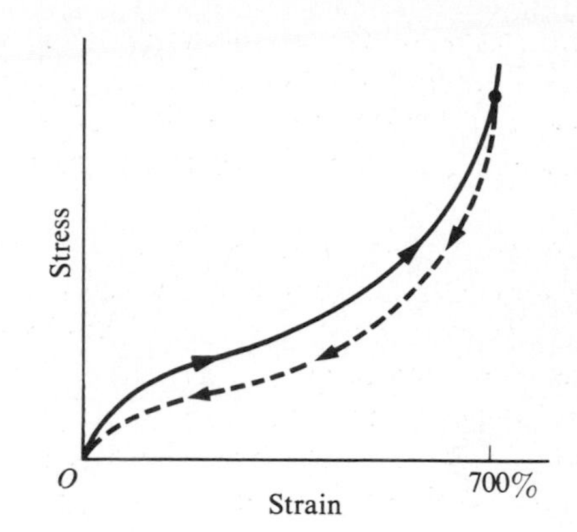

Fig. 10–8 *Typical stress–strain diagram for vulcanized rubber, showing elastic hysteresis.*

or its engineering counterpart, *strength of materials.* When any stress is plotted against the appropriate strain, the resulting stress–strain diagram is found to have several different shapes, depending on the kind of material. Two of the most important materials of present-day science and technology are metal and vulcanized rubber.

There are wide variations in the elastic behavior of metals. A typical stress–strain diagram for a ductile metal is shown in Fig. 10–7. The stress is a simple tensile stress and the strain is the percentage elongation. During the first portion of the curve (up to a strain of less than 1 percent), the stress and strain are proportional until the point *a*, the *proportional limit,* is reached. The fact that there is a region in which stress and strain are proportional is called *Hooke's Law.* (Robert Hooke (1635–1703) was a contemporary of Newton.)

From *a* to *b*, stress and strain are not proportional, but nevertheless, if the load is removed at any point between *O* and *b*, the curve will be retraced and the material will return to its original length. In the region *Ob*, the material is said to be *elastic* or to exhibit *elastic behavior,* and the point *b* is called the *elastic limit,* or the *yield point.* Up to this point, the forces exerted by the material are *conservative*; when the material returns to its original shape, work done in producing the deformation is recovered.

If the material is loaded further, the strain increases rapidly, but when the load is removed at some point beyond *b*, say *c*, the material does not come back to its original length but traverses the dashed line in Fig. 10–7. The length at zero stress is now greater than the original length and the material is said to have a *permanent set.* Further increase of load beyond *c* produces a large increase in strain until a point *d* is reached at which *fracture* takes place. From *b* to *d*, the metal is said to undergo *plastic flow* or *plastic deformation.* If large plastic deformation takes place between the elastic limit and the fracture point, the metal is said to be *ductile.* If, however, fracture occurs soon after the elastic limit is passed, the metal is said to be *brittle.*

Figure 10–8 shows a stress–strain curve for a typical sample of vulcanized rubber which has been stretched to over seven times its original length. During no portion of this curve is the stress proportional to the strain. The substance, however, is elastic, in the sense that when the load is removed, the rubber is restored to its original length. On decreasing the load, the stress–strain curve is *not* retraced but follows the dashed curve of Fig. 10–8.

The lack of coincidence of the curves for increasing and decreasing stress is known as *elastic hysteresis* (accent on the *third* syllable). (An analogous phenomenon observed with magnetic materials is called *magnetic hysteresis.*) When the stress–strain relation exhibits this behavior, the associated forces are *not* conservative, since the work done by the material in returning to its original shape is *less* than the work required to deform it. It can be shown that the area bounded by the two curves, that is, the area of the *hysteresis loop*, is equal to the energy dissipated within the elastic or magnetic material.

The large elastic hysteresis of some types of rubber makes these materials very valuable as vibration absorbers. If a block of such material is placed between a piece of vibrating machinery, and, let us say, the floor, elastic hysteresis takes place during every cycle of vibration. Mechanical energy is converted to a form known as internal energy which evidences itself by a rise in temperature. As a result, only a small amount of energy of vibration is transmitted to the floor.

The stress required to cause actual fracture of a material is called the *breaking stress* or the *ultimate strength.* Two materials, such as two steels, may have very similar elastic constants but vastly different breaking stresses.

10–4 ELASTIC MODULUS

The stress required to produce a given strain depends on the nature of the material under stress. The ratio of stress to strain, or the *stress per unit strain*, is called an *elastic modulus* of the material. The larger the elastic modulus, the greater the stress needed for a given strain.

Consider first longitudinal (tensile or compressive) stresses and strains. Experiment shows that up to the proportional limit, a given longitudinal stress produces a strain of the same magnitude whether the stress is a tension or a compression. Hence, the ratio of tensile stress to tensile strain, for a given material, equals the ratio of compressive stress to compressive strain. This ratio is called the *stretch modulus* or *Young's modulus* of the material, and will be denoted by Y:

$$Y = \frac{\text{tensile stress}}{\text{tensile strain}} = \frac{\text{compressive stress}}{\text{compressive strain}} = \frac{F_\perp/A}{\Delta\ell/\ell_0} = \frac{\ell_0}{A}\frac{F_\perp}{\Delta\ell}. \tag{10–7}$$

If the proportional limit is not exceeded, the ratio of stress to strain is constant, and Hooke's law is therefore equivalent to the statement that *within the proportional limit, the elastic modulus of a given material is constant*, depending only on the nature of the material.

Since a strain is a pure number, the units of Young's modulus are the same as those of stress, namely, force per unit area. Some typical values are listed in Table 10–1.

Table 10–1 APPROXIMATE ELASTIC CONSTANTS

Material	Young's modulus, Y		Shear modulus, S		Bulk modulus, B		Poisson's ratio, σ
	10^{11} N m^{-2}	10^6 lb in^{-2}	10^{11} N m^{-2}	10^6 lb in^{-2}	10^{11} N m^{-2}	10^6 lb in^{-2}	
Aluminum	0.70	10	0.30	3.4	0.70	10	0.16
Brass	0.91	13	0.36	5.1	0.61	8.5	0.26
Copper	1.1	16	0.42	6.0	1.4	20	0.32
Glass	0.55	7.8	0.23	3.3	0.37	5.2	0.19
Iron	1.9	26	0.70	10	1.0	14	0.27
Lead	0.16	2.3	0.056	0.8	0.077	1.1	0.43
Nickel	2.1	30	0.77	11	2.6	34	0.36
Steel	2.0	29	0.84	12	1.6	23	0.19
Tungsten	3.6	51	1.5	21	2.0	29	0.20

When a material elongates under tensile stress, the dimensions *perpendicular* to the direction of stress become shorter by an amount proportional to the fractional change in length. If w is the original width and Δw the change in width, then it is found that

$$\frac{\Delta w}{w} = -\sigma \frac{\Delta \ell}{\ell}, \tag{10–8}$$

where σ is a dimensionless constant characteristic of the material, called *Poisson's ratio.* For many common materials σ has a value between 0.1 and 0.3. Similarly, a material under compressive stress "bulges" at the sides, and again the fractional change in width is given by Eq. (10–8).

The *shear modulus* S of a material, within the Hooke's law region, is defined as the ratio of a shearing stress to the shearing strain it produces:

$$S = \frac{\text{shearing stress}}{\text{shearing strain}} = \frac{F_{\parallel}/A}{x/h} = \frac{h}{A}\frac{F_{\parallel}}{x}. \tag{10–9}$$

(Refer to Fig. 10–6 for the meaning of x and of h.)

The shear modulus of a material is also expressed as force per unit area. For most materials it is one-half to one-third as great as Young's modulus. The shear modulus is also called the *modulus of rigidity* or the *torsion modulus.*

The more general definition of shear modulus is

$$S = \frac{dF_{\parallel}/A}{dx/h} = \frac{h}{A}\frac{dF_{\parallel}}{dx}, \tag{10–10}$$

where dx is the increase in x when the shearing force increases by $dF_{\parallel}$.

The shear modulus has a significance for *solid* materials only. A liquid or gas flows under the influence of a shearing stress, and cannot *permanently* support such a stress.

The modulus relating a hydrostatic pressure to the volume strain it produces is called the *bulk modulus,* and we shall represent it by B. The general definition of bulk modulus is the (negative) ratio of a change in pressure dp to the volume strain dV/V (fractional change in volume) produced by it:

$$B = -\frac{dp}{dV/V} = -V\frac{dp}{dV}. \tag{10–11}$$

The minus sign is included in the definition of B because an *increase* of pressure always causes a *decrease* in volume. That is, if dp is positive, dV is negative. By including a minus sign in its definition, we make the bulk modulus itself a positive quantity.

The change in volume of a *solid* or *liquid* under pressure is so small that the volume V in Eq. (10–11) can be considered constant. Provided the pressure is not too great, the ratio dp/dV is constant also, the bulk modulus is constant, and we can replace dp and dV by finite changes in pressure and volume. The volume of a *gas,* however, changes markedly with pressure, and the general definition of B must be used for gases.

The reciprocal of the bulk modulus is called the *compressibility* k. From its definition,

$$k = \frac{1}{B} = -\frac{dV/V}{dp} = -\frac{1}{V}\frac{dV}{dp}. \tag{10–12}$$

The compressibility of a material thus equals the *fractional decrease in volume,* $-dV/V$, *per unit increase* dp *in pressure.*

The units of a bulk modulus are the same as those of pressure, and the units of compressibility are those of *reciprocal pressure.* Thus the statement that the compressibility of water (see Table 10–2) is 50×10^{-6} atm^{-1} means that the volume decreases by

Table 10–2 COMPRESSIBILITIES OF LIQUIDS

Liquid	Compressibility, k		
	$(\text{N m}^{-2})^{-1}$	$(\text{lb in}^{-2})^{-1}$	atm^{-1}
Carbon disulfide	64×10^{-11}	45×10^{-7}	66×10^{-6}
Ethyl alcohol	110	78	115
Glycerine	21	15	22
Mercury	3.7	2.6	3.8
Water	49	34	50

Table 10–3 STRESSES AND STRAINS

Type of stress	Stress	Strain	Elastic modulus	Name of modulus
Tension or compression	$\dfrac{F_\perp}{A}$	$\dfrac{\Delta \ell}{\ell_0}$	$Y = \dfrac{F_\perp/A}{\Delta\ell/\ell_0}$	Young's modulus
Shear	$\dfrac{F_\parallel}{A}$	$\tan\phi \approx \phi$	$S = \dfrac{F_\parallel/A}{\phi}$	Shear modulus
Hydrostatic pressure	$p\left(=\dfrac{F_\perp}{A}\right)$	$\dfrac{\Delta V}{V_0}$	$B = -\dfrac{p}{\Delta V/V_0}$	Bulk modulus

50 millionths of the original volume for each atmosphere increase in pressure. (1 atm = 1.02×10^5 N m^{-2} = 14.7 lb in^{-2}.)

The various types of stress, strain, and elastic modulus are summarized in Table 10–3.

Example 1 In an experiment to measure Young's modulus, a load of 500 kg, hanging from a steel wire of length 3 m and cross section 0.20 cm^2, was found to stretch the wire 0.4 cm above its no-load length. What were the stress, the strain, and the value of Young's modulus for the steel of which the wire was composed?

$$\text{Stress} = \frac{F_\perp}{A} = \frac{(500\text{ kg})(10\text{ m s}^{-2})}{0.00002\text{ m}^2} = 250 \times 10^6\text{ N m}^{-2};$$

$$\text{Strain} = \frac{\Delta\ell}{\ell_0} = \frac{0.004\text{ m}}{3\text{ m}} = 0.00133;$$

$$Y = \frac{\text{stress}}{\text{strain}} = \frac{250 \times 10^6\text{ N m}^{-2}}{0.00133} = 1.9 \times 10^{11}\text{ N m}^{-2}.$$

Example 2 Suppose the object in Fig. 10–6 is a brass plate 1.0 m square and 0.5 cm thick. How large a force F must be exerted on each of its edges if the displacement x in Fig. 10–6(b) is 0.02 cm? The shear modulus of brass is 0.36×10^{11} N m^{-2}.

The shearing stress on each edge is

$$\text{Shearing stress} = \frac{F_\parallel}{A} = \frac{F}{(1.0\text{ m})(0.005\text{ m})} = 200\, F\text{ N m}^{-2}.$$

The shearing strain is

$$\text{Shearing strain} = \frac{x}{h} = \frac{2 \times 10^{-4}\text{ m}}{1.0\text{ m}} = 2.0 \times 10^{-4}.$$

$$\text{Shear modulus } S = \frac{\text{stress}}{\text{strain}} = 0.36 \times 10^{11}\text{ N m}^{-2} = \frac{200\, F\text{ N m}^{-2}}{2.0 \times 10^{-4}},$$

$$F = 3.6 \times 10^4\text{ N}.$$

Example 3 The volume of oil contained in a certain hydraulic press is 0.2 m^3. Find the decrease in volume of the oil when subjected to a pressure of 2.04×10^7 N m^{-2}. The compressibility of the oil is 20×10^{-6} atm^{-1}.

The volume decreases by 20 parts per million for a pressure increase of 1 atm. Since 2.04×10^7 N m^{-2} = 200 atm, the volume decrease is $200 \times 20 = 4000$ parts per million. Since the original volume is 0.2 m^3, the actual decrease is

$$\left(\frac{4000}{1{,}000{,}000}\right)(0.2\text{ m}^3) = 8 \times 10^{-4}\text{ m}^3 = 800\text{ cm}^3;$$

or, from Eq. (10–12),

$$\Delta V = -kV\,\Delta p = -(20 \times 10^{-6}\text{ atm}^{-1})(0.2\text{ m}^3)(200\text{ atm}) = -800 \times 10^{-6}\text{ m}^3.$$

Although the three elastic moduli and Poisson's ratio have been discussed separately, they are not completely independent. For materials having no distinction between various directions (i.e., *isotropic* materials), only two of these are really independent. For example, the bulk and shear moduli may be expressed in terms of Young's modulus and Poisson's ratio:

$$B = \frac{Y}{3(1 - 2\sigma)}, \qquad S = \frac{Y}{2(1 + \sigma)}.$$

For materials having directional properties, such as wood (which has a grain direction) and single crystals of materials, these relations do not hold, and the elastic behavior is more complex.

10–5 THE FORCE CONSTANT

The various elastic moduli are quantities which describe the elastic properties of a particular *material* and do not directly indicate how much a given rod, cable, or spring constructed of the material will distort under load. If Eq. (10–7) is solved for $F_\perp$, one obtains

$$F_\perp = \frac{YA}{\ell_0}\Delta\ell;$$

or, if YA/ℓ_0 is replaced by a single constant k, and the elongation $\Delta\ell$ is represented by x, then

$$F_\perp = kx. \qquad (10\text{–}13)$$

In other words, the elongation of a body in tension above its no-load length is directly proportional to the stretching force. Hooke's law was originally stated in this form, rather than in terms of stress and strain.

When a helical wire spring is stretched, the stress in the wire is practically pure shear. The elongation of the spring as a whole is directly proportional to the stretching force. That is, an equation of the form $F = kx$ still applies, the constant k depending on the shear modulus of the wire, its radius, the radius of the coils, and the number of coils.

The constant k, or the ratio of the force to the elongation, is called the *force constant* or the *stiffness* of the spring, and is expressed in newtons per meter or pounds per foot. It is equal numerically to the force required to produce unit elongation.

The ratio of elongation to the force, or the elongation per unit force, is called the *compliance* of the spring. The compliance equals the reciprocal of the force constant and is expressed in meters per newton or feet per pound. It is numerically equal to the elongation produced by unit force.

PROBLEMS

10–1 A steel rod 4 m long and 0.5 cm^2 in cross section is found to stretch 0.2 cm under a tension of 12,000 N. What is Young's modulus for this steel?

10–2 The elastic limit of a steel elevator cable is 40,000 lb in^{-2}. Find the maximum upward acceleration which can be given a 2-ton elevator when supported by a cable whose cross section is $\frac{1}{2}$ in^2 if the stress is not to exceed $\frac{1}{4}$ of the elastic limit.

10–3 A copper wire 12 ft long and 0.036 in. in diameter was given the test below. A load of 4.5 lb was originally hung from the wire to keep it taut. The position of the lower end of the wire was read on a scale.

Added load, lb	Scale reading, in.
0	3.02
2	3.04
4	3.06
6	3.08
8	3.10
10	3.12
12	3.14
14	3.65

a) Make a graph of these values, plotting the increase in length horizontally and the added load vertically.

b) Calculate the value of Young's modulus.

c) What was the stress at the proportional limit?

10–4 A steel wire has the following properties:

Length = 10 ft
Cross section = 0.01 in^2
Young's modulus = 30,000,000 lb in^{-2}
Shear modulus = 10,000,000 lb in^{-2}
Proportional limit = 60,000 lb in^{-2}
Breaking stress = 120,000 lb in^{-2}

The wire is fastened at its upper end and hangs vertically. (a) How great a load can be supported without exceeding the proportional limit? (b) How much will the wire stretch under this load? (c) What is the maximum load that can be supported?

10–5 A nylon rope used by mountaineers elongates 1.5 m under the weight of an 80-kg climber. If the rope is 50 m in length and 9 mm in diameter, what is Young's modulus for this material? If Poisson's ratio for nylon is 0.2, find the change in diameter under this stress.

10–6 A relaxed biceps muscle requires a force of 25 N for an elongation of 5 cm, and the same muscle under maximum tension requires a force of 500 N for the same elongation. Regarding the muscle as a uniform cylinder of length 0.2 m and cross-sectional area 50 cm^2, find Young's modulus for the muscle tissue under each of these conditions.

10–7

a) What is the maximum load that can be supported by an aluminum wire 0.1 cm in diameter without exceeding the proportional limit of 8×10^7 N m^{-2}?

b) If the wire was originally 5 m long, how much will it elongate under this load?

c) How much does the *diameter* change under this load?

10–8 A 5-kg mass hangs on a vertical steel wire 0.5 m long and 0.004 cm^2 in cross section. Hanging from the bottom of this weight is a similar steel wire which supports a 10-kg mass. Compute (a) the longitudinal strain, and (b) the elongation of each wire.

10–9 A 15-kg mass, fastened to the end of a steel wire of unstretched length 0.5 m, is whirled in a vertical circle with an angular velocity of 2 rev s^{-1} at the bottom of the circle. The cross section of the wire is 0.02 cm^2. Calculate the elongation of the wire when the weight is at the lowest point of its path.

10–10 A copper wire 8 m long and a steel wire 4 m long, each of cross section 0.5 cm^2, are fastened end-to-end and stretched with a tension of 500 N.

a) What is the change in length of each wire?

b) What is the elastic potential energy of the system?

10–11 A copper rod of length 2 m and cross-sectional area 2.0 cm^2 is fastened end-to-end to a steel rod of length L and cross-sectional area 1.0 cm^2. The compound rod is subjected to equal and opposite pulls of magnitude 3×10^4 N at its ends.

a) Find the length L of the steel rod if the elongations of the two rods are equal.

b) What is the stress in each rod?

c) What is the strain in each rod?

10–12 A rod 1.05 m long, whose weight is negligible, is supported at its ends by wires A and B of equal length, as shown in Fig. 10–9. The cross section of A is 1 mm^2, that of B is 2 mm^2. Young's modulus for wire A is 2.4×10^{11} N m^{-2}, and for B it is 1.6×10^{11} N m^{-2}. At what point along the bar should a weight w be suspended in order to produce (a) equal stresses in A and B, (b) equal strains in A and B?

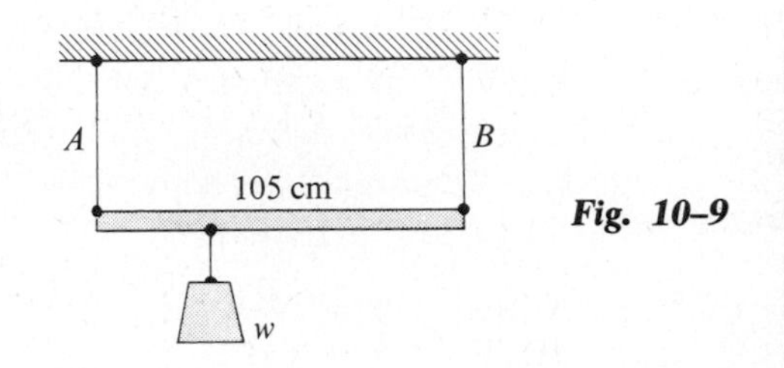

Fig. 10–9

10–13 A bar of length L, cross-sectional area A, and Young's modulus Y, is subjected to a tension F. Represent the stress in the bar by Q and the strain by P. Derive the expression for the elastic potential energy, per unit volume, of the bar in terms of Q and P.

10–14 A certain elevator cable is to have a maximum stress of 10,000 lb in^{-2}, to allow for appropriate safety factors. If it is to support a loaded elevator with total weight of 4000 lb, which has a maximum upward acceleration of 5 ft s^{-2}, what should be the diameter of the cable?

10–15 A steel bar 0.2 cm square and 5 m long is stretched with a force of 400 N at each end. Find the stress, the strain, the total elongation, and the fractional change in thickness of the bar.

10–16 Two round rods, one of steel, the other of brass, are joined end to end. Each rod is 0.5 m long and 2 cm in diameter. The combination is subjected to tensile forces of 5000 N.

a) What is the strain in each rod?

b) What is the elongation of each rod?

c) What is the change in diameter of each rod?

10–17 A specimen of oil having an initial volume of 1000 cm^3 is subjected to a pressure of 12×10^5 N m^{-2}, and the volume is found to decrease by 0.3 cm^3. What is the bulk modulus for the material? The compressibility?

10–18 The compressibility of sodium is to be measured by observing the displacement of the piston in Fig. 10–4 when a force is applied. The sodium is immersed in an oil which fills the cylinder below the piston. Assume that the piston and walls of the cylinder are perfectly rigid, that there is no friction, and no oil leak. Compute the compressibility of the sodium in terms of the applied force F, the piston displacement x, the piston area A, the initial volume of the oil V_0, the initial volume of the sodium v_0, and the compressibility of the oil k_0.

10–19 Two strips of metal are riveted together at their ends by four rivets, each of diameter 0.5 cm. What is the maximum tension that can be exerted by the riveted strip if the shearing stress on the rivets is not to exceed 6×10^8 N m^{-2}? Assume each rivet to carry one-quarter of the load.

10–20 Find the density of ocean water at a depth of 500 m where the pressure is about 5.0×10^6 N m^{-2}. The density at the surface is 1.03×10^3 kg m^{-3}.

10–21 Compute the compressibility of steel, in reciprocal atmospheres, and compare with that of water. Which material is the more readily compressed?

10–22 A steel post 6 in. in diameter and 10 ft long is placed vertically and is required to support a load of 20,000 lb.

a) What is the stress in the post?

b) What is the strain in the post?

c) What is the change in length of the post?

10–23 A hollow cylindrical steel column 10 ft high shortens 0.01 in under a compression load of 72,000 lb. If the inner radius of the cylinder is 0.80 of the outer one, what is the outer radius?

10–24 A bar of cross section A is subjected to equal and opposite tensile forces $\mathbf{F}$ at its ends. Consider a plane through the bar making an angle θ with a plane at right angles to the bar (Fig. 10–10).

a) What is the tensile (normal) stress at this plane, in terms of F, A, θ?

b) What is the shearing (tangential) stress at the plane, in terms of F, A, and θ?

c) For what value of θ is the tensile stress a maximum?

d) For what value of θ is the shearing stress a maximum?

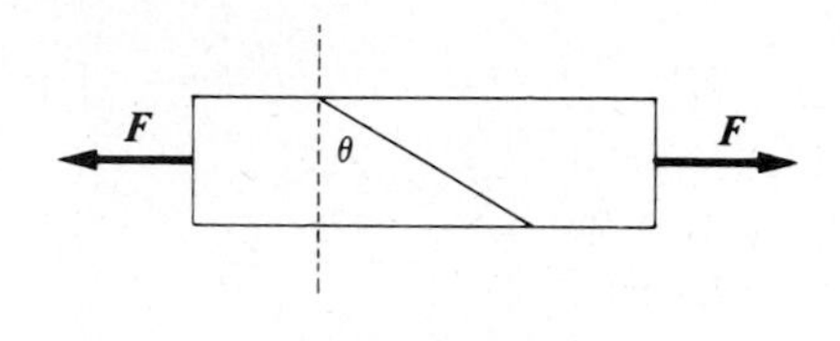

Fig. 10–10

10–25 Suppose the block in Fig. 10–3 is rectangular instead of square, but is in equilibrium under the action of shearing stresses on those outside faces perpendicular to the plane of the diagram. (Note that then $F_x \neq F_y$.)

a) Show that the shearing stress is the same on all outside faces perpendicular to the plane of the diagram.

b) Show that on all sections perpendicular to the plane of the diagram and making an angle of 45° with an end face, the stress is still a pure tension or compression.

Chapter 11

Harmonic Motion

11–1 INTRODUCTION

The motion of a body when acted upon by a constant force was considered in detail in Chapters 4 and 5. The motion is one of constant acceleration, and it was found useful to derive expressions for the position and velocity of the body at any time, and for its velocity at any position. In the present chapter, we are to study the motion of a body when the resultant force on it is *not* constant, but varies during the motion. Naturally, there is an infinite number of ways in which a force may vary; hence no general expressions can be given for the motion of a body when acted on by a variable force, except that the acceleration at each instant must equal the force at that instant divided by the mass of the body.

There is however a particular class of problems that occur so often in practical situations that it is useful to develop formulas for this special case. Such problems involve a situation where a body has an equilibrium position such that when displaced from equilibrium it experiences a restoring force, which causes it to undergo a back-and-forth motion past the equilibrium position. A familiar example is the pendulum, a mass suspended from a string. In the equilibrium position, the mass hangs straight down; when displaced from this position the mass does not simply return to the equilibrium position, but instead swings back and forth in a regular, repetitive manner.

Such a repetitive motion is said to be *oscillatory* or *periodic*. Many other examples might be cited; the balance wheel of a watch, the pistons in a gasoline engine, the strings in a musical instrument. The molecules of a solid body vibrate with oscillatory motion about their equilibrium positions in the crystal lattice, although of course this motion cannot be observed directly. The beating of the human heart is a periodic motion; Galileo is alleged to have used his own heartbeat for timing observations of motion.

In many kinds of *wave motion*, the particles of the medium in which the wave is traveling oscillate with periodic motion. This is true even for light waves and radio waves in empty space, except that instead of material particles the quantities that oscillate are the electric and magnetic fields associated with the wave. As a final example, an electrical circuit in which there is an alternating current is described in terms of voltages, currents, and electrical charges that oscillate with time. It can be seen that a study of periodic motion will lay the foundation for future work in many different fields of physics.

11–2 ELASTIC RESTORING FORCES

It has been shown in Chapter 10 that when a body is caused to change its shape, the distorting force is proportional to the amount of change, provided the proportional limit of elasticity is not exceeded. The change may be in the nature of an increase in length, as of a rubber band or a coil spring, or a decrease in length, or a bending as of a flat spring, or a twisting of a rod about its axis, or of many other forms. The term "force" is to be interpreted liberally as the force, or torque, or pressure, or whatever may be producing the distortion. If we restrict the discussion to the case of a push or a pull, where the distortion is simply the displacement of the point of application of the force, the force and displacement are related by Hooke's law,

$$F = kx,$$

where k is a proportionality constant called the *force constant* and x is the displacement from the equilibrium position.

In this equation, F stands for the force which must be exerted *on* an elastic body to produce the displacement x. The force with which the elastic body pulls back on an object to which it is attached is called the *restoring force* and is equal to $-kx$. The negative sign indicates that the force *on* the body is always opposite in sign to the displacement, so that no matter which way the body is displaced, the force is always toward the equilibrium position.

11–3 DEFINITIONS

To fix our ideas, suppose that a flat strip of steel such as a hacksaw blade is clamped vertically in a vise and a small body is attached to its upper end, as in Fig. 11–1. Assume that the strip is sufficiently long and the displacement sufficiently small so that the motion is essentially along a straight line. The mass of the strip itself is negligible.

Let the top of the strip be pulled to the right a distance A, as in Fig. 11–1, and released. The attached body is then acted on by a restoring force exerted by the steel strip and directed toward the equilibrium position O. It therefore accelerates in the direction of this force, and moves in toward the center with increasing speed. The *rate* of increase (i.e., the acceleration) is not constant, however, since the accelerating force becomes smaller as the body approaches the center.

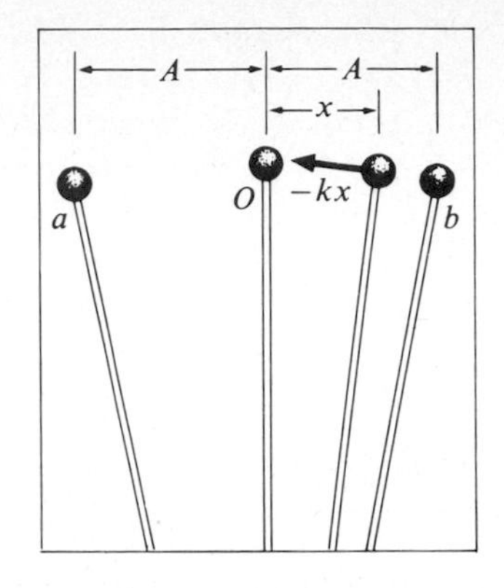

Fig. 11–1 *Motion under a restoring force.*

When the body reaches the center the restoring force has decreased to zero, but because of the velocity which has been acquired, the body "overshoots" the equilibrium position and continues to move toward the left. As soon as the equilibrium position is passed, the restoring force again comes into play, directed now toward the right. The body therefore decelerates, and at a rate which increases with increasing distance from O. It will therefore be brought to rest at some point to the left of O, and repeat its motion in the opposite direction.

Both experiment and theory show that the motion will be confined to a range $\pm A$ on either side of the equilibrium position, each to-and-fro movement taking place in the same interval of time. If there were no loss of energy by friction, the motion would continue indefinitely, once it had been started. This type of motion, under the influence of an elastic restoring force proportional to displacement and in the absence of all friction, is called *simple harmonic motion*, abbreviated SHM.

A *complete vibration* or *complete cycle* means one round trip, say from a to b and back to a, or from O to b to O to a and back to O.

The *periodic time*, or simply the *period* of the motion, represented by T, is the time required for one complete vibration.

The *frequency*, f, is the number of complete vibrations per unit time. Evidently the frequency is the reciprocal of the period, or $T = 1/f$. The mks unit, one cycle per second, is called one *hertz* ($1\ \text{Hz} = 1\ \text{s}^{-1}$).

The *coordinate*, x, at any instant, is the distance away from the equilibrium position or center of the path at that instant. Its sign indicates the direction of the displacement.

The *amplitude*, A, is the maximum value of $|x|$. The total range of the motion is therefore $2A$.

11–4 EQUATIONS OF SIMPLE HARMONIC MOTION

The motion of a point mass m under a restoring force given by $-kx$, where k is the force constant and x the displacement from equilibrium, is an idealized model constituting the simplest possible example of periodic motion. In more complex examples the force may depend on displacement in a more complicated way, but when the force is *directly proportional* to displacement, the resulting motion is called *simple harmonic motion*, abbreviated SHM. Since many more complex motions are approximately simple harmonic and may therefore be described approximately by this model, it is useful to analyze simple harmonic motion in some detail. We shall obtain expressions for the coordinate, velocity, and acceleration of a body moving with simple harmonic motion, just as we found those for a body moving with constant acceleration. It must be emphasized that the equations of motion with *constant* acceleration cannot be applied, since the acceleration is continually changing.

Figure 11–2 represents the vibrating body of Fig. 11–1 at some instant when its displacement from the equilibrium position O is described by the coordinate x. The particle's mass is m, and the resultant force acting on it is simply the elastic restoring force $-kx$. From Newton's second law,

$$F = -kx = ma,$$

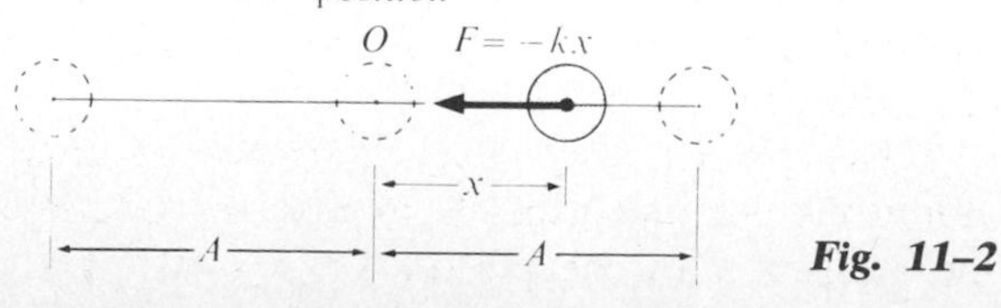

Fig. 11–2

or

$$\boxed{a = -\frac{k}{m}x.} \qquad (11\text{–}1)$$

Thus, an alternate description of the essential feature of simple harmonic motion is that the acceleration at each instant is proportional to the negative of the displacement at that instant. When x has its maximum positive value A, the acceleration has its maximum negative value $-kA/m$; and at the instant the particle passes the equilibrium position ($x = 0$), the acceleration is zero. Its velocity is, of course, *not* zero at this point.

The energy principle provides a convenient basis for analysis of some aspects of simple harmonic motion. The elastic restoring force is a *conservative* force, and the work done by this force can be represented in terms of a potential energy $E_p = \frac{1}{2}kx^2$, just as in Section 7–5. The kinetic energy is $E_k = \frac{1}{2}mv^2$ and, according to the principle of conservation of energy, the total energy $E = E_k + E_p$ is constant. That is,

$$E = \tfrac{1}{2}mv^2 + \tfrac{1}{2}kx^2 = \text{constant}. \qquad (11\text{–}2)$$

The total energy E is also closely related to the amplitude A of the motion. When the particle reaches its maximum displacement $\pm A$, it stops and turns back toward equilibrium. At this instant, $v = 0$, so there is no kinetic energy; the (constant) total energy is thus equal to the potential energy at this point, which is $\frac{1}{2}kA^2 = E$. Thus,

$$\tfrac{1}{2}mv^2 + \tfrac{1}{2}kx^2 = \tfrac{1}{2}kA^2,$$

or

$$\boxed{v = \pm\sqrt{\frac{k}{m}}\sqrt{A^2 - x^2}.} \qquad (11\text{–}3)$$

This relation permits us to obtain the velocity (apart from a sign) for any given position, and Eq. (11–3) is analogous to the relation $v^2 - v_0^2 = 2a(x - x_0)$ for straight-line motion with constant acceleration.

The significance of Eq. (11–2) can be brought out by the graph shown in Fig. 11–3, in which energy

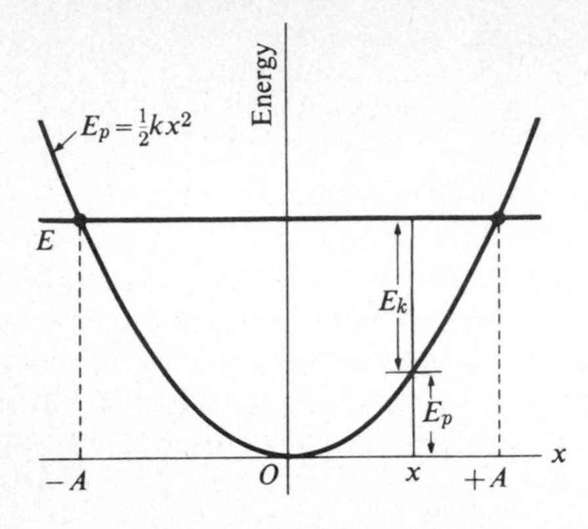

Fig. 11–3 *Relation between total energy E, potential energy E_p, and kinetic energy E_k, for a body oscillating with SHM.*

is plotted vertically and the coordinate x horizontally. First, the curve representing the potential energy, $E_p = \frac{1}{2}kx^2$, is constructed. (This curve is a parabola.) Next, a horizontal line is drawn at a height equal to the total energy E. We see at once that the motion is restricted to values of x lying between the points at which the horizontal line intersects the parabola since, if x were outside this range, the potential energy would exceed the total energy, and that is impossible. The motion of the vibrating body is analogous to that of a particle released at a height E on a frictionless track shaped like the potential energy curve, and the motion is said to take place in a "potential energy well."

If a vertical line is constructed at any value of x within the permitted range, the length of the segment between the x-axis and the parabola represents the potential energy E_p at that value of x, and the length of the segment between the parabola and the horizontal line at height E represents the corresponding kinetic energy E_k. At the endpoints, therefore, the energy is all potential and at the midpoint it is all kinetic. The velocity at the midpoint has its maximum (absolute) value $v_{\max}$:

$$\frac{1}{2}mv_{\max}^2 = E,$$
$$v_{\max} = \pm\sqrt{2E/m}, \tag{11–4}$$

the sign of v being positive or negative depending on the direction of motion. The fact that both the maximum potential energy and maximum kinetic energy are equal to the total energy (and thus to each other) may be used to relate the maximum velocity to the amplitude:

$$\frac{1}{2}kA^2 = \frac{1}{2}mv_{\max}^2,$$
$$v_{\max} = \sqrt{\frac{k}{m}}\,A. \tag{11–5}$$

The position–velocity relation given by Eq. (11–3) is very useful, but it does not tell us where the particle is at any given time. To have a complete description of the motion we need to know the position, velocity, and acceleration for any given time. The expression for the coordinate x as a function of time can be obtained by replacing v with dx/dt, in Eq. (11–3), and integrating. This gives

$$\int \frac{dx}{\sqrt{A^2 - x^2}} = \sqrt{\frac{k}{m}} \int dt,$$
$$\sin^{-1}\frac{x}{A} = \sqrt{\frac{k}{m}}\,t + C. \tag{11–6}$$

Let x_0 be the value of x when $t = 0$. The integration constant C is then

$$C = \sin^{-1}\frac{x_0}{A}.$$

That is, C is the *angle* (in radians) whose sine equals x_0/A. Let us represent this angle by θ_0:

$$\sin\theta_0 = \frac{x_0}{A}, \qquad \theta_0 = \sin^{-1}\frac{x_0}{A}. \tag{11–7}$$

Equation (11–6) can now be written

$$\sin^{-1}\frac{x}{A} = \sqrt{\frac{k}{m}}\,t + \theta_0,$$
$$x = A\sin\left(\sqrt{\frac{k}{m}}\,t + \theta_0\right). \tag{11–8}$$

The coordinate x is therefore a *sinusoidal* function of the time t. The term in parentheses is an *angle*, in radians. It is called the *phase angle*, or simply the *phase* of the motion. The angle θ_0 is the *initial phase angle*.

The period T is the time required for one complete oscillation. That is, the coordinate x has the same value at the times t and $t + T$. In other words, the phase angle $((\sqrt{k/m})t + \theta_0)$ increases by 2π radians in the time T:

$$\sqrt{k/m}\,(t + T) + \theta_0 = ((\sqrt{k/m})t + \theta_0) + 2\pi,$$

$$T = 2\pi\sqrt{\frac{m}{k}}. \qquad (11\text{–}9)$$

The period T therefore depends only on the mass m and the force constant k. It does not depend on the amplitude (or on the total energy). For given values of m and k, the time of one complete oscillation is the same whether the amplitude is large or small. A motion that has this property is said to be *isochronous* (equal time).

The frequency f, or the number of complete oscillations per unit time, is the reciprocal of the period T:

$$f = \frac{1}{T} = \frac{1}{2\pi}\sqrt{\frac{k}{m}}.$$

The *angular frequency* ω is defined as $\omega = 2\pi f$ and is expressed in radians per second. It follows from the two preceding equations that

$$\omega = \sqrt{k/m},$$

and Eqs. (11–8) and (11–3) can be written more compactly as

$$x = A \sin(\omega t + \theta_0), \qquad (11\text{–}10)$$

$$v = \omega\sqrt{A^2 - x^2}. \qquad (11\text{–}11)$$

Expressions for the velocity and acceleration as functions of time can now be obtained by differentiating Eq. (11–10):

$$v = \frac{dx}{dt} = \omega A \cos(\omega t + \theta_0), \qquad (11\text{–}12)$$

$$a = \frac{dv}{dt} = -\omega^2 A \sin(\omega t + \theta_0). \qquad (11\text{–}13)$$

Since $A \sin(\omega t + \theta_0) = x$, Eq. (11–13) may also be written

$$a = -\omega^2 x. \qquad (11\text{–}14)$$

We note that this is consistent with Eq. (11–1).

It follows from Eq. (11–12) that if v_0 is the velocity when $t = 0$, then

$$\cos\theta_0 = \frac{v_0}{\omega A}. \qquad (11\text{–}15)$$

The equation, together with Eq. (11–7), $\sin\theta_0 = x_0/A$, completely determines the initial phase angle θ_0. That is, the angle θ_0 depends both on the initial position x_0 and the initial velocity v_0.

Figure 11–4 shows how the initial phase angle depends on the initial position and initial velocity. It follows from this triangle that

$$A = \sqrt{x_0^2 + (v_0/\omega)^2}, \qquad (11\text{–}16)$$

a relation that can also be obtained by setting the initial total energy, $\frac{1}{2}kx_0^2 + \frac{1}{2}mv_0^2$, equal to the potential energy at maximum displacement. The motion is therefore completely determined, for given values of m and k, when the initial position and initial velocity are known.

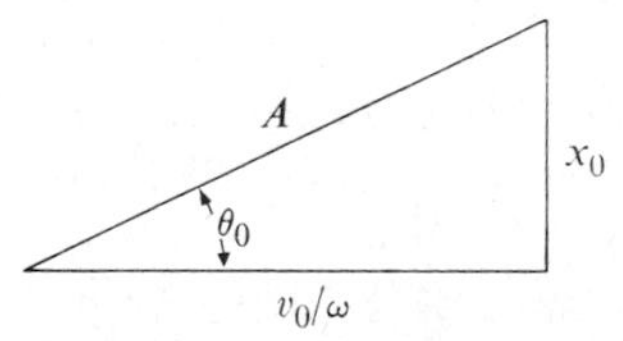

Fig. 11–4 *Relation between the initial phase angle θ_0, initial position x_0, and initial velocity v_0.*

The equations of simple harmonic motion may be summarized by comparing them with similar equations for rectilinear motion with constant acceleration (Table 11–1).

Figure 11–5 shows corresponding graphs of the coordinate x, velocity v, and acceleration a of a body oscillating with simple harmonic motion, plotted as functions of the time t (or of the angle ωt). The initial

TABLE 11–1

Rectilinear motion with constant acceleration	Simple harmonic motion (in terms of ω and θ_0)
$a = \text{constant}$	$a = -\omega^2 x$ $a = -\omega^2 A \sin(\omega t + \theta_0)$
$v^2 = v_0^2 + 2a(x - x_0)$ $v = v_0 + at$	$v = \pm\omega\sqrt{A^2 - x^2}$ $v = \omega A \cos(\omega t + \theta_0)$
$x = x_0 + v_0 t + \frac{1}{2}at^2$	$x = A \sin(\omega t + \theta_0)$
$\omega = 2\pi/T = 2\pi f = \sqrt{k/m}$, $\sin\theta_0 = x_0/A$,	$A = \sqrt{x_0^2 + (v_0/\omega)^2}$, $\cos\theta_0 = v_0/\omega A$.

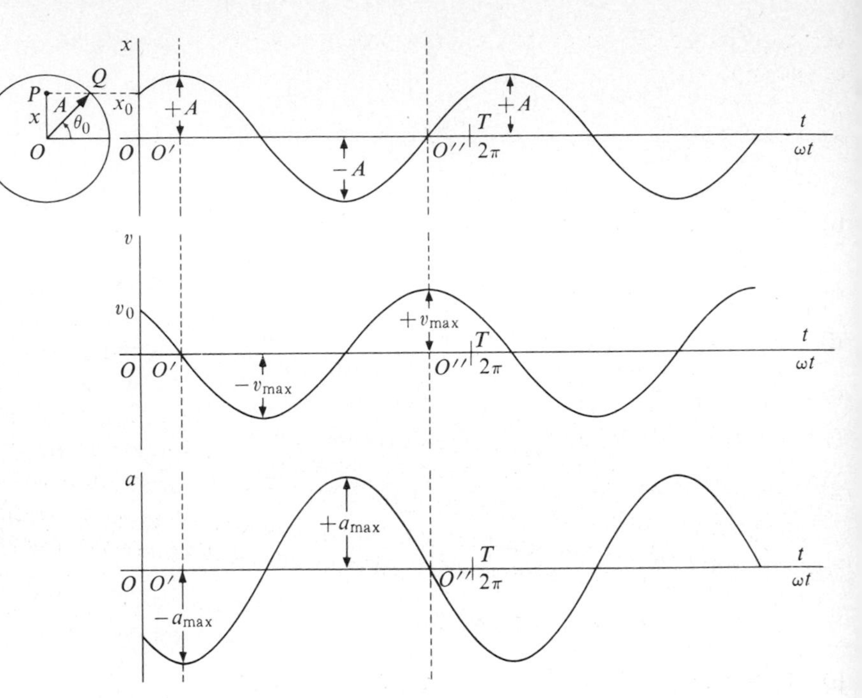

Fig. 11–5 *Graphs of coordinate x, velocity v, and acceleration a of a body oscillating with simple harmonic motion.*

coordinate is x_0 and the initial velocity is v_0. The angle θ_0 has been taken as $\pi/4$ rad. Each curve repeats itself in a time interval equal to the period T, during which the angle ωt increases by 2π rad. Note that when the body is at either end of its path, i.e., when the coordinate x has its maximum positive or negative value ($\pm A$), the velocity is zero and the acceleration has its maximum negative or positive value ($\mp a_{max}$). Note also that when the body passes through its equilibrium position ($x = 0$), the velocity has its maximum positive or negative value($\pm v_{max}$) and the acceleration is zero.

The equations of motion take a simpler form if we set $t = 0$ when the body is at the midpoint, or is at one end of its path. For example, suppose we let $t = 0$ when the body has its *maximum positive* displacement. Then $x_0 = +A$, $\sin\theta_0 = 1$, $\theta_0 = \pi/2$, and

$$\begin{aligned} x &= A\sin(\omega t + \pi/2) = A\cos\omega t, \\ v &= -\omega A\sin\omega t, \\ a &= -\omega^2 A\cos\omega t. \end{aligned} \tag{11–17}$$

This corresponds to moving the origin from the point O, in Fig. 11–5, to the point O'. The graph of x versus t becomes a cosine curve, that of v versus t a negative sine curve, and that of a a negative cosine curve.

If we set $t = 0$ when the body is at the midpoint and moving toward the right, then

$$x_0 = 0, \qquad \sin\theta_0 = 0, \qquad \theta_0 = 0$$

and

$$\begin{aligned} x &= A\sin\omega t, \\ v &= \omega A\cos\omega t, \\ a &= -\omega^2 A\sin\omega t. \end{aligned} \tag{11–18}$$

This corresponds to displacing the origin to the point O'' in Fig. 11–5.

Example Let the mass of the body in Fig. 11–2 be 0.025 kg, the force constant k be 0.4 N m^{-1}, and let the motion be started by displacing the body 0.1 m to the right of its equilibrium position and imparting to it a velocity toward the right of 0.4 m s^{-1}. Compute (a) the period T, (b) the frequency f, (c) the angular frequency ω, (d) the total energy E, (e) the amplitude A, (f) the angle θ_0, (g) the maximum velocity v_{max}, (h) the maximum acceleration a_{max}, (i) the coordinate,

Example A body of mass 1 kg is suspended from a coil spring whose mass is 0.09 kg and whose force constant is 66 N m^{-1}. Find the frequency and amplitude of the ensuing motion if the body is displaced 0.03 m below its equilibrium position and given a downward velocity of 0.4 m s^{-1}.

The angular frequency is

$$\omega = \sqrt{\frac{k}{m + m_s/3}} = \sqrt{\frac{66\,\mathrm{N\,m^{-1}}}{1.03\,\mathrm{kg}}} = 8.00\,\mathrm{rad\,s^{-1}}.$$

The amplitude is expressed in terms of the initial coordinate and velocity by means of Eq. (11–16). Thus,

$$\begin{aligned} A &= \sqrt{x_0^2 + (v_0/\omega)^2} \\ &= \sqrt{(0.03\,\mathrm{m})^2 + (0.4/8\,\mathrm{m})^2} \\ &= 0.0582. \end{aligned}$$

11–6 THE SIMPLE PENDULUM

A simple pendulum, an idealized model for a more complex system, consists of a point mass suspended by an inextensible weightless string. When pulled to one side of its equilibrium position and released, the pendulum bob vibrates about this position. We wish to analyze this motion, asking in particular whether it is simple harmonic.

The necessary condition for simple harmonic motion is that the restoring force F shall be directly proportional to the coordinate x and oppositely directed. The path of the bob is not a straight line, but the arc of a circle of radius L, where L is the length of the supporting cord. The coordinate x refers to distances measured along this arc. (See Fig. 11–8.) Hence, if $F = -kx$, the motion will be simple harmonic; since $x = L\theta$, the requirement may also be written $F = -kL\theta$.

Figure 11–8 shows the forces on the bob at an instant when its coordinate is x. Choose axes tangent to the circle and along the radius, and resolve the weight into components. The restoring force F is

$$F = -mg \sin \theta. \qquad (11\text{–}19)$$

The restoring force is therefore proportional *not* to θ but to $\sin \theta$, so the motion is *not* simple harmonic. However, *if the angle θ is small*, $\sin \theta$ is very nearly equal to θ. For example, when $\theta = 0.1$ rad, about 6°, $\sin \theta = 0.0998$, a difference of only 0.2 percent. With this approximation, Eq. (11–19) becomes

$$F = -mg\theta = -mg\frac{x}{L},$$

or

$$F = -\frac{mg}{L}x. \qquad (11\text{–}20)$$

The restoring force is then proportional to the coordinate *for small displacements*, and the constant mg/L represents the force constant k. The period of a simple pendulum when its amplitude is small is therefore

$$T = 2\pi\sqrt{m/k} = 2\pi\sqrt{m/(mg/L)},$$

or

$$T = 2\pi\sqrt{L/g}. \qquad (11\text{–}21)$$

The corresponding frequency relations are

$$\begin{aligned} f &= \frac{1}{2\pi}\sqrt{\frac{g}{L}}, \\ \omega &= \sqrt{\frac{g}{L}}. \end{aligned} \qquad (11\text{–}22)$$

We note that these expressions do not contain the *mass* of the particle; this is because the restoring force, a component of the particle's weight, is proportional to m. Thus the mass appears on both sides

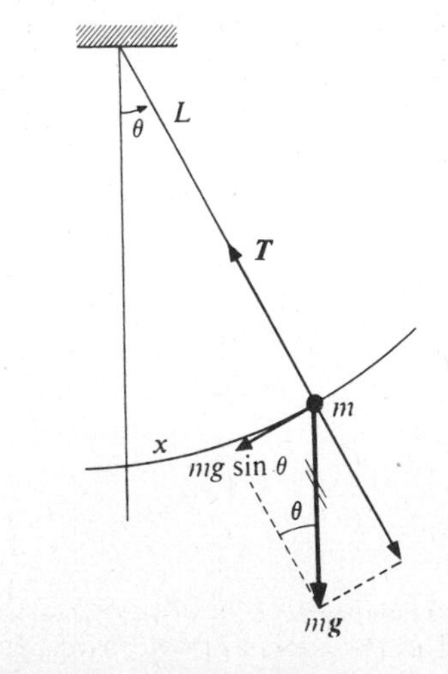

***Fig. 11–8** Forces on the bob of a simple pendulum.*

of $F = ma$ and may be cancelled out. For small oscillations the period of a pendulum for a given value of g is determined entirely by its length.

We emphasize again that the motion of a pendulum is only *approximately* simple harmonic, and when the amplitude is not small the departures from simple harmonic motion can be substantial. What constitutes a "small" amplitude? It can be shown that the general equation for the time of swing, when the maximum angular displacement is α, is

$$T = 2\pi\sqrt{L/g}\left(1 + \frac{1^2}{2^2}\sin^2\frac{\alpha}{2} + \frac{1^2 \cdot 3^2}{2^2 \cdot 4^2}\sin^4\frac{\alpha}{2} + \cdots\right). \tag{11–23}$$

The time may be computed to any desired degree of precision by taking enough terms in the infinite series. When $\alpha = 15°$ (on either side of the central position), the true period differs from that given by the approximate Eq. (11–21) by less than 0.5%.

The utility of the pendulum as a timekeeper is based on the fact that the period is practically independent of the amplitude. Thus, as a clock runs down and the amplitude of the swings becomes slightly smaller, the clock will still keep very nearly correct time.

The simple pendulum is also a precise and convenient method of measuring the acceleration of gravity g without actually resorting to free fall, since L and T may readily be measured. More complicated pendulums find considerable application in the field of geophysics. Local deposits of ore or oil, if their density differs from that of their surroundings, affect the local value of g, and precise measurements of this quantity over an area which is being prospected often furnish valuable information regarding the nature of underlying deposits.

11–7 ANGULAR HARMONIC MOTION

Angular harmonic motion occurs when a body which is pivoted about an axis experiences a restoring *torque* proportional to the angular displacement from its equilibrium position. This type of vibration is very similar to linear harmonic motion, and the corresponding equations may be written down immediately from the analogies between linear and angular quantities. The oscillatory motion of the balance wheel of a watch is an example of angular harmonic motion.

A restoring torque proportional to angular displacement is expressed by

$$\Gamma = -k'\theta, \tag{11–24}$$

where k' is a proportionality constant called the *torque constant.* The moment of inertia of the pivoted body corresponds to the mass of a body in linear motion. Hence, the formula for the period of angular harmonic motion is

$$T = 2\pi\sqrt{I/k'}, \tag{11–25}$$

where k' is the constant in Eq. (11–24).

The equations for angular displacement, angular velocity, and angular acceleration can be obtained by comparison with the corresponding equations in Section 11–4, replacing x by θ, v by ω, and a by α.

The balance wheel of a watch is a familiar example of angular harmonic motion. If the hairspring behaves according to Eq. (11–24), the motion is *isochronous*; the period is constant even though the amplitude decreases somewhat as the mainspring unwinds.

11–8 THE PHYSICAL PENDULUM

A *physical* pendulum is any real pendulum, as contrasted with a simple pendulum in which all the mass is assumed to be concentrated at a point. In Fig. 11–9, a body of irregular shape is pivoted about a horizontal frictionless axis O and displaced from the vertical by an angle θ. The distance from the pivot to the center of gravity is h, the moment of inertia of the pendulum about an axis through the pivot is I, and the mass of the pendulum is m. The weight mg causes a restoring torque

$$\Gamma = -mgh \sin\theta. \tag{11–26}$$

When released, the body will oscillate about its equilibrium position; as in the case of the simple pendulum, the motion is not simple harmonic, since the torque Γ is proportional not to θ but to $\sin\theta$. However, if θ is small, we can again approximate $\sin\theta$ by θ, and the motion is approximately harmon-

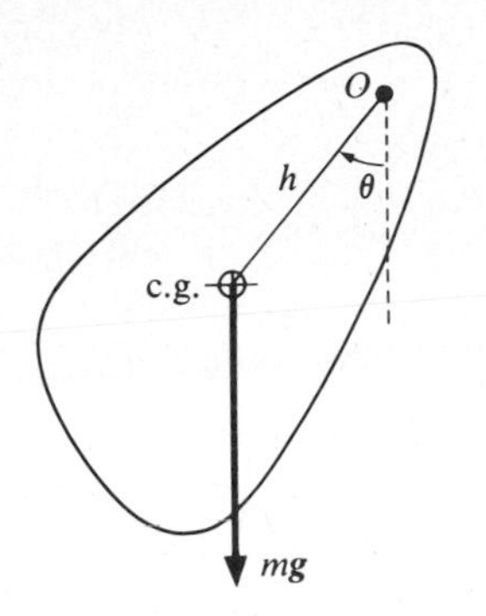

Fig. 11–9 A physical pendulum.

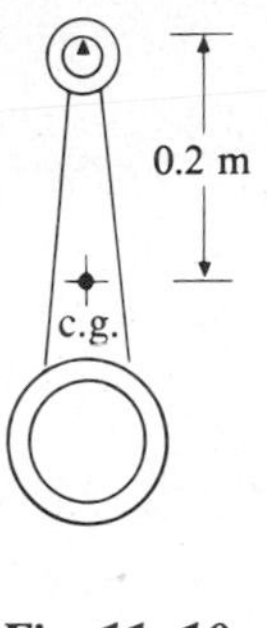

Fig. 11–10

ic. With this approximation,

$$\Gamma \approx -(mgh)\theta,$$

and the effective torque constant is

$$k' = -\frac{\Gamma}{\theta} = mgh.$$

The period T is

$$T = 2\pi\sqrt{\frac{I}{k'}} = 2\pi\sqrt{\frac{I}{mgh}}. \quad (11\text{–}27)$$

Example 1 Let the body in Fig. 11–9 be a meter-stick pivoted at one end. Then if L stands for the total length (1 m),

$$I = \tfrac{1}{3}mL^2, \qquad h = \frac{L}{2}, \qquad g = 9.8 \text{ m s}^{-2},$$

$$T = 2\pi\sqrt{\tfrac{1}{3}mL^2/(mgL/2)} = 2\pi\sqrt{(2/3)(L/g)}$$

$$= 2\pi\sqrt{(2/3)(1 \text{ m})/9.8 \text{ m s}^{-2}} = 1.65 \text{ s}.$$

Example 2 Equation (11–27) may be solved for the moment of inertia I, giving

$$I = \frac{T^2 mgh}{4\pi^2}.$$

The quantities on the right of the equation are all directly measurable. Hence the moment of inertia of a body of any complex shape may be found by suspending the body as a physical pendulum and measuring its period of vibration. The location of the center of gravity can be found by balancing. Since T, m, g, and h are known, I can be computed. For example, Fig. 11–10 illustrates a connecting rod pivoted about a horizontal knife-edge. The connecting rod has a mass of 2 kg, and its center of gravity has been found by balancing to be 0.2 m below the knife-edge. When set into oscillation, it is found to make 100 complete vibrations in 120 s, so that T = 120 s/100 = 1.2 s. Therefore,

$$I = \frac{(1.2 \text{ s})^2(2 \text{ kg})(9.8 \text{ m s}^{-2})(0.2 \text{ m})}{4\pi^2} = 0.143 \text{ kg m}^2.$$

11–9 CENTER OF OSCILLATION

It is always possible to find an *equivalent* simple pendulum whose period is equal to that of a given physical pendulum. If L is the length of the equivalent simple pendulum,

$$T = 2\pi\sqrt{\frac{L}{g}} = 2\pi\sqrt{\frac{I}{mgh}}$$

or

$$L = \frac{I}{mh}. \quad (11\text{–}28)$$

Thus, so far as its period of vibration is concerned, the mass of a physical pendulum may be considered to be concentrated at a point whose distance from the pivot is $L = I/mh$. This point is called the *center of oscillation* of the pendulum.

Example A slender uniform rod of length a is pivoted at one end and swings as a physical pendulum. Find the center of oscillation of the pendulum.

The moment of inertia of the rod about an axis through one end is

$$I = \tfrac{1}{3}ma^2.$$

The distance from the pivot to the center of gravity is $h = a/2$. The length of the equivalent simple pendulum is

$$L = \frac{I}{mh} = \frac{\tfrac{1}{3}ma^2}{m(a/2)} = \tfrac{2}{3}a,$$

and the center of oscillation is at a distance $2a/3$ from the pivot.

Figure 11-11 shows a body, whose center of oscillation is at point C, pivoted about an axis through O. The center of oscillation and the point of support have the following interesting property: If the pendulum is pivoted about a new axis through point C, its period is unchanged and point O becomes the new center of oscillation. The point of support and the center of oscillation are said to be *conjugate* to each other.

The center of oscillation has another important property. Figure 11-12 shows a baseball bat pivoted at O. If a ball strikes the bat at its center of oscillation, no impulsive force is exerted on the pivot and hence no "sting" is felt if the bat is held at that point. Because of this property, the center of oscillation is called the *center of percussion.*

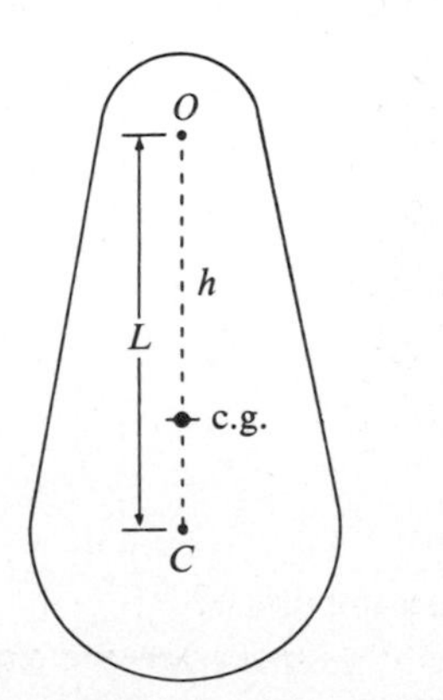

Fig. 11-11 Center of oscillation. The length L equals that of the equivalent simple pendulum.

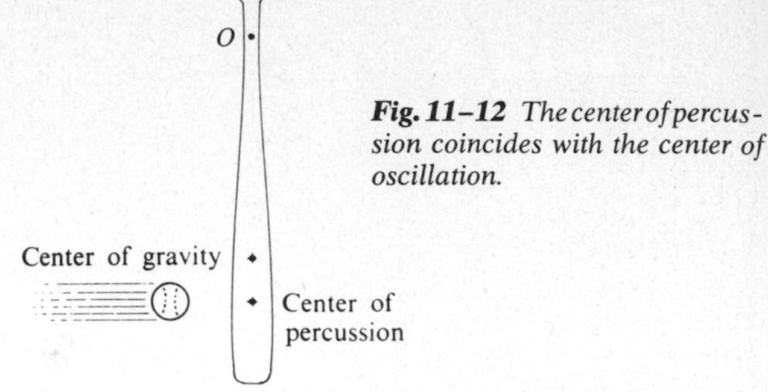

Fig. 11-12 *The center of percussion coincides with the center of oscillation.*

In Fig. 11-13, a series of multiflash photographs, the *center of gravity* is marked by a black band. In (a), the body is struck at its center of percussion relative to a pivot at the upper end of the cord, and it starts to swing smoothly about this pivot. In (b), the body is struck at its center of gravity. Note that it does not start to rotate about the pivot, but that its initial motion is one of pure translation. That is, the center of percussion does not coincide with the center of gravity. In (c), the body is struck above, and in (d) below its center of percussion.

PROBLEMS

11-1 The general equation of simple harmonic motion,

$$x = A\sin(\omega t + \theta_0),$$

can be written in the equivalent form

$$x = B\sin\omega t + C\cos\omega t.$$

a) Find the expressions for the amplitudes B and C in terms of the amplitude A and the initial phase angle θ_0.
b) Interpret these expressions in terms of a rotating vector diagram.

11-2 A body of mass 0.25 kg is acted on by an elastic restoring force of force constant $k = 25\ \mathrm{N\,m^{-1}}$.

a) Construct the graph of elastic potential energy E_p as a function of displacement x, over a range of x from -0.3 m to $+0.3$ m. Let 1 in. = 0.25 J vertically, and 1 in. = 0.1 m horizontally.

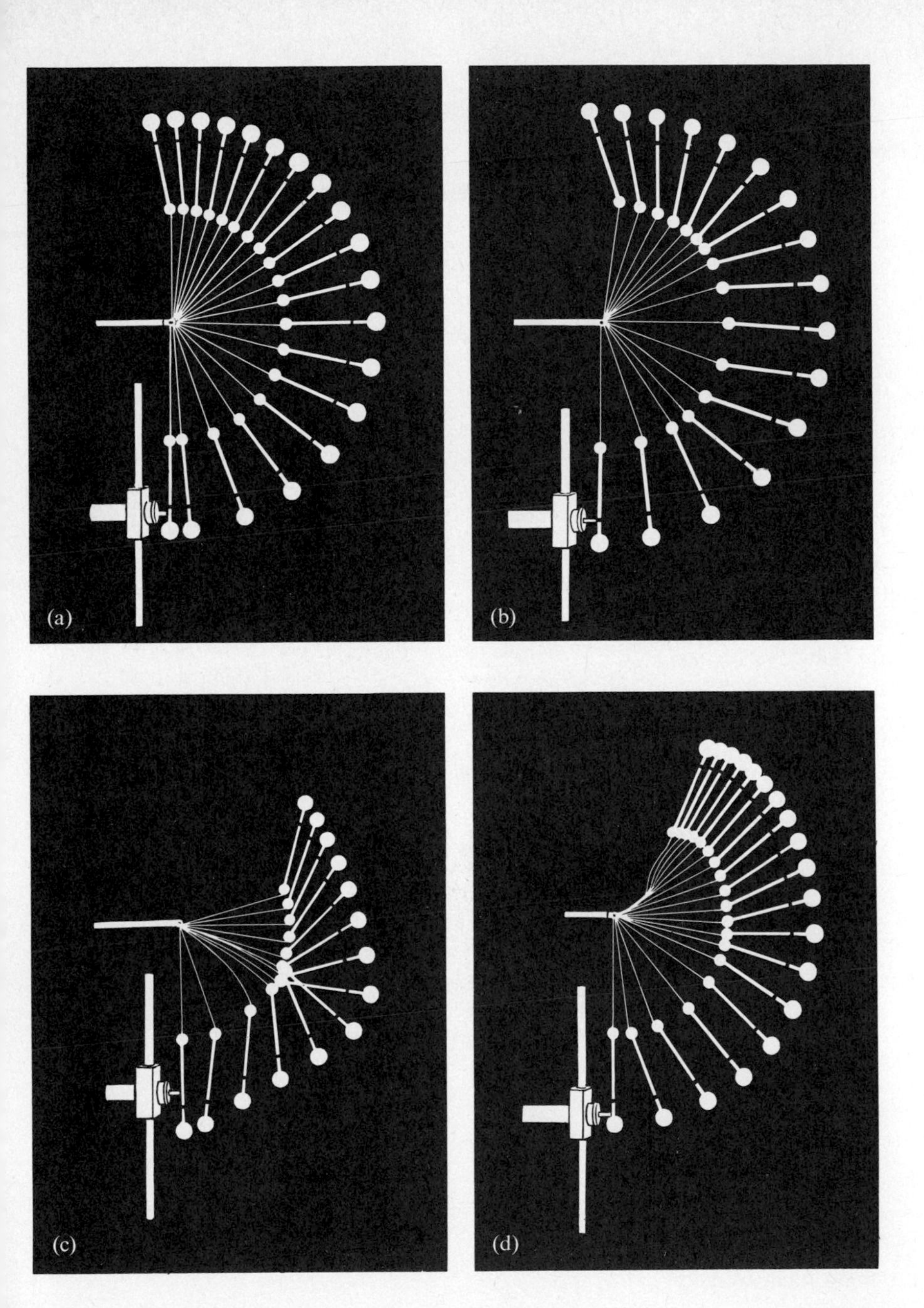

Fig. 11–13 *The motion of a body suspended by a cord when the body is struck a horizontal blow.*

The body is set into oscillation with an initial potential energy of 0.6 J and an initial kinetic energy of 0.2 J. Answer the following questions by reference to the graph.

b) What is the amplitude of oscillation?

c) What is the potential energy when the displacement is one-half the amplitude?

d) At what displacement are the kinetic and potential energies equal?

e) What is the speed of the body at the midpoint of its path?

Find

f) the period T,

g) the frequency f, and

h) the angular frequency ω.

i) What is the initial phase angle θ_0 if the amplitude A = 15 cm, the initial displacement x_0 = 7.5 cm, and the initial velocity v_0 is negative?

11–3 A body is vibrating with simple harmonic motion of amplitude 15 cm and frequency 4 Hz. Compute (a) the maximum values of the acceleration and velocity, (b) the acceleration and velocity when the coordinate is 9 cm, and (c) the time required to move from the equilibrium position to a point 12 cm distant from it.

11–4 A body of mass 10 g moves with simple harmonic motion of amplitude 24 cm and period 4 s. The coordinate is +24 cm when t = 0. Compute (a) the position of the body when t = 0.5 s, (b) the magnitude and direction of the force acting on the body when t = 0.5 s, (c) the minimum time required for the body to move from its initial position to the point where $x = -12$ cm, (d) the velocity of the body when $x = -12$ cm.

11–5 The motion of the piston of an automobile engine is approximately simple harmonic.

a) If the stroke of an engine (twice the amplitude) is 4 in., and the angular velocity is 3600 rev min^{-1}, compute the acceleration of the piston at the end of its stroke.

b) If the piston weighs 1 lb, what resultant force must be exerted on it at this point?

c) What is the velocity of the piston, in miles per hour, at the midpoint of its stroke?

11–6 A body of mass 2 kg is suspended from a spring of negligible mass, and is found to stretch the spring 20 cm.

a) What is the force constant of the spring?

b) What is the period of oscillation of the body, if pulled down and released?

c) What would be the period of a body of mass 4 kg hanging from the same spring?

11–7 The scale of a spring balance reading from zero to 180 N is 9 cm long. A body suspended from the balance is observed to oscillate vertically at 1.5 Hz. What is the mass of the body? Neglect the mass of the spring.

11–8 A body weighing 8 lb is attached to a coil spring and oscillates vertically in simple harmonic motion. The amplitude is 2 ft, and at the highest point of the motion the spring has its natural unstretched length. Calculate the elastic potential energy of the spring, the kinetic energy of the body, its gravitational potential energy relative to the lowest point of the motion, and the sum of these three energies, when the body is (a) at its lowest point, (b) at its equilibrium position, and (c) at its highest point.

11–9 A mass of 100 kg suspended from a wire whose unstretched length ℓ_0 is 4 m is found to stretch the wire by 0.004 m. The cross-sectional area of the wire, which can be assumed constant, is 0.1 cm^2.

a) If the load is pulled down a small additional distance and released, find the frequency at which it will vibrate.

b) Compute Young's modulus for the wire.

11–10 A small block is executing simple harmonic motion in a horizontal plane with an amplitude of 10 cm. At a point 6 cm away from equilibrium, the velocity is 24 cm s^{-1}.

a) What is the period?

b) What is the displacement when the velocity is ±12 cm s^{-1}?

c) If a small object placed on the oscillating block is just on the verge of slipping at the endpoint of the path, what is the coefficient of friction?

11–11 A force of 30 N stretches a vertical spring 15 cm.

a) What weight must be suspended from the spring so that the system will oscillate with a period of $(\pi/4)$ s?

b) If the amplitude of the motion is 5 cm, where is the body and in what direction is it moving $(\pi/12)$ s after it has passed the equilibrium position, moving downward?

c) What force does the spring exert on the body when it is 3 cm below the equilibrium position, moving upward?

11–12 A body of mass m is suspended from a coil spring and the time for 100 complete oscillations is measured for the following values of m:

m(g)	100	200	400	1000
Time of 100 oscillations (s)	23.4	30.6	41.8	64.7

Plot graphs of the measured values of (a) T vs. m, and (b) T^2 vs. m.

c) Are the experimental results in agreement with theory?

d) Is either graph a straight line?

e) Does the straight line pass through the origin?

f) What is the force constant of the spring?

g) What is the mass of the spring?

11–13 A body of mass 100 g hangs from a long spiral spring. When pulled down 10 cm below its equilibrium position and released, it vibrates with a period of 2 s.

a) What is its velocity as it passes through the equilibrium position?

b) What is its acceleration when it is 5 cm above the equilibrium position?

c) When it is moving upward, how long a time is required for it to move from a point 5 cm below its equilibrium position to a point 5 cm above it?

d) How much will the spring shorten if the body is removed?

11–14 A body of mass 5 kg hangs from a spring and oscillates with a period of 0.5 s. How much will the spring shorten when the body is removed?

11–15 Four passengers whose combined weight is 600 lb are observed to compress the springs of an automobile by 2 in. when they enter the automobile. If the total load supported by the springs is 1800 lb, find the period of vibration of the loaded automobile.

11–16

a) A block suspended from a spring vibrates with simple harmonic motion. At an instant when the displacement of the block is equal to one-half the amplitude, what fraction of the total energy of the system is kinetic and what fraction is potential?

b) When the block is in equilibrium, the length of the spring is an amount s greater than in the unstretched state. Prove that $T = 2\pi\sqrt{s/g}$.

11–17

a) With what additional force must a vertical spring carrying an 8-lb body in equilibrium be stretched so that, when released, it will perform 48 complete oscillations in 32 s with an amplitude of 3 in.?

b) What force is exerted by the spring on the body when it is at the lowest point, the middle, and the highest point of the path?

c) What is the kinetic energy of the system when the body is 1 in. below the middle of the path? its potential energy?

11–18 A force of 60 N stretches a certain spring 30 cm. A body of mass 4 kg is hung from this spring and allowed to come to rest. It is then pulled down 10 cm and released.

a) What is the period of the motion?

b) What are the magnitude and direction of the acceleration of the body when it is 5 cm above the equilibrium position, moving upward?

c) What is the tension in the spring when the body is 5 cm above the equilibrium position?

d) What is the shortest time required to go from the equilibrium position to the point 5 cm above?

e) If a small object were placed on the oscillating body, would it remain or leave?

f) If a small object were placed on the oscillating body and its amplitude doubled, where would the object and the oscillating body begin to separate?

11–19 Two springs with the same unstretched length but different force constants k_1 and k_2 are attached to a block of mass m on a level frictionless surface. Calculate the effective force constant in each of the three cases (a), (b), and (c) shown in Fig. 11–14.

d) A body of mass m, suspended from a spring with a force constant k, vibrates with a frequency f_1. When the spring is cut in half and the same body is suspended from one of the halves, the frequency is f_2. What is the ratio f_2/f_1?

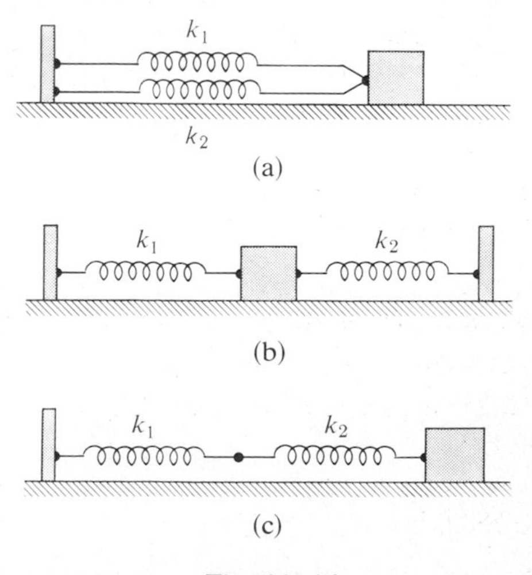

Fig. 11–14

11–20 Two springs, each of unstretched length 0.2 m but having different force constants k_1 and k_2, are attached to opposite ends of a block of mass m on a level frictionless surface. The outer ends of the springs are now attached to two pins P_1 and P_2, 10 cm from the original positions of the springs. Let

$$k_1 = 1\ \mathrm{N\,m^{-1}}, \qquad k_2 = 3\ \mathrm{N\,m^{-1}}, \qquad m = 0.1\ \mathrm{kg}.$$

(See Fig. 11–15.)

a) Find the length of each spring when the block is in its new equilibrium position, after the springs have been attached to the pins.

b) Find the period of vibration of the block if it is slightly displaced from its new equilibrium position and released.

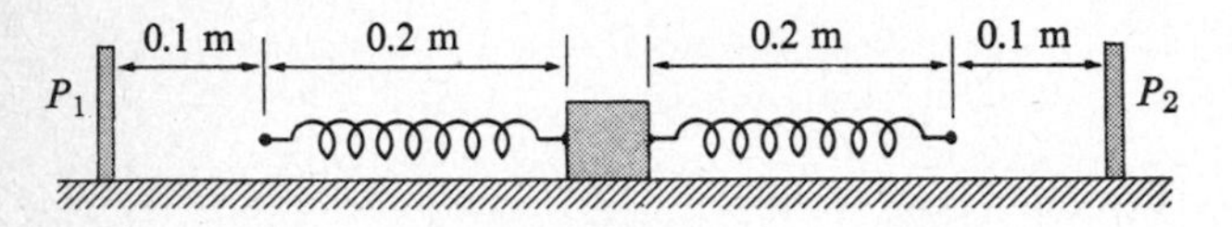

Fig. 11–15

11–21 The block in Problem 11–20 and Fig. 11–15 is oscillating with an amplitude of 0.05 m. At the instant it passes through its equilibrium position, a lump of putty of mass 0.1 kg is dropped vertically onto the block and sticks to it.

a) Find the new period and amplitude.

b) Was there a loss of energy and, if so, where did it go?

c) Would the answers be the same if the putty had been dropped on the block when it was at one end of its path?

11–22 A simple pendulum 8 ft long swings with an amplitude of 1 ft.

a) Compute the velocity of the pendulum at its lowest point.

b) Compute its acceleration at the ends of its path.

11–23 Find the length of a simple pendulum whose period is exactly 1 s at a point where $g = 9.80\ \mathrm{m\,s^{-2}}$.

11–24

a) What is the change ΔT in the period of a simple pendulum when the acceleration of gravity g changes by Δg? (*Hint*: The new period $T + \Delta T$ is obtained by substituting $g + \Delta g$ for g:

$$T + \Delta T = 2\pi\sqrt{\frac{L}{g + \Delta g}}.$$

To obtain an approximate expression, expand the $(g + \Delta g)^{1/2}$ using the binomial theorem and keeping only the first two terms:

$$(g + \Delta g)^{-1/2} = g^{-1/2} - \tfrac{1}{2}g^{-3/2}\Delta g + \cdots$$

The other terms contain higher powers of Δg and are very small if Δg is small.)

b) Find the *fractional* change in period $\Delta T/T$ in terms of the fractional change $\Delta g/g$.

c) The result of part (b) may also be obtained by taking differentials of both sides of Eq. (11–21). Use this procedure to derive the required relationship. The correct result is $dT/T = -\frac{1}{2}dg/g$.

d) A pendulum clock, which keeps correct time at a point where $g = 9.8000\ \mathrm{m\,s^{-1}}$, is found to lose 10 seconds each day at a higher elevation. Use the above results to find approximately the value of g at this new location.

11–25 A certain simple pendulum has a period on earth of 2.0 s. What is its period on the surface of the moon, where $g = 1.7\ \mathrm{m\,s^{-2}}$? (*Note*: The approximate method of Problem 11–24 is *not* adequate here because Δg is not small compared to g.)

11–26 The balance wheel of a watch vibrates with an angular amplitude of π radians and with a period of 0.5 s.

a) Find its maximum angular velocity.

b) Find its angular velocity when its displacement is one-half its amplitude.

c) Find its angular acceleration when its displacement is 45°.

11–27 A certain alarm clock ticks four times each second, each tick representing half a period. The balance wheel consists of a thin rim of radius 1.5 cm, connected to the balance staff by thin spokes of negligible mass. The total mass is 0.8 g.

a) What is the torque constant of the hairspring?

b) If the balance wheel is brass, what temperature increase is required for the clock to lose 10 s per day due to expansion of the balance wheel, assuming the torque constant does not change?

11–28 At one time it was proposed to define the second as the time required for a simple pendulum exactly one meter long to complete one side-to-side swing (one-half period).

a) By how much is this definition in error if $g = 9.8000\ \mathrm{m\,s^{-2}}$?

b) For what value of g would this definition be exactly correct?

11–29 Show that the moment of inertia of a thin uniform rod about a perpendicular axis through the rod, at a distance s from the center of mass, is equal to the moment of inertia about a parallel axis through the center of mass, plus ms^2, where m is the total mass. (This result is a special case of the theorem cited in Problem 11–32.)

11–30 A thin uniform rod is pivoted on a perpendicular axis through the rod, and its period as a physical pendulum is measured. By trial and error, a second pivot point is found so that the period is the same as for the first point. Show that, in this case, the period depends only on the distance L between the two points and is given by $T = 2\pi(L/g)^{1/2}$. (The result of Problem 11–29 will be useful.)

11–31 A monkey wrench is pivoted at one end and allowed to swing as a physical pendulum. The period is 0.9 s and the pivot is 20 cm from the center of gravity.

a) What is the radius of gyration of the wrench about an axis through the pivot?
b) If the wrench was initially displaced 0.1 rad from its equilibrium position, what is the angular velocity of the wrench as it passes through the equilibrium position?

11–32 It is shown in textbooks on mechanics that the moment of inertia I of a body about *any* axis through any point is given by

$$I = I_G + mh^2,$$

where I_G is the moment of inertia about a *parallel axis* through the center of gravity, m is the mass, and h is the perpendicular distance between the two parallel axes. A solid disk of radius $R = 12$ cm oscillates as a physical pendulum about an axis perpendicular to the plane of the disk at a distance r from its center. (See Fig. 11–16.)

a) Calculate the period of oscillation (for small amplitudes) for the following values of r: 0, $R/4$, $R/2$, $3R/4$, R.
b) Let T_0 represent the period when $r = R$, and T the period at any other value of r. Construct a graph of the dimensionless ratio T/T_0 as a function of the dimensionless ratio r/R. (Note that the graph then describes the behavior of *any* solid disk, whatever its radius.)
c) Prove by the methods of calculus that the period is a minimum when $r = R/\sqrt{2}$. Does this result agree with your graph?

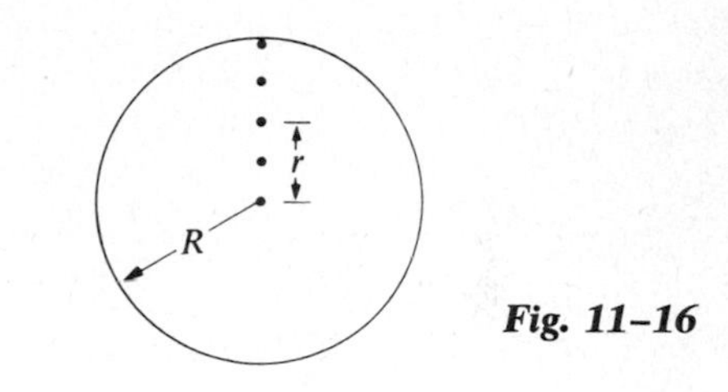

Fig. 11–16

11–33 It is desired to construct a pendulum of period 10 s.

a) What is the length of a *simple* pendulum having this period?
b) Suppose the pendulum must be mounted in a case not over 0.5 m high. Can you devise a pendulum, having a period of 10 s, that will satisfy this requirement?

11–34 A meter stick hangs from a horizontal axis at one end and oscillates as a physical pendulum. A body of small dimensions, and of mass equal to that of the meter stick, can be clamped to the stick at a distance h below the axis. Let T represent the period of the system with the body attached, and T_0 the period of the meter stick alone. Find the ratio T/T_0, (a) when $h = 0.5$ m, (b) when $h = 1.0$ m.

c) Is there any value of h for which $T = T_0$? If so, find it and explain why the period is unchanged when h has this value.

11–35 A meter stick is pivoted at one end. At what distance below the pivot should it be struck in order that it start swinging smoothly about the pivot?

Chapter 12

Hydrostatics

12–1 INTRODUCTION

The term "hydrostatics" is applied to the study of fluids at rest, and "hydrodynamics" to fluids in motion. The special branch of hydrodynamics relating to the flow of gases and of air in particular is called "aerodynamics."

A fluid is a substance which can flow. Hence the term includes both liquids and gases. Liquids and gases differ markedly in their compressibilities; a gas is easily compressed, while a liquid is practically incompressible. The small volume changes of a liquid under pressure can usually be neglected in this part of the subject.

The density of a homogeneous material is defined as its mass per unit volume. Units of density are one kilogram per cubic meter (1 kg m^{-3}) in the mks system, and one gram per cubic centimeter (1 g cm^{-3}) in the cgs system. We shall represent density by the greek letter ρ (rho):

$$\rho = \frac{m}{V}, \qquad m = \rho V. \tag{12–1}$$

Typical values of density at room temperature are given in Table 12–1. The conversion factor

$$1 \text{ g cm}^{-3} = 1000 \text{ kg m}^{-3}$$

is useful in connection with this table.

The *specific gravity* of a material is the ratio of its density to that of water and is therefore a pure number. "Specific gravity" is an exceedingly poor term, since it has nothing to do with gravity. "Relative density" would describe the concept more precisely.

Table 12–1 DENSITIES

Material	Density, g cm^{-3}	Material	Density, g cm^{-3}
Aluminum	2.7	Silver	10.5
Brass	8.6	Steel	7.8
Copper	8.9	Mercury	13.6
Gold	19.3	Ethyl alcohol	0.81
Ice	0.92	Benzene	0.90
Iron	7.8	Glycerin	1.26
Lead	11.3	Water	1.00
Platinum	21.4	Sea water	1.03

Density measurements are an important analytical technique in a wide variety of circumstances. The condition of an automobile storage battery can be judged by measuring the density of the electrolyte, a sulfuric acid solution. As the battery discharges, the sulfuric acid (H_2SO_4) combines with lead in the battery plates to form insoluble lead sulfate ($PbSO_4$), decreasing the concentration of the solution. The density varies from about 1.30 g cm^{-3} for a fully charged battery to 1.15 g cm^{-3} for a discharged battery. Similarly, permanent-type antifreeze is usually a solution of ethylene glycol (density 1.12 g cm^{-3}) in water, with small quantities of additives to retard corrosion. The glycol concentration, which determines the freezing point of the solution, can be determined from a simple density measurement. Both these measurements are performed routinely in service stations with the aid of a simple hydrometer, which measures density by observation of the level at which a calibrated body floats in a sample of the solution. The hydrometer is discussed in Section 12-6.

Medical science finds many uses for density measurements; examples include tests on body fluids such as blood and urine. The density of normal human blood is about 1.04 to 1.06 g cm^{-3}. Since density increases with the concentration of red cells, an unusually low density may indicate anemia. Similarly, the usual density of urine is about 1.02 g cm^{-3}. Certain diseases lead to increased excretion of salts and a corresponding increase in urine density.

12-2 PRESSURE IN A FLUID

When the concept of hydrostatic pressure was introduced in Section 10-1, the weight of the fluid was neglected and the pressure was assumed the same at all points. It is a familiar fact, however, that atmospheric pressure decreases with increasing altitude and that the pressure in a lake or in the ocean decreases with increasing distance from the bottom. We therefore generalize the definition of pressure and define the pressure *at any point* as the ratio of the normal force dF exerted on a small area dA including the point, to the area dA:

$$p = \frac{dF}{dA}, \qquad dF = p\,dA. \tag{12-2}$$

If the pressure is the same at all points of a finite plane surface of area A, these equations reduce to Eq. (10-3):

$$p = \frac{F}{A}, \qquad F = pA.$$

Let us find the general relation between the pressure p at any point in a fluid and the elevation of the point y. If the fluid is in equilibrium, every volume element is in equilibrium. Consider an element in the form of a thin slab, shown in Fig. 12-1, whose thickness is dy and whose faces have an area A. If ρ is the density of the fluid, the mass of the element is $\rho A\, dy$ and its weight dw is $\rho g A\, dy$. The force exerted on the element by the surrounding fluid is everywhere normal to its surface. By symmetry, the resultant horizontal force on its rim is zero. The upward force on its lower face is pA, and the downward force on its upper face is $(p + dp)A$. Since it is in equilibrium,

$$\sum F_y = 0,$$

$$pA - (p + dp)A - \rho g A\, dy = 0,$$

and therefore

$$\frac{dp}{dy} = -\rho g. \tag{12-3}$$

Since ρ and g are both positive quantities, it follows that a positive dy (an increase of elevation) is

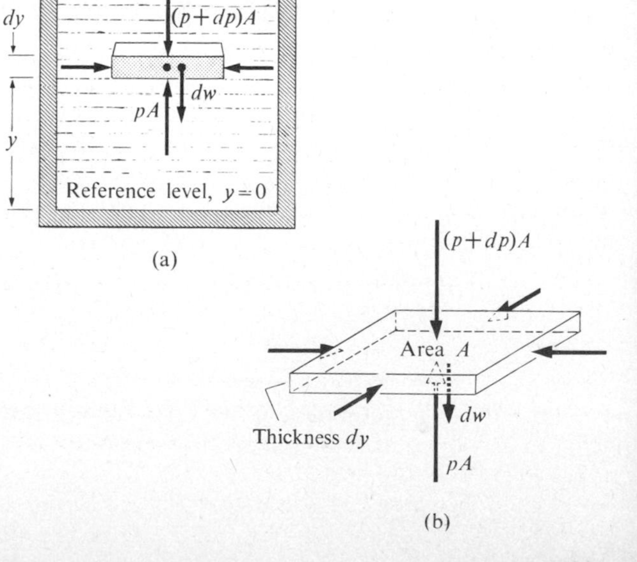

Fig. 12-1 Forces on an element of fluid in equilibrium.

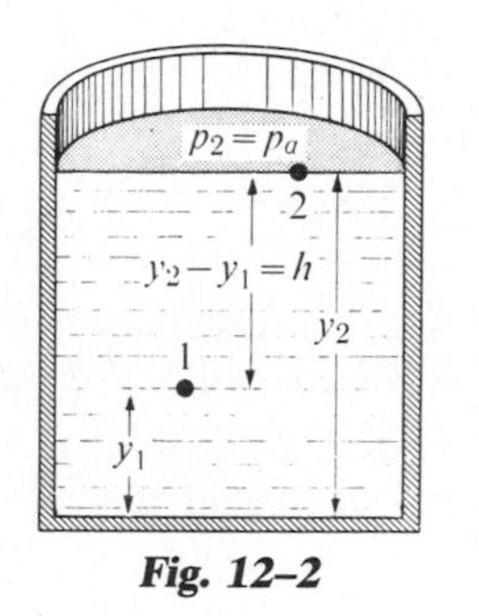

Fig. 12–2

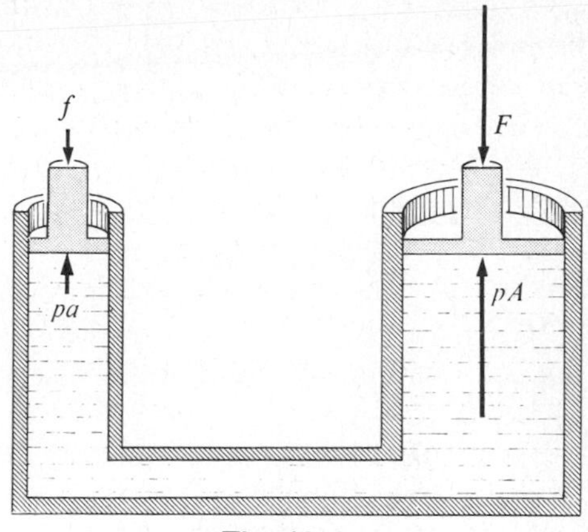

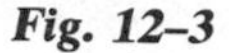

Fig. 12–3

accompanied by a negative dp (decrease of pressure). If p_1 and p_2 are the pressures at elevations y_1 and y_2 above some reference level, then integration of Eq. (12–3), when ρ and g are constant, gives

$$p_2 - p_1 = -\rho g(y_2 - y_1).$$

Let us apply this equation to a liquid in an open vessel, such as that shown in Fig. 12–2. Take point 1 at any level and let p represent the pressure at this point. Take point 2 at the top where the pressure is atmospheric pressure p_a. Then

$$p_a - p = -\rho g(y_2 - y_1),$$

$$p = p_a + \rho g h. \qquad (12\text{–}4)$$

Note that the shape of the containing vessel does not affect the pressure, and that the pressure is the same at all points at the same depth. It also follows from Eq. (12–4) that if the pressure p_a is increased in any way, say by inserting a piston on the top surface and pressing down on it, the pressure p at any depth must increase by exactly the same amount. This fact was recognized in 1653 by the French scientist Blaise Pascal (1623–1662) and is called "Pascal's law." It is often stated: "*Pressure applied to an enclosed fluid is transmitted undiminished to every portion of the fluid and the walls of the containing vessel.*" We can see now that it is not an independent principle but a necessary consequence of the laws of mechanics.

Pascal's law is illustrated by the operation of a hydraulic press, shown in Fig. 12–3. A piston of small cross-sectional area a is used to exert a small force f directly on a liquid such as oil. The pressure $p = f/a$ is transmitted through the connecting pipe to a larger cylinder equipped with a larger piston of area A. Since the pressure is the same in both cylinders,

$$p = \frac{f}{a} = \frac{F}{A} \quad \text{and} \quad F = \frac{A}{a}f.$$

It follows that the hydraulic press is a force-multiplying device with a multiplication factor equal to the ratio of the areas of the two pistons. Barber chairs, dentist chairs, car lifts and jacks, and hydraulic brakes are all devices that make use of the principle of the hydraulic press.

12–3 THE HYDROSTATIC "PARADOX"

If a number of vessels of different shapes are interconnected, as in Fig. 12–4(a), it is found that a liquid poured into them stands at the same level in each. Before the principles of hydrostatics were completely understood, this seemed a very puzzling phenomenon and was called the "hydrostatic paradox." It would appear at first sight, for example, that vessel C should develop a greater pressure at its base than should B, and hence that liquid would be forced from C into B.

However, Eq. (12–4) states that the pressure depends only on the depth below the liquid surface and not at all on the shape of the containing vessel. Since the depth of the liquid is the same in each vessel, the pressure at the base of each is the same

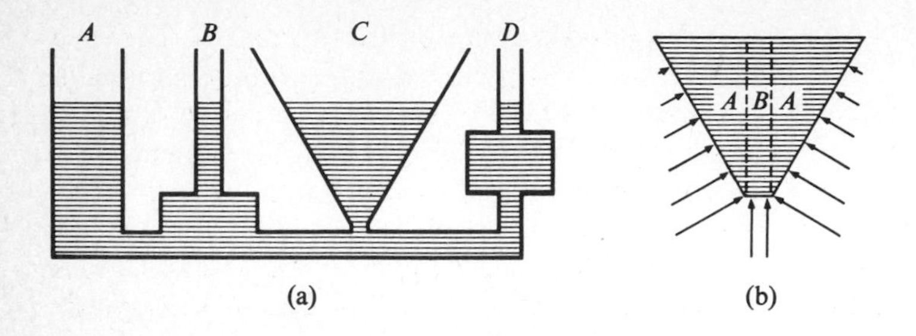

Fig. 12–4 *(a) The hydrostatic paradox. The top of the liquid stands at the same level in each vessel. (b) Forces on the liquid in vessel C.*

and hence the system is in equilibrium. Thus there is no real paradox.

A more detailed explanation may be helpful in understanding the situation. Consider vessel C in Fig. 12–4(b). The forces exerted against the liquid by the walls are shown by arrows, the force being everywhere perpendicular to the walls of the vessel. The inclined forces at the sloping walls may be resolved into horizontal and vertical components. The weight of the liquid in the sections lettered A is supported by the vertical components of these forces. Hence the pressure at the base of the vessel is due only to the weight of the liquid in the cylindrical column B. Any vessel, regardless of its shape, may be treated in the same way.

12–4 PRESSURE GAUGES

The simplest type of pressure gauge is the open-tube manometer, illustrated in Fig. 12–5(a). It consists of a U-shaped tube containing a liquid, one end of the tube being at the pressure p which it is desired to measure, while the other end is open to the atmosphere.

The pressure at the bottom of the left column is $p + \rho g y_1$, while that at the bottom of the right column (the same point) is $p_a + \rho g y_2$, where ρ is the density of the manometric liquid. Since these pressures must be equal,

$$p + \rho g y_1 = p_a + \rho g y_2,$$

and

$$p - p_a = \rho g(y_2 - y_1) = \rho g h. \qquad (12\text{–}5)$$

The pressure p is called the *absolute pressure*, whereas the difference $p - p_a$ between this and the atmospheric pressure is called the *gauge pressure*. It is seen that the gauge pressure is proportional to the difference in height of the liquid columns.

The mercury barometer is a long glass tube that has been filled with mercury and then inverted in a dish of mercury, as shown in Fig. 12–5(b). The space above the mercury column contains only mercury vapor, whose pressure, at room temperature, is so small that it may be neglected. It is easily seen that

$$p_a = \rho g(y_2 - y_1) = \rho g h. \qquad (12\text{–}6)$$

Because mercury manometers and barometers are used so frequently in laboratories, it is customary to express atmospheric pressure and other pressures as so many "inches of mercury," "centimeters of mercury," or "millimeters of mercury." Although these are not real units of pressure, they are so descriptive that they are widely used. The pressure exerted by a column of mercury one millimeter high is commonly called *one Torr,* after the Italian physi-

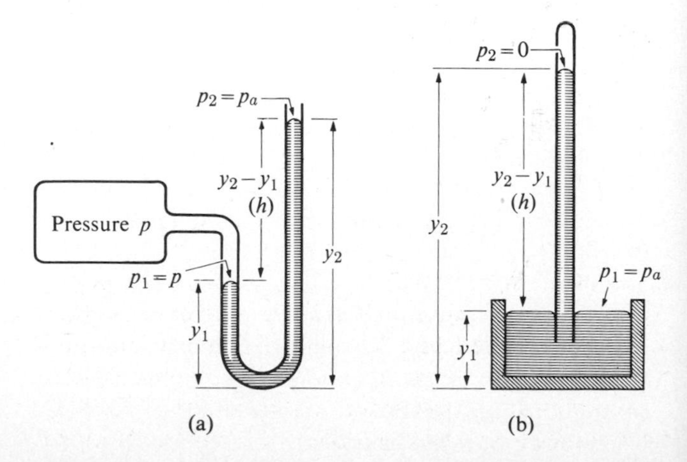

Fig. 12–5 *(a) The open-tube manometer. (b) The barometer.*

cist Torricelli (1608–1647), who first investigated the mercury barometric column.

One type of blood pressure gauge commonly used by physicians, called a *sphygmomanometer*, includes a form of manometer similar to Fig. 12–5(a). Blood-pressure readings, such as 130/80, refer to the maximum and minimum pressures, measured in millimeters of mercury or *Torrs*. Because of height differences, the hydrostatic pressure varies at different points in the body; the standard reference point is the upper arm, level with the heart. Pressure is also affected by the viscous nature of blood flow and by valves throughout the vascular system which act as pressure regulators.

Example Compute the atmospheric pressure on a day when the height of the barometer is 76.0 cm.

The height of the mercury column depends on ρ and g as well as on the atmospheric pressure. Hence both the density of mercury and the local acceleration of gravity must be known. The density varies with the temperature, and g with the latitude and elevation above sea level. All accurate barometers are provided with a thermometer and with a table or chart from which corrections for temperature and elevation can be found. If we assume $g = 9.8 \text{ m s}^{-2}$ and $\rho = 13.6 \times 10^3 \text{ kg m}^{-3}$

$$p_a = \rho g h = (13.6 \times 10^3 \text{ kg m}^{-3})(9.8 \text{ m s}^{-2})(0.76 \text{ m})$$

$$= 101{,}300 \text{ N m}^{-2} = 1.013 \times 10^5 \text{ Pa}.$$

(One pascal, abbreviated Pa, is defined as one newton per square meter; $1 \text{ Pa} = 1 \text{ N m}^{-2}$.)

In English units,

$$0.76 \text{ m} = 30 \text{ in.} = 2.5 \text{ ft},$$

$$\rho g = 850 \text{ lb ft}^{-3},$$

$$p_a = 2120 \text{ lb ft}^{-2} = 14.7 \text{ lb in}^{-2}.$$

A pressure of $1.013 \times 10^5 \text{ Pa} = 14.7 \text{ lb in}^{-2}$, is called *one atmosphere* (1 atm). A pressure of exactly 10^5 Pa is called one *bar*, and a pressure one one-thousandth as great is one *millibar*. Atmospheric pressures are of the order of 1000 millibars, and are now stated in terms of this unit by the National Weather Service.

The Bourdon-type pressure gauge is more convenient for most purposes than a liquid manometer. It consists of a flattened brass tube closed at one end and bent into a circular form. The closed end of the tube is connected by a gear and pinion to a pointer which moves over a scale. The open end of the tube is connected to the apparatus, the pressure within which is to be measured. When pressure is exerted within the flattened tube, it straightens slightly, just as a bent rubber hose straightens when water is admitted. The resulting motion of the closed end of the tube is transmitted to the pointer.

12–5 PUMPS

Many devices of modern life include glass or metal containers from which nearly all the air has been exhausted; examples include electric light bulbs, TV picture tubes, thermos bottles, and many others. Successful operation of these devices requires air pressures as low as 10^{-8} mm of mercury, and special pumps are used to produce these "high" vacuums. Two commonly used types are the *rotary oil pump* for pressures as low as 10^{-4} mm of mercury, and the *mercury or oil diffusion pump* for pressures as low as 10^{-8} mm Hg.

Figure 12–6 depicts schematically a common type of rotary oil pump. The vessel to be exhausted

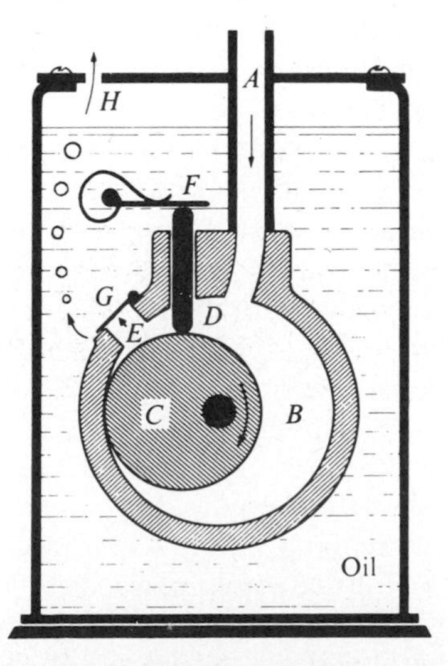

Fig. 12–6 *The rotary oil pump.*

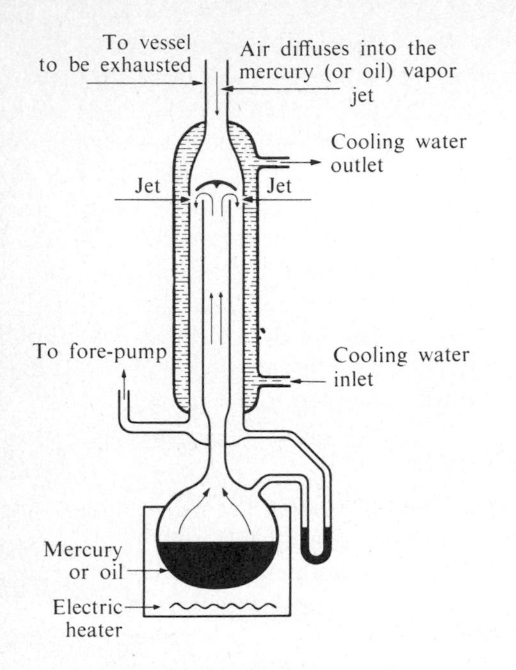

Fig. 12–7 *The diffusion pump.*

is connected to the tube *A*, which communicates directly with the space marked *B*. As an eccentric cylinder *C* rotates in the direction shown, the point of contact between it and the inner walls of the stationary cylinder moves around in a clockwise direction, thereby trapping some air in the space marked *E*. The sliding vane *D* is kept in contact with the rotating cylinder by the pressure of the rod *F*. When the air in *E* is compressed enough to increase the pressure slightly above atmospheric, the valve *G* opens and the air bubbles through the oil and leaves through an opening *H* in the upper plate. The cylinder is caused to rotate by means of a small electric motor.

For pressures below about 10^{-3} or 10^{-4} mm of mercury, a *diffusion pump* is usually employed. In this type of pump, oil or mercury is heated to produce vapor, which is formed by the shape of the pump into a jet of rapidly moving vapor. As air molecules from the vessel to be exhausted diffuse into the jet, they are swept away from this vessel. A common type of diffusion pump is shown in Fig. 12–7. A rotary oil pump is used to reduce the pressure within the diffusion pump to the low value necessary to ensure a well-defined jet, to prevent air molecules from diffusing from the jet back to the vessel. The air molecules swept away by the jet are removed by the rotary oil pump, which is called the *forepump*, and the mercury or oil condenses on the cool walls of the pump and returns to the well at the bottom.

Pumps are also used to compress gases and circulate liquids. Compressors for home refrigerators and air conditioners frequently use rotary pumps similar in principle to that shown in Fig. 12–6; reciprocating-piston pumps, equipped with pistons similar to those in gasoline engines, are also used for refrigerator compressors, and as air-compressors. In these pumps, the essential principle is a varying volume and a valve arrangement to ensure one-way flow of the fluid being pumped.

In situations where it is important to prevent contamination of the fluid, the varying volume is often provided by a flexible diaphragm, as shown in Fig. 12–8. The center portion of the diaphragm moves up and down, driven by a motor. When it moves up, fluid is pushed into the chamber from the left by gravity or external air pressure; when it moves down, the fluid in the chamber is forced out the right side. Backflow is prevented by the two valves. Many automobile-engine fuel pumps operate on this principle, and the human heart comprises four such pumps, with the expansion and contraction of the four chambers by the heart muscles playing the role of the diaphragm in Fig. 12–8. Some artificial heart pumps use diaphragm pumps.

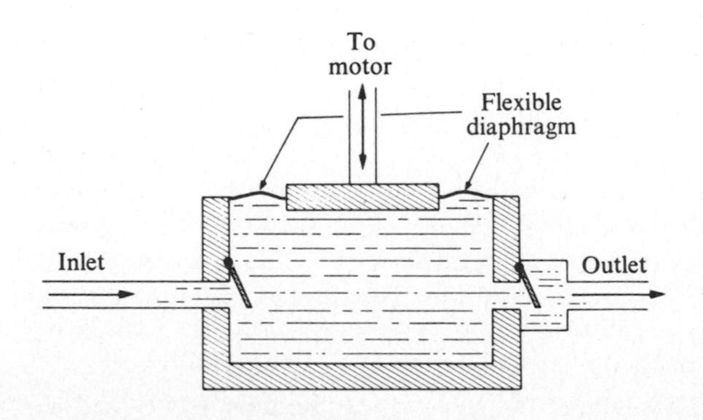

Fig. 12–8 *A diaphragm pump.*

12–6 ARCHIMEDES' PRINCIPLE

Buoyancy is a familiar phenomenon; a body immersed in water seems to have less weight than when immersed in air, and a body whose average density is less than that of the fluid in which it is immersed can float in that fluid. Examples are the human body in water, or a helium-filled balloon in air.

Archimedes' principle states that *when a body is immersed in a fluid, the fluid exerts an upward force on the body equal to the weight of the fluid which is displaced by the body.* To prove this principle, we consider an arbitrary portion of fluid at rest. In Fig. 12–9, the irregular outline is the surface bounding this portion of fluid. The short arrows represent the forces exerted by the surrounding fluid against small elements of the boundary surface.

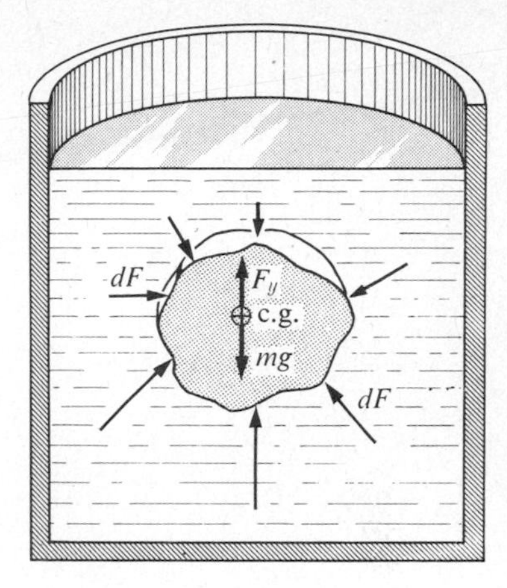

Fig. 12–9 *Archimedes' principle. The buoyant force F_y equals the weight of the displaced fluid.*

Since the entire fluid is at rest, the x-component of the resultant of these surface forces is zero. The y-component of the resultant, F_y, must equal the weight mg of the fluid inside the arbitrary surface, and its line of action must pass through the center of gravity of this fluid.

Now suppose that the fluid inside the surface is removed and replaced by a solid body having exactly the same shape. The pressure at every point is exactly the same as before; the force exerted on the body by the surrounding fluid is unaltered, and is therefore equal to the weight mg of fluid displaced. The line of action of this force again passes through the center of gravity of the displaced fluid.

The submerged body need not be in equilibrium. Its weight may be greater or less than F_y, and if it is not homogeneous, its center of gravity may not lie on the line of F_y. Therefore, in general, it will be acted on by a resultant force through its own center of gravity and by a couple, and will rise or fall and also rotate.

The weight of a balloon floating in air, or of a submarine floating at some depth below the surface of the water, is just equal to the weight of a volume of air, or water, that is equal to the volume of the balloon or submarine. That is, the average density of the balloon equals that of air, and the average density of the submarine equals the density of water.

A body whose average density is less than that of a liquid can float partially submerged at the free upper surface of the liquid. A familiar example is the hydrometer, shown in Fig. 12–10. The instrument sinks in the fluid until the weight of fluid it displaces is exactly equal to its own weight. In a fluid of greater density, less fluid is displaced for the same weight, so less of the instrument is immersed; i.e., in denser fluids it floats *higher.* This corresponds to the familiar fact that when swimming in sea water (density 1.03 g cm^{-3}), one's body floats higher than in fresh water.

The hydrometer is weighted at its bottom end so the upright position is stable, and a scale in the stem at the top permits direct density readings. Fig. 12–10(b) shows a hydrometer commonly used to measure density of battery acid or anti-freeze, as discussed in Section 12–1. When the tube at the bottom is immersed in the fluid and the bulb squeezed to expel air and then released (like a giant medicine dropper) the resulting pressure difference causes the fluid to rise into the outer tube, and the hydrometer floats in this sample.

When making precise "weighings" with a sensitive analytical balance, correction must be made for the buoyant force of the air if the density of the body being "weighed" is very different from that of the standard "weights," which are usually of brass. For example, suppose a block of wood of density 0.4 g cm^{-3} is balanced on an equal-arm balance by brass "weights" of mass 20 g, density 8.0 g cm^{-3}. The

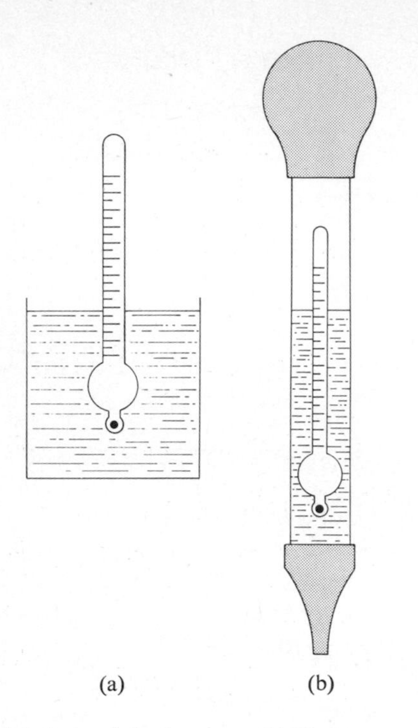

Fig. 12–10 (*a*) *A simple hydrometer.* (*b*) *Hydrometer used as a tester for battery acid or antifreeze.*

apparent weight of each body is the difference between its true weight and the buoyant force of the air. If ρ_w, ρ_b, and ρ_a are the densities of the wood, brass, and air, and V_w and V_b are the volumes of the wood and brass, the apparent weights, which are equal, are

$$\rho_w V_w g - \rho_a V_w g = \rho_b V_b g - \rho_a V_b g.$$

The true mass of the wood is $\rho_w V_w$, and the true mass of the standard is $\rho_b V_b$. Hence,

$$\text{True mass} = \rho_w V_w = \rho_b V_b + \rho_a (V_w - V_b)$$
$$= \text{mass of standard} + \rho_a (V_w - V_b).$$

In the specific example cited,

$$V_w = \frac{20}{0.4} = 50 \text{ cm}^3 \qquad \text{(very nearly)},$$
$$V_b = \frac{20}{8} = 2.5 \text{ cm}^3,$$
$$\rho_a = 0.0013 \text{ g cm}^{-3}.$$

Hence,

$$\rho_a (V_w - V_b) = 0.0013 \times 47.5$$
$$= 0.062 \text{ g}.$$

Therefore,

$$\text{True mass} = 20.062 \text{ g}.$$

If measurements are being made to one one-thousandth of a gram (0.001 g), the correction of 0.062 g is essential.

Example A tank containing water is placed on a spring scale, which registers a total weight W. A stone of weight w is hung from a string and lowered into the water without touching the sides or bottom of the tank (Fig. 12–11(a)). What will be the reading on the spring scale?

First, for the stone alone, the forces are as shown in Fig. 12–11(b), where B is the buoyant force and T is the tension in the string. Since $\sum F_y = 0$,

$$T + B = w.$$

Next, for the tank with the water and stone in it, the forces are as shown in Fig. 12–11(c), where S is the force exerted by the spring scale on the isolated system and, by Newton's third law, is equal in

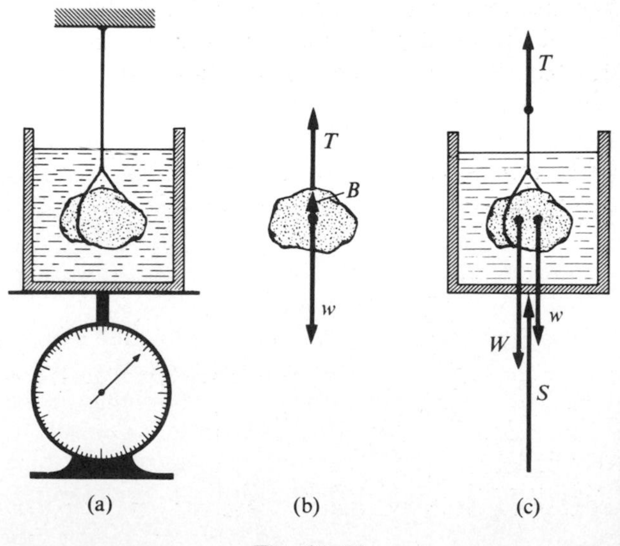

Fig. 12–11

magnitude and opposite in direction to the force exerted on the scale. The condition for equilibrium yields the equation

$$T + S = w + W.$$

Subtracting the first equation from the second, we get

$$S = W + B.$$

That is, the reading of the spring scale has been *increased* by an amount equal to the buoyant force.

An interesting medical application of buoyancy is found in the technique of *hydrotherapy*. A patient may be unable to lift a limb because of disease or injury in the associated muscles or joints. When the body is immersed in water it becomes, in effect, nearly weightless because its average density is only a little greater than that of water. As a result, the forces required to move the limb are greatly reduced, and therapeutic exercise becomes possible.

12-7 FORCES AGAINST A DAM

Water stands at a depth H behind the vertical upstream face of a dam (Fig. 12-12). It exerts a certain resultant horizontal force on the dam, tending to slide it along its foundation, and a certain moment tending to overturn the dam about the point O. We wish to find the horizontal force and its moment.

Figure 12-12(b) is a view of the upstream face of the dam. The pressure at an elevation y is

$$p = \rho g(H - y).$$

(Atmospheric pressure can be omitted, since it also acts upstream against the other face of the dam.) The force against the shaded strip is

$$dF = p\,dA$$
$$= \rho g(H - y)L\,dy.$$

The total force is

$$F = \int dF = \int_0^H \rho g L(H - y)dy$$
$$= \tfrac{1}{2}\rho g L H^2.$$

The moment of the force dF about an axis through O is

$$d\Gamma = y\,dF = \rho g L y(H - y)\,dy.$$

The total torque about O is

$$\Gamma = \int d\Gamma = \int_0^H \rho g L y(H - y)\,dy$$
$$= \tfrac{1}{6}\rho g L H^3.$$

If $\bar{H}$ is the height above O at which the total force F would have to act to produce this torque,

$$F\bar{H} = \tfrac{1}{2}\rho g L H^2 \times \bar{H} = \tfrac{1}{6}\rho g L H^3,$$
$$\bar{H} = \tfrac{1}{3}H.$$

Hence the line of action of the resultant force is at 1/3 of the depth above O, or at 2/3 of the depth below the surface.

12-8 SURFACE TENSION

A liquid flowing slowly from the tip of a medicine dropper emerges not as a continuous stream but as a succession of drops. A sewing needle, if placed carefully on a water surface, makes a small depression in the surface and rests there without sinking,

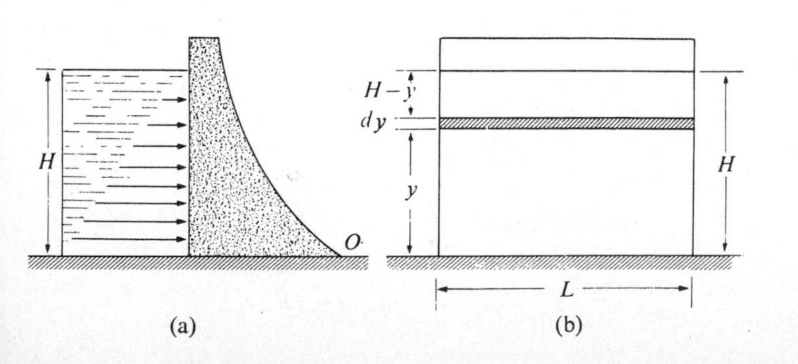

Fig. 12-12 *Forces on a dam.*

even though its density may be as much as 10 times that of water. When a clean glass tube of small bore is dipped into water, the water rises in the tube, but if the tube is dipped in mercury, the mercury is depressed. All these phenomena, and many others of a similar nature, are associated with the existence of a *boundary surface* between a liquid and some other substance.

All surface phenomena indicate that the surface of a liquid can be considered to be in a state of stress such that, for any line lying in or bounding the surface, the material on either side of the line exerts a *pull* on the material on the other side. This pull lies in the plane of the surface and is perpendicular to the line. In Fig. 12–13, a wire ring has a loop of thread attached to it as shown. When the ring and thread are dipped in a soap solution and removed, a thin film of liquid is formed in which the thread "floats" freely, as shown in part (a). If the film inside the loop of thread is punctured, the thread springs out into a circular shape as in part (b), as if the surfaces of the liquid were pulling radially outward on it, as shown by the arrows. Presumably, the same forces were acting before the film was punctured, but since there was film on *both* sides of the thread the net force exerted by the film on every portion of the thread was zero.

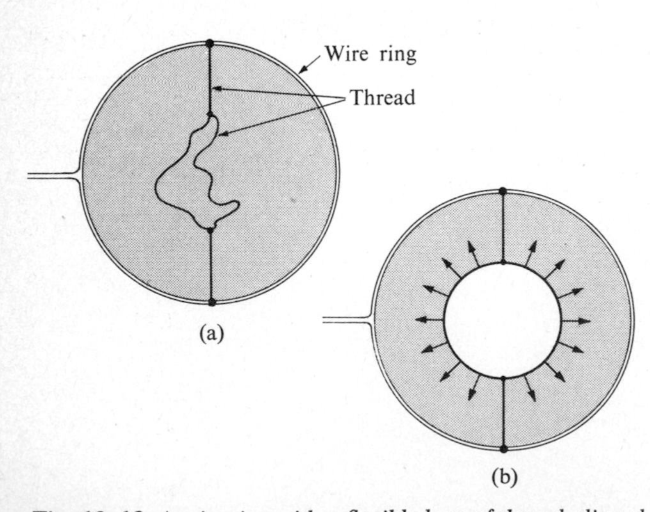

Fig. 12–13 *A wire ring with a flexible loop of thread, dipped in a soap solution, (a) before and (b) after puncturing the surface films inside the loop.*

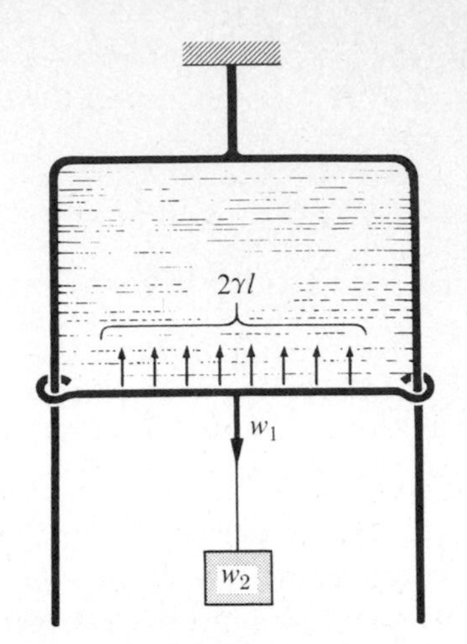

Fig. 12–14 *The horizontal slide wire is in equilibrium under the action of the upward surface force $2\gamma\ell$ and the downward pull $w_1 + w_2$.*

Another simple apparatus for demonstrating surface effects is shown in Fig. 12–14. A piece of wire is bent into the shape of a U and a second piece of wire is used as a slider. When the apparatus is dipped in a soap solution and removed, the slider (if its weight w_1 is not too great) is quickly pulled up to the top of the U. It may be held in equilibrium by adding a second weight w_2. Surprisingly, the same total force $F = w_1 + w_2$ will hold the slider at rest in *any* position, regardless of the area of the liquid film, provided the film remains at constant temperature. That is, the force does not increase as the surface is stretched farther. This is very different from the elastic behavior of a sheet of rubber, for which the force would be greater as the sheet was stretched.

Although a soap film like that in Fig. 12–14 is very thin, its thickness is still enormous compared with the size of a molecule. Hence it can be considered as made up chiefly of bulk liquid, bounded by two surface layers a few molecules thick. When the crossbar in Fig. 12–14 is pulled down and the area of the film is increased, molecules formerly in the main body of the liquid move into the surface layers. That is, these layers are not "stretched" as a rubber sheet would be; more surface is created by molecules moving from the bulk liquid.

Let ℓ be the length of the wire slider. Since the film has two surfaces, the total length along which the surface force acts is 2ℓ. The *surface tension* in the film, γ, is defined as *the ratio of the surface force to the length* (perpendicular to the force) *along which the force acts.* Hence, in this case,

$$\boxed{\gamma = \frac{F}{2\ell}.} \qquad (12\text{–}8)$$

The mks unit of surface tension is the newton per meter (1 N m^{-1}). This unit is not in common use; the usual unit is the cgs unit, the dyne per centimeter (1 dyn cm^{-1}). The conversion factor is

$$1 \text{ N m}^{-1} = 1000 \text{ dyn cm}^{-1}.$$

Another example is shown in Fig. 12–15. A circular wire of circumference ℓ is lifted out from the body of a liquid. The additional force F needed to balance the surface forces $2\gamma\ell$ due to the two surface films on each side is measured either by the stretch of a delicate spring or by the twist of a torsion wire. The surface tension is then given by

$$\gamma = \frac{F}{2\ell}.$$

Other methods of measuring surface tension will be apparent in what is to follow. Some typical values are shown in Table 12–2.

Fig. 12–15 *Lifting a circular wire of length ℓ out of a liquid requires an additional force* **F** *to balance the surface forces $2\gamma\ell$. This method is commonly used to measure surface tension.*

Fig. 12–16 *A drop of milk splashes on a hard surface. (Reproduced from* Flash, *courtesy of Ralph S. Hale & Co.)*

The surface tension of a liquid surface in contact with its own vapor or with air is found to depend *only* on the nature of the liquid and on the temperature. Surface tension usually decreases as temperature increases; Table 12–2 illustrates this behavior for water.

A surface under tension tends to contract until it has the minimum area consistent with any fixed boundaries that may be present and with pressure

Table 12–2 EXPERIMENTAL VALUES OF SURFACE TENSION

Liquid in contact with air	*T*, °C	Surface tension, dyn cm^{-1}
Benzene	20	28.9
Carbon tetrachloride	20	26.8
Ethyl alcohol	20	22.3
Glycerin	20	63.1
Mercury	20	465
Olive oil	20	32.0
Soap solution	20	25.0
Water	0	75.6
Water	20	72.8
Water	60	66.2
Water	100	58.9
Oxygen	−193	15.7
Neon	−247	5.15
Helium	−269	0.12

the surface. For example, a drop of liquid under no external forces, or in free fall in vacuum, is always spherical in shape because the sphere has smaller surface area for a given volume than has any other geometric shape. Figure 12–16 is a beautiful example of formation of spherical droplets in a very complex phenomenon, the impact of a drop on a rigid surface. This was taken by Dr. Edgerton of M.I.T., one of the pioneers in the development of high-speed photographic technique.

12-9 PRESSURE DIFFERENCE ACROSS A SURFACE FILM

A soap bubble consists of two spherical surface films very close together, with liquid between. Surface tension makes the films tend to contract, but as the bubble contracts it compresses the inside air, increasing the interior pressure to a point which prevents further contraction. To obtain a relation between surface tension and this excess pressure, we apply the principles of statics to one-half of the bubble.

As a preliminary, consider first a small element ΔA of the surface, shown in Fig. 12–17. Suppose the air pressure on the left of this element is p and that on the right is p_a. The force normal to the element is therefore $(p - p_a)\Delta A$. The component of this force in the x-direction is

$$(p - p_a)\Delta A \cos \theta.$$

But $\Delta A \cos \theta$ is the area projected on a plane perpendicular to the x-axis. The force in the x-direction is therefore the difference of pressure multiplied by the *projected area* in the x-direction.

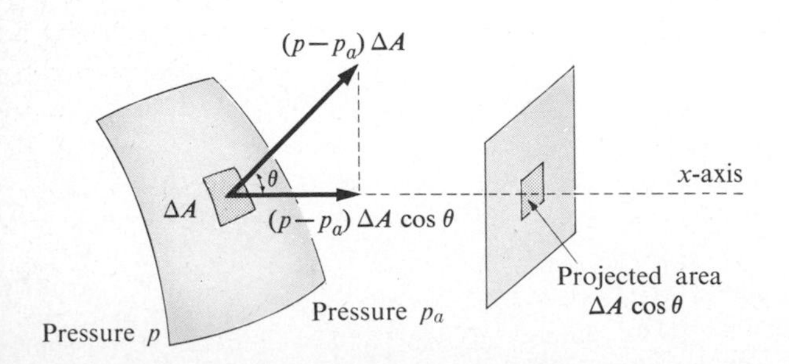

Fig. 12–17 *The force in the x-direction is the difference of pressure multiplied by the projected area in the x-direction.*

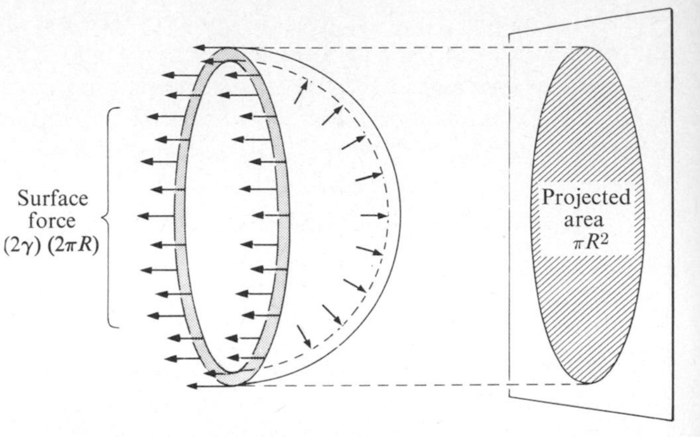

Fig. 12–18 *Equilibrium of half a soap bubble. The force exerted by the outer half is $(2\gamma)(2\pi R)$, and the net force exerted by the air inside and outside the bubble is the pressure difference times projected area, or $(p - p_a)\pi R^2$.*

Now consider the half-bubble shown in Fig. 12–18. The other half exerts a force to the left equal to twice the surface tension times the perimeter or

$$F(\text{to the left}) = (2\gamma)(2\pi R).$$

The force to the right on this half-bubble is equal to the pressure difference $p - p_a$ multiplied by the area obtained by projecting the half-bubble on a plane perpendicular to the direction in question. Since this projected area is πR^2,

$$F(\text{to the right}) = (p - p_a)\pi R^2.$$

Since the half-bubble is in equilibrium,

$$(p - p_a)\pi R^2 = 4\pi R\gamma,$$

or

$$\boxed{p - p_a = \frac{4\gamma}{R} \quad \text{(soap bubble).}} \qquad (12\text{–}9)$$

If the surface tension remains constant (this means constant temperature), the pressure difference is greater for small bubbles than for large ones. If two bubbles are blown at opposite ends of a pipe, the smaller of the two will force air into the larger; the smaller one will get still smaller, and the larger will increase in size.

For a liquid drop, which has only *one* surface film, the difference between pressure of the liquid and that of the outside air is half that for a soap bubble:

$$p - p_a = \frac{2\gamma}{R} \quad \text{(liquid drop).} \qquad (12\text{–}10)$$

Example Calculate the excess pressure inside a drop of mercury of diameter 4 mm at 20°C.

$$\begin{aligned} p - p_a &= \frac{2\gamma}{R} \\ &= \frac{(2)(465 \times 10^{-3}\ \text{N m}^{-1})}{0.002\ \text{m}} \\ &= 465\ \text{N m}^{-2} \cong 0.005\ \text{atm.} \end{aligned}$$

12–10 CONTACT ANGLE AND CAPILLARITY

The preceding sections have been concerned with surface films lying in the boundary between a liquid and a gas. Surface films also exist between a solid wall and a liquid, and between a solid and a vapor. The three boundaries and their accompanying films are shown schematically in Fig. 12–19. The films are only a few molecules thick. Associated with each film is an appropriate surface tension. Thus,

γ_{SL} = surface tension of the solid-liquid film,

γ_{SV} = surface tension of the solid-vapor film,

γ_{LV} = surface tension of the liquid-vapor film.

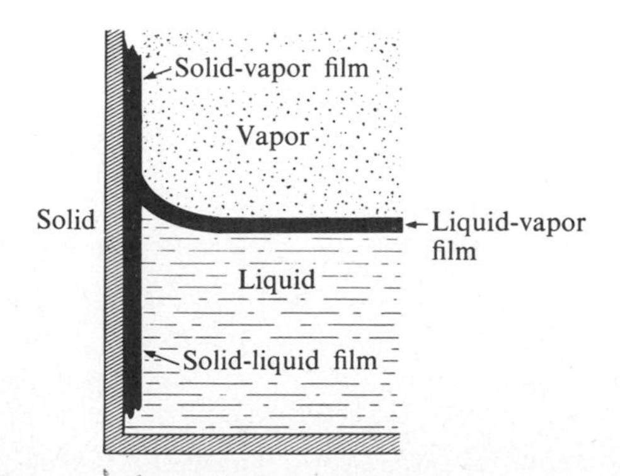

Fig. 12–19 Surface films exist at the solid–vapor boundary as well as the liquid–vapor boundary.

Table 12–3 CONTACT ANGLES

Liquid	Wall	Contact angle
α-Bromonaphthalene ($C_{10}H_7Br$)	Soda-lime glass	5°
	Lead glass	6°45′
	Pyrex	20°30′
	Fused quartz	21°
Methylene iodide (CH_2I_2)	Soda-lime glass	29°
	Lead glass	30°
	Pyrex	29°
	Fused quartz	33°
Water	Paraffin	107°
Mercury	Soda-lime glass	140°

The symbol γ without subscripts, defined and used in the preceding sections, now appears as γ_{LV}.

The curvature of the surface of a liquid near a solid wall depends upon the difference between γ_{SV} and γ_{SL}. Consider a portion of a glass wall in contact with methylene iodide, as shown in Fig. 12–20(a). When γ_{SV} is greater than γ_{SL}, the line along which the three films meet is pulled upward, as in Figs. 12–19 and 12–20(a). Angle θ in Fig. 12–20 is called the *contact angle*, and we see that when γ_{SV} is greater than γ_{SL}, the angle of contact is between zero and 90°. Figure 12–20(b) shows a case where γ_{SV} is *less* than γ_{SL}, the line of intersection of the films is pulled *downward*, and the contact angle is between 90° and 180°. In Fig. 12–20(c), γ_{SV} and γ_{SL} are very nearly equal, and the contact angle is 90°. See Table 12–3 for contact angles of representative liquids in containers of several materials.

Impurities and adulterants present in or added to a liquid may alter the contact angle considerably. Wetting agents or detergents change the contact angle from a large value, greater than 90°, to a value much *smaller* than 90°. Conversely, waterproofing agents applied to cloth cause the contact angle of water in contact with the cloth to be larger than 90°. The effect of a detergent on a drop of water resting on a block of paraffin is shown in Fig. 12–21.

An important phenomenon resulting from surface tension is the elevation of a liquid in an open tube of small cross section. The term *capillarity*, used to describe effects of this sort, originates from the

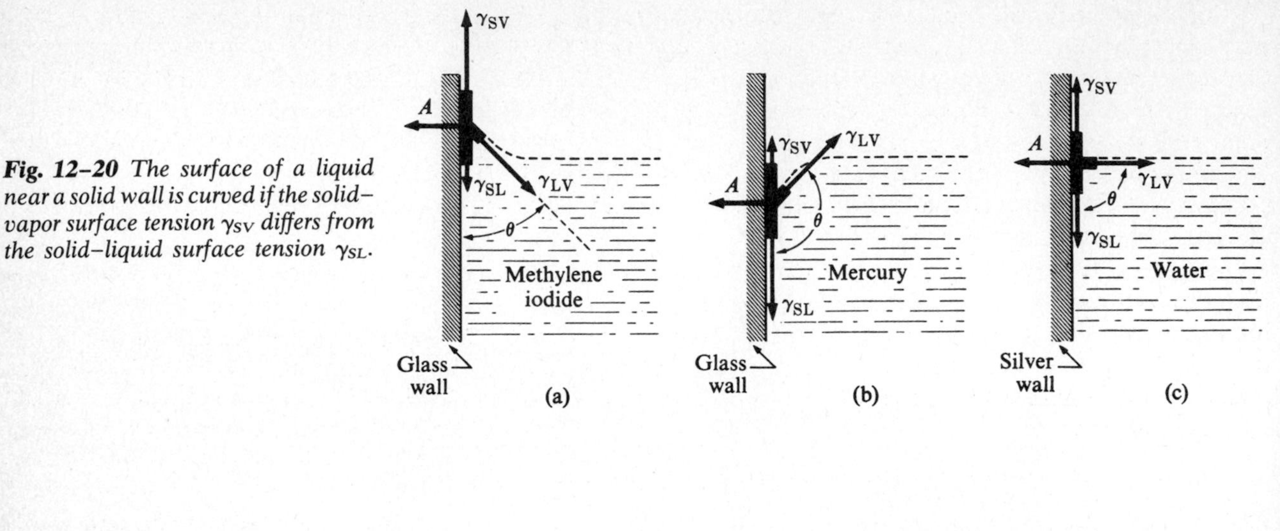

Fig. 12–20 *The surface of a liquid near a solid wall is curved if the solid–vapor surface tension γ_{SV} differs from the solid–liquid surface tension γ_{SL}.*

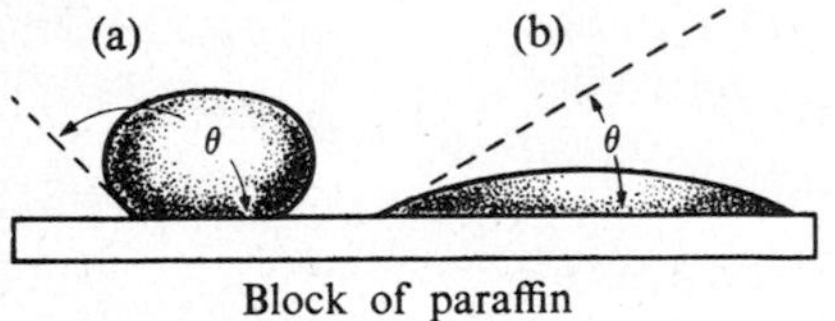

Fig. 12–21 *Effect of decreasing contact angle by a wetting agent.*

description of such tubes as "capillary" or "hairlike." In the case of a liquid that wets the tube, the contact angle is less than 90° and the liquid rises until an equilibrium height y is reached, as shown in Fig. 12–22(a). (The curved liquid surface in the tube is called a *meniscus.*)

If the tube is a cylinder of radius r, the liquid makes contact with the tube along a line of length $2\pi r$. When we isolate the cylinder of liquid of height y and radius r, along with its liquid-vapor film, the total upward force is

$$F = 2\pi r \gamma_{LV} \cos\theta.$$

The downward force is the weight w of the cylinder; this is equal to the weight-density ρg times the volume, which is approximately $\pi r^2 y$, neglecting the small volume of the meniscus. Thus

$$w = \rho g \pi r^2 y.$$

Since the cylinder is in equilibrium,

$$\rho g \pi r^2 y = 2\pi r \gamma_{LV} \cos\theta,$$

or

$$\boxed{y = \frac{2\gamma_{LV}\cos\theta}{\rho g r}.} \tag{12–11}$$

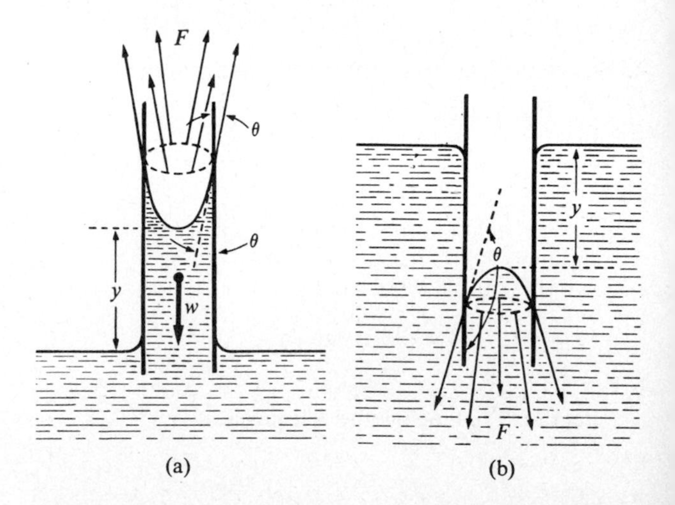

Fig. 12–22 *Surface tension forces on a liquid in a capillary tube. The liquid rises if $\theta < 90°$ and is depressed if $\theta > 90°$.*

The same equation holds for capillary depression, shown in Fig. 12–22(b). Capillarity accounts for the rise of ink in blotting paper, the rise of lighter fluid in the wick of a cigarette lighter, and many other common phenomena.

Capillarity is very important in a variety of life processes. A familiar example is the rising of water (actually a dilute aqueous solution) from the roots of a plant to its foliage, due partly to capillarity and partly to osmotic pressure developed in the roots. In the higher animals, including man, blood is pumped through the arteries and veins, but capillarity is still important in the smallest blood vessels, which indeed are called capillaries.

PROBLEMS

12–1 The piston of a hydraulic automobile lift is 30 cm in diameter. What pressure in pascals (newtons per square meter) is required to lift a car of mass 150 kg? Also express this pressure in atmospheres.

12–2 The expansion tank of a household hot-water heating system is open to the atmosphere and is 10 m above a pressure gauge attached to the furnace. What is the gauge pressure at the furnace, in newtons per square meter? in atmospheres?

12–3 Why can't a skin diver obtain an air supply at any desired depth by breathing through a "snorkel," a tube connected to his face mask and having its upper end above the water surface?

12–4 Suppose the door of a room makes an airtight but frictionless fit in its frame. Do you think you could open the door if the air pressure on one side were standard atmospheric pressure and that on the other side differed from standard by 1%?

12–5 The liquid in the open-tube manometer in Fig. 12–5(a) is mercury, and $y_1 = 3$ cm, $y_2 = 8$ cm. Atmospheric pressure is 970 millibars.

a) What is the absolute pressure at the bottom of the U-tube?
b) What is the absolute pressure in the open tube, at a depth of 5 cm below the free surface?
c) What is the absolute pressure of the gas in the tank?
d) What is the gauge pressure of the gas, in "cm of mercury"?
e) What is the gauge pressure in "cm of water"?

12–6 What gauge pressure must a pump produce in order to pump water from the bottom of Grand Canyon (2400-ft. elevation) to Indian Gardens (4500-ft.)? Express your results in pounds per square inch and in atmospheres.

12–7 A diving bell is to be designed to withstand the pressure of water at a depth of 2000 ft.

a) If sea water weighs 64 lb per cubic foot, what is the pressure at this depth?
b) What is the total force on a glass window 6 inches in diameter?

12–8 A common method for testing gold for purity is to measure its density by weighing it in air and then in water, as discussed in Section 12–6. In an era of rising gold prices, a swindler proposed to make a fake gold ingot by using a hollow slab of iridium (density 22.5 g cm^{-3}) and plating it with a thin layer of gold (density 19.3 g cm^{-3}).

a) To make a fake ingot of total mass 0.5 kg, what should be the total volume?
b) What is the volume of the interior air space?
c) Is iridium really any cheaper than gold?

12–9 According to advertising claims, a certain small car will float in water.

a) If the mass of the car is 1000 kg and its interior volume 4.0 m^3, what fraction of the car is immersed when it floats? The buoyancy of steel and other materials may be neglected.
b) As water gradually leaks in and displaces the air in the car, what fraction of the interior volume is water when the car sinks?

12–10

a) A small test tube, partially filled with water, is inverted in a large jar of water and floats, as shown in Fig. 12–23. The lower end of the test tube is open and the top of the large jar is covered by a tightly fitting rubber membrane. When the membrane is pressed down the test tube sinks, and when the membrane is released it rises again. Explain. (A hollow glass figure in human form is often used instead of the test tube, and is called a "Cartesian diver.")

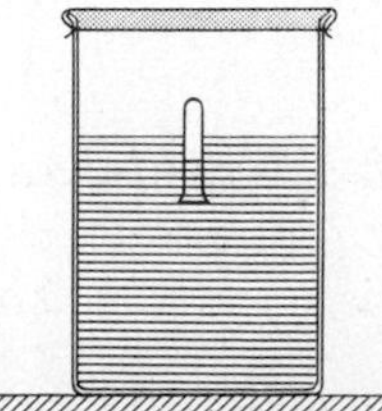

Fig. 12–23

b) A torpedoed ship sinks below the surface of the ocean. If the depth is sufficiently great, is it possible for the ship to remain suspended in equilibrium at some point above the ocean bottom?

12–11 A tube 1 cm^2 in cross section is attached to the top of a vessel 1 cm high and of cross section 100 cm^2. Water is poured into the system, filling it to a depth of 100 cm above the bottom of the vessel, as in Fig. 12–24.

a) What is the force exerted by the water against the bottom of the vessel?

b) What is the weight of the water in the system?

c) Explain why (a) and (b) are not equal.

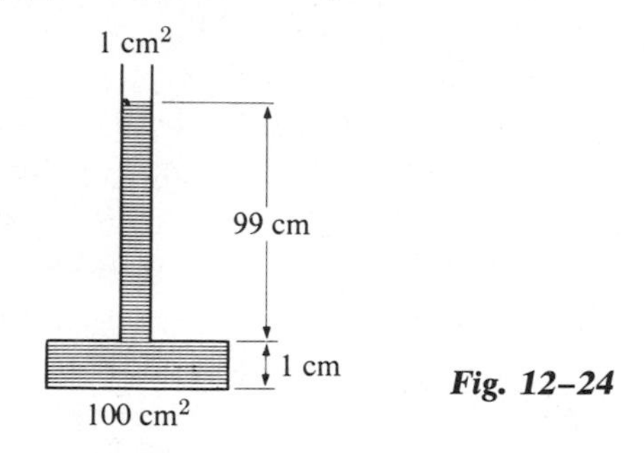

Fig. 12–24

12–12 A piece of gold-aluminum alloy weighs 10 lb. When suspended from a spring balance and submerged in water, the balance reads 8 lb. What is the weight of gold in the alloy if the specific gravity of gold is 19.3 and the specific gravity of aluminum is 2.5?

12–13 What is the area of the smallest block of ice 1 ft thick that will just support a man weighing 180 lb? The specific gravity of the ice is 0.917, and it is floating in fresh water.

12–14 A cubical block of wood 10 cm on a side floats at the interface between oil and water as in Fig. 12–25, with its lower surface 2 cm below the interface. The density of the oil is 0.6 g cm^{-3}.

a) What is the mass of the block?

b) What is the gauge pressure at the lower face of the block?

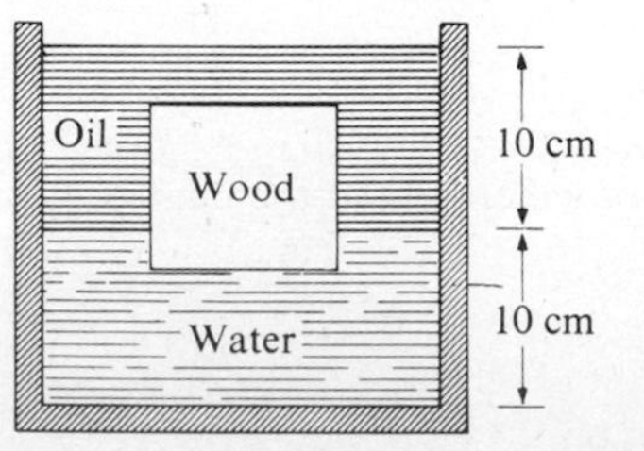

Fig. 12–25

12–15 The densities of air, helium, and hydrogen (at standard conditions) are, respectively, 1.29 kg m^{-3}, 0.178 kg m^{-3}, and 0.0899 kg m^{-3}. What is the volume in cubic meters displaced by a hydrogen-filled dirigible which has a total "lift" of 10,000 kg? What would be the "lift" if helium were used instead of hydrogen?

12–16 A piece of wood is 2 ft long, 1 ft wide, and 2 in. thick. Its specific gravity is 0.6. What volume of lead must be fastened underneath to sink the wood in calm water so that its top is just even with the water level?

12–17 A cubical block of wood 10 cm on a side and of density 0.5 g cm^{-3} floats in a jar of water. Oil of density 0.8 g cm^{-3} is poured on the water until the top of the oil layer is 4 cm below the top of the block.

a) How deep is the oil layer?

b) What is the gauge pressure at the lower face of the block?

12–18 A cubical block of steel (density = 7.8 g cm^{-3}) floats on mercury (density = 13.6 g cm^{-3}).

a) What fraction of the block is above the mercury surface?

b) If water is poured on the mercury surface, how deep must the water layer be so that the water surface just rises to the top of the steel block?

12–19 Block A in Fig. 12–26 hangs by a cord from spring balance D and is submerged in a liquid C contained in beaker B. The weight of the beaker is 2 lb, the weight of the liquid is 3 lb. Balance D reads 5 lb and balance E reads 15 lb. The volume of block A is 0.1 ft^3.

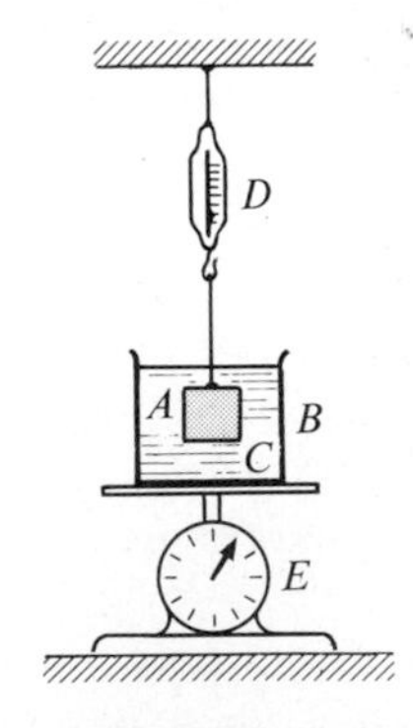

Fig. 12–26

a) What is the mass per unit volume of the liquid?

b) What will each balance read if block A is pulled up out of the liquid?

12–20 A hollow sphere of inner radius 9 cm and outer radius 10 cm floats half submerged in a liquid of specific gravity 0.8.

a) Calculate the density of the material of which the sphere is made.

b) What would be the density of a liquid in which the hollow sphere would just float completely submerged?

12–21 Two spherical bodies having the same diameter are released simultaneously from the same height. If the mass of one is 10 times that of the other and if the air resistance on each is the same, show that the heavier body will arrive at the ground first.

12–22 When a life preserver having a volume of 0.03 m^3 is immersed in sea water (specific gravity 1.03), it will just support a clothed 80-kg man (specific gravity 1.2) with 2/10 of his volume above water. What is the mass per unit volume of the material composing the life preserver?

12–23 A block of balsa wood placed in one scale pan of an equal-arm balance is found to be exactly balanced by a 100-g brass "weight" in the other scale pan. Find the true mass of the balsa wood, if its specific gravity is 0.15.

12–24 A 1500-kg cylindrical can buoy floats vertically in salt water (specific gravity = 1.03). The diameter of the buoy is 1.0 m. Calculate (a) the additional distance the buoy will sink when a 100-kg man stands on top, (b) the period of the resulting vertical simple harmonic motion when the man dives off.

12–25 A cubical block of wood 1 ft on a side is weighted so that its center of gravity is at the point shown in Fig. 12–27(a), and it floats in water with one-half its volume submerged. Compute the restoring torque when the block is "heeled" at an angle of 45° as in Fig. 12–27(b).

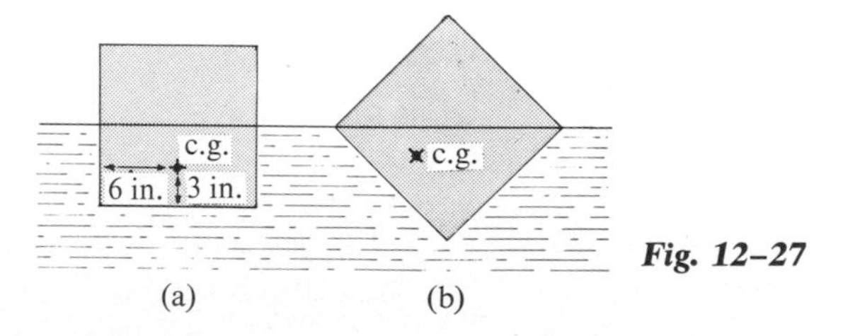

Fig. 12–27

12–26 A hydrometer consists of a spherical bulb and a cylindrical stem of cross section 0.4 cm^2. The total volume of bulb and stem is 13.2 cm^3. When immersed in water the hydrometer floats with 8 cm of the stem above the water surface. In alcohol, 1 cm of the stem is above the surface. Find the density of the alcohol.

12–27 The following is quoted from a letter. How would you reply?

It is the practice of carpenters hereabouts, when laying out and leveling up the foundations of relatively long buildings, to use a garden hose filled with water, into the ends of the hose being thrust glass tubes 10 to 12 inches long.

The theory is that the water, seeking a common level, will be of the same height in both the tubes and thus effect a level. Now the question rises as to what happens if a bubble of air is left in the hose. Our greybeards contend the air will not affect the reading from one end to the other. Others say that it will cause important inaccuracies.

Can you give a relatively simple answer to this question, together with an explanation? I include a rough sketch (Fig. 12–28) of the situation that caused the dispute.

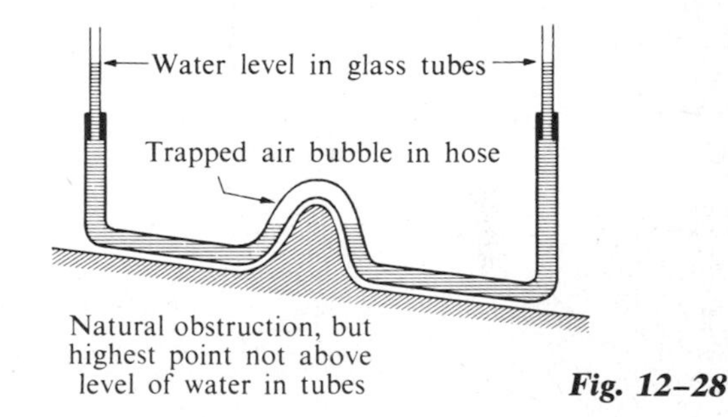

Fig. 12–28

12–28 A swimming pool measures 25 × 8 × 3 m deep. Compute the force exerted by the water against either end and against the bottom.

12–29 The upper edge of a vertical gate in a dam lies along the water surface. The gate is 6 ft wide and is hinged along the bottom edge, which is 10 ft below the water surface. What is the torque about the hinge?

12–30 The upper edge of a gate in a dam runs along the water surface. The gate is 6 ft high and 10 ft wide and is hinged along a horizontal line through its center. Calculate the torque about the hinge.

12–31 Figure 12–29 is a cross-sectional view of a masonry dam whose length perpendicular to the diagram is 100 ft. The depth of water behind the dam is 30 ft. The masonry of which the dam is constructed weighs 150 lb ft^{-3}.

a) Find the dimensions x and $2x$, if the weight of the dam is to be 10 times as great as the horizontal force exerted on it by the water.

b) Is the dam then stable with respect to its overturning about the edge through point O?

c) How does the size of the reservoir behind the dam affect the answers above?

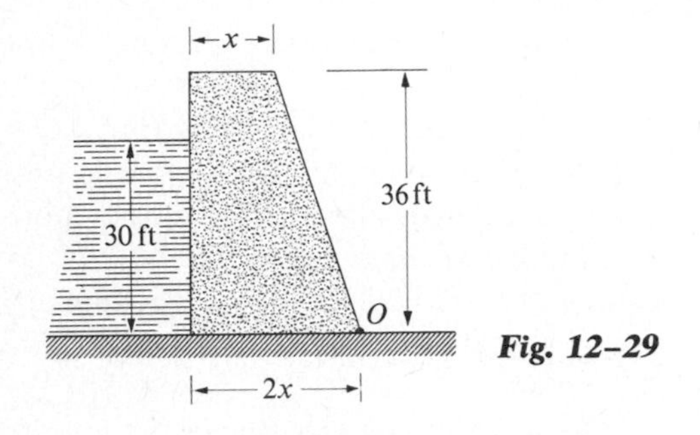

Fig. 12–29

12–32 A U-tube of length ℓ (see Fig. 12–30) contains a liquid. What is the difference in height between the liquid columns in the vertical arms (a) if the tube has an acceleration a toward the right, and (b) if the tube is mounted on a horizontal turntable rotating with an angular velocity ω, with one of the vertical arms on the axis of rotation?

c) Explain why the difference in height does not depend on the density of the liquid, or on the cross-sectional area of the tube. Would it be the same if the vertical tubes did not have equal cross-sections? Would it be the same if the horizontal portion were tapered from one end to the other?

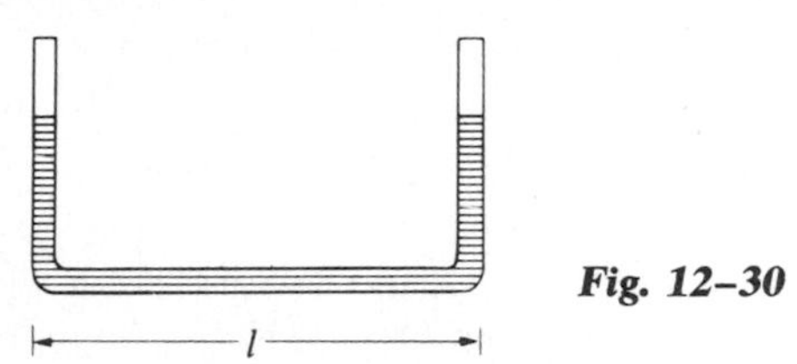

Fig. 12–30

12–33 Compare the tension of a soap bubble with that of a rubber balloon in the following respects:

a) Has each a surface tension?

b) Does the surface tension depend on area?

c) Is Hooke's law applicable?

12–34 Water can rise to a height y in a certain capillary. Suppose that this tube is immersed in water so that only a length $y/2$ is above the surface. Will you have a fountain or not? Explain your reasoning.

12–35 A capillary tube is dipped in water with its lower end 10 cm below the water surface. Water rises in the tube to a height of 4 cm above that of the surrounding liquid, and the angle of contact is zero. What gauge pressure is required to blow a hemispherical bubble at the lower end of the tube?

12–36 A glass tube of inside diameter 1 mm is dipped vertically into a container of mercury, with its lower end 1 cm below the mercury surface.

a) What must be the gauge pressure of air in the tube to blow a hemispherical bubble at its lower end?

b) To what height will mercury rise in the tube if the air pressure in the tube is 3000 N m^{-2} below atmospheric? The angle of contact between mercury and glass is 140°.

12–37 On a day when the atmospheric pressure is 950 millibars, (a) what would be the height of the mercury column in a barometric tube of inside diameter 2 mm?

b) What would be the height in the absence of any surface tension effects?

c) What is the minimum diameter a barometric tube may have in order that the correction for capillary depression shall be less than 0.01 cm of mercury?

12–38

a) Derive the expression for the height of capillary rise in the space between two parallel plates dipping in a liquid.

b) Two glass plates, parallel to each other and separated by 0.5 mm, are dipped in water. To what height will the water rise between them? Assume zero angle of contact.

12–39 A tube of circular cross section and outer radius 0.14 cm is closed at one end. This end is weighted and the tube floats vertically in water, heavy end down. The total mass of the tube and weights is 0.20 g. If the angle of contact is zero, how far below the water surface is the bottom of the tube?

12–40 Find the gauge pressure, in dynes per square centimeter, in a soap bubble 5 cm in diameter. The surface tension is 25 dyn cm^{-1} = 25×10^{-3} N m^{-1}.

12–41 Two large glass plates are clamped together along one edge and separated by spacers a few millimeters thick along the opposite edge to form a wedge-shaped air film. These plates are then placed vertically in a dish of colored liquid. Show that the edge of the liquid forms an equilateral hyperbola.

12–42 When mercury is poured onto a flat glass surface which it does not wet, it spreads out into a pool of uniform thickness regardless of the size of the pool. Find the thickness of the pool.

12–43 Suppose the xylem tubes in the actively growing outer layer of a tree are uniform cylinders, and that the rising of sap is due entirely to capillarity, with a contact angle of 45° and surface tension of 0.05 N m^{-1}. What is the maximum radius of the tubes for a tree 20 m tall?

12–44 A tree 50 m tall has xylem tubes (sap-carrying pipes) in the form of uniform cylinders of radius 2×10^{-4} mm. If the surface tension is 0.05 N m^{-1} and the contact angle 45°, what minimum pressure must be developed in the roots in order for the sap to reach the top of the tree?

Chapter 13

Hydrodynamics and Viscosity

13–1 INTRODUCTION

Hydrodynamics is the study of fluids in motion. It is one of the most complex branches of mechanics, as illustrated by such familiar examples of fluid flow as a river in flood or a swirling cloud of cigarette smoke. While each drop of water or each smoke particle is governed by Newton's laws of motion, the resulting equations can be exceedingly complex. Fortunately, many situations of practical importance can be represented by idealized models which are simple enough to permit detailed analysis.

To begin, we shall consider only a so-called *ideal fluid*, that is, one which is incompressible and which has no internal friction or viscosity. The assumption of incompressibility is usually a good approximation for liquids. A gas can also be treated as incompressible provided the flow is such that pressure differences are not too great. Internal friction in a fluid gives rise to shearing stresses when two adjacent layers of fluid move relative to each other, or when the fluid flows inside a tube or around an obstacle. In some cases these shearing forces can be neglected in comparison with gravitational forces and forces arising from pressure differences.

The path followed by an element of a moving fluid is called a *line of flow*. In general, the velocity of the element changes in both magnitude and direction along its line of flow. If every element passing through a given point follows the same line of flow as that of preceding elements, the flow is said to be *steady* or *stationary*. When any given flow is first started, it passes through a nonsteady state, but in many instances the flow becomes steady after a certain period of time has elapsed. In steady flow, the velocity at each point of space remains constant in time, although the velocity of a particular particle of the fluid may change as it moves from one point to another.

A *streamline* is defined as a curve whose tangent, at any point, is in the direction of the fluid velocity at that point. In steady flow, the streamlines coincide with the lines of flow.

If we construct all of the streamlines passing through the periphery of an element of area, such as the area A in Fig. 13–1, these lines enclose a tube called a *flow tube* or *tube of flow*. From the definition of a streamline, no fluid can cross the side walls of a tube of flow; in steady flow there can be no mixing of the fluids in different flow tubes.

Figure 13–2 illustrates the nature of the flow around a number of obstacles, and in a channel of varying cross section. The photographs were made using an apparatus designed by Pohl, in which

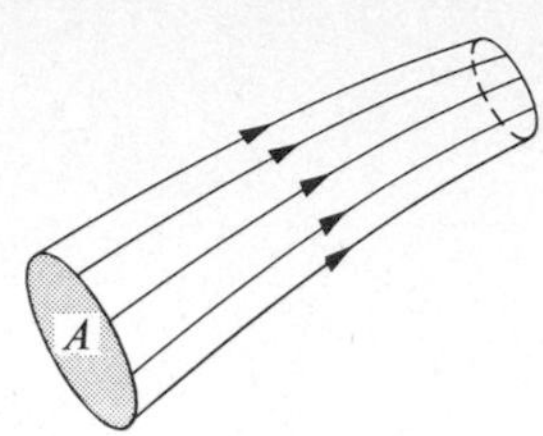

Fig. 13–1 A flow tube bounded by streamlines.

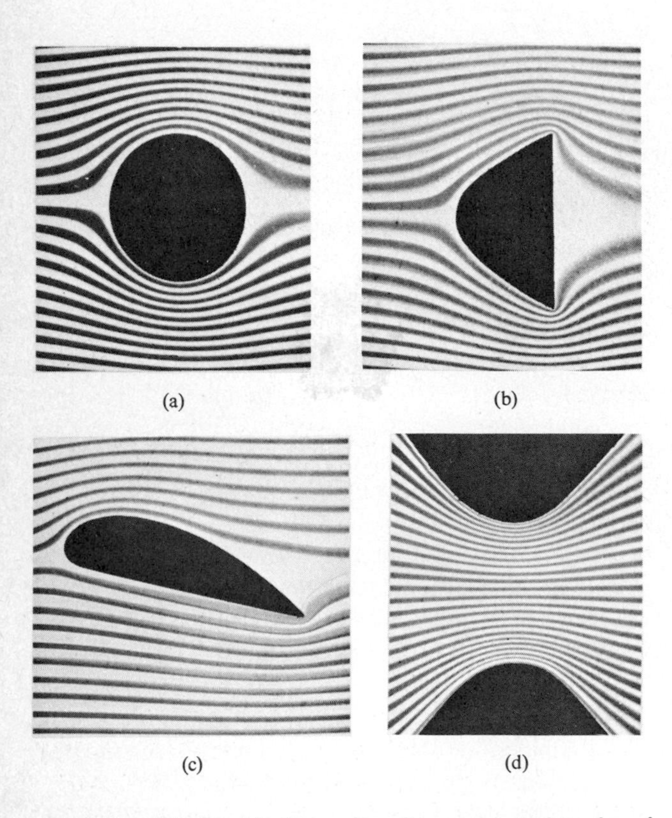

Fig. 13–2 (a), (b), (c): Streamline flow around obstacles of various shapes. (d) Flow in a channel of varying cross-sectional area.

alternate streams of clear and colored water flow between two closely spaced glass plates. The obstacles and the channel walls are opaque flat plates which fit between the glass plates. It will be noted that each obstacle is completely surrounded by a tube of flow. The tube splits into two portions at a so-called *stagnation point* on the upstream side of the obstacle. These portions rejoin at a second stagnation point on the downstream side. The velocity at the stagnation points is zero. It will also be noted that the cross sections of all flow tubes decrease at a constriction and increase again when the channel widens.

The patterns of Fig. 13–2 are typical of *streamline* or *laminar* flow, in which adjacent layers of fluid slide smoothly past each other. At sufficiently high flow rates, or when boundary surfaces cause abrupt changes in velocity, the flow becomes irregular and much more complex and is called *turbulent* flow. In turbulent flow there is, strictly speaking, no steady-state pattern, since the flow pattern continuously changes.

13–2 THE EQUATION OF CONTINUITY

Let us consider any stationary, closed surface in a moving fluid; in general, fluid flows into the volume enclosed by the surface at some points and flows out at other points. The *equation of continuity* is a mathematical statement of the fact that the *net* rate of flow of mass *inward* across any closed surface is equal to the rate of increase of the mass within the surface.

For an incompressible fluid in steady flow, the equation takes the following form. Figure 13–3 represents a portion of a tube of flow, between two fixed cross sections of areas A_1 and A_2. Let v_1 and v_2 be the speeds at these sections. There is no flow across the side walls of the tube. The volume of fluid that flows into the tube across A_1 in a time interval dt is that contained in the short cylindrical element of base A_1 and height $v_1\,dt$, or is $A_1v_1\,dt$. If the density of the fluid is ρ, the *mass* flowing in is $\rho A_1 v_1\,dt$. Similarly, the mass

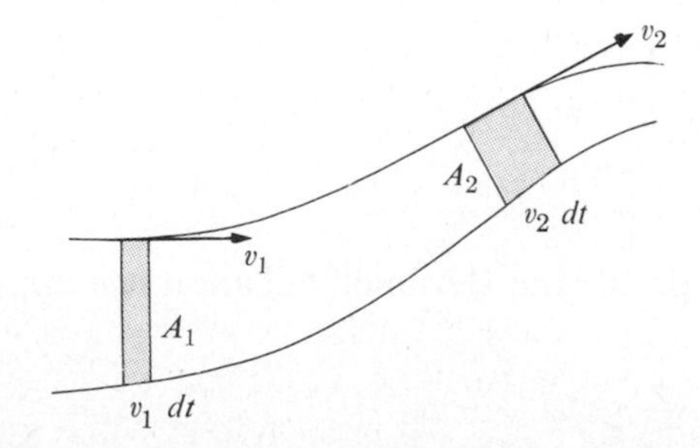

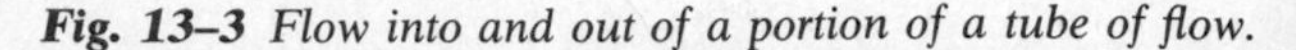

Fig. 13–3 Flow into and out of a portion of a tube of flow.

that flows out across A_2 in the same time is $\rho A_2 v_2\, dt$. The volume between A_1 and A_2 is constant, and since the flow is steady, the mass flowing out equals that flowing in. Hence

$$\rho A_1 v_1\, dt = \rho A_2 v_2\, dt,$$

or

$$\boxed{A_1 v_1 = A_2 v_2,} \qquad (13\text{–}1)$$

and the product Av is constant along any given tube of flow. It follows that when the cross section of a flow tube decreases, as in the constriction in Fig. 13–2(d), the velocity increases. This can readily be shown by introducing small particles into the fluid and observing their motion.

13–3 BERNOULLI'S EQUATION

When an incompressible fluid flows along a horizontal flow tube of varying cross section, its velocity changes; that is, it accelerates or decelerates. It must therefore be acted on by a resultant force, and this means that the pressure must vary *along* the flow tube even though the elevation does not change. For two points at *different* elevations, the pressure difference depends not only on the difference in level but also on the difference between the velocities at the points. The general expression for the pressure difference can be obtained directly from Newton's second law, but it is simpler to make use of the work-energy theorem. The problem was first solved by Daniel Bernoulli in 1738.

Figure 13–4 represents a portion of a tube of flow. We are to follow a small element of the fluid, indicated by shading, as it moves from one point to another along the tube. Let y_1 be the elevation of the first point above some reference level, v_1 the speed at that point, A_1 the cross-sectional area of the tube, and p_1 the pressure. All these quantities may vary from point to point: y_2, v_2, A_2, and p_2 are their values at the second point.

To derive the Bernoulli equation, we apply the work-energy theory to the fluid in a section of a flow tube. At some initial time, this element of fluid lies between two cross-sections a and c of the flow tube, in Fig. 13–4. The width of the tube is exaggerated for

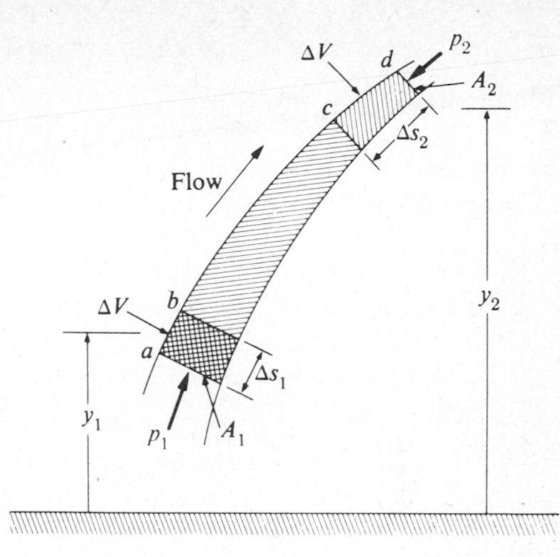

Fig. 13–4 The net work done on the shaded element equals the increase in its kinetic and potential energy.

clarity. In a time interval Δt, the element moves, the ends undergoing displacements Δs_1 and Δs_2 as shown. Because of the continuity relation, $\Delta V = A_1 \Delta s_1 = A_2 \Delta s_2$. The force on the cross-section at a is $p_1 A_1$, and that at b is $p_2 A_2$. The net *work* done on the element during this displacement is therefore

$$W = p_1 A_1 \Delta s_1 - p_2 A_2 \Delta s_2 = (p_1 - p_2)\,\Delta V. \quad (13\text{–}2)$$

The negative sign in the second term arises because the force at c is opposite in direction to the displacement.

We now equate this work to the total change in energy, kinetic and potential, of the element. During Δt, a volume of fluid $\Delta V = A_1 \Delta s_1$ with a mass $\Delta m = \rho\,\Delta V$ enters the tube past section a, bringing with it kinetic energy $\frac{1}{2}\Delta m v^2 = \frac{1}{2}\rho\,\Delta V v_1^2$. Similarly, during this interval, an equal mass leaves the tube past section c, taking with it kinetic energy $\frac{1}{2}\rho\,\Delta V v_2^2$. Thus the net change in kinetic energy is

$$\Delta E_k = \tfrac{1}{2}\rho\,\Delta V(v_2^2 - v_1^2). \qquad (13\text{–}3)$$

The change in potential energy is obtained similarly. The potential energy of the mass entering at a in time Δt is $\Delta m g y_1 = \rho\,\Delta V g y_1$, and that of the mass leaving at c is $\Delta m g y_2 = \rho\,\Delta V g y_2$.

The net change in potential energy is

$$\Delta E_p = \rho \Delta V g(y_2 - y_1). \qquad (13\text{–}4)$$

Combining Eqs. (13–2), (13–3), and (13–4) in the work-energy theorem $W = \Delta E_k + \Delta E_p$, we obtain

$$(p_1 - p_2)\Delta V = \tfrac{1}{2}\rho \Delta V(v_2^2 - v_1^2) + \rho \Delta V g(y_2 - y_1)$$

or

$$p_1 - p_2 = \tfrac{1}{2}\rho(v_2^2 - v_1^2) + \rho g(y_2 - y_1). \qquad (13\text{–}5)$$

In this form, Bernoulli's equation represents the equality of the work per unit volume of fluid $(p_1 - p_2)$ to the sum of the changes in kinetic and potential energies per unit volume that occur during the flow. Or we may interpret Eq. (13–5) in terms of pressures. The second term on the right is the pressure difference arising from the weight of the fluid and the difference in elevation of the two ends of the fluid element. The first term is the additional pressure difference associated with the change of velocity of the fluid.

Equation (13–5) can also be written

$$p_1 + \rho g y_1 + \tfrac{1}{2}\rho v_1^2 = p_2 + \rho g y_2 + \tfrac{1}{2}\rho v_2^2, \qquad (13\text{–}6)$$

and since the subscripts 1 and 2 refer to *any* two points along the tube of flow, Bernoulli's equation may also be written

$$p + \rho g y + \tfrac{1}{2}\rho v^2 = \text{constant}. \qquad (13\text{–}7)$$

Note carefully that p is the *absolute* (not gauge) pressure, and that a consistent set of units must be used. In the mks system, pressure is expressed in N m^{-2}, density in kg m^{-3}, and velocity in m s^{-1}.

Example Water enters a house through a pipe 2.0 cm in inside diameter, at an absolute pressure of 4×10^5 Pa (about 4 atm). The pipe leading to the second floor bathroom 5 m above is 1.0 cm in diameter. When the flow velocity at the inlet pipe is 4 m s^{-1}, find the flow velocity and pressure in the bathroom.

Solution. The flow velocity is obtained from the continuity equation:

$$v_2 = \frac{A_1}{A_2} v_1 = \frac{(2.0\text{ cm})^2}{(1.0\text{ cm})^2}(4\text{ m s}^{-1}) = 16\text{ m s}^{-1}.$$

The pressure is now obtained from Bernoulli's equation:

$$\begin{aligned} p_2 &= p_1 - \tfrac{1}{2}\rho(v_2^2 - v_1^2) - \rho g(y_2 - y_1) \\ &= 4 \times 10^5\text{ Pa} - \tfrac{1}{2}(1.0 \times 10^3\text{ kg m}^{-3}) \\ &\quad \times(256\text{ m}^2\text{ s}^{-2} - 16\text{ m}^2\text{ s}^{-2}) \\ &\quad -(1.0 \times 10^3\text{ kg m}^{-3})(9.8\text{ m s}^{-2})(5\text{ m}) \\ &= 2.3 \times 10^5\text{ Pa} \simeq 2.3\text{ atm}. \end{aligned}$$

We note that when the water is turned off, the second term on the right vanishes and the pressure rises to 3.5×10^5 Pa.

13–4 APPLICATIONS OF BERNOULLI'S EQUATION

1. The equations of hydrostatics are special cases of Bernoulli's equation, when the velocity is everywhere zero. Thus, when v_1 and v_2 are zero, Eq. (13–5) reduces to

$$p_1 - p_2 = \rho g(y_2 - y_1),$$

which is the same as Eq. (12–3).

2. *Speed of efflux. Torricelli's theorem.* Figure 13–5 represents a tank of cross-sectional area A_1, filled to a depth h with a liquid of density ρ. The space above the top of the liquid contains air at pressure p, and the liquid flows out of an orifice of area A_2. Let us consider the entire volume of moving fluid as a single tube of flow, and let v_1 and v_2 be the speeds at points 1 and 2. The quantity v_2 is called the *speed of efflux.* The pressure at point 2 is atmospheric, p_a. Applying Bernoulli's equation to points 1 and 2, and taking the bottom of the tank as our reference level, we get

$$p + \tfrac{1}{2}\rho v_1^2 + \rho g h = p_a + \tfrac{1}{2}\rho v_2^2,$$

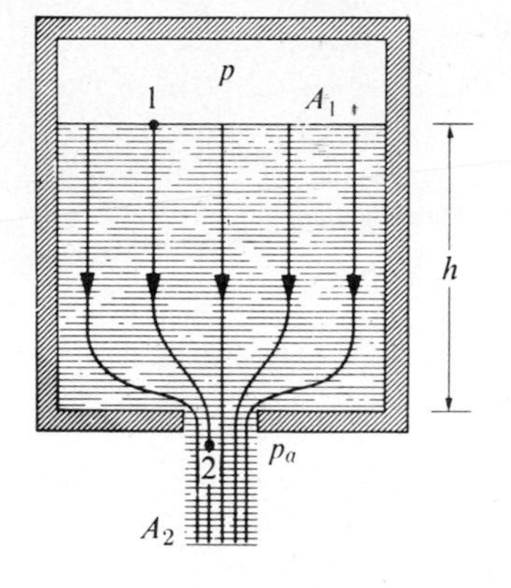

Fig. 13–5 *Flow of a liquid out of an orifice.*

or

$$v_2^2 = v_1^2 + 2\frac{p - p_a}{\rho} + 2gh. \qquad (13\text{–}8)$$

From the equation of continuity,

$$v_2 = \frac{A_1}{A_2}v_1. \qquad (13\text{–}9)$$

Because of the converging of the streamlines as they approach the orifice, the cross section of the stream continues to diminish for a short distance outside the tank. It is the area of smallest cross section, known as the *vena contracta*, which should be used in Eq. (13–9). For a sharp-edged circular opening, the area of the *vena contracta* is about 65 percent as great as the area of the orifice.

Now let us consider some special cases. Suppose the tank is open to the atmosphere, so that

$$p = p_a \qquad \text{and} \qquad p - p_a = 0.$$

Suppose also that $A_1 \gg A_2$. Then v_1^2 is very much less than v_2^2 and can be neglected, and from Eq. (13–8),

$$v_2 = \sqrt{2gh}. \qquad (13\text{–}10)$$

That is, *the speed of efflux is the same as that acquired by any body in falling freely through a height h*. This is *Torricelli's theorem*. It is not restricted to an opening in the bottom of a vessel, but applies also to a hole in the side walls at a depth h below the surface.

Now suppose again that the ratio of areas is such that v_1^2 is negligible and that the pressure p (in a closed vessel) is so large that the term $2gh$ in Eq. (13–8) can be neglected, compared with $2(p - p_a)/\rho$. The speed of efflux is then

$$v_2 = \sqrt{2(p - p_a)/\rho}. \qquad (13\text{–}11)$$

The density ρ is that of the fluid escaping from the orifice. If the vessel is partly filled with a liquid, as in Fig. 13–5, ρ is the density of the liquid. On the other hand, if the vessel contains only a gas, ρ is the density of the gas. The efflux speed of a gas may be very great, even for small pressures, since its density is small. However, if the pressure is too great, our idealized model is no longer adequate. Compressibility of the gas must be considered, and if the speed is too great, the motion may become turbulent. Bernoulli's equation can no longer be applied to the motion under these conditions.

A flow of fluid out of an orifice in a vessel gives rise to a *thrust* or *reaction force* on the remainder of the system. The mechanics of the problem are the same as those involved in rocket propulsion. The thrust can be computed as follows, provided conditions are such that Bernoulli's equation is applicable. If A is the area of the orifice, ρ the density of the escaping fluid, and v the speed of efflux, the mass of fluid flowing out in time Δt is $\rho A v\,\Delta t$, and its momentum (mass × velocity) is $\rho A v^2\,\Delta t$. Since we are neglecting the relatively small velocity of the fluid in the container, we can say that the escaping fluid started from rest and acquired the momentum above in time Δt. Its *rate of change* of momentum was therefore $\rho A v^2$, and from Newton's second law this equals the force acting on it. By Newton's third law, an equal and opposite reaction force acts on the remainder of the system. Taking the expression for v^2 from Eq. (13–11), the reaction force can be written

$$F = \rho A v^2 = \rho A \frac{2(p - p_a)}{\rho},$$

or

$$F = 2A(p - p_a). \qquad (13\text{–}12)$$

Thus, while the *speed* of efflux is inversely proportional to the density, the *thrust* is independent of

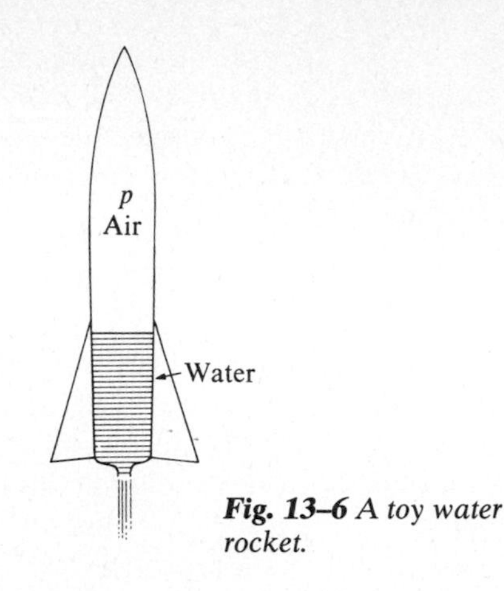

Fig. 13–6 *A toy water rocket.*

the density and depends only on the area of the orifice and the gauge pressure $p - p_a$.

Example Figure 13–6 shows a toy "water rocket." Above the water is air at a pressure $p = 2$ atm. (a) If the rocket is held at rest, what is the speed of efflux out of an opening in the base of the rocket? (b) What is the upward thrust if the area of the opening is 0.5 cm^2?

a) The gauge pressure $p - p_a = 1 \text{ atm} \approx 10^5$ Pa. The efflux speed is therefore

$$v \approx \sqrt{\frac{2 \times 10^5 \text{ Pa}}{10^3 \text{ kg m}^{-3}}} \approx 14 \text{ m s}^{-1}.$$

The upward thrust, or reaction force, is

$$F \approx (2)(0.5 \times 10^{-4} \text{ m}^2)(10^5 \text{ Pa}) \approx 10 \text{ N}.$$

This force is much larger than the weight of the rocket and contents. Note that the reaction force, for the same gauge pressure, would be the same if the rocket initially contained air only. What is the reason for partially filling it with water?

It is interesting to treat this problem in the same way that the rocket was treated earlier. We have the rocket equation, Eq. (8–13)

$$m\frac{dv}{dt} = v_r\frac{dm}{dt} - mg.$$

Each term has a simple interpretation. The left-hand term is the resultant force which is seen to be the difference between the upward reaction force $v_r\, dm/dt$ and the weight mg downward. For the water rocket, $v_r \approx 14 \text{ m s}^{-1}$; $dm/dt = \rho v_r A \approx 0.7 \text{ kg s}^{-1}$. Hence the reaction force is

$$F = v_r\, dm/dt \approx (0.7 \text{ kg s}^{-1})(14 \text{ m s}^{-1}) \simeq 10 \text{ N}.$$

3. The *Venturi tube*, illustrated in Fig. 13–7, consists of a constriction or throat inserted in a pipeline and having properly designed tapers at inlet and outlet to avoid turbulence. Bernoulli's equation, applied to the wide and to the constricted portions of the pipe, becomes

$$p_1 + \tfrac{1}{2}\rho v_1^2 = p_2 + \tfrac{1}{2}\rho v_2^2.$$

From the equation of continuity, the speed v_2 is greater than the speed v_1, and hence the pressure p_2 in the throat is *less* than the pressure p_1. Thus a net force to the right acts to accelerate the fluid as it enters the throat, and a net force to the left decelerates it as it leaves. The pressures p_1 and p_2 can be measured by attaching vertical side tubes as shown in the diagram. From a knowledge of these pressures and of the cross-sectional areas A_1 and A_2, the velocities and the mass rate of flow can be computed. When used for this purpose, the device is called a *Venturi meter.*

The reduced pressure at a constriction finds a number of technical applications. Gasoline vapor is drawn into the manifold of an internal combustion engine by the low pressure produced in a Venturi throat to which the carburetor is connected. The *aspirator pump* is a Venturi throat through which

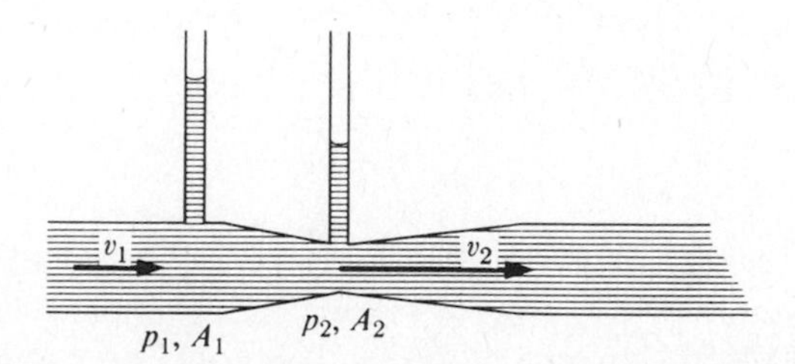

Fig. 13–7 *The Venturi tube.*

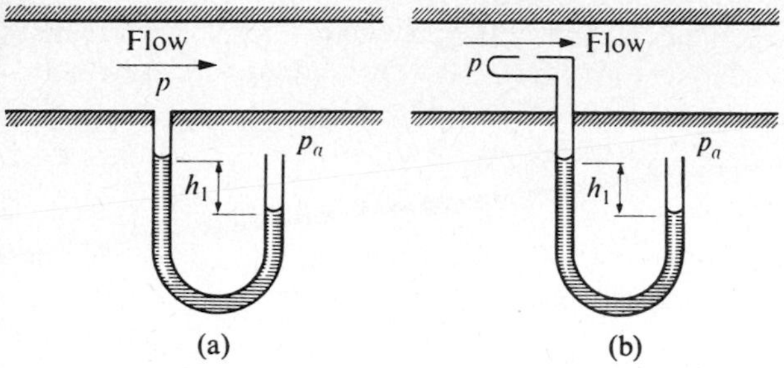

Fig. 13–8 Pressure gauges for measuring the static pressure p in a fluid flowing in an enclosed channel.

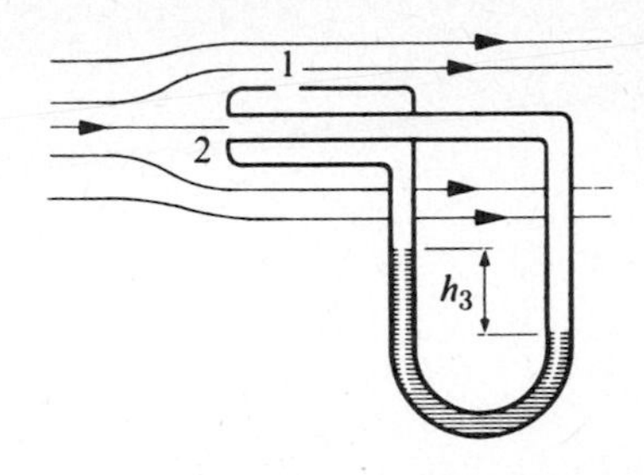

Fig. 13–10 The Prandtl tube.

water is forced. Air is drawn into the low-pressure water rushing through the constricted portion.

4. *Measurement of pressure in a moving fluid.* The pressure p in a fluid flowing in an enclosed channel can be measured with an open-tube manometer, as shown in Fig. 13–8. In (a), one arm of the manometer is connected to an opening in the channel wall. In (b), a *probe* is inserted in the stream. The probe should be small enough so that the flow is not appreciably disturbed and should be shaped so as to avoid turbulence. The difference h_1 in height of the liquid in the arms of the manometer is proportional to the difference between atmospheric pressure p_a and the fluid pressure p. That is,

$$p_a - p = \rho_m g h_1, \qquad p = p_a - \rho_m g h_1, \tag{13–13}$$

where ρ_m is the density of the manometer liquid.

The *Pitot tube*, shown in Fig. 13–9, is a probe with an opening at its upstream end. A stagnation point forms at the opening, where the pressure is p_2

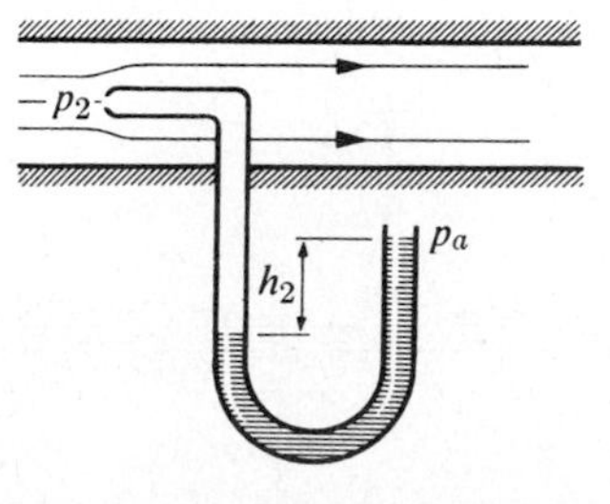

Fig. 13–9 The Pitot tube.

and the speed is zero. Applying Bernoulli's equation to the stagnation point, and to a point at a large distance from the probe where the pressure is p and the speed is v, we get

$$p_2 = p + \tfrac{1}{2}\rho v^2 \tag{13–14}$$

(ρ is the density of the flowing fluid). The pressure p_2 at the stagnation point is therefore the sum of the pressure p and the quantity $\frac{1}{2}\rho v^2$.

The quantity $\frac{1}{2}\rho v^2$ is sometimes called the *dynamic pressure* and p the *static pressure.* These terms are somewhat misleading, inasmuch as p is always the true pressure (force per unit area) in both static and dynamic situations.

The instrument shown in Fig. 13–10 combines in a single device the functions of the instruments in Figs. 13–8 and 13–9. This device is called a *Prandtl tube*; the term Pitot tube is also used. The pressure at opening 1 is the "static pressure," corresponding to the pressure p measured in Fig. 13–8, and that at opening 2 is the quantity $p_2 = p + \frac{1}{2}\rho v^2$ measured in Fig. 13–9. The manometric height h_3 is proportional to the difference of these, or to $\frac{1}{2}\rho v^2$. Hence

$$\tfrac{1}{2}\rho v^2 = \rho_m g h_3. \tag{13–15}$$

This instrument is self-contained, and its reading does not depend on atmospheric pressure. If held at rest, it can be used to measure the velocity of a stream of fluid flowing past it. If mounted on an aircraft, it indicates the velocity of the aircraft relative to the surrounding air and is known as an *airspeed indicator.*

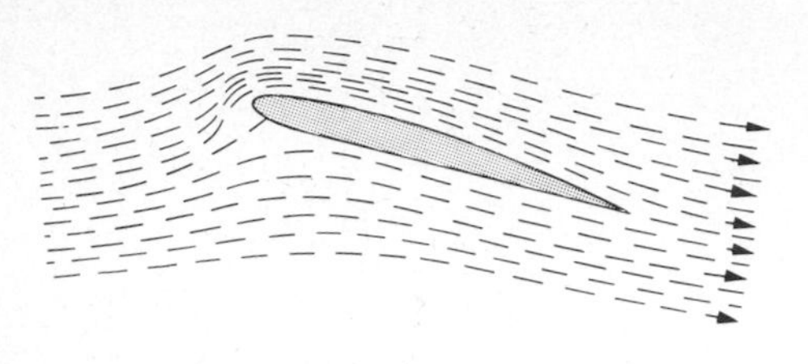

Fig. 13–11 Flow lines around an airfoil.

5. *Lift on an aircraft wing.* Figure 13–11 shows flow lines around a section of an aircraft wing or an airfoil. The orientation of the wing relative to the flow direction causes the flow lines to crowd together above the wing, corresponding to increased flow velocity in this region, much as in the throat of a Venturi. Hence the region above the wing is one of increased velocity and reduced pressure, while below the wing the pressure remains nearly atmospheric. Because the upward force on the under side of the wing is greater than the downward force on the top side, there is a net upward force or *lift.*

This phenomenon can also be understood qualitatively on the basis of Newton's laws. In order for the fluid to exert a net upward force on the wing, the wing must exert a *downward* reaction force on the fluid, deflecting the stream downward. This effect is shown in Fig. 13–11; the fluid suffers a net change in the vertical component of momentum as it passes the wing, corresponding to the downward force the wing exerts on it. The force *on* the wing is thus *upward,* in agreement with the above discussion.

As the angle of the wing relative to the flow increases, turbulent flow occurs in a larger and larger region above the wing, and the pressure drop is no longer as great as that predicted by Bernoulli's principle. The lift on the wing decreases, and in extreme cases the airplane stalls.

6. *The curved flight of a spinning ball.* This effect is somewhat more subtle than the lift on an airplane wing, but the same basic principles are involved. Figure 13–12(a) represents a stationary ball in a blast of air moving from right to left. The motion of the air stream around and past the ball is the same as though the ball were moving through still air from left to right. Because of the large velocities ordinarily involved, there is a region of turbulent flow behind the ball, as shown.

When the ball is spinning, as in Fig. 13–12(b), the viscosity of air causes layers of air near the ball's surface to be pulled around in the direction of spin. The velocity of air relative to the ball surface is greater at the top than at the bottom. The region of turbulence becomes asymmetric, with turbulence occurring farther forward on the top side than on the bottom. This asymmetry gives rise to a pressure difference, the average pressure at the top becoming greater than that at the bottom. The corresponding net downward force deflects the ball as shown. In a baseball curve pitch, the actual deflection is sideways, and in that case Fig. 13–12 shows a *top* view of the situation.

Corresponding to the downward force on the ball in Fig. 13–12(b), the ball must exert an equal and opposite reaction force *upward* on the air. This deflects the air stream upward near the ball, as the figure shows, just as the lift force on an airplane wing corresponds to a downward deflection of the air stream.

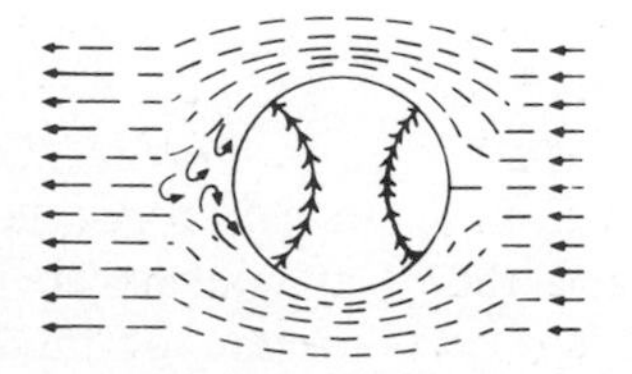

(a)

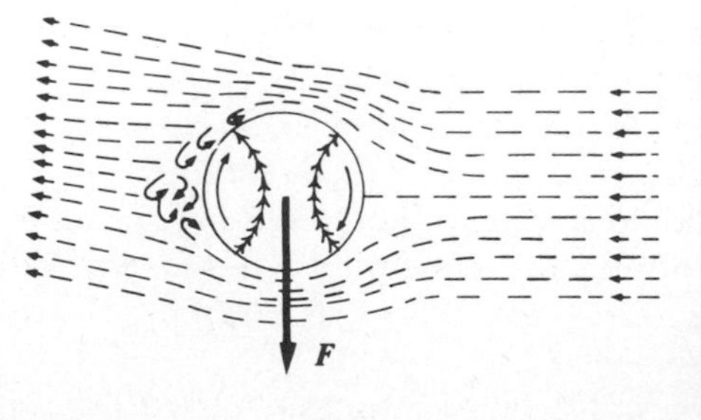

(b)

Fig. 13–12 (a) Air flow past a stationary ball, showing symmetric region of turbulence behind ball. (b) Air flow past a spinning ball, showing asymmetric region of turbulence and deflection of air stream. The net force on the ball is in the direction shown.

A similar effect occurs with golf balls, which always have "backspin" from impact with the slanted club face. The resulting pressure difference between top and bottom of the ball causes a "lift" force which keeps the ball in the air considerably longer than would be possible without spin. A well-hit drive appears from the tee to "float" or even curve *upward* during the initial portion of its flight, and this is a real effect, not an illusion. The dimples on the ball play an essential role; an undimpled ball has a much shorter trajectory than a dimpled one given the same initial velocity and spin. Some manufacturers even claim that the *shape* of the dimples is significant; one feels that polygonal dimples are superior to round ones!

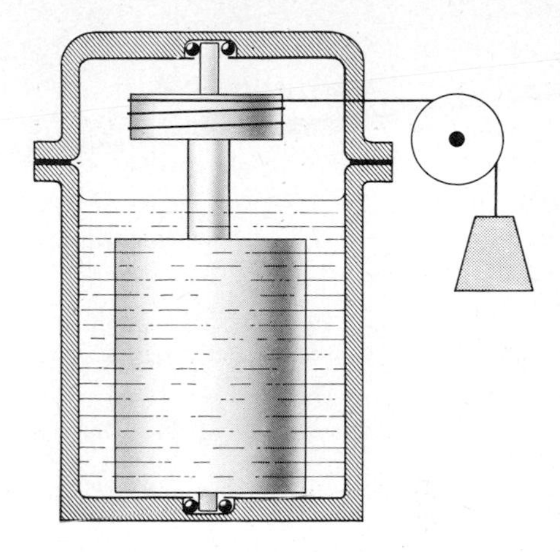

Fig. 13–13 *Schematic diagram of one type of viscosimeter.*

13-5 VISCOSITY

Viscosity may be thought of as the internal friction of a fluid. Because of viscosity, a force must be exerted to cause one layer of a fluid to slide past another, or to cause one surface to slide past another if there is a layer of fluid between the surfaces. Both liquids and gases exhibit viscosity, although liquids are much more viscous than gases. In developing the fundamental equations of viscous flow, it will be seen that the problem is very similar to that of the shearing stress and strain in a solid.

Figure 13–13 illustrates one type of apparatus for measuring the viscosity of a liquid. A cylinder is pivoted on nearly frictionless bearings so as to rotate coaxially within a cylindrical vessel. The liquid whose viscosity is to be measured is poured into the annular space between the cylinders. A torque can be applied to the inner cylinder by the weight-pulley system. When the weight is released, the inner cylinder accelerates momentarily but very quickly reaches a constant angular velocity and continues to rotate at that velocity so long as the torque acts. It is obvious that this velocity will be smaller with a liquid such as glycerine in the annular space than it will be if the liquid is water or kerosene. From a knowledge of the torque, the dimensions of the apparatus, and the angular velocity, the viscosity of the liquid may be computed.

To reduce the problem to its essential terms, we replace the concentric cylinders by two parallel plates, as in Fig. 13–14. The bottom plate is stationary, while the top plate moves with constant velocity v. It is found that the fluid in contact with each surface has the same velocity as that surface; thus at the top surface the fluid has velocity v, while the fluid adjacent to the bottom surface is at rest. The velocities of intermediate layers of fluid increase uniformly from one surface to the other, as shown by the arrows.

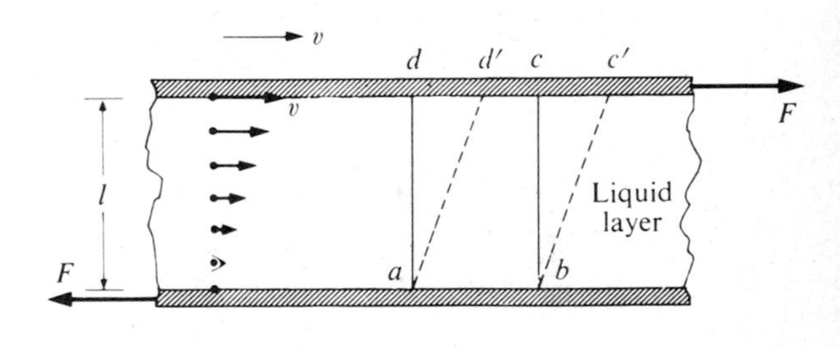

Fig. 13–14 *Laminar flow of a viscous fluid.*

Flow of this type is called *laminar.* (A lamina is a thin sheet.) The layers of liquid slide over one another much as do the leaves of a book when it is placed flat on a table and a horizontal force applied to the top cover. As a consequence of this motion, a portion of the liquid which at some instant has the shape *abcd*, will a moment later take the shape *abc′d′*, and will become more and more distorted as the motion continues. That is, the liquid is in a state of continually increasing shearing strain.

In order to maintain the motion, a constant force must be exerted to the right on the upper, moving plate, and hence indirectly on the upper liquid surface. This force tends to drag the fluid and the lower plate as well to the right. Therefore an equal force must be exerted toward the left on the lower plate to hold it stationary. These forces are lettered F in Fig. 13–14. If A is the area of the fluid over which these forces are applied, the ratio F/A is the shearing stress exerted on the fluid.

When a shearing stress is applied to a *solid*, the effect of the stress is to produce a certain displacement of the solid, such as dd'. The shearing strain is defined as the ratio of this displacement to the transverse dimension ℓ, and within the elastic limit the shearing stress is proportional to the shearing strain. With a fluid, on the other hand, the shearing strain increases without limit so long as the stress is applied, and the stress is found by experiment to depend not on the shearing strain, but on its *rate of change*. The strain in Fig. 13–14, at the instant when the volume of fluid has the shape $abc'd'$, is dd'/ad, or dd'/ℓ. Since ℓ is constant, the rate of change of strain equals $1/\ell$ times the rate of change of dd'. But the rate of change of dd' is simply the velocity of point d', or the velocity v of the moving wall. Hence

$$\text{Rate of change of shearing strain} = \frac{v}{\ell}.$$

The *coefficient of viscosity* of the fluid, or simply its viscosity η, is defined as the ratio of the shearing stress, F/A, to the rate of change of shearing strain:

$$\eta = \frac{\text{shearing stress}}{\text{rate of change of shearing strain}} = \frac{F/A}{v/\ell},$$

or

$$\boxed{F = \eta A \frac{v}{\ell}} \tag{13–16}$$

For a liquid which flows readily, such as water or kerosene, the shearing stress is relatively small for a given rate of change of shearing strain, and the viscosity also is relatively small. For a liquid such as molasses or glycerine, a greater shearing stress is necessary for the same rate of change of shearing strain, and the viscosity is correspondingly greater. Viscosities of gases at ordinary pressures and temperatures are very much smaller than those of common liquids. Viscosities of all fluids are markedly dependent on temperature, increasing for gases and decreasing for liquids as the temperature is increased, hence the expression "as slow as molasses in January." An important consideration in the design of oils for engine lubrication is to reduce the temperature variation of viscosity as much as possible.

Equation (13–16) was derived for the special case in which the velocity increased at a uniform rate with increasing distance from the lower plate. The general term for the *space* rate of change of velocity, in a direction at right angles to the flow, is the *velocity gradient* in this direction. In this special case it is equal to v/ℓ. In the general case, the velocity gradient is not uniform and its value at any point can be written as dv/dy, where dv is the small difference in velocity between two points separated by a distance dy measured at right angles to the direction of flow. Hence the general form of Eq. (13–16) is

$$\boxed{F = \eta A \frac{dv}{dy}.} \tag{13–17}$$

From Eq. (13–16), the unit of viscosity is that of force times distance, divided by area times velocity. In the mks system the unit is

$$1\ \text{N m m}^{-2}\,(\text{m s}^{-1})^{-1} = 1\ \text{N s m}^{-2}.$$

The corresponding cgs unit, 1 dyn s cm^{-2}, is the only viscosity unit in common use, and is called 1 *poise* in honor of the French scientist Poiseuille. Thus,

$$1\ \text{poise} = 1\ \text{dyn s cm}^{-2} = 10^{-1}\ \text{N s m}^{-2}.$$

Small viscosities are expressed in *centipoises* (1 cp = 10^{-2} poise) or *micropoises* (1 μp = 10^{-6} poise). A few typical values are given in Table 13–1.

Not all fluids behave according to the direct proportionality of force and velocity predicted by Eq. (13–16). An interesting exception is blood, for which velocity increases more rapidly than force. Thus doubling the force in Fig. 13–14 produces *more*

Table 13–1 TYPICAL VALUES OF VISCOSITY

Temperature, °C	Viscosity of castor oil, poise	Viscosity of water, centipoise	Viscosity of air, micropoise
0	53	1.792	171
20	9.86	1.005	181
40	2.31	0.656	190
60	0.80	0.469	200
80	0.30	0.357	209
100	0.17	0.284	218

than a twofold increase in velocity. This behavior may be understood on the basis that, on a microscopic scale, blood is not a homogeneous fluid but rather a suspension of solid particles in a liquid. The suspended particles have characteristic shapes; for example, red cells are roughly disk-shaped. At small velocities their orientations are random, but as velocity increases they tend to become oriented so as to facilitate flow. The fluids which provide lubrication in human joints exhibit similar behavior.

Fluids for which Eq. (13–16) holds are called *newtonian fluids*; we see that this description is an idealized model which not all fluids obey. In general, fluids which are suspensions or dispersions are often nonnewtonian in their viscous behavior. Nevertheless, Eq. (13–16) provides a useful model to describe approximately the properties of many pure substances.

13–6 POISEUILLE'S LAW

It is evident from the general nature of viscous effects that the velocity of a viscous fluid flowing through a tube is not the same at all points of a cross section. The outermost layer of fluid clings to the walls of the tube, and its velocity is zero. The tube walls exert a backward drag on this layer, which in turn drags backward on the next layer beyond it, and so on. Provided the velocity is not too great, the flow is laminar, with a velocity which is a maximum at the center of the tube and which decreases to zero at the walls. The flow is like that of a number of telescoping tubes sliding relative to one another, the central tube advancing most rapidly and the outer tube remaining at rest.

Let us consider the variation of velocity with radius for a cylindrical pipe of inner radius R. We consider the flow of a cylindrical element of fluid coaxial with the pipe, of radius r and length L, as shown in Fig. 13–15(a). The force on the left end is $p_1\pi r^2$, and that on the right end $p_2\pi r^2$, as shown. The net force is thus

$$F = (p_1 - p_2)\pi r^2.$$

This force must just balance the viscous retarding force at the surface of this element. This force is given by Eq. (13–17). The area over which the viscous force acts is $A = 2\pi rL$. Thus the viscous force is

$$F = \eta 2\pi rL\frac{dv}{dr}.$$

Equating this to the net force due to pressure on the ends and rearranging,

$$-\frac{dv}{dr} = \frac{(p_1 - p_2)r}{2\eta L}.$$

This shows that the velocity changes more and more rapidly as we go from the center ($r = 0$) to the pipe wall ($r = R$). The negative sign must be introduced because v decreases as r increases. Integrating, we get

$$-\int_v^0 dv = \frac{p_1 - p_2}{2\eta L}\int_r^R r\,dr,$$

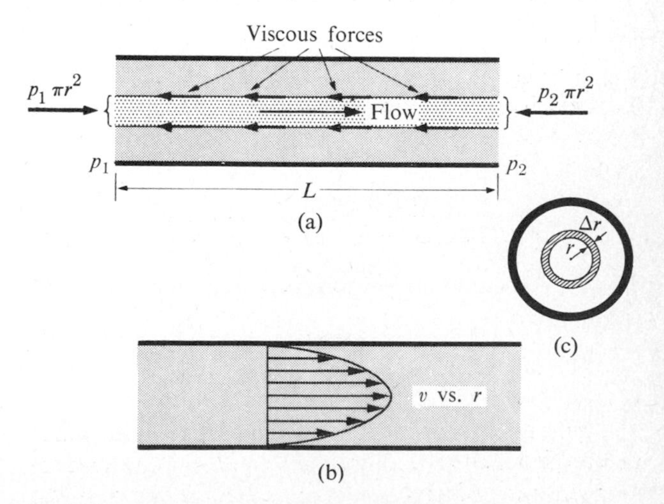

Fig. 13–15 (a) Forces on a cylindrical element of a viscous fluid. (b) Velocity distribution for viscous flow.

and therefore

$$v = \frac{p_1 - p_2}{4\eta L}(R^2 - r^2). \tag{13–18}$$

We see that the velocity decreases from a maximum value $(p_1 - p_2)R^2/4\eta L$ at the center to zero at the wall. Thus the maximum velocity is proportional to the *square* of the pipe radius and is also proportional to the pressure change per unit length $(p_1 - p_2)/L$, called the *pressure gradient.* The curve in Fig. 13–15(b) is a graph of Eq. (13–18) with the v-axis horizontal and the r-axis vertical.

Equation (13–18) may be used to find the total rate of flow of fluid through the pipe. The velocity at each point is proportional to the pressure gradient $(p_1 - p_2)/L$, so the total flow rate must also be proportional to this quantity. Let us consider the thin-walled element in Fig. 13–15(c). The volume of fluid dV crossing the ends of this element in a time dt is $v\,dA\,dt$, where v is the velocity at the radius r and dA is the shaded area, equal to $2\pi r\,dr$. Taking the expression for v from Eq. (13–18), we get

$$dV = \frac{p_1 - p_2}{4\eta L}(R^2 - r^2)\,2\pi r\,dr\,dt.$$

The volume flowing across the entire cross section is obtained by integrating over all elements between $r = 0$ and $r = R$:

$$dV = \frac{\pi(p_1 - p_2)}{2\eta L}\int_0^R (R^2 - r^2)r\,dr\,dt$$

$$= \frac{\pi}{8}\frac{R^4}{\eta}\frac{p_1 - p_2}{L}\,dt.$$

The total volume of flow per unit time, denoted by Q, is given by

$$Q = \frac{dV}{dt} = \frac{\pi}{8}\frac{R^4}{\eta}\frac{p_1 - p_2}{L}. \tag{13–19}$$

This relation was first derived by Poiseuille* and is called *Poiseuille's law.* The volume rate of flow is inversely proportional to viscosity, as might be expected. It is proportional to the pressure gradient along the pipe, and it varies as the fourth power of the radius. For example, if the radius is halved, the flow rate is reduced by a factor of 16. This relation is familiar to physicians in connection with the selection of needles for hypodermic syringes. Needle size is much more important than thumb pressure in determining the flow rate from the needle; doubling the needle diameter has the same effect as increasing the thumb force sixteenfold.

The difference between the flow of an ideal nonviscous fluid and one having viscosity is illustrated in Fig. 13–16, where fluid is flowing along a horizontal tube of varying cross section. The height of the fluid in the small vertical tubes is proportional to the gauge pressure.

In part (a), the fluid is assumed to have no viscosity. The pressure at b is very nearly the static pressure ρgy, since the velocity is small in the large tank. The pressure at c is less than at b because the fluid must accelerate between these points. The pressures at c and d are equal, however, since the velocity and elevation at these points are the same. There is a further pressure drop between d and e, and between

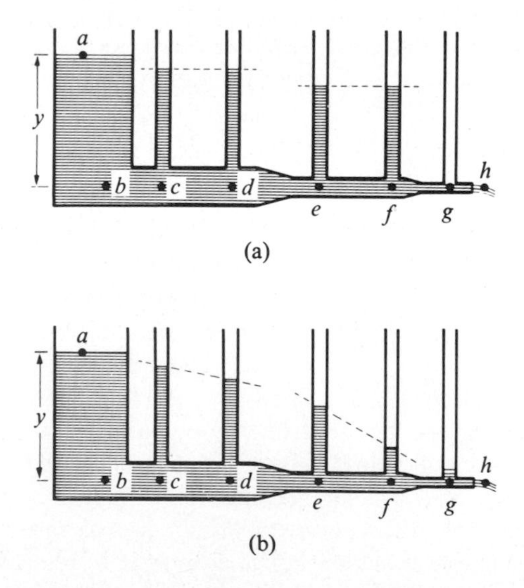

Fig. 13–16 *Pressures along a horizontal tube in which is flowing (a) an ideal fluid, (b) a viscous fluid.*

* Approximate pronounciation: Pwah-zoy′.

f and g. The pressure at g is atmospheric and the gauge pressure at this point is zero.

Part (b) of the diagram illustrates the effect of viscosity. Again, the pressure at b is nearly the static pressure ρgy. There is a pressure drop from b to c, due now in part to viscous effects, and also a further drop from c to d. The pressure gradient in this part of the tube is represented by the slope of the dotted line. The drop from d to e results in part from acceleration and in part from viscosity. The pressure gradient between e and f is greater than between c and d because of the smaller radius in this portion. Finally, the pressure at g is somewhat above atmospheric, since there is now a pressure gradient between this point and the cnd of the tube.

13-7 STOKES' LAW

When an ideal fluid of zero viscosity flows past a sphere, or when a sphere moves through a stationary fluid, the streamlines form a perfectly symmetrical pattern around the sphere, as shown in Fig. 13-2(a). The pressure at any point on the upstream hemispherical surface is exactly the same as that at the corresponding point on the downstream face, and the resultant force on the sphere is zero. If the fluid has viscosity, however, there will be a viscous drag on the sphere. (A viscous drag will of course be experienced by a body of any shape, but only for a sphere is the drag readily calculable.)

We shall not attempt to derive the expression for the viscous force directly from the laws of flow of a viscous fluid. The only quantities on which the force can depend are the viscosity η of the fluid, the radius r of the sphere, and its velocity v relative to the fluid. A complete analysis shows that the force F is given by

$$\boxed{F = 6\pi\eta rv.} \qquad (13\text{-}20)$$

This equation was first deduced by Sir George Stokes in 1845 and is called *Stokes' law*. We have already used it in Section 5-6 (Example 10) to study the motion of a sphere falling in a viscous fluid, although at that point it was necessary to know only that the viscous force on a given sphere in a given fluid is proportional to the relative velocity.

It will be recalled that a sphere falling in a viscous fluid reaches a *terminal velocity* v_T at which the viscous retarding force plus the buoyant force equals the weight of the sphere. Let ρ be the density of the sphere and ρ' the density of the fluid. The weight of the sphere is then $(4/3)\pi r^3 \rho g$, and the buoyant force is $(4/3)\pi r^3 \rho' g$; when the terminal velocity is reached, the total force is zero, and

$$\tfrac{4}{3}\pi r^3 \rho' g + 6\pi\eta r v_T = \tfrac{4}{3}\pi r^3 \rho g,$$

or

$$v_T = \frac{2}{9}\frac{r^2 g}{\eta}(\rho - \rho'). \qquad (13\text{-}21)$$

By measuring the terminal velocity of a sphere of known radius and density, the viscosity of the fluid in which it is falling can be found from the equation above. Conversely, if the viscosity is known, the radius of the sphere can be determined by measuring the terminal velocity. This method was used by Millikan to determine the radius of very small electrically charged oil drops (used to measure the electrical charge of an individual electron) by observing free fall in air.

Even for nonspherical bodies, a relation of the form of Eq. (13-19) holds, with a different numerical coefficient. Biologists call the terminal velocity the *sedimentation velocity*, and experiments with sedimentation can give useful information concerning very small particles. It is often useful to increase the terminal velocity by spinning the sample in a centrifuge, which greatly increases the effective acceleration of gravity.

13-8 REYNOLDS NUMBER

When the velocity of a fluid flowing in a tube exceeds a certain critical value (which depends on the properties of the fluid and the diameter of the tube) the nature of the flow becomes extremely complicated. Within a very thin layer adjacent to the tube walls, called the *boundary layer*, the flow is still laminar. The flow velocity in the boundary layer is zero at the tube walls and increases uniformly throughout the layer. The properties of the boundary layer are of the greatest importance in determining the resistance to flow, and the transfer of heat to or from the moving fluid.

Beyond the boundary layer, the motion is highly irregular. Random local circular currents called *vortices* develop within the fluid, with a large increase in the resistance to flow. Flow of this sort is called *turbulent.*

Experiment indicates that a combination of four factors determines whether the flow of a fluid through a tube or pipe is laminar or turbulent. This combination is known as the *Reynolds number*, N_R, and is defined as

$$N_R = \frac{\rho v D}{\eta}, \tag{13–22}$$

where ρ is the density of the fluid, v the average forward velocity, η the viscosity, and D the diameter of the tube. (The average velocity is defined as the uniform velocity over the entire cross section of the tube, which would result in the same volume rate of flow.) The Reynolds number, $\rho v D/\eta$, is a *dimensionless* quantity and has the same numerical value in any consistent system of units. For example, for water at 20°C flowing in a tube of diameter 1 cm with an average velocity of 10 cm s^{-1}, the Reynolds number is

$$N_R = \frac{\rho v D}{\eta} = \frac{(1\,\text{g cm}^{-3})(10\,\text{cm s}^{-1})(1\,\text{cm})}{0.01\,\text{dyn s cm}^{-2}} = 1000.$$

Had the four quantities been expressed originally in mks units, the same value of 1000 would have been obtained.

All the experiments show that when the Reynolds number is less than about 2000 the flow is laminar, whereas above about 3000 the flow is turbulent. In the transition region between 2000 and 3000

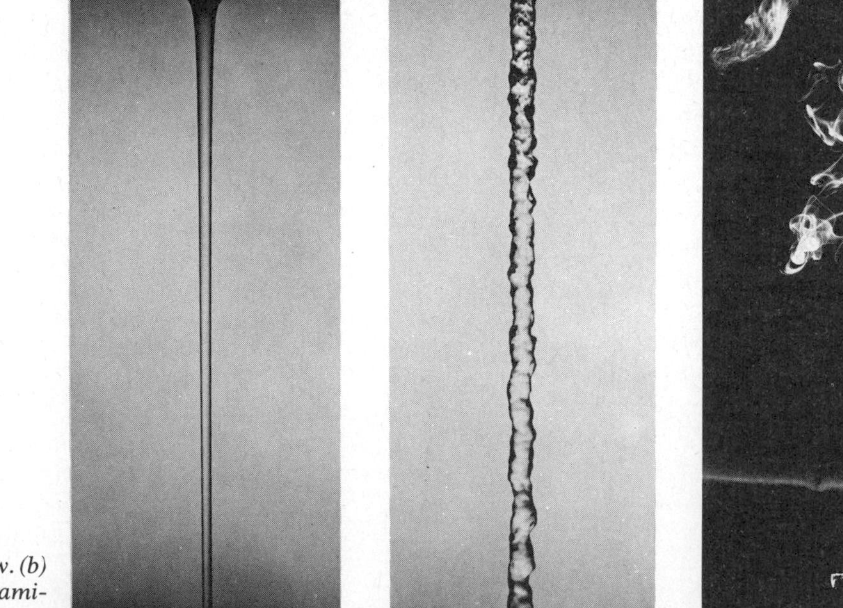

Fig. 13–7 *(a) Laminar flow. (b) Turbulent flow. (c) First laminar, then turbulent.*

(a) (b) (c)

the flow is unstable and may change from one type to the other. Thus for water at 20°C flowing in a tube 1 cm in diameter, the flow is laminar when

$$\frac{\rho v D}{\eta} \leq 2000,$$

or when

$$v \leq \frac{(2000)(0.01 \text{ dyn s cm}^{-2})}{(1 \text{ g cm}^{-3})(1 \text{ cm})} = 20 \text{ cm s}^{-1}.$$

Above about 30 cm s^{-1} the flow is turbulent. If air at the same temperature were flowing at 30 cm s^{-1} in the same tube, the Reynolds number would be

$$N_R = \frac{(0.0013 \text{ g cm}^{-3})(30 \text{ cm s}^{-1})(1 \text{ cm})}{181 \times 10^{-6} \text{ dyn s cm}^{-2}} = 216.$$

Since this is much less than 3000, the flow would be laminar and would not become turbulent unless the velocity were as great as 420 cm s^{-1}.

The distinction between laminar and turbulent flow is shown in the photographs of Fig. 13–17. In (a) and (b) the fluid is water and in (c) air and smoke particles.

The Reynolds number of a system forms the basis for the study of the behavior of real systems through the use of small scale models. A common example is the wind tunnel, in which one measures the aerodynamic forces on a scale model of an aircraft wing. The forces on a full-size wing are then deduced from these measurements.

Two systems are said to be *dynamically similar* if the Reynolds number, $\rho v D/\eta$, is the same for both. The letter D may refer, in general, to any dimension of a system, such as the span or chord of an aircraft wing. Thus the flow of a fluid of given density ρ and viscosity η, about a half-scale model, is dynamically similar to that around the full-size object if the velocity v is twice as great.

PROBLEMS

13–1 A circular hole 2 cm in diameter is cut in the side of a large standpipe, 10 m below the water level in the standpipe. Find (a) the velocity of efflux, and (b) the volume discharged per unit time. Neglect the contraction of the streamlines after emerging from the hole.

13–2 Water stands at a depth H in a large open tank whose side walls are vertical (Fig. 13–18). A hole is made in one of the walls at a depth h below the water surface.

a) At what distance R from the foot of the wall does the emerging stream of water strike the floor?
b) At what height above the bottom of the tank could a second hole be cut so that the stream emerging from it would have the same range?

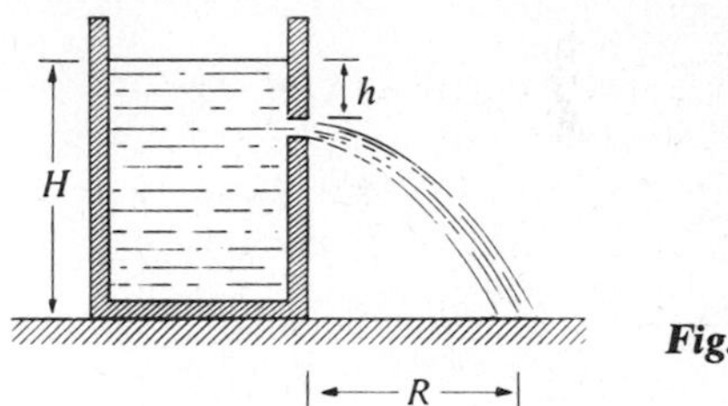

Fig. 13–18

13–3 A cylindrical vessel, open at the top, is 20 cm high and 10 cm in diameter. A circular hole whose cross-sectional area is 1 cm^2 is cut in the center of the bottom of the vessel. Water flows into the vessel from a tube above it at the rate of 140 cm^3 s^{-1}. How high will the water in the vessel rise?

13–4 At a certain point in a pipeline the velocity is 2 m s^{-1} and the gauge pressure is 10^4 Pa above atmospheric. Find the gauge pressure at a second point in the line 1 m lower than the first, if the cross section at the second point is one-half that at the first. The liquid in the pipe is water.

13–5 Water in an enclosed tank is subjected to a gauge pressure of 4 lb in^{-2}, applied by compressed air introduced into the top of the tank. There is a small hole in the side of the tank 16 ft below the level of the water. Calculate the speed with which water escapes from this hole.

13–6 What gauge pressure is required in the city mains in order that a stream from a fire hose connected to the mains may reach a vertical height of 60 ft?

13–7 A tank of large area is filled with water to a depth of 0.3 m. A hole of 5 cm^2 cross section in the bottom allows water to drain out in a continuous stream.

a) What is the rate at which water flows out of the tank, in m^3 s^{-1}?
b) At what distance below the bottom of the tank is the cross-sectional area of the stream equal to one-half the area of the hole?

13–8 A sealed tank containing sea water to a height of 2 m also contains air above the water at a gauge pressure of 40 atm. Water flows out from a hole at the bottom. The cross-sectional area of the hole is 10 cm^2.

a) Calculate the efflux velocity of the water.

b) Calculate the reaction force on the tank exerted by the water in the emergent stream.

13–9 A pipeline 6 in. in diameter, flowing full of water, has a constriction of diameter 3 in. If the velocity in the 6-in. portion is 4 ft s^{-1}, find (a) the velocity in the constriction, and (b) the discharge rate in cubic feet per second.

13–10 A horizontal pipe of 6 in^2 cross section tapers to a cross section of 2 in^2. If water is flowing with a velocity of 180 ft min^{-1} in the large pipe where a pressure gauge reads 10.5 lb in^{-2}, what is the gauge pressure in the adjoining part of the small pipe? The barometer reads 30 in. of mercury.

13–11 At a certain point in a pipeline the velocity is 1 m s^{-1} and the gauge pressure is 3×10^5 Pa. Find the gauge pressure at a second point in the line 20 m lower than the first, if the cross section at the second point is one-half that at the first. The liquid in the pipe is water.

13–12 Water stands at a depth of 1 m in an enclosed tank whose side walls are vertical. The space above the water surface contains air at a gauge pressure of 8×10^5 Pa. The tank rests on a platform 2 m above the floor. A hole of cross-sectional area 1 cm^2 is made in one of the side walls just above the bottom of the tank.

a) Where does the stream of water from the hole strike the floor?

b) What is the vertical force exerted on the floor by the stream?

c) What is the horizontal force exerted on the tank? Assume the water level and the pressure in the tank to remain constant, and neglect any effect of viscosity.

13–13 Water flows steadily from a reservoir, as in Fig. 13–19. The elevation of point 1 is 10 m; of points 2 and 3 it is 1 m. The cross section at point 2 is 0.04 m^2 and at point 3 it is 0.02 m^2. The area of the reservoir is very large compared with the cross sections of the pipe.

a) Compute the gauge pressure at point 2.

b) Compute the discharge rate in cubic feet per second.

13–14 Water flows steadily in a pipeline of constant cross section leading out of an elevated tank. At a point 2 m below the water level in the tank the gauge pressure in the flowing stream is 10^4 Pa.

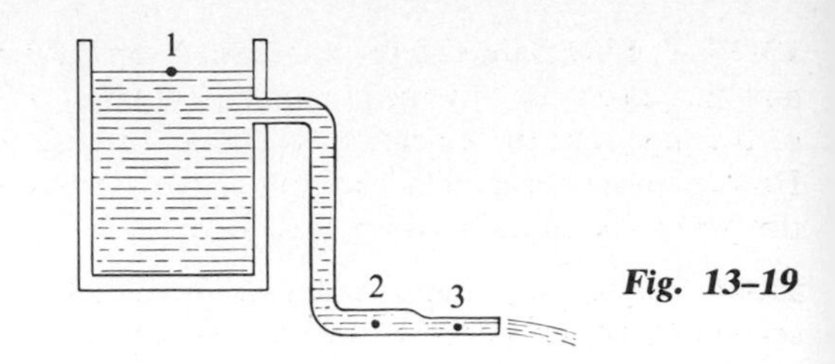

Fig. 13–19

a) What is the velocity of the water at this point?

b) If the pipe rises to a point 3 m above the level of the water in the tank, what are the velocity and the pressure at the latter point?

13–15 Water flows through a horizontal pipe of cross-sectional area 10 cm^2. At one section the cross-sectional area is 5 cm^2. The pressure difference between the two sections is 300 Pa. How many cubic meters of water will flow out of the pipe in 1 min?

13–16 Two very large open tanks, *A* and *F* (Fig. 13–20), both contain the same liquid. A horizontal pipe *BCD*, having a constriction at *C*, leads out of the bottom of tank *A*, and a vertical pipe *E* opens into the constriction at *C* and dips into the liquid in tank *F*. Assume streamline flow and no viscosity. If the cross section at *C* is one-half that at *D*, and if *D* is at distance h_1 below the level of the liquid in *A*, to what height h_2 will liquid rise in pipe *E*? Express your answer in terms of h_1. Neglect changes in atmospheric pressure with elevation.

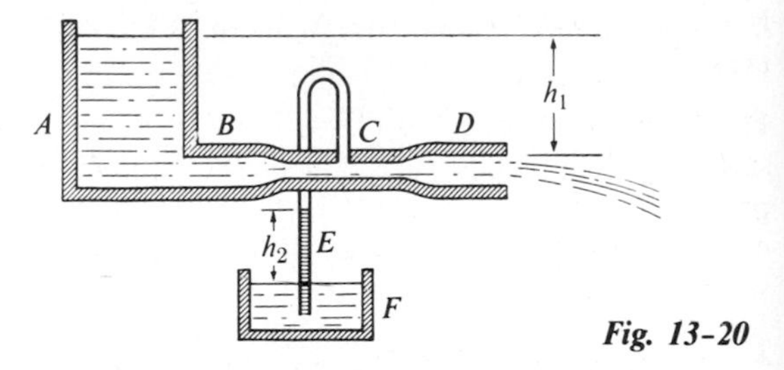

Fig. 13–20

13–17 At a certain point in a horizontal pipeline the gauge pressure is 6.24 lb in^{-2}. At another point the gauge pressure is 4.37 lb in^{-2}. If the areas of the pipe at these two points are 3 in^2 and 1.5 in^2, respectively, compute the number of cubic feet of water which flow across any cross section of the pipe per minute.

13–18 Water flowing in a horizontal pipe discharges at the rate of 0.004 m$^3 \cdot$s^{-1}. At a point in the pipe where the cross section is 0.001 m^2, the absolute pressure is 1.2×10^5 Pa. What must be the cross section of a constriction in the pipe such that the pressure there is reduced to 1.0×10^5 Pa?

13–19 The pressure difference between the main pipeline and the throat of a Venturi meter is 10^5 Pa. The areas of the pipe and the constriction are 0.1 m^2 and 0.05 m^2. How many cubic meters per second are flowing through the pipe? The liquid in the pipe is water.

13–20 The section of pipe shown in Fig. 13–21 has a cross section of 40 cm^2 at the wider portions and 10 cm^2 at the constriction. The discharge of water from the pipe is 3000 $cm^3\ s^{-1}$.

a) Find the velocities at the wide and the narrow portions.

b) Find the pressure difference between these portions.

c) Find the difference in height between the mercury columns in the U-tube.

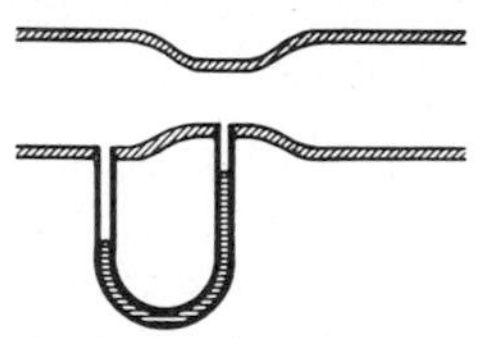

Fig. 13–21

13–21 Water is used as the manometric liquid in a Prandtl tube mounted in an aircraft to measure airspeed. If the maximum difference in height between the liquid columns is 0.1 m, what is the maximum airspeed that can be measured? The density of air is 1.3 kg m^{-3}.

13–22 In Problem 13–21, suppose the manometric liquid is mercury. What is the maximum airspeed that can be measured?

13–23 In a wind-tunnel experiment, the pressure on the top surface of an airplane wing was 0.90×10^5 Pa, and the pressure on the bottom surface 0.91×10^5 Pa. If the area of each surface is 40 m^2, what is the net lift force on the wing?

13–24 An airplane of mass 6000 kg has a wing area of 60 m^2. If the pressure on the lower wing surface is 0.60×10^5 Pa during level flight at an elevation of 4000 m, what is the pressure on the upper wing surface?

13–25 Water at 20° C flows through a pipe of radius 1.0 cm. If the flow velocity at the center is 10 cm s^{-1}, find the pressure drop along a 2-m section of pipe due to viscosity.

13–26 The inner, rotating cylinder of the viscosimeter in Fig. 13–12 is 5 cm in diameter. The inner diameter of the outer, fixed cylinder is 5.4 cm, and the diameter of the pulley attached to the inner cylinder is 4 cm. A liquid whose viscosity is 6 poise fills the space between inner and outer cylinders to a depth of 8 cm. A body of mass 30 g is supported by a thread wrapped around the pulley attached to the inner cylinder, and hangs vertically, as shown in Fig. 13–12. Find the speed of descent of the body after it has reached its terminal velocity.

13–27 A viscous liquid flows through a tube with laminar flow, as in Fig. 13–14(b). Prove that the volume rate of flow is the same as if the velocity were uniform at all points of a cross section and equal to half the velocity at the axis.

13–28

a) With what terminal velocity will an air bubble 1 mm in diameter rise in a liquid of viscosity 150 cp and density 0.90 g cm^{-3}?

b) What is the terminal velocity of the same bubble in water?

13–29

a) With what velocity is a steel ball 1 mm in radius falling in a tank of glycerine at an instant when its acceleration is one-half that of a freely falling body?

b) What is the terminal velocity of the ball? The densities of steel and of glycerine are 8.5 g cm^{-3} and 1.32 g cm^{-3}, respectively. Assume η = 8.3 poise.

13–30 Assume that air is streaming horizontally past an aircraft wing such that the velocity is 40 m s^{-1} over the top surface and 30 m s^{-1} past the bottom surface. If the wing has a mass of 300 kg and an area of 5 m^2, what is the net force on the wing? The density of air is 1.3 kg m^{-3}.

13–31 Modern airplane design calls for a "lift" of about 20 lb per square foot of wing area. Assume that air flows past the wing of an aircraft with streamline flow. If the velocity of flow past the lower wing surface is 300 ft s^{-1}, what is the required velocity over the upper surface to give a "lift" of 20 lb ft^{-2}? The density of air is 0.0025 slug ft^{-3}.

13–32 Water at 20°C flows with a speed of 50 cm s^{-1} through a pipe of diameter 3 mm.

a) What is the Reynolds number?

b) What is the nature of the flow?

13–33 Water at 20°C is pumped through a horizontal smooth pipe 15 cm in diameter and discharges into the air. If the pump maintains a flow velocity of 30 cm s^{-1},

a) What is the nature of the flow?

b) What is the discharge rate in liters per second?

13–34 The tank at the left of Fig. 13–15(a) has a very large cross section and is open to the atmosphere. The depth y = 40 cm. The cross sections of the horizontal tubes leading out of the tank are respectively 1 cm^2, 0.5 cm^2, and 0.2 cm^2. The liquid is ideal, having zero viscosity.

a) What is the volume rate of flow out of the tank?

b) What is the velocity in each portion of the horizontal tube?

c) What are the heights of the liquid in the vertical side tubes?

Suppose that the liquid in Fig. 13–15(b) has a viscosity of 0.5 poise, a density of 0.8 g cm^{-3}, and that the depth of liquid in the large tank is such that the volume rate of flow is the same as in part (a) above. The distance between the side tubes at c and d, and between those at e and f, is 20 cm. The cross sections of the horizontal tubes are the same in both diagrams.

d) What is the difference in level between the tops of the liquid columns in tubes c and d?

e) In tubes e and f?

f) What is the flow velocity on the axis of each part of the horizontal tube?

13–35

a) Is it reasonable to assume that the flow in the second part of Problem 13–34 is laminar?

b) Would the flow be laminar if the liquid were water?

13–36 Oil having a viscosity of 300 centipoise and a density of 0.90 g cm^{-3} is to be pumped from one large open tank to another through 1 km of smooth steel pipe 15 cm in diameter. The line discharges into the air at a point 30 m above the level of the oil in the supply tank.

a) What gauge pressure, in atmospheres, must the pump exert in order to maintain a flow of 0.05 $m^3\ s^{-1}$?

b) What is the power consumed by the pump?

Chapter 14

Relativistic Mechanics

14–1 INVARIANCE OF PHYSICAL LAWS

In previous chapters we have stressed the importance of inertial frames of reference. Newton's laws of motion are valid only in inertial frames, but they are valid in *all* inertial frames. Any frame moving with constant velocity with respect to an inertial frame is itself an inertial frame, and all such frames are equivalent with respect to expressing the basic principles of mechanics. The *laws of mechanics are the same in every inertial frame of reference.*

Einstein proposed in 1905 that this principle should be extended to include *all* the basic laws of physics. This innocent-sounding proposition has far-reaching and startling consequences, a few of which have already been mentioned. If the principle of conservation of momentum is to be valid in all inertial systems, for example, the definition of momentum, for particles moving at speeds comparable to the speed of light, must be changed from $m\mathbf{v}$ to $m\mathbf{v}/(1 - v^2/c^2)^{1/2}$, as discussed in Section 8–8. Even more fundamental are the modifications needed in the *kinematic* aspects of motion, as we shall see. Nevertheless, Einstein's *principle of relativity*, as it has come to be called, is now accepted as an essential requirement for a physical theory. The principle of relativity states that *the laws of physics are the same in every inertial frame of reference.*

The speed of light plays a special role in relativity theory. Light is an electromagnetic wave; it travels in vacuum with a speed c which is independent of the motion of the source. Its numerical value is known very precisely; to six significant figures it is

$$\boxed{c = 2.99793 \times 10^8 \text{ m s}^{-1}.}$$

The approximate value $c = 3.00 \times 10^8$ m s^{-1} is in error by less than one part in 1000 and is often used when greater precision is not required.

At one time it was thought that light traveled through a hypothetical medium called *the ether*, just as sound waves travel through air. Intensive experimental efforts to find direct evidence for its existence yielded consistently negative results, and it is now known that there is no ether. With this result in mind, let us suppose the speed of light is measured by two observers, one at rest with respect to the source, the other moving away from it. Both are in inertial frames of reference, and according to Einstein's principle of relativity, the laws of physics, and in particular the speed of light, must be the same in both frames.

For example, suppose a light source is located in a spaceship moving with respect to earth. An observer moving with the spaceship measures the speed of the light and obtains the value c. But after the light has left the source its motion cannot be influenced by the motion of the source, so an observer on earth measuring the speed of this same light must also obtain the value c, despite the fact that there is relative motion between the two observers. This conclusion may appear not to be consistent with common sense; but it is important to recognize that "common sense" is intuition based on everyday experience, and this does not usually include measurements of the speed of light. We must be prepared to accept results which seem to be in conflict with common sense when they involve realms far removed from everyday observation.

Thus the speed of light (in vacuum) is independent of the motion of the source and is the same in all frames of reference. To explore the consequences of this statement, we consider the newtonian relationship between two inertial frames, labeled S and S' in Fig. 14–1. Let the x axes of the two systems lie along the same line, but let the origin O' of S' move relative to the origin O of S with constant velocity u along the common x-axis. If the two origins coincide at time $t = 0$, then their separation at a later time t is ut.

A point P may be described by the coordinates (x,y,z) in S or by coordinates (x',y',z') in S'. Reference to the figure shows that these are related by

$$x = x' + ut, \qquad y = y', \qquad z = z'. \tag{14–1}$$

These equations are called the *galilean coordinate transformation.*

If the point P moves in the x-direction, its velocity v relative to S is given by $v = dx/dt$, and its velocity v' relative to S' is $v' = dx'/dt$. Intuitively it is clear that these are related by

$$v = v' + u. \tag{14–2}$$

This relation may also be obtained formally from Eqs. (14–1). Suppose the particle is at a point described by coordinate x_1 or x_1' at time t_1, and at x_2 or x_2' at time t_2. Then $\Delta t = t_2 - t_1$, and, from Eq. (14–1),

$$\Delta x = x_2 - x_1 = (x_2' - x_1') + u(t_2 - t_1)$$
$$= \Delta x' + u\,\Delta t,$$
$$\frac{\Delta x}{\Delta t} = \frac{\Delta x'}{\Delta t} + u,$$

and in the limit as $\Delta t \to 0$,

$$v = v' + u,$$

in agreement with Eq. (14–2).

A fundamental problem now appears. Applied to the speed of light, Eq. (14–2) says $c = c' + u$. Einstein's principle of relativity, supported by experimental observations, says $c = c'$. This is a genuine inconsistency, not an illusion, and it demands resolution. If we accept the principle of relativity, we are forced to conclude that Eqs. (14–1) and (14–2), intuitively appealing as they are, cannot be correct, but need to be modified to bring them into harmony with this principle.

The resolution involves modifications of our basic kinematic concepts. The first of these involves an assumption so fundamental that it might seem unnecessary, namely the assumption that the same *time scale* is used in frames S and S'. This may be stated formally by adding to Eqs. (14–1) a fourth equation $t = t'$. Obvious though this assumption may seem, it is not strictly correct when the relative speed u of the two frames of reference is comparable

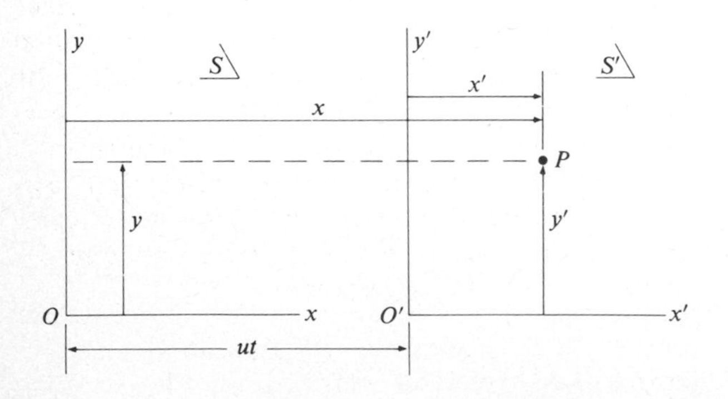

Fig. 14–1 *The position of point P can be described by the coordinates x and y in frame of reference S, or by x' and y' in S'. S' moves relative to S with constant velocity u along the common x-x' axis. The two origins O and O' coincide at time t = t' = 0.*

to the speed of light. The difficulty lies in the concept of *simultaneity*, which we examine next.

14–2 RELATIVE NATURE OF SIMULTANEITY

Measuring times and time intervals involves the concept of *simultaneity*. When a person says he awoke at seven o'clock he means that two *events*, his awakening and the arrival of the hour hand of his clock at the number seven, occurred *simultaneously*. The fundamental problem in measuring time intervals is that in general two events that appear simultaneous in one frame of reference *do not* appear simultaneous in a second frame which is moving relative to the first, even if both are inertial frames.

The following thought experiment, devised by Einstein, illustrates this point. Consider a long train moving with uniform velocity, as shown in Fig. 14–2(a). Two lightning bolts strike the train, one at each end. Each bolt leaves a mark on the train and one on the ground at the same instant. The points on the ground are labeled A and B in the figure, and the corresponding points on the train A' and B'. An observer on the ground is located at O, midway between A and B; another observer is at O', moving with the train and midway between A' and B'. Both these observers use the light signals from the lightning to observe the events.

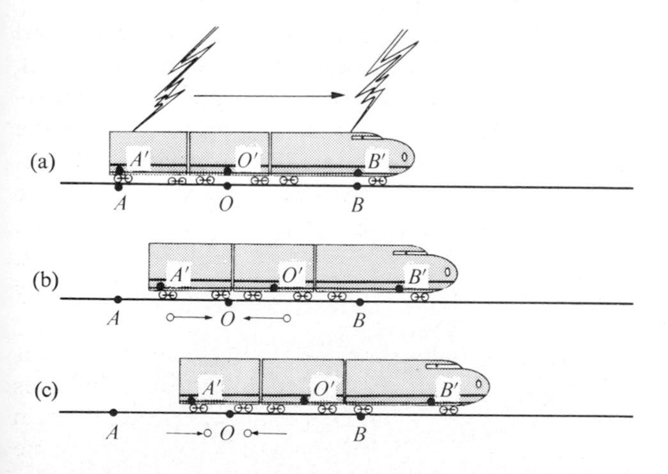

Fig. 14–2 *(a) To the stationary observer at point O, two lightning bolts appear to strike simultaneously. (b) The moving observer at point O′ sees the light from the front of the train first and thinks that the bolt at the front struck first. (c) The two light pulses arrive at O simultaneously.*

Suppose the two light signals reach the observer at O simultaneously; he concludes that the two events took place at A and B simultaneously. But the observer at O' is moving with the train, and the light pulse from B' reaches him before the light pulse from A' does; he concludes that the event at the front of the train happened *earlier* than that at the rear. Thus the two events appear simultaneous to one observer, not to the other. *Whether or not two events at different space points are simultaneous depends on the state of motion of the observer.* It follows that *the time interval between two events at different space points is in general different for two observers in relative motion.*

It might be argued that, in this example, the lightning bolts really *are* simultaneous, and that if the observer at O' could communicate with the distant points without time delay, he would realize this. But the finite speed of information transmission is not the problem. If O' is really midway between A' and B', then, in his frame of reference, the time for a signal to travel from A' to O' is the same as from B' to O'. Two signals arrive simultaneously at O' only if they were emitted simultaneously at A' and B'; in this example they *do not* arrive simultaneously at O', and so O' must conclude that the events at A' and B' were *not* simultaneous.

Furthermore, there is no basis for saying either that O is right and O' is wrong, or the reverse, since according to the principle of relativity, no inertial frame of reference is preferred over any other in the formulation of physical laws. Each observer is correct *in his own frame of reference*, but simultaneity is not an absolute concept. Whether or not two events are simultaneous depends on the frame of reference, and the time interval between two events also depends on the frame of reference.

14–3 RELATIVITY OF TIME

To derive a quantitative relation between time intervals in different coordinate systems, we consider another thought experiment. As before, a frame of reference S' moves with velocity u relative to a frame

S. An observer in S' directs a source of light at a mirror a distance d away, as shown in Fig. 14–3, and measures the time interval $\Delta t'$ for light to make the "round trip" to the mirror and back. The total distance is $2d$, so the time interval is

$$\Delta t' = \frac{2d}{c}. \tag{14–3}$$

As measured in frame S, the time for the round trip is a different interval Δt. During this time, the source moves relative to S a distance ut, and the total round-trip distance is not just $2d$ but is 2ℓ, where

$$\ell = \sqrt{d^2 + (u\,\Delta t/2)^2}\,.$$

The speed of light is the same for both observers, so the relation in S analogous to Eq. (14–3) is

$$\Delta t = \frac{2\ell}{c} = \frac{2}{c}\sqrt{d^2 + (u\,\Delta t/2)^2}\,. \tag{14–4}$$

To obtain a relation between Δt and $\Delta t'$ which does not contain d, we solve Eq. (14–3) for d and substitute the result into Eq. (14–4), obtaining

$$\Delta t = \frac{2}{c}\sqrt{(c\,\Delta t'/2)^2 + (u\,\Delta t/2)^2}\,.$$

This may now be squared and solved for Δt; the result is

$$\boxed{\Delta t = \frac{\Delta t'}{\sqrt{1 - u^2/c^2}}.} \tag{14–5}$$

We may generalize this important result: If a time interval $\Delta t'$ separates two events occurring at the same space point in a frame of reference S' (in this case, the departure and arrival of the light signal at O'), then the time interval Δt between these two events as observed in S is *larger* than $\Delta t'$, and the two are related by Eq. (14–5). Thus when the rate of a clock at rest in S' is measured by an observer in S, the rate measured in S is *slower* than the rate observed in S'. This effect is called *time dilation.*

It is important to note that the observer in S measuring the time interval Δt cannot do so with a single clock. In Fig. 14–3 the points of departure and return of the light pulse are different space points in S, although they are the same point in S'. If S tries to use a single clock, the finite time of communication between two points will becloud the issue. To avoid this, S may use two clocks at the two relevant points. There is no difficulty in synchronizing two clocks in the same frame of reference; one procedure is to send a light pulse simultaneously to two clocks from a point midway between them, with the two operators setting their clocks to a predetermined time when the pulses arrive. In thought experiments it is often helpful to imagine a large number of synchronized clocks distributed conveniently in a single frame of reference. Only when a clock is moving relative to a given frame of reference do ambiguities of synchronization or simultaneity arise.

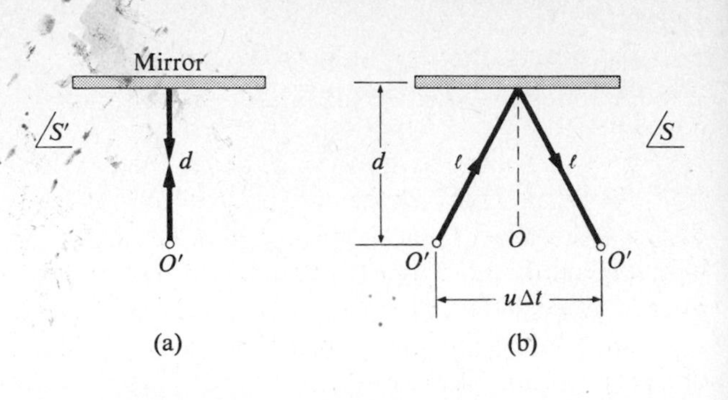

Fig. 14–3 *(a) Light pulse emitted from source at O′ and reflected back along the same line, as observed in S′. (b) Path of the same light pulse, as observed in S. The positions of O′ at the times of departure and return of the pulse are shown. The speed of the pulse is the same in S as in S′, but the path is longer in S.*

Example A spaceship flies past earth with a speed of $0.99c$ (about 2.97×10^8 m s^{-1}). A high-intensity signal light (perhaps a pulsed laser) blinks on and off, each pulse lasting 2×10^{-6} s. At a certain instant the ship appears to an earthling observer to be directly overhead at an altitude of 1000 km, and to be traveling perpendicular to the line of sight. What is the duration of each light pulse, as measured by this observer, and how far does the ship travel relative to earth during each pulse?

The observer does not see the pulse at the instant it is emitted, because the light signal requires a time equal to $(1000 \times 10^3 \text{ m})/(3 \times 10^8 \text{ m s}^{-1})$, or (1/300) s, to travel from the ship to earth. But if the distance from the spaceship to observer is essentially constant during the emission of a pulse, the time delays at beginning and end of the pulse are equal and the time *interval* is not affected.

Let S be the earth's frame of reference, S' that of the spaceship. Then, in the notation of Eq. (14–5), $\Delta t' = 2 \times 10^{-6}$s. This interval refers to two events occurring at the same point relative to S', namely, the starting and stopping of the pulse. The corresponding interval in S is given by Eq. (14–5):

$$\Delta t = \frac{\Delta t'}{\sqrt{1 - u^2/c^2}} = \frac{2 \times 10^{-6}\text{ s}}{\sqrt{1 - (0.99)^2}}$$

$$= 14.1 \times 10^{-6}\text{s}.$$

Thus the time dilation in S is about a factor of seven. The distance D traveled in S during this interval is

$$D = u\,\Delta t = (0.99)(3 \times 10^8 \text{ m s}^{-1})(14.1 \times 10^{-6}\text{ s})$$

$$= 4190 \text{ m} = 4.19 \text{ km}.$$

If the spaceship is traveling directly *toward* the observer, the time interval cannot be measured directly by a single observer because the time delay is not the same at the beginning and end of the pulse. One possible scheme, at least in principle, is to use *two* observers at rest in S, with synchronized clocks, one at the position of the ship when the pulse starts, the other at its position at the end of the pulse. These observers will again measure a time interval in S of 14.1×10^{-6} s.

From the derivation of Eq. (14–5) and the spaceship example, it can be seen that a time interval between two events occurring *at the same point* in a given frame of reference is a more fundamental quantity than an interval between events at different points. The term *proper time* is used to denote an interval between two events occurring at the same space point. Thus Eq. (14–5) may be used *only* when $\Delta t'$ is a proper time interval in S', in which case Δt is *not* a proper time interval in S. If, instead, Δt is proper in S, then Δt and $\Delta t'$ must be interchanged in Eq. (14–5).

When the relative velocity u of S and S' is very small, the factor $(1 - u^2/c^2)$ is very nearly equal to unity, and Eq. (14–5) approaches the newtonian relation $\Delta t = \Delta t'$ (i.e., the same time scale for all frames of reference). This assumption, therefore, retains its validity in the limit of small relative velocities.

14–4 RELATIVITY OF LENGTH

Just as the time interval between two events depends on the frame of reference, the *distance* between two points also depends on the frame of reference. To measure a distance one must, in principle, observe the positions of two points, such as the two ends of a ruler, simultaneously; but what is simultaneous in one frame is not in another.

To develop a relation between lengths in various coordinate systems we consider another thought experiment. We attach a source of light pulses to one end of a ruler and a mirror to the other, as shown in Fig. 14–4. Let the ruler be at rest in S' and the length in this frame be ℓ'. Then the time $\Delta t'$ required for a light pulse to make the round trip from source to

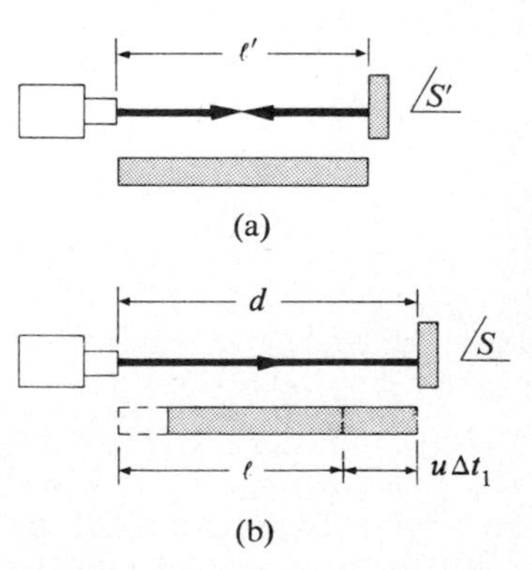

Fig. 14–4 *(a) A light pulse is emitted from a source at one end of a ruler, reflected from a mirror at the opposite end, and returned to the source position. (b) Motion of the light pulse as seen by an observer in S. The distance traveled from source to mirror is greater than the length ℓ measured in S, by the amount $u\,\Delta t_1$, as shown.*

mirror and back is given by

$$\Delta t' = \frac{2\ell'}{c}. \qquad (14\text{–}6)$$

This is a proper time interval, since departure and return occur at the same point in S'.

In S the ruler is displaced during this travel of the light pulse. Let the length of the ruler in S be ℓ, and let the time of travel from source to mirror, as measured in S, be Δt_1. During this interval the mirror moves a distance $u\,\Delta t_1$, and the total length of path d from source to mirror is not ℓ but

$$d = \ell + u\,\Delta t_1. \qquad (14\text{–}7)$$

But since the pulse travels with speed c, it is also true that

$$d = c\,\Delta t_1. \qquad (14\text{–}8)$$

Combining Eqs. (14–7) and (14–8) to eliminate d,

$$c\,\Delta t_1 = \ell + u\,\Delta t_1,$$

or

$$\Delta t_1 = \frac{\ell}{c - u}. \qquad (14\text{–}9)$$

In the same way it can be shown that the time Δt_2 for the return trip from mirror to source is

$$\Delta t_2 = \frac{\ell}{c + u}. \qquad (14\text{–}10)$$

The *total* time $\Delta t = \Delta t_1 + \Delta t_2$ for the round trip, as measured in S, is

$$\Delta t = \frac{\ell}{c - u} + \frac{\ell}{c + u} = \frac{2\ell}{c(1 - u^2/c^2)}. \qquad (14\text{–}11)$$

It is also known that Δt are $\Delta t'$ are related by Eq. (14–5), since $\Delta t'$ is proper in S'. Thus Eq. (14–6) becomes

$$\Delta t\sqrt{1 - u^2/c^2} = \frac{2\ell'}{c}. \qquad (14\text{–}12)$$

Finally, combining this with Eq. (14–11) to eliminate Δt, and simplifying, we obtain

$$\boxed{\ell = \ell'\sqrt{1 - u^2/c^2}.} \qquad (14\text{–}13)$$

Thus the length measured in S, in which the ruler is moving, is *shorter* than in S', where it is at rest. A length measured in the rest frame of the body is called a *proper length*; thus, ℓ' above is a proper length in S', and the length measured in any other frame is less than ℓ'. This effect is called *contraction of length.*

Example In the spaceship example of Section 14–3, what distance does the spaceship travel during emission of a pulse, as measured in its rest frame?

The question is somewhat ambiguous since, of course, in its own frame of reference the ship is at rest. But suppose it leaves markers in space, such as small smoke bombs, at the instants when the pulse starts and stops, and measures the distance between these markers, with the aid of observers behind the ship but moving with it, each with a clock synchronized with that of the ship. The distance d between the markers is a proper length in the earth's frame S. In the spaceship's frame S', the distance d' is contracted by the factor given in Eq. (14–13):

$$\begin{aligned} d' &= d\sqrt{1 - u^2/c^2} = (4190 \text{ m})\sqrt{1 - (0.99)^2} \\ &= 594 \text{ m}. \end{aligned}$$

(Note that because d, not d', is a proper length, we must reverse the roles of ℓ and ℓ'.) An observer in the spaceship can calculate his speed relative to earth from this data:

$$u = \frac{d'}{\Delta t'} = \frac{594 \text{ m}}{2 \times 16^{-6} \text{ s}} = 2.97 \times 10^8 \text{ m s}^{-1},$$

which agrees with the initial data.

When u is very small compared to c, the contraction factor in Eq. (14–13) approaches unity, and in the limit of small speeds we recover the newtonian relation $\ell = \ell'$. This and the corresponding result for time dilation show that Eqs. (14–1) retain their validity in the limit of small speeds; only at speeds comparable to c are modifications needed.

Lengths measured perpendicular to the direction of motion are *not* contracted; this may be verified by constructing a thought experiment for measuring in S and S' the length of a ruler oriented perpendicular to the direction of relative motion. The details of this discussion are not essential for our further work and will not be given here.

14–5 THE LORENTZ TRANSFORMATION

The galilean coordinate transformation given by Eqs. (14–1) is valid only in the limit when u is much smaller than c, but we are now in position to derive a more general transformation not subject to this limitation. The more general relations are called the *Lorentz transformation.* When u is small they reduce to the galilean transformation, but they may also be used when u is comparable to c.

The basic problem is this: When an event occurs at point (x, y, z) at time t, as observed in a frame of reference S, what are the coordinates (x', y', z') and time t' of the event as observed in a second frame S' moving relative to S with constant velocity u along the x-direction?

To derive the transformation equations, we return to Fig. 14–1. As before, we assume that the origins coincide at the initial time $t = t' = 0$. Then in S the distance from O to O' at time t is still ut. The coordinate x' is a proper length in S', so in S it appears contracted by the factor given in Eq. (14–13). Thus the distance x from O to P in S is given not simply by $x = ut + x'$ as in the galilean transformation, but by

$$x = ut + x'\sqrt{1 - u^2/c^2}. \tag{14–14}$$

Solving this equation for x', we obtain

$$x' = \frac{x - ut}{\sqrt{1 - u^2/c^2}}. \tag{14–15}$$

This is half of the Lorentz transformations; the other half is the equation giving t' in terms of x and t. To obtain this we note that the principle of relativity requires that the *form* of the transformation from S to S' must be identical to that from S' to S, the only difference being a change in the sign of the relative velocity u. Thus, from Eq. (14–14), it must be true that

$$x' = -ut' + x\sqrt{1 - u^2/c^2}. \tag{14–16}$$

We may now equate Eqs. (14–15) and (14–16) to eliminate x' from this expression, obtaining the desired relation for t' in terms of x and t. The algebraic details will be omitted; the result is

$$t' = \frac{t - ux/c^2}{\sqrt{1 - u^2/c^2}}. \tag{14–17}$$

As remarked previously, lengths perpendicular to the direction of relative motion are not affected by the motion, so $y' = y$ and $z' = z$. Collecting all the transformation equations, we have

$$\boxed{\begin{aligned} x' &= \frac{x - ut}{\sqrt{1 - u^2/c^2}}, \\ y' &= y, \\ z' &= z, \\ t' &= \frac{t - ux/c^2}{\sqrt{1 - u^2/c^2}}. \end{aligned}} \tag{14–18}$$

These are the *Lorentz transformation equations*, the relativistic generalization of the galilean transformation, Eqs. (14–1). When u is much smaller than c, the two transformations become identical.

Next we consider the relativistic generalization of the velocity transformation relation, Eq. (14–2), which, as previously noted, is valid only when u is very small. The relativistic expression can easily be obtained from the Lorentz transformation. Suppose that a body observed in S' is at point x_1' at time t_1' and point x_2' at time t_2'. Then its speed v' in S' is given by

$$v' = \frac{x_2' - x_1'}{t_2' - t_1'} = \frac{\Delta x'}{\Delta t'}. \tag{14–19}$$

To obtain the speed in S we use Eqs. (14–18) to translate this expression into terms of the corresponding positions x_1 and x_2 and times t_1 and t_2

observed in S. We find

$$x_2' - x_1' = \frac{x_2 - x_1 - u(t_2 - t_1)}{\sqrt{1 - u^2/c^2}} = \frac{\Delta x - u\,\Delta t}{\sqrt{1 - u^2/c^2}},$$

$$t_2' - t_1' = \frac{t_2 - t_1 - u(x_2 - x_1)/c^2}{\sqrt{1 - u^2/c^2}} = \frac{\Delta t - u\,\Delta x/c^2}{\sqrt{1 - u^2/c^2}}.$$

Using these results in Eq. (14–19), we find

$$v' = \frac{\Delta x - u\,\Delta t}{\Delta t - u\,\Delta x/c^2} = \frac{\dfrac{\Delta x}{\Delta t} - u}{1 - \dfrac{u}{c^2}\dfrac{\Delta x}{\Delta t}}.$$

Now $\Delta x/\Delta t$ is just the velocity v measured in S, so we finally obtain

$$\boxed{v' = \frac{v - u}{1 - uv/c^2}.} \qquad (14\text{–}20)$$

We note that when u and v are much smaller than c, the denominator becomes equal to unity, and we obtain the nonrelativistic result $v' = v - u$. The opposite extreme is the case $v = c$; then we find

$$v' = \frac{c - u}{1 - uc/c^2} = c.$$

That is, anything moving with speed c relative to S also has speed c relative to S', despite the relative motion of the two frames. This result demonstrates the consistency of Eq. (14–20) with our initial assumption that the speed of light is the same in all frames of reference.

Equation (14–20) may also be rearranged to give v in terms of v'. The algebraic details are left as a problem; the result is

$$\boxed{v = \frac{v' + u}{1 + uv'/c^2}.} \qquad (14\text{–}21)$$

Example A spaceship moving away from earth with a speed $0.9c$ fires a missile in the same direction as its motion, with a speed $0.9c$ relative to the spaceship. What is the missile's speed relative to earth?

Let the earth's frame of reference be S, the spaceship's S'. Then $v' = 0.9c$ and $u = 0.9c$. The nonrelativistic velocity addition formula would give a velocity relative to earth of $1.8c$. The correct relativistic result, obtained from Eq. (14–21), is

$$v = \frac{0.9c + 0.9c}{1 + (0.9c)(0.9c)/c^2} = 0.994c.$$

When u is less than c, a body moving with a speed less than c in one frame of reference also has a speed less than c in *every other* frame of reference. This is one reason for thinking that no material body may travel with a speed greater than that of light, relative to any frame of reference. The relativistic generalizations of energy and momentum, to be considered next, give further support to this hypothesis.

14-6 MOMENTUM

We have discussed the fact that Newton's laws of motion are *invariant* under the galilean coordinate transformation, but that to satisfy the principle of relativity this transformation must be replaced by the more general Lorentz transformation. This requires corresponding generalizations in the laws of motion and the definitions of momentum and energy.

The principle of conservation of momentum states that *when two bodies collide, the total momentum is constant,* provided there is no interaction except that of the two bodies with each other. However, when one considers a collision in one coordinate system S, in which momentum is conserved, and then uses the Lorentz transformation to obtain the velocities in a second system S', it is found that if the newtonian definition of momentum ($\boldsymbol{p} = m\boldsymbol{v}$) is used, momentum is not conserved in the second system. Thus if the Lorentz transformation is correct, and if we believe in the principle of relativity (i.e., momentum conservation must hold in *all* systems), the only choice remaining is to modify the *definition* of momentum.

Deriving the correct relativistic generalization is beyond our scope, and we simply quote the result, which has already been mentioned in Section 8–8:

$$\mathbf{p} = \frac{m\mathbf{v}}{\sqrt{1 - v^2/c^2}}. \qquad (14\text{–}22)$$

We note that, as usual when the particle's speed v is much less than c, this reduces to the Newtonian expression $\mathbf{p} = m\mathbf{v}$, but that in general the momentum is greater in magnitude than mv.

In newtonian mechanics the second law of motion can be stated in the form

$$\mathbf{F} = \frac{d\mathbf{p}}{dt}. \qquad (14\text{–}23)$$

That is, force equals time rate of change of momentum. Experiment shows that this result is still valid in relativistic mechanics, provided we use the relativistic momentum given by Eq. (14–22). This has the effect that a body under the action of a constant force *does not* experience a constant acceleration; as the particle's speed increases the acceleration for a given force continuously *decreases.* As the speed approaches c, the acceleration approaches zero, no matter how great the force. Thus it is impossible to accelerate a particle from a state of rest to a speed equal to or greater than c, and the speed of light is sometimes referred to as "the ultimate speed."

14–7 WORK AND ENERGY

The work-energy relation developed in Chapter 7 made use of Newton's laws of motion. Since these must be generalized to bring them into accord with the principle of relativity, it is not surprising that the work-energy relation also requires generalization.

Because a constant force on a body no longer causes a constant acceleration (except at very small velocities), even the simplest dynamics problems require the use of the calculus. We may, however, follow the same pattern as that of Section 7–3 in deriving the relativistic generalization of the work–energy principle. We begin with the definition of work: $W = \int F\,dx$. According to Eq. (14–23), $F = dp/dt$. By repeated application of the chain rule for derivatives, we obtain

$$F = \frac{dp}{dt} = \frac{dp}{dv}\frac{dv}{dx}\frac{dx}{dt} = \frac{dp}{dv}\frac{dv}{dx}v. \qquad (14\text{–}24)$$

Thus if the particle has speed v_1 at point x_1 and v_2 at x_2, the work done by F during the motion from x_1 to x_2 may be expressed as

$$W = \int_{x_2}^{x_1} F\,dx = \int_{v_1}^{v_2} \frac{dp}{dv}\,v\,dv. \qquad (14\text{–}25)$$

By differentiating Eq. (14–22), we obtain

$$\frac{dp}{dv} = \frac{m}{(1 - v^2/c^2)^{3/2}}. \qquad (14\text{–}26)$$

The reader should verify this calculation. Substituting this result in Eq. (14–25) and integrating, we find

$$W = \int_{v_1}^{v_2} \frac{mv\,dv}{(1 - v^2/c^2)^{3/2}}$$

$$= \frac{mc^2}{\sqrt{1 - v_2^2/c^2}} - \frac{mc^2}{\sqrt{1 - v_1^2/c^2}}, \qquad (14\text{–}27)$$

where v_1 and v_2 are the initial and final velocities of the particle.

This result suggests defining kinetic energy as

$$E = \frac{mc^2}{\sqrt{1 - v^2/c^2}}. \qquad (14\text{–}28)$$

But this expression is not zero when $v = 0$; instead it becomes equal to mc^2. Thus the correct relativistic generalization of kinetic energy E_k is:

$$\boxed{E_k = \frac{mc^2}{\sqrt{1 - v^2/c^2}} - mc^2.} \qquad (14\text{–}29)$$

This expression, if correct, must reduce to the newtonian expression $E_k = \frac{1}{2}mv^2$ when v is much smaller than c. It is not obvious that this is the case; to demonstrate that it is so, we can expand the radical using the binomial theorem:

$$\left(1-\frac{v^2}{c^2}\right)^{-1/2}=1+\frac{1}{2}\frac{v^2}{c^2}+\frac{3}{8}\frac{v^4}{c^4}+\frac{5}{16}\frac{v^6}{c^6}+\cdots$$

Combining this with Eq. (14–25),

$$\begin{aligned}E_k&=mc^2\left(1+\frac{1}{2}\frac{v^2}{c^2}+\frac{3}{8}\frac{v^4}{c^4}+\cdots\right)-mc^2\\&=\frac{1}{2}mv^2+\frac{3}{8}m\frac{v^4}{c^2}+\cdots\end{aligned}\tag{14–30}$$

In each expression the dots stand for omitted terms. When v is much smaller than c, all terms in the series except the first are negligibly small, and we obtain the classical $\frac{1}{2}mv^2$.

But what is the significance of the term mc^2 that had to be subtracted in Eq. (14–29)? Although Eq. (14–28) does not give the *kinetic energy* of the particle, perhaps it represents some kind of *total* energy, including both the kinetic energy and an additional energy mc^2 which the particle possesses even when it is not moving. This hypothetical energy associated with mass rather than motion may be called the *rest energy* of the particle. This speculation does not prove that the concept of rest energy is meaningful, but it points the way toward further investigation.

There is in fact direct experimental evidence of the existence of rest energy. The simplest example is the decay of the π° meson, an unstable particle that "decays"; in the decay process, the particle disappears and electromagnetic radiation appears. When the particle is at rest (and therefore with no kinetic energy) before its decay, the total energy of the radiation produced is found to be exactly equal to mc^2. There are many other examples of fundamental particle transformations in which the total mass of the system changes, and in every case there is a corresponding energy change consistent with the assumption of a rest energy mc^2 associated with a mass m.

Although the principles of conservation of mass and of energy originally developed quite independently, the theory of relativity shows that they are but two special cases of a single broader conservation principle, the *principle of conservation of mass and energy*. There are physical phenomena where neither mass nor energy is separately conserved, but where the changes in these quantities are governed by the more general relation that a change m in the mass of the system must be accompanied by an opposite change mc^2 in its energy.

The term *mass* as used here always means the rest mass of a particle, the inertial mass m measured through Eq. (14–22). For a given particle, m is a constant, independent of the state of motion of the particle. The concept of a variable, velocity-dependent relativistic mass is unnecessary and is not used in this discussion.

The possibility of conversion of mass into energy is the fundamental principle involved in the generation of power through nuclear reactions, a subject to be discussed in later chapters. When a uranium nucleus undergoes fission in a nuclear reactor, the total mass of the resulting fragments is *less* than that of the parent nucleus, and the total kinetic energy of the fragments is equal to this mass deficit times c^2. This kinetic energy can be used to produce steam to operate turbines for electric power generators or in a variety of other ways.

The total energy (kinetic plus rest) of a particle is related simply to its momentum, as shown by combining Eqs. (14–22) and (14–28) to eliminate the particle's velocity. This is most easily accomplished by rewriting these equations in the following forms:

$$\left(\frac{E}{mc^2}\right)^2=\frac{1}{1-v^2/c^2};\qquad\left(\frac{p}{mc}\right)^2=\frac{v^2/c^2}{1-v^2/c^2}.$$

Subtracting the second of these from the first and rearranging, we find

$$\boxed{E^2=(mc^2)^2+(pc)^2.}\tag{14–31}$$

Again we see that for a particle at rest ($p=0$), $E=mc^2$. Equation (14–31) also suggests that a particle may have energy and momentum even when it has no rest mass. In such a case, $m=0$ and

$$E=pc.\tag{14–32}$$

Massless particles, including photons, the quanta of electromagnetic radiation, and others, were mentioned in Section 8-8. The existence of such particles is well established, and they will be discussed in greater detail in later chapters. These particles always travel with the speed of light; they are emitted and absorbed during changes of state of atomic systems, accompanied by corresponding changes in the energy and momentum of these systems.

14-8 RELATIVITY AND NEWTONIAN MECHANICS

The sweeping changes required by the principle of relativity go to the very roots of newtonian mechanics, including the concepts of length and time, the equations of motion, and the conservation principles. Thus it may appear that foundations on which Newton's mechanics are built have been destroyed. While this is true in one sense, it is essential to keep in mind that the Newtonian formulation still retains its validity whenever speeds are small compared with the speed of light. In such cases time dilation, length contraction, and the modifications of the laws of motion do not appear. In fact, every one of the principles of newtonian mechanics survives as a special case of the more general relativistic formulation.

Relativity does not *contradict* the older mechanics but *generalizes* it. After all, Newton's laws rest on a very solid base of experimental evidence, and it would be very strange indeed to advance a new theory inconsistent with this evidence. So it always is with the development of physical theory. Whenever a new theory is in partial conflict with an older, established theory, it nevertheless must yield the same predictions as the old in areas where the old theory is supported by experimental evidence. Every new physical theory must pass this test, called the *correspondence principle*, which has come to be regarded as a fundamental procedural rule in all physical theory. There are many problems for which newtonian mechanics is clearly inadequate, including all situations where particle speeds approach that of light or there is direct conversion of mass to energy. But there is still a large area, including nearly all the behavior of macroscopic bodies in mechanical systems, in which Newtonian mechanics is still perfectly adequate.

At this point it is legitimate to ask whether the relativistic mechanics just discussed is the final word on this subject or whether *further* generalizations are possible or necessary. For example, inertial frames of reference have occupied a privileged position in all our discussion thus far. Should the principle of relativity be extended to noninertial frames as well?

Here is an example to illustrate some implications of this question. A man decides to go over Niagara Falls while enclosed in a large wooden box. During his free fall over the falls he can in principle perform experiments inside the box. An object released inside the box does not fall to the floor because both the box and the object are in free fall with a downward acceleration of $9.8\ \mathrm{m\ s^{-2}}$. But an alternative interpretation, from this man's point of view, is that the force of gravity has suddenly been turned off. Provided he remains in the box and it remains in free fall, he cannot tell whether he is indeed in free fall or whether the force of gravity has vanished. A similar problem appears in a space station in orbit around the earth. Objects in the spaceship appear weightless, but without going outside the ship there is no way to determine whether gravity has disappeared or the spaceship is in an accelerated (i.e., noninertial) frame of reference.

These considerations form the basis of Einstein's *general theory of relativity.* If one cannot distinguish experimentally between a gravitational field and an accelerated reference system, then there can be no real distinction between the two. Pursuing this concept, we may try to represent *any* gravitational field in terms of special characteristics of the coordinate system. This turns out to require even more sweeping revisions of our space-time concepts than the special theory of relativity did, and we find that, in general, the geometric properties of the space are noneuclidean.

The basic ideas of the general theory of relativity are now well established, but some of the details remain speculative in nature. Its chief application is in cosmological investigations of the structure of the universe, the formation and evolution of stars, and related matters. It is not believed to have any relevance for atomic or nuclear phenomena or macroscopic mechanical problems of less than astronomical dimensions.

PROBLEMS

14–1 The π^+ meson, an unstable particle, lives, on the average, about 2.6×10^{-8}s (measured in its own frame of reference) before decaying.

a) If such a particle is moving with respect to the laboratory with a speed of $0.8c$, what lifetime is measured in the laboratory?
b) What distance, measured in the laboratory, does the particle move before decaying?

14–2 The μ^+ meson (or positive muon) is an unstable particle with a lifetime of about 2.3×10^{-6}s (measured in the rest frame of the muon).

a) If a muon travels with a speed $0.99c$ relative to a laboratory, what is the lifetime as measured in the laboratory?
b) What distance, measured in the laboratory, does the particle travel during its lifetime?

14–3 For the two trains discussed in Section 14–2, suppose the two lightning bolts appear simultaneous to an observer on the train. Show that they *do not* appear simultaneous to an observer on the ground. Which appears to come first?

14–4 Solve Eqs. (14–18) to obtain x and t in terms of x' and t', and show that the resulting transformation has the same form as the original one except for a change of sign for u.

14–5 A light pulse is emitted at the origin of a frame of reference S' at time $t' = 0$. Its distance x' from the origin after a time t' is given by $x'^2 = c^2t'^2$. Use the Lorentz transformation to transform this equation to an equation in x and t, and show that the result is $x^2 = c^2t^2$; that is, the motion appears exactly the same in the frame of reference S of x and t, as it does in S'.

14–6 Two events observed in a frame of reference S have positions and times given by (x_1,t_1) and (x_2,t_2), respectively. Show that in a frame S' moving just fast enough so the two events occur at the same point in S', the time interval $\Delta t'$ between the two events is given by

$$\Delta t' = \sqrt{(\Delta t)^2 - (\Delta x/c)^2},$$

where $\Delta x = x_2 - x_1$, and $\Delta t = t_2 - t_1$. Hence show that, if $\Delta x \geq c\,\Delta t$, there is *no* frame S' in which the two events occur at the same point. The interval $\Delta t'$ is sometimes called the *proper time interval* for the events; is this term appropriate?

14–7 For the two events in Problem 14–6, show that if $\Delta x > c\,\Delta t$ there is a frame of reference S' in which the two events occur *simultaneously*. Find the distance between the two events in S'. This distance is sometimes called a *proper length*; is this term appropriate?

14–8 Two events are observed in a frame of reference S to occur at the same space point, the second occurring 2 s after the first. In a second frame S' moving relative to S, the second event is observed to occur 3 s after the first. What is the distance between the positions of the two events as measured in S'?

14–9 Two events are observed in a frame of reference S to occur simultaneously, at points separated by a distance of 1 m. In a second frame S' moving relative to S along the line joining the two points in S, the two events appear to be separated by 2 m. What is the time interval between the events, as measured in S'?

14–10 A particle is said to be in the *extreme relativistic range* when its kinetic energy is much larger than its rest energy.

a) What is the speed of a particle (expressed as a fraction of c) such that the total energy is ten times the rest energy?
b) For such a particle, what percent error in the energy–momentum relation of Eq.(14–31) results if the term $(mc^2)^2$ is neglected?

14–11 A photon of energy E is emitted by an atom of mass m, which recoils in the opposite direction. Assuming the atom can be treated nonrelativistically, compute the recoil velocity of the atom. Hence, show that the recoil velocity is much smaller than c whenever E is much smaller than the rest energy mc^2 of the atom.

14–12 Two particles emerge from a high-energy accelerator in opposite directions, each with a speed $0.6c$. What is the relative velocity of the particles?

14–13 At what speed is the momentum of a particle twice as great as the result obtained from the nonrelativistic expression mv?

14–14 At what speed does the momentum of a particle differ from the value obtained using the nonrelativistic expression mv by 1 percent? Is the correct relativistic value greater or less than that obtained from the nonrelativistic expression?

14–15 The mass of an electron is 9.11×10^{-31} kg. Comparing the classical definition of momentum with its relativistic generalization, by how much is the classical expression in error if (a) $v = 0.01c$; (b) $v = 0.5c$; (c) $v = 0.9c$?

14–16 A radioactive isotope of cobalt, ^{60}Co, emits an electromagnetic photon (γ ray) of wavelength 0.932×10^{-12} m. The cobalt nucleus contains 27 protons and 33 neutrons, each with a mass of about 1.66×10^{-27} kg. If the nucleus is at rest before emission, what is the speed afterward? Is it necessary to use the relativistic generalization of momentum?

14–17 In Problem 14–16, suppose the cobalt atom is in a metallic crystal containing 0.01 moles of cobalt (about 6.02×10^{21} atoms) and that the entire crystal recoils as a unit, rather than just the single nucleus. Find the recoil velocity. (This recoil of the entire crystal rather than a single nucleus is called the *Mössbauer effect*, in honor of its discoverer, who first observed it in 1958.)

14–18 What is the speed of a particle whose kinetic energy is equal to its rest energy?

14–19 At what speed is the kinetic energy of a particle equal to 10 mc^2?

14–20 How much work must be done to accelerate a particle from rest to a speed $0.1c$? From a speed $0.9c$ to a speed $0.99c$?

14–21 In *positron annihilation*, an electron and a positron (a positively-charged electron) collide and disappear, producing electromagnetic radiation. If each particle has a mass of 9.1×10^{-31} kg and they are at rest just before the annihilation, find the total energy of the radiation.

14–22 The total consumption of electrical energy per year in the United States is of the order 10^{19} joules. If matter could be converted completely into energy, how many kilograms of matter would have to be converted to produce this much energy?

14–23 Compute the kinetic energy of an electron (mass 9.11×10^{-31} kg) using both the nonrelativistic and relativistic expressions, and compare the two results, for speeds of

a) 1.0×10^8 m s^{-1};
b) 2.0×10^8 m s^{-1}.

14–24 In a hypothetical nuclear-fusion reactor, two deuterium nuclei combine or "fuse" to form one helium nucleus. The mass of a deuterium nucleus, expressed in atomic mass units (u), is 2.0147 u; that of a helium nucleus is 4.0039 u. (1 u = 1.66×10^{-27} kg.)

a) How much energy is released when 1 kg of deuterium undergoes fusion?
b) The annual consumption of electrical energy in the United States is of the order of 10^{19} J. How much deuterium must react to produce this much energy?

14–25 A nuclear bomb containing 20 kg of plutonium explodes. The rest mass of the products of the explosion is less than the original rest mass by one part in 10^4.

a) How much energy is released in the explosion?
b) If the explosion takes place in 1 μs, what is the average power developed by the bomb?
c) How much water could the released energy lift to a height of 1 km?

14–26 Construct a right triangle in which one of the angles is α, where $\sin \alpha = v/c$. (v is the speed of a particle, c the speed of light.) If the base of the triangle (the side adjacent to α) is the rest energy mc^2, show that (a) the hypotenuse is the total energy and (b) the side opposite α is c times the relativistic momentum.

c) Describe a simple graphical procedure for finding the kinetic energy E_k.

Chapter 15

Temperature and Expansion

15–1 CONCEPT OF TEMPERATURE

To describe the equilibrium states of mechanical systems, as well as to study and predict the motions of rigid bodies and fluids, only three fundamental indefinables were needed: length, mass, and time. All other physical quantities of importance in mechanics could be expressed in terms of these three indefinables. We come, now, however, to a series of phenomena, called *thermal effects* or *heat phenomena*, which involve aspects that are essentially nonmechanical and which require for their description a fourth fundamental indefinable, the *temperature*.

The familiar sensations of hotness and coldness are described with adjectives such as cold, cool, tepid, warm, hot, etc. When we touch an object, we use our *temperature sense* to ascribe *to the object* a property called *temperature*, which determines whether it will feel hot or cold to the touch. The hotter it feels, the higher the temperature. This procedure plays the same role in "qualitative science" that hefting a body does in determining its weight or that kicking an object does in estimating its mass. To determine the mass of an object *quantitatively*, we must first arrive at the concept of mass by means of *quantitative* operations such as measuring the acceleration imparted to the object by measured force, and then taking the ratio of F to a. Similarly, the quantitative determination of temperature requires a set of operations that are independent of our sense perceptions of hotness or coldness, and which involve quantities that can be measured objectively. How this is done will be explained in the following paragraphs.

Even before treating the concept of *temperature* in a precise, quantitative manner, we can note that there are numerous simple systems in which a quantity characterizing the state of the system varies with the hotness or coldness of the system. A simple example is a liquid such as mercury or alcohol in a bulb attached to a very thin tube, as in Fig. 15–1(a). The significant quantity characterizing the state of this system is the length L of the liquid column, measured from some arbitrary fixed point. Another simple system is a quantity of gas in a constant-volume container, shown in Fig. 15–1(b). Here the varying quantity, which we may refer to in these examples as a *state coordinate*, is the pressure, which varies as the gas becomes hotter or colder. A third example is the electrical resistance of a wire, which also varies with hotness and coldness.

Let A stand for the liquid-in-capillary system, with state coordinate L, and let B stand for the gas at constant volume, with state coordinate p. If A and

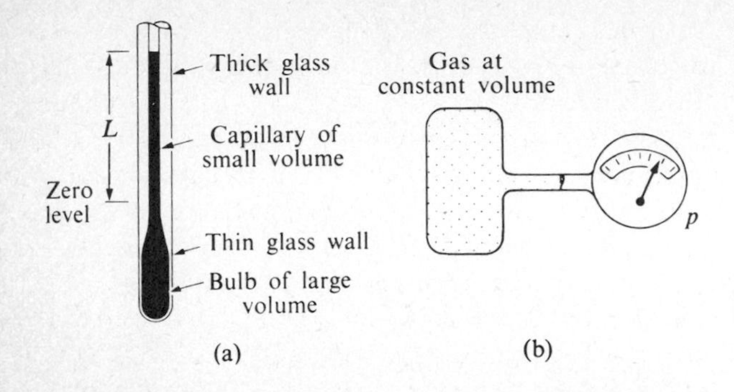

Fig. 15–1 *(a) A system whose state is specified by the value of L. (b) A system whose state is given by the value of p.*

B are brought into contact, their state coordinates, in general, are found to change. When A and B are separated, however, the change is slower, and when thick walls of various materials such as wood, plaster, felt, asbestos, etc., are used to separate A and B, the values of the respective state coordinates L and p are almost independent of each other. Generalizing from these observations, we postulate the existence of an ideal partition, called an ***adiabatic wall,*** *which, when used to separate two systems, allows their state coordinates to vary over a large range of values* ***independently.*** An adiabatic wall is an idealization that cannot be realized perfectly but may be approximated closely. In Fig. 15–2(a), such a wall is represented as a thick cross-shaded region.

When systems A and B are first put into actual contact or are separated by a thin metallic partition, their state coordinates may or may not change. *A wall which enables a state coordinate of one system to influence that of another is called a* ***diathermic wall.*** A thin sheet of copper is an example of a diathermic wall. In Fig. 15–2(b), a diathermic wall is depicted as a thin, darkly shaded region. Eventually, a time will be reached when no further change in the coordinates of A and B takes place. *The joint state of both systems that exists when all changes in the coordinates have ceased is called* ***thermal equilibrium.***

Imagine two systems A and B separated from each other by an adiabatic wall but each in contact with a third system C through diathermic walls, the whole assembly being surrounded by an adiabatic wall as shown in Fig. 15–3(a). Experiment shows that the two systems will come to thermal equilibrium with the third and that no further change will occur if the adiabatic wall separating them is then replaced by a diathermic wall (Fig. 15–3(b)). If, instead of allowing both systems A and B to come to equilibrium with C at the same time, we first have equilibrium between A and C and then equilibrium between B and C (the state of system C being the same in both cases), then, when A and B are brought into communication through a diathermic wall, they will be found to be in thermal equilibrium. We shall use the expression "two systems are in thermal equilibrium"

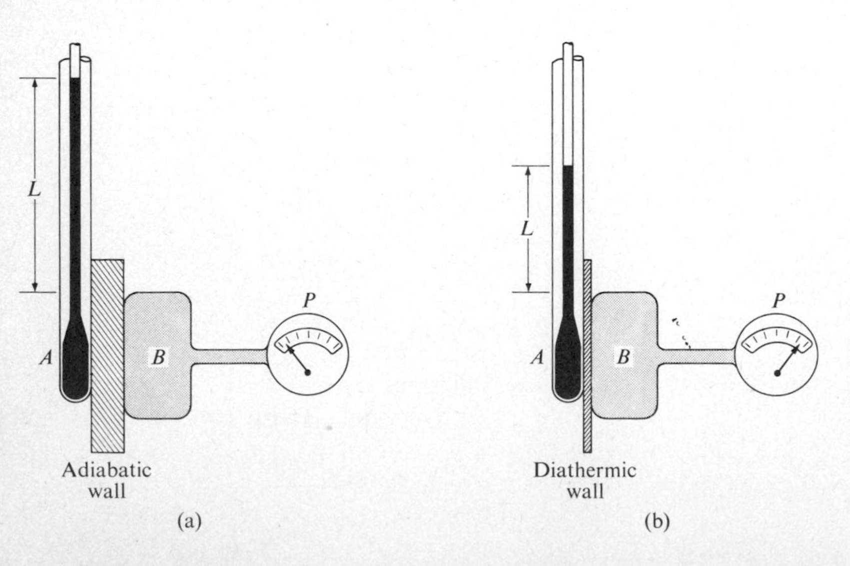

Fig. 15–2 *System A, a liquid column, and system B, a gas at constant volume, separated by (a) an adiabatic wall, p and L independent, and (b) a diathermic wall, p and L dependent.*

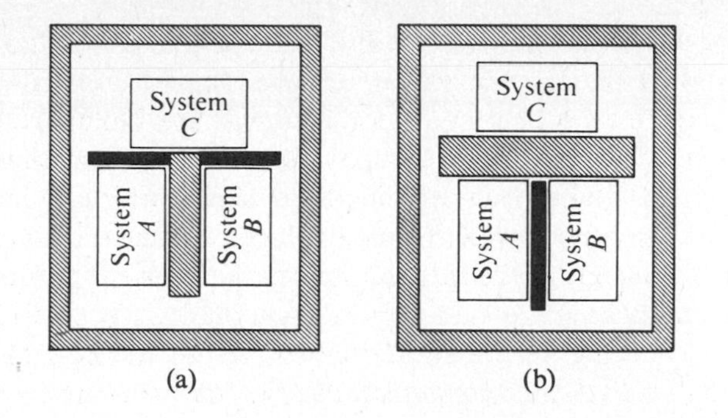

Fig. 15–3 *The zeroth law of thermodynamics. (a) If A and B are each in thermal equilibrium with C, then (b) A and B are in thermal equilibrium with each other.*

to mean that the two systems are in states such that if the two *were* to be connected through a diathermic wall, the combined system *would be* in thermal equilibrium.

These experimental facts may then be stated concisely in the following form: *Two systems in thermal equilibrium with a third are in thermal equilibrium with each other.* Following R. H. Fowler, we shall call this postulate the *zeroth law of thermodynamics.* At first thought it might seem that the zeroth law is obvious, but its truth must be verified by experiment.

When two systems A and B are first put in contact through a diathermic wall, they may or may not be in thermal equilibrium. One is entitled to ask, "What is there about A and B that determines whether or not they are in thermal equilibrium?" *We are led to infer the existence of a new property called the* ***temperature.*** *The temperature of a system is that property which determines whether or not it will be in thermal equilibrium with other systems.* When two or more systems are in thermal equilibrium, they are said to have the same temperature.

The temperature of all systems in thermal equilibrium may be represented by a number. The establishment of a temperature scale is merely the adoption of a set of rules for assigning numbers to temperatures. Once this is done, the condition for thermal equilibrium between two systems is that they have the same temperature. When the temperatures of two systems are different, we may be sure that they are *not* in thermal equilibrium.

The temperature of a material is directly related to the energies of its molecules; as temperature increases, molecular motion increases. The relation of temperature to microscopic mechanical energy will be explored in detail in Chapter 20. It is important to understand, however, that temperature can be defined *without* reference to molecular considerations. Indeed, temperature is inherently a macroscopic concept that has no meaning for an individual molecule. Temperature can be related to molecular motion only by considering the *average* energy of a large number of molecules.

15–2 THERMOMETERS

In defining a temperature scale, the simplest procedure is to choose a system such as one of those described above, arbitrarily assigning a numerical value of temperature to each value of the state coordinate of the system. This then defines quantitatively the temperature of this system, and of all systems in thermal equilibrium with it.

Although the system of Fig. 15–1(a) was one of the earliest thermometers, various other systems are now used. Important characteristics of a thermometer include *sensitivity* (an appreciable change in the state coordinate produced by a small change in temperature), *accuracy* in the measurement of the state coordinate, and *reproducibility.* Another often desirable property is *speed* in coming to thermal equilibrium with other systems. The thermometers which satisfy these requirements best will be described in the following paragraphs.

A thermometer widely used in research and engineering laboratories is the *thermocouple,* which consists of a junction of two different metals or alloys, such as A and B, labeled "test junction" in Fig. 15–4. The test junction is usually embedded in the material whose temperature is to be measured. Since the test junction is small and has a small mass, it can follow temperature changes rapidly and come to equilibrium quickly. The reference junction consists of two junctions: one of A and copper and the other of B and copper. These two junctions are maintained at any desired constant temperature, called the *reference temperature.* The state coordinate of this thermometer is an electrical quantity called

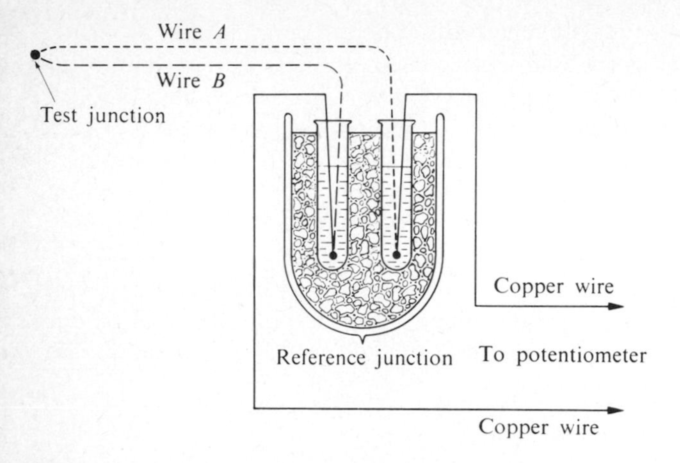

Fig. 15–4 *Thermocouple, showing the test junction and the reference junction.*

the emf (electromotive force) which is measured with an instrument known as a potentiometer. A thermocouple with one junction of pure platinum and the other of 90% platinum and 10% rhodium is often used. Copper and an alloy called constantan are also frequently used.

A different electrical thermometer, the *resistance thermometer*, consists of a fine wire, often enclosed in a thin-walled silver tube for protection. Copper wires lead from the thermometer unit to a resistance-measuring device such as a Wheatstone bridge. Since resistance may be measured with great precision, the resistance thermometer is one of the most precise instruments for the measurement of temperature. In the region of extremely low temperatures, a small carbon cylinder or a small piece of germanium crystal is used instead of a coil of platinum wire.

To measure temperatures above the range of thermocouples and resistance thermometers, an *optical pyrometer* is used. As shown in Fig. 15–5, it consists essentially of a telescope T, in the tube of which is mounted a filter F of red glass and a small electric lamp bulb L. When the pyrometer is directed toward a furnace, an observer looking through the telescope sees the dark lamp filament against the bright background of the furnace. The lamp filament is connected to a battery B and a rheostat R. By turning the rheostat knob, we may gradually increase the current in the filament, and hence the brightness, until the brightness of the filament just matches the brightness of the background. From previous calibration of the instrument at known temperatures, the scale of the ammeter A in the circuit may be marked to read the unknown temperature directly. Since no part of the instrument needs to come into contact with the hot body, the optical pyrometer may be used at very high temperatures, above the melting points of metals.

Among all the state coordinates or, as they are often called, *thermometric properties*, the pressure of a gas whose volume is maintained constant is outstanding in its sensitivity, accuracy of measurement, and reproducibility. The constant-volume gas thermometer is illustrated schematically in Fig. 15–6. The

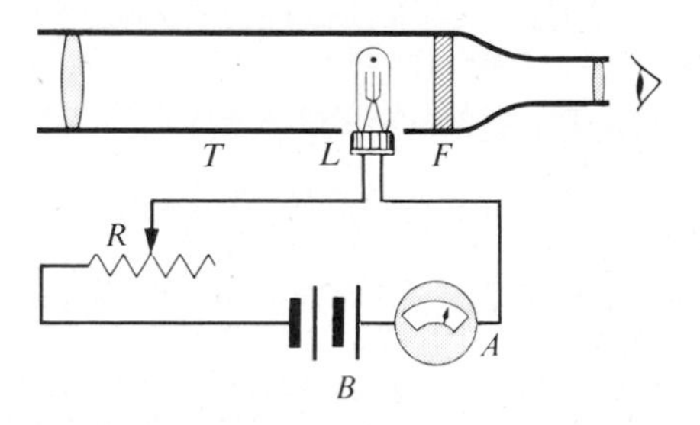

Fig. 15–5 *Principle of the optical pyrometer.*

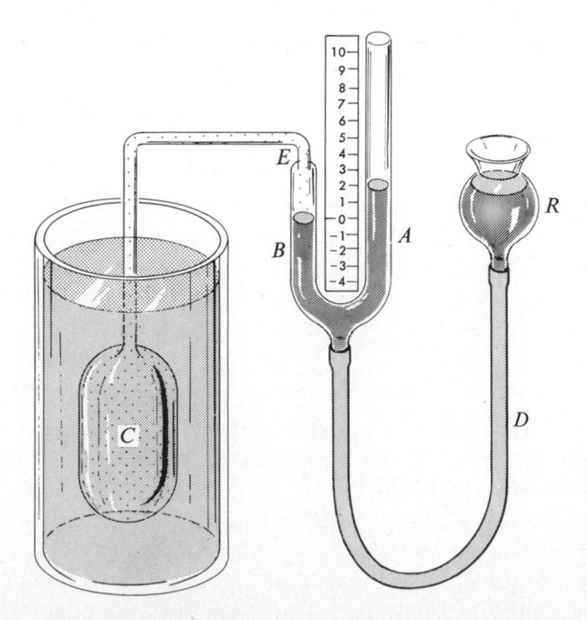

Fig. 15–6 *The constant-volume gas thermometer.*

gas, usually helium, is contained in bulb C and the pressure exerted by it can be measured by the open-tube mercury manometer. As the temperature of the gas increases, the gas expands, forcing the mercury down in tube B and up in tube A. Tubes A and B communicate through a rubber tube D with a mercury reservoir R. By raising R, the mercury level in B may be brought back to reference mark E. The gas is thus kept at constant volume.

Gas thermometers are used mainly in bureaus of standards and in some university research laboratories. They are usually large, bulky, and slow in coming to thermal equilibrium.

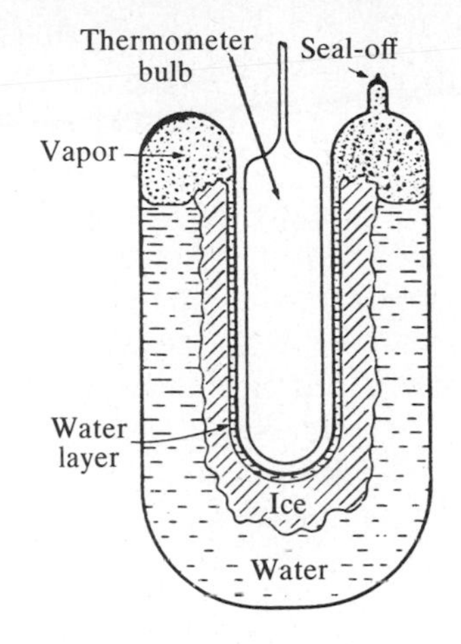

Fig. 15-7 *Triple-point cell with a thermometer in the well, which melts a thin layer of ice nearby.*

15-3 THE ESTABLISHMENT OF A TEMPERATURE SCALE

Any one of the thermometers described in the preceding section may be used to indicate the constancy of a temperature if its state coordinate or thermometric property remains constant. By this means, it has been found that a system composed of a solid and a liquid of the same material maintained at constant pressure will remain in *phase equilibrium* (that is, the liquid and solid exist together without the liquid changing into solid or the solid changing into liquid) only at one definite temperature. Similarly, a liquid will remain in phase equilibrium with its vapor at only one definite temperature when the pressure is maintained constant.

The temperature at which a solid and liquid of the same material coexist in phase equilibrium *at atmospheric pressure* is called the *normal melting point*, abbreviated NMP. The temperature at which a liquid and its vapor exist in phase equilibrium at atmospheric pressure is called the *normal boiling point*, abbreviated NBP.

Phase equilibrium between a solid and its vapor is sometimes possible at atmospheric pressure. The temperature at which this takes place is the *normal sublimation point*, NSP.

It is possible for all three phases—solid, liquid, and vapor—to coexist in equilibrium, but only at one definite pressure and temperature known as the *triple point*, abbreviated TP. The triple-point of water occurs at 4.58 mm of mercury and 0.01°C.

The NMP, NBP, NSP, or TP of any material can be chosen as a standard reference point for the purpose of setting up a temperature scale. Any temperature so chosen is called a *fixed point. The standard fixed point used in modern thermometry is the triple point of water*, to which is given the arbitrary* number

$$273.16\ \text{K},$$

read 273.16 kelvins.†

To achieve the triple point, water of the highest purity is distilled into a vessel like that shown schematically in Fig. 15-7. When all air has been removed, the vessel is sealed off. With the aid of a freezing mixture in the inner well, a layer of ice is formed around the well. When the freezing mixture

* By extrapolating backwards with the aid of the gas law formulas, the "absolute zero" point was "determined" to be −273.15°C. Since the kelvin has the same magnitude as one degree on the Celsius scale, the triple point on the Kelvin scale becomes 273.16 K.

† At a meeting of the Thirteenth General Conference on Weights and Measures on Oct. 13, 1967, the name of the unit of temperature was changed from *degree Kelvin* (symbol °K) to *kelvin* (symbol K). The kelvin, now the unit of temperature, is the fraction 1/273.16 of the thermodynamic temperature of the triple point of water.

is replaced by a thermometer bulb, a thin layer of ice is melted nearby. So long as the solid, liquid, and vapor phases coexist in equilibrium, the system is at the triple point.

We start our program of setting up a temperature scale by denoting with the letter X *any one* of the thermometric properties mentioned previously: the emf of a thermocouple, $\mathscr{E}$, the resistance of a wire, R, the pressure of a gas at constant volume, p, etc. *We define the ratio of two temperatures to be the same as the ratio of the two corresponding values of* X. Thus, if a thermometer with thermometric property X is put in thermal equilibrium with a system and registers a value X, and is then put in thermal equilibrium with another system and registers a value X_3, the ratio of the temperatures of those two systems is given by

$$\boxed{\frac{T(X)}{T(X_3)} = \frac{X}{X_3}.} \tag{15–1}$$

If now, we let the subscript 3 stand for the standard fixed point, the triple point of water, then $T(X_3) = 273.16$ K. Hence

$$T(X) = 273.16\text{ K}\left(\frac{X}{X_3}\right). \tag{15–2}$$

Example Suppose a gas thermometer registers a pressure of 15.0×10^4 Pa at the triple point of water and a pressure of 20.5×10^4 Pa at the normal boiling point; what is the temperature at the normal boiling point?

From Eq. (15–2),

$$T_B = 273.16\text{ K}\frac{20.5 \times 10^4\text{ Pa}}{15.0 \times 10^4\text{ Pa}} = 373\text{ K}.$$

The remaining problem in temperature measurement is that when different thermometers are used to measure the same temperature, the results do not always agree. Best agreement is found with gas thermometers; there is some variation with pressure for any gas, but it is found that, in the limit as pressure becomes smaller and smaller, all gas thermometers at the same temperature approach the same reading, independent of the nature of the gas. Thus the usual procedure is to use the constant-volume thermometer (in the limit of very low pressure) to *define* a temperature scale, and then use this device to calibrate the other thermometers in terms of this scale.

To measure a low temperature, a gas that does not liquefy at the low temperature must be used. The lowest temperature that can be measured with a gas thermometer is about 1 K, provided low-pressure helium is used as the gas. *The temperature $T = 0$ remains as yet undefined.* In Chapter 19 we shall discuss the Kelvin temperature scale, which is independent of the properties of any particular substance. It can be shown that in the temperature region in which a gas thermometer can be used, the gas scale and the Kelvin scale are identical. In anticipation of this result, we write K after a gas temperature.

It will also be shown in Chapter 19 how the absolute zero of temperature is defined on the Kelvin scale. Until then, the term "absolute zero" will have no meaning. The statement made so often that all molecular activity ceases at the temperature $T = 0$ is entirely erroneous. When it is necessary in statistical mechanics to correlate temperature with molecular activity, it is found that classical statistical mechanics must be modified with the aid of quantum mechanics. When this modification is carried out, the molecules of a substance at absolute zero have a *finite* amount of kinetic energy known as the *zero-point* energy.

The body temperatures of warm-blooded animals are held constant to within a few tenths of one celsius degree by an elaborate temperature-control system. The thermometric property which senses the temperature of the blood is provided by a chemical equilibrium condition in a part of the brain called the hypothalamus, which then activates appropriate temperature-controlling mechanisms. The most important mechanisms are dilation or contraction of blood vessels near the surface, increasing or decreasing loss of body heat by conduction, and activation or deactivation of sweat glands to increase or decrease evaporative cooling.

Table 15–1 TEMPERATURES OF FIXED POINTS

Basic fixed points	T, K	t, °C	T_R, °R	t_F, °F
Standard: Triple point of water	**273.16**	**0.01**	**491.688**	**32.018**
NBP of oxygen	90.18	−182.97	162.32	−297.35
Equilibrium of ice and air-saturated water (ice point)	273.15	0.00	491.67	32.00
NBP of water (steam point)	373.15	100.00	671.67	212.00
NMP of zinc	692.66	419.51	1246.78	787.11
NMP of antimony	903.65	630.50	1626.57	1166.90
NMP of silver	1233.95	960.80	2221.11	1761.44
NMP of gold	1336.15	1063.00	2405.07	1945.40

15–4 THE CELSIUS, RANKINE, AND FAHRENHEIT SCALES*

The Celsius temperature scale (formerly called the *centigrade* scale in the United States and Great Britain) employs a degree of the same magnitude as that of the Kelvin scale, but its zero point is shifted so that *the Celsius temperature of the triple point of water is* 0.01 *degree Celsius,* abbreviated 0.01°C. Thus, if t denotes the Celsius temperature

$$t = T - 273.15\ \text{K}. \qquad (15\text{–}3)$$

The Celsius temperature t_S at which steam condenses at 1 atm pressure is

$$t_S = T_S - 273.15\ \text{K};$$

T_S is found to be 373.15 K, so

$$t_S = 373.15° - 273.15° \qquad \text{or } t_S = 100.00°\text{C}.$$

There are two other scales in common use in engineering and in everyday life in the United States and in Great Britain. The Rankine temperature T_R (written °R) is proportional to the Kelvin temperature according to the relation

$$T_R = \frac{9}{5}T. \qquad (15\text{–}4)$$

A degree of the same size is used in the Fahrenheit scale t_F (written °F), but with the zero point shifted according to the relation

$$t_F = T_R - 459.67°\text{R}. \qquad (15\text{–}5)$$

Substituting Eqs. (15–3) and (15–4) into Eq. (15–5), we get

$$t_F = \frac{9}{5}t + 32°\text{F}, \qquad (15\text{–}6)$$

from which it follows that the Fahrenheit temperature of the ice point (t = 0°C) is 32°F and of the steam point (t = 100°C) is 212°F. The 100 Celsius or Kelvin degrees between the ice point and the steam point correspond to 180 Fahrenheit or Rankine degrees, as shown in Fig. 15–8, where the four scales are compared.

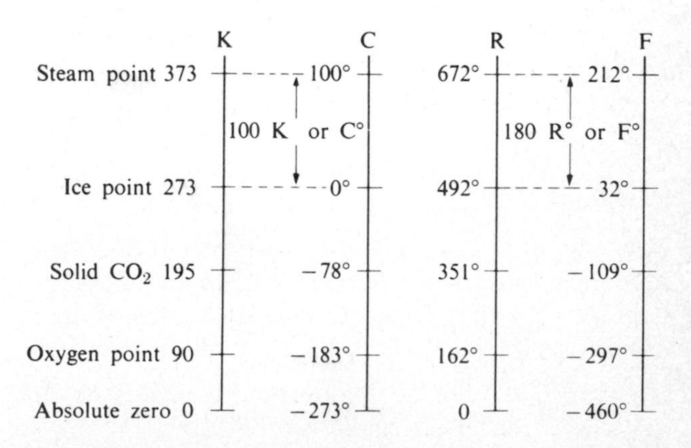

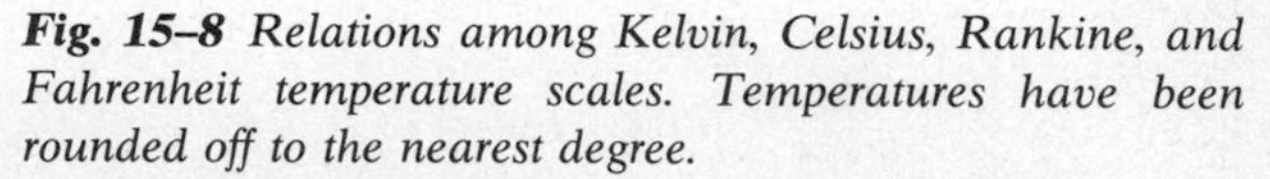
Fig. 15–8 *Relations among Kelvin, Celsius, Rankine, and Fahrenheit temperature scales. Temperatures have been rounded off to the nearest degree.*

* Named after Anders Celsius (1701–1744). William John MacQuorn Rankine (1820–1872). Gabriel Fahrenheit (1686–1736).

The accurate measurement of a boiling point or melting point with the aid of a gas thermometer requires very careful laboratory work. Fortunately, this has been done for a large number of substances which are obtainable with high purity. Some of these results are shown in Table 15–1. With the aid of these basic fixed points other thermometers may be calibrated.

Suppose the temperature of a beaker of water is raised from 20°C to 30°C, through a temperature interval of 10 Celsius degrees. It is desirable to distinguish between such a temperature *interval* and the actual temperature of 10 degrees above the Celsius zero. Hence we shall use the phrase "10 degrees Celsius," or "10°C," when referring to an *actual temperature*, and "10 Celsius degrees," or "10C°" to mean a temperature *interval*. Thus there is an interval of 10 Celsius degrees between 20 degrees Celsius and 30 degrees Celsius.

15–5 EXPANSION OF SOLIDS AND LIQUIDS

With a few exceptions, the volumes of all bodies increase with increasing temperature if the external pressure on the body remains constant. Suppose that a solid or a liquid undergoes a small change of volume dV when the temperature is changed a small amount dT (or dt, since the kelvin and the Celsius degree are temperature intervals of equal magnitude). The *coefficient of volume expansion* β is defined as *the fractional change in volume* dV/V *divided by the change of temperature* dT, or

$$\beta = \frac{1}{V}\frac{dV}{dT} \text{(at constant pressure).} \qquad (15\text{–}7)$$

The unit β is 1 *reciprocal degree*, or 1 $(\text{C}°)^{-1}$. The numerical value depends, of course, on the size of the degree. Since the kelvin and the Celsius degree are $\frac{9}{5}$ as large as the Rankine and Fahrenheit degrees, the fractional volume change per kelvin or Celsius degree is $\frac{9}{5}$ as great as that per Rankine or Fahrenheit degree.

The coefficient of volume expansion is insensitive to a change of pressure, but varies markedly with

Table 15–2 COEFFICIENT OF VOLUME EXPANSION (APPROXIMATE)

Solids	$\bar{\beta}$, $(\text{C}°)^{-1}$	Liquids	$\bar{\beta}$, $(\text{C}°)^{-1}$
Aluminum	7.2×10^{-5}	Alcohol, ethyl	75×10^{-5}
Brass	6.0	Carbon disulfide	115
Copper	4.2	Glycerin	49
Glass	1.2–2.7	Mercury	18
Steel	3.6		
Invar	0.27		
Quartz (fused)	0.12		

temperature. For many substances β decreases as the temperature is lowered, approaching zero as the Kelvin temperature approaches zero. It is also a peculiar circumstance that the higher the melting point of a metal, the lower the coefficient of volume expansion.

Calling $\bar{\beta}$ the average value of β within a moderate temperature interval ΔT, we may rewrite Eq. (15–7):

$$\Delta V = \bar{\beta} V_0 \, \Delta T \qquad (15\text{–}8)$$

where V_0 is the original volume.

Some values of $\bar{\beta}$ in the neighborhood of room temperature are listed in Table 15–2. Note that the values for liquids are much larger than those for solids.

If there is a hole in a solid body, the volume of the hole increases when the body expands, just as if the hole were a solid of the same material as the body. Thus the volume enclosed by a thin-walled glass flask or thermometer bulb increases just as would a solid body of glass of the same size.

Example A glass flask of volume 200 cm³ is just filled with mercury at 20°C. How much mercury will overflow when the temperature of the system is raised to 100°C? The coefficient of volume expansion of the glass is $1.2 \times 10^{-5}(\text{C}°)^{-1}$.

The increase in the volume of the flask is

$$\Delta V = (1.2 \times 10^{-5}(\mathrm{C}°)^{-1})(200\ \mathrm{cm}^3)(100° - 20°)$$
$$= 0.192\ \mathrm{cm}^3.$$

The increase in the volume of the mercury is

$$\Delta V = (18 \times 10^{-5}(\mathrm{C}°)^{-1})(200\ \mathrm{cm}^3)(100° - 20°)$$
$$= 2.88\ \mathrm{cm}^3.$$

The volume of mercury overflowing is therefore

$$2.88 - 0.19 = 2.69\ \mathrm{cm}^3.$$

Water, in the temperature range from 0°C to 4°C, *decreases* in volume with increasing temperature, contrary to the behavior of most substances. That is, between 0°C and 4°C the coefficient of expansion of water is *negative.* Abôve 4°C, water expands when heated. Since the volume of a given mass of water is smaller at 4°C than at any other temperature, the density of water is a maximum at 4°C. Water also expands upon freezing, unlike most materials.

This anomalous behavior has an important effect on plant and animal life in lakes. When a lake cools, the cooled water at the surface flows to the bottom because of its greater density. But when the temperature reaches 4°C this flow ceases and the water near the surface remains colder (and less dense) than that at the bottom. As the surface freezes, the ice floats because it is less dense than water. The water at the bottom remains at 4°C until nearly the entire body is frozen. If water behaved like most substances, contracting continuously on cooling and freezing, lakes would freeze from the bottom up, circulation due to density differences would continuously carry warmer water to the surface for efficient cooling, and lakes would freeze solid much more easily, thus destroying all plant and animal life that can withstand cold water but not freezing.

The anomalous expansion of water in the temperature range 0°C to 10°C is shown in Fig. 15–9. Table 15–3 covers a wider range of temperatures.

For a body in the form of a rod or cable, we are often interested only in the change of *length* with temperature, and we define a *coefficient of linear expansion* α. If L is the length,

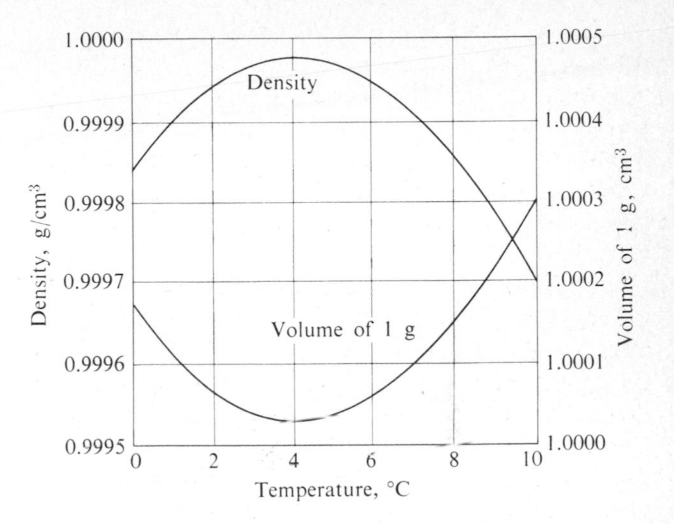

Fig. 15–9 *Density of water, and volume of 1 gram, in the temperature range from 0°C to 10°C.*

$$\alpha = \frac{1}{L}\frac{dL}{dT}, \tag{15–9}$$

and, over a moderate temperature change ΔT,

$$\Delta L = L_0 \bar{\alpha}\, \Delta T, \tag{15–10}$$

and

$$L = L_0 + \Delta L = L_0(1 + \bar{\alpha}\, \Delta T), \tag{15–11}$$

where $\bar{\alpha}$ is the average coefficient within the temperature interval, and L_0 is the original length.

Table 15–3 DENSITY AND VOLUME OF WATER

t,°C	density, g cm^{-3}	Volume of 1 g, cm^3
0	0.9998	1.0002
4	1.0000	1.0000
10	0.9997	1.0003
20	0.9982	1.0018
50	0.9881	1.0121
75	0.9749	1.0258
100	0.9584	1.0434

The volume coefficient is related to the linear coefficient. To obtain the relation, we consider a solid body in the form of a rectangular parallelepiped with dimensions L_1, L_2, and L_3. Then the volume is

$$V = L_1 L_2 L_3,$$

and

$$\frac{dV}{dT} = L_2 L_3 \frac{dL_1}{dT} + L_1 L_3 \frac{dL_2}{dT} + L_1 L_2 \frac{dL_3}{dT}.$$

Dividing by $L_1 L_2 L_3$, we obtain

$$\frac{1}{V}\frac{dV}{dT} = \frac{1}{L_1}\frac{dL_1}{dT} + \frac{1}{L_2}\frac{dL_2}{dT} + \frac{1}{L_3}\frac{dL_3}{dT}.$$

If the solid has the same properties in each of the three directions, each of the three expressions on the right is the coefficient of linear expansion α, and hence

$$\boxed{\beta = 3\alpha.} \qquad (15\text{–}12)$$

15–6 THERMAL STRESSES

If the ends of a rod are rigidly fixed so as to prevent expansion or contraction and the temperature of the rod is changed, tensile or compressive stresses, called *thermal stresses*, will be set up in the rod. These stresses may become very large, sufficiently so to stress the rod beyond its elastic limit or even beyond its breaking strength. Hence in the design of any structure which is subject to changes in temperature, some provision must, in general, be made for expansion. In a long steam pipe this is accomplished by the insertion of expansion joints or a section of pipe in the form of a U. In bridges, one end may be rigidly fastened to its abutment while the other rests on rollers.

It is a simple matter to compute the thermal stress set up in a rod which is not free to expand or contract. Suppose that a rod at a temperature T has its ends rigidly fastened, and that while they are thus held the temperature is reduced to a lower value, T_0.

The fractional change in length if the rod were free to contract would be

$$\frac{\Delta L}{L_0} = \bar{\alpha}(T - T_0) = \bar{\alpha}\,\Delta T. \qquad (15\text{–}13)$$

Since the rod is *not* free to contract, the tension must increase by a sufficient amount to produce the same fractional change in length. But from the definition of Young's modulus,

$$Y = \frac{F/A}{\Delta L/L_0}, \qquad F = AY\frac{\Delta L}{L_0}.$$

Introducing the expression for $\Delta L/L_0$ from Eq. (15–13), we have

$$F = AY\bar{\alpha}\,\Delta T, \qquad (15\text{–}14)$$

which gives the tension F in the rod. The *stress* in the rod is

$$\frac{F}{A} = Y\bar{\alpha}\,\Delta T. \qquad (15\text{–}15)$$

If, instead, Δt represents an *increase* in temperature, F/A is a *compressive* stress of the same magnitude.

Thermal stresses can also be induced by nonuniform heating. Even if a solid body at uniform temperature has no internal stresses, stress may be induced by nonuniform expansion if it is heated unevenly. The breaking of a thick glass container when hot water is poured into it is a familiar phenomenon. Heat-resistant glasses such as Pyrex have exceptionally low expansion coefficients, and usually also have high strength to permit thin-wall construction to minimize temperature differences.

PROBLEMS

15–1 The limiting value of the ratio of the pressures of a gas at the melting point of lead and at the triple point of water, when the gas is kept at constant volume, is found to be 2.19816. What is the Kelvin temperature of the melting point of lead?

15–2

a) If you feel sick in France and are told you have a fever of 40°C, should you be concerned?

b) What is normal body temperature on the Celsius scale?

c) The normal boiling point of liquid oxygen is −182.97°C. What is this temperature on the Kelvin and Rankine scales?

d) At what temperature do the Fahrenheit and Celsius scales coincide?

15–3 When the United States finally converts officially to metric units, the Celsius temperature scale will very likely replace the Fahrenheit scale in everyday use. As a familiarization exercise, find the Celsius temperatures corresponding to (a) a cool room (68°F); (b) a hot summer day (95°F); (c) a cold winter day (5°F).

15–4 In a rather primitive experiment with a constant-volume gas thermometer, the pressure at the triple point of water was found to be 4.0×10^4 Pa. and the pressure at the normal boiling point 5.4×10^4 Pa. According to these data, what is the temperature of absolute zero on the Celsius scale?

15–5 A gas thermometer of the type shown in Fig. 15–6 registered a pressure corresponding to 5.0 cm of mercury when in contact with water at the triple point. What pressure will it read when in contact with water at the normal boiling point?

15–6 The electrical resistance of some metals varies with temperature (as measured by a gas thermometer) approximately according to $R = R_0[1 + \beta(T - T_0)]$, where R_0 is the resistance at temperature T_0. For a certain metal, β is found to be 0.004 K^{-1}.

a) If the resistance at 0°C is 100 ohms, what is the resistance at 20°C?

b) At what temperature is the resistance 200 ohms?

15–7 The pressure p, volume V, number of moles n, and Kelvin temperature T of an ideal gas are related by the equation $pV = nRT$. Prove that the coefficient of volume expansion is equal to the reciprocal of the Kelvin temperature.

15–8

a) For any material, density ρ, mass m, and volume V are related by $\rho = m/V$. Prove that

$$\beta = -\frac{1}{\rho}\frac{\partial \rho}{\partial T}.$$

b) The density of rock salt between −193°C and −13°C is given by the empirical formula

$$\rho = 2.1680(1 - 11.2 \times 10^{-5}T - 0.5 \times 10^{-7}T^2),$$

with T measured on the Celsius scale. Calculate β at −100°C.

15–9 The total change of volume ΔV of a metal, when the temperature changes from room temperature (T_r = 300 K) to the melting point T_m, is still only a small fraction of the original volume, so that

$$\frac{\Delta V}{V} = \int_{T_r}^{T_m} \beta\, dT.$$

Given the data in Table 15–4, calculate the average value of $\Delta V/V$ for these three metals.

Table 15–4

	T_m, K	β, 10^{-6} K^{-1}
Copper	1360	$43 + 0.022T$
Palladium	1830	$30 + 0.015T$
Platinum	2050	$24 + 0.0086T$

15–10 A glass flask whose volume is exactly 1000 cm^3 at 0°C is filled level full of mercury at this temperature. When flask and mercury are heated to 100°C, 15.2 cm^3 of mercury overflow. If the coefficient of volume expansion of mercury is 0.000182 per Celsius degree, compute the coefficient of linear expansion of the glass.

15–11 At a temperature of 20°C, the volume of a certain glass flask, up to a reference mark on the stem of the flask, is exactly 100 cm^3. The flask is filled to this point with a liquid whose cubical coefficient of expansion is 120×10^{-5} $(C°)^{-1}$, with both flask and liquid at 20°C. The linear coefficient of expansion of the glass is 8×10^{-6} $(C°)^{-1}$. The cross section of the stem is 1 mm^2 and can be considered constant. How far will the liquid rise or fall in the stem when the temperature is raised to 40°C?

15–12 To ensure a tight fit, the aluminum rivets used in airplane construction are made slightly larger than the rivet holes and cooled by "dry ice" (solid CO_2) before being driven. If the diameter of a hole is 0.2500 in., what should be the diameter of a rivet at 20°C if its diameter is to equal that of the hole when the rivet is cooled to −78°C, the temperature of dry ice? Assume the expansion coefficient to remain constant at the value given in Table 15–2.

15–13 A metal rod 30.0 cm long expands by 0.075 cm when its temperature is raised from 0°C to 100°C. A rod of a different metal of the same length expands by 0.045 cm for the same rise in temperature. A third rod, also 30.0 cm long, is made up of pieces of each of the above metals placed end-to-end and expands 0.065 cm between 0°C and 100°C. Find the length of each portion of the composite bar.

15–14 A hole 1.000 in. in diameter is bored in a brass plate at a temperature of 20°C. What is the diameter of the hole when the temperature of the plate is increased to 200°C? Assume the expansion coefficient to remain constant.

15–15 Suppose that a steel hoop could be constructed around the earth's equator. just fitting it at a temperature of 20°C. What would be the space between the hoop and the earth if the temperature of the hoop were increased by 1 C°?

15–16 A clock whose pendulum makes one vibration in 2 s is correct at 25°C. The pendulum shaft is of steel and its mass may be neglected compared with that of the bob. (a) What is the fractional change in length of the shaft when it is cooled to 15°C? (b) How many seconds per day will the clock gain or lose at 15°C? [*Hint:* Use differentials.]

15–17 A clock with a brass pendulum shaft keeps correct time at a certain temperature.

a) How closely must the temperature be controlled if the clock is not to gain or lose more than 1 second a day? Does the answer depend on the period of the pendulum?
b) Will an increase of temperature cause the clock to gain or lose?

15–18 The length of a bridge is 2000 ft. (a) If it were a continuous span, fixed at one end and free to move at the other, about what would be the range of motion of the free end between a cold winter day (−20°F) and a hot summer day (100°F)? (b) If both ends were rigidly fixed on the summer day, what would be the stress on the winter day?

15–19 The cross section of a steel rod is 1.5 in^2. What is the least force that will prevent it from contracting while cooling from 520°C to 20°C?

15–20 A steel wire which is 10 ft long at 20°C is found to increase in length by $\frac{3}{4}$ in. when heated to 520°C. Compute its average coefficient of linear expansion. (b) Find the stress in the wire if it is stretched taut at 520°C and cooled to 20°C without being allowed to contract.

15–21 A steel rod of length 40 cm and a copper rod of length 36 cm, both of the same diameter, are placed end-to-end between two rigid supports, with no initial stress in the rods. The temperature of the rods is now raised by 50 C°. What is the stress in each rod?

15–22 A heavy brass bar has projections at its ends, as in Fig. 15–10. Two fine steel wires fastened between the projections are just taut (zero tension) when the whole system is at 0°C. What is the tensile stress in the steel wires when the temperature of the system is raised to 300°C? Make any simplifying assumptions you think are justified, but state what they are.

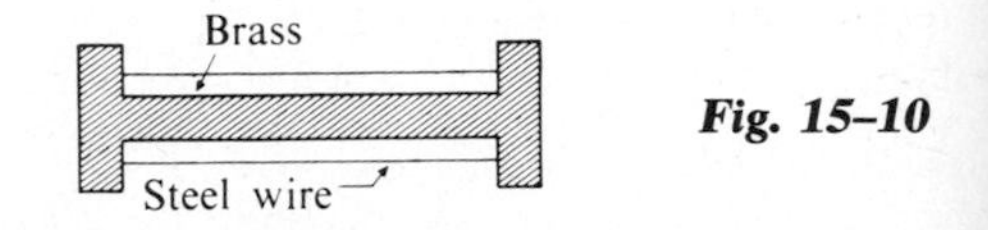

Fig. 15–10

15–23 The steel rails of the missile testing track at Tularosa Basin, New Mexico, are welded into 10,000-ft lengths and each length is prestressed by stretching it 3 ft before it is fastened to the concrete base slab.

a) What is the strain in the rails?
b) If Young's modulus is 30×10^6 lb in^{-2}, what is the stress?
c) What stretching force is required if the cross-sectional area of a rail is 50 in^2?
d) If the stretched rails are fastened to the base slab at a temperature of 60°F, what is the stress when the temperature decreases to 0°F?
e) At what temperature will the stress change from tension to compression?

15–24 Prove that if a body under hydrostatic pressure is raised in temperature but not allowed to expand, the increase in pressure is

$$\Delta p = B\bar{\beta}\,\Delta t.$$

where the bulk modulus B and the average coefficient of volume expansion $\bar{\beta}$ are both assumed positive and constant.

15–25

a) A block of steel at a pressure of 1 atm and a temperature of 20°C is kept at constant volume. If the temperature is raised to 32°C, what will be the final pressure?
b) If the block is maintained at constant volume by rigid walls that can withstand a maximum pressure of 1200 atm, what is the highest temperature to which the sys-

tem may be raised? Assume B and $\bar{\beta}$ to remain practically constant at the values 1.5×10^{11} Pa and $5.0 \times 10^{-5}(\text{C}°)^{-1}$, respectively.

15–26 What hydrostatic pressure is necessary to prevent a copper block from expanding when its temperature is increased from 20°C to 30°C?

15–27 Table 15–3 lists the density of water, and the volume of 1 g. at atmospheric pressure. A steel bomb is filled with water at 10°C and atmospheric pressure, and the system is heated to 75°C. What is then the pressure in the bomb? Assume the bomb to be sufficiently rigid so that its volume is not affected by the increased pressure.

15–28 A liquid is enclosed in a metal cylinder provided with a piston of the same metal. The system is originally at atmospheric pressure and at a temperature of 80°C. The piston is forced down until the pressure on the liquid is increased by 100 atm, and it is then clamped in this position. Find the new temperature at which the pressure of the liquid is again 1 atm. Assume that the cylinder is sufficiently strong so that its volume is not altered by changes in pressure, but only by changes in temperature.

Compressibility of liquid	$k = 50 \times 10^{-6}\ \text{atm}^{-1}$.
Cubical coefficient of expansion of liquid	$\beta = 5.3 \times 10^{-4}\ (\text{C}°)^{-1}$.
Linear coefficient of expansion of metal	$\alpha = 10 \times 10^{-6}\ (\text{C}°)^{-1}$.

Chapter 16

Heat and Heat Measurements

16–1 HEAT TRANSFER

Suppose that system A, at a higher temperature than system B, is put in contact with B. When thermal equilibrium has been reached, A will be found to have undergone a temperature decrease and B a temperature increase. It was therefore quite natural for the early investigators in this field to assume that A lost something and that this "something" flowed into B. While the temperature changes are taking place, it is customary to refer to a *heat flow* or a *heat transfer* from A to B. The process of heat transfer was formerly thought to be a flow of an invisible weightless fluid called *caloric*, which was produced when a substance burned and which could flow from a region rich in caloric (where the temperature was high) to a region where there was less caloric (and a lower temperature). The abandonment of the caloric theory was a part of the general development of physics during the eighteenth and nineteenth centuries. Due to the experimental skill and physical insight of Count Rumford (1753–1814) and Sir James Prescott Joule (1818–1889), the idea slowly emerged that heat flow is an *energy transfer.*

When an energy transfer takes place by virtue of a temperature difference exclusively, it is called a heat flow. Thus, in Fig. 16–1(a), the hot flame of a Bunsen burner is in contact with a system consisting of water and water vapor at a lower temperature. Proceeding from left to right, water is converted to steam at a higher temperature and pressure. Under these conditions the steam is capable of doing more work (by pushing against a turbine blade, for example) than before, and therefore has received energy in a process involving a heat flow.

An energy transfer to a system may take place, however, without a flow of heat. Consider the three diagrams (from left to right) of Fig. 16–1(b). As an object hanging from a string attached to a cam descends, a piston is caused to descend and a gas is compressed. According to the definition given in Chapter 7, this compression is the result of the performance of *work*, since the point of application of a force has moved in the direction of the force. In its compressed state the gas is capable of doing more work than before and hence the gas has received energy in a process that involves the performance of work.

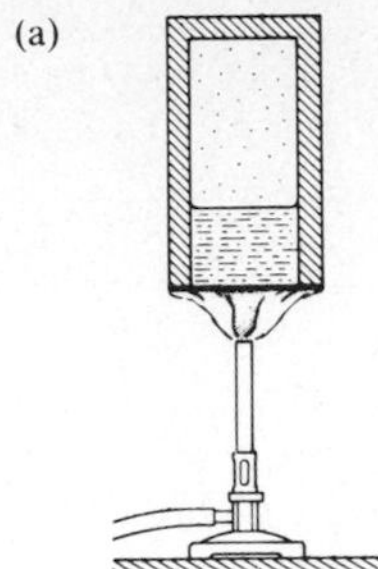

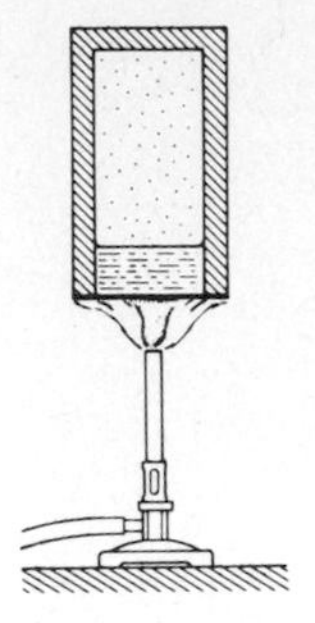
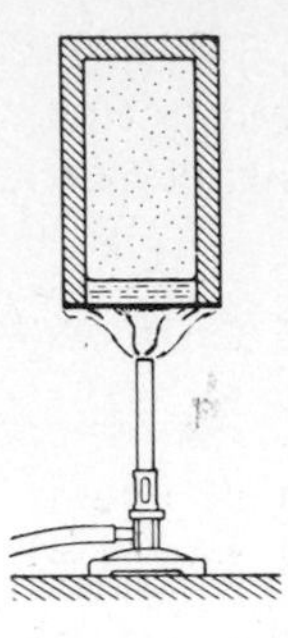
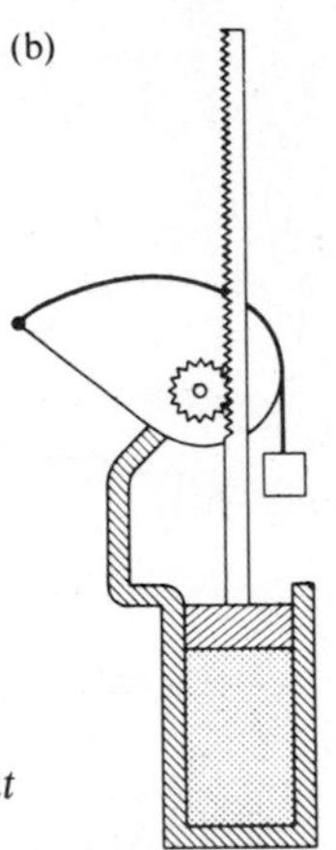

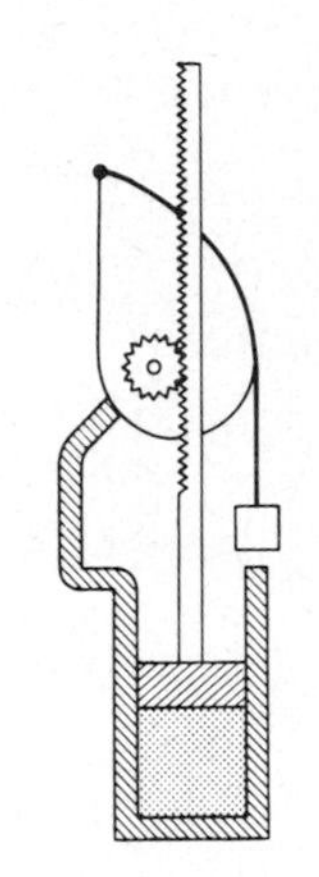
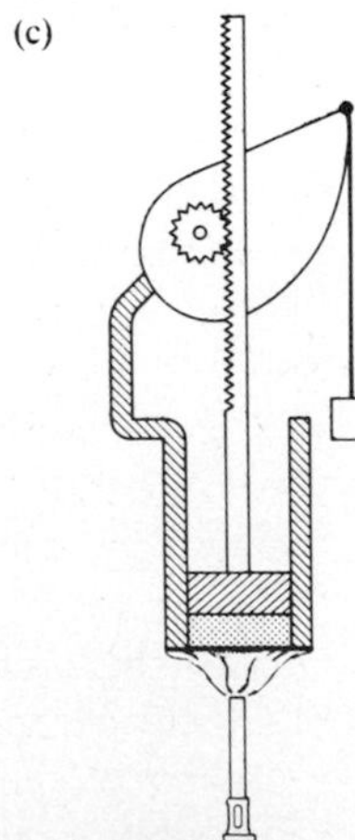

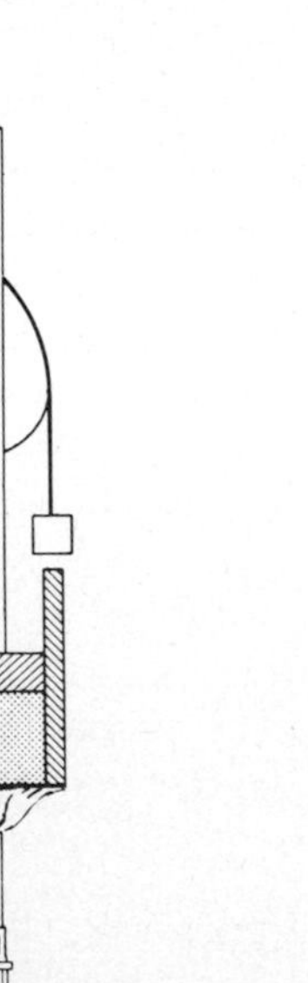
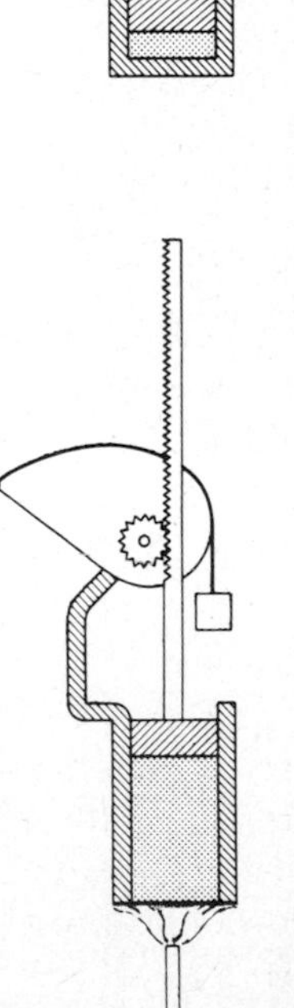

***Fig. 16–1** Distinction between flow of heat and performance of work.*

The three diagrams of Fig. 16–1(c) show how a process may take place involving the simultaneous flow of heat and performance of work.

16–2 QUANTITY OF HEAT

A system consisting of a quantity of water and a small piece of resistance wire is depicted in Fig. 16–2. In part (a), the system is caused to undergo a temperature rise ΔT in a process involving a heat flow through a diathermic wall (Section 15–1) at the bottom of the enclosure. In (b) the system is surrounded by adiabatic walls and there is no heat flow. The same temperature change ΔT is produced by permitting a falling body to turn an electric generator, causing a current in the resistance wire. Thus in (b) the method of energy transfer to the system is the performance of mechanical work.

Suppose, however, that we regard the water alone as our system in part (b). Then the flow of energy *into the water* is due to the presence of a resistance wire whose temperature is maintained a trifle higher than that of the water by means of the electric current in the wire. This process involves a heat flow from the wire to the water.

Two conclusions may be drawn from these considerations:

1. Whether a process is regarded as involving a flow of heat or a performance of work depends on the *choice of the system.*

2. Whether the energy transfer depicted in Fig. 16–2(b) is regarded as a flow of heat or as a performance of work has no bearing on the quantity of energy delivered to the water.

We may describe the results of the two processes depicted in Fig. 16–2 by saying that a heat flow and a performance of work are *equivalent.* Once either process is completed, the energy of the system is greater than before, and no experiment can tell whether this energy increase has been caused by a flow of heat or a performance of work. They are both methods of energy transfer; in each case the *time rate* of energy transfer is expressed in power units, such as watts (joules per second) or foot-pounds per second.

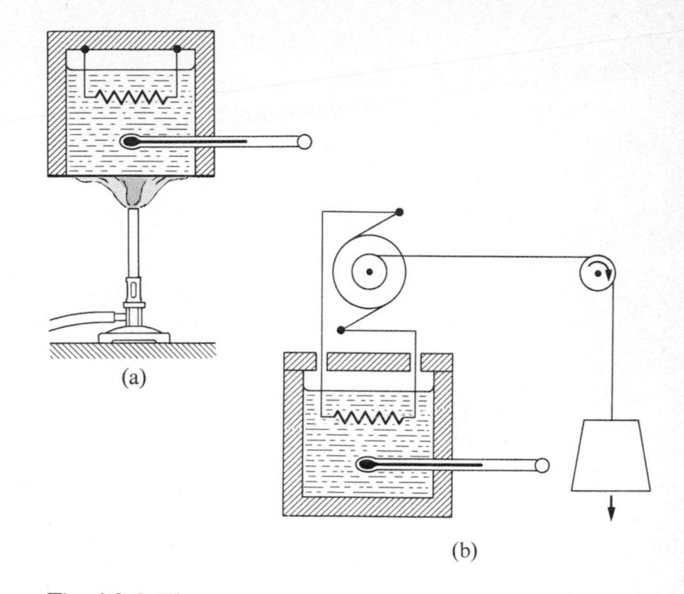

Fig. 16–2 *The same temperature change of the same system may be accomplished by either (a) a heat flow or (b) the performance of work.*

Up to this point, the word *heat* has been used only in conjunction with *flow* or *transfer.* Thus, a heat flow is an energy transfer brought about by a temperature difference only. Ideally, the word *heat* should be used *only* when referring to a *method* of energy *transfer* and, when the transfer is completed, to refer to the total amount of energy so transferred. The expression "quantity of heat," however, has played such a huge role in the development of the subject and is to be found in so many books and tables, that it is almost impossible to avoid.

It is essential to bear in mind that the concept "quantity of heat" has meaning *only* in the context of an interaction in which energy is *transferred* from one system to another as a result of a temperature difference. As we shall see later, to say that a given system *contains* a certain quantity of heat is meaningless. A careful discussion of this point leads to the concept of *internal energy,* to be discussed in detail in Chapter 19.

In the eighteenth century, a unit of quantity of heat, the *calorie,* was defined as that amount of heat required to raise the temperature of one gram of water one kelvin or one Celsius degree. It was found later that more heat was required to raise the temperature of 1 g of water from, say, 90°C to 91°C than from 30°C to 31°C. The definition was then refined, and the chosen calorie became known as the "15° calorie," that is, the quantity of heat required to change the temperature of 1 g of water from 14.5°C to 15.5°C. A corresponding unit defined in terms of Fahrenheit degrees and British units is the *British thermal unit*, or Btu. By definition, 1 Btu is the quantity of heat required to raise the temperature of one pound of water from 63°F to 64°F. A third unit in common use, especially in measuring food energy, is the kilocalorie (kcal). The relations among these three units are

$$1 \text{ Btu} = 252 \text{ cal} = 0.252 \text{ kcal}.$$

It should also be noted that the unit of quantity of heat is fundamentally a unit of energy; thus there must be a relation between the above units and the familiar mechanical energy units such as the joule. It has been found experimentally that

$$1 \text{ cal} = 4.186 \text{ joules} = 4.186 \text{ J}.$$

The International Committee on Weights and Measures no longer recognizes the calorie as a fundamental unit, but recommends instead that the joule be used for quantity of heat as well as all other forms of energy. Although the calorie is convenient in problems involving water, it is awkward when both heat and other forms of energy are involved. There seems little doubt that ultimately the joule will become the universal unit of energy; many of the examples and problems of this and the following chapters will anticipate this usage. The relation between heat and mechanical energy will be explored in detail in Chapters 19 and 20.

16–3 HEAT CAPACITY

Suppose that a small quantity of heat dQ is transferred between a system and its surroundings. If the system undergoes a small temperature change dT. the *specific heat capacity c of the system is defined as the ratio of the heat dQ to the product of the mass m and temperature change dT*; thus

$$c = \frac{dQ}{m\,dT}. \qquad (16\text{–}1)$$

The specific heat capacity of water can be taken to be 4.19 J g^{-1} $(C°)^{-1}$, 1 cal g^{-1} $(C°)^{-1}$, or 1 Btu lb^{-1} $(F°)^{-1}$ for most practical purposes.

It is often convenient to use the *mole* to describe the amount of substance. By definition, one mole (1 mol) of any substance is a quantity of matter such that its mass in grams is numerically equal to the *molecular mass M* (often called *molecular weight*).* To calculate the number of moles n, we divide the mass in grams by the molecular mass; thus $n = m/M$. Replacing the mass m in Eq. (16–1) by the product nM, we get

$$Mc = \frac{dQ}{n\,dT}.$$

The product Mc is called the *molar heat capacity* and is represented by the symbol C. Hence, by definition

$$C = Mc = \frac{dQ}{n\,dT}. \qquad (16\text{–}2)$$

The molar heat capacity of water is approximately 75.3 J mol^{-1} $(C°)^{-1}$ or 18 cal mol^{-1} $(C°)^{-1}$.

The quantity defined by Eq. (16–1) is sometimes called simply *specific heat*, and the molar heat capacity defined by Eq. (16–2) is often called the *molar specific heat*. However, the terms *specific heat capacity* and *molar heat capacity* will be used throughout this book. Representative values of specific and molar heat capacity are given in Table 16–1.

From Eq. (16–1) the *total* quantity of heat Q that must be supplied to a body of mass m to change its temperature from T_1 to T_2 is

$$Q = m\int_{T_1}^{T_2} c\,dT. \qquad (16\text{–}3)$$

The specific heat capacities of all materials vary somewhat with temperature, and of course c must be expressed as a function of T in order to carry out the

* Although the term molecular *weight* is in common use, the term measures the *mass*, not the *weight* of a molecule. Thus the term *molecular mass* is preferable. This term is coming into more common use and will be used throughout this book.

Table 16-1 MEAN SPECIFIC AND MOLAR HEAT CAPACITIES OF METALS

Metal	(Specific) $\bar{c}_p$, J g^{-1} $(C°)^{-1}$	Temperature range °C	M, g mol^{-1}	(Molar) $\bar{C}_p = M\bar{c}_p$, J mol^{-1} $(C°)^{-1}$
Beryllium	1.97	20–100	9.01	17.7
Aluminum	0.91	17–100	27.0	24.5
Iron	0.47	18–100	55.9	26.4
Copper	0.39	15–100	63.5	24.7
Silver	0.234	15–100	108	25.3
Mercury	0.138	0–100	201	27.8
Lead	0.130	20–100	207	26.9

integration. However, when the temperature interval is not too great, this variation may often be ignored, and c may be treated as a constant. In that case, the integral is trivial and Eq. (16-3) becomes

$$Q = mc(T_2 - T_1). \tag{16-4}$$

Similarly, when the molar heat capacity C may be treated as constant, Eq. (16-2) may be integrated to give

$$Q = nC(T_2 - T_1). \tag{16-5}$$

At very low temperatures, specific heat capacities always *decrease*, and the variation of heat capacity with temperature provides useful information concerning the microscopic structure of materials.

The specific or molar heat capacity of a substance is not the only physical property whose experimental determination requires the measurement of a quantity of heat. Heat conductivity, heat of fusion, heat of vaporization, heat of combustion, heat of solution, and heat of reaction are examples of other such properties which are called *thermal properties* of matter. The field of physics and physical chemistry concerned with the measurement of thermal properties is called *calorimetry*.

16-4 THE MEASUREMENT OF HEAT CAPACITY

To measure a heat capacity we need to add a measured quantity of heat to a measured quantity of a material and observe the resulting temperature change. For maximum precision the thermal measurements are usually made electrically. The heat input is provided by passing a current through a resistance wire in contact with (often wound around) the material and measuring the input of electrical energy. The thermometer is usually a small resistance thermometer or thermocouple embedded in the sample and is chosen for its rapidity of response and its sensitivity.

Readings are taken before the switch in the heater circuit is closed; these are shown on the left of Fig. 16-3. The fact that the temperature rises before energy is supplied to the heater indicates that the surroundings are at a temperature higher than that of the sample. The current is maintained in the heater for a time interval Δt, during which we do not measure the temperature. After the switch in the heater circuit is opened, temperature measurements are resumed, giving rise to the righthand part of Fig. 16-3, where the positive slope indicates that the surroundings are at still higher temperature than the sample.

If the potential difference across the heater is V and an electric current I flows, the power delivered to the heater is VI, and if the current is maintained for a time Δt, the total energy (heat) transferred to the sample is $VI\Delta t$. The molar heat capacity is therefore

$$C = \frac{VI\Delta t}{n\Delta T}, \tag{16-6}$$

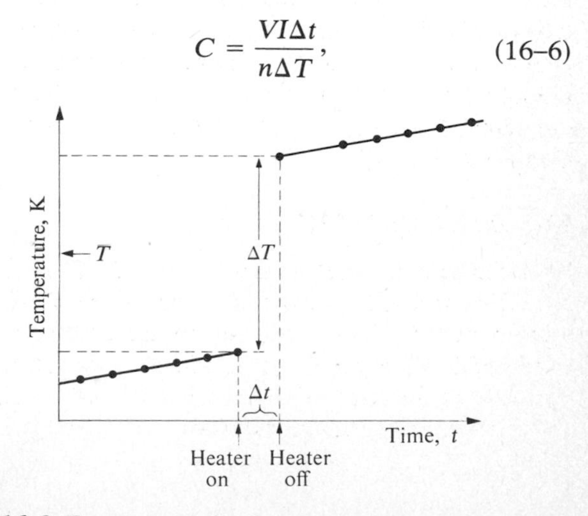

Fig. 16-3 *Temperature-time graph of the data taken during a heat-capacity measurement.*

where C is expressed in joules per mole-degree when I is expressed in amperes, V in volts, t in seconds, and ΔT is read from the graph, as in Fig. 16–3. The resulting molar heat capacity C is the value at the temperature T which is in the middle of the interval ΔT. In careful experiments, ΔT may be as small as 0.01 K.

Precise measurements of specific heats require great experimental skill, partly because of the difficulty of avoiding (and compensating for) unwanted heat transfer between the sample and its surroundings. Still, such measurements are very important. The temperature variation of specific heats provides the most direct approach to the understanding of molecular energies in matter. Low-temperature thermal properties in particular are of great interest in contemporary physics.

Figure 16–4 shows the variation of the specific heat capacity of water with temperature. It may be seen that the quantity of heat necessary to raise the temperature of 1 g of water from 14.5°C to 15.5°C is

$$1\ 15^\circ\ \text{cal} = 4.186\ \text{J}.$$

Two other calories are frequently used. The *international table calorie* (IT cal) is *defined* to be

$$1\ \text{IT cal} = \frac{1\ \text{W hr}}{860} = \frac{3600\ \text{J}}{860} = 4.186\ \text{J},$$

and is almost identical with the 15° cal. The *thermochemical calorie*, however, is equal to 4.1840 J and may be seen from Fig. 16–4 to correspond to about a 17° calorie. In the absence of any additional information, the word *calorie* should be taken to mean 4.186 J. This variety of definitions of the calorie is an additional argument in favor of *eliminating* the calorie completely and using the joule as the fundamental unit of quantity of heat.

Mechanical engineers frequently use the British thermal unit (Btu), defined as the quantity of heat required to raise the temperature of 1 lb (mass) of water from 63°F to 64°F. The following relations hold:

$$1\ \text{Btu} = 778.3\ \text{ft lb} = 252.0\ \text{cal} = 1055\ \text{J}.$$

16–5 EXPERIMENTAL VALUES OF HEAT CAPACITIES

The amount of heat transferred to or from a system depends on the manner in which the system is controlled or constrained during the transfer, that is, whether the system is kept at *constant pressure* or at *constant volume.* The two corresponding specific (or molar) heat capacities are denoted by the respective symbols c_P (or C_P) and c_V (or C_V). In the electrical method of measuring heat capacities, the sample under investigation is maintained at constant pressure. As a matter of fact, it would be almost impossible to make a precise determination of c_V because there is no effective way of keeping the volume of a system constant and of correcting for the heat transferred to the containing walls.

A few average values of heat capacities of metals are listed in Table 16–1. The specific heat capacities are seen to be less than that of water and to decrease with increasing molecular mass. The last column reveals an interesting regularity, first noted in 1819

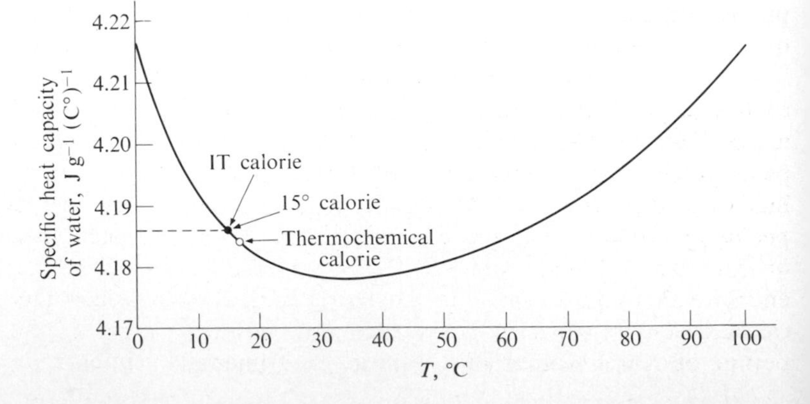

Fig. 16–4 *Specific heat capacity of water as a function of temperature.*

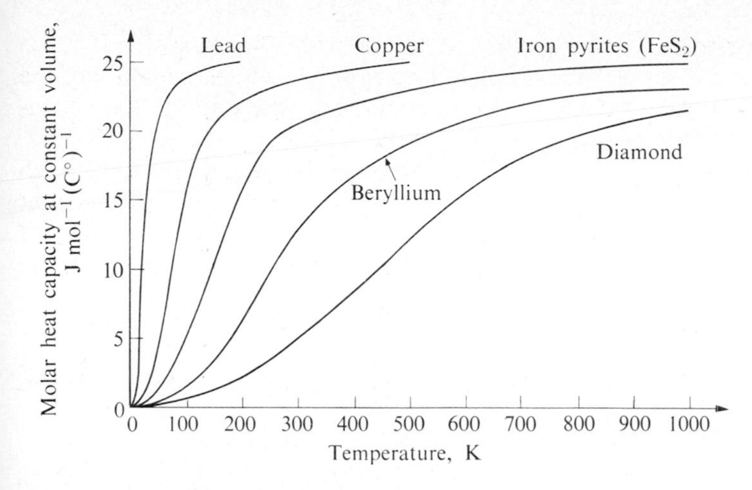

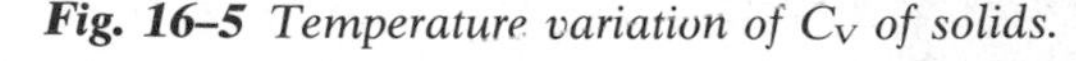
***Fig. 16–5** Temperature variation of C_V of solids.*

by two French physicists, Dulong and Petit. The average molar heat capacities (at constant pressure) for all metals except the very lightest are approximately the same and equal to about 25 J mol^{-1} (C°)$^{-1}$. This is known as the *Dulong and Petit law*; although only an approximate rule, it contains the germ of a very important idea. We know from chemistry that the number of molecules in one mole is the same for all substances. It follows that (nearly) the same amount of heat is required *per molecule* to raise the temperature of each of these metals by a given amount, even though the *mass* of a molecule of lead (for example) is nearly 10 times as great as that of a molecule of aluminum. To put it in a different way, the heat required to raise the temperature of a sample of metal depends only on *how many* molecules the sample contains, and not on the *mass* of an individual molecule. This is the first time in our study of physics that we have met a property of matter so directly related to its molecular structure.

The really important theoretical interpretation of heat-capacity measurements can be made only when the theoretician has at his disposal the complete temperature dependence, from the lowest to the highest possible temperatures, of the *molar heat capacity at constant volume,* C_V, because it is this quantity that is directly connected with the molecular energy which, in turn, may be calculated by statistical methods. Fortunately, there are relations which permit obtaining values of C_V from experimental values of C_P. Figure 16–5 shows the resulting values of C_V and their temperature variation from about 4 K to 1000 K. These curves represent the extremes of behavior. All other metals and nonmetals lie within the boundaries of lead and diamond. Except for certain anomalies that are too complicated to dwell upon, all such curves approach about 6 cal mole^{-1} K^{-1} as the temperature approaches infinity. The Dulong and Petit value, therefore, is approached by all solids, lead arriving at this value at about 200 K (below room temperature) and diamond requiring well over 2000 K.

Each curve in Fig. 16–5 has the same shape; it starts at zero, rises rapidly at first, then bends over and approaches the value 25 J mol^{-1} K^{-1}. It was shown by Debye that the behavior of the molar heat capacity of *nonmetals* over the entire temperature range can be accounted for by the *vibrational motion of the molecules occupying regular positions in the crystal lattice.* There are, however, many substances whose temperature variation of heat capacity cannot be explained entirely on the basis of lattice vibrations. The low-temperature behavior of four "abnormal" types of substance is depicted in Fig. 16–6. There are many more different types of heat capacity curves; each of them indicates some special molecular or atomic or ionic property of the particles that occupy the regular positions within the crystal lattice. The theoretical understanding of heat capacity curves is among the most interesting areas of modern solid-state physics.

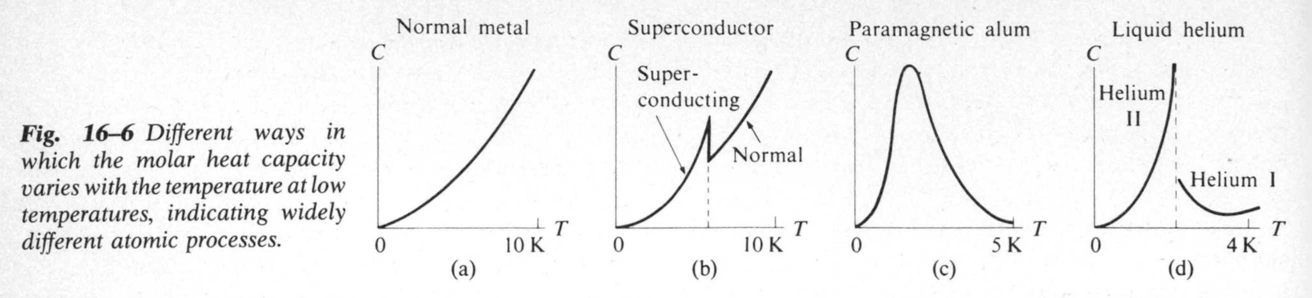

Fig. 16–6 Different ways in which the molar heat capacity varies with the temperature at low temperatures, indicating widely different atomic processes.

16–6 CHANGE OF PHASE

The term *phase* as used here relates to the fact that matter exists either as a solid, liquid, or gas. Thus the chemical substance H_2O exists in the *solid phase* as ice, in the *liquid phase* as water, and in the *gaseous phase* as steam. Provided they do not decompose at high temperatures, all substances can exist in any of the three phases under the proper conditions of temperature and pressure. Transitions from one phase to another are accompanied by the absorption or liberation of heat and usually by a change in volume, even when the transition occurs at constant temperature.

As an illustration, suppose that ice is taken from a freezer where its temperature was, say, −25°C. Let the ice be crushed quickly, placed in a container, and a thermometer inserted in the mass. Imagine the container to be surrounded by a heating coil which supplies heat to the ice at a uniform rate, and suppose that no other heat reaches the ice. The temperature of the ice would be observed to increase steadily, as shown by the portion of the graph (Fig. 16–7) from *a* to *b*, or until the temperature has risen to 0°C. In this temperature range the specific heat capacity of ice is approximately 2.3 J g^{-1} $(C°)^{-1}$ or 0.55 cal g^{-1} $(C°)^{-1}$. As soon as this temperature is reached, some liquid water will be observed in the container. In other words, the ice begins to *melt*. The melting process is a *change of phase*, from the solid phase to the liquid phase. The thermometer, however, will show no *increase in temperature*, and even though heat is being supplied at the same rate as before, the temperature remains at 0°C until all the ice is melted (point *c*, Fig. 16–7), if the pressure is maintained constant at one atmosphere.

As soon as the last of the ice has melted, the temperature begins to rise again at a uniform rate (from *c* to *d*, Fig. 16–7) although this rate will be slower than that from *a* to *b* because the specific heat of water is greater than that of ice. When a temperature of 100°C is reached (point *d*), bubbles of steam (gaseous water or water vapor) start to escape from the liquid surface; the water begins to *boil*. The temperature remains constant at 100°C (at constant atmospheric pressure) until all the water has boiled away. Another change of phase has therefore taken place, from the liquid phase to the gaseous phase.

If all the water vapor had been trapped and not allowed to diffuse away (a very large container would be needed), the heating process could be continued as from *e* to *f*. The gas would now be called "superheated steam."

Although water was chosen as an example in the process just described, the same type of curve as in Fig. 16–7 is obtained for many other substances. Some, of course, decompose before reaching a melting or boiling point, and others, such as glass or

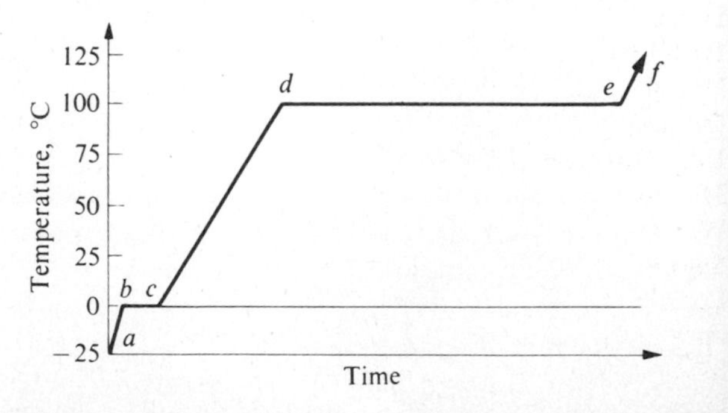

Fig. 16–7 The temperature remains constant during each change of phase, provided the pressure remains constant.

pitch, do not change phase at a definite temperature but become gradually softer as their temperature is raised. Crystalline substances, such as ice, or a metal, melt at a definite temperature. Glass and pitch behave like supercooled liquids of very high viscosity.

The quantity of heat per unit mass that must be supplied to a material at its melting point to convert it completely to a liquid *at the same temperature* is called the *heat of fusion* of the material. The quantity of heat per unit mass that must be supplied to a material at its boiling point to convert it completely to a gas at the same temperature is called the *heat of vaporization* of the material. Heats of fusion and vaporization are expressed in units of energy per unit mass, such as joules per kilogram, calories per gram, or Btu per pound-mass. Thus, the heat of fusion of ice is about 335,000 J kg^{-1}, 80 cal g^{-1}, or 144 Btu lb^{-1}. The heat of vaporization of water at 100°C is 2.26×10^6 J kg^{-1}, 539 cal g^{-1}, or 970 Btu lb^{-1}.

Some heats of fusion and vaporization are listed in Table 16–2. These quantities of heat are sometimes called *latent* heats because they change the phase of the material but not its temperature. This term is somewhat redundant and will not be used here.

When heat is removed from a gas, its temperature falls; at the same temperature at which it boiled, it returns to the liquid phase, or *condenses.* In so doing it gives up to its surroundings the same quantity of heat which was required to vaporize it. The heat so given up, per unit mass, is called the *heat of condensation* and is equal to the heat of vaporization. Similarly, a liquid returns to the solid phase, or freezes, when cooled to the temperature at which it melted, and gives up heat called *heat of solidification* exactly equal to the heat of fusion. Thus the melting point and the freezing point are at the same temperature, and the boiling point and condensation point are at the same temperature.

Under some conditions a material can be cooled below the normal phase-change temperature without a phase change occurring. The resulting state is unstable and is described as *supercooled.* Very pure water can be cooled several degrees below the normal freezing point under ideal conditions: when a small ice crystal is dropped in or the water is agitated, it crystallizes very quickly. Similarly, supercooled water vapor condenses quickly into fog droplets when a disturbance, such as dust particles or ionizing radiation, is introduced. This phenomenon is used in the *cloud chamber,* where charged particles such as protons and electrons induce condensation of supercooled vapor along their path, thus making the path visible.

Table 16–2 HEATS OF FUSION AND VAPORIZATION

Substance	Normal melting point		Heat of fusion, J g^{-1}	Normal boiling point		Heat of vaporization, J g^{-1}
	K	°C		K	°C	
Helium	3.5	−269.65	5.23	4.216	−268.93	20.9
Hydrogen	13.84	−259.31	58.6	20.26	−252.89	452
Nitrogen	63.18	−209.97	25.5	77.34	−195.81	201
Oxygen	54.36	−218.79	13.8	90.18	−182.97	213
Ethyl alcohol	159	−114	104.2	351	78	854
Mercury	234	−39	11.8	630	357	272
Water	273.15	0.00	333	373.15	100.00	2256
Sulfur	392	119	38.1	717.75	444.60	326
Lead	600.5	327.3	24.5	2023	1750	871
Antimony	903.65	630.50	165	1713	1440	561
Silver	1233.95	960.80	88.3	2466	2193	2336
Gold	1336.15	1063.00	64.5	2933	2660	1578
Copper	1356	1083	134	1460	1187	5069

Whether a substance, at its melting point, is freezing or melting depends on whether heat is being supplied or removed. That is, if heat is supplied to a beaker containing both ice and water at 0°C, some of the ice will melt; if heat is removed, some of the water will freeze; the temperature in either case remains at 0°C so long as both ice and water are present. If heat is *neither supplied nor removed*, no change at all takes place and the relative amounts of ice and water, and the temperature, all remain constant. Such a system is said to be in *phase equilibrium.* This furnishes another point of view regarding the melting point. That is, the melting (or freezing) point of a substance is *that temperature at which both the liquid and solid phases can exist together.* At any higher temperature, the substance can be only a liquid; at any lower temperature, it can be only a solid.

The general term *heat of transformation* is applied to both heats of fusion and heats of vaporization, and both are designated by the letter L. Since L represents the heat absorbed or liberated in the change of phase of unit mass, the heat Q absorbed or liberated in the change of phase of a mass m is

$$\boxed{Q = mL.} \tag{16–5}$$

The household steam-heating system makes use of a boiling–condensing process to transfer heat from the furnace to the radiators. Each kilogram of water which is turned to steam in the furnace absorbs 2.26×10^6 J (the heat of vaporization of water) from the furnace, and gives up this amount when it condenses in the radiators. (This figure is correct if the steam pressure is 1 atm. It is slightly smaller at higher pressures.) Thus the steam-heating system does not need to circulate as much water as a hot-water heating system. If water leaves a hot-water furnace at 60°C and returns at 40°C, dropping 20°C, about 27 kg of water must circulate to carry the same heat as is carried in the form of heat of vaporization by 1 kg of steam.

The temperature-control mechanisms of many warm-blooded animals operate on a similar principle. When the hypothalamus detects a slight rise in blood temperature, the sweat glands are activated. As sweat (chiefly water) evaporates from the surface of the body, it removes heat from the body as heat of vaporization. The heat is brought to the surface of the skin by the blood in the vascular system. This plays the same role as the pipes connecting the furnace to the radiators, with the heart playing the role of the circulating pump in a forced-circulation hot-water or steam system.

Under proper conditions of temperature and pressure, a substance can change directly from the solid to the gaseous phase without passing through the liquid phase. The transfer from solid to vapor is called *sublimation*, and the solid is said to *sublime.* "Dry ice" (solid carbon dioxide) sublimes at atmospheric pressure. Liquid carbon dioxide cannot exist at a pressure lower than about 5×10^5 Pa (about 5 atm). Heat is absorbed in the process of sublimation, and liberated in the reverse process. The quantity of heat per unit mass is called the *heat of sublimation.*

Definite quantities of heat are involved in chemical reactions. Perhaps the most familiar are those associated with combustion; complete combustion of one gram of gasoline produces about 46,000 J or about 11,000 cal, and the *heat of combustion* of gasoline is said to be 46,000 J g^{-1} or 46×10^6 J kg^{-1}. Energy values of foods are defined similarly; the unit of food energy, although called a calorie, is really a *kilocalorie*, equal to 1000 cal, or 4186 J. When we say that a gram of peanut butter "contains" 12 calories, we mean that, when it reacts with oxygen, with the help of enzymes, to convert the carbon and hydrogen completely to CO_2 and H_2O, the total energy liberated as heat is 12,000 cal, or 50,200 J. Not all of this energy is directly useful for mechanical work; the matter of efficiency of utilization of energy will be discussed in detail in Chapter 19.

16–7 EXAMPLES

As indicated above, the basic principle in calculations involving quantity of heat is that, when heat flow occurs between two bodies in thermal contact, the amount of heat lost by one body must equal that gained by the other. The following examples illustrate this principle in the context of phenomena discussed in this chapter.

Example 1 A copper cup of mass 0.1 kg, initially at 20°C, is filled with 0.2 kg of coffee initially at 70°C. What is the final temperature after the coffee and cup attain thermal equilibrium?

Let the final temperature be T. The heat lost by the coffee (which is assumed to have a specific heat capacity equal to that of water) is

$$Q = (0.2\ \text{kg})(4186\ \text{J kg}^{-1}\ (\text{C}°)^{-1})(70°\text{C} - T).$$

Similarly, the heat gained by the copper cup is

$$Q = (0.1\ \text{kg})(390\ \text{J kg}^{-1}\ (\text{C}°)^{-1})(T - 20°).$$

Equating these two quantities of heat yields an algebraic equation for T; solution of this equation gives $T = 67.8°\text{C}$. The final temperature is much closer to the initial temperature of the coffee than that of the cup because of the much larger specific heat capacity of water.

Example 2 How much ice at −20°C must be dropped into 0.25 kg of water, initially at 20°C, in order for the final temperature to be 0°C with the ice all melted? The heat capacity of the container may be neglected.

The heat lost by the water is

$$\begin{aligned} Q &= (0.25\ \text{kg})(4186\ \text{J kg}^{-1}\ (\text{C}°)^{-1})(20°\text{C} - 0°\text{C}) \\ &= 20{,}930\ \text{J}. \end{aligned}$$

The specific heat of ice is 2302 J kg^{-1} (C°)$^{-1}$. Let the mass of ice be m; then the heat needed to heat it from −20°C to 0°C is

$$\begin{aligned} Q &= m(2302\ \text{J kg}^{-1}\ (\text{C}°)^{-1})(0°\text{C} - (-20°\text{C})) \\ &= m(46{,}046\ \text{J kg}^{-1}). \end{aligned}$$

The additional heat needed to melt the ice is the latent heat of fusion times the mass:

$$Q = m(335{,}000\ \text{J kg}^{-1}).$$

The sum of these two quantities must equal the heat lost by the water:

$$m(381{,}000\ \text{J kg}^{-1}) = 20{,}930\ \text{J},$$

from which $m = 0.055$ kg $= 55$ g. This is roughly equivalent to two medium-size ice cubes.

Example 3 A certain gasoline camping lantern emits as much light as a 25-watt electric light bulb. Assuming the efficiency of conversion of heat into light is the same for the lantern and the bulb (which is not actually correct), how much gasoline does the lantern burn in 10 hours?

The rate of energy conversion in the lantern is 25 W, so in 10 hours (36,000 s), the total energy needed is

$$(25\ \text{J s}^{-1})(36{,}000\ \text{s}) = 0.9 \times 10^6\ \text{J}.$$

As mentioned in Section 16–6, combustion of one gram of gasoline produces 46,000 J, so the mass of gasoline required is

$$\frac{0.9 \times 10^6\ \text{J}}{46{,}000\ \text{J g}^{-1}} = 19.5\ \text{g}.$$

Actual lanterns are much less efficient than this, and typically require of the order of 300 to 400 g of gasoline (roughly one pint) to operate for ten hours.

PROBLEMS

[*Note.* When no information is given about the temperature variation of specific or molar heat capacity, assume it to be constant and use the average value given in Table 16–1.]

16–1 A combustion experiment is performed by burning a mixture of fuel and oxygen in a constant-volume "bomb" surrounded by a water bath. During the experiment the temperature of the water is observed to rise. Regarding the mixture of fuel and oxygen as the system, (a) has heat been transferred? (b) has work been done?

16–2 A liquid is irregularly stirred in a well-insulated container and thereby undergoes a rise in temperature. Regarding the liquid as the system, (a) has heat been transferred? (b) has work been done?

16–3 An automobile of mass 1500 kg is travelling at 5 m s^{-1}. How many calories are transferred in the brake mechanism when it is brought to rest?

16–4 A copper vessel of mass 200 g contains 400 g of water. The water is heated by a friction device which dissipates mechanical energy, and it is observed that the temperature of the system rises at the rate of 3 C° min^{-1}.

Neglect heat losses to surroundings. What power in watts is dissipated in the water?

16–5 How long could a 2000-hp motor be operated on the heat energy liberated by 1 mi^3 of ocean water when the temperature of the water is lowered by 1 C° if all this heat were converted to mechanical energy? Why do we not utilize this tremendous reservoir of energy?

16–6

a) A certain house burns 10 tons of coal in a heating season. The heat of combustion of the coal is 11,000 Btu lb^{-1}. If stack losses are 15%, how many Btu were actually used to heat the house?

b) In some localities large tanks of water are heated by solar radiation during the summer and the stored energy is used for heating during the winter. Find the required dimensions of the storage tank, assuming it to be a cube, to store a quantity of energy equal to that computed in part (a). Assume that the water is raised to 120°F in summer and cooled to 80°F in the winter.

16–7 An artificial satellite, constructed of aluminum, encircles the earth at a speed of 9000 m s^{-1}.

a) Find the ratio of its kinetic energy to the energy required to raise its temperature by 600 C°. (The melting point of aluminum is 660°C.) Assume a constant specific heat capacity of 0.20 cal g^{-1} $(C°)^{-1}$.

b) Discuss the bearing of your answer on the problem of the re-entry of a satellite into the earth's atmosphere.

16–8 A calorimeter contains 100 g of water at 0°C. A 1000-g copper cylinder and a 1000-g lead cylinder, both at 100°C, are placed in the calorimeter. Find the final temperature if there is no loss of heat to the surroundings.

16–9

a) Compare the heat capacities (heat capacity is the heat per unit temperature change) of equal *masses* of water, copper, and lead.

b) Compare the heat capacities of equal *volumes* of water, copper, and lead.

16–10 An aluminum can of mass 500 g contains 117.5 g of water at a temperature of 20°C. A 200-g block of iron at 75°C is dropped into the can. Find the final temperature, assuming no heat loss to the surroundings.

16–11 A casting weighing 100 lb is taken from an annealing furnace where its temperature was 900°F and plunged into a tank containing 800 lb of oil at a temperature of 80°F. The final temperature is 100°F, and the specific heat capacity of the oil is 0.5 Btu lb^{-1} $(F°)^{-1}$. What was the specific heat capacity of the casting? Neglect the heat capacity of the tank itself and any heat losses.

16–12 A lead bullet, traveling at 350 m s^{-1}, strikes a target and is brought to rest. What would be the rise in temperature of the bullet if there were no heat loss to the surroundings?

16–13 A copper calorimeter (mass 300 g) contains 500 g of water at a temperature of 15°C. A 560-g block of copper, at a temperature of 100°C, is dropped into the calorimeter and the temperature is observed to increase to 22.5°C. Neglect heat losses to the surroundings. Find the specific heat capacity of copper.

16–14 A 50-g sample of a material, at a temperature of 100°C, is dropped into a calorimeter containing 200 g of water initially at 20°C. The calorimeter is of copper and its mass is 100 g. The final temperature of the calorimeter is 22°C. Compute the specific heat capacity of the sample.

16–15 In Fig. 16–8 an electric heater is shown whose purpose is to provide a continuous supply of hot water. Water is flowing at the rate of 300 g min^{-1}, the inlet thermometer registers 15°C, the voltmeter reads 120 V and the ammeter 10 amp.

a) When a steady state is finally reached, what is the reading of the outlet thermometer?

b) Why is it unnecessary to take into account the heat capacity (mc or nC) of the apparatus itself?

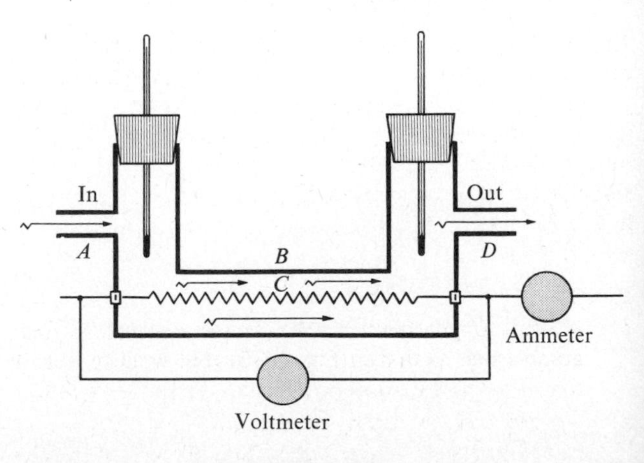

Fig. 16–8

16–16 An automobile engine whose output is 40 hp uses 4.5 gallons of gasoline per hour. The heat of combustion is 12×10^7 joules per gallon. What is the efficiency of the engine? That is, what fraction of the heat of combustion is converted to mechanical work?

16–17 The electric power input to a certain electric motor is 0.5 kW, and the mechanical power output is 0.54 hp.

a) What is the efficiency of the motor? (That is, what fraction of the electrical power is converted to mechanical power?)

b) How many calories are developed in the motor in one hour of operation, assuming all electrical energy not converted to mechanical energy is converted to heat?

16–18 The molar heat capacity at constant pressure of a certain substance varies with temperature according to the empirical equation

$$C_p = 27.2\ \text{J mol}^{-1}\ \text{K}^{-1} + (4 \times 10^{-3}\ \text{J mol}^{-1}\ \text{K}^{-2})T.$$

How much heat is necessary to change the temperature of 10 moles from 27°C to 527°C?

16–19 At very low temperatures, the molar heat capacity of rock salt varies with the temperature according to "Debye's T^3 law"; thus

$$C = k\frac{T^3}{\Theta^3},$$

where $k = 1940\ \text{J mol}^{-1}\ \text{K}^{-1}$ and $\Theta = 281$ K.

a) How much heat is required to raise the temperature of 2 moles of rock salt from 10 K to 50 K?

b) What is the mean molar heat capacity in this range?

c) What is the true molar heat capacity at 50 K?

16–20 Figure 16–9 shows a sketch of a continuous-flow calorimeter used to measure the heat of combustion of a gaseous fuel. Water is supplied at the rate of 12.5 lb min^{-1} and natural gas at 0.020 ft^3 min^{-1}. In the steady state, the inlet and outlet thermometers register 60°F and 76°F, respectively. What is the heat of combustion of natural gas in Btu ft^{-3}? Why should the gas flow be made as small as possible?

16–21

a) Make a rough sketch of the type of temperature–time graph (such as that in Fig. 16–3) that would result if the surroundings were much *cooler* than the sample.

b) An electrical resistor is immersed in a liquid and electrical energy is dissipated for 100 s at a constant rate of 50 W. The mass of the liquid is 530 g, and its temperature increases from 17.64°C to 20.77°C. Find the mean specific heat capacity of the liquid in this temperature range.

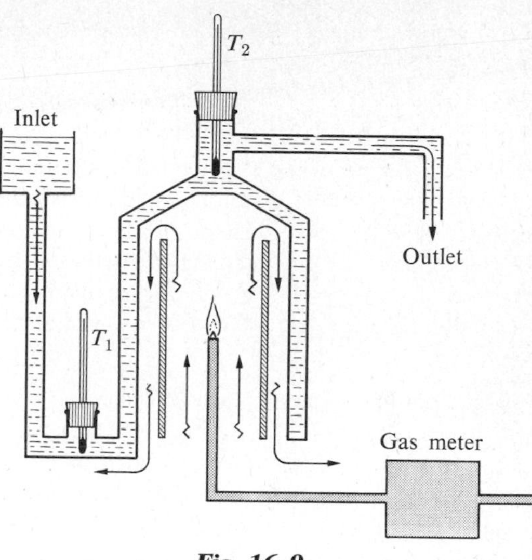

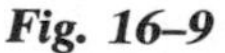

Fig. 16–9

16–22 How much heat is required to convert 1 g of ice at −10°C to steam at 100°C?

16–23 A beaker of very small mass contains 500 g of water at a temperature of 80°C. How many grams of ice at a temperature of −20°C must be dropped in the water so that the final temperature of the system will be 50°C?

16–24 An open vessel contains 500 g of ice at −20°C. The mass of the container can be neglected. Heat is supplied to the vessel at the constant rate of 1000 cal min^{-1} for 100 min. Plot a curve showing the elapsed time as abscissa and the temperature as ordinate.

16–25 A copper calorimeter of mass 100 g contains 150 g of water and 8 g of ice in thermal equilibrium at atmospheric pressure. 100 g of lead at a temperature of 200°C are dropped into the calorimeter. Find the final temperature if no heat is lost to the surroundings.

16–26 500 g of ice at −16°C are dropped into a calorimeter containing 1000 g of water at 20°C. The calorimeter can is of copper and has a mass of 278 g. Compute the final temperature of the system, assuming no heat losses.

16–27 A tube leads from a flask in which water is boiling under atmospheric pressure to a calorimeter. The mass of the calorimeter is 150 g, its heat capacity is 15 cal (C°)$^{-1}$ and it contains originally 340 g of water at 15°C. Steam is allowed to condense in the calorimeter until its temperature increases to 71°C, after which the total mass of calorimeter and contents is found to be 525 g. Compute the heat of condensation of steam from these data.

16–28 An aluminum canteen whose mass is 500 g contains 750 g of water and 100 g of ice. The canteen is dropped from an aircraft to the ground. After landing, the temperature of the canteen is found to be 25°C. Assuming that no energy is given to the ground in the impact, what was the velocity of the canteen just before it landed? State any additional assumptions needed.

16–29 A calorimeter contains 500 g of water and 300 g of ice, all at a temperature of 0°C. A block of metal of mass 1000 g is taken from a furnace where its temperature was 240°C and is dropped quickly into the calorimeter. As a result, all the ice is just melted. What would the final temperature of the system have been if the mass of the block had been twice as great? Neglect heat loss from and the heat capacity of the calorimeter.

16–30 An ice cube whose mass is 50 g is taken from a refrigerator where its temperature was −10°C and is dropped into a glass of water at 0°C. If no heat is gained or lost from outside, how much water will freeze onto the cube?

16–31 A copper calorimeter can (mc = 30 cal deg^{-1}) contains 50 g of ice. The system is initially at 0°C. 12 g of steam at 100°C and 1 atm pressure are run into the calorimeter. What is the final temperature of the calorimeter and its contents?

16–32 A vessel whose walls are thermally insulated contains 2100 g of water and 200 g of ice, all at a temperature of 0°C. The outlet of a tube leading from a boiler, in which water is boiling at atmospheric pressure, is inserted in the water. How many grams of steam must condense to raise the temperature of the system to 20°C? Neglect the heat capacity of the container.

16–33 A 2-kg iron block is taken from a furnace where its temperature was 650°C and placed on a large block of ice at 0°C. Assuming that all the heat given up by the iron is used to melt the ice, how much ice is melted?

16–34 In a household hot-water heating system, water is delivered to the radiators at 140°F and leaves at 100°F. The system is to be replaced by a steam system in which steam at atmospheric pressure condenses in the radiators, the condensed steam leaving the radiators at 180°F. How many pounds of steam will supply the same heat as was supplied by 1 lb of hot water in the first installation?

16–35 A "solar house" has storage facilities for 4 million Btu. Compare the space requirements for this storage on the assumption (a) that the heat is stored in water heated from a minimum temperature of 80°F to a maximum at 120°F, and (b) that the heat is stored in Glauber salt ($Na_2SO_4 \cdot 10H_2O$) heated in the same temperature range.

PROPERTIES OF GLAUBER SALT

Specific heat capacity	
Solid	0.46 Btu lb^{-1} °)$^{-1}$
Liquid	0.68 Btu lb^{-1} °)$^{-1}$
Specific gravity	1.6
Melting point	90°F
Heat of fusion	104 Btu lb^{-1}

16–36 The nominal food-energy value of butter is about 6 kcal g^{-1}, where 1 kcal = 1000 cal = 41 6 J. If all this energy could be converted completely to mechanical energy, how much butter would be required to power an 80-kg mountaineer on his journey from Lupine Meadows (elevation 2070 m) to the summit of Grand Teton (4196 m)?

16–37 The capacity of air conditioners is typically expressed in Btu hr^{-1} or "tons," the latter being the number of tons of ice that can be frozen from water at 0° C in 24 hr by the unit.

a) What is the rating in Btu/hr of a one-ton air conditioner?

b) Express the capacity of a one-ton air conditioner in watts.

16–38 A 6000-Btu-hr^{-1} air conditioner typically consumes about 800 W of electrical power. What is the ratio of the rate of removal of heat to the rate of consumption of electrical energy? Express the two rates in the same units to obtain a dimensionless ratio.

16–39 Suppose 1 liter of gasoline will propel a car a distance of 10 km. The density of gasoline is about 0.7 g cm^{-3}, and its heat of combustion is about 4.6×10^4 J g^{-1}.

a) If the engine is 25% efficient, that is, if 1/4 of the heat of combustion is converted into useful mechanical work, what total work does the engine do during the 10-km displacement?

b) If this work is assumed to be done in opposing a constant resisting force F, find the magnitude of F.

Chapter 17

Transfer of Heat

17–1 CONDUCTION

If one end of a metal rod is placed in a flame while the other is held in the hand, that part of the rod one is holding will be felt to become hotter and hotter, although it was not itself in direct contact with the flame. Heat is said to reach the cooler end of the rod by *conduction* along or through the material of the rod. The molecules at the hot end of the rod increase the violence of their vibration as the temperature of the hot end increases. Then, as they collide with their more slowly moving neighbors farther out on the rod, some of their energy of motion is shared with these neighbors and they in turn pass it along to those still farther out from the flame. Hence energy of thermal motion is passed along from one molecule to the next, while each individual molecule remains at its original position.

It is well known that metals are good conductors of electricity and also good conductors of heat. The ability of a metal to conduct an electric current is due to the fact that there are within it so-called "free" electrons, that is, electrons that have become detached from their parent molecules. The free electrons also play a part in the conduction of heat, and the reason metals are such good heat conductors is that the free electrons provide an effective mechanism for carrying thermal energy from the hotter to the cooler portions of the metal.

Conduction of heat can take place in a body only when different parts of the body are at different temperatures, and the direction of heat flow is always from points of higher to points of lower temperature.

Figure 17–1 represents a rod of material of cross section A and thickness L. Let the whole of the left face of the rod be kept at a temperature T_2, and the whole of the right face at a lower temperature T_1. The direction of the heat current is then from left to right through the rod. The sides of the rod are covered by an insulating material, so no heat transfer occurs at the sides. This is of course an idealized model; all real materials, even the best thermal insulators, conduct heat to some extent.

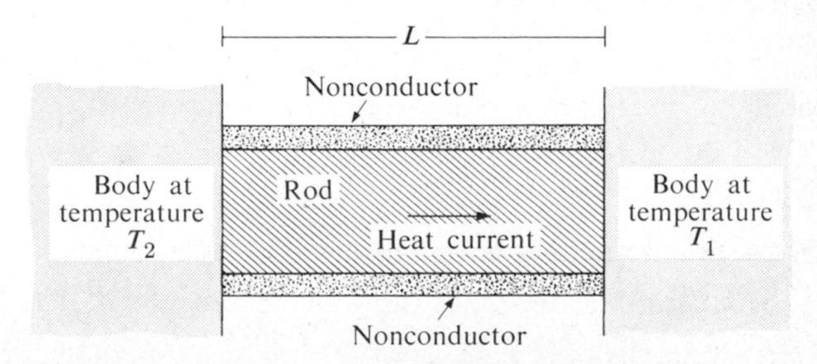

Fig. 17–1 *Steady-state heat flow in a uniform rod.*

After the ends of the rod have been kept at the temperatures T_1 and T_2 for a sufficiently long time, the temperature at points within the rod is found to decrease uniformly with distance from the hot to the cold face. At each point, however, the temperature remains constant with time. This condition is called "steady-state" heat flow.

Experiment shows that the rate of flow of heat through the rod in the steady state is proportional to the area A, proportional to the temperature difference $(T_2 - T_1)$, and inversely proportional to the length L. These proportions may be converted to an equation by introducing a constant k whose numerical value depends on the material of the rod. The quantity k is called the *thermal conductivity* of the material,

$$\boxed{H = \frac{kA(T_2 - T_1)}{L},} \qquad (17\text{–}1)$$

where H is the quantity of heat flowing through the rod per unit time, also called the *heat current.*

Equation (17–1) may also be used to compute the rate of heat flow through a slab, or *any* homogeneous body having a uniform cross-section perpendicular to the direction of flow, provided the flow has attained steady-state conditions and the ends are kept at constant temperatures.

When the cross-section is not uniform, or when steady state conditions are not present, the temperature does not necessarily change uniformly along the direction of heat flow. Equation (17–1) may still be applied to a thin layer of material perpendicular to the direction of flow. If x is the coordinate measured along the flow path, dx the thickness of the layer, and A the cross section area perpendicular to that path, then Eq. (17–1) may be rewritten

$$\boxed{H = -kA\frac{dT}{dx},} \qquad (17\text{–}2)$$

where dT is the temperature change across the distance dx. The negative sign is included because if the temperature *increases* in the direction of increasing x (dx and dT both positive) the direction of heat flow is the direction of *decreasing* x, and conversely. The temperature change per unit length dT/dx is called the *temperature gradient*; Eq. (17–1) refers to the special case where the temperature gradient is constant, and equal to $(T_2 - T_1)/L$.

The mks unit of rate of heat flow is the joule per second, although other energy units such as the calorie are sometimes used. The units of k are revealed by solving Eq. (17–2) for k:

$$k = -\frac{H}{A}\frac{dx}{dT}.$$

This shows that in mks units, the unit of k is

$$\frac{(1\ \text{J s}^{-1})(1\ \text{m})}{(1\ \text{m})^2(1\ \text{C}°)} = 1\ \text{J (s m C°)}^{-1}.$$

Values of thermal conductivity are often tabulated using cgs units, with the calorie as the energy unit. The unit of k is then 1 cal (s cm C°)$^{-1}$. The conversion is

$$1\ \text{cal (s cm C°)}^{-1} = 419\ \text{J (s m C°)}^{-1}.$$

The thermal conductivity of most materials is a function of temperature, increasing slightly with increasing temperature, but the variation is small and can often be neglected. Some numerical values of k, at temperatures near room temperature, are given in Table 17–1. The properties of materials used commercially as heat insulators are expressed in a mixed system in which the unit of heat current is 1 Btu hr^{-1}, the unit of area is 1 ft^2, and the unit of temperature gradient is one fahrenheit degree per *inch* (1 F° in^{-1})!

It is evident from Eq. (17–1) that the larger the thermal conductivity k, the larger the heat current, other factors being equal. A material for which k is large is therefore a good heat conductor, while if k is small, the material is a poor conductor or a good insulator. A "perfect heat conductor" ($k = \infty$) or a "perfect heat insulator" ($k = 0$) does not exist. However, it will be seen from Table 17–1 that the metals as a group have much greater thermal conductivities than the nonmetals, and that those of gases are extremely small.

Before steady-state conditions are reached, the temperature at each point may change with time. For

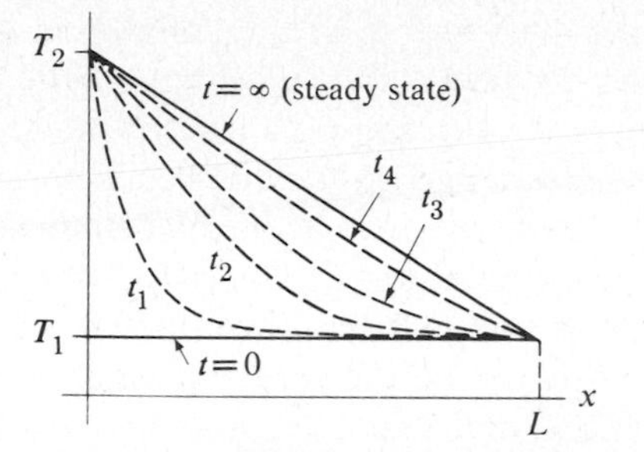

Fig. 17–2 Transient and steady-state temperature distribution along a rod initially at temperature T_1. The transient distributions were computed for $t_2 = 5t_1$, $t_3 = 10t_1$, $t_4 = 20t_1$.

example, suppose the rod in Fig. 17–1 is initially at the temperature T_1 throughout, and then at time $t = 0$ the ends are placed in contact with the bodies of temperature T_1 and T_2. The resulting temperature distributions at various times are shown in Fig. 17–2.

Table 17–1 THERMAL CONDUCTIVITIES

	k, J s^{-1} m^{-1} (C°)$^{-1}$	k, cal s^{-1} cm^{-1} (C°)$^{-1}$
Metals		
Aluminum	205	0.49
Brass	109	0.26
Copper	385	0.92
Lead	34.7	0.083
Mercury	8.3	0.020
Silver	406	0.97
Steel	50.2	0.12
Various solids (Representative values)		
Insulating brick	0.15	0.00035
Red brick	0.6	0.0015
Concrete	0.8	0.002
Cork	0.04	0.0001
Felt	0.04	0.0001
Glass	0.8	0.002
Ice	1.6	0.004
Rock wool	0.04	0.0001
Wood	0.12–0.04	0.0003–0.0001
Gases		
Air	0.024	0.000057
Argon	0.016	0.000039
Helium	0.14	0.00034
Hydrogen	0.14	0.00033
Oxygen	0.023	0.000056

At time t_1 the temperature changes much more rapidly with x near the left end than near the right, corresponding to the fact that the heat flow is much greater near that end to provide a net flow of heat into the middle region and raise its temperature. With increasing time the differences become smaller and smaller, until after a very long time ($t = \infty$) the steady-state uniform temperature distribution is attained. It is a remarkable fact that the final steady state is completely independent of the initial conditions at $t = 0$, and depends only on the temperatures at the ends.

In the steady state the heat current in the rod is the same in all cross sections; otherwise there would be a net flow of heat into certain regions and out of others and their temperatures would change, contradicting the assumption of a steady state. Thus steady-state heat flow is comparable to the flow of an incompressible fluid.

Since there are no restrictions on the relative dimensions of the rod in Fig. 17–1, Eq. (17–1) applies also when the length L becomes relatively small and the area A relatively large, as in the case of a sheet or slab of material such as a portion of the wall of a house, a refrigerator, or a furnace. (The dimension L would then ordinarily be described as the "thickness" rather than the "length.")

17–2 RADIAL HEAT FLOW IN A SPHERE OR CYLINDER

We next consider two examples of heat flow in which the temperature gradient is not uniform along the direction of flow, even in the steady state. Figure 17–3 represents a steam pipe surrounded by a layer

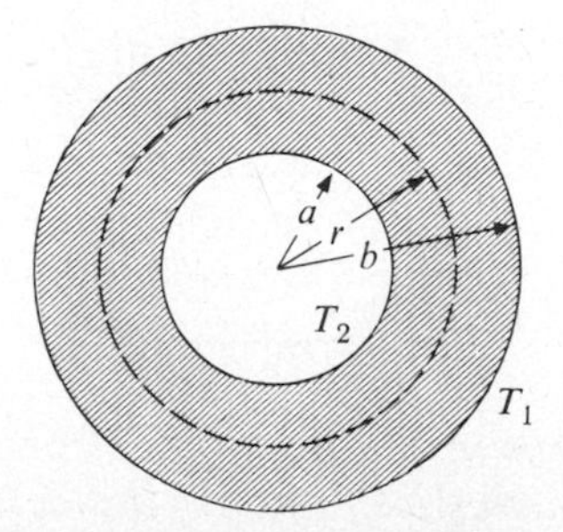

Fig. 17–3 Radial heat flow in a cylinder or sphere.

Table 17–2

Cylinder (length L)	Sphere
$A = 2\pi rL$	$A = 4\pi r^2$
$H = -k(2\pi rL)\dfrac{dT}{dr}$	$H = -k(4\pi r^2)\dfrac{dT}{dr}$
$H\dfrac{dr}{r} = -2\pi kL\,dT$	$H\dfrac{dr}{r^2} = -4\pi k\,dT$
$H\ln r = -2\pi kLT + C$	$-\dfrac{H}{r} = -4\pi kT + C$
$r = a, \quad T = T_2;$	$r = b, \quad T = T_1$
$H = \dfrac{2\pi k(T_2 - T_1)}{[\ln(b/a)]/L}$	$H = \dfrac{4\pi k(T_2 - T_1)}{(b - a)/ab}$

of insulating material, or a sphere surrounded by a spherical shell of insulation. Let T_2 and T_1 be the temperatures of the inner and outer surfaces of the insulation, and a and b be the inner and outer radii. If $T_2 > T_1$, heat flows outward, and in the steady state the heat current H is the same across all surfaces within the insulation, such as that of radius r, shown by a dashed circle. If A is the area of this surface and dT/dr the corresponding temperature gradient,

$$H = -kA\frac{dT}{dr} = \text{constant.}$$

The heat current can be expressed in terms of T_2, T_1, a, and b, by integrating this equation and inserting the appropriate boundary conditions. The calculations for the cylinder and the sphere have been carried out in parallel in Table 17–2.

17–3 CONVECTION

The term *convection* is applied to the transfer of heat from one place to another by the actual motion of hot material. The hot-air furnace, the hot-water heating system, and the flow of blood in the body are examples. If the heated material is forced to move by a blower or pump, the process is called *forced convection*; if the material flows due to differences in density, the process is called *natural* or *free convection.* To understand the latter, consider a U-tube, as illustrated in Fig. 17–4.

In (a), the water is at the same temperature in both arms of the U and hence stands at the same level in each. In (b), the right side of the U has been heated. The water in this side expands and therefore has smaller density; thus a longer column is needed to balance the pressure produced by the cold water in the left column. When the stopcock is opened, water flows from the top of the warmer column into the colder column. This increases the pressure at the bottom of the U produced by the cold column, and decreases the pressure at this point due to the hot column. Hence at the bottom of the U, water is forced from the cold to the hot side. If heat is continually applied to the hot side and removed from the cold side, the circulation continues of itself. The net result is a continual transfer of heat from the hot to the cold side of the column. In the common household hot-water heating system, the "cold" side corresponds to the radiators and the "hot" side to the furnace.

The anomalous expansion of water which was mentioned in Chapter 15 has an important effect on the way in which lakes and ponds freeze in winter.

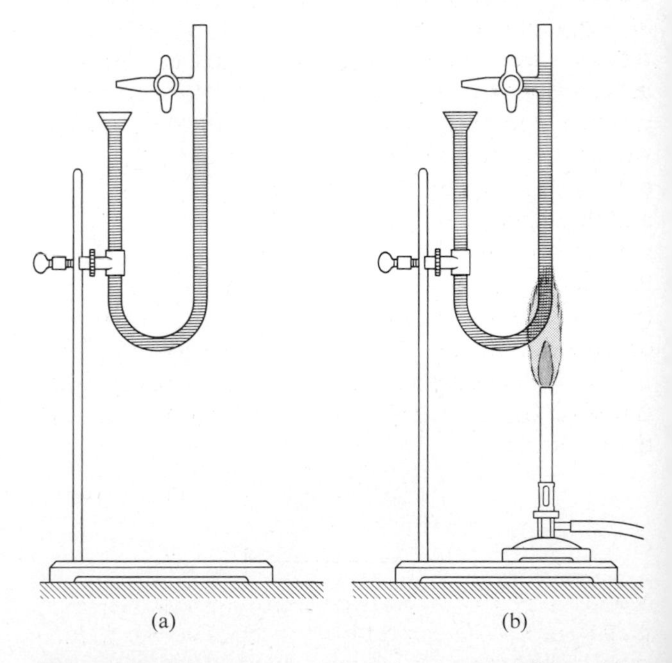

Fig. 17–4 *Convection is brought about by differences in density.*

Consider a pond at a temperature of, say 20°C throughout, and suppose the air temperature at its surface falls to −10°C. The water at the surface becomes cooled to, say, 19°C. It therefore contracts, becomes more dense than the warmer water below it, and sinks in this less dense water, its place being taken by water at 20°C. The sinking of the cooled water causes a mixing process, which continues until all of the water has been cooled to 4°C. Now, however, when the surface water cools to 3°C, it expands, is less dense than the water below it, and hence floats on the surface. Convection and mixing then cease, and the remainder of the water can lose heat only by *conduction.* Since water is an extremely poor heat conductor, cooling takes place very slowly after 4°C is reached, with the result that the pond freezes first at its surface. Then, since the density of ice is even smaller than that of water at 0°C, the ice floats on the water below it, and further freezing can result only from heat flow upward by conduction.

The mathematical theory of heat convection is quite involved. There is no simple equation for convection, as there is for conduction. This arises from the fact that the heat lost or gained by a surface at one temperature in contact with a fluid at another temperature depends on many circumstances, such as (1) whether the surface is flat or curved, (2) whether the surface is horizontal or vertical, (3) whether the fluid in contact with the surface is a gas or a liquid, (4) the density, viscosity, specific heat, and thermal conductivity of the fluid, (5) whether the velocity of the fluid is small enough to give rise to laminar flow or large enough to cause turbulent flow, (6) whether evaporation, condensation, or formation of scale takes place.

The procedure adopted in practical calculations is first to define a *convection coefficient h* by means of the equation

$$\boxed{H = hA\,\Delta T,} \tag{17–3}$$

where H is the heat convection current (the heat gained or lost by convection by a surface per unit of time), A is the area of the surface, and ΔT is the temperature difference between the surface and the main body of fluid. The next step is the determination of numerical values of h that are appropriate to a given piece of equipment. Such a determination is accomplished partly by dimensional analysis and partly by an elaborate series of experiments. An enormous amount of research in this field has been done in recent years so that, by now, there are in existence fairly complete tables and graphs from which the physicist or engineer may obtain the convection coefficient appropriate to certain standard types of apparatus.

A case of common occurrence is that of natural convection from a wall or a pipe which is at a constant temperature and is surrounded by air at atmospheric pressure whose temperature differs from that of the wall or pipe by an amount ΔT. The convection coefficients applicable in this situation are given in Table 17–3.

Example The air in a room is at a temperature of 25°C, and the outside air is at −15°C. How much heat is transferred per unit area of a glass windowpane of thermal conductivity 2.5×10^{-3} cgs units and of thickness 2 mm?

To assume that the inner surface of the glass is at 25°C and the outer surface is at −15°C is entirely erroneous, as anyone can verify by touching the inner surface of a glass windowpane on a cold day.

Table 17–3 COEFFICIENTS OF NATURAL CONVECTION IN AIR AT ATMOSPHERIC PRESSURE

Equipment	Convection coefficient h, cal s^{-1} $cm^{-2}(C°)^{-1}$
Horizontal plate, facing downward	$0.595 \times 10^{-4}(\Delta T)^{1/4}$
Horizontal plate, facing downward	$0.314 \times 10^{-4}(\Delta T)^{1/4}$
Vertical plate	$0.424 \times 10^{-4}(\Delta T)^{1/4}$
Horizontal or vertical pipe $\left(\begin{matrix}\text{diameter}\\ D\end{matrix}\right)$	$1.00 \times 10^{-4}\left(\dfrac{\Delta T}{D}\right)^{1/4}$

One must expect a much smaller temperature difference across the windowpane, so that in the steady state the rates of transfer of heat (1) by convection in the room, (2) by conduction through the glass, and (3) by convection in the outside air, are all equal.

As a first approximation in the solution of this problem, let us assume that the window is at a uniform temperature T. If $T = 5°\text{C}$, then the temperature difference between the inside air and the glass is the same as that between the glass and the outside air, or 20°C. Hence the convection coefficient in both cases is

$$\begin{aligned} h &= 0.424 \times 10^{-4}(20)^{1/4} \text{ cal s}^{-1}\text{ cm}^{-2}\,(\text{C}°)^{-1} \\ &= 0.895 \times 10^{-4} \text{ cal s}^{-1}\text{ cm}^{-2}\,(\text{C}°)^{-1}, \end{aligned}$$

and, from Eq. (17–3), the heat transferred per unit area is

$$\begin{aligned} \frac{H}{A} &= 0.895 \times 10^{-4} \times 20 \\ &= 17.9 \times 10^{-4} \text{ cal s}^{-1}\text{ cm}^{-2}. \end{aligned}$$

The glass, however, is *not* at uniform temperature; there must be a temperature difference ΔT across the glass sufficient to provide heat conduction at the rate of 17.9×10^{-4} cal s^{-1} cm^{-2}. Using the conduction equation, Eq. (17–2), we obtain

$$\begin{aligned} \Delta T &= \frac{L}{k} \times \frac{H}{A} = \frac{0.2}{2.5 \times 10^{-3}} \times 17.9 \times 10^{-4}\ \text{C}° \\ &= 0.14\ \text{C}°. \end{aligned}$$

With sufficient accuracy we may therefore say that the inner surface is at 5.07°C and the outer surface is at 4.93°C.

Heat transfer in the human body involves a combination of mechanisms, which together maintain a remarkably constant and uniform temperature despite large changes in environmental conditions. As mentioned above, the chief *internal* mechanism is forced convection, with the heart serving as the pump and the blood as the circulating fluid. Heat transfer with the surroundings involves conduction, convection, and radiation, in proportions depending on circumstances. Total heat loss from the body is on the order of 2000 to 5000 kcal per day, depending on amount of activity. A dry nude body in still air loses about half its heat by radiation, but under conditions of vigorous activity and copious perspiration, the dominant mechanism is evaporative cooling. Radiation will be discussed in the following sections.

17–4 RADIATION

When one's hand is placed in direct contact with the surface of a hot-water or steam radiator, heat reaches the hand by *conduction* through the radiator walls. If the hand is held above the radiator but not in contact with it, heat reaches the hand by way of the upward-moving *convection* currents of warm air. If the hand is held at one side of the radiator it still becomes warm, even though conduction through the air is negligible and the hand is not in the path of the convection currents. Energy now reaches the hand by *radiation.*

The term radiation refers to the continual emission of energy from the surface of all bodies. This energy is called *radiant energy* and is in the form of electromagnetic waves. These waves travel with the velocity of light and are transmitted through a vacuum as well as through air. (Better, in fact, since they are absorbed by air to some extent.) When they fall on a body which is not transparent to them, such as the surface of one's hand or the walls of the room, they are absorbed, resulting in a transfer of heat to the absorbing material.

The radiant energy emitted by a surface, per unit time and per unit area, depends on the nature of the surface and on its temperature. At low temperatures the rate of radiation is small and the radiant energy is chiefly of relatively long wavelength. As the temperature is increased, the rate of radiation increases very rapidly, in proportion to the fourth power of the absolute temperature. For example, a copper block at a temperature of 100°C (373 K) radiates about 0.03 J s^{-1} or 0.03 W from each square centimeter of its surface. At a temperature of 500°C (773 K), it radiates about 0.54 W from each square centimeter, and at 1000°C (1273 K), it radiates about 4 W per square centimeter. This rate is 130 times as great as that at a temperature of 100°C.

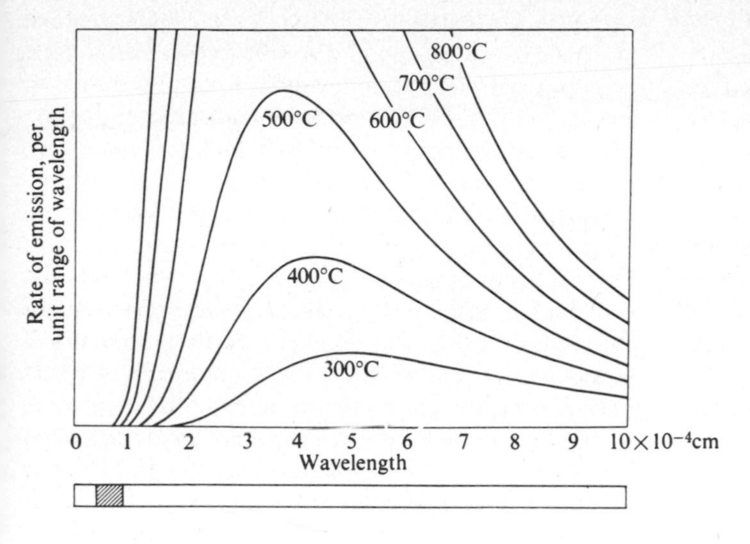

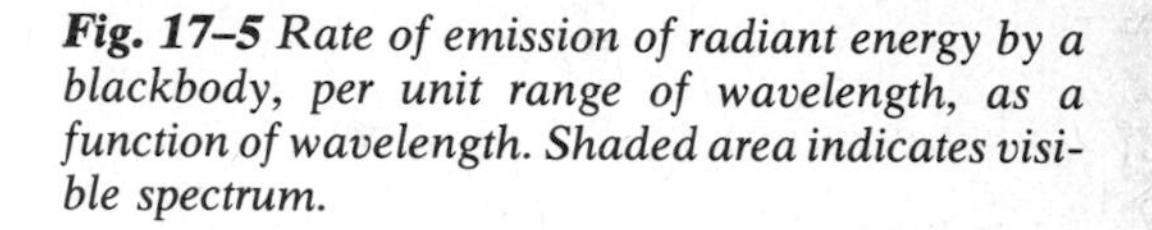
Fig. 17-5 *Rate of emission of radiant energy by a blackbody, per unit range of wavelength, as a function of wavelength. Shaded area indicates visible spectrum.*

At each of these temperatures the radiant energy emitted is a mixture of waves of different wavelengths. At a temperature of 300°C the most intense of these waves has a wavelength of about 5×10^{-4} cm; for wavelengths either greater or less than this value the intensity decreases, as shown by the curve in Fig. 17-5. The corresponding distribution of energy at higher temperatures is also shown in the figure. The area between each curve and the horizontal axis represents the *total* rate of radiation at that temperature. It is evident that this rate increases rapidly with increasing temperature, and also that the wavelength of the most intense wave shifts toward the left, or toward shorter wavelengths, with increasing temperature.

At a temperature of 300°C, practically all of the radiant energy emitted by a body is carried by waves longer than those corresponding to red light. Such waves are called *infrared*, meaning "beyond the red." At a temperature of 800°C a body emits enough visible radiant energy to be self-luminous and appears "red hot." By far the larger part of the energy emitted, however, is still carried by infrared waves. At 3000°C, which is about the temperature of an incandescent lamp filament, the radiant energy contains enough of the shorter wavelengths so that the body appears "white hot."

17-5 STEFAN'S LAW

Experimental measurements of the rate of emission of radiant energy from the surface of a body were made by John Tyndall (1820–1893), and on the basis of these Josef Stefan (1835–1893) concluded in 1879 that the rate of emission could be expressed by the relation

$$\boxed{R = e\sigma T^4} \tag{17-4}$$

which is *Stefan's law*. The quantity R, known as the *radiant emittance*, is equal to the rate of emission of radiant energy per unit area. In the mks system, R is expressed in watts per square meter; in the cgs system it is expressed in ergs per square centimeter per second. The constant σ is a universal physical constant called the *Stefan–Boltzmann constant*; its numerical value in mks units is

$$\boxed{\sigma = 5.6696 \times 10^{-8}\ \mathrm{W\,m^{-2}\,K^{-4}}.}$$

The quantity T is the Kelvin temperature of the surface, and e is a quantity called the emissivity of the surface. The emissivity lies between zero and unity, depending on the nature of the surface. The emissivity of copper, for example, is about 0.3. (Strictly speaking, the emissivity varies somewhat with temperature even for the same surface.) In general, the emissivity is larger for rough and smaller for smooth, polished surfaces.

It may be wondered why it is, if the surfaces of all bodies are continually emitting radiant energy, that all bodies do not eventually radiate away all of their internal energy and cool down to a temperature of absolute zero, where $R = 0$ by Eq. (17–4). The answer is that they *would* do so if energy were not supplied to them in some way. In the case of an electric heater element or the filament of an electric lamp, energy is supplied electrically to make up for the energy radiated. As soon as this energy supply is cut off, these bodies do, in fact, cool down very quickly to room temperature. The reason that they do not cool further is that their surroundings (the walls, and other objects in the room) are also radiating, and some of this radiant energy is intercepted, absorbed, and converted into internal energy. The same thing is true of all other objects in the room: each is both emitting and absorbing radiant energy simultaneously. When any object is hotter than its surroundings, the rate of emission exceeds its rate of absorption. There is thus a net loss of energy and the body cools down, unless heated by some other method. If a body is at a *lower* temperature than its surroundings, its rate of absorption is larger than its rate of emission and its temperature rises. When the body is at the same temperature as its surroundings, the two rates become equal, there is no net gain or loss of energy, and no change in temperature.

If a small body of emissivity e is completely surrounded by walls at a temperature T, the rate of *absorption* of radiant energy per unit area by the body is

$$R = e\sigma T^4.$$

Hence for such a body at a temperature T_1, surrounded by walls at a temperature T_2, the *net* rate of loss (or gain) of energy per unit area by radiation is

$$R_{\text{net}} = e\sigma T_1^4 - e\sigma T_2^4 = e\sigma(T_1^4 - T_2^4). \quad (17\text{–}5)$$

Infrared emission from a body can be studied using a camera equipped with infrared-sensitive film, or better with a special device similar in principle to a television camera, sensitive to infrared radiation. The resulting picture is called a *thermograph*; since emission depends on temperature, thermography permits detailed study of temperature distributions. Some present-day instrumentation is sensitive to differences as small as 0.1 C°.

Thermography has a variety of important medical applications. Local temperature variations in the body are associated with various tumors, such as breast cancers, and growths as small as 1 cm in diameter can be detected. Vascular disorders leading to local temperature anomalies can be studied, and many other examples could be cited.

17–6 THE IDEAL RADIATOR

Suppose that the walls of the enclosure in Fig. 17–6 are kept at temperature T_2 and a number of different bodies having different emissivities are suspended one after another within the enclosure. Regardless of their temperature when they are inserted, it is found that eventually each comes to the same temperature T_2 as that of the walls, even if the air in the enclosure is evacuated. If the bodies are small compared with the size of the enclosure, radiant energy from the walls strikes the surface of each body at the same rate. Of this energy, a part is reflected and the remainder absorbed. In the absence of any other process, the energy absorbed will raise the tempera-

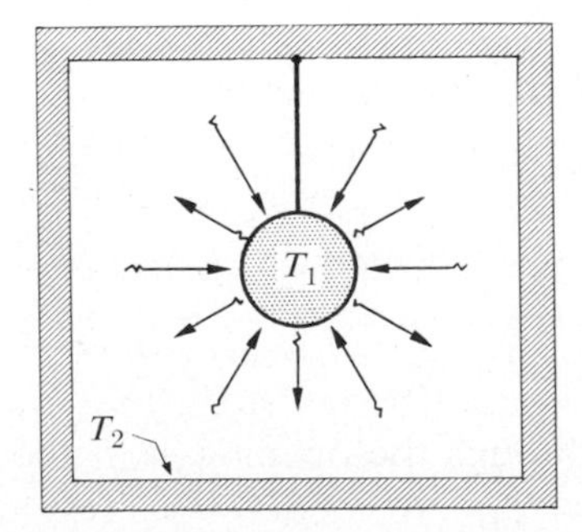

Fig. 17–6 *In thermal equilibrium, the rate of emission of radiant energy equals the rate of absorption. Hence a good absorber is a good emitter.*

ture of the absorbing body, but the temperature is observed *not* to change (once it has reached T_2), and we conclude that each body must *emit* radiant energy at the same rate as it *absorbs* it. Hence a good absorber is a good emitter, and a poor absorber is a poor emitter. But since each body must either absorb or reflect the radiant energy reaching it, a poor absorber must also be a good reflector. Hence a *good reflector* is a *poor emitter.*

This is the reason for silvering the walls of vacuum ("Thermos") bottles. A vacuum bottle is constructed with double glass walls, the space between the walls being evacuated so that heat flow by conduction and convection is practically eliminated. To reduce the radiant emission to as low a value as possible, the walls are covered with a coating of silver, which is highly reflecting and hence is a very poor emitter.

Since a good absorber is a good emitter, the *best* emitter is that surface which is the best absorber. But no surface can absorb more than all of the radiant energy which strikes it. Any surface which *does* absorb all of the incident energy is the best emitting surface possible. Such a surface would reflect no radiant energy, and hence would appear black in color (provided its temperature is not so high that it is self-luminous). It is called an *ideally black surface*, a body having such a surface is called an ideal blackbody, an ideal radiator, or simply a *blackbody.*

No actual surface is ideally black, the closest approach being lampblack, which reflects only about 1%. Blackbody conditions can be closely realized, however, by a small opening in the walls of a closed container. Radiant energy entering the opening is in part absorbed by the interior walls. Of the part reflected, only a very little escapes through the opening, the remainder being eventually absorbed by the walls. Hence the *opening* behaves like an ideal absorber.

Conversely, the radiant energy emitted by the walls or by any body within the enclosure, and escaping through the opening, is, if the walls are of uniform temperature, of the same nature as that emitted by an ideal radiator. This fact is of importance when using an optical pyrometer, described in Section 15–2. The readings of such an instrument are correct only when it is sighted on a blackbody. If used to measure the temperature of a red-hot ingot of iron in the open, its readings will be too low, since iron is a poorer emitter than a blackbody. If, however, the pyrometer is sighted on the iron while still in the furnace, where it is surrounded by walls at the same temperature, "blackbody conditions" are fulfilled and the reading is correct. The failure of the iron to emit as effectively as a blackbody is then just compensated for by the radiant energy which it reflects.

The emissivity e of an ideally black surface is equal to unity. For all real surfaces it is a fraction less than one.

Example Assuming the total surface of the human body is 1.2 m^2 and the surface temperature is 30°C = 303 K, find the total rate of radiation of energy from the body.

Surprisingly, for infrared radiation the body is a very good approximation to an ideal blackbody, irrespective of skin pigmentation. The rate of energy loss per unit area is given by Eq. (17–4). Taking e = 1, we find

$$R = (1)(5.67 \times 10^{-8}\ \mathrm{W\,m^{-2}\,K^{-4}})(303\ \mathrm{K})^4$$

$$= 482\ \mathrm{W\,m^{-2}}.$$

The total rate of energy loss is then

$$P = (482\ \mathrm{W\,m^{-2}})(1.2\ \mathrm{m^2}) = 578\ \mathrm{W}.$$

Of course this loss is partially balanced by *absorption* of radiation, which depends on the temperature of the surroundings. The *net* rate of radiative energy transfer is given by Eq. (17–5).

PROBLEMS

17–1 A slab of a thermal insulator is 100 cm^2 in cross section and 2 cm thick. Its thermal conductivity is 0.1 $\mathrm{J\,s^{-1}\,m^{-1}\,(C°)^{-1}}$. If the temperature difference between opposite faces is 100 C°, how much heat flows through the slab in one day?

17–2 Suppose that the rod in Fig. 17–1 is of copper, of length 10 cm and of cross-sectional area 1 cm^2. Let T_2 = 100°C and T_1 = 0°C.

a) What is the final steady-state temperature gradient along the rod?

b) What is the heat current in the rod, in the final steady state?

c) What is the final steady-state temperature at a point in the rod 2 cm from its left end?

17–3 A long rod, insulated to prevent heat losses, has one end immersed in boiling water (at atmospheric pressure) and the other end in a water–ice mixture. The rod consists of 100 cm of copper (one end in steam) and a length, L_2, of steel (one end in ice). Both rods are of cross-sectional area 5 cm^2. The temperature of the copper–iron junction is 60°C, after a steady state has been set up.

a) How much heat per second flows from the steam bath to the ice-water mixture?

b) How long is L_2?

17–4 A rod is initially at a uniform temperature of 0°C throughout. One end is kept at 0°C and the other is brought into contact with a steam bath at 100°C. The surface of the rod is insulated so that heat can flow only lengthwise along the rod. The cross-sectional area of the rod is 2 cm^2, its length is 100 cm, its thermal conductivity is 0.8 cgs units, its density is 10 g cm^{-3}, and its specific heat capacity is 0.10 cal gm^{-1} (C°)$^{-1}$. Consider a short cylindrical element of the rod 1 cm in length.

a) If the temperature gradient at one end of this element is 200 C° cm^{-1}, how many calories flow across this end per second?

b) If the average temperature of the element is increasing at the rate of 5 C°s^{-1}, what is the temperature gradient at the other end of the element?

17–5 One experimental method of measuring the thermal conductivity of an insulating material is to construct a box of the material and measure the power input to an electric heater, inside the box, which maintains the interior at a measured temperature above the outside surface. Suppose that in such an apparatus a power input of 120 W is required to keep the interior surface of the box 120 F° above the temperature of the outer surface. The total area of the box is 25 ft^2 and the wall thickness is 1.5 in. Find the thermal conductivity of the material in the commercial system of units.

17–6 A container of wall area 5000 cm^2 and thickness 2 cm is filled with water in which there is a stirrer. The outer surface of the walls is kept at a constant temperature of 0°C. The thermal conductivity of the walls is 0.000478 cgs units, and the effect of edges and corners can be neglected. The power required to run the stirrer at an angular velocity of 1800 rpm is found to be 100 W. What will be the final steady-state temperature of the water in the container? Assume that the stirrer keeps the entire mass of water at a uniform temperature.

17–7 A boiler with a steel bottom 1.5 cm thick rests on a hot stove. The area of the bottom of the boiler is 1500 cm^2. The water inside the boiler is at 100°C, and 750 g are evaporated every 5 min. Find the temperature of the lower surface of the boiler, which is in contact with the stove.

17–8 A camping icebox, having wall area of 2 m^2 and thickness 5 cm, is constructed of insulating material having a thermal conductivity of 0.05 J s^{-1} m^{-1} (C°)$^{-1}$. The outside temperature is 20° C, and the inside of the box is to be maintained at 5° C by ice. The melted ice leaves the box at a temperature of 15° C. If ice costs 1 cent per kilogram, what will it cost to run the icebox for 1 hr?

17–9 Rods of copper, brass, and steel are welded together to form a Y-shaped figure. The cross-sectional area of each rod is 2 cm^2. The end of the copper rod is maintained at 100°C and the ends of the brass and steel rods at 0°C. Assume there is no heat loss from the surface of the rods. The lengths of the rods are: copper, 46 cm; brass, 13 cm; steel, 12 cm.

a) What is the temperature of the junction point?

b) What is the heat current in the copper rod?

17–10 A compound bar 2 meters long is constructed of a solid steel core 1 cm in diameter surrounded by a copper casing whose outside diameter is 2 cm. The outer surface of the bar is thermally insulated and one end is maintained at 100°C, the other at 0°C.

a) Find the total heat current in the bar.

b) What fraction is carried by each material?

17–11 Heat flows radially outward through a cylindrical insulator of outside radius R_2 surrounding a steam pipe of outside radius R_1. The temperature of the inner surface of the insulator is T_1, that of the outer surface is T_2. (a) At what radial distance from the center of the pipe is the steady-state temperature exactly halfway between temperatures T_1 and T_2? (b) Sketch a graph of T versus r.

17–12 A steam pipe 2 cm in radius is surrounded by a cylindrical jacket of insulating material 2 cm thick. The temperature of the steam pipe is 100° C, and that of the outer surface of the jacket is 20° C. The thermal conductivity of the insulating material is 0.1 J s^{-1} $m^{-1}(C°)^{-1}$. Compute the temperature gradient, dT/dr, at the inner and outer surfaces of the jacket, and sketch the graph of T versus r.

17–13 A steam pipe of radius 2 cm, carrying steam at 120° C, is surrounded by a cylindrical jacket of cork with inner and outer radii 2 cm and 4 cm, and this in turn is surrounded by a cylindrical jacket of styrofoam having inner and outer radii 4 cm and 6 cm. The thermal conductivity of styrofoam is about 0.01 J s^{-1} m^{-1} $(C°)^{-1}$ The outer surface of the styrofoam is in contact with air at 20° C.

a) What is the temperature at a radius of 4 cm, where the two insulating layers meet?
b) What is the total rate of transfer of heat out of a 2-m length of pipe?

17–14 A spherical shell has inner and outer radii a and b, respectively, and the temperatures at the inner and outer surfaces are T_a and T_b.

a) Find the temperature in the shell as a function of radius.
b) Find the total rate of heat transfer through the shell.

17–15 An electric transformer is in a cylindrical tank 60 cm in diameter and 1 m high, with flat top and bottom. If the tank transfers heat to the air only by natural convection, and the electrical losses are to be dissipated at the rate of 1 kw, how many degrees will the tank surface rise above room temperature?

17–16

a) What would be the difference in height between the columns in the U-tube in Fig. 17–4 if the liquid is water and the left arm is 1 m high at 4°C while the other is at 75°C?
b) What is the difference between the pressures at the foot of two columns of water each 10 m high if the temperature of one is 4°C and that of the other is 75°C?

17–17 A flat wall is maintained at constant temperature of 100°C, and the air on both sides is at atmospheric pressure and at 20°C. How much heat is lost by natural convection from 1 m^2 of wall (both sides) in 1 hr if (a) the wall is vertical and (b) the wall is horizontal?

17–18 A vertical steam pipe of outside diameter 7.5 cm and height 4 m has its outer surface at the constant temperature of 95°C. The surrounding air is at atmospheric pressure and at 20°C. How much heat is delivered to the air by natural convection in 1 hr?

17–19 What is the radiant emittance of a blackbody at a temperature of (a) 300 K, (b) 3000 K?

17–20 The radiant emittance of tungsten is approximately 0.35 that of a blackbody at the same temperature. A tungsten sphere 1 cm in radius is suspended within a large evacuated enclosure whose walls are at 300 K. What power input is required to maintain the sphere at a temperature of 3000 K if heat conduction along the supports is neglected?

17–21 A small blackened solid copper sphere of radius 2 cm is placed in an evacuated enclosure whose walls are kept at 100°C. At what rate must energy be supplied to the sphere to keep its temperature constant at 127°C?

17–22 A cylindrical metal can 10 cm high and 5 cm in diameter contains liquid helium at 4K, at which temperature its heat of vaporization is 5 cal g^{-1}. Completely surrounding the helium can are walls maintained at the temperature of liquid nitrogen, 80 K, the intervening space being evacuated. How much helium is lost per hour? Assume the radiant emittance of the helium can to be 0.2 that of a blackbody at 4 K.

17–23 A solid cylindrical copper rod 10 cm long has one end maintained at a temperature of 20.00 K. The other end is blackened and exposed to thermal radiation from a body at 300 K, no energy being lost or gained elsewhere. When equilibrium is reached, what is the temperature of the blackened end? [*Hint.* Since copper is a very good conductor of heat at low temperature, $k = 4$ cal s^{-1} cm^{-1} $(C°)^{-1}$, the temperature of the blackened end is only slightly greater than 20 K.]

17–24 The operating temperature of a tungsten filament in an incandescent lamp is 2450 K and its emissivity is 0.30. Find the surface area of the filament of a 25-watt lamp.

17–25 A certain incandescent lightbulb operates at 3000 K. The total surface area of the filament is 0.05 cm^2, and the emissivity is 0.3. What electric power must be supplied to the filament?

17–26 The rate at which radiant energy reaches the surface of the earth from the sun is about 1.4 kW m^{-2}. The distance from earth to sun is about 1.5×10^{11} m, and the radius of the sun is about 0.7×10^9 m.

a) What is the rate of radiation of energy, per unit area, from the sun's surface?

b) If the sun radiates as an ideal blackbody, what is the temperature of its surface?

17–27 Suppose that both ends of the rod in Fig. 17–1 are kept at a temperature of 0° C, and that the initial temperature distribution along the rod is given by $T = 100 \sin \pi x/L$, where T is in °C. Let the rod be of copper, of length $L = 10$ cm and of cross section 1 cm^2.

a) Show the initial temperature distribution in a diagram.

b) What is the final temperature distribution after a very long time has elapsed?

c) Sketch curves which you think would represent the temperature distribution at intermediate times.

d) What is the initial temperature gradient at the ends of the rod?

e) What is the initial heat current from the ends of the rod into the bodies making contact with its ends?

f) What is the initial heat current at the center of the rod? Explain. What is the heat current at this point at any later time?

g) What is the value of $k/\rho C$ for copper, and in what unit is it expressed? (This quantity is called the *diffusivity*.)

h) What is the initial rate of change of temperature at the center of the rod?

i) How long a time would be required for the rod to reach its final temperature, if the temperature continued to decrease at this rate? (This time can be described as the *relaxation time* of the rod.)

j) From the graphs in part (c), would you expect the rate of change of temperature at the midpoint to remain constant, increase, or decrease?

k) What is the initial rate of change of temperature at a point in the rod, 2.5 cm from its left end?

Chapter 18

Thermal Properties of Matter

18–1 EQUATIONS OF STATE

The volume V occupied by a definite mass m of any substance depends on the pressure p to which the substance is subjected, and on its temperature T. For every pure substance there is a definite relation between these quantities, called the *equation* of state of the substance. In formal mathematical language, the equation can be written

$$f(m, V, p, T) = 0.$$

The exact form of the function is usually very complicated. It often suffices to know only how some one of the quantities changes when some other is varied, the rest being kept constant. Thus the compressibility k describes the change in volume when the pressure is changed, for a constant mass at a constant temperature, and the coefficient of volume expansion gives the change in volume when the temperature is changed, for a constant mass at a constant pressure.

The term "state" as used here implies an *equilibrium* state. This means that the temperature and pressure are the same at all points. Hence if heat is added at some point to a system in an equilibrium state, we must wait until the processes of heat transfer within the system have brought about a new uniform temperature before the system is again in an equilibrium state.

18–2 THE IDEAL GAS

The simplest equation of state is that of a gas at low pressure. Consider a container whose volume can be varied, such as a cylinder provided with a movable piston. Provision is made for pumping any desired mass of any kind of gas into or out of the cylinder, and the cylinder is provided with a pressure gauge and with a thermometer for determining the Kelvin (absolute) temperature T. Then corresponding values of m, p, V, and T can all be measured. Instead of the mass m, let us express the results in terms of the number of moles, n. Since the molecular mass M is the mass per mole, the total mass m is given by

$$m = nM.$$

In calculations with gases, the number of moles is often the most convenient way to specify the quantity of material.

From measurements of the pressure, volume, temperature, and number of moles, a number of conclusions emerge. First, the volume is proportional to the number of moles. If we double the quantity of gas, keeping pressure and temperature constant, the volume doubles. Second, the volume varies inversely with pressure; if we double the pressure, holding the temperature and quantity of material constant, the gas is compressed to one-half its initial volume. Third, the pressure is proportional to the absolute temperature; if we double the absolute temperature, holding the volume and quantity of material constant, the pressure doubles.

These conclusions can be summarized neatly in a single equation of state:

$$\boxed{pV = nRT.} \tag{18–1}$$

The constant of proportionality R might be expected to have different values for different gases, but instead it turns out to have the same value for *all* gases, at least at sufficiently high temperature and low pressure. This quantity is called the *universal gas constant.* The numerical value of R depends, of course, on the units in which p, V, n, and T are expressed. The adjective "universal" means that in any one system of units R has the same value for *all* gases. In the mks system, where the unit of p is 1 Pa or 1 N m^{-2} and the unit of V is 1 m^3, the numerical value of R is found to be

$$\begin{aligned} R &= 8.314\,(\text{N m}^{-2})\text{m}^3\,\text{mol}^{-1}\,\text{K}^{-1} \\ &= 8.314\,\text{J mol}^{-1}\,\text{K}^{-1}. \end{aligned}$$

In the cgs system, where the unit of p is 1 dyn cm^{-2} and the unit of V is 1 cm^3,

$$\begin{aligned} R &= 8.314 \times 10^7\,(\text{dyn cm}^{-2})\,\text{cm}^3\,\text{mol}^{-1}\,\text{K}^{-1} \\ &= 8.314 \times 10^7\,\text{erg mol}^{-1}\,\text{K}^{-1}. \end{aligned}$$

We note that the units of pressure times volume are the same as units of energy, so, in *all* systems of units, R has units of energy per mole, per unit of absolute temperature. In terms of calories,

$$R = 1.99\,\text{cal mol}^{-1}\,\text{K}^{-1}.$$

In chemical calculations volumes are commonly expressed in liters (ℓ), pressures in atmospheres, and temperatures in kelvins. In this system,

$$R = 0.08207\,\ell\,\text{atm mol}^{-1}\,\text{K}^{-1}.$$

We now define an *ideal gas* as one for which Eq. (18–1) holds precisely for *all* pressures and temperatures. As the term suggests, the ideal gas is an idealized model which represents the behavior of gases very well in some circumstances, less well in others. Generally, gas behavior approximates the ideal-gas model most closely at very low pressures, when the gas molecules are far apart. However, the deviations are not very great at moderate pressures and at temperatures not too near those at which the gas liquefies.

Deviations from ideal-gas behavior are shown in Fig. 18–1, where the ratio pV/nT is plotted as a function of p and T. For an ideal gas this quantity is constant, but for real gases it varies increasingly at lower and lower temperatures. At sufficiently high temperature and low pressure this ratio approaches the value R for an ideal gas.

It follows from Eq. (18–1), and the value of the universal gas constant R, that one mole of an ideal gas occupies a volume of 0.0224 m^3 or 22.4 ℓ at "standard conditions," or at "normal temperature and pressure" (NTP), that is, at a temperature of 0°C = 273 K, and a pressure of 1 atm = 1.01×10^5 N m^{-2}. Thus, from Eq. (18–1),

$$\begin{aligned} V &= \frac{nRT}{p} \\ &= \frac{(1\,\text{mol})(8.31\,\text{J mol}^{-1}\,\text{K}^{-1})(273\,\text{K})}{1.01 \times 10^5\,\text{N m}^{-2}} \\ &= 0.0224\,\text{m}^3 = 22{,}400\,\text{cm}^3 = 22.4\,\ell. \end{aligned}$$

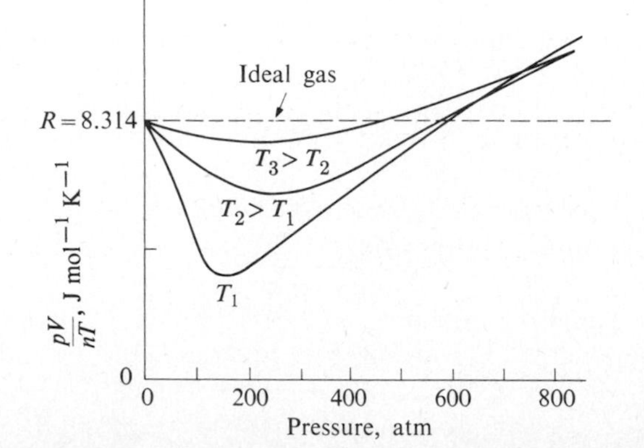

Fig. 18–1 *The limiting value of pV/nT is independent of T for all gases. For an ideal gas, pV/nT is constant.*

For a *fixed mass* (or fixed number of moles) of an ideal gas, the product nR is constant and hence pV/T is constant also. Thus if the subscripts 1 and 2 refer to two states of the same mass of a gas, but at different pressures, volumes, and temperatures,

$$\frac{p_1 V_1}{T_1} = \frac{p_2 V_2}{T_2} = \text{constant.} \qquad (18\text{–}2)$$

If the temperatures T_1 and T_2 are the same, then

$$p_1 V_1 = p_2 V_2 = \text{constant.} \qquad (18\text{–}3)$$

The fact that at constant temperature the product of the pressure and volume of a fixed mass of gas is very nearly constant was discovered experimentally by Robert Boyle in 1660, and Eq. (18–3) is called *Boyle's law.* Although exactly true (by definition) for an *ideal gas,* it is obeyed only approximately by real gases and is not a fundamental law like Newton's laws or the law of conservation of energy.

Example 1 The volume of an oxygen tank is 50 ℓ. As oxygen is withdrawn from the tank, the reading of a pressure gauge drops from 300 lb in^{-2} to 100 lb in^{-2}, and the temperature of the gas remaining in the tank drops from 30° C to 10° C.

a) How many kilograms of oxygen were in the tank originally?
b) How many kilograms were withdrawn?
c) What volume would be occupied by the oxygen withdrawn from the tank at a pressure of 1 atm and a temperature of 20° C?

a) Let us express pressures in pascals, volumes in cubic meters, and temperatures in kelvins. The pressure conversion can be carried out by use of the relations

$$1 \text{ atm} = 14.7 \text{ lb in}^{-2} = 1.01 \times 10^5 \text{ Pa}.$$

A *gauge* pressure of 300 lb in^{-2} is a total pressure of (300 + 14.7) lb in^{-2}, and we have

$$314.7 \text{ lb in}^{-2} = \frac{314.7}{14.7} \text{ atm} = 21.4 \text{ atm}$$
$$= (21.4)(1.01 \times 10^5 \text{ Pa})$$
$$= 2.17 \times 10^6 \text{ Pa}.$$

Similarly, 100 lb in^{-2} (gauge) = 7.90×10^5 Pa. We also have 50 ℓ = 0.05 m^3, 30° C = 303 K, and 10° C = 283 K.

The initial number of moles is

$$n_1 = \frac{p_1 V}{RT_1} = \frac{(2.17 \times 10^6 \text{ Pa})(0.05 \text{ m}^3)}{(8.314 \text{ J mol}^{-1} \text{ K}^{-1})(303 \text{ K})}$$
$$= 43.2 \text{ mol}.$$

The original mass was therefore

$$m_1 = (43.1 \text{ mol})(32 \text{ g mol}^{-1})$$
$$= 1378 \text{ g} = 1.378 \text{ kg}.$$

b) The number of moles remaining in the tank is

$$n_2 = \frac{p_2 V}{RT_2} = \frac{(7.90 \times 10^5 \text{ Pa})(0.05 \text{ m}^3)}{(8.314 \text{ J mol}^{-1} \text{ K}^{-1})(283 \text{ K})}$$
$$= 16.8 \text{ mol},$$

and the mass remaining is

$$(16.8 \text{ mol})(32 \text{ g mol}^{-1}) = 537 \text{ g}$$
$$= 0.537 \text{ kg}.$$

The mass withdrawn is, therefore,

$$1.378 \text{ kg} - 0.537 \text{ kg} = 0.841 \text{ kg}.$$

c) The number of moles withdrawn is

$$43.1 - 16.8 = 26.3,$$

and the volume occupied would be

$$V = \frac{nRT}{p}$$
$$= \frac{(26.3 \text{ mol})(8.314 \text{ J mol}^{-1} \text{ K}^{-1})(293 \text{ K})}{1.01 \times 10^5 \text{ Pa}}$$
$$= 0.634 \text{ m}^3.$$

Example 2 The *McLeod gauge* (pronounced "McLoud"), illustrated in Fig. 18–2, can be used to measure pressures as low as 5×10^{-6} mm of mercury. In part (a), the entire space above point A is occupied by the gas at the low pressure p which is to be measured. When the mercury container B is raised as in (b), the gas in bulb C, whose volume V might be 500 cm^3, is trapped and eventually compressed into a much smaller volume V' above a reference mark on the capillary tube D. Assuming the temperature

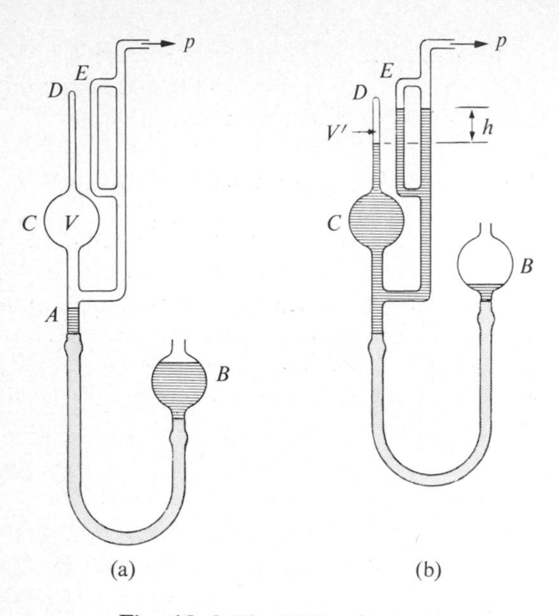

Fig. 18–2 The McLeod gauge.

constant, the pressure p' of the compressed gas is given by applying Boyle's law,

$$p' = \frac{pV}{V'}.$$

The pressure at the upper surface of the mercury in capillary E remains at the value p, so that if h is the difference in elevation between the tops of the mercury columns in E and D,

$$p' = p + \rho gh,$$

where ρ is the density of mercury. Elimination of p' between these equations gives

$$p = \frac{\rho g V'}{V - V'} h \approx \frac{V'}{V} \rho gh,$$

to a good approximation, since $V' \ll V$.

For example, if $V'/V = 10^{-4}$ and $h = 4$ mm, $p = 4 \times 10^{-4}$ torr.

Since ρ, g, V, and V' are constants, the pressure p is directly proportional to h, and a uniform pressure scale can be mounted beside tube E. Corrections for capillary depression (due to surface tension) are eliminated if capillaries D and E have the same diameter.

Example 3 In the upper part of the atmosphere (the stratosphere) the temperature varies only slightly with changes in elevation. Find the law of variation of pressure with elevation.

The rate of change of pressure with elevation in a fluid is $dp/dy = -\rho g$.

Let us replace n by m/M in the ideal gas law, where m is the mass and M the molecular weight. Then

$$pV = \frac{m}{M} RT,$$

and the density ρ of an ideal gas is

$$\rho = \frac{m}{V} = \frac{pM}{RT}.$$

Hence

$$\frac{dp}{dy} = -\frac{pMg}{RT}, \quad \frac{dp}{p} = -\frac{Mg}{RT} dy.$$

If g and T are constant, and if p_1 and p_2 are the pressures at two elevations y_1 and y_2, integration of the preceding equation gives

$$\ln \frac{p_2}{p_1} = -\frac{Mg}{RT} (y_2 - y_1). \tag{18–4}$$

This is known as the *barometric equation.*

18–3 *pVT*-SURFACE FOR AN IDEAL GAS

Since the equation of state for a fixed mass of a substance is a relation among the three variables p, V, and T, it defines a *surface* in a rectangular coordinate system in which p, V, and T are plotted along the three axes. Figure 18–3 shows the pVT-surface of an ideal gas. The solid lines on the surface show the relation between p and V when T is constant (Boyle's law), the dashed lines the relation between V and T when p is constant (Gay-Lussac's law), and the dotted lines the relation between p and T when V is constant. When viewed perpendicular to the pV-plane, the surface appears as in Fig. 18–4(a), and Fig. 18–4(b) is its appearance when viewed perpendicular to the pT-plane.

Every possible equilibrium state of a given quantity of gas corresponds to a point on the surface,

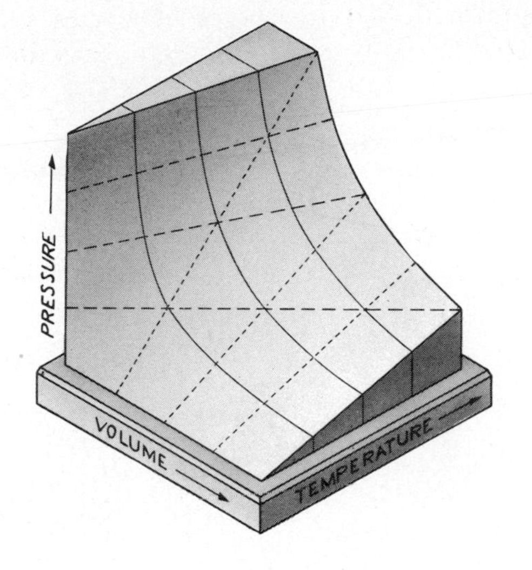

Fig. 18–3 *pVT-surface for an ideal gas.*

and every point of the surface corresponds to a possible equilibrium state. The gas cannot exist in a state that is not on the surface. For example, if the volume and temperature are given, thus locating a point in the base plane of Fig. 18–3, the pressure is then determined by the nature of the gas, and it can have only the value represented by the height of the surface above this point.

In any process in which the gas passes through a succession of equlilbrium states, the point representing its state moves along a curve lying in the pVT-surface. Such a process must be carried out very slowly to give the temperature and pressure time to remain uniform at all points of the gas.

18–4 *pVT*-SURFACE FOR A REAL SUBSTANCE

While all real substances approximate ideal gases at sufficiently low pressures, their behavior departs more and more from that of an ideal gas at high pressures and low temperatures. As the temperature is lowered and the pressure increased, all substances change from the *gas phase* to the *liquid phase* or the *solid phase.* For a fixed mass of a substance, however, there is still a definite relation between the pressure, temperature, and total volume. In other words, the substance has an equation of state in *any* circumstances, and although the general form of the equation is much too complicated to express mathematically, we can represent it *graphically* by a pVT-surface. Figure 18–5 is a schematic diagram (to a greatly distorted scale) of the pVT-surface of a substance that expands on melting (the most common case) and Fig. 18–6 is the corresponding diagram for a substance which, like water, contracts on melting. We see that the substance can exist in either the solid, liquid, or gas phase, or in two phases simultaneously, or, along the triple line, in all three phases. (The distinction between the terms "gas" and "vapor" will be explained later. For the present they may be considered synonymous.)

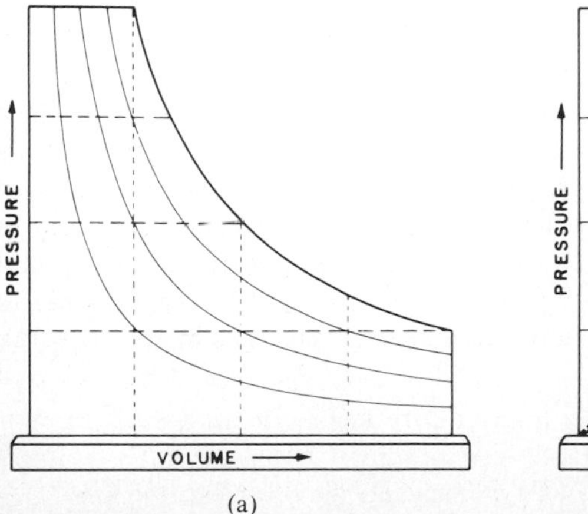

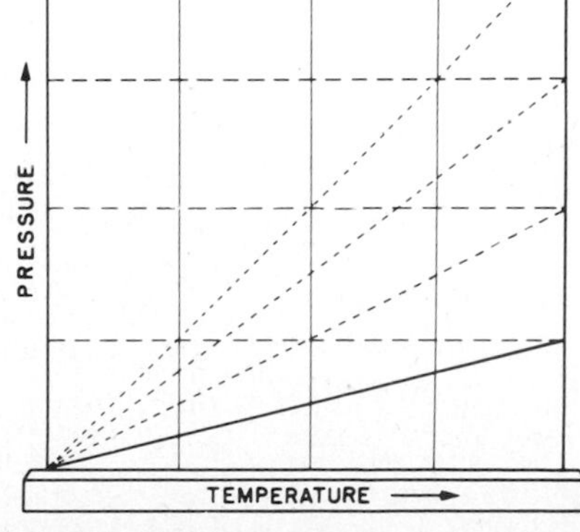

Fig. 18–4 *Projections of ideal gas pVT-surface on (a) the pV-plane, (b) the pT-plane.*

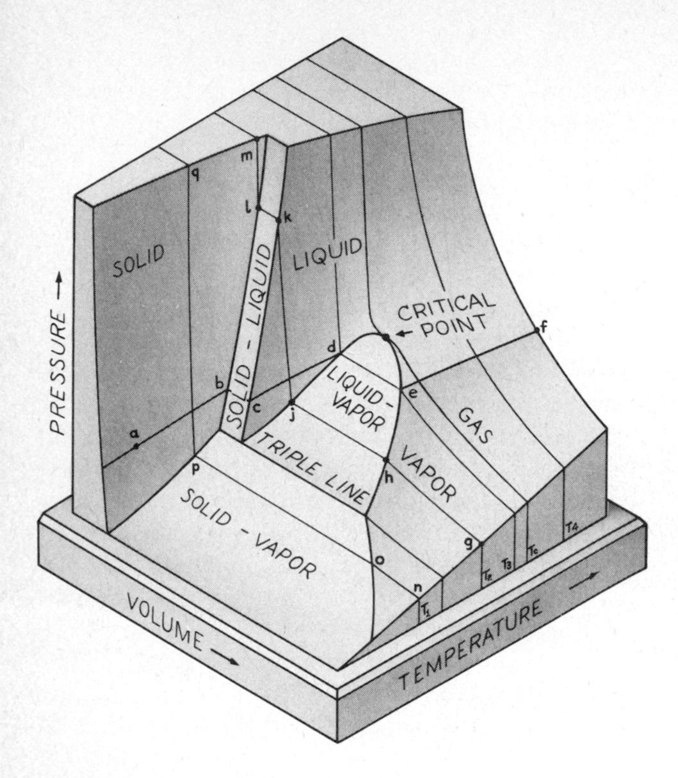

Fig. 18–5 *pVT-surface for a substance that expands on melting.*

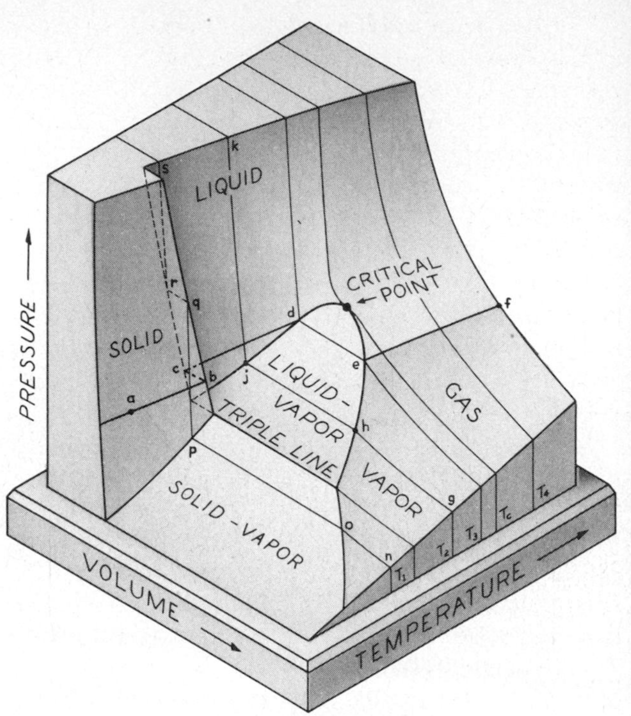

Fig. 18–6 *pVT-surface for a substance that contracts on melting.*

In order that the diagram shall represent the properties of a particular *substance*, but not depend on *how much* of the substance is present, we plot along the volume axis not the actual volume V but the *specific volume* v, the volume per unit mass. Thus for a system of mass m,

$$v = \frac{V}{m}.$$

The actual volume of a unit mass of a substance is numerically equal to its specific volume. The specific volume is the reciprocal of the density ρ,

$$\rho = \frac{m}{V} = \frac{1}{v}.$$

To correlate these diagrams with our familiar experiences regarding the behavior of solids, liquids, and gases, let us start with a substance in the solid phase at point a in Fig. 18–5 or 18–6. Let the substance be contained in a cylinder as in Fig. 18–7, and let a constant force F be exerted on the piston so that the pressure remains constant as the substance expands or contracts. A constant-pressure process is called an *isobaric* process, and lines representing constant-pressure processes are called *isobars.* We now bring the cylinder in contact with some body such as an electric stove, whose temperature can be slowly increased and kept slightly higher than that of the substance, so that heat flows into the substance.

At the start of the process, the temperature of the substance rises as heat is added, at a rate determined by the specific heat capacity of the solid, and the volume increases slightly at a rate determined by its coefficient of expansion. When point b is reached in either diagram the substance starts to melt, that is, to change from the solid to the liquid phase. The temperature ceases to rise even though heat is being continuously supplied. The volume increases in Fig. 18–5 and decreases in Fig. 18–6.

When point c is reached, the substance is wholly in the liquid phase. The temperature now starts to rise again, at a rate determined by the specific heat capacity of the liquid, and the volume increases at a rate determined by its coefficient of expansion. (For water at atmospheric pressure, there would be a small *decrease* in volume at first.)

When point d is reached, the temperature again ceases to rise, although heat is still supplied, and bubbles of vapor start to form in the cylinder and rise to its upper surface. Since the density of the vapor phase is smaller than that of the liquid, the volume increases.

At point e, the substance is wholly in the vapor phase. (The relative volume increase from d to e, except at very high pressures, is enormously greater than the volume change between liquid and solid, so Figs. 18–5 and 18–6 are not constructed on a uniform scale.) With further addition of heat the temperature rises again, the rate now being determined by the specific heat capacity of the vapor phase (at constant pressure). The volume also increases much more rapidly than did that of the solid or liquid.

If the cylinder is now brought in contact with a body whose temperature is kept slightly *lower* than that of the substance, heat flows out of the substance and all the changes in the original process take place in reversed order.

As a second illustration to show how the pVT-surface describes the behavior of a substance, suppose we start with a cylinder containing a gas, at a pressure, volume, and temperature corresponding to point g in Fig. 18–5 or 18–6. Let the external pressure be adjusted so that it is always slightly greater than the pressure exerted by the substance, and let the cylinder be kept in contact with a body at a constant

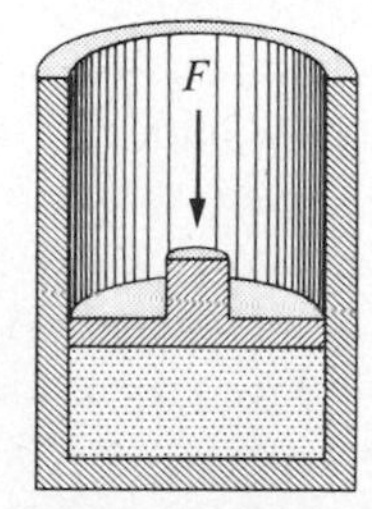

Fig. 18–7

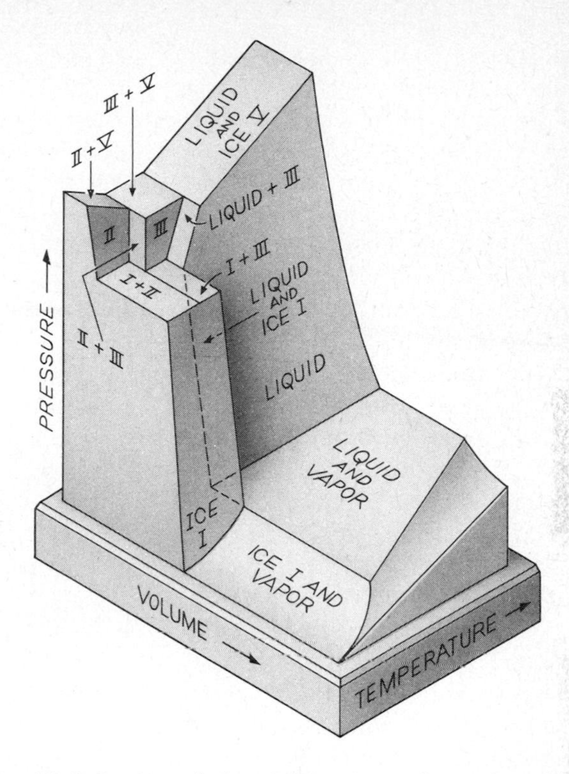

Fig. 18–8 *Portion of the pVT-surface of water at high pressure.*

temperature. A constant-temperature process is called *isothermal*, and lines representing constant-temperature processes are called *isotherms.*

As the pressure increases, the volume decreases along the line gh; simultaneously, heat flows from the gas to the body with which it is in contact; otherwise the temperature of the gas would rise. When point h is reached, drops of liquid begin to form in the cylinder, and the volume continues to decrease without further increase of pressure.

At point j, all of the substance has condensed to the liquid phase. With further increase of pressure the volume decreases, but only slightly because of the small compressibility of liquids. For a substance like that in Fig. 18–6, no other change of phase occurs as the pressure is increased to point k and beyond (unless other forms of the solid state exist; see Fig. 18–8 and the accompanying discussion). In Fig. 18–5, however, another break in the curve takes place at point k. Crystals of the solid begin to appear, and the volume again decreases without an increase of external pressure. At point ℓ, in Fig. 18–5, the substance has been completely converted to the solid phase. Further increase of pressure reduces the volume of the solid only slightly, and unless the substance can exist in more than one modification of the solid phase, no further changes in phase result.

As a final example, suppose we again start with a gas in the cylinder but at a lower temperature, corresponding to point n in Figs. 18–5 and 18–6. If the pressure is increased isothermally, crystals of the solid begin to appear at point o and the gas changes directly to a solid without passing through the liquid phase. The pressure remains constant along the line op, and at point p the substance is all in the solid phase. In Fig. 18–5 no further phase change takes place but in Fig. 18–6 the solid starts to melt at point q and has completely melted at point r.

In all the processes described above, heat must be removed or added continuously in order to keep the pressure or temperature constant. If at any stage of a process the system is thermally insulated so that there can be no flow of heat in or out, and if the external pressure is kept constant, the system remains in equilibrium. Thus at any point on the surfaces lettered solid–liquid, solid–vapor, or liquid–vapor, *two* phases can coexist in equilibrium, and along the *triple line* all three phases can coexist. A vapor at the pressure and temperature at which it can exist in equilibrium with its liquid is called a *saturated vapor*, and the liquid is called a *saturated liquid*. Thus points e and h represent saturated vapor, and points j and d, saturated liquid. (The term "saturated" is poorly chosen. It does not have the same meaning as a "saturated solution" in chemistry. There is no question here of one substance being dissolved in another.)

18–5 CRITICAL POINT AND TRIPLE POINT

A study of Figs. 18–5 and 18–6 will show that the liquid and gas (or vapor) phases can exist together only if the temperature and pressure are less than those at the point lying at the top of the tongue-shaped surface lettered liquid–vapor. This point is called the *critical point*, and the corresponding values of T, p, and v are the *critical temperature, pressure*, and *specific volume*. A gas at a temperature above the critical temperature, such as T_4, does not separate into two phases when compressed isothermally, but its properties change gradually and continuously from those we ordinarily associate with a gas (low density, large compressibility) to those of a liquid (high density, small compressibility). Table 18–1 lists the critical constants for a few substances. The very low critical temperatures of hydrogen and helium make it evident why these gases defied attempts to liquefy them for many years.

Table 18–1 CRITICAL CONSTANTS

Substance	Critical temperature, K	Critical pressure, atm	Critical volume, $m^3\ mol^{-1}$	Critical density, $kg\ m^{-3}$
Helium (4)	5.3	2.26	0.057	69.3
Helium (3)	3.34	1.15	0.0726	41.3
Hydrogen (normal)	33.3	12.80	0.065	31.0
Deuterium (normal)	38.4	16.4	0.0603	66.3
Nitrogen	126.2	33.5	0.0901	311
Oxygen	154.8	50.1	0.078	410
Ammonia	405.5	111.3	0.0725	235
Freon 12	384.7	39.6	0.218	555
Carbon dioxide	304.2	72.9	0.094	468
Sulfur dioxide	430.7	77.8	0.122	524
Water	647.4	218.3	0.056	320
Carbon disulfide	552	78	0.170	440

The term *vapor* is sometimes used to mean a gas at any temperature below its critical temperature, and sometimes is restricted to mean a gas in equilibrium with the liquid phase, that is, a saturated vapor. The term is really unnecessary; no sudden change takes place in the properties of a substance when the critical isotherm is crossed either on the portion of the surface "gas and vapor," or on the portion "liquid."

Many substances can exist in more than one modification of the solid phase. Transitions from one modification to another occur at definite temperatures and pressures, like the phase changes from liquid to solid, etc. Water is one such substance, and at least eight types of ice have been observed at very high pressures Figure 18–8 shows a portion of the pVT-surface of water at high pressure. Note that ordinary ice (ice I) is the only form whose specific volume is greater than that of the liquid phase.

Because of the difficulty of drawing three-dimensional diagrams, it is customary to represent the pVT-surface by its projections onto the pT- and pV-planes. Figure 18–9 shows the two projections of Fig. 18–5, and Fig. 18–10 shows those of Fig. 18–6. The reader should follow through on these diagrams the isobaric and isothermal processes indicated in Figs. 18–5 and 18–6. The pT projection exhibits most clearly the ranges of temperature and pressure in which each phase is stable, and it is often called a *phase diagram.*

The curves in Figs. 18–9(a) and 18–10(b) are loci of corresponding values of pressure and temperature at which the two phases can coexist if a substance is isolated, or at which one phase will transform to the other if heat is supplied or removed. Thus the S-L curve is also a graph of the melting-point temperature or freezing-point temperature of the substance as a function of pressure, the curve S-V is a graph of the *sublimation point* versus *pressure*, and the curve L-V is a graph of the *boiling point* versus *pressure*. The S-V and L-V curves always slope upward to the right. The S-L curve slopes upward to the right for a substance that expands on melting (Fig. 18–9) but upward to the left for a substance such as water that contracts on melting (Fig. 18–10). Thus an increase

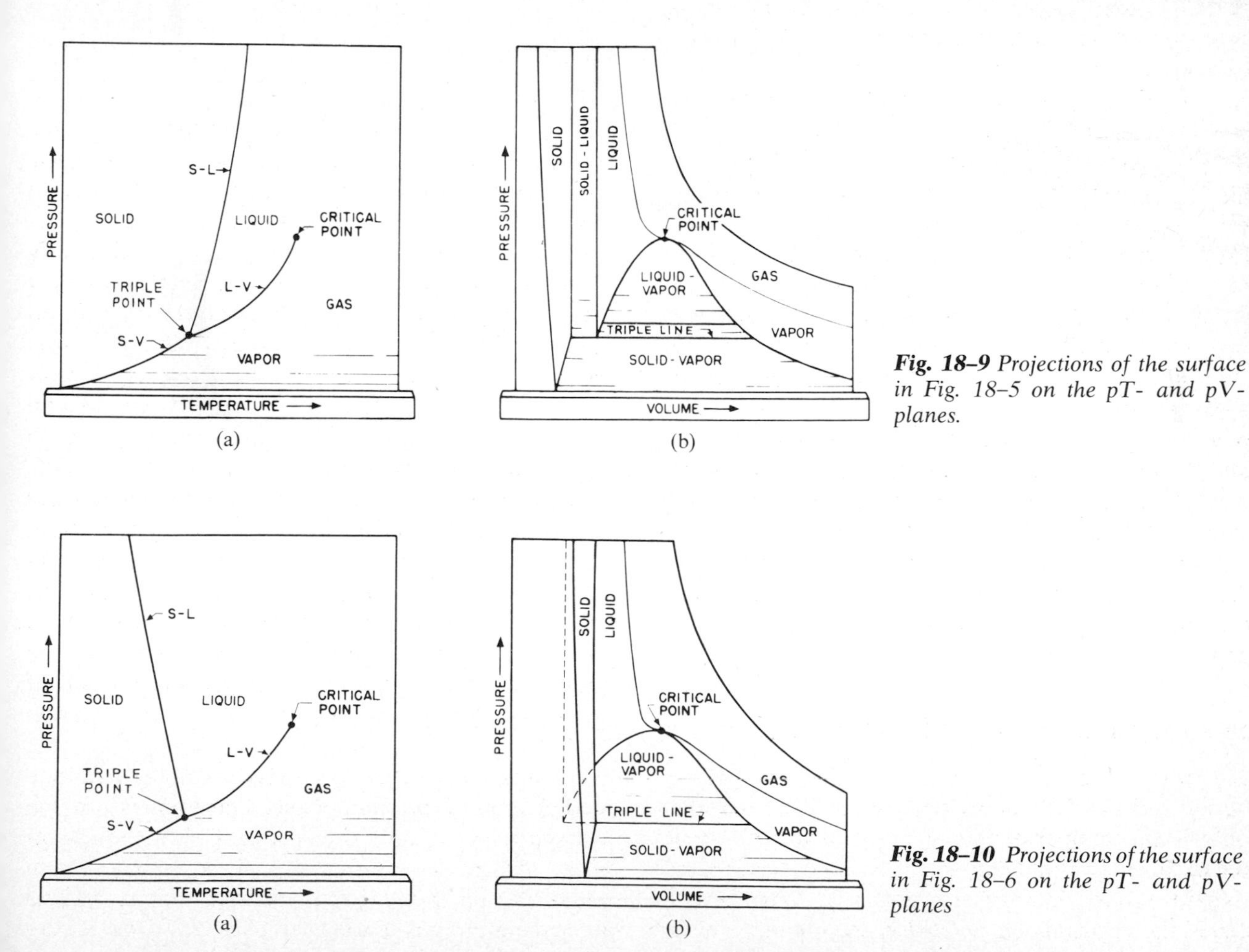

Fig. 18–9 *Projections of the surface in Fig. 18–5 on the pT- and pV-planes.*

Fig. 18–10 *Projections of the surface in Fig. 18–6 on the pT- and pV-planes*

of pressure always increases the temperature of the sublimation point or boiling point, but the temperature of the freezing point may be raised (Fig. 18–9) or lowered (Fig. 18–10) by an increase in pressure.

The pressure of a vapor in equilibrium with the liquid or solid at any temperature is called the *vapor pressure* of the substance at that temperature. Thus the curves S-V and L-V in Figs. 18–9(a) and 18–10(a) are graphs of vapor pressure versus temperature. The vapor pressure of a substance is a function of *temperature only*, not of volume. That is, in a vessel containing a liquid (or solid) and vapor in equilibrium at a *fixed temperature*, the pressure does not depend on the relative amounts of liquid and vapor present. If the volume is decreased some of the vapor condenses, and vice versa, but if the temperature is kept constant by removing or adding heat the pressure does not change.

The boiling-point temperature of a liquid is the temperature at which its vapor pressure equals the external pressure. Table 18–2 gives the vapor pressure of water as a function of temperature; we see that the vapor pressure is 1 atm at a temperature of 100°C. If the external pressure is reduced to 17.5 mm of mercury, water will boil at room temperature (20°C), while under a pressure of 90 lb in^{-2} (about 6 atm) the boiling point is 160°C.

Table 18–2 VAPOR PRESSURE OF WATER

T, °C	Vapor pressure, torr (mm Hg)	Vapor pressure, lb in^{-2}	*T*, °F
0	4.58	0.0886	32
5	6.51	0.126	41
10	8.94	0.173	50
15	12.67	0.245	59
20	17.5	0.339	68
40	55.1	1.07	104
60	149	2.89	140
80	355	6.87	176
100	**760**	**14.7**	**212**
120	1490	28.8	248
140	2710	52.4	284
160	4630	89.6	320
180	7510	145	356
200	11650	225	392
220	17390	336	428

Table 18–3 TRIPLE-POINT DATA

Substance	Temperature, K	Pressure, torr (mm Hg)
Helium (4) (λ point)	2.186	38.3
Helium (3)	None	None
Hydrogen (normal)	13.84	52.8
Deuterium (normal)	18.63	128
Neon	24.57	324
Nitrogen	63.18	94
Oxygen	54.36	1.14
Ammonia	195.40	45.57
Carbon dioxide	216.55	3880
Sulfur dioxide	197.68	1.256
Water	**273.16**	**4.58**

The point of intersection of the three equilibrium lines in Figs. 18–9(a) and 18–10(a), which is an end view of the triple line in Figs. 18–9(b) and 18–10(b), is called the *triple point*. There is only one pressure and temperature at which all three phases can coexist. Triple-point data for a few substances are given in Table 18–3.

As numerical examples, consider the pT-diagrams of water and carbon dioxide, in Fig. 18–11. In (a), a horizontal line at a pressure of 1 atm intersects the freezing-point curve at 0°C and the boiling-point curve at 100°C. The boiling point increases with increasing pressure up to the critical temperature of 374°C. Solid, liquid, and vapor can remain in equilibrium only at the triple point, where the vapor pressure is 4.5 mm of mercury and the temperature is 0.01°C.

The freezing point of a substance such as water, which expands on solidifying, is *lowered* by an increase in pressure. The reverse is true for substances which contract on solidifying. The change in the freezing-point temperature is much smaller than is that of the boiling point; an increase of 1 atm lowers the freezing point of water by only about 0.007°C.

The lowering of the freezing point of water (or the melting point of ice) can be demonstrated by passing a loop of fine wire over a block of ice and hanging a weight of a few pounds from each end of the loop. Suppose that the main body of the ice is at 0°C and at atmospheric pressure. The temperature of the small amount of ice directly under the wire

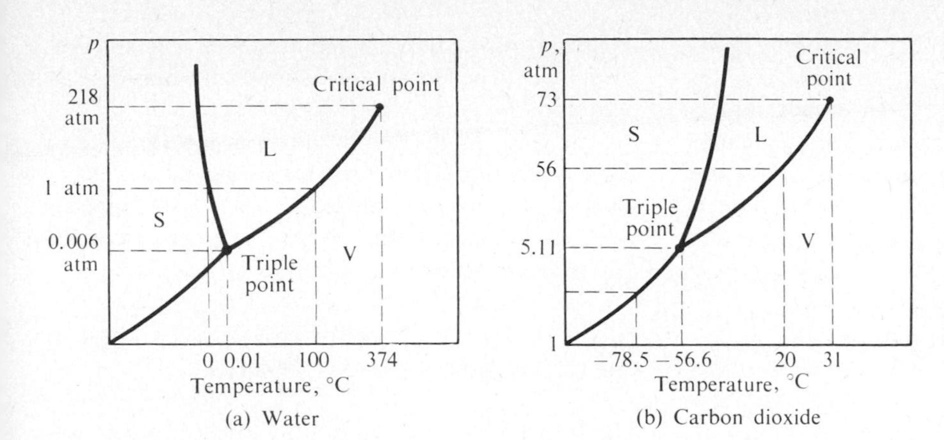

Fig. 18–11 Pressure-temperature diagrams (not to a uniform scale).

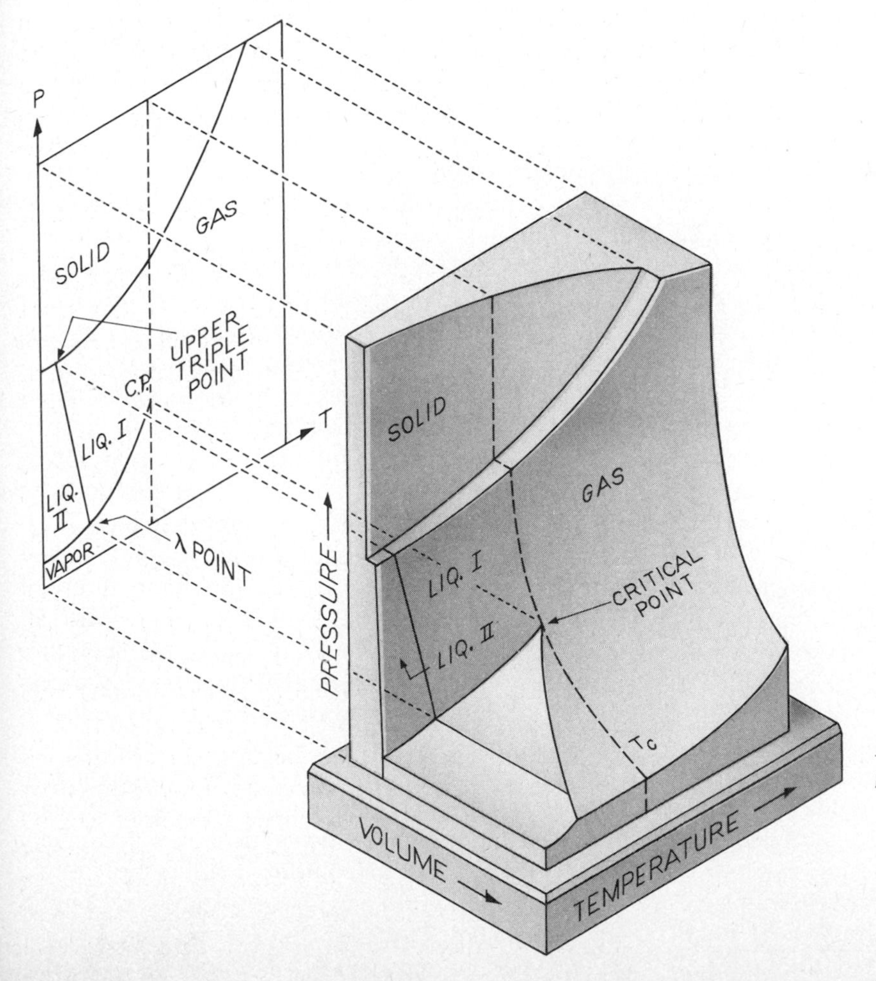

Fig. 18–12 pVT-surface for helium, and p-T phase diagram.

decreases until it achieves the melting point appropriate to the pressure under the wire. During this increase of pressure and decrease of temperature, a small amount of melting takes place. The water thus formed is squeezed out from under the wire and coming to the top of the wire where the pressure is atmospheric, it refreezes and liberates heat which passes through the wire and serves to melt the next bit of ice below the wire.

The wire thus sinks farther and farther into the block, eventually cutting its way completely through, but leaving a solid block of ice behind it. This phenomenon is known as *regelation* (refreezing). Since heat is conducted from the top to the bottom of the wire while the wire is cutting through the ice, the greater the thermal conductivity of the wire, the faster will the wire cut through the ice. Even a perfectly conducting wire would not cut through the ice very rapidly, however, because of the very low thermal conductivity of the water film which is always present beneath the wire.

For carbon dioxide, the triple-point temperature is $-56.6°C$ and the corresponding pressure is 5.11 atm. Hence at atmospheric pressure CO_2 can exist only as a solid or vapor. When heat is supplied to solid CO_2, open to the atmosphere, it changes directly to a vapor, without passing through the liquid phase, whence the name "dry ice." Liquid CO_2 can exist only at a pressure greater than 5.11 atm. The steel tanks in which CO_2 is commonly stored contain liquid and vapor (both saturated). The pressure in these tanks is the vapor pressure of CO_2 at the temperature of the tank. If the latter is 20°C the vapor pressure is 56 atm or about 830 lb in^{-2}.

The pVT-surface for ordinary helium (of mass number 4) is shown in Fig. 18–12. Two remarkable properties may be seen. First, helium has no triple point at which solid, liquid, and gas coexist in equilibrium. Instead, it has two triple points, the lower one (called the "lambda point") representing the temperature and pressure at which two liquid phases I and II coexist with vapor, and the upper one at which the two liquid phases coexist with solid. The existence of two liquid phases is unique to helium; II is the *superfluid* phase, which has a variety of unusual properties, including very low viscosity, very large thermal conductivity, and the ability of the material to form a surface film which creeps up the sides of a container. (The isotope of helium with mass number 3 exhibits no superfluid phase.)

Second, as the temperature is lowered, helium does not solidify, but remains liquid all the way to absolute zero. To obtain solid helium, the pressure must be raised to at least 25 atm, at which the helium atoms can coalesce into a crystal lattice.

18–6 EFFECT OF DISSOLVED SUBSTANCES ON FREEZING AND BOILING POINTS

The freezing point of a liquid is lowered when some other substance is dissolved in the liquid. A common example is the use of an "antifreeze" to lower the freezing point of the water in the radiator of an automobile engine.

The freezing point of a saturated solution of common salt in water is about $-20°C$. To understand why a mixture of ice and salt may be used as a freezing mixture we recall that the freezing point is the only temperature at which the liquid and solid phases can exist in equilibrium. When a saturated salt solution is cooled, it freezes at $-20°C$, and crystals of ice (pure H_2O) separate from the solution. In other words, ice crystals and a salt solution can exist in equilibrium only at $-20°C$, just as ice crystals and pure water can exist together only at 0°C.

When ice at 0°C is mixed with a salt solution at 20°C, some of the ice melts, abstracting its heat of fusion from the solution until the temperature falls to 0°C. But ice and salt solution cannot remain in equilibrium at 0°C, so the ice continues to melt. Heat is now supplied by both the ice and the solution, and both cool down until the equilibrium temperature of $-20°C$ is reached. If the melted ice dilutes the salt solution appreciably, the equilibrium temperature rises, but this can be prevented by supplying excess salt so the solution remains saturated. If no heat is supplied from outside, the mixture remains unchanged at $-20°C$. If the mixture is brought into contact with a warmer body, say an ice cream mixture at 20°C, heat flows from the ice cream mixture to the cold salt solution, melting more of the ice but producing no rise in temperature as long as any ice remains. The flow of heat from the ice cream lowers its temperature to its freezing point (below

0°C, since it is itself a solution). Further loss of heat to the ice-salt mixture causes the ice cream to freeze.

The boiling point of a liquid is also affected by dissolved substances but may be either increased or decreased. Thus the boiling point of a water-alcohol solution is *lower* than that of pure water, while the boiling point of a water-salt solution is *higher* than that of pure water.

18–7 HUMIDITY

Atmospheric air is a mixture of gases consisting of about 80 percent nitrogen, 18 percent oxygen, and small amounts of carbon dioxide, water vapor, and other gases. The mass of water vapor per unit volume is called the *absolute humidity.* The total pressure exerted by the atmosphere is the sum of the pressures exerted by its component gases. These pressures are called the *partial pressures* of the components. It is found that the partial pressure of each of the component gases of a gas mixture is very nearly the same as would be the actual pressure of that component alone if it occupied the same volume as does the mixture, a fact known as *Dalton's law.* That is, each of the gases of a gas mixture behaves independently of the others. The partial pressure of water vapor in the atmosphere is ordinarily equal to a few millimeters of mercury.

It should be evident that the partial pressure of water vapor at any given air temperature can never exceed the vapor pressure of water at that particular temperature. Thus at 10°C, from Table 18–2, the partial pressure cannot exceed 8.94 torr, or at 15°C it cannot exceed 12.67 torr. If the concentration of water vapor, or the absolute humidity, is such that the partial pressure equals the vapor pressure, the vapor is *saturated.* If the partial pressure is less than the vapor pressure, the vapor is *unsaturated.* The ratio of the partial pressure to the vapor pressure at the same temperature is called the *relative humidity,* and is usually expressed as a percentage:

$$\text{Relative humidity (\%)} = 100 \times \frac{\text{partial pressure of water vapor}}{\text{vapor pressure at same temperature}}.$$

The relative humidity is 100% if the vapor is saturated and zero if no water vapor at all is present.

Example The partial pressure of water vapor in the atmosphere is 10 torr and the temperature is 20°C. Find the relative humidity.

From Table 18–2, the vapor pressure at 20°C is 17.5 torr. Hence,

$$\text{Relative humidity} = (10/17.5)(100\%) = 57\%.$$

Since the water vapor in the atmosphere is saturated when its partial pressure equals the vapor pressure at the air temperature, saturation can be brought about either by increasing the water vapor content or by lowering the temperature. For example, let the partial pressure of water vapor be 10 torr when the air temperature is 20°C, as in the preceding example. Saturation or 100% relative humidity could be attained either by introducing enough more water vapor (keeping the temperature constant) to increase the partial pressure to 17.5 torr, *or by lowering the temperature* to 11.4°C, at which, by interpolation from Table 18–2, the vapor pressure is 10 torr.

If the temperature were to be lowered *below* 11.4°C, the vapor pressure would be less than 10 torr. The partial pressure would then be higher than the vapor pressure and enough vapor would condense to reduce the partial pressure to the vapor pressure at the lower temperature. It is this process which brings about the formation of clouds, fog, and rain. The phenomenon is also of frequent occurrence at night when the earth's surface becomes cooled by radiation. The condensed moisture is called *dew.* If the partial pressure is so low that the temperature must fall below 0°C before saturation occurs, the vapor condenses into ice crystals in the form of frost or snow.

The temperature at which the water vapor in a given sample of air becomes saturated is called the *dew point.* Measuring the temperature of the dew point is the most accurate method of determining relative humidity. The usual method is to cool a metal container having a bright polished surface, and to observe its temperature when the surface becomes clouded with condensed moisture. Suppose the dew point is observed in this way to be 10°C, when the air temperature is 20°C. We then know that the water vapor in the air is saturated at 10°C; hence its partial pressure is 8.94 torr, equal to the vapor pressure at 10°C. The pressure necessary for saturation at 20°C is 17.5 torr. The relative humidity is therefore

$$\frac{8.94}{17.5} \times 100 = 51\%.$$

A simpler but less accurate method of determining relative humidity employs a *wet-* and *dry-bulb thermometer.* Two thermometers are placed side by side, the bulb of one being kept moist by a wick dipping in water. The lower the relative humidity, the more rapid will be the evaporation from the wet bulb, and the lower will be its temperature below that of the dry bulb. The relative humidity corresponding to any pair of wet- and dry-bulb temperatures is read from tables.

The rate of evaporation of water from a water–air surface depends on the relative humidity. Evaporation is most rapid when the vapor pressure is low, but when the relative humidity is high evaporation is slower. At 100% relative humidity, *no* further evaporation can take place. The added discomfort which high relative humidity brings to a hot summer day is a familiar effect; the reason is that one of the body's important temperature-control mechanisms, cooling by evaporation of sweat from the skin, is inhibited by high relative humidity.

18–8 THE CLOUD CHAMBER AND THE BUBBLE CHAMBER

The cloud chamber is an apparatus for securing information about elementary particles such as electrons and α-particles. In principle (Fig. 18–13), it consists of a cylindrical enclosure having a glass top B and provided with a movable piston C. The enclosed space contains air, water vapor, and sufficient excess water so that the vapor is saturated. (Other liquids such as alcohol are often used in place of water.) When the piston is suddenly pulled down a short distance, the mixture is lowered in temperature below the dew point. If the air is perfectly clean, the cooled vapor does not immediately condense, and it is said to be *supersaturated.* The point representing its state then lies *above* the pVT-surface, but

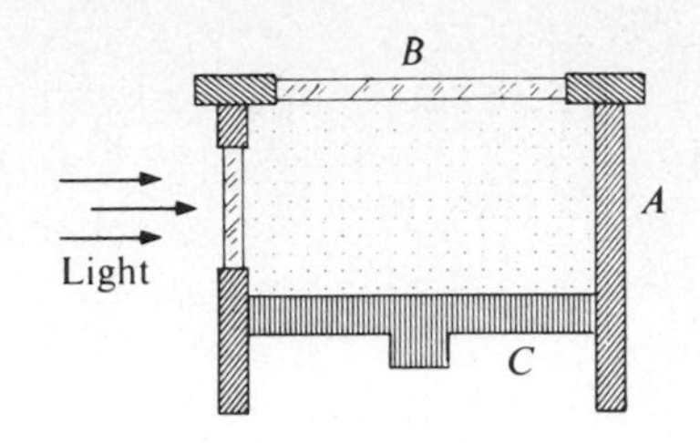

Fig. 18–13 *The Wilson cloud chamber.*

this does not contradict a statement made earlier that all equilibrium states of a substance must lie on its pVT-surface, since the supersaturated state is *not* an equilibrium state. It is found that any ions which may be present serve as nuclei upon which the supersaturated vapor condenses to form liquid droplets. Hence if any ions were present just before the expansion, their presence is made evident by the appearance of tiny droplets immediately after the expansion.

Electrons, protons, and α-particles are all capable of traveling several centimeters through air, but as they collide with or pass near the air molecules they may knock off one or more electrons and hence leave behind them a trail of ions. Therefore if such a particle passed through the chamber just before the expansion, a trail of droplets appears after the expansion, indicating the path the particle followed. For photographing the tracks, an intense beam of light is projected transversely through the chamber and a camera mounted above it. Figure 8–10 is a photograph, made with a Wilson cloud chamber, of the paths of the particles resulting from the fission of a uranium nucleus.

The *bubble chamber,* a more recently developed apparatus for studying ionizing particles, makes use of a *superheated liquid* instead of a supersaturated vapor. A superheated liquid is a liquid at a temperature higher than that of its boiling point at the pressure to which it is subjected. It also is a nonequilibrium state. When ions are produced in a superheated liquid, it "boils" in the vicinity of the ion and forms a tiny bubble of vapor. The track of an ionizing particle through it is thus marked by a line of vapor bubbles rather than of liquid droplets. The advantage of the bubble chamber is that the molecules of a liquid are much closer together than those of a gas, so that there is a greater chance that a particle passing through the liquid will collide with a molecule and produce an ion.

PROBLEMS

(*Assume all gases to be ideal.*)

18–1 A tank contains 0.5 m^3 of nitrogen at an absolute pressure of $1.5 \times 10^5 N\ m^{-2}$ and a temperature of 27°C. What will be the pressure if the volume is increased to 5.0 m^3 and the temperature is raised to 327°C?

18–2 A tank having a capacity of 2 ft^3 is filled with oxygen which has a gauge pressure of 60 lb in^{-2} when the temperature is 47°C. At a later time it is found that because of a leak the gauge pressure has dropped to 50 lb in^{-2} and the temperature has decreased to 27°C. Find (a) the mass of the oxygen in the tank under the first set of conditions, (b) the amount of oxygen that has leaked out.

18–3 A flask of volume 2 ℓ, provided with a stopcock, contains oxygen at 300 K and atmospheric pressure. The system is heated to a temperature of 400 K, with the stopcock open to the atmosphere. The stopcock is then closed and the flask cooled to its original temperature.

a) What is the final pressure of the oxygen in the flask?
b) How many grams of oxygen remain in the flask?

18–4 A balloon whose volume is 20,000 ft^3 is to be filled with hydrogen at atmospheric pressure.

a) If the hydrogen is stored in cylinders of volume 2 ft^3 at an absolute pressure of 200 lb in^{-2}, how many cylinders are required?
b) What is the total weight that can be supported by the balloon, in air at standard conditions?
c) What weight could be supported if the balloon were filled with helium instead of hydrogen?

18–5 Derive from the equation of state of an ideal gas an equation for the density of an ideal gas in terms of pressure, temperature, and appropriate constants.

18–6 At the beginning of the compression stroke, a cylinder of a diesel engine contains 48 in^3 of air at atmospheric pressure and a temperature of 27°C. At the end of the stroke, the air has been compressed to a volume of 3 in^3 and the gauge pressure has increased to 600 lb in^{-2}. Compute the temperature.

18–7 A bubble of air rises from the bottom of a lake, where the pressure is 3 atm, to the surface, where the pressure is 1 atm. The temperature at the bottom of the lake is 7°C and the temperature at the surface is 27°C. What is the ratio of the volume of the bubble as it reaches the surface to its volume at the bottom?

18–8 A liter of helium under a pressure of 2 atm and at a temperature of 27°C is heated until both pressure and volume are doubled.

a) What is the final temperature?
b) How many grams of helium are there?

18–9 A flask contains 1 g of oxygen at an absolute pressure of 10 atm and at a temperature of 47°C. At a later time it is found that because of a leak the pressure has dropped to 5/8 of its original value and the temperature has decreased to 27°C.

a) What is the volume of the flask?
b) How many grams of oxygen leaked out between the two observations?

18–10 The submarine *Squalus* sank at a point where the depth of water was 240 ft. The temperature at the surface is 27°C and at the bottom it is 7°C. The density of sea water may be taken as 2 slugs ft^{-3}.

a) If a diving bell in the form of a circular cylinder 8 ft high, open at the bottom and closed at the top, is lowered to this depth, to what height will the water rise within it when it reaches the bottom?
b) At what gauge pressure must compressed air be supplied to the bell while on the bottom to expel all the water from it?

18–11 A bicycle pump is full of air at an absolute pressure of 15 lb in^{-2}. The length of stroke of the pump is 18 in. At what part of the stroke does air begin to enter a tire in which the gauge pressure is 40 lb in^{-2}? Assume the compression to be isothermal.

18–12 A vertical cylindrical tank 1 m high has its top end closed by a tightly fitting frictionless piston of negligible weight. The air inside the cylinder is at an absolute pressure of 1 atm. The piston is depressed by pouring mercury on it slowly. How far will the piston descend before mercury spills over the top of the cylinder? The temperature of the air is maintained constant.

18–13 A barometer is made of a tube 90 cm long and of cross section 1.5 cm^2. Mercury stands in this tube to a height of 75 cm. The room temperature is 27°C. A small amount of nitrogen is introduced into the evacuated space above the mercury and the column drops to a height of 70 cm. How many grams of nitrogen were introduced?

18–14 A large tank of water has a hose connected to it, as shown in Fig. 18–14. The tank is sealed at the top and has compressed air between the water surface and the top. When the water height h_2 is 3 m, the gauge pressure p_1 is $1.0 \times 10^5 \text{N m}^{-2}$. Assume that the air above the water surface expands isothermally.

a) What is the velocity of flow out of the hose when $h_2 = 3$ m?
b) What is the velocity of flow when h_2 has decreased to 2 m? Neglect friction.

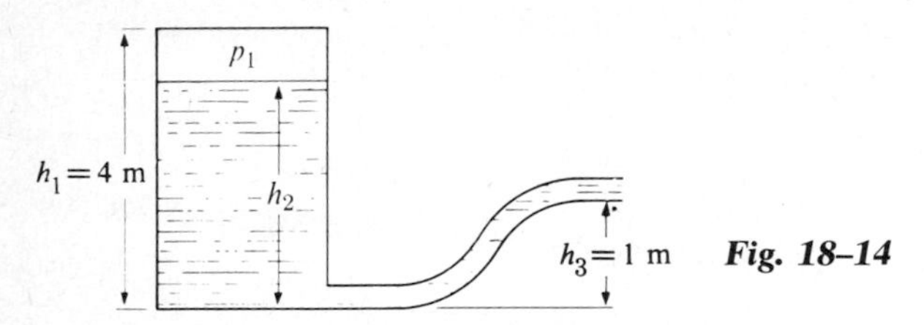

Fig. 18–14

18–15 The volume of an ideal gas is 4 ℓ, the pressure is 2 atm, and the temperature is 300 K. The gas first expands at constant pressure to twice its original volume; it is then compressed isothermally to its original volume, and finally cooled at constant volume to its original pressure.

a) Show the process in a pV-diagram.
b) Compute the temperature during the isothermal compression.
c) Compute the maximum pressure.

18–16

a) Show that the density of an ideal gas is given by pM/RT.
b) What is the density of air at normal atmospheric pressure and 20° C?
c) Under these conditions, what is the total mass of air in a dormitory room 14 m × 3 m × 3 m?

18–17 If air pressure at sea level is 1.00×10^5 Pa, what is air pressure at an elevation of 10,000 m, a common altitude for jet aircraft flight, if the temperature of the air is a uniform 0° C? (This is somewhat higher than the summit of Mt. Everest, 8882 m; this effect illustrates why it is difficult to breath on the summit of Everest.)

18–18 Compare the density of water at the critical point with its density at 0° C and 1 atm; what is the ratio of the two densities?

18–19 Compare the density of carbon dioxide at the critical point to the density of solid carbon dioxide ("dry ice") at 1 atm, which is about 1.56 g cm^{-3}.

18–20 Construct two graphs for a real substance, one showing pressure as a function of volume, and the other showing pressure as a function of temperature. Show on each graph the region in which the substance exists as (a) a gas or vapor, (b) a liquid, (c) a solid. Show also the triple point and the critical point.

18–21 A small amount of liquid is introduced into a glass tube, all air is removed, and the tube is sealed off. Describe the behavior of the meniscus when the temperature of the system is raised:

a) If the volume of the tube is much greater than the critical volume.
b) If the volume of the tube is much less than the critical volume.
c) If the volume of the tube is only slightly different from the critical volume.

18–22 A piece of ice at 0°C is placed alongside a beaker of water at 0°C in a glass vessel, from which all air has been removed. If the ice, water, and vessel are all maintained at a temperature of 0°C by a suitable thermostat, describe the final equilibrium state inside the vessel.

18–23

a) What is the relative humidity on a day when the temperature is 68°F and the dew point is 41°F?
b) What is the partial pressure of water vapor in the atmosphere?
c) What is the absolute humidity, in grams per cubic meter?

18–24 The temperature in a room is 40°C. A can is gradually cooled by adding cold water. At 10°C the surface of the can clouds over. What is the relative humidity in the room?

18–25 A pan of water is placed in a sealed room of volume 60 m^3 at a temperature of 27°C and initial relative humidity of 60 percent.

a) How many grams of water will evaporate?
b) What is the absolute humidity in g m^{-3} after equilibrium has been reached?
c) If the temperature of the room is then increased 1 C°, how many more grams of water will evaporate?

18–26

a) What is the dew-point temperature on a day when the air temperature is 20°C and the relative humidity is 60 percent?

b) What is the absolute humidity, expressed in grams per cubic meter?

18–27 An air-conditioning system is required to increase the relative humidity of 10 ft^3 of air per second from 30 percent to 65 percent. The air temperature is 68°F. How many pounds of water are needed by the system per hour?

18–28 In the lower part of the atmosphere (the troposphere) the temperature is not uniform but decreases with increasing elevation. Show that if the temperature variation is approximated by the linear relation

$$T = T_0 - \alpha y,$$

where T_0 is the temperature at the earth's surface and T is the temperature at a height y, the pressure is given by

$$\ln \frac{p_0}{p} = \frac{Mg}{R\alpha} \ln \frac{T_0}{T_0 - \alpha y}$$

where p_0 is the pressure at the earth's surface and M is the molecular mass.

The coefficient α is called the "lapse rate of temperature." While it varies with atmospheric conditions, an average value is about 0.6 C°/100 m.

18–29 The volume of a closed room, kept at a constant temperature of 20° C, is 60 m^3. The relative humidity in the room is initially 10%. If a pan of water is brought into the room, how many grams will evaporate?

Chapter 19

The Laws of Thermodynamics

19–1 ENERGY AND WORK IN THERMODYNAMICS

Thermodynamics is concerned with energy relationships. The principles of thermodynamics are usually stated with reference to some well-defined *system*, usually a specified quantity of matter. A *thermodynamic system* is one that can interact with its surroundings in at least two ways, one of which must be transfer of heat. A familiar example is a quantity of a gas confined in a cylinder with a piston. Energy can be added to the system by conduction of heat, and it is also possible to do mechanical *work* on the system, since the piston exerts a force which can move through a displacement.

Thermodynamics has its roots in intensely practical problems. A steam engine or steam turbine, for example, makes use of the heat of combustion of coal or other fuel to perform mechanical work to drive an electric generator, pull a train, or perform some other useful function. The gasoline engine in an automobile has a similar function. Because of these practical applications, it is customary to discuss energy relations for a thermodynamic system not in terms of the work done *on* the system *by* its surroundings, but rather in terms of the work done *by the system* on its surroundings, which, in view of Newton's third law, is the negative of the work done *on* the system. This convention will be clarified further in the next section.

Heat can be understood on the basis of microscopic mechanical energy, i.e., the kinetic and potential energies of individual molecules in a material; and it is also possible to develop the principles of thermodynamics from a microscopic viewpoint. In the present chapter, that development is deliberately avoided, in order to emphasize that the central principles and concepts of thermodynamics can be treated in a wholly *macroscopic* way, without reference to microscopic models. Indeed, part of the great power and generality of thermodynamics springs from the fact that it is *not* dependent on details of the structure of matter. In Chapter 20 we return to microscopic considerations and examine their relation to the principles of thermodynamics.

19–2 WORK IN VOLUME CHANGES

When a gas expands, it pushes out on its boundary surfaces as they move outward; hence an expanding gas always does a positive quantity of work. To calculate the work done by a thermodynamic system during a volume change, let us consider a solid or fluid contained in a cylinder equipped with a movable piston, as shown in Fig. 19–1(a). Suppose that the cylinder has a cross-sectional area A and that the pressure exerted by the system at the piston face is p. The force exerted by the system is therefore pA. If the piston moves out an infinitesimal distance dx, the

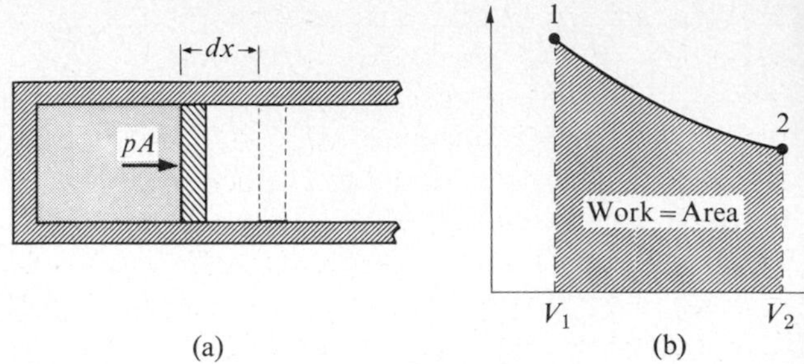

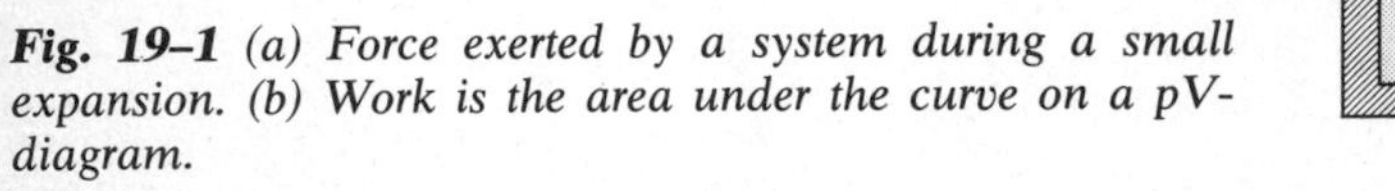
Fig. 19–1 (a) Force exerted by a system during a small expansion. (b) Work is the area under the curve on a *pV*-diagram.

work dW of this force is equal to

$$dW = pA\,dx.$$

But

$$A\,dx = dV,$$

where dV is the change of volume of the system. Therefore

$$dW = p\,dV, \tag{19–1}$$

and in a finite change of volume from V_1 to V_2,

$$\boxed{W = \int_{V_1}^{V_2} p\,dV.} \tag{19–2}$$

In general the pressure of the system will vary during the volume change, and the integral can be evaluated only if the pressure is known as a function of volume. It is customary to represent this relationship graphically by plotting p as a function of V, as in Fig. 19–1(b). Then Eq. (19–2) may be interpreted graphically as the area under the curve between the limits V_1 and V_2. If the system *expands*, then V_2 is greater than V_1 and the work (and the area) are *positive*. *Compression* occurs when the final volume is *less* than the initial volume; the work and the area are then *negative.*

If the pressure remains constant while the volume changes, then the work is

$$W = p(V_2 - V_1) \quad \text{(constant pressure only).}$$

Example An ideal gas is kept in thermal contact with a very large body of constant temperature T and undergoes an *isothermal expansion* in which its volume changes from V_1 to V_2. How much work is done?

From Eq. (19–2)

$$W = \int_{V_1}^{V_2} p\,dV.$$

For an ideal gas

$$p = \frac{nRT}{V}.$$

Since n, R, and T are constant,

$$W = nRT\int_{V_1}^{V_2} \frac{dV}{V} = nRT \ln \frac{V_2}{V_1}. \tag{19–3}$$

In an expansion, $V_2 > V_1$ and W is positive. At constant T,

$$p_1V_1 = p_2V_2, \quad \text{or} \quad \frac{V_2}{V_1} = \frac{p_1}{p_2},$$

and the isothermal work may be expressed also in the form

$$W = nRT \ln \frac{p_1}{p_2}. \tag{19–4}$$

On the pV-diagram in Fig. 19–2 an initial state 1 (characterized by pressure p_1 and volume V_1) and a

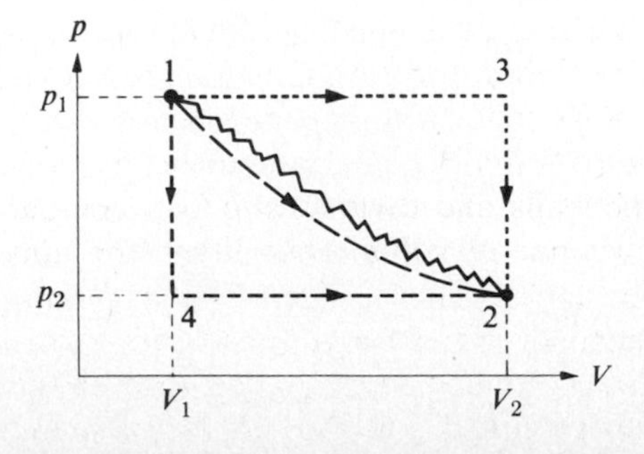

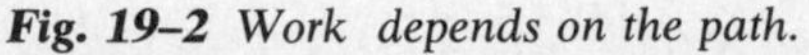
Fig. 19–2 Work depends on the path.

final state 2 (characterized by pressure p_2 and volume V_2) are represented by the two points 1 and 2. There are many ways in which the system may be taken from 1 to 2. For example, the pressure may be kept constant from 1 to 3 (*isobaric* process), and then the volume kept constant from 3 to 2 (*isochoric* process), in which case the work is equal to the area under the line $1 \rightarrow 3$. Another possibility is the path $1 \rightarrow 4 \rightarrow 2$, in which case the work is the area under the line $4 \rightarrow 2$. The jagged line and the continuous curve from 1 to 2 represent other possibilities, in each of which the work is different. We can see, therefore, that *the work depends not only on the initial and final states but also on the intermediate states, i.e., on the path.*

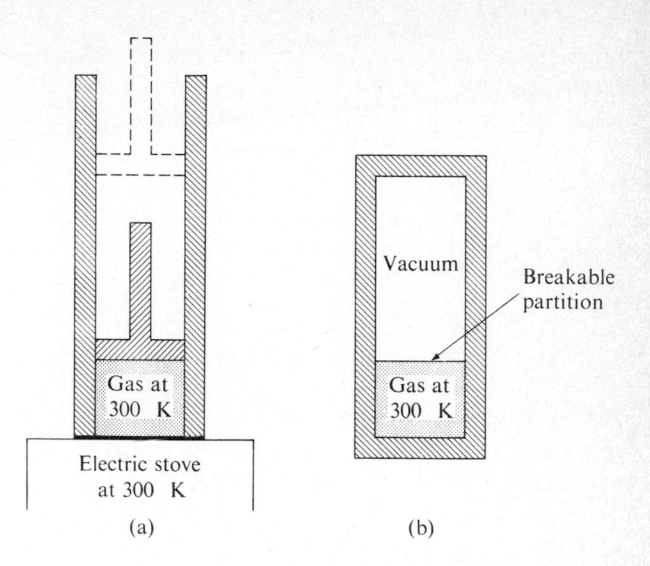

Fig. 19–3 *(a) Slow, controlled, isothermal expansion of a gas from an initial state 1 to a final state 2. (b) Rapid, uncontrolled expansion of the same gas starting at the same state 1 and ending at the same state 2.*

19–3 HEAT IN VOLUME CHANGES

As we have seen in Chapter 16, heat is energy transferred to or from a system by virtue of a temperature difference between the system and its surroundings. Heat is regarded as positive when it enters a system and negative when it leaves. The performance of work and the transfer of heat are *methods of energy transfer*, that is, methods whereby the energy of a system may be increased or decreased.

An important aspect of energy relations is illustrated by the situation of Fig. 19–3. In Fig. 19–3(a), a quantity of gas is contained in a cylinder with a piston, with initial volume of two liters, and maintained at a temperature of 300 K by an electric stove. If the pressure exerted on the piston is only a trifle smaller than the gas pressure, the gas will expand slowly, and heat will flow from the stove to the gas, thereby maintaining the temperature of the gas at the constant value 300 K. Suppose that the gas expands in this slow, controlled isothermal manner until its volume becomes, say, seven liters. A finite amount of heat is absorbed by the gas during this process.

Figure 19–3(b) shows a vessel surrounded by adiabatic walls and divided into two compartments (the lower one of volume two liters, the upper, five liters) by a thin, breakable partition. In the lower compartment the same gas has the same initial volume and temperature as in (a). Suppose the partition is broken and the gas in (b) undergoes a rapid, uncontrolled expansion into the vacuum above, where it encounters no movable piston and therefore exerts no external force. Not only is no work done in this expansion, but also no heat passes through the adiabatic walls. The final volume is seven liters as in (a). The rapid, uncontrolled expansion of a gas into a vacuum is called a *free expansion* and will be discussed later in this chapter. Experiments have indicated that very little temperature change takes place, provided the initial gas pressure is less than a few atmospheres, and therefore the final state of the gas is also the same as the final state in (a). But the states traversed by the gas (the pressures and volumes) while proceeding from state 1 to state 2 are entirely different, so that (a) and (b) represent *two different paths* connecting the *same states* 1 and 2. During path (a) heat is transferred; during path (b) *no* heat is transferred. *Heat*, like work, *depends not only on the initial and final states but also on the path.*

It would be just as incorrect to refer to the "heat in a body" as it would be to speak about the "work in a body." For suppose we assigned an arbitrary value to "the heat in a body" in some standard reference state. The "heat in the body" in some other state would then equal the "heat" in the reference state plus the heat added when the body is carried to the second state. But the heat added depends entirely on the path by which we go from one state to the other, and since there are an infinite number of paths which might be followed, there are an infinite number of values which might equally well be assigned to the "heat in the body" in the second state. Since it is not possible to assign any one value to the "heat in the body," we conclude that this concept is meaningless.

19–4 THE FIRST LAW OF THERMODYNAMICS

The transfer of heat and the performance of work constitute two methods of adding energy to or subtracting energy from a system. Once the transfer of energy has occurred, the system is said to have undergone a change in *internal energy*.

Suppose a system is caused to change from state 1 to state 2 along a definite path and that the heat Q absorbed by the system and the work W done by it are measured. Expressing both Q and W either in thermal units or in mechanical units, we may then calculate the difference $Q - W$. If now we do the same thing over again for many different paths (between the same states 1 and 2), we find experimentally that *$Q - W$ is the same for all paths connecting 1 and 2.* But Q is the energy that has been added to a system by the transfer of heat and W is equal to the energy that has been extracted from the system by the performance of work. The difference $Q - W$, therefore, must represent the internal energy change of the system. It follows that *the internal energy change of a system is independent of the path,* and is therefore equal to the energy of the system in state 2 minus the energy in state 1, or $\Delta U = U_2 - U_1$:

$$\Delta U = U_2 - U_1 = Q - W. \tag{19–5}$$

If some arbitrary value is assigned to the internal energy in some standard reference state, its value in any other state is uniquely defined, since $Q - W$ is the same for all processes connecting the states. Equation (19–5) is known as *the first law of thermodynamics.* In applications of the law in this form, it must be remembered that (1) all quantities must be expressed in the same units, (2) Q is positive when heat goes *into* the system, (3) W is positive when the force exerted *by* the system and the displacement have the same sign, and the system does positive work.

Rewriting Eq. (19–5) in the form $Q = U_2 - U_1 + W$, we may say that when a quantity of heat Q enters a system during a process, some of it ($U_2 - U_1$) remains in the system as increased internal energy, while the remainder (W) leaves the system again in the form of work done by the system against its surroundings.

Internal energy can be interpreted in terms of microscopic mechanical energy, that is, kinetic and potential energies of individual molecules in a material. From the thermodynamic standpoint, however, this is not necessary. Equation (19–5) is the *definition* of the internal energy of a system or, more precisely, of the *change* in its internal energy in any process. As with other forms of energy, only *differences* in internal energy are defined, and not absolute values.

If a system is carried through a process which eventually returns it to its initial state (a *cyclic* process), then

$$U_2 = U_1 \quad \text{and} \quad Q = W.$$

Thus, although net work W may be done by the system in the process, energy has not been created, since an equal amount of energy must have flowed into the system as heat Q.

An *isolated* system is one which does no external work and into which there is no flow of heat. Then, for any process taking place in such a system,

$$W = Q = 0$$

and

$$U_2 - U_1 = 0 \quad \text{or} \quad \Delta U = 0.$$

That is, *the internal energy of an isolated system remains constant.* This is the most general statement of the *principle of conservation of energy.* The internal energy of an *isolated* system cannot be changed by any process (mechanical, electrical, chemical, nuclear, or biological) taking place *within* the system. The energy of a system can be changed only by a flow of heat across its boundary, or by the performance of work. If either of these takes place, the system is no longer isolated. The increase in energy of the system is then equal to the energy flowing in as heat, minus the energy flowing out as work.

19-5 ADIABATIC PROCESS

A process that takes place in such a manner that no heat enters or leaves a system is called an *adiabatic process.* For every adiabatic process, $Q = 0$. This may be accomplished either by surrounding the system with a thick layer of heat-insulating material (such as cork, asbestos, firebrick, or styrofoam), or by performing the process quickly. The flow of heat requires finite time, so any process performed quickly enough will be practically adiabatic. Applying the first law to an adiabatic process, we get

$$U_2 - U_1 = -W \quad \text{(adiabatic process).} \tag{19-6}$$

Thus, the change in the internal energy of a system, in an adiabatic process, is equal in absolute magnitude to the work. If the work W is negative, as when a system is compressed, then $-W$ is positive, U_2 is greater than U_1, and the internal energy of the system increases. If W is positive, as when a system expands, the internal energy of the system decreases. An increase of internal energy is usually accompanied by a rise in temperature, and a decrease in internal energy by a temperature drop.

The compression of the mixture of gasoline vapor and air that takes place during the compression stroke of a gasoline engine is an example of an approximately adiabatic process involving a temperature rise. The expansion of the combustion products during the power stroke of the engine is an approximately adiabatic process involving a temperature decrease. Adiabatic processes, therefore, play a very important role in mechanical engineering.

19-6 ISOCHORIC PROCESS

When a substance undergoes a process in which the volume remains unchanged, the process is called *isochoric.* The rise of pressure and temperature produced by a flow of heat into a substance contained in a rigid container of fixed volume is an example of an isochoric process. When the volume does not change, no work is done and therefore, from the first law

$$U_2 - U_1 = Q \quad \text{(isochoric process).} \tag{19-7}$$

All the added heat has served to increase the internal energy. The very sudden increase of temperature and pressure accompanying the explosion of gasoline vapor and air in a gasoline engine may be treated mathematically as though it were an isochoric addition of heat.

19-7 ISOTHERMAL PROCESS

A process taking place at constant temperature is said to be *isothermal.* In order for the temperature to remain constant, the changes in the pressure and volume must be carried out very slowly, so that the state approximates thermal equilibrium very closely at every stage of the process. In general, *none* of the quantities Q, W, or $U_2 - U_1$ is zero.

There are a few special cases in which the internal energy of the system depends only on temperature, not on pressure or volume. An ideal gas and an ideal paramagnetic crystal are two examples. When such a system undergoes an isothermal process, its internal energy does not change, and therefore $Q = W$. This relation does *not* hold for systems other than those mentioned.

19-8 ISOBARIC PROCESS

A process taking place at constant pressure is called an *isobaric process.* When water enters the boiler of a steam engine and is heated to its boiling point, vaporized, and then the steam is superheated, all these processes take place isobarically. Such processes play an important role in mechanical engineering and also in chemistry.

As discussed in Section 19-2,

$$W = p(V_2 - V_1). \text{ (Isobaric process).} \tag{19-8}$$

A simple example is the vaporization of a mass m of liquid to vapor at constant pressure and temperature. If V_L is the volume of liquid and V_V is the volume of vapor, the work done in expanding from V_L to V_V at constant pressure p is

$$W = p(V_V - V_L).$$

The heat absorbed by each unit of mass is the heat of vaporization L. Hence,

$$Q = mL.$$

From the first law,

$$U_V - U_L = mL - p(V_V - V_L). \qquad (19\text{–}9)$$

Example One gram of water (1 cm^3) becomes 1671 cm^3 of steam when boiled at a pressure of 1 atm. The heat of vaporization at this pressure is 2256 J g^{-1}. Compute the external work and the increase in internal energy.

$$\begin{aligned} W &= p(V_V - V_L) \\ &= (1.013 \times 10^5\ \text{N m}^{-2})(1671 - 1) \times 10^{-6}\ \text{m}^3 \\ &= 169\ \text{J}. \end{aligned}$$

From Eq. (19–5),

$$\begin{aligned} U_V - U_L &= mL - W \\ &= 2256\ \text{J} - 169\ \text{J} \\ &= 2087\ \text{J}. \end{aligned}$$

Hence the external work, or the external part of the heat of vaporization, equals 169 J, and the increase in internal energy, or the *internal* part of the heat of vaporization, is 2087 J.

19–9 THROTTLING PROCESS

A *throttling process* is one in which a fluid, originally at a constant high pressure, seeps through a porous wall or a narrow opening (needle valve or "throttling" valve) into a region of constant lower pressure without a transfer of heat taking place. In Fig. 19–4(a), a fluid is discharged from a pump at a high pressure, then passes through a throttling valve into a pipe which leads directly to the intake or low-pressure side of the pump. Every successive element of fluid undergoes the throttling process in a continuous stream.

Consider any element of fluid enclosed between the piston and throttling valve of Fig. 19–4(b). Suppose this piston moves toward the right and another piston on the other side of the valve moves to the right also, at such rates that the pressure on the left remains at a constant high value, p_1, and that on the right at a constant lower value p_2. After all the fluid has been forced through the valve, the final state is that of Fig. 19–4(c).

The net work done in this process is the difference between the work done in forcing the righthand piston out and the work done in forcing the lefthand piston in. Let V_1 and V_2 be the initial and final volumes. Since the low-pressure fluid changes in volume from zero to V_2 at the constant pressure p_2,

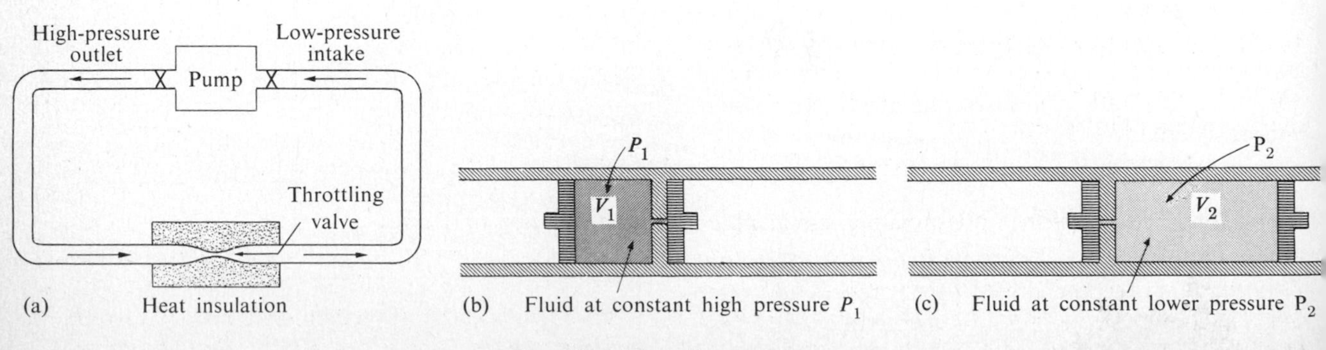

Fig. 19–4 *Throttling process.*

the work is

$$p_2(V_2 - 0),$$

and since the high-pressure fluid changes in volume from V_1 to zero at the constant high pressure p_1, the work is

$$p_1(0 - V_1).$$

The net work W done by the system is, therefore,

$$W = p_2 V_2 - p_1 V_1 .$$

Since the process is adiabatic, $Q = 0$; hence, from the first law,

$$U_2 - U_1 = 0 - (p_2 V_2 - p_1 V_1)$$

or

$$\boxed{U_2 + p_2 V_2 = U_1 + p_1 V_1 .} \qquad (19\text{–}10)$$

This result is of great importance in steam engineering and in refrigeration. The sum $U + pV$, called the *enthalpy*, is tabulated for steam and for many refrigerants. The throttling process plays the main role in the action of a refrigerator, since this is the process that brings about the drop in temperature needed for refrigeration. Liquids that are about to evaporate (saturated liquids) always undergo a drop in temperature and partial vaporization as a result of a throttling process. Gases, however, may undergo either a temperature rise or a drop, depending on the initial temperature and pressure and on the final pressure.

19–10 DIFFERENTIAL FORM OF THE FIRST LAW

Up to this point we have used the first law of thermodynamics only in its finite form,

$$U_2 - U_1 = Q - W.$$

In this form the equation applies to a process in which states 1 and 2 differ in pressure, volume, and temperature by a finite amount. Suppose states 1 and 2 differ only slightly. Then if only a small amount of heat dQ is transferred, and only a small amount of work dW is done, the energy change dU is also very small. In these circumstances, the first law becomes

$$\boxed{dU = dQ - dW.} \qquad (19\text{–}11)$$

If the system is of such a character that the only work possible is by means of expansion or compression, then $dW = p\,dV$, and

$$\boxed{dU = dQ - p\,dV} \qquad (19\text{–}12)$$

is the *differential form of the first law*, applicable to solids, liquids, and gases.

19–11 INTERNAL ENERGY OF AN IDEAL GAS

Imagine a thermally insulated vessel with rigid walls, divided into two compartments by a partition. Suppose that there is a gas in one compartment and that the other is empty. If the partition is removed, the gas will undergo what is known as a *free expansion* in which no work is done and no heat is transferred. From the first law, since both Q and W are zero, it follows that *the internal energy remains unchanged during a free expansion.* The question as to whether or not the temperature of a gas changes during a free expansion has engaged the attention of physicists for over a hundred years.

This is an important question, for it provides information concerning the dependence of internal energy on other quantities. For if the temperature should change while the internal energy stays the same, one would have to conclude that the internal energy depends on *both* the temperature and the volume, or both the temperature and the pressure, but certainly not on the temperature alone. If, on the other hand, T remains unchanged during a free expansion in which we know U remains unchanged, then the only conclusion that is admissible is that U *is a function of* T *only.*

Experiment has shown that the internal energy of a real gas does depend to some extent on the pressure or volume as well as on the temperature. At sufficiently low pressures and high temperatures, however, this dependence disappears and U becomes a function only of T. Thus the assumption that U depends only on T becomes an additional part of the definition of an ideal gas, and in a free expansion an ideal gas undergoes no temperature change.

19–12 HEAT CAPACITIES OF AN IDEAL GAS

The temperature of a substance may be changed under a variety of conditions. The volume may be kept constant, or the pressure may be kept constant, or both may be allowed to vary in some definite manner. The amount of heat per mole necessary to cause unit rise of temperature is different in each case. In other words, a substance has many different molar heat capacities. Two, however, are particularly useful, namely, those at *constant volume* and at *constant pressure.* There is a simple and important relation between these two molar heat capacities of an ideal gas.

Figure 19–5 shows two isotherms of an ideal gas at temperatures T and $T + dT$. Since the internal energy of an ideal gas depends only on the temperature, it is constant if the temperature is constant and the isotherms are also curves of *constant internal energy.* The internal energy therefore has a constant value U at every point of the isotherm at temperature T, and a constant value $U + dU$ at every point of the isotherm at $T + dT$. It follows that the *change* in internal energy, dU, is the same in all processes in which the gas is taken from *any* point on one isotherm to *any* point on the other. Thus dU is the same for all the processes ab, ac, ad, and ef, in Fig. 19–5.

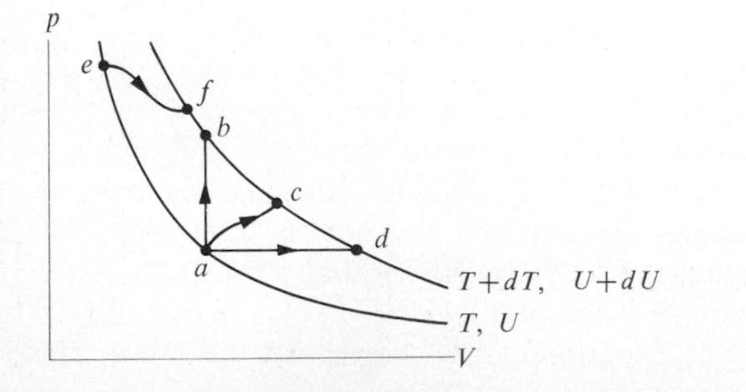

Fig. 19–5 *The change in internal energy of an ideal gas is the same in all processes between two given temperatures.*

Consider first the process ab, at constant volume. To carry out such a process, the gas at temperature T is enclosed in a rigid container and brought in contact with a body at a slightly higher temperature $T + dT$. There will be a flow of heat dQ into the gas, and by definition of the molar heat capacity at constant volume, C_v,

$$dQ = nC_v\,dT. \tag{19–13}$$

The pressure of the gas increases during this process, but no work is done, since the volume is constant. Hence, from the first law (in differential form),

$$dU = dQ - dW,$$

we have

$$\boxed{dU = nC_v\,dT.} \tag{19–14}$$

But dU is the same for *all* processes between the isotherms in Fig. 19–5, so Eq. (19–14) gives the change in internal energy in *all* the processes in Fig. 19–5, *even if they are not at constant volume.*

Now consider the process ad in Fig. 19–5, at constant pressure. To carry out such a process the gas might be enclosed in a cylinder with a piston, acted on by a constant external pressure, and brought in contact with a body at a temperature $T + dT$. As heat flows into the gas it expands at constant pressure and does work. By definition of the molar heat capacity at constant pressure, C_p, the heat dQ flowing into the gas is

$$dQ = nC_p\,dT.$$

The work dW is

$$dW = p\,dV.$$

But from the equation of state, since p is constant,

$$pV = nRT, \qquad p\,dV = nR\,dT,$$

so

$$dW = nR\,dT.$$

Then from the first law,

$$dU = dQ - dW,$$

$$nC_v\,dT = nC_p\,dT - nR\,dT,$$

or

$$\boxed{C_p - C_v = R.} \qquad (19\text{–}15)$$

The molar heat capacity of an ideal gas at constant pressure is therefore *greater* than that at constant volume; the difference is the universal gas constant R. Of course, R must be expressed in the same units as C_p and C_v, usually J mol^{-1} (C°)$^{-1}$ or cal mole^{-1} (C°)$^{-1}$. We have shown in Section 18–2 that, in these units, $R = 8.314$ J mol^{-1} (C°)$^{-1}$ = 1.99 cal mole^{-1} (C°)$^{-1}$.

Although Eq. (19–15) was derived for an ideal gas, it is very nearly true for *real* gases at moderate pressures. Measured values of C_p and C_v are given in Table 19–1 for some real gases at low pressures, and the difference is seen to be very nearly 1.99 cal mol^{-1} (C°)$^{-1}$.

In the last column of Table 19–1 are listed the values of the ratio C_p/C_v, denoted by the Greek letter γ (gamma). It is seen that γ is 1.67 for monatomic gases, and is very nearly 1.40 for the so-called permanent diatomic gases. There is no simple regularity for polyatomic gases.

Solids and liquids also expand when heated, if free to do so, and hence perform work. The coefficients of volume expansion of solids and liquids are, however, so much smaller than those of gases that the work is small. The internal energy of a solid or liquid *does* depend on its volume as well as its temperature, and this must be considered when evaluating the difference between specific heats of solids or liquids. It turns out that here also $C_p > C_v$, but the difference is small and is not expressible as simply as that for a gas. Because of the large stresses set up when solids or liquids are heated and *not* allowed to expand, most heating processes involving them take place at constant pressure, and hence C_p is the quantity usually measured for a solid or liquid.

Table 19–1 MOLAR HEAT CAPACITIES OF GASES AT LOW PRESSURE

Type of gas	Gas	C_p, cal mol^{-1} (C°)$^{-1}$	C_v, cal mol^{-1} (C°)$^{-1}$	$C_p - C_v$	$\gamma = \frac{C_p}{C_v}$
Monatomic	He	4.97	2.98	1.99	1.67
	A	4.97	2.98	1.99	1.67
Diatomic	H_2	6.87	4.88	1.99	1.41
	N_2	6.95	4.96	1.99	1.40
	O_2	7.03	5.04	1.99	1.40
	CO	6.97	4.98	1.99	1.40
Polyatomic	CO_2	8.83	6.80	2.03	1.30
	SO_2	9.65	7.50	2.15	1.29
	H_2S	8.27	6.2	2.1	1.34

19–13 ADIABATIC PROCESS OF AN IDEAL GAS

Any process in which there is no flow of heat into or out of a system is called *adiabatic.* To perform a truly adiabatic process it would be necessary that the system be surrounded by a perfect heat insulator, or that the surroundings of the system be kept always at the same temperature as the system. However, if a process such as compression or expansion of a gas is carried out rapidly, it will be nearly adiabatic, since the flow of heat into or out of the system is slow even under favorable conditions. Thus the compression stroke of a gasoline or diesel engine is approximately adiabatic.

Note that external work may be done *on* or *by* a system in an adiabatic process, and that the temperature usually changes in such a process.

Let an ideal gas undergo an infinitesimal adiabatic process. Then $dQ = 0$, $dU = nC_v\,dT$, $dW = p\,dV$, and, from the first law,

$$nC_v\,dT = -p\,dV. \qquad (19\text{–}16)$$

From the equation of state,

$$p\,dV + V\,dp = nR\,dT.$$

Eliminating dT between these equations and making use of the fact that $C_p - C_v = R$, we obtain the relation

$$\frac{dp}{p} + \frac{C_p}{C_v}\frac{dV}{V} = 0,$$

or, if C_p/C_v is denoted by γ,

$$\frac{dp}{p} + \gamma\frac{dV}{V} = 0.$$

To obtain the relation between p and V in a *finite* adiabatic change we may integrate the preceding equation. This gives

$$\ln p + \gamma \ln V = \ln (\text{constant})$$

or

$$\boxed{p V^{\gamma} = \text{constant}} \qquad \text{(Adiabatic process).}$$

If subscripts 1 and 2 refer to any two points of the process,

$$p_1 V_1^{\gamma} = p_2 V_2^{\gamma}. \qquad (19\text{–}17)$$

It is left as a problem to show that by combining Eq. (19–17) with the equation of state one obtains the alternate forms

$$T_1 V_1^{\gamma-1} = T_2 V_2^{\gamma-1}, \qquad (19\text{–}18)$$

$$T_1 p_1^{(1-\gamma)/\gamma} = T_2 p_2^{(1-\gamma)/\gamma}. \qquad (19\text{–}19)$$

Values of the specific heat ratio γ are listed in Table 19–1 for some common gases.

An adiabatic expansion or compression of an ideal gas may be represented graphically by a plot of Eq. (19–17), as in Fig. 19–6, in which a number of isothermal curves are shown for comparison. The adiabatic curves, at any point, have a somewhat steeper slope than the isothermal curve passing through the same point. That is, as we follow along an adiabatic from right to left (compression process) the curve continually cuts across isotherms of higher and higher temperatures, in agreement with the fact that the temperature generally increases in an adiabatic compression.

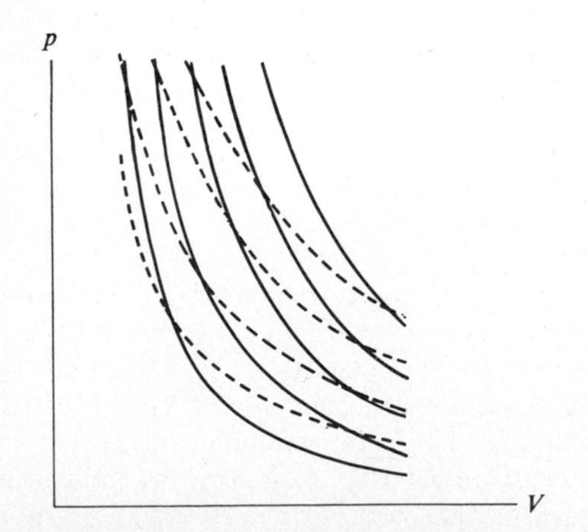

Fig. 19–6 Adiabatic curves (full lines) versus isothermal curves (dotted lines).

The work done by an ideal gas in an adiabatic expansion is computed as follows. We have, from Eq. (19–16),

$$W = \int_{V_1}^{V_2} p\, dV = \int_{T_1}^{T_2} - nC_v\, dT. \qquad (19\text{–}20)$$

Hence the work may be found from either integral. Consider first $\int p\, dV$. Since $p V^{\gamma} = p_1 V_1^{\gamma} = p_2 V_2^{\gamma} =$ a constant, K, we may write

$$W = \int_{V_1}^{V_2} p\, dV = K \int_{V_1}^{V_2} \frac{dV}{V^{\gamma}}$$

$$= \frac{1}{1-\gamma}(K V_2^{1-\gamma} - K V_1^{1-\gamma}).$$

In the first term let $K = p_2 V_2^{\gamma}$ and in the second term let $K = p_1 V_1^{\gamma}$. Then

$$W = \frac{p_2 V_2 - p_1 V_1}{1 - \gamma}. \qquad (19\text{–}21)$$

This expresses the work in terms of initial and final pressures and volumes. If the initial and final temperatures are known, we may return to Eq. (19–20) and write

$$W = \int_{T_1}^{T_2} - nC_v\, dT = nC_v(T_1 - T_2). \qquad (19\text{–}22)$$

Example The compression ratio of a diesel engine, V_1/V_2, is about 15. If the cylinder contains air at 15 lb in^{-2} (absolute) and 60°F (= 520°R) at the start of the compression stroke, compute the pressure and temperature at the end of this stroke. Assume that air behaves like an ideal gas and that the compression is adiabatic. The value of γ for air is 1.40.

From Eq. (19–17),

$$p_2 = p_1 \left(\frac{V_1}{V_2}\right)^{\gamma}$$

$$= (15 \text{ lb in}^{-2})(15)^{1.4}$$

$$= 665 \text{ lb in}^{-2}.$$

The temperature may now be found from Eqs. (19–18) or (19–19), or by the ideal gas law. Thus, from Eq. (19–18), we have

$$T_2 = T_1 \left(\frac{V_1}{V_2}\right)^{\gamma-1}$$
$$= 520°\text{R}\,(15)^{1.4-1}$$
$$= 1536°\text{R} = 1076°\text{F}.$$

Or,

$$T_2 = T_1 \frac{p_2 V_2}{p_1 V_1}$$
$$= (520°\text{R})\frac{(665\text{ lb in}^{-2})(1)}{(15\text{ lb in}^{-2})(15)}$$
$$= 1536°\text{R}.$$

The work done in the compression stroke is found as follows: Let the initial volume V_1 be 60 in^3; then $V_2 = 4$ in^3. From Eq. (19–21),

$$W = \frac{(665\text{ lb in}^{-2})(4\text{ in}^3) - (15\text{ lb in}^{-2})(60\text{ in}^3)}{1 - 1.40}$$
$$= -4400\text{ in. lb}$$
$$= -367\text{ ft lb}.$$

In the above analysis we have used the ideal-gas equation of state, which holds only when the state changes slowly enough so that, at each step, the pressure and temperature are *uniform* throughout the gas. Thus, the validity of our results is limited to situations where the process is rapid enough to prevent appreciable heat exchange with the surroundings, yet slow enough so the system does not depart very far from thermal and mechanical equilibrium.

19–14 HEAT ENGINES

The dominating feature of an industrial society is its ability to utilize, whether for wise or unwise ends, sources of energy other than the muscles of men or animals. Except for waterpower, where mechanical energy is directly available, most energy supplies are in the form of potential energy of molecular or nuclear aggregations. In chemical or nuclear reactions, some of this potential energy is released and converted to random molecular kinetic energy. Heat can be withdrawn and utilized for heating buildings, for cooking, or for maintaining a furnace at a high temperature in order to carry out other chemical or physical processes. But to operate a machine, or to propel a vehicle or projectile, *mechanical* energy is required, and one of the problems of the mechanical engineer is to withdraw heat from a high-temperature source and convert as large a fraction as possible to mechanical energy.

This transformation always requires the services of a *heat engine*, such as a steam engine, gasoline engine, diesel engine, or jet engine. We consider for simplicity an engine in which the so-called "working substance" is carried through a *cyclic* process, that is, a sequence of processes in which it eventually returns to its original state. In the condensing type of steam engine used in marine propulsion, the "working substance," in this case pure water, is actually used over and over again. Water is evaporated in the boilers at high pressure and temperature, does work in expanding against a piston or in a turbine, is condensed by cooling water from the ocean, and pumped back into the boilers. The refrigerant in a household refrigerator also undergoes a cyclic process. Internal combustion engines and steam locomotives do not carry a system through a closed cycle, but they can be analyzed in terms of cyclic processes which approximate their actual operations.

All these devices absorb heat from a source at a high temperature, perform some mechanical work, and reject heat at a lower temperature. When a system is carried through a cyclic process, its initial and final internal energies are equal, and from the first law, for any number of complete cycles,

$$U_2 - U_1 = 0 = Q - W,$$
$$Q = W.$$

That is, the ***net*** *heat flowing into the engine in a cyclic process equals the* ***net*** *work done by the engine.*

The work is represented by the area enclosed by the curve representing the process in the pV-plane. Thus in Fig. 19–7, for example, where the closed curve shows an arbitrary cyclic process, the area under the upper curve from a to b represents work done *by* the system (positive work) in the expansion from a to b, while the area under the lower curve from b to a

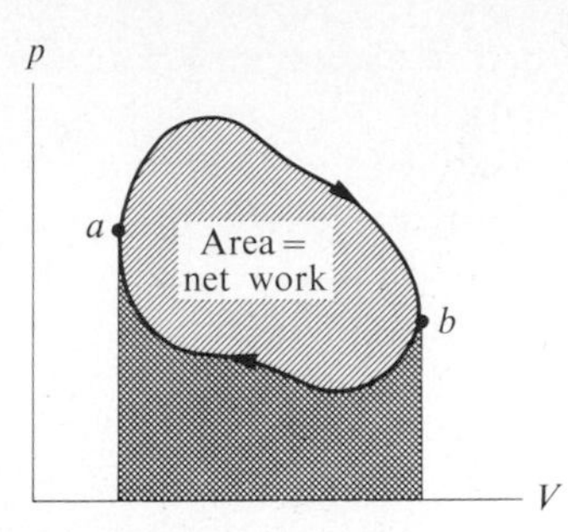

Fig. 19–7 *The area enclosed by the curve representing a cyclic process equals the net work.*

represents work done *on* the system (negative work) in the compression from b to a. Since the average pressure during the compression process is less than that during the expansion, the positive work exceeds the negative work and the area bounded by the closed curve is the net positive work done by the system. If the same process were traversed counterclockwise the net work done by the system would be negative.

In the operation of a heat engine or of a refrigerator there are always two bodies capable of either providing or absorbing large quantities of heat without appreciable changes of temperature. Thus the flames and hot gases surrounding the boilers of a marine steam installation can give up large quantities of heat at a high temperature, and constitute therefore what may be called the *hot reservoir.* Heat transferred between the hot reservoir and the working substance in a heat engine will be represented by the symbol Q_H, where it is understood that a positive value of Q_H means heat entering the working substance. The ocean water used to cool the condenser of the marine installation constitutes the *cold reservoir.* (The words "hot" and "cold" are, of course, relative.) The heat transferred between the working substance and the cold reservoir will be denoted by Q_C. A negative value of Q_C means heat rejected by the working substance.

The energy transformations in a heat engine are conveniently represented schematically by the *flow diagram* of Fig. 19–8. The engine itself is represented by the circle. The heat Q_H supplied to the engine by the hot reservoir is proportional to the cross section of the incoming "pipeline" at the top of the diagram. The cross section of the outgoing pipeline at the bottom is proportional to the heat Q_C which is rejected as heat in the exhaust. The branch line to the right represents that portion of the heat supplied which the engine converts to mechanical work, W.

Consider a heat engine operating in a cycle over and over again and let Q_H and Q_C stand for the heats absorbed and rejected by the working substance *per cycle.* The net heat absorbed is

$$Q = Q_H + Q_C, \tag{19–23}$$

where Q_C is a negative number. The useful output of the engine is the net work W done by the working substance, and from the first law,

$$W = Q = Q_H + Q_C. \tag{19–24}$$

The heat *absorbed* is usually obtained from the combustion of fuel. The heat *rejected* ordinarily has no economic value. The *thermal efficiency* of a cycle is defined as the ratio of the useful work to the heat absorbed ("what you get" divided by "what you pay for"):

$$\boxed{\text{Thermal efficiency} = \frac{W}{Q_H} = \frac{Q_H + Q_C}{Q_H}.} \tag{19–25}$$

Because of friction losses, the *useful* work delivered by an engine is less than the work W, and the overall efficiency is less than the thermal efficiency.

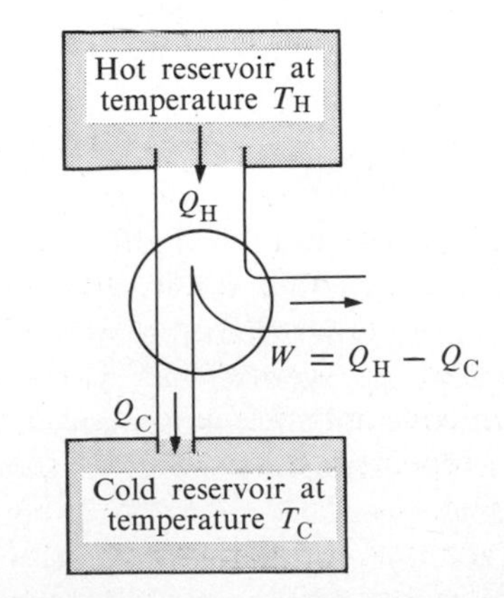

Fig. 19–8 *Schematic flow diagram of a heat engine.*

In terms of the flow diagram in Fig. 19–8, the most efficient engine is the one for which the branch pipeline representing the work obtained is as large as possible, and the exhaust pipeline representing the heat rejected is as small as possible, for a given incoming pipeline or quantity of heat supplied.

We shall now consider, without going into the mechanical details of their construction, the gasoline engine, the diesel engine, and the steam engine.

19–15 THE GASOLINE ENGINE

The common gasoline engine is of the four-cycle type, so called because four processes take place in each cycle. Starting with the piston at the top of its stroke, an explosive mixture of air and gasoline vapor is drawn into the cylinder on the downstroke, the intake valve being open and the exhaust valve closed. This is the *intake* stroke. At the end of this stroke the intake valve closes and the piston rises, performing an approximately adiabatic compression of the air–gasoline mixture. This is the *compression* stroke. At or near the top of this stroke a spark ignites the mixture of air and gasoline vapor, and combustion takes place very rapidly. The pressure and temperature increase at nearly constant volume.

The piston is now forced down, the burned gases expanding approximately adiabatically. This is the *power stroke* or *working stroke*. At the end of the power stroke the exhaust valve opens. The pressure in the cylinder drops rapidly to atmospheric, and the rising piston on the *exhaust stroke* forces out most of the remaining gas. The exhaust valve now closes, the intake valve opens, and the cycle is repeated.

For purposes of computation, the gasoline engine cycle is approximated by the *Otto* cycle illustrated in Fig. 19–9. Starting at point *a*, air at atmospheric pressure is compressed adiabatically in a cylinder to point *b*, heated at constant volume to point *c*, allowed to expand adiabatically to point *d*, and cooled at constant volume to point *a*, after which the cycle is repeated. Line *ab* corresponds to the compression stroke, *bc* to the explosion, *cd* to the working stroke, and *da* to the exhaust of a gasoline engine. V_1 and V_2 in Fig. 19–9 are respectively the maximum and minimum volumes of the air in the cylinder. The ratio V_1/V_2 is called the *compression ratio*, and is about 10 for a modern internal combustion engine.

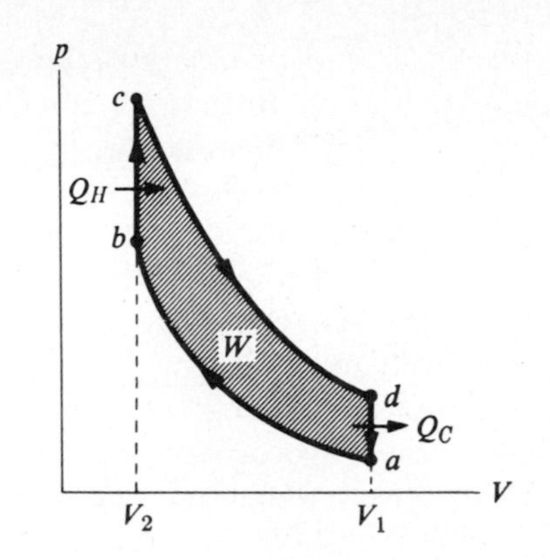

Fig. 19–9 *pV-diagram of the Otto cycle*

The work output in Fig. 19–9 is represented by the shaded area enclosed by the figure *abcd*. The heat *input* is the heat supplied at constant volume along the line *bc*. The exhaust heat is removed along *da*. No heat is supplied or removed in the adiabatic processes *ab* and *cd*.

The heat input and the work output can be computed in terms of the compression ratio, assuming air to behave like an ideal gas. The result is

$$\text{Eff}(\%) = 100\left(1 - \frac{1}{(V_1/V_2)^{\gamma-1}}\right),$$

where γ is the ratio of the specific heat capacity at constant pressure to the specific heat capacity at constant volume, C_p/C_v. For a compression ratio of 10 and a value of $\gamma = 1.4$, the efficiency is about 60%. It will be seen that the higher the compression ratio, the higher the efficiency. However, increasing the compression ratio also increases the temperature at the end of the adiabatic compression of the gas–air mixture. If it is too high, the mixture explodes spontaneously and prematurely, instead of burning evenly after ignition by the spark plug. The mechanical strength and wear of engine parts also pose problems; so the maximum practical compression ratio for ordinary gasoline is about 10. Higher ratios can be used with more exotic fuels.

The Otto cycle is of course a highly idealized model; it neglects friction, turbulence, loss of heat to cylinder walls, and many other effects which combine to reduce the efficiency of a real engine. Another source of inefficiency is incomplete combustion. A mixture of gasoline vapor with just enough air for complete combustion of the hydrocarbons to H_2O and CO_2 does not ignite readily. Reliable ignition requires a mixture "richer" in gasoline, but this leads to CO and unburned hydrocarbons in the exhaust. The heat obtained from the gasoline is then less than the total heat of combustion; the difference is wasted, and the exhaust contributes to air pollution. One attack on this problem is the stratified-charge engine, in which the concentration of gasoline vapor near the spark plug is greater than in the remainder of the combustion chamber.

19–16 THE DIESEL ENGINE

In the Diesel cycle, air is drawn into the cylinder on the intake stroke and compressed adiabatically on the compression stroke, to a sufficiently high temperature so that fuel oil injected at the end of this stroke burns in the cylinder without requiring ignition by a spark. The combustion is not as rapid as in the gasoline engine, and the first part of the power stroke proceeds at essentially constant pressure. The remainder of the power stroke is an adiabatic expansion. This is followed by an exhaust stroke which completes the cycle.

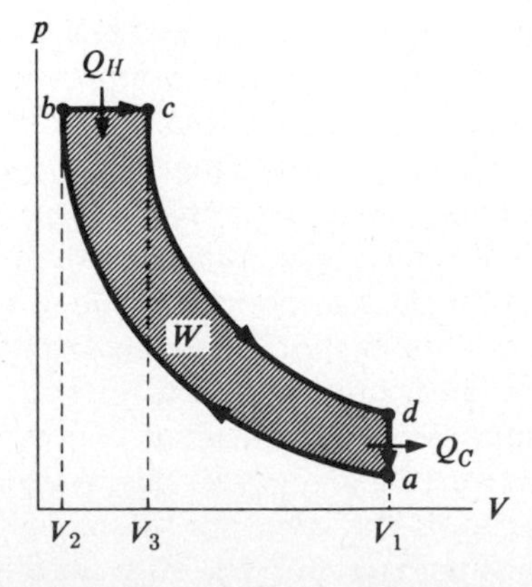

Fig. 19–10 *pV-diagram of the Diesel cycle.*

The idealized air–Diesel cycle is shown in Fig. 19–10. Starting at point *a*, air is compressed adiabatically to point *b*, heated at constant pressure to point *c*, expanded adiabatically to point *d* and cooled at constant volume to point *a*.

Since there is no fuel in the cylinder of a diesel engine on the compression stroke, pre-ignition cannot occur and the compression ratio V_1/V_2 may be much higher than that of an internal combustion engine. A value of 15 is typical. The *expansion* ratio V_1/V_3 may be about 5. Using these values, and taking $\gamma = 1.4$, we compute the efficiency of the air–Diesel cycle at about 56%. As with the Otto cycle, this description is an idealized model; efficiencies of real engines are considerably less than this.

19–17 THE STEAM ENGINE

The condensing type of steam engine performs the following sequence of operations. Water is converted to steam in the boiler, and the steam thus formed is superheated above the boiler temperature. Superheated steam is admitted to the cylinder, where it expands against a piston; connection is maintained to the boiler for the first part of the working stroke, which thus takes place at constant pressure. The inlet valve is then closed and the steam expands adiabatically for the rest of the working stroke. The adiabatic cooling causes some of the steam to condense. The mixture of water droplets and steam (known as "wet" steam) is forced out of the cylinder on the return stroke and into the condenser where the remaining steam is condensed into water. This water is forced into the boiler by the feed pump, and the cycle is repeated.

An idealized cycle (called the Rankine cycle) which approximates the actual steam cycle is shown in Fig. 19–11. Starting with liquid water at low pressure and temperature (point *a*), the water is compressed adiabatically to point *b* at boiler pressure. It is then heated at constant pressure to its boiling point (line *bc*), converted to steam (line *cd*), superheated (line *de*), expanded adiabatically with some condensation (line *ef*), and cooled and condensed (along *fa*) to its initial condition.

The efficiency of such a cycle may be computed in the same way as in the previous examples, by

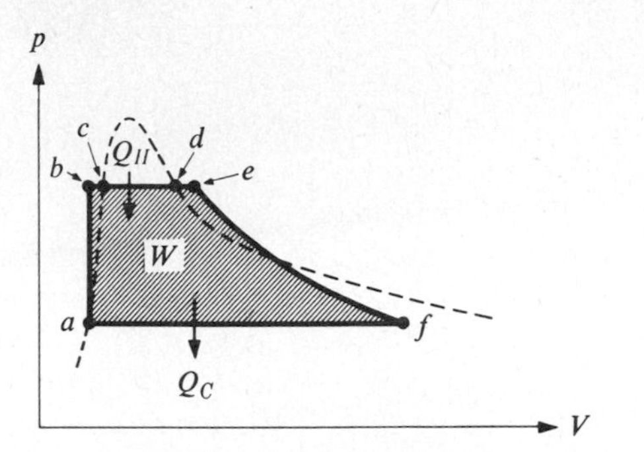

Fig. 19–11 *The Rankine cycle.*

finding the quantities of heat taken in and rejected along the lines *be* and *fa*. Assuming a boiler temperature of 417°F (corresponding to a pressure of 300 lb in^{-2}), a superheat of 63°F above this temperature (480°F) and a condenser temperature of 102°F, the efficiency of a Rankine cycle is about 32%. Efficiencies of actual steam engines are somewhat lower.

The Rankine cycle may also be applied to analysis of the *steam turbine,* in which the steam expands against rotating turbine blades instead of against a piston moving in a cylinder. Nearly all large electric-power plants that burn coal or oil or use nuclear reactors contain steam turbines. The heat obtained from combustion or from the nuclear reaction is used to boil water; the steam drives a steam turbine connected directly to an electric generator, usually on the same shaft.

19–18 THE SECOND LAW OF THERMODYNAMICS

Although their efficiencies differ from one another, none of the heat engines described above has a thermal efficiency of 100%. That is, none of them absorbs heat and converts it *completely* into work. There is nothing in the *first* law of thermodynamics that precludes this possibility. The first law requires only that the energy output of an engine, in the form of mechanical work, shall equal the difference between the energies absorbed and rejected in the form of heat. An engine which rejected no heat and which converted all the heat absorbed to mechanical work would therefore be perfectly consistent with the first law.

We know now that there is another principle, independent of the first law and not derivable from it, which determines the maximum fraction of the energy absorbed by an engine as heat that can be converted to mechanical work. The basis of this principle lies in the difference between the natures of internal energy and of mechanical energy. The former is the energy of *random* molecular motion, while the latter represents *ordered* molecular motion. Superposed on their random motion, the molecules of a moving body have an ordered motion in the direction of the velocity of the body. The total molecular kinetic energy associated with the ordered motion is what we call in mechanics the kinetic energy of the moving body. The kinetic (and potential) energy associated with the *random* motion constitutes the internal energy. When the moving body makes an inelastic collision and comes to rest, the *ordered* portion of the molecular kinetic energy becomes converted to *random* motion. Since we cannot control the motions of individual molecules, it is impossible to reconvert the random motion *completely* to ordered motion. We can, however, convert *a portion* of it; this is what is accomplished by a heat engine.

The impossibility of converting heat *completely* into mechanical energy forms the basis of one form of the *second law of thermodynamics.* Although we have introduced it with a discussion of molecular motion, the second law (as with all of thermodynamics) deals with directly measurable quantities such as heat and work, and can be stated entirely apart from any molecular theory. One statement of the second law is as follows. *No process is possible whose sole result is the absorption of heat from a reservoir at a single temperature and the conversion of this heat completely into mechanical work.*

If the second law were not true, it would be possible to drive a steamship across the ocean by extracting heat from the ocean or to run a power plant by extracting heat from the surrounding air. It should be noted again that neither of these "impossibilities" violates the first law of thermodynamics. After all, both the ocean and the surrounding air contain an enormous store of internal energy which, in principle, may be extracted in the form of a flow of heat. The second law, therefore, is not a deduction from the first but stands by itself as a separate law of nature, referring to an aspect of nature different from that contemplated by the first law. The first law denies the possibility of creating or destroying energy; the second denies the possibility of utilizing energy in a particular way.

The fact that work may be dissipated completely into heat, whereas heat may *not* be converted entirely into work, expresses an essential one-sidedness of nature. All natural, spontaneous processes may be studied in the light of the second law, and in all such cases, this peculiar one-sidedness is found. Thus, heat always flows spontaneously from a hotter to a colder body; gases always seep through an opening spontaneously from a region of high pressure to a region of low pressure; gases and liquids left by themselves always tend to mix, not to unmix. Salt dissolves in water but a salt solution does not separate by itself into pure salt and pure water. Rocks weather and crumble; iron rusts; people grow old. These are all examples of *irreversible* processes that take place naturally in only one direction and, by their one-sidedness, express the second law of thermodynamics.

19-19 THE REFRIGERATOR

A refrigerator may be considered to be a heat engine operated in reverse. A heat engine takes in heat from a hot reservoir, converts a part of the heat into mechanical work output, and rejects the diference as heat to a cold reservoir. A refrigerator, however, takes in heat from a cold reservoir, the compressor supplies mechanical work *input*, and heat is rejected to a hot reservoir. With reference to the ordinary home refrigerator, the food and ice cubes constitute the cold reservoir, work is done by the electric motor, and the hot reservoir is the air in the kitchen.

The flow diagram of a refrigerator is given in Fig. 19-12. In one cycle, heat Q_C enters the refrigerator at low temperature T_C, work W is done on the refrigerator and heat Q_H leaves at a higher temperature T_H. Both W and Q_H are negative quantities, while Q_C is positive. It follows from the first law that

$$-Q_H = Q_C - W,$$

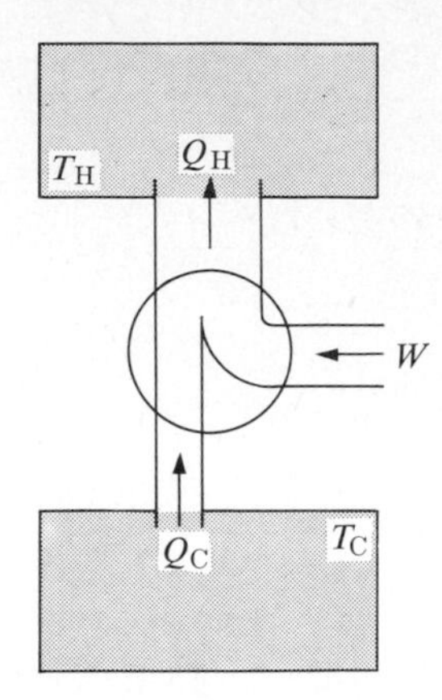

Fig. 19-12 *Schematic flows diagram of a refrigerator.*

and the heat rejected to the hot reservoir is the sum of the heat taken from the cold reservoir and the heat equivalent of the work done by the motor.

From an economic point of view, the best refrigeration cycle is one that removes the greatest amount of heat Q_C from the refrigerator, for the least expenditure of mechanical work W. We therefore define *the coefficient of performance* (rather than the efficiency) of a refrigerator as the ratio $-Q_C/W$, and since $W = Q_H + Q_C$,

$$\text{Coefficient of performance} = -\frac{Q_C}{Q_H + Q_C}. \tag{19-26}$$

The principles of the common refrigeration cycle are illustrated schematically in Fig. 19-13. Compressor A delivers gas (CCl_2F_2, NH_3, etc.) at high temperature and pressure to coils B. Heat is removed from the gas in B by water or air cooling, resulting in condensation of the gas to a liquid still under high pressure. The liquid passes through the throttling valve or expansion valve C, emerging as a mixture of liquid and vapor at a lower temperature. In coils D, heat is supplied that converts the remaining liquid into vapor which enters compressor A to repeat the cycle. In a domestic refrigerator, coils D are placed in the ice compartment, where they cool the refrigerator directly. In a larger refrigerating

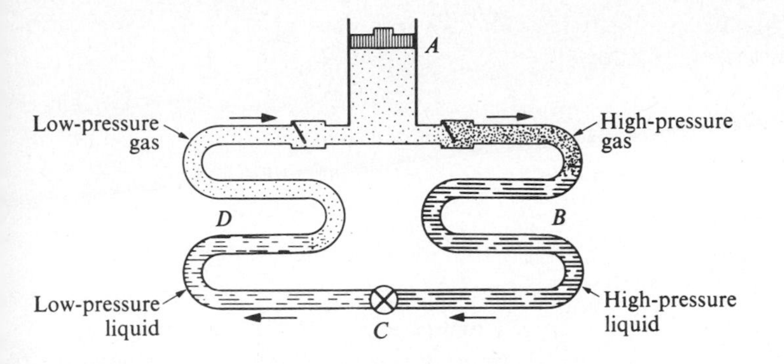

Fig. 19–13 *Principle of the mechanical refrigeration cycle.*

plant, these coils are usually immersed in a brine tank and cool the brine, which is then pumped to the refrigerating rooms.

If no work were needed to operate a refrigerator, the coefficient of performance (heat extracted divided by work done) would be infinite. Coefficients of performance of actual refrigerators vary from about 2 to about 6. Experience shows that work is always needed to transfer heat from a colder to a hotter body. This negative statement leads to another statement of the second law of thermodynamics, namely:

No process is possible whose sole result is the transfer of heat from a cooler to a hotter body.

At first sight this and the previous statement of the second law appear to be quite unconnected, but it can be shown that they are in all respects equivalent. Any device that would violate one statement would violate the other. For example, a "workless" refrigerator which violates this "refrigerator" statement of the second law could be used in conjunction with a real heat engine to violate our first statement of the second law by pumping the rejected heat back to the hot reservoir to be reused.

19-20 THE CARNOT CYCLE

Although their efficiencies differ from one another, none of the heat engines which have been described has an efficiency of 100%. An important question remains to be discussed. What is the maximum attainable efficiency, given a supply of heat at one temperature and a reservoir at a lower temperature for cooling the exhaust? An idealized engine which can be shown to have the maximum efficiency under these conditions was invented by the French engineer Saki Carnot in 1824 and is called a *Carnot engine.* The *Carnot cycle*, shown in a p-V diagram in Fig. 19–14, differs from the Otto and Diesel cycles in that it is bounded by two *isothermals* and two *adiabatics.* Thus all the heat input is supplied at a *single* high temperature and all the heat output is rejected at a *single* lower temperature. (Compare with Figs. 19–9 and 19–10, in which the temperature is different at all points of the lines bc and da.)

This, however, is not the only feature of the Carnot cycle. There are no "one-way" processes in the Carnot cycle, such as explosions or throttling processes. The isothermal and adiabatic processes of

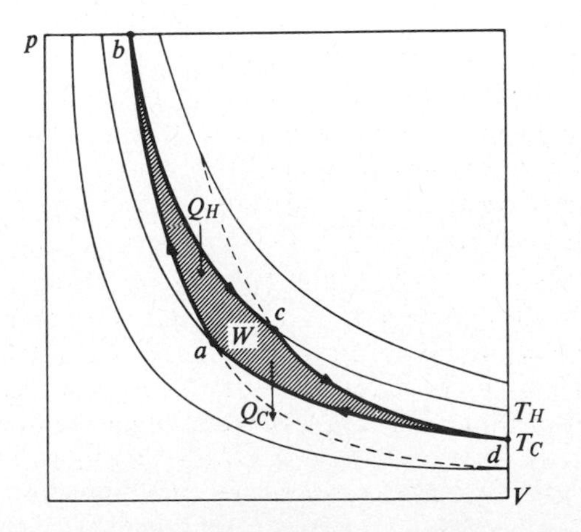

Fig. 19–14 *The Carnot cycle.*

the Carnot cycle are idealizations of actual processes. The direction of either process may be *reversed* by only a slight change in the external pressure; there is no friction, and the working substance is always very close to equilibrium.

A process that is attended by no equilibrium-disturbing effect (such as an unbalanced force or a finite temperature difference) *and no dissipative effect* (such as friction or electrical resistance) *is said to be* ***reversible.*** (A reversible process represents the same sort of idealization in thermodynamics that a point mass, a weightless inextensible cord, or a frictionless, massless pulley represents in mechanics. It may be approximated but not perfectly achieved.)

The isothermal and adiabatic processes of the Carnot cycle may be imagined to proceed in either direction. In the direction shown in Fig. 19–14, heat Q_H goes in, heat Q_C goes out, and work W is done by the engine. If the arrows in the figure are reversed, the cycle becomes a Carnot refrigeration cycle and, *what is most important, the same amount* of heat Q_H which formerly was taken in from the hot reservoir now goes out, *the same amount* of work W, which formerly was delivered to the outside is now required from the surroundings, and *the same amount* of heat Q_C which formerly was rejected to the cold reservoir is now taken in. These numerical equalities would *not* exist if any ordinary engine cycle were reversed.

Suppose an engine (not a Carnot engine) were to operate between a source of heat at some temperature and a reservoir of heat at a lower temperature, thereby delivering to the outside an amount of work W. Suppose this work W were used to operate a Carnot refrigerator which extracted heat from the colder reservoir and delivered it to the warmer source. It can be shown that if the first engine were more efficient than the engine that would result by operating the Carnot refrigerator backward, then the net effect would be a violation of the second law of thermodynamics. Proceeding along these lines, it has been proved that:

No engine operating between two given temperatures can be more efficient than a Carnot engine operating between the same two temperatures.

The following, known as Carnot's theorem, also follows:

All Carnot engines operating between the same two temperatures have the same efficiency, irrespective of the nature of the working substance.

19–21 THE KELVIN TEMPERATURE SCALE

It follows from Carnot's theorem that the efficiency of a Carnot engine operating between reservoirs at two given temperatures *is independent of the nature of the working substance and is a function only of the temperatures.* If we consider a number of Carnot engines using different working substances and absorbing and rejecting heat to the same two reservoirs, the thermal efficiency is the same for all, or

$$\text{Thermal efficiency} = \frac{Q_H + Q_C}{Q_H} = 1 + \frac{Q_C}{Q_H} = \text{constant.}$$

Hence, the ratio Q_H/Q_C is a constant for all the engines. Kelvin proposed that the ratio of the temperatures of the reservoirs be *defined* as equal to this constant ratio of the absolute magnitudes of the quantities of heat absorbed and rejected or, since Q_C is a negative quantity, as equal to the negative of the ratio Q_H/Q_C. Thus,

$$\boxed{\frac{T_H}{T_C} = \frac{|Q_H|}{|Q_C|} = -\frac{Q_H}{Q_C}.} \qquad (19\text{–}27)$$

Since the ratio of temperatures does not involve the properties of any particular substance, the Kelvin temperature scale is truly *absolute.*

To complete the definition of the Kelvin scale we proceed, as in Chapter 15, to assign the arbitrary value of 273.16 K to the temperature of the triple point of water, T_3. For a Carnot engine operating between reservoirs at the temperatures T and T_3, we have

$$\frac{|Q|}{|Q_3|} = \frac{T}{T_3}$$

or

$$T = 273.16\ \text{K}\ \frac{|Q|}{|Q_3|}. \qquad (19\text{–}28)$$

Comparing this with the corresponding expression for the ideal gas temperature, namely,

$$273.16\ \text{K}\left[\lim_{p_3 \to 0}\left(\frac{p}{p_3}\right)_{\text{constant volume}}\right],$$

it is seen that, in the Kelvin scale, Q plays the role of a "thermometric property." This does not, however, have the objection attached to a coordinate of an arbitrarily chosen thermometer, inasmuch as the behavior of a Carnot engine is independent of the nature of the working substance.

It is a simple matter to show that if an ideal gas is taken around a Carnot cycle, the ratio of the heats absorbed and rejected, $|Q_H|/|Q_C|$, is equal to the ratio of the temperatures of the reservoirs *as expressed on the gas scale,* defined in Chapter 15. (The proof is somewhat lengthy, and will not be given here.) Since, in both scales, the triple point of water is chosen to be 273.16 K, it follows that *the Kelvin and the ideal gas scales are identical.*

The efficiency of a Carnot engine rejecting heat Q_C to a reservoir at Kelvin temperature T_C and absorbing heat Q_H from a source at Kelvin temperature T_H is, as usual,

$$\text{Efficiency} = 1 + \frac{Q_C}{Q_H} = 1 - \frac{|Q_C|}{|Q_H|}.$$

But, by definition of the Kelvin scale,

$$\frac{|Q_C|}{|Q_H|} = \frac{T_C}{T_H}.$$

Therefore, the efficiency of a Carnot engine is

$$\text{Efficiency (Carnot)} = 1 - \frac{T_C}{T_H}. \qquad (19\text{–}29)$$

Equation (19–29) points the way to the conditions which a real engine, such as a steam engine, must fulfill to approach as closely as possible the maximum attainable efficiency. These conditions are that the intake temperature T_H must be made as high as possible and the exhaust temperature T_C as low as possible.

The exhaust temperature cannot be lower than the lowest temperature available for cooling the exhaust. This is usually the temperature of the air, or perhaps of river water if this is available at the plant. The only recourse then is to raise the boiler temperature T_H. Since the vapor pressure of all liquids increases rapidly with increasing temperature, a limit is set by the mechanical strength of the boiler. Another possibility is to use, instead of water, some liquid with a lower vapor pressure. Successful experiments in this direction have been made with mercury vapor replacing steam. At a boiler temperature of 200°C, at which the pressure in a steam boiler would be 225 lb in^{-2}, the pressure in a mercury boiler is only 0.35 lb in^{-2}. Liquid sodium is sometimes used for heat transfer in nuclear reactors for similar reasons.

The unavoidable exhaust heat loss in electric power plants creates a serious environmental problem. When a lake or river is used for cooling, the temperature of the body of water may be raised several degrees. Such a temperature change has a severely disruptive effect on the overall ecological balance, inasmuch as relatively small temperature changes can have significant effects on metabolic rates in plants and animals. Since *thermal pollution,* as this effect is called, is an inevitable consequence of the second law of thermodynamics, careful planning is essential to minimize the ecological impact of new power plants.

19–22 ABSOLUTE ZERO

It follows from Eq. (19–28) that the heat transferred isothermally between two given adiabatics decreases as the temperature decreases. Conversely, the smaller the value of Q, the lower the corresponding T. The smallest possible value of Q is zero, and the corresponding T is absolute zero. *Thus, if a system undergoes a reversible isothermal process without transfer of heat, the temperature at which this process takes place is called absolute zero.* In other words, at absolute zero, an isotherm and an adiabatic are identical.

It should be noted that the definition of absolute zero holds for all substances and is therefore independent of the peculiar properties of any one arbitrarily chosen substance. Furthermore, the definition is in terms of purely large-scale concepts. No reference is made to molecules or to molecular energy. Whether absolute zero may be achieved experimentally is a question of some interest and importance. To achieve temperatures below 4.2 K, at which ordinary helium (mass number 4) liquefies, it is necessary to lower the vapor pressure by pumping away the vapor as fast as possible. The lowest temperature that has ever been reached in this way is 0.7 K, and this required larger pumps and larger pumping tubes than are usually employed in low-temperature laboratories. With the aid of the light isotope of helium (mass number 3) which liquefies at 3.2 K, vigorous pumping yields a temperature of about 0.3 K. Still lower temperatures may be achieved magnetically, but it becomes apparent that the closer one approaches absolute zero, the more difficult it is to go further. It is generally accepted as a law of nature that, although absolute zero may be approached as closely as we please, it is impossible actually to reach the zero of temperature This is known as the "*unattainability statement of the third law of thermodynamics.*"

19–23 ENTROPY

The first law of thermodynamics is the law of energy, the second law of thermodynamics is the law of entropy, and every process that takes place in nature, whether it be mechanical, electrical, chemical, or biological, must proceed in conformity with these two laws.

A book of this nature is not the place for a thorough exposition of entropy and the second law of thermodynamics. We shall content ourselves with defining entropy, computing its changes in a few instances, and stating some of its properties.

A re-reading of Section 19–4 will recall to mind that when a system is carried from one state to another it is found by experiment that the difference between the heat added and the work done by the system, $Q - W$, has the same value for all paths. The fact that this difference does have the same value makes it possible to introduce the concept of internal energy, the change in internal energy being defined and measured by the quantity $Q - W$.

Entropy, or rather a change in entropy, may be defined in a similar way. Consider two states of a system and a number of *reversible* paths connecting them. (The restriction to reversible paths was not necessary for internal energy changes.) Although the heat added to the system is different along different paths, it may be proved that if the heat ΔQ added in each small interval of the path is divided by the *absolute* temperature T of the system in the interval, and the resulting ratios summed for the entire path, this sum has the same value for all (reversible) paths between the same endpoints. In mathematical symbols,

$$\int_1^2 \frac{dQ}{T} = \text{constant for all reversible paths between states 1 and 2.}$$

It is therefore *possible* (whether it is of any use or not one can only tell later) to introduce a function whose difference between two states, 1 and 2, is defined by the integral above. This function may be assigned any arbitrary value in some standard reference state, and its value in any other state will be a definite quantity. The function is called the *entropy* of the system, and is denoted by S. We then have

$$S_2 - S_1 = \int_1^2 \frac{dQ}{T} \quad \text{(along any reversible path).} \tag{19–30}$$

If the change is infinitesimal, $dS = dQ/T$. The unit of entropy is $1\ \mathrm{J\ K^{-1}}$, $1\ \mathrm{cal\ K^{-1}}$, $1\ \mathrm{Btu\ (°R)^{-1}}$, etc.

Example 1 One kilogram of ice at 0°C is melted and converted to water at 0°C. Compute its change in entropy.

Since the temperature remains constant at 273 K, T may be taken outside the integral sign. Then

$$S_2 - S_1 = \frac{1}{T}\int dQ = \frac{Q}{T}.$$

But Q is simply the total heat which must be supplied to melt the ice, or 80,000 cal. Hence

$$S_2 - S_1 = \frac{80{,}000}{273} = 293 \text{ cal K}^{-1},$$

and the increase in entropy of the system is 293 cal K^{-1}. In any *isothermal* reversible process, the entropy change equals the heat added divided by the absolute temperature.

Example 2 One kilogram of water at 0°C is heated to 100°C. Compute its change in entropy.

The temperature is not constant and dQ and T must be expressed in terms of a single variable in order to carry out the integration. This may readily be done, since

$$dQ = mc\,dT.$$

Hence

$$\begin{aligned} S_2 - S_1 &= \int_{273\text{K}}^{373\text{K}} mc\frac{dT}{T} = mc \ln\frac{373\text{ K}}{273\text{ K}} \\ &= 312 \text{ cal K}^{-1} \\ &= 1306 \text{ J K}^{-1}. \end{aligned}$$

Example 3 A gas is allowed to expand adiabatically and reversibly. What is its change in entropy?

In an adiabatic process no heat is allowed to enter or leave the system. Hence $Q = 0$ and there is no change in entropy. It follows that every *reversible* adiabatic process is one of constant entropy, and may be described as *isentropic.*

19–24 THE PRINCIPLE OF THE INCREASE OF ENTROPY

One of the features that distinguish entropy from such concepts as energy, momentum, and angular momentum is that *there is no principle of conservation of entropy.* In fact, the reverse is true. Entropy can be created at will, and there is an increase in entropy in every natural process, if all systems taking part in the process are considered.

Consider the process of mixing 1 kg of water at 100°C with 1 kg of water at 0°C. Let us arbitrarily call the entropy of water zero when it is in the liquid state at 0°C. (This is the reference state often used in engineering work.) Then from Example 2 in the previous section, the entropy of 1 kg of water at 100°C is 312 cal K^{-1} and the entropy of 1 kg at 0°C is zero. The entropy of the system, before mixing, is therefore 312 cal K^{-1}.

After the hot and cold water have been mixed, we have 2 kg of water at a temperature of 50°C or 323 K. From the results of Example 2, we see that the entropy of the system is

$$mc \ln\frac{323}{273} = (2000) \ln\frac{373}{273} = 336 \text{ cal K}^{-1}.$$

There has therefore been an *increase* in entropy of

$$(336 - 312) \text{ cal K}^{-1} = 24 \text{ cal K}^{-1}.$$

Physical mixing of the hot and cold water is, of course, not essential in bringing about the final equilibrium state. We might simply have let heat flow by conduction, or be transferred by radiation, from the hot to the cold water. The same increase in entropy would have resulted.

These examples of the mixing of substances at different temperatures, or the flow of heat from a higher to a lower temperature, are illustrative of all natural (i.e., irreversible) processes. When all the entropy changes in the process are included, the increases in entropy are always greater than the decreases. In the special case of a reversible process, the increases and decreases are equal. Hence we can formulate the general principle, which is considered a part of the second law of thermodynamics, that *when all systems taking part in a process are included, the entropy either remains constant or increases.* In other words, *no process is possible in which the entropy decreases,* when all systems taking part in the process are included.

What is the significance of the increase of entropy that accompanies every natural process? The answer, or one answer, is that it represents the extent to which the Universe "runs down" in that process. Consider again the example of the mixing of hot and cold water. We *might* have used the hot and cold water as the high- and low-temperature reservoirs of a heat engine, and in the course of removing heat from the hot water and giving heat to the cold water we could have obtained some mechanical work. But once the hot and cold water have been mixed and have come to a uniform temperature, this opportunity of converting heat to mechanical work is lost and, moreover, it is lost irretrievably. The lukewarm water will never *unmix* itself and separate into a hotter and a colder portion.* Of course, there is no decrease in *energy* when the hot and cold water are mixed, and what has been "lost" in the mixing process is not *energy*, but *opportunity;* the opportunity to convert a portion of the heat flowing out of the hot water to mechanical work. Hence when entropy increases, energy becomes more unavailable, and we say that the Universe has "run down" to that extent. This is the true significance of the term "irreversible."

The tendency of all natural processes such as heat flow, mixing, diffusion, etc., is to bring about a uniformity of temperature, pressure, composition, etc., at all points. One may visualize a distant future in which, as a consequence of these processes, the entire Universe has attained a state of absolute uniformity throughout. When and if such a state is reached, although there would have been no change in the energy of the Universe, all physical, chemical, and presumably biological processes would have to cease. This goal toward which we appear headed has been described as the "heat death" of the Universe.

19–25 ENERGY CONVERSION

We have seen in this chapter that the laws of thermodynamics place very general limitations on conversion of energy from one form to another. In this day of increasing energy demand and diminishing resources, these matters are of the utmost practical importance. We conclude this chapter with a brief discussion of a few energy-conversion systems, present and proposed.

Most (over 80 percent) of the electric power generated in the United States is obtained from coal-fired steam-turbine generating plants. Modern boilers can transfer about 80 to 90 percent of the heat of combustion of coal into steam. The theoretical thermodynamic efficiency of the turbine, given by Eq. (19–29), is usually limited to about 0.55, and the actual efficiency is typically 90 percent of the theoretical value, or about 0.50. The efficiency of large electrical generators in converting mechanical power to electrical is very large, typically 99 percent. Thus the overall thermal efficiency of such a plant is roughly (0.85)(0.50)(0.99), or about 40 percent.

In 1970, a generator with a capacity of about 1 GW (= 1000 MW = 10^9 W) went into operation at the Tennessee Valley Authority's Paradise power plant. The steam is heated to 1003°F, at a pressure of 3,650 lb in^{-2}. The plant consumes 10,500 tons of coal per day, and the overall thermal efficiency is 39.3 percent.

Nuclear power plants have the same theoretical efficiency limit as coal-fired plants. Because at present it is not practical to run nuclear reactors at as high temperatures as coal boilers, the theoretical thermal efficiency is usually lower. The overall thermal efficiency of a nuclear plant is typically 30 percent.

In both coal-fired and nuclear plants, the energy not converted to electrical energy is wasted and must be disposed of. A common practice is to locate such a plant near a lake or river and use the water for disposal of excess heat; this can raise the water temperature several degrees, often with serious ecological consequences.

Solar energy is an interesting possibility. The power in the sun's radiation is about 1.4 kW per square meter of area at the surface of the earth. This radiation could be collected and focused with mirrors and used to generate steam. A more exotic proposal is to use large banks of photocells for direct conversion to electricity; this conversion, since it does not use heat as an intermediate step, is not limited by the Carnot efficiency but could in principle be carried out with 100-percent efficiency.

An indirect scheme for conversion of solar energy would use the temperature gradient in the

* The branch of physics called "statistical mechanics" would modify this statement to read, "It is highly improbable that the water will separate spontaneously into a hotter and a colder portion, but it is not impossible."

ocean. In the Caribbean, for example, the water temperature near the surface is about 25°C, while at a depth of a few hundred meters it may be 10°C. While the Second Law of Thermodynamics forbids taking heat from the ocean and converting it completely into work, there is nothing to forbid running a heat engine between these two temperatures. The thermodynamic efficiency would be very low, but with such a vast reservoir of energy available this would not be a serious problem.

This is only a small sample of present-day activity in energy-conversion research. Many other processes are under discussion or development, and the principles of thermodynamics outlined in this chapter are of central importance in all of them.

PROBLEMS

19–1 A combustion experiment is performed by burning a mixture of fuel and oxygen in a constant-volume "bomb" surrounded by a water bath. During the experiment the temperature of the water is observed to rise. Regarding the mixture of fuel and oxygen as the system; (a) has heat been transferred? (b) has work been done? (c) what is the sign of ΔU?

19–2 A liquid is irregularly stirred in a well-insulated container and thereby undergoes a rise in temperature. Regarding the liquid as the system; (a) has heat been transferred? (b) has work been done? (c) what is the sign of ΔU?

19–3 A resistor immersed in running water carries an electric current. Consider the resistor as the system under consideration.

a) Is there a flow of heat into the resistor?
b) Is there a flow of heat into the water?
c) Is work done?
d) Assuming the state of the resistor to remain unchanged, apply the first law to this process.

19–4 In certain process, 500 cal of heat are supplied to a system, and at the same time 100 J of work are done on the system. What is the increase in the internal energy of the system?

19–5 A piece of ice falls from rest into a lake at 0°C, and one-half of one percent of the ice melts. Compute the minimum height from which the ice falls.

19–6 What must be the initial velocity of a lead bullet at a temperature of 25°C, so that the heat developed when it is brought to rest shall be just sufficient to melt it?

19–7 In a certain process, 200 Btu are supplied to a system and at the same time the system expands against a constant external pressure of 100 lb in^{-2}. The internal energy of the system is the same at the beginning and end of the process. Find the increase in volume of the system.

19–8 An inventor claims to have developed an engine which takes in 100,000 Btu from its fuel supply, rejects 25,000 Btu in the exhaust, and delivers 25 kWh of mechanical work. Do you advise investing money to put this engine on the market?

19–9 A vessel with rigid walls and covered with asbestos is divided into two parts by an insulating partition. One part contains a gas at temperature T and pressure p. The other part contains a gas at temperature T' and pressure p'. The partition is removed. What conclusion may be drawn by applying the first law of thermodynamics?

19–10 A mixture of hydrogen and oxygen is enclosed in a rigid insulating container and exploded by a spark. The temperature and pressure both increase considerably. Neglecting the small amount of energy provided by the spark itself, what conclusion may be drawn by applying the first law of thermodynamics?

19–11 When water is boiled under a pressure of 2 atm, the heat of vaporization is 2.20×10^6 J kg^{-1} and the boiling point is 120°C. At this pressure, one kilogram of water has a volume of 10^{-3} m^3, and one kilogram of steam a volume of 0.824 m^3.

a) Compute the work done when one kilogram of steam is formed at this temperature.
b) Compute the increase in internal energy.

19–12 When a system is taken from state a to state b, in Fig. 19–15, along the path acb, 80 J of heat flow into the system, and 30 J of work are done.

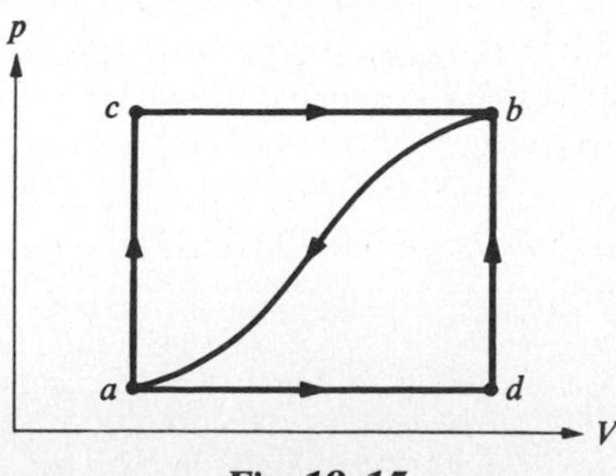

Fig. 19–15

a) How much heat flows into the system along path *adb* if the work is 10 J?

b) When the system is returned from b to a along the curved path, the work is 20 J. Does the system absorb or liberate heat, and how much?

c) If $U_a = 0$ and $U_d = 40$ J, find the heat adsorbed in the processes *ad* and *db*.

19–13 A steel cylinder of cross-sectional area 0.1 ft^2 contains 0.4 ft^3 of glycerin. The cylinder is equipped with a tightly fitting piston which supports a load of 6000 lb. The temperature of the system is increased from 60°F to 160°F. Neglect the expansion of the steel cylinder. Find (a) the increase in volume of the glycerin, (b) the mechanical work of the 6000 lb force, (c) the amount of heat added to the glycerin [specific heat of glycerin = 0.58 Btu $lb^{-1}(F°)^{-1}$], (d) the change in internal energy of the glycerin.

19–14 The volume of 1 mole of an ideal gas is increased isothermally (T = constant) from 1 to 20 liters at 0°C. The pressure of the gas at any moment is given by the equation $pV = RT$, where $R = 8.31 \text{ J mole}^{-1} \text{ K}^{-1}$ and T is the kelvin temperature. How many joules of work are done?

19–15 Calculate the work done when a gas expands from volume V_1 to V_2, the relation between pressure and volume being

$$\left(p + \frac{a}{V^2}\right)(V - b) = K,$$

where a, b, and K are constants.

19–16

a) The change in internal energy, dU, of a system consisting of n moles of a pure substance, in an infinitesimal process at constant volume, is equal to $nC_v\, dT$. Explain why the internal energy change in a process at constant pressure is *not* equal to $nC_p\, dT$.

b) Explain why the change in internal energy of an ideal gas, in *any* infinitesimal process, is given by $nC_v\, dT$.

19–17 A cylinder contains 1 mole of oxygen gas at a temperature of 27°C. The cylinder is provided with a frictionless piston which maintains a constant pressure of 1 atm on the gas. The gas is heated until its temperature increases to 127°C.

a) Draw a diagram representing the process in the pV-plane.

b) How much work is done by the gas in this process?

c) On what is this work done?

d) What is the change in internal energy of the gas?

e) How much heat was supplied to the gas?

f) How much work would have been done if the pressure had been 0.5 atm?

19–18

a) Compare the quantity of heat required to raise the temperature of 1 g of hydrogen through 1 C°, at constant pressure, with that required to raise the temperature of 1 g of water by the same amount.

b) Of the substances listed in Table 19–1, which has the largest specific heat capacity in cal g^{-1} $(C°)^{-1}$?

19–19 Ten liters of air at atmospheric pressure are compressed isothermally to a volume of 2 ℓ, and are then allowed to expand adiabatically to a volume of 10 ℓ. show the process in a pV-diagram.

19–20 An ideal gas is contained in a cylinder closed with a movable piston. The initial pressure is 1 atm and the initial volume is 1 ℓ. The gas is heated at constant pressure until the volume is doubled, then heated at constant volume until the pressure is doubled, and finally expanded adiabatically until the temperature drops to its initial value. Show the process in a pV-diagram.

19–21 An ideal gas initially at 10 atm and 300 K is permitted to expand adiabatically until its volume doubles. Find the final pressure and temperature if the gas is (a) monatomic; (b) diatomic.

19–22 A gasoline engine takes in air at 20° C and 1 atm, and compresses it adiabatically to 1/10 the original volume. Find the final temperature and pressure.

19–23 An ideal gas undergoes an adiabatic expansion, during which its temperature drops from T_1 to T_2. Show that the work done by the gas is given by $nC_v(T_1 - T_2)$, where n is the number of moles.

19–24 A certain ideal gas has $\gamma = 1.33$. Determine the molar heat capacities at constant volume and at constant pressure.

19–25 Compressed air at a pressure of 2×10^6 Pa is used to drive an air engine which exhausts at a pressure of 2×10^5 Pa. What must be the temperature of the compressed air in order that there may be no possibility of frost forming in the exhaust ports of the engine? Assume the expansion to be adiabatic. (*Note*: Frost frequently forms in the exhaust ports of an air-driven engine. This happens when the moist air is cooled below 0°C by the expansion which takes place in the engine.)

19–26 Initially at a temperature of 140°F, 10 ft^3 of air expands at a constant gauge pressure of 20 lb in^{-2} to a volume of 50 ft^3, and then expands further adiabatically to a final volume of 80 ft^3 and a final gauge pressure of 3 lb in^{-2}. Sketch the process in the pV-plane and compute the work done by the air.

19–27 The cylinder of a pump compressing air from atmospheric pressure into a very large tank at 60 lb in^{-2} gauge pressure is 10 in. long.

a) At what position in the stroke will air begin to enter the tank? Assume the compression to be adiabatic.
b) If the air is taken into the pump at 27°C, what is the temperature of the compressed air?

19–28 Two moles of helium are initially at a temperature of 27° C and occupy a volume of 20 ℓ. The helium is first expanded at constant pressure until the volume has doubled, and then adiabatically until the temperature returns to its initial value.

a) Draw a diagram of the process in the pV-plane.
b) What is the total heat supplied in the process?
c) What is the total change in internal energy of the helium?
d) What is the total work done by the helium?
e) What is the final volume?

19–29 A heat engine carries 0.1 mole of an ideal gas around the cycle shown in the pV diagram of Fig. 19–16. Process 1–2 is at constant volume, process 2–3 is adiabatic, and process 3–1 is at a constant pressure of 1 atm. The value of γ for this gas is 5/3.

a) Find the pressure and volume at points 1, 2, and 3.
b) Find the net work done by the gas in the cycle.

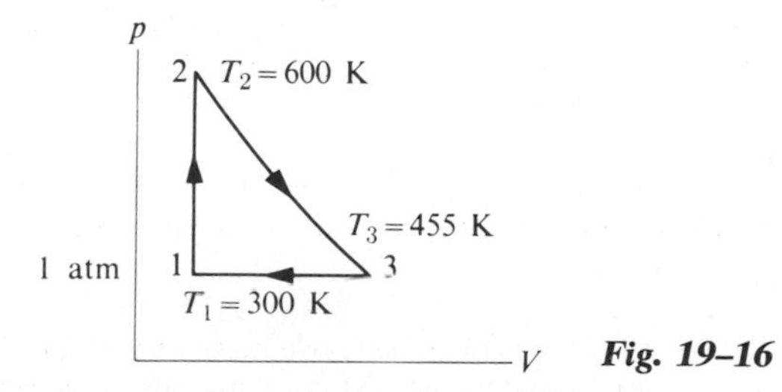

Fig. 19–16

19–30 A cylinder contains oxygen at a pressure of 2 atm. The volume is 3 ℓ and the temperature is 300 K. The oxygen is carried through the following processes:

1) Heated at constant pressure to 500 K.
2) Cooled at constant volume to 250 K.
3) Cooled at constant pressure to 150 K.
4) Heated at constant volume to 300 K.

a) Show the four processes above in a pV-diagram, giving the numerical values of p and V at the end of each process.
b) Calculate the net work done by the oxygen.
c) Find the net heat flowing into the oxygen.
d) What is the efficiency of this device as a heat engine?

19–31 What is the thermal efficiency of an engine which operates by taking an ideal gas through the following cycle? Let $C_v = 3$ cal mole^{-1}(C°)$^{-1}$.

1) Start with n moles at P_0, V_0, T_0.
2) Change to $2P_0$, V_0, at constant volume.
3) Change to $2P_0$, $2V_0$, at constant pressure.
4) Change to P_0, $2V_0$, at constant volume.
5) Change to P_0, V_0, at constant pressure.

19–32 A Carnot engine whose low-temperature reservoir is at 280 K has an efficiency of 40%. It is desired to increase this to 50%.

a) By how many degrees must the temperature of the high-temperature reservoir be increased if the temperature of the low-temperature reservoir remains constant?
b) By how many degrees must the temperature of the low-temperature reservoir be decreased if that of the high-temperature reservoir remains constant?

19–33 A Carnot engine whose high-temperature reservoir is at 400 K takes in 100 cal of heat at this temperature in each cycle, and gives up 80 cal to the low-temperature reservoir.

a) What is the temperature of the latter reservoir?
b) What is the thermal efficiency of the cycle?

19–34 A Carnot refrigerator takes heat from water at 0°C and rejects heat to a room at 27°C. Suppose that 50 kg of water at 0°C are converted to ice at 0°C.

a) How much heat is rejected to the room?
b) How much energy must be supplied to the refrigerator?

19–35 A Carnot engine is operated between two heat reservoirs at temperatures of 400 K and 300 K.

a) If the engine receives 1200 cal from the reservoir at 400 K in each cycle, how many calories does it reject to the reservoir at 300 K?

b) If the engine is operated in reverse, as a refrigerator, and receives 1200 cal from the reservoir at 300 K, how many calories does it deliver to the reservoir at 400 K?

c) How many calories would be produced if the mechanical work required to operate the refrigerator in part (b) were converted directly to heat?

19–36 Is it possible to cool a closed, insulated room by operating an electric refrigerator in the room, leaving the refrigerator door open?

19–37 Consider the entropy of water to be zero when it is in the liquid phase at 0°C and atmospheric pressure.

a) How much heat must be supplied to 1 kg of water to raise its temperature from 0°C to 100°C?

b) What is the entropy of 1 kg of water at 100°C?

c) 1 kg of water at 0°C is mixed with 1 kg at 100°C. What is the final temperature?

d) Find the entropy of the hot and cold water before mixing, and the entropy of the system after mixing. (This is another example of the increase of entropy in an irreversible process.)

19–38

a) Draw a graph of a Carnot cycle, plotting Kelvin temperature vertically and entropy horizontally (a temperature–entropy diagram).

b) Show that the area under any curve in a temperature–entropy diagram represents the heat absorbed by the system.

c) Derive from your diagram the expression for the thermal efficiency of a Carnot cycle.

19–39 Heat is added to 0.5 kg of ice at 0° C until it is all melted.

a) What is the change in entropy of the water?

b) If the source of heat is a very massive body at a temperature of 20° C, what is the change in entropy of this body?

c) What is the total change in entropy of the water and the heat source?

19–40 A block of aluminum of mass 1 kg, initially at 100° C, is dropped into 1 kg of water initially at 0° C.

a) What is the final temperature?

b) What is the total change in entropy of the system?

19–41 Two moles of an ideal gas undergo a reversible isothermal expansion from 0.02 m^3 to 0.04 m^3 at a temperature of 300 K. What is the change in entropy of the gas?

19–42 A solar power plant is to be built with a power output capacity of 1000 MW. What land area must the solar energy collectors occupy if they are: (a) photocells with 90-percent efficiency? (b) mirrors that generate steam for a turbine-generator unit with overall efficiency of 30 percent?

19–43 An engine is to be built to extract power from the temperature gradient of the ocean. If the surface and deep-water temperatures are 25° C and 10° C, respectively, what is the maximum theoretical efficiency of such an engine?

19–44 A coal-fired steam-turbine power plant that produces 1000 MW of electrical power is located beside a river. The overall thermal efficiency of the plant is 40 percent.

a) What is the total thermal power input to the plant?

b) At what rate is waste heat discharged from the plant?

c) If the waste heat is delivered to the river, and if the temperature rise must be no greater than 5 C°, how much water must be available per second?

d) In part (c), if the river is 100 m wide and 5 m deep, what must be the minimum flow velocity of the water?

Chapter 20

Molecular Properties of Matter

20–1 MOLECULAR THEORY OF MATTER

Thousands of physical and chemical observations support the hypothesis that matter in all phases is composed of tiny particles called *molecules*. The molecules of any one substance are identical. They have the same structure, the same mass, and the same mechanical and electrical properties. Many of the large-scale properties of matter which have been discussed heretofore, such as elasticity, surface tension, condensation, and vaporization, can be comprehended with deeper understanding in terms of the molecular theory.

The simplest model of a molecule is that of a rigid sphere, like a small billiard ball, capable of moving, of colliding with other molecules or with a wall, and of exerting attractive or repulsive forces on neighboring molecules. In other parts of physics and chemistry it is important to consider the structure of the molecule, but it is not necessary at this point. The smallest molecules are of the order of 10^{-10} m in size; the largest are at least 10,000 times this large.

One essential characteristic of a molecule is the force that exists between it and a neighboring molecule. There is, of course, a force of *gravitational* attraction between every pair of molecules, but it turns out that this is negligible in comparison with the forces to be considered now. The forces that hold the molecules of a liquid (or solid) together are chiefly of *electrical* origin and do not follow a simple inverse-square law.

When the separation of the molecules is large, as in a gas, the force is extremely small and attractive. The attractive force increases as a gas is compressed and its molecules are brought closer together. But since tremendous pressures are needed to compress a liquid (i.e., to force its molecules closer together than their normal spacing in the liquid state), we conclude that at separations only slightly less than the dimensions of a molecule the force is repulsive and relatively large. Thus the force must vary with separation in somewhat the fashion shown in Fig. 20–1. At large separations the force is small and attractive. As the molecules are brought closer together, the force of attraction becomes larger, passes through a maximum, and then decreases to zero at a separation r_0. When the distance between the molecules is less than r_0, the force becomes repulsive.

A single pair of molecules could remain in equilibrium at a center-to-center spacing equal to r_0 in Fig. 20–1. If they were separated slightly, the force between them would be attractive and they would be drawn together. If they were forced closer together than the distance r_0, the force would be one of

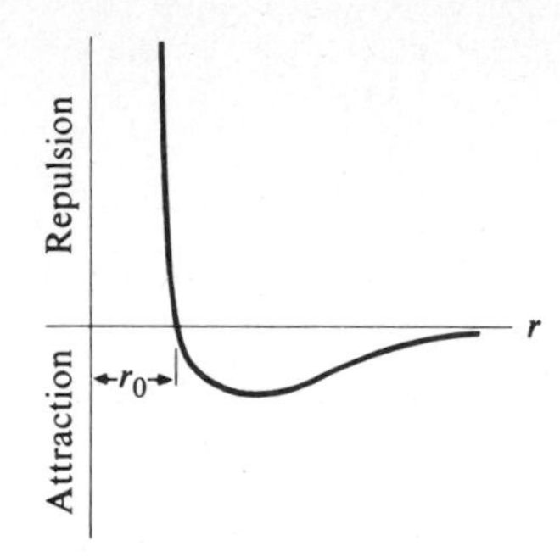

Fig. 20–1 *The force between two molecules changes from an attraction when the separation is large, to a repulsion when the separation is small.*

repulsion and they would spring apart. If they were either pulled apart or pushed together, and then released, they would oscillate about their equilibrium separation r_0.

If the only property of a molecule were the force of attraction between it and its neighbors, all matter would eventually coalesce into the liquid or solid phase. The existence of gases points to another property which enables molecules to stay apart, namely, molecular motion. The more vigorous the motion, the less chance there is for condensation into the liquid or solid phase. In solids the molecules execute vibratory motion about more or less fixed centers. The vibratory motion is relatively weak, and the centers remain fixed at regularly spaced positions which comprise a *space lattice.* This gives rise to the extraordinary regularity and symmetry of crystals. A photograph of the individual, very large molecules of *necrosis virus protein* is shown in Fig. 20–2. It was taken with an electron microscope at a magnification of about 80,000. The molecules look like neatly stacked oranges.

In liquids the intermolecular distances are usually only a trifle greater than those of a solid. The molecules execute vibratory motion of greater energy about centers which are free to move but which remain at approximately the same distances from one another. Liquids show a certain regularity of structure only in the immediate neighborhood of a few molecules. This is called *short-range order,* in contrast with the *long-range order* of a solid crystal.

The molecules of a gas have the greatest kinetic energy, so great in fact that they are, on the average, far away from one another, where only very small attractive forces exist. A molecule of a gas therefore moves with linear motion until a collision takes place either with another molecule or with a wall. In molecular terms, an *ideal gas* is a gas whose molecules exert no forces of attraction at all. The mathematical analysis of a collection of such idealized molecules in random linear motion between collisions is called the *kinetic theory of gases,* and was perfected by Clausius, Maxwell, and Boltzmann in the latter part of the nineteenth century. The analysis of an ideal gas given in this chapter is the simplest example of kinetic theory.

Most common substances exist in the solid phase at low temperatures. When the temperature is raised beyond a definite value, the liquid phase results, and when the temperature of the liquid is

Fig. 20–2 *Crystal of necrosis virus protein. The actual size of the entire crystal is about two thousandths of a millimeter (0.002 mm). By permission of Dr. Ralph Wyckoff, University of Arizona.*

raised further, the substance exists in the gaseous phase; i.e., from a large-scale or *macroscopic* point of view, the transition from solid to liquid to gas is in the direction of increasing temperature. From a *molecular* point of view, this transition is in the direction of increasing molecular kinetic energy. Evidently temperature and molecular kinetic energy are related.

20–2 AVOGADRO'S NUMBER

Surprising as it may seem, it was not until 1811 that the suggestion was first made, by Dalton, that the molecules of a given chemical substance are all alike. Before this, even by those who accepted a molecular theory, it had been assumed that the ultimate particles of the same substance varied in size and shape, just as one might find pebbles of different sizes, all made of the same material.

We now know that a mole of any pure substance contains a definite number of identical molecules. The number of molecules in a mole is called *Avogadro's number*, denoted N_A. This number can be determined in a variety of ways; the first reasonably precise determination was made by the French physicist Jean Perrin in the early 1900's. Before describing Perrin's experiment, we first recall an expression previously derived for the variation of atmospheric pressure with elevation. In Example 3 at the end of Section 18–2 we derived from the laws of hydrostatics and the equation of state of an ideal gas an expression for the ratio of the pressures at two different elevations in an ideal gas at a uniform temperature:

$$\ln \frac{p_2}{p_1} = -\frac{Mg}{RT}(y_2 - y_1). \qquad (20\text{–}1)$$

The molecular mass M equals the mass of one mole, and can be expressed as the product of Avogadro's number N_A and the mass m of a single molecule:

$$M = N_A m \qquad (20\text{–}2)$$

Also, the number of moles n in a given sample of a pure substance equals the total number of molecules N divided by Avogadro's number N_A:

$$n = \frac{N}{N_A}.$$

The ideal gas law can therefore be written

$$pV = nRT = \frac{N}{N_A}RT,$$

or

$$\boldsymbol{n} = \frac{N}{V} = p\frac{N_A}{RT}, \qquad (20\text{–}3)$$

where $\boldsymbol{n}$ is the number of molecules per unit volume, N/V. Since N_A and R are universal constants, it follows that at constant temperature T the number of molecules per unit volume is directly proportional to the pressure. Combining Eqs. (20–2) and (20–3) with Eq. (20–1), we get

$$\ln \frac{\boldsymbol{n}_2}{\boldsymbol{n}_1} = -\frac{N_A mg}{RT}(y_2 - y_1), \qquad (20\text{–}4)$$

where $\boldsymbol{n}_2$ and $\boldsymbol{n}_1$ are the numbers of molecules per unit volume at elevations y_2 and y_1.

One might well ask the following question: If the earth's atmosphere consists of widely separated molecules, not in contact with each other, and if each one is attracted by the earth, then why do the molecules not all settle down to the earth's surface like drops of rain? The explanation is found in the random motions of the molecules, which, in the absence of any gravitational force, would eventually result in their dispersal throughout interstellar space. There are thus two competing tendencies, the gravitational force drawing all molecules toward the earth and their random motions tending to disperse them. The actual distribution represents the compromise between these two tendencies.

It is well known that extremely small particles, called *colloidal* particles, in a liquid of density less than that of the particles, do not all settle to the bottom but remain permanently suspended. These particles are large enough to be observed with a microscope, and the number per unit volume at different heights in the liquid can be counted. Perrin found that the number per unit volume decreased with height, following the same law as predicted by Eq. (20–4) for the number of molecules per unit volume in the atmosphere. Furthermore, the particles are observed to be in a state of continuous motion, called "Brownian motion" after the English botanist Robert Brown, who first observed this effect in 1809 when studying with a microscope small pollen grains suspended in a liquid.

Perrin reasoned that the colloidal particles were analogous in their behavior to the molecules of a "gas," molecules which were actually large enough to be seen and counted. If a colloidal suspension could be made in which all the particles were exactly alike, and if the mass m of each could be determined, and if Eq. (20–4) really describes the vertical distribution of particles, then Avogadro's number could be found by counting the number of particles per unit volume at two known heights in the suspension.

It would occupy too much space to describe the ingenious methods used by Perrin to obtain uniform particles, to count the number per unit volume, and to measure their masses. Suffice it to say that the difficulties were surmounted and that Perrin finally obtained for N_A a value between 6.5 and 7.2×10^{23} molecules per mole. The most precise value of N_A to date, obtained by using x-rays to measure the distance between the layers of molecules in a crystal, is

$$N_A = 6.02217 \times 10^{23} \text{ molecules mol}^{-1}.$$

Once Avogadro's number has been determined, it can be used to compute the mass of a molecule. The mass of 1 mole of atomic hydrogen is 1.008 g. Since by definition the number of molecules in 1 mole is Avogadro's number, it follows that the mass of a single atom of hydrogen is

$$m_H = \frac{1.008 \text{ g mol}^{-1}}{6.022 \times 10^{23}} = 1.673 \times 10^{-24} \text{ g}.$$

The mass of an atom of atomic mass 1 is $1/N_A$ or 1.660×10^{-24} g. Thus, for an oxygen molecule of molecular mass 32,

$$m_{O_2} = (32)(1.660 \times 10^{-24} \text{ g}) = 53.12 \times 10^{-24} \text{ g}.$$

At standard conditions, 1 mole of an ideal gas occupies 22,400 cm^3. The number of molecules per cubic centimeter in an ideal gas at standard conditions is therefore

$$\frac{6.022 \times 10^{23}}{22{,}400 \text{ cm}^3} = 2.69 \times 10^{19} \text{ molecules cm}^{-3}.$$

This is known as *Loschmidt's number*, and of course it is the same for *all* gases. The number of molecules per unit volume can also be computed from Eq. (20–3), at any pressure and temperature.

Numbers of this enormous magnitude are very difficult to comprehend. Sir James Jeans, the eminent British astrophysicist, offered a striking illustration. Each human breath contains the order of 400 cm^3, and thus the order of 10^{22} molecules. Let us assume that George Washington's dying breath has by now become uniformly distributed through the atmosphere, which Jeans estimated to contain 10^{44} molecules. Thus the number of molecules in a breath is roughly equal to the number of breaths in the atmosphere, and since the total volume of the lungs (typically 2000 to 3000 cm^3) is several times the volume of a single breath, each of us at this moment probably has several molecules of Washington's last breath in his lungs!

20–3 PROPERTIES OF MATTER

A molecular theory of matter obviously accomplishes nothing if it simply endows the molecules of a substance with all the properties of that substance. The blue color of copper sulfate is not "explained" by postulating that it consists of molecules, each of which is blue. The fact that a gas is capable of expanding indefinitely is not explained by assuming that it consists of molecules each one of which can expand indefinitely.

Instead, what we must do is to attempt to account for the *complex* properties of matter in bulk as a consequence of *simple* properties ascribed to its molecules. Thus by assuming that a monatomic gas consists of a large number of particles having no properties other than mass and velocity, we can explain the observed equation of state of a gas at low pressure and can show that C_V for the gas should equal 2.99 cal mol^{-1}(C°)$^{-1}$ (see Table 19–1). If we add the hypothesis that the "particles" are not simply geometrical points but have a finite size, then the general features of the viscosity, thermal conductivity, and coefficient of diffusion of a gas can be understood, as well as the fact that polyatomic gases have larger values of C_V than monatomic gases. By assuming that there are forces between the particles, the equation of state can be brought into better agreement with that of a real gas, and the phenomena of liquefaction and solidification at low temperatures can be explained.

The electrical and magnetic properties of matter, and the emission and absorption of light by matter, call for a molecular model that is itself an aggregate of subatomic particles. Some of these are electrically charged, and the forces between molecules originate in these electric charges. To pursue this development of molecular theory further would take us into the area of atomic and nuclear structure. In the next section we return to the starting point and see what properties of a gas at low pressure can be explained by the simplest possible molecular model, an aggregate of particles having mass and velocity.

20–4 KINETIC THEORY OF AN IDEAL GAS

We know that a gas in a cylinder such as that in Fig. 20–3 exerts a pressure against the piston. The molecular theory interprets the pressure not as a static push but as the average effect of many tiny impulsive blows resulting from the collisions of the molecules with the piston. Thus if the piston is forced to the left by a spring, we would expect it not to remain absolutely at rest but to oscillate slightly about some average position, as if it were being bombarded by a helter-skelter rain of shot fired against it. (It is just these random molecular impacts, which do not exactly balance at all instants, that are responsible for the Brownian motion of suspended particles.)

We assume that the gas is composed of a large number of molecules, all having the same mass m and flying about in the cylinder with randomly directed velocities having magnitudes that differ from one molecule to another. Consider first the collision of a single molecule with the piston. Let the mass of the molecule be m, and let the x-component of its velocity be v_{x1}. Figure 20–3 illustrates the collision process. When all the molecules are considered, the *average* y-components of their velocities are not altered by collisions with the piston, since on the average the gas does not gain or lose any vertical motion as a whole. We shall therefore assume that the y-component of the velocity of each molecule is unchanged in the collision. Also, we identify the internal energy of the gas as the sum of the kinetic energies of its molecules. Since the internal energy remains constant if the gas is isolated, the average kinetic energy of the molecules is constant. We therefore assume also that the kinetic energy of any one molecule is constant in a collision with the walls, that is, the collisions are perfectly elastic. Therefore the magnitude of the velocity v is the same before and after the collision, and since the y-component is unchanged, the *magnitude* of the x-component is unchanged also, although it is reversed in direction.

The average force exerted on the piston by one molecule can now be computed by setting the average force equal to the average rate of change of momentum. Since the x-velocity of the molecule reverses from $+v_{x1}$ to $-v_{x1}$ in each collision, the magnitude of the change in momentum in each

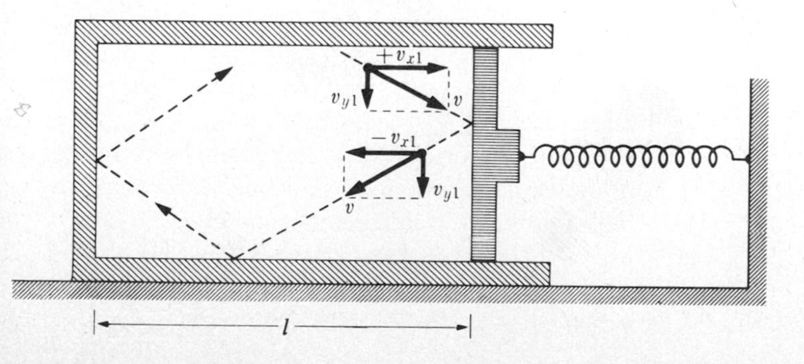

Fig. 20–3 *Collision of a single molecule with a piston.*

collision is $2mv_{x1}$. To find the average force *during the time of one collision* we would divide the change of momentum by this time. However, after having made one collision with the piston, the molecule must travel to the other end of the cylinder and back before colliding again, and during all this time it exerts no force on the piston. To get the correct time average force we must average over the *total* time between collisions. As the molecule travels to the other end of the cylinder and back (see Fig. 20–3), the magnitude of its x-velocity remains constant. If ℓ is the length of the cylinder, the time for the round trip is

$$t = \frac{2\ell}{v_{x1}},$$

and the average force exerted on the piston by one molecule is

$$\text{Average force} = \text{Average rate of change of momentum} = \frac{2mv_{x1}}{2\ell/v_{x1}} = \frac{mv_{x1}^2}{\ell}. \qquad (20\text{–}5)$$

Now consider other molecules 2, 3, etc., with x-velocities v_{x2}, v_{x3}, etc. The average force exerted by each is given by an expression like that in Eq. (20–5), and the *total* average force is

$$\text{Average force} = \frac{m}{\ell}(v_{x1}^2 + v_{x2}^2 + v_{x3}^2 + \cdots).$$

The average pressure p equals the total average force divided by the piston area A, so

$$p = \frac{m}{V}(v_{x1}^2 + v_{x2}^2 + v_{x3}^2 + \cdots), \qquad (20\text{–}6)$$

where $V = \ell A$ is the total volume of the cylinder.

Let N be the total number of molecules. The average value of the square of the x-velocity of all molecules is

$$\overline{v_x^2} = \frac{v_{x1}^2 + v_{x2}^2 + v_{x3}^2 + \cdots}{N}. \qquad (20\text{–}7)$$

Hence, Eq. (20–6) can be written

$$p = \frac{Nm\overline{v_x^2}}{V}. \qquad (20\text{–}8)$$

The *magnitude* of the resultant velocity v, or the *speed* of any molecule, is given by

$$v^2 = v_x^2 + v_y^2 + v_z^2,$$

and averaging over all molecules,

$$\overline{v^2} = \overline{v_x^2} + \overline{v_y^2} + \overline{v_z^2}.$$

But since the x-, y-, and z-directions are all equivalent,

$$\overline{v_x^2} = \overline{v_y^2} = \overline{v_z^2}.$$

Hence

$$\overline{v^2} = 3\overline{v_x^2},$$

and Eq. (20–8) becomes

$$pV = \tfrac{1}{3}Nm\overline{v^2} = (\tfrac{2}{3}N)(\tfrac{1}{2}m\overline{v^2}) \qquad (20\text{–}9)$$

But $\frac{1}{2}m\overline{v^2}$ is the average kinetic energy of a single molecule, and the product of this and the total number of molecules N equals the total random kinetic energy, or internal energy U. Hence the product pV equals two-thirds of the internal energy:

$$pV = \tfrac{2}{3}U. \qquad (20\text{–}10)$$

By experiment, at low pressures, the equation of state of a gas is

$$pV = nRT.$$

The theoretical and experimental laws will therefore be in complete agreement if we set

$$U = \tfrac{3}{2}nRT. \qquad (20\text{–}11)$$

The average kinetic energy of a single molecule is then

$$\frac{U}{N} = \tfrac{1}{2}m\overline{v^2} = \frac{3nRT}{2N}.$$

But the number of moles, n, equals the total number of molecules, N, divided by Avogadro's number, N_A, the number of molecules per mole:

$$n = \frac{N}{N_A}, \qquad \frac{n}{N} = \frac{1}{N_A}.$$

Hence

$$\tfrac{1}{2}m\overline{v^2} = \frac{3}{2}\frac{R}{N_A}T.$$

The ratio R/N_A occurs frequently in molecular theory. It is called the *Boltzmann constant*, k:

$$k = \frac{R}{N_A} = \frac{8.31 \text{ J mol}^{-1}\text{ K}^{-1}}{6.02 \times 10^{23} \text{ molecules mol}^{-1}}$$
$$= 1.38 \times 10^{-23} \text{ J molecule}^{-1}\text{ K}^{-1}.$$

Since R and N_A are universal constants, the same is true of k. Then

$$\boxed{\tfrac{1}{2}m\overline{v^2} = \tfrac{3}{2}kT.} \tag{20–12}$$

The average kinetic energy per molecule, therefore, *depends only on the temperature* and not on the pressure, volume, or species of molecule.

The average value of the square of the speed is

$$\overline{v^2} = \frac{3kT}{m},$$

and the square root of this, or the root-mean-square speed v_{rms}, is

$$v_{rms} = \sqrt{\overline{v^2}} = \sqrt{\frac{3kT}{m}} = \sqrt{\frac{3RT}{M}}. \tag{20–13}$$

Example 1 What is the average kinetic energy of a molecule of a gas at a temperature of 300 K?

$$\tfrac{1}{2}m\overline{v^2} = \tfrac{3}{2}kT = (\tfrac{3}{2})(1.38 \times 10^{-23}\text{ J K}^{-1})(300\text{ K})$$
$$= 6.20 \times 10^{-21}\text{ J}.$$

Example 2 What is the total random kinetic energy of the molecules in 1 mole of a gas at a temperature of 300 K?

$$U = (N_A)(\tfrac{3}{2}kT) = \tfrac{3}{2}RT$$
$$= (\tfrac{3}{2})(8.31\text{ J mol}^{-1}\text{ K}^{-1})(1\text{ mol})(300\text{ K})$$
$$= 3750\text{ J} = 900\text{ cal}.$$

Example 3 What is the root-mean-square speed of a hydrogen molecule at 300 K?

The mass of a hydrogen molecule (see Section 20–2) is

$$m_{H_2} = (2)(1.66 \times 10^{-27}\text{ kg}) = 3.32 \times 10^{-27}\text{ kg}.$$

Hence

$$v_{rms} = \sqrt{\frac{3kT}{m}} = \sqrt{\frac{3(1.38 \times 10^{-23}\text{ J K}^{-1})(300\text{ K})}{3.32 \times 10^{-27}\text{ kg}}}$$
$$= 1934\text{ m s}^{-1}.$$

Alternatively,

$$v_{rms} = \sqrt{\frac{3RT}{M}} = \sqrt{\frac{(3)(8.31\text{ J mol}^{-1}\text{ K}^{-1})(300\text{ K})}{2 \times 10^{-3}\text{ kg mol}^{-1}}}$$
$$= 1934\text{ m s}^{-1}.$$

Example 4 What is the root-mean-square speed of a molecule of mercury vapor at 300 K?

The *kinetic energy* of a mercury molecule (or atom, since mercury vapor is monatomic) is the same as that of a hydrogen molecule at the same temperature. The mass of a mercury atom is

$$m_{Hg} = (201)(1.66 \times 10^{-27}\text{ kg}) = 334 \times 10^{-27}\text{ kg}.$$

Therefore,

$$v_{rms} = \sqrt{\frac{3(1.38 \times 10^{-23}\text{ J K}^{-1})(300\text{ K})}{334 \times 10^{-27}\text{ kg}}}$$
$$= 193\text{ m s}^{-1}.$$

When a gas expands against a moving piston it does work. The source of this work is the random kinetic energy of the gas molecules. Conversely, when work is done in compressing a gas, the random kinetic energy of its molecules increases. But if the collisions with the walls are perfectly elastic, as we have assumed, how can a molecule gain or lose energy in a collision with a piston? To understand this, we must consider the collision of a molecule with a *moving* wall.

When a molecule makes a collision with a *stationary* wall, it exerts a momentary force on the wall but does no work, since the wall does not move. But if the wall is in motion, work *is* done in a collision. Thus, if the piston in Fig. 20–3 is moving to the right, work is done on it by the molecules that strike it, and their velocities (and kinetic energies) after colliding are smaller than they were before a collision. The collision is still completely elastic, since the work done on the moving piston is just equal to the decrease in the kinetic energy of the molecule. Similarly, if the piston is moving toward the left, the kinetic energy of a colliding molecule is *increased*, the increase in kinetic energy being equal to the work done on the molecule.

20–5 MOLAR HEAT CAPACITY OF A GAS

When heat flows into a system in a process at constant volume, no work is done and all of the energy inflow goes into an increase in the internal energy U of the system. From the molecular viewpoint, the internal energy of a system is the sum of the kinetic and potential energies of its molecules. If this sum can be computed as a function of temperature, then, from its rate of change with temperature, we can derive a theoretical expression for the molar heat capacity.

The simplest system is a monatomic ideal gas, for which the molecular energy is wholly kinetic and is given by Eq. (20–11):

$$\text{Random kinetic energy} = U = \tfrac{3}{2}nRT.$$

If the temperature is increased by dT, the random kinetic energy increases by

$$dU = \tfrac{3}{2}nR\,dT.$$

In a process in which the temperature increases by dT at constant volume, the energy flowing into a system is, by definition of C_v,

$$dQ = nC_v\,dT = dU.$$

Hence,

$$\boxed{C_v = \tfrac{3}{2}R.}$$

We showed in Section 19–12 and Table 19–1 that the experimental values of C_v for monatomic gases are, in fact, almost exactly equal to $\frac{3}{2}R$. This agreement is a striking confirmation of the basic correctness of the kinetic model of a gas and did much to establish this theory at a time when many scientists still refused to accept it.

Since $C_p = C_v + R$, it follows that the theoretical ratio of molar heat capacities for a monatomic ideal gas is

$$\frac{C_p}{C_v} = \gamma = \frac{\frac{3}{2}R + R}{\frac{3}{2}R} = \frac{5}{3} = 1.67.$$

This also is in good agreement with the experimental values in Table 19–1.

Polyatomic gases present a more complicated problem. We might expect that a molecule composed of two or more atoms could have additional "internal" energy associated with vibratory motion of the atoms relative to each other and of rotation of the molecule about its center of mass. Also, since every atom consists of a positive nucleus and one or more electrons, there may be additional energy associated with position and motion of these electric charges.

The *temperature* of a gas depends on the average random *translational* kinetic energy of its molecules. When heat flows into a *monatomic* gas, at constant volume, all of this energy goes into an increase in random translational molecular kinetic energy, as evidenced by the agreement between the measured values of C_v and the values computed from the increase in translational kinetic energy. But when heat flows into a *polyatomic* gas, we would expect that a part of the energy would go toward increasing "molecular internal energy." Hence, when equal amounts of heat flow into a monatomic and a polyatomic gas containing the same number of molecules, it is to be expected that the temperature rise in the polyatomic gas will be *less* than in the monatomic, because only a part of the energy is available for increasing *translational* molecular kinetic energy. In other words, to produce *equal* increases of temperature, *more* heat would have to be supplied to the polyatomic gas. This is a qualitative explanation, then, of the larger values of C_v for polyatomic gases listed in Table 19–1.

20–6 MEASUREMENT OF MOLECULAR SPEEDS

Direct measurements of the distribution of molecular speeds have been made by a number of methods. Figure 20–4 is a diagram of the apparatus used by Zartman and Ko in 1930–1934, a modification of a technique developed by Stern in 1920. Metallic silver is melted and evaporated in the oven O. A beam of silver atoms escapes through a small opening in the oven and passes through the slits S_1 and S_2 into an evacuated region. The cylinder C can be rotated at approximately 6000 rpm about the axis A. If the cylinder is at rest, the molecular beam enters the cylinder through a slit S_3 and strikes a curved glass plate G. The molecules stick to the glass plate, and the number arriving at any portion can be determined by removing the plate and measuring with a recording microphotometer the darkening that has resulted.

Now suppose the cylinder is rotated. Molecules can enter it only during the short time intervals during which the slit S_3 crosses the molecular beam. If the rotation is clockwise, as indicated, the glass plate moves toward the right while the molecules cross the diameter of the cylinder. They therefore strike the plate to the left of the point of impact when the cylinder is at rest, and the more slowly they travel, the farther to the left is this point of impact. The blackening of the plate is therefore a measure of the "velocity spectrum" of the molecular beam.

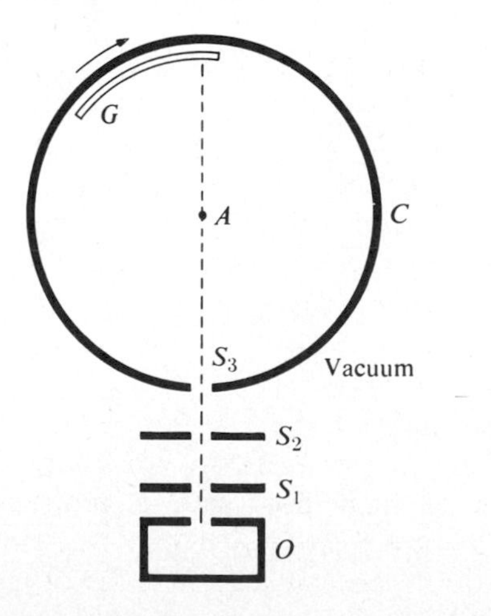

Fig. 20–4 Apparatus used by Zartman and Ko in studying distribution of velocities.

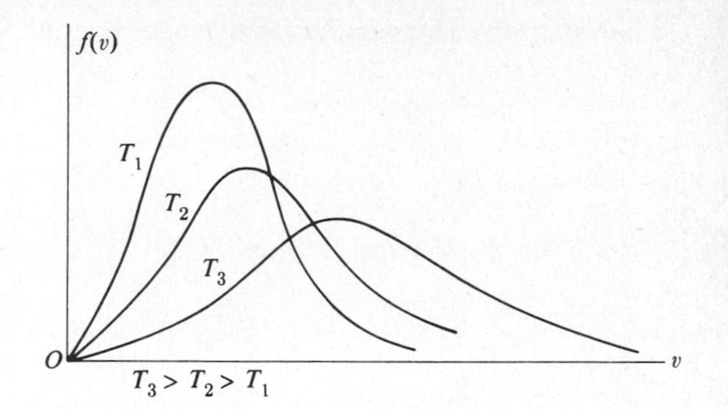

Fig. 20–5 Maxwell–Boltzmann distribution curves for various temperatures. As the temperature increases, the curve becomes flatter, and its maximum shifts to higher temperature.

Several other techniques have been devised in more recent years for measuring distributions of molecular speeds with greater precision. Results of these experiments can be compared with a predicted speed distribution derived from a more detailed kinetic theory of gases and called the *Maxwell–Boltzmann distribution.* This theoretical distribution is shown in Fig. 20–5 for several different temperatures. At any temperature the curve shows a peak corresponding to the most probable speed; very small and very large values of v occur infrequently. As the temperature increases, the peak shifts to higher and higher speeds, corresponding to the increase in average molecular kinetic energy with temperature. Measured molecular speed distributions are generally in excellent agreement with the theoretical prediction of the Maxwell–Boltzmann distribution.

20–7 CRYSTALS

Many materials can exist in a variety of solid forms; a familiar example is the element *carbon.* The black

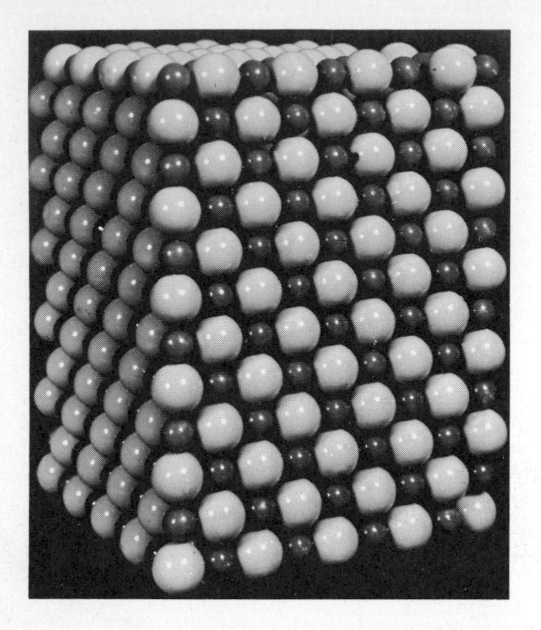

Fig. 20–6 *Close packing of the sodium ions (black spheres) and chlorine ions (white spheres) in a sodium chloride crystal. (Courtesy of Alan Holden; reprinted with permission of Educational Services, Inc., from* Crystals and Crystal Growing, *Doubleday and Co., New York, 1960.)*

soot deposited on a kettle by a smoky campfire is nearly pure carbon. The so-called "lead" in a pencil is not lead at all, but chiefly a different form of carbon called graphite. Diamond is a third form of solid carbon.

The remarkably dissimilar mechanical, thermal, electrical, and optical properties of these three forms of carbon can be understood in terms of the arrangement of the carbon atoms in the solid structure. In soot the atoms have no regular arrangement and are said to constitute an *amorphous* solid. Graphite and diamond are both crystals, but the orderly arrangement of atoms is different in the two crystal lattices. The study of the spatial arrangements of atoms (or molecules or ions) in the various types of crystal lattices is a fascinating area of physics. We shall be concerned here with only a few of the fundamental ideas needed to understand some of the mechanical and thermal properties of single crystals or solids composed of an aggregate of crystals, that is, *polycrystalline solids.*

The particles comprising a crystal lattice are *closely packed,* as seen in Fig. 20–2 and as suggested by the model in Fig. 20–6, where the sodium and chlorine ions of a sodium chloride crystal are represented as spheres touching one another. Strictly speaking, each ion consists of a nucleus surrounded by electrons in motion, so that the boundary of an ion must be thought of as the average positions of the outermost electrons. It follows that the outer electrons of one particle in a crystal lattice come close to and even at times interpenetrate with the outer electrons of a neighboring particle.

To make crystal structures simpler to comprehend, it is customary to exaggerate the distances between neighboring particles, as shown in Fig. 20–7, where a *face-centered cubic* lattice structure is depicted. Much can be learned by representing crystal lattices in this way, with the aid of a supply of gum drops and toothpicks.

Suppose that the chemical composition and density of a very small volume element of a substance are measured at many different places. If the composition and density are the same at all points, the substance is said to be *homogeneous.* If the physical properties determining the transport of heat, electricity, light, etc., are the *same in all directions,* the substance is said to be *isotropic.* Cubic crystals are both homogeneous and isotropic, whereas all other crystals, although homogeneous, are not isotropic (i.e., they are *anisotropic*).

The *anisotropy* (accent on the syllable "ot") of noncubic crystals can be understood when the forces

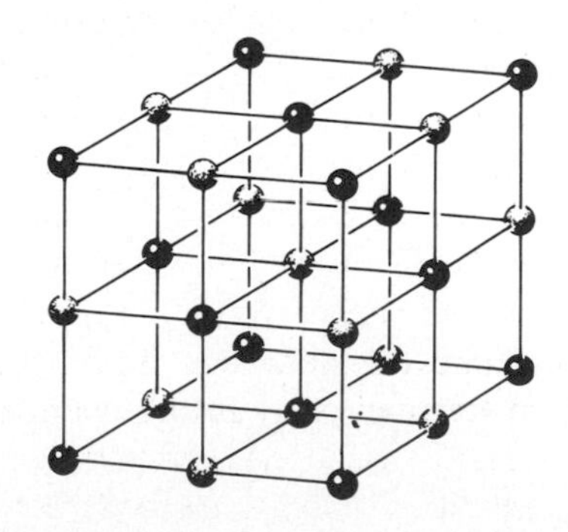

Fig. 20–7 *Symbolic representation of a sodium chloride crystal, with exaggerated distances between ions.*

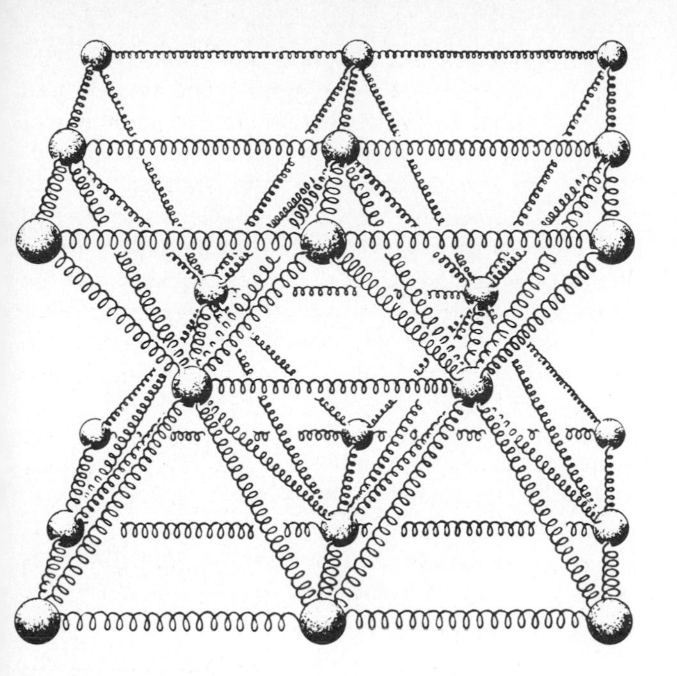

Fig. 20–8 The forces between neighboring particles in a crystal may be visualized by imagining every particle to be connected to its neighbors by springs. In the case of a cubic crystal, all springs are assumed to have the same spring constant. Anisotropy is connected with differing spring constants in different directions.

between neighboring particles in a lattice are taken into account. A convenient way of visualizing these forces is shown in Fig. 20–8, an idealized model which represents a crystal lattice as a large number of spheres, each connected to its neighbors by springs. The force constants of parallel springs are imagined equal, but the force constant of each spring pointing in one direction may be different from that of springs pointing in different directions. If each of the particles represented by spheres in Fig. 20–8 is imagined to be vibrating about its equilibrium position, a fairly accurate picture of the dynamic character of a crystal lattice will result. The mathematical calculations of the vibration frequencies of such a complex mechanical system was first attempted by Born and Von Karman early in the twentieth century. It is, of course, a very laborious and complicated calculation, but it has been carried out in detail for sodium chloride, potassium chloride, diamond, and silver.

The elastic properties of crystals can be partially understood from a diagram such as that in Fig. 20–8. Young's modulus is simply a measure of the stiffness of the springs in the direction in which the crystal is pulled or pushed. The shear modulus depends on the springs which are stretched and those which are compressed when the crystal is twisted. One of the first tests of the correctness of a lattice structure representing a particular material is to compare a calculated value of an elastic modulus with a measured value.

20–8 HEAT CAPACITY OF A CRYSTAL

In Section 20–4 a gas was assumed to consist of N particles in rapid linear motion, so far apart on the average that no force of interaction existed except upon collision with one another or with a wall. By applying simple mechanical principles, it was shown that the product of the pressure p and the volume V is equal to

$$pV = (\tfrac{2}{3}N)(\tfrac{1}{2}m\overline{v^2}) = \tfrac{2}{3}U,$$

where U is the total energy of the particles, in this case, entirely translational kinetic energy. In order for this result to agree with the equation of state of an ideal gas,

$$pV = nRT,$$

the average total energy per particle U/N, under equilibrium conditions, must be equal to

$$\frac{U}{N} = \tfrac{3}{2}kT.$$

Since there are three translational degrees of freedom (i.e., three components of velocity), $\frac{1}{2}kT$ of energy on the average is attributed to each degree of freedom. With some limitations, this is true also of rotational degrees of freedom. The statement that with every degree of freedom of motion of a molecule, there is associated an average energy, at equilibrium, equal to $\frac{1}{2}kT$, is termed the principle of *equipartition of energy*.

The methods of calculation just referred to represent a simplified version of the branch of physics called the *kinetic theory of gases*. There is, however, another way in which the behavior of large numbers of particles may be described and calculated, which, in some respects, is more general and more powerful. This more general method is called *statistical mechanics.* In statistical mechanics it is not necessary to specify whether the particles are molecules of a gas, ions occupying sites in a crystal lattice, electrons, or other entities. The important thing is whether they are almost independent, and if so, whether the weak interaction among the particles is sufficient to enable the particles to exchange energy and finally come to equilibrium. If this is the case, the methods of statistical mechanics enable us to find the equilibrium value of the *average* energy per particle from a knowledge of the types of energy possessed by each individual particle. If the energy of an individual particle is expressed as the sum of a number of squared terms, such as $\frac{1}{2}mv_x^2$ or $\frac{1}{2}I\omega^2$, or $\frac{1}{2}kx^2$, then it can be proved *rigorously* that, at equilibrium, the average energy per particle *associated with each squared* term is $\frac{1}{2}kT$. This is the rigorous statement of the principle of the equipartition of energy.

If we consider an assemblage of N vibrators, each vibrating along the x-axis with very small amplitude almost independently (but not enough to prevent a weak interaction with neighbors), we may consider the energy of each vibrator to consist of two squared terms,

$$\tfrac{1}{2}mv_x^2 + \tfrac{1}{2}kx^2.$$

The principle of equipartition then gives us, for the average energy per particle at equilibrium, the sum of two terms each equal to $\frac{1}{2}kT$, or a total of kT.

Now a crystal is composed of N particles, each of which interacts strongly with its neighbors and vibrates in a very complicated way. In spite of the complications, it may be shown quite generally that the actual situation is *equivalent* to $3N$ oscillators, almost independent of one another, each vibrating with its own frequency. It follows therefore that the total energy at equilibrium is

$$U = 3NkT,$$

or, in terms of the number of moles n and the universal gas constant R,

$$U = 3nRT,$$

and finally

$$C_v = \frac{1}{n}\frac{dU}{dT} = 3R = 25.0\ \text{J mol}^{-1}\ \text{K}^{-1}.$$

This is the law of Dulong and Petit, which we encountered in Chapter 16. As Fig. 16–5 shows, many solids approach this value of C_v at high temperatures, but at low temperatures C_v invariably decreases; it is *not* constant, as the law of Dulong and Petit suggests, but approaches zero as T approaches zero.

A very careful analysis shows that there is nothing wrong with the methods of statistical mechanics or with the equivalence between the actual vibrations of a crystal lattice and $3N$ almost independent oscillators. What is wrong is the assumption that a vibrator can assume *any value* of the energy given by $\frac{1}{2}mv_x^2 + \frac{1}{2}kx^2$. In classical newtonian mechanics, this is certainly correct; but any treatment of mechanical systems on an atomic scale should include considerations of *quantum mechanics,* which we shall discuss in later chapters. A fundamental result of quantum mechanics is that a vibrator of frequency f may *not* take on *any* value of the energy, but only *discrete energy values* given by whole-number multiples of the product hf:

$$1hf,\ 2hf,\ 3hf,\ \ldots,$$

where h is a universal constant of nature, called *Planck's constant.* When this postulate is used in conjunction with the classical methods of statistical mechanics, it is found that the average energy per particle at equilibrium is always *less* than kT, but that at high temperatures it is approximately equal to kT.

This behavior may be understood qualitatively by noting that at very low temperatures the quantity kT is much *smaller* than the smallest energy increment hf that the vibrators can accommodate. As a result, at low T most of the systems remain in their lowest energy states because the energy hf required to make transitions to the next higher energy levels is not available. Thus, at low temperatures the average energy per vibrator is less than kT, and correspondingly the heat capacity per molecule is less than k. But in the other limit, where kT is *large* compared to hf, the classical equipartition theorem holds, and the total heat capacity is $3k$ per molecule, or $3R$ per mole, as the Dulong and Petit relation predicts. This behavior is illustrated by Fig. 16–5. Quantitative understanding of the temperature variation of specific heats was one of the triumphs of quantum mechanics during the early years of its development, early in the twentieth century.

PROBLEMS

Molecular Data

N_A = Avogadro's number
$= 6.02 \times 10^{23}$ molecules mol^{-1},
Mass of a hydrogen atom $= 1.67 \times 10^{-27}$ kg,
Mass of a nitrogen molecule $= 28 \times 1.66 \times 10^{-27}$ kg,
Mass of an oxygen molecule $= 32 \times 1.66 \times 10^{-27}$ kg.

20–1 Consider an ideal gas at 0°C and at 1 atm pressure. Imagine each molecule to be, on the average, at the center of a small cube.

a) What is the length of an edge of this small cube?
b) How does this distance compare with the diameter of a molecule?

20–2 A mole of liquid water occupies a volume of 18 cm^3. Imagine each molecule to be, on the average, at the center of a small cube.

a) What is the length of an edge of this small cube?
b) How does this distance compare with the diameter of a molecule?

20–3 What is the length of the side of a cube, in a gas at standard conditions, that contains a number of molecules equal to the population of the United States (about 200 million)?

20–4 The lowest pressures readily attainable in the laboratory are the order of 10^{-10} torr (the order of 10^{-13} atm). At this pressure and ordinary temperature (say $T = 300$ K) how many molecules are present in a volume of 1 cm^3?

20–5 Experiment shows that the size of an oxygen molecule is the order of 2×10^{-10} m. Make a rough estimate of the pressure at which the finite volume of the molecules should cause noticeable deviations from ideal-gas behavior at ordinary temperatures ($T = 300$ K).

20–6

a) What is the average translational kinetic energy of a molecule of oxygen at a temperature of 300 K?
b) What is the average value of the square of its speed?
c) What is the root-mean-square speed?
d) What is the momentum of an oxygen molecule traveling at this speed?
e) Suppose a molecule traveling at this speed bounces back and forth between opposite sides of a cubical vessel 10 cm on a side. What is the average force it exerts on the walls of the container?
f) What is the average force per unit area?
g) How many molecules traveling at this speed are necessary to produce an average pressure of 1 atm?
h) Compare with the number of oxygen molecules actually contained in a vessel of this size, at 300 K and atmospheric pressure.

20–7 The speed of propagation of a sound wave in air at 27°C is about 350 m s^{-1}. Compare this with the root-mean-square speed of nitrogen molecules at this temperature.

20–8 At what temperature is the rms speed of oxygen molecules equal to the rms speed of hydrogen molecules at 0° C?

20–9 Isotopes of uranium are sometimes separated by gaseous diffusion, using the fact that the rms speeds of their molecules in vapor are slightly different, and hence the vapors diffuse at slightly different rates. Assuming the atomic masses for ^{235}U and ^{238}U are 235 g mol^{-1} and 238 g mol^{-1}, respectively, what is the ratio of the rms speed of the ^{235}U atoms in the vapor to that of the ^{238}U atoms, assuming the temperature is uniform?

20–10 Compute the specific heat capacity at constant volume of hydrogen gas, and compare it with the specific heat capacity of water.

20–11 The molar heat capacities of many diatomic gases can be understood on the assumption that there is an additional average kinetic energy of kT per molecule associated with rotational motion. On this assumption, calculate the molar heat capacity at constant volume (C_v) for an ideal diatomic gas, and compare with the experimental values in Table 19–1.

20–12 What is the total random kinetic energy of the molecules in 1 mole of helium at a temperature of (a) 300 K? (b) 301 K? (c) Compare the difference between these with the change in internal energy of 1 mole of helium when its temperature is increased by 1 K, as computed from the relation $\Delta U = nC_V \, \Delta T$.

20–13 A flask contains a mixture of mercury vapor, neon, and helium. Compare (a) the average kinetic energies of the three types of atoms, and (b) the root-mean-square speeds.

20–14

a) At what temperature is the rms speed of hydrogen molecules equal to the speed of the first earth satellite (about 18,000 mi hr^{-1})?

b) At what temperature is the rms speed equal to the escape speed from the gravitational field of the earth?

20–15 Smoke particles in the air typically have masses the order of 10^{-16} kg. The Brownian motion of these particles resulting from collisions with air molecules can be observed with a microscope.

a) Find the root-mean-square speed of Brownian motion for such a particle in air at 300 K.

b) Would the speed be different if the particle were in hydrogen gas at the same temperature? Explain.

20–16 In the kinetic–molecular model of an ideal gas discussed in Section 20–4, the effect of gravity on molecular motion was neglected. How can this omission be justified?

20–17 The oven in Fig. 20–4 contains bismuth at a temperature of 840 K; the drum is 10 cm in diameter and rotates at 6000 rev min^{-1}. Find the displacement on the glass plate G, measured from a point directly opposite the slit, of the points of impact of the molecules Bi and Bi_2. Assume that all molecules of each species travel with the rms speed appropriate to that species.

20–18 Suppose the two atoms in an oxygen molecule are represented as two point masses 2×10^{-10} m apart.

a) Calculate the moment of inertia of a molecule about an axis through the midpoint, perpendicular to the line joining the atoms.

b) If the rotational kinetic energy of the molecule is $\frac{1}{2}kT$, what is the root-mean-square angular velocity of rotation at $T = 300$ K?

20–19 The density of crystalline sodium chloride is 2.16 g cm^{-3}, and its crystal structure is shown in Fig. 20–7. Find the spacing of adjacent atoms in the crystal lattice. The molecular "weight" of sodium is 23.0 g mol^{-1} and that of chlorine is 35.5 g mol^{-1}.

Chapter 21

Traveling Waves

21–1 INTRODUCTION

A wave is any disturbance from an equilibrium condition which travels or *propagates* with time from one region of space to another. There are many examples of waves in everyday experience, and examples of wave phenomena are found in all branches of physics. The concept of waves, in fact, is one of the most important unifying elements in all of physics.

Waves on the surface of a body of water, produced by the wind or by some other disturbance, are a familiar sight. A sound is carried by means of traveling waves in the intervening atmosphere between source and listener. Many of the observed properties of light are best explained by a wave theory, and we believe that light waves are of the same fundamental nature as radio waves, infrared and ultraviolet waves, x-rays, and gamma rays. One of the outstanding developments of twentieth-century physics has been the discovery that all matter is endowed with wave properties and that a beam of electrons, for example, is reflected by a crystal in much the same way as is a beam of x-rays.

The waves most easily comprehended are *mechanical waves*, in which the wave travels through some material medium as it is displaced from an equilibrium state. Imagine a medium consisting of a large number of particles, each connected or coupled to its neighbors by *elastic* material. If one end of the medium is disturbed or displaced in any way, the displacement does not occur immediately at all other parts of the medium. The original displacement gives rise to an elastic force in the material adjacent to it; then the next particle is displaced, and then the next, and so on. In other words, *the displacement is propagated along the medium with a definite speed.*

In Fig. 21–1(a) the medium is a spring, or even just a wire under tension. If the left end is given a small displacement in a direction perpendicular to the medium, this transverse displacement will occur at successive intervals of time at each coil of the spring, resulting in the propagation of a *transverse pulse* along the spring.

In Fig. 21–1(b) the medium is a liquid or a gas contained in a tube closed at the right end with a rigid wall and at the left end with a movable piston. If the piston is moved slightly toward the right, a *longitudinal pulse* is propagated through the medium in the tube. Longitudinal waves can also travel in solid materials.

In Fig. 21–1(c) the medium is a liquid contained in a trough. The horizontal motion of a flat piece of wood at the left end will create a displacement of the liquid which is *both* longitudinal and transverse, and this disturbance travels along the medium.

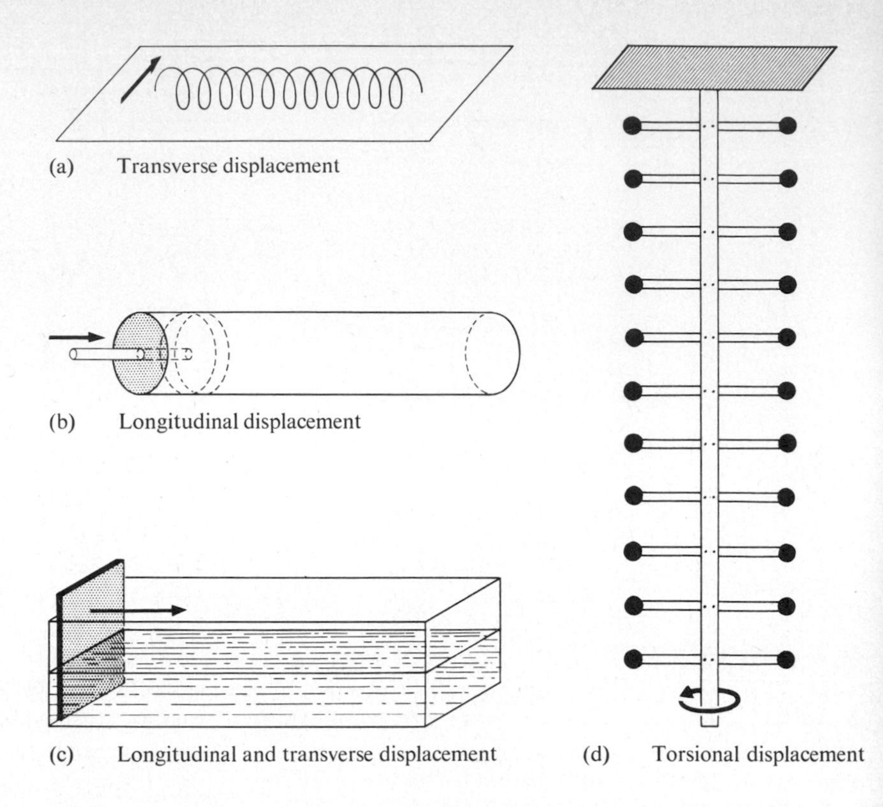

Fig. 21–1 *Propagation of disturbances.*

In Fig. 21–1(d) the medium is a set of "dumbbells" connected to a steel strip. A slight rotation of the lowest dumbbell constitutes a *torsional displacement* which is propagated up the medium with a finite speed.

21–2 PERIODIC WAVES

When the motions of the particles are at right angles to the direction of travel of the wave, the wave is called *transverse.* If the particles move in the direction of propagation, the wave is called *longitudinal.* Suppose that one end of a medium is forced to vibrate periodically, the displacement y (either transverse or longitudinal) varying with time according to the equation of simple harmonic motion:

$$y = \begin{cases} Y \sin \omega t, \\ \quad \text{or} \\ Y \cos \omega t. \end{cases}$$

The resulting continuous train of disturbances, traveling with a speed depending on the properties of the medium, is called a *wave.*

To fix our ideas, suppose that one end of a stretched string is forced to vibrate periodically in a transverse direction with simple harmonic motion of amplitude Y, frequency f, and period $\tau = 1/f$. For the present we shall assume the string to be long enough so that any effects at the far end need not be considered. A *continuous train* of transverse sinusoidal waves then advances along the string. The shape of a portion of the string near the end, at intervals of $\frac{1}{8}$ of a period, is shown in Fig. 21–2 for a total time of one period. The string is assumed to have been vibrating for a sufficiently long time so that the shape of the string is sinusoidal for an indefinite distance from the driven end. It will be seen from the figure that the waveform advances steadily toward the right, as indicated by the short arrow pointing to one particular wave crest, while any one point on the string (see the black dot) oscillates about its equilibrium position with simple harmonic motion. It is important to distinguish between the motion of the

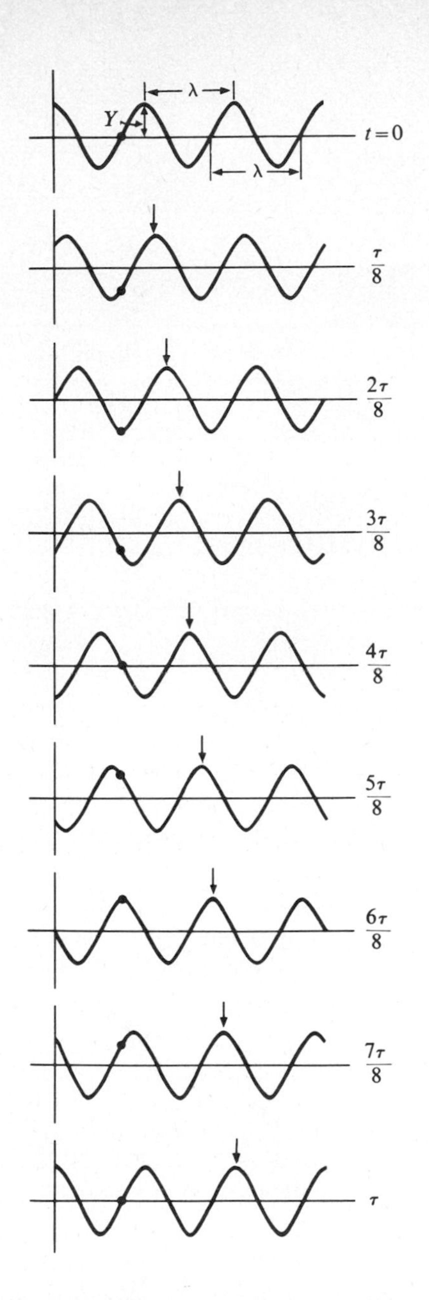

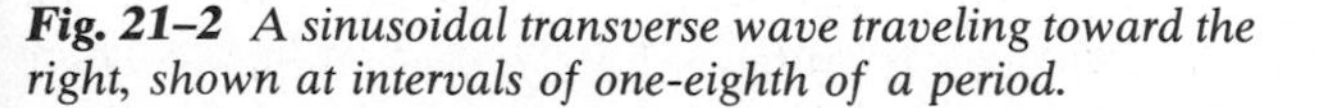
Fig. 21–2 *A sinusoidal transverse wave traveling toward the right, shown at intervals of one-eighth of a period.*

waveform, which moves with constant velocity c along the string, and the motion of *a particle of the string,* which is simple harmonic and transverse to the string.

The distance between two successive maxima (or between any two successive points in the same phase) is the *wavelength* of the wave and is denoted by λ. Since the waveform, traveling with constant velocity c, advances a distance of one wavelength in a time interval of one period, it follows that $c = \lambda/\tau$, or

$$\boxed{c = f\lambda.} \tag{21–1}$$

That is, *the velocity of propagation equals the product of frequency and wavelength.*

To understand the mechanics of a *longitudinal* wave, consider a long tube filled with a fluid and provided with a plunger at the left end, as shown in Fig. 21–3. The dots represent particles of the fluid. Suppose the plunger is forced to undergo a simple harmonic vibration parallel to the direction of the tube. During a part of each oscillation, a region whose pressure is above the equilibrium pressure is formed. Such a region is called a *condensation* and is represented by closely spaced dots. Following the production of a condensation, a region is formed in which the pressure is lower than the equilibrium value. This is called a *rarefaction* and is represented by widely spaced dots. The condensations and rarefactions move to the right with constant velocity c, as indicated by successive positions of the small vertical arrow. The velocity of longitudinal waves in air (sound) at 20°C is 344 m s^{-1} or 1130 ft s^{-1}. The motion of a single particle of the medium, shown by a heavy black dot, is simple harmonic, parallel to the direction of propagation.

The wavelength is the distance between two successive condensations or two successive rarefactions, and the same fundamental equation, $c = f\lambda$, holds in this, as in all types of periodic or repetitive waves.

21–3 MATHEMATICAL REPRESENTATION OF A TRAVELING WAVE

Suppose that a wave of any sort travels from left to right in a medium. Let us compare the motion of any one particle of the medium with that of a second particle to the right of the first. We find that the

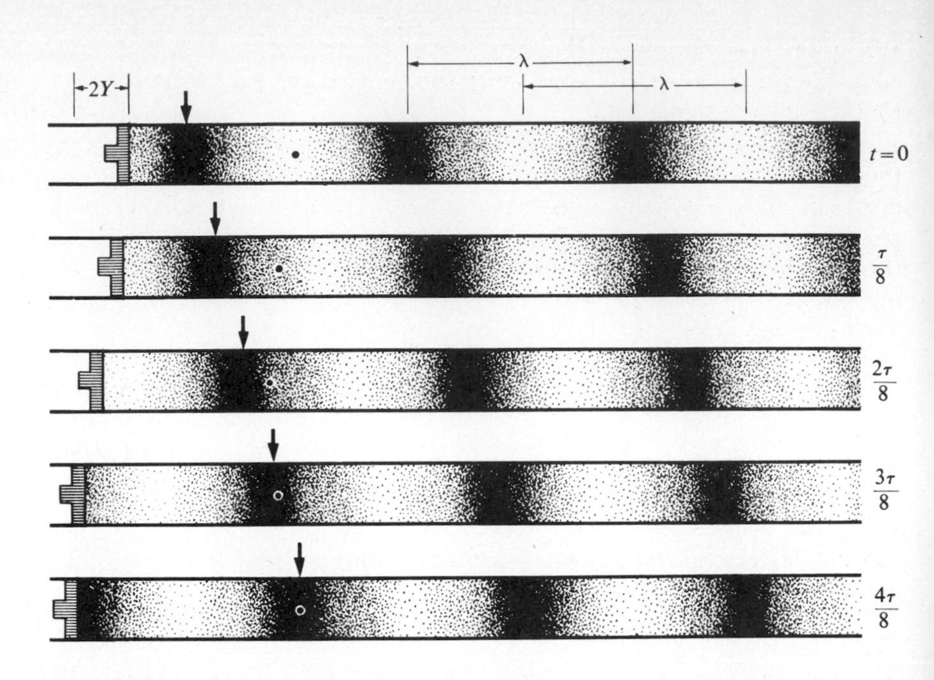

Fig. 21–3 *A sinusoidal longitudinal wave traveling toward the right, shown at intervals of one-eighth of a period.*

second particle moves in the same manner as the first, but after a lapse of time that is proportional to the distance of the second particle from the first. Hence if one end of a stretched string oscillates with simple harmonic motion, all other points oscillate with simple harmonic motion of the same amplitude and frequency. The *phase angle* of the motion, however, is different for different points.

Let the displacement of a particle at the origin ($x = 0$) be given by

$$y = Y \sin \omega t = Y \sin 2\pi f t. \tag{21–2}$$

The time required for the wave disturbance to travel from $x = 0$ to some point x to the right of the origin is given by x/c, where c is the wave speed, as usual. The motion of point x at time t is the same as the motion of point $x = 0$ at the earlier time $t - x/c$. Thus the displacement of point x at time t is obtained simply by replacing t in Eq. (21–2) by $(t - x/c)$, and we find

$$\begin{aligned} y(x, t) &= Y \sin \omega\left(t - \frac{x}{c}\right) \\ &= Y \sin 2\pi f\left(t - \frac{x}{c}\right). \end{aligned} \tag{21–3}$$

The notation $y(x,t)$ is a reminder that the displacement y is a function of two variables x and t, corresponding to the fact that the displacement of a point depends on both the location of the point and the time.

Equation (21–3) can be rewritten in several alternate forms, conveying the same information in different ways. In terms of the period τ and wavelength λ, we find

$$\boxed{y(x, t) = Y \sin 2\pi\left(\frac{t}{\tau} - \frac{x}{\lambda}\right).} \tag{21–4}$$

Another convenient form is obtained by defining a quantity k, called the *propagation constant* or the *wave number*:

$$k = \frac{2\pi}{\lambda}. \tag{21–5}$$

In terms of k and ω, the wavelength–frequency relation $c = \lambda f$ becomes

$$\boxed{\omega = ck,} \tag{21–6}$$

and Eq. (21–4) can be written

$$\boxed{y(x, t) = Y \sin(\omega t - kx).} \tag{21–7}$$

Which of these various forms is used is a matter of convenience in a specific problem; fundamentally they all say the same thing.

At any given *time* t, Eq. (21–4) or (21–7) gives the displacement y of a particle from its equilibrium position, as a function of the *coordinate* x of the particle. If the wave is a transverse wave in a string, the equation represents the *shape* of the string at that instant, as if we had taken a photograph of the string. Thus at time $t = 0$,

$$y = Y \sin(-kx) = -Y \sin kx = -Y \sin 2\pi \frac{x}{\lambda}\cdot$$

This curve is plotted in Fig. 21–4.

At any given *coordinate* x, Eq. (21–4) or (21–7) gives the displacement y of the particle at that coordinate, as a function of *time*. Thus at the position $x = 0$,

$$y = Y \sin \omega t = Y \sin 2\pi \frac{t}{\tau}\cdot$$

This curve is plotted in Fig. 21–5.

The above formulas may be used to represent a wave traveling in the negative x-direction by making a simple modification. In this case the displacement of point x at time t is the same as the motion of point $x = 0$ at the *later* time $(t + x/c)$, and in Eq. (21–2) we must replace t by $(t + x/c)$. Thus, for a wave traveling in the negative x-direction,

$$\begin{aligned} y &= Y \sin 2\pi f\left(t + \frac{x}{c}\right) \\ &= Y \sin 2\pi\left(\frac{t}{\tau} + \frac{x}{\lambda}\right) \\ &= Y \sin(\omega t + kx). \end{aligned} \tag{21–8}$$

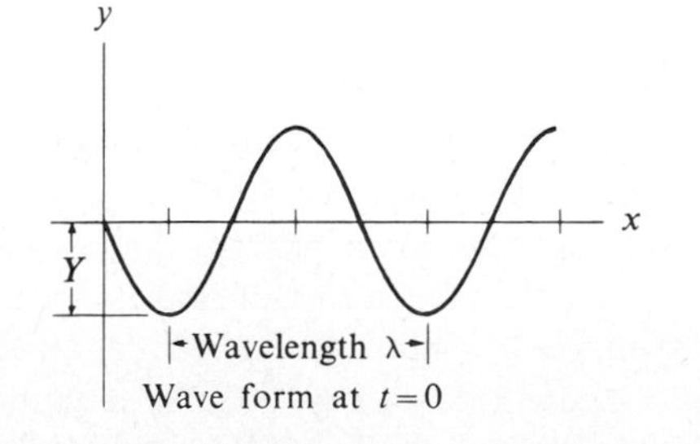

Fig. 21–4

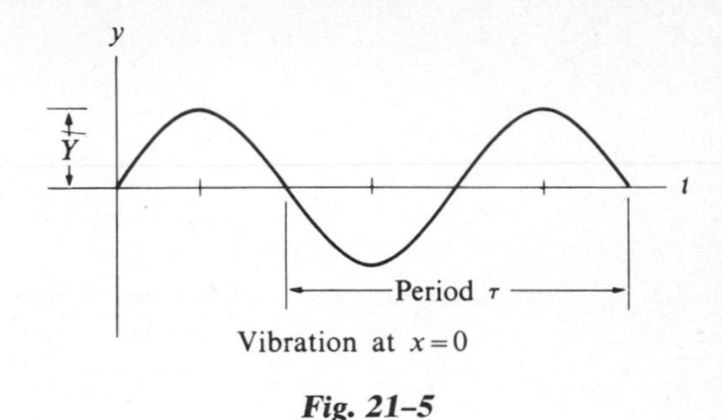

Fig. 21–5

It is essential to distinguish carefully between the *speed of propagation* c of the waveform and the *particle speed* v of a particle of the medium in which the wave is traveling. The wave speed c is given by

$$c = \lambda f = \frac{\omega}{k}\cdot \tag{21–9}$$

The particle speed v for any point in a transverse wave, that is, at a fixed value of x, is obtained by taking the derivative of y with respect to t, holding x constant. Such a derivative is called a *partial derivative* and is written $\partial y/\partial t$. Thus for a sinusoidal wave given by

$$y = Y \sin(\omega t - kx), \tag{21–10}$$

we have

$$v = \frac{\partial y}{\partial t} = \omega Y \cos(\omega t - kx). \tag{21–11}$$

The *acceleration* of the point is the *second* partial derivative:

$$a = \frac{\partial^2 y}{\partial t^2} = -\omega^2 Y \sin(\omega t - kx). \tag{21–12}$$

One may also compute partial derivatives with respect to x, holding t constant. The first derivative $\partial y/\partial x$ is the *slope* of the string at any point. The second partial derivative with respect to x is

$$\frac{\partial^2 y}{\partial x^2} = -k^2 Y \sin(\omega t - kx). \tag{21–13}$$

It follows from Eqs. (21–12) and (21–13) that

$$\frac{\partial^2 y/\partial t^2}{\partial^2 y/\partial x^2} = \frac{\omega^2}{k^2} = c^2,$$

since $c = \omega/k$. The *partial differential equation*

$$\frac{\partial^2 y}{\partial t^2} = c^2 \frac{\partial^2 y}{\partial x^2} \qquad (21\text{-}14)$$

is one of the most important in all of physics. It is called the *wave equation*, and whenever it occurs, the conclusion is made immediately that y is propagated as a traveling wave along the x-axis with a wave speed c.

21-4 SPEED OF A TRANSVERSE WAVE

We shall now derive a relation between the speed of propagation c of a wave on a stretched string and its mechanical properties, *mass* per unit length and *tension*. We consider a particularly simple wave motion, a transverse *pulse*. If a string stretched between two fixed supports is struck a transverse blow at some point, or if a small portion is displaced sideways and released, disturbances will be observed to travel outward in both directions from the displaced portion. Each of these is called a *transverse pulse*, the direction of motion of particles of the string being at right angles to the direction of propagation of the pulse. Each pulse retains its shape as it travels, and each travels with a constant speed which we shall represent by c.

Consider the string depicted in Fig. 21-6 under a tension S and with linear density (mass per unit length) μ. In Fig. 21-6(a) the string is at rest. At time $t = 0$, a constant transverse force F is applied at the left end of the string. As a result, this end moves up with a constant transverse speed v. Figure 21-6(b) shows the shape of the string after a time t has elapsed. All points of the string at the left of the point P are moving with speed v, whereas all points at the right of P are still at rest. The boundary between the moving and the stationary portions is traveling to the right with the *speed of propagation c*. The left end of the string has moved up a distance vt and the boundary point P has advanced a distance ct.

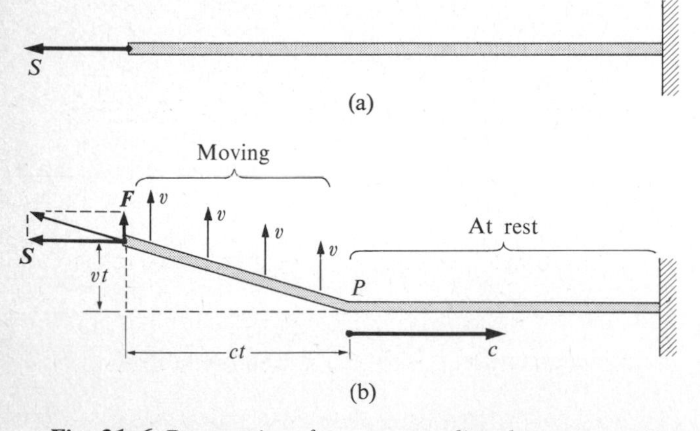

Fig. 21-6 *Propagation of a transverse disturbance in a string.*

The speed of propagation c can be calculated by setting the transverse impulse (transverse force × time) equal to the change of transverse momentum of the moving portion (mass × transverse velocity). The impulse of the transverse force F in time t is Ft. By similar triangles,

$$\frac{F}{S} = \frac{vt}{ct}, \qquad F = S\frac{v}{c}.$$

Hence

$$\text{Transverse impulse} = S\frac{v}{c}\,t.$$

The mass of the moving portion is the product of the mass per unit length μ and the length ct. Hence

$$\text{Transverse momentum} = \mu ctv.$$

Applying the impulse-momentum theorem, we obtain

$$S\frac{v}{c}\,t = \mu ctv,$$

and therefore

$$\boxed{c = \sqrt{S/\mu} \qquad \text{(transverse).}} \qquad (21\text{-}15)$$

Thus it is seen that the velocity of propagation of a transverse pulse in a string depends only on the tension (a force) and the mass per unit length. Although the above calculation of wave speed considered only a very special kind of pulse, a little thought shows that *any* shape of wave disturbance can be considered as a series of pulses with different rates of transverse displacement. Thus, although derived for a special case, Eq. (21-15) is valid for *any* transverse wave motion on a string, including, in particular, the sinusoidal and other periodic waves discussed in Section 21–3.

Example 1 One end of a rubber tube is fastened to a fixed support. The other end passes over a pulley at a distance of 8 m from the fixed end, and carries a load of 2 kg. The mass of the tube, between the fixed end and the pulley, is 600 g. What is the speed of a transverse wave in the tube?

The tension in the tube equals the weight of the 2-kg load, or

$$S = (2\ \text{kg})(9.8\ \text{m s}^{-2}) = 19.6\ \text{N}.$$

The mass of the tube, per unit length, is

$$\mu = \frac{m}{L} = \frac{0.60\ \text{kg}}{8\ \text{m}} = 0.075\ \text{kg m}^{-1}.$$

The speed of propagation is therefore

$$c = \sqrt{\frac{S}{\mu}} = \sqrt{\frac{19.6\ \text{N}}{0.075\ \text{kg m}^{-1}}} = 16\ \text{m s}^{-1}.$$

Example 2 Suppose that a sine wave of amplitude $Y = 10$ cm and wavelength $\lambda = 3$ m travels along the tube from left to right. What is the maximum *transverse* speed of a point of the tube?

The equation of the wave is

$$y = Y \sin(\omega t - kx).$$

The transverse speed is

$$v = \frac{\partial y}{\partial t} = \omega Y \cos(\omega t - kx).$$

The maximum transverse speed is

$$v_{\text{max}} = \omega Y = 2\pi f Y = 2\pi \frac{c}{\lambda} Y$$

$$= (2\pi)\left(\frac{16\ \text{m s}^{-1}}{3\ \text{m}}\right)(0.10\ \text{m})$$

$$= 3.35\ \text{m s}^{-1}.$$

21–5 SPEED OF A LONGITUDINAL WAVE

The propagation speed of longitudinal as well as transverse waves is determined by the mechanical properties of the medium. Figure 21–7 shows a fluid (liquid or gas) of density ρ in a tube of cross-sectional area A, under a pressure p. In Fig. 21–7(a) the fluid is at rest. At time $t = 0$, the piston at the left end of the tube is set in motion toward the right with a speed v. Figure 21–7(b) shows the fluid after a time t has elapsed. All portions of the fluid at the left of point P are moving with speed v, whereas all portions at the right of P are still at rest. The boundary between the moving and the stationary portions travels to the right with the speed of propagation c. The piston has moved a distance vt and the boundary has advanced a distance ct. As in the case of a transverse disturbance in a string, the speed of propagation can be computed from the impulse–momentum theorem.

The quantity of fluid set in motion in time t is the amount that originally occupied a volume of length ct and of cross-sectional area A. The mass of this fluid is therefore ρctA and the longitudinal momentum it has acquired is

$$\text{Longitudinal momentum} = \rho ctAv.$$

We next compute the increase of pressure, Δp, in the moving fluid. The original volume of the moving fluid, Act, has been decreased by an amount Avt. From the definition of bulk modulus B (see Chapter 10),

$$B = \frac{\text{change in pressure}}{\text{fractional change in volume}} = \frac{\Delta p}{Avt/Act}.$$

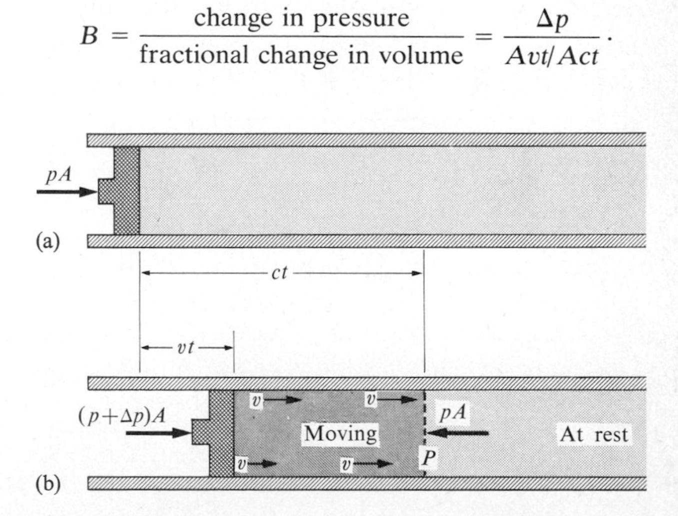

Fig. 21–7 Propagation of a longitudinal disturbance in a fluid confined in a tube.

Therefore,

$$\Delta p = B\frac{v}{c}.$$

The pressure in the moving fluid is therefore $p + \Delta p$, and the force exerted on it by the piston is $(p + \Delta p)A$. The net force on the moving fluid (see Fig. 21–7(b)) is ΔpA, and the longitudinal impulse is

$$\text{Longitudinal impulse} = \Delta pAt = B\frac{v}{c}At.$$

Applying the impulse-momentum theorem, we obtain

$$B\frac{v}{c}At = \rho ctAv,$$

and therefore

$$c = \sqrt{B/\rho} \qquad \text{(longitudinal).} \qquad (21\text{–}16)$$

The speed of propagation of a longitudinal pulse in a fluid therefore depends only on the bulk modulus and the density of the medium.

When a solid bar is struck a blow at one end, the situation is somewhat different from that of a fluid confined in a tube of constant cross section, since the bar expands slightly sidewise when it is compressed longitudinally. It can be shown, by the same type of reasoning as that just given, that the velocity of a longitudinal pulse in the bar is given by

$$c = \sqrt{Y/\rho} \qquad \text{(longitudinal),} \qquad (21\text{–}17)$$

where Y is Young's modulus, defined in Chapter 10.

As with the calculation for the transverse wave on a string, Eqs. (21–16) and (21–17) are valid for all wave motions, not just the special case discussed here. In particular, they are valid for sinusoidal and other periodic waves, discussed in Section 21–3.

21–6 ADIABATIC CHARACTER OF A LONGITUDINAL WAVE

It is a familiar fact that compression of a fluid causes a rise in its temperature unless heat is withdrawn in some way. Conversely, an expansion is accompanied by a temperature decrease unless heat is added. As a longitudinal wave advances through a fluid, the regions which are compressed at any instant are slightly warmer than those that are expanded. The condition is present, therefore, for the conduction of heat from a condensation to a rarefaction. The quantity of heat conducted per unit time and per unit area depends on the thermal conductivity of the fluid and upon the distance between a condensation and its adjacent rarefaction (half a wavelength). Now for ordinary frequencies, say from 20 vibrations per second to 20,000 vibrations per second (the frequency range in which the human ear is sensitive) and for even the best known heat conductors, the wavelength is too large and the thermal conductivity too small for an appreciable amount of heat to flow. The compressions and rarefactions are therefore *adiabatic* rather than isothermal.

In the expression for the speed of a longitudinal wave in a fluid, $c = \sqrt{B/\rho}$, the bulk modulus B is defined by the relation

$$B = \frac{\text{change in pressure}}{\text{fractional change in volume}}.$$

The change in volume produced by a given change of pressure depends upon whether the compression (or expansion) is adiabatic or isothermal. Thus there are two bulk moduli, the adiabatic bulk modulus B_{ad} and the isothermal bulk modulus. The expression for the speed of a longitudinal wave should therefore be written

$$c = \sqrt{\frac{B_{\text{ad}}}{\rho}}. \qquad (21\text{–}18)$$

In the case of an ideal gas, the relation between pressure p and volume V during an adiabatic process is given by

$$pV^{\gamma} = \text{constant}, \qquad (21\text{–}19)$$

where γ is the ratio of the heat capacity at constant pressure to the heat capacity at constant volume.

The definition of the adiabatic bulk modulus is

$$B_{ad} = -\left(\frac{dp}{dV/V}\right)_{ad} = -V\left(\frac{dp}{dV}\right)_{ad}.$$

To calculate the adiabatic bulk modulus B_{ad}, we must evaluate the derivative $(dp/dV)_{ad}$ with the aid of the adiabatic equation. Taking logarithms of both sides of Eq. (21–19), we obtain

$$\ln p + \gamma \ln V = \ln \text{constant},$$

and, taking the differential of this equation, we get

$$\frac{dp}{p} + \gamma \frac{dV}{V} = 0,$$

whence

$$\left(\frac{dp}{dV}\right)_{ad} = -\gamma\frac{p}{V}.$$

Thus for an ideal gas the adiabatic bulk modulus is given by

$$B_{ad} = \gamma p, \qquad (21\text{–}20)$$

while the isothermal bulk modulus B is given simply by

$$B = p. \qquad (21\text{–}21)$$

In each case the bulk modulus, characterizing the material's resistance to being compressed, is proportional to the pressure, but the adiabatic modulus is larger than the isothermal by a factor $\gamma = C_p/C_v$, the ratio of the heat capacities at constant pressure and constant volume.

Combining Eqs. (21–18) and (21–20), we obtain

$$c = \sqrt{\gamma p/\rho} \quad \text{(ideal gas)}. \qquad (21\text{–}22)$$

But, for an ideal gas,

$$p/\rho = RT/M,$$

where R is the universal gas constant, M the molecular mass, and T the Kelvin temperature. Therefore

$$\boxed{c = \sqrt{\gamma RT/M} \quad \text{(ideal gas)},} \qquad (21\text{–}23)$$

and since for a given gas, γ, R, and M are constants, we see that the velocity of propagation is proportional to the square root of the Kelvin temperature.

Let us use Eq. (21–23) to compute the velocity of longitudinal waves in air. The mean molecular mass of air is 28.8 g mol^{-1} = 28.8 × 10^{-3} kg mol^{-1}. Also, $\gamma = 1.40$ and $R = 8.31$ J mol^{-1} K^{-1}. At T = 300 K we obtain

$$c = \sqrt{\frac{(1.40)(8.31\ \text{J mol}^{-1}\ \text{K}^{-1})(300\ \text{K})}{28.8 \times 10^{-3}\ \text{kg mol}^{-1}}}$$

$$\doteq 348\ \text{m s}^{-1} = 1140\ \text{ft s}^{-1} = 778\ \text{mi hr}^{-1}.$$

This agrees with the measured speed at this temperature to within 0.3 percent.

Longitudinal waves in air give rise to the sensation of *sound.* The ear is sensitive to a range of sound frequencies from about 20 to about 20,00 hertz (one hertz = 1 Hz = 1 cycle s^{-1}). From the relation $c = f\lambda$, thc corresponding wavelength range is from about 17 m, corresponding to a 20-Hz note, to about 1.7 cm, corresponding to 20,000 Hz.

The *molecular* nature of a gas has been ignored in the preceding discussion, and a gas has been treated as though it were a continuous medium. Actually, we know that a gas is composed of molecules in random motion, separated by distances which are large compared with their diameters. The vibrations which constitute a wave in a gas are superposed on the random thermal motion. At atmospheric pressure, the mean free path is about 10^{-5} cm, while the displacement amplitude of a faint sound may be only one ten-thousandth of this amount. An element of gas in which a sound wave is traveling can be compared to a swarm of gnats, where the swarm as a whole can be seen to oscillate slightly while individual insects move about through the swarm, apparently at random.

Since the *shape* of a fluid does not change when a longitudinal wave passes through it, it is not as easy to visualize the relation between particle motion and wave motion as it is for transverse waves in a string. Figure 21–8 may help in correlating these motions To use this figure, cut a slit about $\frac{1}{16}$ in. wide and $3\frac{1}{4}$ in. long in a 3-in. × 5-in. card (or fasten two cards edge-to-edge with a $\frac{1}{16}$-in. gap), place the card over

Fig. 21–8 *Diagram for illustrating longitudinal traveling waves.*

the figure with the slit at the top of the diagram, and move the card downward with constant velocity. The portions of the sine curves that are visible through the slit correspond to a row of particles along which there is traveling a longitudinal, sinusoidal wave. Notice that each particle executes simple harmonic motion about its equilibrium position, with a phase that increases continuously along the slit, while the regions of maximum condensation and rarefraction move from left to right with constant speed. Moving the card upward produces a wave traveling from right to left.

21–7 WATER WAVES

Probably the most familiar type of wave motion is that observed at the surface of a body of water and produced by the winds or some other disturbance. The oscillations of the water particles in these waves are not confined to the surface, however, but extend with diminishing amplitude to the very bottom. Furthermore, the oscillations have both a longitudinal and a transverse component. For this reason, and also because the motion is governed by the laws of hydrodynamics, a complete analysis of this wave motion calls for mathematical methods beyond the scope of this book. The basic physical approach, however, is perfectly straightforward.

Consider a long canal of rectangular cross section and with frictionless walls, containing a depth h of an ideal incompressible liquid of density ρ and surface tension γ. When a train of waves travels along the canal, each element of the liquid is displaced from its equilibrium position both horizontally and vertically. The restoring force on the element results in part from pressure differences brought about by the variations in depth from point to point, and in part from surface tension effects arising from the curvature of the free surface. The solution of the resulting equations of motion must satisfy the conditions that the pressure at the upper surface is constant and equal to atmospheric pressure, and that the vertical displacement at the bottom of the canal is always zero. In addition, the motion of the fluid must satisfy the equation of continuity.

For simple harmonic waves, in which the x- and y-components of displacement are sine or cosine functions, the speed c is found to be given by

$$c^2 = \left(\frac{g\lambda}{2\pi} + \frac{2\pi\gamma}{\rho\lambda}\right) \tanh \frac{2\pi h}{\lambda}, \qquad (21\text{–}24)$$

where g is the acceleration of gravity and λ is the wavelength.

We immediately recognize one very significant difference between this expression and that for the speed of propagation of transverse waves in a string, or compressional waves in a fluid. Here, the speed depends not only on the properties of the medium *but on the wavelength λ as well.* When this is the case, the medium is said to be *dispersive.* Such a dependence of speed on wavelength occurs in many other types of waves. For example, a prism separates a parallel beam of white light into a spectrum, with different wavelengths emerging in different directions, because glass is a dispersive medium for light waves.

In general, the particles of liquid move in ellipses in vertical planes parallel to the length of the canal, the long axis of the ellipse being horizontal. This motion can be considered as the superposition of two simple harmonic oscillations of the same frequency and different amplitudes, one in the horizontal and one in the vertical direction, and with a phase difference of 90°. The wave can therefore be considered as the superposition of a longitudinal and a transverse wave 90° out of phase with each other and having different amplitudes. If the wavelength is as small or smaller than the depth, the two amplitudes are very nearly equal at the surface and the surface particles move in circles. The amplitudes of both components decrease with increasing depth, but the vertical component decreases more rapidly than does the horizontal. At the bottom, the vertical component becomes zero and the oscillation is wholly longitudinal.

Figure 21–9 illustrates the paths of motion of particles in the surface layer and at some depth below the surface. The upper horizontal dotted line represents the free surface of the liquid at rest. The circles are the paths of particles whose equilibrium position is at the center of the circle. When a train of waves travels from left to right, the particles revolve clockwise in these paths. The full line gives the shape of the surface at the instant shown.

The lower dotted line passes through the equilibrium positions of particles at some depth below the surface. Their paths are elliptical, as indicated, and the dashed line is their locus at the instant when the free surface has the shape shown by the full line.

Similar waves can occur at the surfaces of elastic solids. The rigidity of the material is an added complication in this case, with the result that the phase difference between the longitudinal and transverse components of displacement changes with depth beneath the surface. Such waves, called *Rayleigh waves,* promise to become important in connection with mechanical devices for processing radar, television, and radio signals. For example, it is possible to make *mechanical* band-pass filters for the megahertz frequency range which are simpler and more compact than their electrical analogs.

There are certain properties possessed by all waves. Propagation, reflection, refraction, absorption, interference, and diffraction are the most common. These wave phenomena may be demonstrated effectively by means of a "ripple tank" consisting of a horizontal sheet of glass or plastic forming the bottom of a shallow tank, filled with water about 1 cm deep. A strong source of light beneath the tank sends light through the water waves, which cast a shadow on a translucent screen (tracing cloth) above the tank. A convenient source of water waves is provided by a smooth sphere about the size of a

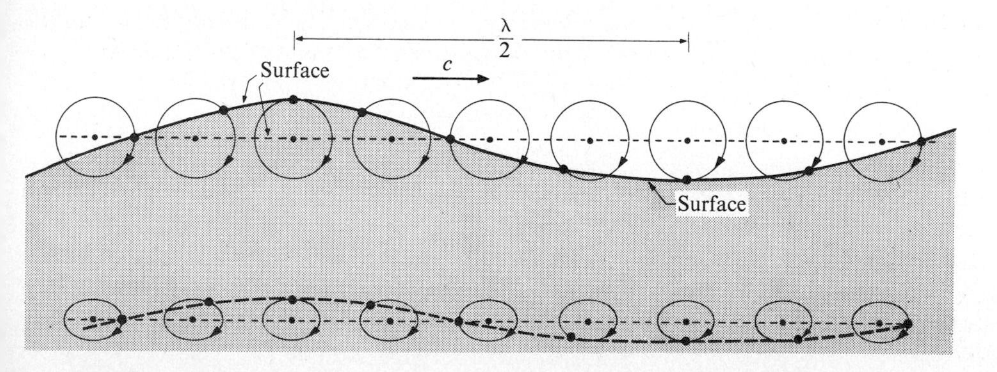

Fig. 21–9 *Particle motion and wave shape for waves in a canal.*

pencil eraser attached to a rod and periodically dipped into the water. At low frequencies the wavelength λ may be made as large as, say, 10 cm. The surface tension of water is about 0.070 N m^{-1}, and the density of water is 10^3 kg m^{-3}. Under these conditions, the second term in the parentheses of Eq. (21–24) turns out to be negligible compared with the first; also, $\tanh 2\pi h/\lambda$ is very nearly equal to $2\pi h/\lambda$. Then

$$c \approx \sqrt{\frac{g\lambda}{2\pi} \cdot \frac{2\pi h}{\lambda}} \approx \sqrt{gh}. \qquad (21\text{–}25)$$

Thus for *long* waves, the speed depends only on the acceleration due to gravity, g, and the depth h, and is independent of the wavelength (no dispersion). Since, according to Eq. (21–25), the shallower the water, the slower the speed of the waves, it is possible to eliminate reflections from the edges of the ripple tank by beveling the edges of the tank. A wave approaching a "shore" is therefore slowed down to zero speed.

PROBLEMS

21–1 The equation of a certain traveling transverse wave is

$$y = 2 \sin 2\pi\left(\frac{t}{0.01} - \frac{x}{30}\right),$$

where x and y are in centimeters and t is in seconds. What are (a) the amplitude, (b) the wavelength, (c) the frequency, and (d) the speed of propagation of the wave?

21–2 Show that Eq. (21–4) may be written

$$y = -Y \sin \frac{2\pi}{\lambda}(x - ct).$$

21–3 A traveling transverse wave on a string is represented by the equation in Problem 21–2. Let $Y = 8$ cm, $\lambda = 16$ cm, and $c = 2$ cm s^{-1}.

a) At time $t = 0$, compute the transverse displacement y at 2-cm intervals of x (i.e., at $x = 0$, $x = 2$ cm, $x = 4$ cm, etc.) from $x = 0$ to $x = 32$ cm. Show the results in a graph. This is the shape of the string at time $t = 0$.
b) Repeat the calculations, for the same values of x, at times $t = 1$ s, $t = 2$ s, $t = 3$ s, and $t = 4$ s. Show on the same graph the shape of the string at these instants. In what direction is the wave traveling?

21–4 The equation of a transverse traveling wave on a string is

$$y = 2 \cos \pi(0.5x - 200t),$$

where x and y are in cm and t is in seconds.

a) Find the amplitude, wavelength, frequency, period, and velocity of propagation.
b) Sketch the shape of the string at the following values of t: 0, 0.0025 s, and 0.005 s.
c) If the mass per unit length of the string is 5 g cm^{-1}, find the tension.

21–5 A steel wire 6 m long has a mass of 60 g and is stretched with a tension of 1000 N. What is the speed of propagation of a transverse wave in the wire?

21–6 One end of a horizontal string is attached to a prong of an electrically driven tuning fork whose frequency of vibration is 240 Hz. The other end passes over a pulley and supports a mass of 5 kg. The linear mass density of the string is 0.02 kg m^{-1}.

a) What is the speed of a transverse wave in the string?
b) What is the wavelength?

21–7 One end of a stretched rope is given a periodic transverse motion with a frequency of 10 Hz. The rope is 50 m long, has total mass 0.5 kg, and is stretched with a tension of 400 N.

a) Find the wave speed and the wavelength.
b) If the tension is doubled, how must the frequency be changed to maintain the same wavelength?

21–8 One end of a rubber tube 20 m long, of total mass 1 kg, is fastened to a fixed support. A cord attached to the other end passes over a pulley and supports a body of mass 10 kg. The tube is struck a transverse blow at one end. Find the time required for the pulse to reach the other end.

21–9 A metal wire has the properties: coefficient of linear expansion $= 1.5 \times 10^{-5}(\text{C}^\circ)^{-1}$, Young's modulus $= 2.0 \times 10^{11}$ N m^{-2}, density $= 9.0 \times 10^3$ kg m^{-3}. At each end are rigid supports. If the tension is zero at 20°C, what will be the speed of a transverse wave at 8°C?

21–10 A metal wire, of density 5×10^3 kg m^{-3}, with a Young's modulus equal to 2.0×10^{11} N m^{-1}, is stretched between rigid supports. At one temperature the speed of a transverse wave is found to be 200 m s^{-1}. When the temperature is raised 100 C°, the speed decreases to 160 m s^{-1}. What is the coefficient of linear expansion?

21–11 A transverse sine wave of amplitude 10 cm and wavelength 200 cm travels from left to right along a long horizontal stretched string with a speed of 100 cm s^{-1}. Take the origin at the left end of the undisturbed string. At time $t = 0$, the left end of the string is at the origin and is moving downward.

a) What is the frequency of the wave?
b) What is the angular frequency?
c) What is the propagation constant?
d) What is the equation of the wave?
e) What is the equation of motion of the left end of the string?
f) What is the equation of motion of a particle 150 cm to the right of the origin?
g) What is the (absolute) maximum transverse velocity of any particle of the string?
h) Find the transverse displacement and the transverse velocity of a particle 150 cm to the right of the origin, at time $t = 3.25$ s.
i) Make a sketch of the shape of the string, for a length of 400 cm, at time $t = 3.25$ s.

21–12 What must be the stress (F/A) in a stretched wire of a material whose Young's modulus is Y, in order that the speed of longitudinal waves shall equal 10 times the speed of transverse waves?

21–13 The speed of longitudinal waves in water is approximately 1450 m s^{-1} at 20°C. Compute the adiabatic compressibility ($1/B_{ad}$) of water and compare with the isothermal compressibility in Table 10–2.

21–14 Provided the amplitude is sufficiently great, the human ear can respond to longitudinal waves over a range of frequencies from about 20 Hz to about 20,000 Hz. Compute the wavelengths corresponding to these frequencies (a) for waves in air, (b) for waves in water. (See Problem 21–13).

21–15 Middle C on the piano corresponds to a frequency of 262 Hz = 262 s^{-1}.

a) Find the wavelength and wave number of the corresponding sound wave in air at 20° C.
b) What is the angular frequency?
c) What is the wavelength of the wave in water at 20° C. if the wavespeed is 1450 m s^{-1}?

21–16 What is the ratio of the speed of sound in a diatomic gas to the rms speed of gas molecules at the same temperature?

21–17

a) If the propagation of sound waves in gases were characterized by isothermal rather than adiabatic expansions and compressions, show that the speed of sound would be given by $(RT/M)^{1/2}$.
b) What would be the speed of sound in air at 20° C in this case?
c) Under what circumstances might the wave propagation be expected to be isothermal?

21–18 A longitudinal wave propagates in a steel bar having density 7.0 g cm^{-3} and Young's modulus 2×10^{11} N m^{-2}.

a) What is the wave speed?
b) By what factor is this speed greater than the speed of sound in air at 20° C?

21–19 At a temperature of 27°C, what is the speed of longitudinal waves in (a) argon, (b) hydrogen? (c) Compare with the speed in air at the same temperature.

21–20 What is the difference between the speeds of longitudinal waves in air at −3°C and at 57°C?

21–21 The sound waves from a loudspeaker spread out nearly uniformly in all directions when their wavelength is large compared with the diameter of the speaker. When the wavelength is small compared with the diameter of the speaker, much of the sound energy is concentrated in the forward direction. For a speaker of diameter 25 cm, compute the frequency for which the wavelength of the sound waves, in air, is (a) 10 times the diameter of the speaker, (b) equal to the diameter of the speaker, (c) 1/10 the diameter of the speaker.

21–22 A steel pipe 100 m long is struck at one end. A person at the other end hears two sounds as a result of two longitudinal waves, one in the pipe and the other in the air. What is the time interval between the two sounds? Take Young's modulus of steel to be 2×10^{11} N m^{-2}.

21–23 Explain why the breakers on a sloping beach are always parallel to the shore line, whatever may be the direction of the waves far from the shore.

21–24 (a) By how many meters per second, at a temperature of 27°C, does the speed of sound in air increase per Celsius degree rise in temperature? [*Hint:* compute dc in terms of dT, and approximate finite changes by differentials.] (b) Is the rate of change of speed with temperature the same at all temperatures?

21–25 A loose coil of flexible rope of length L and mass M rests on a frictionless table. A force S is applied at one end of the rope and more and more rope is pulled from the coil with a constant velocity v.

a) What is the relation between S and v?

b) A transverse pulse is produced while the rope is being pulled from the coil. How does the pulse behave?

c) What is the work of the force S at the moment when the last bit of rope leaves the coil?

d) What is the kinetic energy of the rope?

e) How come?

21–26 The ends of a flexible rope of length L and mass M are joined so as to form a circular loop of radius R. The loop rotates about an axis through its center on a horizontal frictionless surface with a constant linear speed v.

a) Apply Newton's second law to a short piece of rope and calculate the tension S in the rope.

b) Explain the effect of imparting to the rope, while it is spinning, a small rapid radial displacement.

Chapter 22

Vibrating Bodies

22–1 BOUNDARY CONDITIONS FOR A STRING

Let us consider what happens when a wave pulse or wave train, advancing along a stretched string, arrives at the end of the string. If the end is fastened to a rigid support, it just remains at rest. The arriving wave exerts a force on the support, and the reaction to this force "kicks back" on the string and sets up a *reflected* pulse or wave train traveling in the reverse direction.

At the opposite extreme from a rigidly fixed end is one which is perfectly free to move in the direction transverse to the length of the string. For example, the string might be tied to a light ring which slides on a smooth rod perpendicular to the length of the string; the ring and rod maintain the tension of the string but exert no transverse force. At a free end the arriving pulse or wave train causes the end to "overshoot," and again a reflected wave is set up. The conditions imposed on the motion of the end of the string, such as attachment to a rigid support or the complete absence of transverse force, are called *boundary conditions.*

The multiflash photograph of Fig. 22–1 shows the reflection of a pulse at a fixed end of a string. (The camera was tipped vertically while the photographs were taken so that successive images lie one under the other. The "string" is a rubber tube and it

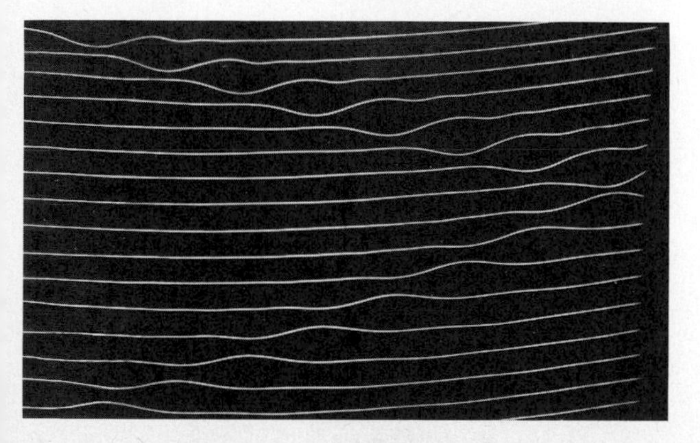

Fig. 22–1 A pulse starts in the upper left corner and is reflected from the fixed end of the string at the right.

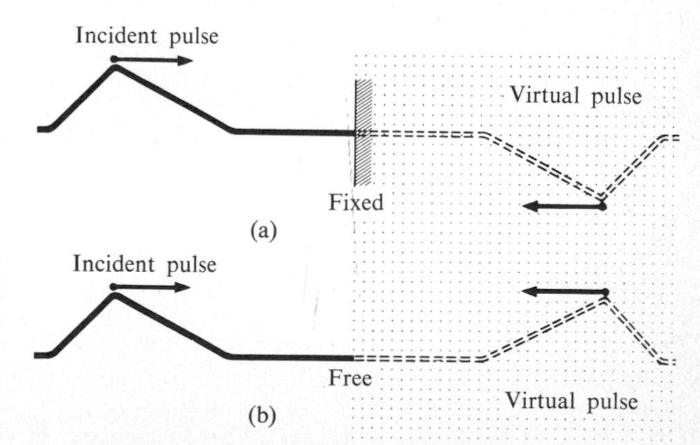

Fig. 22–2 Description of reflection of a pulse (a) at a fixed end of a string, and (b) at a free end, in terms of an imaginary "virtual" pulse.

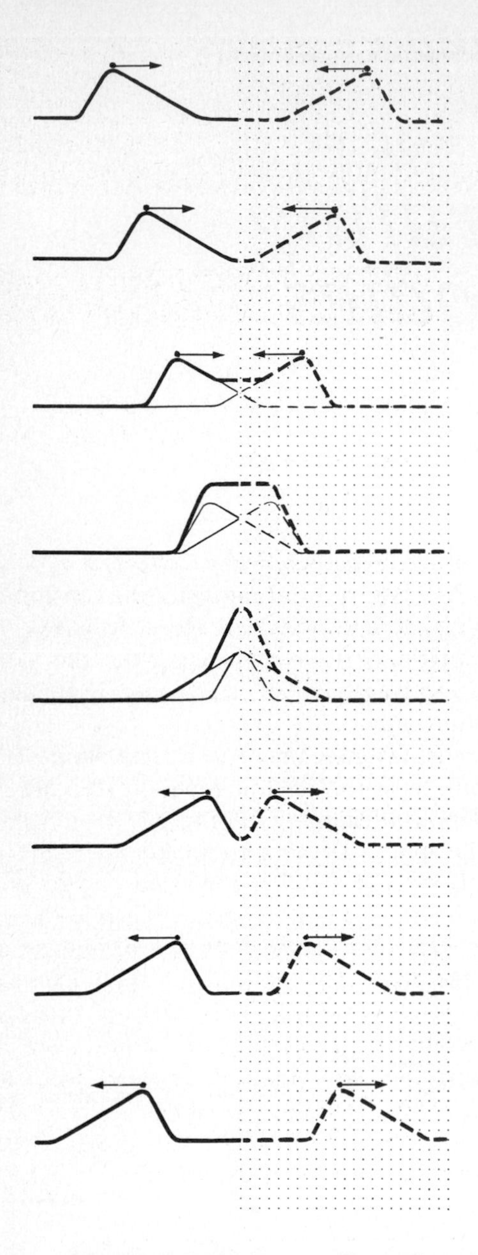

Fig. 22–3 Reflection at a free end.

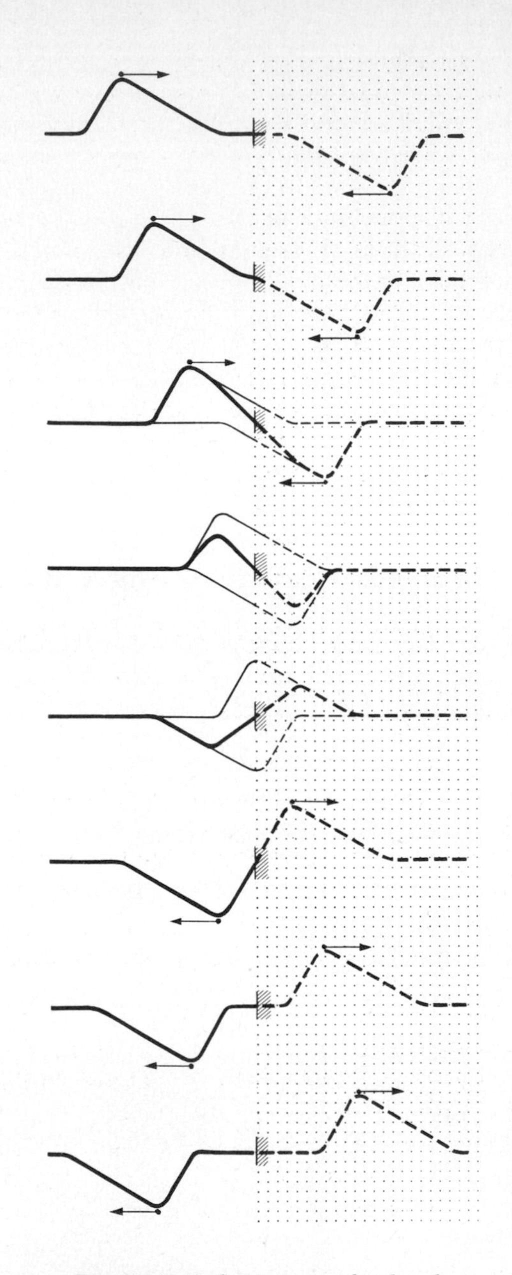

Fig. 22–4 Reflection at a fixed end.

sags somewhat.) It will be seen that the pulse is reflected with its displacement and its direction of propagation both reversed. When reflection takes place at a free end, the direction of propagation is reversed but the direction of the displacement is unchanged.

It is helpful to think of the process of reflection in the following way. Imagine the string to be extended indefinitely beyond its actual terminus. The actual pulse can be considered to continue on into the imaginary portion as though the support were not there, while at the same time a "virtual" pulse, which has been traveling in the imaginary portion, moves out into the real string and forms the reflected pulse. The nature of the reflected pulse depends on whether the end is fixed or free. The two cases are shown in Fig. 22–2.

The displacement at a point where the actual and virtual pulses cross each other is the algebraic sum of the displacements in the individual pulses. Figures 22–3 and 22–4 show the shape of the region near the end of the string for both types of reflected pulses. It will be seen that Fig. 22–3 corresponds to a free end and Fig. 22–4 to a fixed end. In the latter case, the incident and reflected pulses combine in such a way that the displacement of the end of the string is always zero.

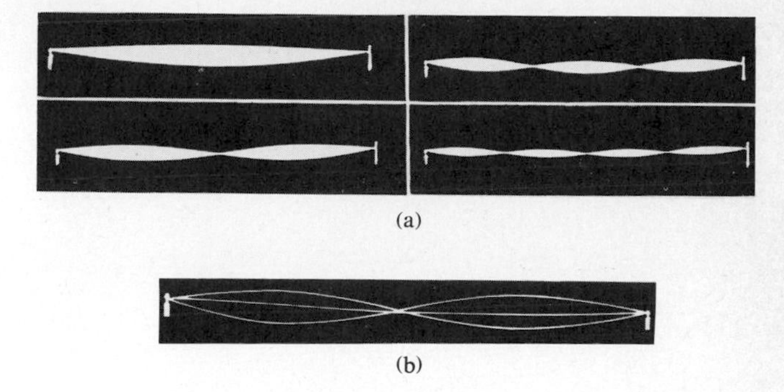

Fig. ***22–5*** *(a) Standing waves in a stretched string (time exposure). (b) Multiflash photograph of a standing wave, with nodes at the center and at the ends.*

22–2 STANDING WAVES IN A STRING

When a continuous train of waves arrives at a fixed end of a string, a continuous train of reflected waves appears to originate at the end and travel in the opposite direction. Provided the elastic limit of the string is not exceeded and the displacements are sufficiently small, the actual displacement of any point of the string is the algebraic sum of the displacements of the incident and reflected waves, a fact called the *principle of superposition.* This principle is extremely important in all types of wave motion and applies not only to waves in a string but to sound waves in air, to light waves, and, in fact, to wave motion of any sort. The general term *interference* is applied to the effect produced by two (or more) wave trains which are simultaneously passing through a given region.

The appearance of the string in these circumstances gives no evidence that two waves are traversing it in opposite directions. If the frequency is sufficiently great so that the eye cannot follow the motion, the string appears subdivided into a number of segments, as in the time exposure photograph of Fig. 22–5(a). A multiflash photograph of the same string, in Fig. 22–5(b), indicates a few of the instantaneous shapes of the string. At any instant (except those when the string is straight) its shape is a sine curve, but whereas in a traveling wave the amplitude remains constant while the wave progresses, here the waveform remains fixed in position (longitudinally) while the amplitude fluctuates. Certain points known as the *nodes* remain always at rest. Midway between these points, at the *loops* or *antinodes*, the fluctuations are a maximum. The vibration as a whole is called a *standing* wave.

To understand the formation of a standing wave, consider the separate graphs of the waveform S at four instants one-tenth of a period apart, shown in Fig. 22–6. The dotted curves represent a wave traveling to the right. The dashed curves represent a wave of the same propagation speed, wavelength, and amplitude traveling to the *left.* The heavy curves represent the resultant waveform, obtained by applying the principle of superposition, that is, by adding displacements. At those places marked N at the bottom of Fig. 22–6, the resultant displacements are *always* zero. These are the ***nodes***. Midway between the nodes, the vibrations have the *largest* amplitude. These are the ***antinodes***. It is evident from the figure that

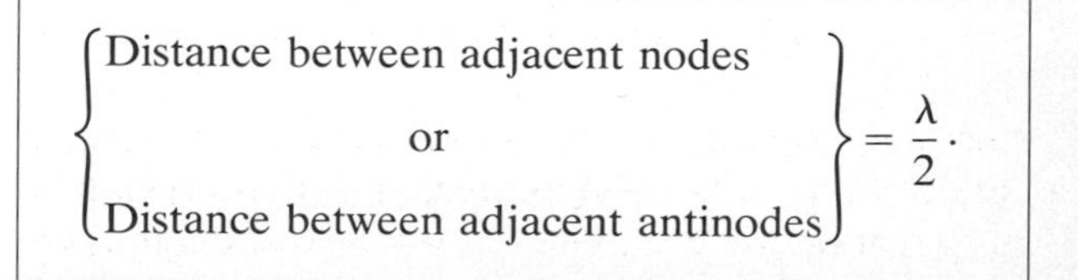

$$\left.\begin{array}{c}\text{Distance between adjacent nodes}\\ \text{or}\\ \text{Distance between adjacent antinodes}\end{array}\right\} = \frac{\lambda}{2}.$$

The equation of a standing wave may be obtained by adding the displacements of two waves of

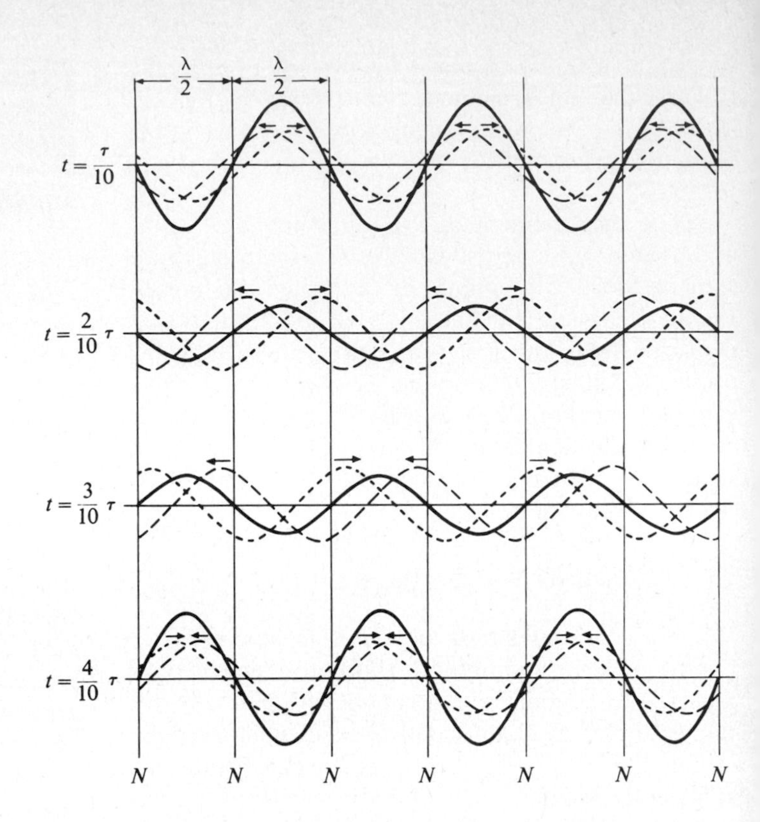

Fig. 22-6 The formation of a standing wave.

equal amplitude, period, and wavelength, but traveling in opposite directions. Thus, if

$$y_1 = A \sin(\omega t - kx) \qquad \text{(positive } x\text{-direction)},$$
$$y_2 = -A \sin(\omega t + kx) \qquad \text{(negative } x\text{-direction)},$$

then

$$y_1 + y_2 = A[\sin(\omega t - kx) - \sin(\omega t + kx)].$$

Introducing the expressions for the sine of the sum and difference of two angles and combining terms, we obtain

$$y_1 + y_2 = -[2A \cos \omega t]\sin kx. \qquad (22\text{-}1)$$

The shape of the string at each instant is, therefore, a sine curve whose amplitude (the expression in brackets) varies with time.

22-3 VIBRATION OF A STRING FIXED AT BOTH ENDS

Thus far we have discussed a long string fixed at one end and have considered the standing waves set up near that end by interference between the incident and reflected waves. Let us next consider the more usual case, that of a string fixed at *both* ends. A continuous train of sine or cosine waves is reflected and re-reflected; and since the string is fixed at both ends, both ends must be nodes. Since adjacent nodes are one-half wavelength apart, the length of the string may be

$$\frac{\lambda}{2}, \quad \frac{2\lambda}{2}, \quad \frac{3\lambda}{2},$$

or, in general, *any* integer number of half-wavelengths. Or, to put it differently, if one considers a particular string of length L, standing waves may be set up in the string by vibrations of a number of different frequencies, namely, those which give rise to waves of wavelengths

$$\lambda = 2L, \frac{2L}{2}, \frac{2L}{3}, \ldots = \frac{2L}{n} \qquad (n = 1, 2, 3, \ldots). \tag{22–2}$$

From the relation $f = c/\lambda$, and since c is the same for all frequencies, the possible frequencies are

$$f = \frac{c}{2L}, \frac{2c}{2L}, \frac{3c}{2L}, \ldots = \frac{nc}{2L} \qquad (n = 1, 2, 3, \ldots). \tag{22–3}$$

The lowest frequency, $c/2L$, is called the *fundamental* frequency f_1 and the others are the *overtones.* The frequencies of the latter are therefore $2f_1$, $3f_1$, $4f_1$, and so on. Overtones whose frequencies are integral multiples of the fundamental are said to form a *harmonic series.* The fundamental is the *first harmonic.* The frequency $2f_1$ is the *first overtone* or the *second harmonic,* the frequency $3f_1$ is the *second overtone* or the *third harmonic,* and so on.

The above results may also be obtained directly from Eq. (22–1). The boundary conditions require that $y_1 + y_2 = 0$ at the ends of the string, that is at $x = 0$ and $x = L$. Since the sine of zero is zero, the first condition is satisfied automatically. The second requires that $\sin kL = 0$, and this is true only when k has certain special values. The sine of an angle is zero only when the angle is zero or an integer multiple of π (180°). Thus we must have

$$kL = n\pi \qquad (n = 1, 2, 3, \ldots). \tag{22–4}$$

We do not include the possibility $n = 0$ because that gives $k = 0$, i.e., a wave with zero displacement *everywhere* (a possible case, to be sure, but not a very interesting one!).

Replacing k above by $2\pi/\lambda$, we obtain

$$\frac{2\pi L}{\lambda} = n\pi \qquad \text{or} \qquad \lambda = \frac{2L}{n},$$

in agreement with Eq. (22–2).

Each of the frequencies given by Eq. (22–3) corresponds to a possible *normal mode* of motion, that is, a motion in which each particle of the string moves sinusoidally, all with the same frequency. As the above analysis shows, there is an infinite number of normal modes, each with its characteristic frequency. This situation is in striking contrast with the simpler harmonic oscillator system, a single mass and a spring. The harmonic oscillator has only one normal mode and one characteristic frequency, while the vibrating string has an infinite number. When a body suspended from a spring is pulled down and released, only one frequency of vibration will ensue. If a string is initially distorted so that its shape is the same as *any one* of the possible harmonics, it will vibrate, when released, at the frequency of that particular harmonic. But when a piano string is struck, not only the fundamental, but many of the overtones are present in the resulting vibration. This motion is therefore a combination or *superposition* of normal modes. Several frequencies and motions are present simultaneously, and the displacement of any point on the string is the sum or superposition of displacements associated with the individual modes.

The fundamental frequency of the vibrating string is $f_1 = c/2L$, where $c = \sqrt{S/\mu}$. It follows that

$$f_1 = \frac{1}{2L}\sqrt{\frac{S}{\mu}}. \tag{22–5}$$

Stringed instruments afford many examples of the implications of this equation. For example, all such instruments are "tuned" by varying the tension S, an increase of tension increasing the frequency or pitch, and vice versa. The inverse dependence of frequency on length L is illustrated by the long strings of the bass section of the piano or the bass viol compared with the shorter strings of the piano treble or the violin. Pressing the strings against the fingerboard of a violin or guitar with the fingers to change the length of the vibrating portion of the string is the usual means of varying pitch when playing these instruments. One reason for winding the bass strings of a piano with wire is to increase the mass per unit length μ, so as to obtain the desired low frequency without resorting to a string which is inconveniently long.

22–4 RESONANCE

Whenever a body capable of oscillating is acted on by a periodic series of impulses having a frequency equal to one of the natural frequencies of oscillation of the body, the body is set into vibration with a relatively large amplitude. This phenomenon is called *resonance*, and the body is said to *resonate* with the applied impulses.

A common example of mechanical resonance is provided by pushing a swing. The swing is a pendulum with a single natural frequency depending on its length. If a series of regularly spaced pushes is given to the swing, with a frequency equal to that of the swing, the motion may be made quite large. If the frequency of the pushes differs from the natural frequency of the swing, or if the pushes occur at irregular intervals, the swing will hardly execute any vibration at all.

Unlike a simple pendulum, which has only one natural frequency, a stretched string (and other systems to be discussed later in this chapter) has a large number of natural frequencies. Suppose that one end of a stretched string is fixed while the other is moved back and forth in a transverse direction. The amplitude at the driven end is fixed by the driving mechanism. Standing waves will be set up in the string, whatever the value of the frequency f. If the frequency is not equal to one of the natural frequencies of the string, the amplitude at the antinodes will be fairly small. However, if the frequency is equal to *any one* of the natural frequencies, the string is in resonance and the amplitude at the antinodes will be very much larger than that at the driven end. In other words, although the driven end is not a node, it lies much closer to a node than to an antinode when the string is in resonance. In Fig. 22–5(a), the right end of the string was held fixed and the left end was forced to oscillate vertically with small amplitude. The photographs show the standing waves of relatively large amplitude which resulted when the frequency of oscillation of the left end was equal to the fundamental frequency or to one of the first three overtones.

A steel bridge or, for that matter, any elastic structure, is capable of vibrating with certain natural frequencies. If the regular footsteps of a column of soldiers were to have a frequency equal to one of the natural frequencies of a bridge which the soldiers are crossing, a vibration of dangerously large amplitude might result. Therefore, in crossing a bridge, a column of soldiers is always ordered to break step.

Tuning a radio or television receiver is an example of electrical resonance. By turning a dial, the natural frequency of an alternating current in the receiving circuit is made equal to the frequency of the waves broadcast by the desired station. Optical resonance may also take place between atoms in a gas at low pressure and light waves from a lamp containing the same atoms. Thus, light from a sodium lamp may cause the sodium atoms in a glass bulb to glow with characteristic yellow sodium light.

The phenomenon of resonance may be demonstrated with the aid of the longitudinal waves set up in air by a vibrating plate or tuning fork. If two identical tuning forks are placed some distance apart and one is struck, the other will be heard when the first is suddenly damped. Should a small piece of wax or modeling clay be put on one of the forks, the frequency of that fork will be altered enough to destroy the resonance.

A similar phenomenon can be demonstrated with a piano. With the damper pedal depressed so the dampers are lifted and the strings free to vibrate, one sings a steady tone into the piano. When the singing stops the piano seems to continue to sing the same note. The sound waves from the voice excite vibrations in those strings with natural frequencies close to those (fundamental or harmonics) of the note sung initially.

22–5 INTERFERENCE OF LONGITUDINAL WAVES

The phenomenon of interference between two longitudinal waves in air may be demonstrated with the aid of the apparatus depicted in Fig. 22–7. A wave emitted by a source S is sent into a metal tube, where it divides into two waves, one following the constant path SAR, the other the path SBR, which may be varied by sliding the tube B to the right. Suppose the frequency of the source is 350 Hz. Then the wavelength is $\lambda = c/f = 1$ m. If both paths are of equal length, the two waves arrive at R at the same time

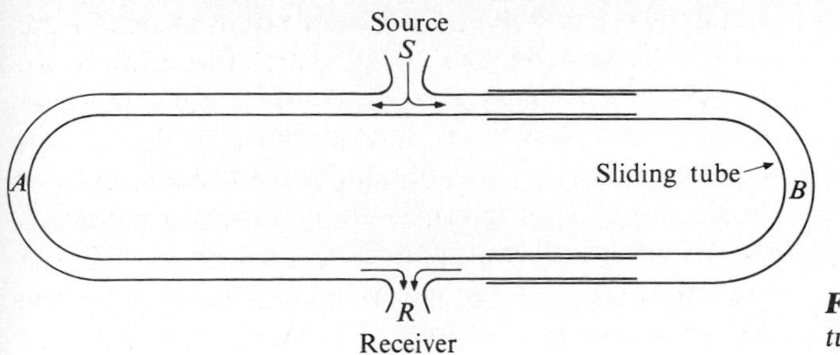

Fig. 22–7 *Apparatus for demonstrating interference of longitudinal waves.*

and the vibrations set up by both waves are in phase. The resulting vibration has an amplitude equal to the sum of the two individual amplitudes and the phenomenon of *reinforcement* may be detected either with the ear at R or with the aid of a microphone, amplifier, and loudspeaker.

Now suppose the tube B is moved out a distance of 0.25 m, thereby making the path SBR 0.5 m longer than the path SAR. The righthand wave travels a distance $\lambda/2$ greater than the lefthand wave and the vibration set up at R by the righthand wave is therefore in opposite phase to that set up by the lefthand wave. The consequent interference is shown by the marked reduction in sound at R.

If the tube B is now pulled out another 0.25 m, so that the *path difference*, SBR minus SAR, is 1 m (1 wavelength), the two vibrations at R again reinforce each other. Thus

$$\left\{\begin{matrix}\text{Reinforcement takes place}\\ \text{when the path difference}\end{matrix}\right\} = 0, \lambda, 2\lambda, \text{etc.}$$

$$\left\{\begin{matrix}\text{Interference takes place}\\ \text{when the path difference}\end{matrix}\right\} = \frac{\lambda}{2}, \frac{3\lambda}{2}, \frac{5\lambda}{2}, \text{etc.}$$

A similar phenomenon can be demonstrated with two loudspeakers driven by the same amplifier, as shown in Fig. 22–8. Suppose the speakers both emit a pure sinusoidal sound wave of constant frequency. When a microphone is placed at point A in the figure, equidistant from the speakers, it receives a strong acoustic signal, corresponding to the fact that two waves emitted from the speakers in phase also arrive at point A in phase and add to each other. But at point B the signal is much *weaker* than when only one speaker is present, corresponding to the fact that the wave from one speaker travels a half-wavelength farther than that from the other, the two arrive a half-cycle out of phase and cancel each other out almost completely. An experiment closely analogous to this one but using light waves, provided conclusive evidence of the wave nature of light.

22–6 LONGITUDINAL STANDING WAVES

Longitudinal waves traveling along a tube of finite length are reflected at the ends of the tube in much the same way that transverse waves in a string are reflected at its ends. Interference between the waves traveling in opposite directions again gives rise to standing waves.

When reflection takes place at a closed end, the displacement of the particles there must always be

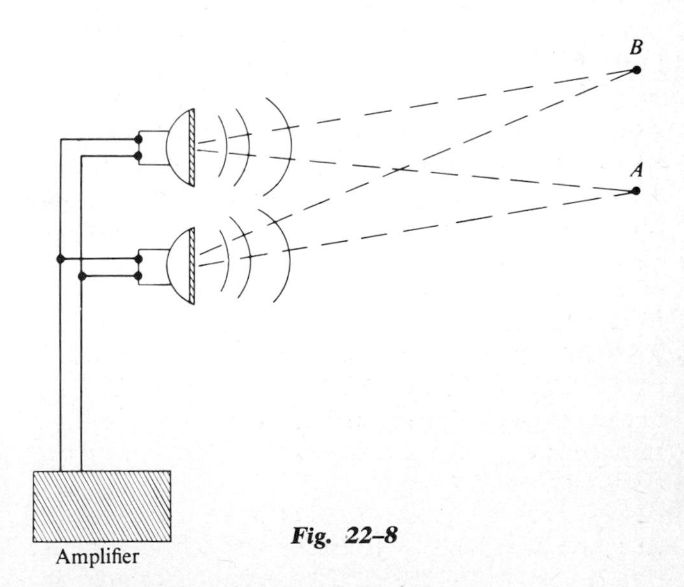

Fig. 22–8

zero. Hence a closed end is a *displacement node.* If the end of the tube is open, the nature of the reflection is more complex and depends on whether the tube is wide or narrow compared with the wavelength. If the tube is narrow compared with the wavelength, which is the case in most musical instruments, the reflection is such that the open end is a *displacement antinode.* Thus the longitudinal waves in a column of fluid are reflected at the closed and open ends of a tube in the same way that transverse waves in a string are reflected at fixed and free ends, respectively.

The reflections at the openings where the instrument is blown are usually characterized by an antinode located at or near the opening. The effective length of the air column of a wind instrument is thus less definite than the length of a string fixed at its ends.

Standing longitudinal waves in a column of gas may be demonstrated conveniently with the aid of the apparatus shown in Fig. 22–9, known as Kundt's tube. A glass tube a few feet long is closed at one end with glass and at the other with a flexible diaphragm. The gas to be studied is admitted to the tube at a known temperature and at atmospheric pressure. A source of longitudinal waves, *S*, whose frequency may be varied, causes vibration of the flexible diaphragm. A small amount of light powder or cork dust is sprinkled uniformly along the tube.

When a frequency is found at which the air column is in resonance, the amplitude of the standing waves becomes large enough for the gas particles to sweep the cork dust along the tube, at all points where the gas is in motion. The powder therefore collects at the displacement nodes, where the gas remains at rest. Sometimes a wire, running along the axis of the tube, is maintained at a dull red heat by an electric current. The nodes show themselves as hot points, compared with the antinodes, which are cooled by moving gas.

With careful manipulation and a good variable-frequency source, a fair determination of the velocity of the wave may be obtained with Kundt's tube. Since, in a standing wave, the distance between two adjacent nodes is one-half (not one!) wavelength, the wavelength λ is obtained by measuring the distance between alternate clumps of powder. When the frequency f is known, the velocity c is then

$$c = f\lambda.$$

At a displacement node, the pressure variations above and below the average are *maximum,* whereas at a displacement antinode, pressure does not vary. This may be understood easily when it is realized that two small masses of gas on opposite sides of a displacement node vibrate in *opposite phase.* When they approach each other, the pressure at the node is maximum, and when they recede from each other, the pressure at the node is minimum. Two small masses of gas, however, on opposite sides of a displacement *antinode* vibrate *in phase,* and hence cause no pressure variations at the antinode.

Figure 22–10 may be helpful in visualizing a longitudinal standing wave. To use this figure, cut a slit about $\frac{1}{16}$ in. wide and $3\frac{1}{2}$ in. long in a card. Place the card over the diagram with the slit horizontal and move it vertically with constant velocity. The portions of the curves that appear in the slit will correspond to the oscillations of the particles in a longitudinal standing wave.

22–7 VIBRATIONS OF ORGAN PIPES

If one end of a pipe is open and a stream of air is directed against an edge, vibrations are set up and

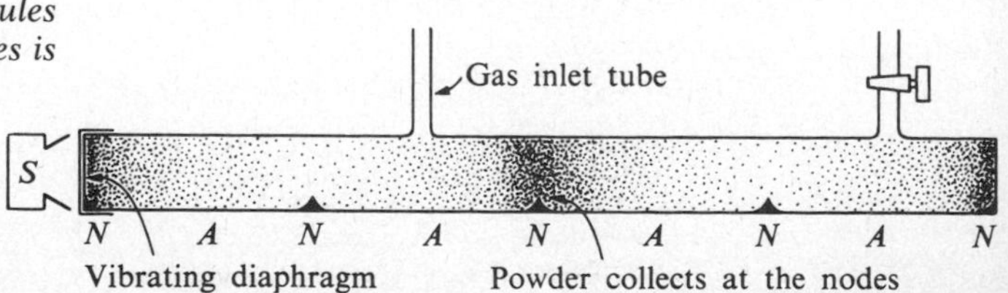

Fig. 22–9 *Kundt's tube for determining the velocity of sound in a gas. The dots represent the density of the gas molecules at an instant when the pressure at the displacement nodes is a maximum or a minimum.*

Fig. 22-10

the tube resonates at its natural frequencies. As in the case of a plucked string, the fundamental and overtones exist at the same time. In the case of a pipe open at both ends, the fundamental frequency f_1 corresponds to an antinode at each end and a node in the middle, as shown at the top of Fig. 22-11. Succeeding diagrams of Fig. 22-11 show two of the overtones, which are seen to be the second and third harmonics. *In an open pipe the fundamental frequency is* $c/2L$ *and all harmonics are present.*

The properties of a *closed* pipe (open at one end but closed at the other) are shown in the diagrams of Fig. 22-12. The fundamental frequency is seen to be $c/4L$, which is one-half that of an open pipe of the same length. In the language of music, the *pitch* of a closed pipe is one octave lower (a factor of two in frequency) than that of an open pipe of equal length. From the remaining diagrams of Fig. 22-12, it may be seen that the second, fourth, etc., harmonics are missing. Hence, *in a closed pipe, the fundamental frequency is* $c/4L$ *and only the odd harmonics are present.*

Almost always in an organ pipe several modes are present at once. The extent to which modes higher than the fundamental are present depends on the cross section of the pipe, the ratio of length to width, whether the pipe is straight or tapered, and to a slight extent on the material. The harmonic content of the tone is an important factor in determining the "tone quality" or timbre. A very thin pipe produces

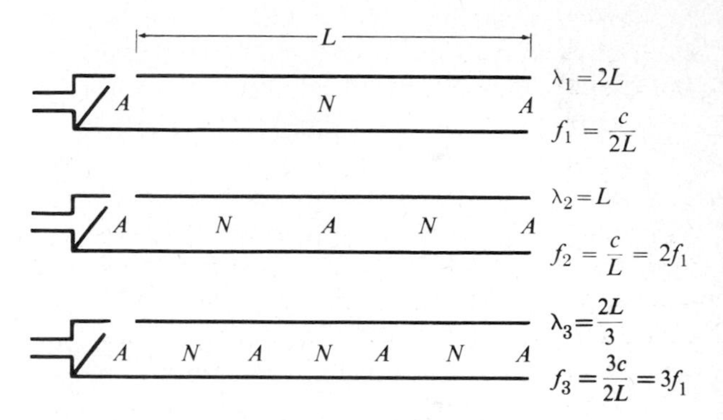

Fig. 22-11 *Modes of vibration of an open organ pipe.*

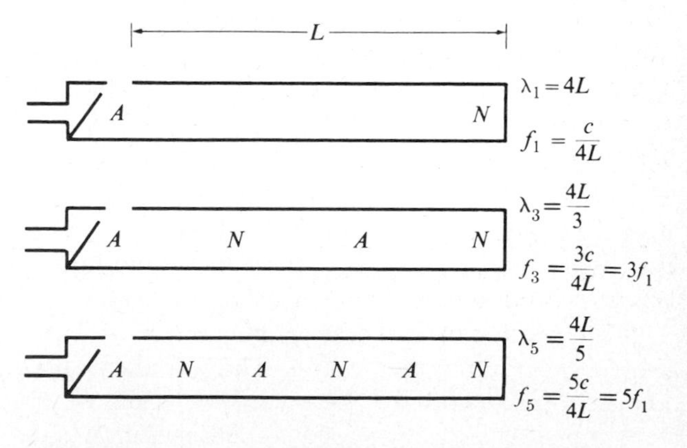

Fig. 22-12 *Modes of vibration of a closed organ pipe.*

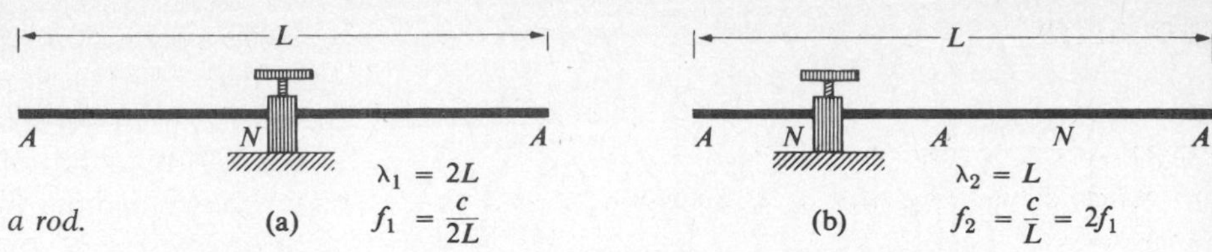

Fig. 22-13 Modes of vibration of a rod.

a tone rich in higher harmonics which the ear perceives as thin and "stringy," while a fatter pipe produces principally the fundamental tone, perceived as a softer, more flute-like tone.

22-8 VIBRATIONS OF RODS AND PLATES

A rod may be set in longitudinal vibration by clamping it at some point and stroking it with a chamois skin that has been sprinkled with rosin. In Fig. 22-13(a) the rod is clamped in the middle; consequently, when it is stroked near the end, a standing wave is set up with a node in the middle and antinodes at each end, exactly the same as the fundamental mode of an open organ pipe. The fundamental frequency of the rod is then $c/2L$, where c is the velocity of a longitudinal wave in the rod. Since the velocity of a longitudinal wave in a solid is much greater than that in air, a rod has a higher fundamental frequency than an open organ pipe of the same length.

By clamping the rod at a point $\frac{1}{4}$ of its length from one end, as shown in Fig. 22-13(b), the second harmonic may be produced.

When a stretched flexible membrane, such as a drumhead, is struck a blow, a two-dimensional pulse travels outward from the struck point and is reflected and re-reflected at the boundary of the membrane. If some point of the membrane is forced to vibrate periodically, continuous trains of waves travel along the membrane. Just as with the stretched string, standing waves can be set up in the membrane; each of these waves has a certain natural frequency. The lowest frequency is the fundamental and the others are overtones. In general, when the membrane is vibrating, a number of overtones are present.

The nodes of a vibrating membrane are *lines* (nodal lines) rather than points. The boundary of the membrane is evidently one such line. Some of the other possible nodal lines of a circular membrane are indicated by arrows in Fig. 22-14. The natural frequency of each mode is given in terms of the fundamental f_1. It will be noted that the frequencies of the overtones are *not* integer multiples of f_1. That is, they are not harmonics.

The restoring force in a vibrating flexible membrane arises from the tension with which it is stretched. A metal plate, if sufficiently thick, vibrates in a similar way, the restoring force being produced by bending stresses in the plate. The study of vibrations of membranes and plates is of importance in connection with the design of loudspeaker diaphragms and the diaphragms of telephone receivers and microphones.

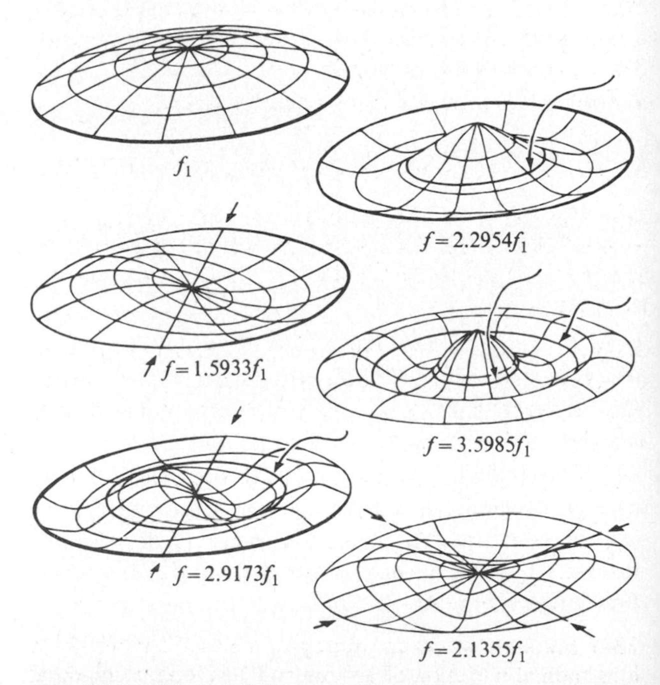

Fig. 22-14 *Modes of vibration of a membrane, showing nodal lines. The frequency of each mode is given in terms of the fundamental frequency, f_1. (Adapted from* Vibration and Sound, *by Philip M. Morse, 2nd Edition, McGraw-Hill Book Company, Inc., 1948. By permission of the publishers.)*

PROBLEMS

22–1 A steel piano wire 50 cm long, of mass 5 g, is stretched with a tension of 400 N.

a) What is the frequency of its fundamental mode of vibration?

b) What is the number of the highest overtone that could be heard by a person who is capable of hearing frequencies up to 10,000 Hz?

22–2 A steel wire of length $L = 100$ cm and density $\rho = 8\text{ g cm}^{-3}$ is stretched tightly between two rigid supports. Vibrating in its fundamental mode, the frequency is $f = 200$ Hz.

a) What is the speed of transverse waves on this wire?

b) What is the longitudinal stress (force per unit area) in the wire?

c) If the maximum acceleration at the midpoint of the wire is 80,000 cm s^{-2}, what is the amplitude of vibration at the midpoint?

22–3 A stretched string is observed to vibrate with a frequency of 30 Hz in its fundamental mode when the supports are 60 cm apart. The amplitude at the antinode is 3 cm. The string has a mass of 30 g.

a) What is the speed of propagation of a transverse wave in the string?

b) Compute the tension in the string.

22–4 The fundamental frequency of the A-string on a cello is 220 Hz. The vibrating portion of the string is 68 cm long and has a mass of 1.29 g. With what tension must it be stretched?

22–5 A standing wave of frequency 1100 Hz in a column of methane at 20°C produces nodes that are 20 cm apart. What is the ratio of the heat capacity at constant pressure to that at constant volume?

22–6 An aluminum object is hung from a steel wire. The fundamental frequency for transverse standing waves on the wire is 300 Hz. The object is then immersed in water so that one-half of its volume is submerged. What is the new fundamental frequency?

22–7 Standing waves are set up in a Kundt's tube by the longitudinal vibration of an iron rod 1 m long, clamped at the center. If the frequency of the iron rod is 2480 Hz and the powder heaps within the tube are 6.9 cm apart, what is the speed of the waves (a) in the iron rod, and (b) in the gas?

22–8 The speed of a longitudinal wave in a mixture of helium and neon at 300 K was found to be 758 m s^{-1}. What is the composition of the mixture?

22–9 The atomic mass of iodine is 127 g mol^{-1}. A standing wave in iodine vapor at 400 K produces nodes that are 6.77 cm apart when the frequency is 1000 Hz. Is iodine vapor monatomic or diatomic?

22–10 A copper rod 1 m long, clamped at the $\frac{1}{4}$ point, is set in longitudinal vibration and is used to produce standing waves in a Kundt's tube containing air at 300 K. Heaps of cork dust within the tube are found to be 4.95 cm apart. What is the speed of the longitudinal waves in copper?

22–11 Find the fundamental frequency and the first four overtones of a 6-in. pipe (a) if the pipe is open at both ends, (b) if the pipe is closed at one end. (c) How many overtones may be heard by a person having normal hearing for each of the above cases?

22–12 A long tube contains air at a pressure of 1 atm and temperature 77°C. The tube is open at one end and closed at the other by a movable piston. A tuning fork near the open end is vibrating with a frequency of 500 Hz. Resonance is produced when the piston is at distances 18.0, 55.5, and 93.0 cm from the open end.

a) From these measurements, what is the speed of sound in air at 77°C?

b) From the above result, what is the ratio of the specific heats γ for air?

22–13 An organ pipe A of length 2 ft, closed at one end, is vibrating in the first overtone. Another organ pipe B of length 1.35 ft, open at both ends, is vibrating in its fundamental mode. Take the speed of sound in air as 1120 ft s^{-1}. Neglect end corrections. (a) What is the frequency of the tone from A? (b) What is the frequency of the tone from B?

22–14 A plate cut from a quartz crystal is often used to control the frequency of an oscillating electrical circuit. Longitudinal standing waves are set up in the plate with displacement antinodes at opposite faces. The fundamental frequency of vibration is given by the equation

$$f_1 = \frac{2.87 \times 10^5}{s},$$

where f_1 is in hertz and s in the thickness of the plate in centimeters.

a) Compute Young's modulus for the quartz plate.

b) Compute the thickness of plate required for a frequency of 1200 kHz. (1 kHz = 1000 Hz.) The density of quartz is 2.66 g cm^{-3}.

22–15 In Problem 22–4, what percent increase in tension is needed to increase the frequency from 220 Hz to 233 Hz, corresponding to a rise in pitch from A to A-sharp?

22–16 The longest pipe found in most medium-size pipe organs is 16 feet long. What is the frequency of the corresponding note if the pipe is (a) open at both ends; (b) open at one end, closed at the other?

22–17 The frequency of middle C is 262 Hz.

a) If an organ pipe is open at both ends, what length must it have to produce this note at 20° C?

b) At what temperature will the frequency be 6 percent higher, corresponding to a rise in pitch from C to C-sharp?

22–18 A certain organ pipe produces a frequency of 440 Hz in air. If the pipe is filled with helium at the same temperature, what frequency does it produce?

Chapter 23

Acoustic Phenomena

23–1 SOUND WAVES

In this chapter we are concerned primarily with longitudinal waves in air which, when striking the ear, give rise to the sensation of *sound*. The human ear is sensitive to waves in the frequency range from about 20 to 20,000 Hz, although the term *sound wave* is sometimes applied also to similar waves with frequencies outside the range of human audibility.

The simplest sound waves are sinusoidal waves with definite frequency, amplitude, and wavelength. When such a wave arrives at the ear, it causes a vibration of the air particles at the eardrum with a definite frequency and amplitude. This vibration may also be described in terms of the variation of *air pressure* at the same point. The air pressure rises above atmospheric pressure, and then drops below atmospheric pressure, with simple harmonic motion of the same frequency as that of an air particle.

The simplest sinusoidal sound wave is described by the wave function

$$y = Y \sin(\omega t - kx), \tag{23–1}$$

where the symbols have the same meaning as in Section 21–3. It is usually much easier to measure the *pressure* variations in sound waves than the displacements, so it is worthwhile to develop a relation between the two. Let p be the instantaneous pressure fluctuation at any point, that is, the amount by which the pressure differs from normal atmospheric pressure. If the displacements of two neighboring points x and $x + \Delta x$ are the same, the gas between these points is neither compressed nor rarefied, there is no volume change, and consequently $p = 0$. Only when y varies from one point to a neighboring point is there a change of volume and therefore of pressure.

The fractional volume change $\Delta V/V$ in an element near point x turns out to be given simply by $\partial y/\partial x$, which is the rate of change of y with x as we go from one point to a neighboring point. From the definition of the bulk modulus B, Eq. (10–11), $p = -B\Delta V/V$, and we find

$$p = -B\left(\frac{\partial y}{\partial x}\right). \tag{23–2}$$

The negative sign arises because, when $\partial y/\partial x$ is positive, the displacement is greater at $x + \Delta x$ than at x, corresponding to an increase in volume and a *decrease* in pressure. For the sinusoidal wave of Eq. (23–1) we find

$$p = -BkY \cos(\omega t - kx). \tag{23–3}$$

The maximum amount by which the pressure differs from atmospheric, that is, the maximum value of p, is called the *pressure amplitude*, denoted P. Equation (23–3) shows that

$$P = BkY. \tag{23–4}$$

We may also develop a relation between p and the particle velocity v of points in the medium, not to be confused with the wave propagation velocity c. The particle velocity is given by $v = \partial y/\partial t$, and for Eq. (23–1) this is

$$v = \frac{\partial y}{\partial t} = -\omega Y \cos(\omega t - kx). \tag{23–5}$$

The maximum velocity, called the velocity amplitude V, is given by $V = \omega Y$. Combining this with Eq. (23–4), we find

$$P = \frac{BkV}{\omega} = \frac{BV}{c}. \tag{23–6}$$

Thus the relation between pressure amplitude and velocity amplitude does not depend on frequency. Recalling that the propagation velocity c is given by

$$c = \sqrt{B/\rho}, \tag{23–7}$$

we may rewrite Eq. (23–6) as follows:

$$P = \sqrt{B\rho}V. \tag{23–8}$$

The quantity $(B\rho)^{\frac{1}{2}}$ is called the *mechanical impedance* of the system, in reference to an analogy with electric circuits, in which pressure is analogous to voltage, velocity to current, and mechanical impedance to resistance or impedance. In terms of this analogy, Eq. (23–8) is the mechanical analog of Ohm's law.

Example Measurements of sound waves show that the maximum pressure variations in the loudest sounds that the ear can tolerate without pain are of the order of 30 N m^{-2} (above and below atmospheric pressure, which is about 100,000 N m^{-2}). Find the corresponding maximum displacement, if the frequency is 100 Hz and $c = 350$ m s^{-1}.

We have $\omega = (2\pi)(1000 \text{ Hz}) = 6283 \text{ s}^{-1}$ and

$$k = \frac{\omega}{c} = \frac{6283 \text{ s}^{-1}}{350 \text{ m s}^{-1}} = 18.0 \text{ m}^{-1}.$$

The adiabatic bulk modulus for air is

$$B = (1.4)(1.01 \times 10^5 \text{ N m}^{-2}) = 1.42 \times 10^5 \text{ N m}^{-2}.$$

From Eq. (23–4) we find

$$Y = \frac{P}{Bk} = \frac{(30 \text{ N m}^{-2})}{(1.42 \times 10^5 \text{ N m}^{-2})(18.0 \text{ m}^{-1})}$$

$$= 1.17 \times 10^{-5} \text{ m}$$

$$= 0.017 \text{ mm}.$$

Thus the displacement amplitude of even the loudest sound is *extremely small.* The maximum pressure variation in the *faintest* audible sound of frequency 1000 Hz is only about 3×10^{-5} N m^{-2}. The corresponding displacement amplitude is about 10^{-9} cm. By way of comparison, the wavelength of yellow light is 6×10^{-5} cm, and the diameter of a molecule is about 10^{-8} cm. The ear is an extremely sensitive organ!

23–2 INTENSITY

An essential aspect of wave propagation of all sorts is transfer of *energy*. A familiar example is the energy supply of the earth, which reaches us from the sun via electromagnetic waves. The intensity I of a traveling wave is defined as *the time average rate at which energy is transported by the wave, per unit area*, across a surface perpendicular to the direction of propagation. More briefly, the intensity is the average *power* transported per unit area.

We have seen that the power developed by a force equals the product of force and velocity. Hence the power per unit area in a sound wave equals the product of the excess pressure (force per unit area) and the *particle* velocity, given by Eqs. (23–3) and (23–5), respectively. We find

$$pv = \omega BkY^2 \cos^2(\omega t - kx). \tag{23–9}$$

The intensity is, by definition, the average value of this quantity. The average value of the function $\cos^2 z$ is $\frac{1}{2}$, so we find

$$I = \tfrac{1}{2}\omega BkY^2. \tag{23–10}$$

It is usually more convenient to express I in terms of

the pressure amplitude P. Using Eq. (23–4), we find

$$I = \frac{P^2}{2\rho c} = \frac{P^2}{2\sqrt{\rho B}}. \qquad (23\text{–}11)$$

The fact that the intensity is proportional to the *square* of the amplitude is a characteristic of all kinds of wave motion.

The intensity of a sound wave of the largest amplitude tolerable to the human ear (about $P = 30\ \mathrm{N\ m^{-2}}$) is

$$\begin{aligned} I &= \frac{(30\ \mathrm{N\ m^{-2}})^2}{2(1.22\ \mathrm{kg\ m^{-3}})(346\ \mathrm{m\ s^{-1}})} \\ &= 1.06\ \mathrm{J\ s^{-1}\ m^{-2}} = 1.06\ \mathrm{W\ m^{-2}} \\ &= 1.06 \times 10^{-4}\ \mathrm{W\ cm^{-2}}. \end{aligned}$$

The unit $1\ \mathrm{W\ cm^{-2}}$ is a mixed unit, neither cgs nor mks. We mention it here because it is unfortunately in general use among acousticians.

The pressure amplitude of the faintest sound wave which can be heard is about $3 \times 10^{-5}\ \mathrm{N\ m^{-2}}$ and the corresponding intensity is about $10^{-12}\ \mathrm{W\ m^{-2}}$ or $10^{-16}\ \mathrm{W\ cm^{-2}}$.

The *total* power carried across a surface by a sound wave equals the product of the intensity at the surface and the surface area, if the intensity over the surface is uniform. The average power developed as sound waves by a person speaking in an ordinary conversational tone is about 10^{-5} W, while a loud shout corresponds to about 3×10^{-2} W. Since the population of the city of New York is about eight million persons, the acoustical power developed, if all were to speak at the same time, would be about 60 W, or enough to operate a moderate-sized electric light. On the other hand, the power required to fill a large auditorium with loud sound is considerable. Suppose the intensity over the surface of a hemisphere 20 m in radius is $10^{-4}\ \mathrm{W\ cm^{-2}}$. The area of the surface is about $25 \times 10^6\ \mathrm{cm^2}$. Hence the acoustic power output of a speaker at the center of the sphere would have to be

$$(10^{-4}\ \mathrm{W\ cm^{-2}})(25 \times 10^6\ \mathrm{cm^2}) = 2500\ \mathrm{W},$$

Table 23–1 NOISE LEVELS DUE TO VARIOUS SOURCES (representative values)

Source or description of noise	Noise level, db	Intensity, $\mathrm{W\ m^{-2}}$
Threshold of pain	120	1
Riveter	95	3.2×10^{3}
Elevated train	90	10^{-3}
Busy street traffic	70	10^{-5}
Ordinary conversation	65	3.2×10^{-6}
Quiet automobile	50	10^{-7}
Quiet radio in home	40	10^{-8}
Average whisper	20	10^{-10}
Rustle of leaves	10	10^{-11}
Threshold of hearing	0	10^{-12}

or 2.5 kW. The electrical power input to the speaker would need to be considerably larger, since the efficiency of such devices is not very high.

23–3 INTENSITY LEVEL AND LOUDNESS

Because of the large range of intensities over which the ear is sensitive, a logarithmic rather than an arithmetic intensity scale is convenient. Accordingly, the *intensity level* β of a sound wave is defined by the equation

$$\beta = 10 \log \frac{I}{I_0}, \qquad (23\text{–}12)$$

where I_0 is an arbitrary reference intensity, taken as $10^{-12}\ \mathrm{W\ m^{-2}}$, corresponding roughly to the faintest sound which can be heard. Intensity levels are expressed in *decibels*, abbreviated db.*

If the intensity of a sound wave equals I_0 or $10^{-12}\ \mathrm{W\ m^{-2}}$, its intensity level is 0 db. The maximum intensity which the ear can tolerate, about $1\ \mathrm{W\ m^{-2}}$, corresponds to an intensity level of 120 db. Table 23–1 gives the intensity levels in decibels of a number

---

* Originally, a scale of intensity levels in *bels* was defined by the relation

$$\text{Intensity level} = \log I/I_0.$$

This unit proved rather large, and hence the *decibel*, one-tenth of a bel, has come into general use. The unit is named in honor of Alexander Graham Bell.

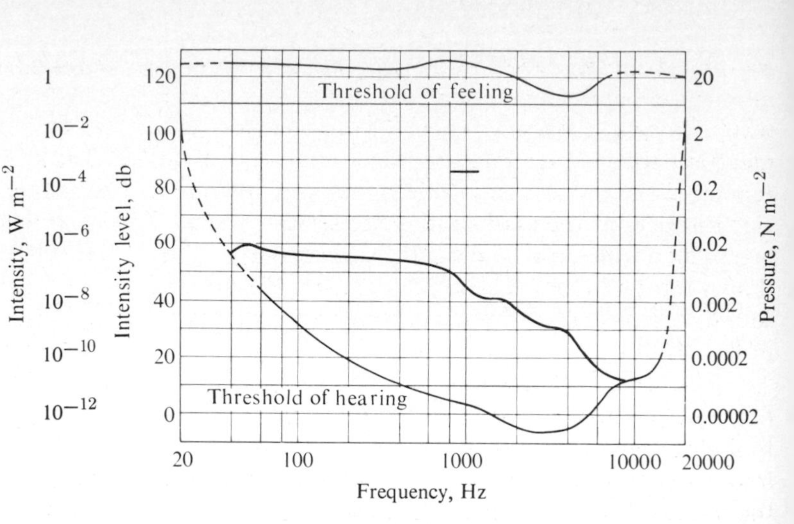

Fig. 23–1 *Auditory area between threshold of hearing and threshold of feeling (light lines), and spectrogram of street noise (heavy line). (Courtesy of Dr. Harvey Fletcher.)*

of familiar noises. It is taken from a survey made by the New York City Noise Abatement Commission.

The range of frequencies and intensities to which the ear is sensitive is conveniently represented by a diagram such as that of Fig. 23–1, which is a graph of the *auditory area* of a person of good hearing. The height of the lower curve at any frequency represents the intensity level of the faintest pure (sinusoidal) tone of that frequency which can be heard. It will be seen from the diagram that the ear is most sensitive to frequencies between 2000 and 3000 Hz where the *threshold of hearing*, as it is called, is about −5 db. The upper curve represents the intensity level of the loudest pure tone which can be tolerated. At intensities above this curve, which is called the *threshold of feeling or pain*, the sensation changes from one of hearing to discomfort or even pain. The height of the upper curve is approximately constant at a level of about 120 db for all frequencies. Every pure tone which can be heard may be represented by a point lying somewhere in the area between these two curves.

Only about 1 percent of the population has a threshold of hearing as low as the bottom curve in Fig. 23–1; 50 percent of the population can hear pure tones of a frequency of 2500 cycles when the intensity level is about 8 db, and 90 percent when the level is 20 db. For a loud tone of intensity level 80 db, the range of audibility is from 20 to 20,000 Hz, but at a level of 20 db it is only from about 200 to about 15,000 Hz. At a frequency of 1000 Hz the range of intensity level is from about 3 db to about 120 db, whereas at 100 Hz it is only from 30 db to 120 db.

The term *loudness* refers to a sensation in the consciousness of a human observer. It is purely subjective, as contrasted with the objective quantity *intensity*, and is not directly measurable with instruments. Loudness increases with intensity but there is no simple linear relationship. Pure tones of the same intensity but different frequencies do not necessarily produce sensations of equal loudness. Thus, for a listener whose auditory area is represented in Fig. 23–1, a pure tone of intensity level 30 db and frequency 60 Hz is completely inaudible, while one of the same intensity level but of frequency 1000 Hz is well above the threshold of audibility. In order for the first tone to appear as loud as the second, its intensity level would have to be raised to about 65 db.

Sounds which are not pure tones do not have a single frequency and hence cannot be represented by a single point on the diagram. A sound such as that from a musical instrument, consisting of a mixture of a relatively few frequencies (the fundamental and overtones), can be represented by a set of points, each point giving the intensity and frequency of one particular overtone. A sound such as street noise, while it cannot be considered as made up of a fundamental and overtones, can still be represented on the diagram. The sound is picked up by a microphone and sent through an electrical network which selects a narrow range of frequencies and measures the average intensity within this range. By repeating the process at a large number of frequencies throughout the audible range, a series of points is obtained which can be plotted. A continuous curve drawn through them is called the *spectrogram* of the sound. A typical spectrogram of street noise is shown in Fig. 23–1.

The term "spectrogram" is borrowed from optics. The process just described is entirely analogous to the optical one of dispersing a beam of light waves into a spectrum by means of a prism and measuring the intensity at a number of wavelengths throughout the spectrum. The light emitted by a gas in an electrical discharge is a mixture of waves of a number of definite frequencies and corresponds to the sound emitted by a musical instrument. Most light beams, however, are a mixture of all frequencies and are therefore the optical analog of noise.

The *total* intensity level of a noise can be found from its spectrogram by an integration process. There are also instruments known as noise meters which measure this level directly. The level of the street noise in Fig. 23–1 is about 85 db and is shown by the short heavy line.

23-4 QUALITY AND PITCH

A string that has been plucked or a plate that has been struck, if allowed to vibrate freely, will vibrate with many frequencies at the same time. It is a rare occurrence for a body to vibrate with only one frequency. A carefully made tuning fork struck lightly on a rubber block may vibrate with only one frequency, but in the case of musical instruments, the fundamental and many harmonics are usually present at the same time. The impulses that are sent from the ear to the brain give rise to one net effect which is characteristic of the instrument. Suppose, for example, the sound spectrum of a tone consisted of a fundamental of 200 Hz and harmonics 2, 3, 4 and 5, all of different intensity, whereas the sound spectrum of another tone consisted of exactly the same frequencies but with a different intensity distribution. The two tones would sound different; they are said to differ in *quality* or *timbre.*

Adjectives used to describe the quality of musical tones are purely subjective in character, such as reedy, golden, round, mellow, tinny, etc. The quality of a sound is determined in part by the number of overtones present and their respective intensity-versus-time curves. The sound spectra of several musical instruments are shown in Fig. 23–2.

Another important factor in determining tone quality is the behavior at the beginning and end of a tone. A piano tone begins percussively with a thump and then dies away gradually; a harpsichord tone, in addition to having different harmonic content, begins much more quickly and incisively with a click, the higher harmonics beginning before the lower ones. The cessation of tone when the key is released is also much prompter with the harpsichord than the piano. Similar effects are present in other musical instru-

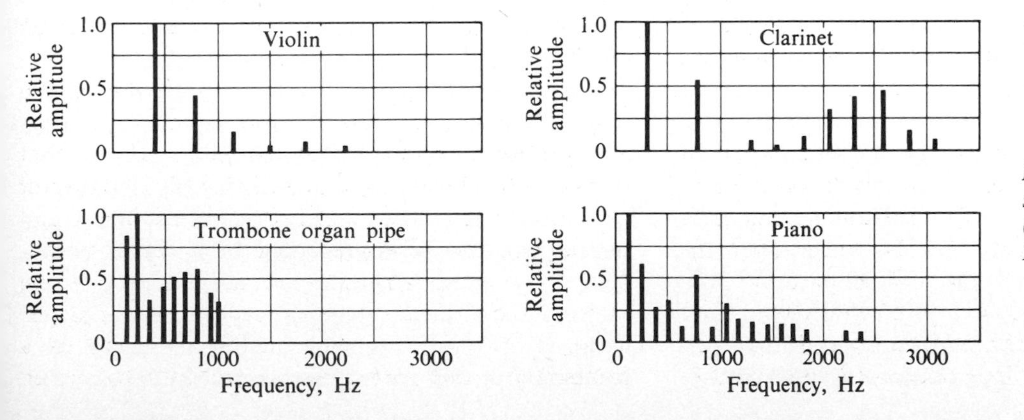

Fig. 23–2 *Sound spectra of some musical instruments (Courtesy of Dr. Harvey Fletcher.)*

ments; with wind and string instruments the player has considerable control over the attack and decay of the tone, and these characteristics help define the unique characteristics of each instrument.

The term *pitch* refers to the attribute of a sound sensation that enables one to classify a note as "high" or "low." Like loudness, it is a subjective quantity and cannot be measured with instruments. Pitch is related to the objective quantity *frequency*, but there is no one-to-one correspondence. For a pure tone of constant intensity, the pitch becomes higher as the frequency is increased, but the pitch of a pure tone of constant frequency becomes lower as the intensity level is raised.

Many of the notes played on musical instruments are rich in harmonics, some of which may be more prominent than the fundamental. Presented with an array of frequencies constituting a harmonic series, the ear still assigns a characteristic pitch to the combination, this pitch being that associated with the fundamental frequency of the series. So definite is this pitch sensation that it is possible to eliminate the fundamental frequency entirely, by means of filters, without changing the pitch! The ear apparently will supply the fundamental, provided the correct harmonics are present. It is this rather surprising property of the ear that enables a small loudspeaker which does not radiate low frequencies well to give nevertheless the impression of good radiation in the low-frequency region. Because the speaker is a fairly efficient radiator for the frequencies of the harmonics, the listener believes he is actually hearing the low frequencies, when instead he is hearing only multiples of these frequencies and his ear is supplying the fundamental. It is possible, by deliberate distortion of the harmonics associated with low musical notes, to make a very small radio set, totally inadequate in the low-frequency range, sound somewhat like a larger, acoustically superior console set. Such synthetic bass is, to the critical ear, inferior to true bass reproduction, where the harmonic content is closer to that of the original sound.

23–5 MUSICAL INTERVALS AND SCALES

When certain musical tones are produced in succession or together, even an untrained listener recognizes a relationship among them. Such relations are described in musical language by words such as octave, major third, minor third, etc. The listener recognizes something basic in these combinations of tones, and experimental measurement of the fundamental frequencies of the separate tones discloses that their relationships, one to the other, may be approximated by simple whole-number ratios. For example, the fundamental frequencies of middle C of the piano and its *octave* above, C′, have a ratio of 1 to 2. Another basic set of tones is obtained by playing C, E, G. These constitute what is known as a *major triad*, and the frequencies are found to be approximately proportional to 4, 5, and 6.

Starting at middle C of the piano, and playing only the white keys toward the right, it is possible to find, within only nine notes, three major triads. These are shown in Table 23–2; they enable one to calculate the frequencies of all the notes, once one of them has been chosen arbitrarily. By international agreement, the frequency of the A above middle C is chosen to be 440 vibrations per second. The frequencies of all the notes, shown in row 5 of Table 23–2, are those that correspond to the *just diatonic scale* of the key of C. The frequency ratios of adjacent notes are seen to be either 9/8, 10/9, or 16/15. The interval between two tones whose frequencies bear the ratio 9/8 or 10/9 are called *whole tones*, whereas the interval between two notes of frequency ratio 16/15 is a *half tone*.

If a major diatonic scale were constructed starting at D instead of at C, four new notes would be needed for a perfect diatonic scale. If all possible musical keys were to be provided for, many additional notes would be needed for each octave. To avoid this tremendous complication, what is known as the *equally tempered scale* has been devised. In this scheme there are 12 half-tone intervals in every range of an octave, adjacent notes a half tone apart bearing the constant ratio of the twelfth root of 2, i.e., 1.05946.

Simple as this scheme is, it results in no one scale being exactly diatonic. Since the ratios of the diatonic scale were originally selected to suit the preferences of the ear (being made up of three sets of major triads, each of which constitutes a harmonious combination), this means that an instrument tuned to

Table 23–2 FREQUENCY RELATIONS IN THE JUST DIATONIC SCALE AND IN THE EQUALLY TEMPERED SCALE

	Do	Re	Mi	Fa	Sol	La	Ti	Do′	Re′
Frequency relations	Middle C	D	E	F	G	A	B	C′	D′
Octave, key of C	1							2	
Major triad, key of C	4		5		6			(8)	
Major triad, key of F				4		5		6	
Major triad, key of G		(3)			4		5		6
Just scale, key of C	264	297	330	352	396	440	495	528	594
Intervals in just scale		9/8 whole	10/9 whole	16/15 half	9/8 whole	10/9 whole	9/8 whole	16/15 half	9/8 whole
Equally tempered scale suitable for all keys	261.6	293.7	329.6	349.2	392.0	440	493.9	523.3	587.4
Intervals in equally tempered scale		$\sqrt[6]{2}$	$\sqrt[6]{2}$	$\sqrt[12]{2}$	$\sqrt[6]{2}$	$\sqrt[6]{2}$	$\sqrt[6]{2}$	$\sqrt[12]{2}$	$\sqrt[6]{2}$

the equally tempered scale is not quite so pleasant to the ear. Thus a piano whose white keys are tuned to the "just" scale sounds better (i.e., more "in tune") than one tuned to equal temperament when played in the key of C, but in some other keys it sounds worse. Hence an instrument to be used for compositions in all keys (as, for example, the *Well-Tempered Clavier* of Bach, containing compositions in all the major and minor keys) is almost always tuned to the equally tempered scale.

23–6 RADIATION FROM A PISTON

As we have seen, sound waves can be produced in a wide variety of ways. One of the simplest to understand is the moving cone of a loudspeaker, and it is instructive to examine the action of such a system. For simplicity, we consider an idealized model in which the cone is represented as a plane moving perpendicularly to its surface. A single surface oscillating this way produces sound waves radiating away from both sides. To simplify the problem further, let us assume that we have an oscillating piston fitting closely in a large wall or *baffle*, as in Fig. 23–3(a). Then only the waves radiated from the right side of the piston need be considered in the space to the right of the baffle.

The directional distribution of radiated energy depends on the shape of the piston and on its dimensions relative to the wavelength of the emitted waves. The mathematics becomes very complicated for any shapes except circles and rectangles. The simplest case is that of a very long rectangle like a long plank, and Fig. 23–3(b) is an end view of such

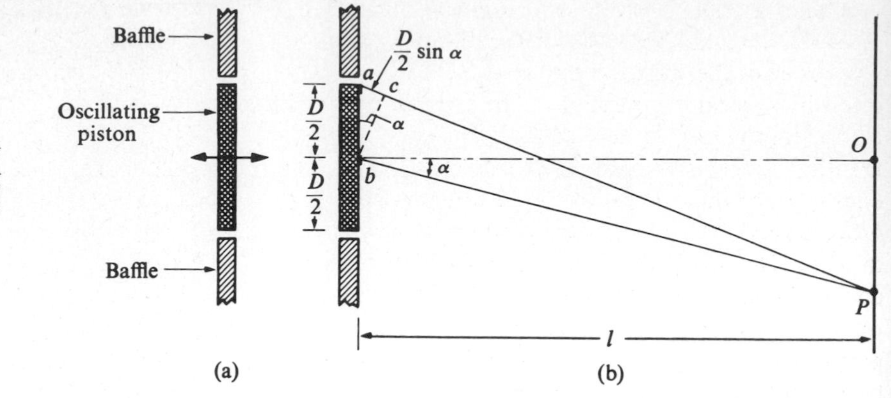

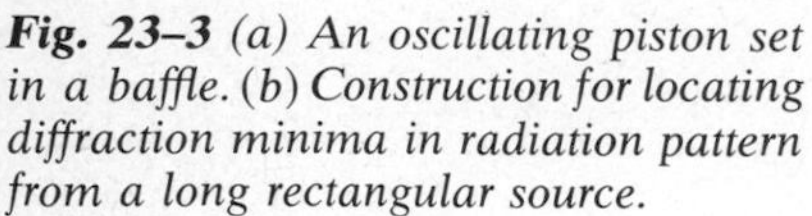
Fig. 23–3 *(a) An oscillating piston set in a baffle. (b) Construction for locating diffraction minima in radiation pattern from a long rectangular source.*

a piston. Imagine the right face of the piston, of width D, to be subdivided into a large number of very narrow strips parallel to the length of the piston and perpendicular to the plane of the diagram. Each of these strips can be treated as a *line source*, sending out waves having *cylindrical* wave surfaces coaxial with the source. We now apply the principle of superposition to these cylindrical waves.

Point P is a point in space at a distance ℓ large compared with the width D of the source, and with the wavelength λ. Consider the two elementary line sources at a and b, one at the top edge of the piston and the other just below its centerline. With P as a center and Pb as radius, strike the arc shown by the dotted line, intersecting the line Pa at c. The distance ac is then the path difference between the waves reaching P from a and b. Since ℓ is large compared with D, the arc bc is very nearly a straight line and abc is very nearly a right triangle with the angle α equal to the angle between Pb and the normal Ob. The path difference ac is then

$$ac = \frac{D}{2}\sin\alpha. \qquad (23\text{–}13)$$

If the angle α has such a value that ac equals a half-wavelength, the waves from a and b will reach P 180° out of phase and will very nearly cancel one another. The cancellation will not be complete because (1) the waves have slightly different distances to travel, and the amplitude decreases with distance, and (2) the directions to P are not exactly the same for both sources, and a line source, unlike a spherical source, does not radiate uniformly in all directions. However, both of these effects will be small if ℓ is large compared with D.

Consider next the pair of line sources just below a and b. Except for the small differences mentioned, the same diagram as in Fig. 23–3(b) can be constructed for them, and the waves from these two sources will *also* cancel at P. Proceeding in this way over the entire surface of the piston, we see that very nearly complete cancellation results in a direction making an angle α with the normal, provided that

$$ac = \frac{\lambda}{2},$$

or, from Eq. (23–5), that

$$\sin\alpha = \frac{\lambda}{D}. \qquad (23\text{–}14)$$

As an example, if $D = 30$ cm and $\lambda = 15$ cm (corresponding to a frequency of about 2000 Hz),

$$\sin\alpha = \frac{15}{30} = 0.50,$$

$$\alpha = 30°,$$

and no energy is radiated at an angle of 30° on either side of the normal.

Other minima occur in directions for which $\sin\alpha = 2\lambda/D,\ 3\lambda/D$, etc. This can be shown by dividing the surface of the piston into quarters, sixths, etc., and pairing off one element against another, as in Fig. 23–3(b).

The relatively simple discussion above, while it gives the angular positions of the *minima*, does not give those of the *maxima* nor does it give the relative intensities in other directions. The complete analysis is too lengthy to give here and we shall only state the results. There is a maximum intensity of radiated energy along the normal to the piston ($\alpha = 0$) and in other directions approximately halfway between the minima. However, the greatest amount of energy is concentrated in the region between the first two minima on either side of the normal, and for most purposes the energy radiated in other directions can be neglected.

The analysis of the radiation pattern from a *circular* piston can be carried out in the same way as that for a long rectangle. The piston is subdivided into narrow circular zones instead of long strips, and the effect of the waves from all the zones is summed at a distant point. There is a maximum of intensity along the axis of the piston, as would be expected. The angle α at which the first minimum occurs is given by

$$\boxed{\sin\alpha = 1.22\frac{\lambda}{D},} \qquad (23\text{–}15)$$

where D is now the piston diameter, and about 85 percent of the radiated energy is concentrated within a cone of this half-angle. If $D = 12$ in., $\lambda = 6$ in.,

$$\sin\alpha = 1.22 \times \frac{6}{12} = 0.61,$$

$$\alpha \approx 37°.$$

Other minima and maxima of rapidly decreasing intensity surround the central maximum. Figure 23–4 is a space diagram of the intensity distribution in front of an oscillating circular piston set in a baffle.

This *directivity* of the sound radiated by a piston has a number of important applications. In a motion picture theater, for example, we wish the sound waves radiated by a speaker behind the screen to spread out over a large angle. If a loudspeaker 12 in. in diameter, approximated by a plane circular piston, is the sound source, then at a frequency of 2000 Hz the directly radiated wave is concentrated mainly in a "beam" of half-angle 37°, centered on the speaker

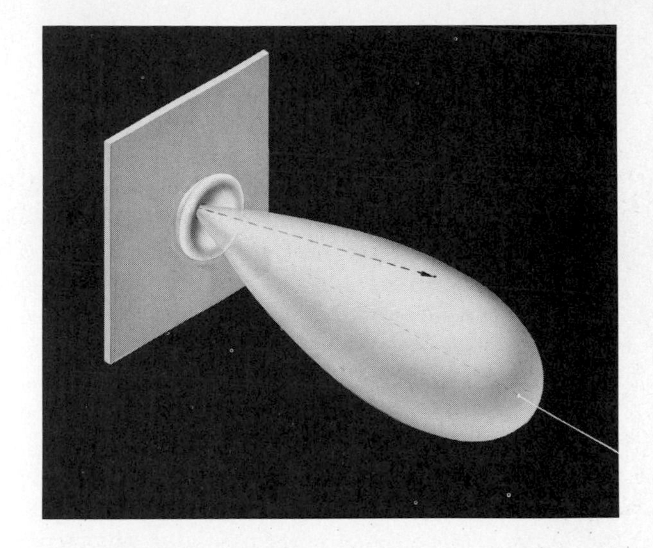

Fig. 23–4 *Space diagram of intensity distribution in front of an oscillating circular piston set in a baffle. A vector drawn from the center of the piston to any point on the surface has a length proportional to the sound intensity in that direction, as observed at a large fixed distance from the piston.*

axis. For a frequency of 10,000 Hz, or a wavelength of about 0.1 ft (about the upper frequency limit for such sound systems), the half-angle is

$$\sin\alpha = 1.22\left(\frac{0.1}{1.0}\right) = 0.122, \qquad \alpha \approx 7°,$$

while at a frequency of 1200 Hz, or a wavelength of about 0.8 ft,

$$\sin\alpha = 1.22\left(\frac{0.8}{1.0}\right) \approx 1.0, \qquad \alpha \approx 90°.$$

This means that for frequencies of 1200 Hz or less the radiated energy spreads out fairly uniformly, while at 10,000 Hz it is concentrated in a narrow beam, like a searchlight beam.

Since the intelligibility of speech depends largely on the high-frequency components, these are usually channeled into a number of speakers directed toward different parts of the auditorium, while the low frequencies can be handled by a single speaker, since the low-frequency radiation pattern has a much wider angular spread. On the other hand, it is sometimes desirable, as in underwater sound signaling or when leading the cheering section, to produce a beam having only a small angular divergence. To accomplish this, the diameter of the source (approximating it by a circular piston) must be large compared with the wavelength of the radiated sound.

If instead of a vibrating piston set in a wall there is merely an aperture in the wall and a train of plane waves is incident from the left, the waves transmitted through the aperture will propagate beyond it in the same way as waves originating at a piston set in the aperture. If the wavelength is small compared with the dimensions of the aperture, the spreading of the waves is small, while if the wavelength is large, the waves spread out in all directions.

When there is an obstacle in the path of a train of waves, the resultant effect at the far side of the obstacle is due to those portions of the advancing wave surface that are *not* obstructed. In very general terms, if the wavelength is relatively small, the spreading is small and the obstacle casts a sharp "shadow." The larger the wavelength, the greater the spreading or bending of the waves. Thus one *can* hear around the corner of a wall but the effect is greater for waves of long wavelength (or low frequency) than for waves of short wavelength (or high frequency). The general term for the phenomena described above, in which one is concerned with the resultant effect of a large number of waves from different parts of a source, is *diffraction.*

23–7 APPLICATIONS OF ACOUSTIC PHENOMENA

The relevance of acoustical principles is by no means limited to sound and hearing. Analysis of mechanical waves is a powerful investigative tool in a wide variety of fields. Elastic waves in water are reflected by solid bodies immersed in the water, such as schools of fish, submarines, or submerged wrecked ships. *Sonar* systems make use of reflections of sound waves in water to determine position and motion of such bodies.

Analysis of elastic waves in the earth provides important information about its structure. The interior of the earth may be pictured crudely as made of concentric spherical shells, with mechanical properties such as density and elastic moduli that are different in different shells. Waves produced by explosions or earthquakes are reflected and refracted at the interfaces between these shells, and analysis of these waves permits partial determination of the dimensions and properties of the shells.

Some of the most interesting recent applications of acoustics lie in the field of medicine, where sound waves are used for both diagnosis and therapy. In diagnosis, *ultrasonic* (i.e., above 20 kHz) frequencies are usually used because their short wavelengths permit study of smaller-scale phenomena than with audible sound. Reflection of such waves from regions in the interior of the body can be used to detect a wide variety of anomalous conditions such as tumors, and to study various phenomena such as heart-valve action. Ultrasound is more sensitive than x-rays in distinguishing various kinds of tissues; it is believed to be less hazardous than x-rays, although possible hazards of ultrasound have not yet been thoroughly explored. At much higher power levels, ultrasound appears to have promise as a selective destroyer of pathological tissue, and may find usefulness in the treatment of certain cancers as well as arthritis and related diseases.

In medical and other applications of ultrasonics, a variety of electronic instrumentation is used. The sound is always produced by first generating an electrical wave, and using this to drive a *transducer* which converts electrical to mechanical (i.e., sound) waves. Thus the *transducer* is similar in action to a loudspeaker except that the frequency ranges may be different; frequencies as high as several megahertz (i.e., several million cycles per second) are routinely used. The detecting instruments include transducers, functioning in the role of microphones, as well as amplifiers and devices for displaying reflected signals. A common technique is to send out pulses, and observe the transmitted and reflected pulse "blips" on an oscilloscope, which permits a direct measurement of the time interval between transmission and reflection.

The applications of acoustic principles to environmental problems are obvious. It is now recognized that noise is an important aspect of environmental degradation. The design of quiet mass-transit vehicles, for example, involves detailed study of

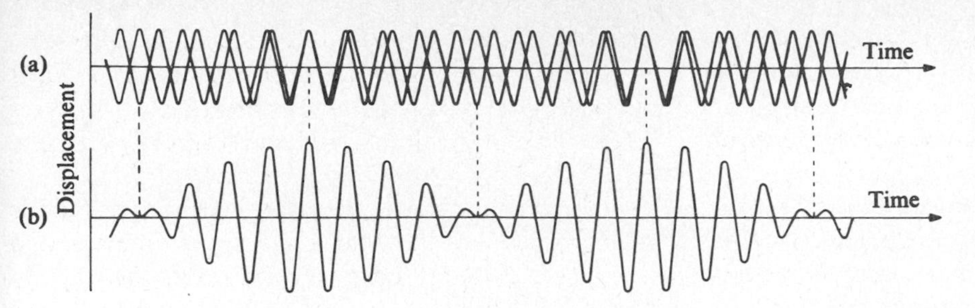

Fig. 23-5 *Beats are fluctuations in amplitude produced by two sound waves of slightly different frequency.*

sound generation and propagation in the motors, wheels, and supporting structures of vehicles. Excessive noise levels can lead to permanent hearing impairment; recent studies have shown that many young "rock" musicians have suffered hearing losses typical of persons 65 years of age, and even prolonged listening to high-level rock music (90 to 100 db) can lead to permanent damage.

23-8 BEATS

Standing waves in an air column have been cited as one example of *interference.* They arise when two wave trains of the same amplitude and frequency are traveling through the same region in opposite directions. We now wish to consider another type of interference which results when two wave trains of equal amplitude but slightly different frequency travel through the same region. Such a condition exists when two tuning forks of slightly different frequency are sounded simultaneously or when two piano strings struck by the same key are slightly "out of tune."

Let us consider some one point of space through which the waves are simultaneously passing. The displacements due to the two waves separately are plotted as a function of time on graph (a) in Fig. 23-5. If the total extent of the time axis represents one second, the graphs correspond to frequencies of 16 Hz and of 18 Hz. Applying the principle of superposition to find the resultant vibration, we get graph (b), where it is seen that the amplitude varies with time. These variations of amplitude give rise to variations of loudness which are called *beats.* Two strings may be tuned to the same frequency by tightening one of them while sounding both until the beats disappear.

The production of beats may be treated mathematically as follows. The displacements due to the two waves passing simultaneously through some one point of space may be written

$$y_1 = A \cos 2\pi f_1 t, \qquad y_2 = A \cos 2\pi f_2 t.$$

(The amplitudes are assumed equal.)

By the principle of superposition, the resultant displacement is

$$y = y_1 + y_2 = A[\cos 2\pi f_1 t + \cos 2\pi f_2 t],$$

and, since

$$\cos a + \cos b = 2 \cos\left(\frac{a+b}{2}\right) \cos\left(\frac{a-b}{2}\right),$$

this may be written

$$y = \left[2A \cos 2\pi\left(\frac{f_1 - f_2}{2}\right)t\right] \cos 2\pi\left(\frac{f_1 + f_2}{2}\right) t. \tag{23-16}$$

The resulting vibration can then be considered to be of frequency $(f_1 + f_2)/2$, or the average frequency of the two tones, and of amplitude given by the expression in brackets. The amplitude therefore varies with time at a frequency $(f_1 - f_2)/2$. If f_1 and f_2 are nearly equal, this term is small and the amplitude fluctuates very slowly. When the amplitude is large, the sound is loud, and vice versa. A beat, or a maximum of amplitude, will occur when

$\cos 2\pi t(f_1 - f_2)/2$ equals 1 or -1. Since *each* of these values occurs once in each cycle, the number of beats per second is *twice* the frequency $(f_1 - f_2)/2$; *the number of beats per second equals the difference of the frequencies.*

Beats between two tones can be detected by the ear up to a beat frequency of 6 or 7 per second. At higher frequencies, individual beats can no longer be distinguished and the sensation merges into one of *consonance* or *dissonance* depending on the frequency ratio of the tones. A beat frequency, even though it lies within the frequency range of the ear, is not necessarily interpreted by the ear as a tone of that frequency. Nevertheless, a tone can be heard of frequency equal to the frequency difference between two others sounded simultaneously. Such a tone is called a *difference* tone. Although not as easy to recognize, a frequency called a *summation* tone and equal to the sum of the frequencies of the two tones can also be heard. The general term applied to both difference tones and summation tones is *combination* tones.

23-9 THE DOPPLER EFFECT

When a source of sound, or a listener, or both, are in motion relative to the air, the pitch of the sound, as heard by the listener, is in general not the same as when source and listener are at rest. The most common example is the sudden drop in pitch of the sound from an automobile horn as one meets and passes a car proceeding in the opposite direction. This phenomenon is called the *Doppler effect.*

Let v_L and v_S represent the velocities of a listener and a source, relative to the air. We shall consider only the special case in which the velocities lie along the line joining listener and source. Since these velocities may be in the same or opposite directions, and the listener may be either ahead of or behind the source, a convention of signs is required. We shall take the positive directions of v_L and v_S as that *from* the position of the listener, toward the position of the source. The velocity of propagation of sound waves, c, will always be considered positive.

In Fig. 23-6, a listener L is at the left of a source S. The positive direction is then from left to right and

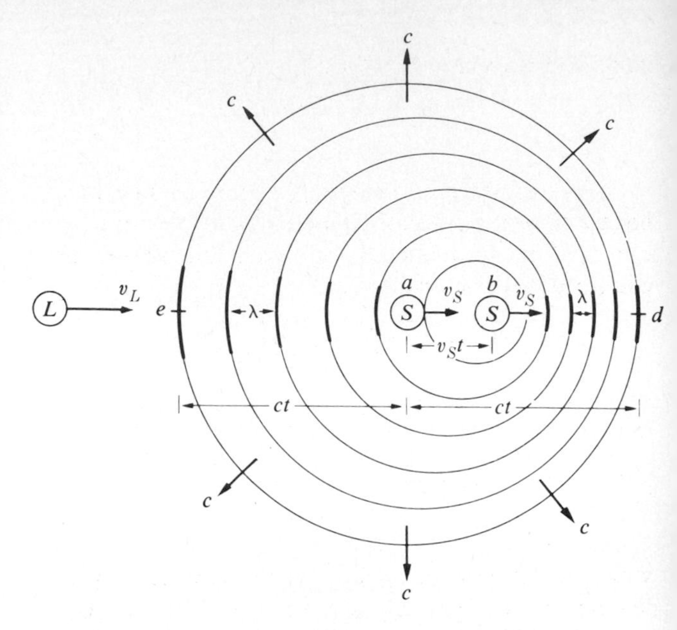

Fig. 23-6 *Wave surfaces emitted by a moving source.*

both v_L and v_S are positive in the diagram. The sound source is at point a at time $t = 0$ and at point b at a later time t. The outer circle represents the wave surface emitted at time $t = 0$. This surface (in free space) is a sphere with center at a, and is traveling radially outward at all points with speed c. (The fact that the wave originated at a *moving* source does not affect its speed after leaving the source. The wave speed c is a property of the *medium* only; the waves forget about the source as soon as they leave it.) The radius of this sphere (the distance ea or ad) is therefore ct. The distance ab equals v_St, so

$$eb = (c + v_S)t, \qquad bd = (c - v_S)t.$$

In the time interval between $t = 0$ and $t = t$, the number of waves emitted by the source is f_St, where f_S is the frequency of the source. In front of the source these waves are crowded into the distance bd, while behind the source they are spread out over the distance eb. The wavelength in front of the source is therefore

$$\lambda = \frac{(c - v_S)t}{f_St} = \frac{c - v_S}{f_S}, \qquad (23\text{-}17)$$

while the wavelength behind the source is

$$\lambda = \frac{(c + v_S)t}{f_S t} = \frac{c + v_S}{f_S}. \tag{23–18}$$

The waves approaching the moving listener L have a speed of propagation *relative to him*, given by $c + v_L$. The frequency f_L at which the listener encounters these waves is

$$f_L = \frac{c + v_L}{\lambda} = \frac{c + v_L}{(c + v_S)/f_S},$$

or

$$\boxed{\frac{f_L}{c + v_L} = \frac{f_S}{c + v_S},} \tag{23–19}$$

which expresses the frequency f_L as heard by the listener in terms of the frequency f_S of the source.

This general relation includes all possibilities for collinear motion of source and listener relative to the medium. If one happens to be at rest in the medium, the corresponding velocity is zero, and of course when both are at rest or have the same velocity relative to the medium, then $f_L = f_S$. Whenever source or listener is moving in the opposite direction to that which we have designated as positive, the corresponding velocity to be used in Eq. (23–19) is negative. The examples illustrate these sign conventions.

Examples Let $f_S = 1000$ Hz, and $c = 1000$ ft s^{-1}. The wavelength of the waves emitted by a stationary source is then $c/f_S = 1.00$ ft.

a) What are the wavelengths ahead of and behind the moving source in Fig. 23–6 if its velocity is 100 ft $^{-1}$?

In front of the source,

$$\lambda = \frac{c - v_S}{f_S} = \frac{1000 - 100}{1000} = 0.90 \text{ ft}.$$

Behind the source.

$$\lambda = \frac{c + v_S}{f_S} = \frac{1000 + 100}{1000} = 1.10 \text{ ft}.$$

b) If the listener L in Fig. 23–6 is at rest and the source is moving away from him at 100 ft s^{-1}, what is the frequency as heard by the listener?

Since

$$v_L = 0 \quad \text{and} \quad v_S = 100 \text{ ft s}^{-1},$$

we have

$$f_L = f_S \frac{c}{c + v_S}$$

$$= 1000\left(\frac{1000}{1000 + 100}\right) = 909 \text{ Hz}.$$

c) If the source in Fig. 23–6 is at rest and the listener is moving toward the left at 100 ft s^{-1}, what is the frequency as heard by the listener?

The positive direction (from listener to source) is still from left to right, so

$$v_L = -100 \text{ ft s}^{-1}, \quad v_S = 0,$$

$$f_L = f_S \frac{c + v_L}{c}$$

$$= 1000\left(\frac{1000 - 100}{1000}\right) = 900 \text{ Hz}.$$

Thus, while the frequency f_L as heard by the listener is less than the frequency f_S both when the source moves away from the listener and when the listener moves away from the source, the decrease in frequency is not the same for the same speed of recession.

In the preceding equations, the velocities v_L, v_S and c are all *relative to the air*, or more generally, to the medium in which the waves are traveling. The Doppler effect exists also for electromagnetic waves in empty space, such as light waves or radio waves. In this case, there is no "medium" relative to which a velocity can be defined, and we can speak only of the *relative* velocity v *of source and receiver.*

In deriving the Doppler relations for light, we must use the relativistic kinematic relations developed in Chapter 14. The wave velocity c is the velocity of light and is the same for both source and listener. In the frame of reference in which the listener is at rest, the source is moving away from the listener with velocity v. (If, instead, the source is *approaching* the listener, v is negative.) The source frequency is again f_S, but this is measured in the rest frame of the source; in the rest frame of the *listener* the corresponding frequency f'_S is less than this by the time-dilation factor $(1 - v^2/c^2)^{1/2}$.

The Doppler relation for light analogous to Eq. (23–19) for sound is obtained by using the time-dilated frequency $f'_S = f_S\sqrt{1 - v^2/c^2}$, in place of f_S. That is, in the rest frame of L,

$$\lambda = \frac{c+v}{f'_S} = \frac{c+v}{f_S\sqrt{1 - v^2/c^2}}.$$

The frequency f_L measured by the listener (i.e., the frequency of arrival of the waves at L) is then given by

$$f_L = \frac{c}{\lambda} = \frac{cf_S\sqrt{1 - v^2/c^2}}{c+v} = \frac{f_S\sqrt{c^2 - v^2}}{c+v}.$$

This can be simplified further by noting that

$$\sqrt{c^2 - v^2} = \sqrt{c-v}\cdot\sqrt{c+v}.$$

The final result is

$$\boxed{f_L = \left(\sqrt{\frac{c-v}{c+v}}\right) f_S.} \qquad (23\text{–}20)$$

When v is positive, the source moves *away* from the listener and f_L is always *less* than f_S; when v is negative, the source moves *toward* the listener and f_L is *greater* than f_S. Thus, the qualitative effect is the same as for sound, although the quantitative relationship is different.

The Doppler effect provides a convenient means of tracking an artificial satellite which is emitting a radio signal of constant frequency f_S. The frequency f_L of the signal that is received on the earth decreases as the satellite is passing, since the velocity component *toward* the earth decreases from position 1 to position 2 in Fig. 23–7, and then points *away* from the earth from 2 to 3. If the received signal is combined with a constant signal generated in the receiver to give rise to *beats*, then the beat frequency may be such as to produce an audible note whose pitch decreases as the satellite passes overhead.

A similar technique is used by law-enforcement officers to measure automobile speeds. An electromagnetic wave is emitted by a source at the side of the road, typically attached to a police car. The wave is reflected from a moving car, which thus acts as a moving source; the reflected wave is Doppler-shifted in frequency. Measurement of the frequency shift using beats, as with satellite tracking, permits simple measurement of the speed.

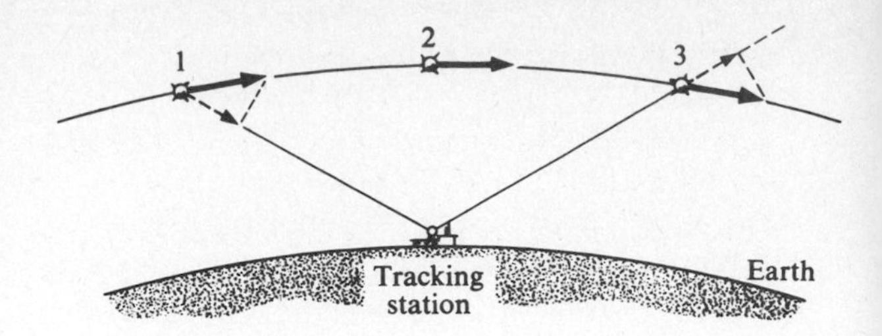

Fig. 23–7 Change of velocity component along the line of sight of a satellite passing a tracking station.

The Doppler effect for *light* is important in astronomy. Analysis of the spectra of light from distant stars shows shifts in wavelength compared to spectra from the same elements on earth. These can be interpreted as Doppler shifts due to motion of the stars. The shift is nearly always toward the longer wavelength or red end of the spectrum, and is therefore called the *red shift.* Such observations have provided practically all the evidence for the "exploding universe" cosmological theories, which represent the universe as having evolved from a great explosion several billion years ago in a relatively small region of space.

PROBLEMS

23–1

a) If the pressure amplitude in a sound wave is tripled, by how many times is the intensity of the wave increased?

b) By how many times must the pressure amplitude of a sound wave be increased in order to increase the intensity by a factor of 16 times?

23–2

a) Two sound waves of the same frequency, one in air and one in water, are equal in intensity. What is the ratio of the pressure amplitude of the wave in water to that of the wave in air?

b) If the pressure amplitudes of the waves are equal, what is the ratio of their intensities?

c) What is the difference between their intensity levels? (The speed of sound in water may be taken as 1490 m s^{-1}.)

23–3

a) Relative to the arbitrary reference intensity of 10^{-16} W cm^{-2}, what is the intensity level in decibels of a sound wave whose intensity is 10^{-10} W cm^{-2}?

b) What is the intensity level of a sound wave in air whose pressure amplitude is 0.2 N m^{-2}?

23–4

a) Show that if β_1 and β_2 are the intensity levels in db of sounds of intensities I_1 and I_2 respectively, the difference in intensity levels of the sound is

$$\beta_2 - \beta_1 = 10 \log \frac{I_2}{I_1}.$$

b) Show that if P_1 and P_2 are the pressure amplitudes of two sound waves, the difference in intensity levels of the waves is

$$\beta_2 - \beta_1 = 20 \log \frac{P_2}{P_1}.$$

23–5

a) Show that if the reference level of intensity is $I_0 = 10^{-16}$ W cm^{-2}, the intensity level of a sound of intensity I (in W cm^{-2}) is

$$\beta = 160 + 10 \log I.$$

b) Is this relation valid if I is expressed in W m^{-2} instead of W cm^{-2}? If not, how should it be modified?

23–6 The intensity due to a number of independent sound sources is the sum of the individual intensities. How many decibels greater is the intensity level when all five quintuplets cry simultaneously than when a single one cries? How many more crying babies would be required to produce a further increase in the intensity level of the same number of decibels?

23–7 A window whose area is 1 m^2 opens on a street where the street noises result in an intensity level, at the window, of 60 db. How much "acoustic power" enters the window via the sound waves?

23–8 A source of sound emits a total power of 10 W, uniformly in all directions. At what distance from the source is the sound level (a) 100 db? (b) 60 db?

23–9 A certain sound source radiates uniformly in all directions in air. At a distance of 5 m the sound level is 80 db. The frequency is 440 Hz.

a) What is the displacement amplitude at this distance?

b) What is the pressure amplitude?

c) At what distance is the sound level 60 db?

23–10

a) What are the upper and lower limits of intensity level of a person whose auditory area is represented by the graph of Fig. 23–3?

b) What are the highest and lowest frequencies he can hear when the intensity level is 40 db?

23–11 Two loudspeakers, A and B, radiate sound uniformly in all directions. The output of acoustic power from A is 8×10^{-4} W, and from B it is 13.5×10^{-4} W. Both loudspeakers are vibrating in phase at a frequency of 173 Hz.

a) Determine the difference in phase of the two signals at a point C along the line joining A and B, 3 m from B and 4 m from A.

b) Determine the intensity at C from speaker A if speaker B is turned off, and the intensity at C from speaker B if speaker A is turned off.

c) With both speakers on, what is the intensity and intensity level at C?

23–12 What should be the diameter of a sound source in the form of a circular piston set in a wall, if the central lobe of the diffraction pattern is to have a half-angle of 45°, for a frequency of 10,000 Hz?

23–13 The sound source of a sonar system operates at a frequency of 50,000 Hz. Approximate the source by a circular disk set in the hull of a destroyer. The velocity of sound in water can be taken as 1450 m s^{-1}.

a) What is the wavelength of the waves emitted by the source?

b) What must be the diameter of the source if the half-angular divergence of the main beam is not to be more than 10°?

c) What is the difference in frequency between the directly radiated waves and the waves reflected from the hull of a submarine traveling directly away from the destroyer at 15 mi hr^{-1}?

23–14 Two identical piano strings when stretched with the same tension have a fundamental frequency of 400 Hz. By what fractional amount must the tension in one string be increased in order that 4 beats s^{-1} shall occur when both strings vibrate simultaneously?

23–15 In Table 23–2, the interval from C to G is called a *fifth.* Compare the frequency ratio for this interval in the just scale to that in the equally tempered scale. Which is larger, and by what percent?

23–16 In Table 23–2, compare the frequency ratio for the interval C to G with the ratio for the interval E to B, in the just scale.

a) By what percent do the two ratios differ?
b) Show that the two ratios are equal in the equally tempered scale.

23–17 The frequency ratio of a half-tone interval on the equally tempered scale is 1.059. Find the velocity of an automobile passing a listener at rest in still air, if the pitch of the car's horn drops a half-tone between the times when the car is coming directly toward him and when it is moving directly away from him.

23–18

a) Refer to Fig. 23–9 and the examples in Section 23–9. Suppose that a wind of velocity 50 ft s^{-1} is blowing in the same direction as that in which the source is moving. Find the wavelengths ahead of and behind the source.
b) Find the frequency heard by a listener at rest when the source is moving away from him.

23–19 A railroad train is traveling at 100 ft s^{-1} in still air. The frequency of the note emitted by the locomotive whistle is 500 Hz. What is the wavelength of the sound waves (a) in front of and (b) behind the locomotive? What would be the frequency of the sound heard by a stationary listener (c) in front of and (d) behind the locomotive? What frequency would be heard by a passenger on a train traveling at 50 ft s^{-1} and (e) approaching the first, (f) receding from the first? (g) How would each of the preceding answers be altered if a wind of velocity 30 ft s^{-1} were blowing in the same direction as that in which the locomotive was traveling?

23–20 A train of plane sound waves of frequency f_0 and wavelength λ_0 travels horizontally toward the right. It strikes and is reflected from a large, rigid, vertical plane surface, perpendicular to the direction of propagation of the wave train and moving toward the left with a velocity v.

a) How many waves strike the surface in a time interval t?
b) At the end of this time interval, how far to the left of the surface is the wave that was reflected at the beginning of the time interval?
c) What is the wavelength of the reflected waves, in terms of λ_0?
d) What is the frequency, in terms of f_0?
e) A listener is at rest at the left of the moving surface. Describe the sensation of sound which he hears as a result of the combined effect of the incident and reflected wave trains.

23–21 Two whistles, A and B, each have a frequency of 500 Hz. A is stationary and B is moving toward the right (away from A) at a velocity of 200 ft s^{-1}. An observer is between the two whistles, moving toward the right with a velocity of 100 ft s^{-1}. Take the velocity of sound in air as 1100 ft s^{-1}.

a) What is the frequency from A as heard by the observer?
b) What is the frequency from B as heard by the observer?
c) What is the beat frequency heard by the observer?

23–22 A man stands at rest in front of a large smooth wall. Directly in front of him, between him and the wall, he holds a vibrating tuning fork of frequency 400 Hz. He now moves the fork toward the wall with a velocity of 4 ft s^{-1}. How many beats per second will he hear between the sound waves reaching him directly from the fork, and those reaching him after being reflected from the wall?

23–23 A source of sound waves, S, emitting waves of frequency 1000 Hz, is traveling toward the right in still air with a velocity of 100 ft s^{-1}. At the right of the source is a large, smooth, reflecting surface moving toward the left with a velocity of 400 ft s^{-1}.

a) How far does an emitted wave travel in 0.01 s?
b) What is the wavelength of the emitted waves in front of (i.e., at the right of) the source?
c) How many waves strike the reflecting surface in 0.01 s?
d) What is the velocity of the reflected waves?
e) What is the wavelength of the reflected waves?

23–24

a) Show that Eq. (23–11) can be written

$$f_L = f_S\left(1 - \frac{v}{c}\right)^{1/2}\left(1 + \frac{v}{c}\right)^{-1/2}$$

b) Use the binomial theorem to show that if $v \ll c$, then, to a good approximation,

$$f_L = f_S\left(1 + \frac{v}{c}\right).$$

c) An earth satellite emits a radio signal of frequency 10^8 Hz. An observer on the ground detects beats between the received signal and a local signal also of frequency 10^8 Hz. At a particular moment, the beat frequency is 2400 Hz. What is the component of the satellite's velocity directed toward the earth at this moment?

PART II

Electricity and Magnetism, Light, and Atomic Physics

Chapter 24

Coulomb's Law

24–1 ELECTRIC CHARGES

It was known to the ancient Greeks as long ago as 600 B.C. that amber, rubbed with wool, acquired the property of attracting light objects. In describing this property today, we say that the amber is *electrified*, or possesses an *electric charge*, or is *electrically charged*. These terms are derived from the Greek word *elektron*, meaning amber. It is possible to impart an electric charge to any solid material by rubbing it with any other material. Thus, an automobile becomes charged by virtue of its motion through the air; an electric charge is developed on a sheet of paper moving through a printing press; a comb is electrified in passing through dry hair. Actually, intimate contact is all that is needed to give rise to an electric charge. Rubbing merely serves to bring many points of the surfaces into good contact.

Hard rubber and fur are commonly used in demonstrations. If, after rubbing with fur, a rubber rod is placed in a dish containing tiny pieces of tissue paper, many of these will at first cling to the rod, but after a few seconds they will fly off. The initial attraction will be explained in Chapter 27; the subsequent repulsion is due to a force that is found to exist whenever two bodies are electrified *in the same way*. Suppose two small, very light pith balls are suspended near each other by fine silk threads. At first they will be attracted to an electrified rubber rod and will cling to it. A moment later, they will be repelled by the rubber and will also repel each other.

A similar experiment performed with a glass rod that has been rubbed with silk gives rise to the same result; pith balls electrified by contact with such a glass rod are repelled not only by the rod but by each other. On the other hand, when a pith ball that has been in contact with electrified rubber is placed near one that has been in contact with electrified glass, the pith balls *attract* each other. We are therefore led to the conclusion that there are *two kinds* of electric charge—that possessed by rubber after being rubbed with fur, called a *negative* charge, and that possessed by glass after being rubbed with silk, called a *positive* charge. The experiments on pith balls described above lead to the fundamental results that (1) *like charges repel*, (2) *unlike charges attract.*

These repulsive or attractive forces of electrical origin are distinct from the gravitational attraction and, in most situations with which we shall deal, are so much larger than the gravitational force that the latter may be completely neglected.

In addition to the forces of attraction or repulsion, other forces are found to exist between electric charges which depend on their relative motion. These forces are responsible for *magnetic* phenomena. For many years, the repulsion or attraction between a pair of bar magnets was explained on the theory that there existed magnetic entities similiar to electric charges and called "magnetic poles." It is a familiar fact, however, that magnetic effects are also observed around a wire carrying an electric current. But a current is simply a motion of electric charge, and it is now known that all magnetic effects come about as a result of the relative motion of electric charges. Hence magnetism and electricity are not two separate subjects, but are related phenomena arising from the properties of electric charges.

Suppose a rubber rod is rubbed with fur and then touched to a suspended pith ball. Both the rubber and the pith ball are negatively charged. If the fur is now brought near the pith ball, the ball will be attracted, indicating that the fur is positively charged. It follows that when rubber is rubbed with fur, opposite charges appear on the two materials. This is found to happen whenever any substance is rubbed with any other substance. Thus, glass becomes positive, while the silk with which the glass was rubbed becomes negative. This suggests strongly that electric charges are not generated or created, but that the process of acquiring an electric charge consists of *transferring* something from one body to another, so that one body has an excess and the other a deficiency of that something. It was not until the end of the nineteenth century that this "something" was found to consist of very small, light pieces of negative electricity, known today as *electrons.*

24–2 ATOMIC STRUCTURE

The word *atom* is derived from the Greek *atomos,* meaning indivisible. It is scarcely necessary to point out that the term is inappropriate. All atoms are more or less complex arrangements of subatomic particles, and there are many methods of splitting off some of these particles, either singly or in groups.

The subatomic particles, the building blocks out of which atoms are constructed, are of three different kinds: the negatively charged *electron,* the positively charged *proton,* and the neutral *neutron.* The negative charge of the electron is of the same magnitude as the positive charge of the proton and no charges of smaller magnitude have ever been observed. The charge of a proton or an electron is the ultimate, natural unit of charge.

The subatomic particles are arranged in the same general way in all atoms. The protons and neutrons always form a closely packed group called the *nucleus,* which has a net positive charge due to the protons. The diameter of the nucleus, if we think of it as roughly spherical, is of the order of 10^{-14} m. Outside the nucleus, but at relatively large distances from it, are the electrons, whose number is equal to the number of protons within the nucleus. If the atom is undisturbed, and no electrons are removed from the space around the nucleus, the atom as a whole is electrically neutral. That is, the equal positive and negative charges of the nucleus and the electrons sum to zero, just as equal positive and negative numbers sum to zero. If one or more electrons are removed, the remaining positively charged structure is called a positive *ion.* A negative ion is an atom which has gained one or more extra electrons. The process of losing or gaining electrons is called *ionization.*

In the atomic model proposed by the Danish physicist Niels Bohr in 1913, the electrons were pictured as whirling about the nucleus in circular or elliptical orbits. More recent research has shown that the electrons are more accurately represented as spread-out distributions of electric charge, governed by the principles of quantum mechanics which will be discussed in Chapter 44. Nevertheless, the Bohr model is still useful for visualizing the structure of an atom. The diameters of the electron charge distributions, which the Bohr model pictures as orbits, determine the overall size of the atom as a whole, and these are of the order of 2 or 3×10^{-10} m, or about ten thousand times as great as the diameter of the nucleus. A Bohr atom is a solar system in miniature, with electrical forces taking the place of gravitational forces. The positively charged central nucleus corresponds to the sun, while the electrons, moving around the nucleus under the electrical force of its attraction for them, correspond to the planets moving around the sun under the influence of its gravitational attraction.

The masses of a proton and a neutron are nearly equal, and the mass of each is about 1840 times as great as that of an electron. Practically all the mass of an atom, therefore, is concentrated in its nucleus. Since one kilomole of monatomic hydrogen consists of 6.02×10^{26} particles (Avogadro's number) and its mass is 1.008 kg, the mass of a single hydrogen atom is

$$\frac{1.008 \text{ kg}}{6.02 \times 10^{26}} = 1.67 \times 10^{-27} \text{ kg}.$$

The hydrogen atom is the sole exception to the rule that all atoms are constructed of three kinds of subatomic particles. The nucleus of a hydrogen atom is a single proton, outside of which there is a single electron. Hence, out of the total mass of the hydrogen atom, 1/1840 part is the mass of the electron and the remainder is the mass of a proton. To three significant figures,

$$\text{Mass of electron} = \frac{1.67 \times 10^{-27} \text{ kg}}{1840} = 9.11 \times 10^{-31} \text{ kg},$$

$$\text{Mass of proton} = 1.67 \times 10^{-27} \text{ kg},$$

and since the masses of a proton and a neutron are nearly equal,

$$\text{Mass of neutron} = 1.67 \times 10^{-27} \text{ kg}.$$

After hydrogen, the atom with the simplest structure is that of helium. Its nucleus consists of two protons and two neutrons, and it has two extranuclear electrons. When these two electrons are absent, the doubly-charged helium ion, which is the helium nucleus itself, is often called an *alpha particle*, or α-particle. The next element, lithium, has three protons in its nucleus and has thus a nuclear charge of three units. In the un-ionized state the lithium atom has three extranuclear electrons. Each element has a different number of nuclear protons and therefore a different positive nuclear charge. In the table of elements listed at the end of this book, known as the *periodic table*, each element occupies a box with which is associated a number, called the *atomic number.*

The atomic number represents the number of nuclear protons or, in the undisturbed state, the number of extranuclear electrons.

Every material body contains a tremendous number of charged particles, positively charged protons in the nuclei of its atoms and negatively charged electrons outside the nuclei. When the total number of protons equals the total number of electrons, the body as a whole is electrically neutral.

To give a body an excess negative charge, we may either *add* a number of *negative* charges to a neutral body, or *remove* a number of *positive* charges from the body. Similarly, either an *addition* of *positive* charge or a *removal* of *negative* charge will result in an excess positive charge. In most instances, it is negative charges (electrons) that are added or removed, and a "positively charged body" is one that has lost some of its normal content of electrons.

The "charge" of a body refers to its *excess* charge only. The excess charge is always a very small fraction of the total positive or negative charge in the body.

24–3 THE LEAF ELECTROSCOPE AND THE ELECTROMETER

A charged pith ball may be used as a test body to determine whether or not a second body is charged. A more sensitive test is afforded by the *leaf electroscope* (Fig. 24–1). Two strips of thin gold leaf or aluminum foil A are fastened at the end of a metal rod B which passes through a support C of rubber,

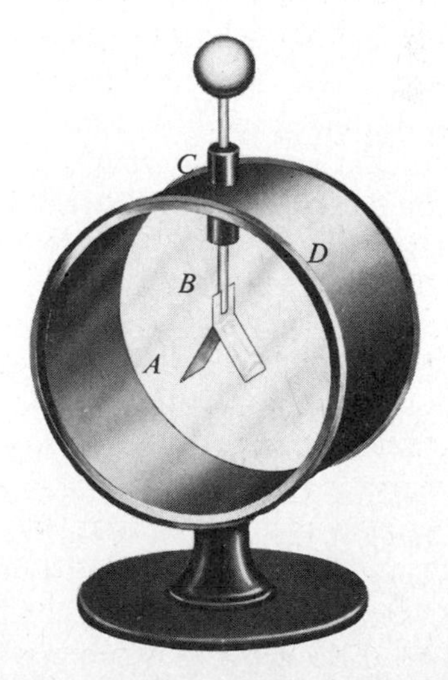

Fig. 24–1 *The leaf electroscope.*

amber, or sulfur. The surrounding case D is provided with windows through which the leaves can be observed and which serve to protect them from air currents. When the knob of the electroscope is touched by a charged body, the leaves acquire charges of the same sign and repel each other, their divergence being a measure of the quantity of charge they have received.

If one terminal of a battery of a few hundred volts potential difference is connected to the knob of an electroscope and the other terminal to the electroscope case, the leaves will diverge just as if they had been charged from a body electrified by contact. There is no difference between the "kinds" of charge given the leaves in these two processes and, in general, there is no distinction between "static electricity" and "current electricity." The term "current" refers to a *flow* of charge, while "electrostatics" is concerned for the most part with interactions between charges at rest. The charges themselves in either case are those of electrons or protons.

Modern electronics has produced instruments called *electrometers* which use electronic amplification to achieve greater sensitivity in charge measurements than is possible with the simple leaf electroscope, and also to make precise *quantitative* measurements of quantity of charge.

24–4 CONDUCTORS AND INSULATORS

Let one end of a copper wire be attached to the knob of an electroscope, its other end being supported by a glass rod, as in Fig. 24–2. If a charged rubber rod is touched to the far end of the wire, the electroscope leaves will immediately diverge. There has, therefore, been a transfer of charge along or through the wire, and the wire is called a *conductor.* If the experiment is repeated, using a silk thread or rubber band in place of the wire, no such deflection of the electroscope occurs and the thread or rubber is called an *insulator* or *dielectric.* The motion of charge through a material substance will be studied in more detail in Chapter 28, but for our present purposes it is sufficient to state that most substances fall into one or the other of the two classes above. Conductors permit the passage of charge through them, while insulators do not.

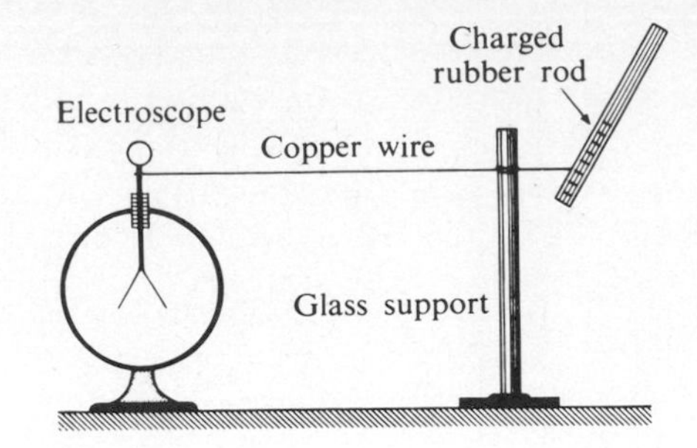

Fig. 24–2 *Copper is a conductor of electricity.*

Metals in general are good conductors, while nonmetals are insulators. The positive valency of metals and the fact that they form positive ions in solution indicate that the atoms of a metal will part readily with one or more of their outer electrons. Within a metallic conductor such as a copper wire, a few outer electrons become detached from each atom and can move freely throughout the metal in much the same way that the molecules of a gas can move through the spaces between grains of sand in a sand-filled container. In fact, these free electrons are often spoken of as an "electron gas." The positive nuclei and the remainder of the electrons remain fixed in position. Within an insulator, on the other hand, there are no, or at most very few, free electrons.

The phenomenon of charging by contact is not limited to rubber and fur, or indeed to insulators in general. Any two dissimilar substances exhibit the effect to a greater or lesser extent, but evidently a conductor must be supported on an insulating handle, or the charges developed on it will at once leak away.

24–5 CHARGING BY INDUCTION

In charging a leaf electroscope by contact with, say, a rubber rod that has been rubbed with fur, some of the extra electrons on the rubber are transferred to the electroscope, leaving the rubber with a smaller negative charge. There is, however, another way to use the rubber rod to charge other bodies, in which the rubber may impart a charge of *opposite* sign and lose none of its own charge. This process, called *charging by induction,* is illustrated in Fig. 24–3.

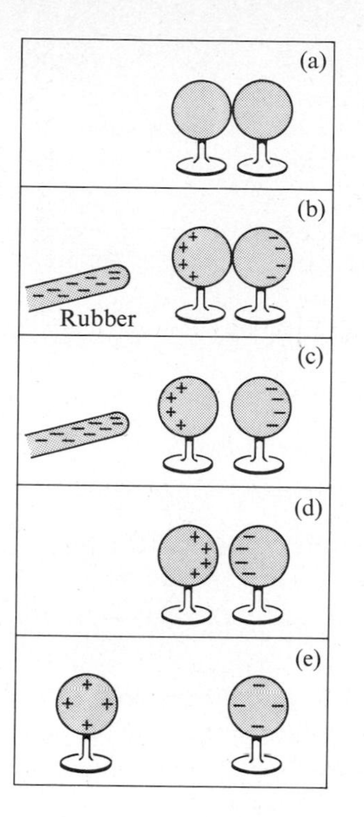

Fig. 24–3 Two metal spheres are oppositely charged by induction.

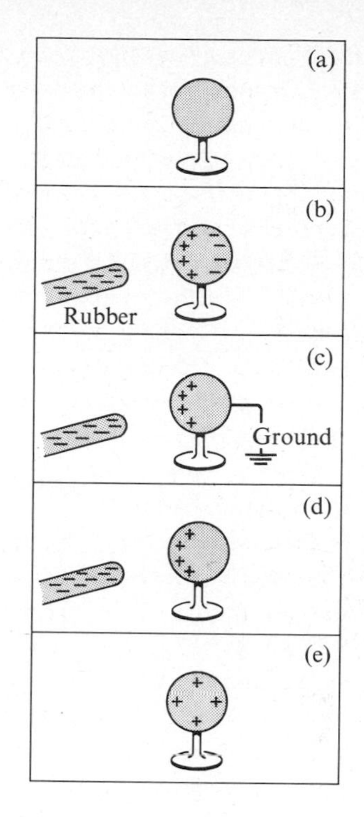

Fig. 24–4 Charging a single metal sphere by induction.

In Fig. 24–3(a), two neutral metal spheres are in contact, both supported on insulating stands. When a negatively charged rubber rod is brought near one of the spheres but without touching it, as in part (b), the free electrons in the metal spheres are repelled and drift slightly away from the rod, toward the right. Since the electrons cannot escape from the spheres, an excess negative charge accumulates at the right surface of the right sphere. This leaves a deficiency of negative charge, or an excess positive charge, at the left surface of the left sphere. These excess charges are called *induced* charges.

It should not be inferred that *all* of the free electrons in the spheres are driven to the surface of the right sphere. As soon as any induced charges develop, they also exert forces on the free electrons within the spheres. In this case this force is toward the left (a repulsion by the negative induced charge and an attraction by the positive induced charge). Within an extremely short time the system reaches an equilibrium state in which, at every point in the interior of the spheres, the force on an electron toward the right, exerted by the charged rod, is just balanced by a force toward the left exerted by the induced charges.

The induced charges remain on the surfaces of the spheres as long as the rubber rod is held nearby. When the rod is removed, the electron cloud in the spheres moves to the left and the original neutral condition is restored.

Suppose that the spheres are separated slightly, as shown in part (c), while the rubber rod is nearby. If the rod is now removed, as in (d), we are left with two oppositely charged metal spheres whose charges attract each other. When the two spheres are separated by a great distance, as in (e), the two charges become uniformly distributed. It should be noticed that the negatively charged rubber rod has lost none of its charge in the steps from (a) to (e).

The steps from (a) to (e) in Fig. 24–4 should be self-explanatory. In this figure, a single metal sphere (on an insulating stand) is charged by induction. The symbol lettered "ground" in part (c) simply means that the sphere is connected to the earth (a conductor). The earth thus takes the place of the second sphere in Fig. 24–3. In step (c), electrons are repelled to ground either through a conducting wire, or along the moist skin of a person who touches the sphere with his finger. The earth thus acquires a negative charge equal to the induced positive charge remaining on the sphere.

The process taking place in Figs. 24–3 and 24–4 could be explained equally well if the mobile charges in the spheres were *positive*, or, in fact, if *both* positive and negative charges were mobile. Although we now know that in a metallic conductor it is actually the *negative* charges that move, it is often convenient to describe a process *as if* the positive charges moved.

24–6 COULOMB'S LAW

The first quantitative investigation of the law of force between charged bodies was carried out by Charles Augustin de Coulomb (1736–1806) in 1784, utilizing for the measurement of forces a torsion balance of the type employed 13 years later by Cavendish in measuring gravitational forces. Coulomb found that the force of attraction or repulsion between two "point charges," that is, charged bodies whose dimensions are small compared with the distance r between them, is inversely proportional to the square of this distance.

The force also depends on the quantity of charge on each body. The net charge of a body might be described by a statement of the excess number of electrons or protons in the body. In practice, however, the charge of a body is expressed in terms of a unit much larger than the charge of an individual electron or proton. We shall use the letter q or Q to represent the charge of a body, postponing for the present the definition of the unit of charge.

In Coulomb's time, no unit of charge had been defined, nor had any method been developed for comparing a given charge with a unit. Despite this, Coulomb devised an ingenious method of showing how the force exerted on or by a charged body depended on its charge. He reasoned that if a charged spherical conductor were brought in contact with a second identical conductor, originally *uncharged*, the charge on the first would, by symmetry, be shared equally between the conductors. He thus had a method for obtaining one-half, one-quarter, and so on, of any given charge. The results of his experiments were consistent with the conclusion that the force between two point charges q and q' is proportional to the product of these charges. The complete expression for the magnitude of the force between two point charges is therefore

$$\boxed{F = k\frac{qq'}{r^2},} \tag{24–1}$$

where k is a proportionality constant whose magnitude depends on the units in which F, q, q', and r are expressed. Equation (24–1) is the mathematical statement of what is known today as *Coulomb's law*:

The force of attraction or repulsion between two point charges is directly proportional to the product of the charges and inversely proportional to the square of the distance between them.

The best verification of Coulomb's law lies in the correctness of many conclusions which have been drawn from it, rather than on direct experiments with point charges, which cannot be made with high precision.

The force on each particle always acts along the line joining the two particles. The charges q and q' are algebraic quantities which may be positive or negative, corresponding to the existence of two kinds of charge, also called positive and negative. Equation (24–1) gives the magnitude of the interaction force in all cases; a positive value of F corresponds to a repulsive interaction between like charges, a negative value to an attractive interaction between unlike charges. In either case, the forces obey Newton's third law; the force which q exerts on q' is the negative of the force q' exerts on q.

Coulomb's law has the same *form* as Newton's law of universal gravitation.

$$F = G\frac{mm'}{r^2}.$$

The electrical constant k corresponds to the gravitational constant G.

If there is matter in the space between the charges, the *net* force acting on each is altered because of redistribution of charge in the molecules of the intervening medium. This effect will be described later on. As a practical matter, the law can be used as it stands for point charges in air, since even at atmospheric pressure the effect of the air is to alter the force from its value in vacuum by only about one part in two thousand.

In the chapters of this book dealing with electrical phenomena, we shall use the mks system of units exclusively. The mks electrical units include all the familiar electrical units such as the volt, the ampere, the ohm, and the watt. The cgs system is also commonly used, more so in scientific work than in commerce and industry, but there is *no* British system of electrical units. This is one of many reasons for abandoning the British system and adopting metric units universally, and there seems little doubt that the mks system will eventually receive worldwide adoption.

To the three basic units in the mks system (the meter, kilogram, and second) we must now add a fourth, the unit of electric charge. This unit is called one *coulomb* (1 C), and the complete system is referred to as the mksc system.* The electrical constant k, in this system, is

$$k = 8.98755 \times 10^9 \text{ N m}^2 \text{ C}^{-2} \approx 9 \times 10^9 \text{ N m}^2 \text{ C}^{-2}.$$

Later, in connection with the study of electromagnetic radiation, we shall show that k is closely related to the speed of light in vacuum,

$$c = 2.998 \times 10^8 \text{ m s}^{-1}.$$

Specifically,

$$k = 10^{-7} c^2.$$

This relationship is not accidental, of course, but results from the definition of the unit of current, which in turn is related to the interaction of electric and magnetic fields, to be studied later.

In the cgs system of electrical units, not used in this book, the constant k is defined to be unity, without units. This defines a unit of electric charge called the *statcoulomb* or the *esu* (electrostatic unit). The conversion factor is

$$1 \text{ C} = 2.998 \times 10^9 \text{ esu}.$$

The "natural" unit of charge is the charge carried by an electron or proton. The most precise measurements up to the present time find this charge e to be

$$e = 1.60219 \times 10^{-19} \text{ C} \approx 1.60 \times 10^{-19} \text{ C}.$$

One coulomb therefore represents the total charge carried by about 6×10^{18} electrons. By way of comparison, the population of the earth is estimated to be about 3×10^9 persons, while on the other hand a cube of copper 1 cm on a side contains about 8×10^{22} free electrons.

Example 1 An α-particle is a nucleus of doubly ionized helium. It has a mass m of 6.68×10^{-27} kg and a charge q of $+2e$ or 3.2×10^{-19} C. Compare the force of electrostatic repulsion between two α-particles with the force of gravitational attraction between them.

The electrostatic force F_e is

$$F_e = k\left(\frac{q^2}{r^2}\right),$$

and the gravitational force F_g is

$$F_g = G\left(\frac{m^2}{r^2}\right).$$

The ratio of the electrostatic to the gravitational force is

$$\frac{F_e}{F_g} = \frac{k}{G}\frac{q^2}{m^2} = 3.1 \times 10^{35}.$$

The gravitational force is evidently negligible compared with the electrostatic force.

* It will be shown in a later chapter that the coulomb is defined as the charge flowing past a point of a circuit in per second (1 s) when there is a current of one ampere (1 A) in the circuit. Hence the system is also referred to as the mksa system.

Example 2 The Bohr model of the hydrogen atom consists of a single electron of charge $-e$ revolving in a circular orbit about a single proton of charge $+e$. The electrostatic force of attraction between electron and proton provides the centripetal force that retains the electron in its orbit. Hence if v is the orbital velocity,

$$k\left(\frac{e^2}{r^2}\right) = m\left(\frac{v^2}{r}\right).$$

In Bohr's theory, the electron may revolve only in some one of a number of specified orbits. The orbit of smallest radius is that for which the angular momentum L of the electron is $h/2\pi$, where h is a universal constant called *Planck's constant*, equal to 6.626×10^{-34} J s. Then,

$$L = mvr = \frac{h}{2\pi}. \qquad (24\text{–}2)$$

When v is eliminated between the preceding equations, we find

$$r = \frac{h^2}{4\pi^2 kme^2};$$

and when numerical values are inserted, we find, for the radius of the *first Bohr orbit,*

$$r = 5.29 \times 10^{-11} \text{ m}$$
$$= 0.529 \times 10^{-8} \text{ cm}.$$

Thus result corresponds reasonably well with other estimates of the "size" of a hydrogen atom obtained from deviations from ideal gas behavior, the density of hydrogen in the liquid and solid states, and other observations.

Example 3 In Fig. 24–5, two equal positive charges $q = 2.0 \times 10^{-6}$ C interact with a third charge $Q = 4.0 \times 10^{-6}$ C. Find the magnitude and direction of the total force on Q.

The key word is *total*; we must compute the force each charge exerts on Q, and then obtain the *vector sum* of the forces. This is most easily accomplished using components. The figure shows the force on Q due to the upper charge q. From Coulomb's law,

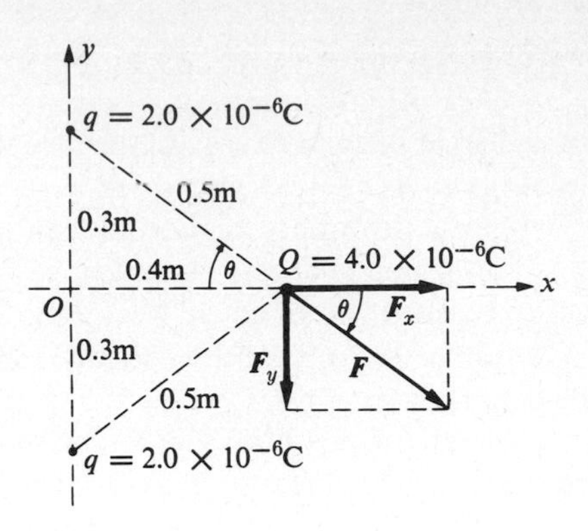

Fig. 24–5 ***F** is the force on Q due to the upper charge q.*

$$F = (9.0 \times 10^9 \text{ N C}^{-2}\text{ m}^{-2})$$
$$\times \frac{(4.0 \times 10^{-6}\text{ C})(2.0 \times 10^{-6}\text{ C})}{(0.5\text{ m})^2}$$
$$= 0.29 \text{ N}.$$

The components of this force are given by:

$$F_x = F\cos\theta = (0.29\text{ N})\left(\frac{0.4\text{ m}}{0.5\text{ m}}\right) = 0.23\text{ N},$$

$$F_y = -F\sin\theta = (0.29\text{ N})\left(\frac{0.3\text{ m}}{0.5\text{ m}}\right) = 0.17\text{ N}.$$

The lower charge q exerts a force of the same magnitude, but in a different direction. From symmetry we see that its x-component is the same as that due to the upper charge, but its y-component is opposite. Hence,

$$\sum F_x = 2(0.23\text{ N})$$
$$= 0.46\text{ N},$$
$$\sum F_y = 0.$$

The total force on Q is horizontal, with magnitude 0.46 N. How would this solution differ if the lower charge were *negative*?

24–7 ELECTRICAL INTERACTIONS

Because matter is made up of charged particles, it is not surprising that electrical interactions play a central and dominant role in all aspects of the structure of matter. The forces that hold atoms together in a molecule or in a solid crystal lattice, the adhesive force of glue, the forces associated with surface tension—all these are basically electrical in nature, arising from the electrical forces between the charged particles making up the interacting atoms. A complete description of the detailed behavior of these forces requires new *mechanical* principles and the introduction of quantum-mechanical concepts, to be discussed in Chapter 44. Nevertheless, Coulomb's law and the additional effects resulting from relative motion of the charges still describe the basic electrical interactions involved.

Electrical interactions alone are *not* sufficient to understand the structure of the *nuclei* of atoms, however. A nucleus is made up of protons, which repel each other, and neutrons, which have no electrical charge; in order for nuclei to be stable there must be additional forces, attractive in nature, to hold them together despite the electrical repulsion. This new kind of interaction, not seen outside the nucleus, is called the *nuclear force*; many phenomena associated with the stability or instability of nuclei pivot around the competition between the repulsive electrical forces and the attractive nuclear forces. These matters will be discussed in greater detail in Chapter 46.

PROBLEMS

24–1 How many excess electrons must be placed on each of two small spheres spaced 3 cm apart if the force of repulsion between the spheres is to be 10^{-19} N?

24–2 Each of two small spheres is positively charged, the combined charge totaling 4×10^{-8} C. What is the charge on each sphere if they are repelled with a force of 27×10^{-5} N when placed 0.1 m apart?

24–3 6.02×10^{23} atoms of monatomic hydrogen have a mass of one gram. How far would the electron of a hydrogen atom have to be removed from the nucleus for the force of attraction to equal the weight of the atom?

24–4 What is the total positive charge, in coulombs, of all the protons in 1 mol of hydrogen atoms?

24–5 If all the positive charges in a mole of hydrogen atoms were lumped into a single charge, and all the negative charges into a single charge, what force would the two lumped charges exert on each other at a distance of (a) 1 m; (b) 10^7 m (comparable to the diameter of the earth)?

24–6 An alpha particle consists of two protons and two neutrons bound together. What is the repulsive force between two alpha particles at a distance of 10^{-15} m, comparable to the sizes of nuclei?

24–7 Two copper spheres, each having mass 1 kg, are separated by 1 m.

a) How many electrons does each sphere contain?
b) How many electrons would have to be removed from one sphere and added to the other to cause an attractive force of 10^4 N (roughly one ton)?
c) What fraction of all the electrons on a sphere does this represent?

24–8 Point charges of 2×10^{-9} C are situated at each of three corners of a square whose side is 0.20 m. What would be the magnitude and direction of the resultant force on a point charge of -1×10^{-9} C if it were placed (a) at the center of the square? (b) at the vacant corner of the square?

24–9 Two charges of $+10^{-9}$ C each are 8 cm apart in air. Find the magnitude and direction of the force exerted by these charges on a third charge of $+5 \times 10^{-11}$ C that is 5 cm distant from each of the first two charges.

24–10 Two positive point charges, each of magnitude q, are located on the y-axis at points $y = +a$ and $y = -a$. A third positive charge of the same magnitude is located at some point on the x-axis.

a) What is the force exerted on the third charge when it is at the origin?
b) What is the magnitude and direction of the force on the third charge when its coordinate is x?
c) Sketch a graph of the force on the third charge as a function of x, for values of x between $+4a$ and $-4a$. Plot forces to the right upward, forces to the left, downward.
d) For what value of x is the force a maximum?

24–11 A negative point charge of magnitude q is located on the y-axis at the point $y = +a$, and a positive charge of the same magnitude is located at $y = -a$. A third positive charge of the same magnitude is located at some point on the x-axis.

a) What is the magnitude and direction of the force exerted on the third charge when it is at the origin?

b) What is the force on the third charge when its coordinate is x?

c) Sketch a graph of the force on the third charge as a function of x, for values of x between $+4a$ and $-4a$.

24–12 Two small balls, each of mass 10 g, are attached to silk threads 1 m long and hung from a common point. When the balls are given equal quantities of negative charge, each thread makes an angle of 4° with the vertical.

a) Draw a diagram showing all of the forces on each ball.

b) Find the magnitude of the charge on each ball.

24–13 A certain metal sphere of volume 1 cm^3 has a mass of 7.5 g and contains 8.2×10^{22} electrons.

a) How many electrons must be removed from each of two such spheres so that the electrostatic force of repulsion between them just balances the force of gravitational attraction? Assume the distance between the spheres is great enough so that the charges on them can be treated as point charges.

b) Express the number of electrons removed as a fraction of the total number of free electrons.

24–14 In the Bohr model of atomic hydrogen, an electron of mass 9.11×10^{-31} kg revolves about a proton in a circular orbit of radius 5.29×10^{-11} m. The proton has a positive charge equal in magnitude to the negative charge on the electron and its mass is 1.67×10^{-27} kg.

a) What is the radial acceleration of the electron?

b) What is its velocity?

c) What is its angular velocity?

24–15 One gram of monatomic hydrogen contains 6.02×10^{23} atoms, each consisting of an electron with charge -1.60×10^{-19} C and a proton with charge $+1.60 \times 10^{-19}$ C.

a) Suppose all these electrons could be located at the north pole of the earth and all the protons at the south pole. What would be the total force of attraction exerted on each group of charges by the other? The diameter of the earth is 12,800 km.

b) What would be the magnitude and direction of the force exerted by the charges in part (a) on a third positive charge, equal in magnitude to the total charge at one of the poles and located at a point on the equator? Draw a diagram.

24–16 The dimensions of atomic nuclei are of the order of 10^{-14} m. Suppose that two α-particles are separated by this distance.

a) What is the force exerted on each α-particle by the other?

b) What is the acceleration of each? (See Example 1 in Section 24–6 for numerical data.)

24–17 The pair of equal and opposite charges in Problem 24–11 is called an *electric dipole.*

a) Show that when the x-coordinate of the third charge in Problem 24–11 is large compared with the distance a, the force on it is inversely proportional to the *cube* of its distance from the midpoint of the dipole.

b) Show that if the third charge is located on the y-axis, at a y-coordinate large compared with the distance a, the force on it is also inversely proportional to the cube of its distance from the midpoint of the dipole.

24–18 Two equal positive point charges are a distance $2a$ apart. Midway between them and normal to the line joining them is a plane. The locus of points where the force on a point charge placed in the plane is a maximum is, by symmetry, a circle. Calculate the radius of this circle.

24–19 A small ball having a positive charge q_1 hangs by an insulating thread. A second ball with a negative charge $q_2 = -q_1$ is kept at a horizontal distance a to the right of the first. (The distance a is large compared with the diameter of the ball.) (a) Show in a diagram all of the forces on the hanging ball in its final equilibrium position. (b) You are given a third ball having a positive charge $q_3 = 2q_1$. Find at least two points at which this ball can be placed so that the first ball will hang vertically.

Chapter 25

The Electric Field. Gauss's Law

25–1 THE ELECTRIC FIELD

Figure 25–1(a) represents two positively charged bodies A and B, between which there is an electrical force of repulsion, $\boldsymbol{F}$. Like the force of gravitational attraction, this force is of the "action-at-a-distance" type, making itself felt without the presence of any material connection between A and B. No one knows "why" this is possible—it is an experimental fact that charged bodies behave in this way. It is useful, however, to think of each of the charged bodies as modifying the state of affairs in the space around it, so that this state is different in some way from whatever it may be when the charged bodies are not present. Thus, let body B be removed. Point P (see Fig. 25–1(b)) is the point of space at which B was formerly located. The charged body A is said to produce or set up an *electric field* at the point P, and if the charged body B is now placed at P, one considers that a force is exerted on B *by the field*, rather than by body A directly. Since a force would be experienced by body B at *any* point of space around body A, the electric field exists everywhere in this space.

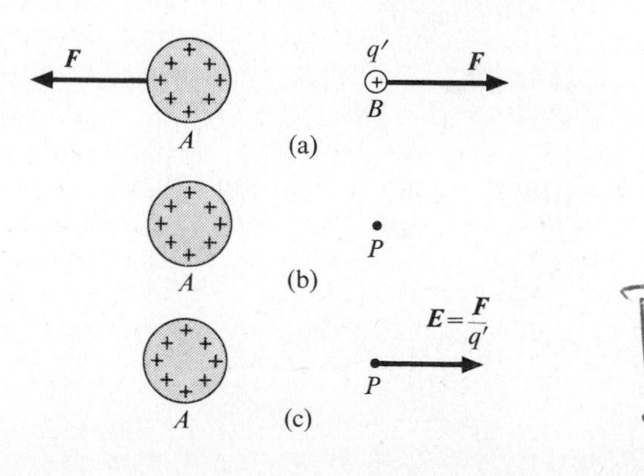

Fig. 25–1 *The space around a charged body is an electric field.*

One can equally well consider that body B sets up a field, and that the force on body A is exerted by the field of B.

The experimental test for the existence of an electric field at any point is simply to place a charged body, which will be called a *test charge*, at a point. If a force (of electrical origin) is exerted on the test charge, then an electric field exists at the point.

An electric field is said to exist at a point if a force of electrical origin is exerted on a charged body placed at the point.

Since force is a vector quantity, the electric field is a *vector field* whose properties are determined when both the magnitude and the direction of an electric force are specified. We define the *electric field* $\mathbf{E}$ at a point as the quotient obtained when the force $\boldsymbol{F}$, acting on a positive test charge, is divided by the magnitude q' of the test charge. Thus,

$$\boldsymbol{E} = \frac{\boldsymbol{F}}{q'}, \tag{25–1}$$

and the direction of $\boldsymbol{E}$ is the direction of $\boldsymbol{F}$. It follows that

$$\boldsymbol{F} = q'\boldsymbol{E}.$$

The force on a *negative* charge, such as an electron, is *opposite* to the direction of the electric field.

The electric field is sometimes called *electric intensity* or *electric field intensity*. In the mksc system, where the unit of force is 1 N and the unit of charge is 1 C, the unit of electric field is 1 newton per coulomb (1 N C^{-1}). Electric field may also be expressed in other units, to be defined later.

One problem with our definition of electric field is that in Fig. 25–1 the force exerted by the test charge q' may change the charge distribution A, especially if the body is a conductor on which charge is free to move, so that the electric field around A when q' is present is not the same as when it is absent. However, when q' is very small, the redistribution of charge on body A is also very small; thus the difficulty can be avoided by refining the definition of electric field to be *the limiting value of the force per unit charge on a test charge q' at the point, as the charge q' approaches zero*:

$$\boldsymbol{E} = \lim_{q' \to 0} \frac{\boldsymbol{F}}{q'}.$$

If an electric field exists within a *conductor*, a force is exerted on every charge in the conductor. The motion of the free charges brought about by this force is called a *current*. Conversely, if there is *no* current in a conductor, and hence no motion of its free charges, the electric field in the conductor must be zero.

In most instances, the magnitude and direction of an electric field vary from point to point. If the magnitude and direction are constant throughout a certain region, the field is said to be *uniform* in this region.

Example 1 When the terminals of a 100-V battery are connected to two large parallel plates 1 cm apart, the field in the region between the plates is very nearly uniform and the electric intensity E is 10^4 N C^{-1}. Suppose we have a field of this intensity whose direction is vertically upward. Compute the force on an electron in this field and compare with the weight of the electron.

$$\text{Electron charge } e = 1.60 \times 10^{-19}\ \text{C},$$

$$\text{Electron mass } m = 9.1 \times 10^{-31}\ \text{kg}.$$

$$F_{\text{elec}} = eE = (1.60 \times 10^{-19}\ \text{C})(10^4\ \text{N C}^{-1})$$
$$= 1.60 \times 10^{-15}\ \text{N}.$$

$$F_{\text{grav}} = mg = (9.1 \times 10^{-31}\ \text{kg})(9.8\ \text{N kg}^{-1})$$
$$= 8.9 \times 10^{-30}\ \text{N}.$$

The ratio of the electrical to the gravitational force is therefore

$$\frac{1.60 \times 10^{-15}\ \text{N}}{8.9 \times 10^{-30}\ \text{N}} = 1.8 \times 10^{14}.$$

It will be seen that the gravitational force is negligible.

Example 2 If released from rest, what velocity will the electron of Example 1 acquire while traveling 1 cm? What will then be its kinetic energy? How long a time is required?

The force is constant, so the electron moves with a constant acceleration of

$$a = \frac{F}{m} = \frac{eE}{m} = \frac{1.60 \times 10^{-15}\ \text{N}}{9.1 \times 10^{-31}\ \text{kg}}$$
$$= 1.8 \times 10^{15}\ \text{m s}^{-2}.$$

Its velocity after traveling 1 cm, or 10^{-2} m, is

$$v = \sqrt{2ax} = 6.0 \times 10^6\ \text{m s}^{-1}.$$

Its kinetic energy is

$$\tfrac{1}{2}mv^2 = 1.6 \times 10^{-17}\ \text{J}.$$

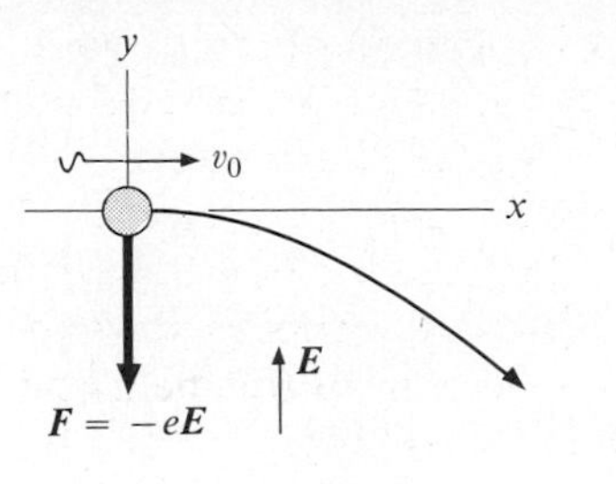

Fig. 25–2 Trajectory of an electron in an electric field.

The time is

$$t = \frac{v}{a} = 3.3 \times 10^{-9}\ \text{s}.$$

Example 3 If the electron of Example 1 is projected into the field with a horizontal velocity, find the equation of its trajectory (Fig. 25–2).

The direction of the field is upward in Fig. 25–2, so the force on the electron is downward. The initial velocity is along the positive x-axis. The x-acceleration is zero, the y-acceleration is $-(eE/m)$. Hence, after a time t,

$$x = v_0 t,$$

$$y = -\frac{1}{2}\left(\frac{eE}{m}\right)t^2.$$

Elimination of t gives

$$y = -\left(\frac{eE}{2mv_0^2}\right)x^2,$$

which is the equation of a parabola. The motion is the same as that of a body projected horizontally in the earth's gravitational field. The deflection of electrons by an electric field is used to control the direction of an electron stream in many electronic devices, such as the cathode-ray oscilloscope.

Example 4 Figure 25–3 represents two point charges of equal magnitude q but of opposite sign, separated by a distance ℓ. Such a pair of charges is called an *electric dipole.* The dipole is in a *uniform* electric field of intensity $\boldsymbol{E}$, whose direction makes an

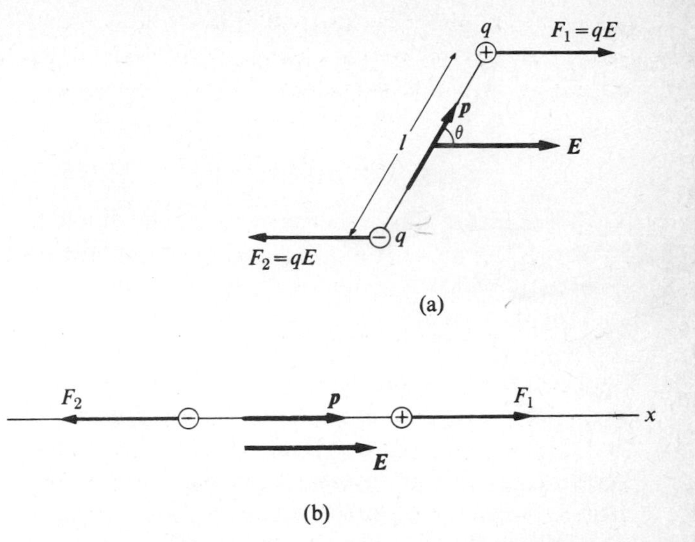

Fig. 25–3 (a) The torque on the dipole is $\Gamma = pE \sin\theta$. (b) The dipole is in equilibrium in a uniform field when **p** *and* **E** *are parallel. If the field is not uniform, the net force on the dipole equals $p(dE/dx)$.*

angle θ with the dipole axis. A force $\boldsymbol{F}_1$, of magnitude qE, in the direction of the field, is exerted on the positive charge, and a force $\boldsymbol{F}_2$, of the same magnitude but in the opposite direction, is exerted on the negative charge. The resultant *force* on the dipole is zero, but since the two forces do not have the same line of action, they constitute a *couple* (see Section 3–4). The moment of the couple is

$$\Gamma = (qE)(\ell \sin\theta),$$

since $\ell \sin\theta$ is the perpendicular distance between the action lines of the forces.

The product $q\ell$ of the charge q and the distance ℓ is called the *electric moment* or the *dipole moment*, and is represented by p:

$$p = q\ell.$$

The torque exerted by the couple is therefore

$$\Gamma = pE \sin\theta. \qquad (25\text{–}2)$$

The *vector dipole moment* of the dipole, **p**, is defined as a vector of magnitude p lying along the dipole axis and pointing from the negative toward the positive charge. The *vector torque* on the dipole is therefore the vector product or cross product of the vectors $\boldsymbol{p}$ and $\boldsymbol{E}$:

$$\boldsymbol{\Gamma} = \boldsymbol{p} \times \boldsymbol{E}. \tag{25–3}$$

The effect of this torque is to rotate the dipole to a position in which the dipole moment $\boldsymbol{p}$ is parallel to the electric vector $\boldsymbol{E}$, as in Fig. 25–3(b). If the field is *uniform*, the dipole is in equilibrium in this position.

Suppose, however, that the field at each charge has the direction of the vector $\boldsymbol{E}$ but that it is *not* uniform and its magnitude is greater at the position of the + charge than at the position of the − charge. The force $\boldsymbol{F}_1$ is then greater than the force $\boldsymbol{F}_2$ and there is a net force on the dipole toward the right, urging it toward a region of *stronger* field.

Let the x-axis be taken in the direction of the field, and let dE/dx be the rate at which E increases with x. The quantity dE/dx is called the *gradient* of the field in the x-direction. Then if E is the magnitude of the field at the negative charge, its magnitude at the positive charge is $E + \ell(dE/dx)$. The resultant force on the dipole is then

$$F = F_1 - F_2 = q\left[E + \ell\frac{dE}{dx}\right] - qE$$

$$= q\ell\frac{dE}{dx} = p\frac{dE}{dx},$$

and the force equals the product of the dipole moment and the gradient of the field.

25–2 CALCULATION OF ELECTRIC FIELD

The preceding section has described an experimental way to measure the electric field at a point. The method consists of placing a small test charge at the point, measuring the force on it, and taking the ratio of the force to the charge. The electric intensity at a point may also be computed from Coulomb's law if the magnitudes and positions of all charges contributing to the field are known. Thus, to find the magnitude of the electric intensity at a point of space P, at a distance r from a point charge q, imagine a test charge q' to be placed at P. The force on the test charge, by Coulomb's law, is

$$F = k\left(\frac{qq'}{r^2}\right),$$

and hence the electric field at P is

$$E = \frac{F}{q'} = k\left(\frac{q}{r^2}\right).$$

The direction of the field is away from the charge q if the latter is positive, toward q if it is negative.

The magnitude and direction of $\boldsymbol{E}$ can both be expressed by a single *vector* equation. Let $\boldsymbol{r}$ be the vector from the charge q to the point P, and $\hat{\boldsymbol{r}}$ a vector of unit magnitude (or a *unit vector*) in the direction of $\boldsymbol{r}$, as in Fig. 25–4(a). Then

$$\boldsymbol{E} = k\frac{q\hat{\boldsymbol{r}}}{r^2}. \tag{25–4}$$

Since $\hat{\boldsymbol{r}}$ is of unit magnitude, the *magnitude* of $\boldsymbol{E}$ is kq/r^2. If q is positive, the direction of $\boldsymbol{E}$ is the *same* as that of the vector $\hat{\boldsymbol{r}}$ (away from q), and if q is negative, it is opposite to $\hat{\boldsymbol{r}}$ (toward q).

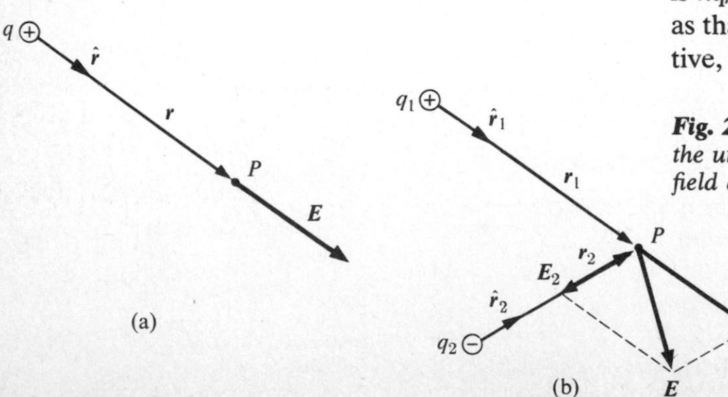

Fig. 25–4 *(a) The electric field* $\boldsymbol{E}$ *is in the same direction as the unit vector* $\hat{\boldsymbol{r}}$ *when q is positive. (b) The resultant electric field at point P is the vector sum of* $\boldsymbol{E}_1$ *and* $\boldsymbol{E}_2$.

If a number of point charges q_1, q_2, etc., are at distances r_1, r_2, etc., from a given point P, as in Fig. 25–4(b), each exerts a force on a test charge q' placed at the point, and the resultant force on the test charge is the vector sum of these forces. The resultant electric intensity is the vector sum of the individual electric intensities, and

$$\boldsymbol{E} = \boldsymbol{E}_1 + \boldsymbol{E}_2 + \cdots = k \sum \frac{q\hat{\boldsymbol{r}}}{r^2}. \tag{25–5}$$

Because each term to be summed is a vector, the sum is a vector sum.

In practice, electric fields are usually set up by charges distributed over the surfaces of conductors of finite size, rather than by point charges. The electric intensity is then calculated by imagining the charge on each conductor to be subdivided into small elements Δq. Not all of the charge in each element will be at the same distance from the point P, but if the elements are small compared with the distance to the point, and r represents the distance from any point within the element to the point P, then approximately

$$\boldsymbol{E} \approx k \sum \frac{\Delta q \hat{\boldsymbol{r}}}{r^2}.$$

The finer the subdivision, the better the approximation, and in the limit as $\Delta q \to 0$,

$$\boldsymbol{E} = k \lim_{\Delta q \to 0} \sum \frac{\Delta q \hat{\boldsymbol{r}}}{r^2}.$$

The limit of the vector sum, however, is the *vector integral*

$$\boldsymbol{E} = k \int \frac{\hat{\boldsymbol{r}}\, dq}{r^2}. \tag{25–6}$$

The limits of integration must be assigned so as to include all charges contributing to the field. Like any vector equation, Eq. (25–6) implies three scalar equations, one for each component of the vectors $\boldsymbol{E}$ and $\hat{\boldsymbol{r}}$. To evaluate the vector integral, we evaluate each of the three scalar integrals.

Example 1 Point charges q_1 and q_2 of $+12 \times 10^{-9}$C and -12×10^{-9}C, respectively, are placed 0.1 m apart, as in Fig. 25–5. Compute the electric fields due to these charges at points a, b, and c.

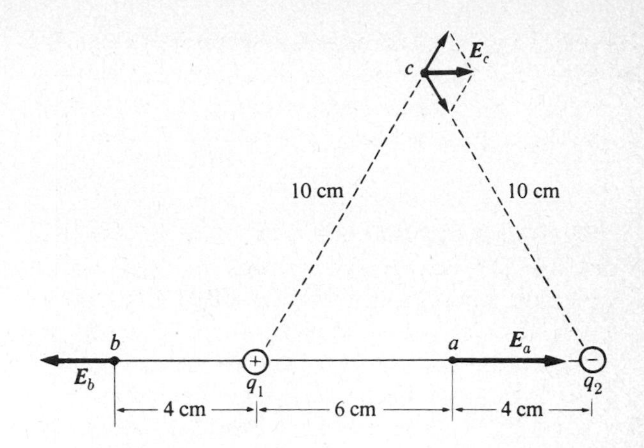

Fig. 25–5 *Electric intensity at three points, a, b, and c, in the field set up by charges q_1 and q_2.*

At point a, the vector due to the positive charge q_1, is directed toward the right, and its magnitude is

$$\begin{aligned} E_1 &= (9 \times 10^9 \text{ N m}^2 \text{ C}^{-2}) \frac{(12 \times 10^{-9}\text{C})}{(0.06 \text{ m})^2} \\ &= 3.00 \times 10^4 \text{ N C}^{-1}. \end{aligned}$$

The vector due to the negative charge q_2 is also directed toward the right; its magnitude is

$$\begin{aligned} E_2 &= (9 \times 10^9 \text{ N m}^2 \text{ C}^{-2}) \frac{(12 \times 10^{-9}\text{ C})}{(0.04 \text{ m})^2} \\ &= 6.75 \times 10^4 \text{ N C}^{-1}. \end{aligned}$$

Hence, at point a,

$$\begin{aligned} E_a &= (3.00 + 6.75) \times 10^4 \text{ N C}^{-1} \\ &= 9.75 \times 10^4 \text{ N C}^{-1}, \qquad \text{toward the right.} \end{aligned}$$

At point b, the vector due to q_1 is directed toward the left,

$$\begin{aligned} E_1 &= (9 \times 10^9 \text{ N m}^2 \text{ C}^{-2}) \frac{(12 \times 10^{-9}\text{ C})}{(0.04 \text{ m})^2} \\ &= 6.75 \times 10^4 \text{ N C}^{-1}. \end{aligned}$$

The vector due to q_2 is directed toward the right, with magnitude

$$E_2 = (9 \times 10^9 \text{ N m}^2 \text{ C}^{-2}) \frac{(12 \times 10^{-9} \text{ C})}{(0.14 \text{ m})^2}$$

$$= 0.55 \times 10^4 \text{ N C}^{-1}.$$

Hence, at point b

$$E_b = (6.75 - 0.55) \times 10^4 \text{ N C}^{-1}$$

$$= 6.20 \times 10^4 \text{ N C}^{-1}, \qquad \text{toward the left.}$$

At point c, the magnitude of each vector is

$$E = (9 \times 10^9 \text{ N m}^2 \text{ C}^{-2}) \frac{(12 \times 10^{-9} \text{ C})}{(0.1 \text{ m})^2}$$

$$= 1.08 \times 10^4 \text{ N C}^{-1}.$$

The directions of these vectors are shown in the figure; their resultant is easily seen to be

$$E_c = 1.08 \times 10^4 \text{ N C}^{-1}, \qquad \text{toward the right.}$$

Example 2 A ring-shaped conductor of radius a carries a total charge Q. Find the electric field at a point a distance x from the center, along the line perpendicular to the plane of the ring, through its center.

The situation is shown in Fig. 25–6. Considering a small segment Δs of the ring, we note that since the total circumference is $2\pi a$, the charge Δq on this segment is

$$\Delta q = \frac{Q\Delta s}{2\pi a}.$$

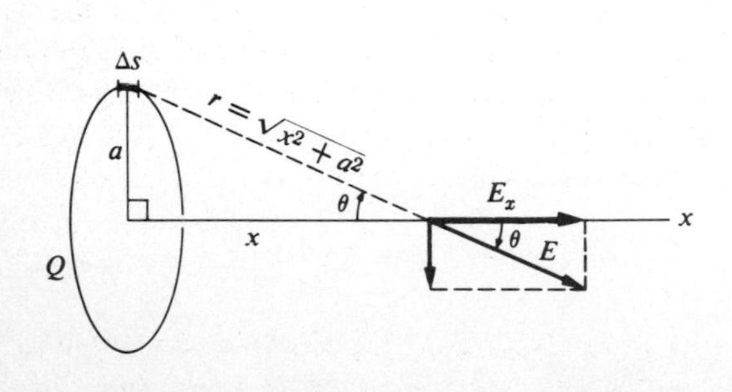

Fig. 25–6 *Electric field due to ring of charge.*

At the point P, this element produces an electric field of magnitude

$$E = k\left(\frac{\Delta q}{r^2}\right) = \frac{kQ\Delta s}{2\pi a} \cdot \frac{1}{x^2 + a^2}.$$

The component of the field along the x-axis is given by

$$E_x = E \cos\theta = \frac{kQ\,\Delta s}{2\pi a(x^2 + a^2)} \cdot \frac{x}{\sqrt{x^2 + a^2}}$$

$$= \frac{kQx\,\Delta s}{2\pi a(x^2 + a^2)^{\frac{3}{2}}}. \qquad (25\text{–}7)$$

To sum the contributions from *all* segments we need only add up their lengths, since the coefficient of Δq in Eq. (25–7) is the same for all segments. Thus, the total field component along the axis is

$$E_x = \frac{kQx}{2\pi a(x^2 + a^2)^{\frac{3}{2}}}(2\pi a)$$

$$= \frac{kQx}{(x^2 + a^2)^{\frac{3}{2}}}. \qquad (25\text{–}8)$$

In principle, this calculation should also be performed for the components perpendicular to the axis, but it is easy to see from symmetry that these add to zero.

Equation (25–8) shows that at the center of the ring ($x = 0$) the total field is zero, as might be expected; charges on opposite sides pull in opposite directions, and their fields cancel. When x is much larger than a, Eq. (25–8) becomes approximately equal to kQ/x^2, corresponding to the fact that at distances much greater than the dimensions of the ring it appears as a point charge.

Example 3 *Long charged wire.* In Fig. 25–7, a fine wire, having a positive charge per unit length, λ, lies on the y-axis. We wish to calculate the electric intensity set up by the wire at point P.

Let the wire be subdivided, in imagination, into short elements of length dy. The charge dq on an element is then $\lambda\ dy$. It will be more convenient in equations to be used later on to let r represent the perpendicular distance from P to the wire and $\mathbf{s}$ the vector from dq to P. The charge dq sets up at P a field $d\mathbf{E}$

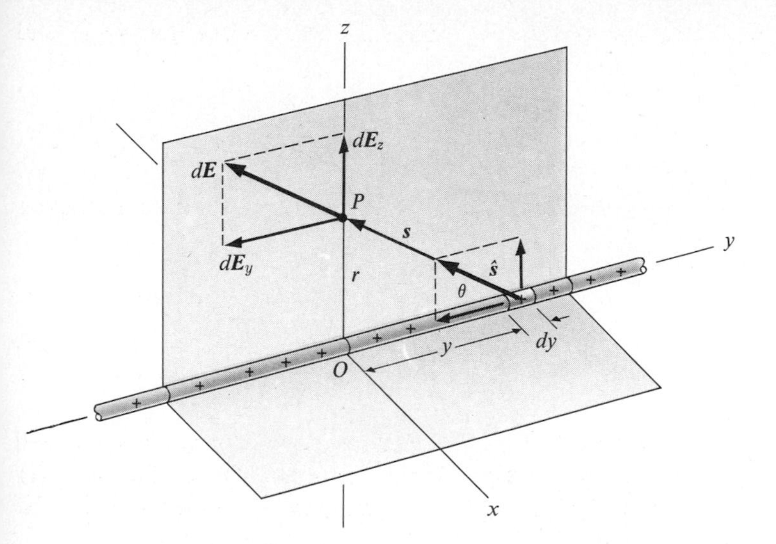

Fig. 25–7

given by

$$d\boldsymbol{E} = k\frac{\hat{\mathbf{s}}\,dq}{s^2} = k\frac{\hat{\mathbf{s}}\lambda\,dy}{s^2},$$

and the resultant intensity $\boldsymbol{E}$ is

$$\boldsymbol{E} = k\int\frac{\hat{\mathbf{s}}\lambda\,dy}{s^2}.$$

The *unit* vector $\hat{\mathbf{s}}$ lies in the yz-plane, so its x-component is zero. The magnitude of its y-component is $\cos\theta$, and that of its z-component is $\sin\theta$. The vector equation above is then equivalent to the three scalar equations

$$E_x = 0, \qquad E_y = k\lambda\int_{-\infty}^{+\infty}\frac{\cos\theta\,dy}{s^2},$$

$$E_z = k\lambda\int_{-\infty}^{+\infty}\frac{\sin\theta\,dy}{s^2}.$$

The wire is considered to be sufficiently long so that the limits of integration are from $-\infty$ to $+\infty$.

To evaluate the integrals, we must either express $\cos\theta$, $\sin\theta$, and s as functions of y, or express all quantities in terms of the same variable. Simplification results if θ is chosen as the independent variable. It will be seen from the diagram that

$$s = r\csc\theta, \qquad y = r\cot\theta.$$

Hence

$$dy = -r\csc^2\theta\,d\theta,$$

and

$$E_y = -k\frac{\lambda}{r}\int_{\pi}^{0}\cos\theta\,d\theta = -k\frac{\lambda}{r}\Big[\sin\theta\Big]_{\pi}^{0} = 0,$$

$$E_z = -k\frac{\lambda}{r}\int_{\pi}^{0}\sin\theta\,d\theta = k\frac{\lambda}{r}\Big[\cos\theta\Big]_{\pi}^{0} = 2k\frac{\lambda}{r}.$$

The y-component of $\boldsymbol{E}$ is zero, as would be expected by symmetry. (For every charge dq at a given positive y, there is an equal charge at the same negative y. The components $d\boldsymbol{E}_y$ set up by these charges are equal and opposite.) The only nonzero component of the field is therefore $\boldsymbol{E}_z$.

Had the point P been taken on the x-axis (the reader should construct his own diagram) the only nonzero component would have been $\boldsymbol{E}_x$. It should be evident, therefore, that the electric intensity at *any* point lies in a plane perpendicular to the wire, is directed radially outward, and is of magnitude

$$E = 2k\frac{\lambda}{r}. \tag{25–9}$$

The resultant field is proportional to the charge per unit length λ, and is inversely proportional to the *first* power of the radial distance r from the wire.

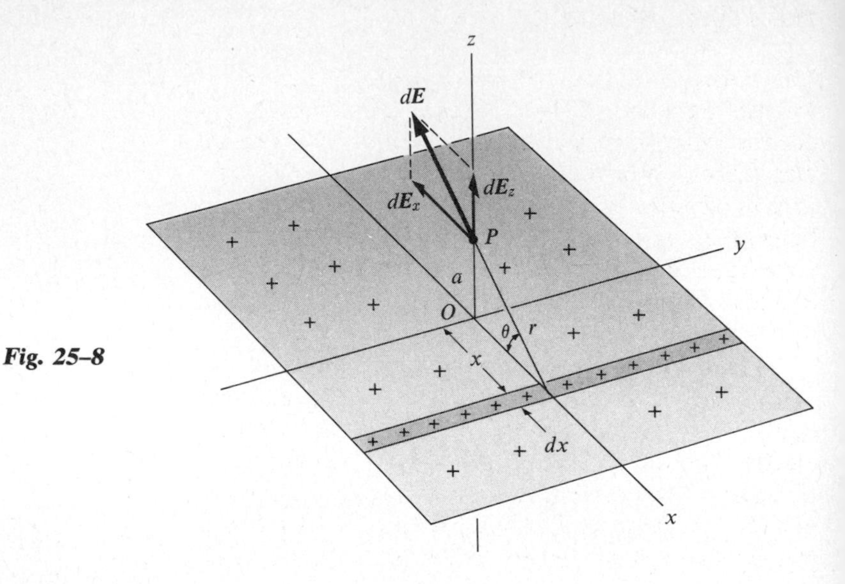

Fig. 25–8

Example 4 *Infinite plane sheet of charge.* In Fig. 25–8, positive charge is distributed uniformly over the entire xy-plane, with a charge per unit area, or *surface density of charge,* σ. We wish to calculate the electric intensity at the point P.

Let the charge be subdivided into narrow strips parallel to the y-axis and of width dx. Each strip can be considered a *line* charge, and we can use the result of the preceding example.

The area of a portion of a strip of length L is $L\,dx$, and the charge dq on the strip is

$$dq = \sigma L\,dx.$$

The charge per unit length, $d\lambda$, is therefore

$$d\lambda = \frac{dq}{L} = \sigma\,dx.$$

From Eq. (25–9), the strip sets up at point P a field $d\mathbf{E}$, lying in the xz-plane and of magnitude

$$dE = 2k\sigma\frac{dx}{r}.$$

The field can be resolved into components $d\mathbf{E}_x$ and $d\mathbf{E}_z$. By symmetry, the components $d\mathbf{E}_x$ will sum to zero when the entire sheet of charge is considered. (Be sure that you understand why.) The resultant field at P is therefore in the z-direction, perpendicular to the sheet of charge. It will be seen from the diagram that

$$dE_z = dE\sin\theta$$

and hence

$$E = \int dE_z = 2k\sigma\int_{-\infty}^{+\infty}\frac{\sin\theta\,dx}{r}.$$

But

$$\sin\theta = \frac{a}{r}, \qquad r^2 = a^2 + x^2,$$

and therefore

$$E = 2k\sigma a\int_{-\infty}^{+\infty}\frac{dx}{a^2+x^2} = 2k\sigma a\left[\frac{1}{a}\tan^{-1}\frac{x}{a}\right]_{-\infty}^{+\infty},$$

$$E = 2\pi k\sigma. \tag{25–10}$$

Note that the distance a from the plane to the point P does not appear in the final result. This means that the intensity of the field set up by an infinite plane sheet of charge is *independent of the distance from the charge.* In other words, the field is *uniform* and *normal* to the plane of charge.

The same result would have been obtained if point P, in Fig. 25–8, had been taken *below* the xy-plane. That is, a field of the same magnitude but in the opposite sense is set up on the opposite side of the plane.

25-3 LINES OF FORCE

The concept of lines of force was introduced by Michael Faraday (1791–1867) as an aid in visualizing electric (and magnetic) fields. A *line of force* (in an electric field) is *an imaginary line drawn in such a way that its direction at any point* (i.e., the direction of its tangent) *is the same as the direction of the field at that point.* (See Fig. 25–9.) Since, in general, the direction of a field varies from point to point, lines of force are usually curves.

Figure 25–10 shows some of the lines of force in two planes containing a single positive charge; two equal charges, one positive and one negative (an electric dipole); and two equal positive charges. The direction of the resultant field at every point in each diagram is along the tangent to the line of force passing through the point. Arrowheads on the lines indicate the direction in which the tangent is to be drawn.

No lines of force originate or terminate in the space surrounding a charge. Every line of force in an *electrostatic* field is a continuous line terminated by a positive charge at one end and a negative charge at the other.* While sometimes for convenience we speak of an "isolated" charge and draw its field as in Fig. 25–10(a), this simply means that the charges on which the lines terminate are at large distances from the charge under consideration. For example, if the charged body in Fig. 25–10(a) is a small sphere suspended by a thread from the laboratory ceiling, the negative charges on which its force lines terminate would be found on the walls, floor, or ceiling, or on other objects in the laboratory.

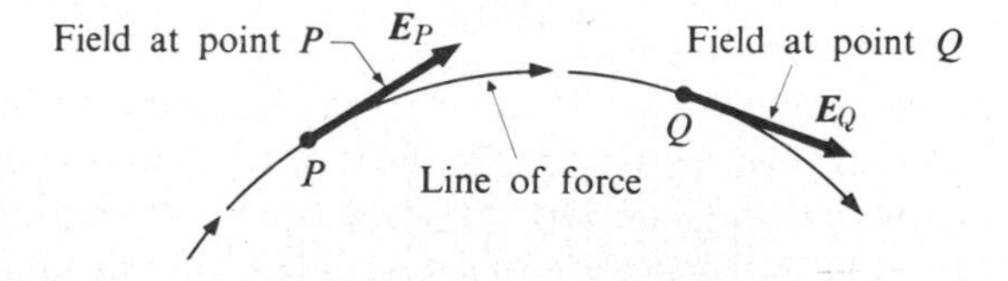

Fig. 25–9 *The direction of the electric field at any point is tangent to the line of force through that point.*

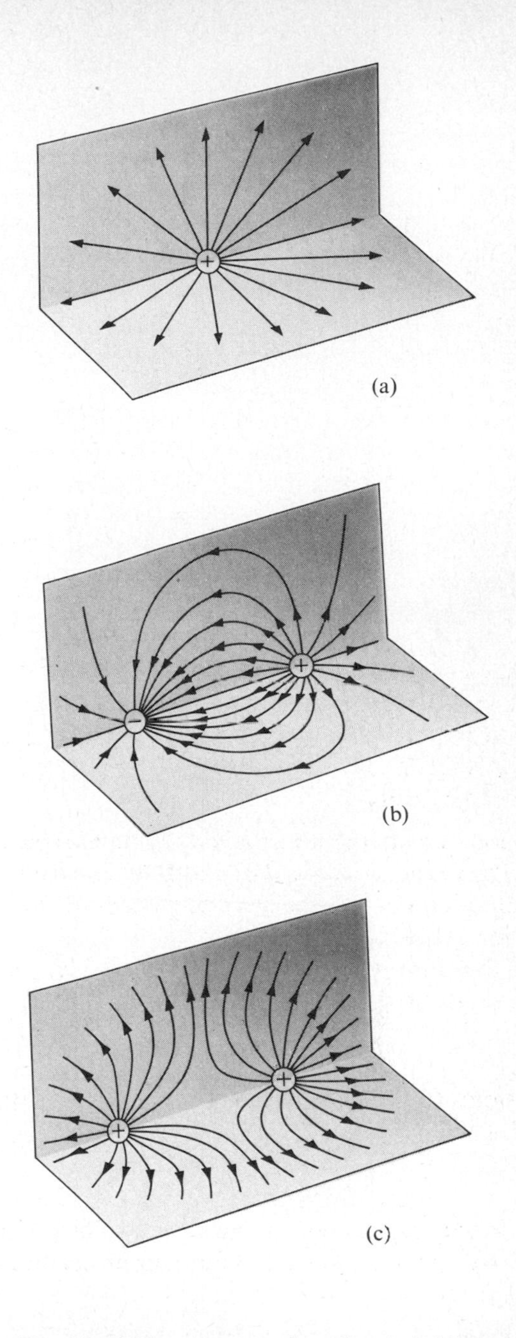

Fig. 25–10 *The mapping of an electric field with the aid of lines of force.*

At any one point, the resultant field can have but one direction. Hence only one line of force can pass through each point of the field. In other words, lines of force never intersect.

* We shall see in a later chapter that a *changing magnetic* field sets up an electric field whose lines do not terminate on electric charges, but close on themselves.

If a line of force were to be drawn through every point of an electric field, the whole of space and the entire surface of a diagram would be filled with lines, and no individual line could be distinguished. By suitably limiting the number of force lines which one draws to represent a field, the lines can be used to indicate the *magnitude* of a field as well as its *direction.* This is accomplished by spacing the lines in such a way that *the number per unit area crossing a surface at right angles to the direction of the field is at every point proportional to the electric intensity.* In a region where the intensity is large, such as that between the positive and negative charges of Fig. 25–10(b), the lines of force are closely spaced, whereas in a region where the intensity is small, such as that between the two positive charges of Fig. 25–10(c), the lines are widely separated. In a *uniform* field, the lines of force are straight, parallel, and uniformly spaced.

25–4 GAUSS'S LAW

Karl Friedrich Gauss (1777–1855) was a German scientist and mathematician who made many contributions to experimental and theoretical physics and to mathematics. The relation known as *Gauss's law* is a statement of an important property of electrostatic fields.

The content of Gauss's law is suggested by consideration of lines of force, discussed in Section 25–3. The field of a positive point charge q is represented by lines of force radiating out in all directions. Suppose we imagine this charge as surrounded by a spherical surface of radius R, with the charge at its center. The area of this imaginary surface is $4\pi R^2$, so if the total number of lines of force emanating from q is N, then the number of lines of force per unit surface area is $N/4\pi R^2$. We imagine a second sphere concentric with the first, but with radius $2R$. Its area is $4\pi(2R)^2 = 16\pi R^2$, and the number of lines of force per unit area on this sphere is $N/16\pi R^2$, one-fourth the density of lines on the first sphere. This corresponds to the fact that at distance $2R$, the field has only one-fourth the magnitude it has at distance R, and verifies our qualitative statement in Section 25–3 that the *density* of lines is proportional to the magnitude of the field.

The fact that the *total* number of lines at distance $2R$ is the same as at R can be expressed another way. The field is inversely proportional to R^2, but the area of the sphere is proportional to R^2, so the *product* of the two is independent of R. For a sphere of arbitrary radius r, the magnitude of E on the surface is

$$E = \frac{kq}{r^2},$$

the surface area is

$$A = 4\pi r^2,$$

and the product of the two is

$$EA = 4\pi kq. \tag{25–11}$$

This is independent of r and depends only on the charge q.

What is true of the entire sphere is also true of any portion of its surface. In the construction of Fig. 25–11, an area ΔA is outlined on a sphere of radius R and then projected onto the sphere of radius $2R$ by drawing lines from the center through points on the boundary of ΔA. The area projected on the larger sphere is clearly $4\ \Delta A$; thus again the product $E\ \Delta A$ is independent of the radius of the sphere.

This projection technique shows how this discussion may be extended to nonspherical surfaces. In-

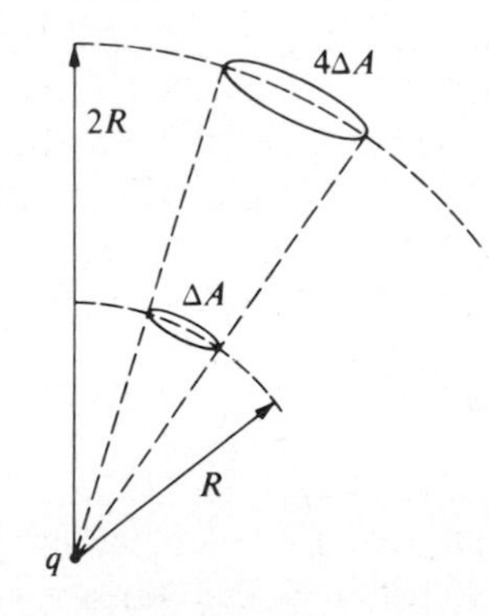

Fig. 25–11 *Projection of an element of area ΔA on a sphere of radius R, onto a sphere of radius $2R$. The projection multiplies each linear dimension by two, so the area element on the larger sphere is $4\ \Delta A$.*

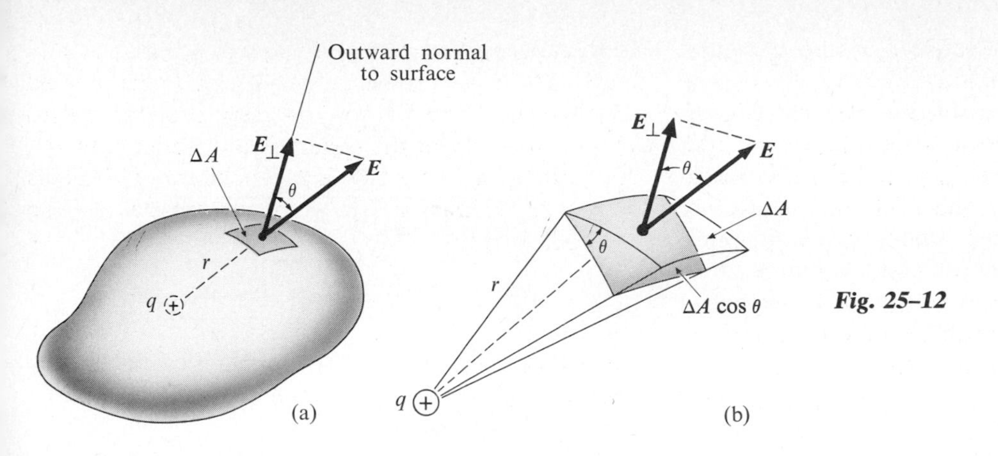

Fig. 25–12

stead of a second sphere, let us surround the sphere of radius R by a surface of irregular shape, as in Fig. 25–12(a). Consider a small areal element ΔA; we note that this area is *larger* than the corresponding element on a spherical surface at the same distance from q. If a normal to the surface makes an angle θ with a radial line from q, two sides of the area projected on the spherical surface are foreshortened by a factor $\cos\theta$, as shown in Fig. 25–12(b). Thus, the quantity corresponding to $E\,\Delta A$ for the spherical surface is $E\,\Delta A\cos\theta$ for the irregular surface.

Now we may divide the entire irregular surface into small elements ΔA, compute the quantity $E\,\Delta A\cos\theta$ for each, and sum the results. Each of these projects onto a corresponding element of area on the sphere, so summing the quantities $E\,\Delta A\cos\theta$ over the irregular surface must yield the same result as summing the quantities $E\,\Delta A$ over the sphere. But we have already performed that calculation; the result, given by Eq. (25–11), depends only on the charge q. Thus, for the irregular surface the result is

$$\sum E\,\Delta A\cos\theta = 4\pi kq, \qquad (25\text{–}12)$$

no matter what the shape of the surface, provided only that it is a *closed* surface enclosing the charge q. Correspondingly, for a closed surface enclosing *no* charge,

$$\sum E\,\Delta A\cos\theta = 0.$$

This is a mathematical statement of the fact that when a region contains no charge, any line of force that enters on one side must leave again at some other point on the boundary surface. Lines of force can begin or end inside a region of space only when there is charge in that region.

Because the field varies from point to point on the irregular surface, Eq. (25–12) is strictly true only in the limit when the area elements become very small. In this limit, the sum becomes an integral called the *surface integral* of $E\cos\theta$, written

$$\oint E\cos\theta\,dA = 4\pi kq. \qquad (25\text{–}13)$$

The circle on the integral sign reminds us that the integral is always taken over a *closed* surface enclosing the charge q.

Since $E\cos\theta$ is the component of $\boldsymbol{E}$ perpendicular to the surface at each point, we may use the notation $E_\perp = E\cos\theta$, and write

$$\boxed{\oint E_\perp\,dA = 4\pi kq.} \qquad (25\text{–}14)$$

The quantity $E_\perp\,dA = E\cos\theta\,dA$ is also called the *electric flux* through the area dA. Equations (25–13) and (25–14) state that the total electric flux out of a closed surface is proportional to the magnitude of charge enclosed.

If the point charge in Fig. 25–12 were negative, the field $\boldsymbol{E}$ would be directed radially *inward*, the angle θ would be greater than 90°, its cosine would be negative, $E_\perp$ would be negative, and the integral in Eq. (25–14) would be negative. But since q would also be negative, Eq. (25–14) still holds.

If a point charge lies *outside* a closed surface (the reader may construct his own diagram), the electric field is *outward* at some points of the surface and *inward* at others. It is not difficult to show that the positive and negative contributions to the integral over the surface exactly cancel, and the sum is zero. But the charge *inside* the closed surface is also zero, so again Eq. (25–14) is obeyed.

Although we have been concerned only with a single point charge, it is easy to generalize the above results to *any* charge distribution. The total electric field $\mathbf{E}$ at a point on the surface is the vector sum of the fields produced by the individual charges, and the quantity $E\,dA\,\cos\theta$ is therefore the sum of the contributions from these charges. Since Eq. (25–14) holds for each point charge, a corresponding relation holds for the *total* $\mathbf{E}$ field and the *total* charge enclosed by the surface. That is,

$$\boxed{\oint E_\perp\,dA = 4\pi k \sum q,} \qquad (25\text{–}15)$$

where $\mathbf{E}$ is now the total electric field and $\sum q$ represents the algebraic sum of all charges enclosed by the surface.

Equation (25–15) is the mathematical statement of *Gauss's law.* It states that when we multiply each element of area of a closed surface by the normal component of $\mathbf{E}$ at the element, and sum over the entire surface, the result is a constant times the total charge inside the surface.

The notation can be simplified in two respects. First, we define the *vector area* $d\mathbf{A}$ as a vector whose magnitude equals dA and whose direction is that of the *outward* normal at dA. The product $E_\perp\,dA = E\cos\theta\,dA$ can then be written as the *scalar product* or *dot product* of the vectors $\mathbf{E}$ and $d\mathbf{A}$:

$$E_\perp\,dA = \mathbf{E}\cdot d\mathbf{A}.$$

Second, to avoid having to write the factor 4π in Eq. (25–15), we define a new constant ϵ_0 by the equation

$$\frac{1}{\epsilon_0} = 4\pi k, \quad \epsilon_0 = \frac{1}{4\pi k}.$$

In the mks system,

$$\epsilon_0 = \frac{1}{4\pi \times 10^{-7}\,c^2} = \frac{1}{(4\pi \times 10^{-7})(2.998\times 10^8)^2};$$

$$\boxed{\epsilon_0 = 8.85\times 10^{-12}\,\mathrm{C^2\,N^{-1}\,m^{-2}}.}$$

Many texts write all the equations of electrostatics in terms of ϵ_0. For example, since $k = 1/4\pi\epsilon_0$, Coulomb's law becomes

$$F = \frac{1}{4\pi\epsilon_0}\left(\frac{qq'}{r^2}\right).$$

Denoting the total charge enclosed by $Q = \sum q$, we may write Gauss's law more compactly as

$$\boxed{\oint \mathbf{E}\cdot d\mathbf{A} = Q/\epsilon_0.} \qquad (25\text{–}16)$$

As mentioned above, the flux of $\mathbf{E}$ across a surface, as well as Gauss's law, can be interpreted graphically in terms of lines of force. If the number of lines per unit area at right angles to their direction is proportional to E, the surface integral of $E_\perp$ over a closed surface is proportional to the total number of lines crossing the surface in an outward direction, and the net charge within the surface is proportional to this number. As an example, consider the field of two equal and opposite point charges shown in Fig. 25–13. Surface A encloses the positive charge only, and 18 lines cross it in an outward direction. Surface

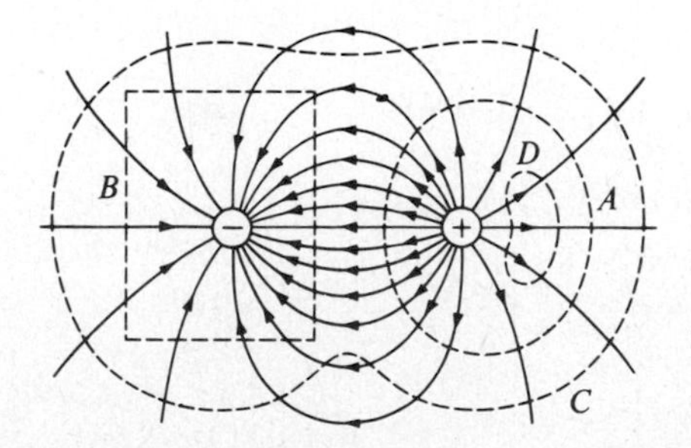

Fig. 25–13

B encloses the negative charge only, and it also is crossed by 18 lines, but in an inward direction. Surface C encloses *both* charges. It is intersected by lines at 16 points, at 8 of which the intersections are outward and at 8 of which they are inward. The *net* number of lines crossing in an outward direction is zero, and the net charge inside the surface is also zero. Surface D is intersected at 6 points, at 3 of which the intersections are outward while at the other 3 they are inward. The net number of lines crossing in an outward direction, and the enclosed charge, are both zero.

In evaluating the surface integral of $E_\perp$ over a closed surface, it is often necessary to divide the surface, in imagination, into a number of portions. The integral over the entire surface is the *sum* of the integrals over each portion. There are many cases of practical importance where symmetry considerations simplify the evaluation of the integral sufficiently that only simple algebraic manipulations are necessary. Several examples will be discussed in the next section. The following observations are also useful:

1. If $\mathbf{E}$ is at right angles to a surface of area A at all points, and has the same magnitude at all points of the surface, then $E_\perp = E =$ constant, and

$$\oint E_\perp \, dA = EA.$$

2. If $\mathbf{E}$ is *parallel* to a surface at all points, $E_\perp = 0$ and the integral is zero.
3. If $\mathbf{E} = \mathbf{0}$ at all points of a surface, the integral is zero.

25–5 APPLICATIONS OF GAUSS'S LAW

1. Location of excess charge on a conductor. It has been explained that the electric intensity $\mathbf{E}$ is zero at all points within a conductor when the charges in the conductor are at rest. (If $\mathbf{E}$ were *not* zero, the charges would move.) We may construct an imaginary surface in the interior of a conductor, such as surface A in Fig. 25–14(a). (In applications of Gauss's law, such a surface is often called a *gaussian surface.*) Because $\mathbf{E} = \mathbf{0}$ everywhere on this surface, Eq. (25–16) requires that the net charge inside the surface be zero.

If we now imagine the surface to shrink to zero like a collapsing balloon, as suggested in Fig. 25–14(a), until it essentially encloses a point, the charge at the point must be zero. Since this process can take place at *any* point in the conductor, there can be no net charge at any point within the conductor. It follows that the entire excess charge on the conductor must be located on the *outer surface* of the conductor, as shown. Gauss's law alone does not enable us to tell precisely *how* the charge is distributed over the surface. To do this, we must make use of another important property of electrostatic fields that will be developed later.

Now suppose there is a cavity in the conductor, as in Fig. 25–14(b), and there are no charges within the cavity. A gaussian surface, such as A, can still be shrunk to zero, so again there is no excess charge at any point within the material of the conductor. Surface B, however, cannot shrink to zero and still remain within the material of the conductor. (Students of topology will recognize this situation.) The

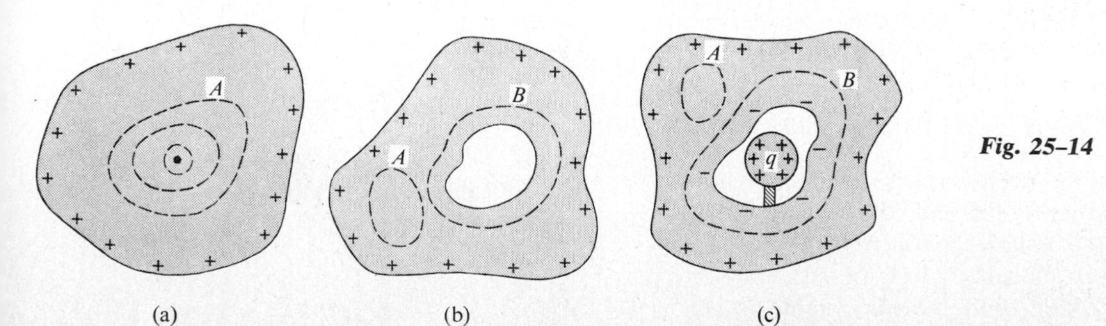

Fig. 25–14

surface encloses the smallest possible volume when it lies just outside the walls of the cavity. But the surface integral of $E_\perp$ over surface B is still zero, and the net charge within it is zero. Since any such charge must lie on the cavity walls, the *net* charge on the wall is zero. This does *not* prove, however, that the entire cavity wall is uncharged; it might be positively charged at some points and negatively charged at others. Again, Gauss's law does not provide a complete answer to the problem: we shall show later that, in fact, the entire cavity wall *is* uncharged. Again, the excess charge on the conductor is confined to its *outer* surface.

Suppose next that there is a charge q on a conductor inside the cavity but insulated from it, as in Fig. 25–14(c). Application of Gauss's law to surface B shows again that the *net* charge inside this surface is zero, so there must be a charge on the cavity wall, equal and opposite in sign to the charge q. If the outer conductor were initially *uncharged* before the charge q was inserted, and is insulated so that the total charge on it cannot change, there must be a charge on its outer surface, equal and opposite to the charge on the cavity wall and therefore equal to, and of the *same* sign, as the charge q. If the outer conductor originally had a charge q', the charge on its outer surface becomes $q + q'$.

It follows that insertion of a charge into a cavity in hollow conductor results in the appearance of an exactly equal charge on the outer surface of the hollow conductor, whether or not this conductor was originally charged.

Finally, suppose the charged conductor inside the cavity is touched to the cavity wall (or connected to the wall by a conducting wire). The excess charge on the cavity wall then neutralizes the charge q, leaving the inner conductor completely uncharged. The charge of magnitude q remains on the outer surface of the hollow conductor, so, in effect, the entire charge q has been transferred to the outer surface of the outer conductor.

This process of transferring charge from one conductor to another by *internal* contact was studied by Faraday. For a hollow conductor, he used the metal pail in which the supply of ice for the laboratory was usually kept, and the experiment is still referred to as the "Faraday ice-pail experiment."

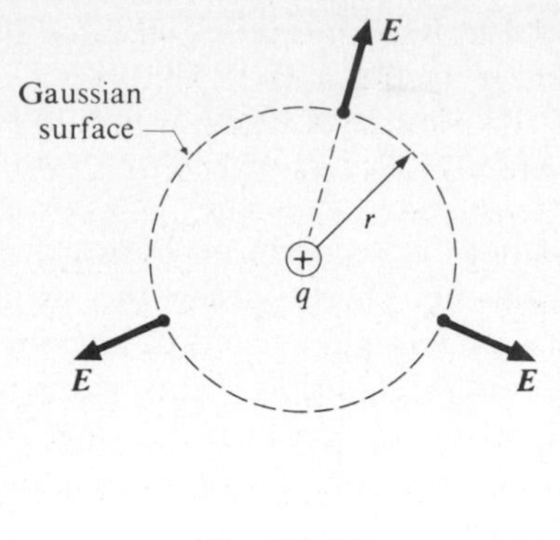

Fig. 25–15

2. Coulomb's law. We have considered Coulomb's law as the fundamental equation of electrostatics and have derived Gauss's law from it. An alternative procedure is to consider Gauss's law as a fundamental experimental relation. Coulomb's law can then be derived from Gauss's law, by using this law to obtain the expression for the electric field $\boldsymbol{E}$ due to a point charge.

Consider the electric field of a single positive point charge q, shown in Fig. 25–15. By *symmetry*, the field is everywhere radial (there is no reason why it should deviate to one side of a radial direction rather than to another) and its magnitude is the same at all points at the same distance r from the charge (any point at this distance is like any other). Hence, if we select as a gaussian surface a spherical surface of radius r, $E_\perp = E = \textit{constant}$ at all points of the surface. Then

$$\oint E_\perp \, dA = E_\perp \oint dA = EA = 4\pi r^2 E.$$

From Gauss's law

$$4\pi r^2 E = \frac{q}{\epsilon_0} \qquad \text{and} \qquad E = \frac{1}{4\pi\epsilon_0}\left(\frac{q}{r^2}\right) = k\left(\frac{q}{r^2}\right).$$

The force on a point charge q' at a distance r from the charge q is then

$$F = q'E = k\left(\frac{qq'}{r^2}\right),$$

which is Coulomb's law.

3. Field of a charged spherical conductor. Any excess charge on an isolated spherical conductor is, by symmetry, distributed *uniformly* over its outer surface. The electric field at any point can be calculated at least in principle by summing the contributions from elements of charge on the surface, but it is much simpler to use Gauss's law. It should be evident that at external points the field has the same symmetry as that of a point charge, so if we construct a gaussian surface of radius r, where r is greater than the radius R of the sphere, and if q is the total charge on the sphere,

$$4\pi r^2 E = q/\epsilon_0,$$

$$E = \frac{1}{4\pi\epsilon_0}\left(\frac{q}{r^2}\right) = k\left(\frac{q}{r^2}\right).$$

The field *outside* the sphere is, therefore, the same as though the entire charge were concentrated at a point at its center. Just outside the surface of the sphere, where $r = R$,

$$E = k\left(\frac{q}{R^2}\right),$$

and inside the sphere, if it is solid, $E = 0$.

The same argument may be used to show that the electric field inside a hollow spherical shell conductor is zero everywhere. This time the radius r of the gaussian sphere is less than R. The field, if it exists, must be spherically symmetric as before, so again $E = kq/r^2$. But this time $q = 0$, so $\mathbf{E}$ must also be zero.

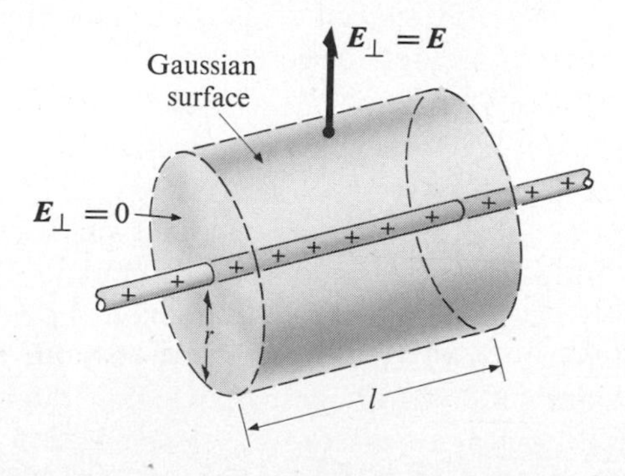

Fig. 25–16 *Cylindrical gaussian surface for calculating the electric intensity due to a long charged wire.*

Because of the relation of Gauss's law to Coulomb's law, this result holds only because the electric field obeys an inverse *square* law. If the field were inversely proportional to r^3, or to $r^{2.147}$, Gauss's law would not hold. Very precise measurements have shown that the internal field of a charged sphere is, in fact, so small that the exponent of r cannot differ from exactly 2 by more than 1 part in 10^7. There seems no reason to doubt that it is exactly 2.

It is left as a problem to find, from Gauss's law, the electric intensity in the interspace between a charged sphere and a concentric hollow sphere surrounding it.

4. Field of a line charge and of a charged cylindrical conductor. We consider next the electric field set up by a long thin uniformly charged wire. We solved this problem in Section 25–2, Example 3, by a straightforward application of Coulomb's law, requiring a somewhat involved integration. Gauss's law makes it possible to find the field by an almost trivial calculation.

If the wire is very long and we are not too near either end, then, by symmetry, the lines of force outside the wire are radial and lie in planes perpendicular to the wire. Also, the field has the same magnitude at all points at the same radial distance from the wire. This suggests that we use as a gaussian surface a *cylinder* of arbitrary radius r and arbitrary length ℓ, with its ends perpendicular to the wire, as in Fig. 25–16. If λ is the charge per unit length on the wire, the charge *within* the gaussian surface is $\lambda\ell$. Since $\mathbf{E}$ is at right angles to the wire, the component of $\mathbf{E}$ normal to the end faces is zero. Thus the end faces make no contribution to the sum in Gauss's law. At all points of the curved surface $E_\perp = E$ = constant, and since the area of this surface is $2\pi r\ell$, we have

$$\lambda\ell = (\epsilon_0 E)(2\pi r\ell), \qquad E = \frac{1}{2\pi\epsilon_0}\left(\frac{\lambda}{r}\right) = 2k\left(\frac{\lambda}{r}\right). \tag{25–17}$$

in agreement with the result obtained in Section 25–2 by a much more laborious method.

It will be noted that although the *entire* charge on the wire contributes to the field $\mathbf{E}$, only that

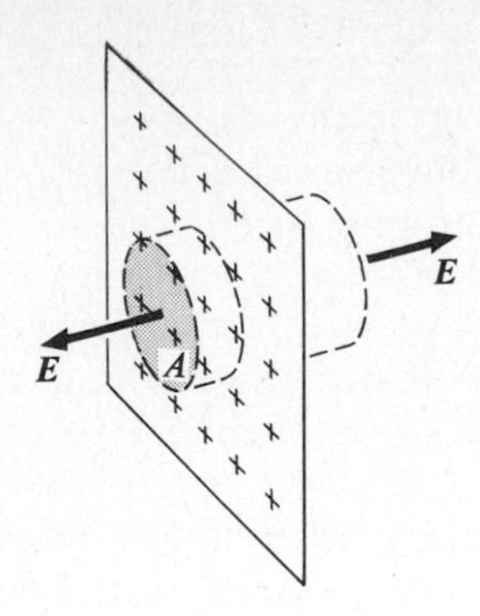

Fig. 25–17 Gaussian surface in the form of a cylinder for finding the field of an infinite plane sheet of charge.

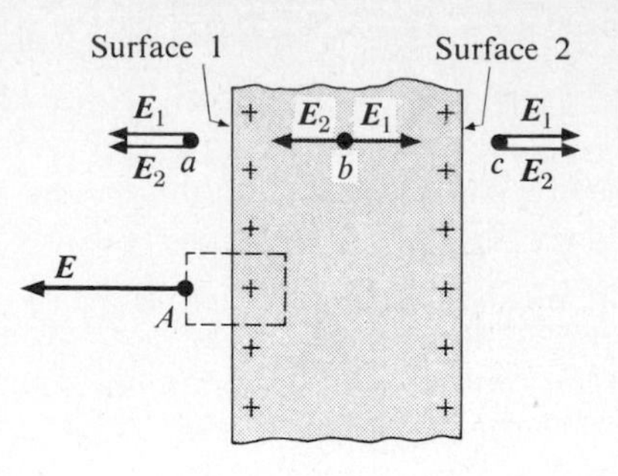

Fig. 25–18 Electric field inside and outside a charged conducting plate.

portion of the total charge lying within the gaussian surface is used when we apply Gauss's law. This feature of the law is puzzling at first; it appears as though we had somehow obtained the right answer by ignoring a part of the charge, and that the field of a *short* wire of length ℓ would be the same as that of a very long wire. The existence of the entire charge on the wire *is*, however, taken into account when we consider the *symmetry* of the problem. Suppose the wire had been a short one, of length ℓ. Then we could *not* conclude by symmetry that the field at one end of the cylinder, say, would equal that at the center, or that the lines of force would everywhere be perpendicular to the wire. So the entire charge on the wire actually *is* taken into account, but in an indirect way.

It is left as a problem (1) to show that the field outside a long charged cylinder is the same as though the charge on the cylinder were concentrated in a line along its axis, and (2) to calculate the electric intensity in the interspace between a charged cylinder and a coaxial hollow cylinder that surrounds it.

5. Field of an infinite plane sheet of charge. To solve this problem, construct the gaussian surface shown by the dotted lines in Fig. 25–17, consisting of a cylinder whose ends have an area A and whose walls are perpendicular to the sheet of charge. By symmetry, since the sheet is infinite, the electric intensity $\boldsymbol{E}$ is the same on both sides of the surface, is uniform, and is directed normally away from the sheet of charge. No lines of force cross the side walls of the cylinder, that is, the component of $\boldsymbol{E}$ normal to these walls is zero. At the ends of the cylinder the normal component of $\boldsymbol{E}$ is equal to $\boldsymbol{E}$. The surface integral of $E_\perp$, calculated over the entire surface of the cylinder, therefore reduces to $2EA$. If σ is the charge per unit area in the plane sheet, the net charge within the gaussian surface is σA. Hence

$$\sigma A = 2\epsilon_0 EA, \qquad E = \frac{\sigma}{2\epsilon_0} = 2\pi k\sigma. \tag{25–18}$$

This agrees with Eq. (25–10), obtained in Section 25–2 (Example 4) by much more laborious means.

We note that the magnitude of the field is *independent* of the distance from the sheet and does *not* decrease inversely with the square of the distance. The lines of force remain everywhere straight, parallel, and uniformly spaced. This is because the sheet was assumed infinitely large.

6. Field of an infinite charged conducting plate. When a metal plate is given a net charge, this charge distributes itself over the entire outer surface of the plate, and if the plate is of uniform thickness and is infinitely large (or if we are not too near the edges of a finite plate), the charge per unit area is uniform and is the same on both surfaces. Hence, the field of such a charged plate arises from the superposition of the fields of *two* sheets of charge, one on each surface of the plate. By symmetry, the field is perpendicular to

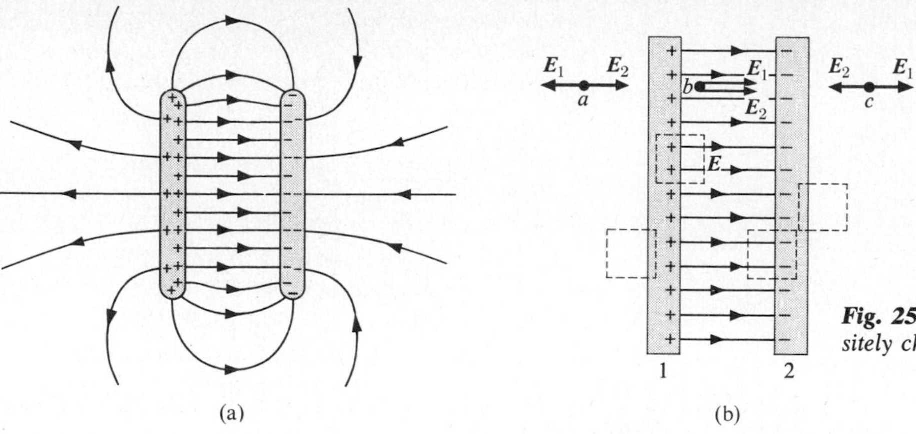

Fig. 25–19 Electric field between oppositely charged parallel plates.

the plate, directed away from it if the plate has a positive charge, and is uniform. The magnitude of the electric intensity at any point can be found from Gauss's law, or by using the results already derived for a sheet of charge.

Figure 25–18 shows a portion of a large charged conducting plate. Let σ represent the charge per unit area in the sheet of charge on *either* surface. At point *a*, outside the plate at the left, the component of electric field $\boldsymbol{E}_1$, due to the sheet of charge on the left face of the plate, is directed toward the left and its magnitude is $\sigma/2\epsilon_0$. The component $\boldsymbol{E}_2$ due to the sheet of charge on the right face of the plate is also toward the left and its magnitude is also $\sigma/2\epsilon_0$. The magnitude of the resultant intensity E is therefore

$$E = E_1 + E_2 = \frac{\sigma}{2\epsilon_0} + \frac{\sigma}{2\epsilon_0} = \frac{\sigma}{\epsilon_0}. \quad (25\text{–}19)$$

At point *b*, inside the plate, the two components of electric field are in opposite directions and their resultant is zero, as it must be in any conductor in which the charges are at rest. At point *c*, the components again add and the magnitude of the resultant is σ/ϵ_0, directed toward the right.

To derive these results directly from Gauss's law, consider the cylinder shown by dotted lines. Its end faces are of area A and one lies inside and one outside the plate. The field inside the conductor is zero. The field outside, by symmetry, is perpendicular to the plate, so the normal component of $\boldsymbol{E}$ is zero over the walls of the cylinder and is equal to $\boldsymbol{E}$ over the outside end face. Hence, from Gauss's law,

$$EA = \frac{\sigma A}{\epsilon_0}, \qquad E = \frac{\sigma}{\epsilon_0}. \quad (25\text{–}20)$$

7. Field between oppositely charged parallel plates. When two plane parallel conducting plates, having the size and spacing shown in Fig. 25–19, are given equal and opposite charges, the field between and around them is approximately as shown in Fig. 25–19(a). While most of the charge accumulates at the opposing faces of the plates and the field is essentially uniform in the space between them, there is a small quantity of charge on the outer surfaces of the plates and a certain spreading or "fringing" of the field at the edges of the plates.

As the plates are made larger and the distance between them diminished, the fringing becomes relatively less. Such an arrangement, two oppositely charged plates separated by a distance small compared with their linear dimensions, is encountered in many pieces of electrical equipment, notably in capacitors. Often the fringing is entirely negligible; even if it is not, neglecting it often provides useful approximations in cases where the work of more detailed calculations is not warranted. We shall therefore assume that the field between two oppositely charged plates is uniform, as in Fig. 25–19(b), and that the charges are distributed uniformly over the opposing surfaces.

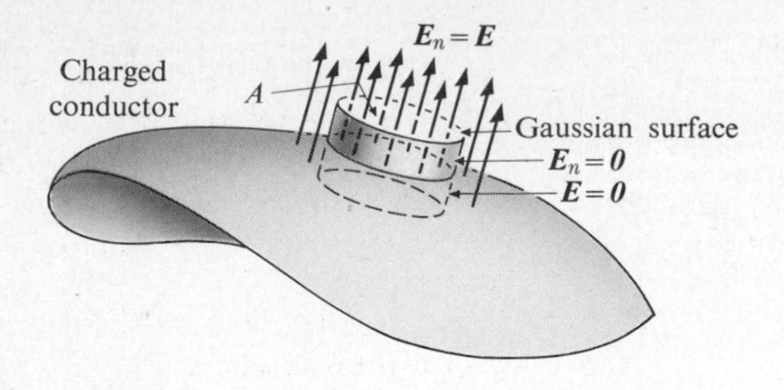

Fig. 25–20 *The field just outside a charged conductor is perpendicular to the surface and is equal to* σ/ϵ_0.

The electric field at any point can be considered as the resultant of that due to two sheets of charge of opposite sign, or it may be found from Gauss's law. Thus at points a and c in Fig. 25–19(b), the components $\mathbf{E}_1$ and $\mathbf{E}_2$ are each of magnitude $\sigma/2\epsilon_0$ but are oppositely directed, so their resultant is zero. At any point b between the plates the components are in the same direction and their resultant is σ/ϵ_0. It is left as an exercise to show that the same results follow from applying Gauss's law to the surfaces shown by dotted lines.

8. Field just outside any charged conductor. Figure 25–20 represents a portion of the surface of a charged conductor of irregular shape. In general, the surface density of charge will vary from point to point of the surface. Let σ represent the surface density at a small area A. We shall show in the next chapter that the electric field *just outside* the surface of any charged conductor is at right angles to the surface.

Let us construct a gaussian surface in the form of a small cylinder, one of whose end faces, of area A, lies within the conductor, while the other lies just outside. The charge within the gaussian surface is σA. The electric field is zero at all points within the conductor. Outside the conductor, the normal com-

Table 25–1 ELECTRIC FIELDS AROUND SIMPLE CHARGE DISTRIBUTIONS

Charge distribution responsible for the electric field	Arbitrary point in the electric field	Magnitude of the electric intensity at this point
Single point charge q	Distance r from q	$E = k\dfrac{q}{r^2}$
Several point charges, $q_1, q_2, \ldots$	Distance r_1 from q_1, r_2 from $q_2, \ldots$	$E = k\left(\dfrac{q_1}{r_1^2} + \dfrac{q_2}{r_2^2} + \cdots\right)$ (vector sum)
Dipole at origin, dipole moment $\mathbf{p}$ on x-axis	(a) Point on x-axis at large distance (b) Point on y-axis at large distance	$E = 2k\dfrac{p}{x^3}$ $E = k\dfrac{p}{y^3}$
Charge q uniformly distributed on the surface of a conducting sphere of radius R	(a) Outside, $r \geq R$ (b) Inside, $r < R$	(a) $E = k\dfrac{q}{r^2}$ (b) $E = 0$
Long cylinder of radius R, with charge per unit length λ	(a) Outside, $r \geq R$ (b) Inside, $r < R$	(a) $E = 2k\dfrac{\lambda}{r}$ (b) $E = 0$
Two oppositely charged conducting plates with charge per until area σ	Any point between plates	$E = \dfrac{\sigma}{\epsilon_0}$
Any charged conductor	Just outside the surface	$E = \dfrac{\sigma}{\epsilon_0}$

ponent of $\boldsymbol{E}$ is zero at the side walls of the cylinder (since $\boldsymbol{E}$ is normal to the conductor), while over the end face the normal component is equal to $\boldsymbol{E}$. Hence from Gauss's law,

$$EA = \frac{\sigma A}{\epsilon_0}, \qquad E = \frac{\sigma}{\epsilon_0}. \qquad (25\text{–}21)$$

This agrees with the results already obtained for spherical, cylindrical, and plane surfaces. Just outside the surface of a sphere of radius R, for example, the electric field is

$$E = k\left(\frac{q}{R^2}\right) = \frac{1}{4\pi\epsilon_0}\left(\frac{q}{R^2}\right).$$

But the surface density of charge on the sphere is $q/4\pi R^2$, so $E = \sigma/\epsilon_0$.

The field outside an infinite charged conducting plate was also shown to equal σ/ϵ_0. In this case, the field is the same at *all* distances from the plate, but in general it decreases with increasing distance from the surface.

The expressions that we have derived for the electric fields set up by a number of simple charge distributions are summarized in Table 25–1.

PROBLEMS

25–1 A small object carrying a charge of -5×10^{-9} C experiences a downward force of 20×10^{-9} N when placed at a certain point in an electric field.

a) What is the electric field at the point?

b) What would be the magnitude and direction of the force acting on an electron placed at the point?

25–2 What must be the charge on a particle of mass 2 g for it to remain stationary in the laboratory when placed in a downward-directed electric field of 500 N C^{-1}?

25–3 A uniform electric field exists in the region between two oppositely charged plane parallel plates. An electron is released from rest at the surface of the negatively charged plate and strikes the surface of the opposite plate, 2 cm distant from the first, in a time interval of 1.5×10^{-8} s.

a) Find the electric field.

b) Find the velocity of the electron when it strikes the second plate.

25–4 An electron is projected with an initial velocity $v_0 = 10^7$ m s^{-1} into the uniform field between the parallel plates in Fig. 25–20. The direction of the field is vertically downward, and the field is zero except in the space between the plates. The electron enters the field at a point midway between the plates. If the electron just misses the upper plate as it emerges from the field, find the magnitude of the electric field.

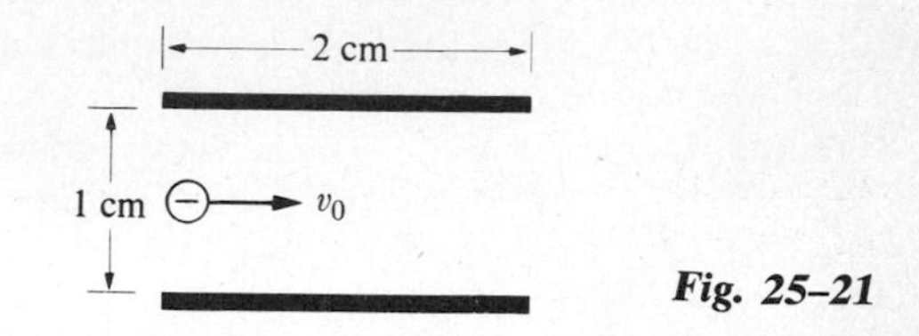

Fig. 25–21

25–5 An electron is projected into a uniform electric field of 5000 N C^{-1}. The direction of the field is vertically upward. The initial velocity of the electron is 10^7 m s^{-1}, at an angle of 30° above the horizontal.

a) Find the maximum distance the electron rises vertically above its initial elevation.

b) After what horizontal distance does the electron return to its original elevation?

c) Sketch the trajectory of the electron.

25–6 In a rectangular coordinate system a charge of 25×10^{-9} C is placed at the origin of coordinates, and a charge of -25×10^{-9} C is placed at the point $x = 6$ m, $y = 0$. What is the electric field at

a) $x = 3$ m, $y = 0$;

b) $x = 3$ m, $y = 4$ m?

25–7 A charge of 16×10^{-9} C is fixed at the origin of coordinates, a second charge of unknown magnitude is at $x = 3$ m, $y = 0$, and a third charge of 12×10^{-9} C is at $x = 6$ m, $y = 0$. What is the magnitude of the unknown charge if the resultant field at $x = 8$ m, $y = 0$ is 20.25 N C^{-1} directed to the right?

25–8 In a rectangular coordinate system, two positive point charges of 10^{-8} C each are fixed at the points $x = +0.1$ m, $y = 0$, and $x = -0.1$ m, $y = 0$. Find the magnitude and direction of the electric field at the following points:

a) the origin;

b) $x = 0.2$ m, $y = 0$;

c) $x = 0.1$ m, $y = 0.15$ m;

d) $x = 0$, $y = 0.1$ m.

25–9 Same as Problem 25–8, except that one of the point charges is positive and the other negative.

25–10

a) What is the electric field at a distance of 10^{-12} cm from a gold nucleus?

b) What is the electric field at a distance of 5.28×10^{-9} cm from a proton?

25–11 A small sphere whose mass is 0.1 g carries a charge of 3×10^{-10} C and is attached to one end of a silk fiber 5 cm long. The other end of the fiber is attached to a large vertical conducting plate which has a surface charge of 25×10^{-6} C m^{-2}. Find the angle which the fiber makes with the vertical.

25–12 How many excess electrons must be added to an isolated spherical conductor 10 cm in diameter to produce a field just outside the surface whose intensity is 1300 N C^{-1}?

25–13 What is the magnitude of an electric field in which the force of an electron is equal in magnitude to the weight of the electron?

25–14 Electric charge is distributed uniformly over a disk of radius a, with total charge Q. Find the electric field at a point on the axis of the disk, distance x from its center. *Hint*: Divide the disk into concentric rings, use the result of Example 2, Section 25–2, to find the field due to each ring, and integrate to find the total field.

25–15 In Example 3 of Section 25–2, suppose that the wire is not infinitely long but has total length $2r$, and that point P lies on the perpendicular bisector of the wire. What is the electric field at point P?

25–16 Electric charge is uniformly distributed around a semicircle of radius a, with total charge Q. What is the electric field at the center of curvature?

25–17 The electric field in the region between a pair of oppositely charged plane parallel plates, each 100 cm^2 in area, is 10^4N C^{-1}. What is the charge on each plate? Neglect edge effects.

25–18 A wire is bent into a ring of radius R and given a charge q.

a) What is the magnitude of the electric field at the center of the ring?

b) Derive the expression for the electric field at a point on a line perpendicular to the plane of the ring and passing through its center, at a distance r from the center of the ring. What is the direction of the **E**-vector at points on this line?

c) Sketch a graph of the magnitude of **E** as a function of r, from $r = 0$ to $r = 2R$.

d) For what value of r/R is the intensity a maximum?

25–19 The electric field **E** in Fig. 25–22 is everywhere parallel to the x-axis, and has the same magnitude at all points of any plane perpendicular to this axis. Its magnitude in the yz-plane equals 400 N C^{-1}.

a) What is the value of $\int E_\perp \, dA$ over surface I in the diagram?

b) What is the value of the surface integral of **E** over surface II?

c) There is a positive charge of 26.6×10^{-9}C within the volume. What is the magnitude and direction of **E** at the face opposite face I?

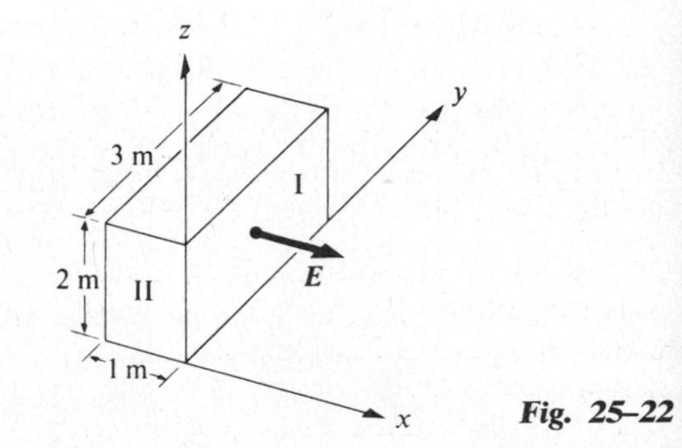

Fig. 25–22

25–20 Apply Gauss's law to the dotted gaussian surfaces in Fig. 25–19(b) to calculate the electric field between and outside the plates.

25–21 A small conducting sphere of radius r_a, mounted on an insulating handle and having a positive charge q, is inserted through a hole in the walls of a hollow conducting sphere of inner radius r_b and outer radius r_c. The hollow sphere is supported on an insulating stand and is initially uncharged, and the small sphere is placed at the center of the hollow sphere. Neglect any effect of the hole.

a) Show that the electric field at a point in the region between the spheres, at a distance r from the center, is equal to

$$E = \frac{kq}{r^2}.$$

b) What is the intensity at a point outside the hollow sphere?

c) Sketch a graph of the magnitude of **E** as a function of r, from $r = 0$ to $r = 2r_c$.

d) Represent the charge on the small sphere by four + signs. Sketch the lines of force of the system, within a spherical volume of radius $2r_c$.

e) The small sphere is moved to a point near the inner wall of the hollow sphere. Sketch the lines of force.

25–22

a) If the charge per unit length λ on the wire in Fig. 25–16 is finite, and the wire is infinitely long, the *total* charge on the wire is infinite. Explain why this infinite charge does not give rise to an infinite electrical field.

b) Draw a diagram showing an end view of an infinitely long charged wire, and the lines of force in a plane perpendicular to the wire, far from either end. Explain in terms of lines of force why the field decreases with $1/r$, although the field of a *point* charge decreases with $1/r^2$.

25–23 Prove that the electric field outside an infinitely long cylindrical conductor with a uniform surface charge is the same as if all the charge were on the axis.

25–24 A long coaxial cable consists of an inner cylindrical conductor of radius r_a and an outer coaxial cylinder of inner radius r_b and outer radius r_c. The outer cylinder is mounted on insulating supports and has no net charge. The inner cylinder has a uniform positive charge λ per unit length. Calculate the electric field

a) at any point between the cylinders, and

b) at any external point.

c) Sketch a graph of the magnitude of $\mathbf{E}$ as a function of the distance r from the axis of the cable, from $r = 0$ to $r = 2r_c$.

25–25 Suppose that positive charge is uniformly distributed throughout a spherical volume of radius R, the charge per unit volume being ρ.

a) Use Gauss's law to prove that the electric field inside the volume, at a distance r from the center, is

$$E = \frac{\rho r}{3\epsilon_0}.$$

b) What is the electric field at a point outside the spherical volume at a distance r from the center? Express your answer in terms of the total charge q within the spherical volume.

c) Compare the answers to (a) and (b) when $r = R$.

d) Sketch a graph of the magnitude of $\mathbf{E}$ as a function of r, from $r = 0$ to $r = 3R$.

25–26 Suppose that positive charge is uniformly distributed throughout a very long cylindrical volume of radius R, the charge per unit volume being ρ.

a) Derive the expression for the electric field inside the volume at a distance r from the axis of the cylinder, in terms of the charge density ρ.

b) What is the electric field at a point outside the volume, in terms of the charge per unit length λ in the cylinder?

c) Compare the answers to (a) and (b) when $r = R$.

d) Sketch a graph of the magnitude of $\mathbf{E}$ as a function of r, from $r = 0$ to $r = 3R$.

Chapter 26

Potential

26–1 ELECTRICAL POTENTIAL ENERGY

In Chapter 7 the concept of a conservative force field was discussed. When a force $\mathbf{F}$, which may vary in magnitude and direction, acts on a body that moves from point a to point b, the work done by the force is given by

$$W_{a\to b} = \int_a^b \mathbf{F}\cdot d\mathbf{s} = \int_a^b F_{\|}\, ds = \int_a^b F\cos\theta\, ds. \qquad (26\text{–}1)$$

This integral is called the *line integral* of $\mathbf{F}$ for the specified path. In general, the work done by the force depends not only on the endpoints but also on the path connecting these points. For a certain class of forces, however, the work is the same for all possible paths connecting points a and b; this property characterizes a *conservative force field.*

As we saw in Chapter 7, the work done by a conservative force can always be represented by means of a potential-energy function E_p. Specifically, if the potential-energy function has the value $(E_p)_a$ when the body is at point a and the value $(E_p)_b$ when the body is at point b, then the work $W_{a\to b}$ done on the body during the displacement from a to b is given by

$$W_{a\to b} = (E_p)_a - (E_p)_b. \qquad (26\text{–}2)$$

If *all* the forces acting on a body are conservative, then the total mechanical energy of the body, kinetic and potential, is *conserved.* As we have seen, this principle simplifies the analysis of many mechanical systems.

We shall now show that the force on a charged particle, in an electric field produced by any combination of charges at rest, is always a conservative force field. We consider first the electric field $\mathbf{E}$ produced by a single point charge q. A test charge q' experiences a force $\mathbf{F}$ given by

$$\mathbf{F} = q'\mathbf{E}. \qquad (26\text{–}3)$$

We wish to calculate the work done on q' during an arbitrary displacement from point a to point b, as shown in Fig. 26–1. From Eqs. (26–1) and (26–3), this work may be expressed as

$$W_{a\to b} = \int_a^b q'\mathbf{E}\cdot d\mathbf{s} = q'\int_a^b E\cos\theta\, ds. \qquad (26\text{–}4)$$

But from the figure, $\cos\theta\, ds = dr$. That is, the work done during a small displacement ds depends only on the change dr in the distance r between the charges. Put differently, the work depends only on the *radial component* of the displacement. In any displacement in which r does not change, no work is done because $\mathbf{E}$ and $d\mathbf{s}$ are perpendicular and hence $\mathbf{E}\cdot d\mathbf{s} = \mathbf{0}$.

In addition, Coulomb's law gives $E = kq/r^2$, so Eq. (26–4) may be written

$$W_{a\to b} = kqq'\int_{r_a}^{r_b}\frac{dr}{r^2} = kqq'\left(\frac{1}{r_a} - \frac{1}{r_b}\right). \qquad (26\text{–}5)$$

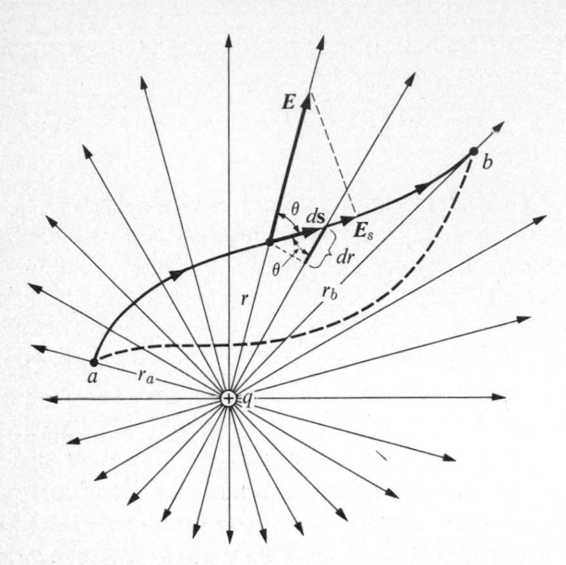

Fig. 26–1 *The work done by the* **E** *field of charge q depends only on the distances r_a and r_b.*

This result shows that the work done on q' by the **E** field depends only on the initial and final distances r_a and r_b and not on the path connecting these points. The work is the same for *any* path between these points, or between any pair of points at distances r_a and r_b from the charge q. Also, if the test charge is returned from b to a along any path, the work done by the **E** field is the negative of that done in the displacement from a to b. Thus, the total work done during a loop displacement returning to the starting point is always zero.

Comparing Eqs. (26–2) and (26–5), we see that the term kqq'/r_a is the potential energy $(E_p)_a$ when q' is at point a, at distance r_a from q, and kqq'/r_b the potential energy $(E_p)_b$ when it is at point b, at distance r_b from q. Thus the potential energy E_p of the test charge q' at *any* distance r from charge q is given by

$$\boxed{E_p = \frac{kqq'}{r}.} \qquad (26\text{–}6)$$

If the field in which charge q' moves is not that of a single point charge but is due to a more general charge distribution, we may always divide the charge distribution into small elements and treat each element as a point charge. Then, since the total field is the vector sum of the fields due to the individual elements, and since the total work on q' is the sum of the contributions from the individual charge elements, we may conclude that *every* electric field due to a static charge distribution is a conservative force field. Furthermore, the potential energy of a test charge q' at point a in Fig. 26–2, due to a collection of charges q_1, q_2, q_3, etc., at distances r_1, r_2, r_3, etc. from the test charge q', is given by

$$E_p = kq'\left(\frac{q_1}{r_1} + \frac{q_2}{r_2} + \frac{q_3}{r_3} + \cdots\right),$$

$$\boxed{E_p = kq'\sum \frac{q}{r}.} \qquad (26\text{–}7)$$

At a second point b, the potential energy is given by the same expression except that $r_1, r_2, \ldots$ now represent the distances from the respective charges to point b. The work of the electric force in moving the test charge from a to b along any path is equal to the difference between its potential energies at a and at b.

We see from Eqs. (26–6) and (26–7) that the "reference level" of electrical potential energy, at which the potential energy is zero, is that at which all the distances $r_1, r_2, \ldots$ are infinite. That is, the potential energy of the test charge is zero when it is very far removed from all the charges setting up the field. This is the most convenient reference level for most electrostatic problems. When dealing with electrical *circuits*, other reference levels are more convenient, which simply means that a constant term is added to the potential energy. This is not of importance, since

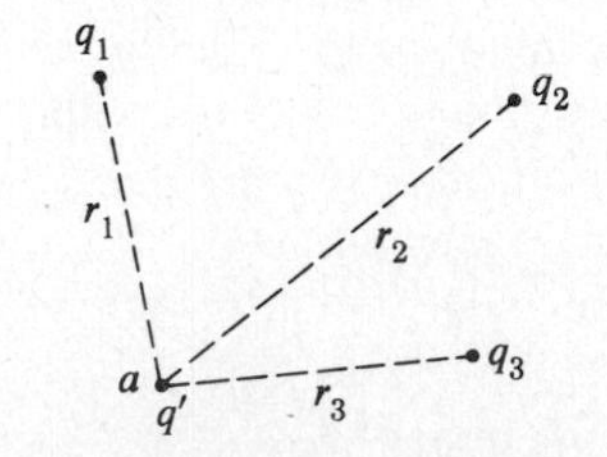

Fig. 26–2 *Potential energy of a charge q' at point a depends on charges q_1, q_2, and q_3, and on their distances r_1, r_2, r_3 from point a.*

it is only *differences* in potential energy that are of practical significance.

From the above definitions, *the potential energy of a test charge at any point in an electric field is equal to the work of the electric force when the test charge is brought from the point in question to a reference level, often taken at infinity.*

26–2 LINE INTEGRAL OF ELECTRIC FIELD

The discussion of the preceding section has shown that the electric field $\mathbf{E}$ of any static electric charge distribution constitutes a conservative force field. This property is expressed succinctly by the equation

$$\oint \mathbf{E} \cdot d\mathbf{s} = \mathbf{0}, \tag{26–8}$$

where the symbol $\oint$ means that the line integral is taken around a closed path. This equation expresses a second property of the electric field, which is as fundamental and useful as that expressed by Gauss's law. It states that *the line integral of the electric field around any closed path in an electrostatic field is zero.*

In some problems of practical interest, this integral can be evaluated without explicitly performing an integration. The following considerations are useful:

1. If $\mathbf{E}$ is *parallel* to a path of length ℓ at all points, and has the same magnitude at all points of the path, then $E_{\|} = E$ = constant, and

$$\int_a^b E_{\|} ds = E\ell.$$

2. If $\mathbf{E}$ is *perpendicular* to a path at all points, $E_{\|} = 0$ and the line integral is zero.
3. If $\mathbf{E} = \mathbf{0}$ at all points of a path, the line integral is zero.

This second general property of an electrostatic field can be used to verify a statement that was made in the preceding chapter, namely, that the electric field just outside the surface of any charged conductor is at right angles to the surface, when the charges in the conductor are at rest. Figure 26–3(a) shows a portion of the surface of the conductor. The dotted rectangle *abcd* is a closed path *(not* a closed *surface).* The line integral around the path is the sum of the integrals along the four sides. Assume first that the $\mathbf{E}$-vector just outside the surface makes some angle θ, less than 90°, with the surface. The line integral along the side *ab* is then the product of $E \cos \theta$ and the length ℓ of this side. We know that $\mathbf{E} = \mathbf{0}$ *inside* the conductor. If $\mathbf{E}$ were different from zero inside the conductor, it would cause the mobile charges in the conductor to move, violating our assumption that we are dealing with a *static* situation. Thus, the line integral along side *cd* is zero. In addition, we assume that the lines *ab* and *cd* are both so close to the surface that sides *bc* and *da* are very short and make no contribution to the line integral. Hence the integral around the entire path is $(E \cos \theta) \cdot \ell$. But the line integral must equal zero, and this can happen only if $\cos \theta = 0$, or if $\theta = 90°$ (or $-90°$ if the surface has a negative charge). Thus when the charges in a conductor are at rest, *the lines of force just outside the surface of the conductor are normal to the surface;* the lines meet the surface at right angles as in Fig. 26–3(b), whatever the shape of the surface. In general, the lines will

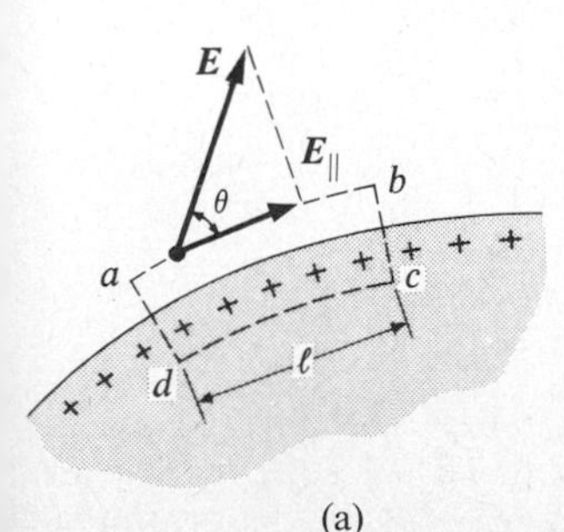

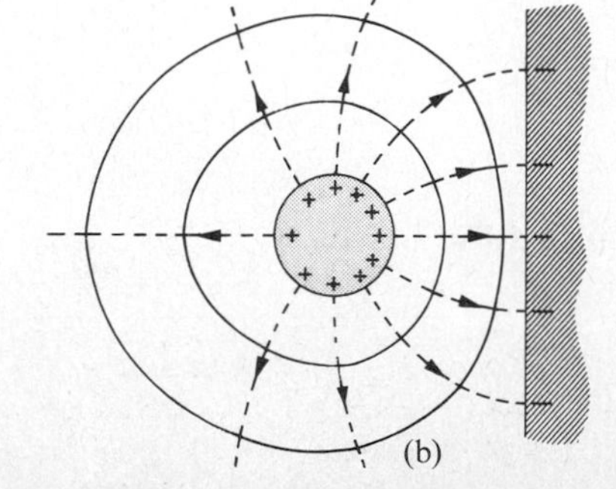

Fig. 26–3 *(a) Construction for finding the direction of* $\mathbf{E}$ *outside the surface of a conductor. (b) Lines of force always meet charged conducting surfaces at right angles.*

change their direction as we move away from the surface, depending on the location of other charges in the vicinity. (If the conductor is carrying a current, the charges within it are not at rest and the lines of force *do not* intersect it at right angles.)

26–3 POTENTIAL

Instead of dealing directly with the potential energy E_p of a charged particle, it is useful to introduce the more general concept of *potential energy per unit charge.* This quantity is called the *potential*, and *the potential at any point of an electrostatic field is defined as the potential energy per unit charge* at the point. Potential is represented by the letter V:

$$\boxed{V = \frac{E_p}{q'},} \qquad \text{or} \qquad E_p = q'V. \quad (26\text{–}9)$$

Potential energy and charge are both *scalars*, so potential is a scalar quantity. Its basic mksc unit is 1 *joule per coulomb* (1 J C^{-1}). For brevity, a potential of 1 J C^{-1} is called 1 *volt* (1 V). The unit is named in honor of the Italian scientist Alessandro Volta (1745–1827). Multiples of the volt are the kilovolt (1 kV = 10^3V), the megavolt (1 MV = 10^6 volts) and the gigavolt (1 GV = 10^9V). Submultiples are the millivolt (1 mV = 10^{-3}V) and the microvolt (1 μV = 10^{-6}V).

To put Eq. (26–2) on a "work per unit charge" basis, we divide both sides by q', obtaining

$$\frac{W_{a\to b}}{q'} = \frac{E_{pa}}{q'} - \frac{E_{pb}}{q'} = V_a - V_b, \quad (26\text{–}10)$$

where $V_a = E_{pa}/q'$ is the potential energy per unit charge at point a, and similarly for V_b. V_a and V_b are called the *potential at point a* and *potential at point b*, respectively. In view of Eq. (26–4), we may also write

$$\boxed{V_a - V_b = \int_a^b \boldsymbol{E} \cdot d\boldsymbol{s}.} \quad (26\text{–}11)$$

The difference $V_a - V_b$ is called the *potential difference* between a and b, the *potential of a with respect to b*, or the *voltage of a with respect to b.* The abbreviation $V_{ab} = V_a - V_b$ is often used. The potential difference between b and a, $V_b - V_a$, is the *negative* of that between a and b:

$$V_{ab} = V_a - V_b = -(V_b - V_a) = -V_{ba}.$$

From Eq. (26–7), the potential V at a point due to an arbitrary collection of point charges is given by

$$\boxed{V = \frac{E_p}{q'} = k\sum\frac{q}{r}.} \quad (26\text{–}12)$$

A continuous distribution of charge can be thought of as divided into small elements of charge, each of which may be treated as a point. The sum in Eq. (26–12) then becomes an integral. Evaluation in particular problems can become complex, but it is often simpler than the evaluation of the *vector* integral, required to obtain the electric field directly from the charge distribution.

A *voltmeter* is an instrument that measures the potential *difference* between the points to which its terminals are connected. The principle of the common type of moving coil voltmeter will be described later. An *electrometer*, which has been considered thus far as an instrument for measuring quantity of charge, can also be used as a voltmeter. Thus, if the knob of a leaf electroscope is connected to a point at one potential and the case to a point at a different potential, the quantity of charge on the leaves is proportional to the potential difference between the points, and the instrument can be calibrated to read this potential difference.

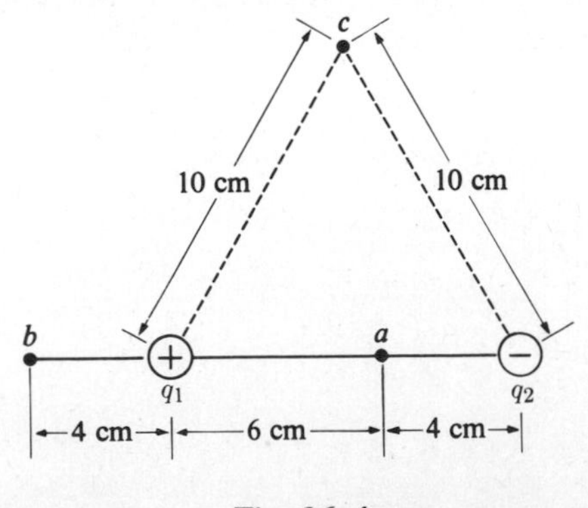

Fig. 26–4

Example 1 Point charges of $+12 \times 10^{-9}$C and -12×10^{-9}C are placed 10 cm apart, as in Fig. 26–4. Compute the potentials at points a, b, and c.

We must evaluate the *algebraic* sum $k\sum(q/r)$ at each point. At point a, the potential due to the positive charge is

$$(9 \times 10^9 \text{ N m}^2 \text{ C}^{-2})\frac{12 \times 10^{-9}\text{ C}}{0.06 \text{ m}} = 1800 \text{ N m C}^{-1}$$
$$= 1800 \text{ J C}^{-1}$$
$$= 1800 \text{ V},$$

and the potential due to the negative charge is

$$(9 \times 10^9 \text{ N m}^2 \text{ C}^{-2})\frac{-12 \times 10^{-9}\text{ C}}{0.04 \text{ m}} = -2700 \text{ J C}^{-1}$$
$$= -2700 \text{ V}.$$

Hence

$$V_a = 1800 \text{ V} - 2700 \text{ V} = -900 \text{ V}$$
$$= -900 \text{ J C}^{-1}.$$

At point b, the potential due to the positive charge is +2700 V and that due to the negative charge is −770 V. Hence,

$$V_b = 2700 \text{ V} - 770 \text{ V} = 1930 \text{ V}$$
$$= 1930 \text{ J C}^{-1}.$$

At point c the potential is

$$V_c = 1080 \text{ V} - 1080 \text{ V} = 0.$$

Example 2 Compute the potential energy of a point charge of $+4 \times 10^{-9}$ C if placed at the points a, b, and c in Fig. 26–4.

First,

$$E_p = qV.$$

Hence at point a,

$$E_p = qV_a = (4 \times 10^{-9}\text{ C})(-900 \text{ J C}^{-1})$$
$$= -36 \times 10^{-7}\text{ J}.$$

At point b,

$$E_p = qV_b = (4 \times 10^{-9}\text{ C})(1930 \text{ J C}^{-1})$$
$$= 77 \times 10^{-7}\text{ J}.$$

At point c,

$$E_p = qV_c = 0.$$

(All relative to a point at infinity.)

26–4 CALCULATION OF POTENTIAL DIFFERENCES

In principle the potential difference between any two points can be calculated by using Eq. (26–12) to find the potential at each point, but in many cases, especially where the electric field is easily obtained, it is easier to use Eq. (26–11), calculating the integral from the known electric field. In some problems a combination of these two approaches is useful. The following problems illustrate these remarks.

1. *Charged spherical conductor.* We consider first a solid conducting sphere of radius R, with total charge q. From Gauss's law, as discussed in Chapter 25, we conclude that, at all points *outside* the sphere, the field is the same as that of a point charge q at the center of the sphere. *Inside* the sphere the field is zero everywhere; otherwise charge would move within the sphere.

As mentioned above, it is convenient to take the reference level of potential (the point where $V = 0$) at a very large distance from all charges. Since the field at any point r greater than R is the same as for a point charge, the work on a test charge moving from a finite value of r to infinity is also the same as for a point charge, so the potential V at a radial distance r, relative to a point at infinity, is

$$V = k\left(\frac{q}{r}\right). \tag{26–13}$$

The potential is positive if q is positive, negative if q is negative.

Equation (26–13) applies to the field of a charged spherical conductor only when r is greater than or equal to the radius R of the sphere. The potential *at* the surface is

$$V = k\left(\frac{q}{R}\right). \tag{26–14}$$

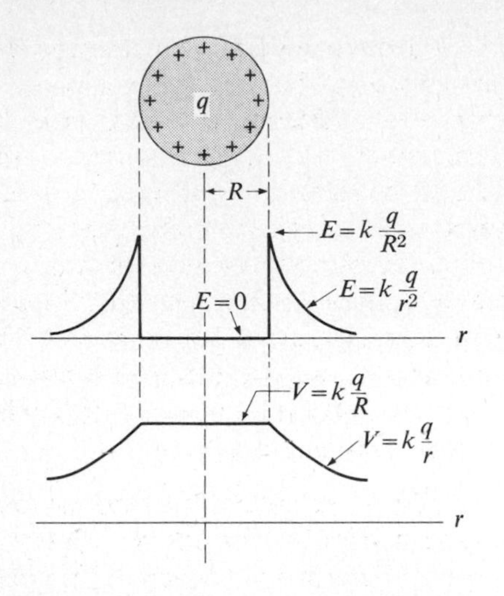

Fig. 26–5 *Electric intensity* **E** *and potential V at points inside and outside a charged spherical conductor.*

Inside the sphere the field is zero everywhere, and no work is done on a test charge displaced from any point to any other in this region. Thus the potential is *the same* at all points inside the sphere, and is equal to its value kq/R at the surface. The field and potential are shown as functions of r in Fig. 26–5.

The electric field $\boldsymbol{E}$ at the surface has a magnitude

$$E = k\left(\frac{q}{R^2}\right). \qquad (26\text{–}15)$$

The maximum potential to which a conductor in air can be raised is limited by the fact that air molecules become ionized, and hence the air becomes a conductor, at an electric field of about $3 \times 10^6\ \mathrm{N\,C^{-1}}$. Comparing Eqs. (26–14) and (26–15), we note that at the surface the field and potential are related by $V = ER$. Thus if E_m represents the upper limit of electric field, known as the *dielectric strength*, the maximum potential to which a spherical conductor can be raised is

$$V_m = RE_m.$$

For a sphere 1 cm in radius, in air,

$$V_m = (10^{-2}\ \mathrm{m})(3 \times 10^6\ \mathrm{N\,C^{-1}}) = 30{,}000\ \mathrm{V},$$

and no amount of "charging" could raise the potential of a sphere of this size, in air, higher than about 30,000 V. It is this fact which necessitates the use of large spherical terminals on high-voltage machines. If we make $R = 2$ m, then

$$V_m = (2\ \mathrm{m})(3 \times 10^6\ \mathrm{N\,C^{-1}}) = 6 \times 10^6\ \mathrm{V} = 6\ \mathrm{MV}.$$

At the other extreme is the effect produced by sharp points, a "point" being a surface of very *small* radius of curvature. Since the maximum potential is proportional to the radius, even relatively small potentials applied to sharp points in air will produce sufficiently high fields just outside the point to result in ionization of the surrounding air.

2. Line charge and charged conducting cylinder. The field of a line charge, and the field outside a charged conducting cylinder, are both given by

$$E = 2k\frac{\lambda}{r},$$

where λ is the charge per unit length.

The potential difference between any two points a and b at radial distances r_a and r_b is

$$V_a - V_b = 2k\lambda \int_{r_a}^{r_b} \frac{dr}{r} = 2k\lambda \ln \frac{r_b}{r_a}. \qquad (26\text{–}16)$$

If we take point b at infinity and set $V_b = 0$, we find for the potential V_a,

$$V_a = 2k\lambda \ln \frac{\infty}{r_a} = \infty.$$

Hence a reference point at infinity is not suitable for this field! We can, however, set $V = 0$ at some arbitrary radius r_0. Then at any radius r,

$$V = 2k\lambda \ln \frac{r_0}{r}. \qquad (26\text{–}17)$$

Equations (26–16) and (26–17) give the potential in the field of a cylinder only for values of r equal to or greater than the radius R of the cylinder. If r_0 is taken as the cylinder radius R, so that the potential of the cylinder is considered zero, the potential at any external point, relative to that of the cylinder, is

$$V = 2k\lambda \ln \frac{R}{r}. \qquad (26\text{–}18)$$

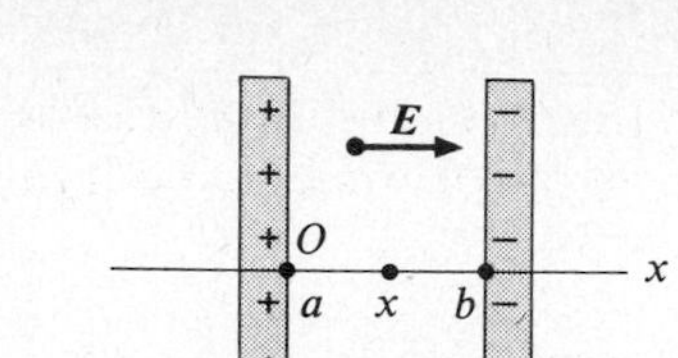

Fig. 26–6

3. Parallel plates. The electric field between oppositely charged parallel plates, as derived in Section 25–5, is

$$E = \frac{\sigma}{\epsilon_0}, \qquad (26\text{–}19)$$

where σ is the magnitude of the surface charge density (charge per unit area) on either plate.

Let us take an x-axis as in Fig. 26–6, perpendicular to the plates and with point a at the origin. Then at any point x,

$$V_a - V_x = \int_0^x E\,dx = Ex,$$

or

$$V_x = V_a - Ex = V_a - \frac{\sigma}{\epsilon_0}x.$$

The potential therefore decreases *linearly* with x. At point b, where $x = \ell$ and $V_x = V_b$,

$$V_b = V_a - E\ell,$$

and hence

$$E = \frac{V_a - V_b}{\ell} = \frac{V_{ab}}{\ell}. \qquad (26\text{–}20)$$

That is, *the electric field equals the potential difference between the plates divided by the distance between them.* This is a more useful expression for E than Eq. (26–19) because the potential difference V_{ab} can readily be measured with a voltmeter, while there are no instruments that read surface density of charge directly.

Equation (26–13) also shows that the unit of electric field can be expressed as 1 *volt per meter* ($1\ \mathrm{V\,m^{-1}}$), as well as $1\ \mathrm{N\,C^{-1}}$. In practice, the volt per meter is most commonly used as the unit of E.

Example A simple type of vacuum tube known as a *diode* consists essentially of two electrodes within a highly evacuated enclosure. One electrode, the cathode, is maintained at a high temperature and emits electrons from its surface. A potential difference of a few hundred volts is maintained between the cathode and the other electrode, known as the *anode* or *plate*, with the anode at the higher potential. Suppose that, in a certain diode, the plate potential is 250 V above that of the cathode, and an electron is emitted from the cathode with no initial velocity. What is its velocity when it reaches the anode?

Let V_c and V_a represent the cathode and anode potentials, respectively, and let the charge on the electron be $-e$. Since this charge is negative, the electron goes from the cathode, where the potential is lower, to the anode, where the potential is higher. Hence, the *decrease* in electrical potential energy is $-e(V_c - V_a)$ or eV_{ac}, and the electron *gains* a corresponding amount of kinetic energy during its travel from cathode to anode. Letting v_c and v_a be the speeds at anode and cathode, respectively, we have

$$eV_{ac} = \tfrac{1}{2}mv_a^2 - \tfrac{1}{2}mv_c^2.$$

Since $v_c = 0$,

$$\begin{aligned} v_a &= \sqrt{2eV_{ac}/m} \\ &= \sqrt{\frac{2(1.6 \times 10^{-19}\ \mathrm{C})(250\ \mathrm{J\,C^{-1}})}{9.1 \times 10^{-31}\ \mathrm{kg}}} \\ &= 9.4 \times 10^6 (\mathrm{J\,kg^{-1}})^{1/2} \\ &= 9.4 \times 10^6\ \mathrm{m\,s^{-1}}. \end{aligned}$$

Note that the shape or separation of the electrodes need not be known. The final velocity depends only on the difference of potential between cathode and anode. Of course, the *time* of transit from cathode to anode depends on the geometry of the tube.

26–5 EQUIPOTENTIAL SURFACES

The potential distribution in an electric field may be represented graphically by *equipotential surfaces.* An equipotential surface is a surface such that the potential has the same value at all points on the surface.

While an equipotential surface may be constructed through every point of an electric field, it is customary to show only a few of the equipotentials in a diagram.

Since the potential energy of a charged body is the same at all points of a given equipotential surface, it follows that no (electrical) work is needed to move a charged body over such a surface. Hence, the equipotential surface through any point must be at right angles to the direction of the field at that point. If this were not so, the field would have a component lying in the surface and work would have to be done against electrical forces to move a charge in the direction of this component. The lines of force and the equipotential surfaces thus form a mutually perpendicular network. In general, the lines of force of a field are curves and the equipotentials are curved surfaces. For the special case of a uniform field, where the lines of force are straight and parallel, the equipotentials are parallel planes perpendicular to the lines of force.

Figure 26–7 shows several arrangements of charges. The lines of force are represented by broken lines, and cross sections of the equipotential surfaces as solid lines. The actual field is of course three-dimensional. At each crossing of an equipotential and a field line, the two are perpendicular.

When charges at rest reside on the surface of a conductor, the electric field just outside the conductor must be everywhere perpendicular to the surface; this was proved in Section 26–2. It follows that when all charges are at rest, *the surface of a conductor is always an equipotential surface.* Figure 26–3(b) illustrates these conclusions. Field lines (broken) and equipotentials (solid) are shown.

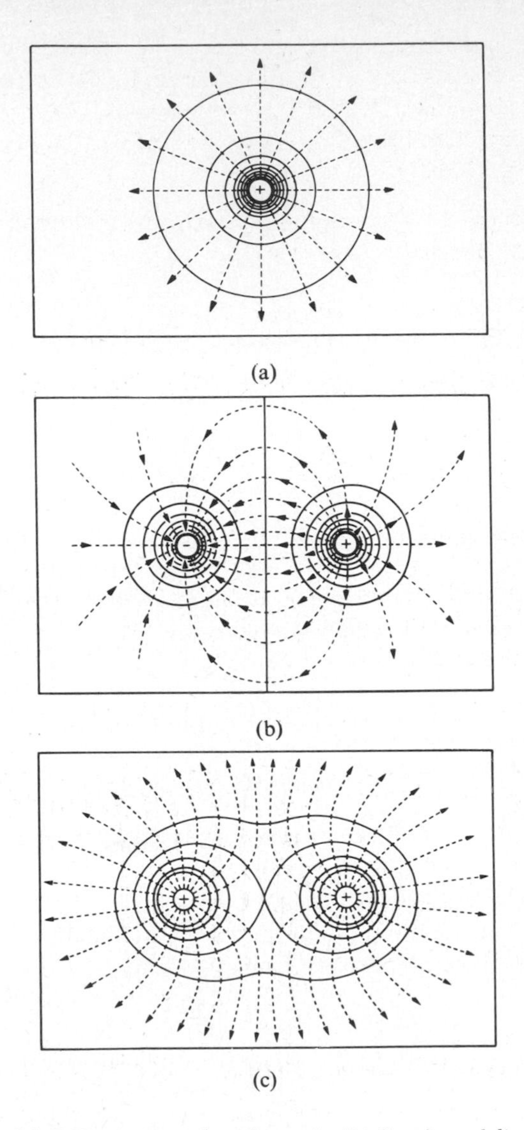

Fig. 26–7 *Equipotential surfaces (solid lines) and lines of force (dotted lines) in the neighborhood of point charges.*

26–6 POTENTIAL GRADIENT

If points a and b in Fig. 26–1 are very close together, the potential difference $V_a - V_b$ becomes simply $-dV$, and the line integral of $\boldsymbol{E}$ from a to b reduces to $E_\parallel \, ds$. The *differential* form of Eq. (26–11) is therefore

$$-dV = E_\parallel \, ds$$

or

$$\boxed{E_\parallel = -\frac{dV}{ds}.} \tag{26–21}$$

The ratio dV/ds, or the rate of change of potential with distance in the direction of ds, is called the *potential gradient*, and $E_\parallel$ is the component of electric field in the direction of ds. Hence we have the important relation: *At any point, the component of electric field in any direction is equal to the negative of the potential gradient in that direction.* In particular, if the direction of ds is the same as that of the electric field, the component of $\boldsymbol{E}$ in the direction of ds is equal to **E** and *the electric field is equal to the negative of the potential gradient in the direction of the field.*

The unit of potential gradient is 1 *volt per meter* (1 V m^{-1}) and Eq. (26–20) is evidently a special case of Eq. (26–21).

Example 1 We have shown that the potential at a radial distance r from a point charge q is

$$V = k\frac{q}{r}.$$

By symmetry, the electric field is in the radial direction, so

$$E = E_r = -\frac{dV}{dr} = -\frac{d}{dr}\left(k\frac{q}{r}\right) = k\frac{q}{r^2},$$

in agreement with Coulomb's law.

Example 2 The potential outside a charged conducting cylinder of radius R and charge per unit length λ (Example 2 in Section 26–4) is

$$V = 2k\lambda \ln\frac{R}{r} = 2k\lambda(\ln R - \ln r).$$

The electric field is radial, and its magnitude is given by

$$E = -\frac{dV}{dr} = \frac{2k\lambda}{r},$$

on agreement with our previous result.

26–7 THE MILLIKAN OIL DROP EXPERIMENT

We have now developed the theory of electrostatics to a point where one of the classical physical experiments of all time can be described. In a brilliant series of investigations carried out at the University of Chicago in the period 1909–1913, Robert Andrews Millikan not only demonstrated conclusively the discrete nature of electric charge, but actually measured the charge of an individual electron.

Millikan's apparatus is shown schematically in Fig. 26–8(a). Two accurately parallel horizontal metal plates, A and B, are insulated from each other and separated by a few millimeters. Oil is sprayed in fine droplets from an atomizer above the upper plate and a few of the droplets are allowed to fall through a

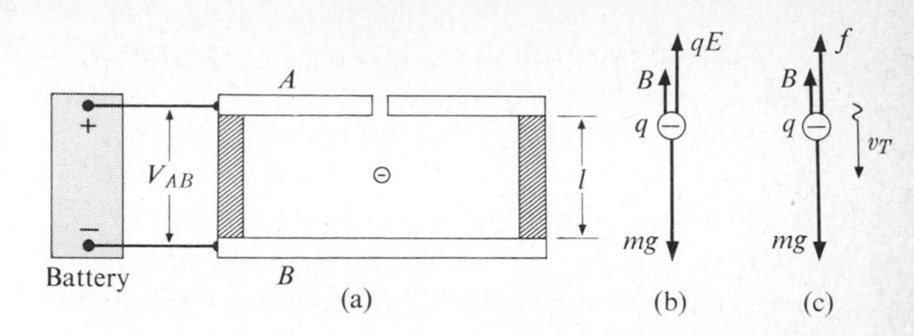

Fig. 26–8 (a) Schematic diagram of Millikan apparatus. (b) Forces on a drop at rest. (c) Forces on a drop falling with its terminal velocity v_T.

small hole in this plate. A beam of light is directed horizontally between the plates, and a telescope is set up with its axis at right angles to the light beam. The oil drops, illuminated by the light beam and viewed through the telescope, appear like tiny bright stars, falling slowly with a terminal velocity determined by their weight and by the viscous air-resistance force opposing their motion.

It is found that some of the oil droplets are electrically charged, presumably because of frictional effects. Charges can also be given the drops if the air in the apparatus is ionized by x-rays or a bit of radioactive material. Some of the electrons or ions then collide with the drops and stick to them. The drops are usually negatively charged, but occasionally one with a positive charge is found.

The simplest method, in principle, for measuring the charge on a drop is as follows. Suppose a drop has a negative charge and the plates are maintained at a potential difference such that a downward electric field of intensity $E(= V_{AB}/\ell)$ is set up between them. The forces on the drop are then its weight mg and the upward force of qE. By adjusting the field E, qE can be made just equal to mg, so that the drop remains at rest, as indicated in Fig. 26–8(b). Under these circumstances.

$$q = \frac{mg}{E}.$$

The mass of the drop equals the product of its density ρ and its volume, $4\pi r^3/3$, and $E = V_{AB}/\ell$, so

$$q = \frac{4\pi}{3}\frac{\rho r^3 g\ell}{V_{AB}}. \qquad (26\text{–}22)$$

All the quantities on the right are readily measurable with the exception of the drop radius r, which is of the order of 10^{-5} cm and is much too small to be measured directly. It can be calculated, however, by cutting off the electric field and measuring the terminal velocity v_T of the drop as it falls through a known distance d defined by reference lines in the ocular of the telescope.

The terminal velocity is that at which the weight mg is just balanced by the viscous force f. The viscous force on a sphere of radius r, moving with a velocity v through a fluid of viscosity η, is given by Stokes' law, discussed in Section 13–7:

$$f = 6\pi\eta r v.$$

If Stokes' law applies, and the drop is falling with its terminal velocity v_T,

$$mg = f,$$

$$\frac{4}{3}\pi r^3 \rho g = 6\pi\eta r v_T,$$

and

$$r = 3\sqrt{\eta v_T / 2\rho g}.$$

When this expression for r is inserted in Eq. (26–22), we have

$$q = 18\pi \frac{\ell}{V_{AB}} \sqrt{\frac{\eta^3 v_T^3}{2\rho g}},$$

which expresses the charge q in terms of measurable quantities.

In actual practice this procedure is modified somewhat. To correct for the buoyant force of the air through which the drop falls, the density ρ of the oil should be replaced by $(\rho - \rho_g)$, where ρ_g is the density of air. A correction to Stokes' law is also required, because air is not a continuous fluid but a collection of molecules separated by distances which are of the same order of magnitude as the dimensions of the drops.

Millikan and his co-workers measured the charges of thousands of drops, and found that, within the limits of their experimental error, every drop had a charge equal to some small integer multiple of a basic charge e. That is, drops were observed with charges of e, $2e$, $3e$, etc., but never with such values as $0.76e$ or $2.49e$. The evidence is conclusive that electric charge is not something which can be divided indefinitely, but that it exists in nature only in units of magnitude e. When a drop is observed with charge e, we conclude it has acquired one extra electron; if its charge is $2e$, it has two extra electrons, and so on.

As stated in Section 24–6, the best experimental value of the charge e is

$$e = 1.60219 \times 10^{-19}\ \text{C} \approx 1.60 \times 10^{-19}\ \text{C}.$$

26–8 THE ELECTRON VOLT

The change in potential energy of a particle having a charge q, when it moves from a point where the potential is V_a to a point where the potential is V_b, is

$$\Delta E_p = qV_{ab}.$$

In particular, if the charge q equals the electronic charge $e = 1.60 \times 10^{-19}$C, and the potential difference $V_{ab} = 1$ V, the change in energy is

$$\Delta E_p = (1.60 \times 10^{-19}\text{C})(1\ \text{V}) = 1.60 \times 10^{-19}\text{J}.$$

This quantity of energy is called 1 *electron volt* (1 eV):

$$\boxed{1\ \text{eV} = 1.60 \times 10^{-19}\text{J}.}$$

Other commonly used units are:

$$1\ \text{keV} = 10^3\ \text{eV},$$
$$1\ \text{MeV} = 10^6\ \text{eV},$$
$$1\ \text{GeV} = 10^9\ \text{eV}.$$

The electron volt is a convenient energy unit when one is dealing with the motions of electrons and ions in electric fields, because the change in potential energy between two points on the path of a particle having a charge e, when expressed in electron volts, is *numerically* equal to the potential difference between the points, in volts. If the charge is some multiple of e, say Ne, the change in potential energy in electron volts is, numerically, N times the potential difference in volts. For example, if a particle having a charge $2e$ moves between two points for which the potential difference is 1000 V, the change in its potential energy is

$$\Delta E_p = qV_{ab} = (2)(1.6 \times 10^{-19}\text{C})(10^3\text{V})$$
$$= 3.2 \times 10^{-6}\text{J}$$
$$= 2000\ \text{eV}.$$

Although the electron volt was defined above in terms of *potential* energy, energy of *any* form, such as the kinetic energy of a moving particle, can be expressed in terms of the electron volt. Thus one may speak of a "one million volt electron," meaning an electron having a kinetic energy of one million electron volts (1 MeV).

It was shown in Section 14–7 that, according to the principles of special relativity, the mass m of a particle is equivalent to a quantity of energy mc^2, where c is the speed of light. The rest mass of an electron is 9.108×10^{-31} kg, and the energy equivalent to this is

$$E_0 = mc^2 = (9.108 \times 10^{-31}\ \text{kg})(3 \times 10^8\ \text{m s}^{-1})^2$$
$$= 82 \times 10^{-15}\ \text{J}.$$

Since 1 eV = 1.60×10^{-19} J, this is equivalent to

$$E_0 = 511{,}000\ \text{eV}$$
$$= 0.511\ \text{MeV}.$$

Suppose that an electron is accelerated from rest through a potential difference of this magnitude, 511,000 V. It then acquires a *kinetic* energy of 511,000 eV, equal to its rest-mass energy, and its *total* energy, rest-mass energy plus kinetic energy, is twice the rest-mass energy:

$$E = E_0 + E_k$$
$$= 2m_0c^2.$$

But according to Eqs. (14–28) and (14–29), its total energy is also given by

$$E = \frac{mc^2}{\sqrt{1 - v^2/c^2}}.$$

When this is equal to $2mc^2$,

$$\sqrt{1 - v^2/c^2} = \frac{1}{2},$$

and

$$v = \sqrt{3/4}\, c \simeq 0.866c.$$

Thus for energies in this range it is essential to use relativistic mechanics.

26–9 THE CATHODE-RAY OSCILLOSCOPE

Figure 26–9 is a schematic diagram of the elements of a cathode-ray oscilloscope tube. The interior of the tube is highly evacuated. The *cathode* at the left is raised to a high temperature by the *heater*, and electrons evaporate from its surface. (Before the nature of this process of electron emission was fully understood, these electrons were given the name "cathode rays.") The *accelerating anode*, which has a

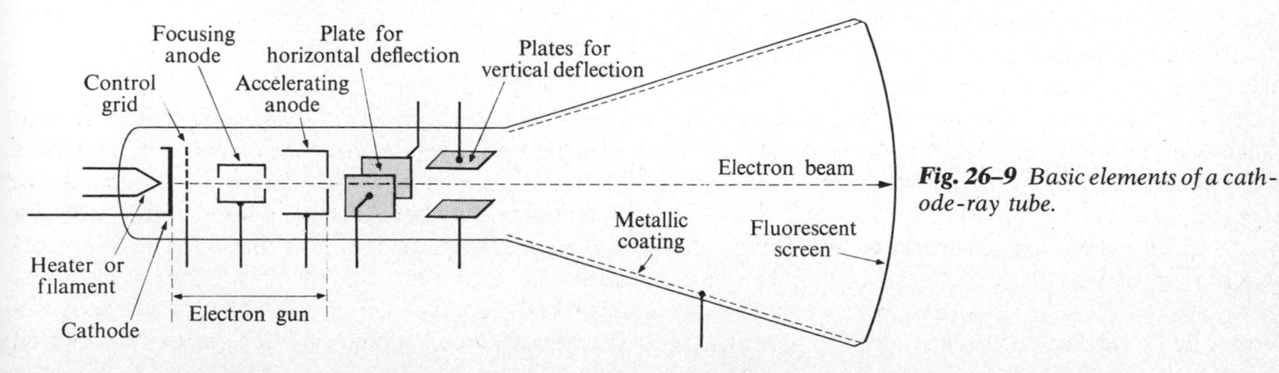

Fig. 26–9 *Basic elements of a cathode-ray tube.*

small hole at its center, is maintained at a high positive potential V_1 relative to the cathode, so that there is an electric field, directed from right to left, between the anode and cathode. This field is confined to the cathode–anode region, and electrons passing through the hole in the anode travel with a *constant* x-velocity from the anode to the *fluorescent screen.*

The function of the *control grid* is to regulate the number of electrons that reach the anode (and hence the brightness of the spot on the screen). The *focusing anode* ensures that electrons leaving the cathode in slightly different directions all arrive at the same spot on the screen. These two electrodes need not be considered in the following analysis. The complete assembly of cathode, control grid, focusing anode, and accelerating electrode is referred to as an *electron gun.*

The accelerated electrons pass between two pairs of *deflecting plates.* An electric field between the first pair of plates deflects them to the right or left, and a field between the second pair deflects them up or down. In the absence of such fields the electrons travel in a straight line from the hole in the accelerating anode to the *fluorescent screen* and produce a bright spot on the screen where they strike it.

We first calculate the velocity imparted to the electrons by the electron gun, just as in the example of Section 26–3. The only force on them is the *conservative* electrical force, so we can use the principle of *conservation of energy.* Let subscripts c and a refer to cathode and anode. Then*

$$\frac{1}{2}mv_c^2 + eV_c = \frac{1}{2}mv_a^2 + eV_a.$$

Although the electrons have a velocity v_c when they evaporate from the cathode, this is very small compared with their final velocity v_a and can be neglected. Therefore,

$$v_a = \sqrt{\frac{2e(V_c - V_a)}{m}} = \sqrt{\frac{2eV_1}{m}}, \tag{26–23}$$

* The potential differences in a typical cathode-ray tube are small enough so that relativistic mechanics need not be used.

where for brevity we have represented the accelerating potential difference V_{ca} by V_1.

The anode is at a *higher* potential than the cathode, so $V_1 = V_{ca}$ is a negative quantity. But the electron charge e is also negative, so the term under the radical is positive.

As a numerical example, if $V_1 = -2000$ V,

$$v_a = \sqrt{\frac{2(-1.6 \times 10^{-19}\text{C})(-2 \times 10^3\text{V})}{9.11 \times 10^{-31}\text{ kg}}}$$
$$= 2.65 \times 10^7 \text{ m s}^{-1}.$$

We can now see one of the advantages of describing electric fields in terms of *potentials.* The kinetic energy of an electron at the anode depends only on the *potential difference* between anode and cathode, and not at all on the details of the fields within the electron gun or on the shape of the electron trajectory within the gun.

If there is no electric field between the plates for horizontal deflection, the electrons enter the region between the other plates with a velocity equal to v_a and represented by v_x in Fig. 26–10. If there is a potential difference v_2 between the plates, and the upper plate is positive, a downward electric field of intensity $E = V_2/\ell$ is set up between the plates. A constant upward force eE then acts on the electrons and their upward acceleration is

$$a_y = \frac{eE}{m}. \tag{26–24}$$

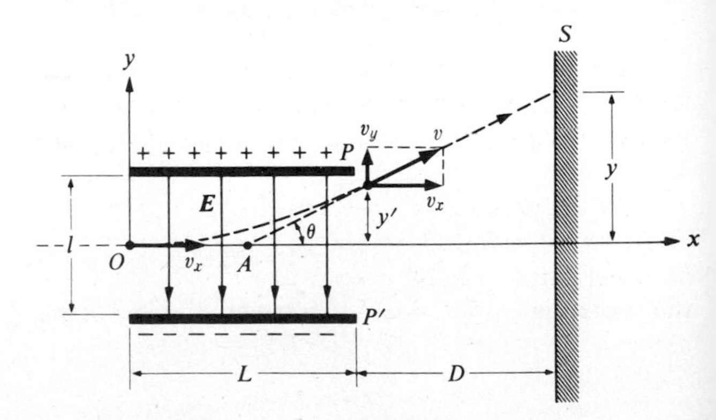

Fig. 26–10 *Electrostatic deflection of cathode rays.*

The *horizontal* velocity remains constant, so the time required for the electrons to travel the length L of the plates is

$$t = \frac{L}{v_x}. \qquad (26\text{–}25)$$

In this time, they acquire an upward velocity given by

$$v_y = a_y t, \qquad (26\text{–}26)$$

and are displaced upward by an amount

$$y' = \frac{1}{2} a_y t^2 .$$

On emerging from the deflecting field, their velocity v makes an angle θ with the x-axis, where

$$\tan\theta = \frac{v_y}{v_x};$$

and from this point on they travel in a straight line to the screen. It is not difficult to show that this straight line, if projected backward, intersects the x-axis at a point A which is midway between the ends of the plates. Then if y is the vertical coordinate of the point of impact with screen S,

$$\tan\theta = \frac{y}{D + (L/2)}.$$

When this is combined with Eqs. (26–24), (26–25), and (26–26), we finally obtain

$$y = \left[\frac{L}{2\ell}\left(D + \frac{L}{2}\right)\right]\frac{V_2}{V_1}. \qquad (26\text{–}27)$$

The term in brackets is a purely geometrical factor. If the accelerating voltage V_1 is held constant, *the deflection* y *is proportional to the deflecting voltage* V_2.

As a numerical illustration, let $L = 2$ cm $= 2 \times 10^{-2}$ m, $\ell = 0.5$ cm $= 5 \times 10^{-3}$ m, and let $v_x = v_a$ have the value computed above, 2.65×10^7 m s^{-1}. If the potential difference V_2 between the deflecting plates is 100 V, the deflecting field E is

$$E = \frac{V_2}{\ell} = 2 \times 10^4\,\mathrm{V\,m^{-1}}.$$

The y-acceleration is

$$a_y = \frac{eE}{m} = 3.52 \times 10^{15}\,\mathrm{m\,s^{-2}}.$$

The time to travel the length of the plates is

$$t = \frac{L}{v_x} = 7.55 \times 10^{-10}\,\mathrm{s}.$$

The upward velocity v_y is

$$v_y = a_y t = 2.66 \times 10^6\,\mathrm{m\,s^{-1}}.$$

The upward displacement y' is

$$y' = \frac{1}{2} a_y t^2 = 10^{-3}\,\mathrm{m}.$$

The tangent of the angle θ is

$$\tan\theta = \frac{v_y}{v_x} = 0.10.$$

If the distance D to the screen is 20 cm, the displacement y on the screen is

$$y = \left(D + \frac{L}{2}\right)\tan\theta = 2.1\ \mathrm{cm}.$$

If a field is also set up between the *horizontal* deflecting plates, the beam is also deflected in the horizontal direction, perpendicular to the plane of Fig. 26–10. The coordinates of the luminous spot on the screen are then proportional, respectively, to the horizontal and vertical deflecting voltages.

The picture tube in a television set is similar in its operation to the oscilloscope tube, except that the beam is deflected by magnetic fields, to be discussed in later chapters, rather than by electric fields. Graphic display devices on some computer terminals also are similar, using a deflected electron beam and a fluorescent screen. The accelerating voltage in TV picture tubes (V_1 in the above discussion) is typically 20 to 25 kV.

26–10 SHARING OF CHARGE BY CONDUCTORS

When a charged conductor is brought into electrical contact with one which is uncharged, the original charge is shared between the two. That this should happen is evident from the mutual forces of repulsion between the component parts of the original charge. The question as to precisely how much charge will be transferred has not yet been answered, but we can now see that it must be such as to bring all points of both conductors to the same potential. Thus if a positively charged body makes *external* contact with an uncharged body, the first will lose some of its charge and its potential will decrease, while the second will gain charge and its potential will increase. The flow of charge will cease when both bodies are at the same potential, but there will still remain some charge on the first body.

When a charged body makes contact with the *interior* of a conductor, however, the situation is quite different. As a consequence of Gauss's law, an induced charge of opposite sign appears on the inner surface of the hollow conductor, and this charge is independent of the position of the charged body within the cavity. Upon touching the charged body to the wall of the cavity, the first body transfers *all* of its charge to the hollow conductor, in spite of the fact that the latter may have originally been charged.

To study this experiment in more detail, consider the large hollow metal sphere B shown in Fig. 26–11(a) with an original positive charge q_B, indicated by positive signs with circles around them, and with an inner radius r_B. Let an opening be made in the walls large enough to admit a small metal sphere A of radius r_A and with a positive charge q_A.

When A is at the center of B and the small effect due to the small opening in B is neglected, the *positive charge on* A and the equal induced negative charge on the interior surface of B are *evenly distributed,* and the electric field between these two charges is symmetrical and radial. By Gauss's law, the field between A and B is due only to the charge of A, and at a distance r from the center of A, is given by

$$E = kq_A\left(\frac{1}{r^2}\right).$$

The potential difference between A and B is therefore

$$V_A - V_B = kq_A\left(\frac{1}{r_A} - \frac{1}{r_B}\right). \qquad (26\text{–}28)$$

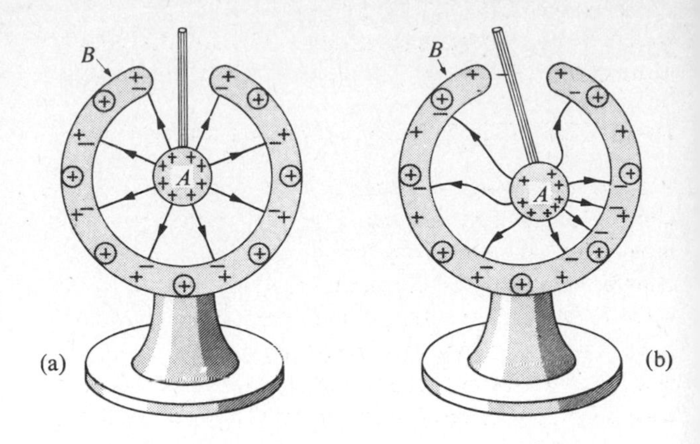

Fig. 26–11 *Body A, with charge* $+q_A$ *introduced into a hollow spherical conductor B with an original charge* $+q_B$ *(indicated by the symbol ⊕). (a) When A is at the center, the field around it is symmetrical and the induced negative charge* $-q_A$ *is distributed evenly on the inner surface of the hollow sphere. (b) When A is off center, the field around it is asymmetrical and the induced negative charge* $-q_A$ *is distributed unevenly on the inner surface of the hollow sphere.*

This equation expresses two important facts:

1. $V_A - V_B$ is positive, or A is at a *higher* potential than B.
2. $V_A - V_B$ depends *only* on q_A, and is thus independent of the original charge residing on B.

If A and B are connected by a conductor, electricity will flow from A to B until $V_A - V_B = 0$, or, from Eq. (26–28), until $q_A = 0$. This leads to the conclusion that *all* of the charge on A is transferred to B, *regardless of the initial value of B's charge and potential.* This is the principle of the Van de Graaff generator, described in the next section.

When body A is off center, as shown in Fig. 26–11(b), the positive charge on A and the equal induced negative charge on the interior wall of B are unevenly distributed. The electric field between A and B is quite asymmetrical and cannot be expressed in simple mathematical form. The difference of potential $V_A - V_B$ is *still positive,* however, but with a value smaller than that which existed when A was at the center, and furthermore, $V_A - V_B$ still depends *only* on q_A, regardless of the original charge and potential of B.

26–11 THE VAN DE GRAAFF GENERATOR

We have already explained that when a charged conductor is inserted in a hollow conductor and makes internal contact with this conductor, all of the charge on the first conductor is transferred to the second, whatever charge the second may already have. Were it not for insulation difficulties, the charge (and hence the potential) of a hollow conductor could be increased without limit by repeating this process. (Of course, as the potential of the conductor is raised, a greater and greater repelling force is exerted on each successive charge brought up to it. Eventually, we might not be strong enough to bring up more charge!)

The generator invented by Robert J. Van de Graaff makes use of the principle above, but instead of inserting charged bodies into a conductor one after another, charge is carried in continuously by a "belt conveyor."

Figure 26–12 is a schematic diagram of a small Van de Graaff generator designed for demonstration purposes. A hollow metal conductor A, approximately spherical, is supported on an insulating tube B mounted on a metal base C which is normally grounded. A nonconducting endless belt D runs over two nonconducting pulleys E and F. Pulley F may be driven by hand or by a small electric motor.

Pulleys E and F are covered with different materials, chosen so that when belt D makes contact with F it acquires a positive charge, while on contact with E it acquires a negative charge. Sharp points G and H are connected electrically to the upper conductor A and the base C, respectively.

The charges developed on the belt, as it makes contact with the pulleys, stick to it and are carried along by it. The left side of the belt, moving upward, carries a continuous stream of positive charge into the upper conductor. As it passes G, it induces a charge on this conductor which, because of the sharp points, results in a sufficiently high field intensity to ionize the air between the point and the belt. The ionized air provides a conducting path along which the positive charge on the belt can flow to conductor A.

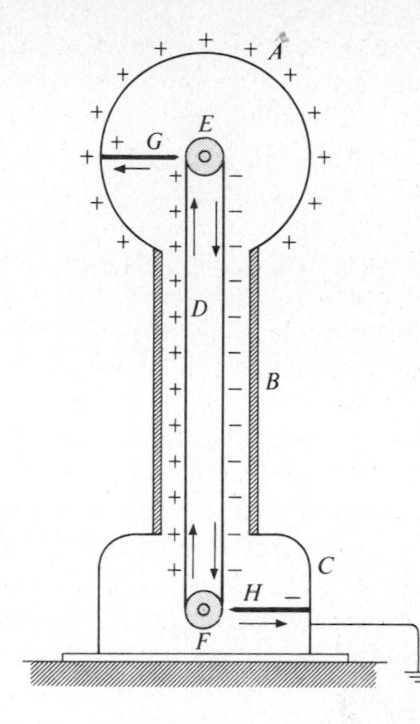

***Fig. 26–12** Simple model of a Van de Graaff generator.*

As the belt leaves the pulley E, it becomes negatively charged and the right side of the belt carries negative charge *out* of the upper terminal. Removal of negative charge is equivalent to addition of positive charge, so both sides of the belt act to increase the net positive charge of terminal A. Negative charge is removed from the belt at the sharp point H, and flows to ground.

PROBLEMS

26–1 A particle of charge $+3 \times 10^{-9}$ C is situated in a uniform electric field directed to the left. In moving to the right a distance of 5 cm, the work of an applied force is 6×10^{-5} J and the change in kinetic energy of the particle is $+4.5 \times 10^{-5}$ J.

a) What is the work of the electrical force?

b) What is the magnitude of the electric field?

26–2 A charge of 2.5×10^{-8} C is placed in an upwardly directed uniform electric field whose intensity is 5

$\times 10^4$ N C^{-1}. What is the work of the electrical force when the charge is moved (a) 45 cm to the right? (b) 80 cm downward? (c) 260 cm at an angle of 45° upward from the horizontal?

26–3

a) Show that 1 N C^{-1} = 1 V m^{-1}.

b) A potential difference of 2000 V is established between parallel plates in air. If the air becomes electrically conducting when the electric intensity exceeds 3×10^6 N C^{-1}, what is the minimum separation of the plates?

26–4 A small sphere of mass 0.2 g hangs by a thread between two parallel vertical plates 5 cm apart. The charge on the sphere is 6×10^{-9} C. What potential difference between the plates will cause the thread to assume an angle of 30° with the vertical?

26–5 Two point charges $q_1 = +40 \times 10^{-9}$C and $q_2 = -30 \times 10^{-9}$ C are 10 cm apart. Point A is midway between them, point B is 8 cm from q_1 and 6 cm from q_2. Find (a) the potential at point A; (b) the potential at point B; (c) the work required to carry a charge of 25×10^{-9} C from point B to point A.

26–6 Three equal point charges of 3×10^{-7} C are placed at the corners of an equilateral triangle whose side is one meter. What is the potential energy of the system? Take as zero potential energy the energy of the three charges when they are infinitely far apart.

26–7 The potential at a certain distance from a point charge is 600 V, and the electric field is 200 N C^{-1}.

a) What is the distance to the point charge?

b) What is the magnitude of the charge?

26–8 Two point charges whose magnitudes are $+20 \times 10^{-9}$ C and -12×10^{-9} C are separated by a distance of 5 cm. An electron is released from rest between the two charges, 1 cm from the negative charge, and moves along the line connecting the two charges. What is its velocity when it is 1 cm from the positive charge?

26–9 Two positive point charges, each of magnitude q, are fixed on the y-axis at the points $y = +a$ and $y = -a$.

a) Draw a diagram showing the positions of the charges.

b) What is the potential V_0 at the origin?

c) Show that the potential at any point on the x-axis is

$$V = k\frac{2q}{\sqrt{a^2 + x^2}}.$$

d) Sketch a graph of the potential on the x-axis as a function of x over the range from $x = +4a$ to $x = -4a$.

e) At what value of x is the potential one-half that at the origin?

26–10 Consider the same distribution of charges as in Problem 26–9.

a) Sketch a graph of the potential on the y-axis as a function of y, over the range from $y = +4a$ to $y = -4a$.

b) Discuss the physical meaning of the graph at the points $+a$ and $-a$.

c) At what point or points on the y-axis is the potential equal to that at the origin?

d) At what points on the y-axis is the potential equal to half its value at the origin?

26–11 Consider the same charge distribution as in Problem 26–9.

a) Suppose a positively charged particle of charge q' and mass m is placed precisely at the origin and released from rest. What happens?

b) What will happen if the charge in part (a) is displaced slightly in the direction of the y-axis?

c) What will happen if it is displaced slightly in the direction of the x-axis?

26–12 Again consider the charge distribution in Problem 26–9. Suppose a positively charged particle of charge q' and mass m is displaced slightly from the origin in the direction of the x-axis.

a) What is its velocity at infinity?

b) Sketch a graph of the velocity of the particle as a function of x.

c) If the particle is projected toward the left along the x-axis from a point at a large distance to the right of the origin, with a velocity half that acquired in part (a), at what distance from the origin will it come to rest?

d) If a negatively charged particle were released from rest on the x-axis, at a very large distance to the left of the origin, what would be its velocity as it passed the origin?

26–13 A positive charge $+q$ is located at the point $x = -a$, $y = -a$, and an equal negative charge $-q$ is located at the point $x = +a$, $y = -a$.

a) Draw a diagram showing the positions of the charges.

b) What is the potential at the origin?

c) What is the expression for the potential at a point on the x-axis, as a function of x?

d) Sketch a graph of the potential as a function of x, in the range from $x = +4a$ to $x = -4a$. Plot positive potentials upward, negative potentials downward.

26–14 A potential difference of 1600 V is established between two parallel plates 4 cm apart. An electron is released from the negative plate at the same instant that a proton is released from the positive plate.

a) How far from the positive plate will they pass each other?

b) How do their velocities compare when they strike the opposite plates?

c) How do their energies compare when they strike the opposite plates?

26–15 Consider the same charge distribution as in Problem 26–9.

a) Construct a graph of the potential energy of a positive point charge on the x-axis, as a function of x.

b) Construct a graph of the potential energy of a negative point charge on the axis, as a function of x.

c) What is the potential gradient at the origin, in the direction of the x-axis?

26–16 In the Bohr model of the hydrogen atom, a single electron revolves around a single proton in a circle of radius R.

a) By equating the electrical force to the electron mass times its acceleration, derive an expression for the electron's speed.

b) Obtain an expression for the electron's kinetic energy, and show that its magnitude is just half that of the electrical potential energy.

c) Obtain an expression for the total energy, and evaluate it using $R = 0.528 \times 10^{-10}$ m.

26–17 A vacuum diode consists of a cylindrical cathode of radius 0.05 cm, mounted coaxially within a cylindrical anode 0.45 cm in radius. The potential of the anode is 300 V above that of the cathode. An electron leaves the surface of the cathode with zero initial velocity. Find its velocity when it strikes the anode.

26–18 A vacuum triode may be idealized as follows. A plane surface (the cathode) emits electrons with negligible initial velocities. Parallel to the cathode and 3 mm away from it is an open grid of fine wire at a potential of 18 V above the cathode. A second plane surface (the anode) is 12 mm beyond the grid and is at a potential of 15 V above the cathode. Assume that the plane of the grid is an equipotential surface, and that the potential gradients between cathode and grid, and between grid and anode, are uniform. Assume also that the structure of the grid is sufficiently open for electrons to pass through it freely.

a) Draw a diagram of potential vs. distance, along a line from cathode to anode.

b) With what velocity will electrons strike the anode?

26–19 A positively charged ring of radius R is placed with its plane perpendicular to the x-axis and with its center at the origin.

a) Construct a graph of the potential V at points on the x-axis, as a function of x.

b) Construct, in the same diagram, a graph of the magnitude of the electric field E.

c) How is the second graph related geometrically to the first?

26–20 A metal sphere of radius r_a is supported on an insulating stand at the center of a hollow metal sphere of inner radius r_b. There is a charge $+q$ on the inner sphere and a charge $-q$ on the outer.

a) Show that the potential difference between the spheres is

$$V_{ab} = kq\left(\frac{1}{r_a} - \frac{1}{r_b}\right).$$

b) Show that the electric intensity at any point between the spheres is

$$E = \frac{V_{ab}}{(1/r_a - 1/r_b)} \cdot \frac{1}{r^2}.$$

c) Find the electric field at a point outside the larger sphere, at a distance r from the center, where $r > r_b$.

d) Suppose the charge on the outer sphere is not $-q$ but a negative charge of different magnitude, say $-Q$. Show that the answers for (a) and (b) are the same as before but (c) is different.

(See Problem 25–21.)

26–21 A long metal cylinder of radius r_a is supported on an insulating stand on the axis of a long hollow metal cylinder of inner radius r_b. The positive charge per unit length on the inner cylinder is λ and there is an equal negative charge per unit length on the outer cylinder.

a) Show that the potential difference between the cylinders is

$$V_{ab} = 2k\lambda \ln \frac{r_b}{r_a}.$$

b) Show that the electric intensity at any point between the cylinders is

$$E = \frac{V_{ab}}{\ln(r_b/r_a)} \cdot \frac{1}{r}$$

c) What is the potential difference if the outer cylinder has no net charge? (See Problem 25–24.)

26–22 Refer to Problem 25–25. (a) Find the expression for the potential V as a function of r, both inside and outside the sphere, relative to a point at infinity. (b) Sketch graphs of V and E as functions of r from $r = 0$ to $r = 3R$, and compare with Fig. 26–5.

26–23 Refer to Problem 25–26. (a) Find the expressions for the potential V as a function of r, both inside and outside the cylinder. Let $V = 0$ at the surface of the cylinder. (b) Sketch graphs of V and E as functions of r, from $r = 0$ to $r = 3R$, and compare with Fig. 26–5.

26–24 In an apparatus for measuring the electronic charge e by Millikan's method, an electric intensity of 6.34×10^4 V m^{-1} is required to maintain a certain charged oil drop at rest. If the plates are 1.5 cm apart, what potential difference between them is required?

26–25 An oil droplet of mass 3×10^{-11} g and of radius 2×10^{-4} cm carries 10 excess electrons. What is its terminal velocity (a) when falling in a region in which there is no electric field? (b) When falling in an electric field whose intensity is 3×10^5 N C^{-1} directed downward? The viscosity of air is 180×10^{-7} N s m^{-2}. (Neglect the buoyant force of the air.)

26–26 A charged oil drop, in a Millikan oil-drop apparatus, is observed to fall through a distance of 1 mm in a time of 27.4 s, in the absence of any external field. The same drop can be held stationary in a field of 2.37×10^4 N C^{-1}. How many excess electrons has the drop acquired? The viscosity of air is 180×10^{-7} N s m^{-2}. The density of oil is 824 kg m^{-3}, and the density of air is 1.29 kg m^{-3}.

26–27

a) Prove that when a particle of constant mass and charge is accelerated from rest in an electric field, its final velocity is proportional to the square root of the potential difference through which it is accelerated.

b) Find the magnitude of the proportionality constant if the particle is an electron, the velocity is in meters per second, and the potential difference is in volts.

c) What is the final velocity of an electron accelerated through a potential difference of 1136 V if it has an initial velocity of 10^7 m s^{-1}?

26–28

a) What is the maximum potential difference through which an electron can be accelerated if its kinetic energy is not to exceed 1% of the rest energy?

b) What is the speed of such an electron, expressed as a fraction of the speed of light, c?

c) Make the same calculations for a *proton.*

26–29 The Cambridge electron accelerator accelerates electrons through a potential difference of 6.5×10^9 V, so that their kinetic energy is 6.5×10^9 eV.

a) What is the ratio of the speed v of an electron having this energy to the speed of light, c?

b) What would the speed be if computed from the principles of classical mechanics?

26–30 An electron in a certain x-ray tube is accelerated from rest through a potential difference of 180,000 V in going from the cathode to the anode. When it arrives at the anode, what is its

a) kinetic energy in eV,

b) its total energy,

c) its velocity.

d) What is the velocity of the electron, calculated classically?

26–31 Calculate, relativistically, the amount of work in MeV that must be done (a) to bring an electron from rest to a velocity of $0.4c$, and (b) to increase its velocity from $0.4c$ to $0.8c$. (c) What is the ratio of the kinetic energy of the electron at the velocity of $0.8c$ to that of $0.4c$ when computed (1) from relativistic values and (2) from classical values?

26–32 Find the potential energy of the interaction of two protons at a distance of 10^{-15} m, typical of the dimensions of nuclei. Express your result in MeV.

26–33 Find the energy equivalent of the rest mass of the proton; express your result in MeV.

26–34 An alpha particle with kinetic energy 10 MeV makes a head-on collision with a gold nucleus at rest. What is the distance of closest approach of the two particles?

26–35 An electron with kinetic energy 100 MeV collides head-on with a gold nucleus at rest. Assuming that the gold nucleus can be treated as a uniform distribution of charge through a sphere of radius 7×10^{-15} m and that the electron can penetrate into the nucleus, what is its kinetic energy when it reaches the center of the nucleus?

26–36 The electric intensity in the region between the deflecting plates of a certain cathode-ray oscilloscope is 30,000 N C^{-1}.

a) What is the force on an electron in this region?

b) What is the acceleration of an electron when acted on by this force?

26–37 In Fig. 26–13, an electron is projected along the axis midway between the plates of a cathode-ray tube with an initial velocity of 2×10^7 m s^{-1}. The uniform electric field between the plates has an intensity of 20,000 N C^{-1} and is upward.

a) How far below the axis has the electron moved when it reaches the end of the plates?

b) At what angle with the axis is it moving as it leaves the plates?

c) How far below the axis will it strike the fluorescent screen S?

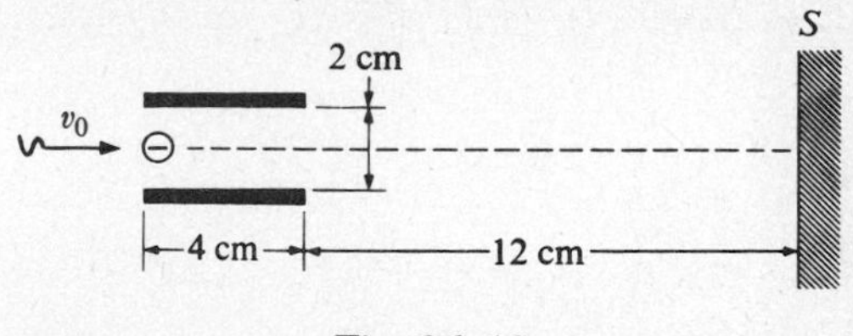

Fig. 26–13

26–38 The maximum charge that can be retained by one of the spherical terminals of a large Van de Graaff generator is about 10^{-3} C. Assume a positive charge of this magnitude, distributed uniformly over the surface of a sphere in otherwise empty space.

a) Compute the magnitude of the electric intensity at a point outside the sphere, 5 m from its center.

b) If an electron were released at this point, what would be the magnitude and direction of its initial acceleration?

26–39 Suppose the potential difference between the spherical terminal of a Van de Graaff generator and the point at which charges are sprayed onto the upward-moving belt is 2×10^6 V. If the belt delivers negative charge to the sphere at the rate of 2×10^{-3} C s^{-1} and removes positive charge at the same rate, what horsepower must be expended to drive the belt against electrical forces?

Chapter 27

Capacitance. Properties of Dielectrics

27–1 CAPACITORS

Any two conductors separated by an insulator are said to form a *capacitor.* In most cases of practical interest the conductors have charges of equal magnitude and opposite sign, so that the *net* charge on the capacitor as a whole is zero. The electric field in the region between the conductors is then proportional to the magnitude of this charge, and it follows that the *potential difference* V_{ab} between the conductors is also proportional to the charge magnitude Q.

The *capacitance* C of a capacitor is defined as the ratio of the magnitude of the charge Q on *either* conductor to the magnitude of the potential difference V_{ab} between the conductors:

$$\boxed{C = \frac{Q}{V_{ab}}.} \tag{27–1}$$

It follows from its definition that the unit of capacitance is one *coulomb per volt* (1 C V^{-1}). A capacitance of one coulomb per volt is called one *farad* (1 F) in honor of Michael Faraday. A capacitor is represented by the symbol

$$—||—$$

Capacitors find many applications in electrical circuits. Capacitors are used for tuning radio circuits and for "smoothing" the rectified current delivered by a power supply. A capacitor is used to eliminate sparking when a circuit containing inductance is suddenly opened. The ignition system of every automobile engine contains a capacitor to eliminate sparking of the "points" when they open and close. The efficiency of alternating current power transmission can often be increased by the use of large capacitors.

27–2 THE PARALLEL-PLATE CAPACITOR

The most common type of capacitor consists in principle of two conducting plates parallel to each other and separated by a distance which is small compared with the linear dimensions of the plates (see Fig. 27–1). Practically the entire field of such a capacitor is localized in the region between the plates as shown. There is a slight "fringing" of the field at its outer boundary, but the fringing becomes relatively less as the plates are brought closer together. If the plates are sufficiently close, the fringing may be neglected, the field between the plates is uniform, and the charges on the plates are uniformly distributed over their opposing surfaces. This arrangement is known as a *parallel-plate capacitor.*

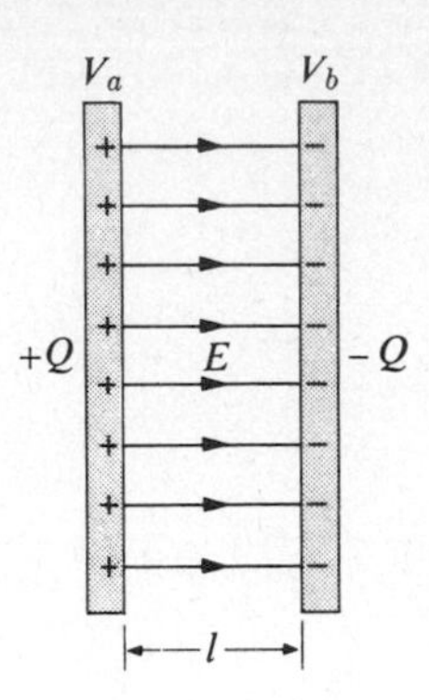

Fig. 27-1 *Parallel-plate capacitor.*

Let us assume first that the plates are in vacuum. It has been shown that the electric intensity between a pair of closely spaced parallel plates in vacuum is

$$E = \frac{\sigma}{\epsilon_0} = \frac{Q}{\epsilon_0 A},$$

where σ is the magnitude of surface charge density on either plate, A is the area of each plate, and Q is the magnitude of total charge on each plate. Since the electric field (potential gradient) between the plates is uniform, the potential difference between the plates is

$$V_{ab} = E\ell = \frac{1}{\epsilon_0}\frac{Q\ell}{A},$$

where ℓ is the separation of the plates. Hence the capacitance of a parallel-plate capacitor in vacuum is

$$\boxed{C = \frac{Q}{V_{ab}} = \epsilon_0 \frac{A}{\ell}.} \qquad (27\text{–}2)$$

Since ϵ_0, A, and ℓ are constants for a given capacitor, the capacitance is a constant independent of the charge on the capacitor, and is directly proportional to the area of the plates and inversely proportional to their separation. If mksc units are used, A is to be expressed in square meters and ℓ in meters. The capacitance C will then be in farads.

As an example, let us compute the area of the plates of a 1-F parallel-plate capacitor if the separation of the plates is 1 mm and the plates are in vacuum:

$$C = \epsilon_0 \frac{A}{\ell},$$

$$A = \frac{C\ell}{\epsilon_0} = \frac{1\text{ F} \times 10^{-3}\text{ m}}{8.85 \times 10^{-12}\text{ C}^2\text{ N}^{-1}\text{ m}^{-2}}$$

$$= 1.13 \times 10^8\text{ m}^2.$$

This corresponds to a square 10,600 m, or 34,800 ft, or about $6\frac{1}{2}$ milies on a side!

Since the farad is such a large unit of capacitance, units of more convenient size are the *microfarad* ($1\ \mu\text{F} = 10^{-6}\text{F}$), and the *picofarad* ($1\text{ pF} = 10^{-12}\text{F}$). For example, a common radio set contains in its power supply several capacitors whose capacitances are of the order of ten or more microfarads, while the capacitances of the tuning capacitors are of the order of ten to one hundred picofarads.

Variable capacitors whose capacitance may be varied at will (between limits) are widely used in the tuning circuits of radio receivers. These are usually air capacitors of relatively small capacitance and are constructed of a number of fixed parallel metal plates connected together and constituting one "plate" of the capacitor, while a second set of movable plates also connected together forms the other "plate" (Fig. 27-2). By rotating a shaft on which the movable plates are mounted, the second set may be caused to interleave the first to a greater or lesser extent. The effective area of the capacitor is that of the interleaved portion of the plates only.

A variable capacitor is represented by the symbol

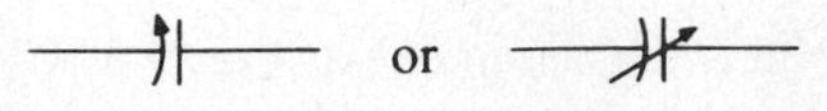

Capacitors consisting of concentric spheres and of coaxial cylinders are sometimes used in standards laboratories, since the corrections for the "fringing" fields can be made more readily and the capacitance

Fig. 27-2 Variable air capacitor. (Courtesy of General Radio Company.)

can be accurately calculated from the dimensions of the apparatus. Problems involving spherical and cylindrical capacitors will be found at the end of the chapter.

Example The plates of a parallel-plate capacitor are 5 mm apart and 2 m^2 in area. The plates are in vacuum. A potential difference of 10,000 volts is applied across the capacitor. Compute (a) the capacitance, (b) the charge on each plate, and (c) the electric intensity in the space between them.

a)
$$C = \epsilon_0 \frac{A}{d} = \frac{(8.85 \times 10^{-12}\ \mathrm{C^2\,N^{-1}\,m^{-2}})(2\ \mathrm{m^2})}{5 \times 10^{-3}\ \mathrm{m}} = 3.54 \times 10^{-9}\ \mathrm{C^2\,N^{-1}\,m^{-1}}.$$

But

$$1\ \mathrm{C^2\,N^{-1}\,m^{-1}} = 1\ \mathrm{C^2\,J^{-1}} = 1\ \mathrm{C(J/C)^{-1}} = 1\ \mathrm{C\,V^{-1}} = 1\ \mathrm{F},$$

so

$$C = 3.54 \times 10^{-9}\mathrm{F} = 0.00354\ \mu\mathrm{F}.$$

b) The charge on the capacitor is

$$Q = CV_{ab} = (3.54 \times 10^{-9}\ \mathrm{C\,V^{-1}})(10^4\mathrm{V}) = 3.54 \times 10^{-5}\mathrm{C}.$$

c) The electric field is

$$E = \frac{\sigma}{\epsilon_0} = \frac{Q}{\epsilon_0 A} = \frac{3.54 \times 10^{-5}\ \mathrm{C}}{(8.85 \times 10^{-12}\ \mathrm{C^2\,N^{-1}\,m^{-2}})(2\ \mathrm{m^2})} = 20 \times 10^5\ \mathrm{N\,C^{-1}};$$

or, since the electric field equals the potential gradient,

$$E = \frac{V_{ab}}{d} = \frac{10^4\mathrm{V}}{5 \times 10^{-3}\ \mathrm{m}} = 20 \times 10^5\mathrm{V\,m^{-1}}.$$

Of course, the newton per coulomb and the volt per meter are equivalent units.

27-3 CAPACITORS IN SERIES AND PARALLEL

In Fig. 27-3(a), two capacitors are connected in *series* between points a and b, maintained at a constant potential difference V_{ab}. The capacitors are both initially uncharged. In this connection, both capacitors always have the same charge Q. One might ask whether the lower plate of C_1 and the upper plate of C_2 might have charges different from those on the remaining two plates, but in this case the net charge on each capacitor would not be zero, and the resulting electric field in the conductor connecting the two capacitors would cause a current to flow until the total charge on each capacitor is zero. Hence, in a series connection the magnitude of charge on all plates is the same.

Referring again to Fig. 27-3(a), we have

$$V_{ac} \equiv V_1 = \frac{Q}{C_1}, \qquad V_{cb} \equiv V_2 = \frac{Q}{C_2},$$

and

$$V_{ab} \equiv V = V_1 + V_2 = Q\left(\frac{1}{C_1} + \frac{1}{C_2}\right),$$

$$\frac{V}{Q} = \frac{1}{C_1} + \frac{1}{C_2}. \qquad (27\text{-}3)$$

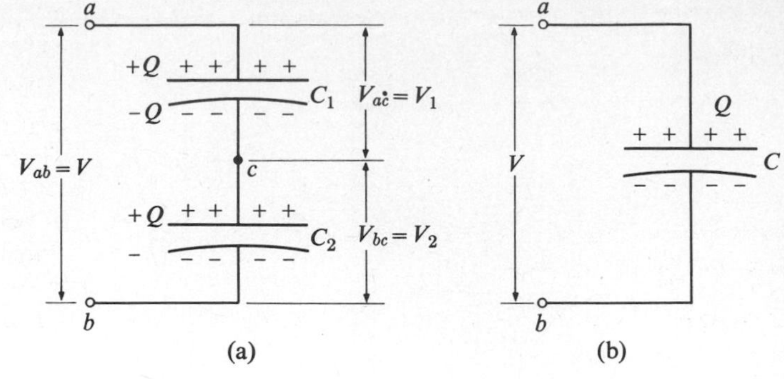

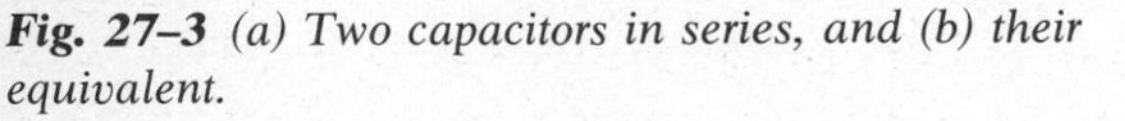

Fig. 27–3 *(a) Two capacitors in series, and (b) their equivalent.*

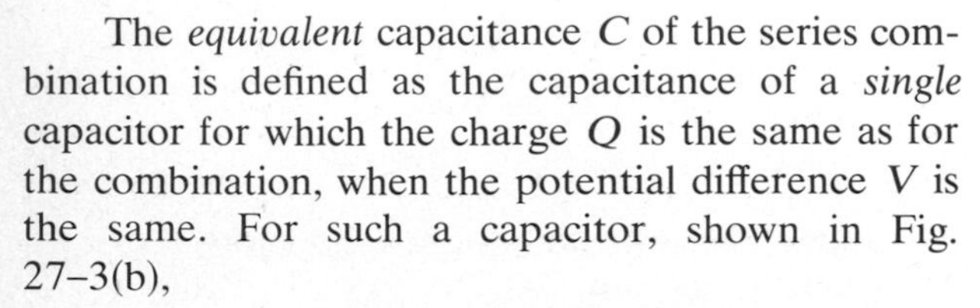

The *equivalent* capacitance C of the series combination is defined as the capacitance of a *single* capacitor for which the charge Q is the same as for the combination, when the potential difference V is the same. For such a capacitor, shown in Fig. 27–3(b),

$$Q = CV, \qquad \frac{V}{Q} = \frac{1}{C}. \tag{27–4}$$

Hence, from Eqs. (27–3) and (27–4),

$$\frac{1}{C} = \frac{1}{C_1} + \frac{1}{C_2}.$$

Similarly, for any number of capacitors in series,

$$\boxed{\frac{1}{C} = \frac{1}{C_1} + \frac{1}{C_2} + \frac{1}{C_3} + \cdots} \tag{27–5}$$

The reciprocal of the equivalent capacitance equals the sum of the reciprocals of the individual capacitances.

In Fig. 27–4(a), two capacitors are conected in *parallel* between points a and b. In this case, the potential difference $V_{ab} \equiv V$ is the same for both, and the charges Q_1 and Q_2, not necessarily equal, are

$$Q_1 = C_1 V, \qquad Q_2 = C_2 V.$$

The *total* charge Q supplied by the source was

$$Q = Q_1 + Q_2 = V(C_1 + C_2),$$

and

$$\frac{Q}{V} = C_1 + C_2. \tag{27–6}$$

The *equivalent* capacitance C of the parallel combination is defined as that of a single capacitor, shown in Fig. 27–4(b), for which the total charge is the same as in part (a). For this capacitor,

$$\frac{Q}{V} = C,$$

and hence

$$C = C_1 + C_2.$$

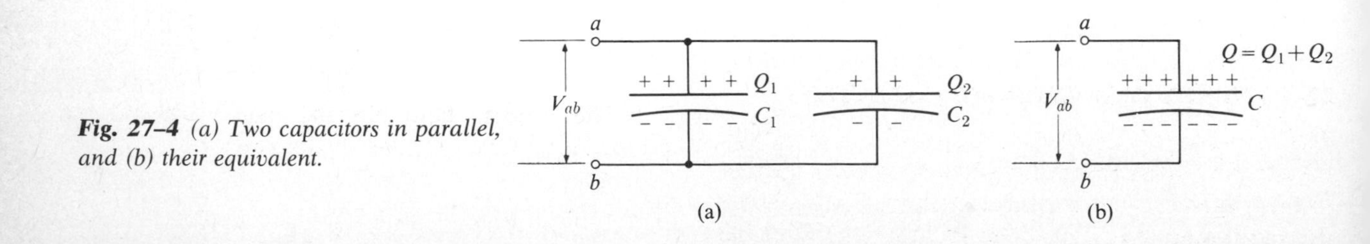

Fig. 27–4 *(a) Two capacitors in parallel, and (b) their equivalent.*

In the same way, for any number of capacitors in parallel

$$C = C_1 + C_2 + C_3 + \cdots \tag{27–7}$$

The equivalent capacitance equals the *sum* of the individual capacitances.

Example In Figs. 27–3 and 27–4, let $C_1 = 6\ \mu\text{F}$, $C_2 = 3\ \mu\text{F}$, $V_{ab} = 18$ V.

The equivalent capacitance of the series combination in Fig. 27–3(a) is given by

$$\frac{1}{C} = \frac{1}{6\ \mu\text{F}} + \frac{1}{3\ \mu\text{F}}, \qquad C = 2\ \mu\text{F}.$$

The charge Q is

$$Q = CV = 2\ \mu\text{F} \times 18\ \text{V} = 36\ \mu\text{C}.$$

The potential differences across the capacitors are

$$V_{ac} \equiv V_1 = \frac{Q}{C_1} = \frac{36\ \mu\text{C}}{6\ \mu\text{F}} = 6\ \text{V},$$

$$V_{cb} \equiv V_2 = \frac{Q}{C_2} = \frac{36\ \mu\text{C}}{3\ \mu\text{F}} = 12\ \text{V}.$$

The *larger* potential difference appears across the *smaller* capacitor.

The equivalent capacitance of the parallel combination in Fig. 27–4(a) is

$$C = C_1 + C_2 = 9\ \mu\text{F}.$$

The charges Q_1 and Q_2 are

$$Q_1 = C_1 V = 6\ \mu\text{F} \times 18\ \text{V} = 108\ \mu\text{C},$$

$$Q_2 = C_2 V = 3\ \mu\text{F} \times 18\ \text{V} = 54\ \mu\text{C}.$$

27–4 ENERGY OF A CHARGED CAPACITOR

The process of charging a capacitor consists of transferring charge from the plate at lower potential to the plate at higher potential. The charging process therefore requires the expenditure of *energy*. Imagine the charging process to be carried out by starting with both plates completely uncharged, and then repeatedly removing small positive charges from one plate and transferring them to the other plate. At a stage of the process at which the magnitude of the net charge on either plate is q, the potential difference v between the plates is $v = q/C$, and, since the work is independent of the path, the work dW to transfer the next charge dq is

$$dW = v\,dq = q\,dq/C.$$

The total work W to increase the charge from zero to a final value Q is

$$W = \int dW = \frac{1}{C}\int_0^Q q\,dq = \frac{Q^2}{2C}.$$

The final potential difference V between the plates is $V = Q/C$ and we can also write

$$W = \frac{Q^2}{2C} = \frac{1}{2}CV^2 = \frac{1}{2}QV. \tag{27–8}$$

The work is expressed in joules when Q is in coulombs and V is in volts.

A charged capacitor is the electrical analog of a stretched spring, whose elastic potential energy equals $\frac{1}{2}kx^2$. The charge Q is analogous to the elongation x, and the *reciprocal* of the capacitance, $1/C$, is analogous to the force constant k. The energy supplied to a capacitor in the charging process is stored by the capacitor and is released when the capacitor discharges.

It is often useful to consider the stored energy to be localized in the *electric field* between the capacitor plates. The capacitance of a parallel-plate capacitor in vacuum is

$$C = \epsilon_0 \frac{A}{\ell}.$$

The electric field fills the space between the plates, of volume $A\ell$, and is given by

$$E = \frac{V}{\ell}.$$

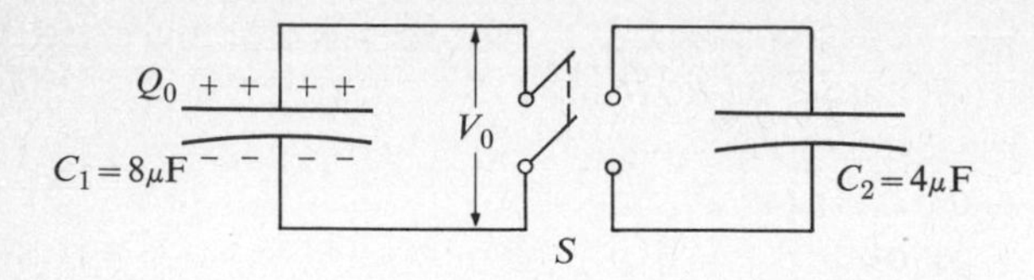

Fig. 27–5 When the switch S is closed, the charged capacitor C_1 is connected to an uncharged capacitor C_2.

The energy per unit volume, or the *energy density* denoted by u, is

$$u = \text{Energy density} = \frac{\frac{1}{2}CV^2}{A\ell}.$$

Making use of the preceding equations, we can express this as

$$u = \tfrac{1}{2}\epsilon_0 E^2. \tag{27–9}$$

Example In Fig. 27–5, let $C_1 = 8\ \mu F$ and $V_0 = 120$ V. The charge Q_0 is

$$Q_0 = C_1 V_0 = 960\ \mu C,$$

and the energy of the capacitor is

$$\frac{1}{2} Q_0 V_0 = 576 \times 10^{-4} J.$$

After the switch S is closed, the positive charge of magnitude Q_0 is distributed over the upper plates of both capacitors, and the negative charge Q_0 is distributed over the lower plates of both. Let Q_1 and Q_2 represent the final charges on the respective capacitors. Then

$$Q_1 + Q_2 = Q_0.$$

When the motion of charges has ceased, both upper plates are at the same potential and both lower plates are at the same potential (different from that of the upper plate). The final potential difference between the plates, V, is therefore the same for both capacitors, and

$$Q_1 = C_1 V, \qquad Q_2 = C_2 V.$$

When this is combined with the preceding equation, we find that

$$V = \frac{Q_0}{C_1 + C_2} = \frac{960\ \mu C}{12\ \mu F} = 80\ V,$$

$$Q_1 = 640\ \mu C, \qquad Q_2 = 320\ \mu C.$$

The final energy of the system is

$$\frac{1}{2} Q_0 V = 384 \times 10^{-4} J.$$

This is less than the original energy of 576×10^{-4} J, the difference being converted to energy of some other form. If the resistance of the connecting wires is large, most of the energy is converted to heat. If the resistance is small, most of the energy is radiated in the form of electromagnetic waves.

The process above is exactly analogous to an inelastic collision of a moving car with a stationary car. In the electrical case, the charge $Q = CV$ is conserved. In the mechanical case, the momentum $p = mv$ is conserved. The electrical energy $\frac{1}{2}CV^2$ is *not* conserved, and the mechanical energy $\frac{1}{2}mv^2$ is *not* conserved.

27–5 EFFECT OF A DIELECTRIC

Most capacitors have a solid, nonconducting material or *dielectric* between their plates. A common type of capacitor incorporates strips of metal foil, forming the plates, separated by strips of wax-impregnated paper or plastic sheet such as Mylar, which serves as the dielectric. A sandwich of these materials is rolled up, forming a compact unit which can provide a capacitance of several microfarads in a relatively small volume.

Electrolytic capacitors have as their dielectric an extremely thin layer of nonconducting oxide between a metal plate and a conducting solution. Because of the thinness of the dielectric electrolytic capacitors of relatively small dimensions may have a capacitance of the order of 100 to 1000 μF.

The function of a solid dielectric between the plates of a capacitor is threefold. First, it solves the mechanical problem of maintaining two large metal sheets at an extremely small separation but without

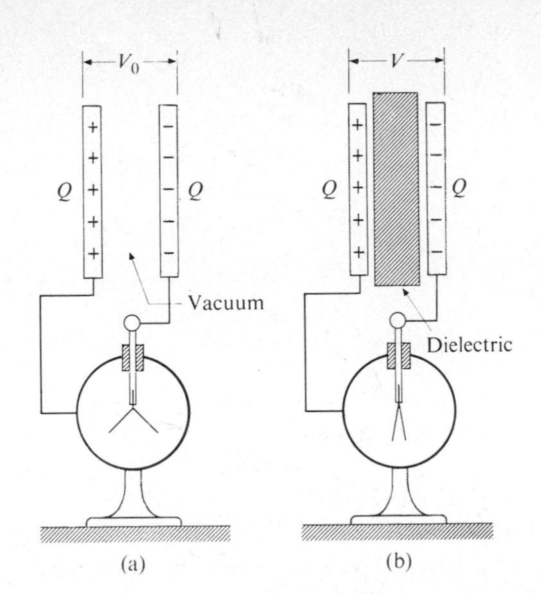

Fig. 27–6 Effect of a dielectric between the plates of a parallel-plate capacitor. (a) With a given charge, the potential difference is V_0. (b) With the same charge, the potential difference V is smaller than V_0.

actual contact. Second, since its dielectric strength is larger than that of air, the maximum potential difference which the capacitor can withstand without breakdown (conduction through the material which is supposed to insulate) is increased. Third, the capacitance of a capacitor of given dimensions is several times larger with a dielectric separating its plates than if the plates were in vacuum.

This effect can be demonstrated as follows. Figure 27–6(a) illustrates a parallel-plate capacitor whose plates have been given equal and opposite charges of magnitude Q. The plates are assumed to be in vacuum, and the potential difference V_0 between the plates is indicated by an electroscope. If a sheet of dielectric, such as glass, bakelite, or hard rubber, is now inserted between the plates, as in Fig. 27–6(b), the potential difference is observed to *decrease* to a smaller value V. If the dielectric is removed, the potential difference returns to its original value, showing that the original charges on the plates have not been affected by insertion of the dielectric.

Table 27–1 DIELECTRIC CONSTANT K

Material	T,° C	K
Vacuum		1
Glass	25	5–10
Mica	25	3–6
Hevea rubber	27	2.94
Neoprene	24	6.70
Bakelite	27	5.50
	57	7.80
	88	18.2
Plexiglas	27	3.40
Polyethylene	23	2.25
Vinylite	20	3.18
	47	3.60
	76	3.92
	96	6.60
	110	9.9
Teflon	22	2.1
Germanium	20	16
Strontium titanate	20	310
Titanium dioxide (rutile)	20	173(⊥), 86(∥)
Water	25	78.54
Glycerin	25	42.5
Liquid ammonia	−77.7	25
Benzene	20	2.284
Air (1 atm)	20	1.00059
Air (100 atm)	20	1.0548

The original capacitance of the capacitor, C_0, was

$$C_0 = \frac{Q}{V_0}.$$

Since Q does not change and V is observed to be less than V_0, it follows that C is greater than C_0. The ratio of C to C_0 is called the *dielectric constant* of the material, K.

$$K = \frac{C}{C_0}. \tag{27–10}$$

Since C is always greater than C_0, the dielectric constants of all dielectrics are greater than unity. Some representative values of K are given in Table 27–1. For a vacuum, of course, $K = 1$, and K for air

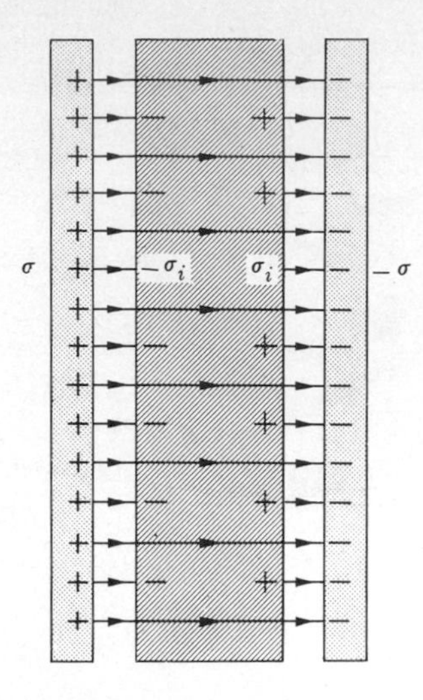

Fig. 27–7 *Induced charges on the faces of a dielectric in an external field.*

is so nearly equal to 1 that for most purposes an air capacitor is equivalent to one in vacuum; the original measurement of V_0 in Fig. 27–6(a) could have been made with the plates in air instead of in vacuum.

With vacuum (or air) between its plates, the electric intensity E_0 in the region between the plates of a parallel-plate capacitor is

$$E_0 = \frac{V_0}{\ell} = \frac{\sigma}{\epsilon_0}.$$

The observed reduction in potential difference, when a dielectric is inserted between the plates implies a reduction in the *electric field*, which in turn implies a reduction in the charge per unit area. Since no charge has leaked off the plates, such a reduction could be caused only by induced charges of opposite sign appearing on the two surfaces of the *dielectric*. That is, the dielectric surface adjacent to the positive plate must have an *induced negative charge* and that adjacent to the negative plate an *induced positive charge of equal magnitude*, as shown in Fig. 27–7.

If σ_i is the magnitude of induced charge per unit area on the surfaces of the dielectric, the electric intensity in the dielectric is

$$E = \frac{V}{\ell} = \frac{\sigma - \sigma_i}{\epsilon_0}. \tag{27–11}$$

But

$$K = \frac{C}{C_0} = \frac{Q/V}{Q/V_0} = \frac{V_0}{V} = \frac{E_0}{E} = \frac{\sigma}{\sigma - \sigma_i}; \tag{27–12}$$

and therefore

$$\sigma - \sigma_i = \frac{\sigma}{K}. \tag{27–13}$$

Substituting Eq. (27–13) into Eq. (27–11), we get

$$E = \frac{\sigma}{K\epsilon_0}. \tag{27–14}$$

The product $K\epsilon_0$ is called the *permittivity* of the dielectric and is represented by ϵ.

$$\boxed{\epsilon = K\epsilon_0.} \tag{27–15}$$

The electric intensity within the dielectric may therefore be written

$$E = \frac{\sigma}{\epsilon}. \tag{27–16}$$

Also,

$$C = KC_0 = K\epsilon_0\frac{A}{\ell},$$

and the capacitance of a parallel plate capacitor with a dielectric between its plates is therefore

$$\boxed{C = \epsilon\frac{A}{\ell}.} \tag{27–17}$$

In empty space, where $K = 1$, $\epsilon = \epsilon_0$, and therefore ϵ_0 may be described as the "permittivity of empty space" or the "permittivity of vacuum." Since K is a pure number, the units of ϵ and ϵ_0 are evidently the same, $C^2\ N^{-1}\ m^{-2}$.

Example The parallel plates in Fig. 27–6 have an area of 2000 cm^2 or $2 \times 10^{-1}\ m^2$, and are 1 cm or 10^{-2} m apart. The original potential difference between them, V_0, is 3000 V, and it decreases to 1000 V when a sheet of dielectric is inserted between the plates. Compute (a) the original capacitance C_0, (b) the charge Q on each plate, (c) the capacitance C after insertion of the dielectric, (d) the dielectric constant K of the dielectric, (e) the permittivity ϵ of the dielectric, (f) the induced charge Q_i on each face of the dielectric, (g) the original electric field E_0 between the plates, and (h) the electric field E after insertion of the dielectric.

a) $$C_0 = \epsilon_0 \frac{A}{\ell}$$

$$= (8.85 \times 10^{-12} \mathrm{C^2\,N^{-1}\,m^{-2}}) \frac{2 \times 10^{-1}\,\mathrm{m}^2}{10^{-2}\,\mathrm{m}}$$

$$= 17.7 \times 10^{-11}\,\mathrm{F} = 177\,\mathrm{pF}.$$

b) $$Q = C_0 V_0$$

$$= (17.7 \times 10^{-11}\,\mathrm{F})(3 \times 10^3\,\mathrm{V})$$

$$= 53.1 \times 10^{-8}\,\mathrm{C}.$$

c) $$C = \frac{Q}{V} = \frac{53.1 \times 10^{-8}\,\mathrm{c}}{10^3\,\mathrm{V}}$$

$$= 53.1 \times 10^{-11}\,\mathrm{F}.$$

d) $$K = \frac{C}{C_0}$$

$$= \frac{53.1 \times 10^{-11}\,\mathrm{F}}{17.7 \times 10^{-11}\,\mathrm{F}} = 3.$$

The dielectric constant could also be found from Eq. (27–12),

$$K = \frac{V_0}{V} = \frac{3000\,\mathrm{V}}{1000\,\mathrm{V}} = 3.$$

e) $$\epsilon = K\epsilon_0 = (3)(8.85 \times 10^{-12}\,\mathrm{C^2\,N^{-1}\,m^{-2}})$$

$$= 26.6 \times 10^{-12}\,\mathrm{C^2\,N^{-1}\,m^{-2}}.$$

f) $$Q_i = A\sigma_i, \qquad Q = A\sigma$$

$$\sigma - \sigma_i = \frac{\sigma}{K}, \qquad \sigma_i = \sigma\left(1 - \frac{1}{K}\right),$$

$$Q_i = \left(1 - \frac{1}{K}\right)Q$$

$$= (53.1 \times 10^{-8}\,\mathrm{C})\left(1 - \frac{1}{3}\right)$$

$$= 35.4 \times 10^{-8}\,\mathrm{C}.$$

g) $$E_0 = \frac{V_0}{\ell} = \frac{3000\,\mathrm{V}}{10^{-2}\,\mathrm{m}} = 3 \times 10^5\,\mathrm{V\,m^{-1}}.$$

h) $$E = \frac{V}{\ell} = \frac{1000\,\mathrm{V}}{10^{-2}\,\mathrm{m}} = 1 \times 10^5\,\mathrm{V\,m^{-1}};$$

or

$$E = \frac{\sigma}{\epsilon} = \frac{Q}{A\epsilon}$$

$$= \frac{53.1 \times 10^{-8}\,\mathrm{C}}{(2 \times 10^{-1}\,\mathrm{m}^2)(26.6 \times 10^{-12}\,\mathrm{C^2\,N^{-1}\,m^{-2}})}$$

$$= 1 \times 10^5\,\mathrm{V\,m^{-1}};$$

or

$$E = \frac{\sigma - \sigma_i}{\epsilon_0} = \frac{Q - Q_i}{A\epsilon_0}$$

$$= \frac{(53.1 - 35.4) \times 10^{-8}\,\mathrm{C}}{(2 \times 10^{-1}\,\mathrm{m}^2)(8.85 \times 10^{-12}\,\mathrm{C^2\,N^{-1}\,m^{-2}})}$$

$$= 1 \times 10^5\,\mathrm{V\,m^{-1}};$$

or, from Eq. (27–12),

$$E = \frac{E_0}{K} = \frac{3 \times 10^5\,\mathrm{V\,m^{-1}}}{3}$$

$$= 1 \times 10^5\,\mathrm{V\,m^{-1}}.$$

Any dielectric material, when subjected to a sufficiently strong electric field, becomes a conductor, a phenomenon known as *dielectric breakdown.* The onset of conduction, associated with cumulative ionization of molecules of the material, is often quite sudden, and may be characterized by spark or arc discharges. When a capacitor is subjected to excessive voltage, an arc may be formed through a layer of dielectric, burning or melting a hole in it, permitting the two metal foils to come in contact, creating a short circuit, and rendering the device permanently useless as a capacitor.

The maximum electric field a material can withstand without the occurrence of breakdown is called the *dielectric strength.* Because dielectric breakdown is affected significantly by impurities in the material, small irregularities in the metal electrodes, and other factors difficult to control, only approximate figures can be given for dielectric strengths. Typical values

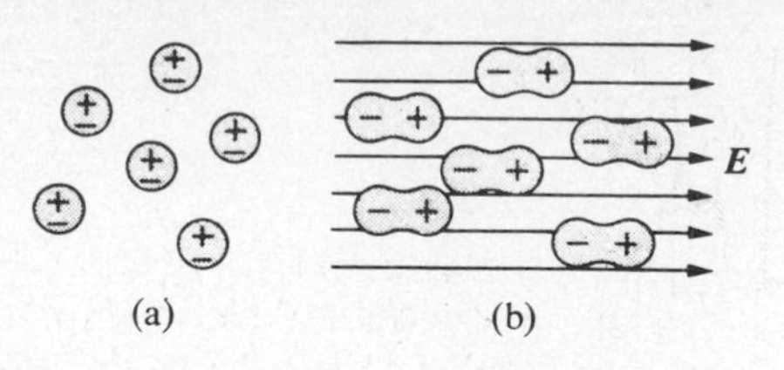

Fig. 27–8 Behavior of nonpolar molecules (a) in the absence and (b) in the presence of an electric field.

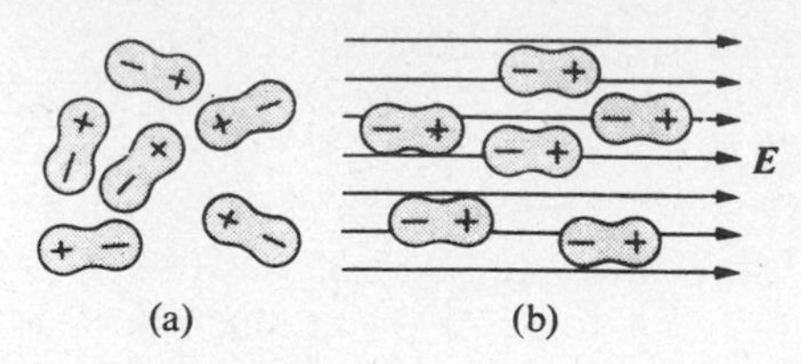

Fig. 27–9 Behavior of polar molecules (a) in the absence and (b) in the presence of an electric field.

for plastic and ceramic materials commonly used to insulate capacitors and current-carrying wires are of the order of 10^7 V m^{-1}. For example, a layer of such a material, 10^{-4} m in thickness, could withstand a maximum voltage of 1000 V, for 1000 V/10^{-4} m = 10^7 V m^{-1}.

27–6 MOLECULAR THEORY OF INDUCED CHARGES ON A DIELECTRIC

When a *conductor* is placed in an electric field, the free charges within it are displaced by the forces exerted on them by the field. In the final steady state, the conductor has an induced charge on its surface, distributed in such a way that the field of this induced charge neutralizes the original field at all internal points and the net electric field within the conductor is reduced to zero. A *dielectric*, however, contains no free charges. How, then, is it possible for an induced charge to appear on the surfaces of a dielectric when it is inserted in the electric field between the plates of a charged capacitor?

The molecules of a dielectric may be classified as either *polar* or *nonpolar*. A nonpolar molecule is one in which the "centers of gravity" of the positive nuclei and the electrons normally coincide, while a polar molecule is one in which they do not. Symmetrical molecules like H_2, N_2, and O_2 are nonpolar. In the molecules N_2O and H_2O, on the other hand, both nitrogen atoms or both hydrogen atoms lie on the same side of the oxygen atom. These molecules are *polar*, and each is a tiny electric dipole.

Under the influence of an electric field, the charges of a nonpolar molecule become displaced, as indicated schematically in Fig. 27–8. The molecules are said to become *polarized* by the field and are

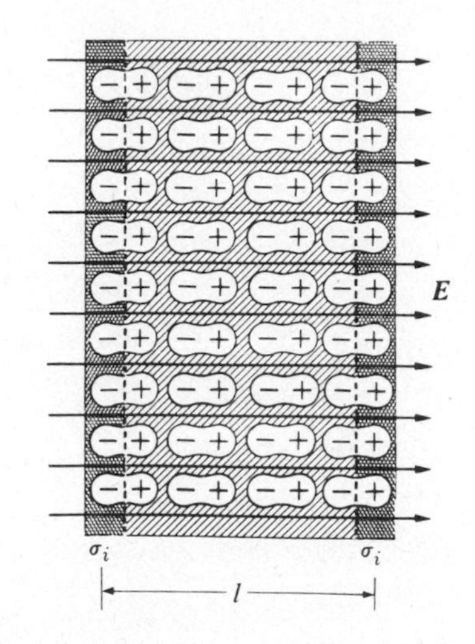

Fig. 27–10 Polarization of a dielectric in an electric field gives rise to thin layers of bound charges on the surfaces.

called *induced* dipoles. When a nonpolar molecule becomes polarized, restoring forces come into play on the displaced charges, pulling them together much as if they were connected by a spring. Under the influence of a given external field, the charges separate until the restoring force is equal and opposite to the force exerted on the charges by the field. Naturally, the restoring forces vary in magnitude from one kind of molecule to another, with corresponding differences in the displacement produced by a given field.

When a dielectric consists of polar molecules or *permanent dipoles*, these dipoles are oriented at random when no electric field is present, as in Fig. 27–9(a). When an electric field is present, as in Fig. 27–9(b), the forces on a dipole give rise to a torque whose effect is to orient the dipole in the same

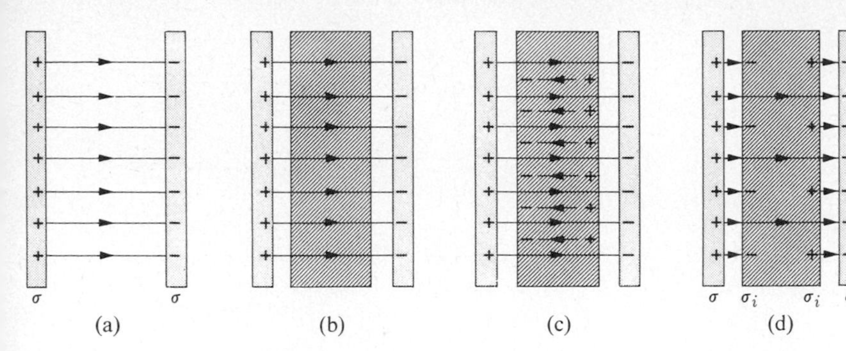

Fig. 27–11 *(a) Electric field between two charged plates. (b) Introduction of a dielectric. (c) Induced surface charges and their field. (d) Resultant field when a dielectric is between charged plates.*

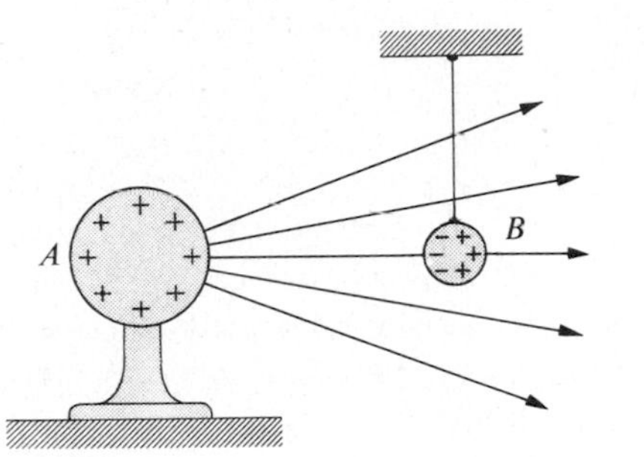

Fig. 27–12 *An uncharged dielectric sphere B in the radial field of a positive charge A.*

direction as the field. The stronger the field, the greater is the aligning effect.

Whether the molecules of a dielectric are polar or nonpolar, the net effect of an external field is substantially the same, as shown in Fig. 27–10. Within the two extremely thin surface layers indicated by dotted lines there is an excess charge, negative in one layer and positive in the other. It is these layers of charge which give rise to the induced charge on the surface of a dielectric. The charges are not free, but each is *bound* to a molecule lying in or near the surface. Within the remainder of the dielectric the net charge per unit volume remains zero.

The four parts of Fig. 27–11 illustrate the behavior of a sheet of dielectric when inserted in the field between a pair of oppositely charged plane parallel plates. Part (a) shows the original field. Part (b) is the situation after the dielectric has been inserted but before any rearrangement of charges has occurred. Part (c) shows by dotted lines the field set up in the dielectric by its induced surface charges. This field is opposite to the original field but, since the charges in the dielectric are not free to move indefinitely, their displacement does not proceed to such an extent that the induced field is equal in magnitude to the original field. The field in the dielectric is therefore *weakened* but not reduced to zero, as it would be in the interior of a conductor.

The resultant field is shown in Fig. 27–11(d). Some of the lines of force leaving the positive plate penetrate the dielectric; others terminate on the induced charges on the faces of the dielectric.

The charges induced on the surface of a dielectric in an external field afford an explanation of the attraction of an *uncharged* pith ball or bit of paper by a charged rod of rubber or glass. Figure 27–12 shows an uncharged dielectric sphere *B* in the radial field of a positive charge *A*. The induced positive charges on *B* experience a force toward the right, while the force on the negative charges is toward the left. Since the negative charges are closer to *A* and therefore in a stronger field than are the positive, the force toward the left exceeds that toward the right, and *B*, although its net charge is zero, experiences a resultant force toward *A*. The sign of *A*'s charge does not affect the conclusion, as may readily be seen. Furthermore, the effect is not limited to dielectrics; a conducting sphere would be similarly attracted.

More general arguments based on energy considerations show that a dielectric body in a nonuniform field *always* experiences a force urging it from a region where the field is weak toward a region where it is stronger, provided the dielectric constant of the body is greater than that of the medium in which it is immersed. If the dielectric constant is less, the reverse is true.

27–7 POLARIZATION AND DISPLACEMENT

The extent to which the molecules of a dielectric become polarized by an electric field, or oriented in the direction of the field, is described by a vector quantity called the *polarization* $\boldsymbol{P}$. If $\boldsymbol{p}$ is the component of the vector dipole moment of each molecule in the direction of the applied field, and there are n molecules per unit volume, the polarization is defined as

$$\boldsymbol{P} = n\boldsymbol{p}. \tag{27–18}$$

Polarization is therefore *dipole moment per unit volume*. The polarization vector has the same direction as the molecular dipole moments, from left to right in Fig. 27–10. For the special case in Fig. 27–10, the magnitude of $\boldsymbol{P}$ is the same at all points of the dielectric. In other cases it can vary from point to point and the quantities n and $\boldsymbol{p}$ then refer to a small volume including the point. The mksc unit of $\boldsymbol{P}$ is one *coulomb meter per cubic meter*, or one *coulomb per square meter* (1 C m^{-2}).

The dipole moment of a dipole is defined as the product of either of the charges making up the dipole, and the charge separation. The polarized dielectric in Fig. 27–10 can be considered a single large dipole, consisting of the induced bound charges Q_i at the opposite faces, separated by the thickness ℓ of the dielectric. The dipole moment of the dielectric is then $Q_i\ell$, and since the volume of the dielectric is the product of its cross-sectional area A and thickness ℓ, the dipole moment per unit volume, or the polarization P, is

$$P = \frac{Q_i\ell}{A\ell} = \frac{Q_i}{A} = \sigma_i, \tag{27–19}$$

where σ_i is the surface density of bound charge. In this special case, the polarization is numerically equal to the surface density of bound charge. More generally, the surface density of bound charge equals the normal component of $\boldsymbol{P}$ at the surface.

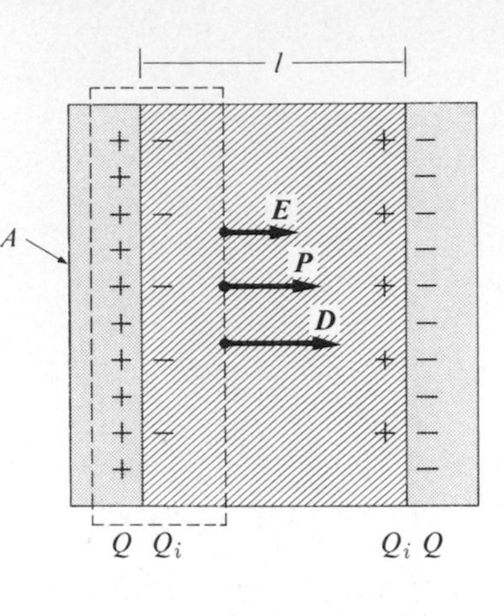

Fig. 27–13

Figure 27–13 shows in cross section a sheet of polarized dielectric between two oppositely charged conducting plates. The thickness is exaggerated; we assume the sheet is very thin compared with its other dimensions, so that the electric field is uniform throughout the dielectric. The broken line is the outline of a Gaussian surface in the form of a cylinder with its ends parallel to the sheet and having cross-sectional area A. We consider the surface integral of the polarization vector $\boldsymbol{P}$ over this closed surface. $\boldsymbol{P}$ is zero at the left face, which is inside the left conducting plate. On the sides of the cylinder, $\boldsymbol{P}$ is parallel to the surface and makes no contribution to the integral. Only the right face, perpendicular to $\boldsymbol{P}$, contributes, and gives simply PA. But, from Eq. (27–19), this is equal to Q_i, the total induced bound charge enclosed by the surface.

In the general case, we have

$$\oint \boldsymbol{P} \cdot d\boldsymbol{A} = -Q_i, \tag{27–20}$$

where the minus sign must be included because as is evident from the diagram the flux of $\boldsymbol{P}$ is *outward* (and hence positive) while the enclosed bound charge is negative. Equation (27–20) is Gauss's law for the polarization vector $\boldsymbol{P}$: *the surface integral of* $\boldsymbol{P}$ *over any closed surface* (the flux of $\boldsymbol{P}$) *equals the negative of the bound charge within the surface.*

The resultant electric field $\boldsymbol{E}$ at any point, when bound charges are present, is due to both the free and bound charges. The general form of Gauss's law for $\boldsymbol{E}$ is therefore

$$\oint \boldsymbol{E} \cdot d\boldsymbol{A} = \frac{1}{\epsilon_0}(Q + Q_i). \qquad (27\text{–}21)$$

We may use Eq. (27–20) to eliminate the induced charge Q_i from Eq. (27–21); the result is

$$\oint \boldsymbol{E} \cdot d\boldsymbol{A} = \frac{1}{\epsilon_0}\left(Q - \oint \boldsymbol{P} \cdot d\boldsymbol{A}\right) \qquad (27\text{–}22)$$

or

$$\oint (\epsilon_0 \boldsymbol{E} + \boldsymbol{P}) \cdot d\boldsymbol{A} = Q. \qquad (27\text{–}23)$$

Let us define a new quantity $\boldsymbol{D}$ called the *displacement* as the vector sum

$$\boldsymbol{D} = \epsilon_0 \boldsymbol{E} + \boldsymbol{P}. \qquad (27\text{–}24)$$

Equation (27–23) then takes the simple form

$$\oint \boldsymbol{D} \cdot d\boldsymbol{A} = Q, \qquad (27\text{–}25)$$

which is Gauss's law for the displacement vector: *the surface integral of* $\boldsymbol{D}$ *over any closed surface* (the flux of $\boldsymbol{D}$) *is equal to the free charge only within the surface.*

In summary, for any closed surface, the flux of $\boldsymbol{E}$ equals the *total* enclosed charge (divided by ϵ_0), the flux of $\boldsymbol{P}$ equals the (negative of the) *bound* charge, and the flux of $\boldsymbol{D}$ equals the *free* charge.

Applying Eq. (27–25) to the Gaussian surface of Fig. 27–13, we see that the magnitude of $\boldsymbol{D}$ is given simply by $D = \sigma$, where σ is the surface density of free charge on the conducting plate. Similarly, applying Gauss's law for $\boldsymbol{E}$ to this same surface gives $E = \sigma - \sigma_i$, where σ_i is the magnitude of the induced surface charge. Thus we find

$$\frac{D}{E} = \frac{\sigma}{E_0(\sigma - \sigma_i)}.$$

Comparing this with Eq. (27–12), we obtain the simple relation

$$D = K\epsilon_0 E = \epsilon E. \qquad (27\text{–}26)$$

More generally, in any dielectric where the polarization is proportional to the electric field (so σ_i is proportional to σ), the vector quantities $\boldsymbol{D}$ and $\boldsymbol{E}$ are related by

$$\boldsymbol{D} = \epsilon \boldsymbol{E}.$$

This relation will be useful in our study of electromagnetic waves in later chapters.

PROBLEMS

27–1 An air capacitor consisting of two closely spaced parallel plates has a capacitance of 1000 pF. The charge on each plate is 1 μC.

a) What is the potential difference between the plates?
b) If the charge is kept constant, what will be the potential difference between the plates if the separation is doubled?
c) How much work is required to double the separation?

27–2 A capacitor has a capacitance of 8.5 μF. How much charge must be removed to lower the potential difference of its plates by 50 V?

27–3 The capacitance of a variable radio capacitor can be changed from 50 pF to 950 pF by turning the dial from 0° to 180°. With the dial set at 180°, the capacitor is connected to a 400-V battery. After charging, the capacitor is disconnected from the battery and the dial is turned to 0°.

a) What is the charge on the capacitor?
b) What is the potential difference across the capacitor when the dial reads 0°?
c) What is the energy of the capacitor in this position?
d) How much work is required to turn the dial, if friction is neglected?

27–4 A 20-μF capacitor is charged to a potential difference of 1000 V The terminals of the charged capacitor are then connected to those of an uncharged 5-μF capacitor. Compute (a) the original charge of the system, (b) the final potential difference across each capacitor, (c) the final energy of the system, and (d) the decrease in energy when the capacitors are connected.

27–5 In Fig. 27–14, each capacitance $C_3 = 3\ \mu$F and each capacitance $C_2 = 2\ \mu$F.

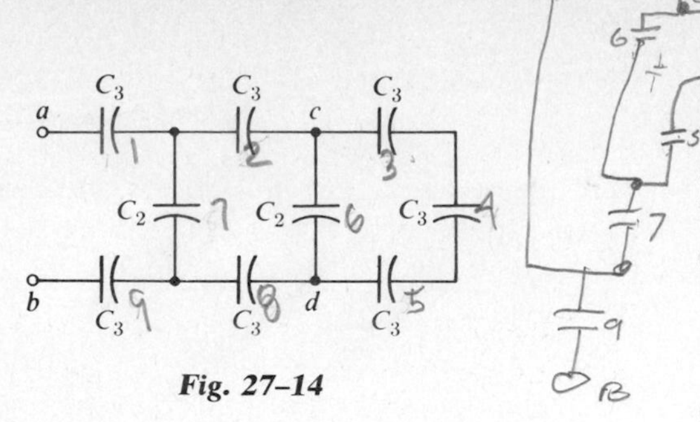

Fig. 27–14

a) Compute the equivalent capacitance of the network between points a and b.

b) Compute the charge on each of the capacitors nearest a and b when $V_{ab} = 900$ V.

c) With 900 V across a and b, compute V_{cd}.

27–6 A number of 0.5-μF capacitors are available. The voltage across each is not to exceed 400 V. A capacitor of capacitance 0.5 μF is required to be connected across a potential difference of 600 V.

a) Show in a diagram how an equivalent capacitor having the desired properties can be obtained.

b) No dielectric is a perfect insulator, of infinite resistance. Suppose that the dielectric in one of the capacitors in your diagram is a moderately good conductor. What will happen?

27–7 A 1-μF capacitor and a 2-μF capacitor are connected in series across a 1200-V supply line.

a) Find the charge on each capacitor and the voltage across each.

b) The charged capacitors are disconnected from the line and from each other, and reconnected with terminals of like sign together. Find the final charge on each and the voltage across each.

27–8 A 1-μF capacitor and a 2-μF capacitor are connected in parallel across a 1200-V supply line.

a) Find the charge on each capacitor and the voltage across each.

b) The charged capacitors are then disconnected from the line and from each other, and reconnected with terminals of unlike sign together. Find the final charge on each and the voltage across each.

27–9 In Fig. 27–4(a), let $C_1 = 6\ \mu$F, $C_2 = 3\ \mu$F, $V_{ab} = 18$ V. Suppose that the charged capacitors are disconnected from the source and from each other, and reconnected with plates of *opposite* sign connected together. By how much does the energy of the system decrease?

27–10 Three capacitors having capacitances of 8, 8, and 4 μF are connected in series across a 12-V line.

a) What is the charge on the 4-μF capacitor?

b) What is the total energy of all three capacitors?

c) The capacitors are disconnected from the line and reconnected in parallel with the positively charged plates connected together. What is the voltage across the parallel combination?

d) What is the energy of the combination?

27–11 A 500-μF capacitor is charged to 120 V. How many calories are produced on discharging the capacitor if all of the energy goes into heating the wire?

27–12 The capacitors in Fig. 27–15 are initially uncharged, and are connected as in the diagram with switch S open. (a) What is the potential difference V_{ab}? (b) What is the potential of point b after switch S is closed? (c) How much charge flowed through the switch when it was closed?

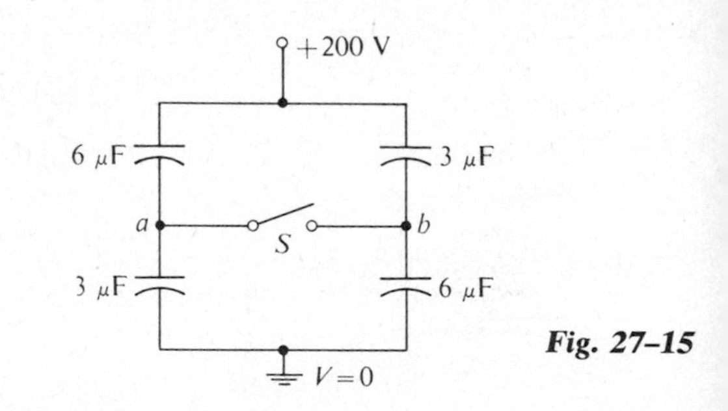

Fig. 27–15

27–13 The plates of a parallel-plate capacitor in vacuum have charges $+Q$ and $-Q$ and the distance between the plates is x. The plates are disconnected from the charging voltage and pulled apart a short distance dx. (a) What is the change dC in the capacitance of the capacitor? (b) What is the change dW in its energy? (c) Equate the work $F\,dx$ to the increase in energy dW and find the force of attraction F between the plates. (d) Explain why F is not equal to QE, where E is the electric intensity between the plates.

27–14 A parallel-plate capacitor is to be constructed using, as a dielectric, rubber having a dielectric coefficient of 3 and a dielectric strength of 2×10^5 V cm^{-1}. The capacitor is to have a capacitance of 0.15 μF and must be able to withstand a maximum potential difference of 6000 V. What is the minimum area the plates of the capacitor may have?

27–15 The paper dielectric in a paper and foil capacitor is 0.005 cm thick. Its dielectric coefficient is 2.5 and its dielectric strength is 50×10^6 V m^{-1}.

a) What area of paper, and of foil, is required for a 0.1-μF capacitor?

b) If the electric intensity in the paper is not to exceed one-half the dielectric strength, what is the maximum potential difference that can be applied across the capacitor?

27–16

a) The permittivity of diamond is $1.46 \times 10^{-10} \text{C}^2 \text{N}^{-1} \text{m}^{-2}$. What is the dielectric constant of diamond?

b) What is its susceptibility?

c) What is the dielectric constant of a metal?

27–17 Three square metal plates, *A*, *B*, and *C*, each 10 cm on a side and 3 mm thick, are arranged as in Fig. 27–16. The plates are separated by sheets of paper 0.5 mm thick and of dielectric constant 5. The outer plates are connected together and connected to point *b*. The inner plate is connected to point *a*.

a) Copy the diagram, and show by + and − signs the charge distribution on the plates when point *a* is maintained at a positive potential relative to point *b*.

b) What is the capacitance between points *a* and *b*?

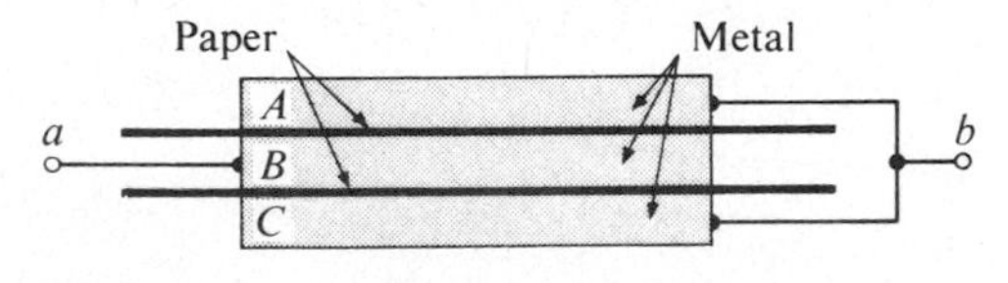

Fig. 27–16

27–18 Two parallel plates have equal and opposite charges. When the space between the plates is evacuated, the electric intensity is $2 \times 10^5 \text{V m}^{-1}$. When the space is filled with dielectric, the electric intensity is $1.2 \times 10^5 \text{V m}^{-1}$.

a) What is the charge density on the surface of the dielectric?

b) What is the permittivity of the dielectric?

c) What is its dielectric constant?

27–19 two oppositely charged conducting plates, having numerically equal quantities of charge per unit area, are separated by a dielectric 5 mm thick, of dielectric coefficient 3. The resultant electric intensity in the dielectric is 10^6 V m^{-1}. Compute (a) the free charge per unit area on the conducting plate, (b) the bound charge per unit area on the surfaces of the dielectric, (c) the polarization *P* in the dielectric, and (d) the displacement *D* in the dielectric.

27–20 A capacitor consists of two parallel plates of area 25 cm^2 separated by a distance of 0.2 cm. The material between the plates has a dielectric constant of 5. The plates of the capacitor are connected to a 300-V battery.

a) What is the capacitance of the capacitor?

b) What is the charge on either plate?

c) What is the energy in the charged capacitor?

d) What is the energy density in the dielectric?

27–21 Two parallel plates of 100 cm^2 area are given equal and opposite charges of 10^{-7}C. The space between the plates is filled with a dielectric material, and the electric intensity within the dielectric is $3.3 \times 10^5 \text{V m}^{-1}$.

a) What is the dielectric constant of the the dielectric?

b) What is the total induced charge on either face of the dielectric?

27–22 A spherical capacitor consists of an inner metal sphere of radius r_a supported on an insulating stand at the center of a hollow metal sphere of inner radius r_b. There is a charge $+Q$ on the inner sphere and a charge $-Q$ on the outer.

a) What is the potential difference V_{ab} between the spheres?

b) Prove that the capacitance is

$$C = \frac{1}{k} \cdot \frac{r_b r_a}{r_b - r_a}.$$

(See Problem 26–20.)

27–23 A coaxial cable consists of an inner solid cylindrical conductor of radius r_a supported by insulating disks on the axis of a conducting tube of inner radius r_b. The two cylinders are oppositely charged with a charge λ per unit length. (a) What is the potential difference between the two cylinders? (b) Prove that the capacitance of a length *L* of the cable is

$$C = \frac{L}{2k \ln (r_b/r_a)}.$$

Neglect any effect of the supporting disks. (See Problem 26–21.)

27–24 A parallel-plate air capacitor is made using two plates 0.2 m square, spaced 1 cm apart. It is connected to a 50-V battery.

a) What is the capacitance?

b) What is the charge on each plate?

c) What is the electric field between the plates?

d) What is the energy stored in the capacitor?

e) If the battery is disconnected and then the plates are pulled apart to a separation of 2 cm, what are the answers to parts (a), (b), (c), and (d)?

27–25 In Problem 27–24, suppose the battery remains connected while the plates are pulled apart. What are the answers to parts (a), (b), (c), and (d)?

27–26 A parallel-plate capacitor with plate area A and separation x is charged to a charge of magnitude q on each plate

a) What is the total energy stored in the capacitor?

b) The plates are now pulled apart an additional distance dx; now what is the total energy?

c) If F is the force with which the plates attract each other, then the difference in the two above energies must equal the work $W = F\,dx$ done in pulling the plates apart. Hence show that $F = q^2/2\,\epsilon_0 A$.

27–27 A parallel-plate capacitor has the space between the plates filled with a slab of dielectric with constant K_1 and one with constant K_2, each of thickness $d/2$, where d is the plate separation. Show that the capacitance is

$$C = \frac{2\epsilon_0 A}{d}\left(\frac{K_1 K_2}{K_1 + K_2}\right).$$

Chapter 28

Current, Resistance, and Electromotive Force

28–1 CURRENT

When there is a net flow of charge across any area, we say there is a *current* across the area. If an isolated conductor is placed in an electrostatic field, the charges in the conductor rearrange themselves so as to make the interior of the conductor a field-free region throughout which the potential is constant. The motion of the charges in the rearranging process constitutes a *transient* current, of short duration only, and the current ceases when the field in the conductor becomes zero. To maintain a *continuous* current, we must in some way maintain a force on the mobile charges in a conductor. The force may result from an electrostatic field or from other causes that will be described later. For the present, we assume that there is maintained within a conductor an electric field $\boldsymbol{E}$ such that a charged particle in the conductor is acted on by a force $\boldsymbol{F} = q\boldsymbol{E}$. We shall refer to this force as the *driving force* on the particle.

The motion of a free charged particle in a conductor is very different from that of a particle in empty space. After a momentary acceleration, the particle makes an inelastic collision with one of the fixed particles in the conductor, loses whatever velocity it has acquired in the direction of the driving force, and makes a fresh start. Thus on the average it moves in the direction of the driving force with an average velocity called its *drift velocity*. The inelastic collisions with the fixed particles result in a transfer of energy to them which increases their energy of vibration and causes a rise in temperature if the conductor is thermally insulated, or results in a flow of heat from the conductor to its surroundings if it is not.

The current across an area is defined quantitatively as *the net charge flowing across the area per unit time*. Thus if a net charge ΔQ flows across an area in a time Δt, the current I across the area is

$$I = \frac{\Delta Q}{\Delta t}. \tag{28–1}$$

Current is a *scalar* quantity.

The mks unit of current, *one coulomb per second*, is called *one ampere* (1 A), in honor of the French scientist André Marie Ampère (1775–1836). Small currents are more conveniently expressed in *milliamperes* (1 mA = 10^{-3} A) or in *microamperes* (1 μA = 10^{-6} A).

Since a current is a flow of charge, the common expression "flow of current" should be avoided, since it means literally "flow of flow of charge." Thus one should say, for example, "The current in a conductor is 10 A," not "The current *flowing* in the conductor is 10 A."

A *galvanometer* is an instrument that indicates, by the deflection of a pointer or a beam of light reflected from a mirror, the existence of a current through it. When the scale is properly calibrated, the galvanometer becomes an *ammeter* (or a *milliammeter* or *microammeter*). The construction of such instruments will be described later.

The current across an area can be expressed in terms of the drift velocity of the moving charges as follows. Consider a portion of a conductor of cross-sectional area A within which there is a resultant electric field $\boldsymbol{E}$ from left to right. For generality, we assume that the conductor contains free charged particles of both signs. The positively charged particles move in the same direction as the field and the negatively charged particles move in the opposite direction. A few positive particles are shown in Fig. 28–1. Suppose there are n_1 such particles per unit volume, all moving with a drift velocity v_1. In a time Δt each advances a distance $v_1\Delta t$. Hence, all of the particles within the shaded cylinder of length $v_1\Delta t$, and only those particles, will flow across the end of the cylinder in time Δt. The volume of the cylinder is $Av_1\Delta t$, the number of particles within it is $n_1Av_1\Delta t$, and if each has a charge q_1, the charge ΔQ_1 flowing across the end of the cylinder in time Δt is

$$\Delta Q_1 = n_1q_1v_1A\Delta t.$$

The current carried by the positively charged particles is therefore

$$I_1 = \frac{\Delta Q_1}{\Delta t} = n_1q_1v_1A.$$

In the same way, if there are n_2 *negative* particles per unit volume, each having a charge q_2 and traveling from right to left with velocity v_2, the current carried by them is

$$I_2 = n_2q_2v_2A.$$

Positive particles crossing from left to right *increase* the *positive* charge at the right of the section, while negative particles crossing from right to left *decrease* the *negative* charge at the right of the section. But a *decrease* of *negative* charge is equivalent to an *increase* of *positive* charge, so the motion of *both* kinds of charge has the same effect, namely, to increase the positive charge at the right of the section. The *total* current I at the section is therefore the *sum* of the currents I_1 and I_2:

$$I = A(n_1q_1v_1 + n_2q_2v_2).$$

In general, if a conductor contains a number of different particles, having different charge densities and moving with different velocities, the current is

$$I = A\sum nqv. \tag{28–2}$$

The *current per unit of cross-sectional area* is called the *current density* J:

$$J = \frac{I}{A} = \sum nqv. \tag{28–3}$$

The *vector current density* $\boldsymbol{J}$ is defined by the equation

$$\boldsymbol{J} = \sum nq\boldsymbol{v}. \tag{28–4}$$

Current, by definition, is a scalar quantity and thus it is not correct to speak of the "direction of a current." This expression is often used, however, for brevity, meaning thereby the direction of the *vector* current density $\boldsymbol{J}$.

The direction of the drift velocity $\boldsymbol{v}$ of a positive charge carrier is the same as that of the electric field $\boldsymbol{E}$, and the direction of the velocity of a negative carrier is opposite to $\boldsymbol{E}$. But since the charge q of such a carrier is negative, each of the vectors $nq\boldsymbol{v}$ is in the same direction as $\boldsymbol{E}$, and hence the *vector current density* $\boldsymbol{J}$ *always has the same direction as the field* $\boldsymbol{E}$. Thus even in a metallic conductor, where the

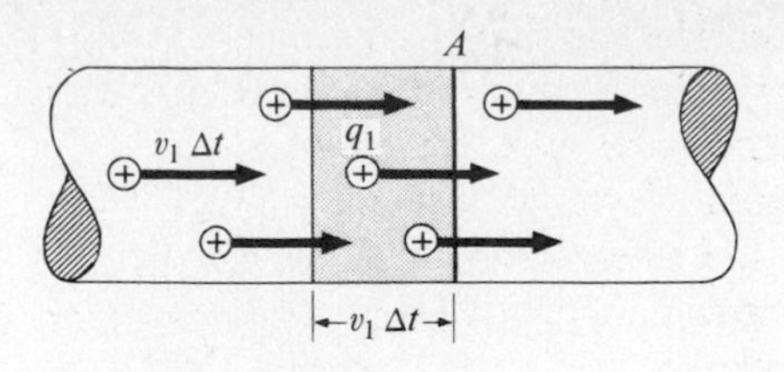

Fig. 28–1 *All of the particles, and only those particles, within the shaded cylinder will cross its base in time* Δt.

charge carriers are negative electrons only and move in the *opposite* direction to $\mathbf{E}$, the *vector* current density $\mathbf{J}$ is in the *same* direction as $\mathbf{E}$.

When there is a steady current in a closed circuit, the total charge in every portion of a conductor remains constant. Hence, if we consider a portion between two fixed cross sections, the rate of flow of charge *out* of the portion at one end equals the rate of flow of charge *into* the portion at the other end. In other words, the current is the same at any two cross sections and hence is *the same at all cross sections.* Current is *not* something that squirts out of the positive terminal of a battery and gets all used up by the time it reaches the negative terminal!

Let us estimate the drift velocity of the electrons in a wire carrying a current. Consider a copper conductor of square cross section 1 mm on a side, carrying a constant current of 20 A. The current density in the wire is

$$J = \frac{I}{A} = 20 \times 10^6 \text{ A m}^{-2}.$$

It was stated at the end of Chapter 24 that there are in copper about 10^{29} free electrons per cubic meter. Then, since $J = nqv$ and $q = e = 1.6 \times 10^{-19}$ C,

$$v = \frac{J}{nq} \approx 10^{-3} \text{ m s}^{-1},$$

or about 1 mm s^{-1}. At this speed, an electron would require 1000 s or about 15 min to travel the length of a wire 1 m long. Thus the drift velocity is very small compared with the velocity of propagation of a current pulse along a wire, about 3×10^8 m s^{-1}.

28–2 RESISTIVITY

The current density $\mathbf{J}$ in a conductor depends on the electric intensity $\mathbf{E}$, and on the nature of the conductor. In general the dependence of $\mathbf{J}$ on $\mathbf{E}$ can be quite complex, but for some materials, especially the metals, it can be represented quite well by a direct proportionality. For such materials the ratio of $\mathbf{E}$ to $\mathbf{J}$ is constant; we define the *resistivity* ρ of a particular material as the ratio of electric field to current density:

$$\rho = \frac{E}{J}. \tag{28–5}$$

That is, the resistivity is the *electric field per unit current density*. The greater the resistivity, the greater the field needed to establish a given current density, or the smaller the current density for a given field. Representative values are given in Table 28–1. (The unit, Ω m, ohm meter, will be explained shortly.) A "perfect" conductor would have zero resistivity and a "perfect" insulator an infinite resistivity. Metals and alloys have the lowest resistivities and are the best conductors. The resistivities of insulators exceed those of the metals by a factor of the order of 10^{22}.

Comparison with Table 17–1 shows that *thermal* insulators have thermal resistivities (the reciprocals of their thermal conductivities) that differ from those of good thermal conductors by factors of only about 10^3. By the use of electrical insulators, electric cur-

Table 28–1 RESISTIVITIES AT ROOM TEMPERATURE

Substance		ρ, Ω m
Conductors		
Metals	Silver	1.47×10^{-8}
	Copper	1.72×10^{-8}
	Aluminum	2.63×10^{-8}
	Tungsten	5.51×10^{-8}
Alloys	Manganin	44×10^{-8}
	Constantan	49×10^{-8}
	Nichrome	100×10^{-8}
Semiconductors		
Pure	Carbon	3.5×10^{-5}
	Germanium	0.60
	Silicon	2300
Insulators		
	Amber	5×10^{14}
	Glass	10^{10}–10^{14}
	Lucite	$> 10^{13}$
	Mica	10^{11}–10^{15}
	Quartz (fused)	75×10^{16}
	Sulfur	10^{15}
	Teflon	$> 10^{13}$
	Wood	10^{8}–10^{11}

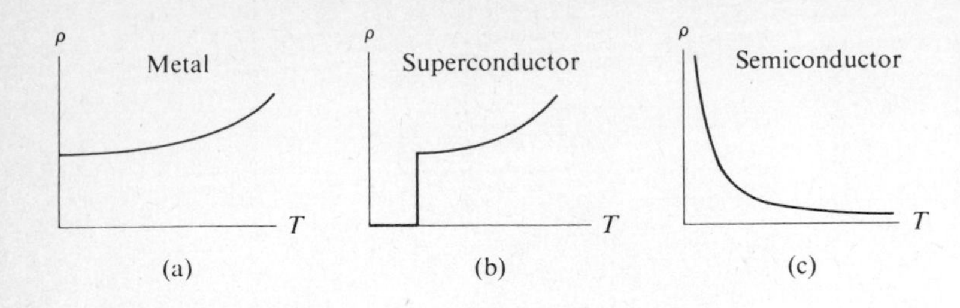

Fig. 28–2 Variation of resistivity with temperature for three conductors: (a) an ordinary metal, (b) a superconducting metal, alloy, or compound, and (c) a semiconductor.

rents can be confined to well-defined paths in good electrical conductors, while it is impossible to confine heat currents to a comparable extent. It is also interesting to note that the metals, as a class, are also the best *thermal* conductors. The free electrons in a metal which carry charge in electrical conduction also play an important role in the conduction of heat; hence, a correlation can be expected between electrical and thermal conductivity. It is a familiar fact that good electrical conductors, such as the metals, are also good conductors of heat, while poor electrical conductors, such as ceramic and plastic materials, are also poor thermal conductors.

The *semiconductors* form a class intermediate between the metals and the insulators. They are of importance not primarily because of their resistivities, but because of the way in which these are affected by temperature and by small amounts of impurities.

It follows from Eq. (28–5) that $E = \rho J$, and since the vectors $\boldsymbol{E}$ and $\boldsymbol{J}$ are in the same direction, we can write this as a vector equation

$$\boldsymbol{E} = \rho \boldsymbol{J}. \qquad (28\text{–}6)$$

The discovery that ρ is a constant for a metallic conductor at constant temperature was made by G. S. Ohm (1789–1854) and is called *Ohm's law.* A material obeying Ohm's law is called an *ohmic* conductor or a *linear* conductor. If Ohm's law is *not* obeyed, the conductor is called *nonlinear.* Thus Ohm's law, like the ideal gas equation, Hooke's law, and many other relations describing the properties of materials, is an *idealized model* which describes the behavior of certain materials reasonably well but is by no means a general property of all matter.

The resistivity of all *metallic* conductors increases with increasing temperature, as shown in Fig. 28–2(a). Over a temperature range that is not too great, the resistivity of a metal can be represented approximately by the equation

$$\rho_T = \rho_0[1 + \alpha(T - T_0)], \qquad (28\text{–}7)$$

where ρ_0 is the resistivity at a reference temperature T_0 and ρ_T the resistivity at temperature $T°$ C. The factor α is called the *temperature coefficient of resistivity.* Some representative values are given in Table 28–2. The resistivity of carbon (a nonmetal) *decreases* with increasing temperature and its temperature coefficient of resistivity is negative. The resistivity of the alloy manganin is practically independent of temperature.

A number of materials have been found to exhibit the property of *superconductivity.* As the tem-

Table 28–2 TEMPERATURE COEFFICIENTS OF RESISTIVITY (approximate values near room temperature)

Material	α, °C^{-1}
Aluminum	0.0039
Brass	0.0020
Carbon	−0.0005
Constantan (Cu 60, Ni 40)	+0.000002
Copper (Commercial annealed)	0.00393
Iron	0.0050
Lead	0.0043
Manganin (Cu 84, Mn 12, Ni 4)	0.000000
Mercury	0.00088
Nichrome	0.0004
Silver	0.0038
Tungsten	0.0045

perature is decreased, the resistivity at first decreases regularly, like that of any metal. At the so-called *critical* temperature, usually in the range 0.1 K to 20 K, the resistivity suddenly drops to zero, as shown in Fig. 28–2(b). A current once established in a superconducting ring will continue of itself, apparently indefinitely, without the presence of any driving field.

The resistivity of a *semiconductor* decreases rapidly with increasing temperature as shown in Fig. 28–2(c). A tiny bead of semiconducting material, called a *thermistor*, serves as a sensitive thermometer.

28–3 THEORY OF METALLIC CONDUCTION

In the simplest microscopic model of metallic conduction, each atom in the crystal lattice is assumed to give up one or more of its outer electrons. These electrons are then free to move through the crystal lattice, colliding at intervals with the stationary positive ions. Their motion is like that of the molecules of gas in a container and they are often referred to as an "electron gas." In the absence of an electric field, the electrons move in straight lines between collisions, but if there is an electric field, the paths are slightly curved, as in Fig. 28–3, which represents schematically a few free paths of an electron in an electric field directed from right to left. At each collision, the electron is assumed to lose any energy it may have acquired from the field and to make a fresh start. The energy given up in these collisions increases the thermal energy of vibration of the positive ions.

A force $F = eE$ is exerted on each electron by the field, and produces an acceleration a in the direction of the force given by

$$a = \frac{F}{m} = \frac{eE}{m},$$

where m is the electron mass. Let u represent the average *random* speed of an electron, and λ the mean free path. The average time t between collisions, called the *mean free time*, is

$$t = \frac{\lambda}{u}.$$

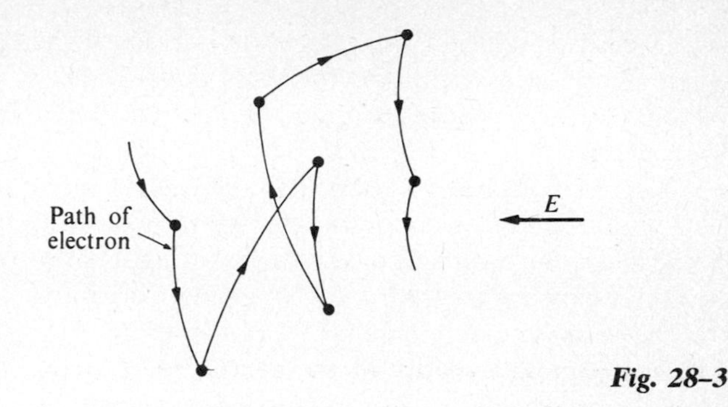

Fig. 28–3

In this time, the electron acquires a final velocity component v_f in the direction of the force, given by

$$v_f = at = \frac{eE}{m}\frac{\lambda}{u}.$$

Its *average* velocity v in the direction of the force, which is superposed on its random velocity and which we interpret as the *drift* velocity, is one-half the final velocity, so

$$v = \frac{1}{2}v_f = \frac{1}{2}\frac{e\lambda}{mu}E.$$

The drift velocity is therefore proportional to the electric field E.

The current density is

$$J = nev = \frac{ne^2\lambda}{2mu}E,$$

and the resistivity is

$$\rho = \frac{E}{J} = \frac{2mu}{ne^2\lambda}. \tag{28–8}$$

This is the theoretical expression for the resistivity, and it is in *qualitative* agreement with experiment.

At a given temperature, the quantities m, u, n, e, and λ are constant. The resistivity is then constant and Ohm's law is obeyed.

If the temperature is increased, the random speed u increases and the theory predicts that the resistivity of a metal increases with increasing temperature.

In a semiconductor, the number of charge carriers per unit volume, n, increases rapidly with increasing temperature. The increase in n far outweighs any increase in u, and the resistivity decreases. At low temperatures, n is very small and the resistivity becomes so large that the material can be considered an insulator.

The modern theory of superconductivity predicts that in effect, at temperatures below the critical temperature the electrons move freely throughout the lattice. The mean free path λ then becomes very large and the resistivity very small.

The complete theory of the conduction process is an active branch of present day physics, but the mathematical methods are beyond the scope of this book.

28–4 RESISTANCE

The current density $\boldsymbol{J}$, at every point within a conductor in which there is a resultant electric field $\boldsymbol{E}$, is given by Eq. (28–6):

$$\boldsymbol{E} = \rho \boldsymbol{J}.$$

It is often difficult to measure $\boldsymbol{E}$ and $\boldsymbol{J}$ directly, and it is useful to put this relation in a form involving readily measured quantities such as total current and potential difference. Toward this end we consider a conductor with uniform cross-section area A and length ℓ, as shown in Fig. 28–4. Assuming a constant current density over a cross section, and a uniform electric field along the length of the conductor, the total current I is given by

$$I = JA,$$

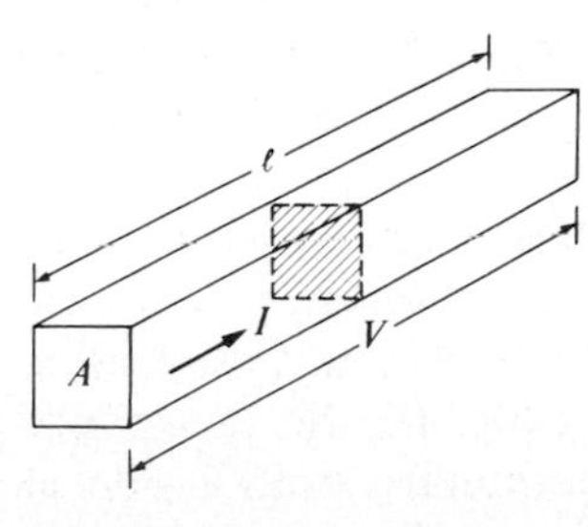

Fig. 28–4 A conductor of uniform cross section. The current density is uniform over any cross section, and the electric field is constant along the length.

and the potential difference V between the ends is

$$V = E\ell. \tag{28–9}$$

Solving these equations for J and E, respectively, and substituting the results in Eq. (28–5), we obtain

$$\frac{V}{\ell} = \frac{\rho I}{A}. \tag{28–10}$$

Thus the total current is proportional to the potential difference.

The quantity $\rho\ell/A$ for a particular specimen of material is called its *resistance* R:

$$\boxed{R = \frac{\rho\ell}{A}.} \tag{28–11}$$

Equation (28–10) then becomes

$$\boxed{V = IR.} \tag{28–12}$$

This relation is often referred to as *Ohm's law*; in this form it refers to a specific piece of material, not to a general property of the material as with Eq. (28–6).

Equation (28–11) shows that the resistance of a wire or other conductor of uniform cross section is directly proportional to its length and inversely proportional to its cross-section area. It is of course also proportional to the resistivity of the material of which the conductor is made.

The mks unit of resistance is one *volt per ampere* ($1\ \mathrm{V\ A^{-1}}$). A resistance of $1\ \mathrm{V\ A^{-1}}$ is called 1 *ohm* ($1\ \Omega$). The unit of resistivity is therefore one *ohm meter* ($1\ \Omega\ \mathrm{m}$).

Large resistances are conveniently expressed in *kilohms* ($1\ \mathrm{k}\Omega = 10^3\ \Omega$) or *megohms* ($1\ \mathrm{M}\Omega = 10^6\ \Omega$), and small resistances in microhms ($1\ \mu\Omega = 10^{-6}\ \Omega$). Resistivities are also expressed in a variety of hybrid units, most common of which is the *ohm-centimeter* ($1\ \Omega\ \mathrm{cm} = 10^{-2}\ \Omega\ \mathrm{m}$).

Because the resistance of any specimen of material is proportional to its resistivity, which varies with temperature, resistance also varies with temperature. For temperature ranges that are not too great, this variation may be represented approximately as a linear relation analogous to Eq. (28–7):

$$R_T = R_0[1 + \alpha(T - T_0)]. \tag{28–13}$$

Here R_T is the resistance at temperature T, and R_0 is the resistance at the temperature T_0, often taken to be 20°C or 0°C. Within the limits of validity of Eq. (28–13), the *change* in resistance resulting from a temperature change $T - T_0$ is given by $\alpha(T - T_0)$, where α is the temperature coefficient of resistivity, given for several common materials in Table 28–2.

Example 1 For the example at the end of Section 28–1, find the electric field, and the potential difference between two points 100 m apart.

From Eq. (28–6), the electric field is given by

$$E = \rho J = (1.72 \times 10^{-8}\ \Omega\ \text{m})(20 \times 10^{6}\ \text{A m}^{-2})$$
$$= 0.344\ \text{V m}^{-1}.$$

The potential difference is given by

$$V = E\ell = (0.344\ \text{V m}^{-1})(100\ \text{m}) = 34.4\ \text{V}.$$

Thus the resistance of a piece of this wire 100 m in length is

$$R = \frac{V}{I} = \frac{34.4\ \text{V}}{20\ \text{A}} = 1.72\ \Omega.$$

This result can also be obtained directly from Eq. (28–11):

$$R = \frac{\rho \ell}{A} = \frac{(1.72 \times 10^{-8}\ \Omega\ \text{m})(100\ \text{m})}{(1 \times 10^{-3}\ \text{m})^2} = 1.72\ \Omega.$$

Example 2 In the previous example, suppose the resistance is 1.72 Ω at a temperature of 20°C. Find the resistance at 0°C and at 100°C.

We use Eq. (28–13). In this instance $T_0 = 20°\text{C}$ and $R_0 = 1.72\ \Omega$. From Table 28–2, the temperature coefficient of resistivity of copper is $\alpha = 0.00393\ (°\text{C})^{-1}$. Thus at $T = 0°\text{C}$,

$$R = 1.72\ \Omega[1 + (0.00393°\text{C}^{-1})(0°\text{C} - 20°\text{C})]$$
$$= 1.58\ \Omega,$$

and at $T = 100°\text{C}$,

$$R = 1.72\ \Omega[1 + (0.00393°\text{C}^{-1})(100°\text{C} - 20°\text{C})]$$
$$= 2.26\ \Omega.$$

Example 3 The space between two metallic coaxial cylinders of radii r_a and r_b is filled with a material of resistivity ρ. What is the resistance between the cylinders?

Equation (28–11) cannot be used directly because the cross section through which the charge travels varies from $2\pi r_a \ell$ at the inner cylinder to $2\pi r_b \ell$ at the outer cylinder. Instead, we consider a cylindrical shell of inner radius r and thickness dr. The area A is then $2\pi r \ell$, and the length of the current path through the shell is dr. Thus, the resistance dR of the shell is

$$dR = \frac{\rho\, dr}{2\pi r \ell}$$

and the total resistance between the cylinders is

$$R = \frac{\rho}{2\pi \ell} \int_{r_a}^{r_b} \frac{dr}{r} = \frac{\rho}{2\pi \ell} \ln \frac{r_b}{r_a}.$$

28–5 ELECTROMOTIVE FORCE

The rectangle in Fig. 28–5 represents schematically a device such as a dry cell, storage battery, or generator, which can supply energy to an electric circuit. Such a device is often called a *source*, but a more appropriate term might be *energy converter.* All such devices have the property that they can maintain a potential difference between conductors a and b, called the *terminals* of the device. In Fig. 28–5, there is no conducting path *external* to the device connecting a and b, and the device is said to be on *open circuit.*

Terminal a, marked +, is maintained by the source at a *higher* potential than terminal b, marked −. Associated with this potential difference is an electrostatic field $\mathbf{E}_e$, at all points between and around the terminals, both inside and outside the source. The lines of force associated with this field are directed from a toward b, as shown by the vector $\mathbf{E}_e$. The source is itself a conductor, however (electrolytic or metallic), and if the *only* force on the free

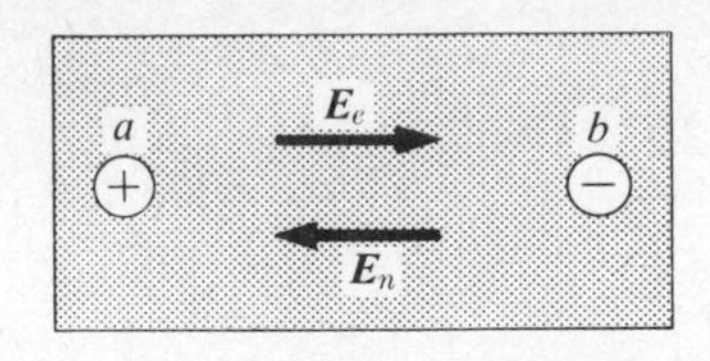

Fig. 28–5 *Schematic diagram showing the general directions of the electrostatic field* $\mathbf{E}_e$ *and the nonelectrostatic field* $\mathbf{E}_n$ *within a source on open circuit. In this case,* $\mathbf{E}_n = -\mathbf{E}_e$ *and* $V_{ab} = \mathcal{E}$.

charges within it were that exerted by the electrostatic field, the positive charges would move from a toward b and the negative charges from b toward a. The excess charges on the terminals would decrease and the potential difference between them would decrease and eventually become zero. Since this is not observed to happen, we conclude that there must also exist, at every point within the source, a force $\boldsymbol{F}_n$ of *nonelectrostatic* origin, which acts on every charged particle and which is equal and opposite to the electrostatic force $\boldsymbol{F}_e = q\boldsymbol{E}_e$.

The origin of this force depends on the nature of the source. In a Van de Graaff generator, the force is exerted by the belt on the charged particles attached to it. In an electrolytic cell, the force is associated with chemical bonding. In the rotor of a generator, the force results from the motion of charged particles transverse to a magnetic field, and in the windings of a transformer it results from the effect of a magnetic field that is changing with time.

Whatever the origin of the nonelectrostatic force, we can define an *equivalent nonelectrostatic field* $\boldsymbol{E}_n$ by the equation

$$\boldsymbol{F}_n = q\boldsymbol{E}_n, \qquad \text{or} \qquad \boldsymbol{E}_n = \frac{\boldsymbol{F}_n}{q}.$$

That is, the nonelectrostatic force is the same *as if* there were a nonelectrostatic field $\boldsymbol{E}_n$ in addition to the purely electrostatic field $\boldsymbol{E}_e$.

When the source is on open circuit, as in Fig. 28–5, the charges are in equilibrium, and the resultant field $\boldsymbol{E}$, the vector sum of $\boldsymbol{E}_e$ and $\boldsymbol{E}_n$, must be zero at every point:

$$\boldsymbol{E} = \boldsymbol{E}_e + \boldsymbol{E}_n = \boldsymbol{0}, \qquad \text{or} \qquad \boldsymbol{E}_e = -\boldsymbol{E}_n.$$

Now the electrostatic potential difference V_{ab} is defined as the work per unit charge performed by the electrostatic field $\boldsymbol{E}_e$ on a charge moving from a to b. That is,

$$V_{ab} = \int_a^b \boldsymbol{E}_e \cdot d\boldsymbol{s}.$$

Similarly, we may consider the work done by the nonelectrostatic field $\boldsymbol{E}_n$. It is customary to speak of the (positive) work of this field during a displacement from b to a rather than the reverse. Specifically, the work performed by $\boldsymbol{E}_n$, per unit charge, when a charge moves from b to a, is called the *electromotive force* $\mathcal{E}$ *of the source.* That is,

$$\mathcal{E} = \int_b^a \boldsymbol{E}_n \cdot d\boldsymbol{s}.$$

When $\boldsymbol{E}_e = -\boldsymbol{E}_n$, we have $V_{ab} = \mathcal{E}$. Hence *for a source on open circuit*, the potential difference V_{ab}, or the *open-circuit terminal voltage*, is equal to the electromotive force:

$$V_{ab} = \mathcal{E} \qquad \text{(Source on open circuit).} \qquad (28\text{–}14)$$

The term *electromotive force*, although widely used, is somewhat unfortunate, in that the concept to which it refers is not a *force* but a *work per unit charge.* the term *electromotance* is sometimes used, but the concept is usually referred to simply as *emf*, pronounced "ee-emm-eff."

The mks unit of $\boldsymbol{E}_n$ is the same as that of $\boldsymbol{E}_e$, namely, one volt per meter, so the unit of emf is the same as that of potential or potential difference, namely, 1 V. However, an electromotive force is not the *same thing* as a potential difference, since the latter is the work of an *electrostatic* field and the former is the work of a *nonelectrostatic* field.

As we shall see later, the electro*static* field within a source, and hence the potential difference between its terminals, depend on the current in the source. The *non*electrostatic field, and hence the emf of the source, is in many cases a *constant* independent of the current, and hence the emf represents a definite property of a source. Unless stated otherwise, we shall assume, in what follows, that the emf of a source is constant.

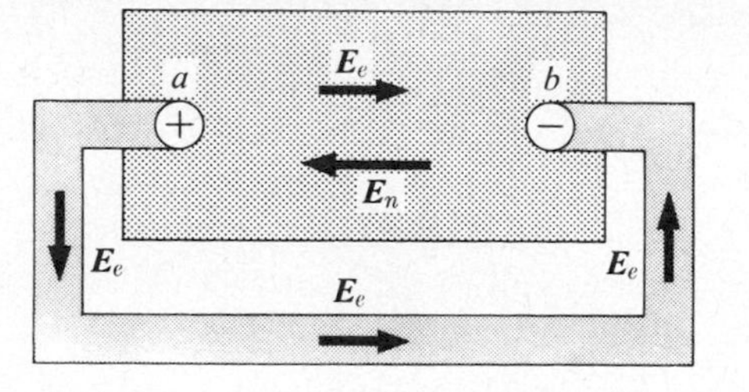

Fig. 28-6 Schematic diagram of a source with a complete circuit. The vectors $\mathbf{E}_n$ *and* $\mathbf{E}_e$ *represent the directions of the corresponding fields. The current is everywhere the same and is in the direction of the resultant field, that is, from a to b in the external circuit and from b to a within the source.* $V_{ab} = IR = \mathcal{E} - Ir$.

Now suppose that the terminals of a source are connected by a wire, as shown schematically in Fig. 28-6. The source and wire are then said to form a *complete circuit.* The driving force on the free charges *in the wire* is due solely to the electro*static* field $\mathbf{E}_e$ set up by the charged terminals a and b of the source. This field sets up a current *in the wire* from a toward b. The charges on the terminals decrease slightly and the electrostatic fields, both within the wire and within the source, decrease also. As a result, the *electrostatic* field within the source becomes smaller than the (constant) *nonelectrostatic* field. Hence positive charges within the source are driven toward the positive terminal, and there is a current within the source from b toward a. The circuit settles down to a steady state in which the current is the same at all cross sections.

If current could travel through the source without impediment (that is, if the source had no *internal* resistance), charge entering the external circuit through terminal a would be replaced immediately by charge flow through the source. In this case the internal electrostatic field in the source would not change under complete-circuit conditions, and the terminal potential difference V_{ab} would still be equal to $\mathcal{E}$. Since V_{ab} is also related to the current and resistance in the external circuit by Eq. (28–12), we would then have

$$\mathcal{E} = IR, \tag{28–15}$$

which determines the current in the circuit, once $\mathcal{E}$ and R are specified.

We say "if" in the above paragraph because no real source has precisely zero internal resistance. Consequently, under closed-circuit conditions the terminal potential difference is *not* precisely $\mathcal{E}$ but is better represented as

$$\boxed{V_{ab} = \mathcal{E} - Ir,} \tag{28–16}$$

where r is the *internal resistance* of the source. The equation determining the current in the complete circuit is then

$$\mathcal{E} - Ir = IR,$$

or

$$\boxed{I = \frac{\mathcal{E}}{R + r}.} \tag{28–17}$$

That is, the current equals the source emf divided by the *total* circuit resistance, external plus internal.

We shall generalize this equation without a formal proof as follows. Let a circuit consist of any number of sources and conductors connected in *series,* that is, in a single closed path. The total resistance of the circuit is defined as the *arithmetic* sum of all resistances, external and internal. For brevity, we write this as $\sum R$. The resultant emf in the circuit is defined as the *algebraic* sum of the emf's of the sources, $\sum \mathcal{E}$. The general form of the circuit equation is then

$$I = \frac{\sum \mathcal{E}}{\sum R}. \tag{28–18}$$

We must adopt a convention of sign for the currents and the emf's. The first step is to decide arbitrarily on a direction around the circuit (clockwise or counterclockwise) that will be considered the *positive direction.* (Either may be chosen.) Then a current in this direction is positive and one in the other direction is negative. An emf is positive if the direction of its associated field $\mathbf{E}_n$ is in the chosen positive direction, negative if in the opposite direction. (Remember that the $\boldsymbol{E}_n$-field within a source is always from the negative toward the positive terminal.)

If the terminals of a source are connected by a conductor of zero (or negligible) resistance, the source is said to be *short circuited.* (This is not an advisable procedure to carry out with the storage battery of your car, or with the terminals of the power line!) Then $R = 0$, and from the circuit equation the *short-circuited current I* is

$$I = \frac{\mathscr{E}}{r}. \tag{28–19}$$

The terminal voltage is then

$$V_{ab} = \mathscr{E} - \left(\frac{\mathscr{E}}{r}\right)r = 0, \tag{28–20}$$

and the terminal voltage drops to zero. The electro*static* field within the source is zero, and the driving force on the charges within it is due to the *non*electrostatic field only.

A source is completely described by its emf $\mathscr{E}$ and its internal resistance r. These properties may be found (at least in principle) from measurements of the open-circuit terminal voltage, which equals $\mathscr{E}$, and the short-circuit current, which enables r to be calculated from Eq. (28–19).

We must consider one more special case. If a source is connected to an external circuit containing other sources, it is possible that the electrostatic field within the given source will be *greater* than the nonelectrostatic field, as in Fig. 28–7. When this is the case, the current *within* the source is from terminal a toward terminal b. This is the case when the storage battery of an automobile is being "charged" by the generator. Equation (28–16) is then written

$$V_{ab} = \mathscr{E} + Ir. \tag{28–21}$$

The terminal voltage is then *greater* than the emf $\mathscr{E}$.

An alternate viewpoint is that V_{ab} is always given by Eq. (28–16) but that I is inherently negative when its direction *in the source* is from + to −. Then $-Ir$ becomes a positive quantity.

The diagrams in the preceding sections have been more or less representational, so as to show the

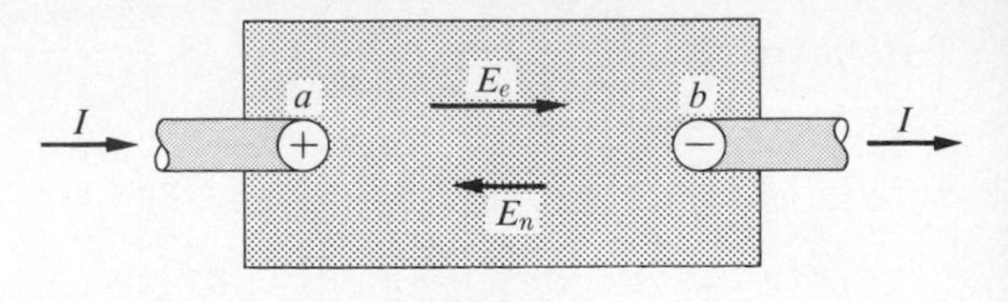

Fig. 28–7

electric fields within sources and conductors. Every conductor (except a superconductor) has resistance, and is therefore a *resistor,* also. In fact, the terms "conductor" and "resistor" may be used interchangeably. Resistance units constructed to introduce into a circuit lumped resistances that are large compared with those of leads and contacts are called *resistors.* A resistor is represented by the symbol

Portions of a circuit of negligible resistance are shown by straight lines.

A variable resistor is called a *rheostat.* A common type consists of a resistor with a sliding contact that can be moved along its length, and is represented by the symbol

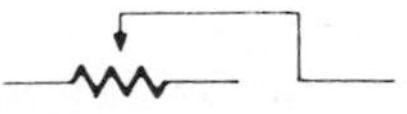

Connections are made to either end of the resistor and to the sliding contact. The symbol

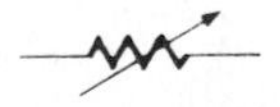

is also used for a variable resistor.

A source is represented by the symbol

The longer vertical line corresponds to the + terminal. We shall modify this in the following examples to

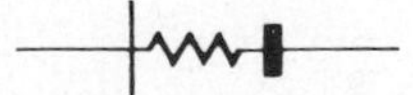

so as to show explicitly that a source has an internal resistance.

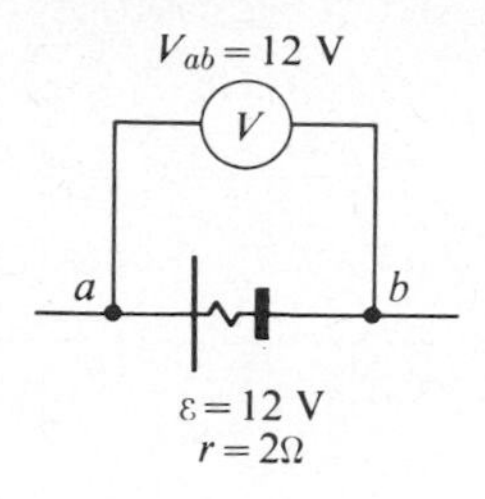

Fig. 28–8 A source on open circuit.

Example 1 Consider a source whose emf $\mathscr{E}$ is constant and equal to 12 V, and whose internal resistance r is 2 Ω. (The internal resistance of a commercial 12-V lead storage battery is only a few thousandths of an ohm.) Fig. 28–8 represents the source with a voltmeter V connected between its terminals a and b. A voltmeter reads the potential difference between its terminals. If it is of the conventional type, the voltmeter provides a conducting path between the terminals and so there is a current in the source (and through the voltmeter). We shall assume, however, that the resistance of the voltmeter is so large (essentially infinite) that it draws no appreciable current. The source is then on *open circuit*, corresponding to the source in Fig. 28–5, and the voltmeter reading V_{ab} equals the emf $\mathscr{E}$ of the source, or 12 V.

Example 2 In Fig. 28–9, an ammeter A and a resistor of resistance $R = 4\ \Omega$ have been connected to the terminals of the source to form a complete circuit. The total resistance of the circuit is the sum of the resistance R, the internal resistance r, and the resistance of the ammeter. The ammeter resistance, however, can be made very small, and we shall assume it so small (essentially zero) that it can be neglected. The ammeter (whatever its resistance) reads the current I through it. The circuit corresponds to that in Fig. 28–6.

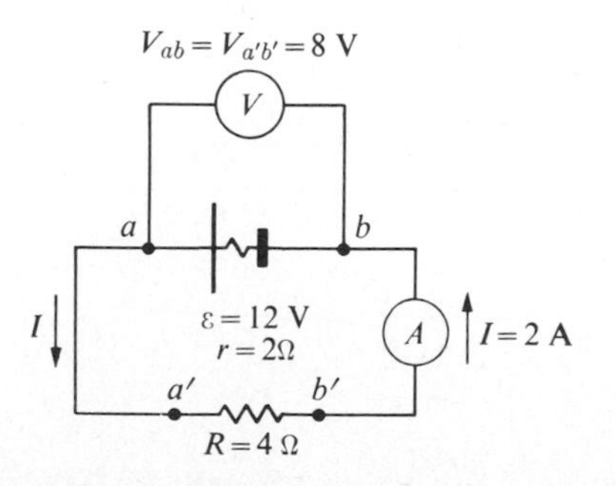

Fig. 28–9 A source in a complete circuit.

The wires connecting the resistor to the source and the ammeter, shown by straight lines, have zero resistance and hence there is no potential difference between their ends. Thus, points a and a' are at the same potential and are electrically equivalent, as are points b and b'. The potential differences V_{ab} and $V_{a'b'}$ are therefore equal. In the future, we shall use the same symbol to represent all points in a circuit that are connected by resistanceless conductors and are at the same potential.

The current I in the resistor (and hence at all points of the circuit) could be found from the relation $I = V_{ab}/R$, if the potential difference V_{ab} were known. However, V_{ab} is the terminal voltage of the source, equal to $\mathscr{E} - Ir$, and since this depends on I it is unknown at the start. We can, however, calculate the current from the circuit equation:

$$I = \frac{\mathscr{E}}{R + r} = \frac{12\ \text{V}}{4\ \Omega + 2\ \Omega} = 2\ \text{A}.$$

The potential difference V_{ab} can now be found by considering a and b either as the terminals of the resistor or as those of the source. If we consider them as the terminals of the resistor,

$$V_{a'b'} = IR = (2\ \text{A})(4\ \Omega) = 8\ \text{V}.$$

If we consider them as the terminals of the source,

$$V_{ab} = \mathscr{E} - Ir = 12\ \text{V} - (2\ \text{A})(2\ \Omega) = 8\ \text{V}.$$

The voltmeter therefore reads 8 V and the ammeter reads 2 A.

Example 3 In Fig. 28–10, the source is short circuited. The current is

$$I = \frac{\mathscr{E}}{r} = \frac{12\ \text{V}}{2\ \Omega} = 6\ \text{A}.$$

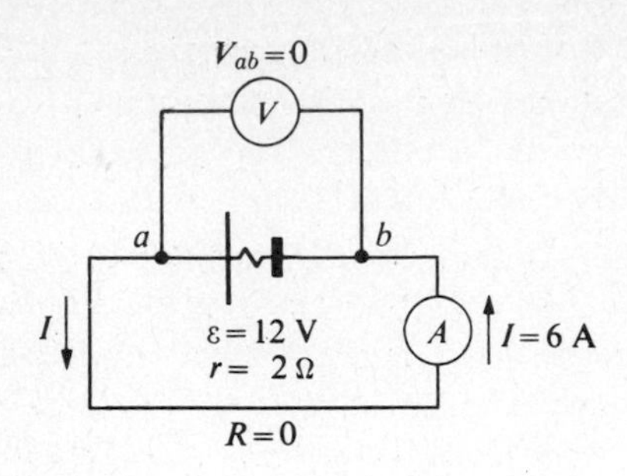

Fig. 28–10 *A source on short circuit.*

The terminal voltage is

$$V_{ab} = \mathscr{E} - Ir = 12\text{ V} - 6\text{ A} \times 2\,\Omega = 0.$$

The ammeter reads 6 A and the voltmeter reads zero.

Example 4 In Fig. 28–11, a second source of emf $\mathscr{E}_2 = 18\text{V}$ and $r_2 = 2\ \Omega$ has been connected to the first, like terminals of the sources being connected together. Let us represent the emf and resistance of the first source by $\mathscr{E}_1$ and r_1. The emf's of the sources have opposite signs because their nonelectrostatic fields are oppositely directed around the circuit, that of the first source being counterclockwise and that of the second clockwise. Hence if we consider the clockwise direction as positive,

$$\mathscr{E}_1 = -12\text{ V}, \qquad \mathscr{E}_2 = 18\text{ V},$$

and

$$\sum \mathscr{E} = 18\text{ V} - 12\text{ V} = 6\text{ V}.$$

From the circuit equation, the current is

$$I = \frac{\sum \mathscr{E}}{\sum R} = \frac{6\text{ V}}{4\,\Omega} = 1.5\text{ A}.$$

Since I has a positive sign, it is in the positive (clockwise) sense, and its direction in source 1 is from left to right. This source therefore corresponds to the source in Fig. 28–7, and is being "charged."

Since the direction of the current in the circuit is opposite to that in Figs. 28–9 and 28–10, the connections to the ammeter terminals would have to be reversed if the zero on the meter scale is at its left end, as is the case in most meters. If the meter is of the "zero-center" type, the connections to it need not be reversed, but its deflection will be in the opposite direction to that in Figs. 28–9 and 28–10.

If we consider a and b as the terminals of source 1, Eq. (28–21) applies and

$$V_{ab} = \mathscr{E}_1 + Ir_1 = 12\text{ V} + (1.5\text{ A})(2\,\Omega) = 15\text{ V}.$$

If we consider a and b as the terminals of source 2, Eq. (28–16) applies and

$$V_{ab} = \mathscr{E} - Ir_2 = 18\text{ V} - (1.5\text{A})(2\,\Omega) = 15\text{ V}.$$

Either viewpoint must lead to the same answer. The potential difference V_{ab} is *greater* than the emf of

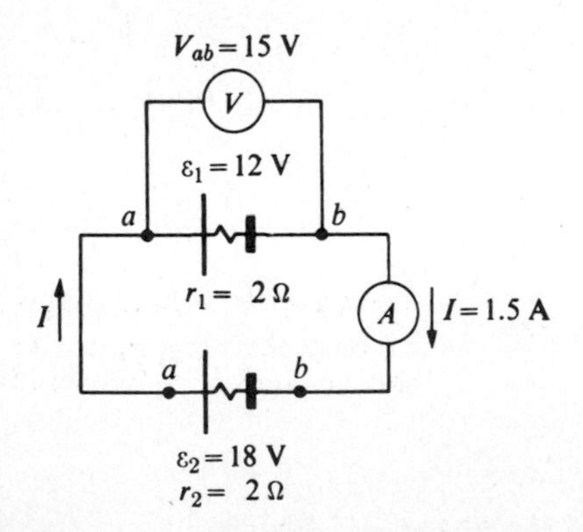

Fig. 28–11 *Two sources with like terminals connected together.*

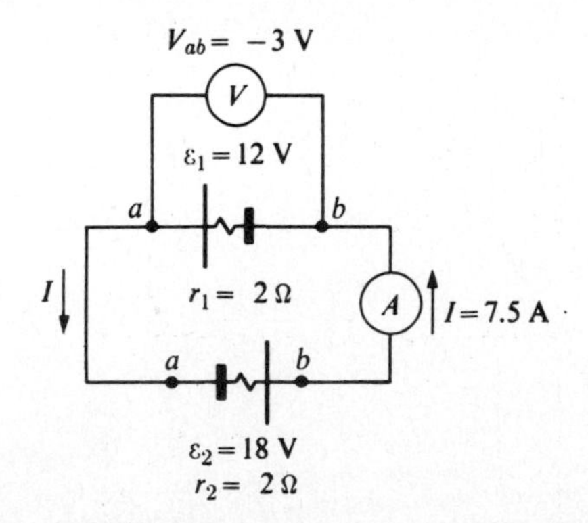

Fig. 28–12 *Two sources with unlike terminals connected together.*

source 1, and is *less* than the emf of source 2. The voltmeter reads 15 V and the ammeter reads 1.5 A.

Example 5 In Fig. 28–12, the polarity of source 2 has been reversed and the sources are connected with unlike terminals together. The fields $\boldsymbol{E}_n$ are now counterclockwise in both sources and, if we consider this as the positive direction, the emf $\mathscr{E}$ of the circuit is

$$\sum \mathscr{E} = \mathscr{E}_1 + \mathscr{E}_2 = 30 \text{ V}.$$

The current is counterclockwise and its magnitude is

$$I = \frac{\sum \mathscr{E}}{\sum R} = \frac{30 \text{ V}}{4 \,\Omega} = 7.5 \text{ A}.$$

The current in source 1 is from right to left, in the same sense as in Fig. 28–10, and it exceeds the short-circuit current of the source (6 A).

Considering a and b as the terminals of source 1, the potential difference V_{ab} is

$$V_{ab} = \mathscr{E}_1 - Ir_1 = 12 \text{ V} - (7.5 \text{ A})(2\,\Omega) = -3 \text{ V}.$$

Since this is negative, terminal b is at a *higher* potential than terminal a.

The subscripts a and b, in Eqs. (28–16) and (28–21), refer respectively to the + and − terminals of a source. In Fig. 28–12, the + terminal of cell 2 is lettered b and the − terminal is lettered a. Equation (28–16) should therefore be written

$$V_{ab} = \mathscr{E}_2 - Ir_2 = 18 \text{ V} - (7.5 \text{ A})(2\,\Omega) = 3 \text{ V}.$$

Point b is therefore at a potential 3 V higher than point a, in agreement with the previous answer.

Unless the voltmeter is of the zero-center type, the connections to its terminals would have to be reversed, relative to the preceding diagrams.

28–6 CURRENT-VOLTAGE RELATIONS

The current through a device such as a resistor depends on the potential difference between its terminals. For a device obeying Ohm's law, the current is directly proportional to voltage, as shown by Eq. (28–12). But as mentioned in Section 28–4, there are many devices for which this simple model is not an adequate description. Current may depend on voltage in a more complicated way, and the current resulting from a given potential difference may depend on the polarity of the potential difference. This is the case with *diodes*, devices constructed deliberately to conduct much better in one direction than the other.

It is convenient to represent the current–voltage relation as a graph, and Fig. 28–13 shows several examples. Part (a) shows the behavior of a resistor that obeys Ohm's law, for which the graph is a straight line. Part (b) shows the relation for a vacuum

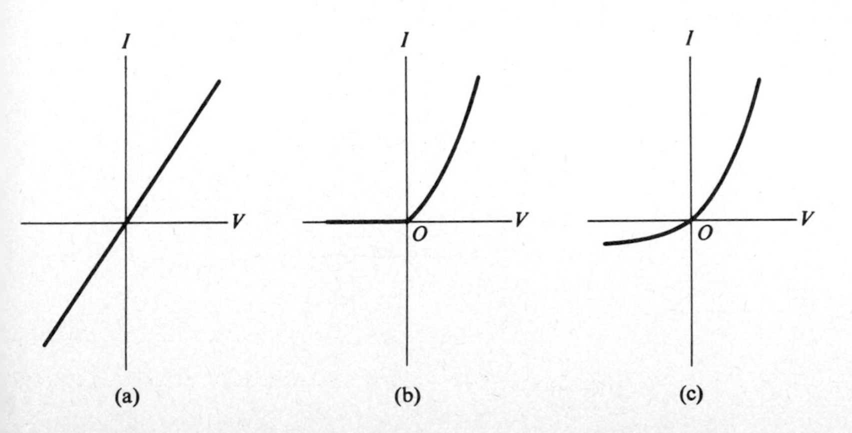

Fig. 28–13 *Current-voltage relations for (a) a resistor obeying Ohm's law; (b) a vacuum diode; (c) a semiconductor diode.*

diode. For positive potentials of anode with respect to cathode, I is approximately proportional to $V^{3/2}$, while for negative potentials the current is several orders of magnitude smaller and for most purposes may be assumed to be zero. Germanium diode behavior (c) is somewhat different but still strongly asymmetric, acting as a one-way valve in a circuit. Diodes are used to convert alternating current to direct and to perform a wide variety of logic functions in computer circuitry. The microscopic basis of diode behavior will be explored in some detail in later chapters.

An additional consideration is that for nearly all materials the current–voltage relation is temperature-dependent. Thus at low temperatures the curve in Fig. 28–13(c) rises more steeply for positive V than at higher temperatures, and at successively higher temperatures the asymmetry in the curve becomes less and less pronounced.

The current–voltage relation for a source may also be represented graphically. For a source represented by Eq. (28–16), that is,

$$V = \mathcal{E} - Ir,$$

the graph appears as in Fig. 28–14. The intercept on the V-axis, corresponding to the open-circuit condition ($I = 0$), is at $V = \mathcal{E}$, and the intercept on the I-axis, corresponding to a short-circuit situation ($V = 0$), is at $I = \mathcal{E}/r$.

This line may be used to find the current in a circuit containing a nonlinear device, as in Fig. 28–14(b). Its current–voltage relation is shown in Fig. 28–14(c), and Eq. (28–16) is also plotted on this graph. Each curve represents a current–voltage relation that must be satisfied, so the intersection represents the only possible values of V and I. This amounts to a graphical solution of two simultaneous equations for V and I, one of which is nonlinear.

Finally, we remark that Eq. (28–16) is not always an adequate representation of the behavior of a source. What we have described as an internal resistance may actually be a more complex voltage–current relation. Nevertheless, the concept of internal resistance frequently provides an adequate description of batteries, generators, and other energy converters. The difference between a fresh flashlight battery and an old one is not in the emf, which decreases only slightly with use, but principally in the internal resistance, which may increase from a few ohms when fresh to as much as 1000 Ω or more after long use. Similarly, the current a car battery can deliver to the starter motor on a cold morning is less than when the battery is warm, not because the emf is appreciably less but because the internal resistance is temperature-dependent, decreasing with increasing temperature.

28–7 WORK AND POWER IN ELECTRICAL CIRCUITS

The rectangle in Fig. 28–15 represents a portion of an electrical circuit within which there is a current I from a toward b. The potentials at terminals a and b are V_a and V_b, respectively. In a time interval Δt, a quantity of charge $\Delta Q = I\Delta t$ enters the portion at terminal a and an equal quantity of charge leaves at terminal b. The change in potential energy of the circulating charge is

$$\Delta W = \Delta Q(V_a - V_b) = V_{ab}I\Delta t. \quad (28\text{–}22)$$

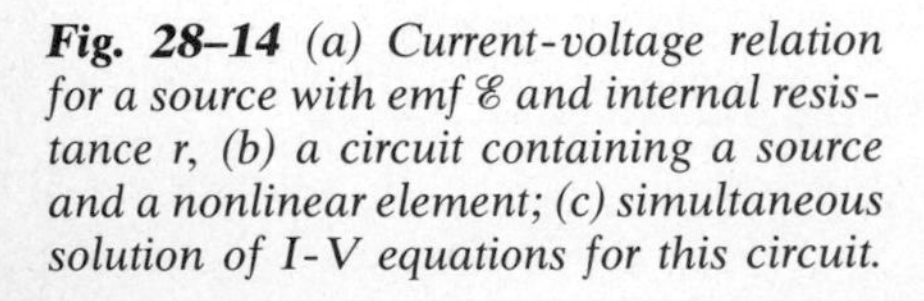

Fig. 28–14 (a) Current-voltage relation for a source with emf $\mathcal{E}$ and internal resistance r, (b) a circuit containing a source and a nonlinear element; (c) simultaneous solution of I-V equations for this circuit.

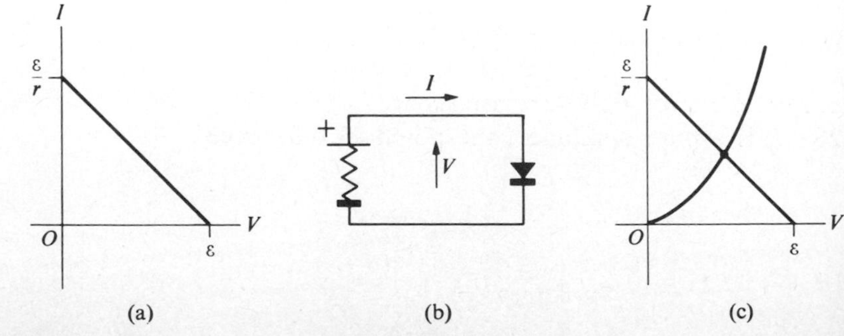

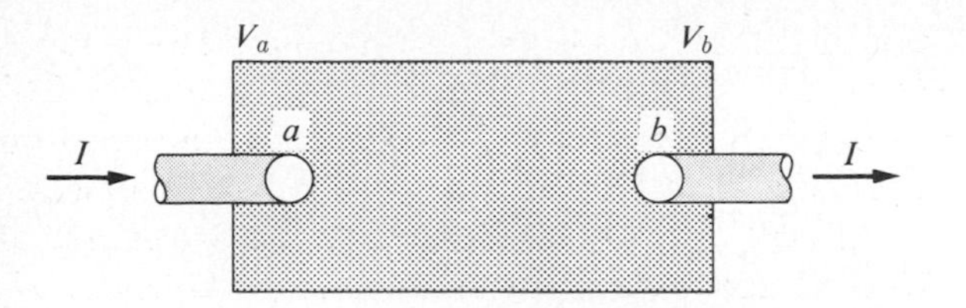

Fig. 28–15 *The power input P to the portion of the circuit between a and b is $P = V_{ab}I$.*

If the potential at a is higher than that at b, the potential energy decreases, corresponding to the fact that the electric field does an amount of work ΔW on a charge ΔQ equal to the potential difference $V_a - V_b$ (work per unit charge) multiplied by the charge ΔQ. That is, the electric field transfers energy *into* the portion of the circuit between a and b.

The *rate* of transfer of energy, or the *power P*, is given by $\Delta W/\Delta t$; hence, from Eq. (28–22),

$$\boxed{P = V_{ab}I.} \qquad (28\text{–}23)$$

If the potential at b is higher than that at a, the charge *gains* potential energy at the expense of some other form of energy, and there is a corresponding transfer of electrical energy out of this portion of the circuit.

Equation (28–23) is the general expression for the magnitude of the electrical power input to (or the power output from) any portion of an electrical circuit. The unit of V_{ab} is one volt, or one joule per coulomb, and the unit of I is one ampere or one coulomb per second. The mksa unit of power is therefore

$$(1\,\mathrm{J\,C^{-1}})(1\,\mathrm{C\,s^{-1}}) = 1\,\mathrm{J\,s^{-1}} = 1\,\mathrm{W} = 1\ \text{watt}.$$

We now consider some special cases.

1. Pure resistance If the portion of the circuit in Fig. 28–15 is a pure resistance, the potential difference $V_{ab} = IR$ and

$$\boxed{P = V_{ab}I = I^2R = V^2/R.} \qquad (28\text{–}24)$$

The potential at a is necessarily higher than at b and there is a power *input* to the resistor. The circulating charges give up energy to the atoms of the resistor when they collide with them, and the temperature of the resistor increases unless there is a flow of heat out of it. We say that energy is *dissipated* in the resistor at a rate I^2R.

2. Power output of a source The upper rectangle in Fig. 28–16 represents a source of emf $\mathcal{E}$ and internal resistance r, connected by resistanceless conductors to an external circuit represented by the lower rectangle and whose precise nature does not matter. We assume only that there is a current I in the circuit in the direction shown, from a to b in the external circuit and from b to a within the source. The letters a and b can be considered to represent either the terminals of the source or of the external circuit, and $V_a > V_b$. The external circuit then corresponds to the rectangle in Fig. 28–15, and the power input to it is

$$P = V_{ab}I.$$

If a and b are considered as the terminals of the source, then as we have shown,

$$V_{ab} = \mathcal{E} - Ir$$

and hence,

$$P = V_{ab}I = \mathcal{E}I - I^2r. \qquad (28\text{–}25)$$

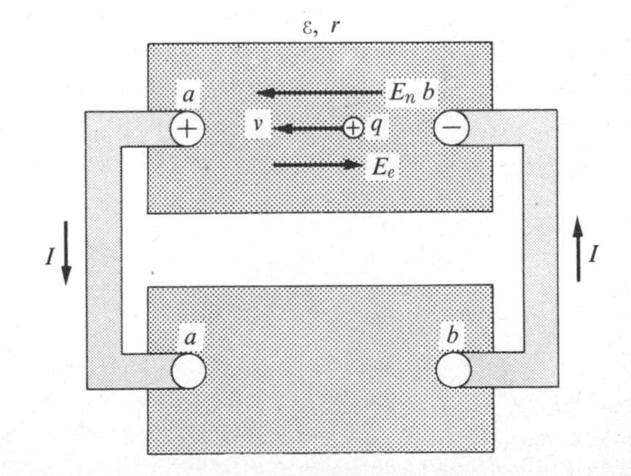

Fig. 28–16 *The rate of conversion of nonelectrical to electrical energy in the source equals $\mathcal{E}I$. The rate of energy dissipation in the source is I^2r. The difference $\mathcal{E}I - I^2r$ is the power output of the source.*

The terms $\mathcal{E}I$ and I^2r have the following significance. The emf $\mathcal{E}$ has been defined as the work per unit charge performed on the charges by the nonelectrostatic field $\boldsymbol{E}_n$ as the charges move from b to a in the source. Hence, if a charge ΔQ flows in time Δt, this field does work $\Delta W = \mathcal{E}\Delta Q$, and its *rate* of doing work, or power, is

$$\frac{\Delta W}{\Delta t} = \mathcal{E}\frac{\Delta Q}{\Delta t} = \mathcal{E}I. \qquad (28\text{–}26)$$

Hence, the product $\mathcal{E}I$ is the rate at which work is done on the circulating charges by the agency that maintains the nonelectrostatic field.

The term I^2r is the rate at which energy is *dissipated* in the internal resistance of the source, and the difference $\mathcal{E}I - I^2r$ is the rate at which energy is delivered by the source to the remainder of the circuit. In other words, the power P in Eq. (28–25) represents the *power output* of the source, or the *power input* to the remainder of the circuit.

3. Power input to a source Suppose that the lower rectangle in Fig. 28–16 is itself a source of emf larger than that of the upper source and with its emf opposite to that of the upper source. The current I in the circuit is then opposite to that shown in Fig. 28–16. The power output of the lower source is

$$P = V_{ab}I.$$

Considering a and b as the terminals of the upper source, we have

$$V_{ab} = \mathcal{E} + Ir,$$

and

$$P = V_{ab}I = \mathcal{E}I + I^2r. \qquad (28\text{–}27)$$

The terms $\mathcal{E}I$ and I^2r have the following interpretation. The charges in the upper source now move from left to right, in a direction opposite the field $\boldsymbol{E}_n$, and work is done by the force $\boldsymbol{F}_n$ on the agent maintaining the nonelectrostatic field. The product $\mathcal{E}I$ equals the rate at which work is done on this agent, and the term I^2r again equals the rate of dissipation of energy in the internal resistance of the source. The sum $\mathcal{E}I + I^2r$ is therefore the *power input* to the upper source. The source converts electrical energy to nonelectrical energy.

We shall illustrate the relations developed in this section by applying them to the circuits in Figs. 28–9 through 28–12.

Example 1 The rate of energy conversion in the source in Fig. 28–9 is

$$\mathcal{E}I = (12\ \text{V})(2\ \text{A}) = 24\ \text{W}.$$

The rate of dissipation of energy in the source is

$$I^2r = (2\ \text{A})^2(2\ \Omega) = 8\ \text{W}.$$

The power *output* of the source is the difference between these, or 16 W.

The power output is also given by

$$IV_{ab} = (2\ \text{A})(8\ \text{V}) = 16\ \text{W}.$$

The power input to the resistor is

$$IV_{a'b'} = (2\ \text{A})(8\ \text{V}) = 16\ \text{W}.$$

This equals the rate of dissipation of energy in the resistor:

$$I^2R = (2\ \text{A})^2(4\ \Omega) = 16\ \text{W}.$$

Example 2 The rate of energy conversion in the source in Fig. 28–10 is

$$\mathcal{E}I = (12\ \text{V})(6\ \text{A}) = 72\ \text{W}.$$

The rate of dissipation of energy in the source is

$$I^2r = (6\ \text{A})^2(2\ \Omega) = 72\ \text{W}.$$

The power *output* of the source (also given by IV_{ab}) equals zero. All of the energy converted is dissipated within the source.

Example 3 The rate of energy conversion in source 1 in Fig. 28–11 is

$$\mathcal{E}_1 I = (12\ \text{V})(1.5\ \text{A}) = 18\ \text{W}.$$

Within this source, electrical energy is converted to nonelectrical energy.

The rate of dissipation of energy in source 1 is

$$I^2 r_1 = (1.5 \text{ A})^2(2\ \Omega) = 4.5 \text{ W}.$$

The total power *input* to source 1 is the *sum* of these, or 22.5 W.

The power input is also given by

$$IV_{ab} = (1.5 \text{ A})(15 \text{ V}) = 22.5 \text{ W}.$$

The same expression gives the power *output* of source 2.

The rate of energy conversion in source 2 is

$$\mathscr{E}_2 I = (18 \text{ V})(1.5 \text{ A}) = 27 \text{ W}.$$

Within this source, nonelectrical energy is converted to electrical energy.

The rate of dissipation of energy in source 2 is

$$I^2 r_2 = (1.5 \text{ A})^2(2\ \Omega) = 4.5 \text{ W}.$$

The *difference* between the rate of energy conversion and the rate of energy dissipation equals the power output of 22.5 W.

Example 4 In the circuit of Fig. 28–12, nonelectrical energy is converted to electrical energy in *both* sources. The rates of conversion are, respectively,

$$\mathscr{E}_1 I = (12 \text{ V})(7.5 \text{ A}) = 90 \text{ W},$$

$$\mathscr{E}_2 I = (18 \text{ V})(7.5 \text{ A}) = 135 \text{ W}.$$

The rates of energy dissipation within the sources are, respectively,

$$I^2 r_1 = (7.5 \text{ A})^2(2\ \Omega) = 112.5 \text{ W}.$$

$$I^2 r_2 = (7.5 \text{A})^2(2\ \Omega) = 112.5 \text{ W}.$$

The sum of these equals the total rate of energy conversion, 225 W.

The power output of source 2 is

$$135 \text{ W} - 112.5 \text{ W} = 22.5 \text{ W}.$$

This equals the power input to source 1 and also equals (in magnitude) the product IV_{ab}.

The rate of energy dissipation in source 1, 112.5 W, equals the sum of the rate of energy conversion in this source, 90 W, and the power input of 22.5 W.

28–8 THERMOELECTRICITY

In 1826 Thomas Johann Seebeck (1770–1831) discovered that an emf could be produced by purely thermal means in a circuit composed of two different metals A and B whose junctions are maintained at different temperatures, as shown schematically in Fig. 28–17. The two metals constitute a *thermocouple*, and the emf in the circuit is called a *thermal* emf, or a *Seebeck* emf. When the temperature of the reference junction T_R is kept constant, the Seebeck emf $\mathscr{E}_{AB}$ is found to be a function of the temperature T of the test junction. This fact enables the thermocouple to be used as a thermometer, as described in Section 15–2, and this is its main use today. The advantage of a thermocouple thermometer is that, because its heat capacity is small, the test junction comes quite rapidly to thermal equilibrium with the system whose temperature is to be measured. It therefore follows (and measures) temperature changes easily.

A *thermopile* is an instrument consisting of many fine thermocouples connected in series, so that the total emf is the sum of the separate emf's. In conjunction with a high-sensitivity galvanometer, a thermopile is an extremely sensitive device for detecting and measuring radiant energy. The reference junctions are covered, while the test junctions are blackened and exposed to the radiant energy. Thermopiles are used to measure the radiation from stars and are commonly employed to investigate the distribution of energy in those portions of a spectrum beyond the limits of a photographic plate.

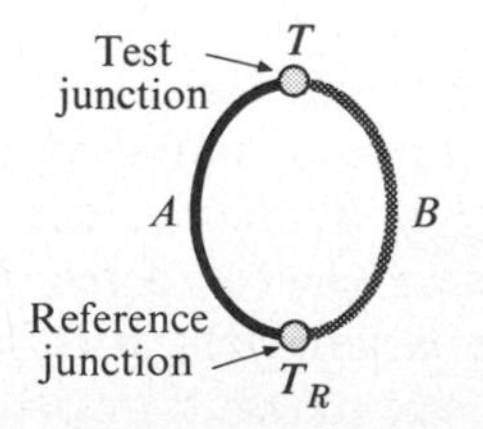

Fig. 28–17 Thermocouple of metals A and B with junctions at T and T_R.

The Seebeck emf arises from the fact that the density of free electrons in a metal differs from one metal to another. Within a metal, the density depends on temperature, and the temperature variation is different for different metals. Each junction thus functions as a temperature-dependent seat of electromotive force. In Fig. 28–17, if the junctions are at the same temperature, the net emf is zero, but when the temperatures are different there is a net emf depending on the temperature, and this is the *Seebeck emf.*

For an individual junction between two metals, transfer of charge ΔQ across the junction, under the action of the nonelectrostatic effects responsible for the emf $\mathcal{E}$, involve a corresponding liberation or absorption of energy $\Delta W = \mathcal{E}\Delta Q$; this energy is ordinarily manifested as a flow of *heat* between the junction and its surroundings. This is called the *Peltier heat,* after its discoverer, Jean C. A. Peltier. Experiment has shown that the Peltier heat transferred at any junction is proportional to the quantity of electricity crossing the junction, and that it reverses its direction of flow when the electric current is reversed. A single junction, therefore, is a *source* within which electrical energy is converted to heat, or heat is converted to electrical energy. The Peltier emf of a junction of metals A and B, π_{AB}, is defined as the heat (energy) absorbed or liberated per unit quantity of electricity crossing the junction. Thus,

$$\pi_{AB} = \frac{\text{Peltier heat}}{\text{charge transferred}}. \qquad (28\text{–}28)$$

It is found that π_{AB} depends not only on the nature of the two metals, but also on the temperature of the junction, and that it is independent of any other junction that may be present.

In a single wire whose ends are maintained at different temperatures, the free-electron density varies from point to point. Each element of a wire of nonuniform temperature is therefore a source; this discovery was made by Sir William Thomson (Lord Kelvin). When a current is maintained in a wire of nonuniform temperature, heat is liberated or absorbed at all points of the wire, and this *Thomson heat* is proportional to the quantity of electricity passing the section of wire and to the temperature difference between the ends of the section. If ΔT is the temperature difference between the ends of a small length of wire of material A, the amount of heat absorbed or liberated in this length of wire per unit quantity of electricity transferred is called the *Thomson* emf, written $\sigma_A \Delta T$. Thus

$$\sigma_A \Delta T = \frac{\text{Thomson heat}}{\text{charge transferred}}. \qquad (28\text{–}29)$$

If σ_A does not depend on temperature, then the *total* Thomson emf in a wire whose ends are at temperatures T_1 and T_2 is given by $\sigma_A(T_2 - T_1)$. In general, σ_A depends on temperature, so this is only approximately correct. Experiment has shown that the Thomson heat is also reversible, and depends on the nature of the wire and on the average temperature of the portion of the wire under consideration. The coefficient σ_A is sometimes called the "specific heat of electricity."

Returning now to the thermocouple depicted in Fig. 28–17, we see that the Seebeck emf $\mathcal{E}_{AB}$ is the resultant of two Peltier emf's, $(\pi_{AB})_T$ and $(\pi_{AB})_{T_R}$, and two Thomson emf's $(T - T_R)\sigma_A$ and $(T - T)\sigma_A$. The relation among these emf's is the *fundamental thermocouple equation*:

$$\mathcal{E}_{AB} = (\pi_{AB})_T - (\pi_{AB})_{T_R} + (T - T_R)(\sigma_A - \sigma_B). \qquad (28\text{–}30)$$

The sign conventions in this equation are:

1. $\mathcal{E}_{AB}$ is positive when the direction of the thermocouple current is from A to B at the test junction, which is taken to be the *warmer* of the two junctions.
2. π_{AB} is positive when the current is from A to B and Peltier heat is *absorbed* by the junction.
3. σ_A is positive when the current is opposite to the direction of the temperature gradient (low to high temperature) and Thomson heat is *absorbed.*

In order to connect a thermocouple to a measuring instrument, it is necessary to break the thermocouple circuit at some point and introduce a third metal, thereby creating two new junctions. It can be shown that no disturbing effects are produced, provided both new junctions are maintained at the same temperature. The correct procedure for connecting a thermocouple to a potentiometer* is shown in Fig. 28–18.

Consider now the two thermocouples depicted in Fig. 28–19, one composed of metals A and C and the other of metals B and C. The fundamental relation of Eq. (28–30) can be used to show that when the junctions are at the temperatures shown in the figure,

$$\mathscr{E}_{AB} = \mathscr{E}_{AC} - \mathscr{E}_{BC}. \tag{28–31}$$

The fact that the emf of a thermocouple AB is the difference between the emf's of two thermocouples AC and BC when the junction temperatures are the same enables us to tabulate convenient numbers from which the thermal emf of *any* thermocouple may be computed. Thus, if M is any metal and L is lead, experiment shows that the thermal emf $\mathscr{E}_{ML}$ depends on the temperature T of the test junction, when $T_R = 0°C$, according to the equation

$$\mathscr{E}_{ML} = aT + \tfrac{1}{2}bT^2, \tag{28–32}$$

provided T is no more than a few hundred degrees. The constants a and b for various metals M are given in Table 28–3.

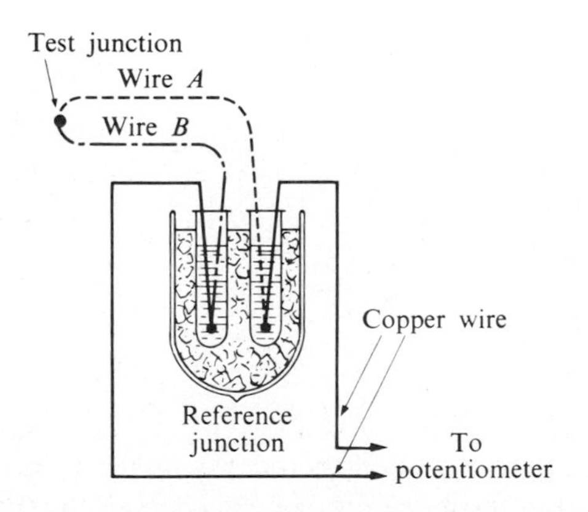

Fig. 28–18 Thermocouple of wires A and B with a reference function consisting of two junctions with copper.

* The potentiometer is described in the next chapter. Essentially, it is a voltmeter that draws no current.

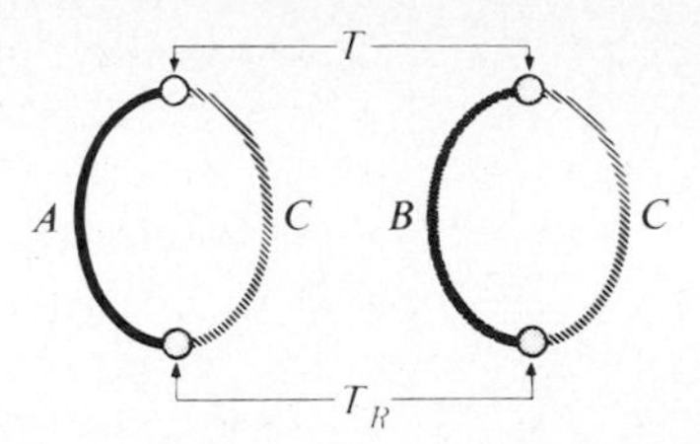

Fig. 28–19 *Two separate thermocouples, each with one wire the same, with functions at the same temperature.* $\mathscr{E}_{AB} = \mathscr{E}_{AC} - \mathscr{E}_{BC}$.

Example Find the emf of a copper–iron thermocouple when $T_R = 0°C$.

We write

$$\mathscr{E}_{CuFe} = \mathscr{E}_{CuPb} - \mathscr{E}_{FePb},$$

and using Eq. (28–32),

$$\mathscr{E}_{CuFe} = (a_{CuPb} - a_{FePb})T + \tfrac{1}{2}(b_{CuPb} - b_{FePb})T^2.$$

From Table 28–3,

$$\begin{aligned}\mathscr{E}_{CuFe} &= (2.76 - 16.6)\ \mu\text{V deg}^{-1}\ \text{T} \\ &\quad + \tfrac{1}{2}(0.012 + 0.030)\ \mu\text{V deg}^{-2}\ \text{T}^2 \\ &= (-13.8\ \mu\text{V deg}^{-1})\text{T} + (0.021\ \mu\text{V deg}^{-2})\text{T}^2.\end{aligned}$$

At $T = 100°C$,

$$\mathscr{E}_{CuFe} = -1.17\ \text{mV},$$

Table 28–3 CONSTANTS IN THE EQUATION $\mathscr{E}_{ML} = aT + \frac{1}{2}bT^2$ ($T_R = 0°C$)

Metal M (L = lead)	a, μV deg^{-1}	b, μV deg^{-2}
Aluminum	−0.47	0.003
Bismuth	−43.7	−0.47
Copper	2.76	0.012
Gold	2.90	0.0093
Iron	16.6	−0.030
Nickel	19.1	−0.030
Platinum	−1.79	−0.035
Silver	2.50	0.012
Steel	10.8	−0.016

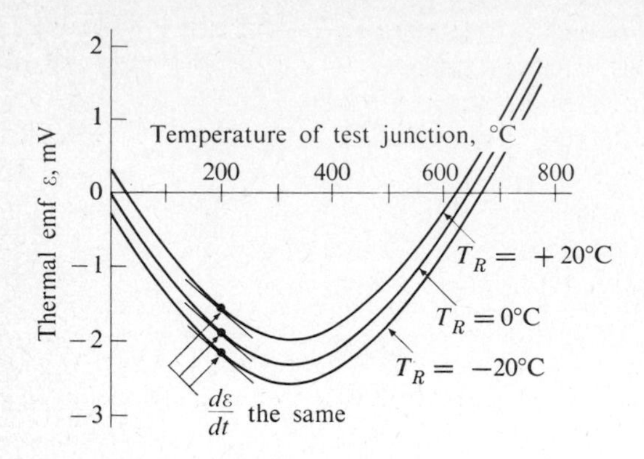

Fig. 28–20 *Thermal emf of a copper-iron thermocouple as a function of the temperature of the test junction, for three different values of the temperature of the reference junction.*

where the minus sign indicates that the direction of the current at the junction is from iron to copper, at 100°C.

The graphs of Fig. 28–20 show the emf of a copper–iron thermocouple as a function of the temperature of the test junction T the three different values of the temperature of the reference junction T_R. It will be seen that at any value of T, the slope is the same for the three curves, and is therefore independent of T_R.

28–9 THE ELECTRIC FIELD OF THE EARTH

If the molecules of the earth's atmosphere were electrically neutral, the atmosphere would be a nonconductor or insulator. Actually, because of the bombardment of the earth by cosmic rays (high-speed protons from outer space), a few ions, both positive and negative, are always present. With increasing elevation, the ionization produced by cosmic rays increases; and since the density of the atmosphere decreases, ions can move more freely and the conductivity increases. Above an elevation of about 50 km, the atmosphere is a relatively good conductor; and since the earth itself is also a relatively good conductor, an extremely oversimplified model of the earth and its lower atmosphere consists of a conducting sphere surrounded by a conducting spherical shell, the two being separated by a poorly conducting layer about 50 km thick.

Near the earth's surface, there is found to be a radial electrostatic field or potential gradient of about 100 V m^{-1}. The field becomes weaker with increasing elevation, and the total potential difference between the earth's surface and the outer conducting layer is about 400,000 V. The direction of the field is downward, so the earth has a negative surface charge and the outer conducting layer has a positive charge. The surface charge density and the electric field are related by Eq. (25–13),

$$\sigma = \epsilon_0 E;$$

and hence, if $E = 100\ \text{V m}^{-1}$, the surface density is, roughly, 10^{-9}C m^{-2}.

Approximating the earth by a sphere of radius 5000 km, we find its surface area is about $3 \times 10^{14}\ \text{m}^2$, so the total surface charge is about $3 \times 10^5\text{C}$.

Because the atmosphere is not a perfect insulator, there is a current in it, directed downward in the same direction as the electrostatic field. (Positive ions drift downward; negative ions drift upward.) The total current over the entire earth's surface is fairly constant and is about 1800 A or 1800 C s^{-1}. The time required to neutralize the entire surface charge of $3 \times 10^5\text{C}$, if it were not replenished in some way, would therefore be only about two or three minutes.

The charge, however, remains constant, so there must be some mechanism that pumps positive charges upward and negative charges downward, opposite to the directions of the forces exerted on them by the electrostatic field. The nature of this mechanism is not fully understood, but it is believed to be the processes that go on in the development of thunderstorms. In some way, positive and negative ions are developed in a thunderstorm. The positive ions are carried upward and the negative ions downward by the air currents in the storm. These currents, in turn, are driven by pressure differences resulting from nonuniform temperatures in the atmosphere, so that the ultimate source of the energy input is the radiant energy reaching the earth from the sun.

The earth's atmosphere can be compared to an enormous Van de Graaff generator immersed in a conducting medium. Instead of a moving belt, it is the motion of the atmosphere that transports charge carriers in a direction opposite to the electrostatic force acting on them. The charges then flow back by conduction through the lower atmosphere. The terminal voltage of the generator is 400,000 V and the current is 1800 A, so the power is about 700 MW. By way of comparison, a modern nuclear power plant may develop nearly 1000 MW, and others to be built are in the GW range.

28–10 PHYSIOLOGICAL EFFECTS OF CURRENTS

Electrical potential differences and currents play a vital role in the nervous systems of animals. Conduction of nerve impulses is basically an electrical process, although the mechanism of conduction is much more complex than in simple materials such as metals. A nerve fiber or *axon*, along which an electrical impulse can travel, includes a cylindrical membrane with one conducting fluid (electrolyte) inside and another outside. By mechanisms similar to those in batteries, a potential difference of the order of 0.1 V is maintained between these fluids.

When a pulse is initiated, the membrane temporarily becomes more permeable to the ions in the fluids, leading to a local drop in potential. As the pulse passes, with a typical speed of the order of 30 m s^{-1}, the membrane recovers and the potential returns to its initial value. There are several aspects of this process that are not yet well understood.

The basically electrical nature of nerve impulse conduction is responsible for the great sensitivity of the body to externally-supplied electrical currents. Currents through the body as small as 0.1 A, much too small to produce significant heating, are fatal because they interfere with nerve processes essential for vital functions such as heartbeat. The *resistance* of the human body is highly variable, principally because of the fairly low conductivity of the skin, but the resistance between two electrodes grasped by dry hands is of the order 5 kΩ to 10 kΩ. For $R = 10 \text{ k}\Omega$, a current of 0.1 A requires a potential difference $V = IR = (0.1 \text{ A})(10 \text{ k}\Omega) = 1000 \text{ V}$.

Even much smaller currents can be very dangerous. A current of 0.01 A causes strong, convulsive muscle action and considerable pain, and with 0.02 A the person typically is unable to release a conductor inflicting the shock. Currents of this magnitude and even as small as 0.001 A can cause ventricular fibrillation, a disorganized twitching of heart muscles which occurs instead of regular beating, and which unfortunately pumps very little blood. Surprisingly, very large currents (over 0.1 A) are somewhat *less* likely to cause fatal fibrillation because the heart muscle is "clamped" in one position and is more likely to resume normal beating when the current is removed. Severe burns are, of course, more likely with large currents.

The moral of this rather morbid story, if there is one, is that under certain conditions voltages as small as 10 V can be dangerous, and should not be regarded with anything but respect and caution.

On the positive side, rapidly alternating currents can have beneficial effects. Alternating currents with frequencies the order of 10^6Hz do not interfere appreciably with nerve processes, and can be used for therapeutic heating for arthritic conditions, sinusitis, and a variety of other disorders. If one electrode is made very small, the resulting concentrated heating can be used for local destruction of tissue, such as tumors, or even for cutting tissue in certain surgical procedures.

Study of particular nerve impulses is also an important *diagnostic* tool in medicine. The most familiar examples are electrocardiography (EKG) and electroencephalography (EEG). Electrocardiograms, obtained by attaching electrodes to the chest and back and recording the regularly-varying potential differences, are used to study heart function. Similarly, electrodes attached to the scalp permit study of potentials in the brain, and the resulting patterns can be helpful in diagnosing epilepsy, brain tumors, and other disorders.

PROBLEMS

28–1 A silver wire 1 mm in diameter carries a charge of 90 C in 1 hr and 15 min. Silver contains 5.8×10^{28} free electrons per m^3. (a) What is the current in the wire? (b) What is the drift velocity of the electrons in the wire?

28–2 When a sufficiently high potential difference is applied between two electrodes in a gas, the gas ionizes, electrons moving toward the positive electrode and positive ions toward the negative electrode.

a) What is the current in a hydrogen discharge if, in each second, 4×10^{18} electrons and 1.5×10^{18} protons move in opposite directions past a cross section of the tube?

b) What is the direction of the current?

28–3 A vacuum diode can be approximated by a plane cathode and a plane anode, parallel to each other and 5 mm apart. The area of both cathode and anode is 2 cm^2. In the region between cathode and anode the current is carried solely by electrons. If the electron current is 50 mA, and the electrons strike the anode surface with a velocity of $1.2 \times 10^7 \text{m s}^{-1}$, find the number of electrons per cubic millimeter in the space just outside the surface of the anode.

28–4 In the Bohr model of the hydrogen atom the electron makes about 6×10^{15} rev s^{-1} around the nucleus. What is the average current at a point on the orbit of the electron?

28–5 The belt of a Van de Graaff generator is 1 m wide and travels with a speed of 25 m s^{-1}.

a) Neglecting leakage, at what rate in coulombs per second must charge be sprayed on one face of the belt to correspond to a current of 10^{-4} A into the collecting sphere?

b) Compute the surface charge per unit area on the belt.

28–6 Refer to the first example in Section 28–4. Assume there are 10^{29} free electrons per cubic meter in the wire, and that the current in the wire is 10 A.

a) What is the current density in the wire?

b) What is the electric intensity?

c) How long a time is required for an electron to travel the length of the wire?

28–7 The current in a wire varies with time according to the relation

$$i = 4 + 2t^2,$$

where i is in amperes and t in seconds.

a) How many coulombs pass a cross section of the wire in the time interval between $t = 5$ s and $t = 10$ s?

b) What constant current would transport the same charge in the same time interval?

28–8 The current in a wire varies with time according to the relation

$$i = 20 \sin 377t,$$

where i is in amperes, t in seconds, and $(377t)$ in radians.

a) How many coulombs pass a cross section of the wire in the time interval between $t = 0$ and $t = 1/120$ s?

b) In the interval between $t = 0$ and $t = 1/60$ s?

c) What constant current would transport the same charge in each of the intervals above?

28–9 A wire 100 m long and 2 mm in diameter has a resistivity of 4.8×10^{-8} Ω m.

a) What is the resistance of the wire?

b) A second wire of the same material has the same weight as the 100-m length, but twice its diameter. What is its resistance?

28–10

a) The following measurements of current and potential difference were made on a resistor constructed of Nichrome wire:

I, A	V_{ab}, V
0.5	2.18
1.0	4.36
2.0	8.72
4.0	17.44

Make a graph of V_{ab} as a function of I. Does the Nichrome obey Ohm's law? What is the resistance of the resistor, in ohms?

b) The following measurements were made on a Thyrite resistor:

I, A	V_{ab}, V
0.5	4.76
1.0	5.81
2.0	7.05
4.0	8.56

Make a graph of V_{ab} as a function of I. Does Thyrite have constant resistance?

c) Construct a graph of the resistance R as a function of I.

28–11 An aluminum bar 2.5 m long has a rectangular cross section 1 cm by 5 cm.

a) What is its resistance?

b) What would be the length of an iron wire 15 mm in diameter having the same resistance?

28–12 A solid cube of brass has a mass of 68.8 g. What is its resistance between opposite faces?

28–13 The two parallel plates of a capacitor have equal and opposite charges Q. The dielectric has a dielectric constant K and a resistivity ρ. Show that the "leakage" current carried by the dielectric is given by the relationship $i = Q/K\epsilon_0 \rho$.

28–14 Refer to the third example in Section 28–4. Let the resistivity of the material between the cylinders be 10 Ω m and let $r_a = 10$ cm, $r_b = 20$ cm, and $\ell = 5$ cm.

a) Find the resistance between the cylinders.

b) Find the current between the cylinders if $V_{ab} = 8$ V.

28–15 The region between two concentric conducting spheres of radii r_a and r_b is filled with a conducting material of resistivity ρ.

a) Show that the resistance between the spheres is given by

$$R = \frac{\rho}{4\pi}\left(\frac{1}{r_a} - \frac{1}{r_b}\right).$$

b) Derive an expression for the current density as a function of radius, if the potential difference between the spheres is V_{ab}.

28–16 In Problem 28–15, let the resistivity be 10 Ω m, and let $r_a = 10$ cm, $r_b = 20$ cm.

a) Find the resistance between the spheres.

b) Find the current between the spheres if $V_{ab} = 8$ V.

c) Find the electric field at $r = 15$ cm, if $V_{ab} = 8$ V.

d) Find the current density at $r = 15$ cm, by (1) using the result of part (c); and (2) using the result of Problem 28–15, part (b).

28–17

a) What is the resistance of a Nichrome wire at 0°C whose resistance is 100.00 Ω at 12°C?

b) What is the resistance of a carbon rod at 30°C whose resistance is 0.0150 Ω at 0°C?

28–18 The resistance of a coil of copper wire is 200 Ω at 20°C. What is its resistance at 50°C?

28–19 A toaster using a Nichrome heating element operates on 120 volts. When it is switched on at 0°C, it carries an initial current of 1.5 A. A few seconds later the current reaches the steady value of 1.33 A. What is the final temperature of the element? The average value of the temperature coefficient of Nichrome over the temperature range is $0.00045(\mathrm{C}°)^{-1}$.

28–20 A certain resistor has a resistance of 150.4 Ω at 20° C and a resistance of 162.4 Ω at 40° C. What is its temperature coefficient of resistivity?

28–21 A resistance thermometer using a platinum wire is used to measure the temperature of a liquid. The resistance is 2.42 Ω at 0° C, and when immersed in the liquid it is 2.98 Ω. The temperature coefficient of resistivity of platinum is $0.0038\ (\mathrm{C}°)^{-1}$. What is the temperature of the liquid?

28–22 A piece of wire has a resistance R. It is cut into three pieces of equal length, and the pieces are twisted together in parallel. What is the resistance of the resulting wire?

28–23 What diameter must an aluminum wire have if it is to have the same resistance as an equal length of copper wire of diameter 2.0 mm?

28–24 When switch S is open, the voltmeter V, connected across the terminals of the dry cell in Fig. 28–21, reads 1.52 V. When the switch is closed, the voltmeter reading drops to 1.37 V and the ammeter A reads 1.5 A. Find the emf and internal resistance of the cell. Neglect meter corrections.

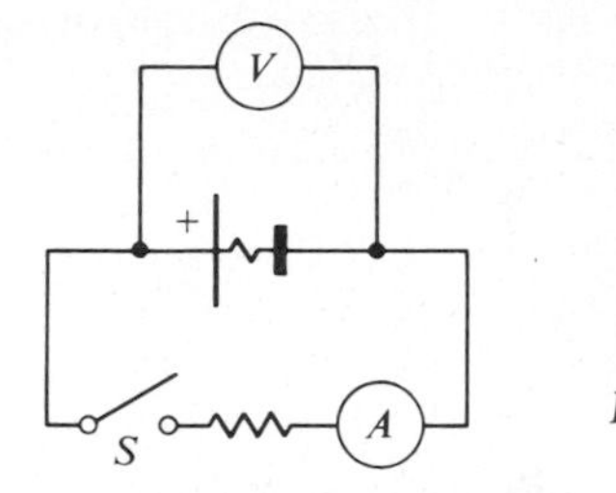

Fig. 28–21

28–25 The potential difference across the terminals of a battery is 8.5 V when there is a current of 3 A in the battery from the negative to the positive terminal. When the current is 2 A in the reverse direction, the potential difference becomes 11 V.

a) What is the internal resistance of the battery?

b) What is the emf of the battery?

28–26

a) What is the potential difference V_{ad} in the circuit of Fig. 28–22?

b) What is the terminal voltage of the 4-V battery?

c) A battery of emf 17 V and internal resistance 1 Ω is inserted in the circuit at d, its positive terminal being

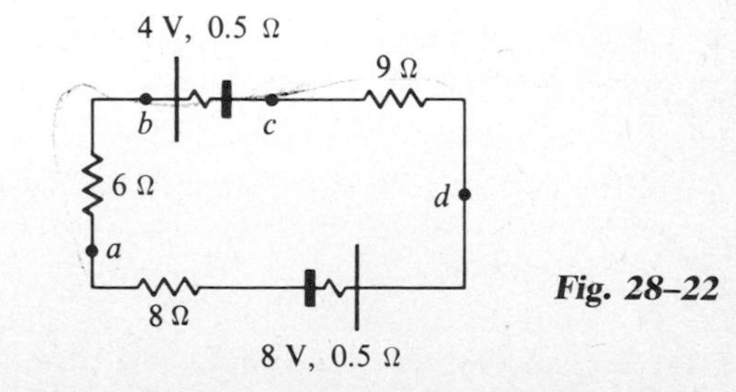

Fig. 28–22

connected to the positive terminal of the 8-V battery. What is now the difference of potential between the terminals of the 4-V battery?

28–27 A closed circuit consists of a 12-V battery, a 3.7-Ω resistor, and a switch. The internal resistance of the battery is 0.3 Ω. The switch is opened. What would a high-resistance voltmeter read when placed (a) across the terminals of the battery, (b) across the resistor, (c) across the switch? Repeat (a), (b), and (c) for the case when the switch is closed.

28–28 The internal resistance of a dry cell increases gradually with age, even though the cell is not used. The emf, however, remains fairly constant at about 1.5 V. Dry cells may be tested for age at the time of purchase by connecting an ammeter directly across the terminals of the cell and reading the current. The resistance of the ammeter is so small that the cell is practically short circuited.

a) The short-circuit current of a fresh No. 6 dry cell is about 30 A. Approximately what is the internal resistance?

b) What is the internal resistance if the short-circuit current is only 10 A?

c) The short-circuit current of a 6-V storage battery may be as great as 1000 A. What is its internal resistance?

28–29 The open-circuit terminal voltage of a source is 10 V and its short-circuit current is 4.0 A.

a) What will be the current when the source is connected to a linear resistor of resistance 2 Ω?

b) What will be the current in the Thyrite resistor of Problem 28–10(b) when connected across the terminals of this source?

c) What is the terminal voltage at this current?

28–30 A "660-W" electric heater is designed to operate from 120-V lines.

a) What is its resistance?

b) What current does it draw?

c) What is the rate of dissipation of energy, in calories per second?

d) If the line voltage drops to 110 V, what power does the heater take, in watts? (Assume the resistance constant. Actually, it will change because of the change in temperature.)

28–31 A resistor develops heat at the rate of 360 W when the potential difference across its ends is 180 V. What is its resistance?

28–32 A motor operating on 120 V draws a current of 2 A. If heat is developed in the motor at the rate of 9 cal s^{-1}, what is its efficiency?

28–33

a) Express the rate of dissipation of energy in a resistor in terms of (1) potential difference and current, (2) resistance and current, (3) potential difference and resistance.

b) Energy is dissipated in a resistor at the rate of 40 W when the potential difference between its terminals is 60 V. What is its resistance?

28–34 A storage battery whose emf is 12 V and whose internal resistance is 0.1 Ω (internal resistances of commercial storage batteries are actually only a few thousandths of an ohm) is to be charged from a 112–V dc supply.

a) Should the + or the − terminal of the battery be connected to the + side of the line?

b) What will be the charging current if the battery is connected directly across the line?

c) Compute the resistance of the series resistor required to limit the current to 10 A.

With this resistor in the circuit, compute (d) the potential difference between the terminals of the battery, (e) the power taken from the line, (f) the power dissipated in the series resistor, and (g) the *useful* power input to the battery.

h) If electrical energy costs 3 cents per kWh, what is the cost of operating the circuit for 2 hr when the current is 10 A?

28–35 In the circuit in Fig. 28–23, find (a) the rate of conversion of internal energy to electrical energy within the battery, (b) the rate of dissipation of energy in the battery, (c) the rate of dissipation of energy in the external resistor.

28–36 A source whose emf is $\mathscr{E}$ and whose internal resistance is r is connected to an external circuit.

a) Show that the power output of the source is maximum when the current in the circuit is one-half the short-circuit current of the source.

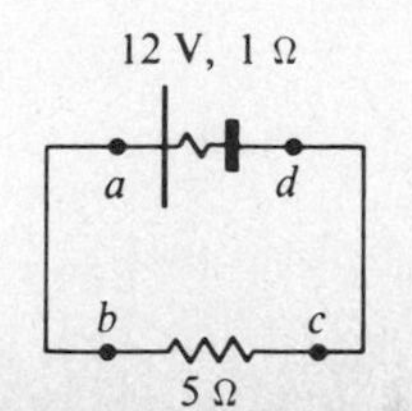

Fig. 28–23

b) If the external circuit consists of a resistance R, show that the power output is maximum when $R = r$, and that the maximum power is $\mathscr{E}^2/4r$.

28–37 Using the data of Table 28–3, compute for a copper-nickel thermocouple (a) the thermal emf when $T = 100°C$, (b) the Peltier emf at 100° C. (c) What Peltier heat is transferred at the junction at 100°C if a current of 1 mA is maintained for 1 hr?

28–38 It is desired to construct a bismuth–silver thermopile that will develop a thermal emf of 1 μV when $T = 0.001°C$, T_R being 0°C. How many separate thermocouples must be connected in series?

Chapter 29

Direct-Current Circuits and Instruments

29–1 RESISTORS IN SERIES AND IN PARALLEL

Most electrical circuits consist not merely of a single source and a single external resistor, but comprise a number of sources, resistors, or other elements such as capacitors, motors, etc., interconnected in a more or less complicated manner. The general term applied to such a circuit is a *network*. We shall next consider a few of the simpler types of network.

Figure 29–1 illustrates four different ways in which three resistors having resistances R_1, R_2, and R_3 might be connected between points a and b. In part (a), the resistors provide only a single path between the points, and are said to be connected in *series* between these points. Any number of circuit elements such as resistors, cells, motors, etc., are similarly said to be in series with one another between two points if connected as in (a) so as to provide only a single path between the points. The *current* is the same in each element.

The resistors in Fig. 29–1(b) are said to be in *parallel* between points a and b. Each resistor provides an alternative path between the points, and any number of circuit elements similarly connected are in

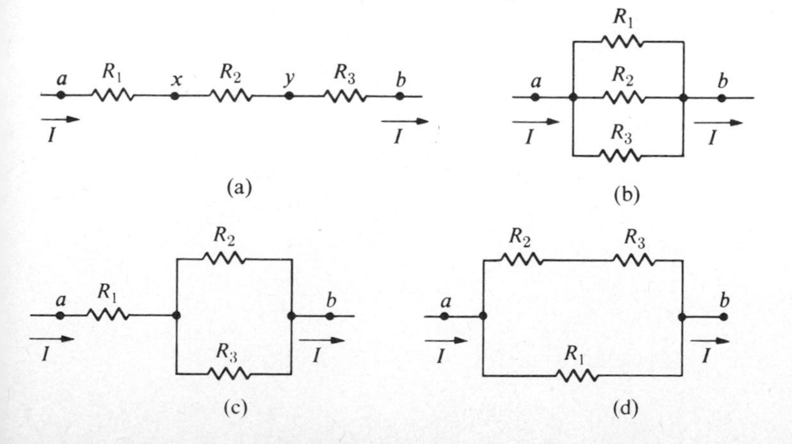

Fig. 29–1 Four different ways of connecting three resistors.

parallel with one another. The *potential difference* is the same across each element.

In Fig. 29–1(c), resistors R_2 and R_3 are in parallel with each other, and this combination is in series with the resistor R_1. In Fig. 29–1(d), R_2 and R_3 are in series, and this combination is in parallel with R_1.

It is always possible to find a single resistor which could replace a combination of resistors in any given circuit and leave unaltered the potential difference between the terminals of the combination and the current in the rest of the circuit. The resistance of this single resistor is called the *equivalent* resistance of the combination. If any one of the networks in Fig. 29–1 were replaced by its equivalent resistance R, we could write

$$V_{ab} = IR \qquad \text{or} \qquad R = \frac{V_{ab}}{I},$$

where V_{ab} is the potential difference between the terminals of the network and I is the current at the point a or b. Hence the method of computing an equivalent resistance is to assume a potential difference V_{ab} across the actual network, compute the corresponding current I (or vice versa), and take the ratio of one to the other. The simple series and parallel connections of resistors are sufficiently common so that it is worthwhile to develop formulas for these two special cases.

If the resistors are in series, as in Fig. 29–1(a), the current in each must be the same and equal to the line current I. Hence

$$V_{ax} = IR_1, \qquad V_{xy} = IR_2, \qquad V_{yb} = IR_3,$$

and

$$\begin{aligned} V_{ab} &= V_{ax} + V_{xy} + V_{yb} \\ &= I(R_1 + R_2 + R_3), \\ \frac{V_{ab}}{I} &= R_1 + R_2 + R_3. \end{aligned}$$

But V_{ab}/I is, by definition, the equivalent resistance R. Therefore

$$\boxed{R = R_1 + R_2 + R_3.} \qquad (29\text{–}1)$$

Evidently the equivalent resistance of any number of resistors in series equals the sum of their individual resistances.

If the resistors are in parallel, as in Fig. 29–1(b), the potential difference between the terminals of each must be the same and equal to V_{ab}. If the currents in each are denoted by I_1, I_2, and I_3, respectively,

$$I_1 = \frac{V_{ab}}{R_1}, \qquad I_2 = \frac{V_{ab}}{R_2}, \qquad I_3 = \frac{V_{ab}}{R_3}.$$

Charge is delivered to point a by the line current I, and removed from a by the currents I_1, I_2, and I_3. Since charge is not accumulating at a, it follows that

$$I = I_1 + I_2 + I_3 = V_{ab}\left(\frac{1}{R_1} + \frac{1}{R_2} + \frac{1}{R_3}\right),$$

or

$$\frac{I}{V_{ab}} = \frac{1}{R_1} + \frac{1}{R_2} + \frac{1}{R_3}.$$

But

$$\frac{I}{V_{ab}} = \frac{1}{R},$$

so

$$\boxed{\frac{1}{R} = \frac{1}{R_1} + \frac{1}{R_2} + \frac{1}{R_3}.} \qquad (29\text{–}2)$$

Evidently, for any number of resistors in parallel, the *reciprocal* of the equivalent resistance equals the *sum of the reciprocals* of their individual resistances.

For the special case of *two* resistors in parallel,

$$\frac{1}{R} = \frac{1}{R_1} + \frac{1}{R_2} = \frac{R_2 + R_1}{R_1 R_2}$$

and

$$R = \frac{R_1 R_2}{R_1 + R_2}.$$

Also, since $V_{ab} = I_1 R_1 = I_2 R_2$,

$$\frac{I_1}{I_2} = \frac{R_2}{R_1}, \qquad (29\text{–}3)$$

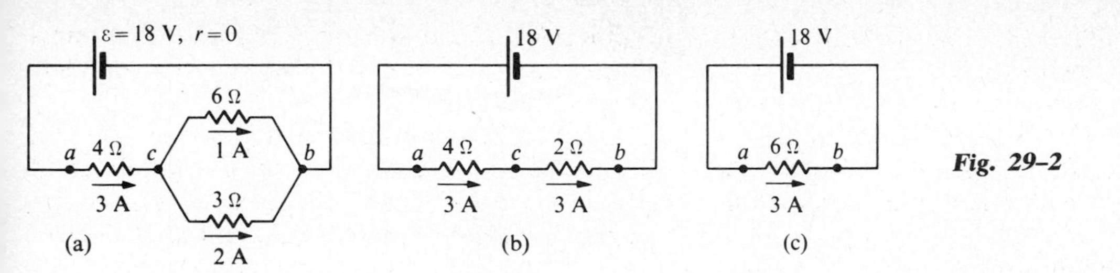

Fig. 29–2

and the currents carried by two resistors in parallel are *inversely proportional* to their resistances.

The equivalent resistances of the networks in Figs. 29–1(c) and 29–1(d) could be found by the same general method, but it is simpler to consider them as combinations of series and parallel arrangements. Thus, in (c) the combination of R_2 and R_3 in parallel is first replaced by its equivalent resistance, which then forms a simple series combination with R_1. In (d), the combination of R_2 and R_3 in series forms a simple parallel combination with R_1. Not all networks, however, can be reduced to simple series–parallel combinations, and special methods must be used for handling such networks.

Example Compute the equivalent resistance of the network in Fig. 29–2, and find the current in each resistor.

Successive stages in the reduction to a single equivalent resistance are shown in parts (b) and (c). The 6-Ω and the 3-Ω resistors in part (a) are equivalent to the single 2-Ω resistor in part (b), and the series combination of this with the 4-Ω resistor results in the single equivalent 6-Ω resistor in part (c).

In the simple series circuit of part (c), the current is 3 A, and hence the current in the 4-Ω and 2-Ω resistors in part (b) is 3 A also. The potential difference V_{cb} is therefore 6 V, and since it must be 6 V in part (a) as well, the currents in the 6-Ω and 3-Ω resistors in part (a) are 1 A and 2 A, respectively.

29–2 KIRCHHOFF'S RULES

Not all networks can be reduced to simple series–parallel combinations. An example is a resistance network with a cross connection, as in Fig. 29–3(a).

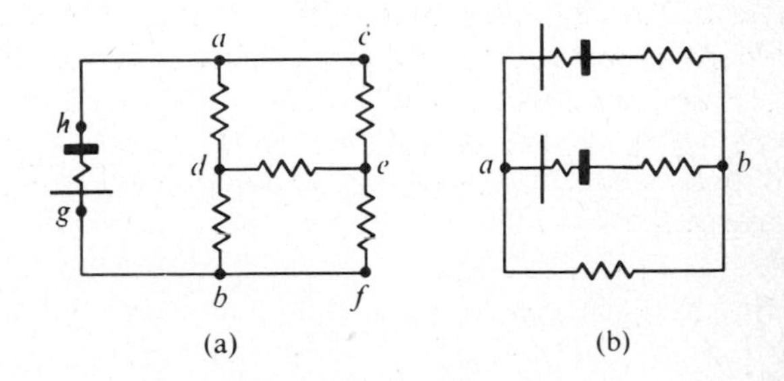

Fig. 29–3 *Two networks that cannot be reduced to simple series–parallel combinations of resistors.*

A circuit like that in Fig. 29–3(b), which contains sources in parallel paths, is another example. No new *principles* are required to compute the currents in these networks, but there are a number of techniques that enable such problems to be handled systematically. We shall describe only one of these, first developed by Gustav Robert Kirchhoff (1824–1887).

We first define two terms. A *branch point* in a network is a point where three or more conductors are joined. A *loop* is any closed conducting path. In Fig. 29–3(a), for example, points *a, d, e,* and *b* are branch points but *c* and *f* are not. In Fig. 29–3(b) there are only two branch points, *a* and *b*.

Possible loops in Fig. 29–3(a) are the closed paths *aceda, defbd, hadbgh,* and *hadefbgh.*

Kirchhoff's rules consist of the following two statements:

Point Rule *The algebraic sum of the currents* **toward** *any branch point is zero:*

$$\boxed{\sum I = 0.} \tag{29–4}$$

Loop Rule *The algebraic sum of the* emf's *in any loop equals the algebraic sum of the IR products in the same loop*:

$$\sum \mathscr{E} = \sum IR. \qquad (29\text{–}5)$$

The first rule merely states formally that no charge accumulates at a branch point. The second rule is a generalization of the circuit equation, Eq. (28–18), and reduces to this equation if the current I is the same in all resistances.

As in so many instances, the chief difficulty encountered in applying Kirchhoff's rules is in keeping track of algebraic signs, not in understanding the physical ideas, which are, in fact, very elementary. The first step is to assign a symbol and direction to each unknown current and emf, and a symbol to each unknown resistance. These, as well as the known quantities, are represented in a diagram with directions carefully shown. The solution is then carried through on the basis of the assumed directions. If, in the numerical solution of the equations, a negative value is found for a current or an emf, its correct direction is opposite to that assumed. The correct *numerical* value is obtained in any case. Hence the rules provide a method for ascertaining the *directions* as well as the magnitudes of currents and emf's, and it is not necessary that these directions be known in advance.

The expressions $\sum I$, $\sum IR$, and $\sum \mathscr{E}$ are *algebraic* sums. When applying the point rule, a current is considered *positive* if its direction is *toward* a branch point, negative if away from the point. (Of course, the opposite convention may also be used.) When applying the loop rule, some direction around the loop (i.e., clockwise or counterclockwise) must be chosen as the positive direction. All currents and emf's in this direction are positive; those in the opposite direction are negative. Note that a current which has a positive sign in the point rule may have a negative sign in the term in which it appears in the loop rule. Note also that the direction around the loop which is considered positive is immaterial, the result of choosing the opposite direction being merely to obtain the same equation with signs reversed.

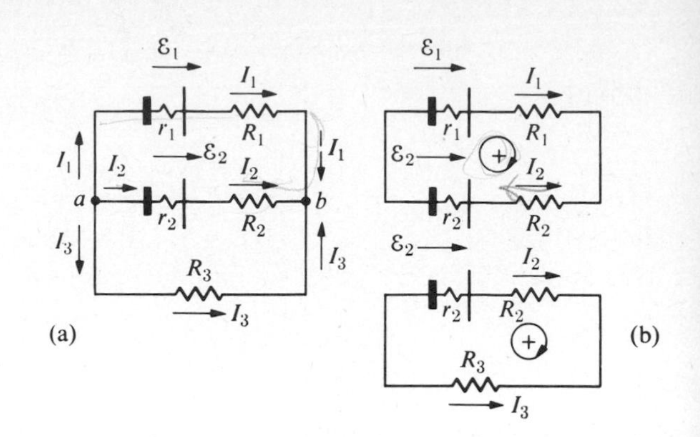

Fig. 29–4 Solution of a network by the application of Kirchhoff's rules.

There is a temptation to assume that the "correct" direction to consider positive is that of the current in the loop, but in general such a choice is not possible, since the currents in some elements of a loop may be clockwise and in other elements, counterclockwise.

In complicated networks, where a large number of unknown quantities may be involved, it is sometimes puzzling to know how to obtain a sufficient number of *independent* equations to solve for the unknowns. The following rules are helpful:

1. If there are n branch points in the network, apply the point rule at $(n - 1)$ of these points. Any points may be chosen. Application of the point rule at the nth point does not lead to an independent equation.
2. Imagine the network to be separated into a number of simple loops, like the pieces of a jigsaw puzzle. Apply the loop rule to each of these loops.

Example Figure 29–4 is the same as Fig. 29–3(b), prepared for solution by Kirchhoff's rules. Let the magnitudes and directions of the emf's and the magnitudes of the resistances be given. We wish to solve for the currents in each branch of the network.

Assign a direction and a letter to each unknown current. The assumed directions are entirely arbitrary. Note that the currents in source 1 and resistor 1 are the same, and require only a single letter, I_1. The same is true for source 2 and resistor 2; the current in both is represented by I_2.

There are only two branch points, a and b. At point b,

$$\sum I = I_1 + I_2 + I_3 = 0.$$

Since there are but two branch points, there is only one independent "point" equation. If the point rule is applied at the other branch point, point a, we get

$$\sum I = -I_1 - I_2 - I_3 = 0,$$

which is the same equation with signs reversed.

In Fig. 29–4(b) the loop is cut up into its "jigsaw" sections. Let us consider the clockwise direction *positive* in each loop. The loop rule then furnishes the following equations:

$$\mathcal{E}_1 - \mathcal{E}_2 = I_1 r_1 + I_1 R_1 - I_2 R_2 - I_2 r_2,$$
$$\mathcal{E}_2 = I_2 r_2 + I_2 R_2 - I_3 R_3,$$

and we have three independent equations to solve for the three unknown currents.

29–3 AMMETERS AND VOLTMETERS

The most common type of ammeter or voltmeter is a modified moving-coil galvanometer, in which a pivoted coil of fine wire carrying a current is deflected by the magnetic interaction between this current and the magnetic field of a permanent magnet. For the present, we are interested only in the instrument as a circuit element. The resistance of the coil of a typical instrument is of the order of 10 to 100 Ω, and a current of the order of a few milliamperes will produce full-scale deflection. The deflection is proportional to the current in the coil, but since the coil is a linear conductor, the current is proportional to the potential difference between the terminals of the coil, and the deflection is also proportional to this potential difference.

As a numerical example, consider a galvanometer whose coil has a resistance of 20 Ω, and which deflects full scale with a current of 1 mA in its coil. The corresponding potential difference is

$$V_{ab} = IR = (10^{-3}\ \text{A})(20\ \Omega) = 0.020\ \text{V} = 20\ \text{mV}.$$

First consider the galvanometer as an ammeter. To measure the current in a circuit, an ammeter must be inserted in *series* in the circuit so that the current to be measured actually passes *through* the meter. If the galvanometer above is inserted in this way, it will measure any current from zero to 1 mA. However, the resistance of the *coil* adds to the total resistance of the circuit, with the result that the current *after* the galvanometer is inserted, although it is correctly indicated by the instrument, may be less than it was *before* insertion of the instrument. It is evidently desirable that the resistance of the instrument should be much *smaller* than that of the remainder of the circuit, so that when the instrument is inserted it does not change the very thing we wish to measure. An *ideal* ammeter would have *zero* resistance.

Furthermore, the *range* of the galvanometer, if it is used without modification, is limited to a maximum current of 1 mA. The range can be extended, and at the same time the equivalent resistance can be reduced, by connecting a low resistance R_{sh} in parallel with the moving coil, as in Fig. 29–5(a). The parallel resistor is called a *shunt.* Coil and shunt are mounted inside a case, with binding posts for external connections at a and b. (In some instruments having interchangeable shunts to cover several different ranges, the shunts are mounted outside the case.)

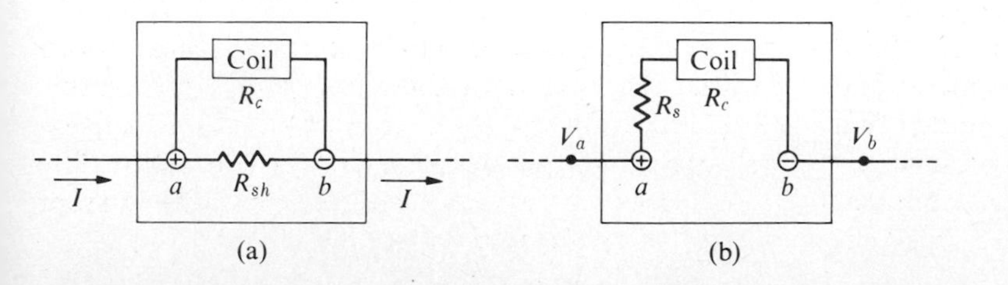

Fig. 29–5 *(a) Internal connections of an ammeter. (b) Internal connections of a voltmeter.*

Suppose we wish to modify the galvanometer above for use as an ammeter with a range of 0 to10 A. That is, the coil is to deflect full scale when the current I in the *circuit* in which the ammeter is inserted equals 10 A. The *coil* current I_c must then be 1 mA, so the current I_{sh} in the shunt is 9.999 A. The potential difference V_{ab} is

$$V_{ab} = I_c R_c = I_{sh} R_{sh}.$$

Hence

$$R_{sh} = R_c\left(\frac{I_c}{I_{sh}}\right) = 20\,\Omega\left(\frac{0.001}{9.999}\right) = 0.00200\,\Omega$$

(to three significant figures).

The equivalent resistance R of the instrument is

$$\frac{1}{R} = \frac{1}{R_c} + \frac{1}{R_{sh}},$$

and

$$R = 0.00200\,\Omega$$

(to three significant figures).

Thus we have a low-resistance instrument with the desired range of 0 to 10 A. Of course, if the current I is *less* than 10 A, the coil current (and the deflection) is correspondingly less, also.

Now consider the construction of a *voltmeter.* A voltmeter measures the potential difference between two points, and its terminals must be connected to these points. Evidently a moving-coil galvanometer cannot be used to measure the potential differences between say, two charged spheres. When the galvanometer terminals are connected to the spheres the galvanometer coil provides a conducting path from one sphere to the other. There will be a *momentary* current in the coil, but the charges on the sphere will change until the entire system is at the *same* potential. Only if the resistance of the instrument is so great that a very long time is required to reach equilibrium can a voltmeter be used for this purpose. An *ideal* voltmeter has an *infinite* resistance, and although an *electrometer* does have a resistance that can be considered infinite, a moving-coil galvanometer can be deflected only by a current in its coil, and its resistance must be finite.

A moving-coil galvanometer can be used to measure the potential difference between the terminals of a *source,* or between *two points of a circuit containing a source,* because the source *maintains* a difference of potential between the points. However, complications arise here also.

We have shown that when a source is an *open* circuit, the potential difference between its terminals equals its emf. It would seem, therefore, that to measure the emf we need only measure this potential difference. But when the terminals of a galvanometer are connected to those of a source, the galvanometer and source form a *complete* circuit in which there is a current. The potential difference *after* the galvanometer is connected, although it is correctly indicated by the instrument, is not equal to $\mathcal{E}$, but to $\mathcal{E} - Ir$, and is less than it was *before* the instrument was connected. Again, the measuring instrument alters the quantity it is intended to measure. It is evidently desirable that the resistance of a voltmeter, even though not infinite, should be as *large* as possible.

Furthermore, the *range* of the galvanometer in our example, if used without modification, is limited to a maximum value of 20 mV. The range can be extended, and at the same time the equivalent resistance can be increased, by connecting a high resistance R_s in *series* with the moving coil, as in Fig. 29-5(b).

Suppose we wish to modify the galvanometer for use as a voltmeter with a range of 0 to 10 V. That is, the coil is to deflect full scale when the potential difference between the terminals of the instrument is 10 V. In other words, the current in the instrument is to be 1 mA when the potential difference between its terminals is 10 V.

The terminal potential difference is

$$V_{ab} = I(R_c + R_s),$$

and the necessary series resistance is

$$R_s = \frac{V_{ab}}{I} - R_c = \frac{10\text{ V}}{0.001\text{ A}} - 20\,\Omega = 9980\,\Omega.$$

The equivalent resistance R is

$$R = R_c + R_s = 10{,}000\,\Omega.$$

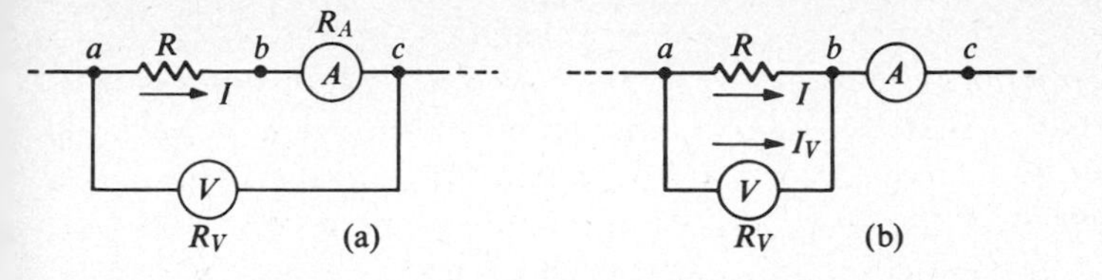

Fig. 29–6 Ammeter-voltmeter method for measuring resistance or power.

Thus we have a high-resistance instrument with the desired range of 0 to 10 V.

The resistance of a resistor equals the potential difference V_{ab} between its terminals, divided by the current I:

$$R = \frac{V_{ab}}{I},$$

and the power input to any portion of a circuit equals the product of the potential difference across this portion and the current:

$$P = V_{ab}I.$$

The most straightforward method of measuring R or P is therefore to measure V_{ab} and I simultaneously.

In Fig. 29–6(a), ammeter A reads correctly the current I in the resistor R. Voltmeter V, however, reads the *sum* of the potential difference V_{ab} across the resistor and the potential difference V_{bc} across the ammeter.

If we transfer the voltmeter terminal from c to b, as in Fig. 29–6(b), the voltmeter reads correctly the potential difference V_{ab} but the ammeter now reads the *sum* of the current I in the resistor and the current I_V in the voltmeter. Thus, whichever connection is used, we must correct the reading of one instrument or the other to obtain the true values of V_{ab} or I (unless, of course, the corrections are small enough to be neglected).

29–4 THE WHEATSTONE BRIDGE

The Wheatstone bridge circuit, shown in Fig. 29–7, is widely used for the rapid and precise measurement of resistance. It was invented in 1843 by the English scientist Charles Wheatstone. M, N, and P are adjustable resistors which have been previously calibrated, and X represents the unknown resistance. To use the bridge, switches K_1 and K_2 are closed and the resistance of P is adjusted until the galvanometer G shows no deflection. Points b and c must then be at the same potential or, in other words, the potential drop from a to b equals that from a to c. Also, the drop from b to d equals that from c to d. Since the galvanometer current is zero, the current in M equals that in N, say I_1, and the current in P equals that in X, say I_2. Then, since $V_{ab} = V_{ac}$, it follows that

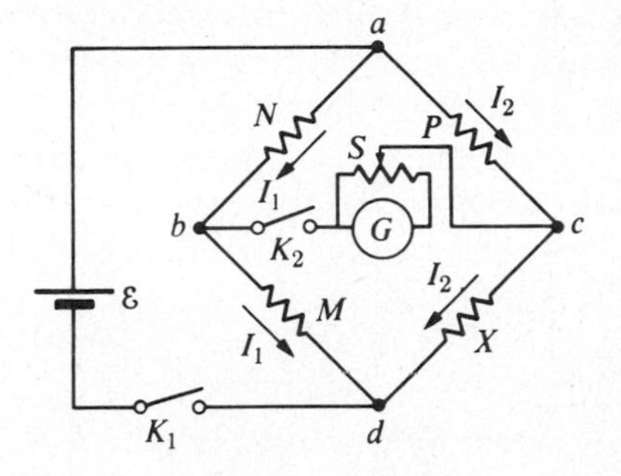

Fig. 29–7 Wheatstone bridge circuit.

$$I_1 N = I_2 P,$$

and since $V_{bd} = V_{cd}$,

$$I_1 M = I_2 X.$$

When the second equation is divided by the first, we find

$$X = \left(\frac{M}{N}\right) P.$$

Hence, if M, N, and P are known, X can be computed. The ratio M/N is usually set at some integral power of 10, such as 0.01, 1, 100, etc., for simplicity in computation.

During preliminary adjustments, when the bridge may be far from balance and V_{bc} large, the galvanometer must be protected from excessive current by the shunt S. A resistor whose resistance is large compared with that of the galvanometer is permanently connected across the galvanometer terminals. When the sliding contact is at the left end of the resistor, none of the current in the path between b and c passes through the galvanometer. In a position such as that shown, that portion of the resistor at the right of the sliding contact is in series with the galvanometer, and this combination is shunted by that portion of the resistor at the left of the contact. Hence only a fraction of the current passes through the galvanometer. With the sliding contact at the right of the resistor, all the current passes through the galvanometer except the small fraction bypassed by the resistor. The galvanometer is, therefore, fully protected when the contact is at the left end of the resistor; and practically full galvanometer sensitivity is attained when the contact is at the right end.

If any of the resistances are inductive, the potentials V_b and V_c may attain their final values at different rates when K_1 is closed; in that case, the galvanometer, if connected between b and c, would show an initial deflection even though the bridge were in balance. Hence K_1 and K_2 are frequently combined in a double key which closes the battery circuit first and the galvanometer circuit a moment later, after the transient currents have died out.

Portable bridges are available which have a self-contained galvanometer and dry cells. The ratio M/N can be set at any integral power of 10 between 0.001 and 1000 by a single dial switch, and the value of P can be adjusted by four dial switches. These instruments are capable of very high precision.

29–5 THE OHMMETER

Although not a precision instrument, the ohmmeter is a useful device for rapid measurement of resistance. It consists of a galvanometer, a resistor, and a source (usually a small dry cell) connected in series, as in Fig. 29–8. The resistance R to be measured is connected between terminals x and y.

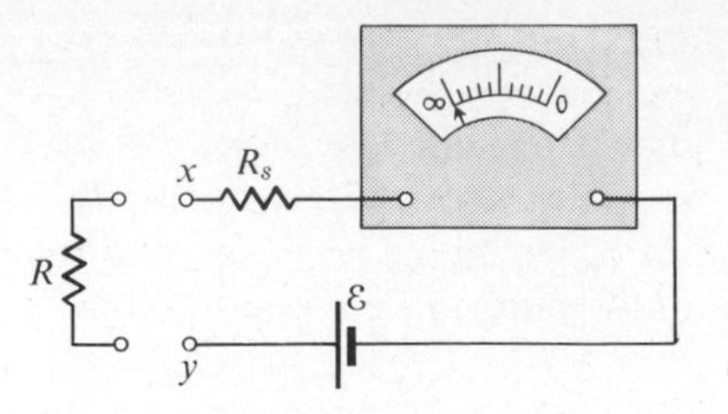

Fig. 29–8

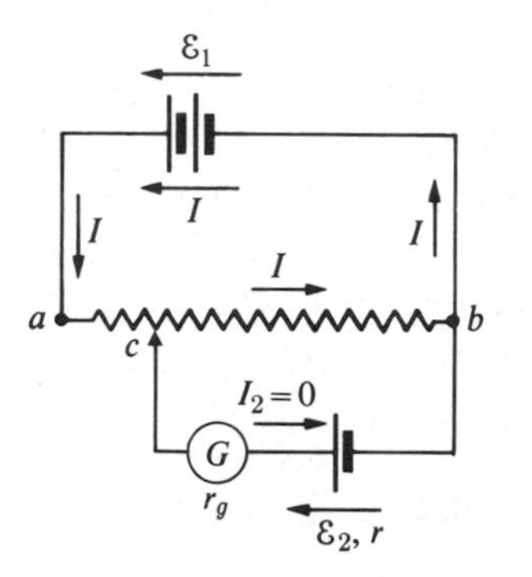

Fig. 29–9 *Principle of potentiometer.*

The series resistance R_s is chosen so that, when terminals x and y are short-circuited (that is, when $R = 0$), the galvanometer deflects full scale. When the circuit between x and y is open (that is, when $R = \infty$), the galvanometer shows no deflection. For a value of R between zero and infinity, the galvanometer deflects to some intermediate point depending on the value of R, and hence the galvanometer scale can be calibrated to read the resistance R.

29–6 THE POTENTIOMETER

The potentiometer is an instrument which can be used to measure the emf of a source without drawing any current from the source; it also has a number of other useful applications. Essentially, it balances an unknown potential difference against an adjustable, measurable potential difference.

The principle of the potentiometer is shown schematically in Fig. 29–9. A resistance wire ab is permanently connected to the terminals of a source of emf $\mathcal{E}_1$. A sliding contact c is connected through the galvanometer G to a second source whose emf $\mathscr{E}_2$ is to be measured. Contact c is moved along the wire until a position is found at which the galvanometer shows no deflection. (This is possible only when $V_{ab} \geq \mathscr{E}_2$.) If we then write the expression for V_{cb} for two paths between these points, we have

$$V_{cb} = IR_{cb} \qquad \text{(resistance wire path)},$$

$$V_{cb} = (r_g + r)I_2 - (-\mathscr{E}_2) = \mathscr{E}_2 \qquad \text{(galvanometer path)},$$

where the last step is possible because *we have adjusted* I_2 *to be equal to zero.*

Hence, IR_{cb} is exactly equal to the emf $\mathscr{E}_2$, and $\mathscr{E}_2$ can be computed if I and R_{cb} are known. No correction need be made for the "Ir" term, since the current in $\mathscr{E}_2$ is zero.

29–7 THE *R–C* SERIES CIRCUIT

In the circuits considered thus far, we have assumed that the emf's and resistances are constant, so that all potentials and currents are constant, independent of time. Figure 29–10 shows a simple example of a circuit in which the current and voltages are *not* constant. When the double-pole, double-throw (dpdt) switch is in the upper position, points a and b are connected to the battery, which charges the capacitor through resistor R.

If the capacitor is initially uncharged, then the initial potential difference across the capacitor is zero, and the entire battery voltage appears across the resistor, causing an initial current $I = V/R$. As the capacitor charges, its voltage increases, and the potential difference across the resistor decreases, corresponding to a decrease in current. After a long time the capacitor has become fully charged, the entire battery voltage V appears across the capacitor, there is no potential difference across the resistor, and the current becomes zero.

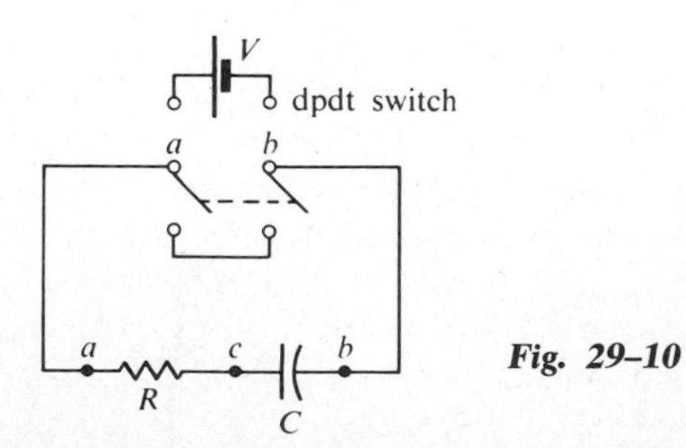

Fig. 29–10

Let q represent the charge on the capacitor and i the charging current at some instant after the switch is thrown in the "up" position. The instantaneous potential differences v_{ac} and v_{cb} are

$$v_{ac} = iR, \qquad v_{cb} = \frac{q}{C}. \tag{29–6}$$

Therefore,

$$V_{ab} = V = v_{ac} + v_{cb} = iR + \frac{q}{C}, \tag{29–7}$$

where V = constant. The current i is then

$$i = \frac{V}{R} - \frac{q}{RC}. \tag{29–8}$$

At the instant connections are made, $q = 0$ and the *initial current* $I_0 = V/R$, which equals the steady current if the capacitor is not present.

As the charge q increases, the term q/RC becomes larger, and the current decreases and eventually becomes zero. When $i = 0$,

$$\frac{V}{R} = \frac{q}{RC}, \qquad q = CV = Q_f,$$

where Q_f is the final charge.

To obtain a quantitative description of the variation with time of the charge and current in this circuit, we replace i in Eq. (29–8) with dq/dt, obtaining

$$\frac{dq}{dt} = \frac{V}{R} - \frac{q}{RC}, \tag{29–9}$$

which may be rearranged as

$$\frac{dq}{VC - q} = \frac{dt}{RC}.$$

Integrating both sides, we obtain

$$-\ln(VC - q) = t/RC + \text{constant}.$$

To evaluate the constant, we note that at $t = 0$, $q = 0$, so

$$-\ln(VC - 0) = 0 + \text{constant}.$$

Rearranging again, we obtain

$$\ln(VC - q) - \ln VC = \ln\frac{VC - q}{VC} = -\frac{t}{RC},$$

$$1 - \frac{q}{VC} = e^{-t/RC},$$

$$q = VC(1 - e^{-t/RC}) = Q_f(1 - e^{-t/RC}). \qquad (29\text{–}10)$$

The time derivative of this expression is the current:

$$i = \frac{dq}{dt} = \frac{V}{R}e^{-t/RC} = I_0 e^{-t/RC}. \qquad (29\text{–}11)$$

The charge and current are therefore both *exponential* functions of time. Figure 29–11(a) is a graph of Eq. (29–11), and Fig. 29–11(b) is a graph of Eq. (29–10). At a time $t = RC$, the current has decreased to $1/e$ of its initial value and the charge has increased to *within* $1/e$ of its final value. The product RC is called the *time constant*, or the *relaxation time*, of the circuit. It is the time in which the current *would* decrease to zero, if it continued to decrease at its initial rate.

The *half-life* of the circuit, t_h, is the time for the current to decrease to half its initial value, or for the capacitor to acquire half its final charge. Setting $i = I_0/2$ in Eq. (29–11), we find

$$t_h = RC \ln 2 = 0.693\ RC.$$

The half-life depends only on the time constant RC, and not on the initial current. Thus if the current decreases from I_0 to $I_0/2$ in a time t_h, as shown in Fig. 29–11(a), it decreases to half this value, or to $I_0/4$, in another half-life, and so on.

Example A resistor of resistance $R = 10\ \text{M}\Omega$ is connected in series with a capacitor of capacitance 1 μF. The time constant is

$$RC = (10 \times 10^6 \Omega)(10^{-6}\text{F}) = 10\ \text{s},$$

and the half-life is

$$t_h = (10\ \text{s})(\ln 2) = 6.9\ \text{s}.$$

On the other hand, if $R = 10\ \Omega$ the time constant is only 10×10^{-6}s, or 10 μs.

Suppose next that the capacitor has acquired a charge Q_0 and that the switch is thrown to the "down" position. The capacitor then *discharges* through the resistor and its charge eventually decreases to zero. (We designate the charge by Q_0 because this is the *initial* charge in the discharge process. It is not necessarily equal to the charge Q_f defined above.)

Again let i and q represent the current and charge at some instant after the switch is thrown. Since V_{ab} is now zero, we have, from Eq. (29–7),

$$0 = v_{ac} + v_{cb}.$$

The direction of the current in the resistor is now

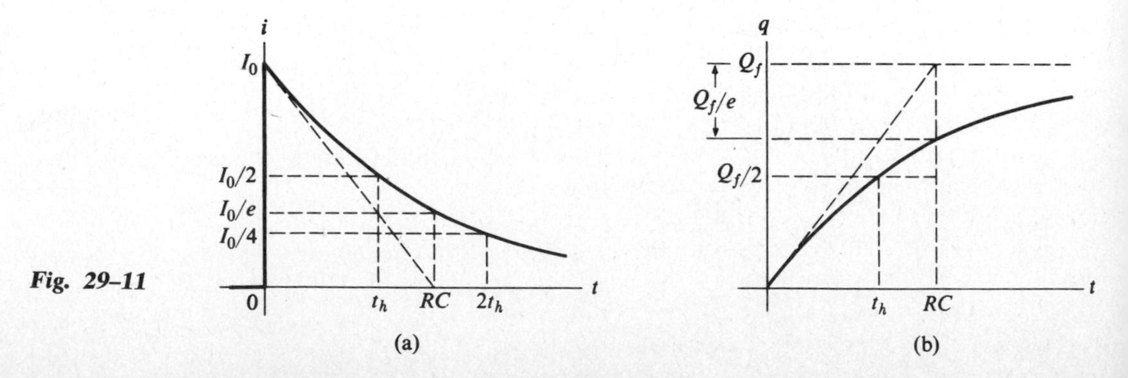

Fig. 29–11

from c to a, so $v_{ca} = iR$ and

$$i = \frac{q}{RC}. \tag{29–12}$$

When $t = 0$, $q = Q_0$ and the initial current I_0 is

$$I_0 = \frac{Q_0}{RC} = \frac{V_0}{R},$$

where V_0 is the initial potential difference across the capacitor. As the capacitor discharges, both q and i decrease.

The same procedures as above can be followed to obtain $i(t)$ and $q(t)$. If we replace i in Eq. (29–12) by $-dq/dt$ (the charge q is now *decreasing*), we get

$$\frac{dq}{dt} = -\frac{q}{RC}. \tag{29–13}$$

Integration of this equation gives $q(t)$, and by differentiation we find $i(t)$.

Alternatively, differentiation of Eq. (29–12) gives

$$\frac{di}{dt} = -\frac{i}{RC}, \tag{29–14}$$

from which we can get $i(t)$ and, by a second integration, get $q(t)$. It is left as a problem to show that

$$i = I_0 e^{-t/RC}, \tag{29–15}$$

$$q = Q_0 e^{-t/RC}. \tag{29–16}$$

Both the current and the charge decrease exponentially with time.

29–8 DISPLACEMENT CURRENT

For a *conducting* circuit, in the steady state, the total current *into* any given portion must be equal to the current *out of* that portion. This statement forms the basis of Kirchhoff's point rule, discussed in Section 29–2. However, this rule is *not* obeyed for a capacitor that is being charged. In Fig. 29–12, there is a conduction current into the left plate but no conduction current out of this plate; similarly, there is conduction current out of the right plate, but none into it.

James Clerk Maxwell (1831–1879) showed that it is possible to generalize the definition of current so that one can still say that the current out of each plate is equal to the current into it. As the capacitor charges, the conduction current increases the charge on each plate, and this in turn increases the electric field between the plates. The *rate* of increase of field is proportional to the conduction current; Maxwell's scheme was to associate an effective current density with this rate of increase of field.

Assuming, for simplicity, that the capacitor consists of two parallel plates with uniform charge density σ, the field E between the plates is given by

$$E = \frac{\sigma}{\epsilon_0} = \frac{Q}{\epsilon_0 A}. \tag{29–17}$$

In a time interval dt, the charge Q increases by $dQ = I\,dt$; the corresponding change in E is

$$dE = \frac{dQ}{\epsilon_0 A} = \frac{I\,dt}{\epsilon_0 A},$$

and the *rate* of change of E is

$$\frac{dE}{dt} = \frac{I}{\epsilon_0 A}. \tag{29–18}$$

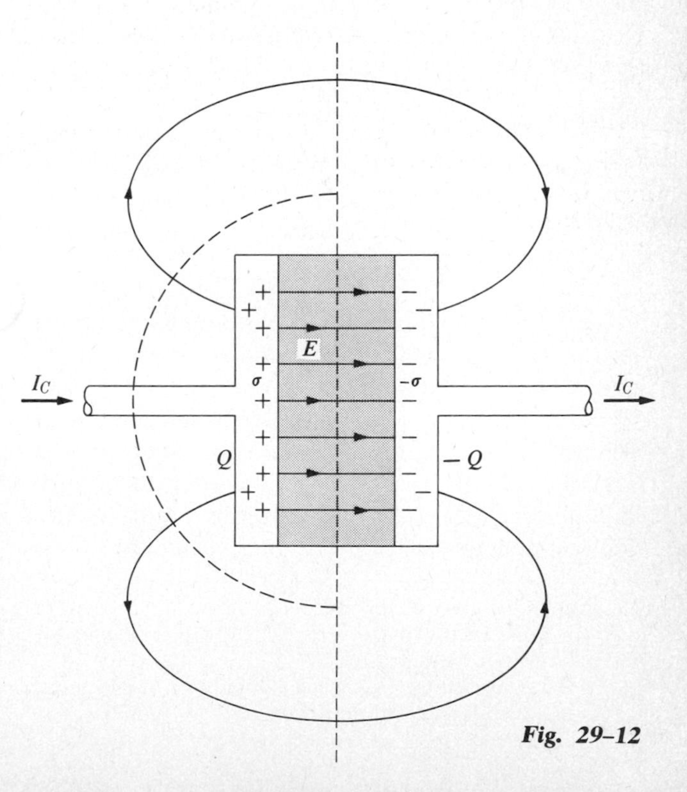

Fig. 29–12

We now define an *effective current density* J_D between the plates as

$$J_D = \epsilon_0 \frac{dE}{dt}. \tag{29–19}$$

Then the total effective current between the plates, which we may call I_D, is

$$I_D = J_D A = \epsilon_0 \left(\frac{dE}{dt}\right) A = I.$$

Thus, if we include the "effective current" as well as the conduction current, the current *into* the region bounded by the broken line in Fig. 29–12 equals the current *out of* this region. The subscript D is chosen because Maxwell originally called this effective current the *displacement current*; this term is still used, although the reasons for its use are dubious.

If the space between capacitor plates contains a dielectric, Eq. (29–17) must be replaced by the more general relation derived in Section (27–5):

$$E = \frac{\sigma}{\epsilon} = \frac{Q}{\epsilon A} = \frac{Q}{K\epsilon_0 A}. \tag{29–20}$$

The corresponding modification of the definition of displacement current density is

$$J_D = \epsilon\left(\frac{dE}{dt}\right) = K\epsilon_0\left(\frac{dE}{dt}\right). \tag{29–21}$$

Maxwell's generalized definition of current may appear simply as an artifice introduced in order to be able to say that the currents into and out of any portion of a circuit are equal, even when the circuit includes a capacitor in which the conduction current is zero.* However, when we consider in a later chapter, the *magnetic* field set up by a current, we shall see that an element of *displacement* current sets up a magnetic field in precisely the same way as does an element of *conduction* current.

* " 'When I use a word,' Humpty Dumpty said, in a rather scornful tone, 'it means just what I choose it to mean—neither more nor less.'

" 'The question is,' said Alice, 'whether you *can* make words mean so many different things.'

" 'The question is,' said Humpty Dumpty, 'which is to be master—that's all.' "

PROBLEMS

29–1 Prove that, when two resistors are connected in parallel, the equivalent resistance of the combination is always smaller than that of either resistor.

29–2

a) A resistance R_2 is connected in parallel with a resistance R_1. What resistance R_3 must be connected in series with the combination of R_1 and R_2 so that the equivalent resistance is equal to the resistance R_1? Draw a diagram.

b) A resistance R_2 is connected in series with a resistance R_1. What resistance R_3 must be connected in parallel with the combination of R_1 and R_2 so that the equivalent resistance is equal to R_1? Draw a diagram.

29–3

a) Calculate the equivalent resistance of the circuit of Fig. 29–13 between x and y.

b) What is the potential difference between x and a if the current in the 8-Ω resistor is 0.5 A?

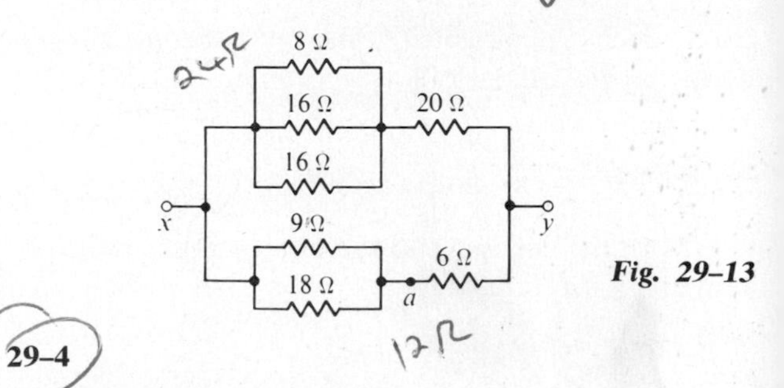

Fig. 29–13

29–4

a) The long resistor between a and b in Fig. 29–14 has a resistance of 300 Ω and is tapped at the one-third points. What is the equivalent resistance between x and y?

b) The potential difference between x and y is 320 V. What is the potential difference between b and c?

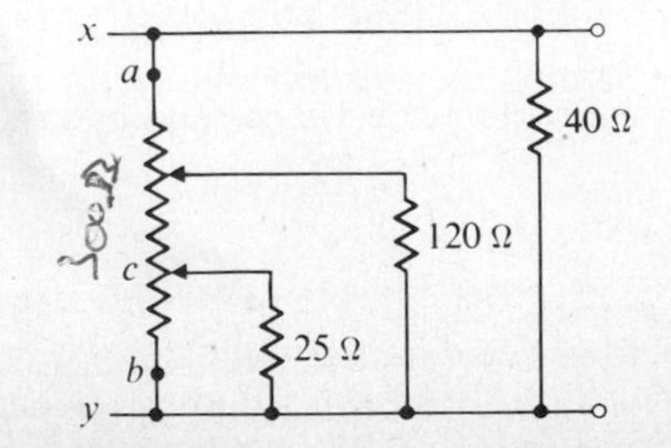

Fig. 29–14

29–5 Each of three resistors in Fig. 29–15 has a resistance of 2 Ω and can dissipate maximum of 18 W without becoming excessively heated. What is the maximum power the circuit can dissipate?

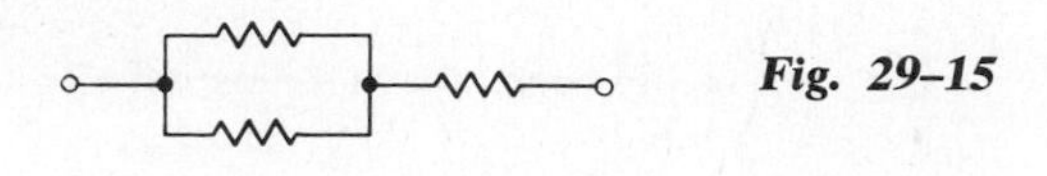

Fig. 29–15

29–6 Two lamps, marked "60 W, 120 V" and "40 W, 120 V," are connected in series across a 120-V line. What power is consumed in each lamp? Assume that the resistance of the filaments does not vary with current.

29–7 Three equal resistors are connected in series. When a certain potential difference is applied across the combination, the total power consumed is 10 W. What power would be consumed if the three resistors were connected in parallel across the same potential difference?

29–8

a) The power rating of a 10,000-Ω resistor is 2 W. (The power rating is the maximum power the resistor can safely dissipate without too great a rise in temperature.) What is the maximum allowable potential difference across the terminals of the resistor?

b) A 20,000-Ω resistor is to be connected across a potential difference of 300 V. What power rating is required?

c) It is desired to connect a resistance of 1000 Ω across a potential difference of 200 V. A number of 10-W, 1000-Ω resistors are available. How should they be connected?

29–9 A 60-Ω resistor and a 90-Ω resistor are connected in parallel and the combination is connected across a 120-V dc line.

a) What is the resistance of the parallel combination?

b) What is the total current through the parallel combination?

c) What is the current through each resistor?

29–10 A 1000-Ω 2-W resistor is needed, but only several 1000-Ω 1-W resistors are available.

a) How can the required resistance and power rating be obtained by a combination of the available units?

b) What power is then dissipated in each resistor?

29–11 A 25-W 120-V lightbulb and a 100-W 120-V lightbulb were connected in series across a 240-V line. Assume the resistance of each bulb does not vary with current.

a) Find the current through the bulbs.

b) Find the power dissipated in each bulb.

c) One bulb burned out very quickly. Which one? Why?

29–12 An electric heating element was designed for a power rating of 550 W when connected to a 110-V line. If the line voltage is 120 V, what power is consumed?

29–13

a) Find the resistance of the network in Fig. 29–16, between the terminals a and b.

b) What potential difference between a and b will result in a current of 1 A in the 4-Ω resistor?

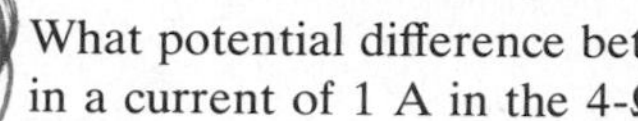

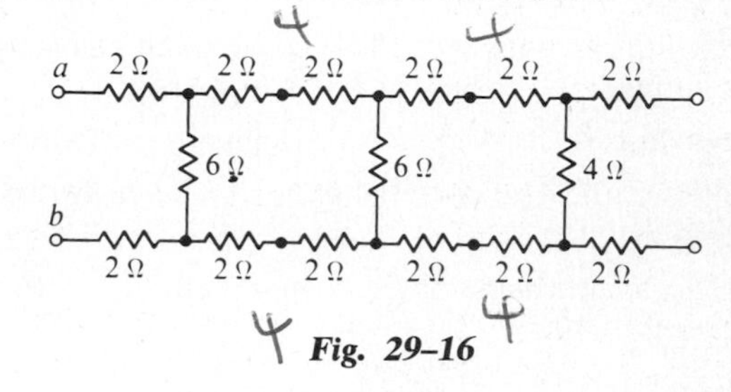

Fig. 29–16

29–14 Prove that the resistance of the infinite network shown in Fig. 29–17 is equal to $(1 + \sqrt{3})r$.

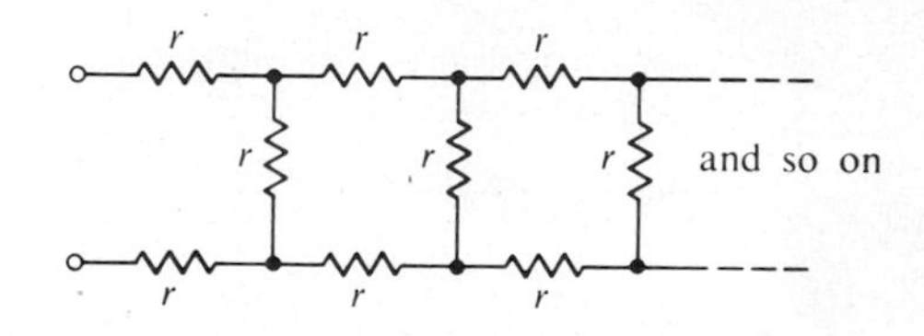

Fig. 29–17

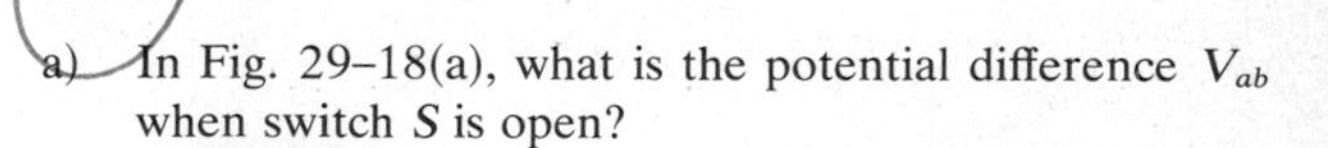

29–15

a) In Fig. 29–18(a), what is the potential difference V_{ab} when switch S is open?

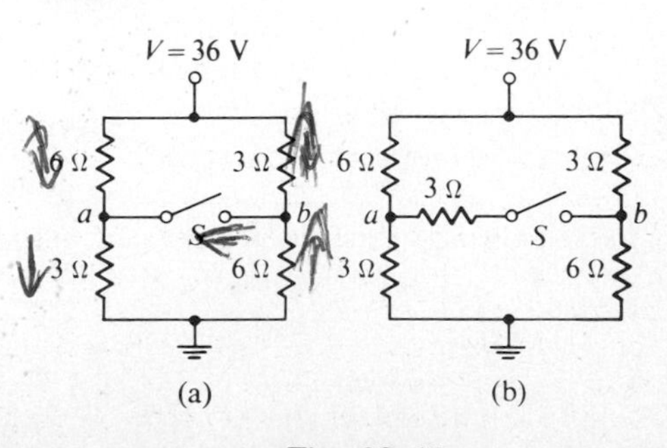

Fig. 29–18

b) What is the current through switch S, when it is closed?

c) In Fig. 29–18(b), what is the potential difference V_{ab} when switch S is open?

d) What is the current through switch S when it is closed?

What is the equivalent resistance of the circuit in Fig. 29–18(b),

e) when switch S is open?

f) when switch S is closed?

29–16

a) What is the potential difference between points a and b in Fig. 29–19 when switch S is open?

b) Which point, a or b, is at the higher potential?

c) What is the final potential of point b when switch S is closed?

d) How much charge flows through switch S when it is closed?

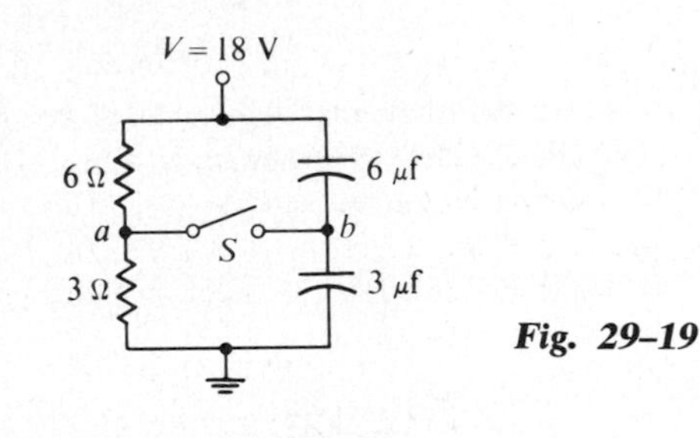

Fig. 29–19

29–17

a) What is the potential difference between points a and b in Fig. 29–20 when switch S is open?

b) Which point, a or b, is at the higher potential?

c) What is the final potential of point b when switch S is closed?

d) How much does the charge on each capacitor change when S is closed?

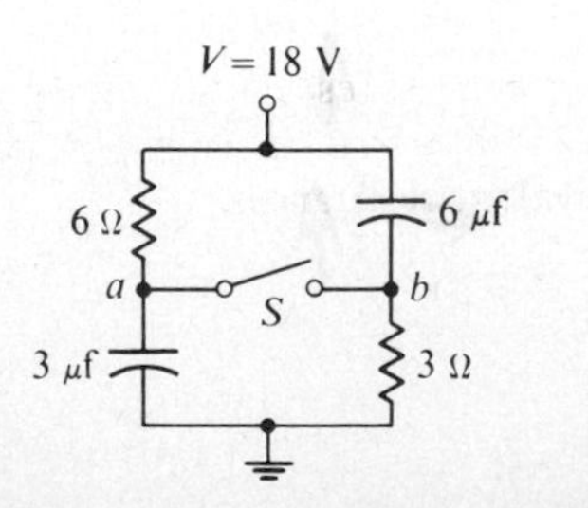

Fig. 29–20

29–18 Calculate the three currents indicated in the circuit diagram of Fig. 29–21.

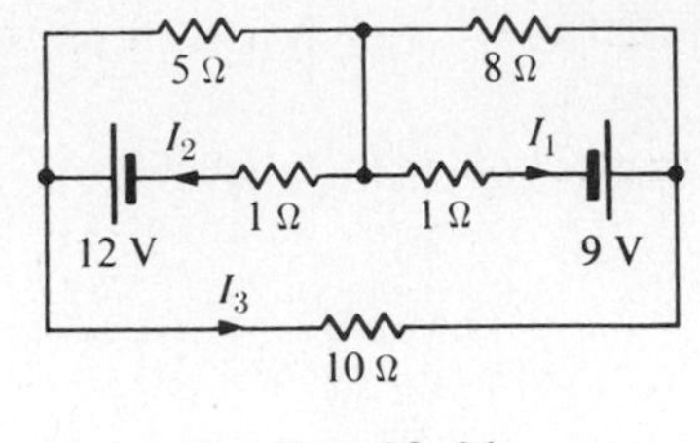

Fig. 29–21

29–19 Find the emf's $\mathcal{E}_1$ and $\mathcal{E}_2$ in the circuit of Fig. 29–22, and the potential difference between points a and b.

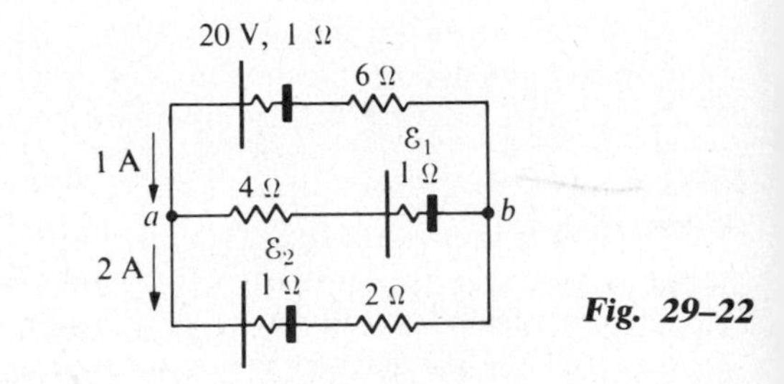

Fig. 29–22

29–20

a) Find the potential difference between points a and b in Fig. 29–23.

b) If a and b are connected, find the current in the 12-V cell.

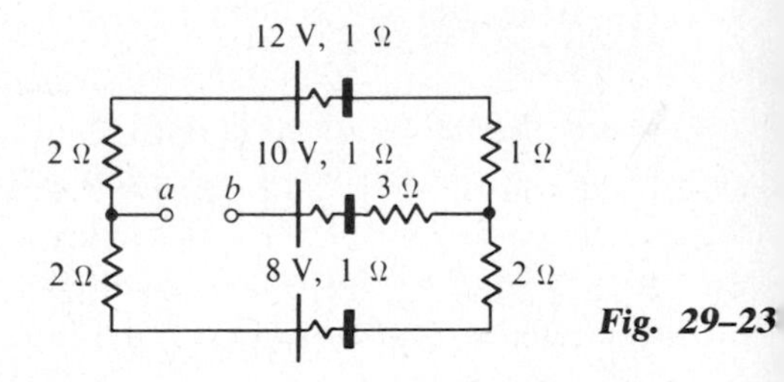

Fig. 29–23

29–21 A 600-Ω resistor and a 400-Ω resistor are connected in series across a 90-V line. A voltmeter across the 600-Ω resistor reads 45 V.

a) Find the voltmeter resistance.

b) Find the reading of the same voltmeter if connected across the 400-Ω resistor.

29–22 Point a in Fig. 29–24 is maintained at a constant potential of 300 V above ground.

a) What is the reading of a voltmeter of the proper range, and of resistance $3 \times 10^4\ \Omega$, when connected between point b and ground?

b) What would be the reading of a voltmeter of resistance $3 \times 10^6\ \Omega$?

c) Of a voltmeter of infinite resistance?

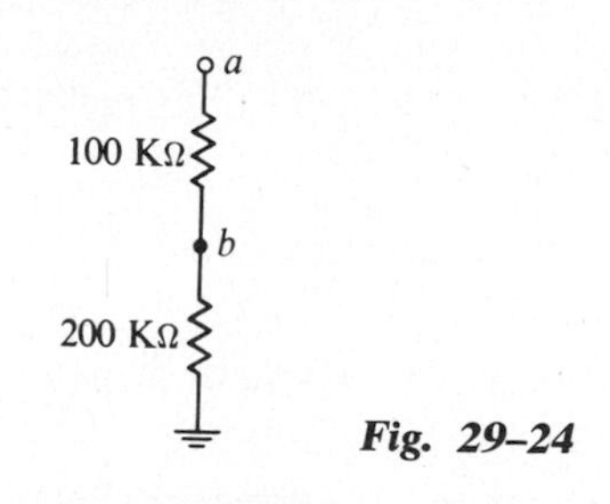

Fig. 29–24

29–23 Two 150-V voltmeters, one of resistance 15,000 Ω and the other of resistance 150,000 Ω, are connected in series across a 120-V dc line. Find the reading of each voltmeter.

29–24 A 150-V voltmeter has a resistance of 20,000 Ω. When connected in series with a large resistance R across a 110-V line, the meter reads 5 V. Find the resistance R. (This problem illustrates one method of measuring large resistances.)

29–25 A 100-V battery has an internal resistance of 5 Ω.

a) What is the reading of a voltmeter having a resistance of 500 Ω when placed across the terminals of the battery?

b) What maximum value may the ratio r/R_V have if the error in the reading of the emf of a battery is not to exceed 5 percent?

29–26 The resistance of a galvanometer coil is 50 Ω and the current required for full-scale deflection is 500 μA.

a) Show in a diagram how to convert the galvanometer to an ammeter reading 5 A full-scale, and compute the shunt resistance.

b) Show how to convert the galvanometer to a voltmeter reading 150 V full-scale, and compute the series resistance.

29–27 The resistance of the coil of a pivoted-coil galvanometer is 10 Ω, and a current of 0.02 A causes it to deflect full scale. It is desired to convert this galvanometer to an ammeter reading 10 A full-scale. The only shunt available has a resistance of 0.03 Ω. What resistance R must be connected in series with the coil? (See Fig. 29–25.)

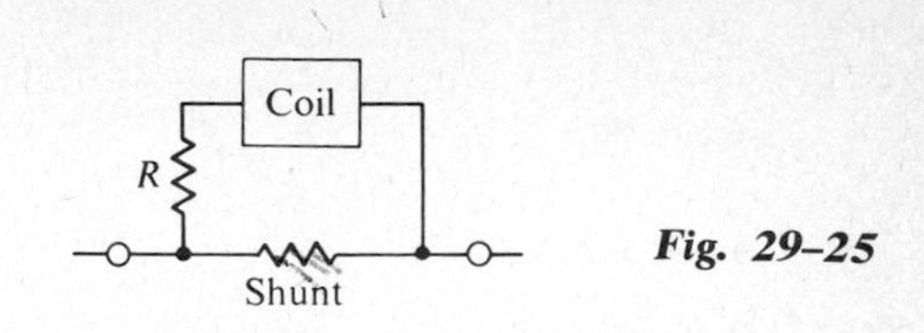

Fig. 29–25

29–28 The resistance of the moving coil of the galvanometer G in Fig. 29–26 is 25 Ω; the meter deflects full scale with a current of 0.010 A. Find the magnitudes of the resistances R_1, R_2, and R_3, to convert the galvanometer to a multirange ammeter deflecting full scale with currents of 10 A, 1 A, and 0.1 A.

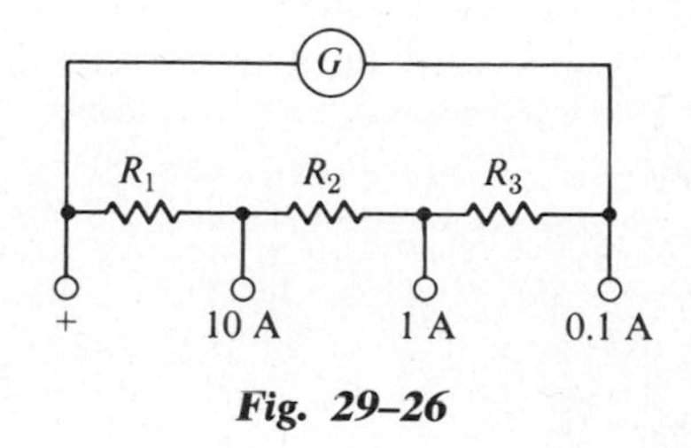

Fig. 29–26

29–29 Figure 29–27 shows the internal wiring of a "three-scale" voltmeter whose binding posts are marked +, 3 V, 15 V, 150 V. The resistance of the moving coil, R_G, is 15 Ω, and a current of 1 mA in the coil causes it to deflect full scale. Find the resistances R_1, R_2, R_3, and the overall resistance of the meter on each of its ranges.

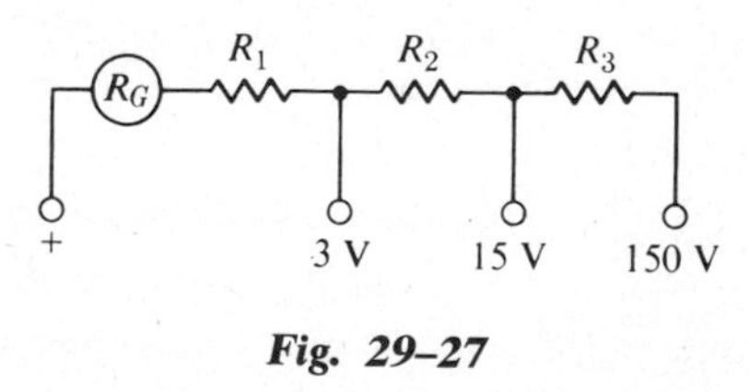

Fig. 29–27

29–30 A certain dc voltmeter is said to have a resistance of "one thousand ohms per volt of full-scale deflection." What current, in milliamperes, is required for full-scale deflection?

29–31 Let V and A represent the readings of the voltmeter and ammeter, respectively, shown in Fig. 29–6, and R_V and R_A their equivalent resistances.

a) When the circuit is connected as in Fig. 29–6(a), show that

$$R = \frac{V}{A} - R_A.$$

b) When the connections are as in Fig. 29–6(b), show that

$$R = \frac{V}{A - (V/R_V)}.$$

c) Show that the power delivered to the resistor in part (a) is $AV - A^2R_A$, and in part (b) is $AV - (V^2/R_V)$.

29–32 A certain galvanometer has a resistance of 200 Ω and deflects full scale with a current of 1 mA in its coil. It is desired to replace this with a second galvanometer whose resistance is 50 Ω and which deflects full scale with a current of 50 μA in its coil. Devise a circuit incorporating the second galvanometer such that (a) the equivalent resistance of the circuit equals the resistance of the first galvanometer, and (b) the second galvanometer will deflect full scale when the current into and out of the circuit equals the full-scale current of the first galvanometer.

29–33 Suppose the galvanometer of the ohmmeter in Fig. 29–8 has a resistance of 50 Ω and deflects full scale with a current of 1 mA in its coil. The emf $\mathscr{E}$ = 1.5 V.

a) What should be the value of the series resistance R_s?

b) What values of R correspond to galvanometer deflections of $\frac{1}{4}$, $\frac{1}{2}$, and $\frac{3}{4}$ full-scale?

c) Does the ohmmeter have a linear scale?

29–34 In the ohmmeter in Fig. 29–28, M is a 1-mA meter having a resistance of 100 Ω. The battery B has an emf of 3 V and negligible internal resistance. R is so chosen that, when the terminals a and b are shorted (R_x = 0), the meter reads full scale. When a and b are open (R_x = ∞), the meter reads zero.

a) What should be the value of the resistor R?

b) What current would indicate a resistance R_x of 600 Ω?

c) What resistances correspond to meter deflections of $\frac{1}{4}$, $\frac{1}{2}$, and $\frac{3}{4}$ full-scale?

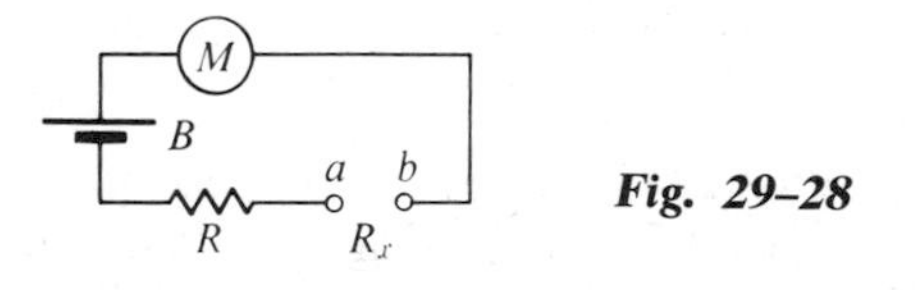

Fig. 29–28

29–35 In Fig. 29–29, a resistor of resistance 75 Ω is connected between points a and b. The resistance of the galvanometer G is 90 Ω. What should be the resistance between b and the sliding contact c, if the galvanometer current I_G is to be $\frac{1}{3}$ of the current I?

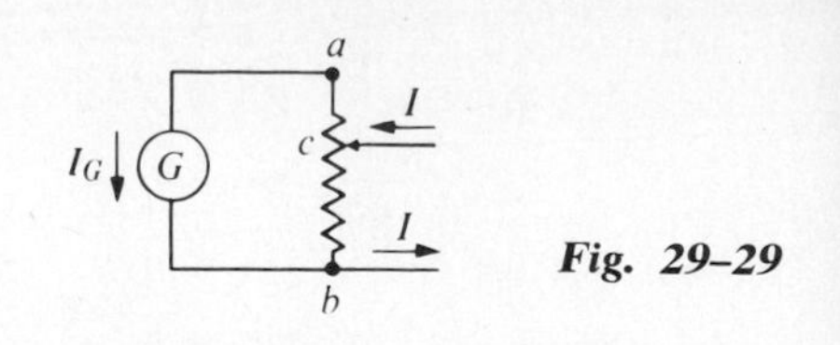

Fig. 29–29

29–36 Figure 29–30 shows a potentiometer set up to measure the emf of cell x. B is a battery whose emf is approximately 3 V and whose internal resistance is unknown. St is a standard cell of 1.0183-V emf. The switch is set at point 2, placing the standard cell in the galvanometer circuit. When the tap b is 0.36 of the distance from a to c, the galvanometer G reads zero.

a) What is the difference of potential across the entire length of resistor ac?

b) The switch is then set at point 1, and a new zero reading of the galvanometer is obtained when b is 0.47 of the distance from a to c. What is the emf of cell x?

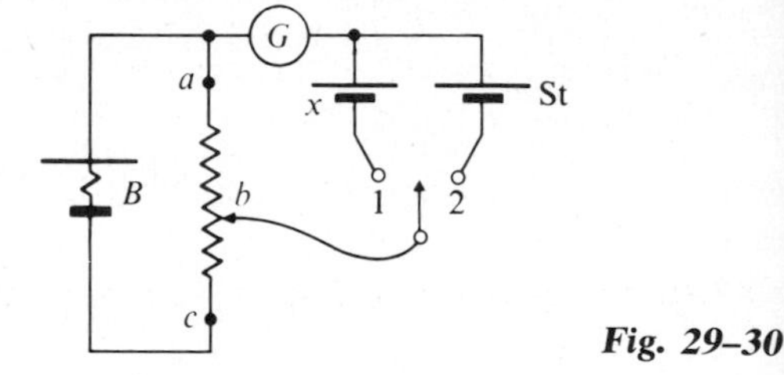

Fig. 29–30

29–37 A 10-μF capacitor is connected through a 1-MΩ resistor to a constant potential difference of 100 V.

a) Compute the charge on the capacitor at the following times after the connections are made: 0, 5 s, 10 s, 20 s, 100 s.

b) Compute the charging current at the same instants.

c) How long a time would be required for the capacitor to acquire its final charge if the charging current remained constant at its initial value?

d) Find the time required for the charge to increase from zero to 5×10^{-4}C.

e) Construct graphs of the results of parts (a) and (b) for a time interval of 20 s.

29–38 Two capacitors are charged in series by a 12-V battery (Fig. 29–31).

a) What is the time constant of the charging circuit?

b) After being closed for the length of time determined in (a), the switch S is opened. What is the voltage across the 6-μF capacitor?

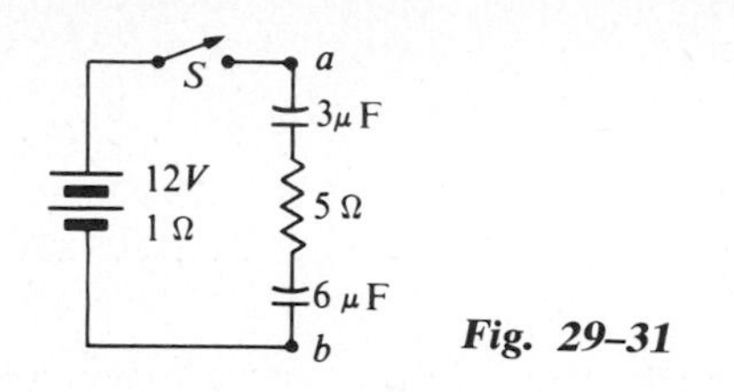

Fig. 29–31

29–39 A capacitor of capacitance C is charged by connecting it through a resistance R to the terminals of a battery of emf $\mathscr{E}$ and of negligible internal resistance.

a) How much energy is supplied by the battery in the charging process?

b) What fraction of this energy appears as heat in the resistor?

29–40 In Fig. 29–10, let $V = 100$ V, $R = 10$ MΩ, $C = 2$ μF. The capacitor is initially uncharged. The switch is thrown to the Up position for 20 s and then quickly reversed to the Down position.

a) Construct graphs of i, q, v_{ac}, and v_{cb} for a time interval of 60 s after the switch is first thrown.

b) How much energy is eventually dissipated in the resistor?

29–41 The current in a discharging capacitor is given by Eq. (29–15).

a) Using Eq. (29–15), derive an expression for the instantaneous power $P = i^2R$ dissipated in the resistor.

b) Integrate the expression for P to find the total energy dissipated in the resistor, and show that this is equal to the total energy initially stored in the capacitor.

29–42 The current in a charging capacitor is given by Eq. (29–11).

a) The instantaneous power supplied by the battery is Vi. Integrate this to find the total energy supplied by the battery.

b) The instantaneous power dissipated in the resistor is i^2R. Integrate this to find the total energy dissipated in the resistor.

c) Find the final energy stored in the capacitor, and show that this equals the total energy supplied by the battery, less the energy dissipated in the resistor, as obtained in parts (a) and (b).

d) What fraction of the energy supplied by the battery is stored in the capacitor? How does this fraction depend on R?

29–43

a) The differential equation for the instantaneous charge q of a capacitor a moment after its terminals have been disconnected from a source and connected to a resistance R is given by Eq. (29–13), namely,

$$\frac{dq}{dt} = -\frac{q}{RC}.$$

Show that

$$q = Q_0 e^{-t/RC}.$$

b) The current in the circuit of part (a) is given by Eq. (29–14), namely,

$$\frac{di}{dt} = -\frac{i}{RC}.$$

Show that

$$i = I_0 e^{-t/RC}.$$

29–44 Suppose that the parallel plates in Fig. 29–12 have an area of 2 m^2 and are separated by a sheet of dielectric 1 mm thick, of dielectric constant 3. (Neglect edge effects.) At a certain instant, the potential difference between the plates is 100 V and the current I_C equals 2 mA.

a) What is the charge Q on each plate?

b) What is the rate of change of charge on the plates?

c) What is the displacement current in the dielectric?

29–45 In a certain copper conductor ($\rho = 2 \times 10^{-8}$ Ω m) carrying a current, the electric field varies sinusoidally with time according to $E = E_0 \sin \omega t$, where $E_0 = 0.1$ V m^{-1} and $\omega = (2\pi)(60$ Hz$)$.

a) Find the magnitude of the maximum conduction current density in the wire.

b) Assuming $\epsilon = \epsilon_0$, find the maximum displacement current density in the wire, and compare with the result of part (a).

29–46

a) Repeat the calculations of Problem 29–45 for a rod of pure silicon having $\rho = 2000$ Ω m.

b) At what frequency would the maximum conduction and displacement densities become equal, if $\epsilon = \epsilon_0$ (which is not actually the case)?

c) At the frequency determined in part (b), what is the relative *phase* of the conduction and displacement currents?

29–47 A capacitor is made from two cylindrical rods of radius R, placed end to end with a gap of width d between them, where d is much smaller than R. There is a current i in each rod, and the charge q on each end surface changes at a rate given by $dq/dt = i$.

a) Find the conduction current density in each rod.

b) Find the electric field between the surfaces, in terms of q.

c) Find the displacement current density in the gap, and show that it is equal to the conduction current in the rods.

Chapter 30

The Magnetic Field

30–1 MAGNETISM

The first magnetic phenomena to be observed were undoubtedly those associated with so-called "natural" magnets, rough fragments of an ore of iron found near the ancient city of Magnesia (whence the term "magnet"). These natural magnets have the property of attracting to themselves unmagnetized iron, the effect being most pronounced at certain regions of the magnet known as its *poles.* It was known to the Chinese as early as 121 A.D. that an iron rod, after being brought near a natural magnet, would acquire and retain this property of the natural magnet, and that such a rod, when freely suspended about a vertical axis, would set itself approximately in the north–south direction. The use of magnets as aids to navigation can be traced back at least to the eleventh century.

The study of magnetic phenomena was confined for many years to magnets made in this way. Not until 1819 was there shown to be any connection between electrical and magnetic phenomena. In that year, the Danish scientist Hans Christian Oersted (1777–1851) observed that a pivoted magnet (a compass needle) was deflected when in the neighborhood of a wire carrying a current. Twelve years later, after attempts extending over a period of several years, Michael Faraday (1791–1867), an English physicist, found that a momentary current existed in a circuit while the current in a nearby circuit was being *started or stopped.* Shortly afterward followed the discovery that the *motion* of a magnet toward or away from the circuit would produce the same effect. Joseph Henry (1797–1878), an American scientist who later became the first director of the Smithsonian Institution, had anticipated Faraday's discoveries by about twelve months but, since Faraday was the first to publish his results, he is usually assigned the credit for them. The work of Oersted thus demonstrated that magnetic effects could be produced by moving electric charges, and that of Faraday and Henry showed that currents could be produced by moving magnets.

It is now known that all so-called magnetic phenomena result from forces between electric charges in motion. That is, charges in motion relative to an observer set up a *magnetic* field as well as an *electrostatic* field, and this magnetic field exerts a force on a second charge in motion relative to the observer. Since the electrons in atoms are in motion about the atomic nuclei, and since each electron has additional motion which can be visualized as rotation about an axis passing through it, all atoms can be expected to exhibit magnetic effects and, in fact, such is found to be the case. The possibility that the magnetic properties of matter were the result of tiny atomic currents was first suggested by Ampère in 1820. Not until recent years has the verification of these ideas been possible.

The medium in which the charges are moving may have a pronounced effect on the observed magnetic forces between them. In the present chapter we shall assume that the charges or conductors are in otherwise empty space. For all practical purposes, the results will apply equally well to other charges and conductors in air.

30–2 THE MAGNETIC FIELD

Instead of dealing directly with the forces exerted on one moving charge by another, it is found more convenient to adopt the point of view that a moving charge sets up in the space around it a *magnetic field*, and that it is this field which exerts a force on another charge moving through it. The magnetic field around a moving charge exists in addition to the electrostatic field which surrounds the charge whether it is in motion or not. A second charged particle in these combined fields experiences a force due to the electric field whether it is in motion or at rest. The magnetic field exerts a force on the particle only if it is in motion.

A magnetic field is said to exist at a point if a force (over and above any electrostatic force) *is exerted on a moving charge at the point.*

The magnetic field, like the electric field, is a *vector field*, and its magnitude and direction at any point are specified by a vector $\boldsymbol{B}$. This vector is sometimes called the *magnetic induction*, the *magnetic induction field*, or the *magnetic flux density*. However, the term *magnetic field* is probably the most commonly used term and will be used in this book.

There are two aspects to the problem of computing the magnetic force between moving charges. The first is that of finding the magnitude and direction of the $\boldsymbol{B}$-vector at a point, given the data on the moving charges that set up the field. The second is to find the magnitude and direction of the *force* on a charge moving in a given field. We shall take up the latter aspect of the problem first. That is, we shall accept, for the present, the fact that moving charges and currents *do* set up magnetic fields, and study the laws that determine the force on a charge moving through the field.

An unknown *electric* field can be "explored" by measuring the magnitude and direction of the force on a test charge at rest relative to the observer. To "explore" an unknown *magnetic* field, we must measure the magnitude and direction of the force on a *moving* test charge. The cathode-ray tube is a convenient experimental device for studying, at least in a qualitative way, the behavior of moving charges in a magnetic field. At one end of this tube is an electron gun which shoots out a narrow electron beam at a speed that can be controlled and calculated. At the other end is a fluorescent screen which emits light at the point where the electron beam strikes it. Let us suppose that our cathode-ray tube is small, that it is not surrounded by iron, and that it can be carried around the room easily. If the spot of light is always in the same place on the screen as we move the tube, we may conclude that there is no detectable magnetic field. If, on the other hand, as we move the tube around the room, the spot of light changes its position by virtue of *deflection* of the electron beam, we conclude that we are in a magnetic field. Since the cathode-ray tube is imagined to be small, rotating the tube about any axis through its center, without shifting the position of its center, will provide information concerning the magnetic field in a small region at the center or, roughly, the magnetic field at a point.

At a given point in a magnetic field, the electron beam will, in general, be deflected. By rotating the cathode-ray tube, however, there can always be found one direction for which no deflection takes place. *The direction of motion of a charge on which a magnetic field exerts* ***no*** *force is defined as the direction of the* $\boldsymbol{B}$*-vector.* (The sense of the vector will be defined later.) Thus, in Fig. 30–1, the electron beam is undeflected when its direction is parallel to the z-axis and hence the $\boldsymbol{B}$-vector points either up or down.

Having determined the direction (but not the sense) of the magnetic field, let us now place the cathode-ray tube so that the electron beam moves in a plane perpendicular to this direction. Experiment shows that a deflection always takes place in such a manner as to indicate a force acting in this plane, but at right angles to the velocity of the electron beam. Thus, in Fig. 30–1, when the electron beam lies in the xy-plane and its direction is initially along the positive x-axis, the beam is deflected in the direction of the positive y-axis. That is, *when the velocity of the*

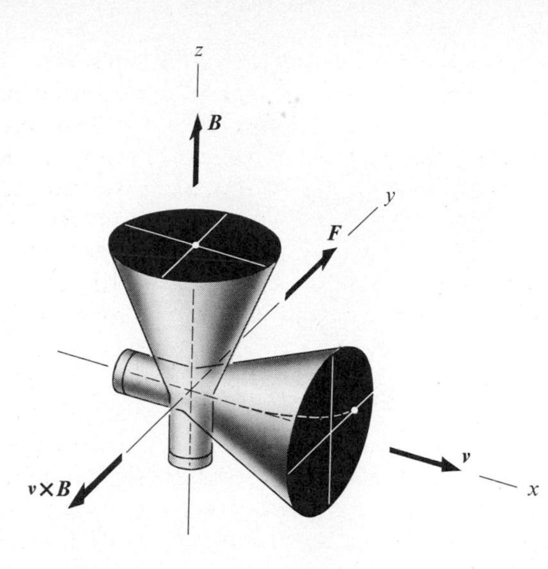

Fig. 30–1 *The electron beam of the cathode-ray tube is undeflected when the beam is parallel to the z-axis. The* ***B****-vector then points either up or down. When the tube axis is parallel to the x-axis, the beam is deflected in the positive y-direction. Then the* ***B****-vector points upward, and the force* ***F*** *on the electrons points along the positive y-axis.*

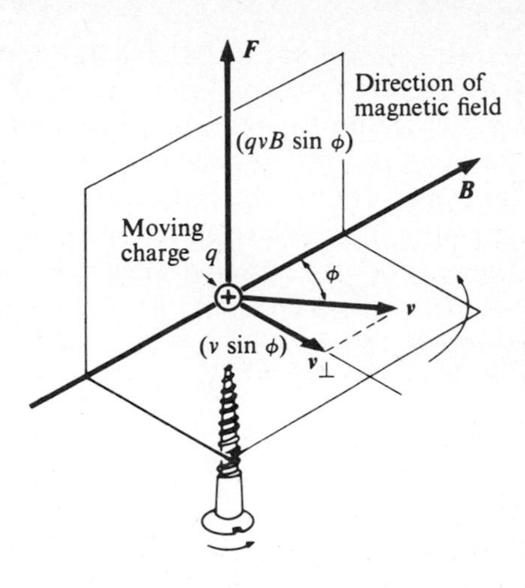

Fig. 30–2 *The magnetic force* ***F*** *acting on a charge q moving with velocity* ***v*** *is perpendicular to both the magnetic field* ***B*** *and to* ***v****.*

moving charge is perpendicular to the magnetic field, the force is perpendicular to both the magnetic field and the velocity. The magnitude of this force is found to be directly proportional to the velocity. If the velocity is not perpendicular to the direction of the magnetic field but makes an angle ϕ with the field, as in Fig. 30–2, then the velocity vector $\boldsymbol{v}$ may be resolved into two components: $v_{\parallel} = v \cos \phi$ in the direction of the field, and $v_{\perp} = v \sin \phi$ perpendicular to the field. In this general case, *the force acting on the moving charge is perpendicular both to the magnetic field and to* $\boldsymbol{v}$ *and has a magnitude proportional to* $v \sin \phi$.

It is possible to conceive of a positive ion tube, with a positive ion source at one end and a detector at the other, so that deflection of the positive ion beam may be observed when the tube is placed in various positions in a magnetic field. By using different positive ions, it is possible to measure the force acting on positive charges of different magnitudes. Thus, with hydrogen ions (protons), the force on singly charged positive ions may be measured, and with other gases under proper conditions, forces on doubly and trebly charged positive ions may be measured. Experiment shows that, in every case, *the force* ***F*** *acting on a charge moving in a magnetic field is proportional to the magnitude of the charge,* and that *the force on a negative charge,* moving in a given direction, *is opposite to that on a positive charge* moving in the same direction.

The magnitude of the ***B***-vector at any point can now be defined by the equation

$$B = \frac{F}{qv \sin \phi}, \quad \text{or} \quad F = qvB \sin \phi, \qquad (30\text{–}1)$$

where q is the magnitude of a moving charge at the point, v is the magnitude of its velocity, and ϕ is the angle between v and the direction of the field.

The mks unit of B is therefore one *newton per (coulomb meter per second)*. But one coulomb per second equals one ampere, so the unit can be expressed as one *newton per ampere · meter* ($1\ \mathrm{NA^{-1}m^{-1}}$). This unit is called one tesla (1 T). Another common equivalent unit will be defined shortly.

The direction of the force $\boldsymbol{F}$ on a positively-charged particle moving with velocity $\boldsymbol{v}$ in a magnetic field $\boldsymbol{B}$ is always perpendicular to the plane determined by $\boldsymbol{v}$ and $\boldsymbol{B}$. Furthermore, the direction of $\boldsymbol{F}$ is the direction of advance of a right hand thread screw when rotated in the sense from $\boldsymbol{v}$ toward $\boldsymbol{B}$, through the smaller of the two possible angles, as shown in Fig. 30–2. This rule completes the definition of the *sense* of the $\boldsymbol{B}$-vector. The direction of force on a particle with *negative* charge is opposite to that shown.

In Chapter 9 (Section 9–11), the *vector product* (or *cross product*) of two vectors $\mathbf{A}$ and $\boldsymbol{B}$ was defined as a vector perpendicular to the plane of $\mathbf{A}$ and $\boldsymbol{B}$, having magnitude $AB \sin \phi$, where ϕ is the angle between $\mathbf{A}$ and $\boldsymbol{B}$, and a sense defined by the right-hand-screw rule. Comparing this definition with the above discussion, we see that the magnetic force on a moving charged particle is given by the vector equation

$$\boxed{\boldsymbol{F} = q\boldsymbol{v} \times \boldsymbol{B}.} \qquad (30\text{–}2)$$

When q is positive, $\boldsymbol{F}$ has the same direction as the vector product $\boldsymbol{v} \times \boldsymbol{B}$; when q is negative, $\boldsymbol{F}$ is opposite to $\boldsymbol{v} \times \boldsymbol{B}$.

Equation (30–1) can also be interpreted as follows. Recalling that ϕ is the angle between the direction of vectors $\boldsymbol{v}$ and $\boldsymbol{B}$, we may interpret ($B \sin \phi$) as the component of $\boldsymbol{B}$ perpendicular to v, that is, $B_\perp$. With this notation, the force expression becomes

$$F = qvB_\perp.$$

Although equivalent to Eq. (30–1), this is sometimes more convenient, especially in problems involving *currents* rather than individual particles. Forces on currents in conductors are discussed in Chapter 31.

30–3 LINES OF INDUCTION. MAGNETIC FLUX

A magnetic field can be represented by lines, such that the direction of a line through a given point is the same as that of the magnetic field vector $\boldsymbol{B}$ at that point. We shall call these *lines of induction*; they are sometimes called magnetic lines of force, but this term is unfortunate because, unlike electric field lines, they *do not* point in the direction of the force on a charge.

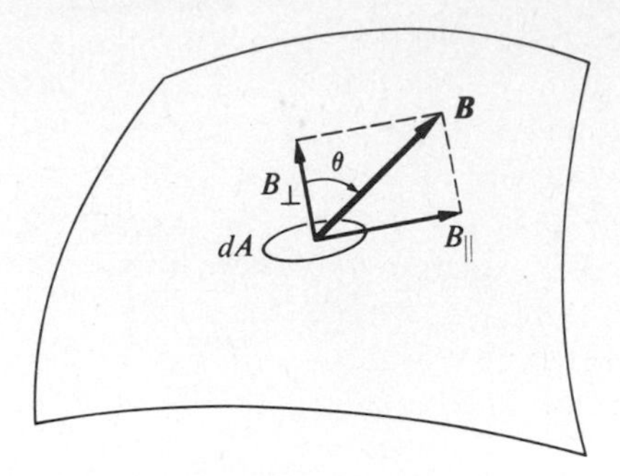

Fig. 30–3 *The magnetic flux through an area element dA is defined to be* $\Phi = B_\perp dA = \mathbf{B}\cdot d\mathbf{A}$.

In a uniform magnetic field, where the $\boldsymbol{B}$ vector has the same magnitude and direction at every point in a region, the lines of induction are straight and parallel. If the poles of an electromagnet are large, flat, and close together, there is a region between them where the magnetic field is approximately uniform.

The *magnetic flux* across a surface is defined in analogy to the electric field flux used with Gauss's law. Any surface can be divided into elements of area, as in Fig. 30–3. For an element dA, we obtain the components of $\mathbf{B}$ normal and tangent to the surface, as shown. From the figure, $B_\perp = B \cos \theta$. The magnetic flux $d\Phi$ through this area is defined as

$$d\Phi = B_\perp \, dA = B \cos \theta \, dA = \mathbf{B}\cdot d\mathbf{A}. \qquad (30\text{–}3)$$

The total magnetic flux through the surface is the sum of the contributions from the individual area elements, given by

$$\Phi = \int B \, dA = \int \mathbf{B}\cdot d\mathbf{A}.$$

In the special case where B is uniform over a plane surface with total area A,

$$\Phi = BA \cos \theta. \qquad (30\text{–}4)$$

If $\boldsymbol{B}$ happens to be perpendicular to the surface, $\cos \theta = 1$, and this expression reduces to $\Phi = BA$. The chief usefulness of the concept of magnetic flux is in the study of electromagnetic induction, to be considered in Chapter 33.

The mks unit of magnetic field B is one tesla = one newton per ampere meter; hence the unit of magnetic flux is *one newton meter per ampere* (1 N m A^{-1}). In honor of Wilhelm Weber (1804–1890), 1 N m A^{-1} is called one *weber* (1 Wb).

If the element of area dA in Eq. (30–3) is at right angles to the lines of induction, $B_\perp = B$, and hence

$$B = \frac{d\Phi}{dA}.$$

That is, the magnetic field equals the *flux per unit area* across an area at right angles to the magnetic field. Since the unit of flux is 1 weber, the unit of field, 1 tesla, is equal to one *weber per square meter* (1 Wb m^{-2}). The magnetic field $\boldsymbol{B}$ is often referred to as the *flux density*.

The total flux across a surface can then be pictured as proportional to the number of lines of induction crossing the surface, and the field (the flux density) as the number of lines per unit area.

In one of a number of systems of electrical and magnetic units based on the cgs mechanical units, the unit of magnetic flux is called one *maxwell* and the corresponding unit of flux density, one maxwell per square centimeter, is called one *gauss* (1 G). (Instruments for measuring flux density are often referred to as *gaussmeters*.) The gauss and the tesla are related by the equation

$$1\,\text{T} = 1\,\text{Wb}\,\text{m}^{-2} = 10^4\text{G}.$$

The largest values of magnetic field that can be produced in the laboratory are of the order of 30 T or 300,000 G, while in the magnetic field of the earth the induction is only a few hundred-thousandths of a tesla, or a few tenths of a gauss.

30–4 MOTION OF CHARGED PARTICLES IN MAGNETIC FIELDS

Let a positively charged particle at point O in a uniform magnetic field $\boldsymbol{B}$ be given a velocity $\boldsymbol{v}$ in a direction at right angles to the field (Fig. 30–4). An upward force $\boldsymbol{F}$, equal in magnitude to qvB, is exerted

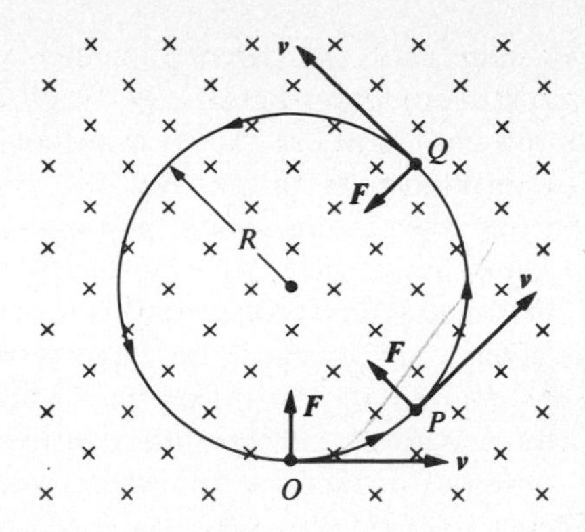

Fig. 30–4 *The orbit of a charged particle in a uniform magnetic field is a circle when the initial velocity is perpendicular to the field. The crosses represent a uniform magnetic field directed away from the reader.*

on the particle at this point. Since the force is at right angles to the velocity, it does not change the *magnitude* of this velocity but merely alters its direction. At points such as P and Q the directions of force and velocity will have changed as shown, the magnitude of the force remaining constant, since the magnitudes of q, $\boldsymbol{v}$, and $\boldsymbol{B}$ are constant. The particle therefore moves under the influence of a force whose *magnitude* is constant but whose *direction* is always at right angles to the velocity of the particle. The orbit of the particle is therefore a *circle* described with constant tangential speed v, the force F being the *centripetal force.* Since

$$\text{Centripetal acceleration} = \frac{v^2}{R},$$

we have, from Newton's second law,

$$qvB = m\left(\frac{v^2}{R}\right),$$

where m is the mass of the particle. The radius of the circular orbit is

$$\boxed{R = \frac{mv}{Bq}.} \qquad (30\text{–}5)$$

If the direction of the initial velocity is *not* perpendicular to the field, the velocity component parallel to the field remains constant and the particle moves in a helix. In this case, v in Eq. (30–5) is the component of velocity perpendicular to the field.

Note that the radius is proportional to the *momentum* of the particle, mv. Note also that the *work* of the magnetic force acting on a charged particle is always *zero*, because the force is always at *right angles* to the motion. The only effect of a magnetic force is to change the *direction* of motion, never to increase or decrease the *magnitude* of the velocity.

Figure 30–5 is a photograph of the circular tracks made in a cloud chamber by charged particles moving in a magnetic field perpendicular to the plane of the paper. The photograph shows three pairs of tracks originating at common points but curving in opposite directions. A study of the density of droplets along the paths shows that both particles ionize like electrons, but since the tracks curve in opposite directions, the charges must be of opposite sign. These tracks are in fact made by *electron-positron pairs*, created at the points from which the tracks originate by the annihilation of a high-energy gamma ray in the process known as *pair production.*

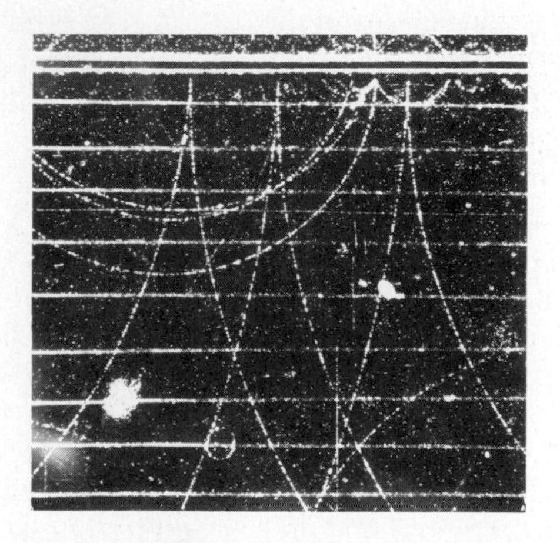

Fig. 30–5 *Cloud-chamber tracks of three electron-positron pairs in a magnetic field. The gamma-ray photons entering at the top materialize into pairs within a lead sheet. The coiled tracks are due to low-energy photoelectrons ejected from the lead. (Courtesy of Radiation Laboratory, University of California.)*

Fig. 30–6 *A 4-MeV electron slowing down in a liquid-hydrogen bubble chamber traversed by a magnetic field. The shape of the path shows how the radius of curvature decreases with velocity. (Courtesy of Radiation Laboratory, University of California.)*

The other tracks at the top of the photograph are portions of the circular paths of photoelectrons ejected from the lead sheet crossing the chamber. One complete circular track appears in the lower part of the photograph.

The spiral in Fig. 30–6 is the track made by a 4-MeV electron as it slows down in a liquid hydrogen bubble chamber in which there is a magnetic field. Many more ions are produced per unit length in a liquid than in a gas, so that although an electron may pass completely through a cloud chamber without appreciable energy loss, as in Fig. 30–5, the electron in Fig. 30-6 loses all its energy and comes to rest at the end of the spiral. The shape of the path shows in a striking way how the radius of curvature decreases as the velocity decreases.

30–5 THOMSON'S MEASUREMENT OF e/m

The charge-to-mass ratio of an electron, e/m, was first measured by Sir J. J. Thomson in 1897 at the Cavendish Laboratory in Cambridge, England. The

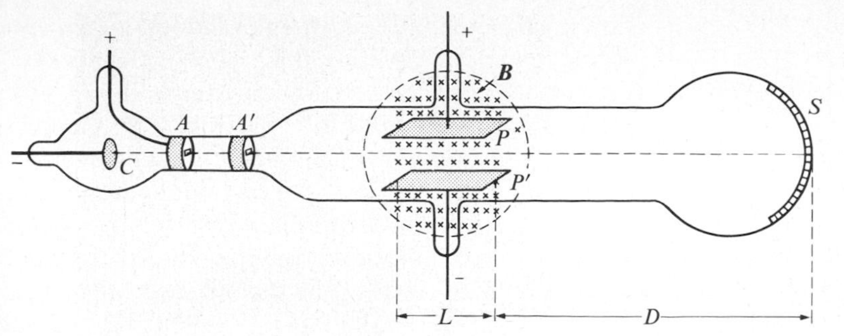

Fig. 30–7 *Thomson's apparatus for measuring the ratio e/m for cathode rays.*

discovery that this ratio is constant provided the best experimental evidence available at that time of the *existence* of electrons, particles with definite mass and charge. Thomson's term for these particles was "cathode corpuscles."

Thomson's apparatus (Fig. 30–7) consisted of a highly evacuated glass tube into which several metal electrodes were sealed. Electrode C is the cathode from which the electrons emerged. Electrode A is the anode, which was maintained at a high positive potential. Most of the electrons hit electrode A, but there was a small hole in A through which some of them passed. They were further restricted by an electrode A' in which there was another hole. Thus a narrow *beam* of electrons passed into the region between the two plates P and P'. After passing between the plates, the electrons struck the end of the tube, where they caused fluorescent material at S to glow. This part of the apparatus was not very different in basic concept from tubes presently used in cathode-ray oscilloscopes and for television picture tubes.

The deflection plates P and P' were separated a known distance, so that when they were at a known difference of potential, the electric field between them could be computed. We shall assume that the field was uniform for a distance L between the plates and zero outside them. When the upper plate P was made positive, the electric field deflected the negative electrons upward. After leaving the region between the plates, the electrons then coasted through the field-free region beyond the plates to the fluorescent screen at S. The deflection was discussed in detail in Chapter 26. The deflection y_E due to the electric field is given by Eq. (26-27). Now the kinetic energy $\frac{1}{2}mv^2$ of an electron is related to the accelerating potential V_1 by $\frac{1}{2}mv^2 = eV_1$, and the transverse deflecting field E is equal to V_2/ℓ, where V_2 is the deflecting potential. Using these relations to eliminate V_1 and V_2 from Eq. (26–27), we obtain

$$y_E = \frac{eEL}{mv^2}\left(D + \frac{L}{2}\right), \tag{30–6}$$

where e is the electron charge, m its mass, v its velocity, E the electric field between the plates, and D and L the dimensions shown in the figure.

If e/m is regarded as a single unknown, then there are two unknowns in this equation. The initial velocity v must be determined before e/m can be found. We need another equation involving the initial velocity v, so that this unknown velocity can be eliminated between the new equation and Eq. (30–6).

Thomson obtained another equation by applying a *magnetic* field perpendicular to both the electron beam and the electric field. It is represented in Fig. 30–7 as being into the page and uniform everywhere within the X-marked area. Thus the electrons experienced electric and magnetic forces in the same geometric space.

Figure 30–8 shows the situation when the magnetic field alone is present. The negative electrons experience an initial force downward, but this force is not constant in direction, and so the electrons move in a circular path. The equation of this path, taking the origin at the center of curvature C, is

$$R = \frac{mv}{qB}.$$

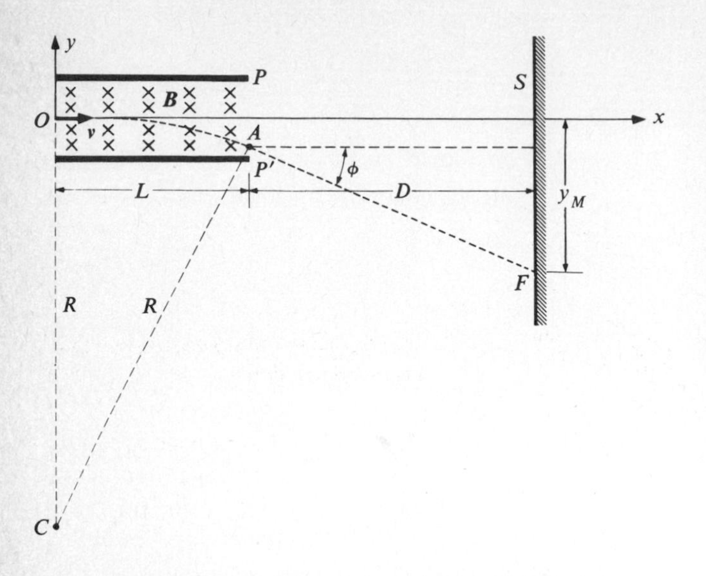

Fig. 30–8 *Magnetic deflection of cathode rays.*

It can be shown, from this equation and the geometry of the apparatus, that the deflection y_M produced by the magnetic field is

$$y_M = -\frac{eBL}{mv}\left(D + \frac{L}{2}\right). \qquad (30\text{–}7)$$

It may be mentioned at this point that the electron beam in all modern television "picture" tubes is deflected by *magnetic* fields, as in Fig. 30–8. To produce a given picture size using electrical deflection, either excessively large deflecting voltages or a much longer tube would be required.

Equation (30–7) is similar to Eq. (30–6). It contains e/m and v, together with measurable quantities, so that v can be eliminated and e/m found. It is interesting, however, to follow Thomson's procedure for determining v by the *simultaneous* application of electric and magnetic fields. If these are adjusted so that there is no deflection on the screen, the force of the electric field on an electron is balanced by that of the magnetic field. Then

$$qE = qvB, \quad \text{or} \quad v = \frac{E}{B}. \qquad (30\text{–}8)$$

For this particular ratio of the fields, the electron goes straight through both fields. It is undeflected, and therefore the measurement of v does not depend on the geometry of the tube. The velocity thus determined may then be substituted into Eq. (30–6).

Thomson measured e/m for his "cathode corpuscles" and found a unique value for this quantity which was independent of the cathode material and the residual gas in the tube. This independence indicated that "cathode corpuscles" are a common constituent of all matter. The modern accepted value of e/m is $(1.758802 \pm 0.000005) \times 10^{11}$C kg^{-1}. Thus Thomson is credited with discovery of the first subatomic particle, the electron. He also found that the velocity of the electrons in the beam was about one-tenth the velocity of light, much larger than any previously measured material particle velocity.

This experiment, performed in 1897, does not determine either the charge or mass of the electron, but only the ratio e/m. The first determination of the electron charge was made by Millikan, fifteen years later, in a classic experiment described in Section 26–7. The value of e is 1.602×10^{-19}C; it follows that the electron mass is

$$m = \frac{1.602 \times 10^{-19}\text{C}}{1.759 \times 10^{11}\text{C kg}^{-1}} = 9.110 \times 10^{-31}\text{kg}.$$

30–6 ISOTOPES

Thomson devised a method similar to the above e/m measurement, for measuring the charge–mass ratio for positive ions. An added difficulty was that in Thomson's day it was difficult to produce a beam of ions all having the same velocity. Because the e/m electron experiment depends on the particles having a common velocity (in order for the electric and magnetic field forces to balance), this method is not directly applicable for a beam of particles having various velocities.

Thomson's method was to make the electric and magnetic fields *parallel*, so that the deflections due to these fields are in perpendicular directions. The net deflection can then never be zero, but it turns out that the relation between the x- and y-deflections for a beam permits determination of the charge–mass ratio of the particles. The field arrangement is shown schematically in Fig. 30–9. The electric field deflects the beam in the y-direction, with the y-deflection

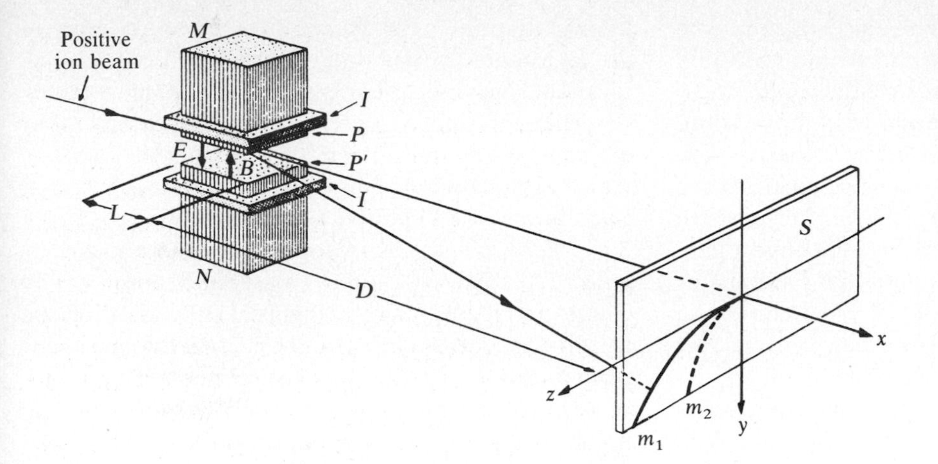

Fig. 30–9 *Formation of positive-ray parabolas.*

given by Eq. (30–6). The magnetic deflection is given by Eq. (30–7), with the change that the deflection is the $+z$-direction rather than the $-y$-direction. Thus, the y- and z-coordinates of the point on the surface S where a particle strikes are given by

$$y = \frac{qEL}{mv^2}\left(D + \frac{L}{2}\right),$$
$$z = \frac{qBL}{mv}\left(D + \frac{L}{2}\right). \qquad (30\text{–}9)$$

Since v is different for different particles, the pattern on the screen S is not a single point but a *curve*, each point of which corresponds to a certain value of v. To obtain the equation of the curve we may eliminate v between Eqs. (30–9), obtaining

$$z^2 = \left(\frac{q}{m}\right)\frac{B^2L}{E}\left(D + \frac{L}{2}\right)y. \qquad (30\text{–}10)$$

This is the equation of a *parabola*, having the general form $z^2 = ay$, where a is a constant. For any parabola experimentally obtained, we can determine the value of a required to "fit" the experimental curve, and then equate this to the coefficient in Eq. (30–10). That is,

$$a = \left(\frac{q}{m}\right)\frac{B^2L}{E}\left(D + \frac{L}{2}\right).$$

Now everything in this equation is known except the ratio q/m, which may therefore be determined.

Figure 30–10 shows the results of a particular experiment. The fact that there are *several* parabolas shows that several values of q/m are represented. Thomson's interpretation was that each particle of the positive "rays" carried a charge equal and opposite to the electronic charge, and he attributed the different parabolas to differences in *mass*. He assumed that the particles were positively charged because each had lost one electron. Thomson could identify particular parabolas with particular ions.

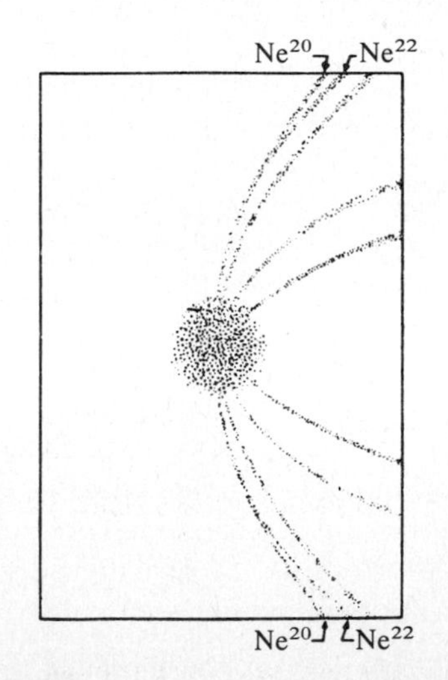

Fig. 30–10 *The parabolas of neon.*

Thus, for atomic hydrogen, he could verify that the q/m he measured was equal to the value one would obtain by dividing the electronic charge by the mass per atom. The reason positive particles move more slowly than electrons and have lower values of q/m than electrons is now clear: the positive particles are much more massive. The *largest* q/m for positive particles is that for the *lightest* element, hydrogen. From the value of q/m it was found that the mass of the *hydrogen ion or proton* is 1836.11 ± 0.01 times the mass of an electron. Electrons contribute only a small amount of the mass of material objects.

The most striking result of these experiments was that certain chemically pure gases had *more than one* value of q/m. Most notable was the case of neon, of atomic mass 20.2. Neon exhibited a parabola corresponding to a particle of atomic mass 20, but it also had a parabola which indicated an atomic mass of 22. Since the next heavier element, sodium, has an atomic mass of 23.0, efforts to explain away the unexpected value of q/m failed at first. Finally, it was concluded that there must be two kinds of neon, with different masses but chemically identical.

This interpretation was confirmed by Aston, one of Thomson's students, using a principle discussed in Chapter 20. The average kinetic energy of a molecule in a gas is $3kT/2$. Different gases mixed together in a container must be at the same temperature, and hence the average kinetic energy of each kind of molecule must be the same. If two gases have different molecular masses, the lighter molecules must have the higher average velocity, and these will make more collisions per unit time with the walls of the container than the heavier molecules. Therefore, if these molecules are allowed to diffuse through a porous plug from one container into another, the number of lighter molecules passing through the plug will be greater than the number of heavier, slower ones.

Aston started with chemically pure neon gas and passed part of it through such a plug. Since one such pass accomplishes only a slight separation, the process had to be repeated many times. He ended with two very small amounts of gas. One fraction had been through the plug many times and the other had been "left behind" many times. He measured the atomic mass of each fraction and found values of 20.15 for the former and 20.28 for the latter. The difference was not great but it was enough to show that there are indeed two kinds of neon. Many other elements have since been shown to exist in forms which are chemically identical but different in mass. Such forms of an element are called *isotopes.*

The discovery of isotopes solved several problems. It explained the two parabolas observed by Thomson. It also gave a logical explanation of the fact that the atomic mass of neon, 20.2, departs so far from an integral value. If chemical neon is a mixture of neon of atomic mass 20 and of neon of atomic mass 22, then there is some proportion of the two which, when mixed, will have an *average* atomic mass of 20.2.

30–7 MASS SPECTROSCOPY

A detailed search for the isotopes of all the elements required a more precise technique. Aston built the first of many instruments called *mass spectrometers* in 1919. His instrument had a precision of one part in 10,000, and he found that many elements have isotopes. Rather than discuss his original instrument, however, we shall describe a more elegant one built

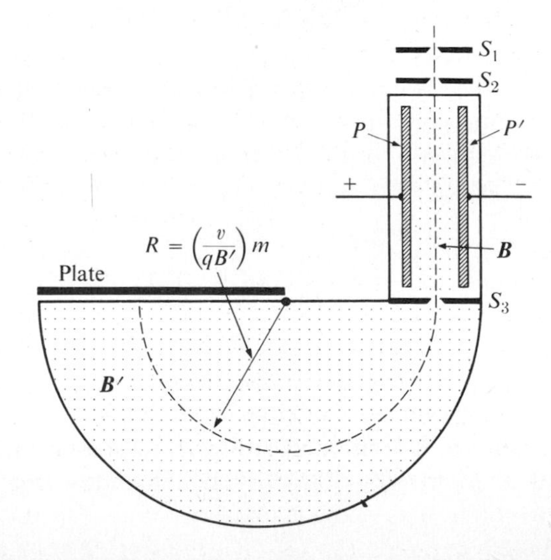

Fig. 30–11 *Bainbridge's mass spectrograph, utilizing a velocity selector.*

by Bainbridge. The Bainbridge mass spectrometer (Fig. 30–11) has a source of ions (not shown) situated above S_1. The ions under study pass through slits S_1 and S_2 and move down into the electric field between the two plates P and P'. In the region of the electric field there is also a magnetic field $\boldsymbol{B}$, perpendicular to the paper. Thus the ions enter a region of crossed electric and magnetic fields like those used by Thomson to measure the velocity of electrons in his determination of e/m. Those ions whose velocity is E/B pass undeviated through this region, but ions with other velocities are stopped by the slit S_3. All ions which emerge from S_3 have the same velocity. The region of crossed fields is called a *velocity selector.* Below S_3 the ions enter a region where there is another magnetic field $\boldsymbol{B}'$, perpendicular to the page, but no electric field. Here the ions move in circular paths of radius R. From Eq. (30–5), we find that

$$m = \frac{qB'R}{v}.$$

Assuming equal charges on each ion, the mass of each ion is proportional to the radius of its path. Ions of different isotopes converge at different points on the photographic plate. The relative abundance of the isotopes is measured from the densities of the photographic images they produce. Figure 30–12 shows the mass spectrum of germanium. The numbers shown beside the isotope images are not the atomic masses of the isotopes but the integers nearest the atomic masses. These integers are called *mass numbers*; isotopes are written with the mass number as a superscript to the chemical symbol. Thus the isotopes shown would be written ^{70}Ge, ^{72}Ge, etc. The mass number is represented by the letter A.

As in the case of neon, the discovery of the isotopes of the various elements largely accounted for the fact that many chemical atomic masses are not integers. If germanium has mass numbers 70, 72, 73, 74, and 76, it is not surprising that a mixture of isotopes of germanium has a chemical atomic mass of 72.6.

Masses of atoms are often expressed in *atomic mass units.* By definition, one atomic mass unit (1 u) is 1/12 the mass of one atom of the most

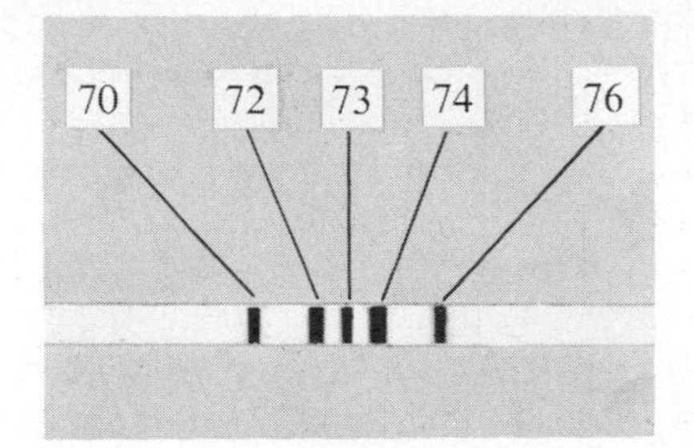

Fig. 30–12 *The mass spectrum of germanium, showing the isotopes of mass numbers 70, 72, 73, 74, 76.*

abundant isotope of carbon, ^{12}C. Since the mass of an atom in grams is equal to its atomic mass divided by Avogadro's number, it follows that

$$\begin{aligned} 1\,\mathrm{u} &= \frac{(1/12)(12\ \mathrm{g\ mol^{-1}})}{6.02 \times 10^{23}\ \mathrm{mol^{-1}}} \\ &= 1.66 \times 10^{-24}\,\mathrm{g} \\ &= 1.66 \times 10^{-27}\,\mathrm{kg}. \end{aligned}$$

30–8 THE CYCLOTRON

The cyclotron, developed in 1931 by Drs. Ernest O. Lawrence and M. Stanley Livingston at the University of California at Berkeley, was the first of many instruments which use electric and magnetic fields to guide particles and accelerate them to form a beam of charged particles with high speed. Despite the size and complexity of the cyclotron, its basic principles of operation are quite simple.

The heart of the cyclotron is a pair of metal chambers shaped like the halves of a pillbox that has been cut along one of its diameters. (See Fig. 30–13). These hollow chambers, referred to as "dees" or "D's" because of their shape, have their diametric edges parallel and slightly separated from each other. A source of ions—the positively charged nuclei of heavy hydrogen (deuterons) are commonly used—is located near the midpoint of the gap between the dees, and the dees are connected to the terminals of a circuit generating an alternating voltage. The potential between the dees is thus caused to alternate

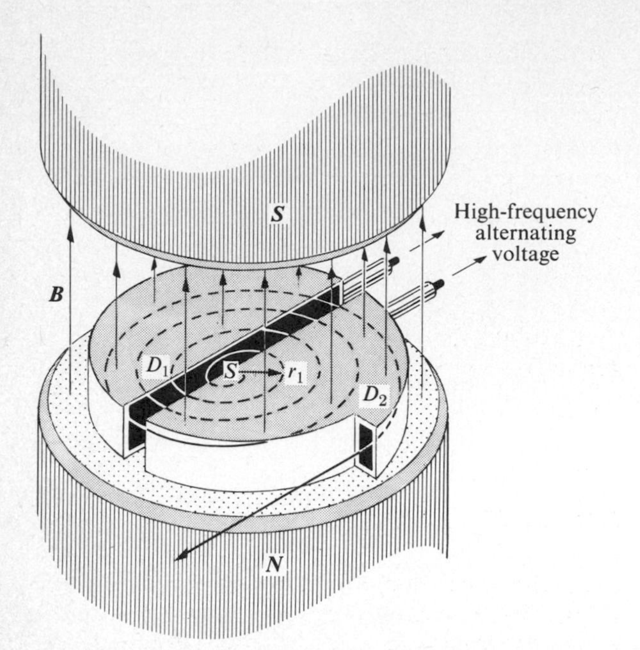

Fig. 30–13 Schematic diagram of a cyclotron.

rapidly, some millions of times per second, so the electric field in the gap between the dees is directed first toward one and then toward the other. But because of the electrical shielding effect of the dees, the space *within* each is a region of zero electric field.

The two dees are enclosed within, but insulated from, a somewhat larger cylindrical metal container from which the air is exhausted, and the whole apparatus is placed between the poles of a powerful electromagnet which provides a uniform magnetic field perpendicular to the ends of the cylindrical container.

Consider an ion of charge $+q$ and mass m, emitted from the ion source S at an instant when D_1 in Fig. 30–13 is positive. The ion is accelerated by the electric field in the gap between the dees and enters the (electric) field-free region within D_2 with a speed, say, of v_1. Since its motion is at right angles to the magnetic field, the ion will travel in a circular path of radius

$$r_1 = \frac{mv_1}{Bq}.$$

If, now, in the time required for the ion to complete a half-circle, the *electric* field has reversed so that its direction is toward D_1, the ion will again be accelerated as it crosses the gap between the dees and will enter D_1 with a greater velocity v_2. It therefore moves in a half circle of larger radius within D_1 to emerge again into the gap.

The angular velocity ω of the ion is

$$\omega = \frac{v}{r} = B\left(\frac{q}{m}\right). \tag{30–10}$$

Hence the angular velocity is *independent of the speed of the ion and of the radius of the circle* in which it travels, depending only on the magnetic induction and the charge-to-mass ratio (q/m) of the ion. If, therefore, the electric field reverses at regular intervals, each equal to the time required for the ion to make a half revolution, the field in the gap will always be in the proper direction to accelerate an ion each time the gap is crossed. It is this feature of the motion, that the time of rotation is independent of the radius, which makes the cyclotron feasible, since the regularly timed reversals are accomplished automatically by the oscillator circuit to which the dees are connected.

The path of an ion is a sort of spiral, composed of semicircular arcs of progressively larger radius connected by short segments along which the radius is increasing. If R represents the outside radius of the dees, and $v_{\max}$ the speed of the ion when traveling in a path of this radius, then

$$v_{\max} = BR\left(\frac{q}{m}\right),$$

and the corresponding kinetic energy of the ion is

$$E_k = \frac{1}{2}mv_{\max}^2 = \frac{1}{2}m\left(\frac{q}{m}\right)^2 B^2R^2.$$

If the ions are deuterons,

$$\frac{q}{m} = \frac{1.6 \times 10^{-19}\text{C}}{(2)(1.66 \times 10^{-27}\text{kg})} = 4.8 \times 10^7\text{C kg}^{-1}.$$

In a typical small cyclotron, $B = 2.0$ T and $R = 0.5$ m.

Hence

$$E_k = (1.66 \times 10^{-27}\ \text{kg})(4.8 \times 10^7\ \text{C kg}^{-1})^2 \times (2.0\ \text{T})^2(0.5\ \text{m})^2 = 3.82 \times 10^{-12}\ \text{J}.$$

Energies of atomic particles are conveniently measured in *electron volts* (introduced in Section 26–8). We recall that $1\ \text{eV} = 1.60 \times 10^{-19}\ \text{J}$, so

$$E_k = \frac{3.82 \times 10^{-12}\ \text{J}}{1.60 \times 10^{-19}\ \text{J eV}^{-1}} = 2.39 \times 10^7\ \text{eV} = 23.9\ \text{MeV}.$$

Since the magnitude of the deuteron charge is the same as that of the electron, the deuterons acquire as much energy as they would if permitted to accelerate through a potential difference of 23.9 MV. In principle this potential could be supplied by a machine such as a Van de Graaff generator, but practical difficulties make such a voltage impractical.

The success of the cyclotron hinges on the ion motion remaining "in step" with the alternating electric field, which is not difficult so long as the angular velocity is constant, as in Eq. (30–10). However, when the particle speeds reach a significant fraction of the speed of light, relativistic effects must be considered. The appropriate relativistic modification of Eq. (30–5) consists in replacing the momentum mv by its relativistic generalization $mv/(1 - v^2/c^2)^{1/2}$. The radius of curvature of the path for a given value of v becomes

$$R = \frac{mv}{Bq\sqrt{1 - v^2/c^2}},$$

and the angular velocity $\omega = v/R$, is no longer given by Eq. (30–10) but instead is

$$\omega = B\left(\frac{q}{m}\right)\sqrt{1 - v^2/c^2}. \tag{30–11}$$

Hence, if the ions are to remain in step with the alternating potential as they are accelerated, the oscillator frequency must *decrease*. This is accomplished in a *synchrocyclotron* by periodically varying the oscillator frequency at just the right rate to keep in step with groups of particles, and the particles are then accelerated in bunches.

In recent years, other high-energy particle accelerators have been constructed in which the magnetic field is varied as the particles speed up, so that they retrace the same trajectory over and over. Such machines are called synchrotrons; the world's highest-energy synchrotron is located at the National Accelerator Laboratory at Batavia, Illinois, and can accelerate protons to an energy of 400 GeV(400×10^9eV).

PROBLEMS

30–1 Each of the lettered circles at the corners of the cube in Fig. 30–14 represents a positive charge q moving with a velocity of magnitude v in the direction indicated. The region in the figure is a uniform magnetic field $\boldsymbol{B}$, parallel to the x-axis and directed toward the right. Copy the figure, find the magnitude and direction of the force on each charge, and show the force in your diagram.

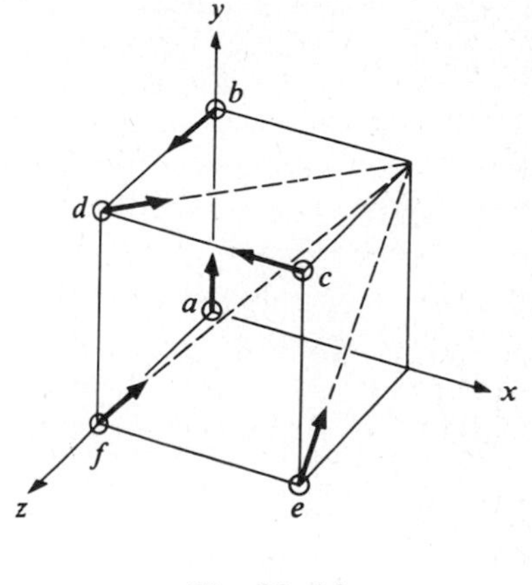

Fig. 30–14

30–2 The magnetic field $\boldsymbol{B}$ in a certain region is 2 T and its direction is that of the positive x-axis in Fig. 30–15.

a) What is the magnetic flux across the surface *abcd* in the figure?
b) What is the magnetic flux across the surface *befc*?
c) What is the magnetic flux across the surface *aefd*?

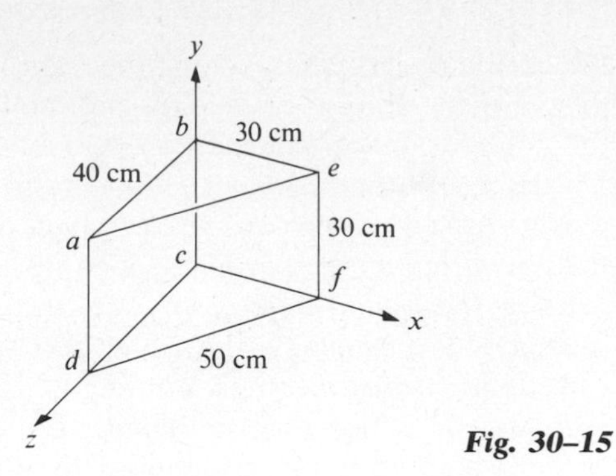

Fig. 30–15

30–3 A particle having a mass of 0.5 g carries a charge of 2.5×10^{-8}C. The particle is given an initial horizontal velocity of 6×10^4m s^{-1}. What is the magnitude and direction of the minimum magnetic field that will keep the particle moving in a horizontal direction?

30–4 A deuteron, an isotope of hydrogen whose mass is very nearly 2 u, travels in a circular path of radius 40 cm in a magnetic field of flux density 1.5 T.

a) Find the speed of the deuteron.
b) Find the time required for it to make one-half a revolution.
c) Through what potential difference would the deuteron have to be accelerated to acquire this velocity?

30–5 An electron at point A in Fig. 30–16 has a velocity v_0 of 10^7m s^{-1}. Find (a) the magnitude and direction of the magnetic field that will cause the electron to follow the semi-circular path from A to B, (b) the time required for the electron to move from A to B.

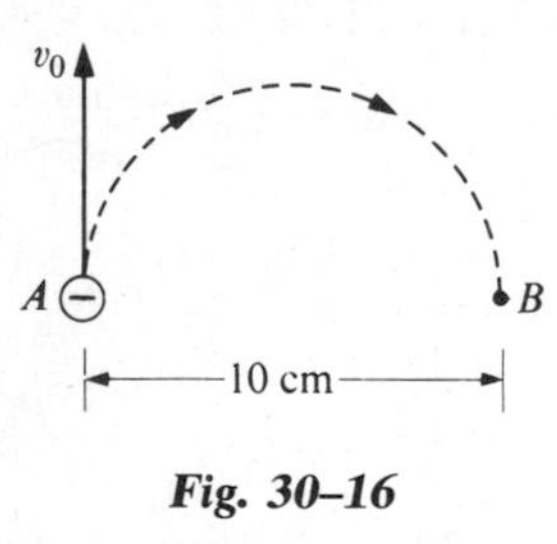

Fig. 30–16

30–6 In Problem 30–5, suppose the particle is a proton rather than an electron. Answer the same questions as in that problem.

30–7 In a magnetic field directed vertically upward, a particle initially moving north is deflected toward the west. What is the sign of the charge on the particle?

30–8 In a TV picture tube, an electron in the beam is accelerated by a potential difference of 20,000 V. Then it passes through a region of transverse magnetic field where it moves in a circular arc with radius 12 cm. What is the magnitude of the field?

30–9 Estimate the effect of the earth's magnetic field on the electron beam in a TV picture tube. Suppose the accelerating voltage is 20,000 V; calculate the approximate deflection of the beam over a distance of 0.4 m from the electron gun to the screen, under the action of a transverse field of magnitude 0.5×10^{-4} T (comparable to the magnitude of the earth's field), assuming there are no other deflecting fields. Is this deflection significant?

30–10 A particle carries a charge of 4×10^{-9}C. When it moves with a velocity v_1 of 3×10^4m s^{-1} at 45° above the x-axis in the xy-plane, a uniform magnetic field exerts a force F_1 along the z-axis. When the particle moves with a velocity v_2 of 2×10^4m s^{-1} along the z-axis, there is a force F_2 of 4×10^{-5}N exerted on it along the x-axis. What are the magnitude and direction of the magnetic field? (See Fig. 30–17).

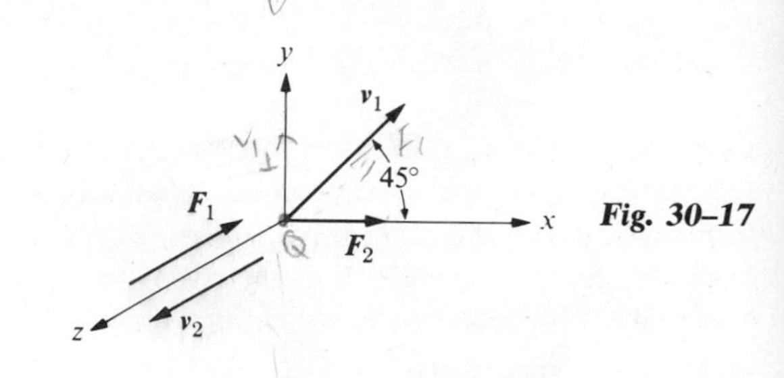

Fig. 30–17

30–11 An electron moves in a circular path of radius 1.2 cm perpendicular to a uniform magnetic field. The velocity of the electron is 10^6m s^{-1}. What is the total magnetic flux encircled by the orbit?

30–12 An electron and an alpha particle (a doubly ionized helium atom) both move in circular paths in a magnetic field with the same tangential velocity. Compare the number of revolutions they make per second. The mass of the alpha particle is 6.68×10^{-27}kg.

30–13 For a particular parabola in Thomson's mass spectrograms, what physical quantity is different for the ions which land close to the origin than for those landing farther away? Why does this difference exist, since the accelerating voltage is the same for all the ions?

30–14 What must be the direction of the electric field $\boldsymbol{E}$ and the magnetic field $\boldsymbol{B}$ in Fig. 30–9 so that the segment of the positive-ion parabola will be in (a) the lower right quadrant, (b) the upper right quadrant, and (c) the upper left quadrant, as viewed from the right of the diagram?

30–15

a) If the ion beam in Fig. 30–9 contains two types of ions having equal charges but different masses, which of the two parabolic segments will have those of greater mass?

b) If the masses are equal but the charges different, which segment will contain those having the larger charge?

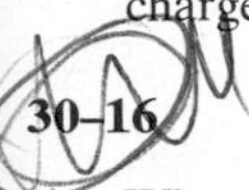

30–16

a) What is the velocity of a beam of electrons when the simultaneous influence of an electric field of $34 \times 10^4 \text{V m}^{-1}$ and a magnetic field of 2×10^{-3} T, both fields being normal to the beam and to each other, produces no deflection of the electrons?

b) Show in a diagram the relative orientation of the vectors $\boldsymbol{v}$, $\boldsymbol{E}$, and $\boldsymbol{B}$.

c) What is the radius of the electron orbit when the electric field is removed?

30–17 A singly charged Li^7 ion has a mass of 1.16×10^{-23} g. It is accelerated through a potential difference of 500 V and then enters a magnetic field of 0.4 T, moving perpendicular to the field. What is the radius of its path in the magnetic field?

30–18 Suppose the electric field between the plates P and P' in Fig. 30–11 is 150 V cm^{-1}, and the magnetic field is 0.5 T. If the source contains the three isotopes of magnesium ${}_{12}Mg^{24}$, ${}_{12}Mg^{25}$, and ${}_{12}Mg^{26}$, and the ions are singly charged, find the distance between the lines formed by the three isotopes on the photographic plate. Assume the atomic masses of the isotopes equal to their mass numbers.

30–19 The electric field between the plates of the velocity selector in a Bainbridge mass spectrograph is 1200 V cm^{-1}, and the magnetic field in both regions is 0.6 T. A stream of singly charged neon moves in a circular path of 7.28-cm radius in the magnetic field. Determine the mass number of the neon isotope.

30–20 A particle of mass m and charge $+q$ starts from rest at the origin in Fig. 30–18. There is a uniform electric field E in the positive y-direction and a uniform magnetic field B directed toward the reader. It is shown in more advanced books that the path is a cycloid whose radius of curvature at the top point is twice the y-coordinate at that point.

a) Explain why the path has a cycloidal shape and why it is repetitive.

b) Prove that the speed at any point is equal to $\sqrt{2qEy/m}$.

c) Applying Newton's second law at the top point, prove that the speed at this point is $2E/B$.

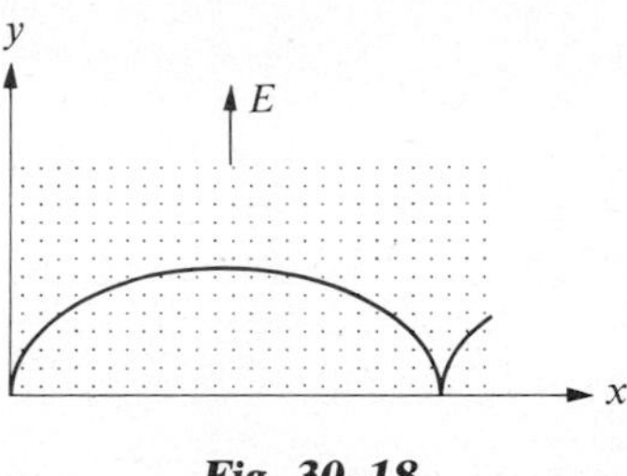

Fig. 30–18

30–21 Two positive ions having the same charge q but different masses, m_1 and m_2, are accelerated horizontally from rest through a potential difference V. They then enter a region where there is uniform magnetic field $\boldsymbol{B}$ normal to the plane of the trajectory.

a) Show that if the beam entered the magnetic field along the x-axis, the value of the y-coordinate for each ion at any time t is approximately

$$y = Bx^2\left(\frac{q}{8mV}\right)^{1/2}.$$

b) Can this arrangement be used for isotope separation?

30–22 The magnetic field in a cyclotron which is accelerating protons is 1.5 Wb m^{-2}.

a) How many times per second should the potential across the dees reverse?

b) The maximum radius of the cyclotron is 0.35 m. What is the maximum velocity of the proton?

c) Through what potential difference would the proton have to be accelerated to give it the maximum cyclotron velocity?

30–23 Deuterons in a cyclotron describe a circle of radius 32.0 cm just before emerging from the dees. The frequency of the applied alternating voltage is 10 MHz. Find (a) the magnetic field, and (b) the energy and speed of the deuterons upon emergence.

30–24 The cyclotron in Problem 30–22 is to be modified to accelerate alpha particles having twice the charge and four times the mass of the proton. The magnetic field is to remain at the same value, but the oscillator frequency may be changed.

a) What frequency should the oscillator have?

b) What is the maximum kinetic energy, in electron volts?

c) How does the maximum kinetic energy compare with the maximum kinetic energy of protons, for which the machine was originally designed?

Chapter 31

Magnetic Forces on Current-Carrying Conductors

31–1 FORCE ON A CONDUCTOR

When a current-carrying conductor lies in a magnetic field, magnetic forces are exerted on the moving charges within the conductor. These forces are transmitted to the material of the conductor, and the conductor as a whole experiences a force distributed along its length. The electric motor and the moving-coil galvanometer both depend for their operation on the magnetic forces on conductors carrying currents.

Figure 31–1 represents a portion of a conducting wire of length ℓ and cross-sectional area A, in which the current density J is from left to right. The wire is in a magnetic field $\boldsymbol{B}$, perpendicular to the plane of the diagram, and directed *into* the plane. A positive charge q_1 within the wire, moving with its drift velocity $\boldsymbol{v}_1$, is acted on by an upward force $\boldsymbol{F}_1$ of magnitude $q_1 v_1 B$. The drift velocity $\boldsymbol{v}_2$ of a negative-charge carrier q_2 is opposite to $\boldsymbol{v}_1$ and the force $\boldsymbol{F}_2$ on it has magnitude $q_2 v_2 B$. Forces $\boldsymbol{F}_1$ and $\boldsymbol{F}_2$ are both upward, because the signs of both q_2 and v_2 are opposite to those of q_1 and v_1, respectively.

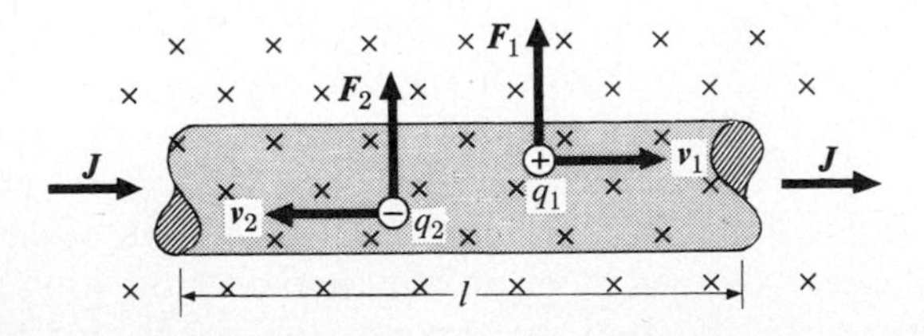

Fig. 31–1 Forces on the moving charges in a current-carrying conductor. The forces on both positive and negative charges are in the same direction.

Let n_1 and n_2 represent the numbers of positive and negative charge carriers per unit volume. The numbers of carriers in the portion are then $n_1 A\ell$ and $n_2 A\ell$; the total force $\boldsymbol{F}$ on all carriers (and hence the total force on the wire) has magnitude

$$\begin{aligned} F &= (n_1 A\ell)(q_1 v_1 B) + (n_2 A\ell)(q_2 v_2 B) \\ &= (n_1 q_1 v_1 + n_2 q_2 v_2) A\ell B. \end{aligned} \tag{31–1}$$

But $n_1 q_1 v_1 + n_2 q_2 v_2$ (or more generally, $\sum nqv$) equals the current density J, and the product JA equals the current I, so finally

$$F = I\ell B.$$

If the $\boldsymbol{B}$ field is not perpendicular to the wire but makes an angle ϕ with it, the situation is just like that discussed in Section 30–2 for a single charge. The component of $\boldsymbol{B}$ parallel to the wire (and thus to the drift velocities of the charges) exerts no force; the component perpendicular to the wire is given by

$B_\perp = B \sin\phi$ so, in general,

$$F = I\ell B_\perp = I\ell B \sin\phi. \tag{31–2}$$

To find the direction of the force on a current-carrying conductor placed in a magnetic field, we may use the same righthand screw rule that was used in the case of a moving positive charge. Rotate a righthand screw from the direction $\boldsymbol{i}$ toward $\boldsymbol{B}$ and the direction of advance will be the direction of $\boldsymbol{F}$. The situation is the same as that shown in Fig. 30–2, but with $\boldsymbol{v}$ replaced by the direction of i.

Comparison of the above discussion with that of Section 30–2 also shows that the force on an element of conductor $d\ell$ is given by

$$d\boldsymbol{F} = I\, d\boldsymbol{\ell} \times \boldsymbol{B}, \tag{31–3}$$

and the force on a straight conductor of length ℓ is

$$\boxed{\boldsymbol{F} = I\boldsymbol{\ell} \times \boldsymbol{B}.} \tag{31–4}$$

The vectors $d\boldsymbol{\ell}$ and $\boldsymbol{\ell}$ are in the direction of the current density $\boldsymbol{J}$, and Eqs. (31–3) and (31–4) describe both the *magnitude* and *direction* of the force.

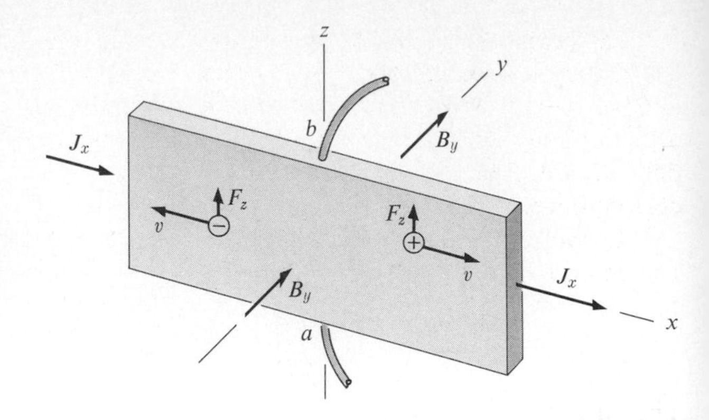

Fig. 31–2 *Forces on charge carriers in a conductor in a magnetic field.*

31–2 THE HALL EFFECT

The reality of the forces on the moving charges in a conductor in a magnetic field is strikingly demonstrated by the *Hall effect*. The conductor in Fig. 31–2 is in the form of a flat strip. The current carriers within it are driven toward the upper edge of the strip by the magnetic force qvB exerted on them. Here $\boldsymbol{v}$ is the drift velocity of the moving charges in the material.

This force is an example of the *nonelectrostatic* forces $\boldsymbol{F}_n$ discussed in Chapter 28, and in this case the equivalent *nonelectrostatic field* $\boldsymbol{E}_n$ (the force per unit charge) has the magnitude

$$E_n = vB.$$

If the charge carriers are electrons, an excess negative charge accumulates at the upper edge of the strip, leaving an excess positive charge at its lower edge, until the transverse electrostatic field $\boldsymbol{E}_e$ within the conductor is equal and opposite to the *nonelectrostatic* field $\boldsymbol{E}_n$. Because the final transverse current is zero, the conductor is on "open circuit" in the transverse direction, and the potential difference between the edges of the strip, which can be measured with a potentiometer, is equal to the *Hall emf* in the strip. The study of this Hall emf has provided much information regarding the conduction process. It is found that, for the metals, the upper edge of the strip in Fig. 31–2 *does* become charged negatively relative to the lower, which justifies our belief that the charge carriers in a metal are negative electrons.

Suppose, however, that the charge carriers are *positive*. Then *positive* charge accumulates at the upper edge, and the potential difference is *opposite* to that resulting from the deflection of negative charges. Soon after the discovery of the Hall effect, in 1879, it was observed that many materials, notably the *semiconductors*, exhibited a Hall emf opposite to that of the metals, *as if* their charge carriers were *positively* charged. We now believe that these materials conduct by a process known as *hole conduction*. There are sites within the material that would normally be occupied by an electron but are actually empty, and a *missing negative* charge is equivalent to a *positive* charge. When an electron moves in one direction to fill a hole, it leaves another hole behind it, and the result is that the *hole* (equivalent to a positive charge) migrates in the direction *opposite* to that of the electron.

In terms of the set of coordinate axes in Fig. 31–2, the electrostatic field E_e is in the z-direction, and we write it as E_z. The magnetic field is in the y-direction, and we write it as B_y. The nonelectrostatic field E_n (also in the z-direction) equals vB_y. The current density, J_x is in the x-direction.

In the final steady state, when the fields E_e and E_n are equal,

$$E_z = vB_y.$$

The current density J_x is

$$J_x = nqv.$$

When v is eliminated, we have

$$nq = \frac{J_x B_y}{E_z}. \qquad (31\text{–}5)$$

Thus, from measurements of J_x, B_y, and E_z, one can compute the product nq. In both metals and semiconductors, q is equal in magnitude to the electron charge, so the Hall effect permits a direct measurement of n, the density of current-carrying charges in the material.

31–3 FORCE AND TORQUE ON A COMPLETE CIRCUIT

The net force and torque on a complete circuit in a magnetic field can be found from Eq. (31–4). Three simple cases will be analyzed.

Rectangular loop. Figure 31–3 shows a rectangular loop of wire with sides of lengths a and b. The normal to the plane of the loop makes an angle α with the direction of a uniform magnetic field, and the loop carries a current i. (Provision must be made for leading the current into and out of the loop, or for inserting a seat of emf. This is omitted from the diagram, for simplicity.)

The force $\boldsymbol{F}$ on the right side of length a is in the direction of the x-axis, toward the right, as shown. In this side $\boldsymbol{B}$ is perpendicular to the current direction, and the total force on this side (the sum of the forces distributed along the side) is

$$F = IaB.$$

A force of the same magnitude but in the opposite direction acts on the opposite side, as shown in the figure.

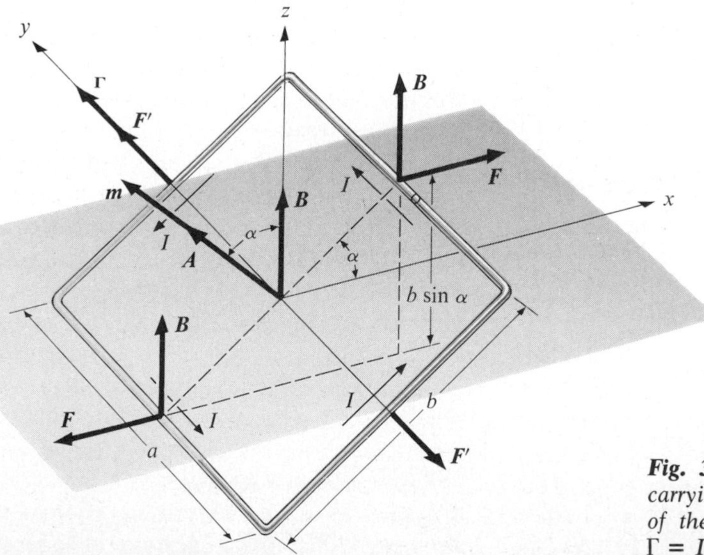

Fig. 31–3 *Forces on the sides of a current-carrying loop in a magnetic field. The resultant of the set of forces is a couple of moment* $\Gamma = I(\boldsymbol{A} \times \boldsymbol{B})$.

The forces on the sides of length b, represented by the vectors $\boldsymbol{F}'$, are of magnitude $IbB \cos \alpha$. The lines of action of both lie along the y-axis.

The resultant *force* on the loop is evidently zero, since the forces on opposite sides are equal and opposite. The resultant *torque*, however, is not zero, since the forces on the sides of length a constitute a couple of moment

$$\Gamma = (IBa)(b \sin \alpha). \quad (31\text{–}6)$$

The couple is maximum when $\alpha = 90°$ (when the plane of the coil is parallel to the field), and it is zero when $\alpha = 0$ and the plane of the coil is perpendicular to the field. What is the position of stable equilibrium?

Since ab is the area A of the coil, Eq. (31–6) may also be written

$$\boxed{\Gamma = IBA \sin \alpha,} \quad (31\text{–}7)$$

or, in vector form,

$$\boxed{\boldsymbol{\Gamma} = I\mathbf{A} \times \boldsymbol{B},} \quad (31\text{–}8)$$

where $\mathbf{A}$ is the vector area of the loop. The torque vector $\boldsymbol{\Gamma}$ is in the direction of the vector product $\mathbf{A} \times \boldsymbol{B}$ and points along the positive y-axis.

The product $I\mathbf{A}$ is called the *magnetic moment* $\boldsymbol{m}$ of the loop. (It is analogous to the *electric moment* of an electric dipole.) Hence the result obtained above can be expressed by the simple equation.

$$\boldsymbol{\Gamma} = \boldsymbol{m} \times \boldsymbol{B}, \quad (31\text{–}9)$$

which is the analog of Eq. (25–3) in Section 25–1 for the torque on a dipole in an electric field,

$$\boldsymbol{\Gamma} = \boldsymbol{p} \times \boldsymbol{E}.$$

The effect of the torque $\boldsymbol{\Gamma}$ is to tend to rotate the loop toward its equilibrium position, in which it lies in the xy-plane with its vector magnetic moment $\boldsymbol{m}$ in the same direction as the field $\boldsymbol{B}$.

When a magnetic dipole changes its orientation in a magnetic field, the field does *work* on it, and there is a corresponding potential energy. As the above discussion suggests, the potential energy is least when $\boldsymbol{m}$ and $\boldsymbol{B}$ are parallel and greatest when they are antiparallel. It can be shown that the potential energy E_p of this interaction is given by

$$E_p = -\boldsymbol{m} \cdot \boldsymbol{B} = -mB \cos \alpha. \quad (31\text{–}10)$$

Circular loop. A circular loop may be approximated by a very large number of small rectangular loops; then the sum of the areas of the rectangular loops may be made to approach the area of the circular one as closely as we please. Furthermore, the boundary of the rectangular loops will approximate the circular loop with any desired accuracy. Currents in the same sense in all the rectangular loops will give rise to forces which will cancel at all points except on the boundary. It can therefore be proved quite rigorously that, *not only for a circular loop, but for a plane loop of* ***any shape whatever,*** carrying a current i in a magnetic field of flux density B, the torque is given by Eq. (31–7) or (31–8).

Solenoid. A helical winding of wire, such as that obtained by winding wire around the surface of a mailing tube, is called a *solenoid.* If the windings are closely spaced, the solenoid can be approximated by a number of circular loops lying in planes at right angles to its long axis. The total torque on a solenoid in a magnetic field is simply the sum of the torques on the individual turns. Hence, for a solenoid of N turns in a uniform field B,

$$\Gamma = NIAB \sin \alpha, \quad (31\text{–}11)$$

where α is the angle between the axis of the solenoid and the direction of the field. The torque is a maximum when the magnetic field is parallel to the planes of the individual turns or perpendicular to the long axis of the solenoid. The effect of this torque, if the solenoid is free to turn, is to rotate it into a position in which each turn is perpendicular to the field and the axis of the solenoid is parallel to the field.

Although little has been said thus far regarding permanent magnets, the behavior of the solenoid as described above is the same as that of a bar magnet or compass needle, in that both the solenoid and the magnet will, if free to turn, set themselves with their axes parallel to a magnetic field. The behavior of a bar magnet or compass is usually explained by ascribing the torque on it to magnetic forces exerted on "poles" at its ends. We see, however, that no such interpretation is demanded in the case of the solenoid. May it not be, therefore, that the whirling electrons in a bar of magnetized iron are equivalent to the current in the windings of a solenoid, and that the observed torque arises from the same cause in both instances? We shall return to this question later.

Example 1 Consider the electron of a hydrogen atom, revolving in the first Bohr orbit, as discussed in Example 2 at the end of Section 24–6. The orbit is equivalent to a current loop of radius r and area πr^2. The average charge per unit time passing a point of the orbit, or the average current I, equals e/T, where T is the time of one revolution, equal to $2\pi r/v$. The magnetic moment of the loop is called 1 *Bohr magneton* and is represented by μ_B. Thus

$$\mu_B = IA = \tfrac{1}{2}evr.$$

But according to Bohr's theory, the angular momentum mvr is equal to $h/2\pi$, or

$$vr = \frac{h}{2\pi m}.$$

Therefore

$$\mu_B = \frac{h}{4\pi}\cdot\frac{e}{m}$$

$$= \frac{6.62\times10^{-34}\text{ J s}\times1.76\times10^{11}\text{ C kg}^{-1}}{12.57}$$

$$= 9.27\times10^{-24}\text{ A m}^2.$$

Example 2 Find the interaction potential energy when the hydrogen atom described above is placed in a magnetic field of 2 T.

According to Eq. (31–10), the interaction energy when $\alpha = 0$ is

$$E_p = -(9.27\times10^{-24}\text{ A m}^2)(2\text{ T}) = -1.85\times10^{-23}\text{ J}$$

$$= -1.15\times10^{-4}\text{ eV}.$$

When $\boldsymbol{m}$ and $\boldsymbol{B}$ are antiparallel, the energy is $+1.15\times10^{-4}$ eV. We note that these energies are much *smaller* than the electrostatic potential energy of the electron in the field of the nucleus, about -27 eV in the ground state. Nevertheless, the energy-level shifts due to magnetic interactions can be measured by precise spectroscopic observations. Such energy shifts are called the *Zeeman effect.*

31–4 THE GALVANOMETER

Any device used for the detection or measurement of current is called a *galvanometer.* The earliest form of galvanometer was simply the apparatus of Oersted, namely a compass needle placed below the wire in which the current was to be measured. Wire and needle were both aligned in the north–south direction, with no current in the wire. The deflection of the needle when a current was sent through the wire was then a measure of the current. The sensitivity of this form of galvanometer was increased by winding the wire into a coil in a vertical plane with the compass needle at its center; instruments of this type were developed by Lord Kelvin in the 1890's to a point where their sensitivity is scarcely exceeded by any available at the present time.

Practically all galvanometers used today, however, are of the D'Arsonval moving-coil or pivoted-coil type, in which the roles of magnet and coil are interchanged. The magnet is made much larger and is stationary, while the moving element is a light coil swinging in the field of the magnet.* The construction of a moving-coil galvanometer is illustrated in Fig. 31–4(a), and a photograph of one type of high-sensitivity galvanometer is shown in Fig. 31–4(b). The magnetic field of a horseshoe magnet whose poles are designated by N and S is concentrated in the vicinity of the coil C by the soft iron cylinder A. The coil consists of from 10 to 20 turns, more or less, of insulated copper wire wound on a rectangular frame and suspended by a fine conducting wire or thin flat strip F which provides a restoring torque when the coil is deflected from its normal position, and which also serves as one current lead to the coil. The other terminal of the coil is connected to the loosely wound helix H which serves as the second lead, but which exerts a negligible torque on the coil.

* The magnetic field surrounding a permanent magnet is discussed more fully in Chapter 34. For our present purposes, we may take it for granted that, in the region between the magnet poles of Fig. 31–4(a), there exists a field whose general direction is from N to S.

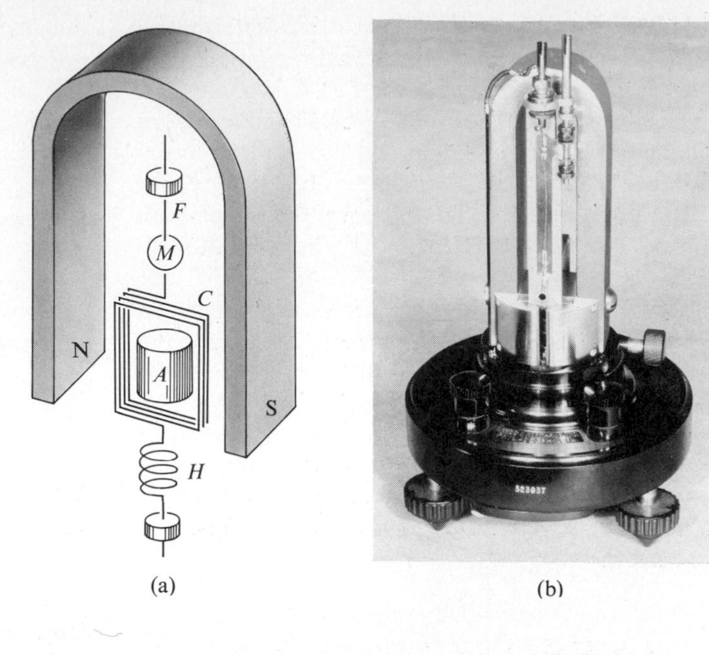

Fig. 31–4 *(a) Principle of the D'Arsonval galvanometer. (b) D'Arsonval galvanometer. (Courtesy of Leeds and Northrup.)*

When a current is sent through the coil, horizontal and oppositely directed side thrusts are exerted on its vertical sides, producing a torque about a vertical axis through its center. The coil rotates in the direction of torque and eventually comes to rest in such a position that the restoring torque exerted by the upper suspension equals the deflecting torque due to the side thrust. The angle of deflection is observed with the aid of a beam of light reflected from a small mirror M cemented to the upper suspension, the light beam serving as a weightless pointer. Since light incident on the mirror is reflected at an angle of reflection equal to the angle of incidence, rotation of the mirror through an angle θ deflects the light beam through an angle 2θ.

Because of the geometry of the field in which the moving coil swings, the deflections of a D'Arsonval galvanometer are not directly proportional to the current in the galvanometer coil except for relatively small angles. Hence these instruments are used chiefly as *null* instruments, that is, in connection with circuits such as those of a Wheatstone bridge or a potentiometer, in which other circuit elements are to be adjusted so that the galvanometer current is zero. High-sensitivity galvanometers can detect currents as small as 10^{-10} A.

31–5 THE PIVOTED-COIL GALVANOMETER

The pivoted-coil galvanometer, while essentially the same in principle as the D'Arsonval instrument, differs from the latter in two respects. One is that the moving coil, instead of being suspended by a fine fiber, is pivoted between two jewel bearings. The instrument may hence be used in any position and is much more rugged and conveniently portable. The second difference is that the permanent magnetic

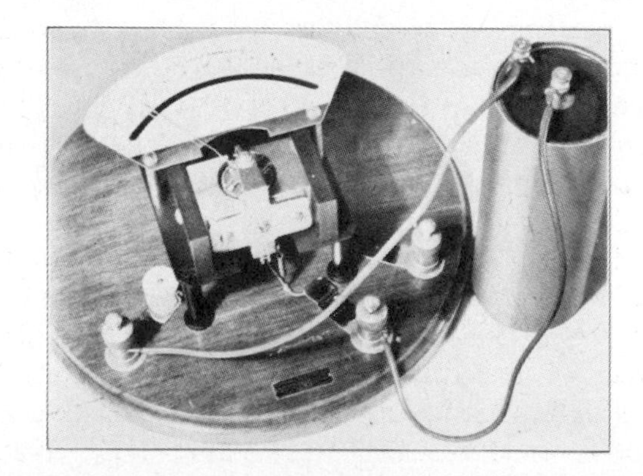

Fig. 31–5 *Pivoted-coil galvanometer, modified for use as an ammeter or a voltmeter. Series resistor may be seen at left. (Courtesy of Houghton Mifflin Co.)*

field is modified by the use of soft iron pole pieces attached to the permanent magnet, as shown in Fig. 31–5, so that the coil swings in a field which is everywhere radial. The side thrusts on the coil are therefore always perpendicular to the plane of the coil, and the angular deflection of the coil is directly proportional to the current in it. The restoring torque is provided by two hairsprings, which serve also as current leads. A length of aluminum tubing, flattened at its tip in a vertical plane, serves as a pointer.

The frictional torque of the jewel bearings, while small, is greater than that of a supporting fiber. Since the deflecting and restoring torques must both be considerably larger than the friction torque, pivoted-coil instruments cannot be made as sensitive as the D'Arsonval type. The smallest currents which can be read on such an instrument are of the order of magnitude of 0.1 μA (10^{-7} A).

31–6 THE BALLISTIC GALVANOMETER

A ballistic galvanometer is used for measuring the *quantity of charge* displaced by a current of short duration, as, for example, in the charging or discharging of a capacitor. While any moving-coil galvanometer can be used ballistically, instruments designed specifically for the purpose have coils with somewhat larger moments of inertia, and suspensions with somewhat smaller torque constants, than are found in instruments designed primarily for current measurement.

Consider a galvanometer coil of N turns, of area A, and moment of inertia I' suspended from a torsion wire whose torque constant (torque per unit angle) is k'. Suppose a current pulse is sent through the coil. The current is at first zero, then rises to a maximum, and then comes down to zero in such a short time that the coil does not move through an appreciable angle *while the current pulse exists*, but rotates only after the current has ceased. The torque at any moment of time *during the current pulse* is given by $\Gamma = NiAB$, and the angular impulse is the average torque Γ_{av} multiplied by the time Δt during which the charge flows. The average torque is given in turn by the average current i_{av} multiplied by NAB. Since the angular impulse is equal to the change of angular momentum, we have

$$\Gamma_{av}\Delta t = Ni_{av}AB\,\Delta t = I'\omega, \tag{31-12}$$

where ω is the angular velocity at the start of the galvanometer deflection.

The initial kinetic energy of the coil is converted to elastic potential energy of the torsion fiber, so that

$$\frac{1}{2}I'\omega^2 = \frac{1}{2}k'\theta_{max}{}^2, \tag{31-13}$$

where θ_{max} is the maximum angle of deflection. When ω is eliminated between these equations, we get

$$NABi_{av}\,\Delta t = \sqrt{k'I'}\,\theta_{max}.$$

But $i_{av}\,\Delta t$ is the total charge Q transferred through the coil. Hence

$$\boxed{Q = \frac{\sqrt{k'I'}}{NAB}\theta_{max},} \tag{31-14}$$

which shows that the total charge sent through a ballistic galvanometer coil is proportional to the maximum angle of deflection. The process is analogous to that which takes place when a ballistic pendulum is struck by a bullet.

31–7 THE DIRECT-CURRENT MOTOR

The direct-current motor is illustrated schematically in Fig. 31–6. The armature, A, is a cylinder of soft steel mounted on a shaft so that it can rotate about its axis. Embedded in longitudinal slots in the surface of the armature are a number of copper conductors C. Current is led into and out of these conductors through graphite brushes making contact with a cylinder on the shaft called the *commutator* (not shown in Fig. 31–6). The commutator is an automatic switching arrangement which maintains the currents in the conductors in the directions shown in the figure, whatever the position of the armature.

The current in the field coils F and F' sets up a magnetic field which, because of the shape of the pole pieces P and P', is essentially radial in the gap between them and the armature. The motor frame M provides a path for the magnetic field. Some of the

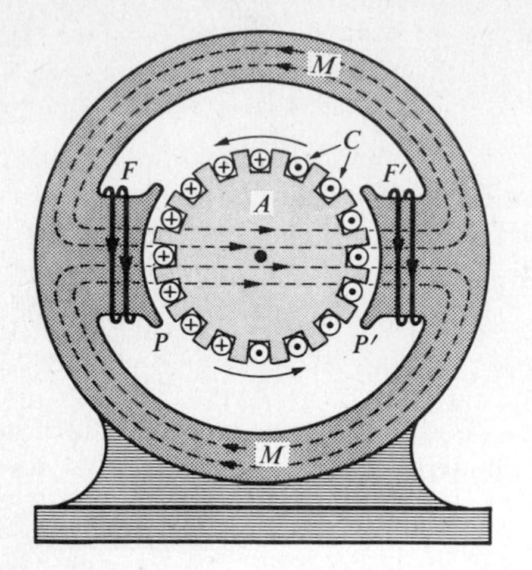

Fig. 31–6 Schematic diagram of a dc motor.

lines of induction are indicated by the dotted lines in the figure.

With the relative directions of field and armature currents as shown, the side thrust on each conductor is such as to produce a counterclockwise torque on the armature. When the motor is running, the armature develops mechanical energy at the expense of electrical energy. It must therefore be a source, in which there is an emf. This is an "induced" emf and is discussed further in Chapter 33. The field windings, however, are static and behave like a pure resistance.

If the armature and the field windings are connected in series, we have a *series* motor; if they are connected in parallel, we have a *shunt* motor. In some motors the field windings are in two parts, one in series with the armature and the other in parallel with it; the motor is then *compound.*

Example A series-wound dc motor has an internal resistance of 2.0 Ω. When running at full load on a 120-V line, it draws a current of 4.0 A.

a) What is the emf in the armature?

$$\begin{aligned} V_{ab} &= \mathscr{E} + Ir, \\ 120\text{ V} &= \mathscr{E} + (2.0\,\Omega)(4.0\text{ A}), \\ \mathscr{E} &= 112\text{ V}. \end{aligned}$$

b) What is the power delivered to the motor?

$$\begin{aligned} P &= IV_{ab} \\ &= (4.0\text{ A})(120\text{ V}) \\ &= 480\text{ W}. \end{aligned}$$

c) What is the rate of dissipation of energy in the motor?

$$\begin{aligned} P &= I^2r \\ &= (4.0\text{ A})^2(2.0\,\Omega) \\ &= 32\text{ W}. \end{aligned}$$

d) What is the mechanical power developed?

Mechanical power

$$\begin{aligned} &= \text{total power} - \text{rate of dissipation of energy} \\ &= 480\text{ W} - 32\text{ W} \\ &= 448\text{ W}. \end{aligned}$$

The mechanical power may also be calculated from the relation

$$\begin{aligned} \text{Mechanical power} &= \mathscr{E}I = (112\text{ V})(4.0\text{ A}) \\ &= 448\text{ W}. \end{aligned}$$

31–8 THE ELECTROMAGNETIC PUMP

Magnetic forces acting on conducting fluids provide a convenient means of *pumping* these fluids. An example is found in nuclear reactor design; in some types of reactors, heat is transferred from the reactor core to the point where it is to be utilized, by a circulating flow of liquid metal. (Sodium, lithium, bismuth, and sodium-potassium alloys have been used.) The flow is maintained by an *electromagnetic pump,* one form of which is shown schematically in Fig. 31–7.

A current is sent transversely through the liquid metal, in a direction at right angles to a transverse magnetic field. The resulting side thrust on the current-carrying metal drives it along the pipe in which it is contained. The system is completely sealed and the only moving part is the metal itself.

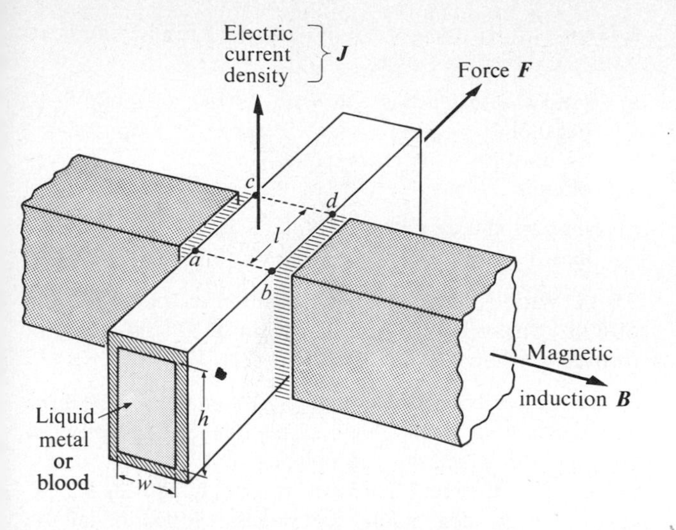

Fig. 31–7 *An electromagnetic pump.*

Magnetic pumps are also finding a variety of applications in medical technology, particularly for pumping blood. Ordinary mechanical pumps with moving parts can damage blood cells. Blood contains ions, making it an electrical conductor, and magnetic pumping is possible. The pump is again completely sealed, reducing danger of contamination, and there are no moving parts except the blood itself. Magnetic pumps are now in use in some heart-lung machines and artificial kidney machines.

PROBLEMS

31–1 The cube in Fig. 31–8, 0.5 m on a side, is in a uniform magnetic field of 0.6 Wb m^{-2}, parallel to the x-axis. The wire *abcdef* carries a current of 4 A in the direction indicated. Determine the magnitude and direction of the force acting on the segments *ab*, *bc*, *cd*, *de*, and *ef*.

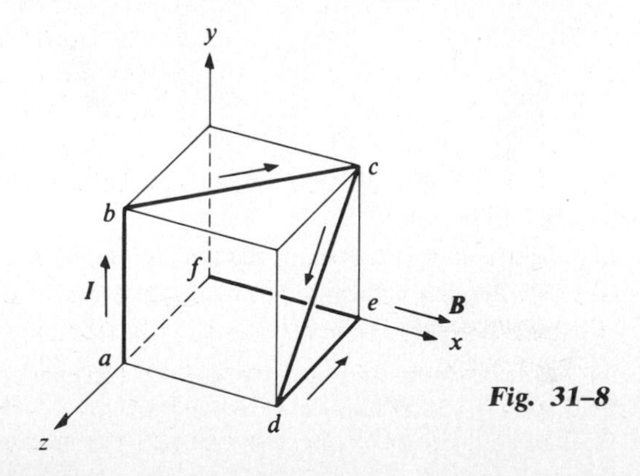

Fig. 31–8

31–2 Figure 31–9 shows a portion of a silver ribbon with z_1 = 2 cm and y_1 = 1 mm, carrying a current of 200 A in the positive x-direction. The ribbon lies in a uniform magnetic field, in the y-direction, of magnetic induction 1.5 Wb m^{-2}. If there are 7.4 × 10^{28} free electrons per m^3, find (a) the drift velocity of the electrons in the x-direction, (b) the magnitude and direction of the electric field in the z-direction due to the Hall effect, (c) the Hall emf.

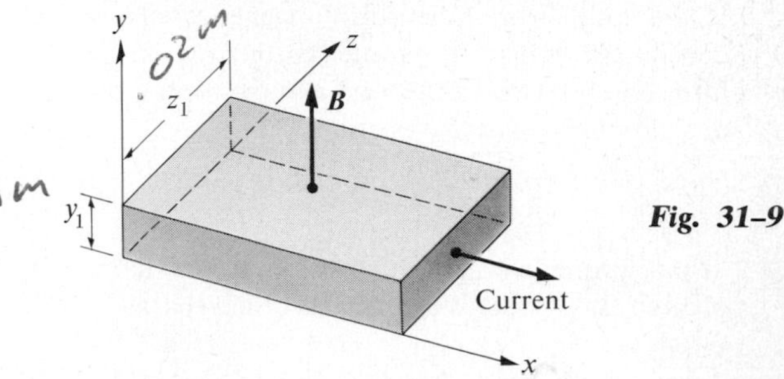

Fig. 31–9

31–3 Let Fig. 31–9 represent a strip of copper of the same dimensions as those of the silver ribbon of the preceding problem. When the magnetic induction is 5 Wb m^{-2} and the current is 100 A, the Hall emf is found to be 45.4 μV. What is the concentration of free electrons?

31–4 The plane of a rectangular loop of wire 5 cm × 8 cm is parallel to a magnetic field whose flux density is 0.15 Wb m^{-2}.

a) If the loop carries a current of 10 A, what torque acts on it?

b) What is the maximum torque that can be obtained with the same total length of wire carrying the same current in this magnetic field?

31–5 An electromagnet produces a magnetic field of 1.2 T in a cylindrical region of radius 5 cm between its poles. A wire carrying a current of 20 A passes through this region, through the axis of the cylinder and perpendicular to it. What force is exerted on the wire?

31–6 A wire 0.5 m long lies along the y-axis and carries a current of 10 A in the $+y$ direction. The magnetic field is uniform and has components B_x = 0.3 T, B_y = − 1.2 T, and B_z = 0.5 T.

a) Find the components of force on the wire.

b) What is the magnitude of the total force on the wire?

31–7 A horizontal rod 0.2 m long is mounted on a balance and carries a current. In the vicinity of the wire is a uniform horizontal magnetic field of magnitude 0.05 T, perpendicular to the wire. The force on the rod is measured by the balance and is found to be 0.24 N. What is the current?

31–8 In Problem 31–7, suppose the magnetic field is horizontal but makes an angle of 30° with the rod. What is the current in the rod?

31–9 The rectangular loop in Fig. 31–10 is pivoted about the y-axis and carries a current of 10 A in the direction indicated.

a) If the loop is in a uniform magnetic field of flux density 0.2 Wb m^{-2}, parallel to the x-axis, find the force on each side of the loop and the torque required to hold the loop in the position shown.

b) Same as (a) except that the field is parallel to the z-axis.

c) What torque would be required if the loop were pivoted about an axis through its center, parallel to the y-axis?

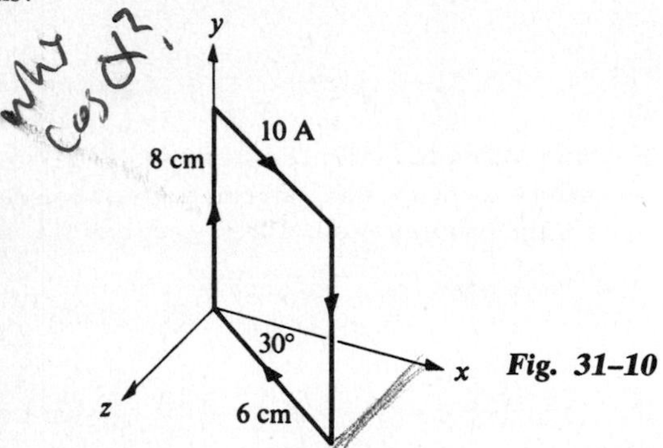

Fig. 31–10

31–10 The rectangular loop of wire in Fig. 31–11 has a mass of 0.1 g per centimeter of length, and is pivoted about side ab as a frictionless axis. The current in the wire is 10 A in the direction shown.

a) Find the magnitude and sense of the magnetic field, parallel to the y-axis, that will cause the loop to swing up until its plane makes an angle of 30° with the yz-plane.

b) Discuss the case where the field is parallel to the x-axis.

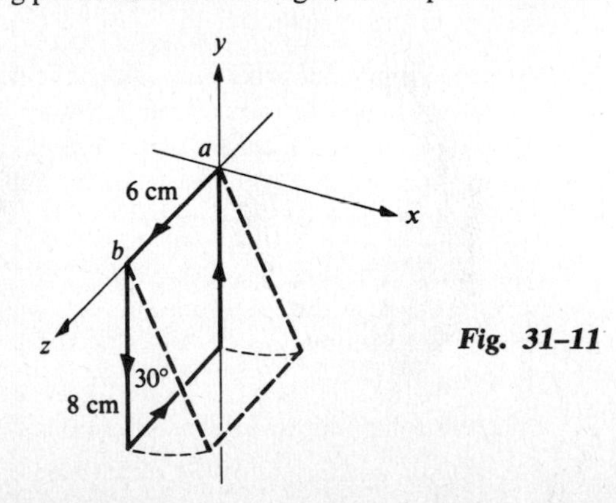

Fig. 31–11

31–11 What is the maximum torque on a coil 5 × 12 cm, of 600 turns, when carrying a current of 10^{-5} A in a uniform field where the flux density is 0.10 Wb m^{-2}?

31–12 The coil of a pivoted-coil galvanometer has 50 turns and encloses an area of 6 cm^2. The magnetic induction in the region in which the coil swings is 0.01 Wb m^{-2} and is radial. The torsional constant of the hairsprings is $k' = 0.1$ dyn cm deg^{-1}. Find the angular deflection of the coil for a current of 1 mA.

31–13 A circular coil of wire 8 cm in diameter has 12 turns and carries a current of 5 A. The coil is in a field where the magnetic induction is 0.60 Wb m^{-2}.

a) What is the maximum torque on the coil?

b) In what position would the torque be one-half as great as in (a)?

31–14 A 1-μF capacitor is connected to a dry cell of emf 1.5 V and is then discharged through a ballistic galvanometer whose coil of 20 turns has an area of 4 cm^2 and swings in a magnetic field of 0.1 Wb m^{-2}. If the maximum angle of deflection is $\frac{1}{2}$ rad, what is the product of the torsion constant of the suspension and the moment of inertia of the coil?

31–15 A potential difference which varies with the time according to the equation

$$v = V \sin \frac{2\pi}{T} t$$

is established across a ballistic galvanometer coil of resistance R for a time interval of half a cycle, $T/2$.

a) What quantity of charge was transferred in the coil in this time interval?

b) Derive an expression for the maximum angle of deflection in terms of the physical constants of the galvanometer.

31–16 In a shunt-wound dc motor, the resistance of the field coils is 150 Ω and the resistance of the armature is 2 Ω. When a difference of potential of 120 V is applied to the brushes, and the motor is running at full speed delivering mechanical power, the current supplied to it is 4.5 A.

a) What is the current in the field coils?

b) What is the current in the armature?

c) What is the emf developed by the motor?

d) How much mechanical power is developed by this motor?

31–17 Figure 31–12 is a diagram of a shunt-wound dc motor, operating from 120-V dc mains. The resistance of the field windings, R_f, is 240 Ω. The resistance of the armature, R_a, is 3 Ω. When the motor is running the armature develops an emf $\mathcal{E}$. The motor draws a current of 4.5 A from the line. Compute (a) the field current, (b) the armature current, (c) the emf $\mathcal{E}$, (d) the rate of development of heat in the field windings, (e) the rate of development of heat in the armature, (f) the power input to the motor, and (g) the efficiency of the motor, if friction and windage losses amount to 50 W.

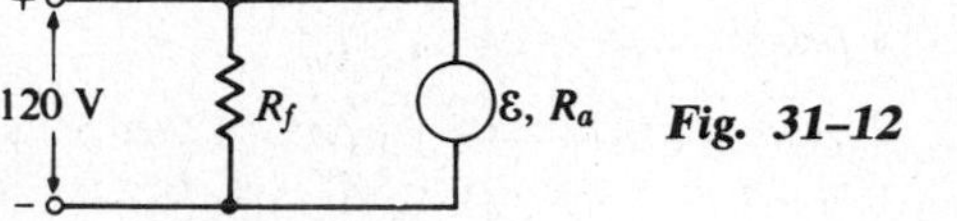

Fig. 31–12

31–18 A horizontal tube of rectangular cross section (height h, width w) is placed at right angles to a uniform magnetic field of induction B, so that a length ℓ is in the field. (See Fig. 31–7) The tube is filled with liquid sodium and an electric current of density J is maintained in the third mutually perpendicular direction.

a) Show that the difference of pressure between a point in the liquid on a vertical plane through ab (Fig. 31–7) and a point in the liquid on another vertical plane through cd, under conditions in which the liquid is prevented from flowing, is

$$\Delta p = J\ell B.$$

b) What current would be needed to provide a pressure difference between these two points of 1 atm if $B = 1 \text{ Wb m}^{-2}$ and $\ell = 0.1$ m?

Chapter 32

Magnetic Field of a Current

32–1 FIELD OF A MOVING CHARGE

In Chapter 30 we discussed one aspect of the magnetic interaction of moving charges; we took the existence of magnetic fields as a given fact and studied the *forces* exerted by the magnetic field on moving charges and on currents in conductors. We now return to the second aspect of this interaction, the problem of *calculating* the magnetic field established by moving charges or by currents in a conductor.

The first recorded observations of magnetic fields set up by currents were those of Oersted, who discovered that a pivoted compass needle, beneath a wire in which there was a current, set itself with its long axis perpendicular to the wire. Later experiments by Biot and Savart, and by Ampère, led to a relation by means of which we can compute the magnetic field at any point of space around a circuit in which there is a current.

We consider first the $\boldsymbol{B}$ field produced by a single moving point charge. Figure 32–1 shows a positive point charge q, moving relative to the observer with a velocity $\boldsymbol{v}$. The $\boldsymbol{B}$-vectors show the directions and relative magnitudes of the magnetic field set up by the charge at a number of points. At every point, such as the point P, the $\boldsymbol{B}$-vector lies in a plane perpendicular to the velocity $\boldsymbol{v}$, and is itself perpendicular to the plane determined by $\boldsymbol{v}$ and the vector $\boldsymbol{r}$ from the charge to the point. The $\boldsymbol{B}$-vector at any point is found experimentally to be

$$\boxed{\boldsymbol{B} = k' \frac{q\boldsymbol{v} \times \hat{\boldsymbol{r}}}{r^2},} \qquad (32\text{–}1)$$

where k' is a constant analogous to the constant k in the expression for the electric vector $\boldsymbol{E}$ set up by a

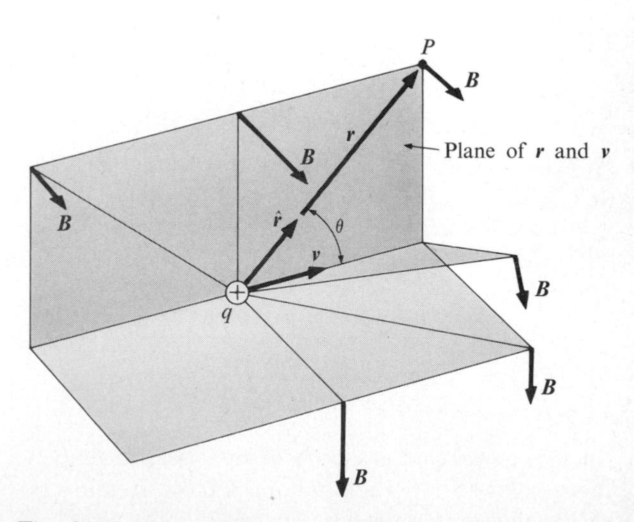

Fig. 32–1 *Magnetic field vectors due to a moving positive point charge q.*

charge q, $\boldsymbol{r}$ is the vector from the charge q to the point, and $\hat{\boldsymbol{r}}$ is a *unit* vector in the direction of $\boldsymbol{r}$. That is,

$$\hat{\boldsymbol{r}} = \boldsymbol{r}/r.$$

If the charge q is positive, the direction of $\boldsymbol{B}$ is the same as that of the vector product $\boldsymbol{v} \times \hat{\boldsymbol{r}}$. If q is negative, the direction of $\boldsymbol{B}$ is opposite to that of $\boldsymbol{v} \times \hat{\boldsymbol{r}}$. The magnitude of $\boldsymbol{B}$ is

$$B = k' \frac{qv \sin \theta}{r^2}. \quad (32\text{–}2)$$

The $\boldsymbol{B}$-vector is zero at all points on a line through the charge in the direction of the velocity vector $\boldsymbol{v}$, since $\sin \theta = 0$ at all such points. At a given distance r from the charge, the $\boldsymbol{B}$-vector is a maximum in the plane through the charge perpendicular to $\boldsymbol{v}$, since $\theta = 90°$ and $\sin \theta = 1$ at all points in this plane.

Unlike the $\boldsymbol{E}$-field of the charge, which is radial, the lines of induction of the $\boldsymbol{B}$-field are *circles* in planes perpendicular to the velocity vector, with centers on a line through this vector.

Since the unit of $\boldsymbol{B}$ is $1\ \mathrm{N\ A^{-1}\ m^{-1}} = 1\ \mathrm{N\ s\ C^{-1}\ m^{-1}}$, the unit of the magnetic constant k' is

$$1\ \mathrm{N\ s^2\ C^{-2}} = 1\ \mathrm{N\ A^{-2}} = 1\ \mathrm{Wb\ A^{-1}\ m^{-1}}$$
$$= 1\ \mathrm{T\ A^{-1}\ m}.$$

In the mks system, the numerical value of k' is arbitrarily assigned to be *exactly* 10^{-7}. Thus

$$k' = 10^{-7}\ \mathrm{N\ s^2\ C^{-2}} = 10^{-7}\ \mathrm{N\ A^{-2}}$$
$$= 10^{-7}\ \mathrm{Wb\ A^{-1}\ m^{-1}} \quad \text{(exactly)}. \quad (32\text{–}3)$$

The reason for this choice of the numerical value of k' will be explained later in the chapter.

It will be recalled that the electrical constant $k = 8.98755 \times 10^9\ \mathrm{N\ m^2\ C^{-2}}$. The ratio k/k' is therefore

$$\frac{k}{k'} = \frac{8.98755 \times 10^9\ \mathrm{N\ m^2\ C^{-2}}}{10^{-7}\ \mathrm{N\ s^2\ C^{-2}}}$$
$$= 8.98755 \times 10^{16}\ \mathrm{m^2\ s^{-2}},$$

which is equal to the square of the speed of light, c. This suggests that there may be a close relation between electricity, magnetism, and light. We shall return to this question in a later chapter.

We can now write the expression for the magnetic force between two point charges, both of which are in motion relative to an observer. Thus a charge q', moving with velocity $\boldsymbol{v}'$ in a magnetic field $\boldsymbol{B}$, experiences a force

$$\boldsymbol{F} = q'\boldsymbol{v}' \times \boldsymbol{B},$$

and if the field $\boldsymbol{B}$ is set up by a charge q moving with velocity $\boldsymbol{v}$,

$$\boldsymbol{F} = k' \frac{qq'\boldsymbol{v}' \times (\boldsymbol{v} \times \hat{\boldsymbol{r}})}{r^2}, \quad (32\text{–}4)$$

which corresponds to Coulomb's law for the electrical force between the charges.

32–2 MAGNETIC FIELD OF A CURRENT ELEMENT. THE BIOT LAW

The magnetic field set up at any point by the current in a circuit is the resultant (vector sum) of the fields due to all of the moving charges in the circuit. The circuit is to be divided, in imagination, into short elements of length $d\ell$, one of which is shown in Fig. 32–2. The volume of the element equals $A\ d\ell$, where A is its cross-sectional area. If there are n charge carriers per unit volume, each of charge q, the total moving charge dQ in the element is

$$dQ = nqA\ d\ell.$$

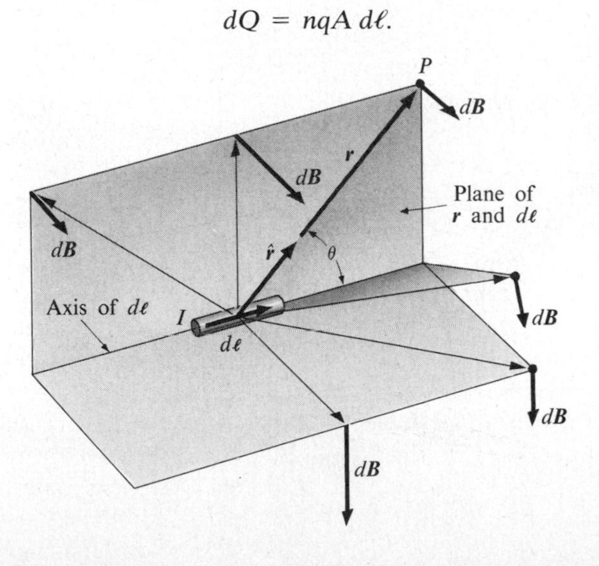

Fig. 32–2 *Magnetic field vectors due to a current element.*

The moving charges are therefore equivalent to a single charge dQ, traveling with the *drift* velocity $\mathbf{v}$. (Fields due to the *random* velocities of the carriers will, on the average, cancel out at every point.) From Eq. (32-2), the magnitude of dB at any point is

$$dB = k' \frac{dQ\, v \sin\theta}{r^2} = k' \frac{nqvA\, d\ell \sin\theta}{r^2}.$$

But $(nqvA)$ equals the current I in the element, so

$$dB = k' \frac{I\, d\ell \sin\theta}{r^2}, \qquad (32\text{-}5)$$

or in vector form,

$$d\mathbf{B} = k' \frac{I\, d\boldsymbol{\ell} \times \hat{\mathbf{r}}}{r^2}, \qquad (32\text{-}6)$$

where $d\boldsymbol{\ell}$ is a vector of length $d\ell$.

As shown in Fig. 32-2, the field vectors $d\mathbf{B}$ are exactly like those set up by a finite positive charge Q, moving in the direction of the drift velocity $\mathbf{v}$.

The resultant flux density $\mathbf{B}$ at any point in space, due to a complete circuit, is the *vector integral* of the values of $d\mathbf{B}$ due to all elements of the circuit. Thus

$$\boxed{\mathbf{B} = k' \int \frac{I\, d\boldsymbol{\ell} \times \hat{\mathbf{r}}}{r^2}.} \qquad (32\text{-}7)$$

It should be pointed out that it is impossible to verify Eq. (32-6) experimentally, since one can never obtain an isolated element of a current-carrying circuit. Only the *resultant* field $\mathbf{B}$, given by the integral in Eq. (32-7), can be measured experimentally. Equation (32-6) was *deduced* originally by the French physicist Jean Biot, in 1820, from experimental studies of the magnetic fields around circuits of various shapes. It is known as the *Biot law* (pronounced "bee-OH").

If there is matter in the space around a circuit, the field at a point will be due not entirely to currents in conductors, but in part to the *magnetization* of this matter. The magnetic properties of matter are discussed more fully in Chapter 35. However, unless iron or some other ferromagnetic material is present, this effect is so small that, while strictly speaking Eq. (32-7) holds only for conductors in vacuum, as a practical matter it can be used without correction for conductors in air, or in the vicinity of any nonferromagnetic material.

32-3 MAGNETIC FIELD OF A LONG STRAIGHT CONDUCTOR

Let us use the Biot law to compute the magnetic field $\mathbf{B}$ at point P in Fig. 32-3 due to a long straight conductor carrying a current I. It will be convenient to change the notation (as we did when evaluating the

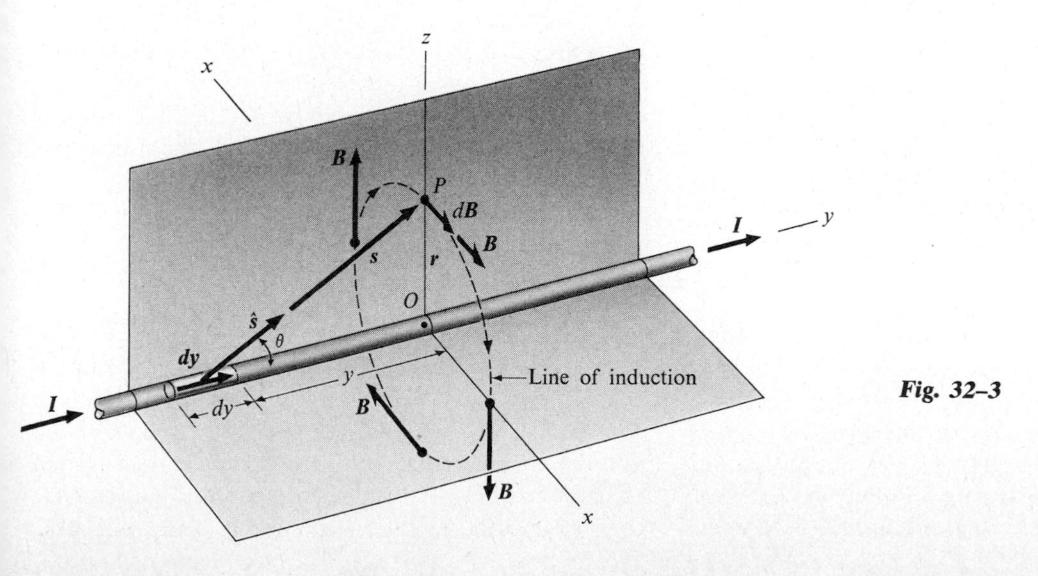

Fig. 32-3

electric field around the charged wire in Fig. 25–7) and let r represent the radial distance of point P from the wire and $\mathbf{s}$ the vector from a current element to P. Then in terms of the symbols in Fig. 32–3,

$$\mathbf{B} = k'I \int \frac{d\mathbf{y} \times \hat{\mathbf{s}}}{s^2}.$$

This is a vector integral, but since the field $d\mathbf{B}$ set up at P by any other element of the conductor is *parallel* to the vector $d\mathbf{B}$ in the diagram, the resultant field is the *algebraic* sum or the sum of the magnitudes dB. Then, since $\hat{\mathbf{s}}$ is a unit vector,

$$B = k'I \int_{-\infty}^{+\infty} \frac{\sin\theta}{s^2}\, dy.$$

Simplification results if θ is chosen as the integration variable. It will be seen from the diagram that

$$s = r \csc\theta, \qquad y = -r\cot\theta.$$

Hence

$$dy = r\csc^2\theta\, d\theta$$

and

$$B = k'\frac{I}{r}\int_0^{\pi} \sin\theta\, d\theta.$$

The integration limits are from 0 to π, since the wire is assumed to be very long. (The remainder of the circuit is considered to be so far away that any contribution it makes to the field can be neglected.) Carrying out the integration, we find

$$\boxed{B = 2k'\frac{I}{r} \quad \text{(long straight wire).}} \qquad (32\text{–}8)$$

This relation was deduced from experimental observations by Biot and Savart before the differential form, Eq. (32–6), had been discovered. It is called the *Biot–Savart law*, although this name is sometimes applied also to Eqs. (32–6) and (32–7).

A portion of the magnetic field around a long, straight conductor is shown in the cutaway view of Fig. 32–4. Unlike the *electric* field around a charged

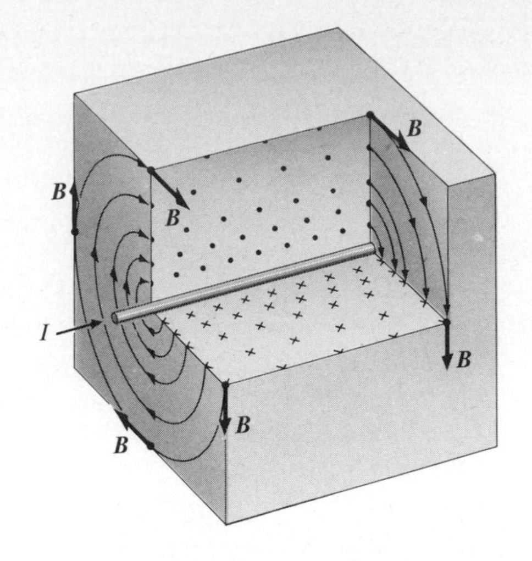

Fig. 32–4 *Magnetic field around a long straight conductor.*

wire, which is radial, the lines of magnetic induction are *circles* concentric with the wire and lying in planes perpendicular to it. It will also be noted that each line of induction is a *closed* line, and that, in this respect, lines of induction differ from the lines of force in an electric field, which terminate on positive or negative charges. This property of lines of induction is true whatever the geometry of the circuit setting up the field—lines of induction never have endpoints.

If a closed surface is constructed in a magnetic field, the number of lines of induction that emerge from the surface must equal the number that enter. We have shown that the number of lines of induction crossing a surface is proportional to the flux Φ across the surface. Hence in a magnetic field, the flux across any *closed* surface is zero, or

$$\oint B_{\perp} dA = \oint \mathbf{B}\cdot d\mathbf{A} = 0. \qquad (32\text{–}9)$$

This should be compared with Gauss's law for electrostatic fields, in which the surface integral of $\mathbf{E}$ over a closed surface equals $1/\epsilon_0$ times the enclosed charge. Thus Eq. (32–9) can be interpreted as saying that there is no such thing as "magnetic charge" to act as a source of $\mathbf{B}$. The sources of $\mathbf{B}$ are moving electric charges, as outlined above.

32–4 FORCE BETWEEN PARALLEL CONDUCTORS. THE AMPERE AND THE COULOMB

Figure 32–5 shows a portion of two long straight parallel conductors separated by a distance r and carrying currents I and I', respectively, in the same direction. Since each conductor lies in the magnetic field set up by the other, each will experience a force. The diagram shows some of the lines of induction set up by the current in the *lower* conductor. From Eq. (32–8), the magnitude of $\boldsymbol{B}$ at the upper conductor is

$$B = 2k'\frac{I}{r}.$$

From Eq. (31–2), the force on a length ℓ of the upper conductor is

$$F = I'\ell B = 2k'\ell\frac{II'}{r}$$

and the force *per unit length* is therefore

$$\boxed{\frac{F}{\ell} = 2k'\frac{II'}{r}.} \qquad (32\text{–}10)$$

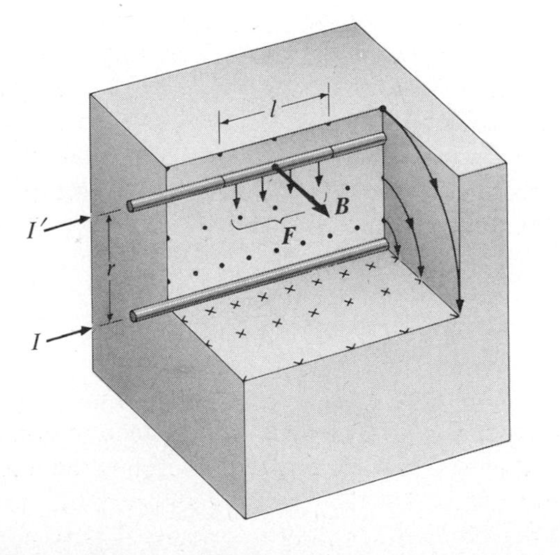

Fig. 32–5 *Parallel conductors carrying currents in the same direction attract each other.*

The direction of the force on the upper conductor is downward. There is an equal and opposite force per unit length on the lower conductor, as may be seen by considering the field set up by the upper conductor. Hence the conductors *attract* each other.

If the direction of either current is reversed, the forces reverse also. Parallel conductors carrying currents in *opposite* directions *repel* each other.

The fact that two straight parallel conductors exert forces of attraction or repulsion on each other is made the basis of the definition of the ampere in the mks system. The ampere is defined as follows:

> *One ampere is that unvarying current which, if present in each of two parallel conductors of infinite length and one meter apart in empty space, causes each conductor to experience a force of exactly* 2×10^{-7} *newton per meter of length.*

It follows from this and the preceding equation that by definition the constant k' is *exactly* 10^{-7} N A^{-2}.

From the definition above, the ampere can be established, in principle, with the help of a meter stick and a spring balance. For the practical standardization of the ampere, coils of wire are used instead of straight wires and their separation is made only a few centimeters. The complete instrument, which is capable of measuring currents with a high degree of precision, is called a *current balance.*

Having defined the ampere, we can now define the coulomb as *the quantity of charge that in one second crosses a section of a circuit in which there is a constant current of one ampere.*

Mutual forces of attraction exist not only between *wires* carrying currents in the same direction, but between each of the longitudinal elements into which a single current-carrying conductor may be subdivided. If the conductor is a liquid or an ionized gas (a plasma), these forces result in a constriction of the conductor as if its surface were acted on by an external, inward, pressure force. The constriction of the conductor is called the *pinch effect,* and attempts are being made to utilize the high temperature produced by the pinch effect in a plasma to bring about nuclear fusion.

32–5 MAGNETIC FIELD OF A CIRCULAR TURN

In many devices in which a current is used to establish a magnetic field, as in an electromagnet or a transformer, the wire carrying the current is wound into a coil of some sort. We therefore consider next the magnetic field set up by a single circular turn of wire carrying a current.

Figure 32–6 shows a circular turn of radius R carrying a current I and lying in the xz-plane. Point P is on the axis of the turn, at a distance y from its center O, and r is the distance from an element of the turn of length $d\ell$ to point P. The plane determined by $d\boldsymbol{\ell}$ and $\boldsymbol{r}$ is shaded, and the field $d\boldsymbol{B}$ set up at P by the current in the element $d\boldsymbol{\ell}$ is at right angles to this plane and lies in the yz-plane. The angle θ between $d\boldsymbol{\ell}$ and $\boldsymbol{r}$ is 90°, so in terms of the symbols used in this diagram, the Biot law becomes

$$dB = k'I\frac{d\ell}{r^2}.$$

Let α represent the angle between r and the y-axis. The vector $d\boldsymbol{B}$ can then be resolved into a component $dB \sin \alpha$ along the y-axis and a component $dB \cos \alpha$ at right angles to it. It will be seen by symmetry that each element contributes an equal component $dB \sin \alpha$, but that the components $dB \cos \alpha$ set up by diametrically opposite elements will cancel.

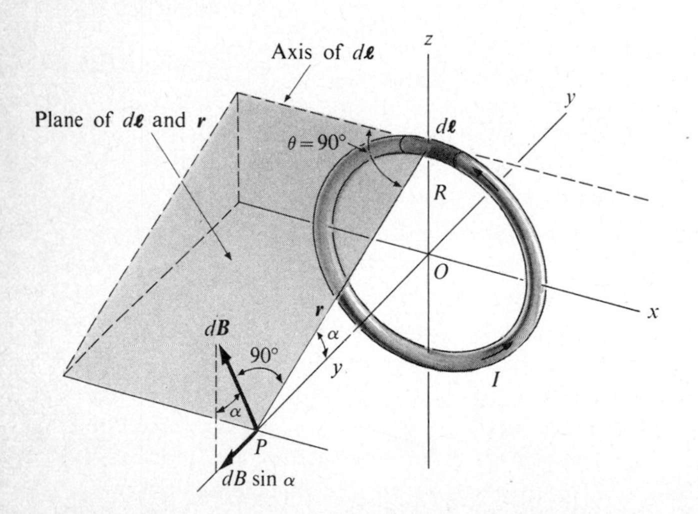

Fig. 32–6 Magnetic field set up by an element $d\ell$ of a circular turn of wire carrying a current.

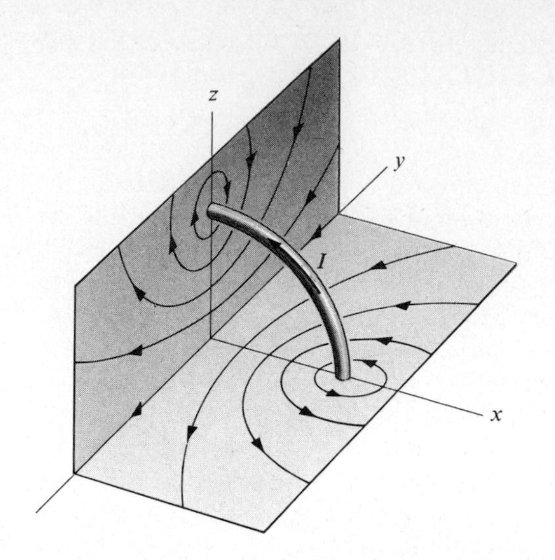

Fig. 32–7 Lines of induction surrounding a circular turn.

Hence the resultant field is along the y-axis, and is found by integration of the components $dB \sin \alpha$:

$$B = \oint dB \sin \alpha = k'\frac{I \sin \alpha}{r^2}\oint d\ell, \quad (32\text{–}11)$$

since $\sin \alpha$ and r are constants and may be taken outside the integral sign. The integral is merely the circumference of the turn, $2\pi R$, so finally

$$\boxed{B = 2\pi k'\frac{IR \sin \alpha}{r^2}.} \quad (32\text{–}12)$$

At the center of the turn, $\alpha = 90°$ and $r = R$, so at the center

$$B = 2\pi k'\frac{I}{R}. \quad (32\text{–}13)$$

If instead of a single turn, as in Fig. 32–7, we have a coil of N closely spaced turns all of essentially the same radius, each turn contributes equally to the field and Eq. (32–13) becomes

$$\boxed{B = 2\pi k'\frac{NI}{R}.}$$

Since $\sin\alpha = R/r$, Eq. (32–12) can also be written

$$B = 2k'\frac{I(\pi R^2)}{r^3}.$$

But πR^2 is the area of the loop and $I(\pi R^2)$ is its magnetic moment m. Hence

$$B = k'\frac{2m}{r^3}.$$

Some of the lines of induction surrounding a circular turn are shown in Fig. 32–7.

It was pointed out in Chapter 31 that a loop of wire of any shape and carrying a current is acted on by a torque when placed in an external magnetic field. The position of stable equilibrium is one in which the plane of the loop is perpendicular to the external field. Now we see that this position is such that *within the area enclosed by the loop, the loop's own **B**-field is in the same direction as that of the external field.* In other words, a loop, if free to turn, will set itself in such a plane that the flux passing through the area enclosed by it has its maximum possible value. This is found to be true in all instances and is a useful general principle. For example, if a current is sent through an irregular loop of *flexible* wire in an external magnetic field, the loop will assume a circular form with its plane perpendicular to the field and with its own flux adding to that of the field. The same conclusion can, of course, be drawn by analyzing the side-thrusts on the elements of the conductor.

32–6 AMPÈRE'S LAW

Ampère's law is a useful relation that is analogous to Gauss's law. The latter, it will be recalled, is a relation between the integral of the normal component of electric field over a closed surface and the net electric charge enclosed by the surface. Ampère's law is a relation between the *line integral* of the *tangential* component of *magnetic* field around a closed curve and the net current through the area bounded by the curve.

We consider first a long, straight conductor carrying a current I, passing through the center of a circle of radius r in a plane perpendicular to the conductor. At every point on the circle, $\mathbf{B}$ has a magnitude $2k'I/r$, and it is tangent to the circle at each point. Thus the value of the line integral of $B_\parallel$ around this circle is simply

$$\oint \mathbf{B}\cdot d\mathbf{l} = B(2\pi r) = 4\pi k'I. \qquad (32\text{–}14)$$

Thus the line integral of $B_\parallel$ equals $4\pi k'$ times the current through the area bounded by the circle.

This result may also be derived for a more general integration path, such as that shown in Fig. 32–8. The $\mathbf{B}$ field at any point is in the plane of the diagram and at right angles to the radius $\mathbf{r}$ from the wire to the point. The magnitude of $\mathbf{B}$ is

$$B = 2k'\frac{I}{r}.$$

The $\mathbf{B}$-vector makes an angle θ with an element ds of a closed path encircling the wire and the component of $\mathbf{B}$ in the direction of ds is $B_\parallel = B\cos\theta$. From the small right "triangle" of which ds is the hypotenuse, we see that

$$r\,d\phi = ds\cos\theta,$$

or

$$ds = r\,d\phi/\cos\theta.$$

Therefore

$$\begin{aligned} B_\parallel\,ds &= \mathbf{B}\cdot d\mathbf{s} = B\cos\theta\,ds \\ &= \left(2k'\frac{I}{r}\right)(\cos\theta)\frac{r\,d\phi}{\cos\theta} \\ &= 2k'I\,d\phi. \end{aligned}$$

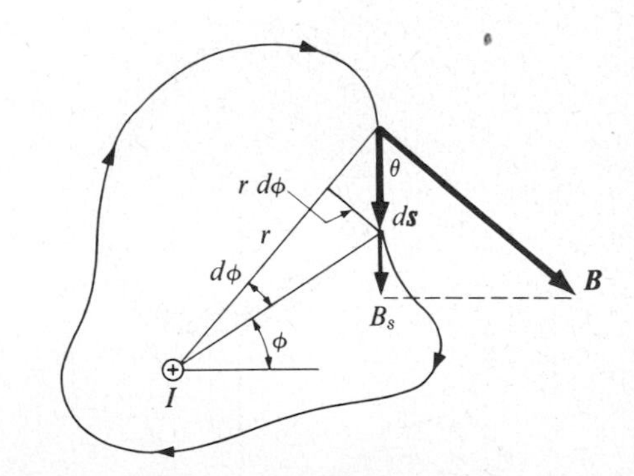

Fig. 32–8 *The line integral of **B** around a closed path equals $4\pi k'$ times the net current through the area bounded by the path.*

The line integral of $\boldsymbol{B}$ around the closed path is

$$\oint \boldsymbol{B} \cdot d\boldsymbol{s} = 2k'I \oint d\phi,$$

and since the angle ϕ increases by 2π as we go once around the path,

$$\oint \boldsymbol{B} \cdot d\boldsymbol{s} = 4\pi k'I. \tag{32–15}$$

The line integral does not depend on the shape of the path, or on the position of the wire within it. If the current in the wire were opposite to that shown, the integral would have the opposite sign. Hence if any number of long straight conductors pass through the surface bounded by the path, the line integral equals $4\pi k'$ times the *algebraic sum* of the currents.

If a closed path does not encircle the wire (or if a wire lies outside the path), the line integral of the $\boldsymbol{B}$-field of that wire is zero, because the angle ϕ then has the same value at the start and finish of any round trip. Hence if there are other conductors present that do not pass through a given path, they may contribute to the value of $\boldsymbol{B}$ at every point but the line integrals of their fields are zero.

It follows that if we interpret I in Eq. (32–15) to mean the algebraic sum of the currents across the area bounded by a closed path, this equation implies all of the statements above and is the analytic form of Ampère's law:

> *The line integral of the magnetic induction* $\boldsymbol{B}$ *around any closed path is equal to* $4\pi k'$ *times the net current across the area bounded by the path.*

Although derived only for the special case of the field of a number of long straight parallel conductors, the law is true for conductors and paths of any shape. The general derivation is no different in principle from that above, but it is complicated geometrically and will not be given.

The appearance of the factor 4π in Eq. (32–15) suggests that the form of certain magnetic equations can be simplified if we define a new constant μ_0 by the equation

$$4\pi k' = \mu_0.$$

Since k' is exactly 10^{-7} Wb A^{-1} m^{-1},

$$\mu_0 = 4\pi \times 10^{-7}\ \text{Wb A}^{-1}\ \text{m}^{-1} = 12.57 \times 10^{-7}\ \text{Wb A}^{-1}\ \text{m}^{-1}.$$

Hence Ampère's law can be written more compactly as

$$\oint \boldsymbol{B} \cdot d\boldsymbol{s} = \mu_0 I. \tag{32–16}$$

In terms of μ_0, Eq. (32–7) becomes

$$\boldsymbol{B} = \frac{\mu_0}{4\pi} \int \frac{I\, d\boldsymbol{\ell} \times \hat{\boldsymbol{r}}}{r^2}. \tag{32–17}$$

The field due to a long straight conductor, Eq. (32–8), becomes

$$B = \frac{\mu_0}{2\pi}\left(\frac{I}{a}\right), \tag{32–18}$$

the force per unit length between two long parallel conductors, Eq. (32–10), is

$$\frac{F}{\ell} = \frac{\mu_0}{4\pi}\left(\frac{2II'}{a}\right), \tag{32–19}$$

and the magnetic field of a circular turn, Eq. (32–12), is

$$B = \frac{\mu_0}{2} \frac{IR \sin \alpha}{r^2}. \tag{32–20}$$

32–7 APPLICATIONS OF AMPÈRE'S LAW

In some instances, *symmetry* considerations make it possible to use Ampère's law to compute the magnetic flux density. As was the case with similar applications of Gauss's law, we must be able to replace the *integral* in Eq. (32–16) by a *product* to obtain an *algebraic* equation that can be solved for B. We consider three examples.

1. Field of a solenoid A solenoid is constructed by winding wire in a helix around the surface of a cylindrical form, usually of circular cross section. The turns of the winding are ordinarily closely spaced and may consist of one or more layers. For simplicity, we have represented a solenoid in Fig. 32–9 by a relatively small number of circular turns, each carrying a current I. The resultant field at any point is the vector sum of the $\boldsymbol{B}$-vectors due to the individual turns. The diagram shows the lines of induction in the xy- and yz-planes. Exact calculations show that for a long, closely-wound solenoid, half of the lines passing through a cross section at the center

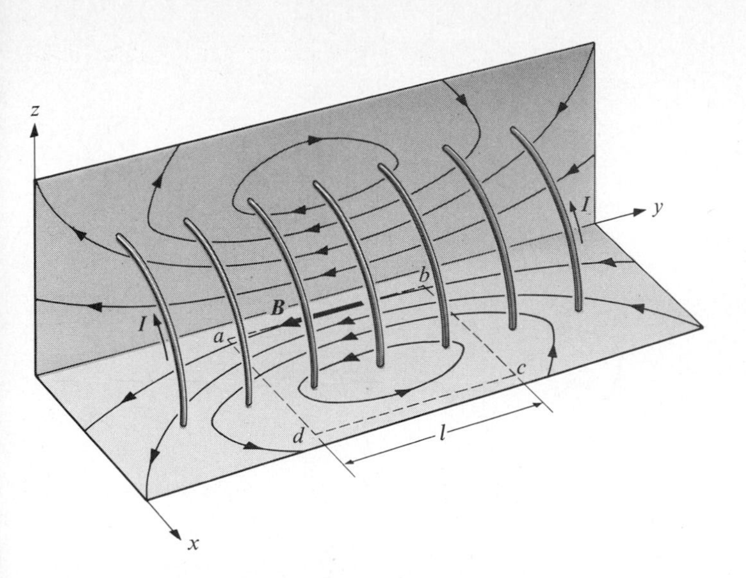

Fig. 32–9 *Lines of induction surrounding a solenoid. The dotted rectangle abcd is used to compute the flux density **B** in the solenoid from Ampère's law.*

emerge from the ends and half "leak out" through the windings between center and end.

If the length of the solenoid is large compared with its cross-sectional diameter, the *internal* field near its center is very nearly uniform and parallel to the axis, and the *external* field near the center is very small. The internal field at or near the center can then be found by use of Ampère's law.

We select as a closed path the dotted rectangle *abcd* in Fig. 32–9. Side *ab*, of length ℓ, is parallel to the axis of the solenoid. Sides *bc* and *da* are to be taken very long so that side *cd* is far from the solenoid and the field at this side is negligibly small.

By symmetry, the $\boldsymbol{B}$ field along side *ab* is parallel to this side and is constant, so that for this side $B_{\|} = B$ and

$$\oint \boldsymbol{B} \cdot d\boldsymbol{\ell} = B\ell.$$

Along sides *bc* and *da*, $B_{\|} = 0$ since $\boldsymbol{B}$ is perpendicular to these sides; and along side *cd*, $B_{\|} = 0$ also since $B = 0$. The sum around the entire closed path therefore reduces to $B\ell$.

Let n be the number of turns *per unit length* in the windings. The number of turns in length ℓ is then $n\ell$. Each of these turns passes once through the rectangle *abcd* and carries a current I, where I is the current in the windings. The total current through the rectangle is then $n\ell I$, and from Ampères law,

$$B\ell = \mu_0 n\ell I,$$

$$\boxed{B = \mu_0 nI \qquad \text{(solenoid).}} \qquad (32\text{–}21)$$

Since side *ab* need not lie on the axis of the solenoid, the field is uniform over the entire cross section.

2. Field of a toroid Figure 32–10(a) represents a toroid, wound with wire carrying a current I. The dotted lines in Fig. 32–10(b) are a number of paths to which we wish to apply Ampère's law. Consider first path 1. By symmetry, if there is any field at all in this region, it will be *tangent* to the path at all points, and $\oint \boldsymbol{B} \cdot d\boldsymbol{\ell}$ will equal the product of B and the circumference of the path. The current through the path, however, is zero, and hence from Ampère's law (since the circumference is *not* zero), the field B must be zero.

Similarly, if there is any field at path 3, it will also be tangent to the path at all points. Each turn of the winding passes *twice* through the area bounded by this path, carrying equal currents in opposite directions. The *net* current through the area is therefore zero, and hence $B = 0$ at all points of the path. *The field of the toroid is therefore confined wholly to the space enclosed by the windings.* The toroid may be

Fig. 32–10 (a) A toroidal winding. (b) Closed paths (dotted circles) used to compute the flux density B set up by a current in a toroidal winding. The field is very nearly zero at all points except those within the space enclosed by the windings.

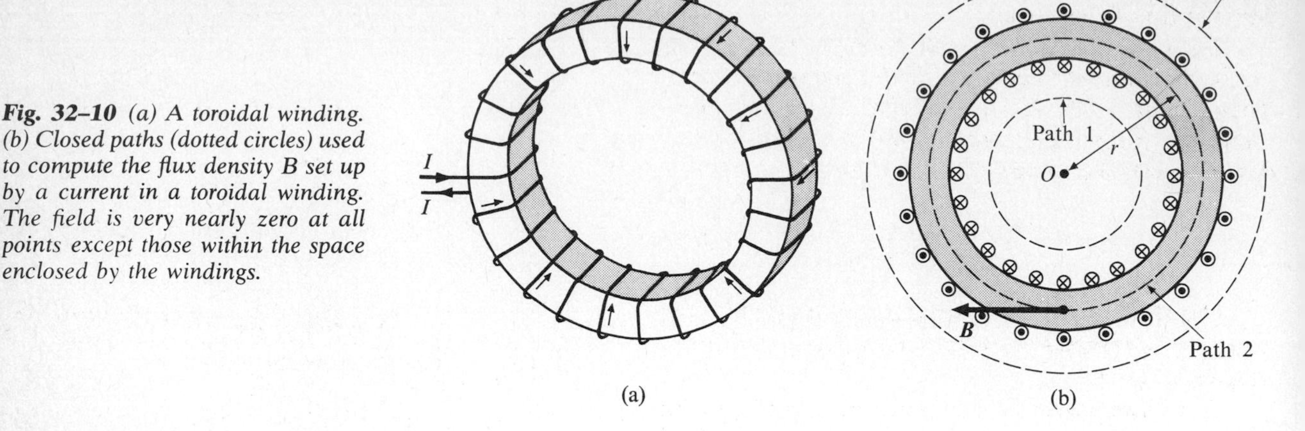

thought of as a solenoid that has been bent into a circle.

Finally, consider path 2, a circle of radius r. Again by symmetry, the $\boldsymbol{B}$ field is tangent to the path, and $\oint \boldsymbol{B} \cdot d\boldsymbol{\ell}$ equals $2\pi rB$. Each turn of the winding passes *once* through the area bounded by path 2, and the total current through the area is NI, where N is the *total* number of turns in the winding. Then, from Ampère's law,

$$2\pi rB = \mu_0 NI,$$

and

$$B = \frac{\mu_0}{2\pi}\left(\frac{NI}{r}\right) \quad \text{(toroid).} \qquad (32\text{–}22)$$

The magnetic field is *not* uniform over a cross section of the core, because the path length ℓ is larger at the outer side of the section than at the inner side. However, if the radial thickness of the core is small compared with the toroid radius r, the field varies only slightly across a section. In that case, considering that $2\pi r$ is the circumferential length of the toroid and that $N/2\pi r$ is the number of turns per unit length n, the field may be written

$$B = \mu_0 nI,$$

just as at the center of a long straight solenoid.

The equations derived above for the field in a closely wound solenoid or toroid are strictly correct only for a winding in *vacuum.* For most practical purposes, however, they can be used for a winding in air, or on a core of any nonferromagnetic material. We shall show in Chapter 35 how they are modified if the core is of iron.

3. Field between parallel plates Figure 32–11 is a sectional view of two long parallel plates of width w. The plate on the left carries a current I toward the reader, and that on the right carries an equal current *away* from the reader.

In the region between the plates and not too near their edges, the $\boldsymbol{B}$-field is uniform. The field

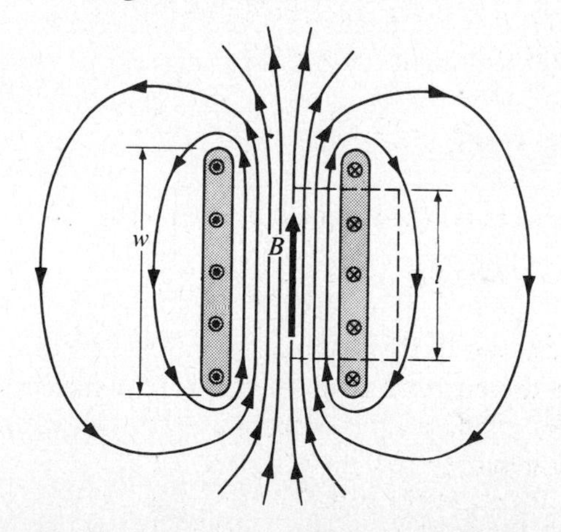

Fig. 32–11

outside the plates is small, and becomes smaller as the width w is increased.

Let us apply Ampère's law to the dotted rectangle, and assume the field outside the plates to be zero. Then $\mathbf{B} \cdot d\ell = B\ell$. If I is the total current in either plate, the current per unit width is I/w and the current through the rectangle is $I\ell/w$. Hence

$$B\ell = \frac{\mu_0 I\ell}{w}, \qquad B = \frac{\mu_0 I}{w}. \tag{32–23}$$

We shall use this relation later, in developing the equation for the velocity of propagation of electromagnetic waves.

It is interesting to compare the nature of the magnetic field between two long, current-carrying plates, with the electric field between two large charged plates. Both fields are approximately uniform, the magnetic lines being parallel to the plates and the electric lines perpendicular. In the magnetic case, $B = \mu_0(I/w)$, whereas in the electric case $E = (1/\epsilon_0)(Q/A)$.

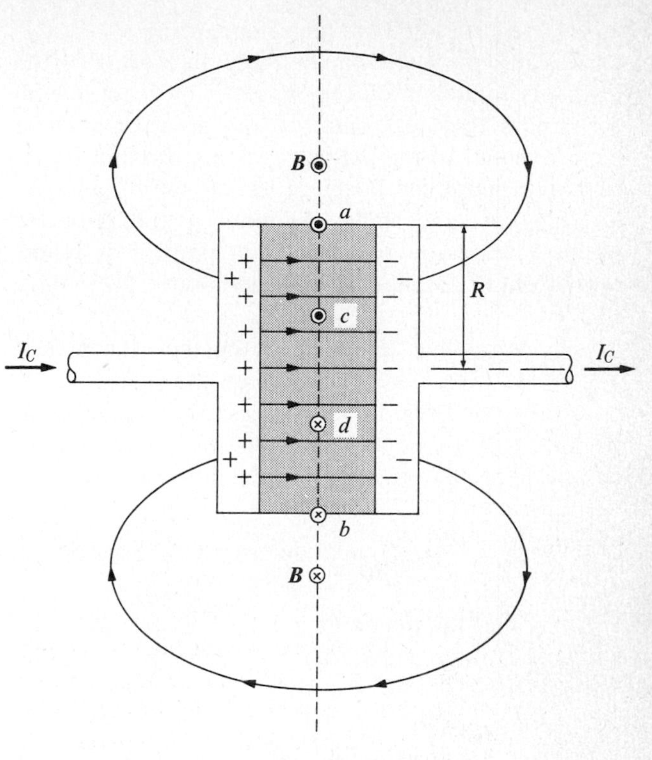

Fig. 32–12

32–8 MAGNETIC FIELDS AND DISPLACEMENT CURRENTS

When the Biot and Savart law was discussed in Section 32–2, it was assumed that the current I in an element $d\ell$ was a *conduction* current. Experiment shows, however, that a *displacement* current I_D (Section 29–8) contributes to a magnetic field in exactly the same way as a conduction current, I_C, so the general form of the Biot law is

$$d\mathbf{B} = \frac{\mu_0}{4\pi} \frac{(I_C + I_D)\, d\ell \times \hat{\mathbf{r}}}{r^2}, \tag{32–24}$$

and the general form of Ampère's law is

$$\oint \mathbf{B} \cdot d = \mu_0(I_C + I_D). \tag{32–25}$$

Note that in the Biot law, the symbols I_C and I_D refer to the currents in an element $d\ell$, while in Ampère's law they refer to the *total* currents across an area bounded by the path of integration.

As an example, Fig. 32–12 is a sectional view of two circular conducting plates of radius R separated by a vacuum (i.e., a parallel-plate capacitor). The diagram shows some of the lines of the $\mathbf{E}$ field, both within and outside the dielectric. Suppose we wish to calculate the $\mathbf{B}$ field at some point in the midplane, shown in the figure as a broken line. In principle, this could be calculated from the Biot law. It would be necessary to subdivide the whole of space (including the wires and the plates) into elementary filaments $d\ell$, find the conduction and displacement currents in each filament, and perform a vector integration over all space. Such a calculation would be extremely complicated, if not impossible. Some idea of the result can be obtained, however, if we use Ampère's law and take advantage of symmetry considerations.

As is evident from the diagram, there is no *conduction* current across the midplane, and the current across this plane is a *displacement* current only. Also, since every line of displacement crosses the plane at some point, the *total* displacement current across the plane is the *displacement* current *out of* the left plate, and hence equals the *conduction* current *into* the plate. Furthermore, by symmetry, the lines of induction of the $\boldsymbol{B}$ field are circles with centers on the axis. Thus, at any point on a circle perpendicular to the axis and passing through points such as a and b, the $\boldsymbol{B}$-vector has the same value and is tangent to the circle. At point a, the $\boldsymbol{B}$-vector points toward the reader and at point b it points away from the reader, as indicated by the symbols $\odot$ and $\otimes$.

If the plates are very close together, the $\boldsymbol{E}$ field is confined to the region between the plates and is uniform in that region. In that case, the displacement current *density* j_D is uniform; the total displacement current is equal to the total conduction current, so we have

$$j_D = \frac{I_C}{\pi R^2}.$$

We may find the magnetic field at points c and d by applying Ampère's law to a circle of radius r passing through these points. The total current enclosed is

$$I_D = \pi r^2 j_D = \left(\frac{r^2}{R^2}\right) I_C.$$

Ampère's law becomes

$$\oint \boldsymbol{B} \cdot d\boldsymbol{\ell} = 2\pi r B = \mu_0 \left(\frac{r^2}{R^2}\right) I_C,$$

or

$$B = \frac{\mu_0}{2\pi}\left(\frac{r}{R^2}\right) I_C,$$

showing that, in the region between the plates, $\boldsymbol{B}$ is zero at the axis, increasing linearly with distance from the axis. A similar calculation shows that outside the region between the plates, $\boldsymbol{B}$ is the same as though the wire were continuous and the plates not present at all.

PROBLEMS

$$1\ \text{T} = 1\ \text{Wb m}^{-2} = 1\ \text{N A}^{-1}\ \text{m}^{-1}$$

$$k' = 10^{-7}\ \text{N A}^{-2} = 10^{-7}\ \text{Wb A}^{-1}\ \text{m}^{-1}$$

$$\mu_0 = 4\pi k' = 12.57 \times 10^{-7}\ \text{Wb A}^{-1}\ \text{m}^{-1}$$

32–1 A long straight wire, carrying a current of 200 A, runs through a cubical wooden box, entering and leaving through holes in the centers of opposite faces, as in Fig. 32–13. The length of each side of the box is 20 cm. Consider an element of the wire 1 cm long at the center of the box. Compute the magnitude of the magnetic induction ΔB produced by this element at the points lettered a, b, c, d, and e in Fig. 32–13. Points a, c, and d are at the centers of the faces of the cube, point b is at the midpoint of one edge, and point e is at a corner. Copy the figure and show by vectors the directions and relative magnitudes of the field vectors.

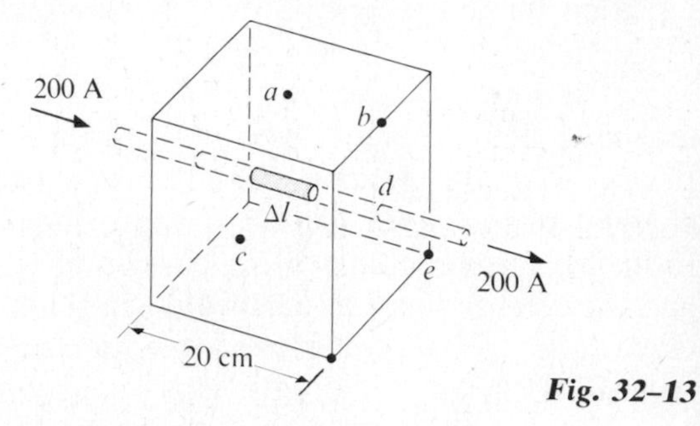

Fig. 32–13

32–2 A long straight wire carries a current of 10 A along the y-axis, as shown in Fig. 32–14. A uniform magnetic field $B_0 = 10^{-6}$ T is directed parallel to the x-axis. What is the resultant magnetic field at the following points: (a) $x = 0$, $z = 2$ m, (b) $x = 2$ m, $z = 0$, (c) $x = 0$, $z = -0.5$ m?

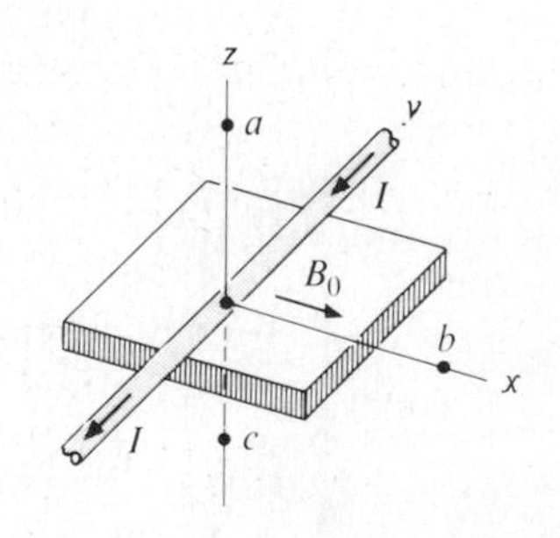

Fig. 32–14

32–3 Figure 32–15 is an end view of two long parallel wires perpendicular to the xy-plane, each carrying a current I, but in opposite directions. (a) Copy the diagram, and show by vectors the $\boldsymbol{B}$-field of each wire, and the resultant $\boldsymbol{B}$-field, at point P. (b) Derive the expression for the magnitude of $\boldsymbol{B}$ at any point on the x-axis, in terms of the coordinate x of the point. (c) Construct a graph of the magnitude of $\boldsymbol{B}$ at any point on the x-axis. (d) At what value of x is B a maximum?

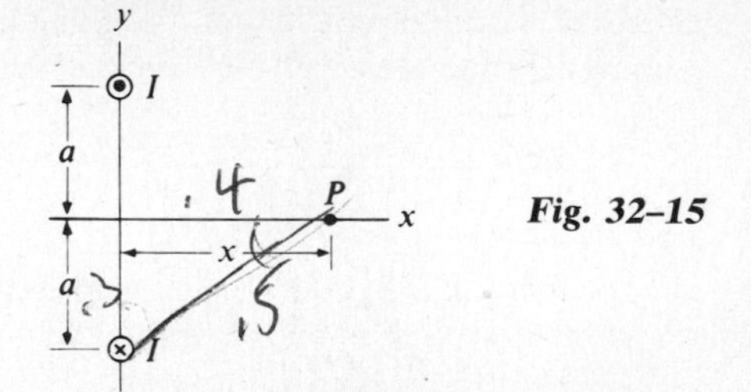

Fig. 32–15

32–4 Same as Problem 32–3, except that the current in both wires is away from the reader.

32–5 Two long, straight, parallel wires are separated by a distance $2a$. If the wires carry equal currents in opposite directions, what is the flux density in the plane of the wires at a point (a) midway between them, and (b) at a distance a above the upper wire? If the wires carry equal currents in the same direction, what is the flux density in the plane of the wires at a point (c) midway between them, and (d) at a distance a above the upper wire?

32–6 Two long, straight, parallel wires are 100 cm apart, as in Fig. 32–16. The upper wire carries a current I_1 of 6 A into the plane of the paper. (a) What must be the magnitude and direction of the current I_2 for the resultant field at point P to be zero? (b) What is then the resultant field at Q? (c) At S?

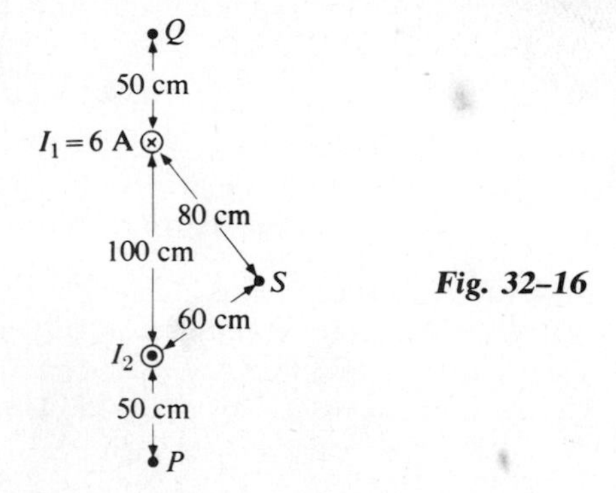

Fig. 32–16

32–7 In Fig. 32–15 suppose a third long straight wire, parallel to the other two, passes through point P, and that each wire carries a current I = 20 A. Let a = 30 cm and x = 40 cm. Find the magnitude and direction of the force per unit length on the third wire, (a) if the current in it is away from the reader, (b) if the current is toward the reader.

32–8 A long straight wire carries a current of 1.5 A. An electron travels with a velocity of 5×10^6 cm s^{-1} parallel to the wire, 10 cm from it, and in the same direction as the current. What force does the magnetic field of the current exert on the moving electron?

32–9 A long horizontal wire AB rests on the surface of a table. (See Fig. 32–17). Another wire CD vertically above the first is 100 cm long and is free to slide up and down on the two vertical metal guides C and D. The two wires are connected through the sliding contacts and carry a current of 50 A. The mass of the wire CD is 0.05 g cm^{-1}. To what equilibrium height will the wire CD rise, assuming the magnetic force on it to be due wholly to the current in the wire AB?

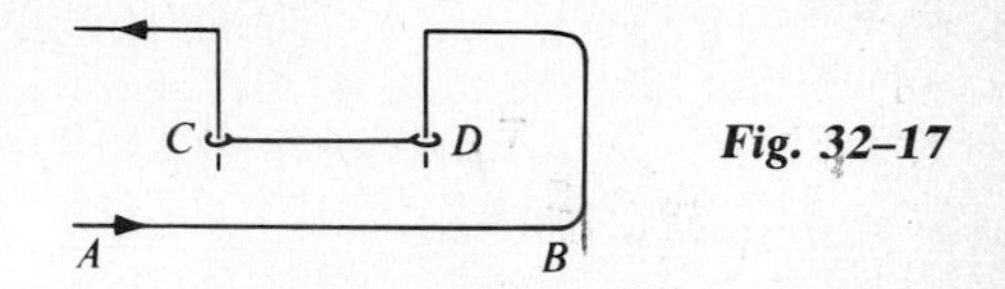

Fig. 32–17

32–10 Two long parallel wires are hung by cords of 4 cm length from a common axis. The wires have a mass of 50 g m^{-1} and carry the same current in opposite directions. What is the current if the cords hang at an angle of 6° with the vertical?

32–11 The long straight wire AB in Fig. 32–18 carries a current of 20 A. The rectangular loop whose long edges are parallel to the wire carries a current of 10 A. Find the magnitude and direction of the resultant force exerted on the loop by the magnetic field of the wire.

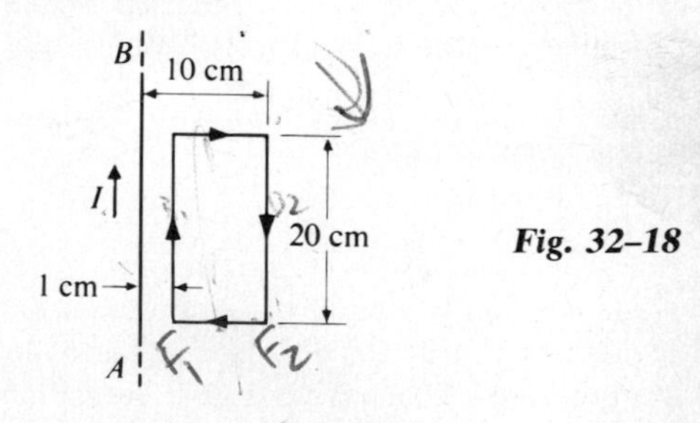

Fig. 32–18

32–12 The long straight wire AB in Fig. 32–19 carries a constant current I. (a) What is the flux density at the

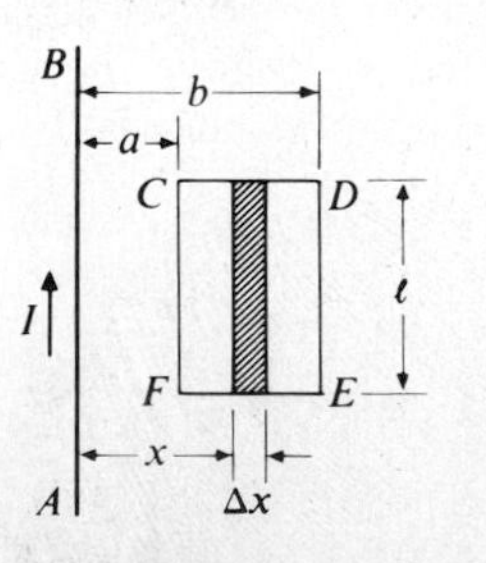

Fig. 32–19

shaded area at a perpendicular distance x from the wire? (b) What is the magnetic flux $d\Phi$ through the shaded area? (c) What is the flux Φ through the rectangular area $CDEF$, in terms of I, ℓ, a, and b?

32–13 Refer to Fig. 32–6. Sketch a graph of the magnitude of the flux density B on the axis of the coil, from $y = -3R$ to $y = +3R$.

32–14 Figure 32–20 is a sectional view of two circular coils of radius a, each wound with N turns of wire carrying a current I, circulating in the same direction in both coils. The coils are separated by a distance a equal to their radii. (a) Derive the expression for the flux density B at point P, midway between the coils. (b) Calculate the magnitude of $\mathbf{B}$ if $N = 100$ turns, $I = 5$ A, and $a = 30$ cm.

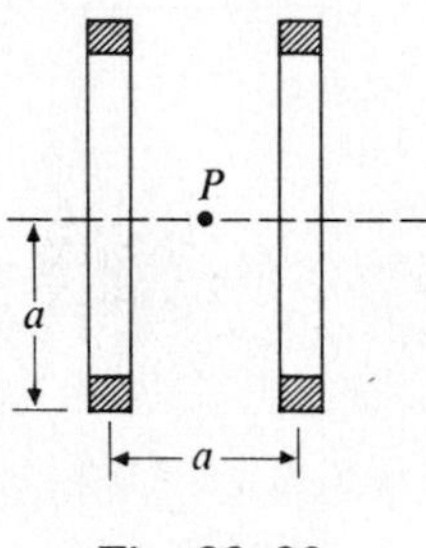

Fig. 32–20

32–15 Considering the magnetic field along the axis of a circular loop of radius R, at what distance from the center of the loop is the field $\frac{1}{10}$ of its value at the center?

32–16 A circular coil of radius 5 cm has 200 turns and carries a current of 0.2 A. What is the magnetic field (a) at the center of the coil? (b) At a point on the axis of the coil, 10 cm from its center?

32–17 A closely wound coil has a diameter of 40 cm and carries a current of 2.5 A. How many turns does it have if the magnetic induction at the center of the coil is 1.26×10^{-4} T?

32–18 A thin disk of dielectric material, having a total charge $+Q$ distributed uniformly over its surface, and of radius a, rotates n times per second about an axis perpendicular to the surface of the disk and passing through its center. Find the magnetic induction at the center of the disk.

32–19 A solenoid is 30 cm long and is wound with two layers of wire. The inner layer consists of 300 turns, the outer layer of 250 turns. The current is 3 A in the same direction in both layers. What is the magnetic induction at a point near the center of the solenoid?

32–20 A wire of circular cross section and radius R carries a current I, uniformly distributed over its cross-sectional area.

a) In terms of I, R, and r_1, what is the current through a circular area of radius r_1, inside the wire?

b) Use Ampère's law to find B inside the wire, at a distance r_1 from the axis.

c) What is B outside the wire, at a distance r_2 from the axis?

d) What would the field be at this distance if the current were concentrated in a very fine wire along the axis?

e) Sketch a graph of the magnitude of $\mathbf{B}$ as a function of r, from $r = 0$ to $r = 2R$.

32–21 A coaxial cable consists of a solid conductor of radius R_1, supported by insulating disks on the axis of a tube of inner radius R_2 and outer radius R_3. If the central conductor and the tube carry equal currents in opposite directions, find the magnetic field (a) at points outside the axial conductor but inside the tube, and (b) at points outside the tube.

32–22 A solenoid of length 20 cm and radius 2 cm is closely wound with 200 turns of wire. The current in the winding is 5 A. Compute the magnetic field at a point near the center of the solenoid.

32–23 A wooden ring whose mean diameter is 10 cm is wound with a closely spaced toroidal winding of 500 turns. Compute the field at a point on the mean circumference of the ring when the current in the windings is 0.3 A.

32–24 A conductor is made in the form of a hollow cylinder with inner and outer radii a and b, respectively. It carries a current I, uniformly distributed over the cross section. Derive expressions for the magnetic field in the regions $r < a$, $a < r < b$, and $r > b$.

32–25 A solenoid is to be designed to produce a magnetic field of 0.1 T at its center. The radius is to be 5 cm and the length 50 cm, and the available wire can carry a maximum current of 10 A.

a) How many turns per unit length should the solenoid have?

b) What total length of wire is required?

Chapter 33

Induced Electromotive Force

33–1 MOTIONAL ELECTROMOTIVE FORCE

The present large-scale production, distribution, and use of electrical energy would not be economically feasible if the only seats of emf available were those of *chemical* nature, such as dry cells. The development of electrical engineering, as we now know it, began with Faraday and Henry who, independently and at nearly the same time, discovered the principles of induced emf's and the methods by which mechanical energy can be converted directly to electrical energy.

Figure 33–1 represents a conductor of length ℓ in a uniform magnetic field, perpendicular to the plane of the diagram and directed away from the reader. If the conductor is set in motion toward the right with a velocity $\boldsymbol{v}$, perpendicular both to its own length and to the magnetic field, a charge q within it is acted on by a force equal to $q\boldsymbol{v} \times \boldsymbol{B}$. The direction of the force on a positive charge is from b toward a in Fig. 33–1. Since this force is of nonelectrostatic origin, we denote it by $\boldsymbol{F}_n$:

$$\boldsymbol{F}_n = q\boldsymbol{v} \times \boldsymbol{B}. \tag{33–1}$$

The state of affairs within the conductor is the same as though it had been inserted in an electric field of intensity $\boldsymbol{v} \times \boldsymbol{B}$, directed from b toward a. Following the discussion of Section 28–5, we define the equivalent nonelectrostatic field $\boldsymbol{E}_n$ as the nonelectrostatic force per unit charge:

$$\boldsymbol{E}_n = \boldsymbol{v} \times \boldsymbol{B}. \tag{33–2}$$

The free charges in the conductor move in the direction of the force acting on them until the accumulation of excess charges at the ends of the conductor establishes an *electrostatic* field $\boldsymbol{E}_e$ such that the *resultant* force on each charge within the conductor is zero. The charges are then in equilibrium.

Suppose now that the moving conductor slides along a stationary U-shaped conductor, as in Fig. 33–2. There is no magnetic force on the charges within the stationary conductor but, since it lies in the electrostatic field surrounding the moving conductor, a current will be established within it; the direction of this current (defined as usual as the direction of positive-charge motion) is counterclockwise, or from b toward a. As a result of this current,

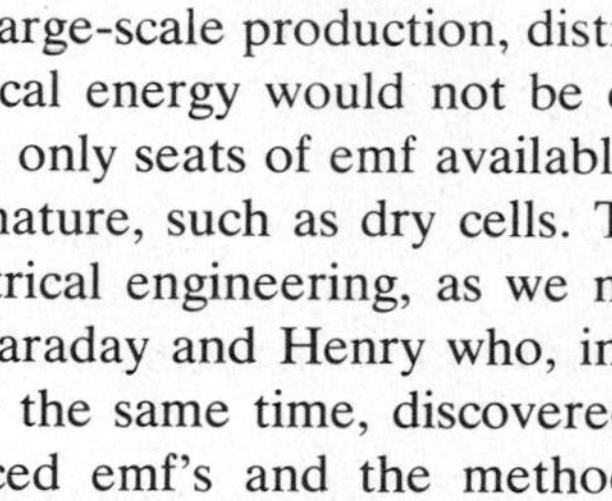

Fig. 33–1 *Conducting rod in uniform magnetic field.*

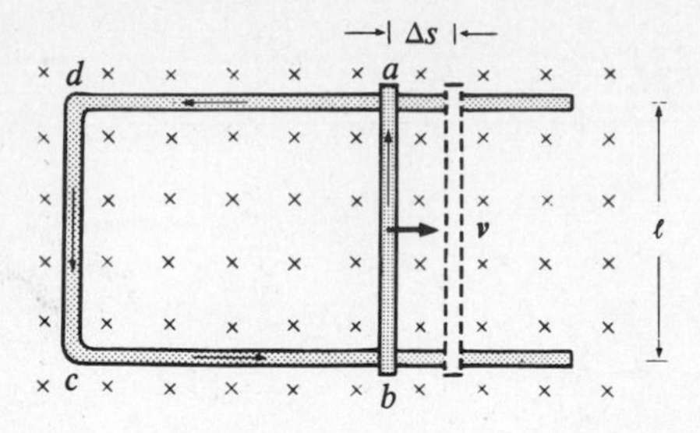

Fig. 33–2 Current produced by the motion of a conductor in a magnetic field.

the excess charges at the ends of the moving conductor are reduced, the electrostatic field within the moving conductor is weakened, and the magnetic forces cause a further displacement of the free electrons within the wire from a toward b. As long as the motion of the conductor is maintained there will, therefore, be a continual current in a counterclockwise direction. The moving conductor corresponds to a seat of electromotive force, and is said to have *induced* within it a *motional electromotive force.*

Following the discussion of Section 28–5, we define the electromotive force $\mathscr{E}$ as the line integral of the nonelectrostatic field:

$$\mathscr{E} = \int_b^a \mathbf{E}_n \cdot d\mathbf{s} = E_n\ell = vB\ell. \qquad (33\text{–}3)$$

Equation (33–3) was derived for the special case in which $\boldsymbol{v}$, $\boldsymbol{B}$, and $\boldsymbol{\ell}$ are mutually perpendicular. The general expression for the motional emf $d\mathscr{E}$ in an element of a conductor of length $d\boldsymbol{\ell}$ is

$$d\mathscr{E} = \mathbf{E}_n \cdot d\boldsymbol{\ell} = (\boldsymbol{v} \times \mathbf{B}) \cdot d\boldsymbol{\ell}. \qquad (33\text{–}4)$$

Thus if the velocity $\boldsymbol{v}$ of the conductor makes an angle ϕ with the field $\boldsymbol{B}$, we must replace vB by $vB \sin \phi$, and the induced emf becomes

$$\boxed{\mathscr{E} = vB\ell \sin \phi.} \qquad (33\text{–}5)$$

If v is expressed in m s^{-1}, B in T (or Wb m^{-2}), and ℓ in meters, the emf is in joules per coulomb or volts, as the reader should verify.

Example 1 Let the length ℓ in Fig. 33–2 be 0.1 m, the velocity v be 0.1 m s^{-1}, the resistance of the loop be 0.01 Ω, and let B = 1 Wb m^{-2}.

The field E_n is

$$E_n = vB = 0.1 \text{ V m}^{-1}.$$

The emf $\mathscr{E}$ is

$$\mathscr{E} = \ell E_n = 0.01 \text{ V}.$$

The current in the loop is

$$I = \frac{\mathscr{E}}{R} = 1 \text{ A}.$$

Because of this current, there is a force F on the loop, in the opposite direction to its motion, and equal to

$$F = IB\ell = 0.1 \text{ N}.$$

The power necessary to move the loop against this force is

$$P = Fv = 0.01 \text{ W}.$$

The product $\mathscr{E}I$ is

$$\mathscr{E}I = 0.01 \text{ W}.$$

Thus the rate of energy conversion, $\mathscr{E}I$, equals the mechanical power input to the system.

Example 2 The rectangular loop in Fig. 33–3, of length a and width b, is rotating with uniform angular velocity ω about the y-axis. The entire loop lies in a uniform, constant $\boldsymbol{B}$-field, parallel to the z-axis. We wish to calculate the motional emf in the loop from Eq. (33–3).

The velocity v of the sides of the loop of length a is

$$v = \omega \frac{b}{2}.$$

The direction of the motional field $\mathbf{E}_n (= \boldsymbol{v} \times \mathbf{B})$ in each of the sides of length a is shown in the diagram. Its magnitude is

$$E_n = vB \sin \theta = \omega \frac{b}{2} B \sin \theta.$$

The motional fields in the other two sides of the loop are transverse to these sides and contribute

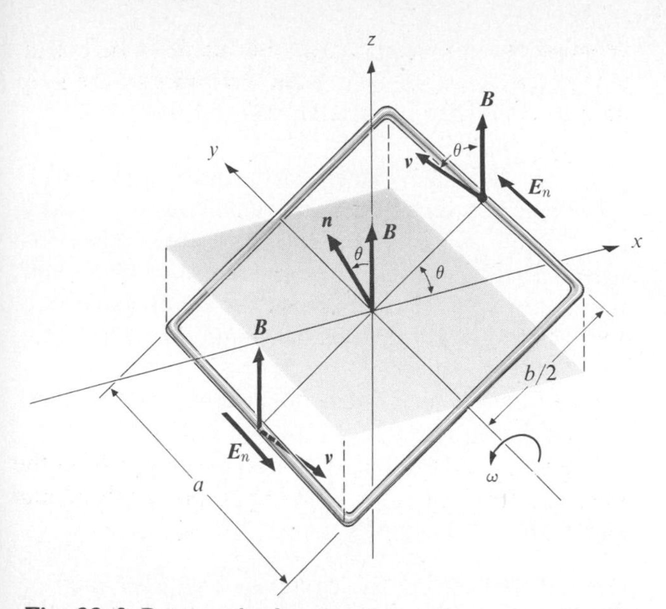

Fig. 33–3 *Rectangular loop rotating with constant angular velocity in a uniform magnetic field.*

nothing to the emf. The line integral of $\boldsymbol{E}_n$ around the loop reduces to $2E_na$, so

$$\mathcal{E} = \oint \boldsymbol{E}_n \cdot ds = 2E_na = \omega(ab)B \sin\theta.$$

The product ab equals the area A of the loop, and if the loop lies in the xy-plane at $t = 0$, then $\theta = \omega t$. Hence

$$\mathcal{E} = \omega AB \sin \omega t. \qquad (33\text{–}6)$$

The emf therefore varies sinusoidally with the time. The *maximum* emf $\mathcal{E}_m$, which occurs when $\sin \omega t = 1$, is

$$\mathcal{E}_m = \omega AB,$$

so we can write Eq. (33–6) as

$$\mathcal{E} = \mathcal{E}_m \sin \omega t. \qquad (33\text{–}7)$$

The rotating loop is the prototype of the *alternating current generator*, or *alternator;* we say it develops a *sinusoidal alternating emf.*

The emf is a maximum (in absolute value) when $\theta = 90°$ or $270°$ and the long sides are moving at right angles to the field. The emf is zero when $\theta = 0$

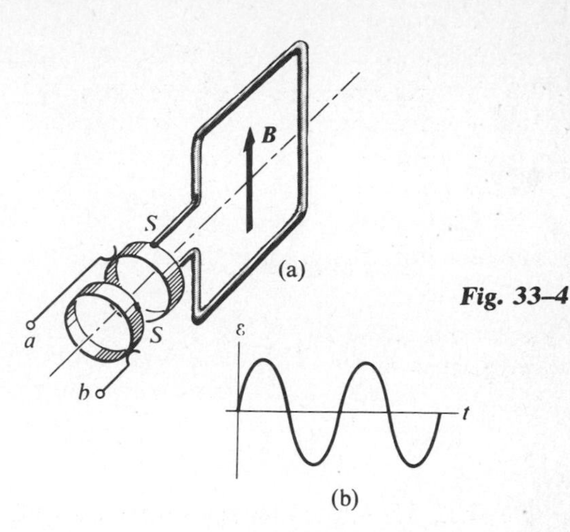

Fig. 33–4

or 180° and the long sides are moving parallel to the field. The emf depends only on the *area* A of the loop and not on its shape. This is most easily verified by use of Faraday's law, to be discussed in the next section.

The rotating loop in Fig. 33–3 can be utilized as the source in an external circuit by making connections to *slip rings S, S,* which rotate with the loop as shown in Fig. 33–4(a). Stationary brushes bearing against the rings are connected to the output terminals a and b. The instantaneous terminal voltage v_{ab}, on open circuit, equals the instantaneous emf. Figure 33–4(b) is a graph of v_{ab} as a function of time.

A terminal voltage that always has the same *sign*, although it fluctuates in *magnitude*, can be obtained by connecting the loop to a split ring or *commutator*, as in Fig. 33–5(a). At the position in which the emf

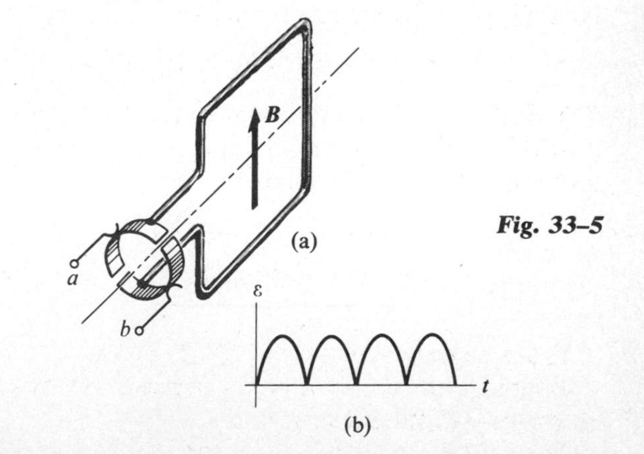

Fig. 33–5

reverses, the connections to the external circuit are interchanged. Figure 33–5(b) is a graph of the terminal voltage, and the device is the prototype of a dc generator.

Commercial dc generators have a large number of coils and commutator segments, and their terminal voltage is not only unidirectional but also practically constant.

Example 3 A disk of radius R, shown in Fig. 33–6, lies in the xz-plane and rotates with uniform angular velocity ω about the y-axis. The disk is in a uniform, constant $\boldsymbol{B}$-field parallel to the y-axis. Consider a short portion of a narrow radial segment of the disk, of length dr. Its velocity is $v = \omega r$, and since $\boldsymbol{v}$ is at right angles to $\boldsymbol{B}$, the motional field E_n in the segment is

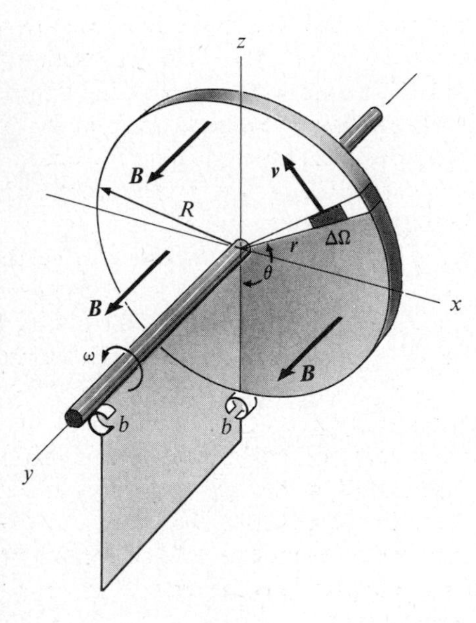

Fig. 33–6

$$E_n = vB = \omega rB.$$

The direction of $\boldsymbol{E}_n$ is radially outward.

The emf between center and rim is

$$\mathcal{E} = \int_0^R \boldsymbol{E}_n \cdot d\boldsymbol{r} = \omega B \int_0^R r\,dr = \omega BR^2/2.$$

All of the radial segments of the disk are in *parallel*, so the emf between center and rim equals that in any radial segment. The entire disk can therefore be considered a *source* for which the emf between center and rim equals $\omega BR^2/2$. The source can be included in a closed circuit by completing the circuit through sliding contacts of brushes b, b.

The emf in such a disk was studied by Faraday, and the device is called a *Faraday disk dynamo.*

33–2 FARADAY'S LAW

The induced emf in the circuit of Fig. 33–2 may be considered from another viewpoint. While the conductor moves toward the right a distance ds, the cross-sectional area of the closed circuit $abcd$ increases by

$$dA = \ell\,ds,$$

and the change in *flux* through the circuit is

$$d\Phi = B\,dA = B\ell\,ds.$$

When both sides are divided by Δt, we obtain

$$\frac{d\Phi}{dt} = \left(\frac{ds}{dt}\right)B\ell = vB\ell.$$

But the product $vB\ell$ equals the induced emf $\mathcal{E}$, so the preceding equation states that *the induced emf in the circuit is numerically equal to the rate of change of the magnetic flux through it.* The righthand screw convention of sign is usually used with this equation. If one faces the circuit, an emf is considered positive if it results in a conventional current in a clockwise direction, and $d\Phi/dt$ is considered positive if there is an increase in the flux directed away from the observer. (Then, a decrease in flux away from the observer is negative, an increase in flux toward the observer is negative, and a decrease in flux toward the observer is positive.) In Fig. 33–2 the current is counterclockwise, so the emf is negative, while the flux is away from the observer and is increasing, so

$d\Phi/dt$ is positive. A study of other possibilities shows that $\mathcal{E}$ and $d\Phi/dt$ always have *opposite* signs. Hence, we write the equation

$$\mathcal{E} = -\frac{d\Phi}{dt}. \tag{33–8}$$

Equation (33–8) is known as *Faraday's law.* As it stands, it appears to be merely an alternative form of Eq. (33–3) for the emf in a moving conductor. It turns out, however, that the relation has a much deeper significance than might be expected from its derivation. That is, it is found to apply to circuits through which the flux is caused to vary, even though there is no *motion* of any part of the circuit and hence no emf directly attributable to a force on a moving charge.

Suppose, for example, that two loops of wire are located as in Fig. 33–7. A current in circuit 1 sets up a magnetic field whose magnitude at all points is proportional to this current. A part of this flux passes through circuit 2, and if the current in circuit 1 is increased or decreased, the flux through circuit 2 will also vary. Circuit 2 is not moving in a magnetic field and hence no "motional" emf is induced in it, but there is a change in the flux through it, and it is found experimentally that an emf appears in circuit 2 of

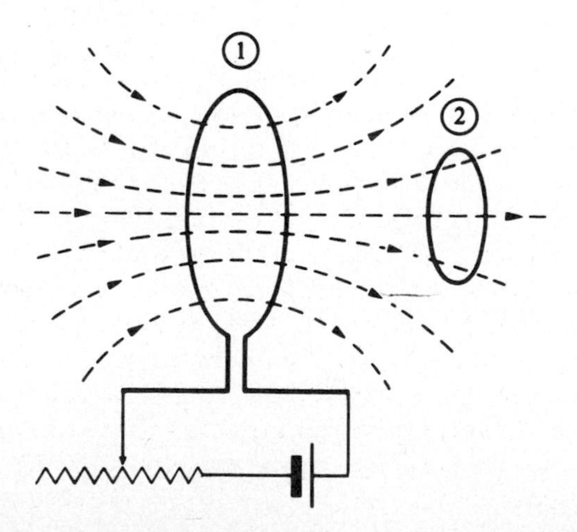

Fig. 33–7 *As the current in circuit 1 is varied, the magnetic flux through circuit 2 changes.*

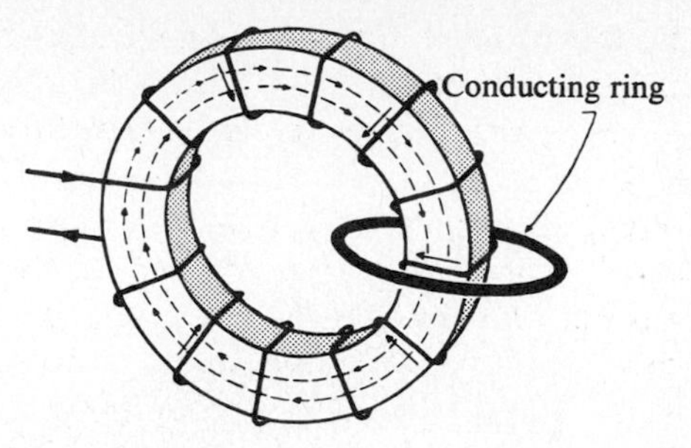

Fig. 33–8 *An emf is induced in the ring when the flux in the toroid varies.*

magnitude $\mathcal{E} = d\Phi/dt$. In such a situation, no one portion of circuit 2 can be considered the seat of emf; the *entire circuit* constitutes the seat.

Here is another example. Suppose we set up a magnetic field within the toroidal winding of Fig. 33–8, link the toroid with a conducting ring, and vary the current in the winding of the toroid. We have shown that the flux lines set up by a steady current in a toroidal winding are wholly confined to the space enclosed by the winding; not only is the ring not *moving* in a magnetic field, but if the current were steady it would not even be *in* a magnetic field! However, lines of induction do pass through the area bounded by the ring, and the flux changes as the current in the windings changes. Equation (33–8) predicts an induced emf in the ring, and we find by experiment that the emf actually exists. In case the reader has not identified the apparatus in Fig. 33–8, it may be pointed out that it is merely a *transformer* with a one-turn secondary, so that the phenomenon we are now discussing is the basis of the operation of every transformer.

To sum up, then, an emf is induced in a circuit whenever the flux through the circuit varies with time. The flux may be caused to vary in two ways: (1) by the *motion* of a conductor in a constant magnetic field where the circuit is not deformed during the motion, as in Fig. 33–2; or (2) by a *change in the magnitude* of the flux through a stationary circuit, as in Fig. 33–7 or 33–8. For case (1), the emf may be computed *either* from

$$\mathcal{E} = vB\ell$$

or from

$$\mathscr{E} = -\frac{d\Phi}{dt}.$$

For case (2), the emf may be computed *only* by

$$\mathscr{E} = -\frac{d\Phi}{dt}.$$

If we have a coil of N turns and the flux varies at the same rate through each, the induced emf's in the turns are in *series*, and the total emf is

$$\boxed{\mathscr{E} = -N\left(\frac{d\Phi}{dt}\right).} \tag{33–9}$$

Example 1 We consider again the rotating rectangular loop in Fig. 33–3, discussed in Example 2 of Section 33–1; this time we compute the emf from Eq. (33–8). The flux through the loop equals that through its projected area on the xy-plane (shaded in Fig. 33–3).

$$\Phi = \boldsymbol{B} \cdot \boldsymbol{A} = BA \cos \theta = BA \cos \omega t.$$

Then

$$\dot{\Phi} = \frac{d\Phi}{dt} = -\omega BA \sin \omega t$$

and

$$\mathscr{E} = -\dot{\Phi} = \omega BA \sin \omega t,$$

in agreement with Eq. (33–6).

Note that the maximum value of $\mathscr{E}$ occurs when $\theta = 90°$ and the flux through the loop is zero, and that $\mathscr{E} = 0$ when $\theta = 0$ and the flux is a maximum. That is, the emf depends not on the flux through the loop, but on its *rate of change*.

Example 2 We consider again the rotating disk in Fig. 33–6, discussed in Example 3 of Section 33–1. To compute the emf from Faraday's law, Eq. (33–8), we consider the circuit to be the periphery of the shaded areas in Fig. 33–6. The rectangular portion in the yz-plane is fixed. The area of the shaded section in the xz-plane is $\frac{1}{2}R^2\theta$, and the flux through it is

$$\Phi = \tfrac{1}{2}BR^2\theta.$$

As the disk rotates, the shaded area increases. In a time dt, the angle θ increases by $d\theta = \omega\, dt$. The flux increases by

$$d\Phi = \tfrac{1}{2}BR^2\, d\theta = \tfrac{1}{2}BR^2\omega\, dt,$$

and the induced emf is

$$\mathscr{E} = \frac{d\Phi}{dt} = \tfrac{1}{2}Br^2\omega,$$

in agreement with the previous result.

33–3 THE SEARCH COIL

A useful experimental method of measuring magnetic field strength uses a ballistic galvanometer connected by flexible leads to the terminals of a small, closely wound coil of N turns called a *search coil* or a *snatch coil.* Assume first, for simplicity, that the search coil is placed with its plane perpendicular to a magnetic field $\boldsymbol{B}$. If the area enclosed by the coil is A, the flux Φ through it is $\Phi = BA$. Now if the coil is quickly given a quarter-turn about one of its diameters so that its plane becomes parallel to the field, or if it is quickly snatched from its position to another where the field is known to be zero, the flux through it decreases rapidly from BA to zero. During the time that the flux is decreasing, an emf of short duration is induced in the coil and a "kick" is imparted to the ballistic galvanometer. The maximum deflection of the galvanometer is noted.

The galvanometer current at any instant is

$$i = \frac{\mathscr{E}}{R},$$

where R is the combined resistance of galvanometer and search coil, $\mathscr{E}$ is the instantaneous induced emf, and i the instantaneous current.

From Faraday's law, we obtain

$$\mathscr{E} = -N\frac{d\Phi}{dt}, \qquad i = -\frac{N d\Phi}{R\, dt}.$$

The total charge Q which flows through the galvanometer is given by

$$Q = \int_0^t i\, dt = -\frac{N}{R}\int_\Phi^0 d\Phi = \frac{N\Phi}{R}.$$

Thus Φ and B are given by

$$\Phi = \frac{RQ}{N} \quad \text{and} \quad B = \frac{\Phi}{A} = \frac{RQ}{NA}.$$

The maximum deflection of a ballistic galvanometer is proportional to the quantity of charge displaced through it. Hence, if this proportionality constant is known, Q may be found, and from Q we can obtain Φ and B.

Strictly speaking, while this method gives correctly the total flux through the coil, it is only the *average* field over the area of the coil which is measured. However, if the area is sufficiently small, this approximates closely the field at, say, the center of the coil.

The preceding discussion assumed the plane of the coil to be initially perpendicular to the direction of the field. If one is "exploring" a field whose direction is not known in advance, the same apparatus may be used to find the direction by performing a series of experiments in which the coil is placed at a given point in the field in various orientations, and snatched out of the field from each orientation. The deflection of the galvanometer will be a maximum for the particular orientation in which the plane of the coil was perpendicular to the field. Thus the magnitude and direction of an unknown field can both be found by this method.

Since the search coil is permanently connected to the galvanometer terminals the galvanometer will be highly damped and must either be calibrated with the search coil connected, or the corrections mentioned in the next section must be applied.

33–4 GALVANOMETER DAMPING

Suppose a ballistic galvanometer is connected as in Fig. 33–9(a) to measure the quantity of charge on a capacitor. Let the switch S be closed momentarily, allowing the capacitor to discharge through the galvanometer, and then immediately opened. The surge of charge starts the galvanometer coil swinging and, since it is rotating in a magnetic field, an emf is induced in it. The current through it is zero, however, since the switch has been opened and there is no closed circuit. The motion of the coil is controlled solely by the suspension and friction. If the latter were entirely absent, the coil would oscillate in-

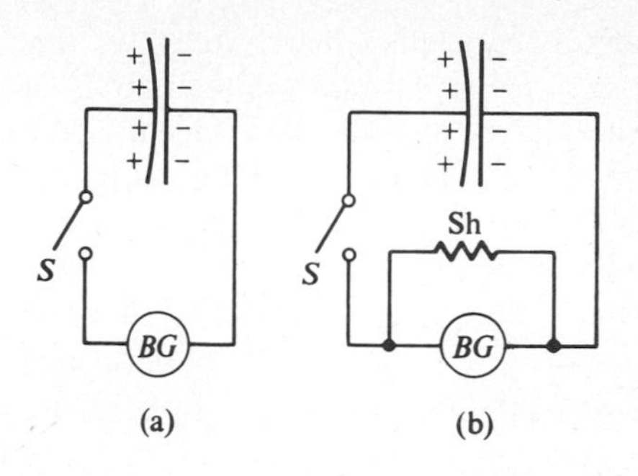

Fig. 33–9 *Discharge of a capacitor through a ballistic galvanometer.*

definitely with angular harmonic motion.

Now let the shunt resistor Sh in Fig. 33–9(b) be connected across the galvanometer terminals and the experiment repeated. The motion of the galvanometer will be affected for two reasons. First, a part of the discharge current of the capacitor will be bypassed by the shunt and the impulse imparted to the coil will be correspondingly less. Second, the galvanometer and shunt now form a closed circuit even when switch S is opened, so that there will now be a current in the swinging coil. The side thrusts on this current give rise to a torque on the coil; the direction of this torque is such as to oppose the motion of the coil and aid in bringing it to rest. From the energy standpoint, a part of the kinetic energy of the swinging coil is dissipated, thereby increasing the internal energy (and also the temperature) of the coil and the shunt resistor. The motion of the coil is *damped harmonic.*

It should be evident that the smaller the shunt resistance, the larger will be the induced current and the greater the damping. With a sufficiently small shunt resistance the motion ceases to be oscillatory; the galvanometer makes but one swing and returns slowly to its zero position. The particular resistance for which the motion just ceases to be oscillatory is called the *critical external damping resistance*, and when shunted by this resistance, the galvanometer is said to be *critically damped.* With more resistance it is *underdamped*, and with less it is *overdamped.*

Since the presence of damping reduces the maximum swing of a ballistic galvanometer, the simple theory in Section 31–6, which assumes that *all* of the initial kinetic energy of the coil is converted to potential energy of the suspension must be extended if damping is present. The complete analysis shows that the quantity of charge displaced is still proportional to the maximum angle of swing, although with a modified proportionality constant. However, if the galvanometer is calibrated with the same external resistance as that with which it is to be used, the modified constant is automatically determined.

In many pivoted-coil instruments, such as portable ammeters and voltmeters, the necessary amount of damping is "built into" the moving coil, so to speak, by winding this coil on a light aluminum frame. The frame itself then forms a closed circuit and the side thrusts on the current induced in this circuit quickly bring the swinging coil to rest.

33–5 INDUCED ELECTRIC FIELDS

The examples of induced emf considered thus far have all involved conductors moving in a magnetic field, but we have also pointed out, in introducing Faraday's law, that induced emf can also occur with stationary conductors. An example is shown in Fig. 33–10, a solenoid encircled by a conducting loop of arbitrary shape. A current I in the windings of the solenoid sets up a magnetic field $\mathbf{B}$ along the solenoid axis and a magnetic flux $\Phi = BA$ passes through any

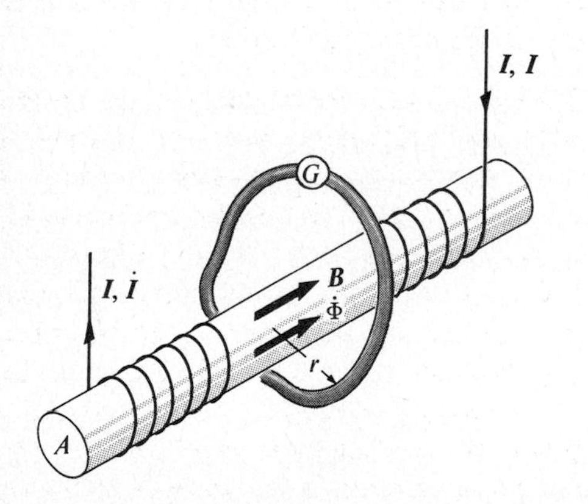

Fig. 33–10 The windings of a long solenoid carry a current I that is increasing at a rate dI/dt. The magnetic flux in the solenoid is increasing at a rate dΦ/dt, and this changing flux passes through a wire loop of arbitrary size and shape. An emf ℰ is induced in the loop, given by ℰ = −dΦ/dt.

surface bounded by the wire. Suppose that a small galvanometer G is inserted in the loop. It is found that if the current I is changed (and hence if the flux Φ is changed), the galvanometer indicates an emf $\mathcal{E}$ in the wire *during the time that the flux is changing*, and that this emf is equal to the *time rate of change of flux* through the surface bounded by the wire. That is,

$$\mathcal{E} = -\frac{d\Phi}{dt}. \qquad (33\text{–}10)$$

The reason for the minus sign is the same as in Section 33–2. This is Faraday's law; the essential difference between this and previous examples is that here it is applied not to a moving conductor but to a stationary one in which there is a changing flux because the magnetic field in the area bounded by the conductor is changing.

Example Suppose the long solenoid in Fig. 33–10 is wound with 1000 turns per meter, and the current in its windings is increasing at the rate of $100\ \text{A s}^{-1}$. The cross-sectional area of the solenoid is $4\ \text{cm}^2 = 4 \times 10^{-4}\ \text{m}^2$.

The flux Φ in the solenoid, not too near its ends, is

$$\Phi = BA = \mu_0 nIA.$$

The rate of change of flux is

$$\begin{aligned} d\Phi/dt &= \mu_0 nA\, dI/dt \\ &= (4\pi \times 10^{-7}\ \text{Wb A}^{-1}\ \text{m}^{-1})(1000\ \text{turns m}^{-1}) \\ &\quad \times (4 \times 10^{-4}\ \text{m}^2)(100\ \text{A s}^{-1}) \\ &= 16\pi \times 10^{-6}\ \text{Wb s}^{-1}. \end{aligned}$$

The induced emf is

$$\begin{aligned} \mathcal{E} &= \frac{d\Phi}{dt} = 16\pi \times 10^{-6}\ \text{V} \\ &= 16\pi\ \mu\text{V}. \end{aligned}$$

The same emf will be induced in a loop of any size and shape encircling the solenoid, since $\mathcal{E}$ depends only on the rate of change of flux through the loop, and not on its size or shape.

We have previously associated electromotive force with the line integral of an electric field $\boldsymbol{E}_n$ of nonelectrostatic origin, and the same point of view must be taken here. The emf is the line integral of $\boldsymbol{E}_n$ around the loop:

$$\mathcal{E} = \oint \boldsymbol{E}_n \cdot d\boldsymbol{s}.$$

Hence the Faraday law states that

$$\oint \boldsymbol{E}_n \cdot d\boldsymbol{s} = -\frac{d\Phi}{dt}. \qquad (33\text{–}11)$$

As an example, suppose the loop in Fig. 33–10 is a circle of radius r. Because of the axial symmetry, the nonelectrostatic field $\boldsymbol{E}_n$ is everywhere tangent to this circle, so the line integral in Eq. (33–11) becomes simply the magnitude E_n times the circumference $2\pi r$ of the circle. Thus, the induced electric field at a distance r from the axis is given by

$$E_n = \frac{1}{2\pi r}\frac{d\Phi}{dt}, \qquad (33\text{–}12)$$

where Φ is the flux through a circle of radius r.

If it should happen that at a particular instant the $\boldsymbol{B}$ field is uniform across this circle, as is *not* the case with the solenoid above, then

$$\Phi = \pi r^2 B,$$

$$\frac{d\Phi}{dt} = \pi r^2 \frac{dB}{dt},$$

and

$$E = \frac{1}{2\pi r}\pi r^2 \frac{dB}{dt} = \frac{r}{2}\frac{dB}{dt}. \qquad (33\text{–}13)$$

In summary, Eq. (33–10) is valid for two rather different situations. In the first, an emf is induced by the magnetic forces on charges in a conductor moving through a magnetic field, while in the second a time-varying magnetic field induces an electric field of nonelectrostatic nature in a stationary conductor and hence induces an emf. The $\boldsymbol{E}_n$ field in the latter case differs from an electro*static* field in two significant ways. First, its line integral around a closed path is *not* zero, so it is not a *conservative* field. In contrast, an electrostatic field is *always* conservative, as discussed in Section 26–1. Second, the nonelectrostatic field is not produced by static charges; it can be shown that the surface integral of $\boldsymbol{E}_n$ over a closed surface (the same integral that appears in Gauss's law) always has the value zero, whether charges are enclosed or not.

Thus, a changing magnetic field acts as a source of electric field, but one that differs from an electrostatic field in two ways: It is nonconservative, and its integral over a closed surface is zero. The reader may also note that this situation is analogous to that of the displacement current discussed in Section 29–8, in which a changing *electric* field acts as a source of *magnetic* field. These relations thus exhibit a kind of symmetry in the behavior of the two fields. We shall return to this relationship in Chapter 37, in connection with the analysis of electromagnetic waves.

33–6 LENZ'S LAW

H. F. E. Lenz (1804–1864) was a German scientist who, without knowledge of the work of Faraday and Henry, duplicated many of their discoveries nearly simultaneously. The law which goes by his name is a useful rule for predicting the direction of an induced current. It states:

> *The direction of an induced current is such as to oppose the cause producing it.*

The "cause" of the current may be the motion of a conductor in a magnetic field, or it may be the change of flux through a stationary circuit. In the first case, the direction of the induced current in the moving conductor is such that the direction of the side-thrust exerted on the conductor by the magnetic field is opposite in direction to its motion. The motion of the conductor is therefore "opposed."

In the second case, the induced current sets up a magnetic field of its own which, within the area bounded by the circuit, is (a) *opposite* to the original field if this is *increasing*, but (b) is in the *same* direction as the original field if the latter is *decreasing*. Thus it is the *change in flux* through the circuit (not the flux itself) which is "opposed" by the induced current.

In order for there to be an induced current, we must have a closed circuit. If a conductor does not form a closed circuit, then we mentally complete the circuit between the ends of the conductor and use Lenz's law to determine the direction of the current. The polarity of the ends of the open-circuit conductor may then be deduced.

Lenz's law is not really a separate principle, but conveys the same information as the sign convention in Faraday's law. Nevertheless, it is a convenient rule to use in many induction problems.

33–7 THE BETATRON

The *magnetic induction accelerator*, or *betatron*, is one of the family of machines designed for the purpose of accelerating charged particles to high energies. It was invented in 1941 by Donald W. Kerst of the University of Illinois. The following description refers to a machine constructed in 1945 by the General Electric Company, which could accelerate electrons to an energy of 100 MeV. Although the betatron has now been superseded by other machines, it is, nevertheless, of interest at this point because it affords an excellent example of the induced electric field set up by a varying magnetic field.

Figure 33–11 is a photograph of the toroidal vacuum tube of the 1945 machine. The tube is placed horizontally between the pole faces of a large electromagnet. (The magnet is not shown in the photograph.) Alternating current at a frequency of 60 Hz is sent through the windings of the electromagnet, so that the magnetic flux through the plane of the toroid reverses from a maximum in one direction to a maximum in the opposite direction in (1/120) s. Electrons accelerated through approximately 50,000 V by an electron gun are shot tangentially into the tube and are caused by the magnetic field to circle around within the tube in an orbit about 5 ft in diameter or about 5 m in circumference.

Figure 33–12 is a top view of the tube at an instant when the magnetic field $\boldsymbol{B}$ is into the plane of the diagram, and is increasing. For reasons that will be explained later, the magnetic field near the center is larger than that near the periphery. The field is symmetrical, however, so the lines of force of the induced electric field are *circles.*

The line integral of $\boldsymbol{E}_n$ around a circular path of radius r within the tube is

$$\mathscr{E} = \oint \boldsymbol{E}_n \cdot d\boldsymbol{s} = 2\pi r E_n.$$

By the Faraday law, this equals the rate of change of flux through the path and, disregarding algebraic

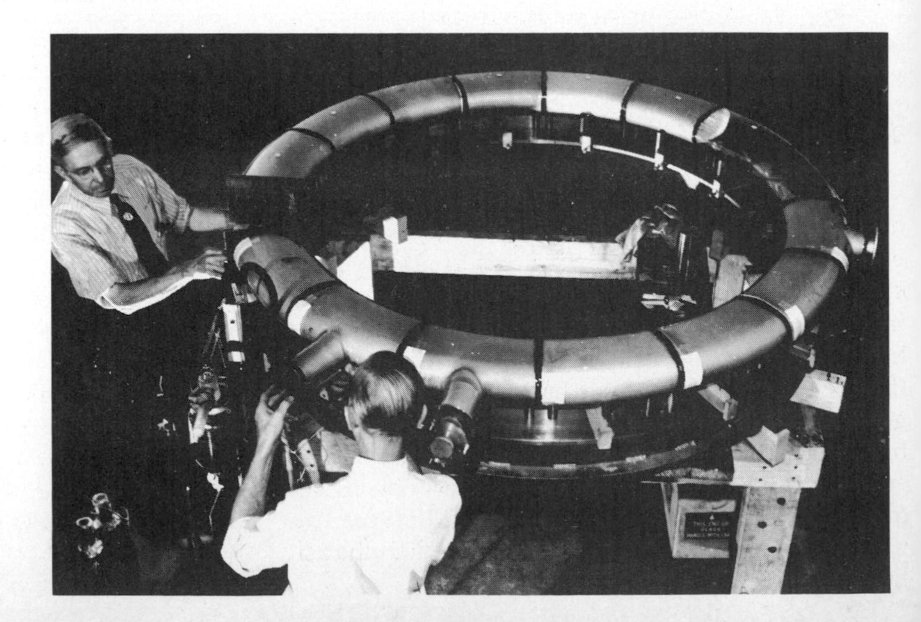

Fig. 33–11 *Assembling the vacuum tube of the betatron. (Courtesy of General Electric Company.)*

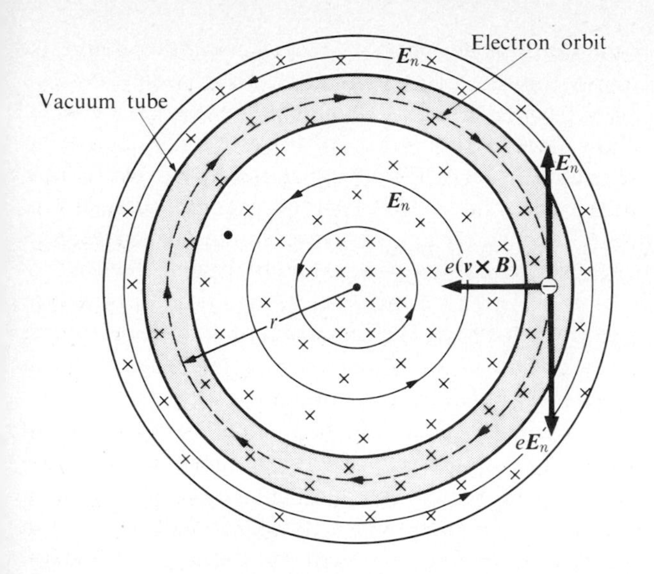

Fig. 33–12 *An electron in the vacuum tube of a betatron is accelerated by the induced electric field* $\mathbf{E}_n$ *and forced to move in a circle by the magnetic field* $\mathbf{B}$.

signs,

$$2\pi r E_n = \dot{\Phi}, \qquad E_n = \frac{\dot{\Phi}}{2\pi r}.$$

Note that there would be an *electric* field at the vacuum tube, even if the *magnetic* field were limited to a small region near the axis. The values of $\dot{\Phi}$ and r are such that the electric field E_n is about 80 N C^{-1} or 80 V m^{-1}. The emf in the path is

$$\mathscr{E} = 2\pi r E_n = (80\,\mathrm{V\,m^{-1}})(5\,\mathrm{m}) = 400\,\mathrm{V} = 400\,\mathrm{J\,C^{-1}}.$$

Because there *is* a magnetic field at the electron orbit, an electron is acted on by a centripetal force $e(\mathbf{v} \times \mathbf{B})$, which constrains it in a circular path. As the electron is accelerated by the E_n-field, its velocity $\mathbf{v}$ increases and a larger $\boldsymbol{B}$-field is necessary to provide the requisite centripetal force. It is not difficult to show that if the value of $\boldsymbol{B}$ at the orbit is always just half as great as the *average* value of $\boldsymbol{B}$ over the area bounded by the orbit, the increasing $\boldsymbol{B}$-field will be just large enough to provide the increasing centripetal force required because of the increasing velocity. Hence the value of $\boldsymbol{B}$ near the axis must be greater than that at the orbit, as mentioned earlier.

If a loop of wire were inserted in the field at the same place as the vacuum tube, the electrons in the wire would be acted on by the same electric field as those in the tube, and in each journey around the loop they would acquire 400 eV of energy. Their motion, however, would be a series of accelerations followed by decelerations as they collided with the fixed particles in the wire. At each collision they would give up the energy acquired in the interval following the preceding collision, and their average velocity would be only the relatively small drift velocity.

The electrons in the vacuum tube of the betatron, however, instead of being confined in a wire, are forced to move in a circle by the magnetic field. In each revolution they also acquire 400 eV of energy, but they can accelerate continuously without making collisions and the energy gain is *cumulative*. During the interval while the flux increases from zero to its maximum value an electron makes 250,000 revolutions and hence gains (250,000)(400 eV) or 100 MeV of energy.

The accelerator may be compared to an ordinary transformer with the secondary replaced by the electrons in the evacuated tube. The electrons are accelerated by the *induced electric field* set up by the *changing magnetic field*, and at the same time are forced to move in a circular orbit by the *existence* of the magnetic field.

33–8 EDDY CURRENTS

Thus far we have considered only instances in which the currents resulting from induced emf's were confined to well-defined paths provided by the wires and apparatus of the external circuit. In many pieces of electrical equipment, however, one finds masses of metal moving in a magnetic field or located in a changing magnetic field, with the result that induced currents circulate throughout the volume of the metal. Because of their general circulatory nature, these are referred to as *eddy currents.*

Consider a disk rotating in a magnetic field perpendicular to the plane of the disk but confined to a limited portion of its area, as in Fig 33–13(a). Element *Ob* is moving across the field and has an emf induced in it. Elements *Oa* and *Oc* are not in the field, but, in common with all other elements located outside the field, provide return conducting paths along which charges displaced along *Ob* can return

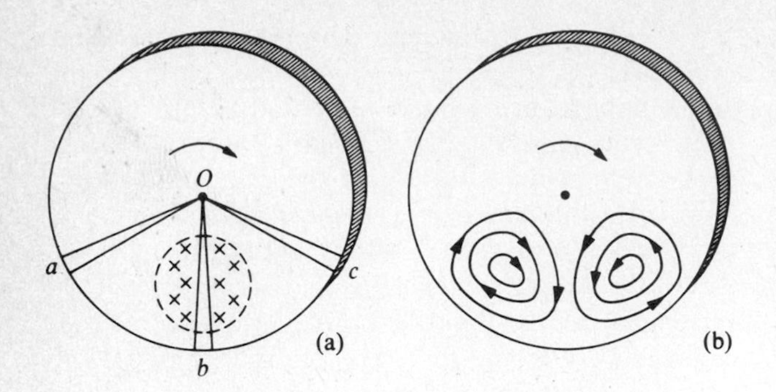

Fig. 33–13 *Eddy currents in a rotating disk.*

from b to O. A general eddy-current circulation is therefore set up in the disk, somewhat as sketched in Fig. 33–13(b).

The currents in the neighborhood of radius Ob experience a side thrust which *opposes* the motion of the disk, while the return currents, since they lie outside the field, do not experience such a thrust. The interaction between the eddy currents and the field therefore results in a braking action on the disk. The apparatus finds some technical applications and is known as an "eddy current brake."

As a second example of eddy currents, consider the core of an alternating current transformer, shown in Fig. 33–14. The alternating current in the transformer windings sets up an alternating flux within the core, and an induced emf develops in the secondary windings because of the continual change in flux through them. The iron core, however, is also a conductor, and any section such as that at AA can be thought of as a number of closed conducting circuits, one within the other. The flux through each of these circuits is continually changing, so that there is an eddy current circulation in the entire volume of the core, the lines of flow lying in planes perpendicular to the flux. These eddy currents are very undesirable both because of the energy which they dissipate, and because of the flux which they themselves set up.

In all actual transformers, the eddy currents are greatly reduced by the use of a *laminated* core, that is, one built up of thin sheets or laminae. The electrical resistance between the surfaces of the laminations (due either to a natural coating of oxide or to an insulating varnish) effectively confines the eddy currents to individual laminae. The resulting length of path is greatly increased, with consequent increase in resistance. Hence, although the induced emf is not altered, the currents and their heating effects are minimized.

In small transformers where eddy-current losses must be kept to an absolute minimum, the cores are sometimes made of *ferrites*, which are complex oxides of iron and other metals. These materials are ferromagnetic but have relatively high resistivity.

PROBLEMS

33–1 In Fig. 33–1, let $\ell = 1.5$ m, $B = 0.5$ T, $v = 4$ m s^{-1}.

a) Find the magnitude and direction of the equivalent nonelectrostatic field $\boldsymbol{E}_n$ in the rod.
b) Find the magnitude and direction of the electrostatic field $\boldsymbol{E}_e$ in the rod.
c) What is the motional emf in the rod?

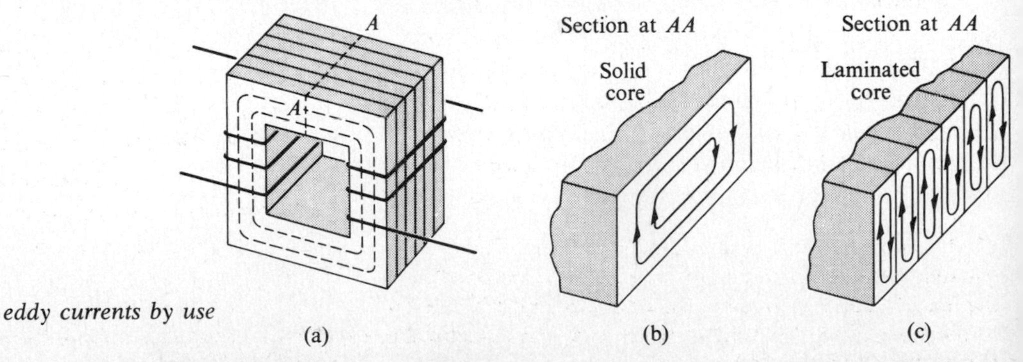

Fig. 33–14 *Reduction of eddy currents by use of a laminated core.*

d) What is the potential difference between its terminals? Which end is at the higher potential?

33–2 The cube in Fig. 33–15, 1 m on a side, is in a uniform magnetic field 0.2 T, directed along the y-axis. Wires A, C, and D move in the directions indicated, each with a velocity of 50 cm s^{-1}.

a) What is the magnitude of the equivalent nonelectrostatic field $\boldsymbol{E}_n$ in each wire?

b) What is the motional emf in each wire?

c) What is the potential difference between the terminals of each?

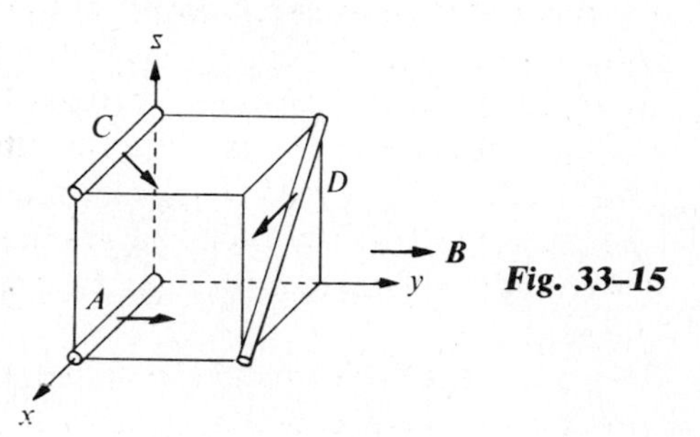

Fig. 33–15

33–3 A conducting rod ab in Fig. 33–2 makes contact with the metal rod $adcb$. The apparatus is in a uniform magnetic field 0.5 Wb m^{-2}, perpendicular to the plane of the diagram, and ℓ = 0.5 m.

a) Find the magnitude and direction of the emf induced in the rod when it is moving toward the right with a velocity of 4 m s^{-1}.

b) If the resistance of the circuit is 0.2 Ω (assumed constant), find the force required to maintain the rod in motion. Neglect friction.

c) Compare the rate at which mechanical work is done by the force (Fv) with the rate of development of heat in the circuit (i^2R).

33–4 A closely wound rectangular coil of 50 turns has dimensions of 12 cm × 25 cm. The plane of the coil is rotated from a position where it makes an angle of 45° with a magnetic field 2 T to a position perpendicular to the field in time t = 0.1 s. What is the average emf induced in the coil?

33–5 A coil of 1000 turns enclosing an area of 20 cm^2 is rotated from a position where its plane is perpendicular to the earth's magnetic field to one where its plane is parallel to the field, in 0.02 s. What average emf is induced if the earth's magnetic field is 6×10^{-5} T?

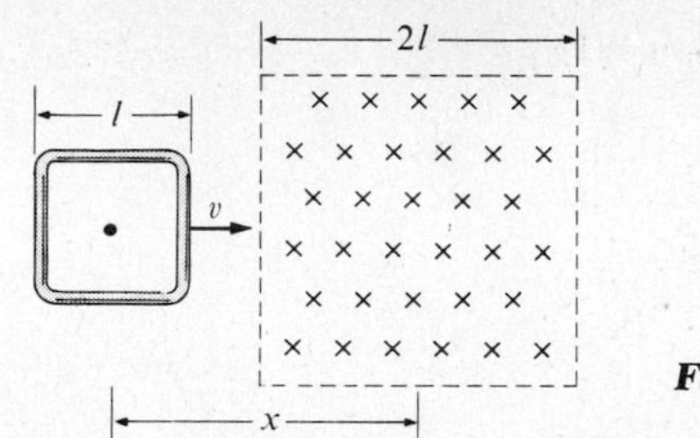

Fig. 33–16

33–6 A square loop of wire is moved at constant velocity v across a uniform magnetic field confined to a square region whose sides are twice the length of those of the square loop. (See Fig. 33–16.)

a) Sketch a graph of the external force F needed to move the loop at constant velocity, as a function of the distance x, from $x = -2\ell$ to $x = +2\ell$.

b) Sketch a graph of the induced current in the loop as a function of x, plotting clockwise currents upward and counterclockwise currents downward.

33–7 The long rectangular loop in Fig. 33–17, of width ℓ, mass m, and resistance R, starts from rest in the position shown and is acted on by a constant force $\boldsymbol{F}$. At all points to the right of the dotted line there is a uniform magnetic field of flux density $\boldsymbol{B}$, perpendicular to the plane of the diagram.

a) Sketch a graph of the velocity of the loop as a function of time.

b) Find the terminal velocity.

c) Derive the equation for the velocity as a function of time.

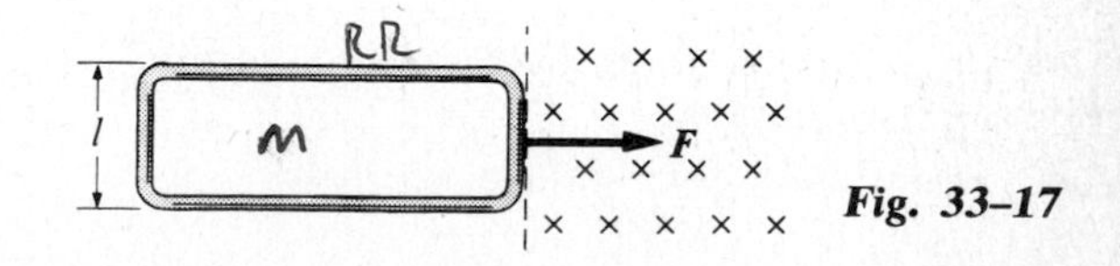

Fig. 33–17

33–8 A slender rod 1 m long rotates about an axis through one end and perpendicular to the rod, with an angular velocity of 2 rev s^{-1}. The plane of rotation of the rod is perpendicular to a uniform magnetic field 0.5 T.

a) What are the magnitude and direction of the equivalent nonelectrostatic field in the rod, at its midpoint?

b) What is the induced emf in the rod?

c) What is the potential difference between its terminals?

33–9 A flat square coil of 10 turns has sides of length 12 cm. The coil rotates in a magnetic field 0.025 T.

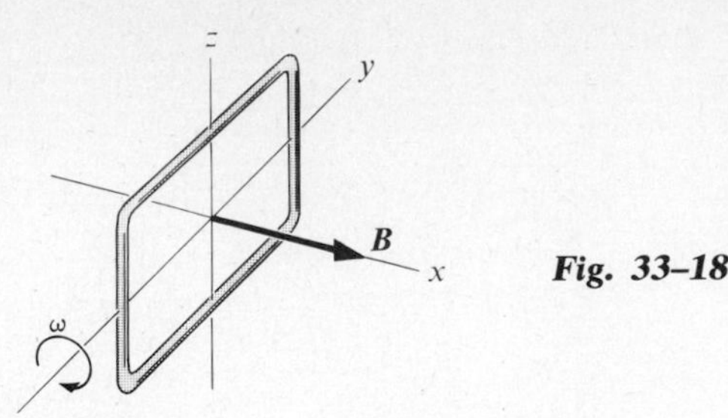

Fig. 33–18

a) What is the angular velocity of the coil if the maximum emf produced is 20 mV?

b) What is the average emf at this velocity?

33–10 A rectangular coil of wire having 10 turns with dimensions of 20 cm × 30 cm rotates at a constant speed of 600 rpm in a magnetic field 0.10 T. The axis of rotation is perpendicular to the field. Find the maximum emf produced.

33–11 The rectangular loop in Fig. 33–18, of area A and resistance R, rotates at uniform angular velocity ω about the y-axis. The loop lies in a uniform magnetic field of flux density $\boldsymbol{B}$ in the direction of the x-axis. Sketch the following graphs:

a) the flux Φ through the loop as a function of time (let $t = 0$ in the position shown in Fig. 33–18);

b) the rate of change of flux, $d\Phi/dt$;

c) the induced emf in the loop;

d) the torque Γ needed to keep the loop rotating at constant angular velocity;

e) the induced emf if the angular velocity is doubled. (Neglect the self-inductance of the loop.)

33–12 In Problem 33–11 and Fig. 33–18, let $A = 400\ \text{cm}^2$, $R = 2\ \Omega$, $\omega = 10\ \text{rad s}^{-1}$, $B = 0.5$ T. Find

a) the maximum flux through the loop,

b) the maximum induced emf,

c) the maximum torque.

d) Show that the work of the external torque in one revolution is equal to energy dissipated in the loop.

33–13 Suppose the loop in Fig. 33–18 is

a) rotated about the z-axis;

b) rotated about the x-axis;

c) rotated about an edge parallel to the y-axis.

What is the maximum induced emf in each case if the angular velocity is the same as in Problem 33–12?

33–14 A flexible circular loop 10 cm in diameter lies in a magnetic field 1.2 T, directed into the plane of the diagram in Fig. 33–19. The loop is pulled at the points indicated by the arrows, forming a loop of zero area in 0.2 s.

a) Find the induced emf in the circuit.

b) What is the direction of the current in R?

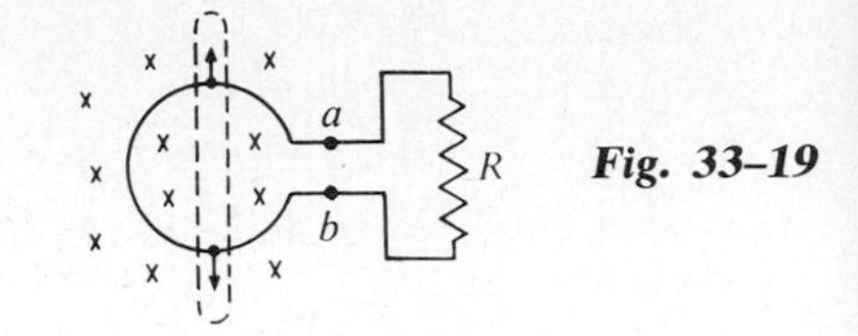

Fig. 33–19

33–15 The magnetic field within a long, straight solenoid of circular cross section and radius R is increasing at a rate dB/dt.

a) What is the rate of change of flux through a circle of radius r_1 inside the solenoid, normal to the axis of the solenoid, and with center on the solenoid axis?

b) Find the induced electric field $\boldsymbol{E}_n$ inside the solenoid, at a distance r_1 from its axis. Show the direction of this field in a diagram.

c) What is the induced electric field *outside* the solenoid, at a distance r_2 from the axis?

d) Sketch a graph of the magnitude of $\boldsymbol{E}_n$ as a function of the distance r from the axis, from $r = 0$ to $r = 2R$. Compare with part (e) of Problem 32–18.

e) What is the induced emf in a circular turn of radius $R/2$?

f) Of radius R?

g) Of radius $2R$?

33–16 A long, straight solenoid of cross-sectional area 6 cm^2 is wound with 10 turns of wire per centimeter, and the windings carry a current of 0.25 A. A secondary winding of 2 turns encircles the solenoid. When the primary circuit is opened, the magnetic field of the solenoid becomes zero in 0.05 s. What is the average induced emf in the secondary?

33–17 The magnetic field B at all points within the dotted circle of Fig. 33–20 is 0.5 T. It is directed into the plane of the diagram and is decreasing at the rate of $0.1\ \text{T s}^{-1}$.

a) What is the shape of the lines of force of the induced $\boldsymbol{E}_n$-field in Fig. 33–20, within the dotted circle?

b) What are the magnitude and direction of this field at any point of the circular conducting ring of radius 10 cm, and what is the emf in the ring?

c) What is the current in the ring, if its resistance is 2 Ω?

d) What is the potential difference between points a and b of the ring?

e) How do you reconcile your answers to (c) and (d)?

f) If the ring is cut at some point and the ends separated slightly, what will be the potential difference between the ends?

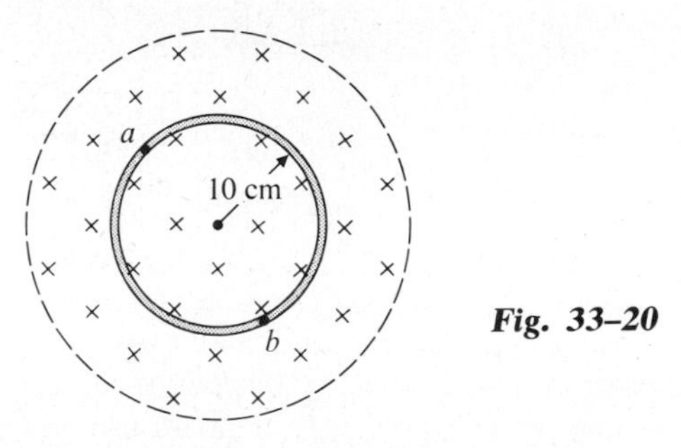

Fig. 33–20

33–18 A square conducting loop, 20 cm on a side, is placed in the same magnetic field as in Problem 33–17. (See Fig. 33–21.)

a) Copy Fig. 33–21, and show by vectors the directions and relative magnitudes of the induced electric field $\boldsymbol{E}_n$ at points a, b, and c.

b) Prove that the component of $\boldsymbol{E}_n$ along the loop has the same value at every point of the loop and is equal to that at the ring of Fig. 33–20.

c) What is the current in the loop, if its resistance is 2 Ω?

d) What is the potential difference between points a and b?

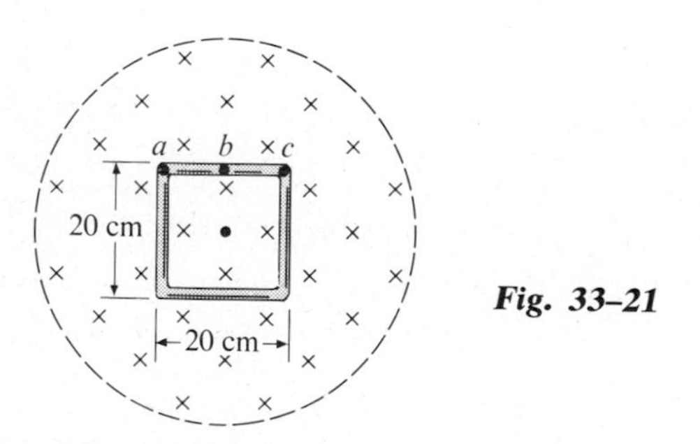

Fig. 33–21

33–19 A square conducting loop, 20 cm on a side, is placed in the same magnetic field as in Problem 33–17 with side ac along a diameter and with point b at the center of the field. (See Fig. 33–22.)

a) Copy Fig. 33–22, and show by vectors the directions and relative magnitudes of the induced electric field $\boldsymbol{E}_n$ at the lettered points.

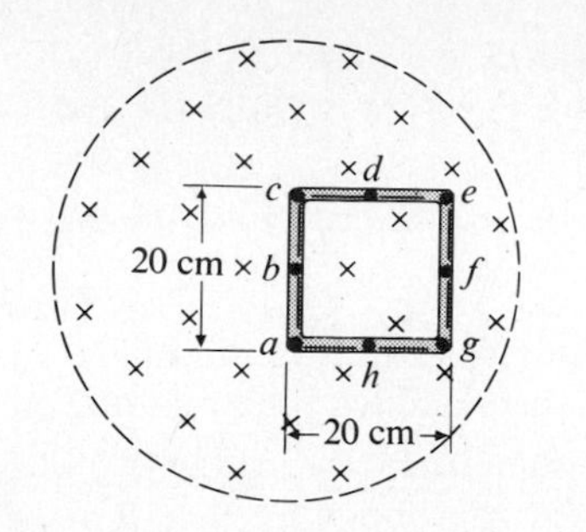

Fig. 33–22

b) What is the induced emf in side ac?

c) What is the induced emf in the loop?

d) What is the current in the loop, if its resistance is 2 Ω?

e) What is the potential difference between points a and c? Which is the higher potential?

33–20

a) Find the induced field $\boldsymbol{E}_n$ in each side of the square loop in Problem 33–19 and Fig. 33–22.

b) Find the induced emf in each side.

c) Find the electro*static* field $\boldsymbol{E}_e$ in each side.

d) Find the potential differences V_{ac}, V_{ce}, V_{eg}, and V_{ga}. What should be the sum of these potential differences?

33–21

a) What is the direction of the drift velocity of an electron at point b in the wire loop of Fig. 33–22?

b) What would be the direction of the drift velocity if the loop were at the left side of the center line instead of at the right side?

c) What would be the direction of the force on an electron if it were at rest at the center of the magnetic field, with no wire loop present?

d) How do you reconcile the answers to (a), (b), and (c)?

33–22 The magnetic flux in a toroid of small cross-sectional area and radius R is increasing at a constant rate $d\Phi/dt$.

a) What are the magnitude and direction of the induced $\boldsymbol{E}_n$-field at a point on the axis of the toroid at a distance x from its center? (See Section 32–5 for the corresponding expression for the $\boldsymbol{B}$-field at a point on the axis of a circular turn of wire carrying a constant current.)

b) Sketch a graph of the magnitude of $\boldsymbol{E}_n$ as a function of x.

c) Evaluate $\int_{-\infty}^{+\infty} E_n \, dx$ to find the induced emf in a wire that extends along the axis of the toroid from $x = -\infty$ to $x = +\infty$.

d) If the ends of the wire are joined by a conductor very far from the toroid, what is the induced emf in this circuit?

e) What is the induced emf in a ring that links the toroid closely?

33–23 The current in the wire AB of Fig. 33–23 is upward, and is increasing steadily at the rate of $2\ \text{A s}^{-1}$.

a) At an instant when the current is i, what are the magnitude and direction of the field $\boldsymbol{B}$ at a distance r from the wire?

b) What is the flux $d\Phi$ through the narrow shaded strip?

c) What is the total flux through the square loop of wire, 20 cm on a side?

d) What is the induced emf in the loop?

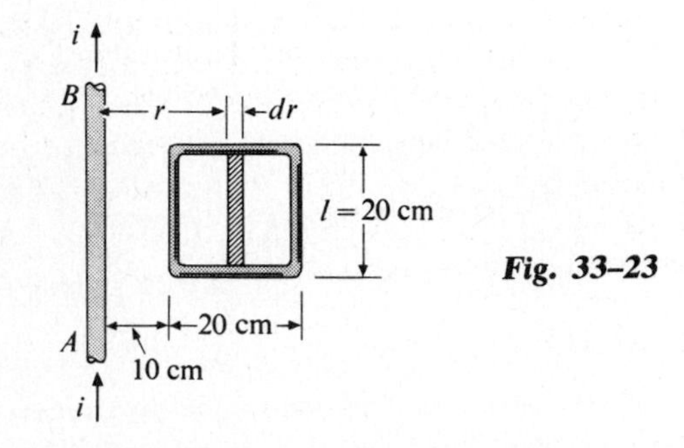

Fig. 33–23

33–24 A cardboard tube is wound with two windings of insulated wire, as in Fig. 33–24. Terminals a and b of winding A may be connected to a battery through a reversing switch. State whether the induced current in the resistor R is from left to right, or from right to left, in the following circumstances:

a) the current in winding A is from a to b and is increasing;

b) the current is from b to a and is decreasing;

c) the current is from b to a and is increasing.

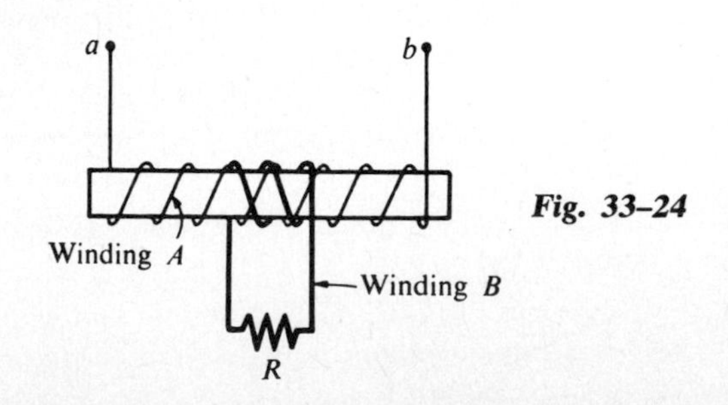

Fig. 33–24

33–25 Using Lenz's law, determine the direction of the current in resistor ab of Fig. 33–25 when

a) switch S is opened,

b) coil B is brought closer to coil A,

c) the resistance of R is decreased.

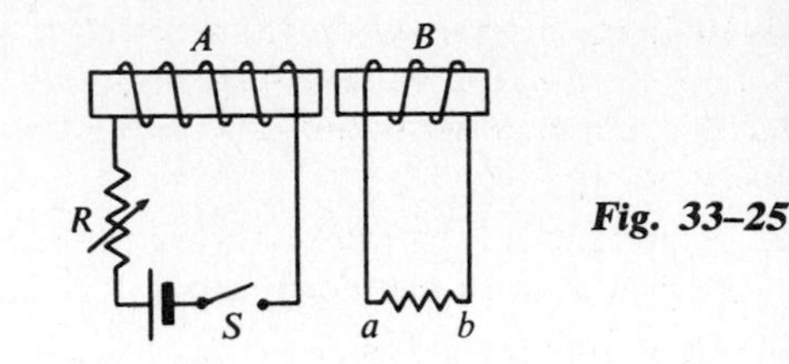

Fig. 33–25

33–26 The cross-sectional area of a closely wound search coil having 20 turns is $1.5\ \text{cm}^2$ and its resistance is $4\ \Omega$. The coil is connected through leads of negligible resistance to a ballistic galvanometer of resistance $16\ \Omega$. Find the quantity of charge displaced through the galvanometer when the coil is pulled quickly out of a region where $B = 1.8$ T to a point where the magnetic field is zero. The plane of the coil, when in the field, made an angle of 90° with the magnetic induction.

33–27 A solenoid 50 cm long and 8 cm in diameter is wound with 500 turns. A closely wound coil of 20 turns of insulated wire surrounds the solenoid at its midpoint, and the terminals of the coil are connected to a ballistic galvanometer. The combined resistance of coil, galvanometer, and leads is $25\ \Omega$.

a) Find the quantity of charge displaced through the galvanometer when the current in the solenoid is quickly decreased from 3 A to 1 A.

b) Draw a sketch of the apparatus, showing clearly the directions of windings of the solenoid and coil, and of the current in the solenoid. What is the direction of the current in the coil when the solenoid current is decreased?

33–28 A closely wound search coil has an area of $4\ \text{cm}^2$, 160 turns, and a resistance of $50\ \Omega$. It is connected to a ballistic galvanometer whose resistance is $30\ \Omega$. When the coil is rotated quickly from a position parallel to a uniform magnetic field to one perpendicular to the field, the galvanometer indicates a charge of 4×10^{-5} C. What is the magnitude of the field?

33–29 The orbit of an electron in a betatron is a circle of radius R. Suppose the electron is revolving in this orbit with a tangential velocity v.

a) What flux density is required to maintain the electron in this orbit if the magnitude of its velocity is constant?

b) If the flux density is uniform over the plane of the orbit, and is increasing at a rate dB/dt, what is the equivalent voltage accelerating the electron in each revolution?

33–30 In the diagram of the toroidal tube of the betatron, Fig. 33–12, a magnetic field is directed into the plane of the diagram and is increasing. An electron is moving clockwise.

a) Show that the emf induced around the electron's orbit is such as to accelerate the electron.

b) Show that the increasing radial force on the electron due to the magnetic field tends to prevent the electron from going to larger orbits.

33–31 A Faraday disk dynamo is to be used to supply current to a large electromagnet which requires 20,000 A at 1.0 V. The disk is to be 0.6 m in radius, and it turns in a magnetic field of 1.2 T, supplied by a smaller electromagnet.

a) How many revolutions per second must the disk turn?

b) What torque is required to turn the disk, assuming all the mechanical energy is dissipated as heat in the electromagnet?

33–32 A coil 4 cm in radius, containing 500 turns, turns with constant angular velocity about an axis along a diameter perpendicular to the earth's magnetic field, which may be taken as 0.5×10^{-4} T. What angular velocity must it have for the induced emf to have a maximum value of 1.0×10^{-3} V?

33–33 The coil described in Problem 33–32 is placed in a magnetic field which varies with time according to $B = 0.01t + (2 \times 10^{-4})t^3$, where B is in teslas and t in seconds. It is connected to a 500-Ω resistor.

a) Find the induced emf in the coil as a function of time.

b) What is the current in the resistor at time $t = 10$ s?

33–34 A search coil used to measure magnetic fields is to be made with a radius of 2 cm. It is to be designed so that flipping it 180° in a field of 0.1 T causes a total charge of 10^{-4} C to flow in a ballistic galvanometer when the total circuit resistance is 50 Ω. How many turns should the coil have?

Chapter 34

Inductance

34–1 MUTUAL INDUCTANCE

An emf is induced in a stationary circuit whenever the magnetic flux through the circuit varies with time. If the variation in flux is brought about by a varying current in a second circuit, it is convenient to express the induced emf in terms of the varying *current*, rather than in terms of the varying *flux*. We shall use the symbol i to represent the instantaneous value of a varying current.

Figure 34–1 is a sectional view of two closely wound coils of wire. A current in coil 1 sets up a magnetic field, as indicated by the dotted lines, and some of these lines pass through coil 2. Let us represent the flux through coil 2, produced by a current i_1 in coil 1, by Φ_{21}. The *mutual inductance* M_{21} of the coils is defined as the ratio of the product $N_2\Phi_{21}$ to the current i_1, where N_2 is the number of turns in coil 2:

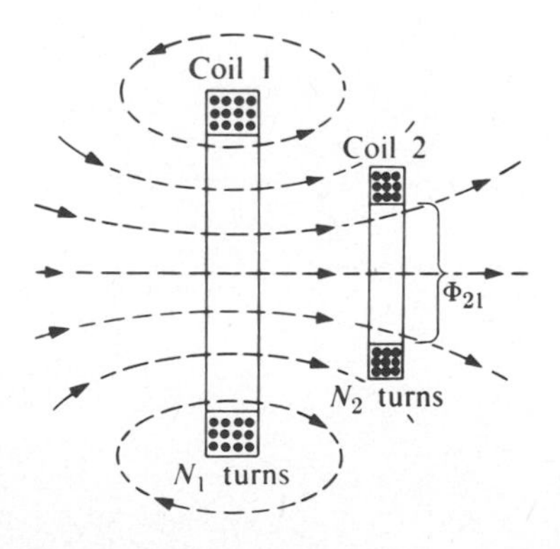

Fig. 34–1 A portion of the flux set up by a current in circuit 1 links with circuit 2.

$$M_{21} = \frac{N_2\Phi_{21}}{i_1}, \qquad \text{or } N_2\Phi_{21} = M_{21}i_1. \tag{34–1}$$

The product $N_2\Phi_{21}$ is called the number of *flux linkages* with coil 2.

If the current i_1 varies with time,

$$N_2\frac{d\Phi_{21}}{dt} = M_{21}\frac{di_1}{dt}.$$

The left side of this equation is the negative of the induced emf $\mathscr{E}_2$ in coil 2, so

$$\mathscr{E}_2 = -M_{21}\frac{di_1}{dt}. \tag{34–2}$$

From this point of view, the mutual inductance can be considered as the *induced emf in coil 2, per unit rate of change of current in coil 1*.

It can be shown that the same emf is induced in either of two circuits between which there is mutual

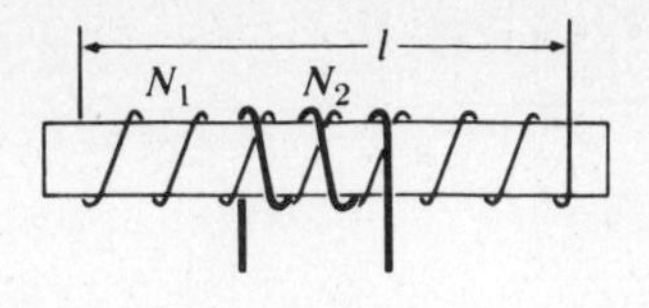

Fig. 34-2

inductance, when the current in the other changes at a given rate. That is,

$$\frac{N_2\Phi_{21}}{i_1} = \frac{N_1\Phi_{12}}{i_2}.$$

We can therefore drop the subscripts from M_{21} and write

$$\mathscr{E}_2 = -M\frac{di_1}{dt},$$

$$\mathscr{E}_1 = -M\frac{di_2}{dt}.$$

The mks unit of mutual inductance is *one volt per ampere per second.* This is called 1 *henry* (1 H), in honor of Joseph Henry.

Example A long solenoid of length ℓ and cross-sectional area A is closely wound with N_1 turns of wire. A small coil of N_2 turns surrounds it at its center, as in Fig. 34–2. A current i_1 in the solenoid sets up a $\boldsymbol{B}$ field at its center, of magnitude

$$B = \frac{\mu_0 N_1 I_1}{\ell}.$$

The flux through the central section is equal to BA, and since all of this flux links with the small coil the mutual inductance is

$$M = \frac{N_2\Phi_{21}}{i_1} = \frac{\mu_0 A N_1 N_2}{\ell}.$$

If $\ell = 0.50$ m, $A = 10\text{ cm}^2 = 10^{-3}\text{ m}^2$, $N_1 = 1000$ turns, $N_2 = 10$ turns,

$$M = \frac{(4\pi \times 10^{-7})(10^{-3})(10^3)(10)}{0.50}$$
$$\approx 25 \times 10^{-6}\text{ H} \approx 25\mu\text{H}.$$

34-2 SELF-INDUCTANCE

In the preceding section, the source of the magnetic field linking a circuit was assumed to be independent of the circuit in which the induced emf appeared. But whenever there is a current in any circuit, this current sets up a magnetic field which itself links with the circuit and which varies when the current varies. Hence any circuit in which there is a varying current has induced in it an emf, because of the variation in *its own* magnetic field. Such an emf is called a *self-induced electromotive force.*

Suppose a circuit has N turns of wire and that a flux Φ passes through each turn, as in Fig. 34–3. The number of flux linkages ($N\Phi$) per unit current is called the *self-inductance* of the circuit, L:

$$\boxed{L = \frac{N\Phi}{i}.} \tag{34–3}$$

The self-inductance of a circuit depends on its size, shape, number of turns, etc. It also depends on the magnetic properties of the material in which a magnetic field exists. For example, the self-inductance of a solenoid of given dimensions is much greater if it has an iron core than if it is in vacuum. If no *ferro*magnetic materials are present, the self-inductance is a constant, independent of the current, since then the flux density at any point is directly proportional to the current. When ferromagnetic

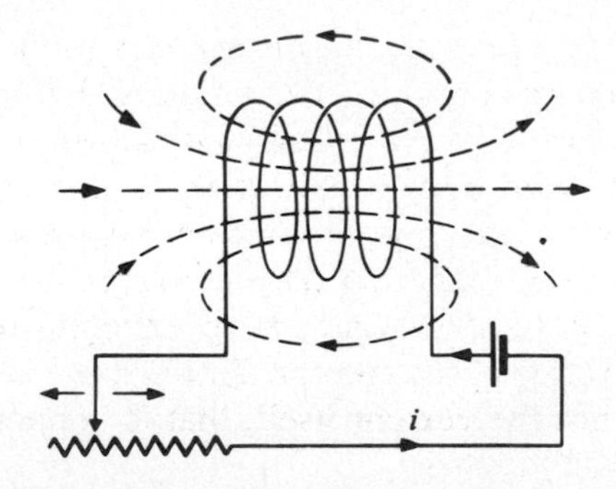

Fig. 34–3 A flux Φ linking a coil of N turns. When the current in the circuit changes, the flux changes also, and a self-induced emf appears in the circuit.

materials are present, the self-inductance varies in a complicated way as the current varies, because of the variations in permeability. For simplicity, we shall consider only circuits of constant self-inductance.

Equation (34-3) can be written as

$$N\Phi = Li.$$

If Φ and i change with time, then

$$N\frac{d\Phi}{dt} = L\frac{di}{dt},$$

and since the self-induced emf $\mathscr{E}$ is

$$\mathscr{E} = -N\frac{d\Phi}{dt},$$

it follows that

$$\boxed{\mathscr{E} = -L\frac{di}{dt}.} \tag{34-4}$$

The self-inductance of a circuit is therefore *the self-induced emf per unit rate of change of current.* The mks unit of self-inductance is 1 henry.

A circuit, or part of a circuit, which has inductance is called an *inductor.* An inductor is represented by the symbol

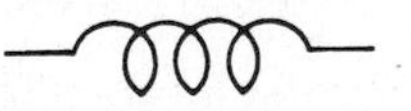

The direction of a self-induced (nonelectrostatic) field can be found from Lenz's law. We consider the "cause" of this field, and hence of the emf associated with it, to be the *changing current* in the conductor. If the current is *increasing,* the direction of the induced field is *opposite* to that of the current. If the current is *decreasing,* the induced field is in the *same* direction as the current. Thus it is the *change* in current, not the current itself, that is "opposed" by the induced field.

Example 1 An air-core toroid of cross-sectional area A and mean circumferential length ℓ is closely wound with N turns of wire. The flux in the toroid is

$$\Phi = BA = \frac{\mu_0 NiA}{\ell}.$$

Since all of the flux links with each turn, the self-inductance is

$$L = \frac{N\Phi}{i} = \frac{\mu_0 N^2 A}{\ell}.$$

Thus, if $N = 100$ turns, $A = 10\ \text{cm}^2 = 10^{-3}\ \text{m}^2$, $\ell = 0.50$ m,

$$L = \frac{(4\pi \times 10^{-7})(100)^2(10^{-3})}{0.50}$$
$$\approx 25 \times 10^{-6}\ \text{H}.$$

Example 2 If the current in the coil above increases uniformly from zero to 1 A in 0.1 s, find the magnitude and direction of the self-induced emf:

$$\mathscr{E} = L\frac{di}{dt} = (25 \times 10^{-6}\ \text{H})\frac{1\ \text{A}}{0.1\ \text{s}}$$
$$= 2.5 \times 10^{-4}\ \text{V}.$$

Since the current is increasing, the direction of this emf is opposite to that of the current.

34-3 ENERGY ASSOCIATED WITH AN INDUCTOR

Consider an inductor carrying a current i which is increasing at the rate di/dt. This changing current results in an emf $L(di/dt)$, so that power P is supplied to the inductor, where

$$P = \mathscr{E}i = Li\frac{di}{dt}.$$

The energy dW supplied in time dt is $P\,dt$, or

$$dW = Li\,di,$$

and the total energy supplied while the current increases from zero to I is

$$W = L\int_0^I i\,di = \tfrac{1}{2}LI^2. \tag{34-5}$$

After the current has reached its final steady value, $di/dt = 0$, and the power input is zero. The energy that has been supplied to the inductor is used to establish the magnetic field around the inductor, where it is "stored" as a form of potential energy so long as the current is maintained. When the circuit is opened the magnetic field collapses and this energy is returned to the circuit. It is this release of energy that maintains the arc often seen when a switch is opened in an inductive circuit.

Consider an inductor in the form of a closely wound toroid. From the preceding example, the self-inductance of the toroid is

$$L = \frac{\mu_0 N^2 A}{\ell},$$

and the energy stored in the toroid when the current in the windings is I is

$$W = \frac{1}{2} LI^2 = \frac{1}{2} \frac{\mu_0 N^2 A}{\ell} I^2.$$

We can think of this energy as localized in the volume enclosed by the windings, equal to ℓA. The energy per unit volume, denoted by u, is then

$$u = \frac{W}{\ell A} = \frac{1}{2} \mu_0 \frac{N^2 I^2}{\ell^2}.$$

But $N^2 I^2/\ell^2 = B^2/\mu_0^2$, so

$$\boxed{u = \frac{1}{2} \frac{B^2}{\mu_0},} \qquad (34\text{–}6)$$

which is the analog of the expression for the energy per unit volume in the electric field of an air capacitor, $\frac{1}{2}\epsilon_0 E^2$, Eq. (27–9).

34-4 THE *R–L* CIRCUIT

Every inductor necessarily has some resistance (unless its windings are superconducting). To distinguish between the effects of the resistance R and the self-inductance L, we represent the inductor as in Fig. 34–4, replacing it with an ideal resistanceless inductor in series with a noninductive resistor. The same diagram can also represent a resistor in series with an inductor, in which case R is the *total* resistance of the combination. By means of the dpdt switch, the $R–L$ circuit may be connected to a source of constant

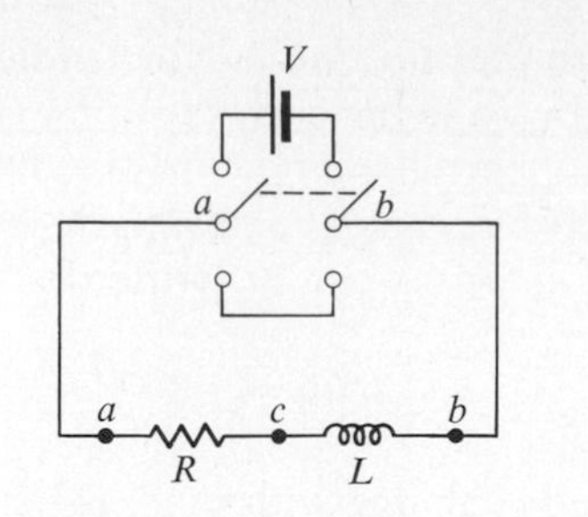

Fig. 34-4

terminal voltage V, or it may be shorted by the conductor across the lower switch terminals.

Suppose the switch in the diagram is suddenly closed in the "up" position. Because of the self-induced emf, the current does not rise to its final value at the instant the circuit is closed, but grows at a rate which depends on the inductance and resistance of the circuit.

At some instant after the switch is closed, let i represent the current in the circuit. The instantaneous potential difference across the inductor is

$$v_{cb} = L\frac{di}{dt},$$

and that across the resistor is

$$v_{ac} = iR.$$

Since $V = v_{ac} + v_{cb}$, it follows that

$$V = L\frac{di}{dt} + iR. \qquad (34\text{–}7)$$

The rate of increase of current is therefore

$$\frac{di}{dt} = \frac{V - iR}{L} = \frac{V}{L} - \frac{R}{L}i. \qquad (34\text{–}8)$$

At the instant the circuit is first closed, $i = 0$ and the current starts to grow at the rate

$$\left(\frac{di}{dt}\right)_{\text{initial}} = \frac{V}{L}.$$

The greater the self-inductance L, the more slowly does the current increase.

As the current increases, the term Ri/L increases also, and hence the *rate* of increase of current di/dt becomes smaller and smaller, as Eq. (34–8) shows.

When the current reaches its final *steady-state* value I, its rate of increase is zero. Then

$$0 = \frac{V}{L} - \left(\frac{R}{L}\right) I$$

and

$$I = \frac{V}{R}.$$

That is, the final current does not depend on the self-inductance and is the same as it would be in a pure resistance R connected to a cell of emf V.

To obtain an expression for the current as a function of time, we proceed just as we did for the problem of the charging capacitor, discussed in Section 29–7. We first rearrange Eq. (34–8):

$$\frac{di}{(V/R) - i} = \frac{R}{L}\,dt.$$

Integrating both sides, we find

$$-\ln(V/R - i) = (R/L)t + \text{constant}.$$

The integration constant is evaluated by noting that the initial current is zero, so $i = 0$ at time $t = 0$.

$$\text{constant} = -\ln V/R.$$

Rearranging again, we obtain

$$\ln\left(\frac{V}{R} - i\right) - \ln\frac{V}{R} = \ln\left(1 - \frac{Ri}{V}\right) = -\left(\frac{R}{L}\right)t,$$

$$\boxed{i = \frac{V}{R}\left(1 - e^{-(R/L)t}\right).} \qquad (34\text{–}9)$$

From this,

$$\frac{di}{dt} = \frac{V}{L} e^{-(R/L)t}. \qquad (34\text{–}10)$$

We note that at time $t = 0$, $i = 0$ and $di/dt = V/L$, and that as $t \to \infty$, $i \to V/R$ and $di/dt \to 0$, as predicted above. Figure 34–5(a) is a graph of Eq. (34–9).

At a time equal to L/R, the current has risen to $(1 - 1/e)$ or about 0.63 of its final value. This time is called the *time constant*, or the *decay constant*, for this circuit. The time required for the current to reach *half* its final value is obtained by setting

$$e^{-(R/L)t} = \tfrac{1}{2},$$

from which $T_{1/2} = (L/R)\ln 2$. For a given value of R, these times increase when the inductance L increases. Thus, although the graph of i vs. t has the same general shape whatever the inductance, the current rises rapidly to its final value if L is small, and slowly if L is large. For example, if $R = 100\ \Omega$ and $L = 10$ H,

$$\frac{L}{R} = \frac{10\text{ H}}{100\ \Omega} = 0.1\text{ s},$$

and the current increases to about 63% of its final value in 0.1 s. On the other hand, if $L = 0.01$ H,

$$\frac{L}{R} = \frac{0.01\text{ H}}{100\ \Omega} = 10^{-4}\text{ s},$$

and only 10^{-4} s is required for the current to increase to 63% of its final value.

If there is a steady current I in the circuit of Fig. 34–4 and the switch is then quickly thrown to the "down" position, the current decays, as shown in

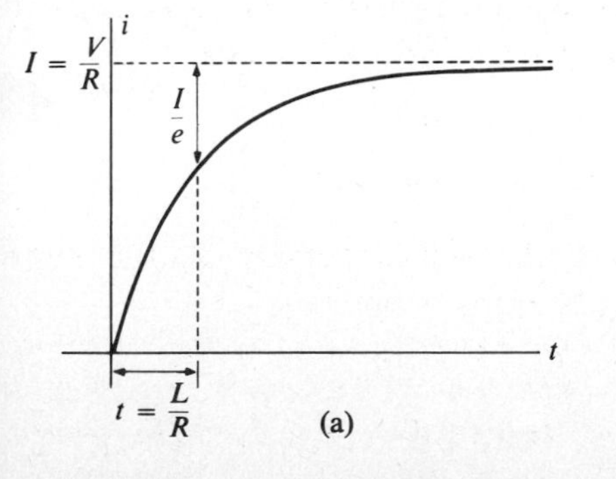

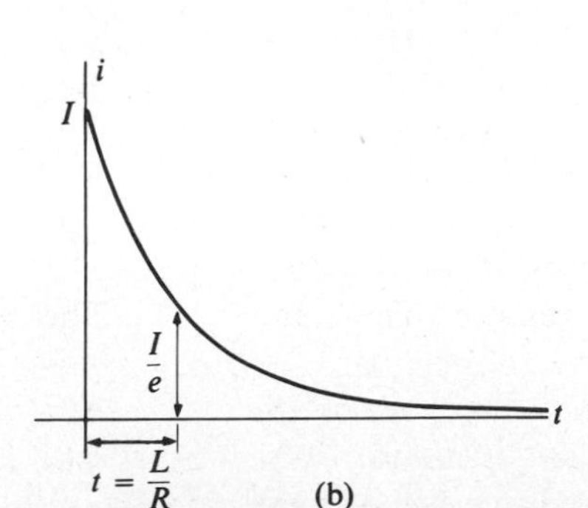

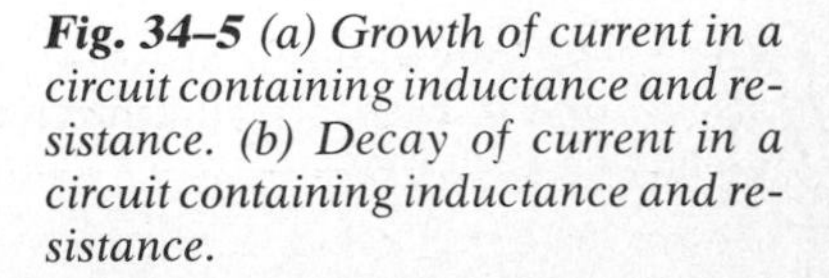

Fig. 34–5 *(a) Growth of current in a circuit containing inductance and resistance. (b) Decay of current in a circuit containing inductance and resistance.*

Fig. 34–5(b). The equation of the decaying current is

$$i = Ie^{-Rt/L}, \tag{34–11}$$

and the time constant, L/R, is the time for the current to decrease to $1/e$ or about 37% of its original value. Derivation of Eq. (34–11) is left as a problem. The energy necessary to maintain the current during this decay is provided by the energy stored in the magnetic field of the inductor.

34–5 THE *L–C* CIRCUIT

The behavior of an R–C circuit was discussed in Section 29–7, and that of an R–L circuit in Section 34–4. We now consider the L–C circuit shown in Fig. 34–6, a resistanceless inductor connected between the terminals of a charged capacitor. At the instant connections are made, in Fig. 34–6(a), the capacitor starts to discharge through the inductor. At a later instant, represented in Fig. 34–6(b), the capacitor has completely discharged and the potential difference between its terminals (and those of the inductor) has decreased to zero. The current in the inductor has meanwhile established a magnetic field in the space around it. This magnetic field now decreases, inducing an emf in the inductor in the same direction as the current. The current therefore persists, although with diminishing magnitude, until the magnetic field has disappeared and the capacitor has been charged in the opposite sense to its initial polarity, as in Fig. 34–6(c). The process now repeats itself in the reversed direction, and, in the absence of energy losses, the charges on the capacitor surge back and forth indefinitely. This process is called an *electrical oscillation.*

From the energy standpoint, the oscillations of an electrical circuit consist of a transfer of energy back and forth from the electric field of the capacitor to the magnetic field of the inductor, the *total* energy associated with the circuit remaining constant. This is analogous to the transfer of energy in an oscillating mechanical system from kinetic to potential, and vice versa.

The frequency of the electrical oscillations of a circuit containing inductance and capacitance only (a so-called L–C circuit) may be calculated in exactly the same way as the frequency of oscillation of a body suspended from a spring, discussed in Chapter 11. In the mechanical problem, a body of mass m is attached to a spring of force constant k. Let the body be displaced a distance A from its equilibrium positions and released from rest at time $t = 0$. Then, as shown in the left column of Table 34–1, the kinetic energy of the system at any later time is $\frac{1}{2}mv^2$, and its elastic potential energy is $\frac{1}{2}kx^2$. Because the system is conservative, the sum of these equals the initial energy of the system, $\frac{1}{2}kA^2$. The velocity v at any coordinate x is therefore

$$v = \pm\sqrt{\frac{k}{m}}\sqrt{A^2 - x^2}.$$

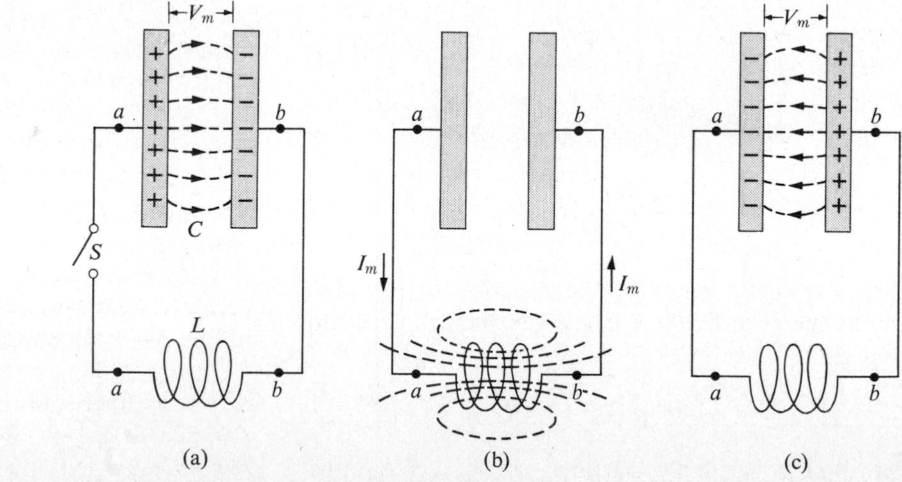

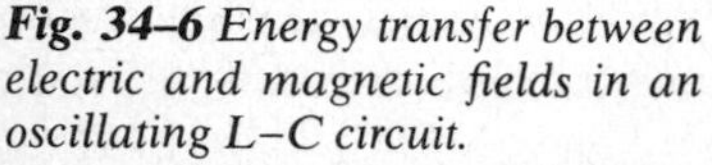
***Fig. 34–6** Energy transfer between electric and magnetic fields in an oscillating L–C circuit.*

Table 34–1 OSCILLATION OF A MASS ON A SPRING COMPARED WITH THE ELECTRICAL OSCILLATION IN AN L–C CIRCUIT

Mass on a spring	Circuit containing inductance and capacitance
Kinetic energy $= \frac{1}{2}mv^2$	Magnetic energy $= \frac{1}{2}Li^2$
Potential energy $= \frac{1}{2}kx^2$	Electrical energy $\frac{1}{2}\frac{q^2}{C}$
$\frac{1}{2}mv^2 + \frac{1}{2}kx^2 = \frac{1}{2}kA^2$	$\frac{1}{2}Li^2 + \frac{1}{2}\frac{q^2}{C} = \frac{1}{2}\frac{Q^2}{C}$
$v = \pm\sqrt{k/m}\sqrt{A^2 - x^2}$	$i = \pm\sqrt{1/LC}\sqrt{Q^2 - q^2}$
$v = \frac{dx}{dt}$	$i = \frac{dq}{dt}$
$x = A\cos\sqrt{k/m}\,t$	$q = Q\cos\sqrt{1/LC}\,t$
$= A\cos\omega t$	$= Q\cos\omega t$
$v = -\omega A\sin\omega t$	$i = -\omega Q\sin\omega t$
$= -v_{\max}\sin\omega t$	$= -I\sin\omega t$

The velocity equals dx/dt, and the coordinate x as a function of t has been shown to be

$$x = A\cos\left(\sqrt{\frac{k}{m}}\right)t = A\cos\omega t.$$

In the electrical problem, also a conservative system, a capacitor of capacitance C is given a charge Q and, at time $t = 0$, is connected to the terminals of an inductor of self-inductance L. The magnetic energy of the inductor at any later time corresponds to the kinetic energy of the vibrating body and is given by $\frac{1}{2}Li^2$. The electrical energy of the capacitor corresponds to the elastic potential energy of the spring and is given by $q^2/2C$, where q is the charge on the capacitor. The sum of these equals the initial energy of the system, $Q^2/2C$. The current, i, when the charge on the capacitor is q, is therefore

$$i = \sqrt{\frac{1}{LC}}\sqrt{Q^2 - q^2}.$$

The current $i = dq/dt$ varies with time in the same way as the coordinate x in the mechanical problem. That is,

$$q = Q\cos\left(\sqrt{\frac{1}{LC}}\right)t = Q\cos\omega t. \quad (34\text{–}12)$$

The angular frequency of the electrical oscillations is therefore

$$\boxed{\omega = \sqrt{\frac{1}{LC}}.} \quad (34\text{–}13)$$

This is called the *natural frequency* of the L–C circuit.

The striking parallelism between the mechanical and electrical systems displayed in Table 34–1 is only one of many such examples in physics. So close is the parallelism between electrical and mechanical (and acoustical) systems that it has been found possible to solve complicated mechanical and acoustical problems by setting up analogous electrical circuits and measuring the currents and voltages which correspond to the desired mechanical and acoustical "unknowns."

34–6 THE R–L–C CIRCUIT

In the above discussion we have assumed that the L–C circuit contains no *resistance.* This is an idealization, of course; for every real inductor there is resistance associated with the windings, and there may be resistance in the connecting wires as well. The effect of resistance is to drain away the electromagnetic energy and convert it to heat; thus, resistance in an electric circuit plays a role analogous to that of friction in a mechanical system.

Suppose an inductor of self-inductance L and resistance R is connected across the terminals of a capacitor. If the capacitor is initially charged, it starts to discharge at the instant the connections are made but, because of i^2R losses in the resistor, the energy of the inductor when the capacitor is completely discharged is *less* than the original energy of the capacitor. In the same way, the energy of the capacitor when the magnetic field has collapsed is still smaller, and so on.

If the resistance R is relatively small, the circuit oscillates, but with *damped harmonic motion* as shown in Fig. 34–7(a). As R is increased, the oscillations die out more rapidly. At a sufficiently large value of R the circuit no longer oscillates and is said to be *critically damped,* as in Fig. 34–7(b). For still larger resistances it is *overdamped,* as in Fig. 34–7(c).

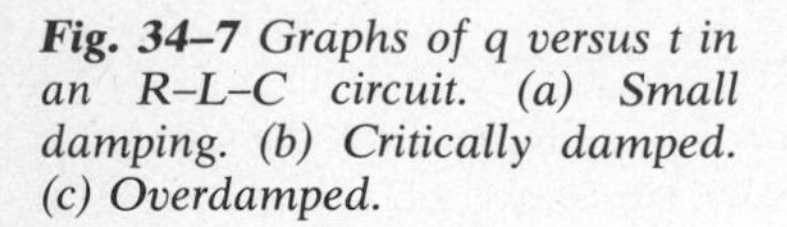

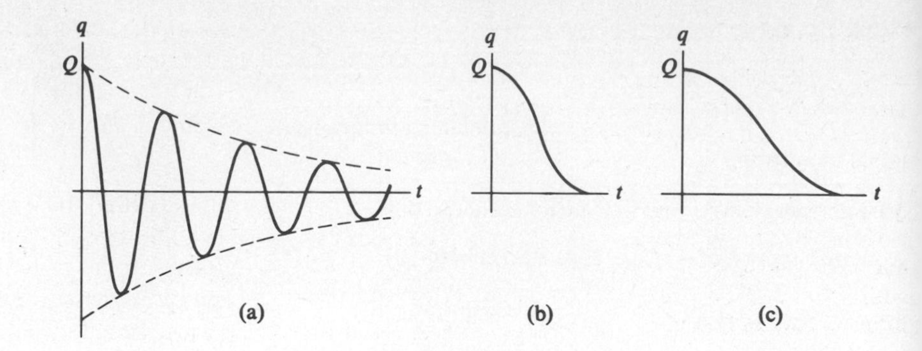

Fig. 34–7 *Graphs of q versus t in an R–L–C circuit. (a) Small damping. (b) Critically damped. (c) Overdamped.*

With the proper electronic circuitry, energy can be fed *into* an R–L–C circuit at the same rate as that at which it is dissipated by i^2R or radiation losses. In effect, a *negative resistance* is inserted in the circuit so that its total resistance is zero. The circuit then oscillates with *sustained* oscillations, as does the idealized circuit discussed in Section 34–5.

As a second example, we consider the process of charging a capacitor, which was discussed in Section 29–7 on the assumption that the circuit was a pure R–C circuit. We now take into account the self-inductance L which, although it may be small, is always present. We shall assume that the resistance R is sufficiently large so that the circuit is *overdamped.* Then instead of the circuit of Fig. 29–10 we have that of Fig. 34–8.

Let i and q represent the current and the charge on the capacitor at some instant after the switch is closed. Then

$$V_{ab} = iR + L\frac{di}{dt} + \frac{q}{C} = \text{constant.}$$

When this equation is differentiated with respect to t, and dq/dt is replaced with i, we obtain

$$\frac{d^2i}{dt^2} + \frac{R}{L}\frac{di}{dt} + \frac{1}{LC}i = 0. \qquad (34\text{–}14)$$

Thus when *all* of the electrical properties of a circuit are taken into account, the current as a function of time is the solution of a *second-order* differential equation. We shall not give the complete solution, but only state that the current is given by the *difference* between two exponential curves such as those shown by dotted lines in Fig. 34–9. It will be seen that the current does not change discontinuously from zero to a finite value, as in Fig. 29–11(a) where the effect of self-inductance was ignored, but increases continuously from zero to a maximum and then decreases. In many practical cases, however, the self-inductance of a circuit is so small that the analysis in Section 29–7 is a satisfactory approximation.

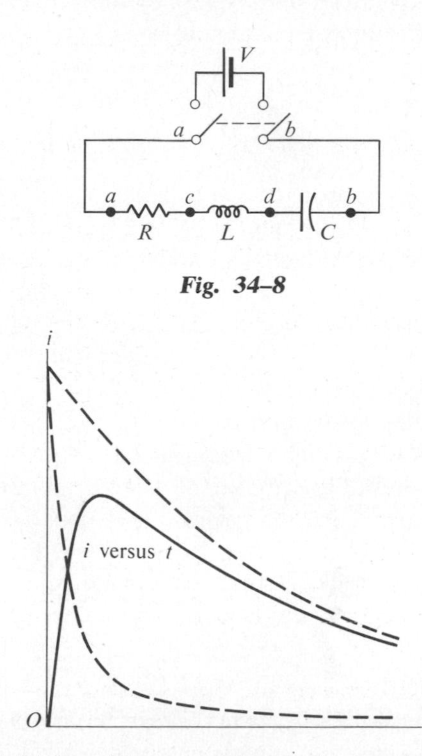

Fig. 34–8

Fig. 34–9 *The solid curve represents the charging current of a capacitor in an R–L–C circuit. It should be compared with Fig. 29–11, which is the graph when the self-inductance L of the circuit is neglected.*

PROBLEMS

34–1 A solenoid of length 10 cm and radius 2 cm is wound uniformly with 1000 turns. A second coil of 50 turns is wound around the solenoid at its center. What is the mutual inductance of the two coils?

34–2 A toroidal solenoid (cf. Section 32–7, Example 2) has a radius of 10 cm and a cross-sectional area of 5 cm^2, and is wound uniformly with 1000 turns. A second coil with 500 turns is wound uniformly on top of the first. What is the mutual inductance?

34–3 Find the self-inductance of the toroidal solenoid in Problem 34–2 if only the 1000-turn coil is used. How would your answer change if the two coils were connected in series?

34–4 A toroidal solenoid has two coils with n_1 and n_2 turns, respectively; it has radius r and cross-sectional area A.

a) Derive an expression for the self-inductance L_1 when only the first coil is used, and that for L_2 when only the second coil is used.

b) Derive an expression for the mutual inductance of the two coils.

c) Show that $M^2 = L_1 L_2$. This result is valid whenever all the flux linked by one coil is also linked by the other.

34–5 Two coils have a mutual inductance $M = 0.01$ H. The current i in the first coil is given by $i = (10 \text{ A}) \sin (120\ \pi \text{ s}^{-1})t$.

a) Find the induced emf in the second coil, as a function of time.

b) Suppose the above expression gives the current in the second coil. What is the induced emf in the first coil?

34–6 An inductor of inductance 5 H carries a current that decreases at a uniform rate, $di/dt = -0.02 \text{ A s}^{-1}$. Find the induced emf; what is its polarity?

34–7 An inductor with $L = 40$ H carries a current i that varies with time according to $i = (0.1 \text{ A}) \sin (120\ \pi \text{ s}^{-1})t$. Find an expression for the induced emf. What is the phase of $\mathcal{E}$ relative to i?

34–8 A coaxial cable consists of a small solid conductor of radius r_a supported by insulating disks on the axis of a thin-walled tube of inner radius r_b. Show that the self-inductance of a length ℓ of the cable is

$$L = \ell \frac{\mu_0}{2\pi} \ln \frac{r_b}{r_a}.$$

Assume the inner and outer conductors to carry equal currents in opposite directions. [*Hint:* Use Ampère's law to find the flux density at any point in the space between the conductors. Write the expression for the flux $d\Phi$ through a narrow strip of length ℓ parallel to the axis, of width dr, at a distance r from the axis of the cable and lying in a plane containing the axis. Integrate to find the total flux linking a current i in the central conductor.]

34–9 An inductor used in a dc power supply has an inductance of 20 H and a resistance of 200 Ω, and carries a current of 0.1 A.

a) What is the energy stored in the magnetic field?

b) At what rate is energy dissipated in the resistor?

34–10 (a) Show that the two expressions for self-inductance, namely,

$$\frac{N\Phi}{i} \quad \text{and} \quad \frac{\mathcal{E}}{di/dt},$$

have the same units. (b) Show that L/R and RC both have the units of time. (c) Show that 1 Wb s^{-1} equals 1 V.

34–11 The current in a resistanceless inductor is caused to vary with time as in the graph of Fig. 34–10.

a) Sketch the pattern that would be observed on the screen of an oscilloscope connected to the terminals of the inductor. (The oscilloscope spot sweeps horizontally across the screen at constant speed, and its vertical deflection is proportional to the potential difference between the inductor terminals.)

b) Explain why the inductor can be described as a "differentiating circuit."

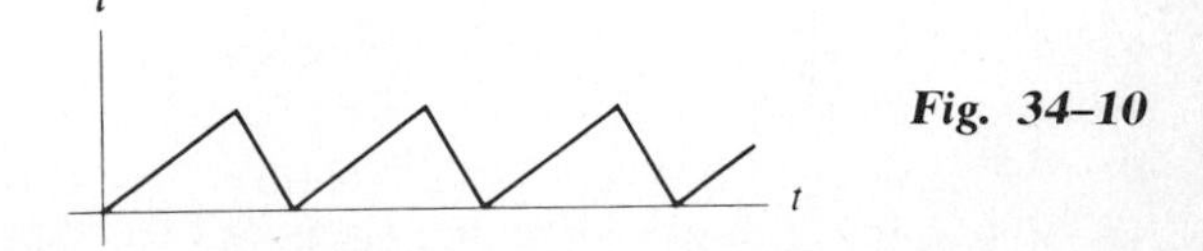

Fig. 34–10

34–12 An inductor of inductance 3 H and resistance 6 Ω is connected to the terminals of a battery of emf 12 V and of negligible internal resistance. Find (a) the initial rate of increase of current in the circuit, (b) the rate of increase of current at the instant when the current is 1 A, (c) the current 0.2 s after the circuit is closed, (d) the final steady-state current.

34–13 The resistance of a 10-H inductor is 200 Ω. The inductor is suddenly connected across a potential difference of 10 V.

a) What is the final steady current in the inductor?

b) What is the initial rate of increase of current?

c) At what rate is the current increasing when its value is one-half the final current?

d) At what time after the circuit is closed does the current equal 99% of its final value?

e) Compute the current at the following times after the circuit is closed: 0, 0.025 s, 0.05 s, 0.075 s, 0.10 s. Show the results in a graph.

34–14 An inductor of resistance R and self-inductance L is connected in series with a noninductive resistor of resistance R_0 to a constant potential difference V (Fig. 34–11). (a) Find the expression for the potential difference v_{cb} across the inductor at any time t after switch S_1 is closed. (b) Let $V = 20\text{ V}$, $R_0 = 50\,\Omega$, $R = 150\,\Omega$, $L = 5\text{ H}$. Compute a few points, and construct graphs of v_{ac} and v_{cb} over a time interval from zero to twice the time constant of the circuit.

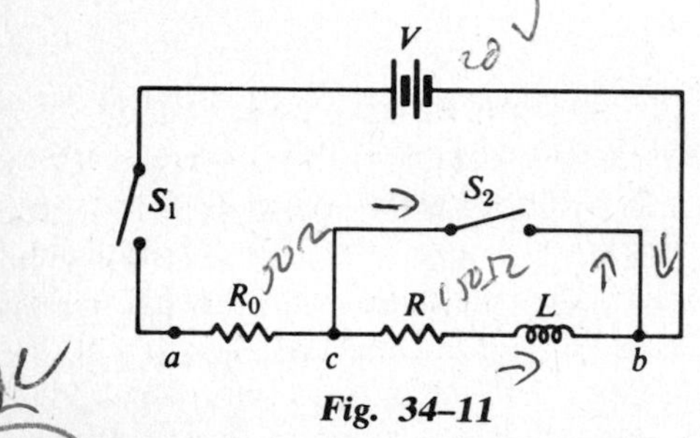

Fig. 34–11

34–15 After the current in the circuit of Fig. 34–11 has reached its final steady value the switch S_2 is closed, thus short-circuiting the inductor. What will be the magnitude and direction of the current in S_2, 0.01 s after S_2 is closed?

34–16 Refer to Problem 34–12. (a) What is the power input to the inductor at the instant when the current in it is 0.5 A? (b) What is the rate of dissipation of energy at this instant? (c) What is the rate at which the energy of the magnetic field is increasing? (d) How much energy is stored in the magnetic field when the current has reached its final steady value?

34–17 An inductor having inductance L and resistance R carries a current I. Show that the time constant is equal to twice the ratio of the energy stored in the magnetic field to the rate of dissipation in the resistance.

34–18 Show that the quantity $(L/C)^{1/2}$ has units of resistance (ohms).

34–19 The maximum capacitance of a variable air capacitor is 35 pF.

a) What should be the self-inductance of a coil to be connected to this capacitor if the natural frequency of the L–C circuit is to be 550×10^3 Hz, corresponding to one end of the broadcast band?

b) The frequency at the other end of the broadcast band is 1550×10^3 Hz. What must be the minimum capacitance of the capacitor if the natural frequency is to be adjustable over the range of the broadcast band?

34–20 Write an equation corresponding to Eq. (34–8) for the current in Fig. 34–4 just after the switch is thrown to the down position, if the initial current is I. Solve the resulting differential equation and verify Eq. (34–11).

34–21 In Fig. 34–6, equate the voltage across the inductor to that across the capacitor, to show that

$$L\frac{d^2q}{dt^2} = -\frac{q}{C}.$$

Show that this differential equation is satisfied by the function $q = Q\cos\omega t$, with ω given by $1/(LC)^{1/2}$.

34–22 An inductor having $L = 40$ mH is to be combined with a capacitor to make an L–C circuit with natural frequency 2×10^6 Hz. What value of capacitance should be used?

34–23 An inductor is made with two coils wound close together on a form, so all the flux linking one coil also links the other. The number of turns is the same in each. If the inductance of one coil is L, what is the inductance when the two coils are connected (a) in series; (b) in parallel? (c) If an L–C circuit using this inductor has natural frequency ω using one coil, what is the natural frequency when the two coils are used in series?

34–24 An L–C circuit in an AM radio tuner uses an inductor with inductance 0.1 mH and a variable capacitor. If the natural frequency of the circuit is to be adjustable over the range 0.5 to 1.5 MHz, corresponding to the AM broadcast band, what range of capacitance must the variable capacitor cover?

34–25 The equation preceding Eq. (34–14) may be converted into an energy relation. Multiply both sides of this equation by $i = dq/dt$. Show that the first term on the right is i^2R, the second can be written $d(\frac{1}{2}Li^2)/dt$, and the third can be written $d(q^2/2C)/dt$. Hence, show that the rate at which the battery delivers energy, $V_{ab}i$, is equal to the rate of energy dissipation in the resistor plus the sum of the rates of change of energy stored in the inductor and the capacitor.

Chapter 35

Magnetic Properties of Matter

35–1 MAGNETIC MATERIALS

In the preceding chapters we discussed the magnetic fields set up by moving charges or by currents in conductors, when the charges or conductors are in air (or, strictly speaking, in a vacuum). However, pieces of technical equipment, such as transformers, motors, generators, or electromagnets, always incorporate iron or an iron alloy in their structures, both for the purpose of increasing the magnetic flux and for confining it to a desired region. The magnetic field of a galvanometer or loudspeaker is set up by a *permanent* magnet, which produces this field without any apparent circulation of charge. The coating on magnetic tape makes a record of information supplied to it by the extent to which it becomes permanently magnetized.

We therefore turn next to a consideration of the magnetic properties which make iron and a few other ferromagnetic materials so useful. We shall find that magnetic properties are not confined to ferromagnetic materials but are exhibited (to a much smaller extent, to be sure) by *all* substances. A study of the magnetic properties of materials affords another means of gaining an insight into the nature of matter in general.

The existence of the magnetic properties of a substance can be demonstrated by supporting a small spherically shaped specimen by a fine thread and placing it near the poles of a powerful electromagnet. If the specimen is of iron or one of the *ferromagnetic* substances, it will be attracted into the strong part of the magnetic field. Not so familiar is the fact that any substance whatever will be influenced by the field, although to an extent which is extremely small compared with a substance like iron. Some substances will, like iron, be forced into the strong part of the field, while others will be urged to move toward the weak part of the field. The first type is called *paramagnetic*; the second, *diamagnetic*. All substances, including liquids and gases, fall into one or the other of these classes. Liquids and gases must, of course, be enclosed within some sort of container, and due allowance must be made for the properties of the container as well as those of the medium (usually air) in which the specimens are immersed.

In our discussion of induced emf's in Section 33–5, it was assumed that the core of the solenoid in Fig. 33–10 was empty. Suppose, however, that the solenoid is wound on a solid rod. It is found that, for a given change of current in the solenoid windings, the induced emf is not the same as when the core is empty, and hence the change in flux is not the same. If the rod is ferromagnetic, the change in flux for a given change in current will be very much larger; if paramagnetic, it will be *slightly* larger; and if diamagnetic, slightly *smaller* than if the core were empty. The differences in the latter two cases are, in fact, so small that this method is not a practical one for investigating these substances, but its principle is simpler than that of other methods and we shall ignore experimental difficulties.

The results are the same as if there were, in addition to the conduction current I_c in the solenoid, an additional current I_s circulating around the *surface* of the rod. This current is in such a direction as to set up a field of its own which *adds* to the field of the solenoid current if the rod is ferromagnetic or paramagnetic, and subtracts from it if the rod is diamagnetic. That is, in the former case, the surface current I_s is in the same direction as I_c and in the latter case it is opposite to I_c.

These *equivalent surface currents* are analogous to *bound surface charges* on a polarized dielectric, but with the difference that while the $\boldsymbol{E}$ field due to bound charges is *always* opposite to the $\boldsymbol{E}$ field of free charges, the $\boldsymbol{B}$ field due to equivalent surface currents may be in the same or the opposite direction as the $\boldsymbol{B}$ field of the conduction current in the solenoid. We return to the subject of surface currents in Section 35-3.

35-2 MAGNETIC PERMEABILITY

In Chapter 27, our formulation of the properties of a dielectric substance was based on a specimen in the form of a flat slab, inserted in the field between oppositely charged parallel plates. The electric field is wholly confined to the region between the plates if their separation is small, and therefore a flat slab between the plates will completely occupy all points of space at which an electric field exists. A specimen of this shape is not as well suited for a study of magnetic effects, however, since magnetic field lines are always *closed* lines and there is no way of producing a magnetic field which is confined to the region between two closely spaced surfaces. The magnetic field within a closely spaced toroidal winding, however, *is* wholly confined to the space enclosed by the winding. We shall accordingly use such a field as a basis for our discussion of magnetic properties of

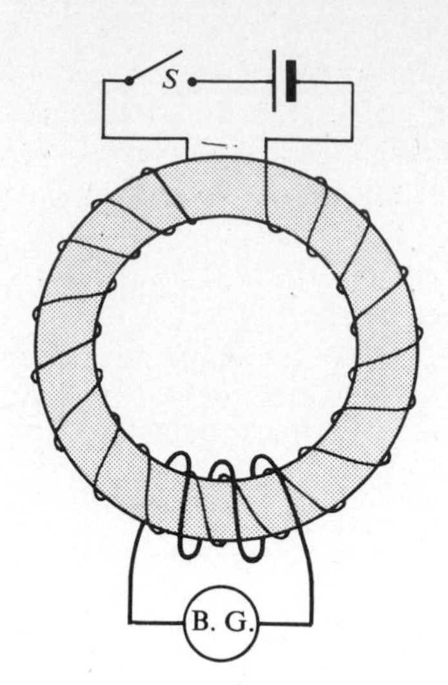

Fig. 35-1 *Magnetic specimen in the form of a ring with a toroidal winding.*

materials, using a specimen in the form of a ring on whose surface the wire is wound. Such a specimen is often called a *Rowland ring*, after J. H. Rowland, who made much use of it in his experimental and theoretical work on electricity and magnetism. The winding of wire around the specimen is called the *magnetizing winding*, and the current in the winding, the *magnetizing current.*

The magnetic field (flux density) within the space enclosed by a toroidal winding *in vacuum* is, from Section 32-7,

$$B_0 = \mu_0\left(\frac{Ni}{\ell}\right), \tag{35-1}$$

where ℓ is the circumference.

Suppose now that the same coil is wound on a Rowland ring, and that a second winding is placed on the ring, as in Fig. 35-1, with its terminals connected to a ballistic galvanometer. The $\boldsymbol{B}$ field within the ring may be measured by a procedure essentially the same as that of the search coil, discussed in Section 33-3. When switch S is opened and the magnetizing current suddenly drops to zero, the deflection of the ballistic galvanometer is proportional to the total flux change, and the initial flux Φ and magnetic field B can be determined.

The results are found *not* to agree with that computed from Eq. (35–1). If the core is made of a ferromagnetic material, the measured flux density will be very much larger; if made of a paramagnetic material, very slightly larger; and if made of a diamagnetic material, it will be very slightly *smaller* than the calculated value.

Let B represent the magnetic field in a material ring and B_0 the field in a "ring of vacuum." The ratio of B to B_0 is called the *relative permeability* of the material and is represented by K_m:

$$\boxed{K_m = \frac{B}{B_0}.} \tag{35–2}$$

$$K_m \text{ is} \begin{cases} \text{equal to 1 for vacuum,} \\ \text{slightly larger than 1 for paramagnetic materials,} \\ \text{slightly smaller than 1 for diamagnetic materials,} \\ \text{often much larger than 1 for ferromagnetic materials.} \end{cases}$$

Substituting for B_0 its value $\mu_0 Ni/\ell$, we get

$$B = K_m \mu_0 \left(\frac{Ni}{\ell}\right).$$

The product $K_m\mu_0$ is called the *permeability* of the material and is denoted by μ:

$$\boxed{\mu = K_m \mu_0.} \tag{35–3}$$

$$\mu \text{ is} \begin{cases} \text{equal to } \mu_0 \text{ for vacuum,} \\ \text{slightly greater than } \mu_0 \text{ for paramagnetic materials,} \\ \text{slightly smaller than } \mu_0 \text{ for diamagnetic materials,} \\ \text{often much larger than } \mu_0 \text{ for ferromagnetic materials.} \end{cases}$$

The unit in which μ is expressed is the same as that for μ_0, namely, webers per ampere·meter.

The expression for the flux density B in a material substance in the form of a Rowland ring may now be written

$$B = \mu\left(\frac{Ni}{\ell}\right). \tag{35–4}$$

35–3 MOLECULAR THEORY OF MAGNETISM

The earliest speculations as to the origin of the magnetic properties of a piece of lodestone, or of a permanent magnet, were that such a substance contained a number of particles each of which was itself a tiny magnet. In an unmagnetized body, the magnets were thought to be oriented at random; the process of magnetizing a body consisted of aligning these elementary magnets.

When it was discovered that magnetic effects could also be produced by *currents*, Ampère proposed the theory that the magnetic properties of a body arose from a multitude of tiny closed current loops within the body. The currents in these loops were assumed to continue indefinitely, as if there were no resistance. At the time the theory was proposed it was highly speculative, since no currents were known that would continue indefinitely of themselves. We now believe the theory to be essentially correct, the elementary current loops consisting of electrons spinning about their axes or revolving in orbits around nuclei. We have also seen that a current loop in a magnetic field experiences a torque, similar to the torque exerted on a bar magnet and attributable to a *magnetic moment* associated with the current loop. Thus these two viewpoints on the basis of magnetic behavior of matter are not inconsistent.

The theory of *diamagnetism* is based on the Faraday law. The electrons in an atom can be thought of as revolving in orbits about their parent nuclei, and hence their paths are equivalent to current loops of zero resistance, since their motion goes on indefinitely. When a magnetic field is established in a material, the increasing flux gives rise to an emf in each loop and the electrons are accelerated or decelerated, depending on their direction of revolution. (The process is like that taking place in a betatron.) The end result is the same as if a current were induced in each loop, *opposite* to the current in

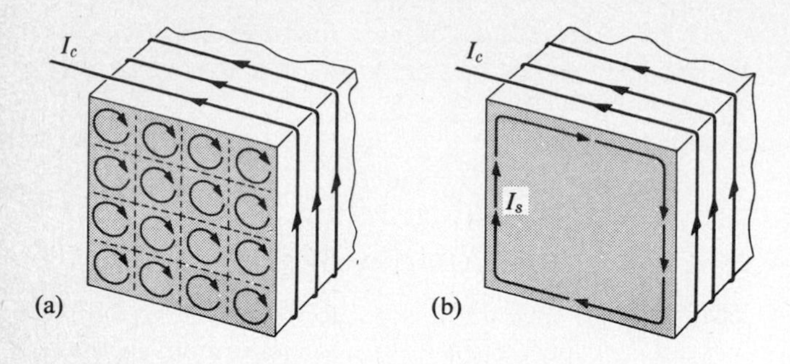

Fig. 35–2 (a) Alignment of induced atomic current loops in a diamagnetic substance. (b) Surface current equivalent to part (a).

the windings setting up the original field. In Fig. 35–2(a), these induced currents are represented by the small circles. At interior points the currents in adjacent loops are in opposite directions and cancel. The outer portions of the outside loops, however, are uncompensated, and the entire assembly of loops is equivalent to a current I_s circulating around the outside of the body, as in Fig. 35–2(b). These equivalent surface currents have essentially the same shape as a magnetizing winding around the body, and since they are opposite to the magnetizing current, their effect is to *decrease* the field set up by the current in the windings.

Since all atoms contain electron current loops, the phenomenon of diamagnetism is present in all substances. In many materials, the molecular currents in one direction are just equal to those in the other, and in the absence of an external field the molecules have no net magnetic moment. The phenomenon of *paramagnetism* results when the molecules of a substance have a *permanent magnetic moment*, corresponding to the molecules of a *polar* dielectric. A molecule with a permanent magnetic moment tends to become *aligned* with the field. We have shown that, in its equilibrium position, the flux through a current loop is in the same direction as that of the external field, so that in this case the molecular field *adds* to the external field.

As with the induced current loops in a diamagnetic material, the aligned loops in a paramagnetic substance are equivalent to a surface current around the periphery of a magnetized body, with the difference that they are now in the *same* direction as the magnetizing current.

Paramagnetic effects are small, but if they exist at all they always outweigh the (even smaller) diamagnetic effect which is present in *every* substance, and the material appears paramagnetic.

35–4 MAGNETIZATION AND MAGNETIC INTENSITY

The extent to which the molecular magnetic moments in a material are aligned in a particular direction, either spontaneously or by an external magnetic field, is described by a vector quantity called *magnetization* $\boldsymbol{M}$. We consider a small volume V; let a typical molecular magnetic moment in this volume be $\boldsymbol{m}$. Then the *total* magnetic moment within V is the vector sum $\sum \boldsymbol{m}$. The magnetization is defined as

$$\boldsymbol{M} = \frac{\sum \boldsymbol{m}}{V}. \tag{35–5}$$

Magnetization is therefore *magnetic moment per unit volume*, or *density of magnetic moment.* If the molecular magnetic moments are randomly oriented, or if there are none, as in a diamagnetic material in the absence of an external field, then the vector sum is zero and $\boldsymbol{M} = \boldsymbol{0}$.

Figure 35–3, which corresponds to Fig. 32–9, represents a portion of a long rod surrounded by a solenoidal winding. If the portion is not too near the ends of the solenoid, the $\boldsymbol{B}$ field set up by the solenoid is uniform and parallel to the solenoid axis and the magnetization $\boldsymbol{M}$ within the rod is uniform, also. We assume the rod to be paramagnetic, so that the equivalent surface currents are in the same direction as the current in the windings.

The magnetic moment of a current loop is defined as the product of the current and the area of the loop. Let I_s represent the total surface current around the periphery of a portion of the rod of length ℓ. (The surface current is actually distributed *continuously* along the surface of the rod.) The magnetic moment of this portion is then I_sA, where A is the cross-sectional area of the rod, and since the volume of the portion is $A\ell$, the magnetic moment per unit

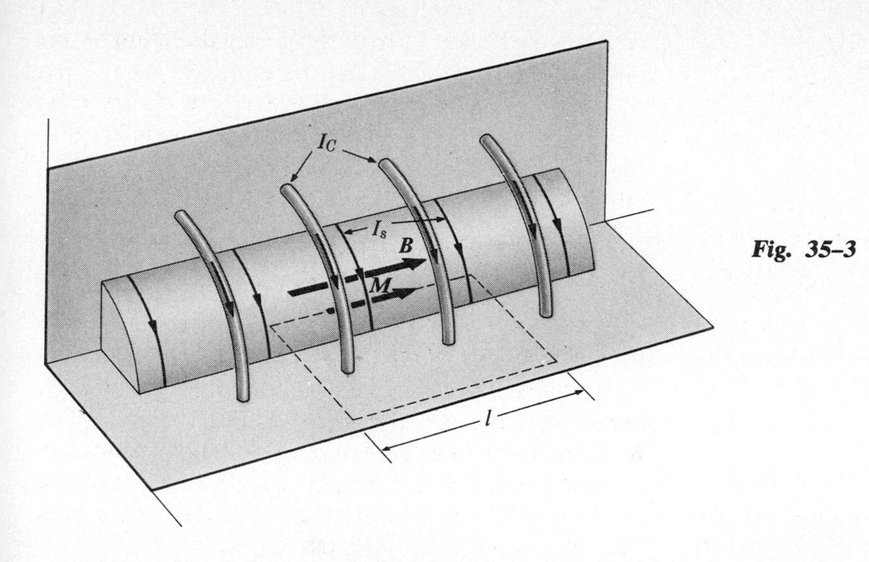

Fig. 35-3

volume, or the magnetization $\boldsymbol{M}$, has magnitude

$$M = \frac{I_s A}{A\ell} = \frac{I_s}{\ell} = j_s, \qquad (35\text{-}6)$$

where j_s is the *surface current per unit length* and is analogous to the bound charge per unit area, σ_b, in a polarized dielectric. In this special case, the magnetization is numerically equal to the surface current per unit length. More generally, the surface current per unit length equals the tangential component of $\boldsymbol{M}$ at the surface.

The broken line in Fig. 35-3 (compare Fig. 32-9) is a closed path of length ℓ. In analogy to Ampère's law, we may consider the integral

$$\oint \boldsymbol{M} \cdot d\boldsymbol{s}$$

around the broken line. Now the magnetization $\boldsymbol{M}$ is uniform within the rod and is zero outside; the sides are perpendicular to $\boldsymbol{M}$ and do not contribute to the integral, which is therefore equal simply to the product $M\ell$. But from Eq. (35-6), $M\ell = I_s$, so the integral is also equal to the total surface current I_s through the area bounded by this path. Thus,

$$\oint \boldsymbol{M} \cdot d\boldsymbol{s} = I_s. \qquad (35\text{-}7)$$

Although we have obtained this relation only for a special case, it is found to be true in general; it may be called Ampère's law for the magnetization vector $\boldsymbol{M}$.

The total $\boldsymbol{B}$ field in the rod is due both to the conduction current in the windings of the solenoid, and to the surface currents. Hence if I_c is the total *conduction current* through the rectangle, Ampère's law for $\boldsymbol{B}$ states that

$$\oint \boldsymbol{B} \cdot d\boldsymbol{s} = \mu_0(I_C + I_s). \qquad (35\text{-}8)$$

When I_s is eliminated between the preceding equations, we have

$$\oint \boldsymbol{B} \cdot d\boldsymbol{s} = \mu_0\left(I_C + \oint \boldsymbol{M} \cdot d\boldsymbol{s}\right)$$

or

$$\oint \left(\frac{\boldsymbol{B}}{\mu_0} - \boldsymbol{M}\right) \cdot d\boldsymbol{s} = I_C.$$

Let us define a new quantity called the *magnetic intensity* $\boldsymbol{H}$ (analogous to the displacement $\boldsymbol{D}$) as the vector difference

$$\boxed{\boldsymbol{H} = \frac{\boldsymbol{B}}{\mu_0} - \boldsymbol{M}.} \qquad (35\text{-}9)$$

Equation (35–8) then takes the simple form,

$$\oint \boldsymbol{H} \cdot d\boldsymbol{s} = I_C, \tag{35–10}$$

which is Ampère's law for the magnetic intensity vector: *the line integral of* **H** *around a closed path is equal to the conduction current only across any surface bounded by the path.*

The preceding equations have the same form if the rod is diamagnetic. If this is the case, the surface currents are *opposite* to the conduction currents, the vector **M** in Fig. 35–3 points to the left, and the magnitude of the vector difference $\boldsymbol{B}/\mu_0 - \boldsymbol{M}$ equals the algebraic sum $B/\mu_0 + M$.

In summary, for any closed path, the line integral of **B** equals the product of μ_0 and the *total* current across any surface bounded by the path, the line integral of **M** equals the *surface* current, and the line integral of **H** equals the *conduction* current. (If displacement currents are also present, they must be added to the conduction current.)

The magnetic intensity **H** has a number of useful properties that we shall discuss later. Like the magnetic field **B**, the magnetic intensity can be represented by lines called *lines of magnetic force.* In free space, these have the same shape as the lines of **B**. In a magnetized body, $\boldsymbol{H} = \boldsymbol{B}/\mu_0 - \boldsymbol{M}$. In free space, where $\boldsymbol{M} = 0$, $\boldsymbol{H} = \boldsymbol{B}/\mu_0$.

Although the symbols **H** and **B** are used nearly universally, these fields are given various names by various authors. In some books **H** is called the *magnetic field* and **B** the *magnetic induction* or the *magnetic flux density*. Other authors avoid confusion by referring simply to "the **H** field" and "the **B** field." In this book we usually call **B** the *magnetic field* and **H** the *magnetic intensity*. In the following discussion of the relation of these fields, we shall sometimes call **B** the *magnetic flux density* to emphasize that the **B** field is always used in computing magnetic flux for applications of Faraday's law.

35-5 MAGNETIC SUSCEPTIBILITY AND PERMEABILITY

The magnetization **M** in a material depends on the magnetic intensity **H** and on the nature of the material. We define a property of a material called its *magnetic susceptibility* χ_m by the equation

$$\boldsymbol{M} = \chi_m \boldsymbol{H}. \tag{35–11}$$

That is, the magnetic susceptibility is the *magnetization per unit magnetic intensity.* (Compare with the corresponding definition of electric susceptibility.) The magnetic susceptibility of vacuum is zero, because only a material substance can become magnetized. Magnetic susceptibility is a pure number, since the unit of both **H** and **M** is 1 A m^{-1}.

The magnetic susceptibility of a paramagnetic material is positive and that of a diamagnetic material is negative. Some representative values are given in Table 35–1.

Table 35–1 MAGNETIC SUSCEPTIBILITIES OF PARAMAGNETIC AND DIAMAGNETIC MATERIALS

Materials	Temperature, °C	$\chi_m = K_m - 1$
Paramagnetic		
Iron ammonium alum	−269	4830×10^{-5}
Iron ammonium alum	−183	213
Oxygen, liquid	−183	152
Iron ammonium alum	+20	66
Uranium	20	40
Platinum	20	26
Aluminum	20	2.2
Sodium	20	0.72
Oxygen gas	20	0.19
Diamagnetic		
Bismuth	20	-16.6×10^{-5}
Mercury	20	−2.9
Silver	20	−2.6
Carbon (diamond)	20	−2.1
Lead	20	−1.8
Rock salt	20	−1.4
Copper	20	−1.0

The orienting influence of a magnetic field on the molecules of a paramagnetic material is opposed by the *disorienting* effect of thermal agitation, which increases with temperature. Hence, the magnetic susceptibility of a paramagnetic material *decreases* with increasing temperature. For many materials, the temperature dependence is satisfactorily represented by the *Curie law*,

$$\chi_m = \frac{C}{T},$$

where C is a constant called the *Curie constant* and T is the Kelvin temperature. Diamagnetic susceptibilities are independent of temperature.

In terms of χ_m, $\mathbf{B}$ *is*

$$\boxed{\mathbf{B} = \mu_0(\mathbf{H} + \mathbf{M}) = \mu_0(1 + \chi_m)\mathbf{H}.}$$

The relative permeability K_m, defined in Section 35–2, is

$$K_m = 1 + \chi_m, \qquad (35\text{–}12)$$

so

$$B = \mu_0 K_m H.$$

The product $\mu_0 K_m$ is called the *permeability* μ:

$$\mu = \mu_0 K_m, \qquad (35\text{–}13)$$

and hence

$$\boxed{\mathbf{B} = \mu\mathbf{H}.} \qquad (35\text{–}14)$$

In vacuum, $K_m = 1$ and $\mu = \mu_0$. For this reason, the magnetic constant μ_0 is often referred to as "the permeability of vacuum" or "the permeability of free space." The "relative permeability of vacuum," where $\chi_m = 0$, is equal to 1. The relative permeability of a paramagnetic substance is slightly greater than 1, and that of a diamagnetic substance is slightly less than 1.

35–6 FERROMAGNETISM

If an iron rod is inserted in a solenoid, or if a toroidal winding surrounds an iron core, the $\mathbf{B}$ field in the iron may be hundreds or even thousands of times as great as that due to the current in the windings alone. Furthermore, B is not a *linear* function of H; in other words, the permeability μ is not a constant. To complicate matters even further, the permeability depends on the past history (magnetically speaking) of the iron, a phenomenon known as *hysteresis*. In fact, a flux may exist in the iron even in the absence of any external field; when in this state the iron is called a *permanent magnet.*

Any substance which exhibits the properties above is called *ferromagnetic*. Iron, nickel, cobalt, and gadolinium are the only ferromagnetic *elements* at room temperature, but a number of elements at low temperatures and several alloys whose components are not ferromagnetic also show these effects.

Because of the complicated relation between the flux density B and the magnetic intensity H in a ferromagnetic material, it is not possible to express B as a simple function of H. Instead, the relation between these quantities is either given in tabular form or represented by a graph of B versus H, called the *magnetization curve* of the material.

The initial magnetization curve of a specimen of annealed iron is shown in Fig. 35–4 in the curve labeled B versus H. The permeability μ, equal to the ratio of B to H, can be found at any point of the curve by dividing the flux density B, at the point, by the corresponding magnetic intensity H. For example, when $H = 150\ \text{A m}^{-1}$, $B = 1.01\ \text{T} = 1.01\ \text{Wb m}^{-2}$, and

$$\mu = \frac{B}{H} = \frac{1.01\ \text{T}}{150\ \text{A m}^{-1}}$$
$$= 67{,}500 \times 10^{-7}\ \text{Wb A}^{-1}\ \text{m}^{-1}.$$

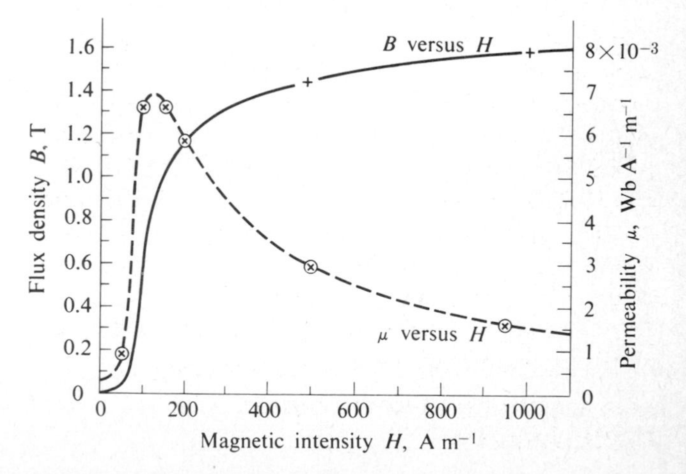

Fig. 35–4 *Magnetization curve and permeability curve of annealed iron.*

Table 35–2 MAGNETIC PROPERTIES OF ANNEALED IRON

H, $\mathrm{A\,m^{-1}}$	B, T	$\mu = B/H$, $\mathrm{Wb\,A^{-1}\,m^{-1}}$	$K_m = \mu/\mu_0$	$\mu_0 H$, T	$\mu_0 M = B - \mu_0 H$, T
0	0	$3{,}100 \times 10^{-7}$	250	0	0
10	0.0042	4,200	330	0.000013	0.0042
20	0.010	5,000	400	0.000025	0.010
40	0.028	7,000	560	0.000050	0.028
50	0.043	8,600	680	0.000063	0.043
60	0.095	16,000	1270	0.000078	0.095
80	0.45	56,000	4500	0.000104	0.45
100	0.67	67,000	5300	0.00013	0.67
150	1.01	67,500	5350	0.00019	1.01
200	1.18	59,000	4700	0.00025	1.18
500	1.44	28,800	2300	0.00063	1.44
1,000	1.58	15,800	1250	0.0013	1.58
10,000	1.72	1,720	137	0.013	1.71
100,000	2.26	226	18	0.13	2.13
800,000	3.15	39	3.1	1.00	2.15

It is evident that the permeability is not a constant. The curve labeled μ versus H in Fig. 35–4 is a graph of μ as a function of H.

Table 35–2 covers a wider range of values for the same specimen, and also includes values of the relative permeability K_m and the magnetization M (multiplied by μ_0). The values of B, $\mu_0 H$, and $\mu_0 M$ are plotted in Fig. 35–5 as functions of H. Because of the wide range of values of H, the horizontal scale has been compressed beyond $H = 1000\ \mathrm{A\,m^{-1}}$. It will be seen that when H is relatively small, practically all of the flux density B is due to the magnetization M (or to the equivalent surface currents). Beyond the point where H is of the order of $100{,}000\ \mathrm{A\,m^{-1}}$ there is very little further increase in the magnetization M and the iron is said to become *saturated*. Further increases in B are due almost wholly to increases in H resulting from increased current in the windings.

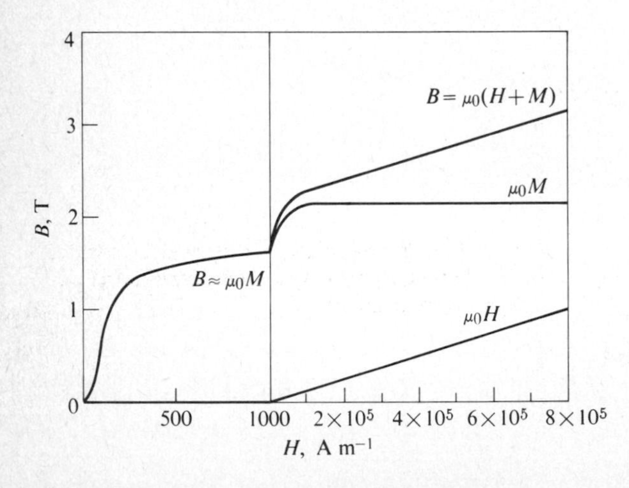

Fig. 35–5 *Graph of the data in Table 35–2.*

35–7 MAGNETIC DOMAINS

The electrons in most of the metallic ions which form the lattice structure of a metal are paired off, with half spinning one way and half the other way. The ion is thus "magnetically neutral." The ions of the ferromagnetic elements, iron, nickel, cobalt, etc., are exceptions. In particular, that of iron has an excess of two uncompensated electrons; the magnetization of iron is due almost entirely to the alignment of the magnetic moments of these electrons. When a sample of iron is magnetically *saturated*, all of the uncompensated electrons are spinning with their axes in the direction of the magnetizing field.

We can now understand the reason for the shape of the graph of $\mu_0 M$ versus H in Fig. 35–5. As H is increased from zero, more and more of the spinning electrons become aligned with the $\boldsymbol{H}$ field and M increases steadily. When H is about 100,000 A m^{-1}, practically all of the electrons have lined up with the $\boldsymbol{H}$ field. No further increase in M can occur, and the iron is saturated.

The magnetic moment of a spinning electron is called *one Bohr magneton*, equal to

$$\mu_B = 0.927 \times 10^{-23} \text{ A m}^2.$$

Reference to Fig. 35–5 shows that in saturated iron the value of $\mu_0 M$ is about 2.15 T, so that the saturation magnetization is

$$M = \frac{2.15 \text{ T}}{12.57 \times 10^{-7} \text{ Wb A}^{-1} \text{ m}^{-1}} = 1.71 \times 10^6 \text{ A m}^{-1}.$$

The density of iron is 7.8 g cm^{-3} and its atomic mass is 56 g $mole^{-1}$. The volume of 1 mole is therefore 7.18 cm^3 and since there are 6.02×10^{23} atoms in a mole, the number of atoms per cubic centimeter is 8.38×10^{22} and the number per cubic meter is 8.38×10^{28}. The magnetic moment *per atom*, in saturated iron, is therefore

$$\frac{1.71 \times 10^6 \text{ A m}^{-1}}{8.38 \times 10^{28} \text{ atoms m}^{-3}} = 2.04 \times 10^{-23} \text{ A m}^2 \text{ atom}^{-1}$$

which is almost exactly equal to two Bohr magnetons!

The above description enables us to understand the phenomenon of *saturation*, the existence of an upper limit on the magnetization of a material, but it does not account for the observation that in small fields ferromagnetic materials typically have much larger susceptibility, and therefore larger permeability, than paramagnetic materials. Ferromagnetism results because of a spontaneous, self-aligning, cooperative interaction among relatively large numbers of iron atoms.

Even a reasonably complete description of this phenomenon would take us beyond the scope of this book. It must suffice to state that there exist in ferromagnetic materials small regions called *domains*, in each of which, as a result of molecular interactions, the molecular magnetic moments are all aligned parallel to one another. In other words, each domain is *spontaneously magnetized to saturation* even in the absence of any external field. The directions of magnetization in different domains are not necessarily parallel to one another, so that in an unmagnetized specimen the *resultant* magnetization is zero. When the specimen is placed in a magnetic field the resultant magnetization may increase in two different ways, either by an increase in the volume of those domains which are favorably oriented with respect to the field at the expense of unfavorably oriented domains, or by the rotation of the direction of

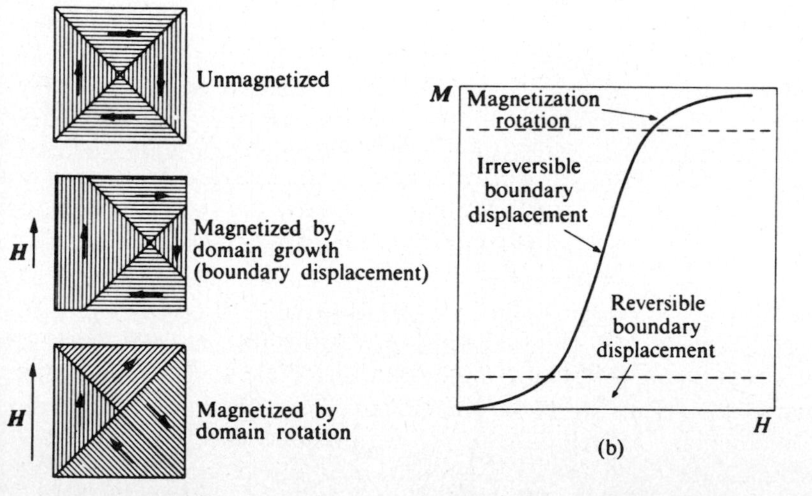

Fig. 35–6 *Schematic diagram showing magnetization by domain growth and domain rotation.*

magnetization toward the direction of the field. These two methods are shown in Fig. 35–6(a).

In weak fields, the magnetization usually changes by means of domain boundary displacements, so that the favorably oriented domains increase in size. In strong fields the magnetization usually changes by rotation of the direction of magnetization. The curve in Fig. 35–6(b) shows the regions in which each process is dominant. In small fields the changes are reversible. That is, the boundaries return to their original positions when the field is removed. In stronger fields the changes are irreversible and the substance remains magnetized when the external field is removed.

The sizes of the domains can be studied by spreading a finely divided magnetic powder on the surface of the specimen, a technique first developed by F. H. Bitter. The powder particles collect along the boundaries between domains and may be examined under a microscope. The size of the domains may vary widely, depending on the size of the specimen and whether it is a single crystal or is polycrystalline. Typical values are from 10^{-6} to 10^{-2} cm^3, which means that a domain may contain from 10^{17} to 10^{21} molecules.

The *Barkhausen effect*, which is most pronounced along the steeply rising portion of the magnetization curve, is believed to be caused by the irregular motion of domain boundaries. If a rod of ferromagnetic material is surrounded by a search coil connected to an audio amplifier, and the rod is placed in a magnetic field that can be steadily increased or decreased, a crackling sound is heard from a speaker connected to the amplifier. As the domain boundaries change in size, each change induces a sudden short rush of current through the search coil and these surges are heard as noise from the speaker.

Every ferromagnetic material has a critical temperature called the *Curie temperature*, above which the material is paramagnetic but not ferromagnetic. As the Curie temperature is reached, the energy associated with thermal motion in the material becomes large enough (compared to the interaction energies responsible for the alignment of magnetic moments in a domain) so that this alignment disappears. Above the Curie temperature there is no spontaneous magnetization. The transition from ferromagnetic to paramagnetic behavior is a *phase change* analogous to the transitions between solid, liquid, and gaseous phases of matter, discussed in Chapter 18.

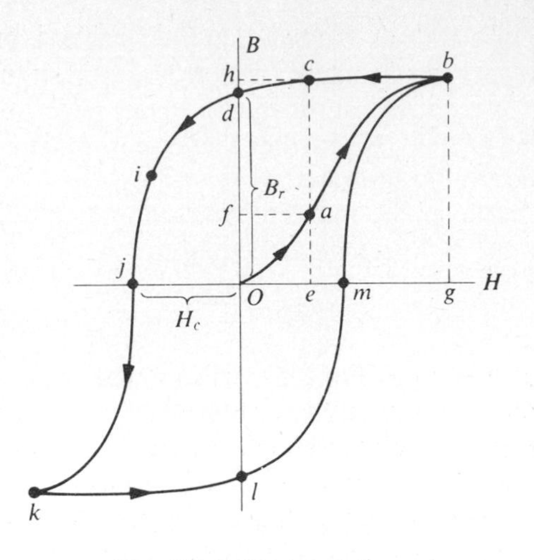

Fig. 35–7 *Hysteresis loop.*

35–8 HYSTERESIS

A magnetization curve such as that in Fig. 35–4 expresses the relation between the magnetic field B in a ferromagnetic material and the corresponding magnetic intensity H, *provided the sample is initially unmagnetized and the magnetic intensity is steadily increased from zero.* Thus, in Fig. 35–7, if the magnetizing current in the windings of an unmagnetized ring sample (as in Section 35–2) is steadily increased from zero until the magnetic intensity H corresponds to the abscissa Oe, the field B is given by the ordinate Of. If, starting from the same unmagnetized state, the magnetic intensity H is first increased from zero to Og and then *decreased* to Oe, the magnetic state of the sample follows the path $Oabc$. The value of B when the magnetic intensity has been reduced to Oe is represented by the ordinate Oh rather than Of. If the magnetizing current is now reduced to zero, the curve continues to point d, where B is Od.

Thus, B in the sample is seen to depend not only on H, but on the *magnetic history* of the sample as well. The sample "remembers" that it has been magnetized to point b even after the magnetizing current has been cut off. At point d it has become a *permanent magnet.* This behavior of the material, as evidenced by the fact that the B–H curve for decreasing H does not coincide with that for increasing H, is called *hysteresis.* The term means literally, "to lag behind."

In many pieces of electrical apparatus, such as transformers and motors, masses of iron are located in magnetic fields whose direction is continually reversing. That is, the magnetic intensity H increases from zero to a certain maximum in one direction, then decreases to zero, increases to the same maximum but in the opposite direction, decreases to zero, and continues to repeat this cycle over and over. The magnetic field B within the iron reverses also, but in the manner indicated in Fig. 35–7, tracing out a closed curve in the B–H plane known as a *hysteresis loop.*

Positive values of H or B in Fig. 35–7 indicate that the respective directions of these quantities are, say, clockwise around the ring, while negative values mean that the directions are counterclockwise. The magnitude and direction of H are determined solely by the current in the winding, while those of B depend on the magnetic properties of the sample and its past history. Note that at points in the second and fourth quadrants, B and H are *opposite* in direction.

The ordinate Od in Fig. 35–7 represents the flux density B remaining in the specimen when the magnetic intensity has been reduced to zero. It is called the *retentivity* or *remanence* of the specimen and is designated by B_r. The abscissa Oj represents the reversed magnetic intensity H needed to reduce the flux density to zero after the specimen has been magnetized to saturation in the opposite direction, and it is called the *coercive force* or the *coercivity*, H_c.

The magnetization curve Oab in Fig. 35–7 shows the sample initially unmagnetized. One may wonder how this can be accomplished, since cutting off the magnetizing current does not reduce the flux density in the material to zero. A sample may be demagnetized by reversing the magnetizing current a number of times, decreasing its magnitude with each reversal.

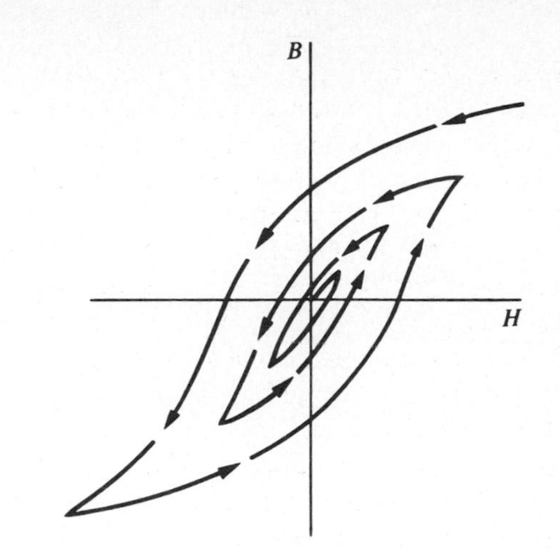

Fig. 35–8 *Successive hysteresis loops during the operation of demagnetizing a ferromagnetic sample.*

The sample is thus carried around a hysteresis curve which winds more and more closely about the origin (see Fig. 35–8).

Hysteresis effects introduce a similar difficulty in measuring the flux in a sample. In the absence of hysteresis, the flux falls to zero when the magnetizing current is cut off, and a ballistic galvanometer connected to a search coil around the specimen indicates the flux previously present in the specimen. But since B does not become zero when H is reduced to zero, ballistic galvanometer deflections indicate only the *changes* in flux corresponding to changes in the magnetizing force. Hence, in practice, the complete hysteresis loop must be traced out in stepwise fashion, measuring the changes in flux that accompany the changes in magnetizing current.

It is evidently desirable that a material for permanent magnets should have both a large retentivity (so that the magnet will be "strong") and a large coercive force (so that the magnetization will not be wiped out by stray external fields). Some typical values are given in Table 35–3. Alnico 5 is widely used for permanent magnets. Its superiority to carbon steel, which was the material used for many years, is evident.

One consequence of the phenomenon of hysteresis is the dissipation of *energy* within a ferromagnetic material each time the material is caused to

Table 35–3 RETENTIVITY AND COERCIVE FORCE OF PERMANENT MAGNET MATERIALS

Material	Composition percent	B_r, Wb m^{-2}	H_c A m^{-1}
Carbon steel	98 Fe, 0.86 C, 0.9 Mn	0.95	3.6×10^3
Cobalt steel	52 Fe, 36 Co, 7 W, 3.5 Cr, 0.5 Mn, 0.7 C	0.95	18×10^3
Alnico 2	55 Fe, 10 Al, 17 Ni, 12 Co, 6 Cu	0.76	42×10^3
Alnico 5	51 Fe, 8 Al, 14 Ni, 24 Co, 3 Cu	1.25	44×10^3

traverse its hysteresis loop. This results from the irreversible motion of domain boundaries. It can be shown that the energy dissipated per unit volume, in each cycle, is proportional to the *area* enclosed by the hysteresis loop. Hence if a ferromagnetic material is to be subjected to a field which is continually reversing its direction (the core of a transformer, for example), it is desirable that the hysteresis loop of the material be narrow, to minimize losses. Fortunately, iron or iron alloys are available which combine high permeability with small hysteresis loss.

35–9 SELF-INDUCTANCE

As we saw in Section 35–2, the presence of a magnetic material in a toroidal coil increases the flux for a given current by a factor of the relative permeability K_m of the material. This effect influences the self-inductance of the coil, which can be defined as the number of flux-linkages per unit current in the coil. If a magnetic material of permeability μ replaces a vacuum at all points of the magnetic field set up by a coil, the flux density at every point is increased by a factor K_m over its value in empty space. The number of flux-linkages, and hence the self-inductance, increases in the same ratio. Another way of stating this is that in any expression for the self-inductance in vacuum, the term μ_0 is to be replaced by $\mu = K_m\mu_0$. Thus the self-inductance of a closely wound toroid on a magnetic core (see example in Section 34–2) is

$$L = \frac{\mu N^2 A}{\ell}. \tag{35–15}$$

The self-inductance of an iron-core coil is therefore very much larger than that of the same winding in vacuum. On the other hand, since the permeability of a ferromagnetic material is not constant, the self-inductance of such a coil is not constant either, but depends on the current in the coil and on its magnetic history.

The expression for the energy density in a magnetic field is modified in the same way. It was shown in Section 34–3 that the energy density in a magnetic field in vacuum equals $B^2/2\mu_0$. In a magnetic material of permeability μ, the energy density is $u = B^2/2\mu$, and since $B = \mu H$, this can also be written as

$$\boxed{u = \frac{1}{2}\frac{B^2}{\mu} = \frac{1}{2}\mu H^2 = \frac{1}{2}BH.} \tag{35–16}$$

35–10 PERMANENT MAGNETS

Magnetic fields can be set up by magnetized matter, as well as by currents in conductors. Figure 35–9 is a sectional view of a thin circular ferromagnetic disk that has been permanently magnetized in a direction perpendicular to its flat end faces. The magnetization $\boldsymbol{M}$ (magnetic moment per unit volume) is uniform within the disk and is zero at points outside the disk.

Part (b) of the diagram shows the $\boldsymbol{B}$ field of the disk. Since there are no conduction currents, the $\boldsymbol{B}$ field is due wholly to the equivalent surface currents circulating around the rim of the disk in the narrow shaded zone. The lines of the $\boldsymbol{B}$ field are continuous and are from left to right through the disk.

Now we consider the $\boldsymbol{H}$ field. The values of $\boldsymbol{H}$, $\boldsymbol{B}$, and $\boldsymbol{M}$, at every point, are related by the equation

$$\boldsymbol{H} = \frac{\boldsymbol{B}}{\mu_0} - \boldsymbol{M}.$$

Outside the disk, where $\boldsymbol{M} = \boldsymbol{0}$, $\boldsymbol{H} = \boldsymbol{B}/\mu_0$. Since this is a *vector* equation, the *direction* of the $\boldsymbol{H}$

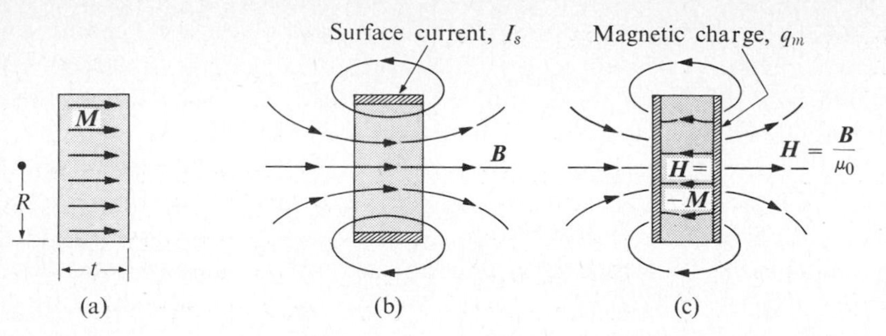

Fig. 35–9 *(a) A uniformly magnetized disk. (b)* ***B****-field of the disk. (c)* ***H****-field of the disk.*

field at external points is the same as that of the $\boldsymbol{B}$ field, and the *magnitude* of $\boldsymbol{H}$ at every point equals B/μ_0.

Inside the disk, where $M \neq 0$, the magnetic intensity $\boldsymbol{H}$ equals the vector *difference* $\boldsymbol{B}/\mu_0 - \boldsymbol{M}$. Let us compute B/μ_0 at the center of the disk. From Eq. (35–6), the surface current per unit length, j_s, is equal to M and the total surface current I_s is therefore

$$I_s = j_s t = Mt.$$

The $\boldsymbol{B}$ field at the center of a circular current loop is

$$B = \frac{\mu_0}{2}\left(\frac{I_s}{R}\right) = \frac{\mu_0}{2}\left(\frac{Mt}{R}\right).$$

Therefore at the center,

$$H = \frac{B}{\mu_0} - M = M\left(\frac{t}{2R} - 1\right).$$

If the thickness t is very small compared with the radius R, the term $t/2R$ is very much less than 1 and can be neglected. Hence, at the center

$$H = -M.$$

The $\boldsymbol{H}$ field at the center is therefore opposite to $\boldsymbol{M}$, or is from right to left, and is numerically equal to M. Thus the lines of the $\boldsymbol{H}$ field must be as shown in Fig. 35–9(c). Because the vector difference $\boldsymbol{B}/\mu_0 - \boldsymbol{H}$ must equal the constant value $\boldsymbol{M}$ at every point, the lines of $\boldsymbol{H}$ within the disk, except on the axis, are curved, as indicated.

A complete calculation shows that the $\boldsymbol{H}$ field is geometrically identical to the $\boldsymbol{E}$ field between a pair of oppositely charged parallel plates, as shown in Fig. 25–19. We might therefore postulate (for the purpose of calculating the $\boldsymbol{H}$ field of a magnetized body) the existence of hypothetical positive and negative *magnetic charges,* which set up a radial inverse-square $\boldsymbol{H}$ field in the same way that *electric charges* set up a radial inverse-square $\boldsymbol{E}$ field. For the disk in Fig. 35–9, these charges are distributed in thin layers over the flat end faces of the disk, as suggested by the shading in part (c). The surface density of magnetic charge is found to be related to the magnetization by the equation

$$\sigma_m = -M.$$

Although the term "magnetic charge" is an appropriate one, the term "magnetic pole" is ordinarily used. A *north magnetic pole* (an N-pole) corresponds to a positive magnetic charge, and a *south magnetic pole* (an S-pole) to a negative magnetic charge. We would say that N-poles are distributed over the right face of the disk in Fig. 35–9, and S-poles over its left face. Unnecessary and often confusing geographical implications can be avoided by speaking of these simply as "enn-poles" and "ess-poles."

It will be seen from Fig. 35–9 that we find *equivalent surface currents* at points where the $\boldsymbol{M}$-vector is *parallel* to the outer surface of a magnetized body, and *magnetic charges* or *poles* where the $\boldsymbol{M}$-vector is *perpendicular* to the surface. For a magnetized body in vacuum, the general relations between $\boldsymbol{M}$ and the surface currents and charges are

$$j_s = M_{\parallel}, \qquad \sigma_m = -M_{\perp},$$

where $M_{\parallel}$ and $M_{\perp}$ are the components of $\boldsymbol{M}$, parallel and perpendicular to the surface.

A few remarks are in order about the concept of magnetic poles. A *single, isolated* magnetic pole has never been found, and many physicists believe that such isolated poles do not exist. The fundamental magnetic entities are magnetic *dipoles*, which arise from molecular current loops but which can also be described as *pairs* of opposite magnetic poles. Nevertheless, the concept of magnetic poles is a useful calculational aid, and magnetic poles do have a physical reality if we define them simply as *those portions of the surface of the body where the **M**-vector has a component perpendicular to the surface.*

The ***H*** field inside the disk in Fig. 35–9 (and, in fact, inside any permanent magnet) is *opposite* to the magnetization ***M*** and to the flux density ***B***. It is called a *demagnetizing* field.

The demagnetizing field in a body depends on the shape of the body and, in general, varies in magnitude and direction from point to point. Within a uniformly magnetized sphere the demagnetizing field equals $-M/3$; at a point within a long, slender rod, not too near its ends, the demagnetizing field is very small. The magnetic state of a bar magnet or a horseshoe magnet is extremely complicated, and in general the problem is not capable of an exact analytical solution.

35–11 THE MAGNETIC FIELD OF THE EARTH

The magnetic field of the earth, at distances up to about five earth radii, is approximately the same as that outside a uniformly magnetized sphere. Figure 35–10(a) represents a section through the earth. The dot-dash line is its axis of rotation, and the *geographic* north and south poles are lettered N_g and S_g. The direction of the (presumed) internal magnetization makes an angle of about 15° with the earth's axis. The dashed line indicates the plane of the magnetic equator, and the letters N_m and S_m represent the so-called *magnetic* north and south poles. Note carefully that lines of induction emerge from the earth's surface over the entire southern magnetic hemisphere and enter its surface over the entire northern magnetic hemisphere. Hence, if we wish to attribute the earth's field to magnetic poles, we must assume, on the basis of this hypothesis about the earth's internal magnetic state, that magnetic N-poles are distributed over the entire *southern* magnetic hemisphere, and magnetic S-poles over the entire *northern* magnetic hemisphere. This can be very confusing. The north and south magnetic poles, considered as points on the earth's surface, are simply those points where the field is vertical. The former is located at latitude 70° N, longitude 96° W.

It is interesting to note that the same field at *external* points would result if the earth's magnetism were due to a short bar magnet near its center, as in Fig. 35–10(b), with the S-pole of the magnet pointing toward the north magnetic pole. The field within the earth is different in the two cases, but for obvious reasons experimental verification of either hypothesis is impossible.

Except at the magnetic equator, the earth's field is not horizontal. The angle which the field makes with the horizontal is called the *angle of dip* or the *inclination.* At Cambridge, Mass. (about 45° N lat.),

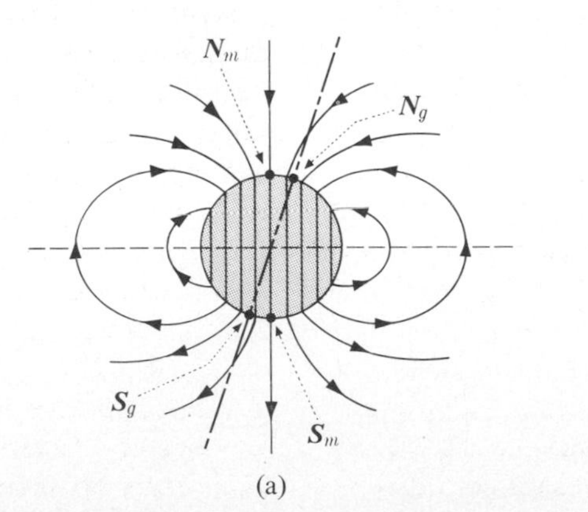

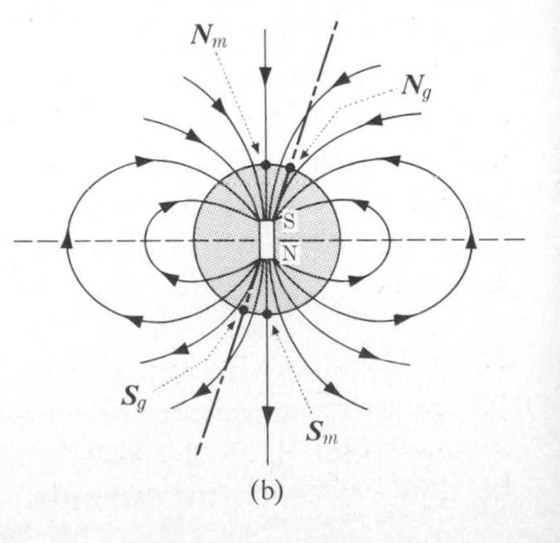

Fig. 35–10 *Simplified diagram of the magnetic field of the earth.*

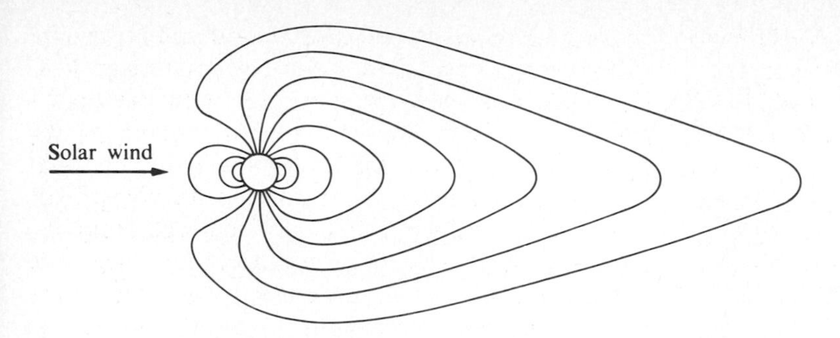

Fig. 35–11 *Probable distortion of the geomagnetic field by the solar wind.*

the magnitude of the earth's field is about 5.8 $\times 10^{-5}$ T and the angle of dip about 73°. Hence, the horizontal component at Cambridge is about 1.7×10^{-5} T and the vertical component about 5.5×10^{-5} T. In northern magnetic latitudes, the vertical component is directed downward; in southern magnetic latitudes it is upward. The angle of dip is, of course, 90° at the magnetic poles.

Out to a distance from the earth of about five earth radii, the magnetic field is governed almost entirely by the earth. At larger distances, the motions of ionized particles play an important role. These motions are strongly influenced by the *solar wind*, a thin hot gas expelled by the sun. Recent studies using satellites and space probes suggest that at large distances the field is distorted as suggested in Fig. 35–11.

The angle between the horizontal component of $\boldsymbol{B}$ and the true north–south direction is called the *variation* or *declination*. At Cambridge, Mass., the declination at present is about 15°W, that is, a compass needle points about 15° to the west of true north.

The magnetic field of the earth is not as symmetrical as one might be led to suspect from the idealized drawing of Fig. 35–10. It is, in reality, very complicated; the inclination and the declination vary irregularly over the earth's surface, and also vary with time.

35–12 MAGNETIC CIRCUITS

As we have seen, every line of induction ($\boldsymbol{B}$ field) is a closed line. Although there is nothing in the nature of a *flow* along these lines, it is useful to draw an analogy between the closed paths of the flux lines and a closed conducting circuit in which there is a current. The region occupied by the magnetic flux is called a *magnetic circuit*; a Rowland ring is the simplest example. When the windings on such a ring are closely spaced over its surface, practically all the flux lines are confined to the ring (Fig. 35–12(a)). Even if the winding is concentrated over only a small portion of the ring, as in Fig. 35–12(b), the permeability of the ring is so much greater than that of the surrounding air that most of the flux is still confined to the material of the ring. The small part which returns via an air path is called the *leakage flux* and is indicated by dotted lines.

If the ring contains an air gap, as in Fig. 35–12(c), there will be a certain amount of spreading or "fringing" of the flux lines at the air gap but, again, most of the flux is confined to a well-defined path. This magnetic circuit may be considered to consist of the iron ring and the air gap "in series."

Figure 35–12(d) shows a section of a common type of transformer core. Here the magnetic circuit is divided, and Sections A and C may be considered to be "in parallel" with each other, and in series with Section B. Figure 35–12(e) is the magnetic circuit of a motor or generator. The two air gaps are in series with the iron portion of the circuit.

An important problem in the design of apparatus in which there is a magnetic circuit is to compute the flux density $\boldsymbol{B}$ which results from a given current in a given winding on a given core or, conversely, to design a core and windings so as to produce a desired flux density. Consider first a closed ring (no air gap) of uniform cross section. We have shown that, within

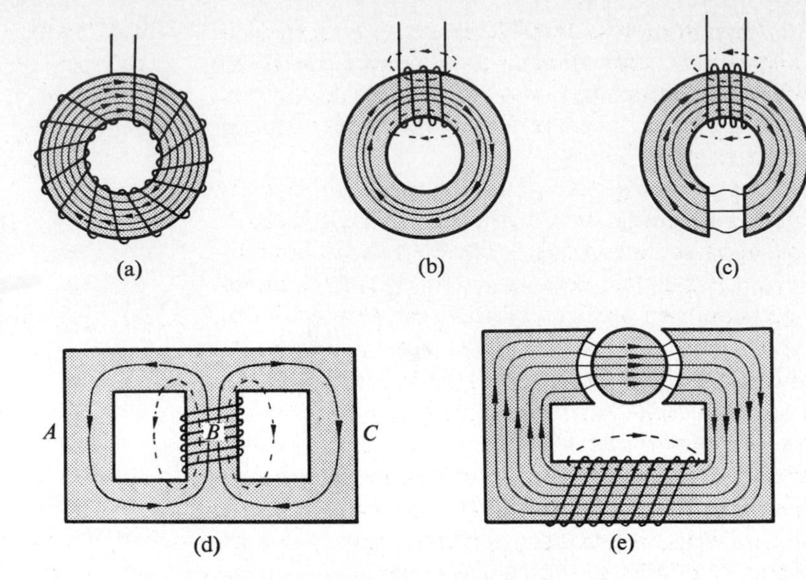

Fig. 35-12 Various magnetic circuits. (a) Rowland ring completely wound. (b) Rowland ring partially wound, showing leakage. (c) Fringing in an air gap. (d) Leakage in a transformer core. (e) Fringing and leakage in the core of a motor or generator.

the ring,

$$B = \mu\left(\frac{Ni}{\ell}\right);$$

and, since $\Phi = BA$,

$$\Phi = \mu\left(\frac{NiA}{\ell}\right),$$

or

$$\Phi = \left(\frac{Ni}{\ell/\mu A}\right). \qquad (35\text{-}17)$$

Now the resistance R of a conductor of uniform cross section A, length ℓ, and resistivity ρ, is given by

$$R = \rho\left(\frac{\ell}{A}\right),$$

or, in terms of the conductivity σ ($\sigma = 1/\rho$),

$$R = \frac{\ell}{\sigma A}.$$

If such a conductor is connected to the terminals of a seat of emf $\mathscr{E}$ of negligible internal resistance, the circuit equation becomes

$$I = \frac{\mathscr{E}}{\ell/\sigma A}.$$

The *form* of this equation is the same as that of Eq. (35-17), with current corresponding to magnetic flux, the quantity Ni corresponding to electromotive force, and $\ell/\mu A$ corresponding to resistance. In view of the close analogy, the numerator in Eq. (35-17) is called the *magnetomotive force*, and the denominator is called the *reluctance* of the magnetic circuit:

$$\boxed{\text{Magnetomotive force} = \mathcal{M} = Ni,} \qquad (35\text{-}18)$$

$$\boxed{\text{Reluctance} = \mathcal{R} = \frac{\ell}{\mu A},} \qquad (35\text{-}19)$$

and Eq. (35-17) may be written

$$\boxed{\Phi = \frac{\mathcal{M}}{\mathcal{R}}.} \qquad (35\text{-}20)$$

Magnetomotive force is evidently expressed in *ampere-turns*, and reluctance in *ampere-turns per weber*. The reluctance of a magnetic circuit is the required number of ampere-turns, per weber of magnetic flux in the circuit.

The advantage of writing the expression for the flux in a magnetic circuit in the form Eq. (35–20) is most apparent when one considers a circuit containing an air gap (or more generally, when the circuit is composed of sections of different permeabilities, lengths, and cross sections). It turns out that the *equivalent reluctance* of such a circuit may be found in the same way that one finds the equivalent resistance of a network of conductors. For example, a ring containing an air gap corresponds to two resistors in series, and the equivalent reluctance of the circuit is the *sum* of the reluctances of ring and gap. The arms A and C of Fig. 35–12(d) are in *parallel*, and the *reciprocal* of their equivalent reluctance is the sum of the reciprocals of the reluctances of the arms individually. For a simple "series" magnetic circuit, one has

$$\Phi = \frac{\mathcal{M}}{\Sigma\mathcal{R}} = \frac{\mathcal{M}}{\Sigma(\ell/\mu A)} = \frac{\mathcal{M}}{\ell_1/\mu_1 A_1 + \ell_2/\mu_2 A_2 + \cdots}, \tag{35–21}$$

where ℓ_1, ℓ_2, etc., are the lengths of the various portions of the circuit, μ_1, μ_2, etc., the corresponding permeabilities, and A_1, A_2, etc., the cross-sectional areas.

The statements above are correct only when the leakage flux is small, i.e., when there is very little flux *outside* the areas A_1, A_2, etc.

Example 1 The mean length of a toroidal ring is 50 cm and its cross section is 4 cm^2. Use the data of Table 35–2 to compute the magnetomotive force needed to establish a flux of 4×10^{-4} Wb in the ring. What current is required if the ring is wound with 200 turns of wire?

The desired flux density B is

$$B = \frac{\Phi}{A} = \frac{4 \times 10^{-4}\ \text{Wb}}{4 \times 10^{-4}\ \text{m}^2} = 1\ \text{Wb m}^{-2} = 1\ \text{T}.$$

From Table 35–2, the permeability at this flux density is about 67×10^{-4} Wb A^{-1} m^{-1}. Hence the reluctance $\mathcal{R}$ is

$$\mathcal{R} = \frac{\ell}{\mu A} = \frac{0.5\ \text{m}}{(67 \times 10^{-4}\ \text{Wb A}^{-1}\ \text{m}^{-1})(4 \times 10^{-4}\ \text{m}^2)} = 1.86 \times 10^5\ \text{A Wb}^{-1},$$

and since $\mathcal{M} = \Phi\mathcal{R}$, the required magnetomotive force is

$$\mathcal{M} = (4 \times 10^{-4}\ \text{Wb})(1.86 \times 10^5\ \text{A Wb}^{-1}) = 74.4\ \text{A-turns}.$$

If the ring is wound with 200 turns, the current required is 0.372 A.

Example 2 If an air gap 1 mm in length is cut in the ring, what current is required to maintain the same flux?

The reluctance of the air gap is

$$\mathcal{R} = \frac{\ell}{\mu_0 A} = \frac{10^{-3}\ \text{m}}{(12.57 \times 10^{-7}\ \text{Wb A}^{-1}\ \text{m}^{-1})(4 \times 10^{-4}\ \text{m}^2)} = 20 \times 10^5\ \text{A Wb}^{-1}.$$

If we neglect the small change in length of the iron, its reluctance is the same as before, or 1.86×10^5 A Wb^{-1}. Thus although the gap is only 1 mm long, its reluctance is 10 times as great as that of the iron portion of the circuit. The reluctance of the entire circuit is now $(20 \times 10^5 + 1.86 \times 10^5)$A Wb^{-1} $= 22 \times 10^5$ A Wb^{-1}. The number of ampere-turns required is 880 and the corresponding current is 4.4 A.

PROBLEMS

$$\mu_0 = 4\pi \times 10^{-7}\ \text{Wb A}^{-1}\ \text{m}^{-1} = 12.57 \times 10^{-7}\ \text{Wb A}^{-1}\ \text{m}^{-1}$$

35–1 Experimental measurements of the magnetic susceptibility of iron ammonium alum are given in Table 35–4.

Table 35–4

T,°C	χ_m
−258.15	129×10^{-4}
−173	19.4×10^{-4}
−73	9.7×10^{-4}
27	6.5×10^{-4}

Make a graph of $1/\chi_m$ against Kelvin temperature, and determine whether Curie's law holds. If so, what is the Curie constant?

35–2 A toroid having 500 turns of wire and a mean circumferential length of 50 cm carries a current of 0.3 A. The relative permeability of the core is 600.

a) What is the flux density in the core?

b) What is the magnetic intensity?

c) What part of the flux density is due to surface currents?

35–3 The current in the windings on a toroid is 2.0 A. There are 400 turns and the mean circumferential length is 40 cm. With the aid of a search coil and ballistic galvanometer, the magnetic field is found to be 1.0 T. Calculate (a) the magnetic intensity, (b) the magnetization, (c) the magnetic susceptibility, (d) the equivalent surface current, and (e) the relative permeability.

35–4 Each of the two preceding coils has a cross-sectional area of 8 cm^2. Calculate the self-inductance L of each coil.

35–5 In 1911, Kamerlingh-Onnes discovered that at low temperatures some metals lose their electrical resistance and become superconductors. Thirty years later, Meissner showed that the magnetic induction inside a superconductor is zero. If the current in the windings of a toroidal superconductor is increased, a critical value H_c may be reached at which the metal suddenly becomes normal with practically zero magnetization.

a) Plot a graph of B/μ_0 against H from $H = 0$ to $H = 2H_c$.

b) Plot a graph of M against H in the same range.

c) Is a superconductor paramagnetic, diamagnetic, or ferromagnetic?

d) If the current in the windings is 10 A, how large are the surface currents on the superconducting ring and in what direction?

e) What happens when you try to place a small permanent magnet on a superconducting plate?

35–6 Table 35–5 lists corresponding values of H and B for a specimen of commercial hot-rolled silicon steel, a material widely used in transformer cores.

a) Construct graphs of B and μ as functions of H, in the range from $H = 0$ to $H = 1000$ A m^{-1}.

b) What is the maximum permeability?

c) What is the initial permeability ($H = 0$)?

d) What is the permeability when $H = 800,000$ A m^{-1}?

Table 35–5 MAGNETIC PROPERTIES OF SILICON STEEL

Magnetic intensity H, A m^{-1}	Flux density B, T
0	0
10	0.050
20	0.15
40	0.43
50	0.54
60	0.62
80	0.74
100	0.83
150	0.98
200	1.07
500	1.27
1,000	1.34
10,000	1.65
100,000	2.02
800,000	2.92

35–7 Construct graphs of $\mu_0 H$, $\mu_0 M$, and K_m against H, like those in Fig. 35–5 for the specimen of silicon steel in Table 35–5.

35–8 Suppose the ordinate of the point b in Fig. 35–7 corresponds to a flux density of 1.6 T, and the abscissa to a magnetic intensity H of 1000 A m^{-1}. Approximately what is the relative permeability at points a, b, c, d, i, and j?

35–9 A bar magnet has a coercivity of 4×10^3 A m^{-1}. It is desired to demagnetize it by inserting it inside a solenoid 12 cm long and having 60 turns. What current should be carried by the solenoid?

35–10 An iron disk 6 cm in diameter and 4 mm thick is uniformly magnetized in a direction perpendicular to its end faces. The magnetization $M = 1.5 \times 10^6$ A m^{-1}.

a) What is the equivalent surface current around the rim of the disk?

b) What is the flux density B at the center of the disk?

c) What is the magnetic intensity H at the center of the disk? What is its direction relative to that of B?

d) What is the relative permeability of the disk?

e) What is the surface density of magnetic charge on the end faces?

f) What is the magnetic moment of the disk?

35–11 The horizontal component of the flux density of the earth's magnetic field at Cambridge, Mass., is 1.7×10^{-5} Wb m^{-2}. What is the horizontal component of the magnetic intensity?

35–12 Show that, for a continuous magnetic circuit, Eq. (35–21) reduces to $B = \mu H$, where $H = NI/\ell$.

35–13 An iron toroid has a mean circumferential length of 40 cm and an area of 5 cm^2. The toroid is wound with 350 turns of wire and carries a current of 0.2 A.

a) What is the magnetomotive force in the toroid?

b) What is the magnetic intensity?

c) Using the permeability curve of Fig. 35–4, determine the permeability of iron at this value of magnetic intensity.

d) What is the reluctance of the magnetic circuit?

e) What is the total flux in the toroid?

35–14 A gap 0.5 mm wide is cut in the toroid of Problem 35–13.

a) What is the reluctance of the magnetic circuit?

b) What is now the flux in the ring?

35–15 A Rowland ring has a cross section of 2 cm^2, a mean length of 30 cm, and is wound with 400 turns. Find the current in the winding that is required to set up a flux density of 0.1 T in the ring

a) if the ring is of annealed iron (Table 35–2);

b) if the ring is of silicon steel (Table 35–5).

c) Repeat the computations above if a flux density of 1.2 T is desired.

Chapter 36

Alternating Currents

36–1 INTRODUCTION

A coil of wire, rotating with constant angular velocity in a magnetic field, develops a sinusoidal alternating emf as explained in Section 33–2. This simple device is the prototype of the commercial alternating current generator, or *alternator.* An L–C circuit, as explained in Section 34–5, oscillates sinusoidally, and with the proper circuitry provides an alternating potential difference between its terminals having a frequency, depending on the purpose for which it is designed, that may range from a few hertz to many millions of hertz.

We consider next a number of circuits connected to an alternator or oscillator which maintains between its terminals a sinusoidal alternating potential difference

$$v = V \cos \omega t,$$

where V is the maximum potential difference or the *voltage amplitude*, v is the *instantaneous* potential difference, and ω the *angular frequency*, equal to 2π times the frequency f. For brevity, the alternator or oscillator will be referred to as an "ac source," although this is not the best English usage, since "ac" is an abbreviation for "*alternating current.*" The circuit-diagram symbol for an ac source is ⊙.

Analysis of alternating-current circuits is facilitated by use of vector diagrams similar to those used in the study of harmonic motion (Section 11–4), in which the instantaneous value of a quantity that varies sinusoidally with time is represented by the projection onto a horizontal axis of a vector of length corresponding to the amplitude of the quantity and rotating counterclockwise with angular velocity ω. In the context of ac-circuit analysis, these rotating vectors are often called *phasors,* and diagrams containing them are called *phasor diagrams.*

Alternating currents are of utmost importance in technology and industry. Transmission of power over long distance is very much easier and more economical with alternating than with direct currents. Circuits used in modern communication equipment, including radio and television, make extensive use of alternating currents. Many life processes involve alternating voltages and currents. The beating of the heart induces alternating currents in the surrounding tissues; the detection and study of these currents, called electrocardiography, provide valuable information concerning the health or pathology of the heart. Electroencephalograms, recordings of alternating currents in the brain, provide analogous information regarding brain function. Both electrocardiograms and electroencephalograms are invaluable diagnostic tools in modern medicine.

36–2 CIRCUITS CONTAINING RESISTANCE, INDUCTANCE, OR CAPACITANCE

Let a resistor of resistance R be connected between the terminals of an ac source, as in Fig. 36–1. The instantaneous potential difference between points a and b is $v_{ab} = V \cos \omega t$, and the instantaneous current in the resistor is

$$i = \frac{v_{ab}}{R} = \frac{V}{R} \cos \omega t.$$

The maximum current I, or the *current amplitude*, is evidently

$$I = \frac{V}{R}, \tag{36–1}$$

and we can therefore write

$$i = I \cos \omega t. \tag{36–2}$$

The current and voltage are both proportional to $\cos \omega t$, so the current is *in phase* with the voltage. The current and voltage amplitudes, from Eq. (36–1), are related in the same way as in a dc circuit.

Figure 36–1(b) shows graphs of i and v as functions of time. The fact that the curve representing the current has the greater amplitude in the diagram is of no significance, because the choice of vertical scales for i and v is arbitrary. The corresponding vector diagram is given in Fig. 36–1(c). Because i and v are *in phase* and have the same frequency, the current and voltage vectors rotate together.

The current density in a wire carrying alternating current is not uniform over the cross section of the wire but is greater near the surface, a phenomenon known as "skin effect." The effective cross section of the wire is therefore reduced, and its resistance is *larger* than when the current is constant. Skin effect results from self-induced emf's set up by the variation in the internal flux in the conductor, and is greater the higher the frequency. Unless the frequency is extremely high, however (of the order of some millions of hertz), the change in resistance is not large; we shall assume the resistance to be independent of frequency.

Next, suppose that a capacitor of capacitance C is connected across the source, as in Fig. 36–2. The instantaneous charge q on the capacitor is

$$q = Cv_{ab} = CV \cos \omega t.$$

In this case the instantaneous current is equal to *the rate of change* of the capacitor charge: and therefore to the rate of change of voltage,

$$i = \frac{dq}{dt} = -\omega CV \sin \omega t.$$

The maximum current is evidently

$$I = \omega CV, \tag{36–3}$$

and we can write

$$i = -I \sin \omega t. \tag{36–4}$$

Thus if the voltage is represented by a *cosine* function, the current is represented by a *negative sine* function, as shown in Fig. 36–2(b). The current is *not* in phase with the voltage; the current is greatest at times when the v curve is rising or falling most steeply, and is zero at times when the v curve "levels off," that is, when it reaches its maximum and minimum values.

Fig. 36–1 *(a) Resistance R connected across an ac source. (b) Graphs of instantaneous voltage and current. (c) Vector diagram; current and voltage in phase.*

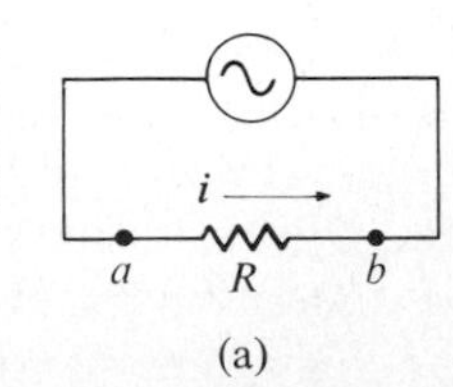

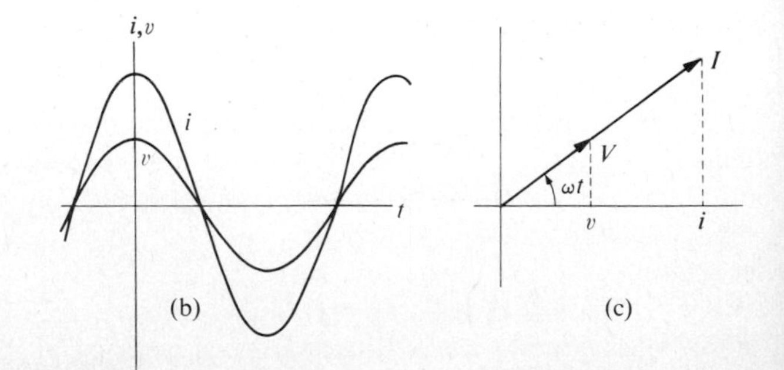

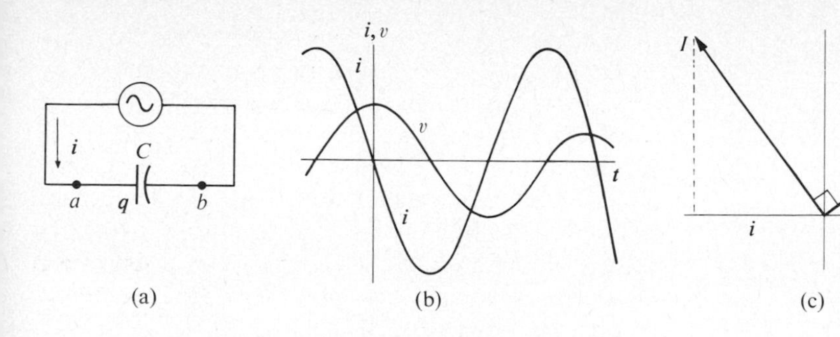

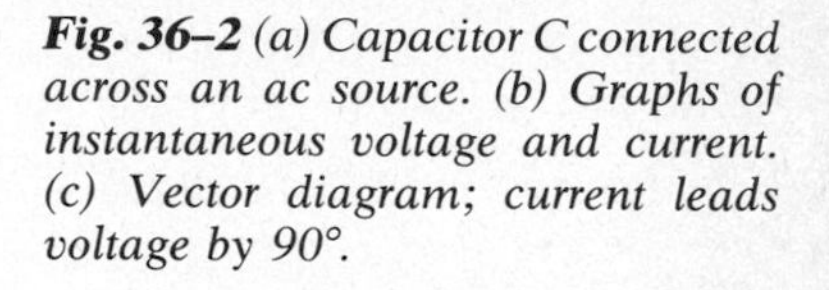

Fig. 36–2 *(a) Capacitor C connected across an ac source. (b) Graphs of instantaneous voltage and current. (c) Vector diagram; current leads voltage by 90°.*

Thus the voltage and current are "out of step" or *out of phase* by a quarter-cycle, with the current a quarter-cycle ahead. The peaks of current occur a quarter-cycle *before* the corresponding voltage peaks. This result is also represented by the vector diagram of Fig. 36–2(c) which shows the current vector ahead of the voltage vector by a quarter-cycle, or 90°. This *phase difference* between current and voltage can also be obtained by rewriting Eq. (36–4), using the trigonometric identity $\cos(A + 90°) = -\sin A$:

$$i = \omega CV \cos(\omega t + 90°). \qquad (36\text{–}5)$$

This shows that the expression for i can be viewed as a cosine function with a "head start" of 90° compared with that for the voltage. We say that the current *leads* the voltage by 90° or that the voltage *lags* the current by 90°.

The expression for the maximum current can be put in the same *form* as that for the maximum current in a resistor ($I = V/R$) if we write Eq. (36–3) as

$$I = \frac{V}{1/\omega C},$$

and define a quantity X_C, called the *capacitive reactance* of the capacitor, as

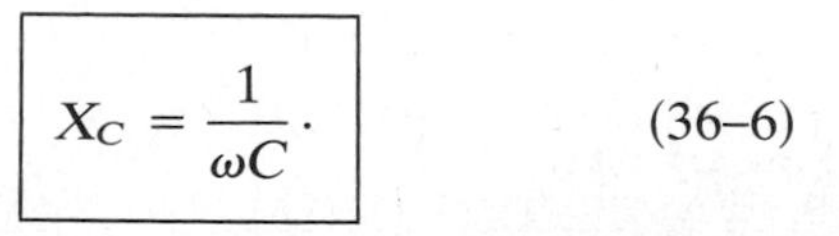

$$\boxed{X_C = \frac{1}{\omega C}}. \qquad (36\text{–}6)$$

Then

$$I = \frac{V}{X_C}. \qquad (36\text{–}7)$$

It will be seen from Eq. (36–7) that the unit of capacitive reactance is one *volt per ampere* ($1\ \mathrm{V\ A^{-1}}$), or one *ohm* ($1\ \Omega$).

The reactance of a capacitor is inversely proportional both to the capacitance C and to the angular frequency ω; the greater the capacitance, and the higher the frequency, the *smaller* is the reactance X_C.

Example At an angular frequency of $1000\ \mathrm{rad\ s^{-1}}$, the reactance of a $1\text{-}\mu\mathrm{F}$ capacitor is

$$X_C = \frac{1}{\omega C} = \frac{1}{(10^3\ \mathrm{rad\ s^{-1}})(10^{-6}\ \mathrm{F})} = 1000\ \Omega.$$

At frequency of $10{,}000\ \mathrm{rad\ s^{-1}}$, the reactance of the same capacitor is only $100\ \Omega$, and at a frequency of $100\ \mathrm{rad\ s^{-1}}$ it is $10{,}000\ \Omega$.

Finally, suppose a pure inductor of zero resistance and having a self-inductance L is connected to an ac source as in Fig. 36–3. Since the potential difference between the terminals of an inductor equals $L\,di/dt$, we have

$$L\frac{di}{dt} = V \cos \omega t$$

and

$$di = \frac{V}{L} \cos \omega t\ dt.$$

Integration of both sides gives

$$i = \frac{V}{\omega L} \sin \omega t.$$

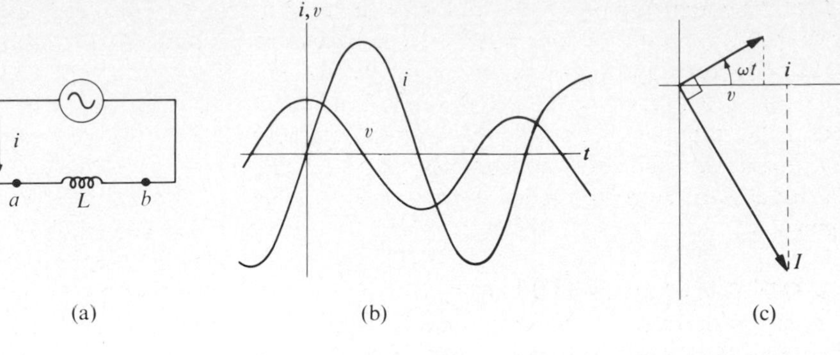

Fig. 36–3 (a) Inductance L connected across an ac source. (b) Graphs of instantaneous voltage and current. (c) Vector diagram; current lags voltage by 90°.

The maximum current is

$$I = \frac{V}{\omega L}, \tag{36–8}$$

and

$$i = I \sin \omega t. \tag{36–9}$$

If the voltage is represented by a *cosine* curve, the current is given by a *sine* curve, as in Fig. 36–3(b).

Again, the voltage and current are a quarter-cycle out of phase, but this time the current *lags* the voltage (or the voltage *leads* the current) by 90°. This may also be seen by rewriting Eq. (36–9) using the trigonometric identity $\cos(A - 90°) = \sin A$:

$$i = \frac{V}{\omega L} \cos(\omega t - 90°). \tag{36–10}$$

This result and the vector diagram of Fig. 36–3(c) shows that the current can be viewed as a cosine function with a "late start" of 90°.

The *inductive reactance* X_L of an inductor is defined as

$$\boxed{X_L = \omega L,} \tag{36–11}$$

and Eq. (36–8) can also be written in the same form as that for a resistor:

$$I = \frac{V}{X_L}. \tag{36–12}$$

The unit of inductive reactance is also 1 *ohm*.

The reactance of an inductor is directly proportional both to its inductance L and to the angular frequency ω; the greater the inductance, and the higher the frequency, the *larger* is the reactance. If the inductor has a ferromagnetic core, self-inductance is not constant, but for simplicity we shall ignore this variation.

Example At an angular frequency of 1000 rad s^{-1}, the reactance of a 1-H inductor is

$$X_L = \omega L = (10^3 \text{ rad s}^{-1})(1 \text{ H}) = 1000\ \Omega.$$

At a frequency of 10,000 rad s^{-1} the reactance of the same inductor is 10,000 Ω, while at a frequency of 100 rad s^{-1} it is only 100 Ω.

The graphs in Fig. 36–4 summarize the variations with frequency of the resistance of a resistor,

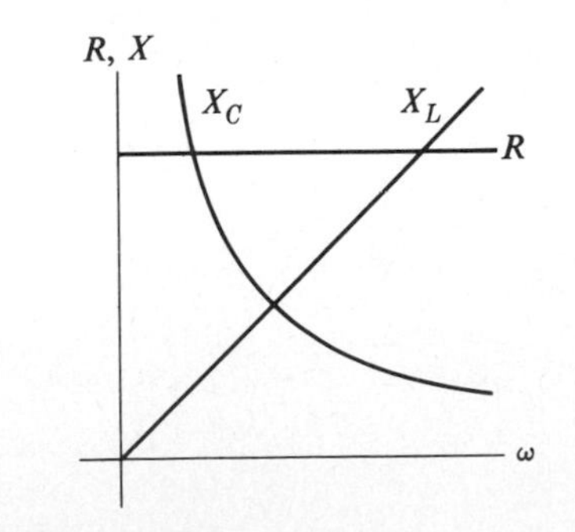

Fig. 36–4 *Graphs of R, X_L, and X_C, as functions of frequency.*

and of the reactances of an inductor and of a capacitor. As the frequency increases, the reactance of the inductor approaches infinity, and that of the capacitor approaches zero. As the frequency decreases, the inductive reactance approaches zero and the capacitive reactance approaches infinity. The limiting case of zero frequency corresponds to a dc circuit.

36–3 THE *R–L–C* SERIES CIRCUIT

In many instances, ac circuits include resistance, inductive reactance, and capacitive reactance. A simple series circuit is shown in Fig. 36–5(a). Analysis of this and similar circuits is facilitated by use of a vector diagram which includes the voltage and current vectors for the various components. In this instance the *total* voltage across all three components is equal to the source voltage, and its vector is the *vector sum* of the vectors for the individual voltages. The complete vector diagram for this circuit is shown in Fig. 36–5(b). This may appear complex, so we shall explain it step by step.

Provided the frequency is not too high, the instantaneous current i has the same value at all points of the circuit. Thus, a *single* vector I, of length proportional to the current amplitude, suffices to represent the current in each circuit element.

Let us use the symbols v_R, v_L, and v_C for the instantaneous voltages across R, L, and C, and V_R, V_L, and V_C for their maximum values. The instantaneous and maximum voltages across the source will be represented by v and V. Then $v = v_{ab}$, $v_R = v_{ac}$, $v_L = v_{cd}$, and $v_C = v_{db}$.

We have shown that the potential difference between the terminals of a resistor is *in phase* with the current in the resistor, and that its maximum value V_R is

$$V_R = IR.$$

Thus the vector V_R in Fig. 36–5(b), in phase with the current vector, represents the voltage across the resistor. Its projection on the horizontal axis, at any instant, gives the instantaneous potential difference v_R.

The current in an inductor *lags* the voltage by 90° or, what is the same thing, the voltage *leads* the current by 90°. The voltage amplitude is

$$V_L = IX_L.$$

The vector V_L in Fig. 36–5(b) represents the voltage across the inductor, and its projection at any instant onto the horizontal axis equals v_L.

The current in a capacitor *leads* the voltage by 90°, or, the voltage *lags* the current by 90°. The voltage amplitude is

$$V_C = IX_C.$$

The vector V_C in Fig. 36–5(b) represents the voltage across the capacitor and its projection at any instant onto the horizontal axis equals v_C.

The instantaneous potential difference v between terminals a and b equals at every instant the (algebraic) sum of the potential differences v_R, v_L,

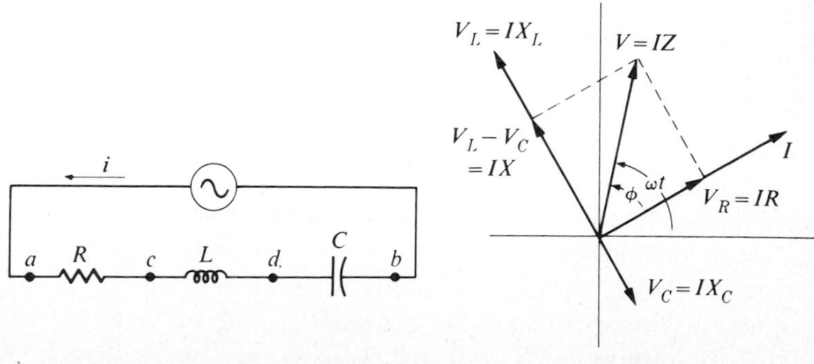

Fig. 36–5 *(a) A series R–L–C circuit. (b) Vector diagram.*

and v_C. That is, it equals the sum of the projections of the vectors V_R, V_L, and V_C. But the *projection* of the *vector sum* of these vectors is equal to the *sum* of their *projections*, so this vector sum V must be the vector that represents the source voltage. To form the vector sum, we first subtract the vector V_C from the vector V_L (since these always lie in the same straight line), giving the vector $V_L - V_C$. Since this is at right angles to the vector V_R, the magnitude of the vector V is

$$\begin{aligned} V &= \sqrt{V_R{}^2 + (V_L - V_C)^2} \\ &= \sqrt{(IR)^2 + (IX_L - IX_C)^2} \\ &= I\sqrt{R^2 + (X_L - X_C)^2}. \end{aligned}$$

Finally, we define the *impedance* Z of the circuit as

$$Z = \sqrt{R^2 + (X_L - X_C)^2}, \qquad (36\text{–}13)$$

so we can write

$$V = IZ \quad \text{or} \quad I = \frac{V}{Z}. \qquad (36\text{–}14)$$

Thus, the *form* of the equation relating current and voltage *amplitudes* is the same as that for a dc circuit, the impedance Z playing the same role as the resistance R of the dc circuit. Note, however, that the impedance is actually a function of R, L, and C, as well as of the frequency ω. The complete expression for Z, for a series circuit, is

$$\begin{aligned} Z &= \sqrt{R^2 + (X_L - X_C)^2} \\ &= \sqrt{R^2 + [\omega L - (1/\omega C)]^2}. \end{aligned} \qquad (36\text{–}15)$$

The unit of impedance, from Eq. (36–15), is evidently one *volt per ampere* ($1\ \mathrm{V\ A^{-1}}$) or one *ohm*.

The expressions for the impedance Z of (a) an R–L series circuit, (b) an R–C series circuit, and (c) an L–C series circuit can be obtained from Eq. (36–15) by letting (a) $X_C = 0$, (b) $X_L = 0$, and (c) $R = 0$.

Equation (36–15) gives the impedance Z only for a *series* R–L–C circuit. But whatever the nature of an R–L–C network, its impedance can be *defined* by Eq. (36–14) as the ratio of the voltage amplitude to the current amplitude.

The angle ϕ in Fig. 36–5(b) is the phase angle of the source voltage V with respect to the current I. It should be evident from the diagram that

$$\tan\phi = \frac{V_L - V_C}{V_R} = \frac{I(X_L - X_C)}{IR} = \frac{X}{R}. \qquad (36\text{–}16)$$

Hence, if the source voltage is represented by a cosine function

$$v = V\cos\omega t,$$

the current lags by an angle ϕ between 0 and 90°, and its equation is

$$i = I\cos(\omega t - \phi).$$

Figure 36–5 has been constructed for a circuit in which $X_L > X_C$. If $X_L < X_C$, vector V lies on the opposite side of the current vector I and the current *leads* the voltage. In this case, $X = X_L - X_C$ is a *negative* quantity, $\tan\phi$ is negative, and ϕ is a negative angle between 0 and −90°.

To summarize, we can say that the *instantaneous* potential differences in an ac series circuit add *algebraically*, just as in a dc circuit, while the voltage *amplitudes* add *vectorially*.

Example In the series circuit of Fig. 36–5, let $R = 300\ \Omega$, $L = 0.9$ H, $C = 2.0\ \mu$F, and $\omega = 1000$ rad s^{-1}. Then

$$X_L = \omega L = 900\ \Omega, \qquad X_C = \frac{1}{\omega C} = 500\ \Omega.$$

The reactance X of the circuit is

$$X = X_L - X_C = 400\ \Omega,$$

and the impedance Z is

$$Z = \sqrt{R^2 + X^2} = 500\ \Omega.$$

If connected across an ac source of voltage amplitude 50 V, the current amplitude is

$$I = \frac{V}{Z} = 0.10 \text{ A}.$$

The lag angle ϕ is

$$\phi = \tan^{-1}\frac{X}{R} = 53°.$$

The voltage amplitude across the resistor is

$$V_R = IR = 30 \text{ V}.$$

The voltage amplitudes across the inductor and capacitor are, respectively,

$$V_L = IX_L = 90 \text{ V}, \qquad V_C = IX_C = 50 \text{ V}.$$

The above analysis describes the *steady-state* condition of a circuit, the condition that prevails after the circuit has been connected to the source for a long time. When the source is first connected, there may be additional voltages and currents, called *transients*, whose nature depends on the time in the cycle when the circuit is initially completed. A detailed analysis of transients is beyond our scope. In any event, they always die out after a sufficiently long time, and therefore do not affect the steady-state behavior of the circuit.

36–4 AVERAGE AND ROOT-MEAN-SQUARE VALUES. AC INSTRUMENTS

The *instantaneous* potential difference between two points of an ac circuit can be measured by connecting a calibrated oscilloscope between the points, and the instantaneous current by connecting an oscilloscope across a resistor in the circuit. The usual moving-coil galvanometer, however, has too large a moment of inertia to follow the instantaneous values of an alternating current. It averages out the fluctuating torque on its coil, and its deflection is proportional to the *average* current.

The average value of any quantity that varies with time, $f(t)$, over a time interval from t_1 to t_2, is defined as

$$f_{\text{av}} = \frac{1}{t_2 - t_1}\int_{t_1}^{t_2} f(t)\,dt, \qquad f_{\text{av}}(t_2 - t_1) = \int_{t_1}^{t_2} f(t)\,dt.$$

The average has the following graphical interpretation. The integral $\int_{t_1}^{t_2} f(t)\,dt$ is the *area* under a graph of $f(t)$ versus t, between vertical lines at t_1 and t_2. The product $f_{\text{av}}(t_2 - t_1)$ is the area of a rectangle of height f_{av} and base $(t_2 - t_1)$. From the definition of f_{av}, these areas are equal.

Let us apply this definition to a sinusoidally varying quantity, for example, a current given by

$$i = I \sin \omega t.$$

The average value of the current for the *half-cycle* from $t = 0$ to $t = \pi/\omega$ is

$$\boxed{I_{\text{av}} = \frac{\omega}{\pi}\int_0^{\pi/\omega} I \sin \omega t\,dt = \frac{2I}{\pi}} \qquad (36\text{–}17)$$

That is, the average current is $2/\pi$ (about $\frac{2}{3}$) times the maximum current, and the area under the rectangle in Fig. 36–6 equals the area under one loop of the sine curve.

The average current for a *complete cycle* (or any number of complete cycles) is

$$I_{\text{av}} = \frac{\omega}{2\pi}\int_0^{2\pi/\omega} I \sin \omega t\,dt = 0,$$

as would be expected, since the *positive* area of the loop between 0 and π/ω is equal to the *negative* area of the loop between π/ω and $2\pi/\omega$. Hence if a sinusoidal current is sent through a moving-coil galvanometer, the meter reads zero!

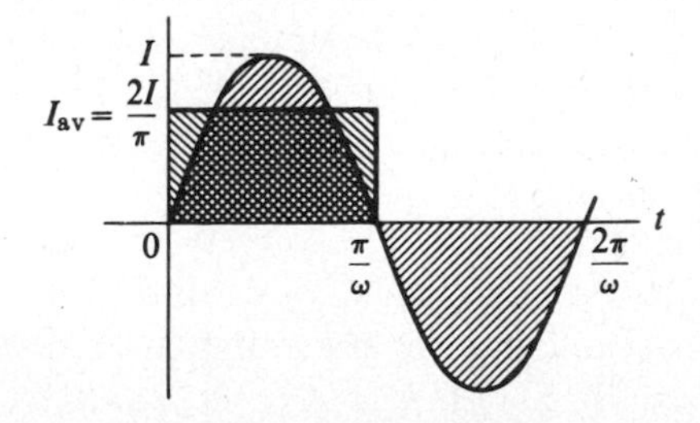

Fig. 36–6 *The average value of a sinusoidal current over a half-cycle is $2I/\pi$. The average over a complete cycle is zero.*

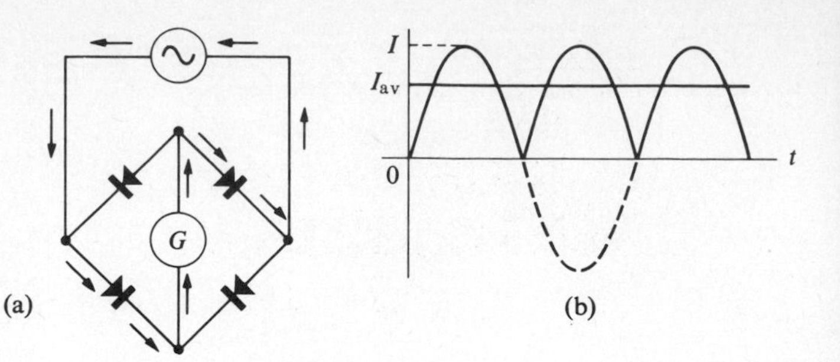

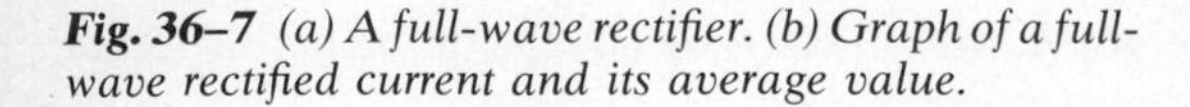

Fig. 36–7 *(a) A full-wave rectifier. (b) Graph of a full-wave rectified current and its average value.*

Such a meter can be used in an ac circuit, however, if it is connected in the *full-wave rectifier circuit* shown in Fig. 36–7(a). As explained in Section 28–6, an ideal rectifier offers a constant finite resistance to current in the forward direction and an infinite resistance to current in the opposite direction. By tracing through the circuit, one can see that, when the line current is in the direction shown, two of the rectifiers are conducting and two are nonconducting, and the current in the galvanometer is upward. When the line current is in the opposite direction, the rectifiers that are nonconducting in Fig. 36–7(a) carry the current, and the current in the galvanometer is still upward. Thus, if the line current alternates sinusoidally, the current in the galvanometer has the waveform shown in Fig. 36–7(b). Although pulsating, it is always in the same direction and its average value is *not* zero. Then, when provided with the necessary series resistance or shunt, as for a dc voltmeter or ammeter, the galvanometer can serve as an ac voltmeter or ammeter.

The average value of the rectified current, in any number of complete cycles, is the same as the average current in the first half-cycle in Fig. 36–6, or it is $2/\pi$ times the maximum current I. Hence, if the meter deflects full-scale with a steady current I_0 through it, it will also deflect full-scale when the average value of the rectified current, $2I/\pi$, is equal to I_0. The current amplitude I, when the meter deflects full scale, is then

$$I = \frac{\pi I_0}{2}.$$

For example, if $I_0 = 1$ A, $I = 1.57$ A.

Most ac meters are calibrated to read not the maximum value of the current or voltage, but the *root-mean-square* value, that is, the *square root* of the *average* value of the *square* of the current or voltage, abbreviated as the rms value. Since i^2 is always positive, the average of i^2 is never zero even when the average of i *is* zero.

Figure 36–8 shows graphs of a sinusoidally varying current and of its square. If $i = I \sin \omega t$, then

$$i^2 = I^2 \sin^2 \omega t = I^2\left(\frac{1}{2}(1 - \cos 2\omega t)\right)$$
$$= \frac{1}{2} I^2 - \frac{1}{2} I^2 \cos 2\omega t.$$

The average value of i^2, or the *mean square current,* is equal to the constant term $\frac{1}{2}I^2$, since the average value of $\cos 2\omega t$, over any number of complete cycles, is zero:

$$(I^2)_{\text{av}} = \frac{I^2}{2}.$$

The root-mean-square current is the square root of this, or

$$\boxed{I_{\text{rms}} = \sqrt{(I^2)_{\text{av}}} = \frac{I}{\sqrt{2}}.} \qquad (36\text{–}18)$$

In the same way, the root-mean-square value of a sinusoidal voltage is

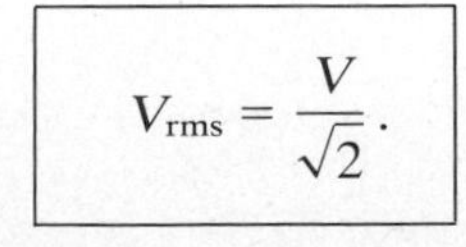

$$\boxed{V_{\text{rms}} = \frac{V}{\sqrt{2}}.}$$

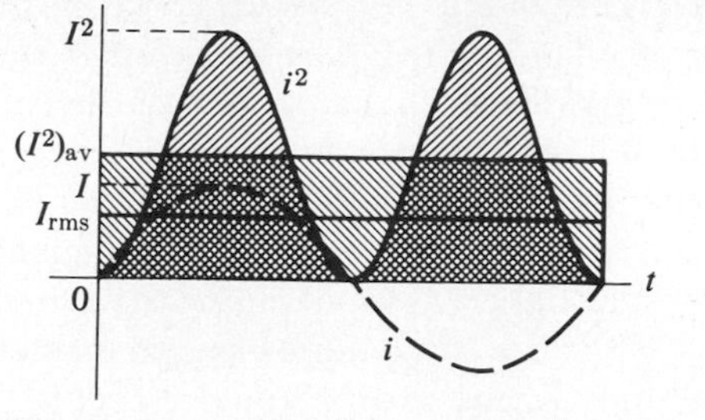

Fig. 36–8 *The average value of the square of a sinusoidally varying current, over any number of half-cycles, is* $I^2/2$. *The root-mean-square value is* $I/\sqrt{2}$.

Voltages and currents in power distribution systems are always referred to in terms of their rms values. Thus, when we speak of our household power supply as "115-volt AC," this means that the rms voltage is 115 V. The voltage amplitude is

$$V = \sqrt{2}\, V_{\text{rms}} = 163 \text{ V}.$$

36–5 POWER IN AC CIRCUITS

The instantaneous power input to an ac circuit is

$$p = vi,$$

where v is the instantaneous source potential difference and i is the instantaneous current. We consider first some special cases.

If the circuit consists of a pure resistance R, as in Fig. 36–1, i and v are *in phase.* The graph representing p is obtained by multiplying together at every instant the ordinates of the graphs of v and i in Fig. 36–1(b), and it is shown by the solid curve in Fig. 36–9(a). (The product vi is positive when v and i are both positive or both negative.) That is, energy is supplied *to* the resistor at all instants, whatever the direction of the current, although the *rate* at which it is supplied is not constant.

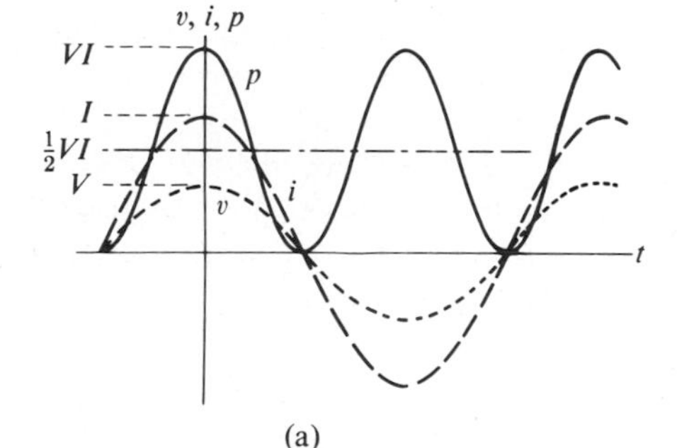

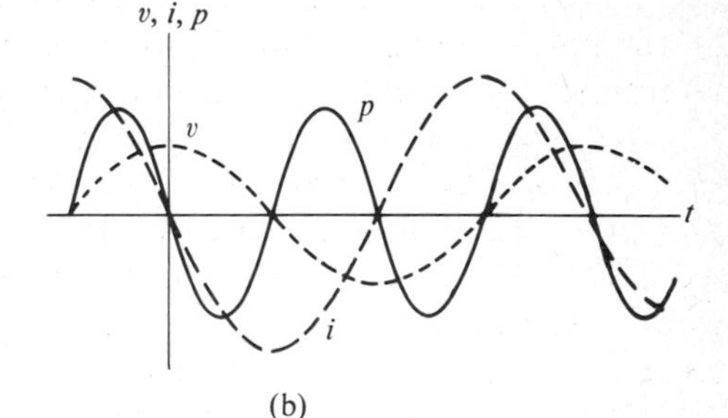

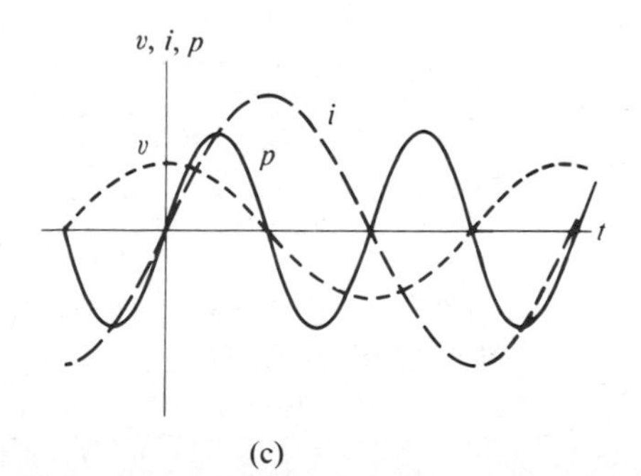

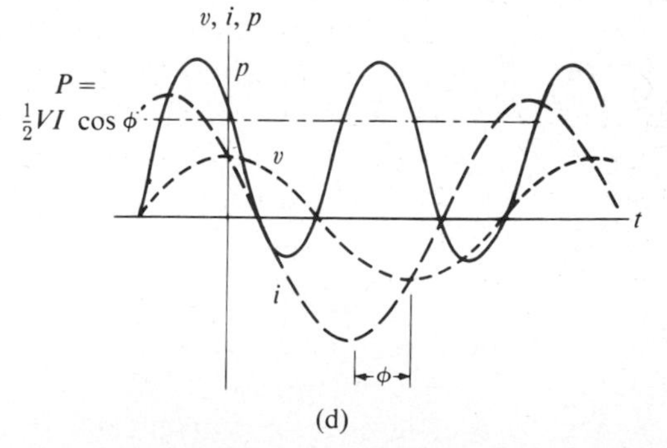

Fig. 36–9 *(a) Instantaneous power input to a resistor. The average power is* $\frac{1}{2}VI$. *(b) Instantaneous power input to a capacitor. The average power is zero. (c) Instantaneous power input to a pure inductor. The average power is zero. (d) Instantaneous power input to an arbitrary ac circuit. The average power is* $\frac{1}{2}VI \cos\phi = V_{\text{rms}}I_{\text{rms}} \cos\phi$.

The power curve is symmetrical about a value equal to one-half its maximum ordinate VI, so the *average* power P is

$$P = \frac{1}{2} VI. \tag{36–19}$$

The average power can also be written

$$P = \frac{V}{\sqrt{2}}\left(\frac{I}{\sqrt{2}}\right) = V_{\mathrm{rms}} I_{\mathrm{rms}}. \tag{36–20}$$

Furthermore, since $V_{\mathrm{rms}} = I_{\mathrm{rms}} R$, we have

$$P = I_{\mathrm{rms}}^2 R. \tag{36–21}$$

Note that Eqs. (36–20) and (36–21) have the same *form* as those for a dc circuit.

Suppose next that the circuit consists of a capacitor, as in Fig. 36–2. The current and voltage are then 90° out of phase. When the curves of v and i are multiplied together (the product vi is *negative* when v and i have *opposite* signs), we get power curve in Fig. 36–9(b), which is symmetrical about the horizontal axis. The average power is therefore *zero.*

To see why this is so, we recall that positive power means that energy is supplied *to* a device and that negative power means that energy is supplied *by* a device. The process we are considering is merely the charge and discharge of a capacitor. During the intervals when p is positive, energy is supplied to charge the capacitor, and when p is negative the capacitor is discharging and returning energy to the source.

Figure 36–9(c) is the power curve for a pure inductor. As with a capacitor, the current and voltage are out of phase by 90°, and the average power is zero. Energy is supplied to establish a magnetic field around the inductor and is returned to the source when the field collapses.

In the most general case, current and voltage differ in phase by an angle ϕ and

$$p = V \sin \omega t \times I \sin(\omega t - \phi). \tag{36–22}$$

The instantaneous power curve has the form shown in Fig. 36–9(d). The area under the positive loops is greater than that under the negative loops and the net average power is positive.

The preceding analyses have shown that when v and i are *in phase*, the average power equals $\frac{1}{2}VI$; and when v and i are 90° *out of phase*, the average power is zero. Hence, in the general case, when v and i differ by an angle ϕ, the average power equals $\frac{1}{2}V$, multiplied by $I \cos \phi$, the component of I that is *in phase* with V. That is,

$$P = \frac{1}{2} VI \cos \phi = V_{\mathrm{rms}} I_{\mathrm{rms}} \cos \phi. \tag{36–23}$$

This is the general expression for the power input to *any* ac circuit. The factor $\cos \phi$ is called the *power factor* of the circuit. For a pure resistance, $\phi = 0$, $\cos \phi = 1$, and $P = V_{\mathrm{rms}} I_{\mathrm{rms}}$. For a capacitor or inductor, $\phi = 90°$, $\cos \phi = 0$, and $P = 0$.

A low power factor (large angle of lag or lead) is usually undesirable in power circuits because, for a given potential difference, a large current is needed to supply a given amount of power, with correspondingly large heat losses in the transmission lines. Since many types of ac machinery draw a lagging current, this situation is likely to arise. It can be corrected by connecting a capacitor in parallel with the load. The leading current drawn by the capacitor compensates for the lagging current in the other branch of the circuit. The capacitor itself takes no net power from the line.

36–6 SERIES RESONANCE

The impedance of an R–L–C series circuit depends on the frequency, since the inductive reactance is directly proportional to frequency, and the capacitive reactance inversely proportional. This dependence is illustrated in Fig. 36–10(a), where a logarithmic frequency scale has been used because of the wide range of frequencies covered. Note that there is one particular frequency at which X_L and X_C are numerically equal. At this frequency, $X_L - X_C$ is zero. Hence the impedance Z, equal to $\sqrt{R^2 + (X_L - X_C)^2}$, is a *minimum* at this frequency and is equal to the resistance R.

Fig. 36–10 (a) Reactance, resistance, and impedance as functions of frequency (logarithmic frequency scale). (b) Impedance, current, and phase angle as functions of frequency logarithmic frequency scale).

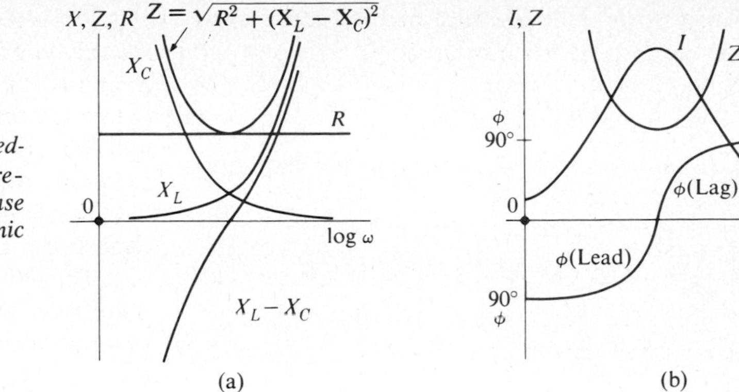

If an ac source of constant voltage amplitude but variable frequency is connected across the circuit, the current amplitude I varies with frequency, as shown in Fig. 36–10(b), and is a *maximum* at the frequency at which the impedance Z is a *minimum.* The same diagram also shows the phase difference ϕ as a function of frequency. At low frequencies, where capacitive reactance X_C predominates, the current *leads* the voltage. At the frequency where $X = 0$, the current and voltage are *in phase,* and at high frequencies, where X_L predominates, the current *lags* the voltage.

The behavior of the current in an ac series circuit, as the source frequency is varied, is exactly analogous to the response of a spring–mass system having a viscous damping force as the frequency of the driving force is varied. The frequency ω_0 at which the current is a maximum is called the *resonant frequency* and is easily computed from the fact that, at this frequency, $X_L = X_C$:

$$X_L = X_C, \quad \omega_0 L = \frac{1}{\omega_0 C}, \quad \omega_0 = \frac{1}{\sqrt{LC}}. \tag{36–24}$$

Note that this is equal to the natural frequency of oscillation of an L-C circuit, as derived in Section 33–12.

If the inductance L or the capacitance C of a circuit can be varied, the resonant frequency can be varied also. This is the procedure by which a radio or television receiving set may be "tuned" to receive the signal from a desired station.

Example The series circuit in Fig. 36–11 is connected to the terminals of an ac source whose frequency is variable but whose rms terminal voltage is constant and equal to 100 V. At the resonant frequency,

$$X_L = X_C = 2000\ \Omega, \qquad \text{and} \qquad R = 500\ \Omega.$$

The impedance Z is then equal to the resistance R, and the rms current is

$$I = \frac{V}{Z} = \frac{V}{R} = 0.20\ \text{A}.$$

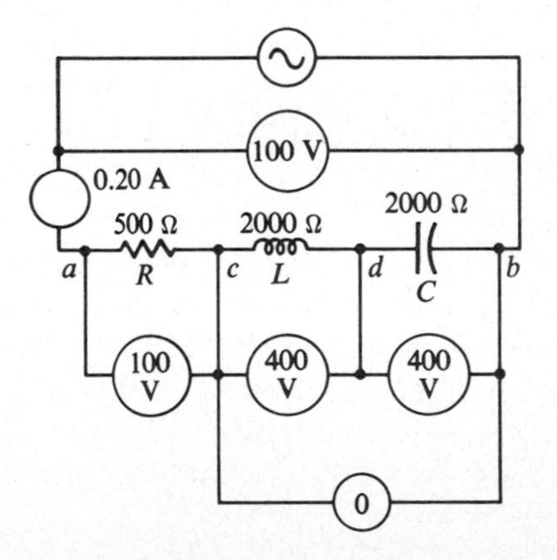

Fig. 36–11 Series resonant circuit.

The rms potential difference across the resistor is

$$V_R = IR = 100\ \text{V}.$$

The potential differences across the inductor and capacitor are, respectively,

$$V_L = IX_L = 400\ \text{V}, \qquad V_C = IX_C = 400\ \text{V}.$$

The potential difference across the inductor–capacitor combination (V_{cb}) is

$$V = IX = I(X_L - X_C) = 0.$$

This is because the instantaneous potential differences across the inductor and the capacitor are 180° out of phase. Although the effective values of each may be large, their resultant at each instant is zero.

36–7 CIRCUITS IN PARALLEL

Circuit elements connected in parallel across an ac source can be analyzed by the same procedure as for elements in series. The vector diagram for the circuit in Fig. 36–12(a) is given in Fig. 36–12(b). In this case, the instantaneous potential difference across each element is the same, and a *single* vector V represents the common terminal voltage. The vector I_R, of amplitude V/R and in phase with V, represents the current in the resistor. Vector I_L, of amplitude V/X_L and lagging V by 90°, represents the current in the inductor, and vector I_C, of amplitude V/X_C and leading V by 90°, represents the current in the capacitor.

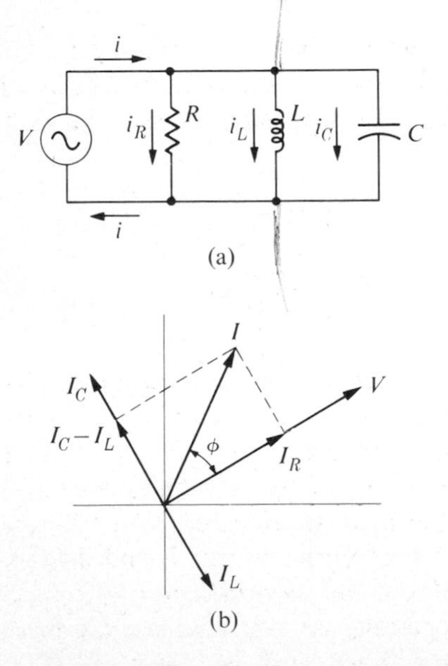

Fig. 36–12

The instantaneous current i, by Kirchhoff's point rule, equals the (algebraic) sum of the instantaneous currents i_R, i_L, and i_C, and is represented by the vector I, the vector sum of vectors I_R, I_L, and I_C. Angle ϕ is the phase angle between current and source voltage.

From Fig. 36–12,

$$I = \sqrt{I_R{}^2 + (I_C - I_L)^2}$$

$$= \sqrt{\left(\frac{V}{R}\right)^2 + \left(\omega CV - \frac{V}{\omega L}\right)^2}$$

$$= V\sqrt{\frac{1}{R^2} + \left(\omega C - \frac{1}{\omega L}\right)^2}. \tag{36–25}$$

The maximum current I is frequency-dependent, as expected. It is minimum when the second factor in the radical is zero; this occurs when the two reactances have equal magnitudes, at the resonant frequency ω_0 given by Eq. (36–24).

Thus at resonance the total current in the parallel R–L–C circuit is *minimum*, in contrast to the R–L–C *series* circuit, which has *maximum* current at resonance. This can be understood by noting that, in the parallel circuit, the currents in L and C are always exactly a half-cycle out of phase; when they also have equal *magnitudes*, they cancel each other completely, and the total current is simply that through R. Indeed, when $\omega C = 1/\omega L$, Eq. (36–25) becomes simply $I = V/R$. This does *not* mean there is *no* current in L or C at resonance, but only that the two currents cancel. If R is large, the equivalent impedance of the circuit near resonance is much *larger* than the individual reactances X_L and X_C.

36–8 THE TRANSFORMER

For reasons of efficiency, it is desirable to transmit electrical power at high voltages and small currents, with consequent reduction of I^2R heating in the transmission line. On the other hand, considerations of safety and of insulation of moving parts require relatively low voltages in generating equipment and in motors and household appliances. One of the most useful features of ac circuits is the ease and efficiency with which voltages (and currents) may be changed from one value to another by *transformers.*

In principle, the transformer consists of two coils electrically insulated from each other and wound on the same iron core. An alternating current in one winding sets up an alternating flux in the core, and the induced electric field produced by this varying flux (see Section 33–5) induces an emf in the other winding. Energy is thus transferred from one winding to another via the core flux and its associated induced electric field. The winding to which power is supplied is called the *primary*; that from which power is delivered is called the *secondary.* The circuit symbol for an iron core transformer is

The power output of a transformer is necessarily less than the power input because of unavoidable losses. These losses consist of I^2R losses in the primary and secondary windings, and hysteresis and eddy-current losses in the core. Hysteresis losses are minimized by the use of iron having a narrow hysteresis loop; and eddy currents are minimized by laminating the core. In spite of these losses, transformer efficiencies are usually well over 90%, and in large installations may reach 99%.

For simplicity, we shall consider only an idealized transformer in which there are *no* losses and in which all of the flux is confined to the iron core, so that the same flux links both primary and secondary. The transformer is shown schematically in Fig. 36–13. A primary winding of N_1 turns and a secondary winding of N_2 turns both encircle the core in the same sense. An ac source voltage amplitude V_1 is

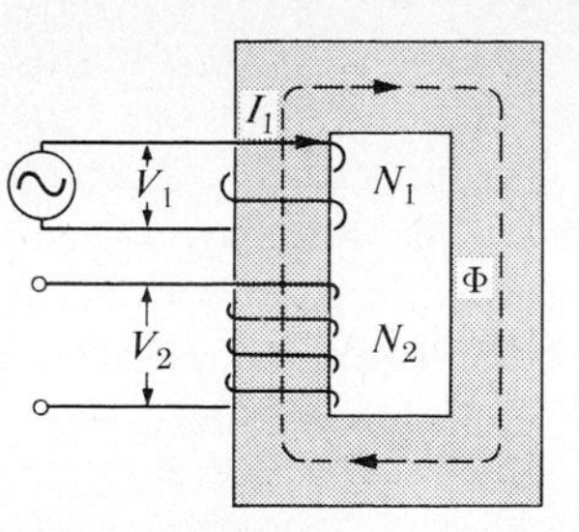

Fig. 36–13 *Schematic diagram of a transformer with secondary open.*

connected to the primary and, to begin with, we assume the secondary to be open, so that there is no secondary current. The primary winding then functions merely as an inductor.

The primary current, which is small, lags the primary voltage by 90° and is called the *magnetizing* current. The power input to the transformer is zero. The core flux is in phase with the primary current. Since the same flux links both primary and secondary, the induced emf *per turn* is the same in each. The ratio of primary to secondary induced emf is therefore equal to the ratio of primary to secondary turns, or

$$\frac{\mathcal{E}_2}{\mathcal{E}_1} = \frac{N_2}{N_1}.$$

Since the windings are assumed to have zero resistance, the induced emf's $\mathcal{E}_1$ and $\mathcal{E}_2$ are numerically equal to the corresponding terminal voltages V_1 and V_2, and

$$\boxed{\frac{V_2}{V_1} = \frac{N_2}{N_1}.} \qquad (36\text{–}26)$$

Hence by properly choosing the turn ratio N_2/N_1, we may obtain any desired secondary voltage from a given primary voltage. If $V_2 > V_1$, we have a *step-up* transformer; if $V_2 < V_1$, a *step-down* transformer.

Consider next the effect of closing the secondary circuit. The secondary current I_2 and its phase angle ϕ_2 will, of course, depend on the nature of the secondary circuit. As soon as the latter is closed, some power must be delivered by the secondary (except when $\phi_2 = 90°$); and, from energy considerations, an equal amount of power must be supplied to the primary. The process by which the transformer is enabled to draw the requisite amount of power is as follows. When the secondary circuit is open, the core flux is produced by the primary current only. But when the secondary circuit is closed, *both* primary and secondary currents set up a flux in the core. The secondary current, by Lenz's law, tends to weaken the core flux and therefore to decrease the back-emf in the primary. But (in the absence of losses) the back-emf in the primary must equal the primary terminal voltage, which is assumed to be fixed. The primary current therefore increases until the core flux is restored to its original no-load magnitude.

If the secondary circuit is completed by a resistance R, $I_2 = V_2/R$. From energy considerations, the power delivered to the primary equals that taken out of the secondary (neglecting losses), so

$$V_1 I_1 = V_2 I_2 . \qquad (36\text{–}27)$$

Combining this with the above expression for I_2 with Eq. (36–26) to eliminate V_2 and I_2, we find

$$\boxed{I_1 = \frac{V_1}{(N_1/N_2)^2 R}.} \qquad (36\text{–}28)$$

Thus, when the secondary circuit is completed through a resistance R, the result is the same as if the *source* had been connected directly to a resistance equal to R multiplied by the reciprocal of the *square* of the turns ratio. In other words, the transformer "transforms" not only voltages and currents, but resistances (more generally, impedances) as well. It can be shown that maximum power is supplied by a source to a resistor when its resistance equals the internal resistance of the source. The same principle applies in ac circuits, with resistance replaced by impedance. When a high-impedance ac source must be connected to a low-impedance circuit, as when an audio amplifier is connected to a loudspeaker, the impedance of the source can be *matched* to that of the circuit by the insertion of a transformer of the correct turns ratio.

PROBLEMS

Angular frequency = 2π (frequency),

ω (rad s^{-1}) = $2\pi f$ (Hz)

36–1

a) At what frequency would a 5-H inductor have a reactance of 4000 Ω?

b) At what frequency would a 5-μF capacitor have the same reactance?

36–2 What is the reactance of an 0.015-μF capacitor at (a) 1 Hz? (b) 5 kHz? (c) 2 MHz?

36–3

a) What is the reactance of a 1-H inductor at a frequency of 60 Hz?

b) What is the inductance of an inductor whose reactance is 1 ohm at 60 Hz?

c) What is the reactance of a 1-μF capacitor at a frequency of 60 Hz?

d) What is the capacitance of a capacitor whose reactance is 1 ohm at 60 Hz?

36–4

a) Compute the reactance of a 10-H inductor at frequencies of 60 Hz and 600 Hz.

b) Compute the reactance of a 10-μF capacitor at the same frequencies.

c) At what frequency is the reactance of a 10-H inductor equal to that of a 10-μF capacitor?

36–5 A 1-μF capacitor is connected across an ac source whose voltage amplitude is kept constant at 50 V, but whose frequency can be varied. Find the current amplitude when the angular frequency is (a) 100 rad s^{-1}, (b) 1000 rad s^{-1}, (c) 10,000 rad s^{-1}. (d) Construct a log-log plot of current amplitude vs. frequency.

36–6 The voltage amplitude of an ac source is 50 V and its angular frequency is 1000 rad s^{-1}. Find the current amplitude if the capacitance of a capacitor connected across the source is (a) 0.01 μF, (b) 1.0 μF, (c) 100 μF.

d) Construct a log-log plot of current amplitude vs. capacitance.

36–7 An inductor of self-inductance 10 H and of negligible resistance is connected across the source in Problem

36–5. Find the current amplitude when the angular frequency is (a) 100 rad s^{-1}, (b) 1000 rad s^{-1}, (c) 10,000 rad s^{-1}.

d) Construct a log-log plot of current amplitude vs. frequency.

36–8 Find the current amplitude if the self-inductance of a resistanceless inductor connected across the source of Problem 36–6 is (a) 0.01 H, (b) 1.0 H, (c) 100 H.

b) Construct a log-log plot of current amplitude versus self-inductance.

36–9 The expression for the impedance Z of an R–L series circuit can be obtained from Eq. (36–15) by setting $X_C = 0$, which corresponds to $C = \infty$. Explain.

36–10 In the series circuit of Fig. 36–5, the source has a constant voltage amplitude of 50 V and a frequency of 1000 rad s^{-1}. $R = 300\ \Omega$, $L = 0.9$ H, and $C = 2.0\ \mu$F. Suppose a series circuit contains only the resistor and the inductor in series.

a) What is the impedance of the circuit?

b) What is the current amplitude?

c) What are the voltage amplitudes across the resistor and across the inductor?

d) What is the phase angle ϕ? Does the current lag or lead?

e) Construct the vector diagram.

36–11 Same as Problem 36–10, except that the circuit consists of the resistor and the *capacitor* of Fig. 36–5 in series.

36–12 Same as Problem 36–10, except that the circuit consists of the *inductor* and capacitor of Fig. 36–5 in series.

36–13

a) Compute the impedance of the circuit in Fig. 36–5 at an angular frequency of 500 rad s^{-1}.

b) Describe how the current amplitude varies as the frequency of the source is slowly reduced from 1000 rad s^{-1} to 500 rad s^{-1}.

c) What is the phase angle when $\omega = 500$ rad s^{-1}? Construct the vector diagram when $\omega = 500$ rad s^{-1}. (See the example at the end of Section 36–3 for data.)

36–14 Five infinite impedance voltmeters, calibrated to read rms value, are connected as shown in Fig. 36–14 to the circuit of Fig. 36–5. What does each voltmeter read? (See the example at the end of Section 36–3 for data.)

36–15 What is the reading of each voltmeter in Fig. 36–14 if the angular frequency $\omega = 500$ rad s^{-1}? (See the example at the end of Section 36–3 for data.)

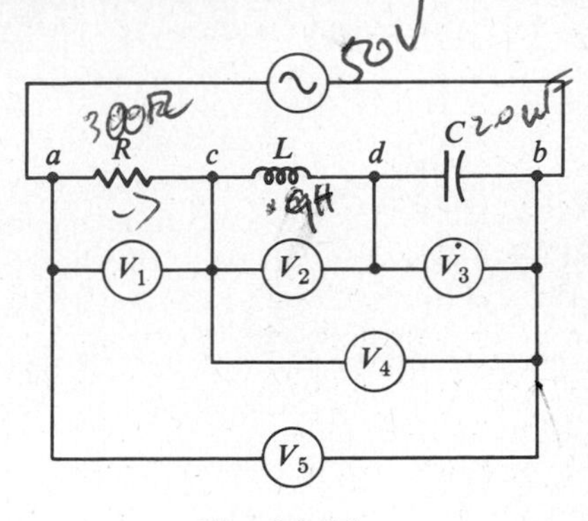

Fig. 36–14

36–16 A 400-Ω resistor is in series with a 0.1-H inductor and a 0.5-μF capacitor. Compute the impedance of the circuit and draw the vector impedance diagram (a) at a frequency of 500 Hz, and (b) at a frequency of 1000 Hz. Compute, in each case, the phase angle between line current and line voltage, and state whether the current lags or leads.

36–17

a) Construct a graph of the current amplitude in the circuit of Fig. 36–5 as the angular frequency of the source is increased from 500 rad s^{-1} to 1000 rad s^{-1}.

b) At what frequency is the circuit in resonance?

c) What is the power factor at resonance?

d) What is the reading of each voltmeter in Fig. 36–14 when the frequency equals the resonant frequency?

e) What would be the resonant frequency if the resistance were reduced to 100 Ω?

f) What would then be the rms current at resonance? See the example at the end of Section 36–3 for data.

36–18

a) At what frequency is the voltage amplitude across the inductor in an R–L–C series circuit a maximum?

b) At what frequency is the voltage amplitude across the capacitor a maximum?

36–19 In Fig. 36–5, the resistance of R is 250 Ω, L has an inductance of 0.5 H and zero resistance, and C has a capacitance of 0.02 μF.

a) What is the resonant frequency of the circuit?

b) The capacitor can withstand a peak voltage of 350 V. What maximum effective terminal voltage may the generator have at resonant frequency?

36–20 A resistor of 500 Ω and a capacitor of 2 μF are connected in parallel to an ac generator supplying a constant voltage amplitude of 282 V and having an angular frequency of 374 rad s^{-1}. Find (a) the current amplitude in the resistor, (b) the current amplitude in the capacitor, (c) the phase angle, and (d) the line current amplitude.

36–21 A coil has a resistance of 20 Ω. At a frequency of 100 Hz, the voltage across the coil leads the current in it by 30°. Determine the inductance of the coil.

36–22 An inductor having a reactance of 25 Ω gives off heat at the rate of 2.39 cal s^{-1} when it carries a current of 0.5 A. What is the impedance of the inductor?

36–23 The circuit of Problem 36–16 carries an rms current of 0.25 A with frequency 100 Hz.

a) What power is consumed in the circuit?

b) In the resistor?

c) In the capacitor?

d) In the inductor?

e) What is the power factor of the circuit?

36–24 A series circuit has a resistance of 75 Ω and an impedance of 150 Ω. What power is consumed in the circuit when a potential difference of 120 V is impressed across it?

36–25 A circuit draws 330 W from a 110-V 60-Hz ac line. The power factor is 0.6 and the current lags the voltage.

a) Find the capacitance of the series capacitor that will result in a power factor of unity.

b) What power will then be drawn from the supply line?

36–26 A series circuit has an impedance of 50 Ω and a power factor of 0.6 at 60 Hz, the voltage lagging the current.

a) Should an inductor or a capacitor be placed in series with the circuit to raise its power factor?

b) What size element will raise the power factor to unity?

36–27 The internal resistance of an ac source is 10,000 Ω.

a) What should be the turns ratio of a transformer to match the source to a load of resistance 10 Ω?

b) If the voltage amplitude of the source is 100 V, what is the voltage amplitude of the secondary on open circuit?

36–28 A 100-Ω resistor, a 0.1-μF capacitor, and a 0.1-H inductor are connected in parallel to a voltage source with amplitude 100 V.

a) What is the resonant frequency? The resonant angular frequency?

b) What is the maximum total current through the parallel combination at the resonant frequency?

c) What is the maximum current in the resistor at resonance?

d) What is the maximum current in the inductor at resonance?

e) What is the maximum energy stored in the inductor at resonance? In the capacitor?

36–29 The same three components as in Problem 36–28 are connected in *series* to a voltage source with amplitude 100 V.

a) What is the resonant frequency? The resonant angular frequency?

b) What is the maximum current in the resistor at resonance?

c) What is the maximum voltage across the capacitor at resonance?

d) What is the maximum energy stored in the capacitor at resonance?

36–30 A transformer connected to a 120-V ac line is to supply 12 V to a low-voltage lighting system for a model-railroad village. The total equivalent resistance of the system is 2 Ω.

a) What should be the the turns ratio of the transformer?

b) What current must the secondary supply?

c) What power is delivered to the load?

d) What resistance connected directly across the 120-V line would draw the same power as the transformer? Show that this is equal to 2 Ω times the square of the turns ratio.

36–31 A step-up transformer connected to a 120-V ac line is to supply 18,000 V for a neon sign. To reduce shock hazard, a fuse is to be inserted in the primary circuit; the fuse is to blow when the secondary circuit exceeds 10 mA.

a) What is the turns ratio of the transformer?

b) What power must be supplied to the transformer when the secondary current is 10 mA?

c) What current rating should the fuse in the primary circuit have?

Chapter 37

Electromagnetic Waves

37–1 INTRODUCTION

Our discussion of electric and magnetic fields in the last several chapters can be classified in two general categories. The first includes fields that do not vary with time. The electrostatic field of a distribution of charges at rest and the magnetic field of a steady current in a conductor are examples of fields which, while they may vary from point to point in space, do not vary with time at any individual point. For such situations we found it possible to treat the electric and magnetic fields independently, without worrying unduly about interactions between the two fields.

The second category includes situations in which the fields *do* vary with time, and in all such cases we have found that it is *not* possible to treat the fields independently. Faraday's law tells us that a time-varying magnetic field acts as a source of electric field. This field is manifested directly in induction accelerators such as the betatron (Section 33–7) and indirectly in the induced emf's in inductances and transformers. Similarly, in developing the general formulation of Ampère's law (Section 32–7), which is valid for charging capacitors and similar situations as well as for ordinary conductors, we found it necessary to regard a changing electric field as a source of magnetic field.

Thus, when either field is changing with time, a field of the other kind is induced in adjacent regions of space, and we are led naturally to consider the possibility of an electromagnetic disturbance, consisting of time-varying electric and magnetic fields, which can propagate through space from one region to another, even in the absence of any matter in the intervening region. Such a disturbance, if it exists, will have the properties of a *wave*, and an appropriate descriptive term is *electromagnetic wave*. Such waves do exist, of course, and it is a familiar fact that radio and television transmission, light, x-rays, and many other phenomena are examples of electromagnetic radiation. In the following pages we shall show how the existence of such waves is related to the principles of electromagnetism studied thus far, and shall examine their properties.

As so often happens in the development of science, the theoretical understanding of electromagnetic waves originally took a considerably more devious path than the one outlined above. In the early days of electromagnetic theory (the early nineteenth century), two different units of electric charge were used, one for electrostatics, the other for magnetic phenomena involving currents. The particular system of units in common use at the time had the property that these two units of charge had different physical dimensions; their ratio turned out to have units of velocity. This in itself is not so astounding, but experimental measurements revealed that the ratio had a numerical value precisely equal to the speed of light, $3.00 \times 10^8 \text{m s}^{-1}$. At the time, physicists regarded this as an extraordinary coincidence, and had no idea how to explain it. It was the search for an explanation of this result that led Maxwell, in 1864, to prove by theoretical reasoning that an electrical disturbance should propagate in free space with a speed equal to that of light, and hence to postulate that light waves were *electromagnetic* waves.

It remained for Heinrich Hertz, in 1887, actually to *produce* electromagnetic waves with the aid of oscillating circuits and to receive and detect these waves with other circuits tuned to the same frequency. Hertz then produced stationary electromagnetic waves and measured the distance between adjacent nodes, to measure the wavelength. Knowing the frequency of his resonators, he then found the velocity of the waves from the fundamental wave equation $c = f\lambda$, and verified Maxwell's theoretical value directly.

The possible use of electromagnetic waves for purposes of long-distance communication does not seem to have occurred to Hertz. It remained for the enthusiasm and energy of Marconi and others to make "wireless telegraphy" a familiar household phenomenon.

The unit of frequency, one cycle per second, is named one *hertz* (1 Hz) in honor of Hertz.

37–2 SPEED OF AN ELECTROMAGNETIC WAVE

In developing the relation of electromagnetic wave propagation to familiar electromagnetic principles, we begin with a particularly simple and somewhat artificial example of an electromagnetic wave. We shall postulate a particular configuration of fields, and then test whether it is consistent with the principles mentioned above, Faraday's law and Ampère's law including displacement current.

Using an x-y-z coordinate system, as shown in Fig. 37–1, we imagine that all space is divided into two regions by a plane perpendicular to the x-axis (parallel to the y-z plane). At all points to the right of this plane there are no electric or magnetic fields; at all points to the left there is a uniform electric field $\boldsymbol{E}$ in the $+y$-direction and a uniform magnetic field $\boldsymbol{B}$ in the $+z$-direction, as shown. Furthermore, we suppose that the boundary surface, which may also be called the *wave front*, moves to the right with a constant speed c, as yet unknown. Thus the $\boldsymbol{E}$ and $\boldsymbol{B}$ fields travel to the right into previously field-free regions, with a definite speed. The situation, in short, does describe a rudimentary electromagnetic wave, provided it is consistent with the laws of electromagnetism.

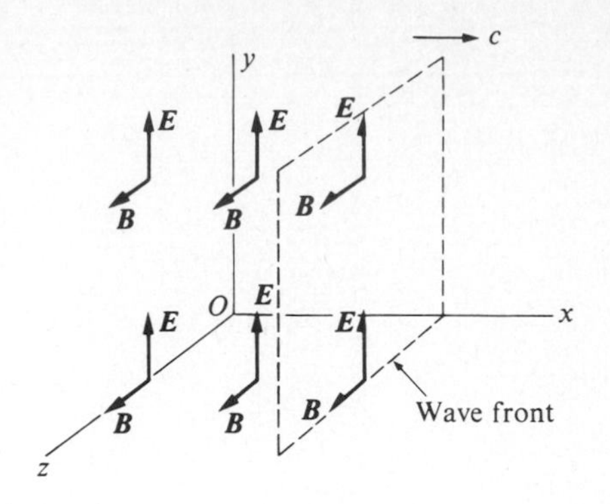

Fig. 37–1 An electromagnetic wave front. The **E** and **B** fields are uniform over the region to the left of the plane, but are zero everywhere to the right of it. The plane representing the wave front moves to the right with speed c.

We shall not worry about how such a field configuration can be produced. It can be shown that, in principle, an infinitely large sheet of charge in the y-z plane, which is initially at rest and suddenly starts moving with constant velocity in the $+y$-direction, gives such fields, but of course there is no practical way to realize an infinitely large charge sheet. At any rate, we now apply Faraday's law to a rectangle in the x-y plane, as in Fig. 37–2, located so that at some instant the wave front has progressed partway through the rectangle, as shown. In a time dt, the boundary surface moves a distance $c\,dt$ to the right, sweeping out an area $ac\,dt$ of the rectangle, and in this time the magnetic flux increases by $B(ac\,dt)$. Thus the rate of change of flux is given by

$$\frac{d\Phi}{dt} = Bac. \tag{37–1}$$

According to Faraday's law, this must equal the line integral of $\boldsymbol{E}$ around the boundary of the area. The

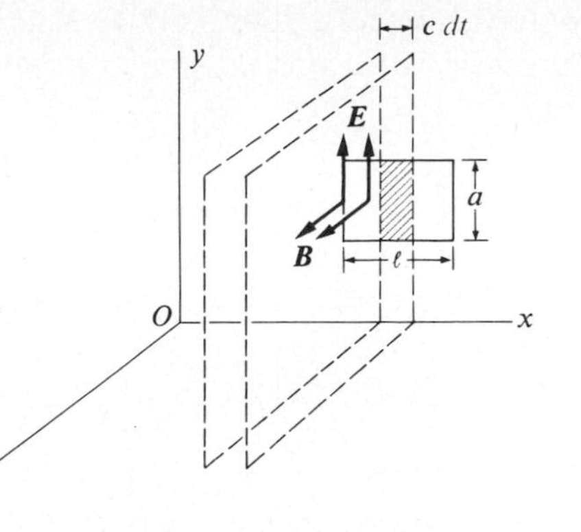

Fig. 37–2 In time dt the wave front moves to the right a distance c dt. The magnetic flux through the rectangle in the xy-plane increases by an amount dΦ equal to the flux through the shaded rectangle of area ac dt, that is, dΦ = Bac dt.

field at the right end is zero, and the top and bottom sides do not contribute because there the component of **E** along **ds** is zero. Thus the integral becomes simply ***Ea***, and Faraday's law gives

$$E = cB. \tag{37–2}$$

Thus the postulated wave is consistent with Faraday's law only if ***E***, ***B***, and c are related as in Eq. (37–2).

Next we consider a rectangle in the y-z plane, as shown in Fig. 37–3. There is a changing electric field flux through this rectangle, and thus a *displacement* current. Ampère's law, with no conduction current, gives

$$\oint \mathbf{B} \cdot d\mathbf{s} = \mu_0 I_D. \tag{37–3}$$

The displacement current I_D, defined in Section 29-8, is equal to ϵ_0 times the rate of change of electric flux through the area bounded by the integration path. In Fig. 37–3, the change in electric flux in time dt is the area $b(c\,dt)$ swept out by the wave front, multiplied by E. In evaluating the line integral of ***B***, we note that ***B*** is zero on the right end, and that it is perpendicular to the path on the front and back sides. Thus only the left end contributes to the integral, and we find $\oint \mathbf{B} \cdot d\mathbf{s} = Bb$. Combining these results with Eq. (37–3), we obtain

$$B = \mu_0 \epsilon_0 c E. \tag{37–4}$$

Thus Ampère's law is obeyed only if ***B***, c, and ***E*** are related as in Eq. (37–4).

Since *both* Ampère's law and Faraday's law must be obeyed simultaneously, Eqs. (37–2) and (37–4) must both be satisfied. This can occur only when $\mu_0 \epsilon_0 c = 1/c$, or

$$c = \frac{1}{\sqrt{\epsilon_0 \mu_0}}. \tag{37–5}$$

Inserting the numerical values of these quantities, we find

$$c = \frac{1}{\sqrt{(8.85 \times 10^{-12})(4\pi \times 10^{-7})}}$$
$$= 3.00 \times 10^8 \text{m s}^{-1}.$$

The postulated field configuration *is* consistent with the laws of electrodynamics, provided the wave front moves with the speed given above, which of course is recognized as the speed of light.

We have chosen a particularly simple wave for study in order to avoid undue mathematical complexity, but this special case nevertheless illustrates several features of *all* electromagnetic waves:

1. The wave is *transverse*; both ***E*** and ***B*** are perpendicular to the direction of propagation of the wave, and to each other.

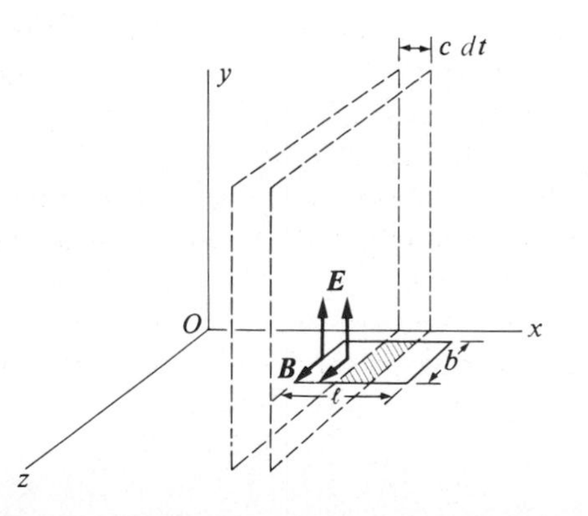

*Fig. 37–3 In time dt the electric flux through the rectangle in the xz-plane increases by an amount equal to **E** times the area (bc dt) of the shaded rectangle.*

2. There is a definite ratio between $\boldsymbol{E}$ and $\boldsymbol{B}$.
3. The waves travel in vacuum with a definite and unchanging speed.

It is not difficult to generalize the above discussion to a more realistic situation. Suppose we have several wave fronts in the form of parallel planes perpendicular to the axis and all moving to the right with speed c. Suppose that, within a single region between two planes, the $\boldsymbol{E}$ and $\boldsymbol{B}$ fields are the same at all points in the region, but that they differ from region to region. An extension of the above development shows that such a situation is also consistent with Ampère's and Faraday's laws, provided the wave fronts all move with the speed c given by Eq. (37–5). From this picture it is only a short additional step to a wave picture in which the $\boldsymbol{E}$ and $\boldsymbol{B}$ fields at any instant vary smoothly, rather than in steps, as we move along the x-axis and the entire field pattern moves to the right with speed c. In Section 37–5 we consider waves in which the dependence of $\boldsymbol{E}$ and $\boldsymbol{B}$ on position and time is *sinusoidal*; first, however, we consider the *energy* associated with an electromagnetic wave.

37–3 ENERGY IN ELECTROMAGNETIC WAVES

Analysis of energy needed to charge a capacitor and to establish a current in an inductor has led us (Sections 27–4 and 34–3) to associate an energy density (energy per unit volume) with electric and magnetic fields. Specifically, Eqs. (27–9) and (34–6) show that the total energy density u in a region of space where $\boldsymbol{E}$ and $\boldsymbol{B}$ fields are present is given by

$$u = \frac{1}{2}\epsilon_0 E^2 + \frac{1}{2\mu_0} B^2. \tag{37–6}$$

This and subsequent expressions become somewhat more symmetrical when expressed in terms of the field $\boldsymbol{H} = \boldsymbol{B}/\mu_0$ introduced in Chapter 35. In terms of $\boldsymbol{E}$ and $\boldsymbol{H}$, the energy density is

$$\boxed{u = \frac{1}{2}\epsilon_0 E^2 + \frac{1}{2}\mu_0 H^2.} \tag{37–7}$$

Now we have found above that $\boldsymbol{E}$ and $\boldsymbol{B}$ in an electromagnetic wave are not independent but are related by

$$B = \frac{E}{c} = \sqrt{{}_0\mu_0}\, E \qquad \text{or} \qquad H = \sqrt{\frac{\epsilon_0}{\mu_0}}\, E. \tag{37–8}$$

Thus the energy density may also be expressed

$$\begin{aligned} u &= \frac{1}{2}\epsilon_0 E^2 + \frac{1}{2}\mu_0 \left(\sqrt{\frac{\epsilon_0}{\mu_0}}\, E\right)^2 \\ &= \frac{1}{2}\epsilon_0 E^2 + \frac{1}{2}\epsilon_0 E^2 = \epsilon_0 E^2, \end{aligned} \tag{37–9}$$

which shows that in a wave, the energy density associated with the $\boldsymbol{E}$ field is equal to that of the $\boldsymbol{H}$ field.

Equation (37–8) may also be written

$$\boxed{\frac{E}{H} = \sqrt{\frac{\mu_0}{\epsilon_0}}.} \tag{37–10}$$

The unit of E is $1\ \mathrm{V\,m^{-1}}$ and that of H is $1\ \mathrm{A\,m^{-1}}$, so the unit of the ratio E/H is $1\ \mathrm{V\,A^{-1}}$ or $1\ \Omega$. When numerical values of ϵ_0 and μ_0 are inserted, we find

$$\frac{E}{H} = 377\ \Omega. \tag{37–11}$$

For reasons which we cannot discuss in detail, this ratio is called the *impedance of free space.*

Because the $\boldsymbol{E}$ and $\boldsymbol{H}$ fields in the simple wave considered above advance with time into regions where originally there were no fields, it is clear that the wave transports energy from one region to another. This energy transfer is conveniently characterized by considering the energy transferred *per unit time, per unit cross-sectional area* for an area perpendicular to the direction of wave travel. This quantity will be denoted by $\boldsymbol{S}$. This is analogous to the concept of current density, the charge per unit time transferred across unit area perpendicular to the direction of flow.

To see how the energy flow is related to the fields, we consider a stationary plane perpendicular to the x-axis, which at a certain time coincides with the wave front. In a time dt after this time, the wave front moves a distance $c\,dt$ to the right. Considering an area A on the stationary plane, we note that the energy in the space to the right of this area must have passed through it to reach its new location. The volume dV of the relevant region is the base area A times the length $c\,dt$, and the total energy dU in this region is the energy density u times this volume:

$$dU = \epsilon_0 E^2 Ac\,dt. \qquad (37\text{–}12)$$

Since this much energy passed through area A in time dt, the energy flow S per unit time, per unit area is

$$S = \frac{dU}{A\,dt} = \epsilon_0 c E^2.$$

Using Eqs. (37–5) and (37–8), we obtain the alternate forms

$$S = \frac{\epsilon_0}{\sqrt{\epsilon_0 \mu_0}} E^2 = \sqrt{\frac{\epsilon_0}{\mu_0}} E^2 = EH. \qquad (37\text{–}13)$$

The quantity S is called the *intensity* of the radiation. The units of S are energy per unit time, per unit area, or power per unit area. In the mks system the unit of S is $1\ \mathrm{J\,s^{-1}\,m^{-2}}$ or $1\ \mathrm{W\,m^{-2}}$.

We can define a *vector* quantity that describes both the magnitude and direction of the energy flow rate. We define

$$\mathbf{S} = \mathbf{E} \times \mathbf{H}. \qquad (37\text{–}14)$$

$\mathbf{S}$ is called the Poynting vector; its magnitude is given by Eq. (37–13) and its direction is the direction of propagation of the wave. The magnitude EH gives the flow of energy across a cross section perpendicular to the propagation direction, per unit area and per unit time.

Example In the wave discussed above, suppose

$$E = 100\ \mathrm{V\,m^{-1}} = 100\ \mathrm{N\,C^{-1}}.$$

Find the values of B and H, the energy density, and the energy flow S.

From Eq. (37–2),

$$B = \frac{E}{c} = \frac{100\ \mathrm{V\,m^{-1}}}{3.0 \times 10^8\ \mathrm{m\,s^{-1}}} = 3.33 \times 10^{-7}\ \mathrm{T};$$

$$H = \frac{B}{\mu_0} = \frac{3.33 \times 10^{-7}\ \mathrm{T}}{4\pi \times 10^{-7}\ \mathrm{Wb\,A^{-1}\,m^{-1}}} = 0.268\ \mathrm{A\,m^{-1}}.$$

From Eq. (37–9),

$$u = \epsilon_0 E^2 = (8.85 \times 10^{-12}\ \mathrm{C^2\,N^{-1}\,m^{-2}}) \times (100\ \mathrm{N\,C^{-1}})^2$$
$$= 8.85 \times 10^{-8}\ \mathrm{N\,m^{-2}} = 8.85 \times 10^{-8}\ \mathrm{J\,m^{-3}};$$
$$S = EH = (100\ \mathrm{V\,m^{-1}})(0.268\ \mathrm{A\,m^{-1}})$$
$$= 26.8\ \mathrm{W\,m^{-2}}.$$

The idea that energy can travel through empty space without the aid of any matter in motion may seem strange, yet this is the very mechanism by which energy reaches us from the sun. The conclusion that electromagnetic waves transport energy is as inescapable as the conclusion that energy is required to establish electric and magnetic fields.

It is also possible to show that electromagnetic waves carry *momentum*, with a corresponding momentum per unit volume of magnitude

$$\frac{EH}{c^2} = \frac{S}{c^2}. \qquad (37\text{–}15)$$

This momentum is a property of the field alone and is not associated with moving mass. There is also a corresponding momentum flow rate; just as the energy density u corresponds to S, the rate of energy flow per unit area, the momentum density given by Eq. (37–15) corresponds to the momentum flow rate

$$\frac{1}{A}\frac{dp}{dt} = \frac{S}{c} = \frac{EH}{c}, \qquad (37\text{–}16)$$

which represents the momentum transferred per unit surface area, per unit time.

This momentum is responsible for the phenomenon of *radiation pressure;* when an electromagnetic wave is absorbed by a surface perpendicular to the propagation direction, the rate of change of momentum is equal to the force on the surface. Thus the force per unit area, or pressure, is equal to S/c. If the wave is totally reflected, the momentum change is twice as great, and the pressure is $2S/c$. For example, the value of S for direct sunlight is about $1.4\ \mathrm{kW\,m^{-2}}$, and the corresponding pressure on a completely absorbing surface is

$$p = \frac{1.4 \times 10^3\ \mathrm{W\,m^{-2}}}{3.0 \times 10^8\ \mathrm{m\,s^{-1}}} = 4.7 \times 10^{-6}\ \mathrm{Pa}$$
$$= 4.7 \times 10^{-6}\ \mathrm{N\,M^{-2}}.$$

37–4 ELECTROMAGNETIC WAVES IN MATTER

The above analysis can be extended to include electromagnetic waves in dielectrics. The wave speed is not the same as in vacuum, and we denote it by w instead of c. Faraday's law is unaltered, but Eq. (37–2) is replaced by $E = wB$. In Ampère's law the displacement current density is given not by $\epsilon_0\, dE/dt$ but by $\epsilon\, dE/dt = K\epsilon_0\, dE/dt$. In addition, the constant μ_0 in Ampère's law must be replaced by $\mu = K_m\mu_0$. Thus, Eq. (37–4) is replaced by

$$B = \mu\epsilon w E,$$

and the wave speed is given by

$$w = \frac{1}{\sqrt{\epsilon\mu}} = \frac{1}{\sqrt{KK_m}}\frac{1}{\sqrt{\epsilon_0\mu_0}}. \tag{37–17}$$

For many dielectrics the relative permeability K_m is very nearly equal to unity; in such cases we can say

$$w \simeq \frac{1}{\sqrt{K}}\frac{1}{\sqrt{\epsilon_0\mu_0}} = \frac{c}{\sqrt{K}}.$$

Because K is always greater than unity, the velocity of electromagnetic waves in a dielectric is always less than the velocity in vacuum, by a factor of $1/\sqrt{K}$. The ratio of the speed in vacuum to the speed in a material is known in optics as the *index of refraction* n of the material; we see that this is given by

$$\frac{c}{w} = n = \sqrt{KK_m} \simeq \sqrt{K}. \tag{37–18}$$

Corresponding modifications are required in the expressions for the energy density and the Poynting vector; these are developed in detail in more advanced texts on electromagnetic theory.

The waves described above cannot propagate in a *conducting* material because the $\boldsymbol{E}$ field leads to currents which provide a mechanism for dissipating the energy of the wave. For an ideal conductor with zero resistivity, $\boldsymbol{E}$ must be zero everywhere inside the material. When an electromagnetic wave strikes such a material, it is totally reflected. Real conductors with finite resistivity permit some penetration of the wave into the material, with partial reflection. A polished metal surface is usually a good *reflector* of electromagnetic waves, but metals are not *transparent* to radiation.

37–5 SINUSOIDAL WAVES

Sinusoidal electromagnetic waves are closely analogous to sinusoidal transverse mechanical waves on a stretched string, as discussed in Chapter 21. At any point in space, the $\boldsymbol{E}$ and $\boldsymbol{H}$ fields are sinusoidal functions of time and, at any instant of time, the spatial variation of the fields is also sinusoidal.

The simplest sinusoidal electromagnetic waves share with the wave of Section 37–2 the property that at any instant the fields are uniform over any plane perpendicular to the direction of propagation. The entire pattern travels to the right with speed c. Since $\boldsymbol{E}$ and $\boldsymbol{H}$ are at right angles to the direction of propagation, the wave is *transverse.*

The frequency f, the wavelength λ, and the speed of propagation c are related by the equation applicable to any sort of periodic wave motion, namely, $c = f\lambda$. If the frequency f is the power-line frequency of 60 Hz, the wavelength is

$$\lambda = \frac{c}{f} = \frac{3 \times 10^8\mathrm{m\,s^{-1}}}{60\ \mathrm{Hz}} = 5 \times 10^6\mathrm{m} = 5000\ \mathrm{km},$$

which is of the order of the earth's radius! Hence at this frequency a distance of many miles includes only a small fraction of a wavelength. On the other hand, if the frequency is 10^8 Hz (100 M Hz), the wavelength is

$$\lambda = \frac{3 \times 10^8\mathrm{m\,s^{-1}}}{10^8\ \mathrm{Hz}} = 3\ \mathrm{m},$$

Fig. 37-4 ***E*-, *H*-, and *S*-vectors in a sinusoidal electromagnetic wave traveling in the x-direction.**

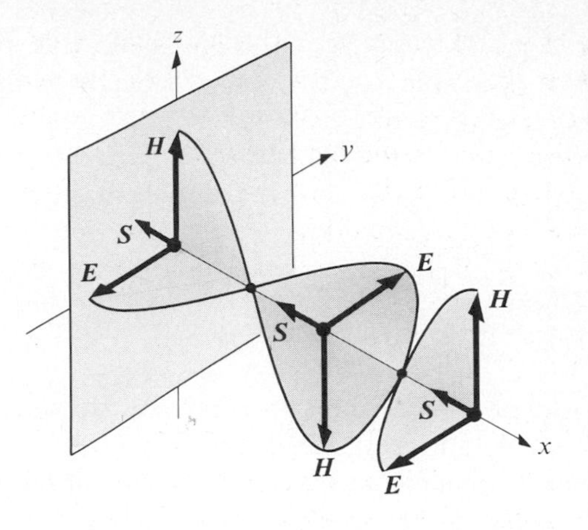

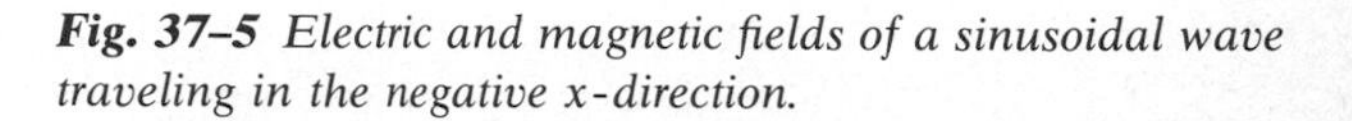

***Fig. 37-5** Electric and magnetic fields of a sinusoidal wave traveling in the negative x-direction.*

and a moderate distance can include a number of complete waves.

Figure 37–4 represents schematically the magnitudes and directions of the $\boldsymbol{E}$ and $\boldsymbol{H}$ fields in planes perpendicular to the direction of propagation. In those planes in which the $\boldsymbol{E}$-vector is in the direction of the positive y-axis, the $\boldsymbol{H}$-vector is in the direction of the positive z-axis. Where $\boldsymbol{E}$ is in the direction of the negative y-axis, $\boldsymbol{H}$ is in the direction of the negative z-axis.

It will be recalled that the equation of a transverse wave traveling to the right along a stretched string is $y = Y \sin(\omega t - kx)$, where y is the transverse displacement from its equilibrium position, at the time t, of a point of the string whose coordinate is x. The quantity Y is the maximum displacement or the *amplitude* of the wave, ω is its *angular frequency*, equal to 2π times the frequency f, and k is the *propagation constant*, equal to $2\pi/\lambda$, where λ is the wavelength.

Let e and h represent the instantaneous values, and E and H the maximum values or amplitudes, of the electric and magnetic fields in Fig. 37–4. The equations of the traveling electromagnetic wave are then

$$e = E \sin(\omega t - kx), \qquad h = H \sin(\omega t - kx). \tag{37–19}$$

The sine curves in Fig. 37–4 represent instantaneous values of e and h, as functions of x at the time $t = 0$. The wave travels to the right with speed c.

The instantaneous value s of the Poynting vector is

$$s = eh = EH \sin^2(\omega t - kx)$$

$$= \frac{1}{2} EH[1 - \cos 2(\omega t - kx)].$$

The time average value of $\cos 2(\omega t - kx)$ is zero, so the average value S_{av} of the Poynting vector, or the average power transmitted per unit area, is

$$S_{av} = \frac{1}{2} EH. \tag{37–20}$$

Now suppose the region in Fig. 37–4 does not extend indefinitely to the right, but is of finite length. Waves traveling to the right may then be *reflected* or *absorbed* at the far end, depending on conditions. The problem is closely analogous to that of a string with a source of waves at one end, where the nature of the reflection at the other end depends on whether the end is held rigidly or is free to move. The superposition of incident and reflected waves gives rise to *standing waves*.

Figure 37–5 represents schematically the electric and magnetic fields of a wave traveling in the *negative* x-direction. At points where $\boldsymbol{E}$ is in the positive

y-direction, $\boldsymbol{H}$ is in the *negative* z-direction, and where $\boldsymbol{E}$ is in the negative y-direction, $\boldsymbol{H}$ is in the positive z-direction. The Poynting vector is in the negative x-direction at all points. (Compare with Fig. 37–4, which represents a wave traveling in the *positive* x-direction.) The equations of the wave are

$$e = -E \sin(\omega t + kx), \qquad h = H \sin(\omega t + kx). \tag{37–21}$$

Assume that we have an *incident* wave traveling to the right, and also a *reflected* wave of the same amplitude traveling to the left. Let the subscripts 1 and 2 refer to the incident and reflected waves, respectively. From the principle of superposition, the resultant $\boldsymbol{E}$ and $\boldsymbol{H}$ fields are

$$e = e_1 + e_2,$$
$$h = h_1 + h_2.$$

Then from Eqs. (37–19) and (37–21),

$$e = 2E \cos \omega t \sin kx,$$
$$h = -2H \sin \omega t \cos kx.$$

The electric field is zero at all times in those planes for which $\sin kx = 0$, or at which

$$x = 0, \quad \frac{\lambda}{2}, \quad 2\frac{\lambda}{2}, \quad 3\frac{\lambda}{2}, \quad \text{etc.}$$

These are the *nodal planes* of the $\boldsymbol{E}$ field. Thus the reflected wave might have been produced by inserting a conducting sheet perpendicular to the x-axis at one of these values of x. By using *two* conducting sheets located at nodal planes, we can set up standing waves; since the distance between planes must be an integer number of half-wavelengths, this provides a means of measuring the wavelength. This technique was used by Hertz in his pioneering investigations of electromagnetic waves.

The magnetic field is zero at all times in those planes for which $\cos kx = 0$, or at which

$$x = \frac{\lambda}{4}, \quad 3\frac{\lambda}{4}, \quad 5\frac{\lambda}{4}, \quad \text{etc.}$$

These are the nodal planes of the $\boldsymbol{H}$ field. The nodal planes of one field are midway between those of the other, and the nodal planes of either field are separated by one-half wavelength.

The electric field is a *cosine* function of t and the magnetic field is a *sine* function of t. The fields are therefore 90° out of phase. At times when $\cos \omega t = 0$, the electric field is zero *everywhere* and the magnetic field is a maximum. When $\sin \omega t = 0$, the magnetic field is zero *everywhere* and the electric field is a maximum.

37–6 RADIATION FROM AN ANTENNA

The waves discussed above are called *plane waves,* referring to the fact that at any instant the fields are uniform over a plane perpendicular to the direction of propagation of the wave. Although these waves are the simplest to describe and analyze, they are by no means the simplest to produce experimentally. *Any* charge or current distribution that oscillates sinusoidally with time produces sinusoidal electromagnetic waves. The simplest example of such a source is an oscillating dipole, a pair of electric charges of equal magnitude and opposite sign, with the charge magnitude varying sinusoidally with time. Such an oscillating dipole can be constructed in a number of ways, the details of which need not concern us.

The radiation from an oscillating dipole is *not* a plane wave, but travels out in all directions from the source. At points far from the source the $\boldsymbol{E}$ and $\boldsymbol{H}$ fields are perpendicular to the direction from the source and to each other; in this sense the wave is still transverse. The value of S drops off as the square of the distance from the source and also depends on the direction from the source; the radiation is most intense in directions perpendicular to the dipole axis, and $S = 0$ in directions parallel to the axis. The radiation pattern from a dipole source is shown schematically in Fig. 37–6.

37–7 WAVES ON TRANSMISSION LINES

The above discussion has included only electromagnetic waves in vacuum or in a dielectric, but wave phenomena also occur in transmission lines. Figure 37–7 represents a region near one end of a long two-wire line of zero resistance in empty space. At time $t = 0$ the switch is closed, connecting the upper wire to the positive terminal of a source of constant terminal voltage V, and the lower wire to its negative terminal. The entire line does not become charged instantaneously, however, nor is the current I set up instantaneously at all points. Instead, as shown by the dotted line in the figure, these exists a boundary plane which travels to the right along the line with a constant speed u. At time t after the source has been connected, this plane has advanced a distance ut. At the left of the plane there is a current I, toward the right in the upper wire and toward the left in the lower. Within the same portion of the line, the upper wire has a **positive** charge and the lower wire has a negative charge. To the right of the plane there is no current and no excess charge. This portion has not yet received the news that the source has been connected. Figure 37–8 shows the electric and magnetic fields in the region near the end of the line.

By reversing the polarity of the battery at regular intervals, perhaps by using a reversing switch as in Fig. 37–9(a), we obtain a periodically varying wave on the transmission line. The figure shows the current and charge distributions and the associated **E** field at one instant. Similarly, if the source voltage varies sinusoidally, the result is a sinusoidal wave, as shown in Fig. 37–9(b).

This system can be analyzed from either a circuit or a wave point of view. We cannot pursue

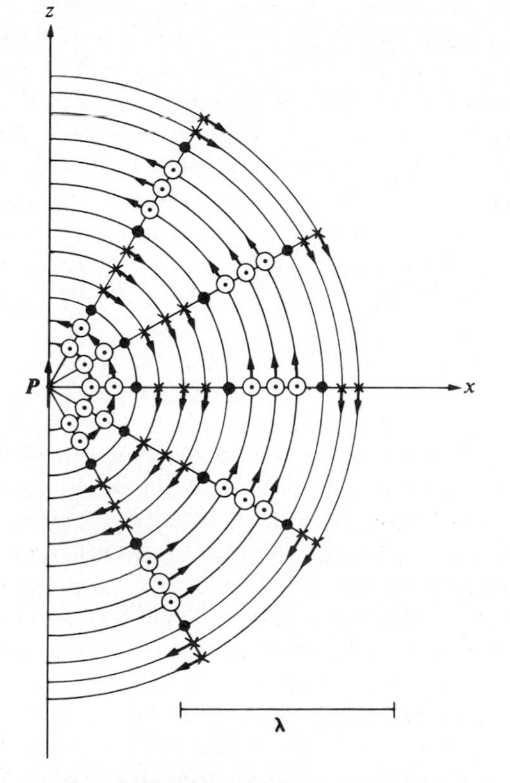

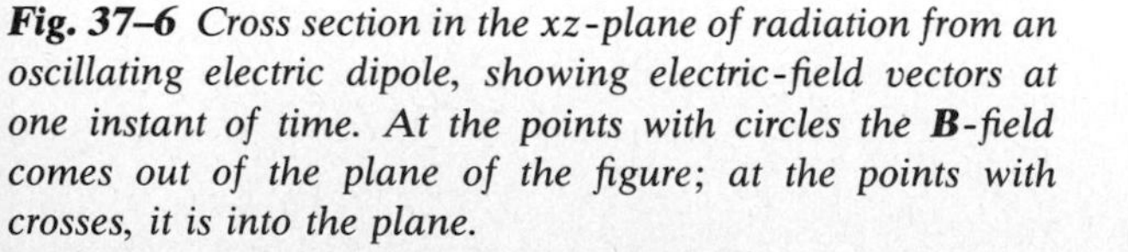
Fig. 37–6 Cross section in the xz-plane of radiation from an oscillating electric dipole, showing electric-field vectors at one instant of time. At the points with circles the **B***-field comes out of the plane of the figure; at the points with crosses, it is into the plane.*

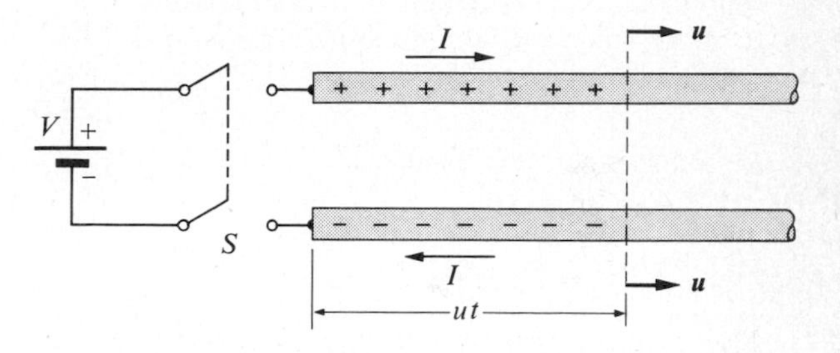

Fig. 37–7 *Propagation of an electric disturbance along a two-wire line of zero resistance in a vacuum.*

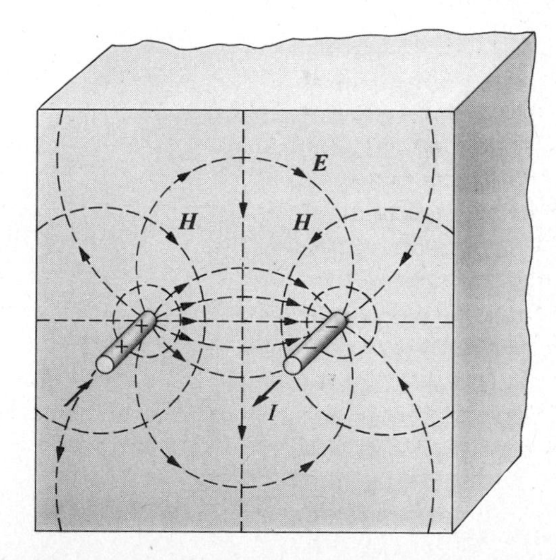

Fig. 37–8 *Electric and magnetic fields near a two-wire line.*

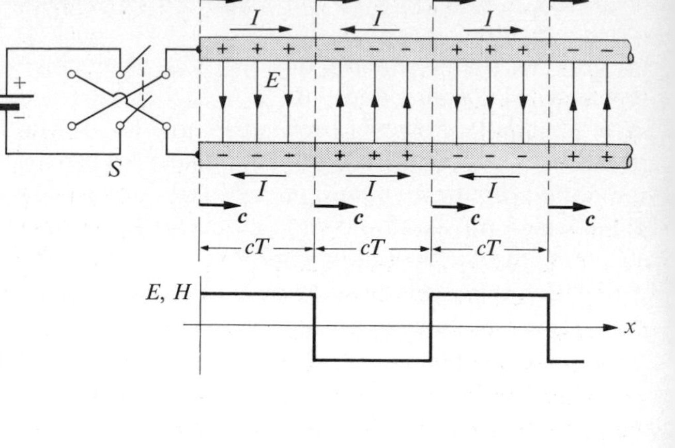

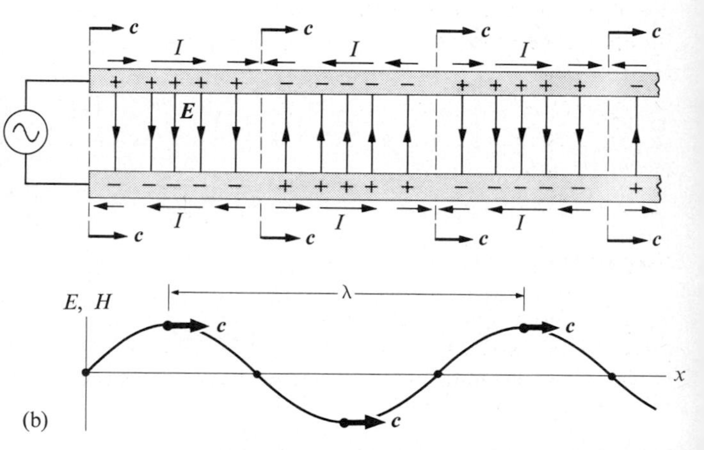

Fig. 37–9 *(a) Propagation of an electromagnetic wave along a two-wire line. (b) Current and charge distribution in a line consisting of a pair of parallel plates at one moment during the propagation of a sinusoidal electromagnetic wave.*

either analysis in detail here, but a few general remarks are in order. From the *circuit* point of view, the two conductors act as a capacitor, with the capacitance distributed along their length. Different portions of the capacitor charge at different times, and so the rate of change of potential difference between conductors at any position is related to the varying currents in various portions of the line. Similarly, there is a distributed self-inductance, and a corresponding relation between the time rate of change of current and the potential difference between conductors at each point.

When the details are worked out, one finds that the speed of propagation of this disturbance along the line is equal to the speed of light, given by Eq. (37–5). If the wires are embedded in a dielectric, then the speed of propagation is given instead by Eq. (37–17). If the line is terminated by a short circuit, the situation is analogous to reflection of an electromagnetic wave by a conducting sheet, and a reflected wave on the line occurs. In fact, it can be shown that a reflected wave originates at the far end of the line no matter how it is terminated, *unless* the termination consists of a pure resistance of the correct value. The resistance required to prevent reflections is called the *characteristic impedance* of the line, and typically has a value of 50 to 500 Ω depending on the dimensions of the conductors and the properties of the dielectric insulation material used.

From the *wave* viewpoint, the electric and magnetic field patterns around the wires propagate along the length of the line in the same manner as the plane waves discussed above, except that the fields are localized in the vicinity of the conductors. The speed of propagation of the waves can be shown again to be given by Eq. (37–5) or (37–17). Transmission lines for high-frequency use are often made in the form of coaxial cylinders; such a line is called a *coaxial line.* The advantage of this geometry is that the fields are confined entirely in the region between cylinders, so there is no danger of interference from adjacent lines.

Finally, at very high frequencies, of the order of 10^{10} Hz (corresponding to wavelengths of a few centimeters), an arrangement called a *waveguide* is used. This is basically a conducting pipe of rectangular cross section. Wave propagation in a waveguide can be pictured roughly in terms of waves traveling at an angle to the axis of the pipe, undergoing successive oblique reflections from opposite walls. For such a wave pattern, the propagation velocity along the waveguide is *not* equal to the speed given by Eq. (37–5), and more detailed analysis is necessary for complete understanding.

PROBLEMS

37–1 An electromagnetic wave propagates in a ferrite material having $K = 10$ and $K_M = 1000$. Find (a) the speed of propagation; (b) the wavelength of a wave having a frequency of 100 MHz.

37–2 The maximum electric field in the vicinity of a certain radio transmitter is 1.0×10^{-3} V m^{-1}. What is the maximum magnitude of the $\boldsymbol{B}$ field? How does this compare in magnitude with the earth's field?

37–3 The energy flow associated with sunlight is about 1.4 kW m^{-2}. (a) Find the maximum values of E and B for a wave of this intensity. (b) The distance from the earth to the sun is about 1.5×10^{11} m. Find the total power radiated by the sun.

37–4 For a 50,000-W radio station, find the maximum magnitudes of $\boldsymbol{E}$ and $\boldsymbol{B}$ at a distance of 100 km from the antenna, assuming that the antenna radiates equally in all directions (which is probably not actually the case).

37–5 The nineteenth-century inventor Nikolai Tesla proposed to transmit electric power via electromagnetic waves. Suppose power is to be transmitted in a beam of cross-sectional area 100 m^2; what $\boldsymbol{E}$ and $\boldsymbol{B}$ strengths are required to transmit an amount of power comparable to that handled by modern transmission lines (of the order of 500 kV and 1000 A)?

37–6 A bar magnet is mounted on an insulating support, as in the side and end views of Fig. 37–10(a) and (b), and is given a positive electric charge. Copy the diagram, sketch the lines of the $\boldsymbol{E}$ and $\boldsymbol{H}$ fields around the magnet, and draw the Poynting vector at a number of points.

a) What is the general nature of the energy flow in the electromagnetic field?
b) What can you say about the momentum flow in this case?

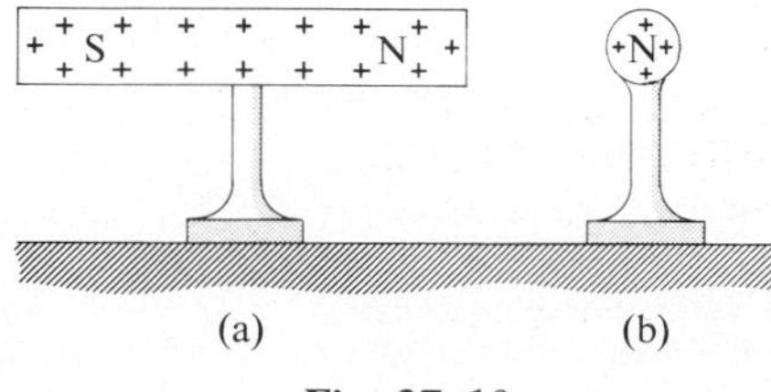

Fig. 37–10

37–7 A cylindrical conductor of circular cross section has a radius a and a resistivity ρ and carries a constant current I.

a) What are the magnitude and direction of the $\boldsymbol{E}$-vector at a point inside the wire, at a distance r from the axis?
b) What are the magnitude and direction of the $\boldsymbol{H}$-vector at the same point?
c) What are the magnitude and direction of the Poynting vector $\boldsymbol{S}$ at the same point?
d) Compare your answer to (c) with the rate of dissipation of energy within a volume of the conductor of length ℓ and radius r.

37–8 A wire of radius 1 mm whose resistance per unit length is $3 \times 10^{-3}\Omega$ m^{-1} carries a current of 25.1 A. At a point very near the surface of the wire, calculate (a) the magnitude of $\boldsymbol{H}$, (b) the component of $\boldsymbol{E}$ parallel to the wire, (c) the component of $\boldsymbol{S}$ perpendicular to the wire.

37–9 Assume that 10% of the power input to a 100-W lamp is radiated uniformly as light of wavelength 500 nm (1 nm $= 10^{-9}$ m). At a distance of 2 m from the source, the electric and magnetic intensities vary sinusoidally according to the equations $e = E \sin(\omega t + \phi)$ and $h = H \sin(\omega t + \phi)$. Calculate E and H.

37–10 A capacitor consists of two circular plates of radius r separated by a distance ℓ. Neglecting fringing, show that while the capacitor is being charged, the rate at which energy flows into the space between the plates is equal to the rate at which the electrostatic energy increases.

37–11 A very long solenoid of n turns per unit length and radius a carries an increasing current i.

a) Calculate the induced electric field at a point at a distance r from the solenoid axis.

b) Compute the magnitude and direction of the Poynting vector at this point.

37–12 An FM radio station antenna radiates a power of 10 kW at a wavelength of 3 m. Assume for simplicity that the radiated power is confined to, and is uniform over, a hemisphere with the antenna at its center. What are the amplitudes of E and H in the radiation field at a distance of 10 km from the antenna?

37–13 In a TV picture, ghost images are formed when the signal from the transmitter travels directly to the receiver and also indirectly after reflection from a building or other large metallic mass. In a 25-inch set the ghost is about 1 cm to the right of the principal image if the reflected signal arrives 1 μs after the principal signal. In this case, what is the difference in path length for the two signals?

37–14 For a sinusoidal electromagnetic wave in vacuum, such as that described by Eqs. (37–19), show that the average density of energy in the electric field is the same as that in the magnetic field.

37–15 A plane electromagnetic wave has a wavelength of 3.0 cm and an $\boldsymbol{E}$-field amplitude of 30 V m^{-1}.

a) What is the frequency?

b) What is the $\boldsymbol{B}$-field amplitude?

c) What is the intensity?

d) What average force does the radiation pressure exert on a totally absorbing surface of area 0.5 m^2 perpendicular to the the direction of propagation?

37–16 If the intensity of direct sunlight is 1.4 kW m^{-2}, find the radiation pressure (in pascals) on (a) a totally absorbing surface; (b) a totally reflecting surface. Also express your results in atmospheres.

Chapter 38

The Nature and Propagation of Light

38–1 THE NATURE OF LIGHT

Until about the middle of the seventeenth century it was generally believed that light consists of a stream of particles or *corpuscles.* These corpuscles were supposedly emitted by light sources, such as the sun or a candle flame, and traveled outward from the source in straight lines. They could penetrate transparent materials and were reflected from the surfaces of opaque materials. When the corpuscles entered the eye, the sense of sight was stimulated.

By the middle of the seventeenth century, while most workers in the field of optics accepted the corpuscular theory, the idea had begun to develop that light might be a *wave* motion of some sort. Christian Huygens, in 1678, showed that the laws of reflection and refraction could be explained on the basis of a wave theory and that such a theory furnished a simple explanation of the recently discovered phenomenon of double refraction. The wave theory did not find immediate acceptance, however. For one thing, the objection was raised that if light were a wave motion, one should be able to see around corners, since waves can bend around obstacles in their path. We know now that the wavelengths of light waves are so short that the bending, while it *does* actually take place, is so small that it is not ordinarily observed. As a matter of fact, the bending of a light wave around the edges of an object, a phenomenon known as *diffraction,* was noted by Grimaldi in a book published in 1665, but the significance of his observations was not understood at the time.

In the first quarter of the nineteenth century the experiments of Thomas Young and Augustin Fresnel on interference, and the measurements of the velocity of light in liquids by Leon Foucault at a somewhat later date, conclusively demonstrated the existence of optical phenomena for whose explanation a corpuscular theory was inadequate. These phenomena, including interference and diffraction, can be understood on the basis of a wave theory and will be discussed further in Chapter 42. Young's experiments enabled him to measure the wavelength of the waves and Fresnel showed that the rectilinear propagation of light, as well as the diffraction effects observed by Grimaldi and others, could be accounted for by the behavior of waves of short wavelength.

The next great forward step in the theory of light was the work of Maxwell, discussed in the preceding chapter. In 1873, Maxwell showed that an oscillating electrical circuit should radiate electromagnetic waves. The speed of propagation of the waves could be computed from purely electrical and magnetic measurements, and it turned out to be very nearly $3 \times 10^8 \mathrm{m\ s^{-1}}$. Within the limits of experimental error, this was equal to the measured speed of propagation of light. The evidence seemed inescapable that light consisted of electromagnetic waves of extremely short wavelength. In 1887 Heinrich Hertz, using an oscillating circuit of small dimensions, succeeded in producing short-wavelength waves (we would speak of them today as *microwaves*) of undoubted electromagnetic origin and showed that they possessed all the properties of light waves. They could be reflected, refracted, focused by a lens, polarized, and so on, just as could waves of light. Maxwell's electromagnetic theory of light and its experimental justification by Hertz constituted one of the triumphs of physical science. By the end of the nineteenth century it was the general belief that little, if anything, would be added in the future to our knowledge of the nature of light. Such was not to be the case!

The classical electromagnetic theory failed to account for several phenomena associated with *emission* and *absorption* of light. One example is the phenomenon of photoelectric emission, that is, the ejection of electrons from a conductor by light incident on its surface. In 1905, Einstein extended an idea proposed 5 years earlier by Planck and postulated that the energy in a light beam was concentrated in packets or *photons.* A vestige of the wave picture was retained, in that a photon was still considered to have a frequency and the energy of a photon was proportional to its frequency. The mechanism of the photoelectric effect consisted in the transfer of a photon's energy to an electron. Experiments by Millikan showed that the kinetic energies of photoelectrons were in excellent agreement with the formula proposed by Einstein. These experiments will be discussed in more detail in Chapter 44.

Another striking confirmation of the photon nature of light is the Compton effect. A. H. Compton, in 1921, succeeded in determining the motion of a photon and a single electron, both before and after a "collision" between them, and found that they behaved like material bodies having kinetic energy and momentum, both of which were conserved in the collision. The photoelectric effect and the Compton effect, then, both seem to demand a return to a corpuscular theory of light.

The reconciliation of these apparently contradictory experiments has come only since 1930 with the development of quantum electrodynamics, a comprehensive theory which includes both wave and particle properties. The phenomena of light *propagation* may best be described by the electromagnetic *wave* theory, while the interaction of light with matter, in the processes of emission and absorption, is a *corpuscular* phenomenon.

38–2 SOURCES OF LIGHT

All bodies emit electromagnetic radiation as a result of thermal motion of their molecules; this radiation, called *thermal radiation,* is a mixture of different wavelengths. At a temperature of 300°C the most intense of these waves has a wavelength of 5000×10^{-9}m or 5000 nm, which is in the *infrared* region. At a temperature of 800°C a body emits enough visible radiant energy to be self-luminous and appears "red hot." By far the larger part of the energy emitted, however, is still carried by infrared waves. At 3000°C, which is about the temperature of an incandescent lamp filament, the radiant energy contains enough of the "visible" wavelengths, between 400 nm and 700 nm, that the body appears nearly "white hot."

In modern incandescent lamps, the filament is a coil of fine tungsten wire. An inert gas such as argon is introduced to reduce evaporation of the filament. Incandescent lamps vary in size from one no larger than a grain of wheat to one with a power input of 5000 W, used for illuminating airfields.

One of the brightest sources of light is the *carbon arc.* Rods of carbon from 10 to 20 cm in length and about 1 cm in diameter are placed either horizontally, as shown in Fig. 38–1(a), or at an angle, as shown in Fig. 38–1(b). Sometimes the carbon rods are copper coated to improve electrical conductivity. To start a carbon arc, the two carbons are connected to a 110-V or 220-V dc source, are allowed to touch momentarily, and are then drawn apart. Intense

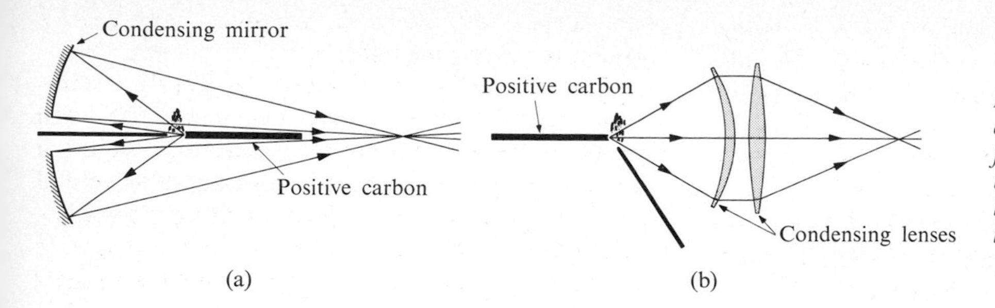

Fig. 38-1 Two types of carbon arcs. (a) With condensing mirror, for moderately sized motion picture theaters. (b) With condensing lenses for large motion picture theaters.

electron bombardment of the positive rod causes an extremely hot crater to form at its end. This end, at a temperature of about 4000°C, is the source of light. An electric motor or a clockwork mechanism is used to keep the carbons the correct distance apart as they burn away. Carbon arcs are used in all motion-picture theaters, where they operate on from 50 to several hundred amperes.

A common laboratory source of light is provided by a mercury arc. A glass or quartz tube has tungsten electrodes sealed in each end. A pool of mercury surrounds the negative electrode. A difference of potential is established across the electrodes and the tube is tilted until the mercury makes contact between the two electrodes. Some mercury is vaporized, and when the tube is restored to its vertical position an electric discharge is maintained by electrons and positive mercury ions. When the mercury is at low pressure, the mercury atoms emit a characteristic light consisting of only yellow, green, blue, and violet. A didymium filter may be used to absorb the yellow, and a yellow glass filter to absorb the blue and violet, leaving an intense green light consisting of a very small band of wavelengths whose average value is 546 nm. Low-pressure mercury arcs containing only one isotope of mercury, of atomic mass 198, are obtainable from the United States National Bureau of Standards. The green light from these lamps consists of an extremely narrow band of wavelengths, and is a close approach to *monochromatic* light.

If the current through a quartz mercury-arc lamp is allowed to increase, the temperature rises greatly and the vapor pressure of the mercury rises to between 50 and 100 atm. Such a lamp requires water cooling, and when operating is a very intense source of white light.

An intense source of yellow light of average wavelength 589.3 nm is provided by a sodium arc lamp. This is usually made of a special kind of glass that is not attacked by sodium and into which electrodes are sealed. Each electrode is a filament for providing electrons to maintain an electric discharge through an inert gas. After the inert gas discharge has taken place for a few minutes, the temperature rises to a value at which the vapor pressure of the sodium is great enough to provide sufficient sodium atoms to emit the characteristic yellow sodium light. Sodium lamps are sometimes used for street lighting because of their economy and because great visual acuity results from the absence of chromatic aberration of the eye when almost monochromatic light is used.

A very important development in illumination engineering is provided by the *fluorescent* lamp. This consists of a glass tube containing argon and a droplet of mercury. The electrodes consist of tungsten filaments. When an electric discharge takes place in the mercury–argon mixture, only a small amount of visible light is emitted by the mercury and argon atoms. There is, however, considerable *ultraviolet* light (light of wavelength shorter than that of visible violet). This ultraviolet light is absorbed in a thin layer of material, called a *phosphor*, with which the interior walls of the glass tube are coated. The phosphor has the property of *fluorescence*, which means that it emits visible light when illuminated by light of shorter wavelength. Lamps may be obtained which will fluoresce with any desired color, depending on the nature of the phosphor. Phosphors commonly used are cadmium borate for pink, zinc silicate for green, calcium tungstate for blue, and a mixture for white.

The brightest source among those described in the preceding pages is the carbon arc, which emits a large amount of radiant power from a relatively small surface. The radiation, however, spreads out in a large solid angle and only a fraction of it may be collected by a lens or a mirror and focused on a piece of equipment. In Section 44–7, a source of light readily available, called a *laser*, will be described and explained. The laser produces a narrow beam of enormously intense radiation, *all* of which may be intercepted by a lens and focused on an object. The power of this beam can be so great that high-intensity lasers have been used to cut through steel, fuse high-melting-point materials, and bring about many other effects that are important in physics, chemistry, biology, and engineering.

38–3 WAVES, WAVE FRONTS, AND RAYS

It is convenient to represent a train of waves of any sort by means of *wave fronts. A wave front is defined as the locus of all points at which the phase of vibration of a physical quantity is the same.* Thus, in the case of sound waves spreading out in all directions from a point source, any spherical surface concentric with the source is a possible wave front. Some of these spherical surfaces are the loci of points at which the pressure is a *maximum*, others where it is a *minimum*, and so on; but the *phase* of the pressure variations is the same over any spherical surface. It is customary to draw only a few wave fronts, usually those which pass through the maxima and minima of the disturbance. Such wave fronts are separated from one another by one-half wavelength.

If the wave is a light wave, the quantity corresponding to the pressure in a sound wave is the electric or magnetic field. It is usually unnecessary to indicate in a diagram either the magnitude or direction of the field, but simply to show the *shape* of the wave by drawing the wave fronts of their intersections with some reference plane. For example, the

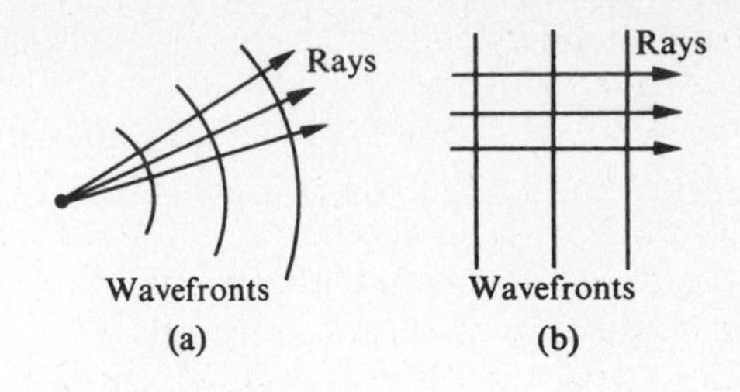

Fig. 38–2 *Wave fronts and rays.*

electromagnetic waves radiated by a small light source may be represented by spherical surfaces concentric with the source or, as in Fig. 38–2(a), by the intersections of these surfaces with the plane of the diagram. At a sufficiently great distance from the source, where the radii of the spheres have become very large, the spherical surfaces can be considered planes and we have a train of plane waves, as in Fig. 38–2(b).

A train of light waves may often be represented more simply by means of *rays* than by wave fronts. In a corpuscular theory, a ray is simply the path followed by a light corpuscle. From the wave viewpoint, *a ray is an imaginary line drawn in the direction in which the wave is traveling.* Thus, in Fig. 38–2(a) the rays are the radii of the spherical wave fronts and in Fig. 38–2(b) they are straight lines perpendicular to the wave fronts. In fact, in every case in which the waves are traveling in a homogeneous isotropic medium, the rays are straight lines normal to the wave fronts. At a boundary surface between two media, such as the surface between a glass plate and the air outside it, the direction of a ray may change suddenly, but it is a straight line both in the air and in the glass. If the medium is *not* homogeneous, for instance, if one is considering the passage of light through the earth's atmosphere, where the density and hence the velocity vary with elevation, the rays are curved but are still normal to the wave fronts. If the medium is anisotropic, as is the case in certain crystals, the direction of rays is not always normal to the wave fronts. The problem will be considered in more detail in Chapter 43.

A narrow cone of rays diverging from a common point is called a *pencil.* The entire group of pencils orginating at all points of a surface of finite extent is called a *beam*.

The wavelength of visible electromagnetic waves (those capable of affecting the sense of sight) lies between 0.00004 cm and 0.00007 cm. Because these wavelengths are so small, it is convenient to express them in terms of a small unit of length.

Three such units are commonly used: the micrometer (1 μm), the nanometer (1 nm) (both accented on the *first* syllable), and the angstrom (1 Å):

$$1\ \mu\text{m} = 10^{-6}\text{m} = 10^{-4}\text{cm},$$

$$1\ \text{nm} = 10^{-9}\text{m} = 10^{-7}\text{cm},$$

$$1\ \text{Å} = 10^{-10}\text{m} = 10^{-8}\text{cm}.$$

In older literature the micrometer is sometimes called the *micron*, and the nanometer is sometimes called the *millimicron*; these terms are passing out of common usage. Most workers in the fields of optical instrument design, color, and physiological optics express wavelengths in *nanometers*. For example, the wavelength of the yellow light from a sodium flame is 589 nm. A spectroscopist, however, would specify the wavelengths present in this same yellow light as 5889.963 Å and 5895.930 Å.

Figure 38–3 is a chart of the electromagnetic spectrum. Although waves of different wavelengths must be produced by different methods, all are alike so far as their fundamental nature is concerned. Note the relatively small portion occupied by the visible spectrum.

Different parts of the visible spectrum evoke the sensations of different colors. Wavelengths for colors in the visible spectrum are (approximately) as follows:

From 400 to 450 nm	Violet
From 450 to 500 nm	Blue
From 500 to 550 nm	Green
From 550 to 600 nm	Yellow
From 600 to 650 nm	Orange
From 650 to 700 nm	Red

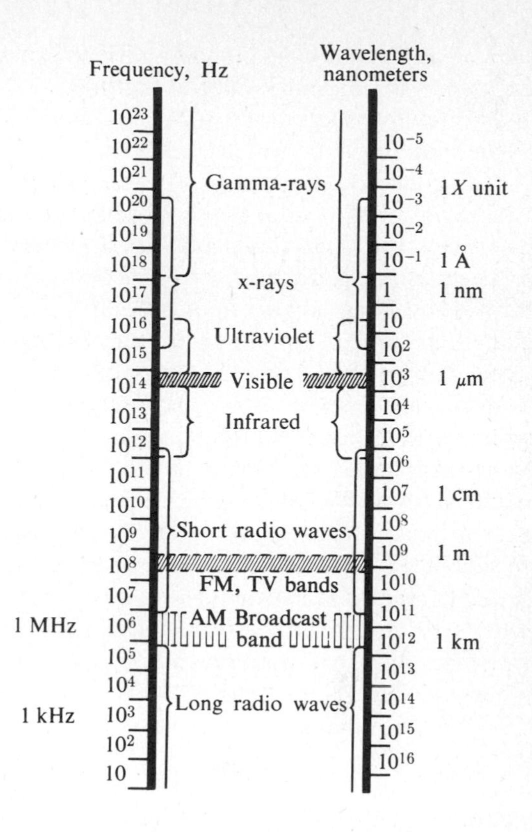

Fig. 38–3 *A chart of the electromagnetic spectrum.*

By the use of special sources or special filters, it is possible to limit the wavelength spread to a small band, say from 1 to 10 nm. Such light is called roughly *monochromatic light*, meaning light of a single color. Light consisting of only one wavelength is an idealization that is useful in theoretical calculations but represents an experimental impossibility. When the expression "monochromatic light of wavelength 550 nm" is used in theoretical discussions, it refers to one wavelength, but in descriptions of laboratory experiments it means a small band of wavelengths *around* 550 nm.

38–4 THE SPEED OF LIGHT

The speed of light in free space is one of the fundamental constants of nature. Its magnitude is so great (about 186,000 mi s^{-1} or 3×10^8m s^{-1}) that it evaded experimental measurement until 1676. Up to that time it was generally believed that light traveled with an infinite speed.

The first attempt to measure the speed of light involved a method proposed by Galileo. Two experimenters were stationed on the tops of two hills about a mile apart. Each was provided with a lantern, the experiment being performed at night. One man was first to uncover his lantern and, observing the light from this lantern, the second was to uncover his. The velocity of light could then be computed from the known distance between the lanterns and the time elapsing between the instant when the first observer uncovered his lantern and when he observed the light from the second. While the experiment was entirely correct in principle, we know that the speed is too great for the time interval to be measured in this way with any degree of precision.

In 1676, the Danish astronomer Olaf Roemer, from astronomical observations made on one of the satellites of the planet Jupiter, obtained the first definite evidence that light is propagated with a finite speed. Jupiter has twelve small satellites or moons, four of which are sufficiently bright to be seen with a moderately good telescope or a pair of field glasses. The satellites appear as tiny bright points at one side or the other of the disk of the planet. These satellites revolve about Jupiter just as does our moon about the earth and, since the plane of their orbits is nearly the same as that in which the earth and Jupiter revolve, each is eclipsed by the planet during a part of every revolution.

Roemer was engaged in measuring the time of revolution of one of the satellites by taking the time interval between consecutive eclipses (about 42 hr). He found, by a comparison of results over a long period of time, that while the earth was receding from Jupiter the periodic times were all somewhat longer than the average, and that while it was approaching Jupiter the times were all somewhat shorter. He concluded rightly that the cause of these variations was the varying distance between Jupiter and the earth.

Figure 38–4, not to scale, illustrates the situation. Let observations be started when the earth and Jupiter are in the positions E_1 and J_1. Since Jupiter

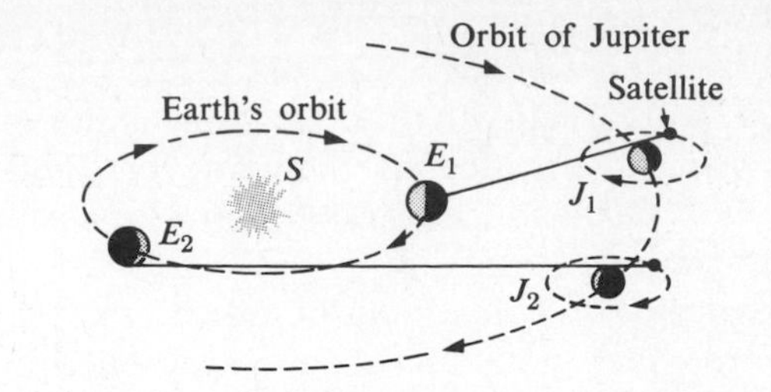

Fig. 38–4 *Roemer's method for deducing the velocity of light.*

requires about 12 years to make one revolution in its orbit, then by the time the earth has moved to E_2 (about 5 months later) Jupiter has moved only to J_2. During this interval the distance between the planets has been continually increasing. Hence at each eclipse, the light from the satellite must travel a slightly greater distance than at the preceding eclipse, and the observed time of revolution is slightly larger than the true time.

Roemer concluded from his observations that a time of about 22 min was required for light to travel a distance equal to the diameter of the earth's orbit. The best figure for this distance, in Roemer's time, was about 172,000,000 mi. Although there is no record that Roemer actually made the computation, if he *had* used the data above he would have found a speed of about 130,000 mi s^{-1} or 2.1×10^8 m s^{-1}.

The first successful determination of the speed of light from purely *terrestrial* measurements was made by the French scientist Fizeau in 1849. A schematic diagram of his apparatus is given in Fig. 38–5. Lens L_1 forms an image of the light source S at a point near the rim of a toothed wheel T, which can be set into rapid rotation. G is an inclined plate of clear glass. Suppose first that the wheel is stationary and the light passes through one of the openings between the teeth. Lenses L_2 and L_3, which are separated by about 8.6 km, form a second image on the mirror M. The light is reflected from M, retraces its path, and is in part reflected from the glass plate G through the lens L_4 into the eye of an observer at E.

If the wheel T is set in rotation, the light from S is "chopped up" into a succession of wave trains of

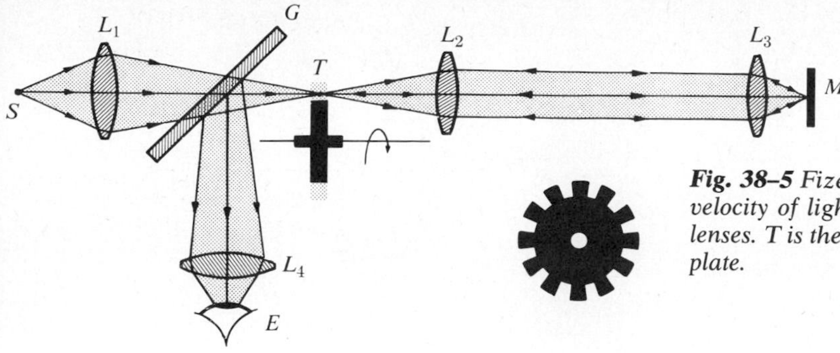

Fig. 38–5 *Fizeau's toothed wheel method for measuring the velocity of light. S is a light source,* L_1, L_2, L_3, *and* L_4 *are lenses. T is the toothed wheel, M is a mirror, and G is a glass plate.*

limited length. If the speed of rotation is such that by the time the front of one wave train has traveled to the mirror and returned, an opaque segment of the wheel has moved into the position formerly occupied by an open portion, no reflected light will reach the observer E. At twice this angular velocity, the light transmitted through any one opening will return through the next and an image of S will again be observed. From a knowledge of the angular velocity and radius of the wheel, the distance between openings, and the distance from wheel to mirror, the speed of light may be computed. Fizeau's measurements were not of high precision. He obtained a value of

$$3.15 \times 10^8 \text{ m s}^{-1}.$$

Fizeau's apparatus was modified by Foucault, who replaced the toothed wheel with a rotating mirror. In addition, Foucault introduced between the wheel and the mirror a tube filled with water and proved that the speed of light in water is less than in air. (The old corpuscular theory demanded that it should be greater, and at the time these measurements were made they were taken as conclusive proof that a corpuscular theory was untenable.)

The most precise measurements by the Foucault method were made by the American physicist Albert A. Michelson (1852-1931). His first experiments were performed in 1878 while he was on the staff of the Naval Academy at Annapolis. The last, under way at the time of his death, were completed in 1935 by Pease and Pearson.

From analysis of all measurements up to 1969, the most probable value is

$$c = 2.9979250 \times 10^8 \text{ m s}^{-1},$$

which is believed to be correct within ± 100 m s^{-1}.

Since we feel sure today that the speed of any electromagnetic wave in free space is given by

$$c = \sqrt{1/\epsilon_0 \mu_0} = \sqrt{k/k'},$$

and since in the mks system k' is assigned a value of *exactly* 10^{-7} N s^2 C^{-2}, the preceding equation provides the most precise means for finding the value of the electrical constant k:

$$k = k'c^2,$$

since the speed of light can be measured with much higher precision than the force between two charged bodies.

38–5 THE LAWS OF REFLECTION AND REFRACTION

Most of the objects we see are visible because they *reflect* light into our eyes. In the most common class of reflection, called *diffuse reflection*, light is reflected in all directions. A book on a table in a room lighted only by a point source of light can be seen from any part of the room. This type of reflection occurs whenever the roughness of the reflecting body has "dimensions" large compared with the wavelength of the reflected wave. In the other class of reflection, called *regular* or *specular reflection*, a narrow beam or

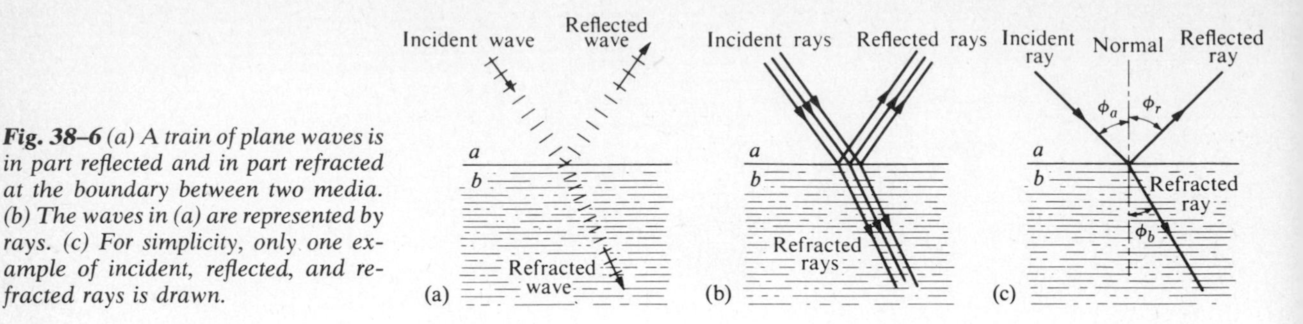

Fig. 38–6 *(a) A train of plane waves is in part reflected and in part refracted at the boundary between two media. (b) The waves in (a) are represented by rays. (c) For simplicity, only one example of incident, reflected, and refracted rays is drawn.*

pencil of light is reflected in one direction only. This type of reflection occurs from smooth surfaces whose irregularities are small compared with the wavelength of the reflected wave. For example, reflection from a blotter is diffuse, whereas reflection from a mirror is specular. Reflection from a polished automobile is intermediate, in that the reflected light is both diffuse and specular. Our main interest is in specular reflection, and in what follows, the word "reflection" will mean "specular reflection."

Figure 38–6(a) shows a narrow train of plane waves of light incident from the upper left on a plane surface separating two transparent substances, *a* and *b*, such as air and glass. A part of the incident light is *reflected* at the surface and a part passes into the lower medium, or is *refracted.*

The directions of the incident, reflected, and refracted beams of light are specified in terms of the angles they make with the normal to the surface at the point of incidence. For this purpose it is sufficient to indicate one ray, as in Fig. 38–6(c), although a single ray of light is a geometrical abstraction. A careful experimental study of the directions of the incident, reflected, and refracted beams leads to the following results:

1. *The incident, reflected, and refracted rays, and the normal to the surface, all lie in the same plane.* Thus, if the incident ray is in the plane of the diagram, and the surface of separation is perpendicular to this plane, the reflected and refracted rays are in the plane of the diagram.

2. *The angle of reflection ϕ_r is equal to the angle of incidence ϕ_a for all colors and any pair of substances.* Thus

$$\boxed{\phi_r = \phi_a.} \qquad (38\text{–}1)$$

The experimental result that $\phi_r = \phi_a$, and that the incident and reflected rays and the normal all lie in the same plane, is known as the *law of reflection.*

3. For monochromatic light and for a given pair of substances, *a* and *b*, on opposite sides of the surface of separation, *the ratio of the sine of the angle ϕ_a* (between the ray in substance *a* and the normal) *and the sine of the angle ϕ_b* (between the ray in substance *b* and the normal) *is a constant.* Thus

$$\frac{\sin \phi_a}{\sin \phi_b} = \text{constant.} \qquad (38\text{–}2)$$

This experimental result, together with the fact that the incident and refracted rays and the normal to the surface all lie in the same plane, is known as the *law of refraction.* The discovery of this law is usually credited to Willebrord Snell (1591–1626), although there appears to be some doubt that it was actually original with him. It is called *Snell's law.*

The laws of reflection and refraction relate only to the *directions* of the corresponding rays but say nothing about an equally important question, namely, the *intensities* of the reflected and refracted rays. These depend on the angle of incidence; for the present we simply state that the fraction reflected is smallest at *normal* incidence, where it is a few percent, and that it increases with increasing angle of incidence to almost 100% at grazing incidence or when $\phi_a = 90°$.

Table 38–1 INDEX OF REFRACTION FOR YELLOW SODIUM LIGHT ($\lambda = 589$ nm)

Substance	Index of refraction	Substance	Index of refraction
Solids		Liquids at 20°C	
Ice (H_2O)	1.309	Methyl alcohol (CH_3OH)	1.3290
Fluorite (CaF_2)	1.434	Water (H_2O)	1.3330
Rock salt (NaCl)	1.544	Ethyl alcohol (C_2H_5OH)	1.3618
Quartz (SiO_2)	1.544	Carbon tetrachloride (CCl_4)	1.4607
Zircon ($ZrO_2 \cdot SiO_2$)	1.923	Turpentine	1.4721
Diamond (C)	2.417	Glycerine	1.4730
Fabulite ($SrTiO_3$)	2.409	Benzene	1.5012
Rutile (TiO_2)	{2.616, 2.903}	Carbon disulfide (CS_2)	1.6276

Suppose next that a ray of light is directed from *below* the surface in Fig. 38–6. Again, we find that reflected and refracted rays exist, and that these, together with the incident ray and the normal, all lie in the same plane. The same law of reflection applies as when the ray is originally traveling in air, and the same law of refraction, Eq. (38–2). *The passage of a ray of light in going from one medium to another is reversible.* It follows the same path in going from b to a as when going from a to b. Also, in going from b to a, when the incident ray is normal to the surface, the same fraction of the incident light is reflected as when normal to the surface from above. As the angle between the ray in water and the normal is increased, the fraction reflected increases, but according to a different law from that applying in the direction from a to b.

38–6 INDEX OF REFRACTION

Let a beam of monochromatic light traveling *in vacuum* make an angle of incidence ϕ_v with the normal to the surface of a substance a, and let ϕ_a be the angle of refraction in the substance. The constant in Snell's law is then called the *index of refraction* of substance a and is designated by n_a:

$$\frac{\sin \phi_v}{\sin \phi_a} = n_a. \qquad (38\text{–}3)$$

The index of refraction depends not only on the substance but on the wavelength of the light. If no wavelength is stated, the index is usually assumed to be that corresponding to the yellow light from a sodium lamp, of wavelength 589 nm.

The index of refraction of most of the common glasses used in optical instruments lies between 1.46 and 1.96. There are only a very few substances having indices larger than this value, diamond being one, with an index of 2.42, and rutile (crystalline titanium dioxide) another, with an index of 2.7. The values for a number of solids and liquids are given in Table 38–1.

The index of refraction of air at standard conditions, for violet light of wavelength 436 nm, is 1.0002957, while for red light of wavelength 656 nm, the index is 1.0002914. It follows that for most purposes the *index of refraction of air can be assumed to be unity.* The index of refraction of a gas increases uniformly as the density of the gas is increased.

It follows from Eq. (38–3) that the angle of refraction ϕ_a is always *less than* the angle of incidence ϕ_v for a ray passing from a vacuum into one of the materials listed in Table 38–1, where all indices are seen to be greater than unity. In such a case, the ray is bent *toward* the normal. If the light is traveling in the opposite direction, the reverse is true and the ray is bent *away from* the normal.

Consider two parallel-sided plates of substances a and b placed parallel to each other with an arbitrary space between them, as shown in Fig. 38–7(a). Let the medium surrounding both plates be vacuum, although the behavior of light would be

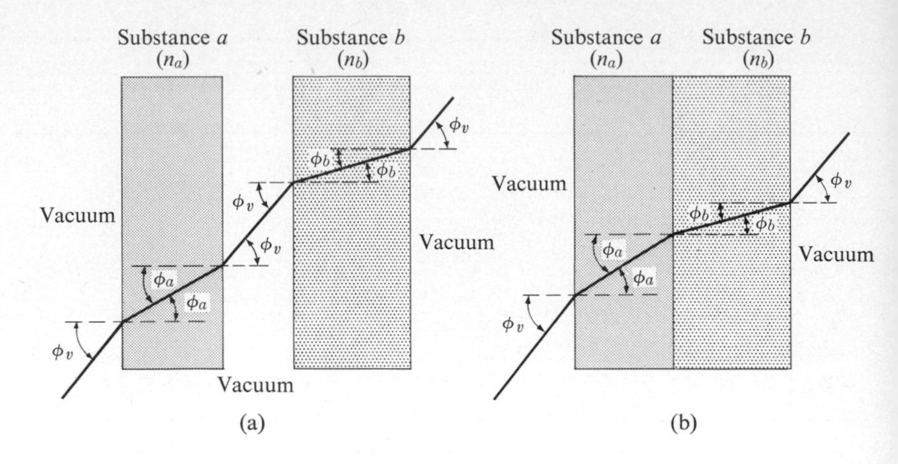

Fig. 38–7 The transmission of light through parallel plates of different substances. The incident and emerging rays are parallel, regardless of direction and regardless of the thickness of the space between adjacent slabs.

practically the same if the plates were surrounded by air. If a ray of monochromatic light starts at the lower left with an angle of incidence ϕ_v, the angle between ray and normal in substance a is ϕ_a, and the light emerges from substance a at an angle ϕ_v equal to its incident angle. The light ray therefore enters plate b with an angle of incidence ϕ_v, makes an angle ϕ_b in substance b, and emerges again at an angle ϕ_v. Exactly the same path would be traversed if the same light ray were to start at the upper right and enter substance b at an angle ϕ_v. Moreover, *the angles are independent of the thickness of the space between the two plates,* and are the same when the space shrinks to nothing, as in Fig. 38–7(b).

Applying Snell's law to the refractions that take place at the surface between vacuum and substance a, and also at the surface between vacuum and substance b, we have

$$\frac{\sin \phi_v}{\sin \phi_a} = n_a, \qquad \frac{\sin \phi_v}{\sin \phi_b} = n_b.$$

Dividing the second equation by the first, we get

$$\frac{\sin \phi_a}{\sin \phi_b} = \frac{n_b}{n_a}, \tag{38–4}$$

which shows that the *constant of Snell's law for the refraction between substances a and b is the ratio of the indices of refraction.* From Eq. (38–4) we see that the simplest and most symmetrical way of writing Snell's law for any two substances a and b, and for any direction, is

$$\boxed{n_a \sin \phi_a = n_b \sin \phi_b.} \tag{38–5}$$

38–7 ABSORPTION

No material is perfectly transparent; as light passes through any optical medium (except vacuum) its energy is partially absorbed, increasing the internal energy in the material, and the intensity (power per unit area) is correspondingly attenuated.

When a beam of light passes through a thin sheet of material, of thickness dx, the decrease dI in its intensity I is found to be proportional to the initial intensity I and to the thickness dx. Thus

$$dI = -\alpha I\, dx. \tag{38–6}$$

The proportionality constant α, which depends on the material, is called the *absorption coefficient.* The intensity after passage through a slab of finite thickness x can be obtained by integrating Eq. (38–6):

$$\boxed{I = I_0\, e^{-\alpha x},} \tag{38–7}$$

where I_0 is the intensity at $x = 0$. Equation (38–7) is called *Lambert's law.*

The absorption coefficient is often strongly wavelength-dependent. A clear optical glass for which $\alpha = 4\ \text{m}^{-1}$ (a typical value) in the middle of the visible spectrum may have $\alpha = 1000\ \text{m}^{-1}$ at $\lambda = 250$ nm (the near ultraviolet) and hence be essentially opaque to ultraviolet radiation. If α varies substantially *within* the visible spectrum, white light incident on the material appears colored when it emerges. For example, if α is larger at 500 nm than at 600 nm, proportionally more green and blue light is absorbed, and the emergent light appears red.

In some materials, absorption depends on the *polarization* of the incident light, that is, the plane in which the electric-field vector oscillates. A familiar example is the Polaroid filter, discussed in Section 43–4, used for sunglasses and many other applications. This material is transparent for light with one plane of polarization but absorbent for light in the perpendicular plane; the two absorption coefficients may differ by as much as a factor of 100.

A beam of light passing through an optical medium may also be attenuated by *scattering*. In contrast to absorption, in which the energy is ordinarily converted to internal energy, scattering simply redirects some of the radiation into directions other than that of the beam. The visibility from the side of a searchlight beam at night results from scattering of light out of the beam by dust particles and water droplets in the air and, to a small extent, by the air molecules themselves.

Scattering of visible light is usually wavelength-dependent, usually increasing with increasing frequency (decreasing wavelength). The sky is blue because the shorter-wavelength (blue) visible light from the sun is scattered more strongly in the earth's atmosphere than is the longer-wavelength (red) light. Similarly, at sunset the sun appears red because proportionally more of the blue light has been scattered out of the beam.

38–8 ILLUMINATION

We have defined the *intensity* of light and other electromagnetic radiation as power per unit area, measured in watts per square meter. Similarly, the *total* rate of radiation of energy from any of the sources of light discussed in Section 38–2 is called the *radiant power* or *radiant flux*, measured in watts. These quantities are not adequate to measure the visual sensation of *brightness*, however, for two reasons: First, not all the radiation from a source lies in the visible spectrum; an ordinary incandescent light bulb radiates more energy in the infrared than in the visible spectrum. Second, the eye is not equally sensitive to all wavelengths; a bulb emitting 1 watt of yellow light appears brighter than one emitting one watt of blue light.

The quantity analogous to radiant power, but compensated to include the above effects, is called *luminous flux*, denoted by F. The unit of luminous flux is the *lumen*, abbreviated lm, defined as that quantity of light emitted by $\frac{1}{60}\ \text{cm}^2$ surface area of pure platinum at its melting temperature (about 1770°C), within a solid angle of 1 steradian (1 sr). As an example, the total light output (luminous flux) of a 40-watt incandescent light bulb is about 500 lm, while that of a 40-watt fluorescent tube is about 2300 lm.

When luminous flux strikes a surface, the surface is said to be *illuminated*. The intensity of illumination, analogous to the intensity of electromagnetic radiation (which is power per unit area) is the *luminous flux per unit area*, called the *illuminance*, denoted by E. The unit of illuminance is the lumen per square meter, also called the *lux*:

$$1\ \text{lux} = 1\ \text{lm m}^{-2}.$$

An older unit, the lumen per square foot, or foot-candle, has become obsolete. If luminous flux F falls at normal incidence on an area A, the illuminance E is given by

$$E = \frac{F}{A}. \qquad (38\text{–}8)$$

Most light sources do not radiate equally in all directions; it is useful to have a quantity that describes the intensity of a source in a specific direction, without using any specific distance from the source. We place the source at the center of an imaginary sphere of radius R. A small area A of the sphere subtends a solid angle Ω given by $\Omega = A/R^2$. If the luminous flux passing through this area is F, we define the *luminous intensity* I in the direction of the area as

$$I = \frac{F}{\Omega}. \tag{38–9}$$

The unit of luminous intensity is one lumen per steradian, also called one *candela*, abbreviated cd:

$$1 \text{ cd} = 1 \text{ lm sr}^{-1}.$$

The term "luminous intensity" is somewhat misleading. The usual usage of *intensity* connotes power per unit area, and the intensity of radiation from a point source decreases as the square of the distance. Luminous intensity, however, is flux per unit *solid angle*, not per unit *area*, and the luminous intensity of a source in a particular direction *does not* decrease with increasing distance.

Example A certain 100-watt light bulb emits a total luminous flux of 1200 lm, distributed uniformly over a hemisphere. Find the illuminance and the luminous intensity at a distance of 1 m, and at 5 m.

The area of a half-sphere of radius 1 m is

$$(2\pi)(1 \text{ m})^2 = 6.28 \text{ m}^2.$$

The illuminance at 1 m is

$$E = \frac{1200 \text{ lm}}{6.28 \text{ m}^2} = 191 \text{ lm m}^{-2} = 191 \text{ lux}.$$

Similarly, the area of a half-sphere of radius 5 m is

$$(2\pi)(5 \text{ m})^2 = 157 \text{ m}^2,$$

and the illuminance at 5 m is

$$E = \frac{1200 \text{ lm}}{157 \text{ m}^2} = 7.64 \text{ lm m}^{-2} = 7.64 \text{ lux}.$$

This is smaller by a factor of 5^2 than the illuminance at 1 m, and illustrates the inverse-square law for illuminance from a point source.

The solid angle subtended by a hemisphere is 2π sr. The luminous intensity is

$$I = \frac{1200 \text{ lm}}{2\pi \text{ sr}} = 191 \text{ lm sr}^{-1} = 191 \text{ cd}.$$

The luminous intensity does not depend on distance.

PROBLEMS

38–1 What is the wavelength in meters, microns, nanometers, and angstrom units of (a) soft x-rays of frequency 2×10^{17} Hz? (b) green light of frequency 5.6×10^{14} Hz?

38–2 The visible spectrum includes a wavelength range from about 400 nm to about 700 nm. Express these wavelengths in inches.

38–3 Assuming the radius of the earth's orbit to be 92,900,000 mi, and taking the best value of the speed of light, compute the time required for light to travel a distance equal to the diameter of the earth's orbit. Compare with Roemer's value of 22 min.

38–4 Fizeau's measurements of the speed of light were continued by Cornu, using Fizeau's apparatus but with the distance between mirrors increased to 22.9 km. One of the toothed wheels used was 40 mm in diameter and had 180 teeth. Find the angular velocity at which it should rotate so that light transmitted through one opening will return through the next.

38–5 The speed of light in matter is less than in vacuum by a factor of the refractive index of the material. When orange light of wavelength 600 nm enters glass having refractive index 1.5, find (a) the wavelength, (b) the frequency, in the material.

38–6 An inside corner of a cube is lined with mirrors. A ray of light is reflected successively from each of three mutually perpendicular mirrors; show that its final direction is always exactly opposite to its initial direction. This principle is used in tail-light lenses and reflecting highway signs.

38–7 The density of the earth's atmosphere increases as the surface of the earth is approached.

a) Draw a diagram showing how the light from a star or planet is bent as it goes through the atmosphere.
b) Indicate the apparent position of the light source.
c) Explain how one can see the sun after it has set.
d) Explain why the setting sun appears flattened.

38–8

a) Why does the surface of a smooth highway under hot sun appear wet when it is seen at a glancing angle?
b) Account for a mirage in the desert.

38–9 A ray of light traveling with speed c leaves point 1 of Fig. 38–8 and is reflected to point 2. Show that the time

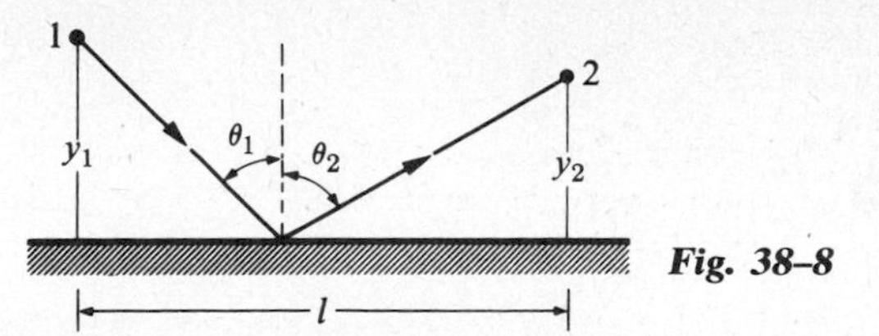

Fig. 38-8

required for the light to travel from 1 to 2 is $(y_1 \sec \theta_1 + y_2 \sec \theta_2)/c$.

38–10 Prove that a ray of light reflected from a plane mirror rotates through an angle 2θ when the mirror rotates through an angle θ about an axis perpendicular to the plane of incidence.

38–11 A parallel beam of light is incident on a prism, as shown in Fig. 38-9, part reflected from one face and part from another. Show that the angle θ between the two reflected beams is twice the angle A between the two reflecting surfaces.

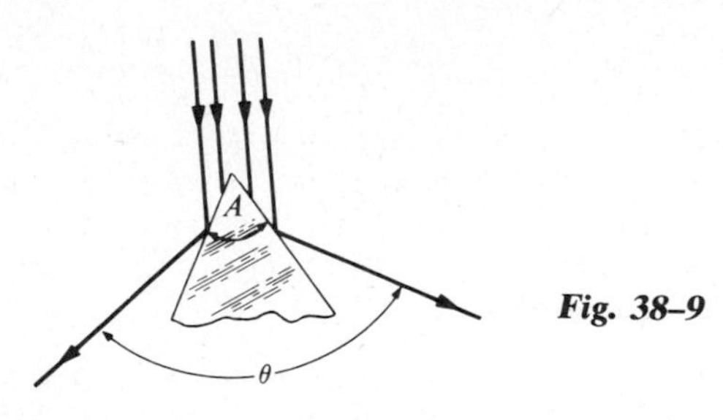

Fig. 38-9

38–12 A parallel beam of light makes an angle of 30° with the surface of a glass plate having a refractive index of 1.50. (a) What is the angle between the refracted beam and the surface of the glass? (b) What should be the angle of incidence ϕ with this plate for the angle of refraction to be $\phi/2$?

38–13 Light strikes a glass plate at an angle of incidence of 60°, part of the beam being reflected and part refracted. It is observed that the reflected and refracted portions make an angle of 90° with each other. What is the index of refraction of the glass?

38–14 A ray of light is incident on a plane surface separating two transparent substances of indices 1.60 and 1.40. The angle of incidence is 30° and the ray originates in the medium of higher index. Compute the angle of refraction.

38–15 A parallel-sided plate of glass having a refractive index of 1.60 is held on the surface of water in a tank. A ray coming from above makes an angle of incidence of 45° with the top surface of the glass. (a) What angle does the ray make with the normal in the water? (b) How does this angle vary with the refractive index of the glass?

38–16 A lens for sunglasses is to be made from a gray-tinted glass having an absorption coefficient of 500 m^{-1}. If the intensity of light passing through the lens is to be decreased by a factor of $\frac{1}{4}$, what should be the thickness of the lens?

38–17 If a swimming pool is 5m deep and the water has an absorption coefficient of 2 m^{-1}, by what factor is the intensity of light attenuated as it travels from the surface to the bottom of the pool?

38–18 If the absorption coefficient of sea water is 2 m^{-1}, and if the eye can perceive light of intensity smaller than that of sunlight by a factor of 10^{-18}, what is the greatest depth below the ocean surface at which light can be seen?

38–19 The illuminance of direct sunlight is about 10^5 lux. If a photoflash lamp has an intensity in a certain direction of 5×10^6 cd, at what distance from a surface should it be placed to produce illuminance equal to that of sunlight?

38–20 The *luminous efficiency* of a lamp is defined as the ratio of luminous flux to electric power input. A certain lamp mounted 3 m above a desk top has a luminous efficiency of 20 lm W^{-1}. What is the power input to the lamp if the illuminance on the desk is equal to that of sunlight, about 10^5 lux? Assume that the lamp radiates uniformly over its lower half-sphere.

38–21 A certain baseball field in the shape of a square 140 m on a side is to be illuminated for night games by six towers supporting banks of 1000-watt incandescent lamps with luminous efficiency (see Problem 38–20) of 30 lm W^{-1}. The illuminance required on the playing field is 200 lux. Assume that 50% of the luminous flux from the lamps reaches the field.

a) How many lamps are required in each tower?
b) What is the electric power input to each tower?
c) If power for all six towers is supplied by a generator driven by a gasoline engine, what must be the power capacity of the engine?

Chapter 39

Reflection and Refraction at Plane Surfaces

39–1 HUYGENS' PRINCIPLE

In Chapters 37 and 38 we have considered electromagnetic waves for which the wavefronts are planes or spheres, the latter originating from a very small source which may be idealized as a point. In the absence of obstructions, such a wave can propagate indefinitely far; a relatively simple expression can be written down which describes the electric or magnetic field at any point and at any time. We have written such expressions for sinusoidal plane electromagnetic waves in Section 37–5.

The absence of obstructions is of course an idealization; in real situations waves almost always *are* obstructed. When light enters a telescope, only that portion of a wave front corresponding to the diameter of the objective lens is admitted. When light passes through a narrow slit, almost the entire wave front is absorbed by the screen and only a small portion goes through. The fundamental problem of wave propagation is to determine the amplitude, the phase, and the polarization of the light arriving at a point when only a portion of a wave front is exposed. This can be an extraordinarily difficult problem when the exposed portion of the wave front has an unusual shape, and when the electrical and magnetic characteristics of the obstructing screen have to be taken into account. If, however, the direction of the $\mathbf{E}$-vector is unimportant, a scalar theory may be used and the mathematical equations become somewhat simpler. The technique of finding the intensity and phase of the light at any point when only a portion of a wave front is exposed is called *Huygens' principle.* (The diphthong "uy" is pronounced like the French sound "euille," somewhat as the "oy" in "boy," and the "g" is guttural.)

The principle, as stated originally by Huygens, is a geometrical method for finding, from the known shape of a portion of a wave front at some instant, what the shape will be at some later instant. It is assumed that *every point of a wave front may be considered the source of secondary wavelets, which spread out in all directions with a speed equal to the speed of propagation of the waves.* The new wave front is then found by constructing a surface *tangent* to the secondary wavelets or, as it is called, the *envelope* of the wavelets.

Huygens' principle is illustrated in Fig. 39–1. The original wave front, AA', is traveling as indicated by the small arrows. We wish to find the shape of the wave front after a time interval t. Let v represent the speed of propagation. Construct a number of circles (traces of spherical wavelets) of radius $r = vt$, with centers along AA'. The trace of the envelope of these

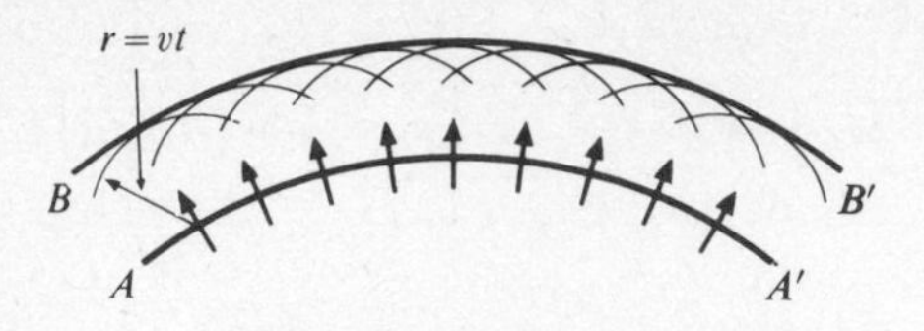

Fig. 39–1 *Huygens' principle.*

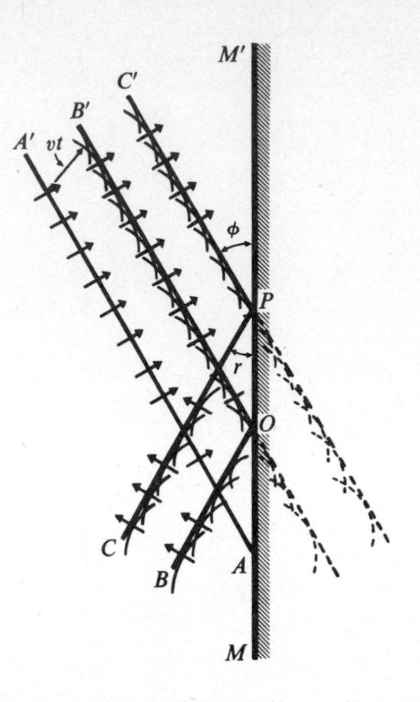

Fig. 39–3 *Successive positions of a plane wave AA′ as it is reflected from a plane surface.*

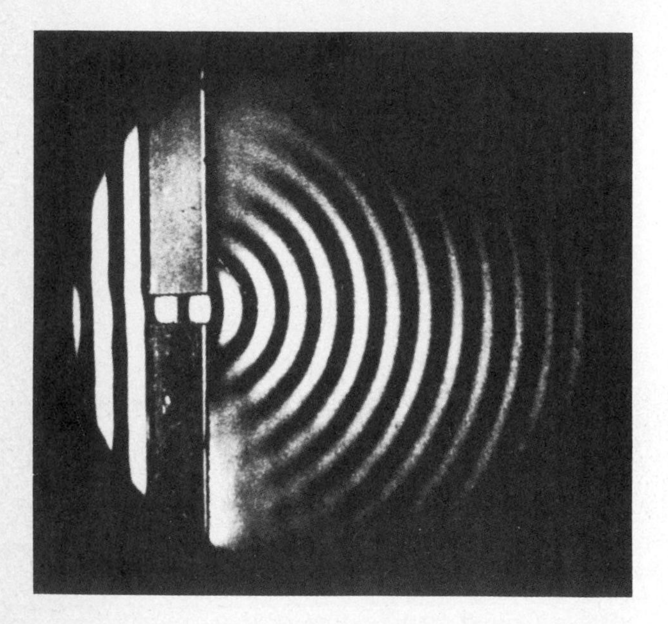

Fig. 39–2 *Demonstration of a Huygens wavelet with the aid of water waves in a shallow tank. (Reproduced from* Ripple Tank Studies of Wave Motion *with permission of W. Llowarch, the Clarendon Press, Oxford.)*

wavelets, which is the new wave front, is the curve BB'. The speed v has been assumed the same at all points and in all directions.

The fact that any point on a wave front may be regarded as a source of a secondary wavelet may be demonstrated convincingly with the aid of a ripple tank. If plane wave fronts are produced in the water in a shallow tank and these wave fronts come to a narrow slit, a complete circular wave is formed, as shown in Fig. 39–2, rather than a narrow parallel beam, as might have been expected.

39–2 DERIVATION OF THE LAW OF REFLECTION FROM HUYGENS' PRINCIPLE

In Fig. 39–3, consider the trace of a plane wave surface AA', called the incident *wave front,* which is just making contact with a reflecting surface MM' along a line through A perpendicular to the plane of the diagram. The planes of the wave front and the reflecting surface are also normal to the plane of the figure. The position of the wave front after a time interval t may be found by applying Huygens' principle. With points on AA' as centers, draw a number of secondary wavelets of radius vt, where v is the speed of propagation in the medium near the surface. Those wavelets originating near the upper end of AA' spread out unhindered, and their envelope gives that portion of the new wave surface OB'. The wavelets originating near the lower end of AA', however, strike the reflecting surface. If the latter had not been there, they would have occupied the positions shown by the dotted circular arcs. The effect of the reflecting surface is to reverse the direction of travel of

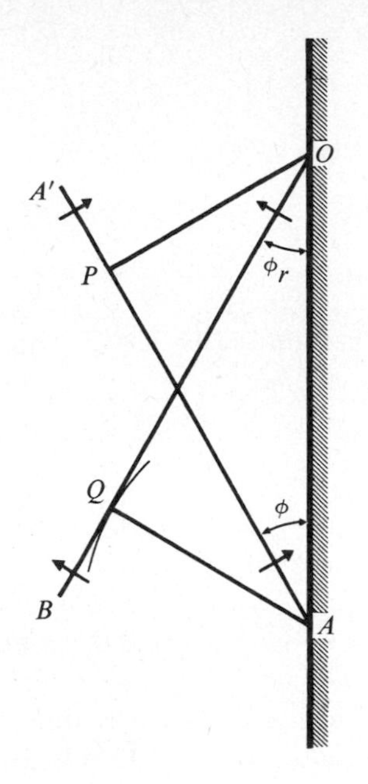

Fig. 39-4 A portion of Fig. 39-3.

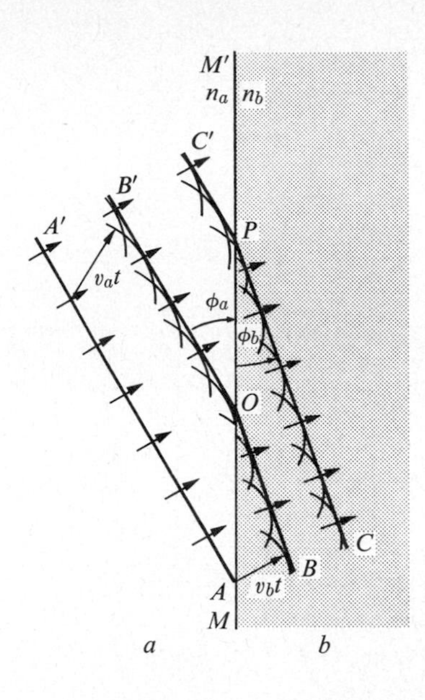

Fig. 39-5 Successive positions of a plane wave front AA' as it is refracted by a plane surface.

those wavelets which strike it, so that that part of a wavelet which would have penetrated the surface actually lies to the left of it, as shown by the full lines. The envelope of these reflected wavelets is then that portion of the wave front OB. The trace of the entire wave front at this instant is the broken line BOB'. A similar construction gives the line CPC' for the wave front after another interval t.

The angle ϕ between the incident *wave front* and the *surface* is the same as that between the incident *ray* and the *normal* to the surface, and is therefore the angle of incidence. Similarly, ϕ_r is the angle of reflection. To find the relation between these angles, consider Fig. 39-4, which is the same as a portion of Fig. 39-3. From O, draw $OP = vt$, perpendicular to AA'. Now OB, by construction, is tangent to a circle of radius vt with center at A. Hence, if AQ is drawn from A to the point of tangency, the triangles APO and AQO are equal (right triangles with the side AO in common and with $AQ = OP$). The angle ϕ therefore equals the angle ϕ_r, and we have the law of reflection.

39-3 DERIVATION OF SNELL'S LAW FROM HUYGENS' PRINCIPLE

Consider the trace of a plane wave front AA' (Fig. 39-5) which is just making contact with the surface MM' along a line through A perpendicular to the plane of the diagram. MM' represents a boundary surface between two transparent media a and b, of indices of refraction n_a and n_b. (The reflected waves are not shown in the figure; they proceed exactly as in Fig. 39-3.) Let us apply Huygens' principle to find the position of the refracted wave front after a time t.

With points on AA' as centers, draw a number of secondary wavelets. Those originating near the upper end of AA' travel with speed v_a and, after a time interval t, are spherical surfaces of radius $v_a t$. The wavelet originating at point A, however, is traveling in the second medium b with speed v_b and at time t is a spherical surface of radius $v_b t$. The envelope of the wavelets from the original wave front is the plane whose trace is the broken line BOB'. A similar construction leads to the trace CPC' after a second interval t.

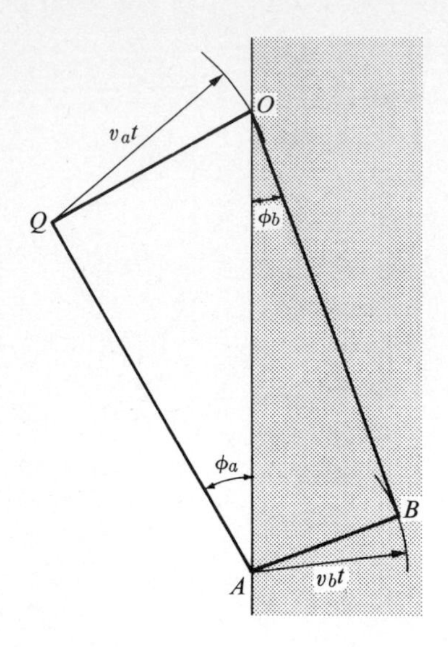

Fig. 39–6 *A portion of Fig. 39–5.*

The angles ϕ_a and ϕ_b between the surface and the incident and refracted wave fronts are, respectively, the angle of incidence and the angle of refraction. To find the relation between these angles, refer to Fig. 39–6 which is the same as a portion of Fig. 39–5. Draw $OQ = v_a t$, perpendicular to AQ, and draw $AB = v_b t$, perpendicular to BO. From the right triangle AOQ,

$$\sin \phi_a = \frac{v_a t}{AO},$$

and from the right triangle AOB,

$$\sin \phi_b = \frac{v_b t}{AO}.$$

Hence,

$$\frac{\sin \phi_a}{\sin \phi_b} = \frac{v_a}{v_b}. \tag{39–1}$$

Since v_a/v_b is a constant, Eq. (39–1) expresses Snell's law, and we have derived Snell's law from a wave theory.

The most general form of Snell's law is given by Eq. (38–4), namely

$$\frac{\sin \phi_a}{\sin \phi_b} = \frac{n_b}{n_a},$$

whence

$$\frac{v_a}{v_b} = \frac{n_b}{n_a},$$

and

$$n_a v_a = n_b v_b .$$

When either medium is a vacuum, the index is 1 and the speed is c. Hence

$$\boxed{n_a = \frac{c}{v_a}, \qquad n_b = \frac{c}{v_b},} \tag{39–2}$$

showing that the *index of refraction of any medium is the ratio of the speed of light in a vacuum to the speed in the medium.*

It was shown in Chapter 37 that the speed of an electromagnetic wave in vacuum is given by

$$c = \sqrt{1/\epsilon_0 \mu_0}\,.$$

For a light wave in a *material* medium of permittivity ϵ and permeability μ, the speed v is

$$v = \sqrt{1/\epsilon\mu} = \sqrt{1/K\epsilon_0 K_m \mu_0} = c\sqrt{1/KK_m}\,.$$

Most transparent substances show a magnetic behavior that differs only slightly from that of a vacuum. The quantity K_m is therefore very close to 1, and $v = c/\sqrt{K}$, or

$$\boxed{n = \sqrt{K}.} \tag{39–3}$$

The values of K listed in Table 27–1 are *static* values, obtained by measurements with nonvarying electric fields. When the electric field varies rapidly, as it does in a light wave, K is found to depend on the frequency. The index of refraction, therefore, cannot be calculated from the values in Table 27–1.

In Fig. 39–5, the spacing of the incident wave fronts is arbitrary. If t is chosen to be the period T of the wave, the spacing is vT, which is the wavelength λ. Figure 39–5 shows that the wavelength in the second medium is smaller, since the wave speed is smaller. When a light wave proceeds from one medium to another, where the speed is different, the wavelength changes *but not the frequency*. Since

$$v_a = f\lambda_a \quad \text{and} \quad v_b = f\lambda_b,$$

$$\frac{\lambda_a}{v_a} = \frac{\lambda_b}{v_b} \quad \text{and} \quad \lambda_a\frac{c}{v_a} = \lambda_b\frac{c}{v_b}.$$

Therefore,

$$\lambda_a n_a = \lambda_b n_b .$$

If either medium is vacuum, the index is 1 and the wavelength in a vacuum is represented by λ_v. Hence

$$\boxed{\lambda_a = \frac{\lambda_v}{n_a}, \qquad \lambda_b = \frac{\lambda_v}{n_b},} \tag{39–4}$$

showing that *the wavelength in any medium is the wavelength in a vacuum divided by the index of refraction of the medium.*

39–4 TOTAL INTERNAL REFLECTION

Figure 39–7 shows a number of rays diverging from a point source P in medium a of index n_a and striking the surface of a second medium b of index n_b, where $n_a > n_b$. From Snell's law,

$$\sin\phi_b = \frac{n_a}{n_b}\sin\phi_a .$$

Since n_a/n_b is greater than unity, $\sin\phi_b$ is larger than $\sin\phi_a$ and evidently equals unity (i.e., $\phi_b = 90°$) for some angle ϕ_a less than 90°. This is illustrated by ray 3 in the diagram, which emerges just grazing the surface at an angle of refraction of 90°. The angle of incidence for which the refracted ray emerges tangent to the surface is called the *critical angle* and is designated by ϕ_{crit} in the diagram. If the angle of incidence is greater than the critical angle, the sine of the angle of refraction, as computed by Snell's law, would have to be greater than unity. This may be interpreted to mean that beyond the critical angle the ray *does not* pass into the upper medium but is *totally internally reflected* at the boundary surface. Total internal reflection can occur only when a ray is incident on the surface of a medium whose index is *smaller* than that of the medium in which the ray is traveling.

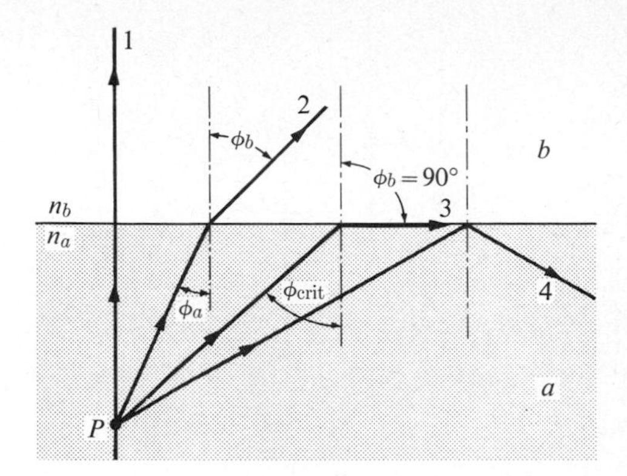

Fig. 39–7 *Total internal reflection. The angle of incidence ϕ_a for which the angle of refraction is 90°, is called the critical angle.*

The critical angle for two given substances may be found by setting $\phi_b = 90°$ or $\sin\phi_b = 1$ in Snell's law. We then have

$$\boxed{\sin\phi_{\text{crit}} = \frac{n_b}{n_a}.} \tag{39–5}$$

The critical angle of a glass–air surface, taking 1.50 as a typical index of refraction of glass, is

$$\sin\phi_{\text{crit}} = \frac{1}{1.50} = 0.67, \qquad \phi_{\text{crit}} = 42°.$$

This angle, very conveniently, is slightly less than 45°, which makes possible the use in many optical instruments of prisms of angles 45°-45°-90° as totally reflecting surfaces. The advantages of totally reflecting prisms over metallic surfaces as reflectors are, first, that the light is *totally* reflected, while no metallic surface reflects 100% of the light incident on it, and second, the reflecting properties are permanent and not affected by tarnishing. Offsetting these is the fact that there is some loss of light by reflection at the surfaces where light enters and leaves the prism, although coating the surfaces with so-called "nonreflecting" films can reduce this loss considerably.

The simplest type of reflecting prism is shown in Fig. 39–8. Its angles are 45°-45°-90°. Light incident normally on one of the shorter faces strikes the inclined face at an angle of incidence of 45°. This is

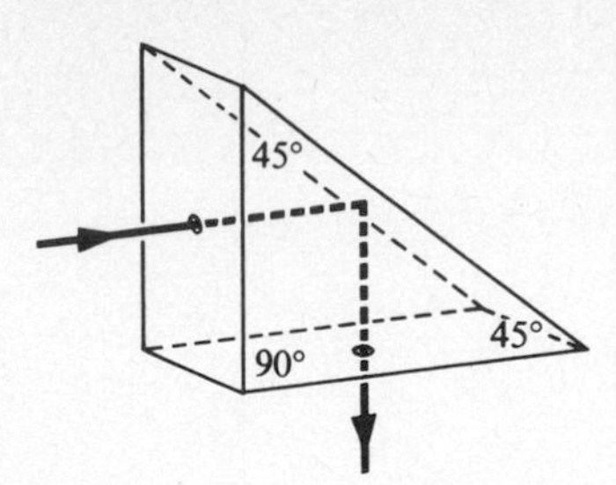

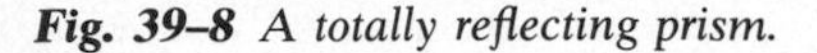

Fig. 39–8 A totally reflecting prism.

greater than the critical angle, so the light is totally internally reflected and emerges from the second of the shorter faces after undergoing a deviation of 90°.

A 45°-45°-90° prism, used as in Fig. 39–9(a), is called a *Porro* prism. Light enters and leaves at right angles to the hypotenuse and is reflected at each of the shorter faces.The deviation is 180°. Two Porro prisms are often combined, as in Fig. 39–9(b).

If a beam of light enters one end of a transparent rod, as in Fig. 39–10, the light is totally reflected internally and is "trapped" within the rod even if it is curved, provided the curvature is not too great. The rod is sometimes referred to as a *light pipe.* A bundle of fine fibers will behave in the same way and has the advantage of being flexible. The study of the properties of such a bundle is an active field of research known as *fiber optics.*

A bundle may consist of thousands of individual fibers, each the order of 0.002 to 0.01 mm in diameter. If the fibers can be assembled in the bundle so that the relative positions of the ends are the same at both ends, the bundle can transmit an image, as shown in Fig. 39–11. Bundles several feet in length have been made. Devices using fiber optics are finding a wide range of applications in medical science. The interior of lungs and other passages in the human body can be viewed by insertion of a fiber bundle. A bundle can be enclosed in a hypodermic needle for the study of tissues and blood vessels far beneath the skin. Despite the enormous technical difficulties of manufacturing fiber-optic components, such systems promise to become an extremely important class of optical systems.

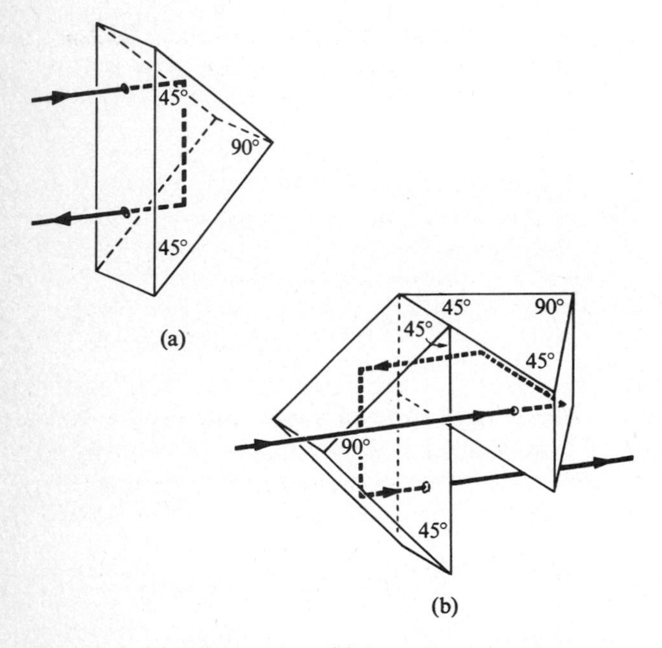

Fig. 39–9 (a) A Porro prism. (b) A combination of two Porro prisms.

39–5 REFRACTION BY A PRISM

The prism, in one or another of its many forms, is second only to the lens in its usefulness as a piece of optical apparatus. Totally reflecting prisms have been mentioned briefly. We consider now the *deviation* and the *dispersion* produced by a prism.

Consider a light ray incident at an angle ϕ on one face of a prism, as in Fig. 39–12(a). Let the index of the prism be n, the included angle at the apex be A, and let the medium on either side of the prism be air. It is desired to find the *angle of deviation,* δ. In

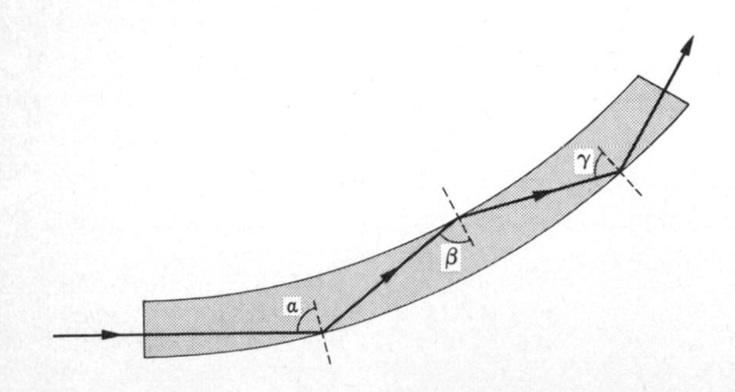

Fig. 39–10 A light ray "trapped" by internal reflections.

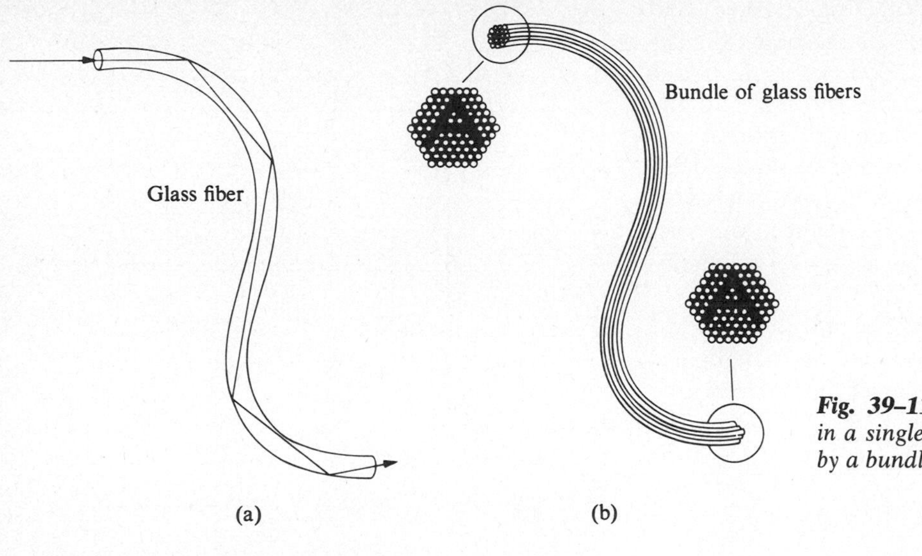

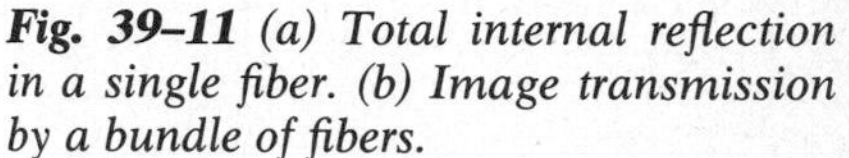

Fig. 39–11 (a) Total internal reflection in a single fiber. (b) Image transmission by a bundle of fibers.

principle, this is a straightforward problem in surveying. One has only to apply Snell's law at the first surface, compute the angle of refraction, then by geometry find the angle of incidence at the second surface, and from a second application of Snell's law find the angle of refraction at the second surface. The direction of the emergent ray is then known, and the angle of deviation may be found.

While the method is simple enough, the expression for the angle δ turns out in the general case to be rather complicated. However, as the angle of incidence is, say, decreased from a large value, the angle of deviation is found to decrease at first and then increase, and to be a *minimum* when the ray passes through the prism symmetrically, as in Fig. 39–12(b). The angle δ_m is then called the angle of *minimum deviation*; it is related to the angle of the prism and its index by the equation

$$n = \frac{\sin[(A + \delta_m)/2]}{\sin A/2}. \tag{39–6}$$

To derive Eq. 39–6, we have, from Fig. 39–12(b),

$$\phi_1' = \frac{A}{2} \quad \text{(sides mutually perpendicular)},$$

$$\delta_1 = \frac{\delta_m}{2} \quad \text{(half the deviation takes place at each surface)},$$

$$\phi_1 = \phi_1' + \delta_1 = \frac{A}{2} + \frac{\delta_m}{2} = \frac{A + \delta_m}{2},$$

$$\sin \phi_1 = n \sin \phi_1', \qquad \text{and} \quad \sin \frac{A + \delta_m}{2} = n \sin \frac{A}{2},$$

which is Eq. (39–6).

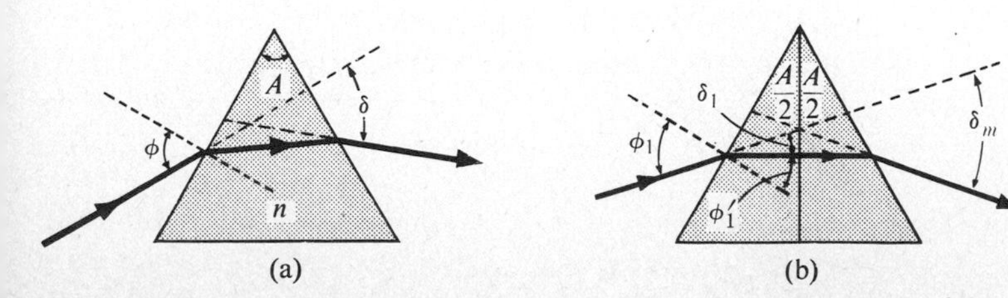

Fig. 39–12 (a) Deviation by a prism. (b) The deviation is a minimum when the ray passes through the prism symmetrically.

The index of refraction of a transparent solid may be measured, making use of the equation derived above. The specimen whose index is desired is ground into the form of a prism. The angle of the prism A and the angle of minimum deviation δ_m are measured with the aid of a spectrometer. Since these angles may be determined with a high degree of precision, this method is an extremely accurate one; indices of refraction may be measured to six significant figures.

If the angle of the prism is *small*, the angle of minimum deviation is small also, and we may replace the sines of the angles by the angles themselves. One then obtains

$$n = \frac{A + \delta_m}{A},$$

or

$$\delta_m = (n - 1)A, \qquad (39\text{–}7)$$

a useful approximate relation.

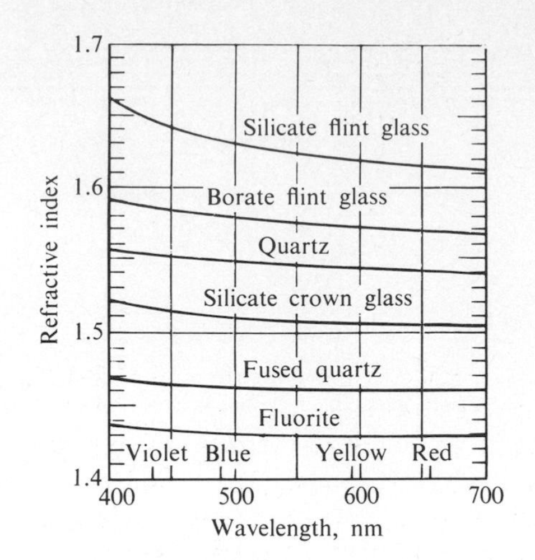

Fig. 39–13 *Variation of index with wavelength.*

39–6 DISPERSION

Most light beams are a mixture of waves whose wavelengths extend throughout the visible spectrum. While the speed of light waves in a vacuum is the same for all wavelengths, the speed in a material substance is different for different wavelengths. Hence the index of refraction of a substance is a function of wavelength. The dependence of wave speed on wavelength was first mentioned in connection with water waves, in Section 21–7. A substance in which the speed of a wave varies with wavelength is said to exhibit *dispersion.* Figure 39–13 is a diagram showing the variation of index of refraction with wavelength for a number of the more common optical materials.

Consider a ray of white light, a mixture of all visible wavelengths, incident on a prism, as in Fig. 39–14. Since the deviation produced by the prism increases with increasing index of refraction, violet light is deviated most and red least, with other colors occupying intermediate positions. On emerging from the prism, the light is spread out into a fan-shaped beam, as shown. The light is said to be *dispersed* into a spectrum.

When white light is dispersed by a prism, it can be seen from Fig. 39–14 that the whole fan-shaped beam is deviated from the incident direction. A convenient measure of this *deviation* is provided by the angle of deviation of yellow light, since yellow is roughly midway between red and violet. A simple measure of the *dispersion* is provided by the angular separation of the red and violet rays. Since deviation and index of refraction are related, the deviation of the entire spectrum is controlled by the index of refraction for yellow light, whereas the dispersion

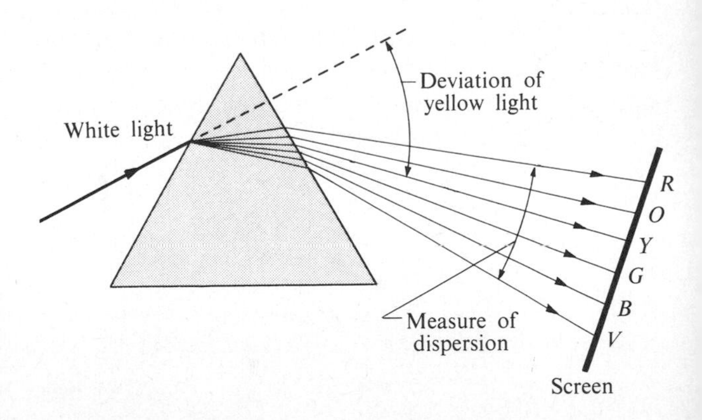

Fig. 39–14 *Dispersion by a prism. The bank of colors on the screen is called a spectrum.*

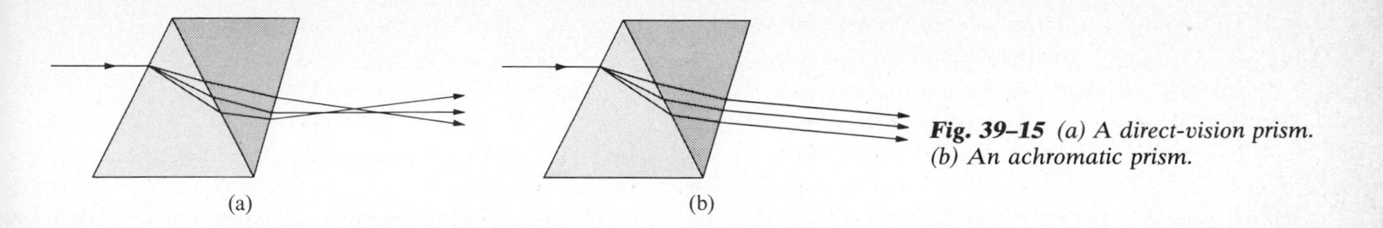

Fig. 39–15 (a) A direct-vision prism. (b) An achromatic prism.

depends on the *difference* between the index for violet light and that for red light. From Fig. 38–13 it can be seen that for a substance such as fluorite, whose index for yellow light is small, the difference between the indices for red and violet is also small. On the other hand, in the case of silicate flint glass, both the index for yellow light and the difference between extreme indices are large. In other words, for most transparent materials the greater the deviation, the greater the dispersion.

The brilliance of diamond is due in part to its large dispersion. In recent years synthetic crystals of titanium dioxide and of strontium titanate, with about eight times the dispersion of diamond, have been produced.

Because of the shape of the curves in Fig. 39–13, the dispersions produced by two prisms of equal angles are not exactly proportional to the mean deviations. It is therefore possible to combine two (or more) prisms of different materials in such a way that there is *no net deviation* of a ray of some chosen wavelength, while there remains an outstanding dispersion of the spectrum as a whole. Such a device, illustrated in Fig. 39–15(a), is known as a *direct-vision prism.* Two prisms may also be designed so that the dispersion of one is offset by the dispersion of the other, although the deviation is not. A compound prism of this sort, shown in Fig. 39–15(b), is called *achromatic* (without color).

39–7 THE RAINBOW

The rainbow is produced by the combined effects of refraction, dispersion, and internal reflection of sunlight by drops of rain. When conditions for its observation are favorable, two bows may be seen, the inner being called the primary bow and the outer the secondary bow. The inner bow, which is the brighter, is red on the outside and violet on the inside, while in the fainter outer bow the colors are reversed.

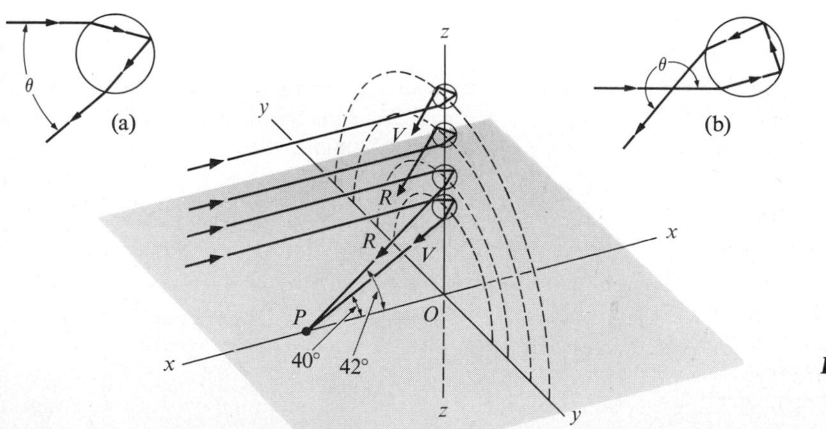

Fig. 39–16 The rainbow.

The primary bow is produced in the following manner. Assume that the sun's rays are horizontal, and consider a ray striking a raindrop, as in Fig. 39–16(a). This ray is refracted at the first surface and is in part reflected at the second surface, passing out again at the front surface as shown. The French scientist Descartes computed the paths of some thousands of rays incident at different points on the surface of a raindrop and showed that, if a ray of any given color were incident at such a point that its deviation was a *maximum*, all other rays of the same color which struck the surface of the drop in the immediate neighborhood of this point would be reflected in a direction very close to that of the first. Hence each color is strongly reflected in the direction of maximum deviation of that particular color. The angle of maximum deviation of red light is 138°, so the angle θ in Fig. 39–16(a) is $180° - 138° = 42°$. The corresponding angle for violet light is 40°, while that for other colors lies between these.

Consider now an observer at P in Fig. 39–16. The xy-plane is horizontal and sunlight is coming from the left parallel to the x-axis. All drops which lie on a circle subtending an angle of 42° at P, and with the center at O, will reflect red light strongly to P. All those on a circle subtending 40° at P will reflect violet light strongly, while those occupying intermediate positions will reflect the intermediate colors of the spectrum.

The point O, the center of the circular arc of the bow, may be considered the shadow of P on the yz-plane. As the sun rises above the horizon, the point O moves down, and hence with increasing elevation of the sun a smaller and smaller part of the bow is visible. Evidently an observer at ground level cannot see the primary bow when the sun is more than 42° above the horizon. If the observer is in an elevated position, however, the point O moves up and more and more of the bow may be seen. In fact, it is not uncommon for a complete *circular* rainbow to be seen from an airplane.

The secondary bow is produced by *two* internal reflections, as shown in Fig. 39–16(b). As before, the light which is reflected in any particular direction depends upon the angle of maximum deviation. Since the angle of deviation is here the angle θ, and since the violet is deviated more than the red, the violet rays in the secondary bow are deflected down at a steeper angle than the red and the secondary bow is red on the *inside* and violet on the outside edge. The corresponding angles are 50.5° for red and 54° for violet.

The preceding discussion applies when the drops producing a rainbow are relatively large. When the drops are small, diffraction plays just as important a role as dispersion and reflection, and red light, for example, is received in appreciable amounts from drops lying on circles other than those subtending an angle of 42°. The bow is then a complicated mixture of colors and its appearance depends on the size of the drops.

PROBLEMS

39–1

a) What is the speed of light of wavelength 500 nm (in vacuum), in glass whose index at this wavelength is 1.50?

b) What is the wavelength of these waves in the glass?

39–2 A glass plate 3 mm thick, of index 1.50, is placed between a point source of light of wavelength 600 nm (in vacuum) and a screen. The distance from source to screen is 3 cm. How many waves are there between source and screen?

39–3 The speed of light of wavelength 656 nm in heavy flint glass is 1.60×10^8 m s^{-1}. What is the index of refraction of this glass?

39–4 Light of a certain frequency has a wavelength in water of 442 nm. What is the wavelength of this light when it passes into carbon disulfide? (See Table 38–1.)

39–5 A ray of light goes from point A in a medium where the velocity of light is v_1 to point B in a medium where the velocity is v_2, as in Fig. 39–17. Show (a) that the time required for the light to go from A to B is

$$t = \frac{h_1 \sec \theta_1}{v_1} + \frac{h_2 \sec \theta_2}{v_2},$$

and (b) that this time is minimum when the relation between the angles is $n_1 \sin \theta_1 = n_2 \sin \theta_2$. (Note that $\ell = h_1 \tan \theta_1 + h_2 \tan \theta_2$.)

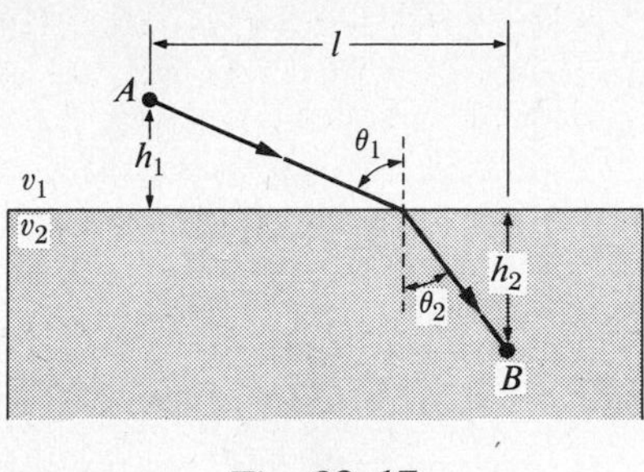

Fig. 39–17

39–6 A point light source is 5 cm below a water-air surface. Compute the angles of refraction of rays from the source making angles with the normal of 10°, 20°, 30°, and 40°, and show these rays in a carefully drawn full-size diagram.

39–7 A glass cube in air has a refractive index of 1.50. Parallel rays of light enter the top obliquely and then strike a side of the cube. Is it possible for the rays to emerge from this side?

39–8 A point source of light is 20 cm below the surface of a body of water. Find the diameter of the largest circle at the surface through which light can emerge from the water.

39–9 The index of refraction of the prism shown in Fig. 39–18 is 1.56. A ray of light enters the prism at point a and follows in the prism the path ab which is parallel to the line cd. (a) Sketch carefully the path of the ray from a point outside the prism at the left, through the glass, and out some distance into the air again. (b) Compute the angle between the original and final directions in air. (Dotted lines are construction lines only.)

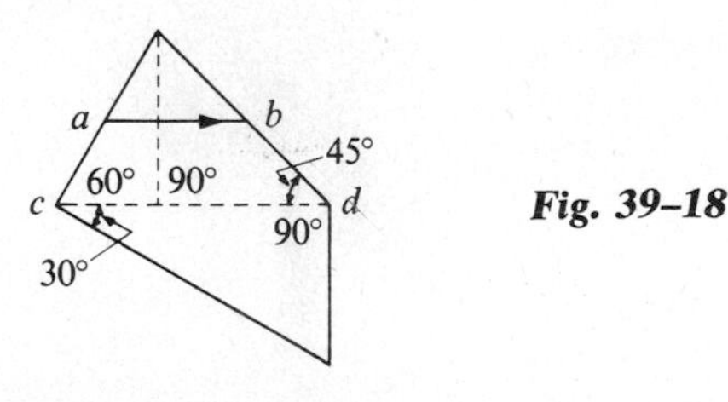

Fig. 39–18

39–10 Light is incident normally on the short face of a 30°-60°-90° prism, as in Fig. 39–19. A drop of liquid is placed on the hypotenuse of the prism. If the index of the prism is 1.50, find the maximum index the liquid may have if the light is to be totally reflected.

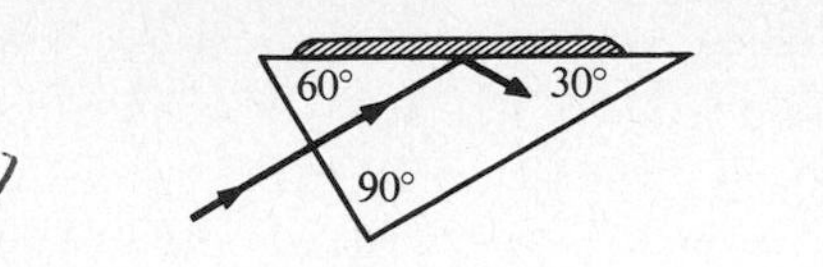

Fig. 39–19

39–11 A 45°-45°-90° prism is immersed in water. What is the minimum index of refraction the prism may have if it is to reflect totally a ray incident normally on one of its shorter faces?

39–12 The velocity of a sound wave is 330 m s^{-1} in air and 1320 m s^{-1} in water.

a) What is the critical angle for a sound wave incident on the surface between air and water?

b) Which medium has the higher "index of refraction" for sound?

39–13 Light is incident at an angle ϕ_1 (as in Fig. 39–20) on the upper surface of a transparent plate, the surfaces of the plate being plane and parallel to each other.

a) Prove that $\phi_1 = \phi_2'$.

b) Show that this is true for any number of different parallel plates.

c) Prove that the lateral displacement of the emergent beam is given by the relation

$$PQ = t\frac{\sin(\phi_1 - \phi_1')}{\cos\phi_1'},$$

where t is the thickness of the plate.

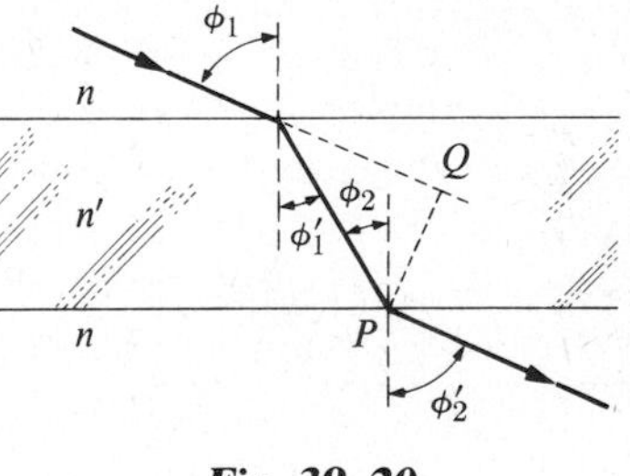

Fig. 39–20

39–14 A parallel beam of light containing wavelengths A and B is incident on the face of a triangular glass prism having a refracting angle of 60°. The indices of refraction are $n_A = 1.40$ and $n_B = 1.60$. If beam A goes through the prism at minimum deviation, find (a) the angle of emergence of each beam and (b) the angle of deviation of each.

39–15 A ray of light is incident at an angle of 60° on one surface of a glass plate 2 cm thick, of index 1.50. The medium on either side of the plate is air. Find the transverse displacement between the incident and emergent rays.

39–16 The prism of Fig. 39–21 has a refractive index of 1.414, and the angles A are 30°. Two light rays m and n are parallel as they enter the prism. What is the angle between them after they emerge?

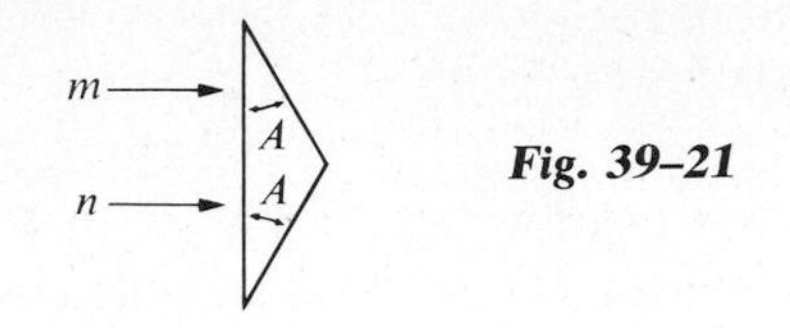

Fig. 39–21

39–17 What is the angle of minimum deviation of an equiangular prism having a refractive index of 1.414?

39–18 An equiangular prism is constructed of the silicate flint glass whose index of refraction is given in Fig. 39–13. Find the angles of minimum deviation for light of wavelength 400 nm and of 700 nm.

39–19 A silicate crown prism of apex angle 15° is to be combined with a prism of silicate flint so as to result in no net deviation of light of wavelength 550 nm. (See Fig. 39–13 for indices of refraction.) Find the angle of the flint prism. Assume that light passes through both prisms at the angle of minimum deviation.

39–20 The glass vessel shown in Fig. 39–22(a) contains a large number of small, irregular pieces of glass and a liquid. The dispersion curves of the glass and of the liquid are shown in Fig. 39–22(b). Explain the behavior of a parallel beam of white light as it traverses the vessel. (This is known as a *Christiansen filter.*)

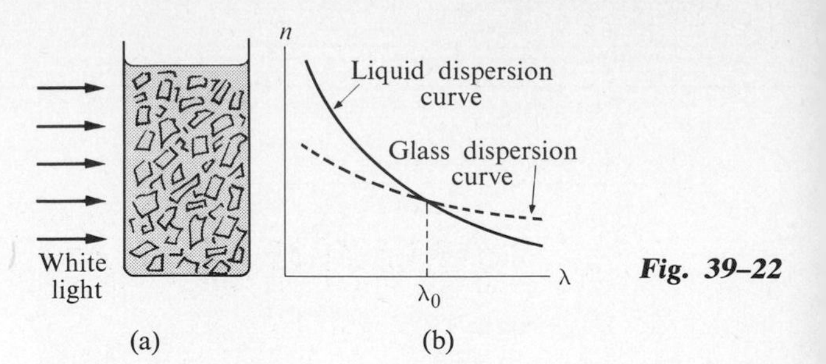

Fig. 39–22

39–21 A slab of glass 5 cm thick, having $n = 1.5$, is immersed in water, for which $n = 1.33$. A ray of light from a source in the water is incident on the surface of the glass slab at an angle of 30° to the normal.

a) What is the angle to the normal of the refracted ray?
b) What is the lateral displacement of the ray after passing through the slab?

39–22 In Problem 39–21, if a ray of light enters the slab through an edge, what is the critical angle (with the normal) for total internal reflection in the glass?

39–23 A prism has an apex angle (angle A in Fig. 39–12) of 60°; its refractive index is 1.54 for red light of wavelength 650 nm and 1.58 for blue light of wavelength 450 nm. If rays of these two colors are directed so that each passes through the prism at minimum deviation, find the angle between the two emerging rays.

39–24 A prism having an apex angle of 5° was found to deviate a ray of light 3° at minimum deviation.

a) What is the refractive index of the prism?
b) What percent error is introduced by using the approximate relation of Eq. (39–7) instead of the exact expression, Eq. (39–6)?

Chapter 40

Images Formed by a Single Reflection or Refraction

40–1 INTRODUCTION

In the preceding chapter we discussed the reflection and refraction rays of light at a surface separating two susbtances of different index, but considered only a single ray or bundles of parallel rays. We now take up the problem of tracing the paths of a large number of rays, all of which diverge from some one point of an object which either is self-luminous or has a rough surface capable of reflecting light in all directions. Such a point is represented by P in Fig. 40–1. At the right of P is a plane surface such as that of a block of glass. A ray incident on any point of the surface is in part transmitted and in part reflected. The direction of the *reflected* ray is given by the law of reflection and that of the transmitted or *refracted* ray by Snell's law. These two principles suffice to determine the directions of all rays after they encounter a surface of any shape. We shall first consider the reflected rays only, from both plane and spherical surfaces, and then discuss refraction at plane and spherical surfaces. A discussion of surfaces other than plane and spherical is beyond the scope of this book.

In Chaper 39 we used the subscripts a and b to refer to two different materials on opposite sides of a reflecting or refracting surface. The notation in the chapters to follow is simplified by the use of unprimed symbols for the material on one side of a surface, and primed symbols for that on the other side. Thus, in Fig. 40–1 the indices of refraction are written n and n'.

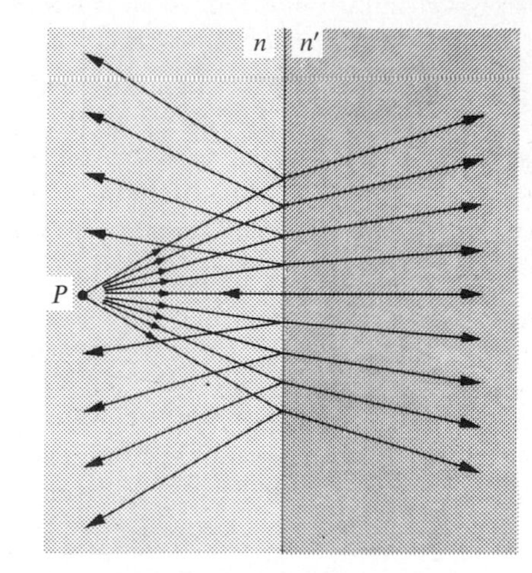

Fig. 40–1 *Reflection and refraction of rays at a plane surface.*

40–2 REFLECTION AT A PLANE MIRROR

Although some light is reflected at any surface separating two substances of different index, it is often

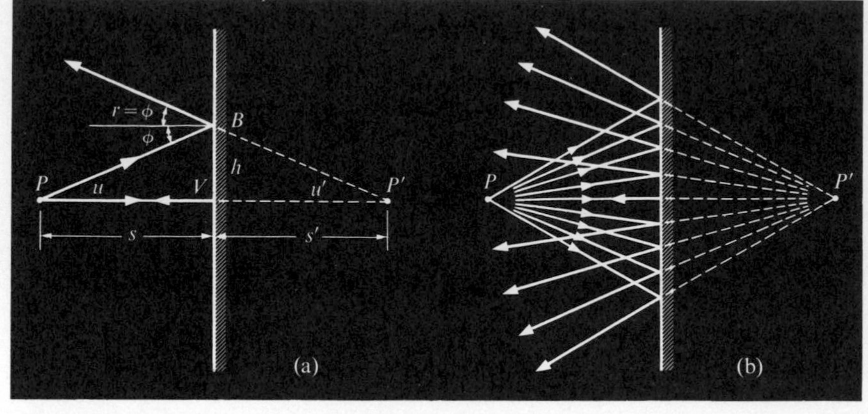

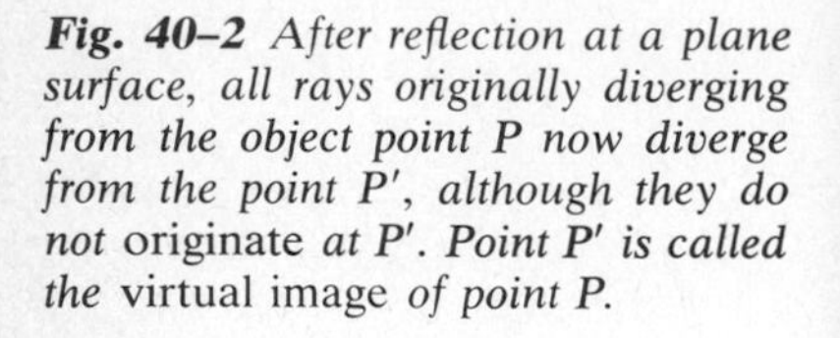
Fig. 40–2 *After reflection at a plane surface, all rays originally diverging from the object point P now diverge from the point P′, although they do not* originate *at P′. Point P′ is called the* virtual image *of point P.*

desirable that the fraction shall be as large as possible. By making the surface of highly polished metal, or by applying a thin metallic coat to a polished surface, we may increase the fraction of the light reflected to nearly 100%. A highly reflecting smooth surface is called a *mirror*, and we shall speak of the reflecting surfaces in the sections to follow as mirrors, although the equations to be derived will apply to smooth surfaces from which the light is only partially reflected, such as the surface of a pane of window glass, or that of a lens.

Figure 40–2(a) shows two rays diverging from a point P at a distance s at the left of a plane mirror. The ray PV, incident normally on the mirror, returns along its original path. The ray PB, making an arbitrary angle u with PV, strikes the mirror at an angle of incidence $\phi = u$, and is reflected at an angle $r = \phi = u$. When extended backward, the reflected ray intersects the axis at point P'. The angle u' is equal to r and hence is equal to u.

In Fig. 40–2(b), a number of rays diverging from P are shown. We shall prove in a moment that *all* rays diverging from P give rise to reflected rays which, when projected backward, intersect at the *same* point P'. We call P' the *image* of the *object* point P. The distance s is the *object distance*, and the distance s' is the *image distance.*

In other problems to be discussed later, we shall see that an object point and its corresponding image point may lie on either side of a reflecting or refracting surface. Hence, as in so many other problems, it is necessary to adopt a convention of sign. Many different conventions are in use; we shall use the following:

1. *If the direction from the object to the reflecting or refracting surface is the same as that of the light* ***incident*** *on the surface, the object distance s is* ***positive.***
2. *If the direction from the reflecting or refracting surface to the image point is the same as that of the light* ***reflected*** *or* ***refracted*** *from the surface, the image distance s′ is* ***positive****.*

It follows that in Fig. 40–2 the object distance s is *positive* (the direction from the object point P to the surface is the *same* as that of the incident light) and the image distance s' is *negative* (the direction from the surface to the image point P' is *opposite* to that of the reflected light).

The image P' is called *virtual*, meaning that the reflected rays, although they diverge from P', do not *originate* at P'. A virtual image corresponds to a negative image distance.

To find the position of P', let h represent the distance VB in Fig. 40–2(a). Then from the triangles PVB and $P'BV$,

$$\tan u = \frac{h}{s}, \qquad \tan u' = \frac{h}{-s'},$$

and since $u = u'$, it follows that

$$\boxed{s = -s'.} \tag{40–1}$$

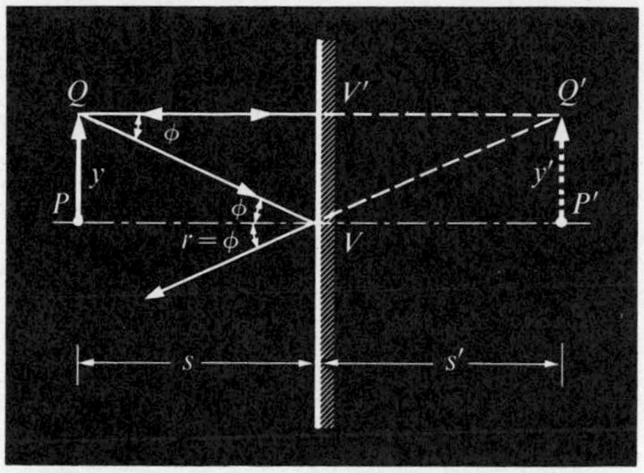

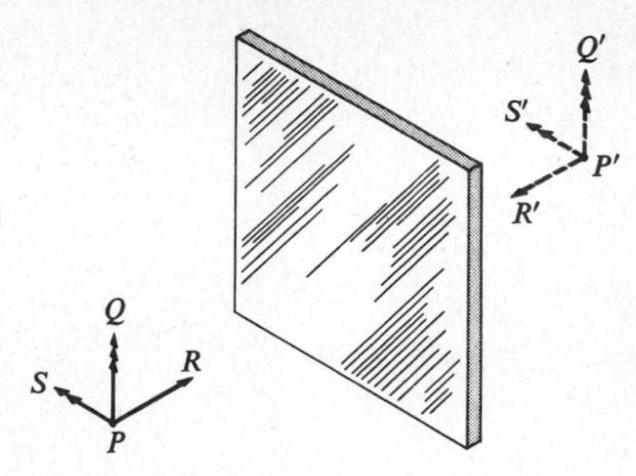

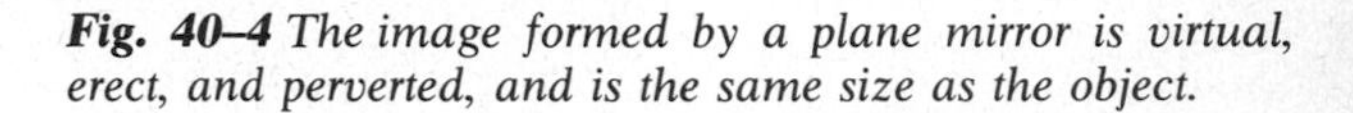

Fig. 40–3 *Construction for determining the height of an image formed by reflection at a plane surface.*

Fig. 40–4 *The image formed by a plane mirror is virtual, erect, and perverted, and is the same size as the object.*

This result is true whatever the value of the angle u, so that *all* rays originally diverging from P diverge from P' after reflection. For the special case of a plane mirror, then, the image of an object point lies on the extension of the normal from the object point to the mirror, and the object and image distances are equal in absolute value.

Now consider an object of finite size and parallel to the mirror, represented by the arrow PQ in Fig. 40–3. Point P', the image of P, is found as in Fig. 40–2. Two of the rays from Q are shown, and *all* rays from Q diverge from its image Q' after reflection. Other points of the object PQ are imaged between P' and Q'.

Let y and y' represent the lengths of object and image, respectively. The ratio y'/y is called the *lateral magnification* m:

$$m = \frac{y'}{y}.$$

From the triangles PQV and $P'Q'V$,

$$\tan\phi = \frac{y}{s} = \frac{y'}{-s'}$$

and since $s' = -s$,

$$m = \frac{y'}{y} = -\frac{s'}{s} = +1. \qquad (40\text{–}2)$$

The lateral magnification for a plane mirror is unity.

In general, if a transverse object is represented by an arrow, its image may point in the same direction as the object, or in the opposite direction. If the directions are the same, the image is called *erect*; if they are opposite, the image is *inverted.* When the lateral magnification m is positive, the object and image arrows point in the same direction and the image is erect.

The three-dimensional virtual image of a three-dimensional object, formed by a plane mirror, is shown in Fig. 40–4. The image of every object point lies on the normal from that point to the mirror, and the distances from object and image to the mirror are equal. The images $P'Q'$ and $P'S'$ are parallel to their objects, but $P'R'$ is reversed relative to PR. We can think of the arrow PR as generated by the displacement of an object point P from P to R, through a distance Δs. Since

$$s' = -s,$$

it follows that the corresponding displacement $\Delta s'$ of the image, from P' to R', is equal to $-\Delta s$. The ratio $\Delta s'/\Delta s$ is called the *longitudinal magnification* m'. In this case,

$$m' = \frac{\Delta s'}{\Delta s} = -1. \qquad (40\text{–}3)$$

The negative sign means that the arrows PR and $P'R'$ point in opposite directions. Hence the image of a three-dimensional object formed by a plane mirror

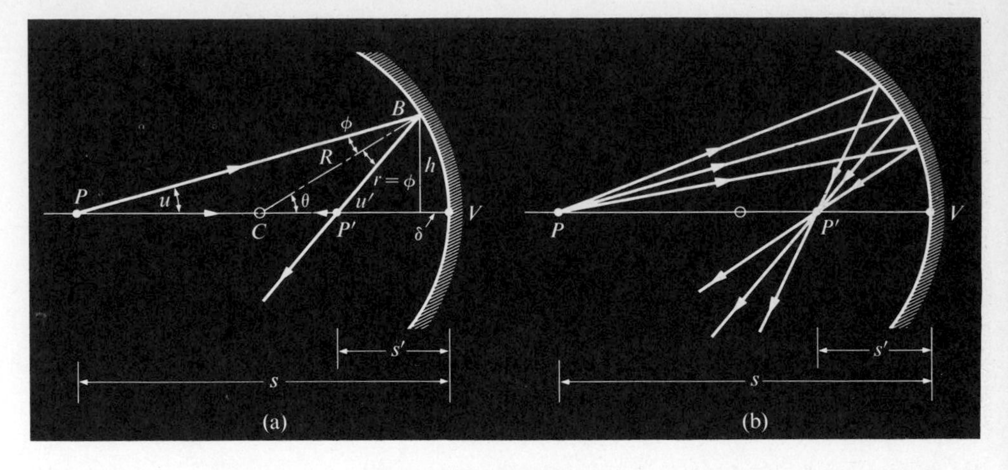

Fig. 40–5 (a) Construction for finding the position of the image P′ of a point object P, formed by a concave spherical mirror. (b) If the angle u is small, all rays from P intersect at P′.

is the same size as the object in both its lateral and transverse dimensions. However, the image and object are not identical in all respects but are related in the same way as are a right hand and a left hand. To verify this, point your thumbs along PR and $P'R'$, your forefingers along PQ and $P'Q'$, and your middle fingers along PS and $P'S'$. When an object and its image are related in this way the image is called *perverted*. When the transverse dimensions of object and image are in the same direction, the image is erect. Thus a plane mirror forms an erect but perverted image.

40–3 REFLECTION AT A SPHERICAL MIRROR

Figure 40–5(a) shows a spherical mirror of radius of curvature R, with its concave side facing the incident light. The center of curvature of the surface is at C. Point P is an object point. The ray PV, passing through C, strikes the mirror normally and is reflected back on itself. Point V is called the *vertex* of the mirror, and the line PCV the *axis*.

Ray PB, at an arbitrary angle u with the axis, strikes the mirror at B, where the angle of incidence is ϕ and the angle of reflection is $r = \phi$. The reflected ray intersects the axis at point P'. We shall show shortly that if the angle u is small, *all* rays from P intersect the axis at the *same* point P', as in Fig. 40–5(b). Point P' is therefore the *image* of the object point P. The object distance, measured from the vertex V, is s, and the image distance is s'.

The direction from the object point P to the surface is the same as that of the incident light, so the object distance s is positive. The direction from the surface to the image point P' is the same as that of the reflected light, so the image distance s' is positive also.

Unlike the reflected rays in Fig. 40–2, the reflected rays in Fig. 40–5(b) actually intersect at point P', and then diverge from P' *as if* they had originated at this point. The image P' is called *real*, and a real image corresponds to a *positive* image distance.

Making use of the fact that an exterior angle of a triangle equals the sum of the two opposite interior angles, and considering the triangles PBC and $P'BC$ in Fig. 40–5(a), we have

$$\theta = u + \phi, \qquad u' = \theta + \phi.$$

Eliminating ϕ between these equations gives

$$u + u' = 2\theta. \tag{40–4}$$

It is now necessary to introduce a sign convention for radii of curvature. We shall use the following:

1. If the direction from a reflecting or refracting surface to the center of curvature is the same as that of the

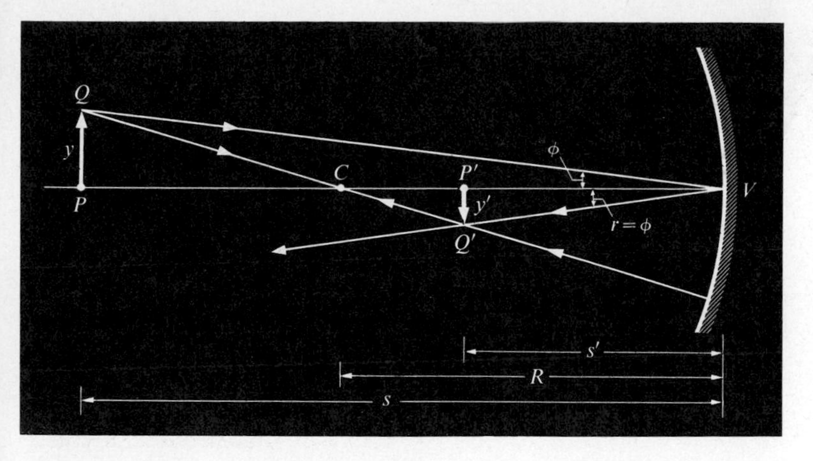

Fig. 40–6 *Construction for determining the height of an image formed by a concave spherical mirror.*

reflected *or* ***refracted*** *light, the radius of curvature is positive.*

Hence in Fig. 40–5, R is positive, since the direction from the mirror to the center of curvature C is the same as that of the reflected light.

We may now compute the image position s'. Let h represent the height of point B above the axis, and δ the short distance from V to the foot of this vertical line. Now write the expressions for the tangents of u, u', and θ, remembering that s, s', and R are all positive quantities:

$$\tan u = \frac{h}{s - \delta}, \qquad \tan u' = \frac{h}{s' - \delta},$$

$$\tan \theta = \frac{h}{R - \delta}.$$

These *trigonometric* equations cannot be solved as simply as the corresponding *algebraic* equations for a plane mirror. However, *if the angle u is small,* the angles u' and θ will be small also. Since the tangent of a small angle is nearly equal to the angle itself (in radians), we can replace $\tan u'$ by u', etc., in the equations above. Also if u is small, the distance δ can be neglected compared with s', s, and R. Hence, *approximately, for small angles,*

$$u = \frac{h}{s}, \qquad u' = \frac{h}{s'}, \qquad \theta = \frac{h}{R}.$$

Substituting in Eq. (40–4) and cancelling h, we get

$$\boxed{\frac{1}{s} + \frac{1}{s'} = \frac{2}{R}} \qquad (40\text{–}5)$$

as a general relation among the three quantities s, s', and R. Since the equation above does not contain the angle u, *all* rays from P making sufficiently small angles with the axis will, after reflection, intersect at P'. Such rays, nearly parallel to the axis, are called *paraxial* rays. As the angle increases, the point P' moves closer to the vertex; a spherical mirror, unlike a plane mirror, does not form precisely a point image of a point object. This property of a spherical mirror is called *spherical aberration.*

If $R = \infty$, the mirror becomes *plane* and Eq. (40–5) reduces to Eq. (40–1), previously derived for this special case.

Now, suppose we have an object of finite size, represented by the arrow PQ in Fig. 40–6, perpendicular to the axis PV. The image of P formed by paraxial rays is at P'. Since the object distance for point Q is a trifle greater than that for the point P, the image $P'Q'$ is not a straight line but is curved, another aberration of spherical surfaces, called *curvature of field.* However, if the height PQ is not too great, the image is nearly straight and is perpendicular to the axis; we shall assume this to be the case.

We now compute the lateral magnification m. The ray QCQ', incident normally, is reflected back on itself. The ray QV makes an angle of incidence ϕ and an angle of reflection $r = \phi$. From the triangles PQV and $P'Q'V$,

$$\tan\phi = \frac{y}{s} = \frac{-y'}{s'}.$$

Hence,

$$m = \frac{y'}{y} = -\frac{s'}{s}. \quad (40\text{–}6)$$

For a plane mirror, $s = -s'$ and hence $y' = y$, as we have already shown.

To find the longitudinal magnification m', imagine point P in Fig. 40–6 to be displaced a short distance Δs toward the mirror. The corresponding displacement of the image is $\Delta s'$, and Eq. (40–5) gives

$$\frac{1}{s + \Delta s} + \frac{1}{s' + \Delta s'} = \frac{2}{R},$$

which may be rewritten,

$$\frac{1}{s}\cdot\frac{1}{1 + \Delta s/s} + \frac{1}{s'}\cdot\frac{1}{1 + \Delta s'/s'} = \frac{2}{R}.$$

We must now make an approximation. When a is a quantity much smaller than unity, $1/(1 + a)$ is very nearly equal to $(1 - a)$. For example, when $a = 0.01$, $1/(1 + 0.01) = 0.99 = 1 - 0.01$. Using this approximation, we rewrite the above equation:

$$\frac{1}{s}\left(1 - \frac{\Delta s}{s}\right) + \frac{1}{s'}\left(1 - \frac{\Delta s'}{s'}\right) = \frac{2}{R}.$$

Now we subtract Eq. (40–5) from this:

$$-\frac{\Delta s}{s^2} - \frac{\Delta s'}{s'^2} = 0,$$

or

$$m' = \frac{\Delta s'}{\Delta s} = -\left(\frac{s'}{s}\right)^2 = -m^2, \quad (40\text{–}7)$$

since the lateral magnification is $m = -s'/s$. Thus the longitudinal magnification equals the *square* of the lateral magnification and is always *negative*, regardless of the sign of m, since m^2 is always positive. The negative sign means that image and object always move in *opposite* directions; if an object point moves *toward* the mirror, its image moves *away from* the mirror.

Although the ratio of image size to object size is referred to as the *magnification*, the image formed by a mirror or lens may be *smaller* than the object. The magnification is then a small fraction and might more appropriately be called the *reduction*. The image formed by an astronomical telescope mirror, or by a camera lens, is much smaller than the object. Since the longitudinal magnification is the square of the lateral magnification, the two are not equal except in the special case of *unit* magnification when $m = 1$. In particular, if m is a small fraction, m^2 is very small and the three-dimensional image of a three-dimensional object is reduced *longitudinally* much more than it is reduced *transversely*. Figure 40–7 illustrates this effect.

Example 1 a) What type of mirror is required to form an image, on a wall 3 m from the mirror, of the filament of a headlight lamp 10 cm in front of the mirror? b) What is the height of the image if the height of the object is 5 mm?

a)

$$s = 10 \text{ cm}, \qquad s' = 300 \text{ cm},$$

$$\frac{1}{10 \text{ cm}} + \frac{1}{300 \text{ cm}} = \frac{2}{R},$$

$$R = 19.4 \text{ cm}.$$

Since the radius is positive, a concave mirror is required.

b)

$$m = \frac{y'}{y} = -\frac{s'}{s} = -\frac{300 \text{ cm}}{10 \text{ cm}} = -30.$$

The image is therefore inverted (m is negative) and is 30 times the height of the object, or $30 \times 5 = 150$ mm.

Example 2 A wire frame in the form of a small cube 3 cm on a side is placed with its center on the axis of a concave mirror of radius of curvature 30 cm. The sides of the cube are parallel or perpendicu-

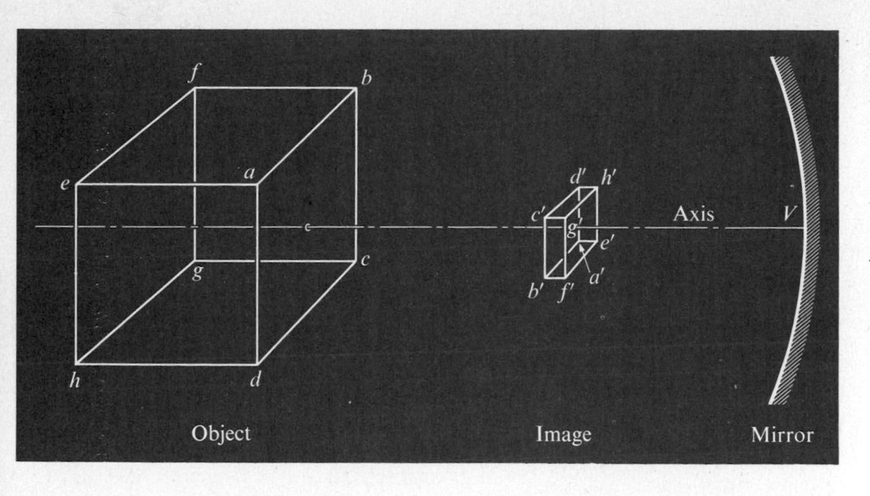

Fig. 40–7 *Schematic diagram of an object and its real, inverted, reduced image formed by a concave mirror.*

lar to the axis. The face toward the mirror is 60 cm to the left of the vertex. Find (a) the position of the image of the cube, (b) the lateral magnification, (c) the longitudinal magnification.

a) Let us calculate the position of the image of a point at the center of the right face of the cube. From the given data, $s = +60$ cm, $R = +30$ cm:

$$\frac{1}{s} + \frac{1}{s'} = \frac{2}{R}, \qquad \frac{1}{60} + \frac{1}{s'} = \frac{2}{30}, \qquad s' = +20 \text{ cm}.$$

The image of this point is therefore 20 cm to the left of the vertex (s' is positive) and is real.

b) $$m = \frac{y'}{y} = -\frac{s'}{s} = -\frac{1}{3}, \qquad y' = -\frac{1}{3}y.$$

The image of the right face is therefore inverted (m is negative) and is a square 1 cm on a side.

c) The object distance for the left face of the cube is 63 cm. To find the position and magnification of the left face of the cube, we could repeat the calculations above with $s = 63$ cm, but since the dimensions of object and image are small compared with the object distance, let us instead find the longitudinal magnification.

$$m' = \frac{\Delta s'}{\Delta s} = -m^2 = -\frac{1}{9}, \qquad \Delta s' = -\frac{1}{9}\Delta s.$$

The longitudinal magnification is negative, so the image of the left face is nearer the mirror than the image of the right face, and the images of the cube sides parallel to the axis are $\frac{1}{3}$ cm in length.

Figure 40–7 (not to scale) represents schematically the object and image, with corresponding points designated by unprimed and primed letters. It has been assumed that the lateral magnification is the same for both transverse faces.

In Fig. 40–8(a), the *convex* side of a spherical mirror faces the incident light so that R is negative. Ray PB is reflected with the angle of reflection r equal to the angle of incidence ϕ, and the reflected ray, projected backward, intersects the axis at P'. As in the case of a concave mirror, *all* rays from P will, after reflection, diverge from the same point P', provided that the angle u is small, so that P' is the image of P. The object distance s is positive, the image distance s' is negative, and the radius of curvature R is negative.

Figure 40–8(b) shows two rays diverging from the head of the arrow PQ, and the virtual image $P'Q'$ of this arrow. It is left to the reader to show, by the same procedure as that used for a concave mirror, that (1)

$$\frac{1}{s} + \frac{1}{s'} = \frac{2}{R};$$

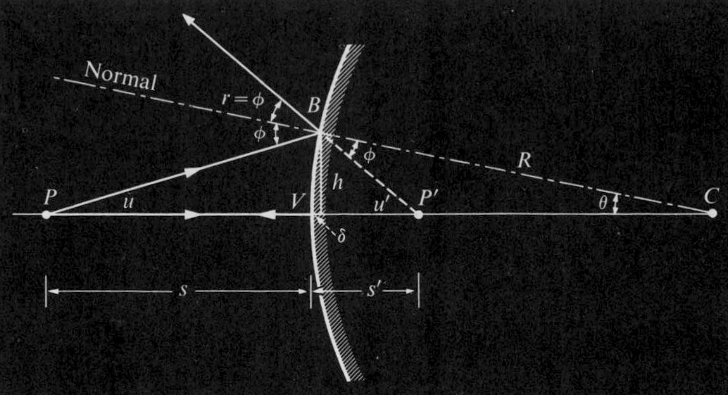

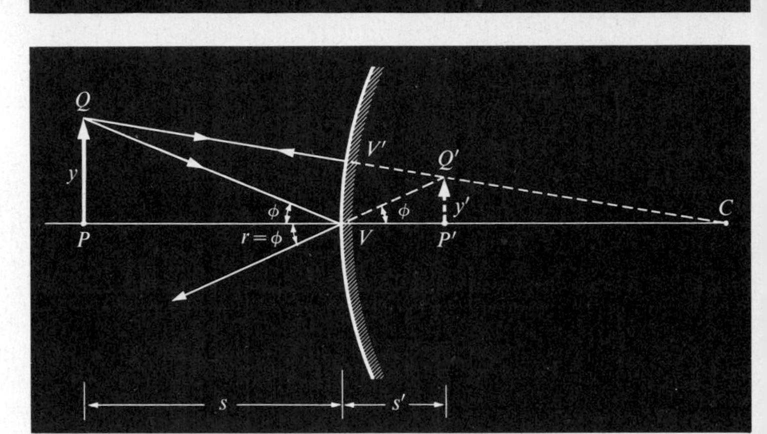

Fig. 40–8 *Construction for finding (a) the position and (b) the magnification of the image formed by a convex mirror.*

2) the lateral magnification is

$$m = \frac{y'}{y} = -\frac{s'}{s};$$

and (3) the longitudinal magnification again equals the negative square of the lateral magnification. These expressions are exactly the same as those for a concave mirror, as they should be when a consistent sign convention is used.

40–4 FOCAL POINT AND FOCAL LENGTH

When an object point is a very large distance from a mirror, all rays from that point that strike the mirror are parallel to one another. The object distance is $s = \infty$ and, from Eq. (40–4),

$$\frac{1}{\infty} + \frac{1}{s'} = \frac{2}{R}, \qquad s' = \frac{R}{2}.$$

The image distance s' then equals one-half the radius of curvature and has the same sign as R. This means that if R is positive, as in Fig. 40–9(a), the image point F lies to the left of the mirror and is real, while if R is negative, as in Fig. 40–9(b), the image point F lies to the right of the mirror and is virtual.

Conversely, if the image distance s' is very large, the object distance s is

$$\frac{1}{s} + \frac{1}{\infty} = \frac{2}{R}, \qquad s = \frac{R}{2}.$$

The object distance then equals half the radius of curvature. If R is positive, as in Fig. 40–10(a), s is positive. If R is negative, as in Fig. 40–10(b), s is negative and the *object is behind* the mirror. That is, rays previously made converging by some other surface are converging toward F.

The point F in Figs. 40–9 and 40–10 is called the *focal point* of the mirror. It may be considered either

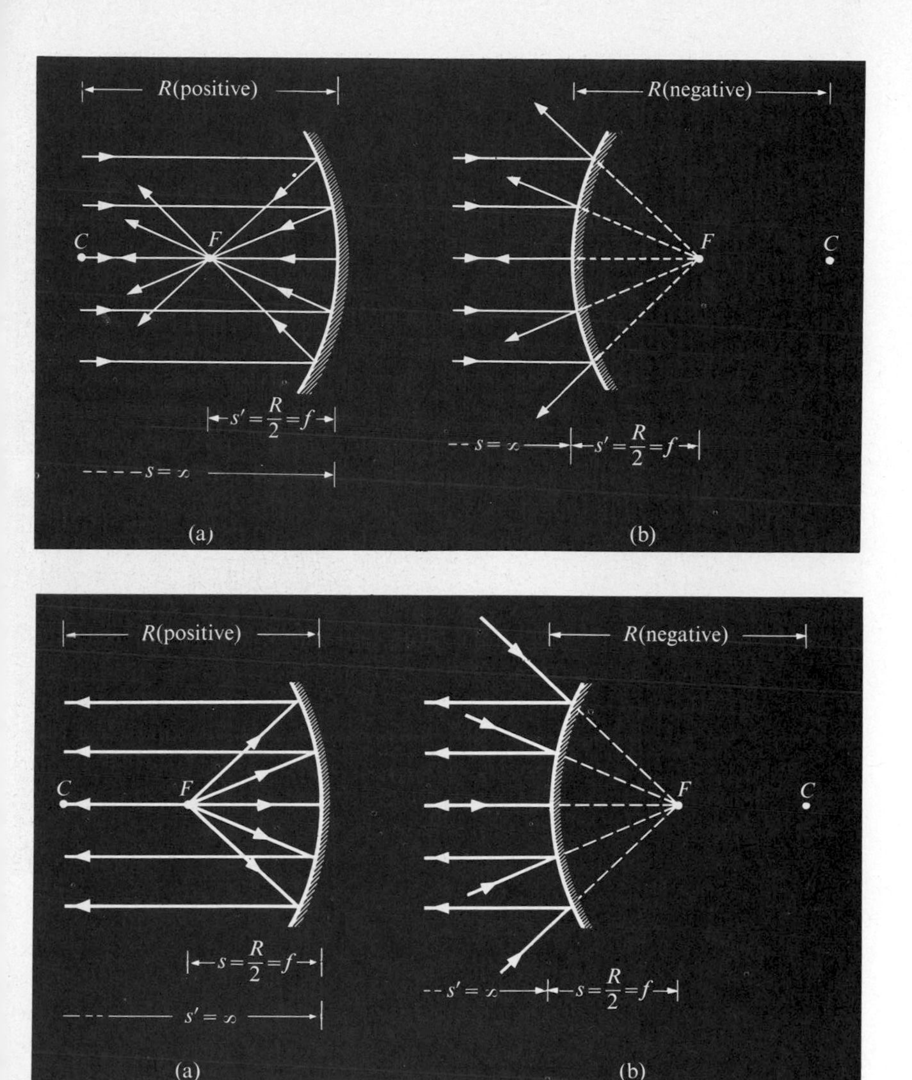

Fig. 40–9 *Incident rays parallel to the axis (a) converge to the focal point F of a concave mirror, (b) diverge as though coming from the focal point F of a convex mirror.*

Fig. 40–10 *Rays from a point object at the focal point of a spherical mirror are parallel to the axis after reflection. The object in part (b) is virtual.*

as the image point of an infinitely distant object point on the mirror axis, or as the object point of an infinitely distant image point. Thus the mirror of an astronomical telescope forms at its focal point an image of a star on the axis of the mirror.

The distance between the vertex of a mirror and the focal point is called the *focal length* of the mirror and is represented by f. As seen from the preceding discussion, the magnitude of the focal length equals one-half the radius of curvature. We shall see in the next chapter that a lens also has a focal length, and that the focal length of a lens which *converges* parallel rays to a real image is a positive quantity. A concave mirror, which behaves like a converging lens, has a positive focal length:

$$f = \frac{R}{2}. \tag{40–8}$$

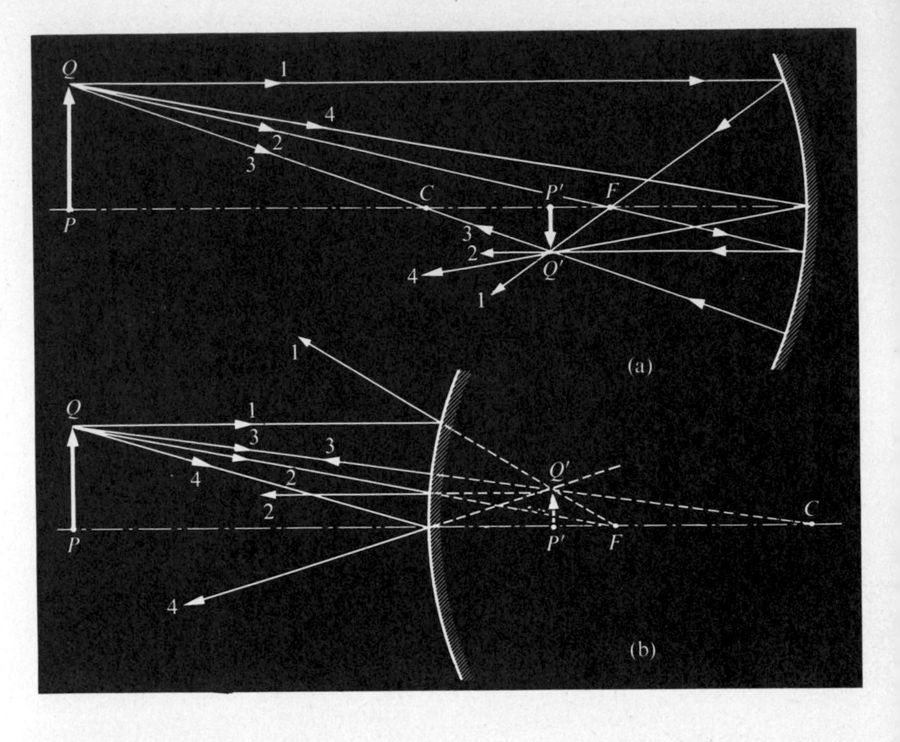

Fig. 40–11 *Rays used in the graphical method of locating an image.*

The relation between object and image distances, Eq. (40–5), for a mirror may now be written as

$$\boxed{\frac{1}{s} + \frac{1}{s'} = \frac{1}{f}.} \quad (40\text{–}9)$$

40–5 GRAPHICAL METHODS

The position and size of the image formed by a mirror may be found by a simple graphical method. This method consists of finding the point of intersection, after reflection from the mirror, of a few particular rays diverging from some point of the object *not* on the mirror axis, such as point Q in Fig. 40–11. Then (neglecting aberrations) *all* rays from this point which strike the mirror will intersect at the same point. Four rays whose paths may readily be traced are shown in Fig. 40–11. These are often called *principal rays.*

1. *A ray parallel to the axis,* after reflection, passes through the focal point of a concave mirror or appears to come from the focal point of a convex mirror.
2. *A ray from (or proceeding toward) the focal point* is reflected parallel to the axis.
3. *A ray along the radius* (extended if necessary) intersects the surface normally and is reflected back along its original path.
4. *A ray to the center* is reflected with equal angles with the optic axis.

Once the position of the image point has been found by means of the intersection of any two of these principal rays (1, 2, 3, 4), the paths of all other rays from the same object point may be drawn.

Example A concave mirror has a radius of curvature of magnitude 20 in. Find graphically the image of an object in the form of an arrow perpendicular to the axis of the mirror and at the following object distances: 30 in., 20 in., 10 in., and 5 in. Check the construction by computing the size and magnification of the image.

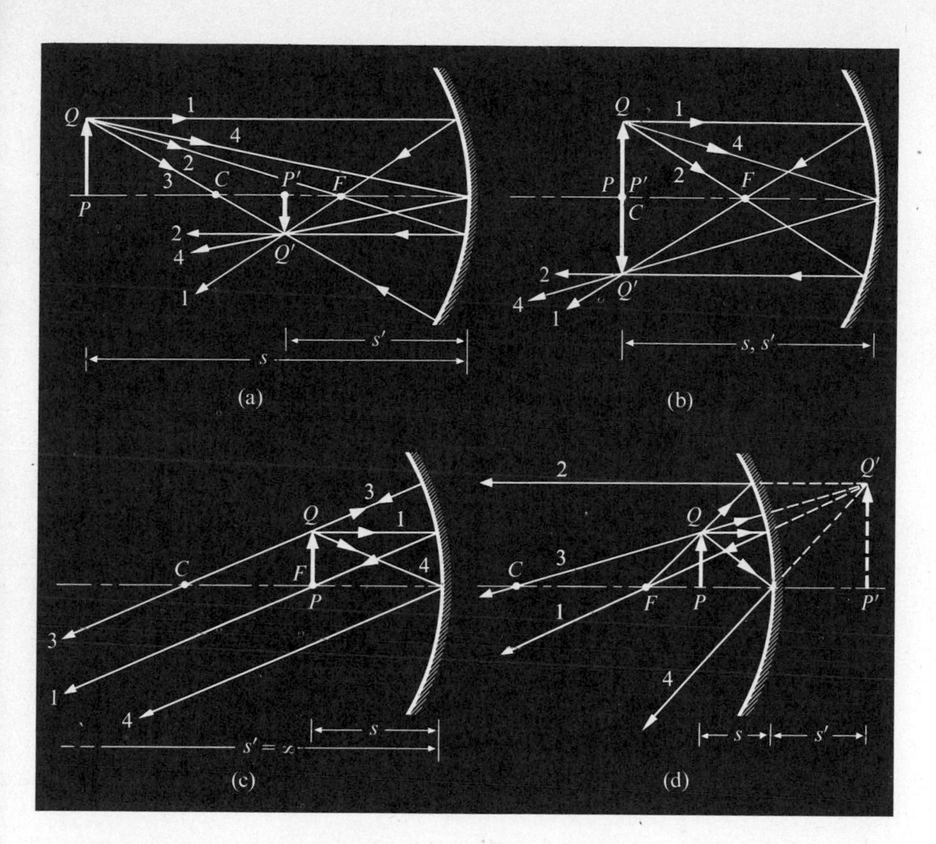

Fig. 40–12 Image of an object at various distances from a concave mirror, showing principal-ray construction.

The graphical construction is indicated in the four parts of Fig. 40–12. [Note that in (b) and (c) only three of the four rays can be used.] The calculations are left as a problem.

40–6 REFRACTION AT A PLANE SURFACE

The method of finding the image of a point object formed by rays refracted at a plane or spherical surface is essentially the same as for reflection, the only difference being that Snell's law replaces the law of reflection. In every case, we let n represent the index of the medium on one side of the surface and n' that of the medium on the other side. The same convention of signs is used as in reflection.

Consider first a plane surface, shown in Fig. 40–13, and assume $n' > n$. A ray from the object point P toward the vertex V is incident normally and passes into the second medium without deviation. A ray making an angle u with the axis is incident at B with an angle of incidence $\phi = u$. The angle of refraction, ϕ', is found from Snell's law,

$$n \sin \phi = n' \sin \phi'.$$

The two rays both appear to come from point P' after refraction. From the triangles PVB and $P'VB$,

$$\tan \phi = \frac{h}{s}, \qquad \tan \phi' = \frac{h}{-s'}. \qquad (40\text{–}10)$$

We must write $-s'$, since the direction from the surface to the image is *opposite* to that of the refracted light.

If the angle u is small, the angles ϕ, u', and ϕ' are small also, and therefore, approximately,

$$\tan \phi = \sin \phi, \qquad \tan \phi' = \sin \phi'.$$

Then Snell's law can be written as

$$n \tan \phi = n' \tan \phi',$$

and from Eq. (40–10), after cancelling h, we have

$$\frac{n}{s} = -\frac{n'}{s'},$$

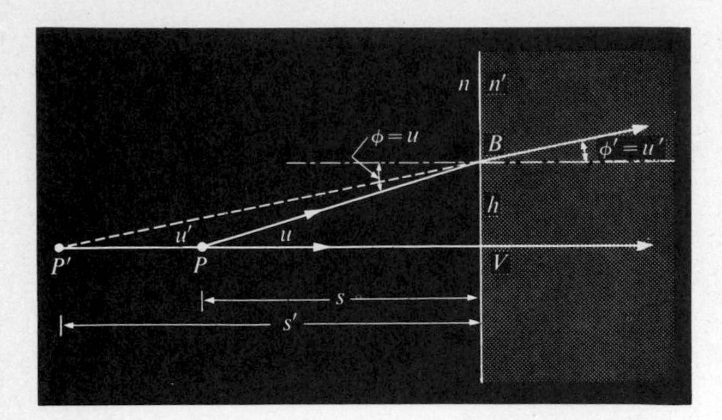

Fig. 40–13 Construction for finding the position of the image P′ of a point object P, formed by refraction at a plane surface.

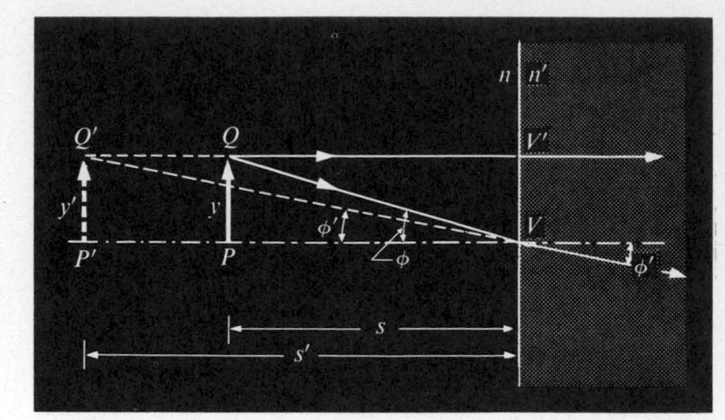

Fig. 40–14 Construction for determining the height of an image formed by refraction at a plane surface.

or

$$\frac{s'}{s} = -\frac{n'}{n}. \tag{40–11}$$

This is an *approximate* relation, *valid for paraxial rays only*. That is, a plane refracting surface does *not* image all rays from a point object at the same image point.

Consider next the image of a finite object, as in Fig. 40–14. The two rays diverging from point Q appear to diverge from its image Q' after refraction. From the triangles PQV and $P'Q'V$,

$$\tan\phi = \frac{y}{s},$$

$$\tan\phi' = \frac{y'}{-s'}.$$

Combining with Snell's law and using the small-angle approximation,

$$\sin\phi = \tan\phi, \ \sin\phi' = \tan\phi',$$

we obtain

$$\frac{ny}{s} = -\frac{n'y'}{s'},$$

and

$$m = \frac{y'}{y} = -\frac{ns'}{n's}. \tag{40–12}$$

However, from Eq. (40–11), $ns' = -n's$, so

$$m = \frac{y'}{y} = 1 \quad \text{(plane refracting surface)},$$

in agreement with Fig. 40–14. The image distance is greater than the object distance, but image and object are the same height.

A common example of refraction at a plane surface is afforded by looking vertically downward into the quiet water of a pond or a swimming pool; the apparent depth is less than the actual depth. Figure 40–15 illustrates this case. Two rays are

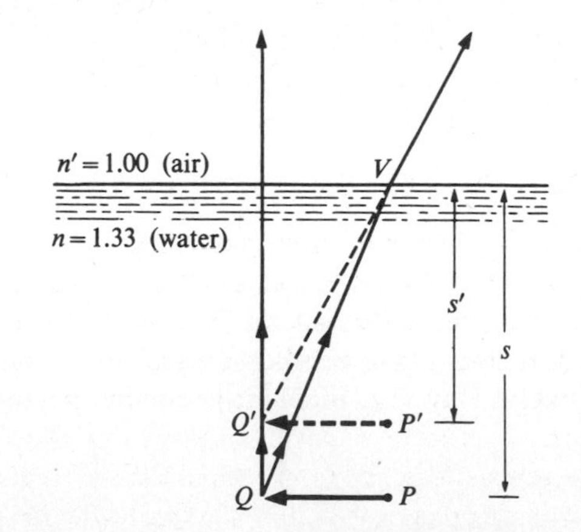

Fig. 40–15 Arrow $P'Q'$ is the image of the underwater object PQ.

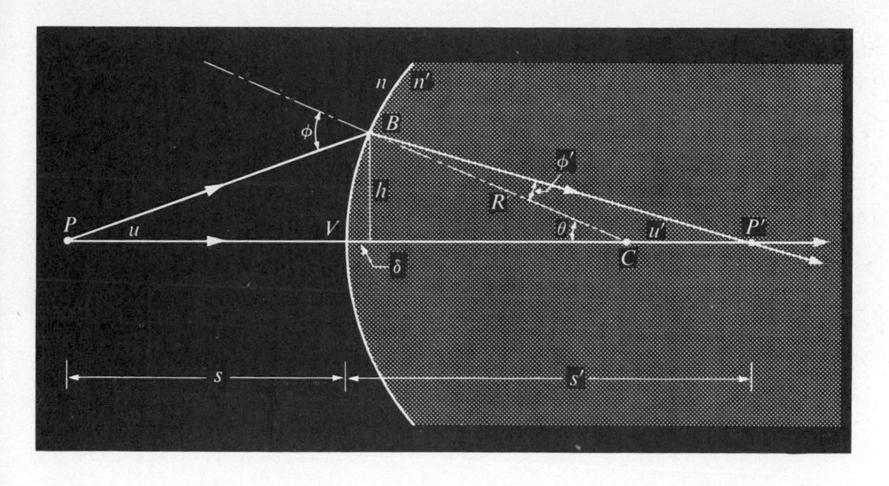

Fig. 40–16 *Construction for finding the position of the image P′ of a point object P, formed by refraction at a spherical surface.*

shown diverging from a point Q at a distance s below the surface. Here, n' (air) is less than n (water) and the ray incident at V is deviated *away from* the normal. The rays after refraction appear to diverge from Q', and the arrow PQ, to an observer looking vertically downward, appears lifted to the position $P'Q'$. From Eq. (40–11),

$$s' = -\frac{n'}{n}s = -\frac{1.00}{4/3}s = -\frac{3}{4}s.$$

The apparent depth s' is therefore only three-fourths of the actual depth s. The same phenomenon accounts for the apparent sharp bend in an oar when a portion of it extends below a water surface. The submerged portion appears lifted above its actual position.

40–7 REFRACTION AT A SPHERICAL SURFACE

Finally, we consider refraction at a spherical surface. In Fig. 40–16, P is an object point at a distance s to the left of a spherical surface of radius R. The indices at the left and right of the surface are n and n', respectively. Ray PV, incident normally, passes into the second medium without deviation. Ray PB, making an angle u with the axis, is incident at an angle ϕ with the normal and is refracted at an angle ϕ'. These rays intersect at P' at a distance s' to the right of the vertex.

We shall show that if the angle u is small, all rays from P intersect at the same point P', so P' is the real image of P. The object and image distances are both positive. The radius of curvature is positive also, because the direction from the surface to the center of curvature is the same as that of the refracted light.

When considering rays from P that are *reflected* at the surface, as in Fig. 40–8, the radius of curvature is negative. Thus the radius of curvature of a given surface has one sign for reflected light and the opposite sign for refracted light. This is the price we must pay in order that the relation connecting s, s', and the focal length, shall have the same algebraic form for both reflection and refraction.

From the triangles PBC and $P'BC$, we have

$$\phi = \theta + u, \qquad \theta = u' + \phi'. \qquad (40\text{–}13)$$

From Snell's law,

$$n \sin\phi = n' \sin\phi'.$$

Also, the tangents of u, u', and θ are

$$\tan u = \frac{h}{s+\delta},$$

$$\tan u' = \frac{h}{s'-\delta},$$

$$\tan\theta = \frac{h}{R-\delta}.$$

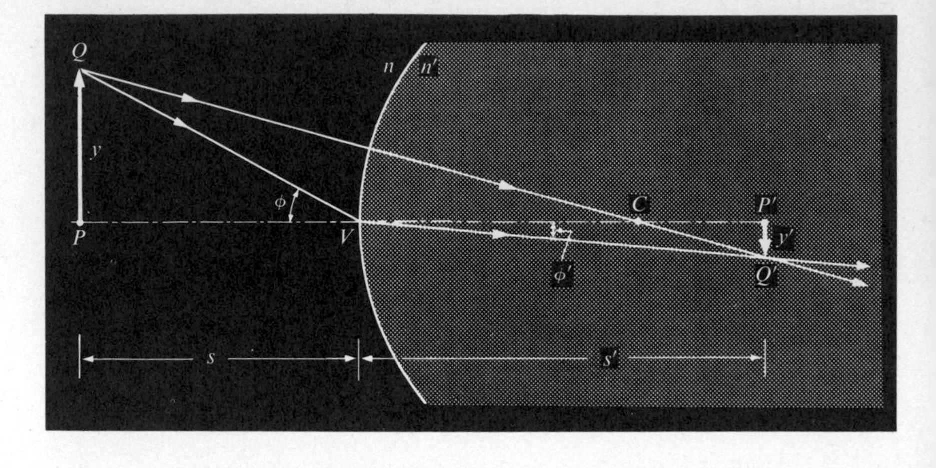

Fig. 40–17 *Construction for determining the height of an image formed by refraction at a spherical surface.*

For paraxial rays we may approximate both the sine and tangent of an angle by the angle itself, and neglect the small distance δ. Snell's law then becomes

$$n\phi = n'\phi',$$

and, combining with the first of Eqs. (40–13), we get

$$\phi' = \frac{n}{n'}(u + \theta).$$

Substituting this in the second of Eqs. (40–13) gives

$$nu + n'u' = (n' - n)\theta.$$

Using the small-angle approximations and cancelling h, we obtain

$$\boxed{\frac{n}{s} + \frac{n'}{s'} = \frac{n' - n}{R}.} \qquad (40\text{–}14)$$

This equation does not contain the angle u, so the image distance is the same for all paraxial rays from P.

If the surface is plane, $R = \infty$ and this equation reduces to Eq. (40–11), already derived for the special case of a plane surface.

The magnification is found from the construction in Fig. 40–17. From point Q draw two rays, one through the center of curvature C and the other incident at the vertex V. From the triangles PQV and $P'Q'V$,

$$\tan\phi = \frac{y}{s}, \qquad \tan\phi' = \frac{-y'}{s'},$$

and from Snell's law,

$$n \sin\phi = n' \sin\phi'.$$

For small angles,

$$\tan\phi = \sin\phi, \qquad \tan\phi' = \sin\phi',$$

and hence

$$\frac{ny}{s} = -\frac{n'y'}{s'},$$

or

$$\boxed{m = \frac{y'}{y} = -\frac{ns'}{n's},} \qquad (40\text{–}15)$$

which agrees with the relation previously derived for a plane surface.

The longitudinal magnification is

$$\boxed{m' = -\frac{n'}{n}m^2.}$$

Table 40–1

	Plane mirror	Spherical mirror	Plane refracting surface	Spherical refracting surface
Object and image distances	$\frac{1}{s}+\frac{1}{s'}=0$	$\frac{1}{s}+\frac{1}{s'}=\frac{2}{R}=\frac{1}{f}$	$\frac{n}{s}+\frac{n'}{s'}=0$	$\frac{n}{s}+\frac{n'}{s'}=\frac{n'-n}{R}$
Lateral magnification	$m=-\frac{s'}{s}=1$	$m=-\frac{s'}{s}$	$m=-\frac{ns'}{n's}=1$	$m=-\frac{ns'}{n's}$
Longitudinal magnification	$m'=\frac{\Delta s'}{\Delta s}=-1$	$m'=-\frac{s'^2}{s^2}$	$m'=-\frac{n'}{n}$	$m'=-\frac{ns'^2}{n's^2}$

Equations (40–14) and (40–15) can be applied to both convex and concave refracting surfaces when a consistent sign convention is used, and they apply whether n' is greater or less than n. The reader should construct diagrams like Figs. 40–16 and 40–17, when R is negative and $n' < n$, and use them to derive Eqs. (40–14) and (40–15).

The concepts of focal point and focal length can also be applied to a refracting surface. Such a surface is found to have *two* focal points. The first is the *object* point when the image is at infinity, the second is the *image* point of an infinitely distant object. These points lie on opposite sides of the surface and at different distances from it, so that a single refracting surface has two focal lengths. The positions of the focal points can readily be found from Eq. (40–14) by setting s or s' equal to infinity.

40–8 SUMMARY

The results of this chapter are summarized in Table 40–1. Note that by letting $R = \infty$, the equation for a plane surface follows immediately from the appropriate equation for a curved surface.

Example 1 One end of a cylindrical glass rod (Fig. 40–18) is ground to a hemispherical surface of radius $R = 20$ mm. Find the image distance of a point object on the axis of the rod, 80 mm to the left of the vertex. The rod is in air.

$$n = 1, \qquad n' = 1.5,$$
$$R = +20\text{ mm}, \qquad s = +80\text{ mm}.$$
$$\frac{1}{80\text{ mm}} + \frac{1.5}{s'} = \frac{1.5-1}{+20\text{ mm}},$$
$$s' = +120\text{ mm}.$$

The image is therefore formed at the right of the vertex (s' is positive) and at a distance of 120 mm from it. Suppose that the object is an arrow 1 mm high, perpendicular to the axis. Then

$$m = -\frac{ns'}{n's} = \frac{(1)(120\text{ mm})}{(1.5)(80\text{ mm})} = -1.$$

That is, the image is the same height as the object, but is inverted.

Example 2 Let the same rod be immersed in water of index 1.33, the other quantities having the same

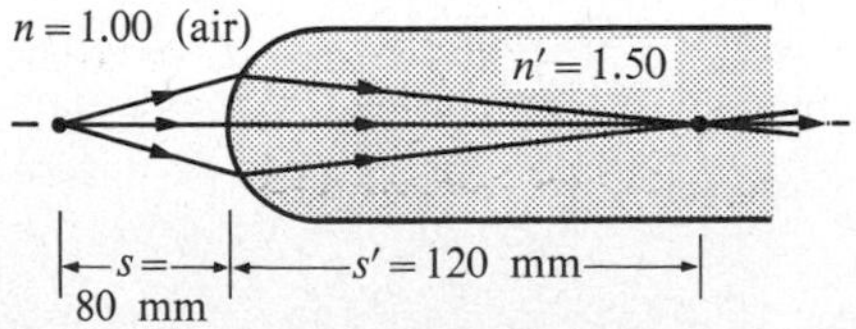

Fig. 40–18

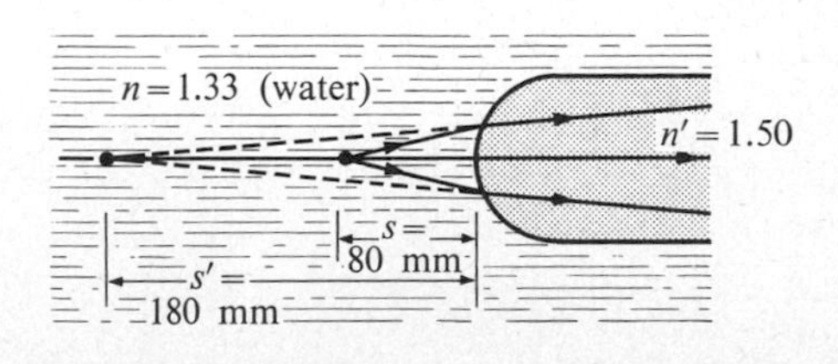

Fig. 40–19

values as before. Find the image distance (Fig. 40–19).

$$\frac{1.33}{80\text{ mm}} + \frac{1.5}{s'} = \frac{1.5 - 1.33}{+20\text{ mm}}, \qquad s' = -180\text{ mm}.$$

The fact that s' is a negative quantity means that the rays, after refraction by the surface, are not converging but appear to diverge from a point 180 mm to the left of the vertex. We have met a similar case before in the refraction of spherical waves by a plane surface, and have called the point a *virtual image*. In this example, then, the surface forms a virtual image 180 mm to the left of the vertex.

PROBLEMS

40–1 What is the size of the smallest vertical plane mirror in which an observer standing erect can see his full-length image?

40–2 The image of a tree just covers the length of a 5-cm plane mirror when the mirror is held 30 cm from the eye. The tree is 100 m from the mirror. What is its height?

40–3 An object is placed between two mirrors arranged at right angles to each other.

a) Locate all of the images of the object.
b) Draw the paths of rays from the object to the eye of an observer.

40–4 An object 1 cm high is 20 cm from the vertex of a concave spherical mirror whose radius of curvature is 50 cm. Compute the position and size of the image. Is it real or virtual? Erect or inverted?

40–5 A concave mirror is to form an image of the filament of a headlight lamp on a screen 4 m from the mirror. The filament is 5 mm high, and the image is to be 40 cm high.

a) What should be the radius of curvature of the mirror?
b) How far in front of the vertex of the mirror should the filament be placed?

40–6 The diameter of the moon is 3480 km and its distance from the earth is 386,000 km. Find the diameter of the image of the moon formed by a spherical concave telescope mirror of focal length 4 m.

40–7 A spherical concave shaving mirror has a radius of curvature of 30 cm.

a) What is the magnification when the face is 10 cm from the vertex of the mirror?
b) Where is the image?

40–8 A concave spherical mirror has a radius of curvature of 10 cm. Make a diagram of the mirror to scale, and show rays incident on it parallel to the axis and at distances of 1, 2, 3, 4, and 5 cm from the axis. Using a protractor, construct the reflected rays and indicate the points at which they cross the axis.

40–9 An object is 16 cm from the center of a silvered spherical glass Christmas tree ornament 8 cm in diameter. What are the position and magnification of its image?

40–10 A concave mirror of radius 5 cm has a radius of curvature of 20 cm.

a) What is the focal length?
b) If the mirror is immersed in water (refractive index 1.33), what is the focal length?

40–11 An object 2 cm high is placed 5 cm away from a concave spherical mirror having radius of curvature of 20 cm.

a) Draw a principal-ray diagram showing formation of the image.
b) Determine the position, size, orientation, and nature of the image.

40–12 Prove that the image formed of a real object by a convex mirror is always virtual, no matter what the object position.

40–13 If light striking a convex mirror does not diverge from an object point but instead is converging toward a point at a (negative) distance s to the right of the mirror, this point is called a *virtual object*

a) For a convex mirror having radius of curvature 10 cm, for what range of virtual-object positions is a real image formed?
b) What is the orientation of the image?
c) Draw a principal-ray diagram showing formation of such an image.

40–14 A tank whose bottom is a mirror is filled with water to a depth of 20 cm. A small object hangs motionless 8 cm under the surface of the water. What is the apparent depth of its image when viewed at normal incidence?

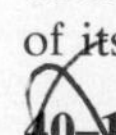

40–15 A ray of light in air makes an angle of incidence of 45° at the surface of a sheet of ice. The ray is refracted within the ice at an angle of 30°.

a) What is the critical angle for the ice?

b) A speck of dirt is embedded 2 cm below the surface of the ice. What is its apparent depth when viewed at normal incidence?

40–16 A microscope is focused on the upper surface of a glass plate. A second plate is then placed over the first. In order to focus on the bottom surface of the second plate, the microscope must be raised 1 mm. In order to focus on the upper surface it must be raised 2 mm *farther.* Find the index of refraction of the second plate. (This problem illustrates one method of measuring index of refraction.)

40–17 A layer of ether ($n = 1.36$) 2 cm deep floats on water ($n = 1.33$) 4 cm deep. What is the apparent distance from the ether surface to the bottom of the water layer, when viewed at normal incidence?

40–18 The end of a long glass rod 8 cm in diameter has a hemispherical surface 4 cm in radius. The refractive index of the glass is 1.50. Determine each position of the image if an object is placed on the axis of the rod at the following distances from its end:

a) infinitely far, b) 16 cm, c) 4 cm.

40–19 The rod of Problem 40–18 is immersed in a liquid. An object 60 cm from the end of the rod and on its axis is imaged at a point 100 cm inside the rod. What is the refractive index of the liquid?

40–20 What should be the index of refraction of a transparent sphere in order that paraxial rays from an infinitely distant object will be brought to a focus at the vertex of the surface opposite the point of incidence?

40–21 The left end of a long glass rod 10 cm in diameter, of index 1.50, is ground and polished to a convex hemispherical surface of radius 5 cm. An object in the form of an arrow 1 mm long, at right angles to the axis of the rod, is located on the axis 20 cm to the left of the vertex of the convex surface. Find the position and magnification of the image of the arrow formed by paraxial rays incident on the convex surface.

40–22 A transparent rod 40 cm long is cut flat at one end and rounded to a hemispherical surface of 12 cm radius at the other end. A small object is embedded within the rod along its axis and halfway between its ends. When viewed from the flat end of the rod the apparent depth of the object is 12.5 cm. What is its apparent depth when viewed from the curved end?

40–23 A solid glass hemisphere having a radius of 10 cm and a refractive index of 1.50 is placed with its flat face downward on a table. A parallel beam of light of circular cross section 1 cm in diameter travels directly downward and enters the hemisphere along its diameter. What is the diameter of the circle of light formed on the table?

40–24 A small tropical fish is at the center of a spherical fish bowl 30 cm in diameter. Find its apparent position and magnification to an observer outside the bowl. The effect of the thin walls of the bowl may be neglected.

40–25 In Fig. 40–20(a) the first focal length f is seen to be the value of s corresponding to

$$s' = \infty;$$

in (b) the second focal length f' is the value of s' when

$$s = \infty.$$

a) Prove that $n/n' = f/f'$.

b) Prove that the general relation between object and image distances is

$$\frac{f}{s} + \frac{f'}{s'} = 1.$$

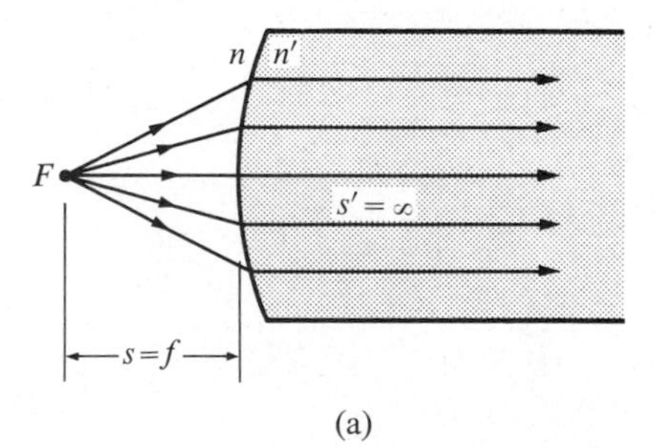

(a)

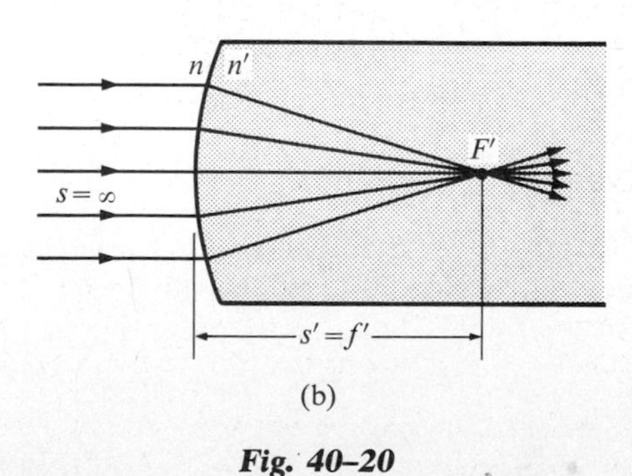

(b)

Fig. 40–20

Chapter 41

Lenses and Optical Instruments

41-1 IMAGES AS OBJECTS

Most optical systems include more than one reflecting or refracting surface. The image formed by the first surface serves as the object for the second; the image formed by the second surface serves as the object for the third; etc. Figure 41–1 illustrates the various situations which may arise, and the following discussion of them should be thoroughly understood.

In Fig. 41–1, the arrow at point O represents a small object at right angles to the axis. A narrow cone of rays diverging from the head of the arrow is traced through the system. Surface 1 forms a real image of the arrow at point P. Distance OV_1 is the object distance for the first surface and distance V_1P is the image distance. Both of these are positive.

The image at P, formed by surface 1, serves as the object for surface 2. The object distance is PV_2 and is positive, since the direction from P to V_2 is the same as that of the incident light. The second surface forms a virtual image at point Q. The image distance is V_2Q and is negative because the direction from V_2 to Q is opposite to that of the refracted light.

The image at Q, formed by surface 2, serves as the object for surface 3. The object distance is QV_3 and is positive. The image at Q, although virtual, constitutes a *real object* so far as surface 3 is concerned. The rays incident on surface 3 are rendered converging and, except for the interposition of surface 4, would converge to a real image at point R. Even though this image is never formed, distance V_3R is the image distance for surface 3 and is positive.

The rays incident on surfaces 1, 2, and 3 have all been diverging, and the object distance has been the distance from the surface to the point from which the

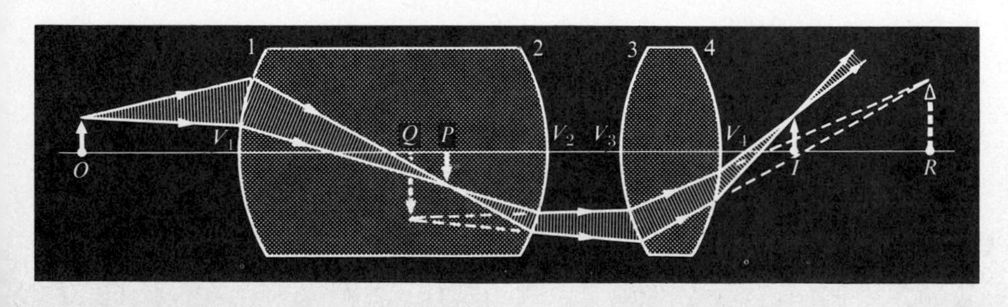

Fig. 41–1 *The object for each surface, after the first, is the image formed by the preceding surface.*

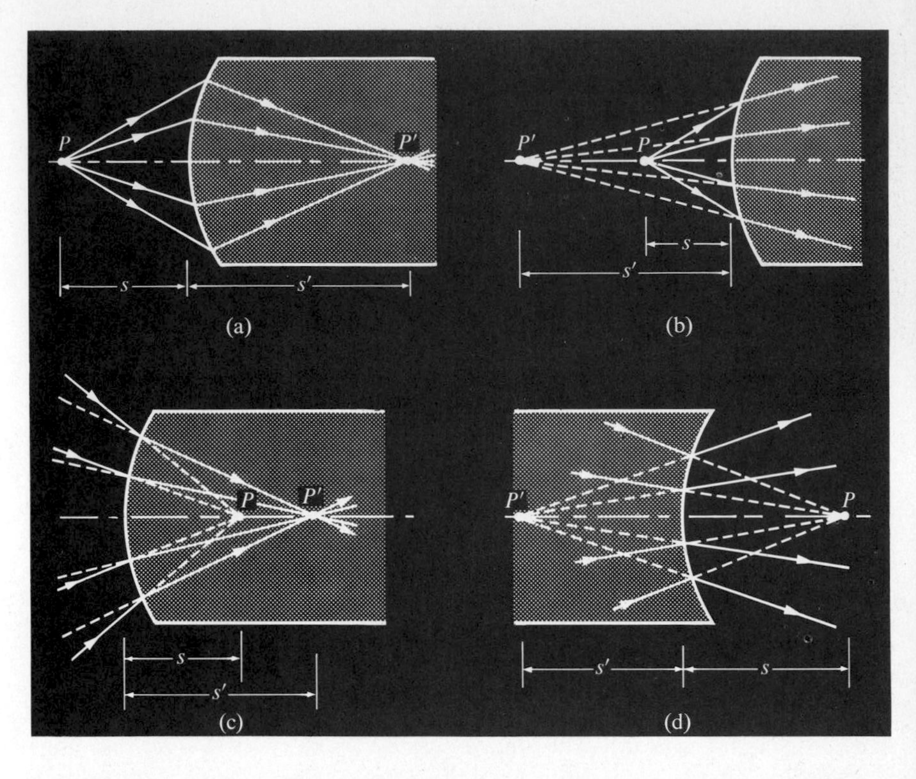

Fig. 41–2 *(a) A real image of a real object. (b) A virtual image of a real object. (c) A real image of a virtual object. (d) A virtual image of a virtual object.*

rays were actually or apparently diverging. The rays incident on surface 4, however, are *converging* and there is no point at the left of the vertex from which they diverge or appear to diverge. The *image at R, toward which the rays are converging,* is the object for surface 4, and since the direction from R to V_4 is opposite to that of the incident light, the object distance RV_4 is negative. The image at R is called a *virtual object* for surface 4. In general, whenever a *converging* cone of rays is incident on a surface, the point toward which the rays are converging serves as the object, the object distance is negative, and the point is called a virtual object.

Finally, surface 4 forms a real image at I, the image distance being V_4I and positive.

The meaning of virtual object and of virtual image are further exemplified by Fig. 41–2.

41–2 THE THIN LENS

A lens is an optical system including two refracting surfaces. The general problem of refraction by a lens is solved by applying the methods of Section 41–1 to each surface in turn, the object for the second surface being the image formed by the first. Figure 41–3 shows a pencil of rays diverging from point Q of an object PQ. The first surface of lens L forms a virtual image of Q at Q'. This virtual image serves as a real object for the second surface of the lens, which forms a real image of Q' at Q''. Distance s_1 is the object distance for the first surface; s_1' is the corresponding image distance. The object distance for the second surface is s_2, equal to the sum of s_1' and the lens thickness t, and s_2' is the image distance for the second surface.

If, as is often the case, the lens is so thin that its thickness t is negligible in comparison with the distances s_1, s_1', s_2, s_2', we may assume that s_1' equals $-s_2$, and measure object and image distances from either vertex of the lens. We shall also assume the medium on both sides of the lens to be air, with index of refraction 1.00. For the first refraction, Eq. (40–14) becomes

$$\frac{1}{s_1} + \frac{n}{s_1'} = \frac{n-1}{R_1}.$$

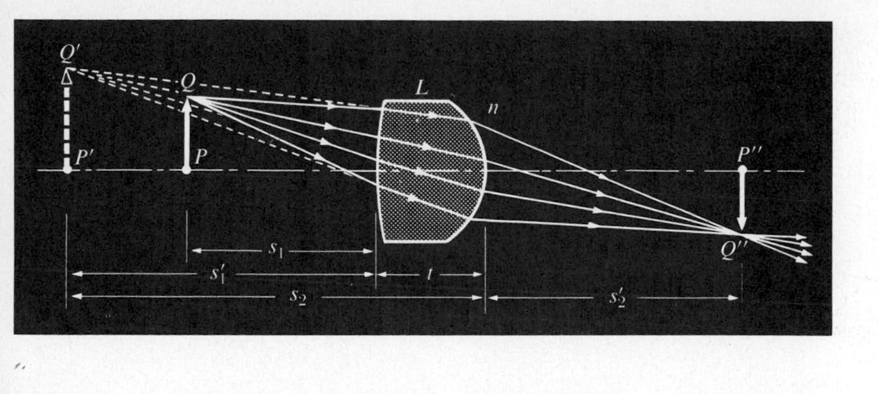

Fig. 41–3 *The image formed by the first surface of a lens serves as the object for the second surface.*

Refraction at the second surface yields the equation

$$\frac{n}{s_2} + \frac{1}{s_2'} = \frac{1 - n}{R_2}.$$

Adding these two equations, and remembering that the lens is so thin that $s_2 = -s_1'$, we get

$$\frac{1}{s_1} + \frac{1}{s_2'} = (n - 1)\left(\frac{1}{R_1} - \frac{1}{R_2}\right).$$

Since s_1 is the object distance for the thin lens and s_2' is the image distance, the subscripts may be omitted, and we get finally

$$\boxed{\frac{1}{s} + \frac{1}{s'} = (n - 1)\left(\frac{1}{R_1} - \frac{1}{R_2}\right).} \qquad (41\text{–}1)$$

The usual sign conventions apply to this equation. Thus, in Fig. 41–4, s, s', and R_1 are positive quantities, but R_2 is negative.

The *focal length f* of a thin lens may be defined either as (a) the object distance of a point object on the lens axis whose image is at infinity, or (b) the image distance of a point object on the lens axis at an infinite distance from the lens. When we set either s or s' equal to infinity in Eq. (41–1) we find, for the focal length,

$$\boxed{\frac{1}{f} = (n - 1)\left(\frac{1}{R_1} - \frac{1}{R_2}\right),} \qquad (41\text{–}2)$$

which is known as the *lensmaker's equation.*

Example In Fig. 41–4, let the absolute magnitudes of the radii of curvature of the lens surfaces be respectively 20 cm and 5 cm. Since the direction from the first surface to its center of curvature is the same as that of the refracted light, $R_1 = +20$ cm, and since the direction from the second surface to its

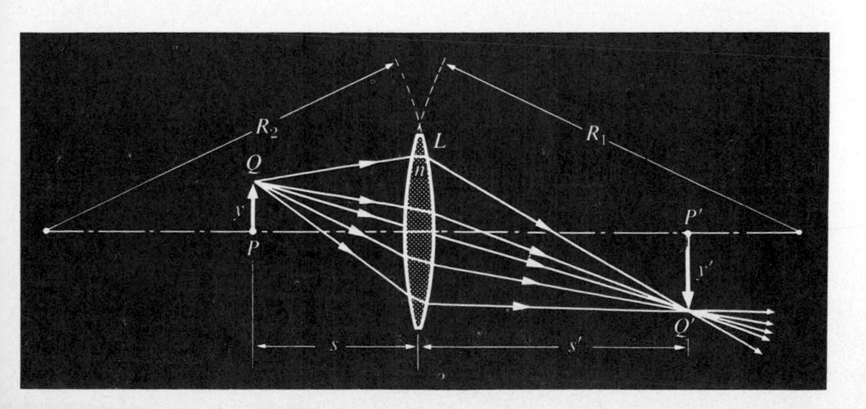

Fig. 41–4 *A thin lens.*

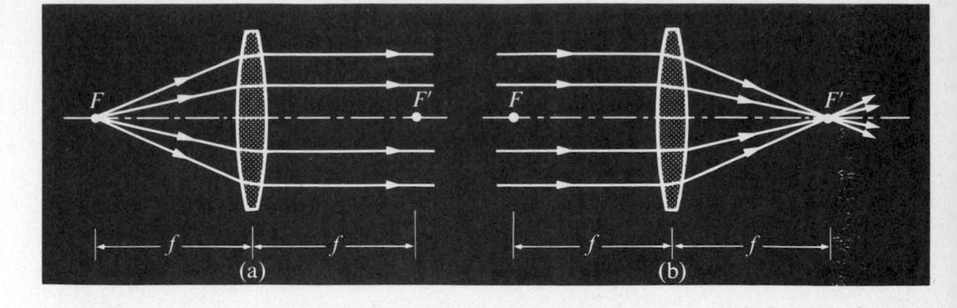

Fig. 41–5 First and second focal points of a thin lens.

center of curvature is opposite to that of the refracted light, $R_2 = -5$ cm. Let $n = 1.50$. Then

$$\frac{1}{f} = (1.50 - 1)\left(\frac{1}{20} - \frac{1}{-5}\right),$$

$$f = +8 \text{ cm}.$$

Substituting Eq. (41–2) in Eq. (41–1), we get the thin-lens equation

$$\boxed{\frac{1}{s} + \frac{1}{s'} = \frac{1}{f}.} \tag{41–3}$$

This is known as the *gaussian* form of the thin-lens equation, after Karl F. Gauss, the same mathematician responsible for the law in electrostatics bearing his name. Note that Eq. (41–3) has exactly the same form as the equation for a spherical mirror.

The object point for which the image is at infinity is called the *first focal point* of the lens and is lettered F in Fig. 41–5(a). The image point for an infinitely distant object is called the *second focal point* and is lettered F' in Fig. 41–5(b). The focal points of a thin lens lie on opposite sides of the lens at distances from it equal to its focal length.

Figure 41–6 corresponds to Fig. 41–5, except that it is drawn for object and image points not on the axis of the lens. Planes through the first and second focal points of a lens, perpendicular to the axis, are called the first and second *focal planes.* Paraxial rays from point Q in Fig. 41–6(a), in the first focal plane of the lens, are parallel to one another after refraction. In other words, they converge to an infinitely distant image of point Q. In Fig. 41–6(b) a bundle of parallel rays from an infinitely distant point, not on the lens axis, converges to an image Q' lying in the second focal plane of the lens.

The lateral magnification produced by a lens may be obtained by inspection of Fig. 41–4; the object PQ, the image $P'Q'$, and the lines PP' and QQ' form two similar triangles. Hence

$$\frac{-y'}{y} = \frac{s'}{s},$$

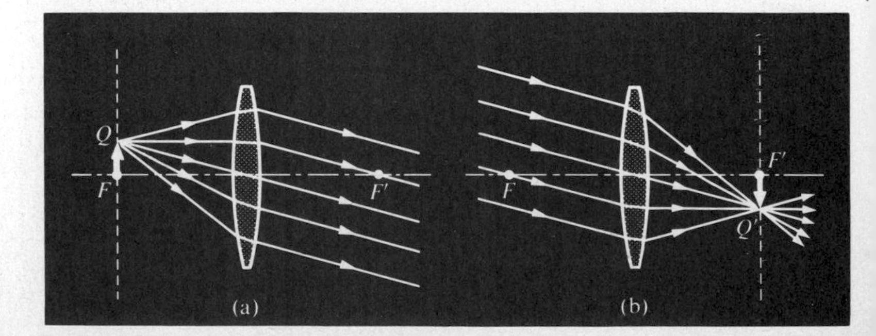

Fig. 41–6 Planes through the focal points of a lens are called focal planes.

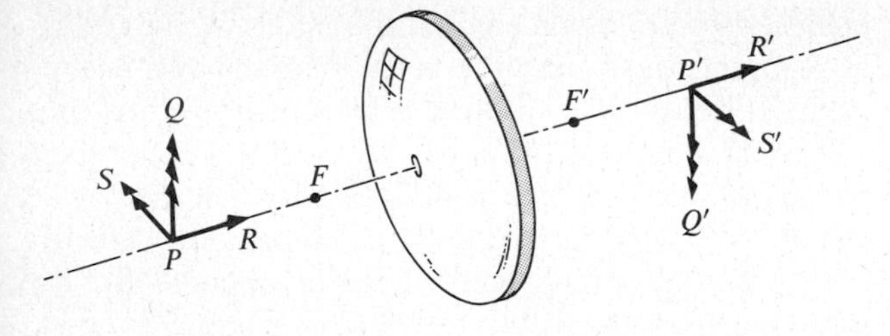

Fig. 41-7 *A lens forms a three-dimensional image of a three-dimensional object*

and since m, as usual, is y'/y, we obtain

$$m = -\frac{s'}{s}. \tag{41-4}$$

It follows from Eq. (41-3) that the *longitudinal* magnification m' is

$$m' = \frac{\Delta s'}{\Delta s} = -\left(\frac{s'}{s}\right)^2 = -m^2.$$

Although Eqs. (41-3) and (41-4) were derived for the special case of rays making small angles with the axis and, in general, do not apply to rays making large angles, they may be used for any lens which has been corrected so that all rays are imaged at the same point. They are therefore two of the most important equations in geometrical optics.

The three-dimensional image of a three-dimensional object, formed by a lens, is shown in Fig. 41-7. Since point R is nearer the lens than point P, its image, from Eq. (41-3), is farther from the lens than is point P', and the image $P'R'$ points in the same direction as the object PR. Arrows $P'S'$ and $P'Q'$ are reversed in space, relative to PS and PQ. Although we speak of the image as "inverted," only its transverse dimensions are reversed.

Figure 41-7 should be compared with Fig. 40-4, showing the image formed by a plane mirror. Note that the image formed by a lens, although it is inverted, is *not* perverted. That is, if the object is a left hand, its image is a left hand also. This may be verified by pointing the left thumb along PR, the left forefinger along PQ, and the left middle finger along PS. A rotation of 180° about the thumb as an axis then brings the fingers into coincidence with $P'Q'$ and $P'S'$. In other words, *inversion* of an image is equivalent to a rotation of 180° about the lens axis.

41-3 DIVERGING LENSES

A bundle of parallel rays incident on the lens shown in Figs. 41-5 and 41-6 converges to a real image after passing through the lens. The lens is called a *converging lens.* Its focal length, as computed from Eq. (41-2), is a positive quantity and therefore the lens is also called a *positive lens.*

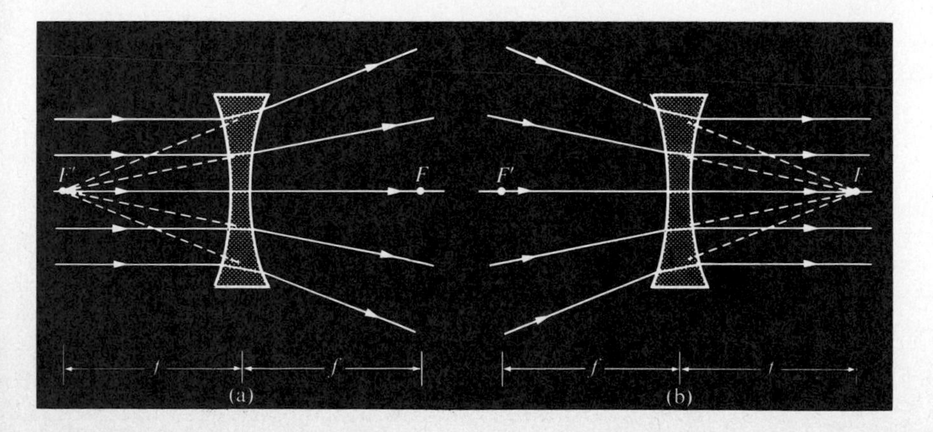

Fig. 41-8 *Focal points of a diverging lens.*

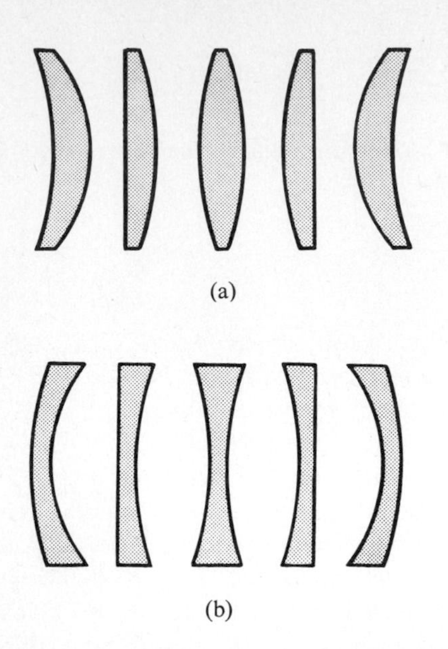

Fig. 41–9 (*a*) *Meniscus, plano-convex, and double-convex converging lenses.* (*b*) *Meniscus, plano-concave, and double-concave diverging lenses.*

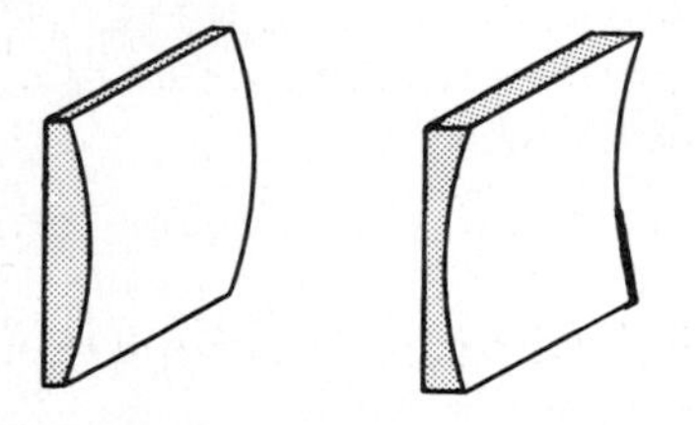

Fig. 41–10 Cylindrical lenses.

A bundle of parallel rays incident on the lens in Fig. 41–8 becomes diverging after refraction, and the lens is called a *diverging lens.* Its focal length, computed by Eq. (41–2), is a negative quantity and therefore the lens is also called a *negative lens.* The focal points of a negative lens are reversed, relative to those of a positive lens. The second focal point, F', of a negative lens is the point from which rays, originally parallel to the axis, appear to diverge after refraction, as in Fig. 41–8(a). Incident rays converging toward the first focal point F, as in Fig. 41–8(b), emerge from the lens parallel to its axis. That is, just as for a positive lens, the second focal point is the (virtual) image of an infinitely distant object on the axis of the lens, while the first focal point is the object point (a virtual object if the lens is diverging) for which an image is formed at infinity. Equations (41–2) through (41–6) apply both to negative and to positive lenses. Various types of lenses, both converging and diverging, are illustrated in Fig. 41–9.

In addition to lenses having spherical surfaces, use is frequently made of *cylindrical* lenses, particularly in spectacles, to correct a defect of vision known as *astigmatism.* One or both surfaces of a cylindrical lens are portions of cylinders (Fig. 41–10). Since defects of vision other than astigmatism are frequently present also, a spectacle lens may be cylindrical at one surface and spherical at the other; the same effect can be achieved by making one surface ellipsoidal.

41–4 GRAPHICAL METHODS

The position and size of the image of an object formed by a thin lens may be found by a simple graphical method. This method consists of finding the point of intersection, after passing through the lens, of a few rays (called *principal rays*) diverging from some chosen point of the object *not* on the lens axis, such as point Q in Fig. 41–11. Then (neglecting lens aberrations) all rays from this point which pass through the lens will intersect at the same point. In using the graphical method, the entire deviation of any ray is assumed to take place at a plane through the center of the lens. Three principal rays whose paths may readily be traced are shown in Fig. 41–11.

1. A *ray parallel to the axis,* after refraction by the lens, passes through the second focal point of a converging lens, or appears to come from the second focal point of a diverging lens.

2. A *ray through the center of the lens* is not appreciably deviated, since the two lens surfaces through which the central ray passes are very nearly parallel if the lens is thin. We have seen that a ray passing through a plate with parallel faces is not deviated, but only displaced. For a thin lens, the displacement may be neglected.

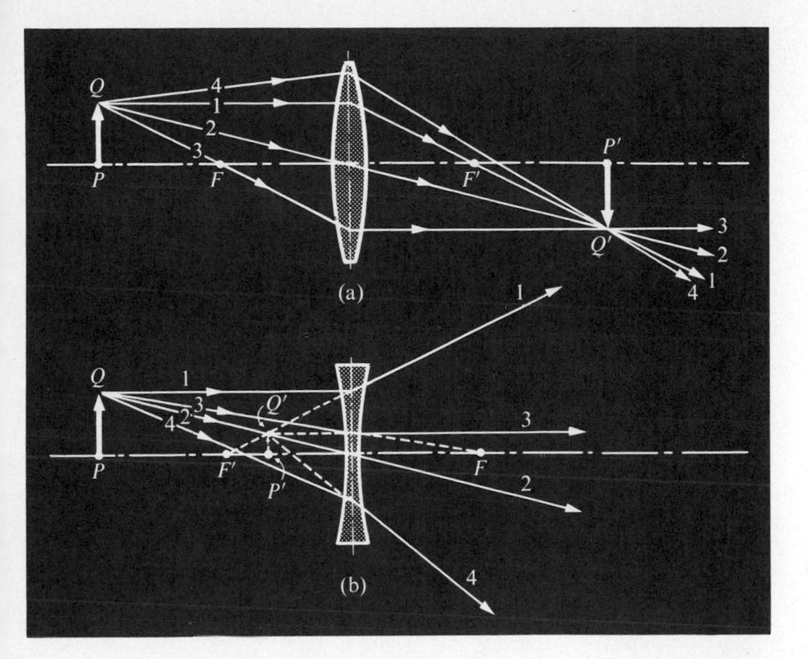

Fig. 41–11 *Graphical method of locating an image.*

3. A *ray through* (*or proceeding toward*) *the first focal point* emerges parallel to the axis.

Once the position of the image point has been found by means of the intersection of any two of rays 1, 2, and 3, the paths of all other rays from the same point, such as ray 4 in Fig. 41–11, may be drawn. A few examples of this procedure are given in Fig. 41–12.

41–5 IMAGES AS OBJECTS FOR LENSES

It was shown in Section 41–1 and Fig. 41–1 that the image formed by any one *surface* in an optical system serves as the object for the next surface. In the majority of optical systems using lenses, more than one lens is used and the image formed by any one *lens* serves as the object for the next lens. Figure 41–13 illustrates the various possibilities. Lens 1 forms a real image at P of a real object at O. This real image serves as a real object for lens 2. The virtual image at Q formed by lens 2 is a real object for lens 3. If lens 4 were not present, lens 3 would form a real image at R. Although this image is never formed, it serves as a virtual object for lens 4, which forms a final real image at I.

41–6 THE NEWTONIAN FORM OF THE LENS EQUATION

Figure 41–14 shows two rays proceeding from point Q of an object PQ to point Q' of its image $P'Q'$. The lengths PQ and Oa are equal to the height y of the object, and the lengths $P'Q'$ and Ob are equal to the height y' of the image. Let x and x' represent object and image distances, *measured from the corresponding focal points* F and F'. From the similar triangles PQF and FOb, we have

$$\frac{y}{x} = \frac{y'}{f}, \qquad \text{or} \qquad \frac{y'}{y} = \frac{f}{x},$$

and from the triangles $P'Q'F'$ and $F'Oa$,

$$\frac{y}{f} = \frac{y'}{x'}, \qquad \text{or} \qquad \frac{y'}{y} = \frac{x'}{f}.$$

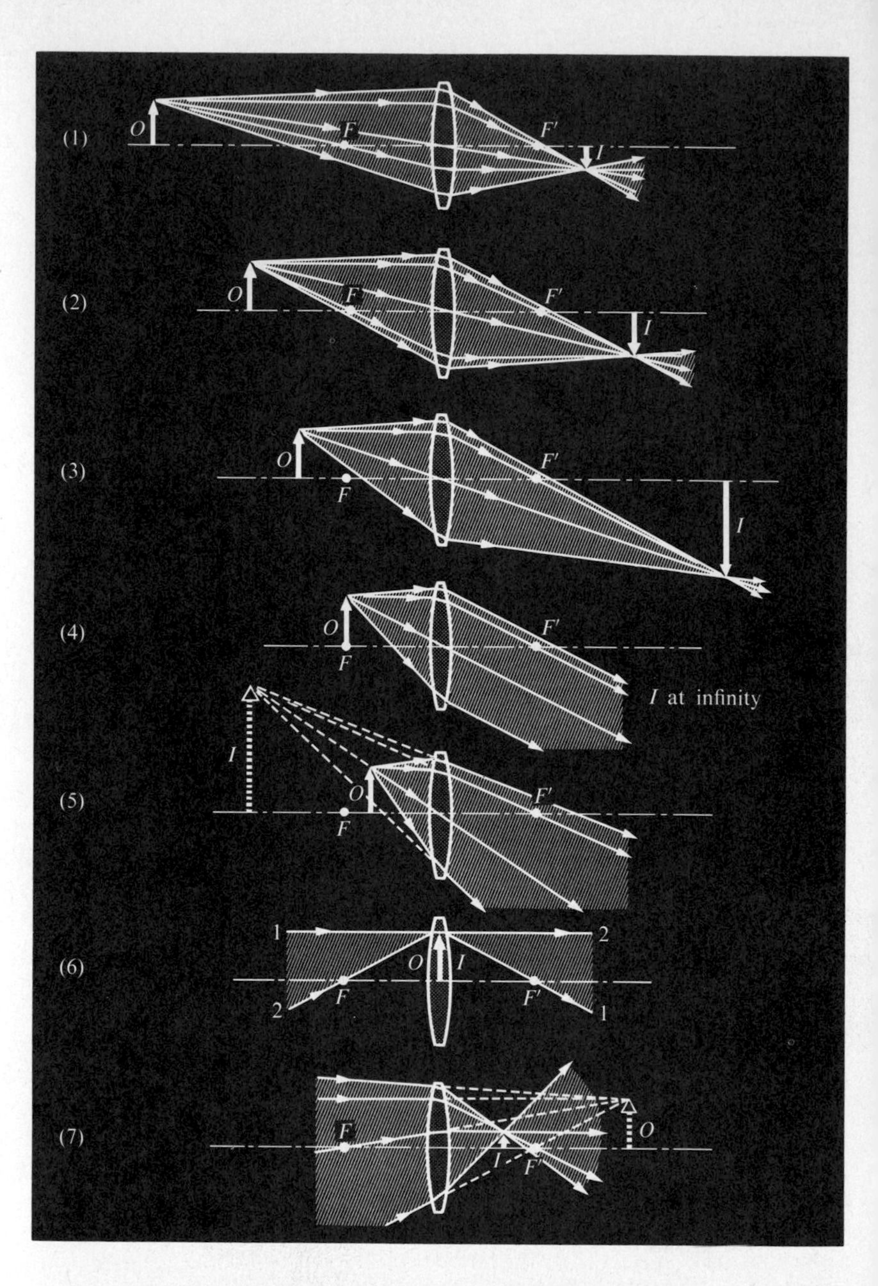

Fig. 41–12 *Formation of an image by a thin lens.*

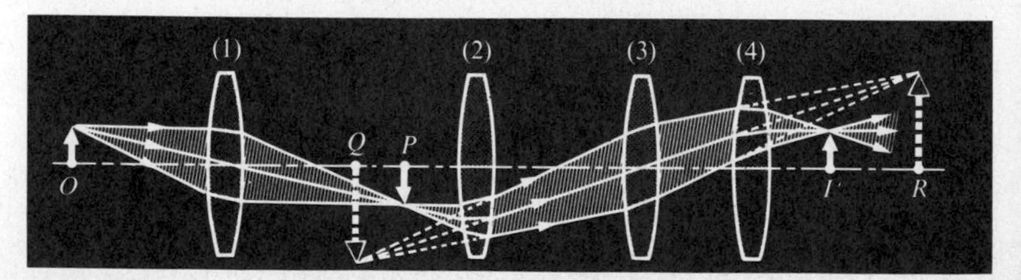

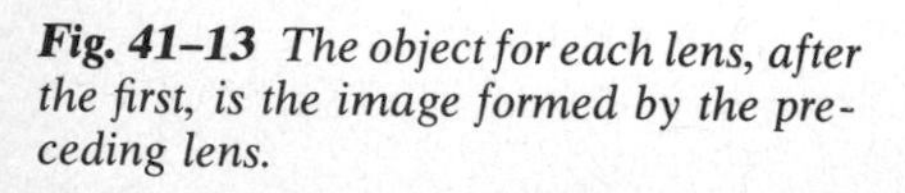

Fig. 41–13 *The object for each lens, after the first, is the image formed by the preceding lens.*

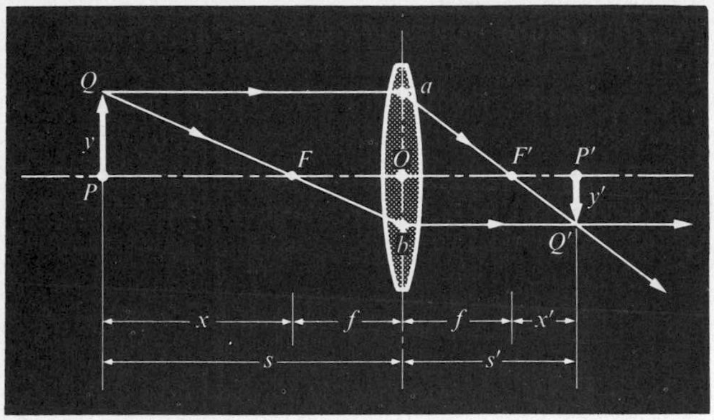

Fig. 41-14 In the newtonian form of the lens equations, object and image distance x and x′ are measured from the respective focal points F and F′ and $xx' = f^2$.

It follows that

$$\frac{f}{x} = \frac{x'}{f} \quad \text{or} \quad \boxed{xx' = f^2.} \qquad (41\text{–}5)$$

This is known as the *newtonian form* of the lens equation. It can also be derived from the gaussian form by substituting $x + f$ for s and $f + x'$ for s'.

The lateral magnification, y'/y, is

$$\boxed{m = \frac{y'}{y} = -\frac{f}{x} = -\frac{x'}{f}.} \qquad (41\text{–}6)$$

The negative sign must be included since y' and y have opposite signs.

41-7 THICK LENSES

The term *thick lens* means either a single lens whose thickness is not negligible, or any combination of lenses such as a corrected camera lens. Since Eqs. (41-3) and (41-6) were derived for a *thin* lens, the question arises whether or not they can also be used for a thick lens.

Figure 41-15 represents a simple thick lens, and shows two rays proceeding from point Q of an object PQ to point Q' of its image $P'Q'$. A ray parallel to the axis is deviated at both lens surfaces and its intersection with the axis determines the second focal point F'. Let the incident and emergent portions of the ray be extended, as shown by dotted lines, to their point of intersection at a. The plane through a, perpendicular to the axis, is called the *second principal plane* of the lens, and the point H' at which it intersects the axis is called the *second principal point.* Thus any incident ray parallel to the axis emerges from the lens as *if* it had undergone a *single* deviation in the second principal plane.

In the same way, the incident ray through the first focal point F emerges from the lens as *if* it had undergone a single deviation at point b. The plane through b is the *first principal plane* and point H is the *first principal point.*

The same procedure can be followed with any combination of lenses. In some instances, the principal points may lie outside the lens combination and may be *crossed,* with H' at the left of H. The principal points of a *thin* lens coincide at the center of the lens.

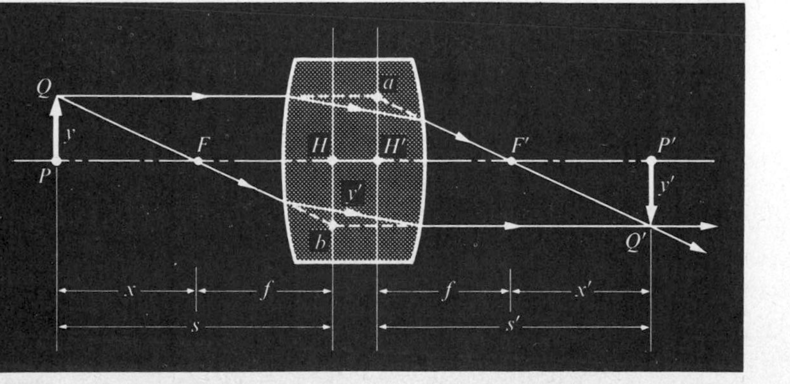

Fig. 41-15 The first principal point H of a thick lens is at a distance f to the right of the first focal point F, and the second principal point H′ is at a distance f to the left of the second focal point F′. Object and image distance s and s′ are measured from the corresponding principal points. The lens equation can be written either as $xx' = f^2$, *or as* $f/x = x'/f$.

The focal length f of a thick lens is now defined as the distance from the first focal point F to the first principal point H, or from the second principal point H' to the second focal point F'.

Let y and y' represent the lengths of object and image. From the similar triangles PQF and FHb (compare the discussion in the preceding section), we have

$$\frac{y}{x} = \frac{y'}{f}, \qquad \text{or} \qquad \frac{y'}{y} = \frac{f}{x},$$

and from the triangles $P'Q'F'$ and $F'H'a$,

$$\frac{y'}{x'} = \frac{y}{f}, \qquad \text{or} \qquad \frac{y'}{y} = \frac{x'}{f}.$$

Therefore,

$$\frac{f}{x} = \frac{x'}{f} \qquad \text{and} \qquad xx' = f^2.$$

Thus the newtonian lens equation applies to a thick lens as well as to a thin lens, provided the focal length is defined as the distance FH or $H'F'$.

The gaussian form of the lens equation can also be used for a thick lens, provided object and image distances are measured, not from the lens *surfaces*, but from the respective *principal points*, as shown in Fig. 41–15. If we substitute $(s - f)$ for x, and $(s' - f)$ for x', in the newtonian equation, we have, after some algebraic manipulation,

$$\frac{1}{s} + \frac{1}{s'} = \frac{1}{f}.$$

The focal points of any lens are easily located experimentally by finding the position of the image of a distant object, but this is not true of the principal points, for which special equipment is required. For this reason, the manufacturer usually measures the focal length of a lens before it leaves the factory and stamps the value on the lens mount. The positions of the principal points can then be determined from the positions of the focal points and the focal length. The position and magnification of the image of a given object are completely determined by the positions of the focal points and principal points, and it is not necessary to refer to the lens *surfaces.*

41–8 LENS ABERRATIONS

The relatively simple equations we have derived connecting object and image distances, focal lengths, radii of curvature, etc., were based upon the approximation that all rays made small angles with the axis. In general, however, a lens is called upon to image not only points on its axis, but points which lie off the axis as well. Furthermore, because of the finite size of the lens, the cone of rays which forms the image of any point is of finite size. Nonparaxial rays proceeding from a given object point do not, in general, all intersect at the same point after refraction by a lens. Consequently, the image formed by these rays is not a sharp one. Furthermore, the focal length of a lens depends upon its index of refraction, which varies with wavelength. Therefore, if the light proceeding from an object is not monochromatic, a lens forms a number of colored images which lie in different positions and are of different sizes, even if formed by paraxial rays.

The departures of an actual image from the predictions of simple theory are called *aberrations.* Those caused by the variation of index with wavelength are the *chromatic aberrations.* The others, which would arise even if the light were monochromatic, are the *monochromatic aberrations.* Lens aberrations are not caused by faulty construction of the lens, such as the failure of its surfaces to conform to a truly spherical shape, but are simply consequences of the laws of refraction at spherical surfaces.

The monochromatic aberrations are all related to the failure of the paraxial-ray approximation for lenses of finite aperture, but it is customary to distinguish various aspects of this difficulty, each with its characteristic effect. *Spherical aberration* is the failure of rays from a point object on the optic axis to converge to a point image; instead, the rays converge to a circle of minimum radius, called *the circle of least confusion,* and then diverge again, as shown in Fig. 41–16. The corresponding effect for points off the axis produces images that are comet-shaped figures rather than circles; this is called *coma.*

Astigmatism is the imaging of a point off the axis as a *line*; in this aberration the rays from a point object converge at some distance from the lens to a line in the plane defined by the optic axis and the

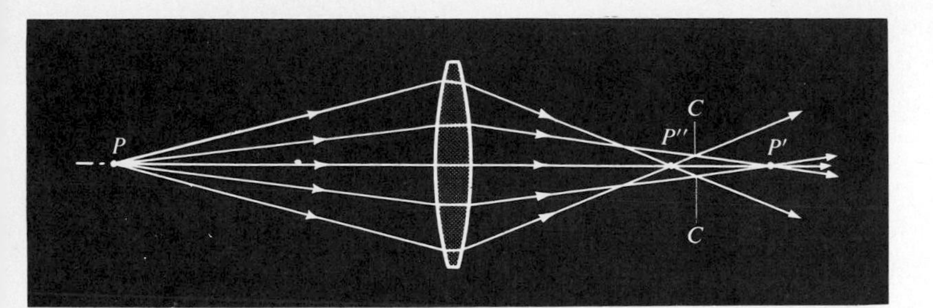

Fig. 41–16 *Spherical aberration. The circle of least confusion is shown by C-C.*

object point and, at a somewhat different distance from the lens, to a line *perpendicular* to this plane. The circle of least confusion appears between these two positions, at a location which depends on the object point's distance from the axis as well as its distance from the lens. As a result, object points lying in a plane are in general imaged not in a plane but in some curved surface; this effect is called *curvature of field.* Finally, the image of a straight line that does not pass through the axis may be curved; as a result the image of a square with the axis through its center may resemble a barrel (sides bent outward) or a pincushion (sides bent inward). This effect, called *distortion,* is not related to lack of sharpness of the image but results from a change in lateral magnification with distance from the axis.

Chromatic aberrations result directly from the variation of index of refraction with wavelength. Even in the absence of all monochromatic aberration, different wavelengths are imaged at different points, and when an object is illuminated with white light containing a mixture of wavelengths, there is no single point at which a point object is imaged. The magnification of a lens also varies with wavelength; this effect is responsible for the rainbow-fringed images seen with inexpensive binoculars or telescopes.

It is impossible to eliminate these aberrations from a single lens, but in a compound lens of several elements, the aberrations of one element may partially cancel those of another element. Design of such lenses is an extremely complex problem which has been aided greatly in recent years by the use of high-speed computers. It is still impossible to eliminate all aberrations, but it *is* possible to decide which ones are most troublesome for a particular application and to design accordingly.

41–9 THE EYE

Since the purpose of most optical instruments is to enable us to see better, the logical place to begin a discussion of such instruments is with the eye. The essential parts of the eye, considered as an optical system, are shown in Fig. 41–17.

The eye is very nearly spherical in shape, and about an inch in diameter. The front portion is somewhat more sharply curved, and is covered by a tough, transparent membrane *C*, called the *cornea.* The region behind the cornea contains a liquid *A* called the *aqueous humor.* Next comes the *crystalline lens, L,* a capsule containing a fibrous jelly, hard at the center and progressively softer at the outer portions. The crystalline lens is held in place by ligaments which attach it to the ciliary muscle *M.* Behind the lens, the eye is filled with a thin jelly *V* consisting largely of water, called the *vitreous humor.* The indices of refraction of both the aqueous humor and the vitreous humor are nearly equal to that of water, about 1.336. The crystalline lens, while not homogeneous, has an "average" index of 1.437. This is not very different from the indices of the aqueous and vitreous humors, so that most of the refraction of light entering the eye occurs at the cornea.

A large part of the inner surface of the eye is covered with a delicate film of nerve fibers, *R,* called the *retina.* A cross section of the retina is shown in Fig. 41–18(a). Nerve fibers branching out from the *optic nerve O* terminate in minute structures called rods and cones. The rods and cones, together with a

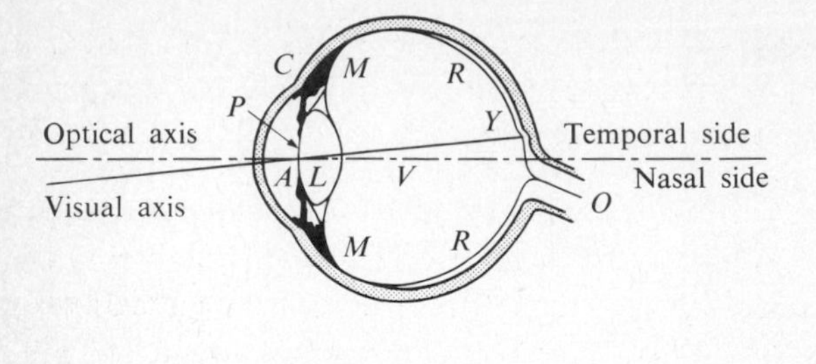

Fig. 41–17 *The eye.*

bluish liquid called the visual purple, which circulates among them, receive the optical image and transmit it along the optic nerve to the brain. There is a slight depression in the retina at Y called the yellow spot or macula. At its center is a minute region, about 0.25 mm in diameter, called the *fovea centralis*, which contains cones exclusively. Vision is much more acute at the fovea than at other portions of the retina, and the muscles controlling the eye always rotate the eyeball until the image of the object toward which attention is directed falls on the fovea. The outer portion of the retina merely serves to give a general picture of the field of view. The fovea is so small that motion of the eye is necessary to focus distinctly two points as close together as the dots in a colon (:).

There are no rods or cones at the point where the optic nerve enters the eye, and an image formed at this point cannot be seen. This region is called the *blind spot*. The existence of the blind spot can be demonstrated by closing the left eye and looking with the right eye at the cross in Fig. 41–18(b). When the diagram is about 25 cm from the eye, the image of the square falls on the blind spot and the square disappears. At a smaller distance, the square reappears while the circle disappears. At a still smaller distance, the circle again appears.

In front of the crystalline lens is the iris, at the center of which is an opening P called the *pupil*, which regulates the quantity of light entering the eye, dilating if the brightness of the field is low, and contracting if the brightness is increased. This process is known as *adaptation*. However, the range of pupillary diameter is only about fourfold (hence, the range in area is about sixteenfold) over a range of brightness which is 100,000-fold. The receptive mechanism of the retina can also adapt itself to large differences in quantity of light.

To see an object distinctly, a sharp image of it must be formed on the retina. If all the elements of the eye were rigidly fixed in position, there would be but one object distance for which a sharp retinal image would be formed, while in fact the normal eye can focus sharply on an object at any distance from infinity up to about 25 cm in front of the eye. This is made possible by the action of the crystalline lens and the ciliary muscle to which it is attached. When

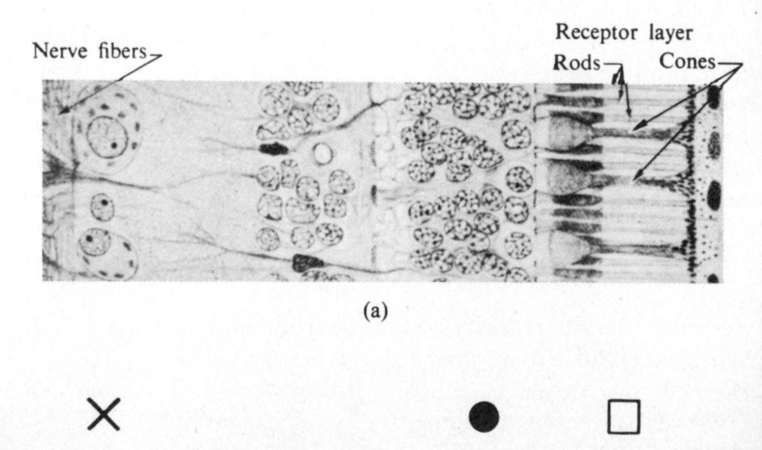

Fig. 41–18 *(a) Section of the human retina (500X). Light is incident from the left. (b) Figure for demonstrating the blind spot.*

relaxed, the normal eye is focused on objects at infinity, i.e., the second focal point is at the retina. When it is desired to view an object nearer than infinity, the ciliary muscle tenses and the crystalline lens assumes a more nearly spherical shape. This process is called *accommodation.*

The extremes of the range over which distinct vision is possible are known as the *far point* and the *near point* of the eye. The far point of a normal eye is at infinity. The position of the near point evidently depends on the extent to which the curvature of the crystalline lens may be increased in accommodation. The range of accommodation gradually diminishes with age as the crystalline lens loses its flexibility. For this reason the near point gradually recedes as one grows older. This recession of the near point with age is called *presbyopia,* and should not be considered a defect of vision, since it proceeds at about the same rate in all normal eyes. The following is a table of the approximate position of the near point at various ages:

Age, years	Near point, cm
10	7
20	10
30	14
40	22
50	40
60	200

41–10 DEFECTS OF VISION

Several common defects of vision result from an incorrect relation between the parts of the optical system of the eye. A normal eye forms an image on the retina of an object at infinity when the eye is relaxed, as in Fig. 41–19(a). In the *myopic* (nearsighted) eye, the eyeball is too long from front to back in comparison with the radius of curvature of the cornea, and rays from an object at infinity are focused in front of the retina. The most distant object for which an image can be formed on the retina is then nearer than infinity. In the *hyperopic* (farsighted) eye, the eyeball is too short and the image of an infinitely distant object would be formed behind the retina. By accommodation, these parallel rays may be made to converge on the retina, but evidently, if the range of accommodation is normal, the near point will be more distant than that of a normal eye. The myopic eye produces too much convergence in a parallel bundle of rays for an image to be formed on the retina; the hyperopic eye, not enough.

Astigmatism refers to a defect in which the surface of the cornea is not spherical, but is more sharply curved in one plane than another. (It should not be confused with the lens aberration of the same name, which applies to the behavior, after passing through a lens having spherical surfaces, of rays making a large angle with the axis.) Astigmatism makes it impossible, for example, to focus clearly on the horizontal and vertical bars of a window at the same time.

These defects can be corrected by the use of spectacles. The near point of either a presbyopic or a hyperopic eye is farther from the eye than normal. To see clearly an object at normal reading distance (usually assumed to be 25 cm) we must place in front of the eye a lens of such focal length that it forms an image of the object at or beyond the near point. Thus the function of the lens is not to make the object appear larger, but in effect to move the object farther away from the eye to a point where a sharp retinal image can be formed.

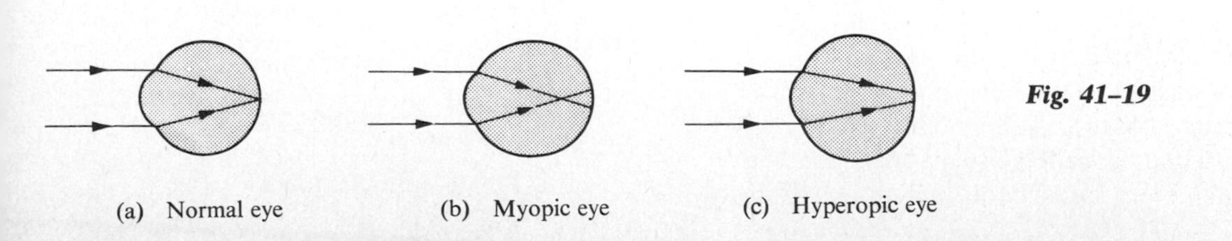

Fig. 41–19

Example The near point of a certain eye is 100 cm in front of the eye. What lens should be used to see clearly an object 25 cm in front of the eye?

We have

$$s = +25 \text{ cm}, \qquad s' = -100 \text{ cm},$$

$$\frac{1}{f} = \frac{1}{s} + \frac{1}{s'} = \frac{1}{+25 \text{ cm}} + \frac{1}{-100 \text{ cm}},$$

$$f = +33 \text{ cm}.$$

That is, a converging lens of focal length 33 cm is required.

The far point of a *myopic* eye is nearer than infinity. To see clearly objects beyond the far point, a lens must be used which will form an image of such objects, not farther from the eye than the far point.

Example The far point of a certain eye is 1 m in front of the eye. What lens should be used to see clearly an object at infinity?

Assume the image to be formed at the far point. Then

$$s = \infty, \qquad s' = -100 \text{ cm},$$

$$\frac{1}{f} = \frac{1}{s} + \frac{1}{s'} = \frac{1}{\infty} + \frac{1}{-100 \text{ cm}},$$

$$f = -100 \text{ cm}.$$

A *diverging* lens of focal length 100 cm is required.

Astigmatism is corrected by means of a *cylindrical* lens. The curvature of the cornea in a horizontal plane may have the proper value such that rays from infinity are focused on the retina. In the vertical plane, however, the curvature may not be sufficient to form a sharp retinal image. When a cylindrical lens with axis horizontal is placed before the eye, the rays in a horizontal plane are unaffected, while the additional convergence of the rays in a vertical plane now causes these to be sharply imaged on the retina.

The optometrist describes the converging or diverging effect of lenses in terms, not of the focal length, but of its *reciprocal*. The reciprocal of the focal length of a lens is called its *power*, and if the focal length is in meters the power is in diopters. Thus the power of a positive lens whose focal length is 1 m is 1 diopter; if the focal length is 2 m the power is 0.5 diopter, and so on. If the focal length is negative, the power is negative also. For example, a lens of power −0.5 diopter is a diverging lens of focal length −2 m.

41–11 THE MAGNIFIER

The apparent size of an object is determined by the size of its retinal image, which, in turn, if the eye is unaided, depends upon the *angle* subtended by the object at the eye. When one wishes to examine a small object in detail one brings it close to the eye, in order that the angle subtended and the retinal image may be as large as possible. Since the eye cannot focus sharply on objects closer than the near point, a given object subtends the maximum possible angle at an unaided eye when placed at this point. (We shall assume hereafter that the near point is 25 cm from the eye.) By placing a converging lens in front of the eye, the accommodation may, in effect, be increased. The object may then be brought closer to the eye than the near point and will subtend a correspondingly larger angle. A lens used for this purpose is called a *magnifying glass, a simple microscope*, or a *magnifier*. The magnifier forms a virtual image of the object and the eye "looks at" this virtual image. Since a (normal) eye can focus sharply on an object anywhere between the near point and infinity, the image can be seen equally clearly if it is formed anywhere within this range. We shall assume that the image is formed at infinity.

The magnifier is illustrated in Fig. 41–20. In (a), the object is at the near point, where it subtends an angle u at the eye. In (b), a magnifier in front of the eye forms an image at infinity, and the angle subtended at the magnifier is u'. The *angular magnification M* (not to be confused with the *lateral magnification m*) is defined as the ratio of the angle u' to the angle u. The value of M may be found as follows.

From Fig. 41–20,

$$u = \frac{y}{25 \text{ cm}} \quad \text{(approximately)},$$

$$u' = \frac{y}{f} \quad \text{(approximately)}.$$

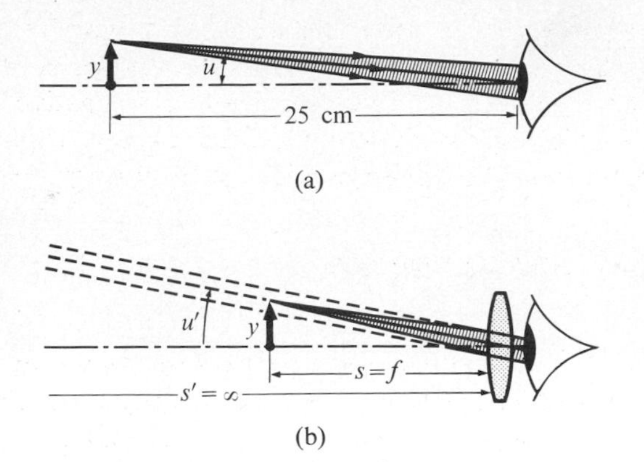

Fig. 41-20 *A simple magnifier.*

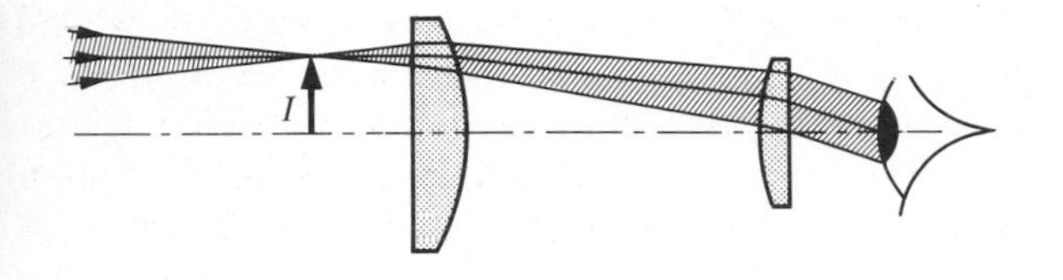

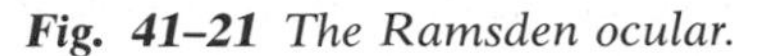

Fig. 41-21 *The Ramsden ocular.*

Hence,

$$M = \frac{u'}{u} = \frac{y/f}{y/25} = \frac{25}{f} \qquad (f \text{ in centimeters}). \tag{41-7}$$

While it appears at first that the angular magnification may be made as large as desired by decreasing the focal length f, the aberrations of a simple double convex lens set a limit to M of about $2X$ or $3X$. If these aberrations are corrected, the magnification may be carried as high as $20X$.

An *ocular* or *eyepiece* is a magnifier used for viewing an image formed by a lens or lenses preceding it in an optical system. The Ramsden ocular is illustrated in Fig. 41-21. It is constructed of two plano-convex lenses of equal focal length, separated by a distance of about $\frac{2}{3}$ of this length. The image to be examined is shown at I; the final image is at infinity. Since four refracting surfaces are available, the aberrations of a simple magnifier can be greatly reduced.

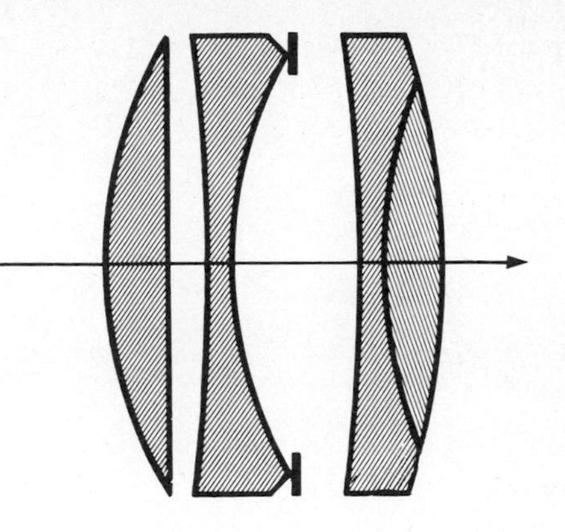

Fig. 41-22 *Zeiss "Tessar" lens.*

41-12 THE CAMERA

The essential elements of a camera are a lens, a light-tight box, and a sensitized plate or film for receiving the image. The "normal" lens usually covers a field of about 45°, less for telephoto lenses and more for wide-angle lenses. Furthermore, the aperture of the lens must be large, in order that it may collect sufficient light to permit short exposures. The combination of wide field and large aperture makes the problem of correcting a photographic lens a difficult one. Nevertheless, even the simplest photographic lenses are corrected for chromatic aberration and curvature of field. Many modern high-speed, short-focal-length lenses are modifications of the Zeiss "Tessar" lens, illustrated in Fig. 41-22.

The light-gathering power of a photographic objective is usually stated in terms of its f/number, which is determined by the focal length of the lens and by its diameter or the diameter of the aperture which effectively determines the lens area. Thus the notation $f/4.5$ means that the focal length of the lens is 4.5 times its effective diameter. The smaller the f/number, the larger the lens diameter for a given focal length, and the greater the light-gathering power or "speed" of the lens. Extremely fast lenses may have f/numbers as small as $f/1.4$ or $f/1.2$. The required time of exposure increases with the square of the f/number.

For a given position of the photographic film only those points lying in the corresponding object plane are sharply focused; objects at greater or lesser

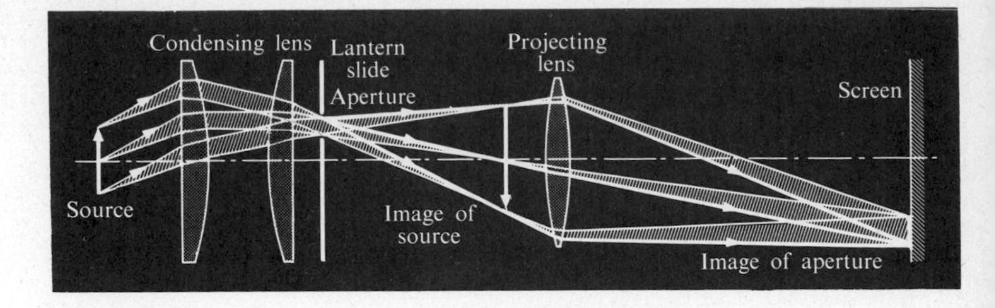

Fig. 41–23 *The projector.*

distance appear somewhat blurred. However, because of lens aberrations, a point of a given object will be imaged as a small circle, called the *circle of confusion*, even with the best focusing. The circles of confusion of points at other distances will be larger. If extremely sharp definition of the image is not essential, there is evidently a certain range of object distances, called the *depth of field*, such that all objects within this range are simultaneously "in focus" on the plate. That is, the circles of confusion of points within this range are not so large that the image is unsatisfactory.

The longitudinal magnification produced by a lens is

$$m^2 = \left(\frac{f}{x}\right)^2.$$

The ratio f/x is small for a camera lens of short focal length, and the longitudinal "magnification" is actually a very small fraction. The images of objects at different distances from the lens are therefore compressed into a very short distance along the lens axis (see Fig. 40–7), and the depth of field is correspondingly large.

The "zoom" lens, widely used in television and motion picture cameras, is a compound lens in which one or more of the individual lenses can be displaced along the lens axis, relative to the other lenses and the lens mounting. The objects for such a lens are at such large distances that they are imaged essentially in the second focal plane. Displacement of the movable lens elements leaves the second focal plane fixed relative to the mounting, but changes the positions of the principal points and hence changes the focal length f. The lateral magnification of an object at a distance x from the first focal point is

$$m = \frac{f}{x}.$$

Hence, an increase in f increases the image *size* without changing its *position*, so that the image remains in satisfactory focus.

41–13 THE PROJECTOR

The optical system of a projector for slides or motion pictures is illustrated in Fig. 41–23. The arrow at the left represents the light source, for example, the filament of a projection lamp. For simplicity, the slide to be projected is represented as opaque except for a single transparent aperture.

The diagram traces the course of three pencils of rays originating at the ends and at the midpoint of the source. The function of the condensing lens is to deviate the light from the source inward, so that it can pass through the projecting lens. If the condensing lens were omitted, light passing through the outer portions of the slide would not strike the projecting lens and only a small portion of the slide near its center would be imaged on the screen.

A study of the figure shows that for the three selected points of the source, only those rays within the shaded pencils can pass through the aperture, all others striking the condensing lens being intercepted by the opaque portions of the slide. Similar pencils of rays could be drawn from all other points of the source.

Each of these pencils converges, after passing through the aperture, to form an image of its point of origin just to the left of the projecting lens. In practice, this image would be formed *at* the projecting lens, but for clarity in the diagram the image and the lens have been displaced slightly. The focal length of the condensing lens should be such that the image of the source just fills the projecting lens. If the image of the source is larger than the projecting lens, some of the light passing through the slide is wasted. If it is smaller, the area of the projecting lens is not being fully utilized. Thus, in the diagram, the outer portions of the projecting lens serve no useful purpose.

Three rays tangent to the upper edge of the aperture have been emphasized in the figure. These rays originate at *different* points of the source. Hence, although they intersect at the edge of the aperture, this point of intersection does not constitute an image of any point of the source; but these three rays diverge from a common point of the *slide*, and therefore this point of the slide is imaged, as shown on the screen. Similarly, rays tangent to any point of the edge of the aperture are imaged at a corresponding point on the screen. Thus, if the aperture is circular, a circular spot of light appears on the screen. Note that light from *all* points of the source illuminates *every* point of the image of the aperture, and would do the same were the aperture at any other point of the slide.

The preceding discussion has explained the conditions that determine the *focal length* of the condensing lens and the *diameter* of the projecting lens (the image of the source formed by the condensing lens should just fill the projecting lens). The *diameter* of the condensing lens must evidently be at least as great as the diagonal of the largest slide to be projected, while the *focal length* of the projecting lens is determined by the magnification desired between the slide and its image, and the distance of the projector from the screen.

41–14 THE COMPOUND MICROSCOPE

When an angular magnification higher than that attainable with a simple magnifier is desired, it is necessary to use a *compound microscope*, usually called merely a *microscope*. The essential elements of a microscope are illustrated in Fig. 41–24. The object O to be examined is placed just beyond the first focal point F_1 of the *objective* lens, which forms a real and enlarged image I. This image lies just within the first focal point F_2 of the ocular, which forms a virtual image of I at I'. As was stated earlier, the position of I' may be anywhere between the near and far points of the eye. Although both the objective and ocular of an actual microscope are highly corrected compound lenses, they are shown as simple thin lenses for simplicity.

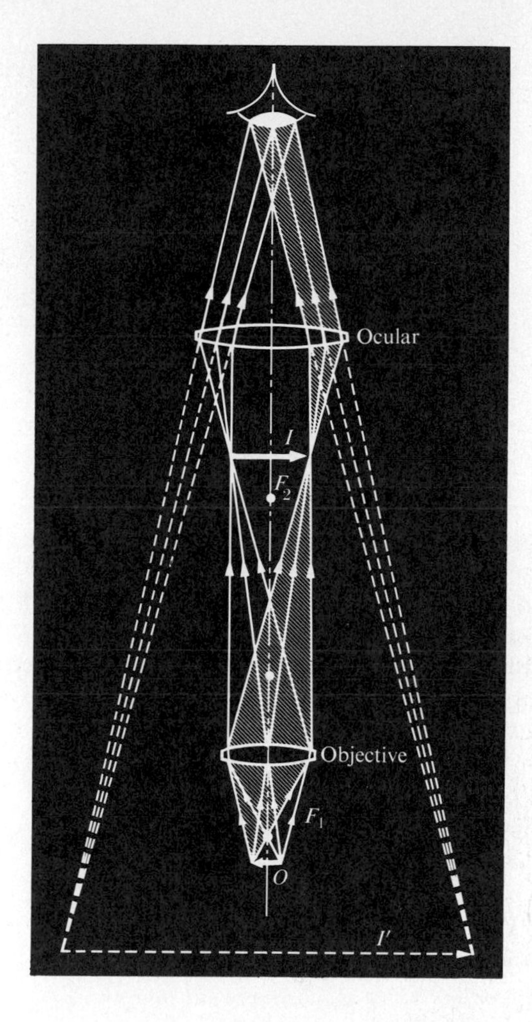

Fig. 41–24 *The microscope.*

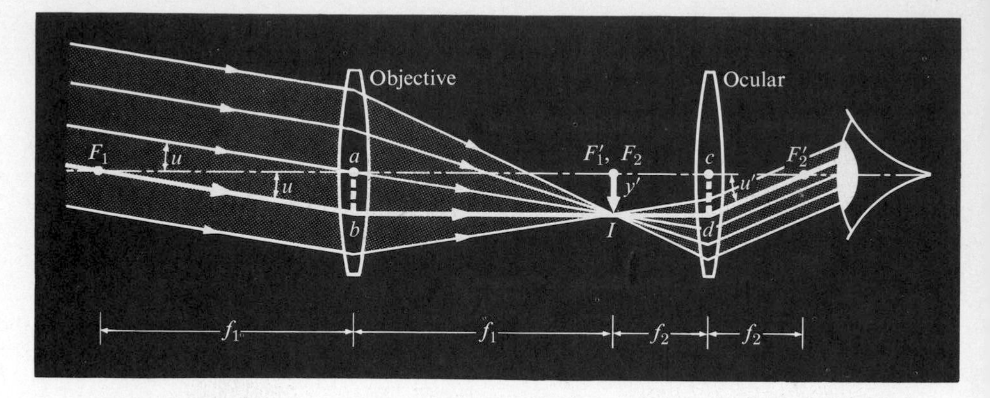

***Fig. 41–25** Telescope; final image at infinity.*

Since the objective merely forms an enlarged real image which is examined by the ocular, the overall magnification M of the compound microscope is the product of the lateral magnification m_1 of the objective and the angular magnification M_2 of the ocular. The former is given by

$$m_1 = -\frac{x'}{f_1},$$

where f_1 is the focal length of the objective and x' is the image distance measured from its second focal point. The angular magnification of the ocular (considered as a simple lens) is $M_2 = 25/f_2$, where f_2 is the focal length of the ocular. Hence, the overall magnification M of the compound microscope is

$$\boxed{M = m_1 \times M_2 = -\frac{x'}{f_1} \times \frac{25}{f_2}} \qquad (41\text{–}8)$$

(where x', f_1, f_2 are measured in cm). It has become customary among microscope manufacturers to specify the values of m_1 and M_2, rather than the focal lengths of objective and ocular.

41–15 TELESCOPES

The optical system of a refracting telescope is essentially the same as that of a compound microscope. In both instruments, the image formed by an objective is viewed through an ocular. The difference is that the telescope is used to examine large objects at large distances and the microscope to examine small objects close at hand.

The *astronomical* telescope is illustrated in Fig. 41–25. The objective lens forms a real, reduced image I of the object O. I' is the virtual image of I formed by the ocular. As with the microscope, the image I' may be formed anywhere between the near and far points of the eye. In practice, the objects examined by a telescope are at such large distances from the instrument that the image I is formed very nearly at the second focal point of the objective. Furthermore, if the image I' is at infinity, the image I is at the first focal point of the ocular. The distance between objective and ocular, or the length of the telescope, is therefore the *sum* of the focal lengths of objective and ocular, $f_1 + f_2$.

The angular magnification of a telescope is defined as the ratio of the angle subtended at the eye by the final image I', to the angle subtended at the (unaided) eye by the object. This ratio may be expressed in terms of the focal lengths of objective and ocular as follows. In Fig. 41–25, the ray passing through F_1, the first focal point of the objective, and through F_2', the second focal point of the ocular, has been emphasized. The object (not shown) subtends an angle u at the objective and would subtend essentially the same angle at the unaided eye. Also, since the observer's eye is placed just to the right of the focal point F_2', the angle subtended at the eye by the final image is very nearly equal to the angle u'. The distances ab and cd are evidently equal to each

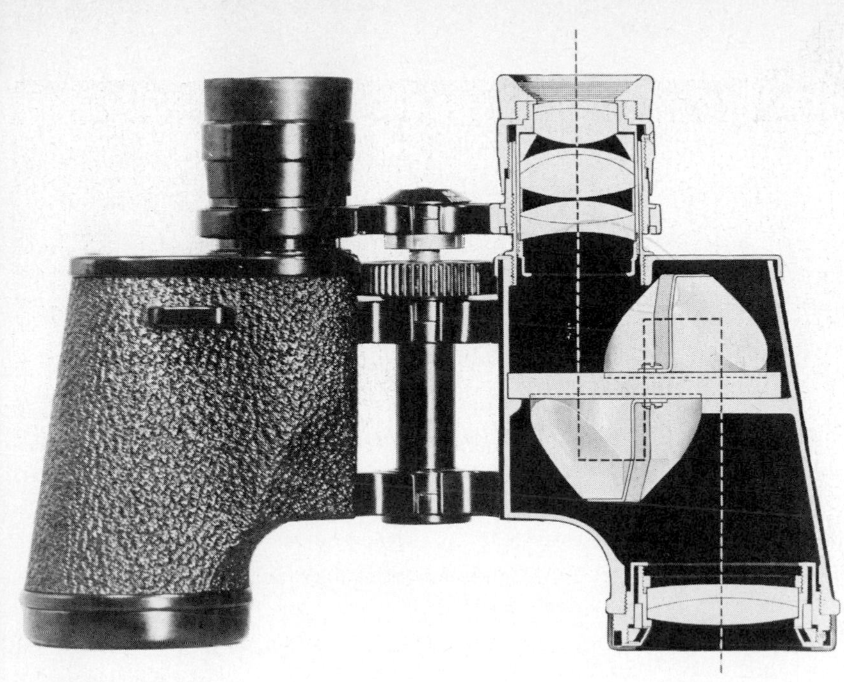

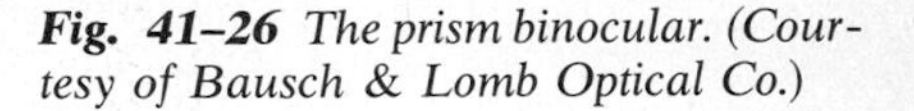

Fig. 41–26 *The prism binocular. (Courtesy of Bausch & Lomb Optical Co.)*

other and to the height y' of the image I. Since u and u' are small, they may be approximated by their tangents. From the right triangles F_1ab and $F_2'cd$,

$$u = \frac{-y'}{f_1}, \qquad u' = \frac{y'}{f_2}.$$

Hence,

$$\boxed{M = \frac{u'}{u} = -\frac{y'/f_2}{y'/f_1} = -\frac{f_1}{f_2}.} \tag{41–9}$$

The angular magnification of a telescope is therefore equal to the ratio of the focal length of the objective to that of the ocular. The minus sign denotes an inverted image.

An inverted image is not a disadvantage if the instrument is to be used for astronomical observations, but it is desirable that a terrestrial telescope form an erect image. This is accomplished in the *prism binocular*, of which Fig. 41–26 is a cutaway view, by a pair of 45°-45°-90° totally reflecting prisms inserted between objective and ocular. The image is inverted by the four reflections from the

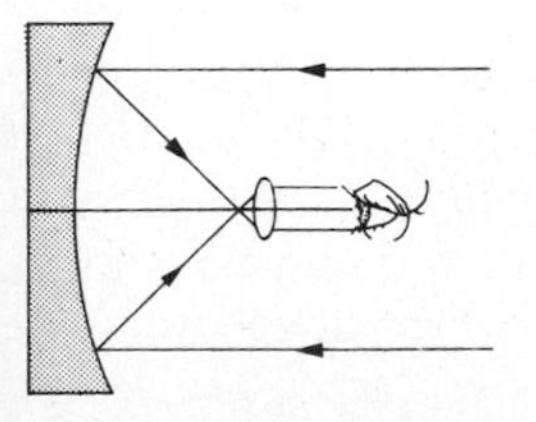

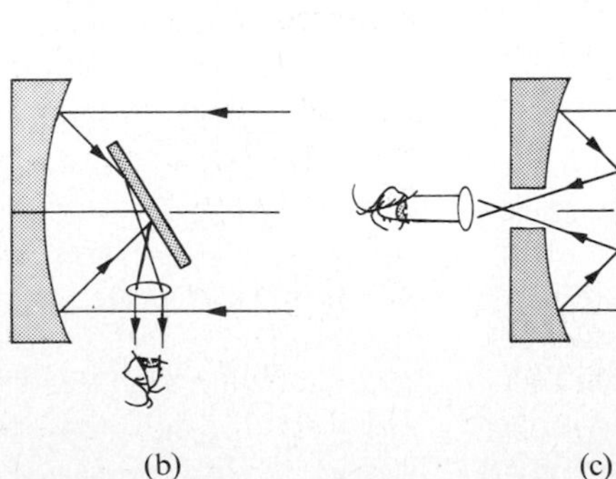

(a) (b) (c)

Fig. 41–27 *The reflecting telescope.*

inclined faces of the prisms. It is customary to stamp on a flat metal surface of a binocular two numbers separated by a multiplication sign, thus, 7 × 50. The first number is the magnification and the second is the diameter of the objective lenses in millimeters.

In the *reflecting telescope* the objective lens is replaced by a concave mirror, as shown in Fig. 41–27. In large telescopes this scheme has many advantages, both theoretical and practical. The mirror is intrinsically free of chromatic aberrations, and spherical aberrations are much easier to correct than with a lens. The material need not be transparent, and the reflector can be made more rigid than a lens, which has to be supported only at its edges. The largest reflecting telescope in the world has a mirror over 5 m in diameter.

Because the image is formed in a region traversed by incoming rays, this image can be observed directly with an ocular only by blocking off part of the incoming beam; this is practical only for the very largest telescopes. Alternate schemes use a mirror to reflect the image out the side or through a hole in the mirror, as shown in Figs. 41–26(b) and 41–26(c).

PROBLEMS

41–1 A thin-walled glass sphere of radius R is filled with water. An object is placed a distance $3R$ from the surface of the sphere. Determine the position of the final image. The effect of the glass wall may be neglected.

41–2 A transparent rod 40 cm long is cut flat at one end and rounded to a hemispherical surface of 12 cm radius at the other end. An object is placed on the axis of the rod, 10 cm from the hemispherical end.

a) What is the position of the final image?
b) What is its magnification? Assume the refractive index to be 1.50.

41–3 Both ends of a glass rod 10 cm in diameter, of index 1.50, are ground and polished to convex hemispherical surfaces of radius 5 cm at the left end and radius 10 cm at the right end. The length of the rod between vertices is 60 cm. An arrow 1 mm long, at right angles to the axis and 20 cm to the left of the first vertex, constitutes the object for the first surface.

a) What constitutes the object for the second surface?
b) What is the object distance for the second surface?
c) Is the object real or virtual?
d) What is the position of the image formed by the second surface?
e) What is the height of the final image?

41–4 The same rod as in Problem 41–3 is now shortened to a distance of 10 cm between its vertices, the curvatures of its ends remaining the same.

a) What is the object distance for the second surface?
b) Is the object real or virtual?
c) What is the position of the image formed by the second surface?
d) Is the image real or virtual, erect, or inverted, with respect to the original object?
e) What is the height of the final image?

41–5 A glass rod of refractive index 1.50 is ground and polished at both ends to hemispherical surfaces of 5 cm radius. When an object is placed on the axis of the rod and 20 cm from one end, the final image is formed 40 cm from the opposite end. What is the length of the rod?

41–6 A solid glass sphere of radius R and index 1.50 is silvered over one hemisphere, as in Fig. 41–28. A small object is located on the axis of the sphere at a distance $2R$ from the pole of the unsilvered hemisphere. Find the position of the final image after all refractions and reflections have taken place.

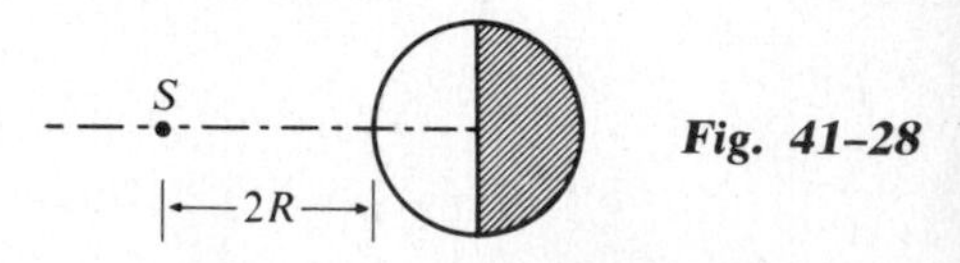

Fig. 41–28

41–7 A narrow beam of parallel rays enters a solid glass sphere in a radial direction. At what point outside the sphere are these rays brought to a focus? The radius of the sphere is 3 cm and its index is 1.50.

41–8 A glass plate 2 cm thick, of index 1.50, having plane parallel faces, is held with its faces horizontal and its lower face 8 cm above a printed page. Find the position of the image of the page, formed by rays making a small angle with the normal to the plate.

41–9 a) Show that the equation

$$\frac{1}{s} + \frac{1}{s'} = \frac{1}{f}$$

is that of an equilateral hyperbola having as asymptotes the lines $x = f$ and $y = f$.

b) Construct a graph with object distance s as abscissa, and image distance s' as ordinate for a lens of focal length f, and for object distances from 0 to ∞.

c) On the same set of axes, construct a graph of magnification (ordinate) *vs.* object distance.

41–10 A converging lens has a focal length of 10 cm. For object distances of 30 cm, 20 cm, 15 cm, and 5 cm determine (a) image position, (b) magnification, (c) whether the image is real or virtual, (d) whether the image is erect or inverted.

41–11 Sketch the various possible thin lenses obtainable by combining two surfaces whose radii of curvature are, in absolute magnitude, 10 cm and 20 cm. Which are converging and which are diverging? Find the focal length of each lens if made of glass of index 1.50.

41–12 The radii of curvature of the surfaces of a thin lens are +10 cm and +30 cm. The index is 1.50.

a) Compute the position and size of the image of an object in the form of an arrow 1 cm high, perpendicular to the lens axis, 40 cm to the left of the lens.

b) A second similar lens is placed 160 cm to the right of the first. Find the position of the final image.

c) Same as (b), except the second lens is 40 cm to the right of the first.

d) Same as (c), except the second lens is diverging, of focal length −40 cm.

41–13 An object is placed 18 cm from a screen.

a) At what points between object and screen may a lens of 4 cm focal length be placed to obtain an image on the screen?

b) What is the magnification of the image for these positions of the lens?

41–14 An object is imaged by a lens on a screen placed 12 cm from the lens. When the lens is moved 2 cm farther from the object, the screen must be moved 2 cm closer to the object to refocus it. What is the focal length of the lens?

41–15 Three thin lenses, each of focal length 20 cm, are aligned on a common axis, and adjacent lenses are separated by 30 cm. Find the position of the image of a small object on the axis, 60 cm to the left of the first lens.

41–16 An equiconvex thin lens made of glass of index 1.50 has a focal length in air of 30 cm. The lens is sealed into an opening in one end of a tank filled with water (index = 1.33). At the end of the tank opposite the lens is a plane mirror, 80 cm distant from the lens. Find the position of the image formed by the lens-water-mirror system, of a small object outside the tank on the lens axis and 90 cm to the left of the lens. Is the image real or virtual, erect or inverted?

41–17 A diverging meniscus lens of 1.48 refractive index has spherical surfaces whose radii are 2.5 and 4 cm. What would be the position of the image if an object were placed 15 cm in front of the lens?

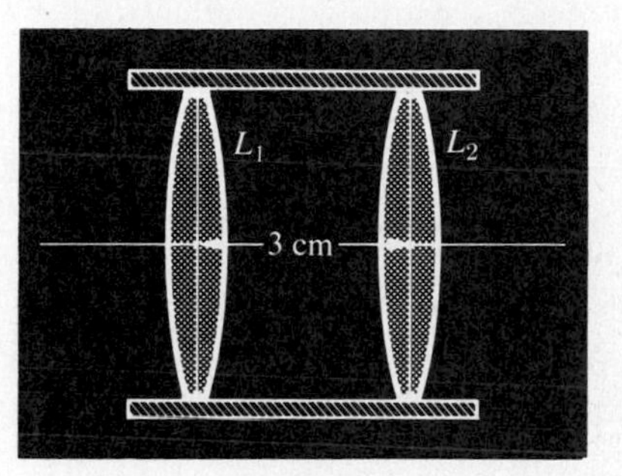

Fig. 41–29

41–18 Figure 41–29 represents a compound lens consisting of two thin lenses L_1 and L_2 each of focal length 6 cm, separated by a distance of 3 cm.

a) Find the position of the image formed by lens L_1 of a point on the axis at an infinite distance to the left of the lens.

b) Find the position of the image of this image formed by lens L_2. This locates the second focal point F' of the compound lens and, by symmetry, the first focal point F lies at the same distance to the left of L_1.

c) Find the position of the image formed by L_1 of a point on the axis at a distance $x = 2$ cm to the left of the first focal point F.

d) Find the distance x' from the second focal point F' to the image of this image formed by lens L_2.

e) What is the focal length f of the compound lens?

f) Construct a full-size diagram showing the positions of the focal points and principal points of the compound lens, and use the method illustrated in Fig. 41–15 to locate graphically the image of an object 4 cm to the left of the first focal point F.

41–19 A plano-convex lens is 2 cm thick along its axis. The refractive index is 1.50 and the radius of curvature of the convex surface is 10 cm. The convex surface faces toward the left.

a) Find the distance from the first focal point to the vertex of the convex surface.

b) Find the distance from the vertex on the plane surface to the second focal point.

41–20 When an object is placed at the proper distance in front of a converging lens, the image falls on a screen 20 cm from the lens. A diverging lens is now placed halfway between the converging lens and the screen, and it is found that the screen must be moved 20 cm farther away from the lens to obtain a sharp image. What is the focal length of the diverging lens?

41–21 a) Prove that when two thin lenses of focal lengths f_1 and f_2 are placed *in contact*, the focal length f of the combination is given by the relation

$$\frac{1}{f} = \frac{1}{f_1} + \frac{1}{f_2}.$$

b) A converging meniscus lens has an index of refraction of 1.50, and the radii of its surfaces are 5 and 10 cm. The concave surface is placed upward and filled with water. What is the focal length of the water–glass combination?

41–22 Two thin lenses, both of 10 cm focal lengths, the first converging, the second diverging, are placed 5 cm apart. An object is placed 20 cm in front of the first (converging) lens.

a) How far from this lens will the image be formed?

b) Is the image real or virtual?

41–23 Rays from a lens are converging toward a point image P, as in Fig. 41–30. What thickness t of glass of index 1.50 must be interposed, as in the figure, in order that the image shall be formed at P'?

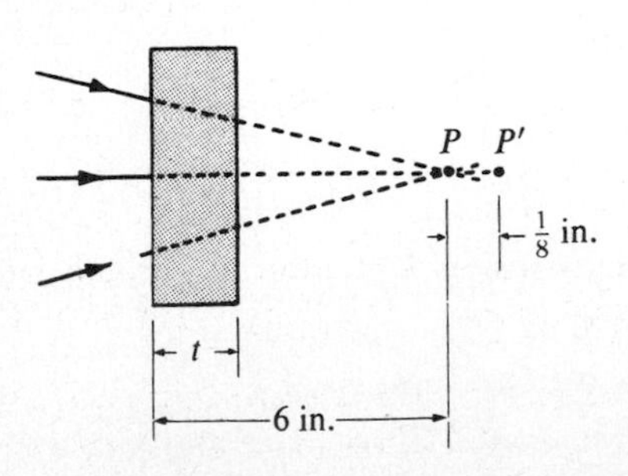

Fig. 41–30

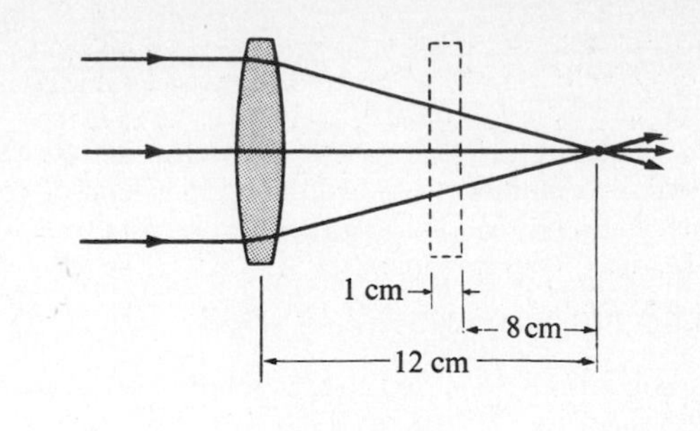

Fig. 41–31

41–24 An object 2.4 m in front of a camera lens is sharply imaged on a photographic film 12 cm behind the lens. A glass plate 1 cm thick, of index 1.50, having plane parallel faces, is interposed between lens and plate, as shown in Fig. 41–31.

a) Find the new position of the image.

b) At what distance in front of the lens will an object be in sharp focus on the film with the plate in place, the distance from lens to film remaining 12 cm? Consider the lens as a simple thin lens.

41–25 The picture size on ordinary 35-mm camera film is 24 × 36 mm. Focal lengths of lenses available for 35-mm cameras typically include 28 mm, 35 mm, 50 mm (the "standard" lens), 85 mm, 100 mm, 135 mm, 200 mm, and 300 mm, among others. Which of these lenses should be used to photograph the following objects, assuming the object is to fill most of the picture area?

a) A cathedral 100 m high and 150 m long, at a distance of 150 m?

b) An eagle with wingspan 2.0 m, at a distance of 15 m?

41–26 During a lunar eclipse, a picture of the moon (diameter 3.48×10^6 m, distance from earth 3.8×10^8 m) is taken with a camera whose lens has focal length 50 mm. What is the diameter of the image on the film?

41–27 The *resolution* of a camera lens can be defined as the maximum number of lines per millimeter in the image that can barely be distinguished as separate lines. A certain lens has a focal length of 50 mm and resolution of 100 lines mm^{-1}. What is the minimum separation of two lines in an object 100 m away if they are to be visible in the image as separate lines?

41–28 Show that when two thin lenses are placed in contact, the *power* of the combination in diopters, as defined in Section 41–10, is the sum of the powers of the separate lenses. Is this relation valid even when one lens has positive power and the other negative?

41–29 An ocular consists of two similar positive thin lenses having focal lengths of 6 cm, separated by a distance of 3 cm. Where are the focal points of the ocular?

41–30 When two thin lenses are closely spaced, the power of the combination is the sum of the powers of the individual lenses. Two thin lenses of 25 cm and 40 cm focal lengths are in contact. What is the power of the combination?

41–31 What is the power of the spectacles required (a) by a hyperopic eye whose near point is 125 cm? (b) by a myopic eye whose far point is 50 cm?

41–32 a) What spectacles are required for reading purposes by a person whose near point is at 200 cm?

b) The far point of a myopic eye is at 30 cm. What spectacles are required for distant vision?

41–33

a) Where is the near point of an eye for which a spectacle lens of power +2 diopters is prescribed?

b) Where is the far point of an eye for which a spectacle lens of power −0.5 diopter is prescribed for distant vision?

41–34 A thin lens of focal length 10 cm is used as a simple magnifier.

a) What angular magnification is obtainable with the lens?

b) When an object is examined through the lens, how close may it be brought to the eye?

41–35 The focal length of a simple magnifier is 10 cm.

a) How far in front of the magnifier should an object to be examined be placed if the image is formed at the observer's near point, 25 cm in front of his eye?

b) If the object is 1 mm high, what is the height of its image formed by the magnifier? Assume the magnifier to be a thin lens.

41–36 A camera lens is focused on a distant point source of light, the image forming on a screen at *a* (Fig. 41–32). When the screen is moved backward a distance of 2 cm to *b*, the circle of light on the screen has a diameter of 4 mm. What is the *f*/number of the lens?

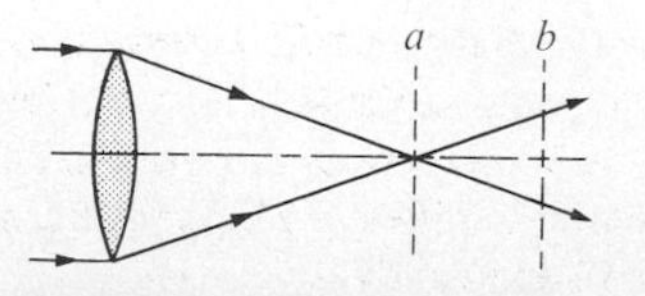

Fig. 41–32

41–37 Camera *A*, having an *f*/8 lens 2.5 cm in diameter, photographs an object using the correct exposure of 1/100 s. What exposure should camera *B* use in photographing the same object if it has an *f*/4 lens 5 cm in diameter?

41–38 The focal length of an *f*/2.8 camera lens is 8 cm.

a) What is the diameter of the lens?

b) If the correct exposure of a certain scene is 1/200 s at *f*/2.8, what would be the correct exposure at *f*/5.6?

41–39 The dimensions of the picture on a 35-mm color slide are 24 mm × 36 mm. It is desired to project an image of the slide, enlarged to 2 m × 3 m, on a screen 10 m from the projection lens.

a) What should be the focal length of the projection lens?

b) Where should the slide be placed?

41–40 The image formed by a microscope objective of focal length 4 mm is 180 mm from its second focal point. The ocular has a focal length of 31.25 mm.

a) What is the magnification of the microscope?

b) The unaided eye can distinguish two points as separate if they are about 0.1 mm apart. What is the minimum separation, using this microscope?

41–41 A certain microscope is provided with objectives of focal lengths 16 mm, 4 mm, and 1.9 mm, and with oculars of angular magnification 5X and 10X. What is (a) the largest, and (b) the least overall magnification obtainable? Each objective forms an image 160 mm beyond its second focal point.

41–42 The focal length of the ocular of a certain microscope is 2.5 cm. The focal length of the objective is 16 mm. The distance between objective and ocular is 22.1 cm. The final image formed by the ocular is at infinity. Treat all lenses as thin.

a) What should be the distance from the objective to the object viewed?

b) What is the linear magnification produced by the objective?

c) What is the overall magnification of the microscope?

41–43 A microscope with an objective of focal length 9 mm and an ocular of focal length 5 cm is used to project an image on a screen 1 m from the ocular. What is the lateral magnification of the image? Let $x' = 18$ cm.

41–44 The moon subtends an angle at the earth of approximately $\frac{1}{2}^\circ$. What is the diameter of the image of the moon produced by the objective of the Lick Observatory telescope, a refractor having a focal length of 18 m?

41–45 The ocular of a telescope has a focal length of 10 cm. The distance between objective and ocular is 2.1 m. What is the angular magnification of the telescope?

41–46 A crude telescope is constructed of two spectacle lenses of focal lengths 100 cm and 20 cm, respectively.

a) Find its angular magnification.
b) Find the height of the image formed by the objective of a building 80 m high and distant 2 km.

41–47 Figure 41–33 is a diagram of a *Galilean telescope*, or *opera glass*, with both the object and its final image at infinity. The image I serves as a virtual object for the ocular. The final image is virtual and erect. Prove that the angular magnification $M = -f_1/f_2$.

41–48 A Galilean telescope is to be constructed, using the same objective as in Problem 41–46.

a) What type lens should be used as an ocular and what focal length should it have, if the telescopes are to have the same magnification?
b) Compare the lengths of the telescopes.

41–49 A reflecting telescope is made using a mirror of radius of curvature 0.50 m and an ocular of focal length 10 cm. What is the lateral magnification? What should be the position of the ocular if both the object and the final image are at infinity?

41–50 A certain reflecting telescope has a mirror 10 cm in diameter, with radius of curvature 1.0 m, and an ocular of focal length 1.0 cm. If the lateral magnification is 40, find the position of the lens and the position and nature of the final image.

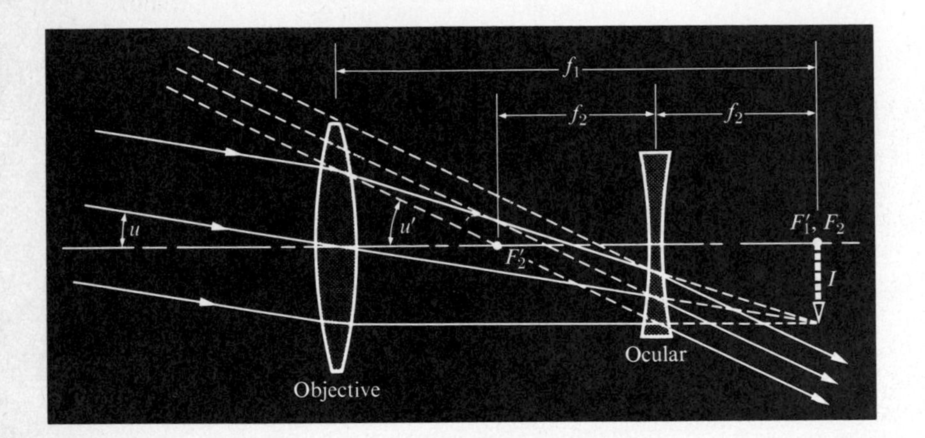

Fig. 41–33

Chapter 42

Interference and Diffraction

42–1 INTERFERENCE AND COHERENT SOURCES

In analyzing the formation of images by lenses and mirrors, we have represented light as *rays* which travel in straight lines in a homogeneous medium and which are deviated in accordance with simple laws at a reflecting surface or an interface between two optical media. This simple model forms the basis of *geometrical optics*, which, as we have seen, is adequate for understanding a wide variety of phenomena involving lenses and mirrors.

In this chapter we shall discuss the phenomena of *interference* and *diffraction*, for whose understanding the principles of geometrical optics *do not* suffice. Instead, we must return to the more fundamental point of view that light is a *wave motion*, and that the total effect of a number of waves arriving at one point depends on the *phases* of the waves as well as upon their amplitudes. This part of the subject is called *physical optics.*

In our discussions of mechanical waves in Chapter 21 and electromagnetic waves in Chapter 37, we have often considered *sinusoidal* waves having a single frequency and a single wavelength. Such a wave is called a *monochromatic* (single-color) wave. Common sources of light, such as an incandescent light bulb or a flame, *do not* emit monochromatic light but rather a continuous distribution of wavelengths. Indeed, a strictly monochromatic light wave is an unattainable idealization, like a frictionless plane, a point mass, an inductor with zero resistance, and many other idealizations in physics. Nevertheless, it remains a useful concept for the analysis of phenomena of light interference.

Monochromatic light can be *approximated* in the laboratory. Continuous-spectrum light can be passed through a filter which blocks all but a narrow range of wavelengths. Gas-discharge lamps, such as the mercury-arc lamp, emit line spectra in which the light consists of a discrete set of colors, each having a narrow band of wavelengths called a *spectrum line.* For example, the bright green line in the mercury spectrum has an average wavelength of 546.1 nm, with a spread of wavelength of the order of ±0.001 nm, depending on the pressure and temperature of the mercury vapor in the lamp. By far the most nearly monochromatic source available at present is the *laser*, to be discussed in Chapter 44. The familiar helium–neon laser, inexpensive and readily available, emits visible light at 632.8 nm with a line width (wavelength range) of the order of ±0.000001 nm, or about one part in 10^9. Laser light also has much greater *coherence* (to be discussed later) than ordinary light.

The term *interference* refers to any situation in which two or more waves overlap in space. This term was introduced in Section 22–2 in connection with standing waves on a stretched string, formed by the superposition of two sinusoidal waves traveling in opposite directions. In such cases, the total displacement at any point at any instant of time is governed by the *principle of linear superposition*, introduced in Section 22–2. This principle, the most important in all of physical optics, states that *when two or more waves overlap, the resultant displacement at any point and at any instant may be found by adding the instantaneous displacements that would be produced at the point by the individual waves if each were present alone.* The term "displacement" as used here is a general one. If one is considering surface ripples on a liquid, the displacement means the actual displacement of the surface above or below its normal level. If the waves are sound waves, the term refers to the excess or deficiency of pressure. If the waves are electromagnetic, the displacement means the magnitude of the electric or magnetic field. When light of extremely high intensity passes through matter, the principle of linear superposition is *not* precisely obeyed, and the resulting phenomena are classified under the heading *nonlinear optics.* (These effects are beyond the scope of this book.)

To introduce the essential ideas of interference, we consider first the problem of two identical sources of monochromatic waves, S_1 and S_2, separated in space by a certain distance. The two sources are permanently *in phase*, so that at every point in space there is a definite and unchanging phase relation between waves from the two sources. They might be, for example, two loudspeakers driven by the same amplifier, or two radio antennas powered by the same transmitter, or two small apertures in an opaque screen, illuminated by the same monochromatic light source.

We locate the two sources S_1 and S_2 along the x-axis, as shown in Fig. 42–1. Let P be a point in a vertical plane for which the path difference $PS_1 - PS_2$ is some whole number of wavelengths $m\lambda$ ($m = 0, 1, 2, 3, \ldots$). Two waves starting from points S_1 and S_2 in phase, will arrive at point P in phase and will reinforce at this point. Reinforcement will also take place at point P' in a horizontal plane if

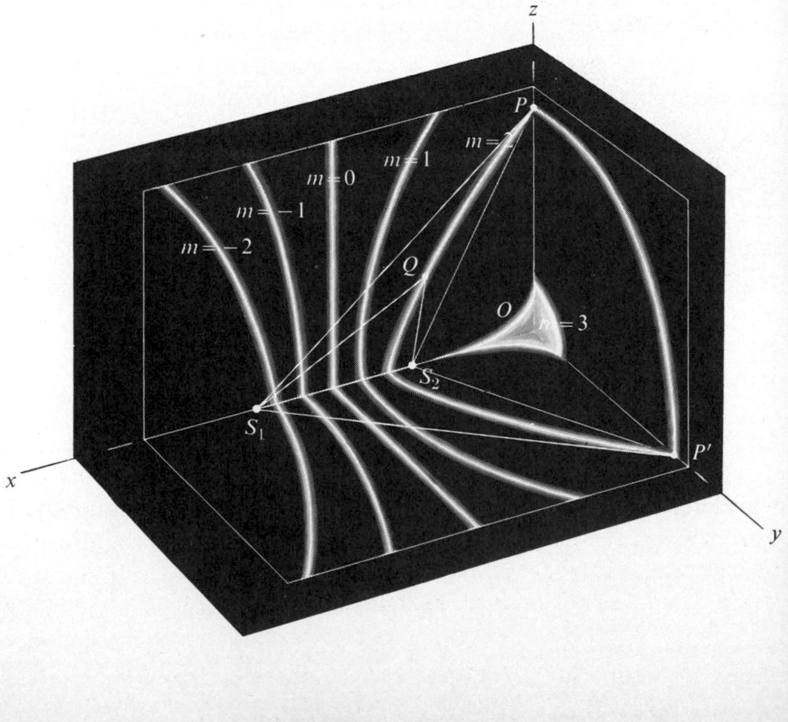

Fig. 42–1 *Curves of maximum intensity in the interference pattern of two monochromatic point sources.*

$P'S_1 = PS_1$ and $P'S_2 = PS_2$. In fact, at all points of the circle through P and P', with center at O, the waves from the two sources will reinforce, and a bright circular region of reinforcement will appear on a vertical screen through P and P', perpendicular to the line xO. Close to this region of maximum reinforcement will be circular regions of lesser intensity where the waves come together neither in phase nor in opposite phase. These regions are terminated by a black circle where there is complete destructive interference because the waves arrive in opposite phase. The intensity distribution on a screen from dark to bright to dark again is called an *interference fringe.*

Point Q in Fig. 42–1 is another point for which the path difference $QS_1 - QS_2$ is also $m\lambda$, and the curve passing through P and Q is the locus of all such points in the vertical plane passing through xO. This curve is a hyperbola, a curve having the property that the difference between the distances from any point on it to two fixed points is a constant. If we imagine this hyperbola to be rotated about the line xO as an axis, it sweeps out a surface called a *hyperboloid,* and we see that the waves from S_1 and S_2 will arrive at *all* points of this surface so as to reinforce each other.

The diagram of Fig. 42–1 has been drawn for the simple case where the distance from S_1 to S_2 is 3λ, and where the path difference for point P is

$$PS_1 - PS_2 = 2\lambda \qquad (m = 2).$$

The hyperbola lettered $m = 1$ is the locus of points in a vertical plane for which the path difference is λ. The locus of all points for which the path difference is zero ($m = 0$) is a line passing through the midpoint of $S_1 S_2$. Hyperbolas for which $m = -1$ and $m = -2$ are also shown. Rotation of these curves about xO gives rise, in this case, to five hyperboloids. Bright lines will appear on a screen in any position along those curves where the hyperboloids intersect the screen.

If the distance between the sources is many wavelengths, there will be a large number of surfaces over which the waves reinforce, and a large number of alternate bright and dark hyperbolic (almost straight) fringes will be formed on a screen parallel to the line joining the sources. Also, a great number of alternate bright and dark circular fringes will be formed on a screen perpendicular to the line joining the sources.

In the above discussion the constant phase relationship between the sources is an essential requirement. If the relative phase of the sources changes, the positions of the maxima and minima in the resulting interference pattern also change. When the radiation is light, it is possible for the two sources to have a definite and constant phase relation *only when they both emit light coming from a single primary source;* it is *not* possible with two separate sources. The reason is a fundamental one associated with the mechanism of light emission.

In ordinary light sources, atoms of the material of the source are given excess energy by thermal agitation or impact with accelerated electrons. An atom thus "excited" begins to radiate and continues until it has lost all the energy it can, typically in a time of the order of 10^{-8} s. A source ordinarily contains a very large number of atoms, which radiate in an unsynchronized and random phase relationship. Thus emission from two such sources has a rapidly varying phase relation; the result is a constantly changing interference pattern which, with ordinary observations, does not reveal a visible interference pattern at all.

However, if the light from a single source is split so that parts of it emerge from two or more regions of space, forming two or more *secondary sources,* any random phase change in the source affects these secondary sources equally and does not change their *relative* phase. Two such sources derived from a single source and having a definite phase relation are said to be *coherent.*

The distinguishing feature of light from a *laser* is that the emission of light from many atoms is *synchronized* in frequency and phase, by mechanisms to be discussed in Chapter 44. As a result, the random phase changes mentioned above occur *much* less frequently. Definite phase relations are preserved over correspondingly much greater lengths in the beam. Accordingly, laser light is said to be much more *coherent* than ordinary light.

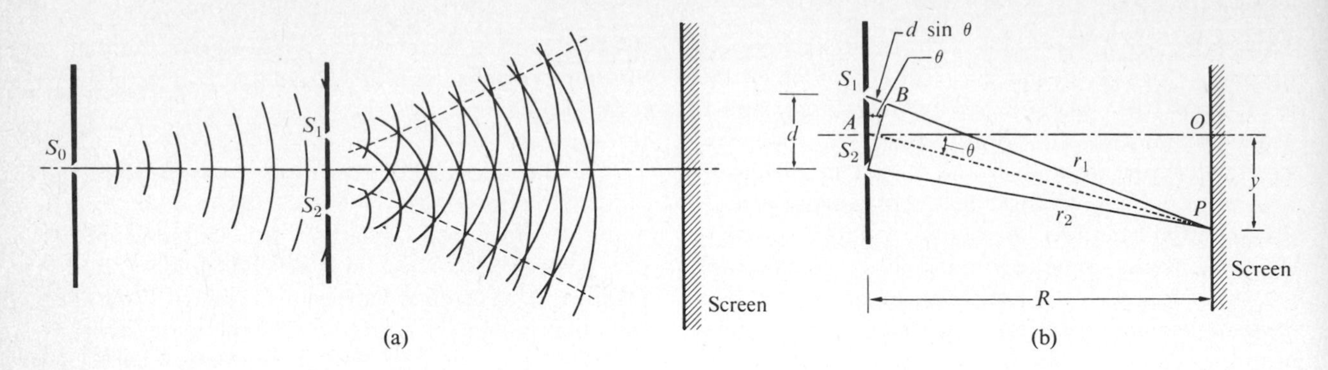

Fig. 42-2 (a) Interference of light waves passing through two slits. (b) Young's experiment.

42–2 YOUNG'S EXPERIMENT AND POHL'S EXPERIMENT

One of the earliest demonstrations of the fact that light can produce interference effects was performed in 1800 by the English scientist Thomas Young. The experiment was a crucial one at the time, since it added further evidence to the growing belief in the wave nature of light. A corpuscular theory was quite inadequate to account for the effects observed.

Young's apparatus is shown in Fig. 42–2(a). Monochromatic light issuing from a narrow slit S_0 is divided into two parts by falling upon a screen in which are cut two other narrow slits S_1 and S_2, very close together. The dimensions in this figure are distorted for clarity. The distance from the source slit S_0 to the screen containing S_1 and S_2 is 20 cm to 100 cm. The distance from the double-slit screen to the final screen is usually from 1 m to 5 m. The slits are 0.1 mm to 0.2 mm wide and the separation of the slits S_1 and S_2 is less than 1 mm. In short, all slit widths and slit separations are fractions of millimeters; all other distances are hundreds or thousands of millimeters.

According to Huygens' principle, cylindrical wavelets such as those shown in Fig. 39–2 spread out from slit S_0 and reach slits S_1 and S_2 at the same instant. A train of Huygens' wavelets diverges from each slit; therefore they act as *coherent* sources. Let d represent the distance between the slits, and consider a point P on the screen, in a direction making an angle θ with the axis of the system (Fig. 42–2(b)). With P as a center and PS_2 as radius, strike an arc intersecting PS_1 at B. If the distance R from slits to screen is large in comparison with the distance d between the slits, the arc S_2B can be considered a straight line at right angles to PS_2, PA, and PS_1.

Fig. 42–3 Interference fringes produced by Young's double-slit inteferometer.

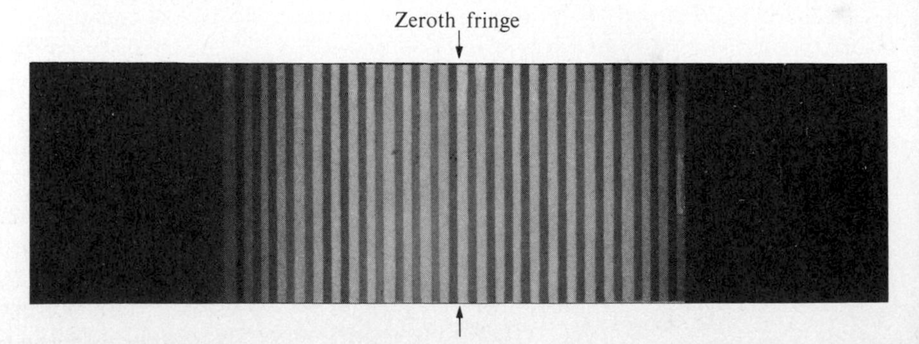

Then the triangle BS_1S_2 is a right triangle, similar to POA, and the distance S_1B equals $d \sin \theta$. This latter distance is the difference in path length $r_1 - r_2$ between the waves reaching P from the two slits. The waves spreading out from S_1 and S_2 necessarily start in phase, but they may not be in phase at P because of this difference in length of path. According to the principles discussed in the preceding section, complete reinforcement will take place at the point P (that is, P will lie at the center of a bright fringe) only when the path difference $d \sin \theta$ is some integral number of wavelengths, say $m\lambda (m = 0, 1, 2, 3, \text{etc.})$. Thus

$$d \sin \theta = m\lambda \qquad \text{or} \qquad \sin \theta = \frac{m\lambda}{d}. \tag{42–1}$$

Now λ is of the order 5×10^{-5} cm, while d cannot be made much smaller than about 10^{-2} cm. As a rule, only the first five to ten fringes are bright enough to be seen, so that m is at most, say, 10. Therefore the very largest value of $\sin \theta$ is

$$\sin \theta \text{ (maximum)} = \frac{10 \times 5 \times 10^{-5} \text{ cm}}{10^{-2} \text{ cm}} = 0.05,$$

which corresponds to an angle of only 3°. Figure 42–3 shows a typical pattern.

The central bright fringe at point O, or *zeroth* fringe ($m = 0$), corresponds to zero path difference, or $\sin \theta = 0$. If point P is at the center of the mth fringe, the distance y_m from the zeroth to the mth fringe is, from Fig. 42–2(b),

$$y_m = R \tan \theta_m.$$

Since, however, the angle θ_m for all values of m is extremely small, $\tan \theta_m \approx \sin \theta_m$ and

$$y_m = R \sin \theta_m.$$

Therefore,

$$y_m = R\frac{m\lambda}{d},$$

and

$$\lambda = \frac{y_m d}{mR}. \tag{42–2}$$

Hence, by measuring the distance d between the slits, the distance R to the screen, and the distance y_m from the center of the zeroth fringe to the center of the mth fringe on either side, one may compute the wavelength of the light producing the interference pattern.

Example With two slits spaced 0.2 mm apart, and a screen at a distance of 1 m, the third bright fringe is found to be displaced 7.5 mm from the central fringe. Find the wavelength of the light used.

Let λ be the unknown wavelength. Then

$$\lambda = \frac{y_m d}{mR} = \frac{0.75 \text{ cm} \times 0.02 \text{ cm}}{3 \times 100 \text{ cm}} = 5 \times 10^{-5} \text{ cm}$$

$$= 500 \times 10^{-9} \text{ m} = 500 \text{ nm}.$$

Circular interference fringes may be produced very easily with the aid of a simple apparatus suggested by Robert Pohl and shown in Fig. 42–4. A small arc lamp S_0 is placed a few centimeters away from a sheet of mica of thickness about 0.05 mm. Some light is reflected from the first surface, as though it were issuing from the virtual image S_1. An approximately equal amount of light is reflected from the back surface, as though it were coming from the virtual image S_2. The circular interference fringes formed by the light issuing from these two mutually coherent sources may be shown on the entire wall of a room, as shown in Fig. 42–5.

42–3 INTENSITY DISTRIBUTION IN INTERFERENCE FRINGES

In Section 42–2 we computed the positions of the maxima (bright fringes) in the two-slit interference pattern; we may also compute the intensity at *any* point in the pattern. In Fig. 42–2(b), each source produces a sinusoidally-varying electric field at point P; and if the sources are in phase, the fields arriving at P differ in phase by an amount proportional to the

Fig. 42-4 Pohl's mica-sheet interferometer.

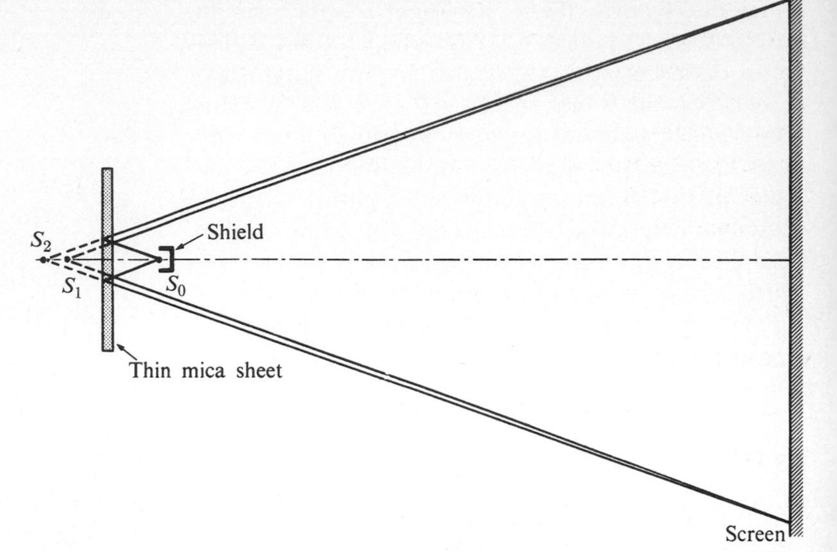

path difference $(r_2 - r_1)$. If the path difference is one wavelength, the phase difference is $\delta = 2\pi$ (or 360°). If $(r_2 - r_1)/\lambda = \frac{1}{2}$, the phase difference is π, and so on. The general relation is that a path difference $r_2 - r_1$ yields a phase difference δ given by

$$\delta = \frac{2\pi}{\lambda}(r_2 - r_1) = k(r_2 - r_1), \tag{42-3}$$

where $k = 2\pi/\lambda$ is the *wave number* introduced in Section 21–3. If the medium in the space between the sources and P is other than vacuum, the wavelength *in the medium* must be used in Eq. (42–3). If the medium has refractive index n, then

$$\lambda = \frac{\lambda_0}{n} \quad \text{and} \quad k = nk_0,$$

where λ_0 and k_0 are the values of λ and k, respectively, in vacuum.

To find the amplitude of the resulting E field, we represent each sinusoidal field as a rotating vector or *phasor*, just as in Chapters 11 and 36. The appropriate diagram is shown in Fig. 42–6; E_1 and E_2 are the amplitudes of the two fields, and E_p is the amplitude of the resultant. Applying the law of cosines, we find

$$E_p^2 = E_1^2 + E_2^2 + 2E_1E_2 \cos \delta.$$

When the two coherent sources S_1 and S_2 are equally intense, $E_1 = E_2 = E$, and

$$E_p^2 = 2E^2 + 2E^2 \cos \delta = 2E^2(1 + \cos \delta) = 4E^2 \cos^2 \frac{\delta}{2}.$$

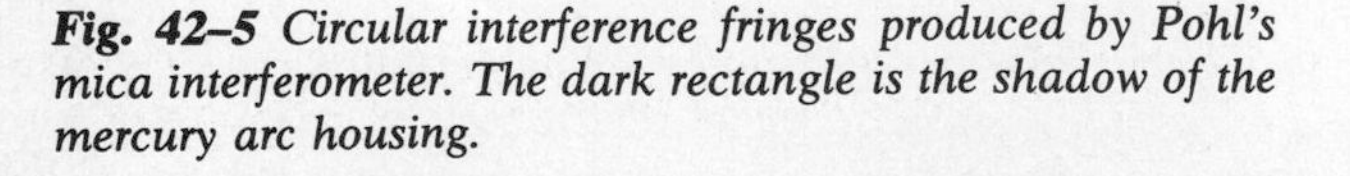

Fig. 42–5 *Circular interference fringes produced by Pohl's mica interferometer. The dark rectangle is the shadow of the mercury arc housing.*

Now the intensity of the resultant wave at P is proportional to the square of the resultant E-field amplitude, as discussed in Section 37–3, so the intensity is proportional to the above expression. The interference fringes therefore consist of gradations of light intensity varying with the cosine squared of *half* the phase difference.

Since

$$\delta = k(r_1 - r_2) = \frac{2\pi}{\lambda} d \sin \theta,$$

and

$$y = R \tan \theta = R \sin \theta,$$

we get

$$\frac{\delta}{2} = \left(\frac{\pi d}{\lambda R}\right) y.$$

If I is the intensity of the light at any value of y and I_0 the intensity at the point where $y = 0$, we get finally

$$I = I_0 \cos^2\left(\frac{\pi d}{\lambda R}\right) y, \qquad (42\text{–}4)$$

which enables us to calculate the intensity of light at any point, and to reproduce the pattern of Fig. 42–3.

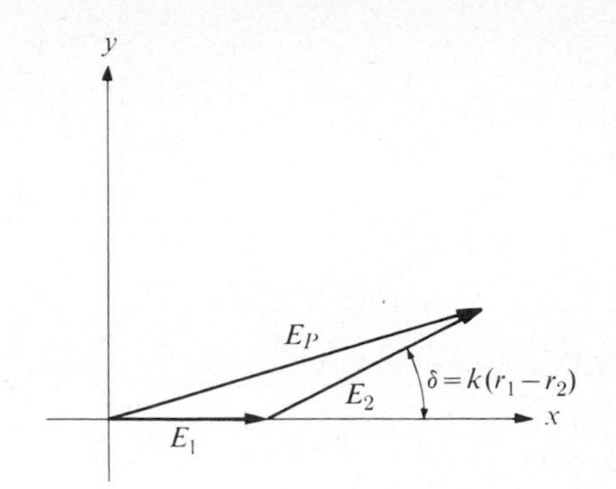

Fig. 42–6 Phasor diagram of the variations of the electric vector at a point P where two waves meet.

42–4 PHASE CHANGES IN REFLECTION

A very important fact concerning the reflection of light from the surface of a material of higher index of refraction than that in which it is originally traveling is demonstrated with the aid of an interferometer known as *Lloyd's mirror,* shown in Fig. 42–7. In this arrangement, the two coherent sources are the actual source slit S_0 and its virtual image S_1. The fringes formed by interference between the light waves from these coherent sources may be viewed on a ground-glass screen placed anywhere beyond the mirror. If, instead, an ocular of high magnification is used to view the fringes that form in space in a plane passing through the edge B, the fringe nearest this edge is seen to be black. This is the fringe corresponding to zero path difference; if the two wave trains giving rise to this fringe had both traveled in air or had both been reflected from glass, this fringe would have been *bright.* The zeroth fringe, however, was formed by two wave trains, one of which had undergone reflection from the glass and one of which had proceeded directly from S_1. The fact that the zeroth fringe is black indicates that *the waves reflected from glass have undergone a phase shift of* 180°. In other words, the wave has gained (or lost) half a wavelength in the process of reflection.

42–5 INTERFERENCE IN THIN FILMS. NEWTON'S RINGS

The brilliant colors that are often seen when light is reflected from a soap bubble or from a thin layer of oil floating on water are produced by interference

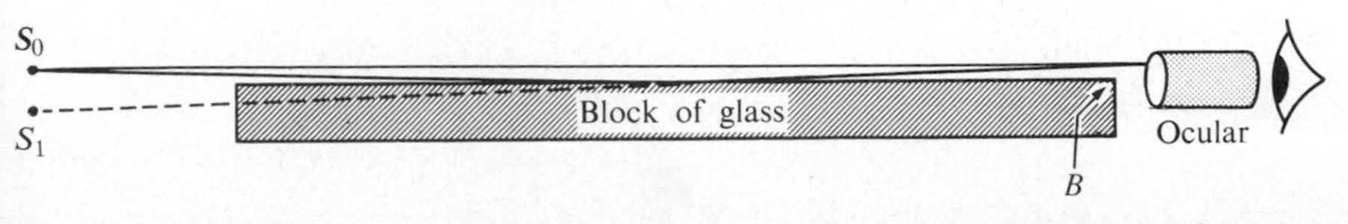

Fig. 42–7 Lloyd's mirror. When the ocular is focused on the edge B of the glass block, the fringe nearest the edge is black.

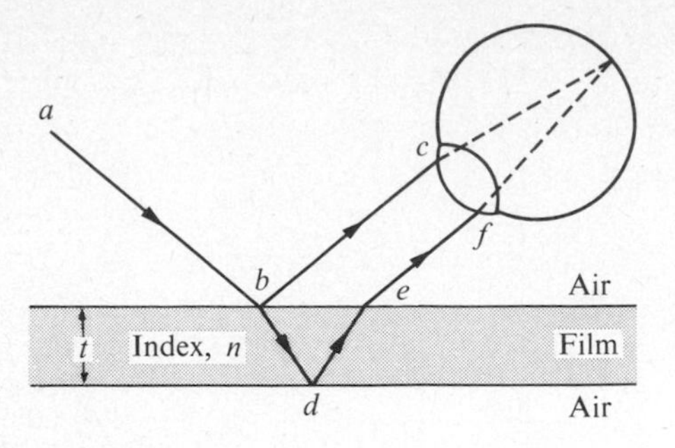

Fig. 42–8 *Interference between rays reflected from the upper and lower surface of a film.*

effects between the two trains of light waves reflected at opposite surfaces of the thin films of soap solution or of oil. In Fig. 42–8, the line *ab* is one ray in a beam of monochromatic light incident on the upper surface of a thin film. A part of the incident light is reflected at the first surface, as indicated by ray *bc*, and a part, represented by *bd*, is transmitted. At the second surface a part is again reflected, and, of this, a part emerges as represented by ray *ef*. The rays *bc* and *ef* come together at a point on the retina of the eye, provided the film is thin enough to allow both rays to enter the pupil of the eye. Needless to say, the thickness of the film and the spacing of the interfering rays are greatly exaggerated in Fig. 42–8.

If an extended monochromatic source is used, there are many rays like *ab* which give rise to two rays that enter the eye. Interference fringes are therefore formed. If the film has parallel surfaces and is illuminated normally, the fringes are circular and appear in space. If, on the other hand, the film is wedge-shaped, the fringes are fairly straight and seem to be localized near the film itself. At the apex of the wedge, where the path difference is zero, destructive interference takes place for all wavelengths because, of the two interfering rays, the one reflected in air from the film undergoes a phase reversal, whereas the other ray does not. At the apex, therefore, the edge of the fringe is black. At a distance from the apex such that the film thickness is $\frac{1}{4}$ wavelength, it will be bright. Where the thickness equals $\frac{1}{2}$ wavelength it will be dark, and so on. If the film is illuminated first by blue, then by red light, the spacing of the red fringes is greater than that of the blue, as is to be expected from the greater wavelength of the red light. The fringes produced by intermediate wavelengths occupy intermediate positions. If the film is illuminated by white light, its color at any point is that due to the mixture of those colors which may be reflected at that point, while the colors for which the thickness is such as to result in destructive interference are absent. Just those colors which are absent in the reflected light, however, are found to predominate in the transmitted light. At any point, the color of the film by reflected light is complementary to its color by transmitted light.

If the convex surface of a lens is placed in contact with a plane glass plate, as in Fig. 42–9, a thin film of air is formed between the two surfaces.

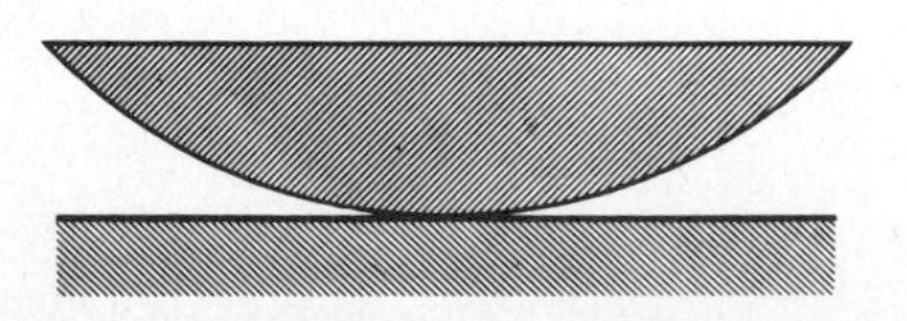

Fig. 42–9 *Air film between a convex and a plane surface.*

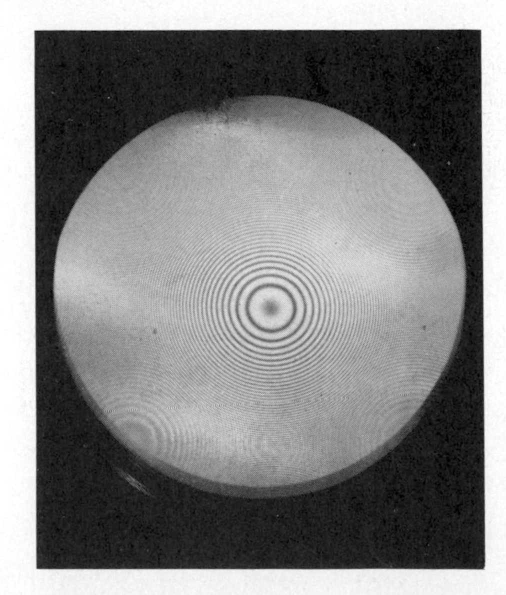

Fig. 42–10 *Newton's rings formed by interference in the air film between a convex and a plane surface. (Courtesy of Bausch & Lomb Optical Co.)*

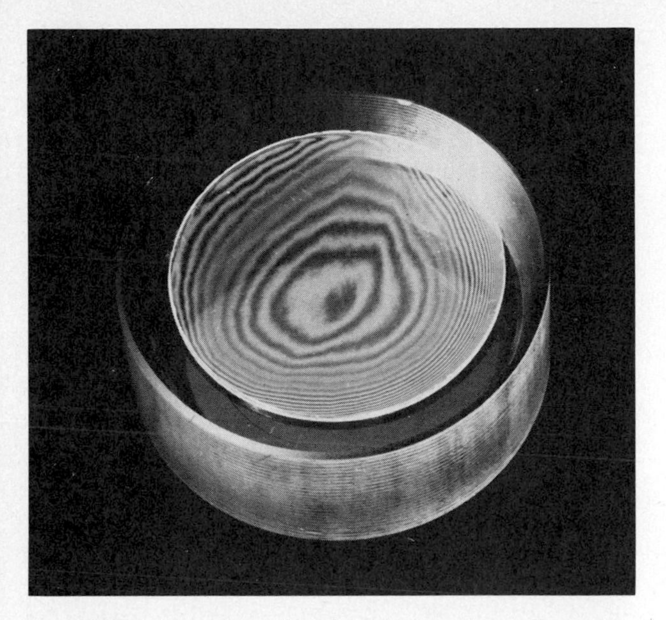

Fig. 42–11 The surface of a telescope objective under inspection during manufacture. (Courtesy of Bausch & Lomb Optical Co.)

The thickness of this film is very small at the point of contact, gradually increasing as one proceeds outward. The loci of points of equal thickness are circles concentric with the point of contact. Such a film is found to exhibit interference colors, produced in the same way as the colors in a thin soap film. The interference bands are circular, concentric with the point of contact. When viewed by reflected light, the center of the pattern is black, as is a thin soap film. Note that in this case there is no phase reversal of the light reflected from the upper surface of the film (which here is of smaller index than that of the medium in which the light is traveling before reflection), but the phase of the wave reflected from the lower surface is reversed. When viewed by transmitted light, the center of the pattern is bright. If white light is used, the color of the light reflected from the film at any point is complementary to the color transmitted.

These interference fringes were studied by Newton, and are called *Newton's rings.* Figure 42–10 is a photograph of Newton's rings formed by the air film between a convex and a plane surface.

The surface of an optical part which is being ground to some desired curvature may be compared with that of another surface, known to be correct, by bringing the two in contact and observing the interference fringes. Figure 42–11 is a photograph made at one stage of the process of manufacturing a telescope objective. The lower, larger-diameter, thicker disk is the master. The smaller upper disk is the objective under test. The "contour lines" are Newton's interference fringes, and each one indicates an additional departure of the specimen from the master of $\frac{1}{2}$ wavelength of light. That is, at 10 lines from the center spot the space between the specimen and master is 5 wavelengths, or about 0.0002 inch. This specimen is very poor; high-quality lenses are routinely ground with a precision of less than a wavelength.

42–6 THIN COATINGS ON GLASS

The phenomenon of interference is utilized in nonreflecting coatings for glass. A thin layer or film of hard transparent material with an index of refraction smaller than that of the glass is deposited on the surface of the glass, as in Fig. 42–12. If the coating has the proper index of refraction, equal quantities of light will be reflected from its outer surface and from the boundary surface between it and the glass. Furthermore, since in both reflections the light is reflected from a medium of greater index than that in which it is traveling, the same phase change occurs in each reflection. It follows that if the film thickness

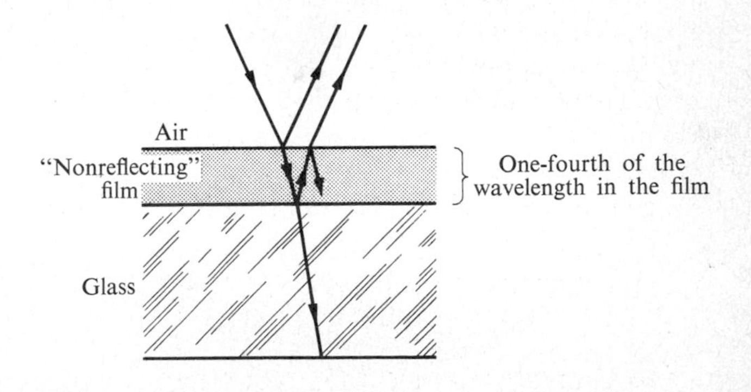

is $\frac{1}{4}$ wavelength *in the film* (normal incidence is assumed), the light reflected from the first surface will be 180° out of phase with that reflected from the second, and complete destructive interference will result.

The thickness can, of course, be $\frac{1}{4}$ wavelength for only one particular wavelength. This is usually chosen in the yellow-green portion of the spectrum, where the eye is most sensitive. Some reflection then takes place at both longer and shorter wavelengths and the reflected light has a purple hue. The overall reflection from a lens or prism surface can be reduced in this way from 4 or 5% to a fraction of 1%. The treatment is extremely effective in eliminating stray reflected light and increasing the contrast in an image formed by highly corrected lenses having a large number of air–glass surfaces.

The electromagnetic theory of light can be used to show that the fraction of the incident electric amplitude reflected from the film will be equal to the fraction reflected from the glass when the index of refraction of the coating n_c is the *geometric mean* of the index of refraction of the air n_a and that of the glass n_g, that is

$$n_c = \sqrt{n_a n_g}. \tag{42–5}$$

Taking the index of refraction of air to be 1 and that of glass to be 1.5, we find $n_c = 1.22$. Unfortunately, it has so far proved impossible to find a hard, transparent material that has this low an index and can easily be evaporated and condensed on glass. The best approximation is magnesium fluoride, MgF_2, with an index of 1.38. With this coating, the wavelength of green light in the coating is

$$\lambda_c = \frac{\lambda_a}{n_c} = \frac{5.5 \times 10^{-5} \text{ cm}}{1.38}$$
$$= 4 \times 10^{-5} \text{ cm},$$

and the thickness of a "nonreflecting" film of MgF_2 is 10^{-5} cm.

If a material whose index of refraction is *greater* than that of glass is deposited on glass with a thickness of $\frac{1}{4}$ wavelength, then the reflectivity is *increased*. For example, a coating of index 2.5 will allow 38% of the incident energy to be reflected, instead of the usual 4% when there is no coating. With the aid of multiple coatings it is possible to achieve an equivalent index of refraction greater than 100, in which case the reflectivity for a particular wavelength is almost 100%.

42–7 THE MICHELSON INTERFEROMETER

Young's double slit and Lloyd's mirror are examples of optical interferometers in which a wavefront from

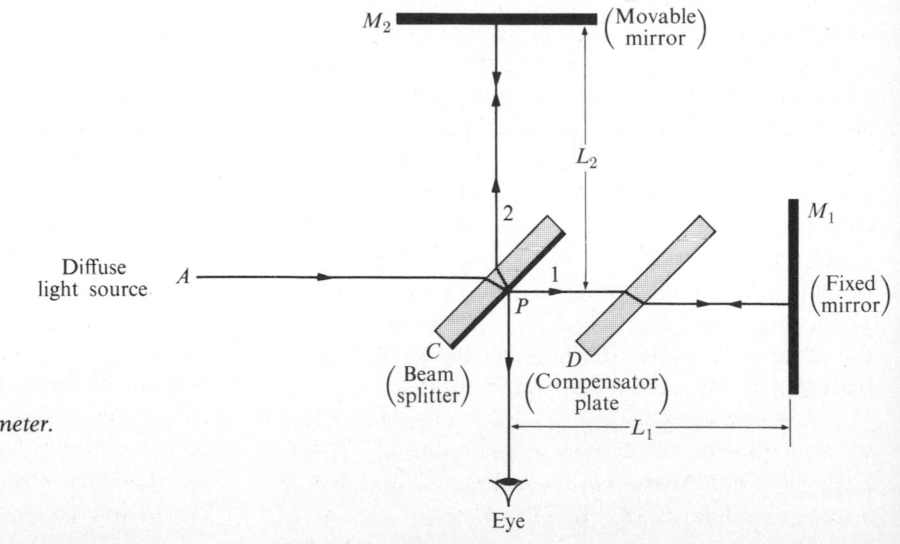

Fig. 42–13 *The Michelson interferometer.*

a very narrow source slit is subdivided into two wavefronts by reflecting or transmitting regions of the interferometer. Some interferometers can be used in conjunction with a large extended source. Of these, the Michelson interferometer has been most important in the past and is still of some significance.

Figure 42–13 is a diagram of the principal features of the Michelson interferometer. The figure shows the path of one ray from a point A of an extended source. Light from the source strikes a glass plate C, the right side of which has a thin coating of silver. Part of the light is reflected from the silvered surface at point P to the mirror M_2 and back through C to the observer's eye. The remainder of the light passes through the silvered surface and the compensator plate D, and is reflected from mirror M_1. It then returns through D and is reflected from the silvered surface of C to the observer. The compensator plate D is cut from the same piece of glass as plate C, so that its thickness will not differ from that of C by more than a fraction of a wavelength. Its purpose is to ensure that rays 1 and 2 pass through the same thickness of glass. Plate C is called a *beam splitter*.

The whole apparatus is mounted on a heavy rigid frame and a fine, very accurate screw thread is used to move the mirror M_2. A common commercial model of the interferometer is shown in Fig. 42–14. The source is placed to the left, and the observer is directly in front of the handle that turns the screw.

If the distances L_1 and L_2 in Fig. 42–13 are exactly equal, and the mirrors M_1 and M_2 are exactly at right angles, the virtual image of M_1 formed by reflection at the silvered surface of plate C will coincide with mirror M_2. If L_1 and L_2 are not exactly equal, the image of M_1 will be displaced slightly from M_2; and if the angle between the mirrors is not exactly a right angle, the image of M_1 will make a slight angle with M_2. Then the mirror M_2, and the virtual image of M_1, play the same roles as the two surfaces of a thin film, discussed in Section 42–5, and the same sort of interference fringes result from the light that is reflected from these surfaces.

Suppose that the extended source in Fig. 42–13 is monochromatic, of wavelength λ, and that the angle between mirror M_2 and the virtual image of M_1 is such that five or six vertical fringes are present in

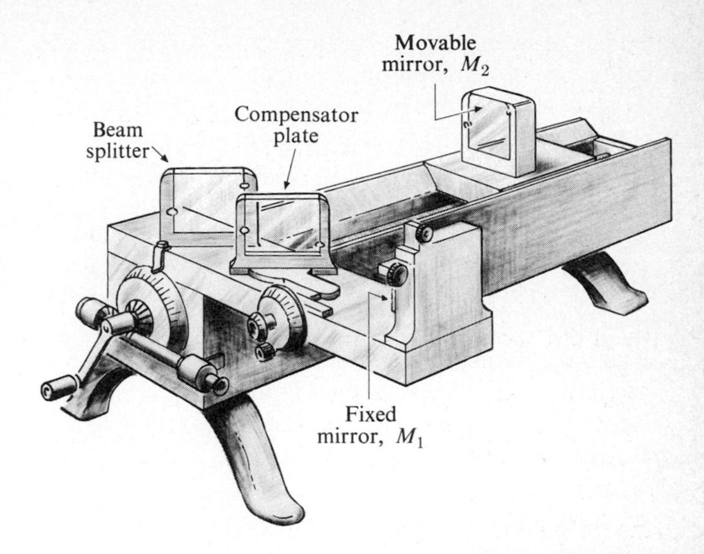

Fig. 42–14 *A common type of Michelson interferometer.*

the field of view. If the mirror M_2 is now moved slowly either backward or forward by a distance $\lambda/2$, the effective film thickness will change by λ and each of the fringes will move either to the right or to the left through a distance equal to the spacing of the fringes. If the fringes are observed through a telescope whose ocular is equipped with a crosshair, and m fringes cross the crosshair when the mirror is moved a distance x, then

$$x = m\,\frac{\lambda}{2} \qquad \text{or} \qquad \lambda = \frac{2x}{m}. \tag{42–6}$$

If m is as large as several thousand, the distance x is sufficiently great so that it can be measured with good precision and hence a precise value of the wavelength λ can be obtained.

It was stated in Chapter 1 that the meter is defined as a length equal to a specified number of wavelengths of the orange-red light of krypton-86. Before this number could be agreed on, it was necessary to measure as accurately as possible the number of these wavelengths in the *former* standard meter, defined as the distance between two scratches on a bar of platinum–iridium. The measurement was made with a modified Michelson interferometer. It would not be practical to make an interferometer whose movable mirror has a range of motion of one meter, so a number of secondary standards called *etalons* were used. The details of this procedure need not concern us. The measurement was made many times under very carefully controlled conditions, with the result that the best value of the number of wavelengths in a distance equal to the old standard meter was 1,650,763.73 wavelengths. The meter is now defined as *exactly* this number of wavelengths.

Another application of the Michelson interferometer of considerable historical interest is the Michelson–Morley experiment. To understand the purpose of this experiment, we must recall that before the electromagnetic theory of light and Einstein's special theory of relativity became established, physicists believed that light waves were propagated in a medium called *the ether.*

Although the ether was assumed to be very rigid, in order to propagate waves with the enormous speed of light, it was assumed also to be very tenuous, in order to allow the planets to move freely through it. A light wave was considered to travel with speed c relative to the ether, just as sound waves in a medium travel with the speed of sound relative to the medium. If a medium is in motion relative to the earth, the velocity of a *sound* wave relative to the earth, v_{WE}, is the vector sum of its velocity relative to the medium, $\mathbf{v}_{WM}$, and the velocity of the medium relative to the earth, $\mathbf{v}_{ME}$:

$$\mathbf{v}_{WE} = \mathbf{v}_{WM} + \mathbf{v}_{ME}. \qquad (42\text{–}7)$$

This point of view has been substantiated so many times with sound waves and water waves that it seemed obvious that the same results should hold for light waves.

In 1887, Michelson and Morley utilized the Michelson interferometer in an attempt to detect the relative motion of the earth and the ether. Suppose the interferometer in Fig. 42–13 is moving from left to right relative to the ether. According to nineteenth-century theory, this would lead to changes in the speed of light in the horizontal portions of the path, and corresponding fringe shifts relative to the positions the fringes would have if the instrument were at rest in the ether. Then when the entire instrument was rotated 90°, the vertical paths would be similarly affected, giving a fringe shift in the opposite direction.

A reasonable guess as to the velocity of the earth relative to the ether is that it is of the same order as the orbital velocity of the earth around the sun, about 3×10^4 m s^{-1}. Michelson and Morley calculated that, for green light, this velocity should cause a fringe shift of about four-tenths of a fringe when the instrument was rotated. The shift actually observed was less than a hundredth of a fringe, and within the limits of experimental uncertainty appeared in fact to be zero. Despite its orbital velocity the earth appeared to be at rest relative to the ether. This negative result baffled physicists of the time, and to this day the Michelson–Morley is the most significant "negative-result" experiment ever performed.

Understanding of this result had to wait for Einstein's special theory of relativity, published in 1905. Einstein realized that the classical equation for combining relative velocities, Eq. (42–7), is only the limiting case of a more general equation, and that this general equation leads to the result that the velocity of a light wave has the same magnitude c relative to *all* reference frames, whatever their velocity may be relative to other frames. The relativistic addition of velocities was discussed in detail in Section 14–5.

Even if there *is* an ether breeze past an interferometer, this breeze has no effect on the velocity of light relative to the interferometer. The result is that rays 1 and 2 in Fig. 42–13 travel, relative to the interferometer, with the same speed c, regardless of the orientation of the interferometer. There is no path difference, and no shift when the interferometer is rotated. The presumed ether then plays no role, and the very concept of an ether has been given up. The underlying principle of Einstein's theory of special relativity may be stated as follows: *there is no preferred frame of reference for the measurement of the speed of light; the speed is the same for every observer, without regard to the magnitude or direction of his velocity relative to other observers.*

42–8 FRESNEL DIFFRACTION

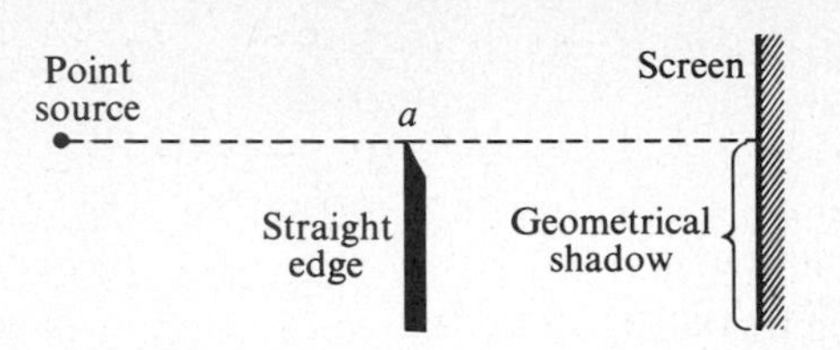

Fig. 42–15 *Geometrical shadow of a straight edge.*

According to geometrical optics, if an opaque object is placed between a point light source and a screen, as in Fig. 42–15, the edges of the object cast a sharp shadow on the screen. No light reaches the screen at points within the geometrical shadow, while outside the shadow the screen is uniformly illuminated. Geometrical optics is, however, an idealized model of the behaviour of light. There are situations in which the representation in terms of straight-line or ray propagation is inadequate; we now proceed to examine some of these situations. An important class of phenomena in which the ray model of geometrical optics is *not* adequate is grouped under the heading *diffraction.*

The photograph reproduced in Fig. 42–16 was made by placing a razor blade halfway between a pinhole illuminated by monochromatic light and a photographic film, so that the film made a record of the shadow cast by the blade. Figure 42–17 is an enlargement of a region near the shadow of an edge of the blade. The boundary of the *geometrical* shadow is indicated by the short arrows. Note that a small amount of light has "bent" around the edge, into the geometrical shadow, which is bordered by alternating bright and dark bands. Note also that in the first bright band, just outside the geometrical shadow, the illumination is actually *greater* than in the region of uniform illumination to the extreme left. This simple experimental setup serves to give some idea of the true complexity of what is often considered the most elementary of optical phenomena, the shadow cast by a small source of light. The distribution of light and dark on any screen after a beam of light has been partially blocked by a perforated diaphragm is called a *diffraction pattern.*

Diffraction patterns such as that in Fig. 42–16 are not commonly observed in everyday life because most ordinary light sources are not point sources of

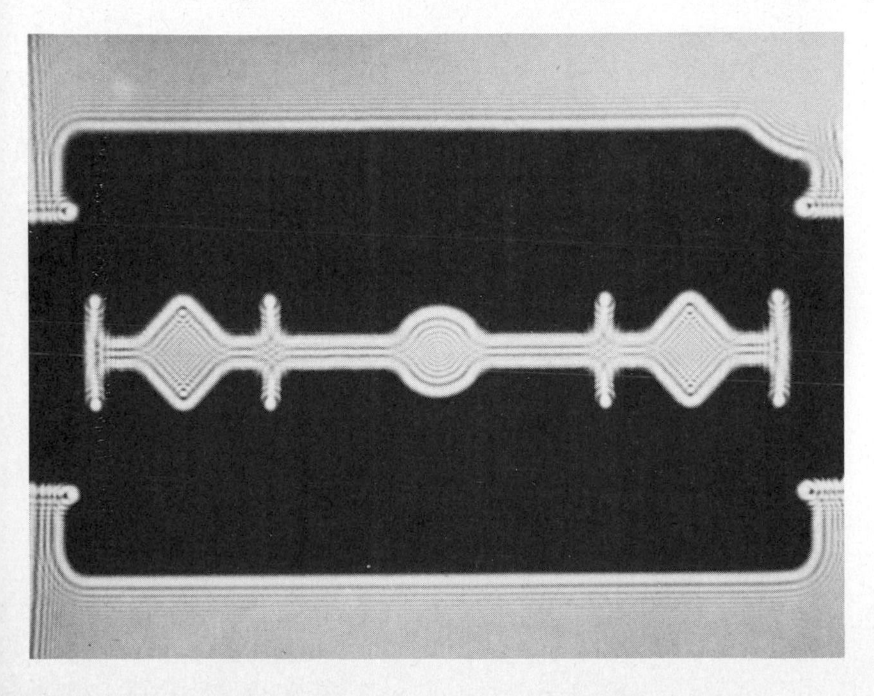

Fig. 42–16 *Shadow of a razor blade.*

Fig. 42–17 *Shadow of a straight edge.*

monochromatic light. If the shadow of a razor blade is cast by a frosted-bulb incandescent lamp, for example, the light from every point of the surface of the lamp forms its own diffraction pattern, but these overlap to such an extent that no individual pattern can be observed.

The term *diffraction* is applied to problems in which one is concerned with *the resultant effect produced by a limited portion of a wave front.* Since in most diffraction problems some light is found within the region of geometrical shadow, diffraction is sometimes defined as "the bending of light around an

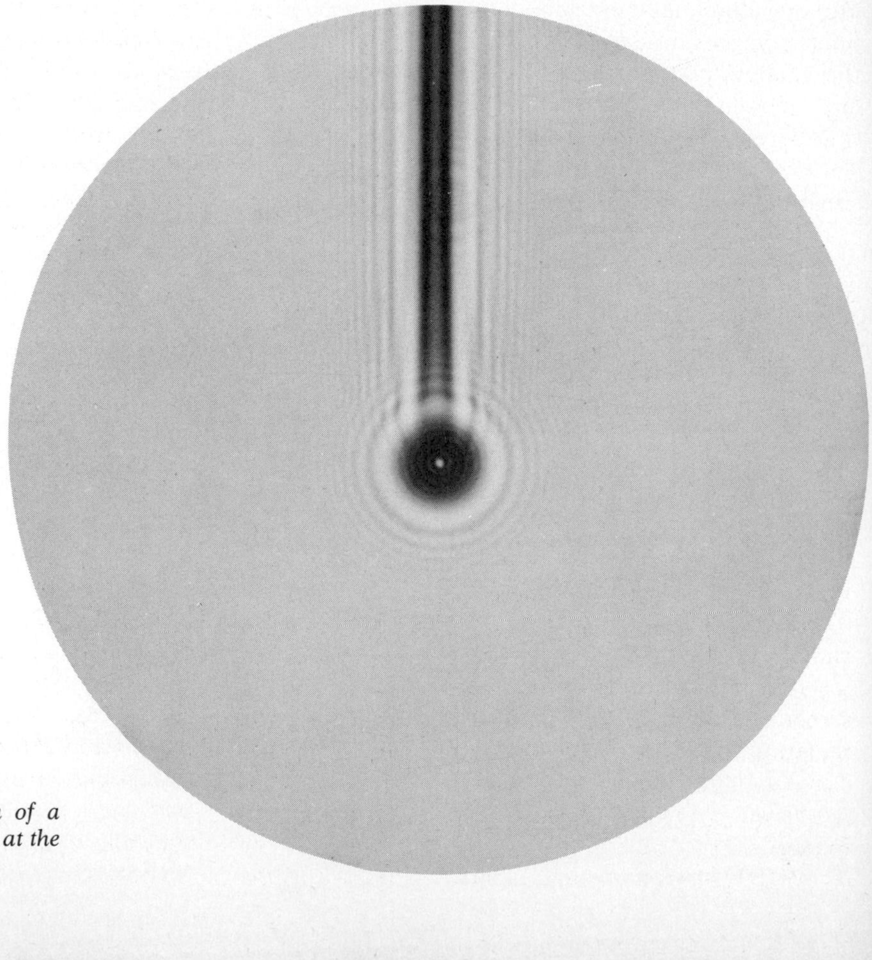

Fig. 42–18 *Fresnel diffraction pattern of a small circular obstacle. A bright spot is at the center of the shadow.*

obstacle." It should be emphasized, however, that the process by which diffraction effects are produced is going on continuously in the propagation of *every* wave. Only if a part of the wave is cut off by some obstacle are the effects commonly called "diffraction effects" observed. But since every optical instrument does, in fact, make use of only a limited portion of a wave (a telescope, for example, utilizes only that portion of a wave admitted by the objective lens), it is evident that a clear comprehension of the nature of diffraction is essential for a complete understanding of practically all optical phenomena.

A circular *opening* in an opaque screen transmits only a small circular patch of a wave surface; the remainder of the wave is obscured. An interesting effect is observed if we reverse this procedure and insert a small circular *obstacle* in the light from a distant point source. A small circular patch of the wave is then obscured, while the remainder is allowed to proceed. Figure 42–18 is a photograph of the shadow of a small steel ball, supported from the tip of a magnetized sewing needle. Constructive interference of the wavelets from the unobstructed portion of the incident wave results in the small bright spot at the center of the geometrical shadow.

The essential features observed in diffraction effects can be predicted with the help of Huygens' principle, according to which every point of a wave surface can be considered the source of a secondary wavelet which spreads out in all directions. However, instead of finding the new wave surface by the simple process of constructing the envelope of all the secondary wavelets, we must combine these wavelets according to the principles of interference. That is, at every point we must combine the displacements that would be produced by the secondary wavelets, taking into account their amplitudes and relative phases. The mathematical operations are often quite complicated.

In Fig. 42–15 both the point source and the screen are at large but finite distances (say several meters) from the obstacle forming the diffraction pattern, and no lenses are used. This situation is described as *Fresnel diffraction* (after Augustin Jean Fresnel, 1788–1827), and the resulting pattern on the screen is called a *Fresnel diffraction pattern.* If the source is far enough away so that the waves striking the obstacle can be considered plane waves, and if a lens is used so that the diffraction pattern appears on a screen in the second focal plane of the lens, the phenomenon is called *Fraunhofer diffraction* (after Joseph von Fraunhofer, 1787–1826). The latter situation is simpler to treat in detail, and we shall consider it first.

42–9 FRAUNHOFER DIFFRACTION FROM A SINGLE SLIT

Suppose a beam of parallel monochromatic light is incident from the left on an opaque plate having a narrow vertical slit. According to geometrical optics, the transmitted beam should have the same cross section as the slit, and a screen in the path of the beam would be illuminated uniformly over an area of the same size and shape as the slit. What is actually observed is the pattern shown in Fig. 42–19. The beam spreads out horizontally after passing through the slit, and the diffraction pattern consists of a central bright band, which may be much wider than the slit width, bordered by alternating dark bands and bright bands of decreasing intensity. A diffraction pattern of this nature can readily be observed by looking at a point source such as a distant street light through a narrow slit formed between two fingers in front of the eye. The retina of the eye then corresponds to the screen.

Let us now apply Huygens' principle to compute the distribution of light on the screen. We consider a plane wave front at the moment it reaches the space between the edges of a slit such as that shown in section in Fig. 42–20(a). Small elements of area are obtained by subdividing the wave front into narrow strips, parallel to the long edges of the slit, perpendicular to the page. From each of these strips, secondary Huygens wavelets spread out in all directions, as shown.

In Fig. 42–20(b) a screen is placed at the right of the slit and P is one point on a line in the screen, the line being parallel to the long edges of the slit and perpendicular to the plane of the diagram. The light reaching a point on the line is calculated by applying the principle of superposition to all the wavelets

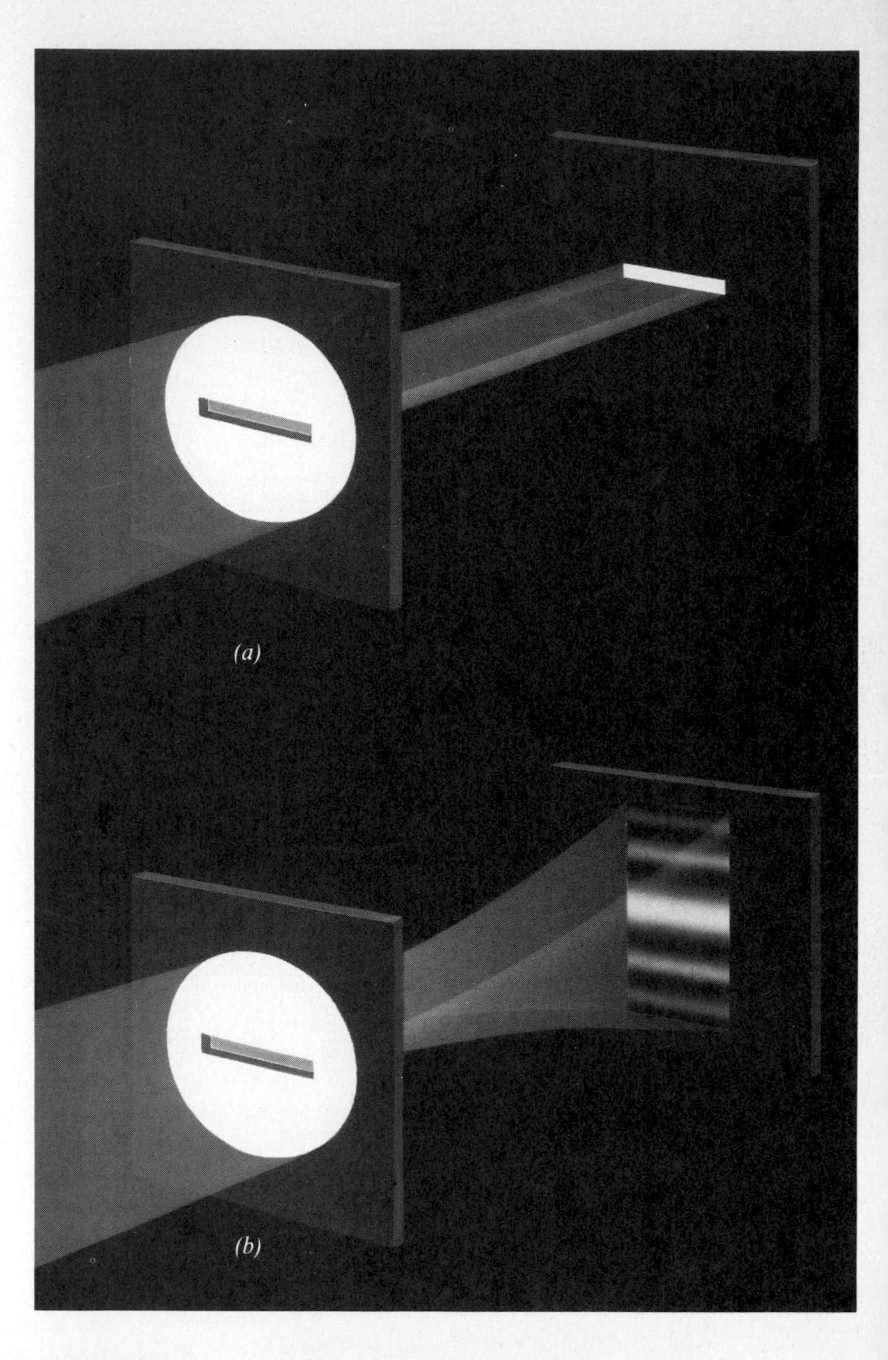

Fig. 42–19 *(a) Geometrical "shadow" of a slit. (b) Diffraction pattern of a slit. The slit width has been greatly exaggerated.*

arriving at the point, from all the elementary strips of the original wave surface. Because of the varying distances to the point, and the varying angles with the original direction of the light, the amplitudes and phases of the wavelets at the point will be different.

The problem is greatly simplified when the screen is sufficiently distant, or the slit sufficiently narrow, so that all rays from the slit to a point on the screen can be considered parallel, as in Fig. 42–20(c). The former case, where the screen is relatively close to the slit (or the slit is relatively wide) is Fresnel diffraction, whereas the latter is Fraunhofer diffraction. There is, of course, no difference in the *nature* of the diffraction process in the two cases, and Fresnel diffraction merges gradually into Fraunhofer diffraction as the screen is moved away from the slit, or as the slit width is decreased.

Fraunhofer diffraction occurs also if a lens is placed just beyond the slit, as in Fig. 42–20(d), since the lens brings to a focus, in its second focal plane, all light traveling in a specified direction. That is, the lens forms in its focal plane a reduced image of the pattern that would appear on an infinitely distant screen in the absence of the lens.

Some aspects of Fraunhofer diffraction from a single slit can be deduced easily. We first consider two narrow strips, one just above the bottom edge of the slit and one at its center. The difference in path length to point P in Fig. 42–21 is $(a/2)\sin\theta$, where a is the slit width. Suppose this path difference happens to be equal to $\lambda/2$; then light from these two strips arrives at point P with a half-cycle phase difference, and cancellation occurs. Similarly, light from two strips just above these two will also arrive a half-cycle out of phase; and, in fact, light from *every* strip in the top half cancels out that from a corresponding strip in the bottom half, resulting in complete cancellation, and giving a dark fringe in the interference

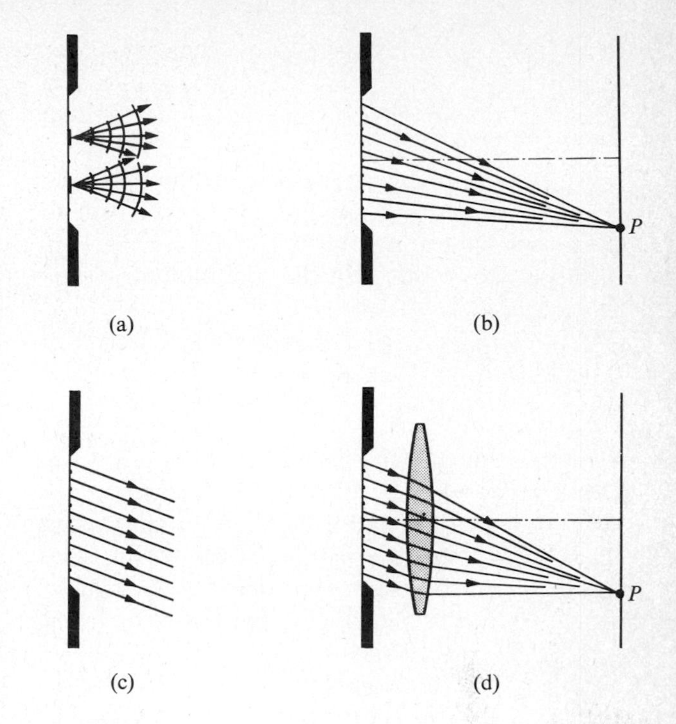

Fig. 42–20 *Transition from Fresnel diffraction to Fraunhofer diffraction by a single slit.*

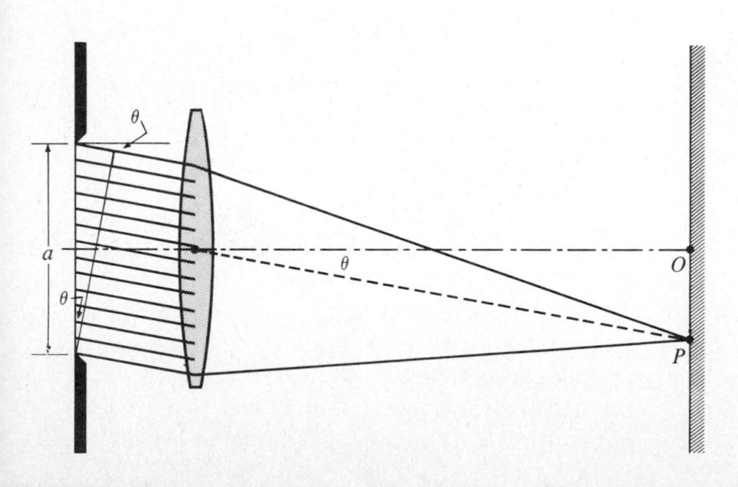

Fig. 42–21 *Subdivision of the wavefront into a large number of narrow strips.*

pattern. Thus, a dark fringe occurs whenever

$$\left(\frac{a}{2}\right)\sin\theta = \frac{\lambda}{2} \qquad \text{or} \qquad \sin\theta = \frac{\lambda}{a}. \quad (42\text{–}8)$$

We may also divide the screen into quarters, sixths, and so on, and use the above argument to show that a dark fringe occurs whenever $\sin\theta = 2\lambda/a$, $3\lambda/a$, and so on. Thus the condition for a *dark* fringe is

$$\boxed{\sin\theta = \frac{n\lambda}{a},} \qquad \text{where } n = 1, 2, 3, \ldots \quad (42\text{–}9)$$

For example, if the slit width is equal to ten wavelengths, dark fringes occur at $\sin\theta = \frac{1}{10}, \frac{2}{10}, \frac{3}{10}, \ldots$ Midway between the dark fringes are bright fringes. We also note that $\sin\theta = 0$ is a *bright* band, since then light from the entire slit arrives at P in phase. Thus the central bright fringe is twice as wide as the others, as Fig. 42–19 shows.

To compute the complete intensity distribution in the single-slit diffraction pattern, we again imagine a plane wavefront at the slit subdivided into a large number of strips each of which sends out rays in *all* directions toward the lens, shown in Fig. 42–21. If we choose an arbitrary point P on a screen in the focal plane of the lens, only those rays making an angle θ with the axis of the lens will arrive at P. Point O on the screen is the special point at which all rays making the angle $\theta = 0$ arrive. Figure 42–22(a) is the vector diagram showing that when the slit is subdivided into 14 sections and each section emits a Huygens wavelet at the angle $\theta = 0$, all wavelets arrive in phase. The resultant amplitude of the electric intensity at O is denoted by S.

Keeping the same subdivision of the wavefront into 14 strips and considering the wavelets that make the angle θ and arrive at P, there will be a slight phase difference between the electric intensity at P due to each succeeding wavelet, and the corresponding vector diagram is shown in Fig. 42–22(b). The sum S is now the perimeter of a portion of a many-sided polygon and E_P, the amplitude of the resultant electric intensity at P, is the chord. The angle δ is the phase difference between the wave from the bottom strip of Fig. 42–21 and the wave from the top strip.

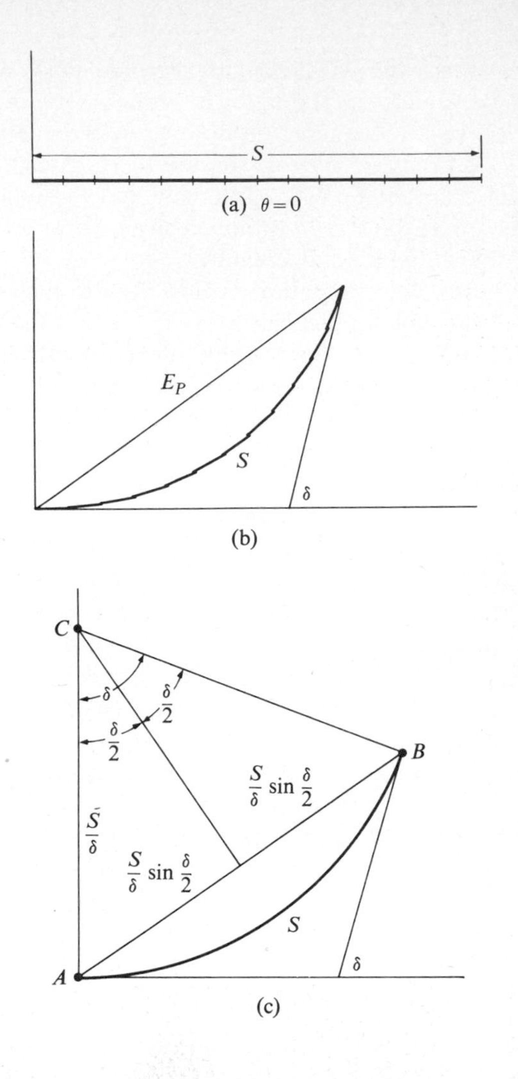

Fig. 42–22 *(a) Vector diagram when all elementary electric intensities are in phase ($\theta = 0$, $\delta = 0$). (b) Vector diagram when each elementary electric intensity differs in phase slightly from the preceding one. (c) Limit reached by the vector diagram when the slit is subdivided infinitely.*

In the limit, as the number of strips into which the slit is subdivided is increased, the vector diagram becomes an *arc of a circle,* as shown in Fig. 42–22(c), with the length of arc S equal to a constant. By constructing perpendiculars at A and B, the center of the circular arc C is found. The radius of the circle of which S is an arc is S/δ, and the resultant amplitude

E_P (distance AB) is $2(S/\delta)\sin\delta/2$. We have then

$$E_P = S\frac{\sin\delta/2}{\delta/2},$$

where δ is the phase difference between the two rays at the extreme edges of the slit.

The phase difference between the vibrations at a point due to the arrival of two waves starting in phase at the same source and traveling different paths is $2\pi/\lambda$ times the path difference. We see from Fig. 42–21 that the path difference between the ray from the top of the slit and that from the bottom is $a\sin\theta$. Therefore

$$\delta = \frac{2\pi}{\lambda}a\sin\theta, \qquad (42\text{–}10)$$

and

$$E_P = S\frac{\sin(\pi a\sin\theta/\lambda)}{\pi a\sin\theta/\lambda}.$$

Since the intensity I is proportional to the *square* of the amplitude,

$$\boxed{I = I_0\left[\frac{\sin(\pi a\sin\theta/\lambda)}{\pi a\sin\theta/\lambda}\right]^2,} \qquad (42\text{–}11)$$

where I_0 is the intensity at O in Fig. 42–21 where $\theta = 0$.

Equation (42–11) is plotted in Fig. 42–23(a), and a photograph of the diffraction pattern is shown di-

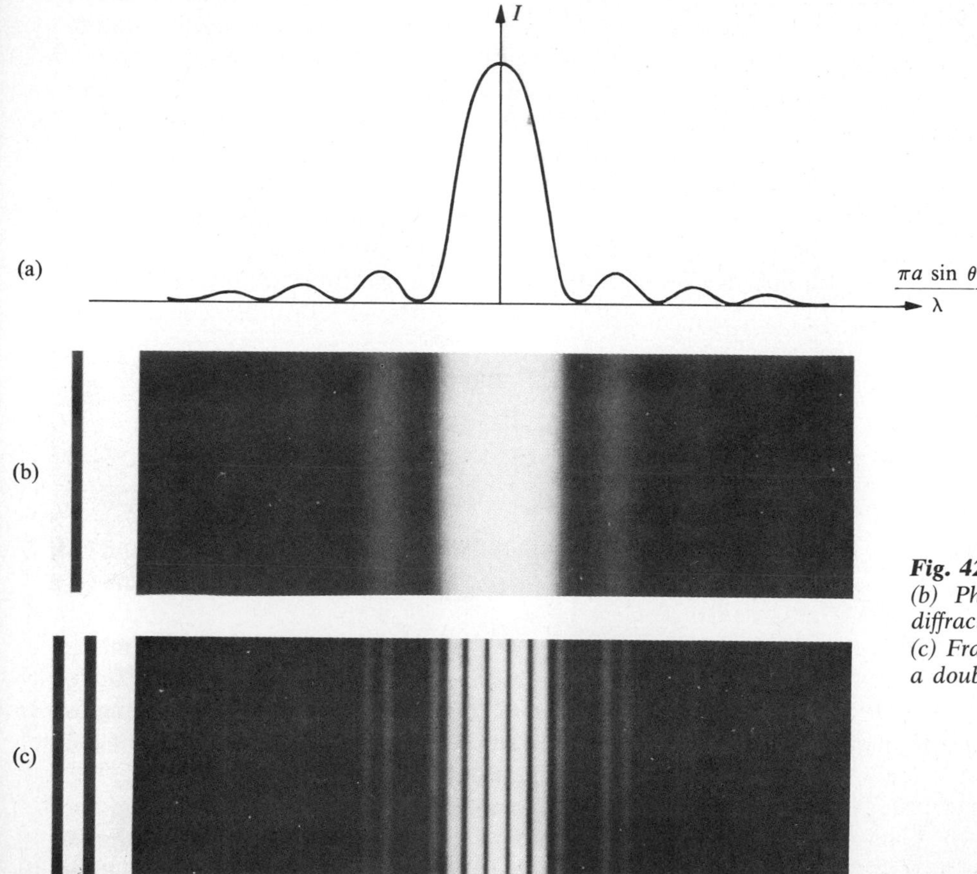

Fig. 42–23 *(a) Intensity distribution. (b) Photograph of the Fraunhofer diffraction pattern of a single slit. (c) Fraunhofer diffraction pattern of a double slit.*

rectly underneath. Note that most of the light exists in the region close to the point where $\theta = 0$, or the geometrical focus. From Eq. (42–11), the smallest value of $\pi a \sin \theta/\lambda$ at which the intensity becomes zero is the value π. This corresponds to a value of θ equal to θ_1 and given by

$$\sin \theta_1 = \frac{\lambda}{a}.$$

Since λ is of the order of 5×10^{-5} cm, and a typical slit width is 10^{-2} cm, $\sin \theta_1$ is ordinarily so small that $\sin \theta_1 = \theta_1$, and

$$\theta_1 = \frac{\lambda}{a}.$$

When a is several centimeters, θ_1 is so small that one can consider practically all the light to be concentrated at the geometrical focus.

The photograph in Fig. 42–23(c) shows the Fraunhofer diffraction pattern of *two* slits each of the same width a as the one above, but separated by a distance $d = 4a$. Note that the interference fringes due to cooperation between the slits have intensities that follow the diffraction pattern of each separate slit. The "cosine-squared" interference fringes are modulated by the shape of the curve shown in part (a) of the figure, because of the finite width of the slits S_1 and S_2.

42–10 THE PLANE DIFFRACTION GRATING

Suppose that instead of a single slit, or two slits side by side as in Young's experiment, we have a very large number of parallel slits, all of the same width and spaced at regular intervals. Such an arrangement, known as a *diffraction grating,* was first constructed by Fraunhofer. The earliest gratings were of fine wires, 0.04 mm to 0.6 mm in diameter. Gratings are now made by using a diamond point to rule a large number of equidistant grooves on a glass or metal surface.

Let GG, in Fig. 42–24, represent the grating, the slits of which are perpendicular to the plane of the paper. While only five slits are shown in the diagram, an actual grating contains several thousand, with a grating spacing d of the order of 0.002 mm. Let a train of plane waves be incident normally on the grating from the left. The problem of finding the intensity of the light transmitted by the grating then combines the principles of interference and diffraction. That is, each slit gives rise to a diffracted beam whose nature, as we have seen, depends on the slit width. These diffracted beams then interfere with one another to produce the final pattern.

Let us assume that the slits are so narrow that the diffracted beam from each spreads out over a sufficiently wide angle for it to interfere with all the other diffracted beams. Consider first the light pro-

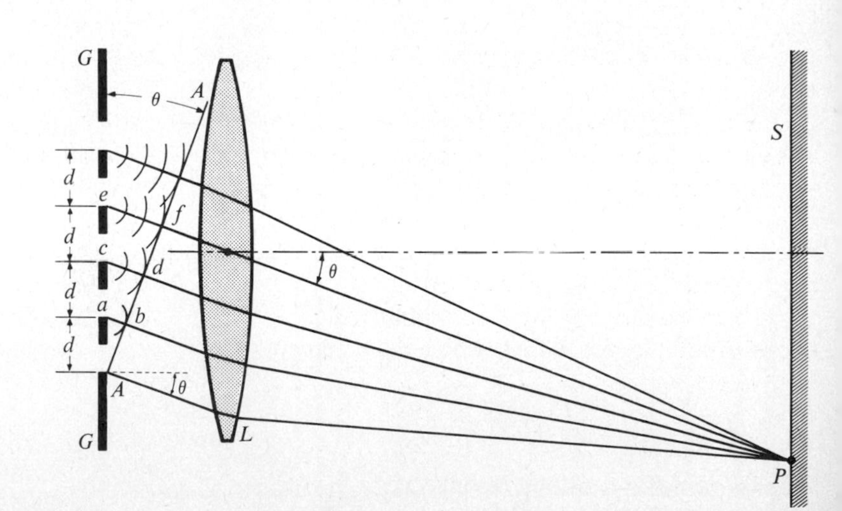

Fig. 42–24 The plane diffraction grating.

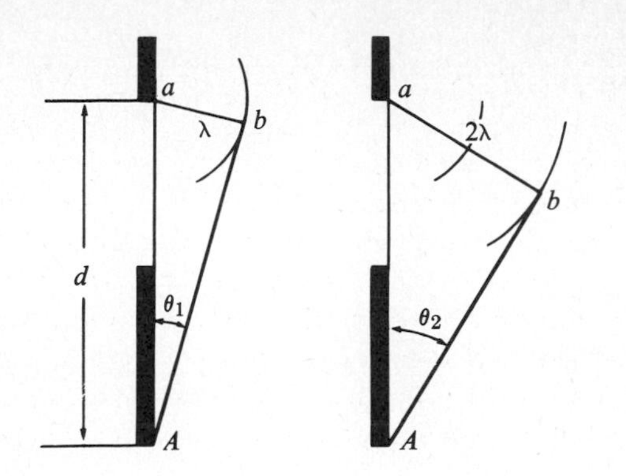

Fig. 42–25 *First-order maximum when ab = λ, second-order maximum when ab = 2λ.*

ceeding from elements of infinitesimal width at the lower edges of each opening, and traveling in a direction making an angle θ with that of the incident beam, as in Fig. 42–24. A lens at the right of the grating forms in its focal plane a diffraction pattern similar to that which would appear on a screen at infinity.

Suppose the angle θ in Fig. 42–24 is taken so that the distance $ab = \lambda$, the wavelength of the incident light. Then $cd = 2\lambda$, $ef = 3\lambda$, etc. The waves from all of these elements, since they are in phase at the plane of the grating, are also in phase along the plane AA and therefore reach the point P in phase. The same holds true for any set of elements in corresponding positions in the various slits.

If the angle θ is increased slightly, the disturbances from the grating elements no longer arrive at AA in phase with one another, and even an extremely small change in angle results in almost complete destructive interference between them, provided there is a large number of slits in the grating. Hence, the maximum at the angle θ is an extremely sharp one, differing from the rather broad maxima and minima which result from interference or diffraction effects with a small number of openings.

As the angle θ is increased still further, a position is eventually reached in which the distance ab in Fig. 42–24 becomes equal to 2λ. Then cd equals 4λ, cf equals 6λ, and so on. The disturbances at AA are again all in phase, the path difference between them now being 2λ, and another maximum results. Evidently still others will appear when $ab = 3\lambda, 4\lambda, \ldots$ Maxima will also be observed at corresponding angles on the opposite side of the grating normal, as well as along the normal itself, since in the latter position the phase difference between disturbances reaching AA is zero.

The angles of deviation for which the maxima occur may readily be found from Fig. 42–25. Consider the right triangle Aba. Let d be the distance between successive grating elements, called the *grating spacing*. The necessary condition for a maximum is that $ab = m\lambda$, where $m = 0, 1, 2, 3$, etc. It follows that

$$\boxed{\sin\theta = m\frac{\lambda}{d}} \qquad (42\text{–}12)$$

is the necessary condition for a maximum. The angle θ is also the angle by which the rays corresponding to the maxima have been *deviated* from the direction of the incident light.

In practice, the parallel beam incident on the grating is usually produced by a lens having a narrow illuminated slit at its first focal point. Each of the maxima is then a sharp image of the slit, of the same color as that of the light illuminating the slit, assumed thus far to be monochromatic. If the slit is illuminated by light consisting of a mixture of wavelengths, the lens will form a number of images of the slit in different positions, every wavelength in the original light giving rise to a set of slit images deviated by the appropriate angles. If the slit is illuminated with white light, a continuous group of images is formed side by side or, in other words, the white light is dispersed into continuous spectra. In contrast with the single spectrum produced by a prism, a grating forms a number of spectra on either side of the normal. Those which correspond to $m = 1$ in Eq. (42–12) are called *first-order*, those which correspond to $m = 2$ are called *second-order*, and so on. Since for $m = 0$ the deviation is zero, all colors combine to produce a white image of the slit in the direction of the incident beam.

In order that an appreciable deviation of the light may be produced, it is necessary that the grating spacing be of the same order of magnitude as the wavelength of light. Gratings for use in or near the visible spectrum are ruled with from about 500 to 1500 lines per millimeter.

The diffraction grating is widely used in spectrometry, instead of a prism, as a means of dispersing a light beam into spectra. If the grating spacing is known, then from a measurement of the angle of deviation of any wavelength, the value of this wavelength may be computed. In the case of a prism this is not so; the angles of deviation are not related in any simple way to the wavelengths but depend on the characteristics of the material of which the prism is constructed. Since the index of refraction of optical glass varies more rapidly at the violet than at the red end of the spectrum, the spectrum formed by a prism is always spread out more at the violet end than it is at the red. Also, while a prism deviates red light the least and violet the most, the reverse is true of a grating, since in the latter case the deviation increases with increasing wavelength.

Example 1 The limits of the visible spectrum are approximately 400 nm to 700 nm. Find the angular breadth of the first-order visible spectrum produced by a plane grating having 6,000 lines per centimeter, when light is incident normally on the grating.

The grating spacing d is

$$d = \frac{1}{6{,}000 \text{ lines cm}^{-1}} = 1.67 \times 10^{-6} \text{ m}.$$

The angular deviation of the violet is

$$\sin\theta = \frac{400 \times 10^{-9} \text{ m}}{1.67 \times 10^{-6} \text{ m}} = 0.240,$$

$$\theta = 13.9°.$$

The angular deviation of the red is

$$\sin\theta = \frac{700 \times 10^{-9} \text{ m}}{1.67 \times 10^{-6} \text{ m}} = 0.420,$$

$$\theta = 24.8°.$$

Hence, the first-order visible spectrum includes an angle of

$$24.8° - 13.9° = 10.9°.$$

Example 2 Show that the violet of the third-order spectrum overlaps the red of the second-order spectrum.

The angular deviation of the third-order violet is

$$\sin\theta = \frac{(3)(400 \times 10^{-9} \text{ m})}{d}$$

and of the second-order red it is

$$\sin\theta = \frac{(2)(700 \times 10^{-9} \text{ m})}{d}.$$

Since the first angle is smaller than the second, whatever the grating spacing, the third order will *always* overlap the second.

42–11 DIFFRACTION OF X-RAYS BY A CRYSTAL

Although x-rays were discovered by Roentgen in 1895, it was not until 1913 that x-ray wavelengths were measured with any degree of precision. Experiments had indicated that these wavelengths might be of the order of 10^{-8} cm, which is about the same as the interatomic spacing in a solid. It occurred to Laue in 1913 that if the atoms in a crystal were arranged in a regular way, a crystal might serve as a three-dimensional diffraction grating for x-rays. The experiment was performed by Friederich and Knipping and it succeeded, thus verifying in a single stroke both the hypothesis that x-rays *are* waves (or, at any rate, wavelike in some of their properties) and

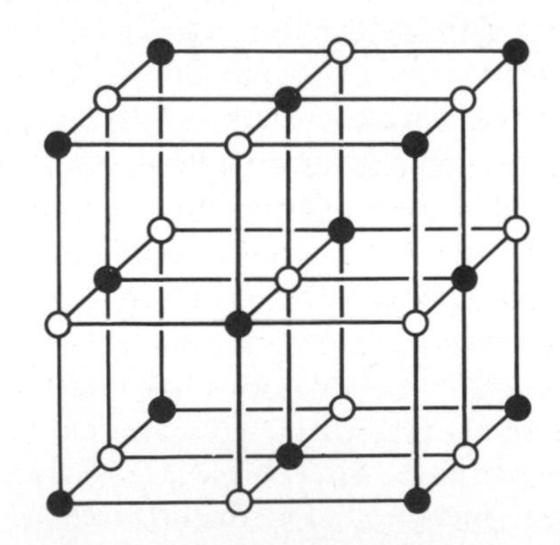

Fig. 42–26 Model of arrangement of ions in a crystal of NaCl. Black circles, Na; open circles, Cl.

that the atoms in a crystal *are* arranged in a regular manner. Since that time, the phenomenon of x-ray diffraction by a crystal has proved an invaluable tool of the physicist, both as a way to measure x-ray wavelengths and a method of studying the structure of crystals.

Figure 42–26 is a diagram of a simple type of crystal, that of sodium chloride (NaCl). The black circles represent the sodium, and the open circles the chlorine ions. Figure 42–27 is a diagram of a section through the crystal. Planes such as those parallel to *aa, bb, cc*, etc., can be constructed through the crystal in such a way that they pass through relatively large numbers of atoms.

Figure 42–28 is a photograph made by directing a narrow beam of x-rays at a thin section of a crystal of quartz and allowing the diffracted beams to strike a photographic plate. Each atom in the crystal scatters some of the incident radiation, and interference occurs between the waves scattered by the various atoms. Just as with the diffraction grating, nearly complete cancellation occurs for all but certain very specific directions for which constructive interference is possible; hence, the spots in Fig. 42–28. The interference pattern is often described in terms of *reflections* from various planes of atoms as in Fig. 42–27, but the basic phenomenon is one of *interference.*

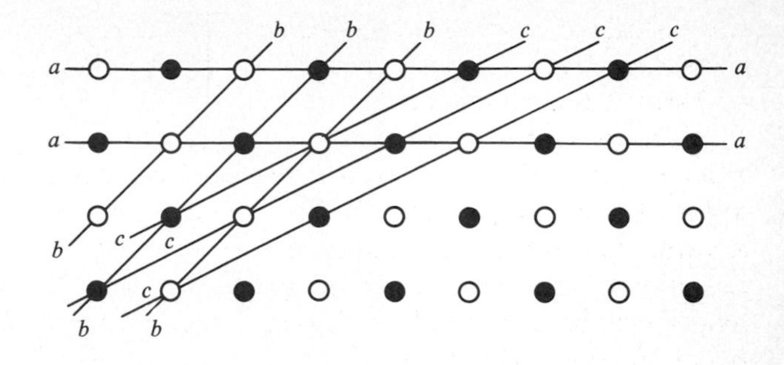

Fig. 42–27 *Crystal planes such as aa, bb, and cc serve as a three-dimensional diffraction grating for x-rays.*

42–12 THE RESOLVING POWER OF OPTICAL INSTRUMENTS

The expressions for the magnification of a telescope or a microscope (except for certain numerical factors) involved only the focal lengths of the lenses making up the optical system of the instrument. It appears, at first glance, as though any desired magnification might be attained by a proper choice of these focal lengths. Beyond a certain point, however, while the image formed by the instrument becomes larger (or subtends a larger angle) it *does not gain in detail,* even though all lens aberrations have been corrected. This limit to the useful magnification is set by the fact that light is a wave motion and the laws of geometrical optics do not hold strictly for a wave surface of limited extent. Physically, the image of a point source is not the intersection of *rays* from the source, but the diffraction pattern of those *waves* from the source, that pass through the lens system.

It is an important experimental fact that the light from a point source, diffracted by a circular opening, is focused by a lens not as a geometrical point, but as a disk of finite radius surrounded by dark and bright rings. The larger the wave surface admitted (i.e., the larger the lenses or diaphragms in

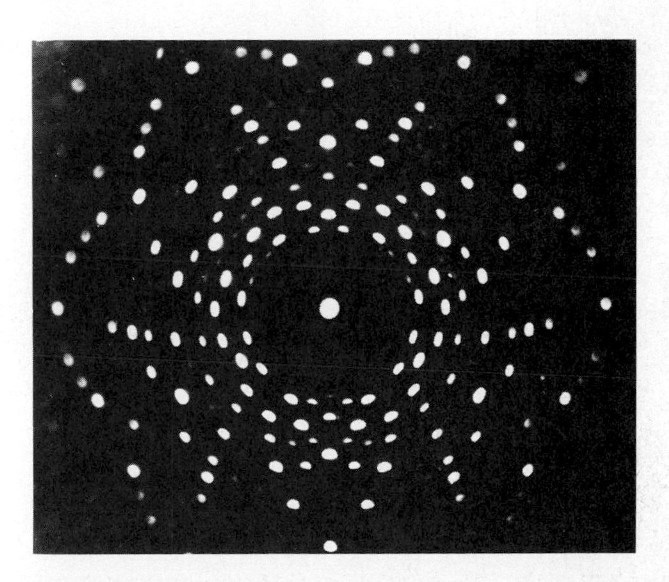

Fig. 42–28 *Laue diffraction pattern formed by directing a beam of x-rays at a thin section of quartz crystal (Courtesy of Dr. B. E. Warren.)*

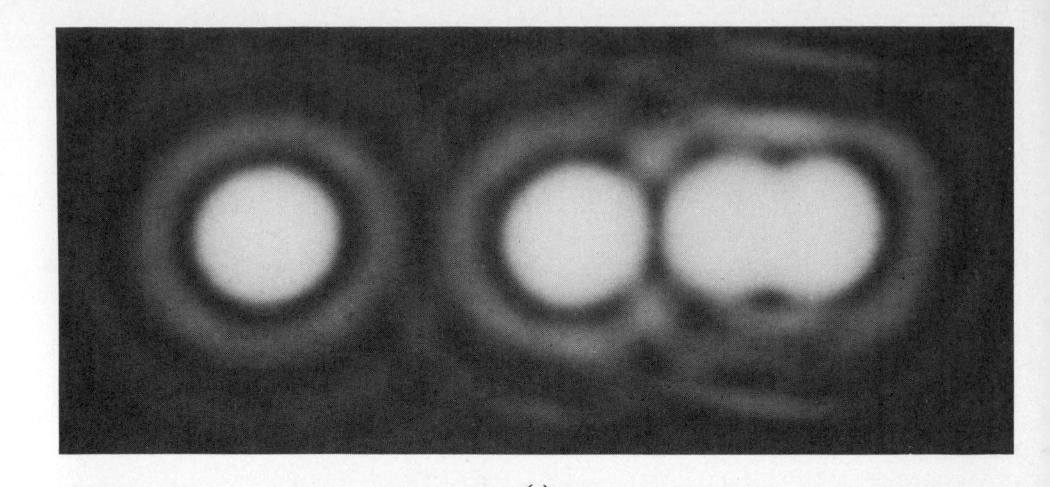

(a)

Fig. 42–29 *Diffraction patterns of four "point" sources, with a circular opening in front of the lens. In (a), the opening is so small that the patterns at the right are just resolved, by Rayleigh's criterion. Increasing the aperture decreases the size of the diffraction patterns, as in (b) and (c).*

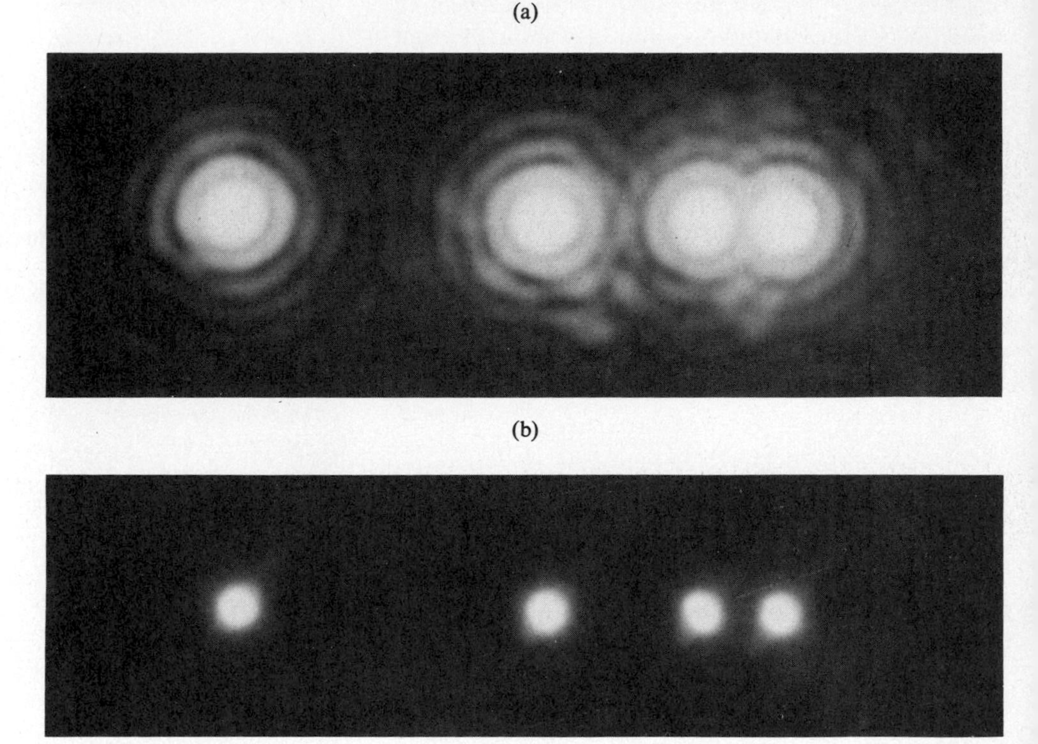

(b)

(c)

an optical system), the smaller the diffraction pattern of a point source and the closer together may two point sources be before their diffraction disks overlap and become indistinguishable. An optical system is said to be able to *resolve* two point sources if the corresponding diffraction patterns are sufficiently small or sufficiently separated to be distinguished. The numerical measure of the ability of the system to resolve two such points is called its *resolving power* or its *resolution*.

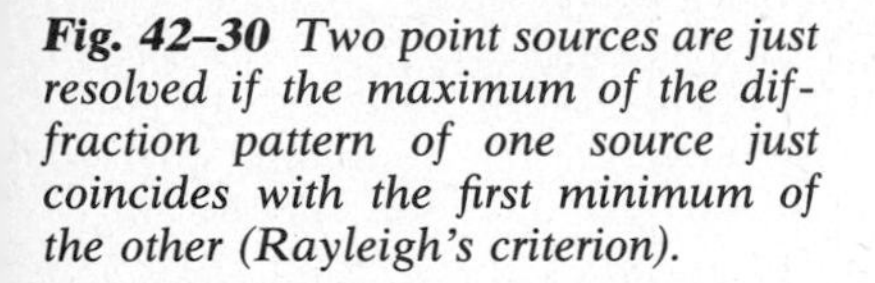

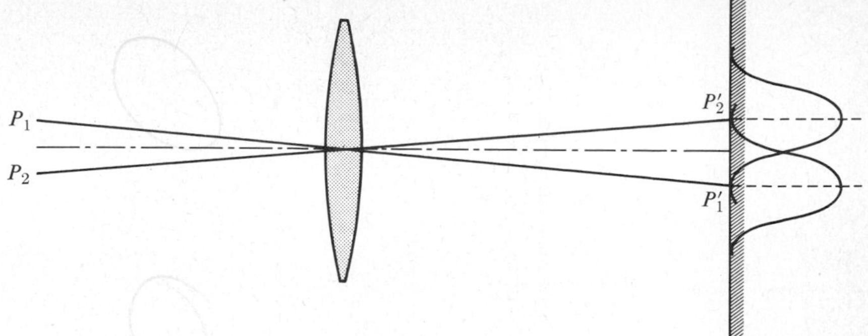

Fig. 42–30 *Two point sources are just resolved if the maximum of the diffraction pattern of one source just coincides with the first minimum of the other (Rayleigh's criterion).*

Figure 42–29(a) is a photograph of four point sources made with the camera lens "stopped down" to an extremely small aperture. The nature of the diffraction patterns is clearly evident, and it is obvious that further magnification of the picture would not aid in resolving the sources. What is necessary is not to make the image *larger,* but to make the diffraction patterns *smaller.* Figures 42–29(b) and 42–29(c) show how the resolving power of the lens is increased by increasing its aperture. In (b) the diffraction patterns are sufficiently small for all four sources to be distinguished. In (c) the full aperture of the lens was utilized.

An arbitrary criterion proposed by Lord Rayleigh is that two point sources are just resolvable if the central maximum of the diffraction pattern of one source just coincides with the first *minimum* of the other. Let P_1 and P_2 in Fig. 42–30 be the central rays in the two parallel beams coming from two very distant point sources. Let P_1' and P_2' be the centers of their diffraction patterns, formed by some optical instrument such as a microscope or telescope, which for simplicity is represented in the diagram as a single lens. If the images are just resolved, that is, if according to Rayleigh's criterion the first minimum of one pattern coincides with the center of the other, the separation of the centers of the patterns equals the radius of the central bright disk. From a knowledge of the focal lengths and separations of the lenses in any particular instrument, one can compute the corresponding distance between the two point objects. This distance, the minimum separation of two points that can just be resolved, is called the *limit of resolution* of the instrument. The smaller this distance, the greater is said to be the *resolving power.* The resolving power increases with the solid angle of the cone of rays intercepted by the instrument, and is inversely proportional to the wavelength of the light used. It is as if the light waves were the "tools" with which our optical system is provided, and the smaller the tools the finer the work the system can do.

42–13 HOLOGRAPHY

Holography is a technique for recording and reproducing an image of an object without the use of lenses or mirrors. Unlike the two-dimensional images recorded by an ordinary photograph or television system, a holographic image is truly three-dimensional. Such an image can be viewed from different directions to reveal different sides, and from various distances to reveal changing perspective.

The basic procedure for making a hologram is very simple in principle. A possible arrangement is shown in Fig. 42–31(a). The object to be holographed is illuminated by monochromatic light, and a photographic film is located so that it is struck by scattered light from the object and also by direct light from the source. In practice the source must be a laser, for reasons to be discussed later. Interference between the direct and scattered light leads to the formation and recording of a complex interference pattern on the film.

To form the images, one simply projects laser light through the developed film, as shown in Fig.

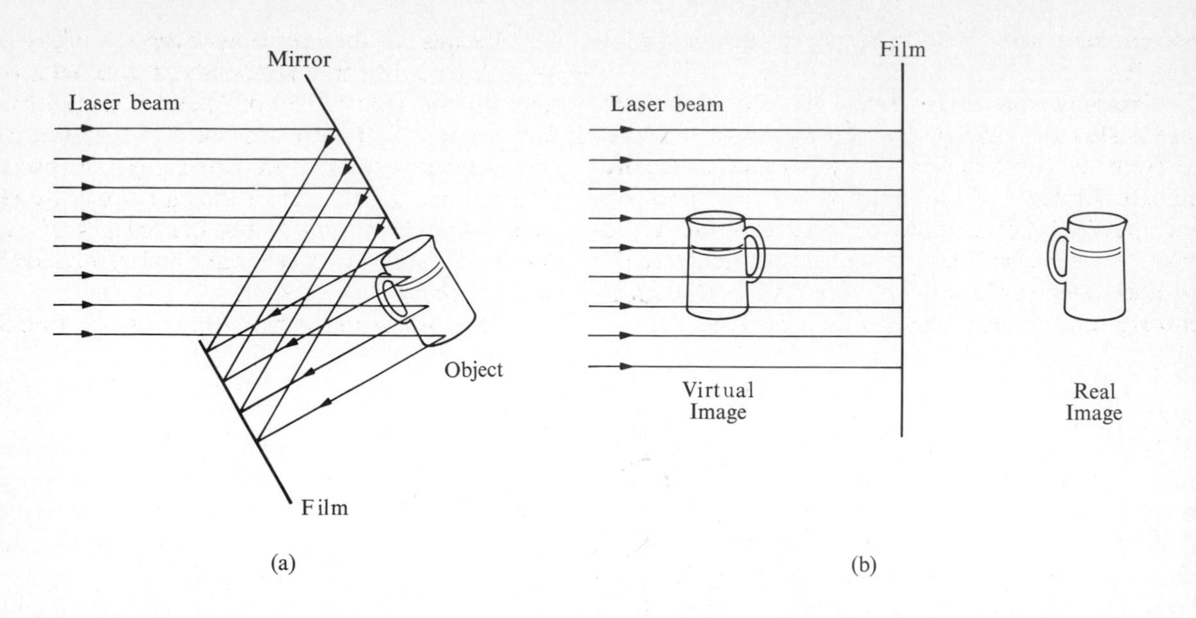

Fig. 42–31 *(a) The hologram is the record on film of the interference pattern formed with light directly from the source and light scattered from the object. (b) Images are formed when light is projected through the hologram.*

42–31(b). Two images are formed, a virtual image on the side of the film nearer the source, and a real image on the opposite side.

A complete analysis of holography is beyond our scope, but we can gain some insight into the process by examining how a single point is holographed and imaged. We consider the interference pattern formed on a screen by the superposition of an incident plane wave and a spherical wave, as shown in Fig. 42–32(a). The spherical wave originates at a point source P a distance d_o from the screen; P may in fact be a small object that *scatters* part of the incident plane wave. In any event, we assume that the two waves are monochromatic and coherent, and that the phase relation is such that constructive interference occurs at point O on the diagram. Then constructive interference will *also* occur at any point Q on the screen that is farther from P than O is, by an integer number of wavelengths. That is, if $d_n - d_o = n\lambda$, where n is an integer, constructive interference occurs. The points where this condition is satisfied form circles centered at O, with radii r_n given by

$$d_n - d_o = \sqrt{d_o^2 + r_n^2} - d_o = n\lambda, \quad (n = 1, 2, 3, \ldots). \tag{42–13}$$

Solving this for r_n^2, we find

$$r_n^2 = \lambda(2nd_o + n^2\lambda).$$

Ordinarily d_o is very much larger than λ, so we neglect the second term in parentheses, obtaining

$$r_n = \sqrt{2n\lambda d_o}. \tag{42–14}$$

Since n may be any integer, the interference pattern consists of a series of concentric bright circular fringes, with the radii of the brightest regions given by Eq. (42–14). Between these bright fringes will be darker fringes. This pattern may be recorded on film by simply placing the film in the position of the screen.

Now the film is developed and a transparent positive print is made, so the bright-fringe areas have the greatest transparency on the film. It is then illuminated with monochromatic plane-wave light of the same wavelength as that used initially. In Fig. 42–32(b), consider a point P' at a distance d_o along the axis from the film; the centers of successive bright fringes differ in their distance from P' by an integer number of wavelengths, and therefore a strong *maximum* in the diffracted wave occurs at P'. That is, light converges to P' and then diverges from it on the opposite side, and P' is therefore a *real image* of point P.

This is not the entire diffracted wave, however; there is also a diverging spherical wave which would represent a continuation of the wave originally emanating from P if the film had not been present. Thus the *total* diffracted wave is a superposition of a converging spherical wave forming a real image at P', and a diverging spherical wave shaped as though it had originated at P, forming a virtual image at P.

Because of the principle of linear superposition, what is true for the imaging of a single point is also true for the imaging of any number of points. The film records the superposed interference pattern from the various points, and when light is projected through the film the various image points are reproduced simultaneously. Thus the images of an extended object can be recorded and reproduced just as for a single point object.

Several practical problems must be overcome.

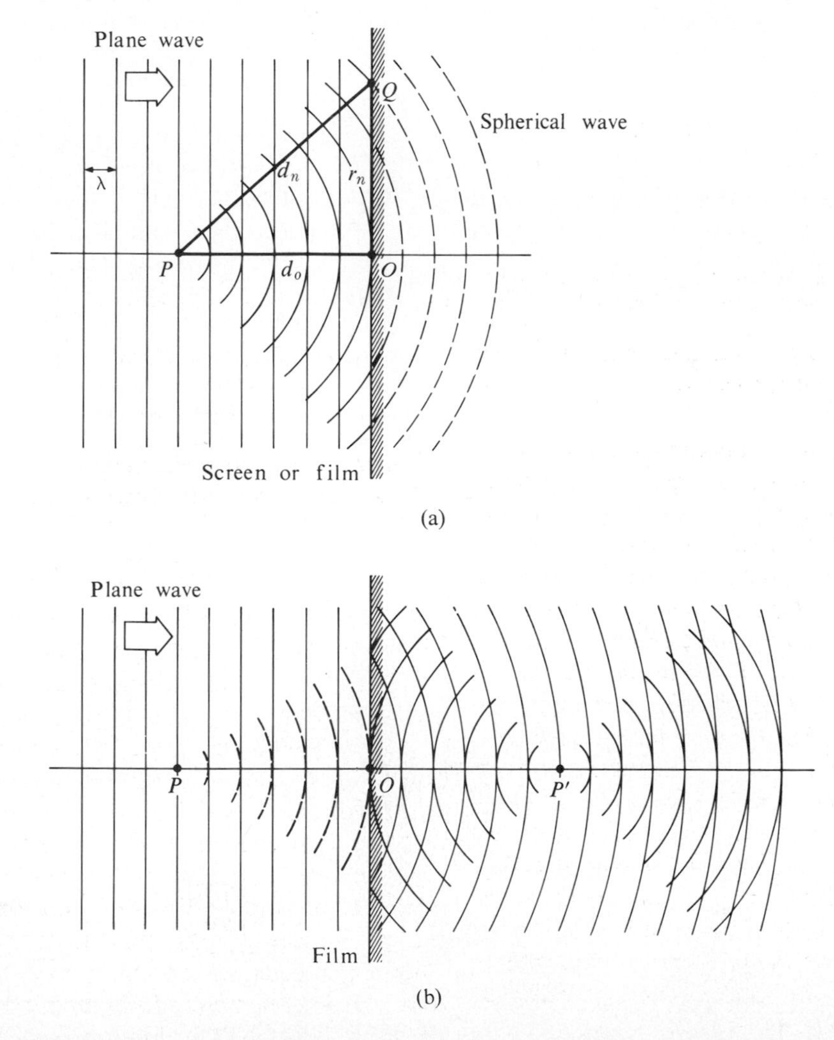

Fig. 42–32 *(a) Constructive interference of the plane and spherical waves occurs in the plane of the film at every point Q for which the distance d_n from P is greater than the distance d_0 from BP to O by an integer number of wavelengths $n\lambda$. For the point shown, $n = 2$. (b) When a plane wave strikes the developed film, the diffracted wave consists of a wave converging to P′ and then diverging again, and a diverging wave that appears to originate at P. These waves form the real and virtual images, respectively.*

First, the light used must be *coherent* over distances large compared to the dimensions of the object and its distance from the film. Ordinary light sources *do not* satisfy this requirement, for reasons discussed in Section 42–1, and laser light is essential. Second, extreme mechanical stability is needed. If there is any relative motion of source, object, or film during exposure, even as much as a wavelength, the interference pattern is blurred on the film enough to prevent satisfactory image formation. These obstacles are not insurmountable, however, and holography promises to become increasingly important in research, entertainment, and a wide variety of technological applications.

PROBLEMS

42–1 Two slits are spaced 0.3 mm apart and are placed 50 cm from a screen. What is the distance between the second and third dark lines of the interference pattern when the slits are illuminated with light of 600-nm wavelength?

42–2 Young's experiment is performed with sodium light ($\lambda = 589$ nm). Fringes are measured carefully on a screen 100 cm away from the double slit, and the center of the twentieth fringe is found to be 11.78 mm from the center of the zeroth fringe. What is the separation of the two slits?

42–3 In the case of the double slit, show that (a) the intensity of light on a screen is given by $I = I_0 \cos^2(\pi d \sin\theta/\lambda)$, and (b) the first minimum occurs at the angle $\theta_1 = \lambda/2d$. If there are two extremely narrow *source* slits S_0 and S_0' subtending an angle α at the center of the double slit, there are two sets of interference fringes whose zeroth fringes also subtend an angle α. (c) Sketch the two sets of fringes when $\alpha \gg \theta_1$, $\alpha = \theta_1$, and $\alpha \ll \theta_1$.

42–4 Light of wavelength 500 nm is incident perpendicularly from air on a film 10^{-4} cm thick and of refractive index 1.375. Part of the light enters the film and is reflected back at the second face. (a) How many waves are contained along the path of this light in the film? (b) What is the phase difference between these waves as they leave the film?

42–5 A glass plate 0.40 μm thick is illuminated by a beam of white light normal to the plate. The index of refraction of the glass is 1.50. What wavelengths within the limits of the visible spectrum ($\lambda = 400$ nm to $\lambda = 700$ nm) will be intensified in the reflected beam?

42–6 Figure 42–33 shows an interferometer known as *Fresnel's biprism.* The magnitude of the prism angle A is extremely small.

a) If S_0 is a very narrow source slit, show that the separation of the two virtual coherent sources S_1 and S_2 is given by $d = 2aA(n-1)$, where n is the index of refraction of the material of the prism.
b) Calculate the spacing of the fringes of green light of wavelength 500 nm on a screen 2 m from the biprism. Take $a = 20$ cm and $n = 1.5$.

42–7 In Lloyd's mirror, the source slit S_0 and its virtual image S_1 lie in a plane 20 cm behind the left edge of the mirror. (See Fig. 42–7.) The mirror is 30 cm long and a ground glass screen is placed at the right edge. Calculate the distance from this edge to the first light maximum, if the perpendicular distance from S_0 to the mirror is 2 mm and $\lambda = 7.2 \times 10^{-5}$ cm.

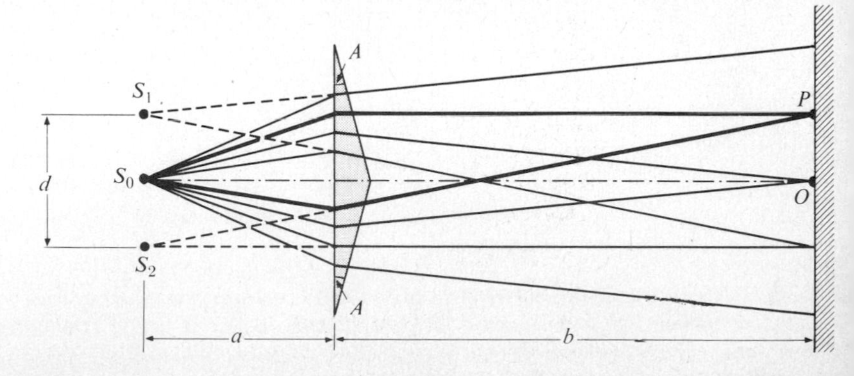

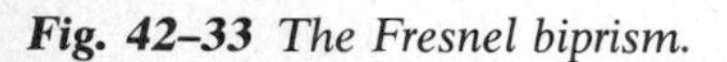
Fig. 42–33 *The Fresnel biprism.*

42–8 How far must the mirror M_2 of the Michelson interferometer be moved in order that 3000 fringes of krypton-86 light (λ = 606 nm) move across a line in the field of view?

42–9 Two rectangular pieces of plane glass are laid one upon the other on a table. A thin strip of paper is placed between them at one edge so that a very thin wedge of air is formed. The plates are illuminated by a beam of sodium light at normal incidence (λ = 589 nm). Interference fringes are formed, there being ten per centimeter length of wedge measured normal to the edges in contact. Find the angle of the wedge.

42–10 A sheet of glass 10 cm long is placed in contact with a second sheet, and is held at a small angle with it by a metal strip 0.1 mm thick placed under one end. The glass is illuminated from above with light of 546-nm wavelength. How many interference fringes are observed per cm in the reflected light?

42–11 The radius of curvature of the convex surface of a plano-convex lens is 120 cm. The lens is placed convex side down on a plane glass plate, and illuminated from above with red light of wavelength 650 nm. Find the diameter of the third bright ring in the interference pattern.

42–12 What is the thinnest film of 1.40 refractive index in which destructive interference of the violet component (400 nm) of an incident white beam in air can take place by reflection? What is then the residual color of the beam?

42–13 The surfaces of a prism of index 1.52 are to be made "nonreflecting" by coating them with a thin layer of transparent material of index 1.30. The thickness of the layer is such that at a wavelength of 550 nm (in vacuum), light reflected from the first surface is $\frac{1}{2}$ wavelength out of phase with that reflected from the second surface. Find the thickness of the layer.

42–14 (a) Is a thin film of quartz suitable as a nonreflecting coating for fabulite? (See Table 38–1.) (b) If so, how thick should the film be?

42–15 Light from a mercury-arc lamp is passed through a filter that blocks everything except for one spectrum line in the green region of the spectrum. It then falls on two slits separated by 0.6 mm. In the resulting interference pattern on a screen 2.5 m away, adjacent bright fringes are separated by 2.27 mm. What is the wavelength?

42–16 In Problem 42–1, suppose the entire apparatus is immersed in water. Then what is the distance between the second and third dark lines?

42–17 Light of wavelength 589 nm from a distant source is incident on a slit 1.0 mm wide, and the resulting diffraction pattern is observed on a screen 2.0 m away. What is the distance between the two dark fringes on either side of the central bright fringe?

42–18 In Problem 42–17, suppose the entire apparatus is immersed in water. Then what is the distance between the two dark fringes?

42–19 Parallel rays of green mercury light of wavelength 546 nm pass through a slit of width 0.437 mm covering a lens of focal length 40 cm. In the focal plane of the lens, what is the distance from the central maximum to the first minimum?

42–20 A slit of width a was placed in front of a lens of focal length 80 cm. The slit was illuminated by parallel light of wavelength 600 nm and the diffraction pattern of Fig. 42–28(b) was formed on a screen in the second focal plane of the lens. If the photograph of Fig. 42–28(b) represents an enlargement to twice the actual size, what was the slit width?

42–21 The intensity of light in the Fraunhofer diffraction pattern of a single slit is

$$I = I_0 (\sin \beta/\beta)^2,$$

where

$$\beta = (\pi a \sin \theta)/\lambda.$$

Show that the equation for the values of β at which I is a maximum is $\tan \beta = \beta$. How can you solve such an equation graphically?

42–22 Plane monochromatic waves of wavelength 600 nm are incident normally on a plane transmission grating having 500 lines mm^{-1}. Find the angles of deviation in the first, second, and third orders.

42–23 A plane transmission grating is ruled with 4000 lines cm^{-1}. Compute the angular separation in degrees, in the second-order spectrum, between the α and δ lines of atomic hydrogen, whose wavelengths are respectively 656 nm and 410 nm. Assume normal incidence.

42–24 (a) What is the wavelength of light which is deviated in the first order through an angle of 20° by a transmission grating having 6000 lines cm^{-1}? (b) What is the second-order deviation of this wavelength? Assume normal incidence.

42–25 What is the longest wavelength that can be observed in the fourth order for a transmission grating having 5000 lines cm^{-1}? Assume normal incidence.

42–26 In Fig. 42–34, two point sources of light, a and b, at a distance of 50 m from lens L and 6 mm apart, produce images at c which are just resolved by Rayleigh's criterion. The focal length of the lens is 20 cm. What is the diameter of the diffraction circles at c?

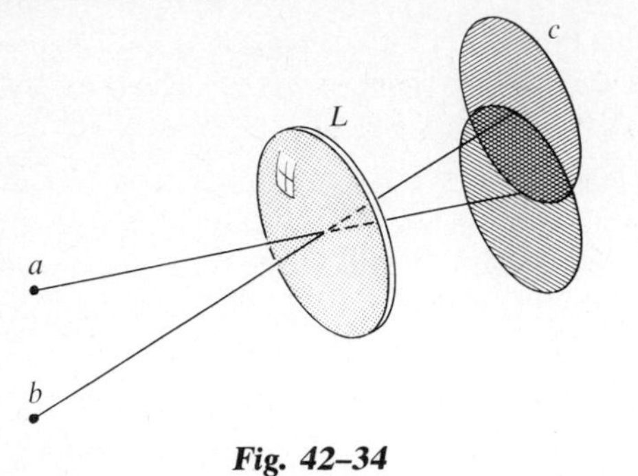

Fig. 42–34

42–27 A telescope is used to observe two distant point sources 1 ft apart. (λ = 500 nm.) The objective of the telescope is covered with a slit of width 1 mm. What is the maximum distance in feet at which the two sources may be distinguished as two?

42–28 A typical lens in a 35-mm camera has a focal length of 50 mm and a diameter of 25 mm (f:2). The resolution of such lenses is expressed as the number of lines per millimeter in the image that can be resolved. If the resolution of this lens is limited by diffraction effects, approximately what is its resolution in lines mm^{-1}?

42–29 If a hologram is made using 600-nm light and then viewed using 500-nm light, how will the images look compared to those observed with 500-nm light?

42–30 A hologram is made using 600-nm light and is then viewed using continuous-spectrum white light from an incandescent bulb. What will be seen?

42–31 Ordinary photographic film reverses black and white, in the sense that the most brightly illuminated areas become blackest upon development (hence the term *negative*). Suppose a hologram negative is viewed directly, without making a positive transparency. How will the resulting images differ from those obtained with the positive hologram?

Chapter 43

Polarization

43–1 POLARIZATION

The phenomena of interference and diffraction can occur with any sort of waves, such as sound waves or surface waves on a liquid. In this chapter we consider some optical phenomena that depend directly on the *transverse* character of light waves. These are called *polarization* effects. They can be observed only with *transverse* waves and cannot be duplicated with sound waves in air because the latter are *longitudinal.*

Let us recall for a moment the nature of the electromagnetic waves radiated by a radio antenna, as described in Section 37–6. We suppose the antenna is vertical, and we consider a portion of a wave front in a vertical plane far away from the antenna, as in Fig. 43–1. The electric field intensity $\boldsymbol{E}$ at all points of this wavefront is in a vertical direction, as indicated. If, in the wave front in the diagram, the electric intensity is maximum in the upward direction, then, in wave fronts $\frac{1}{2}$-wavelength ahead of or behind this one, the intensity is maximum in a downward direction. At all points of any plane fixed in space, the electric vector oscillates up and down along a vertical line and the wave is said to be *linearly* polarized along this direction. (Waves of this sort are also described as *plane polarized*, or merely as *polarized.*)

(To avoid confusion, the magnetic intensity $\boldsymbol{H}$ is not shown in Fig. 43–1. It is always at right angles to the electric field $\boldsymbol{E}$.)

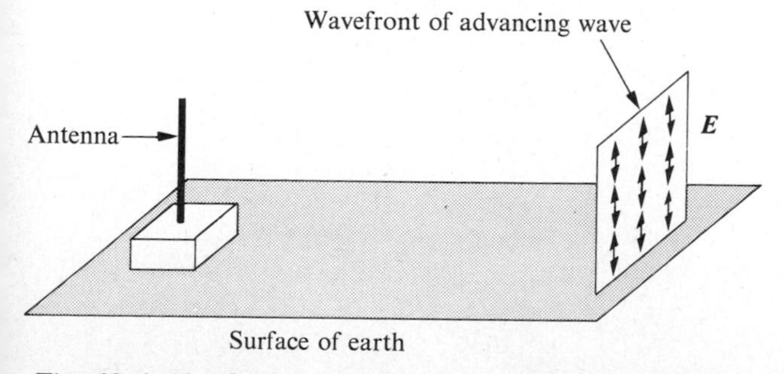

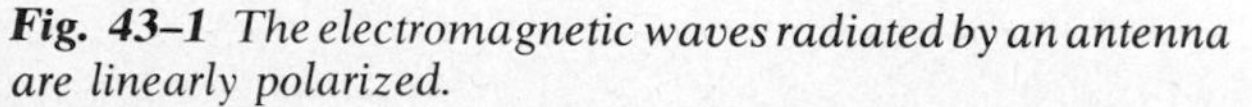

Fig. 43–1 *The electromagnetic waves radiated by an antenna are linearly polarized.*

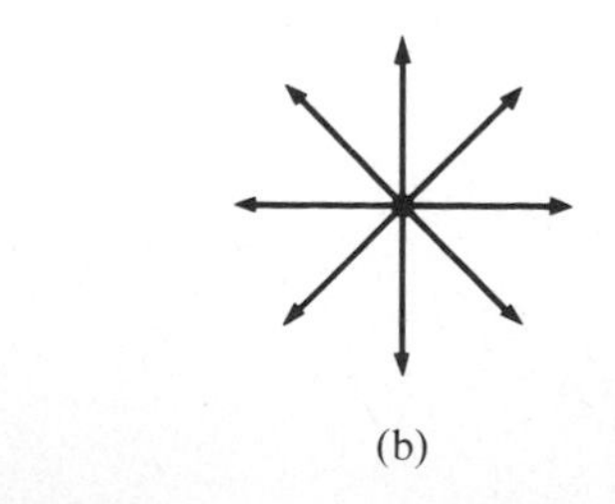

Fig. 43–2 *Schematic diagrams of (a) linearly polarized light, and (b) ordinary light.*

The "antennas" that radiate light waves are the molecules of which light sources are composed. The electrically charged particles in the molecules acquire energy in some way, and radiate this energy as electromagnetic waves of short wavelength. The waves from any one molecule may be linearly polarized, like those from a radio antenna; but since any actual light source contains a tremendous number of molecules, oriented at random, the light emitted is a random mixture of waves linearly polarized in all possible transverse directions. Let the plane of the diagram in Fig. 43–2 represent a wavefront in a beam of light advancing toward the reader, and the dot an end view of one ray in this beam. A linearly polarized light wave is represented schematically by the double arrow, which indicates that the electric field oscillates in the vertical direction only. A beam of natural light is represented as in part (b), in which the arrows indicate a mixture of waves, linearly polarized in all possible transverse directions.

We could equally well characterize the state of polarization of a light wave or any other electromagnetic wave by specifying the direction of the *magnetic* field $\boldsymbol{B}$ or $\boldsymbol{H}$ rather than the electric field. The latter is chosen in practice because most mechanisms for detecting radiation employ principally the electric-field forces on electrons in materials. The most common manifestations of electromagnetic radiation are due chiefly to the electric-field force, not the magnetic-field force.

43–2 POLARIZATION BY REFLECTION

There are a number of methods by which the vibrations in one particular direction can be "sorted out," in whole or in part, from a beam of natural light. One of these is the familiar process of reflection. When natural light strikes a reflecting surface, there is found to be a preferential reflection for those waves in which the electric vector is perpendicular to the plane of incidence. (The plane of incidence is the plane containing the incident ray and the normal to the surface. See Fig. 43–3.) An exception is that at *normal* incidence all directions of polarization are reflected equally. At one particular angle of incidence, known as the *polarizing angle*, ϕ_p, no light whatever is reflected except that in which the electric vector is perpendicular to the plane of incidence. This case is illustrated in Fig. 43–3.

The situation depicted in Fig. 43–3 calls for somewhat more explanation. The heavy double arrow lettered E in Fig. 43–4 represents the amplitude

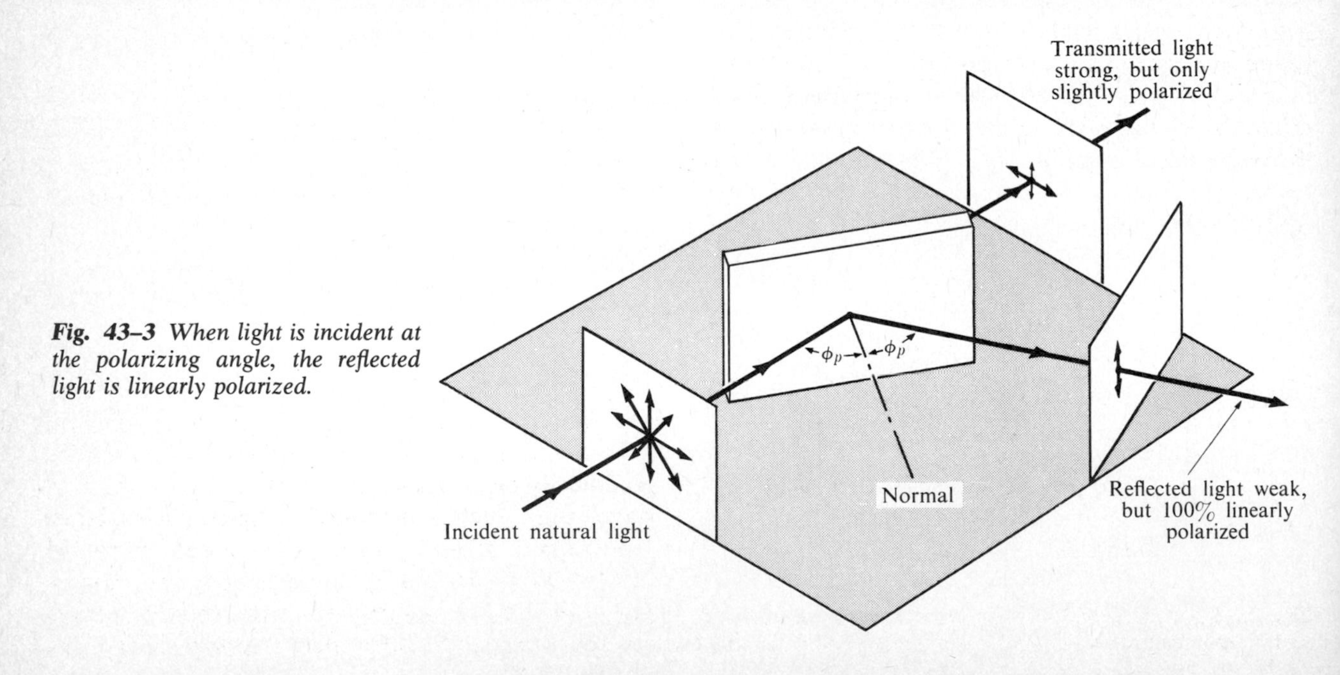

Fig. 43–3 *When light is incident at the polarizing angle, the reflected light is linearly polarized.*

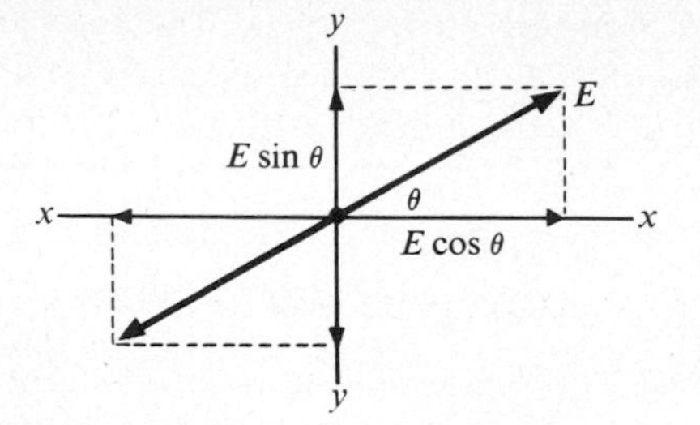

Fig. 43–4 Linearly polarized light resolved into two linearly polarized components.

of the electric field in a linearly polarized wave advancing toward the reader, the direction of vibration making an angle θ with the x-axis. This wave can be resolved into two component waves (that is, it is equivalent to these two waves) linearly polarized along the x- and y-axes, and of amplitudes $E \cos \theta$ and $E \sin \theta$. In the same way, each linearly polarized component in the incident beam of natural light in Fig. 43–3, as represented by the "star" of vectors, can be resolved into two components, one perpendicular and the other parallel to the plane of incidence.

We can describe the reflection of light at the surface by stating what happens to each component of an arbitrary linearly polarized wave in the incident light. When incident at the polarizing angle, *none* of the components parallel to the plane of incidence is reflected; that is each is 100% transmitted in the *refracted* beam. Of the components perpendicular to the plane of incidence, about 15% are reflected if the

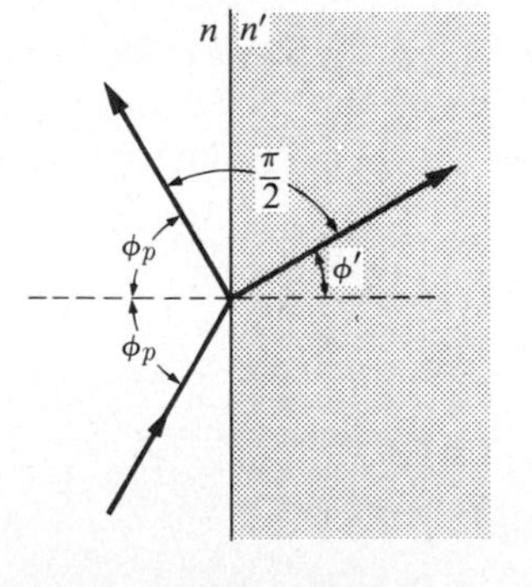

Fig. 43–5 At the polarizing angle the reflected and transmitted rays are perpendicular to each other.

reflecting surface is glass. The fraction reflected depends on the index of the reflecting material. Hence the *reflected* light is weak and *completely* linearly polarized. The *refracted* light is a mixture of the parallel components, all of which are refracted, and the remaining 85% of the perpendicular component. It is therefore strong, but only *partially* polarized.

At angles of incidence other than the polarizing angle some of the components parallel to the plane of incidence are reflected, so that, except at the polarizing angle, the reflected light is not completely linearly polarized.

To increase the intensity of the reflected light, a pile of thin glass plates is often used. Due to the many rays reflected at the polarizing angle from the various surfaces, there is not only an increase in intensity of the reflected light, but also an increase in the polarization of the transmitted light, which now contains much less of the perpendicular component.

In 1812, Sir David Brewster noticed that when the angle of incidence is equal to ϕ_p, the reflected ray and the refracted ray are perpendicular to each other, as shown in Fig. 43–5. When this is the case, the angle of refraction ϕ' becomes the complement of ϕ_p, so that $\sin \phi' = \cos \phi_p$. Since

$$n \sin \phi_p = n' \sin \phi',$$

we get $n \sin \phi_p = n' \cos \phi_p$, and

$$\boxed{\tan \phi_p = \frac{n'}{n},} \qquad (43\text{–}1)$$

a relation known as *Brewster's law.*

43–3 DOUBLE REFRACTION

The progress of a wave train through a homogeneous isotropic medium, such as glass, may be determined graphically by Huygens' construction. The secondary wavelets in such a medium are spherical surfaces. There exist, however, many transparent crystalline substances which, while homogeneous, are *anisotropic*. That is, the velocity of a light wave in them is not the same in all directions. Crystals having this

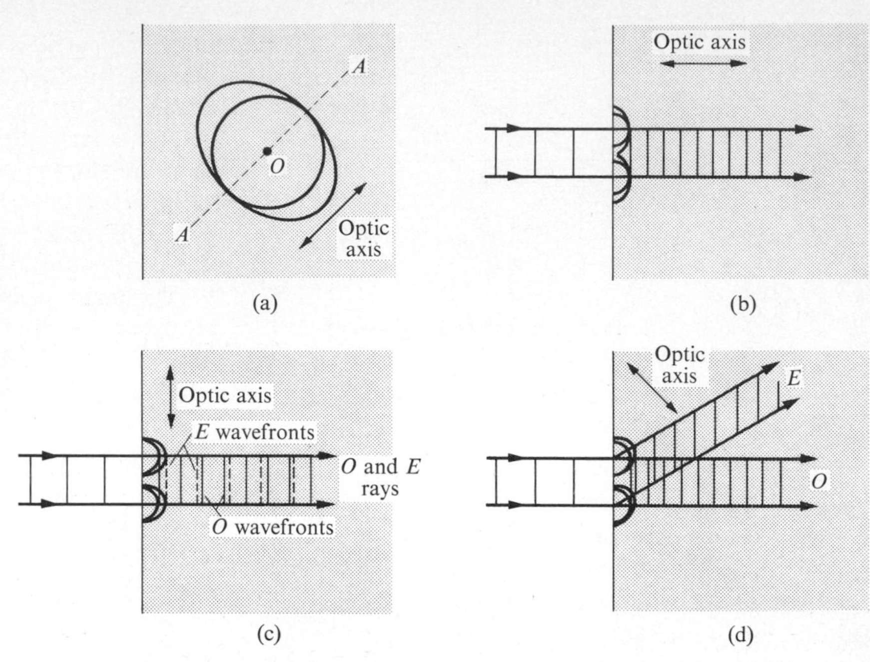

Fig. 43–6 *(a) Spherical and ellipsoidal waves diverge from a point source in a uniaxial crystal. (b) In the direction of the optic axis there is no distinction between O- and E-beams. (c) Perpendicular to the optic axis there is the maximum difference in the speed and the wavelength of the O- and E-beams. (d) At an arbitrary angle to the optic axis the O- and E-beams differ in direction, speed, and wavelength.*

property are said to be *doubly refracting*, or *birefringent.* Two sets of Huygens wavelets propagate from every wave surface in such a crystal, one set being spherical and the other ellipsoidal. The two sets are tangent to each other in one direction, called the optic axis of the crystal.

Figure 43–6(a) shows the traces of the Huygens wavelets from a point source O within a doubly refracting crystal. The complete wave surfaces are obtained by rotating the diagram about axis AA. The direction of line AOA is the optic axis. (The optic axis is a *direction* in the crystal, not just one line. Any other line parallel to AOA is also an optic axis.)

Parts (b), (c) and (d) of Fig. 43–6 show the wave fronts in three sections cut from the crystal in different directions, when light is incident normally on the surface of the section. It will be seen that two sets of wave fronts travel through the crystal, one formed by the tangents to the spheres and the other by the tangents to the ellipsoids.

It may be seen in Fig. 43–6 that a ray incident normally is broken up into two rays in traversing the crystal. The ray which corresponds to wave surfaces tangent to the spherical wavelets is undeviated and is called the *ordinary ray*. The ray corresponding to the wave surfaces tangent to the ellipsoids is deviated even though the incident ray is normal to the surface, and is called the *extraordinary ray*. If the crystal is rotated about the incident ray as an axis, the ordinary ray remains fixed but the extraordinary ray revolves around it, as shown in Fig. 43–7. Furthermore, for angles of incidence other than 0°, Snell's law (i.e., $\sin\phi/\sin\phi' =$ constant) holds for the ordinary but *not* for the extraordinary ray, as the velocity of the latter is different in different directions.

The index of refraction for the extraordinary ray is therefore a function of direction. It is customary to state the index for the direction at right angles to the

Table 43–1 INDICES OF REFRACTION OF DOUBLY REFRACTING CRYSTALS (For light wavelength 589 nm)

Material	n_O	n_E
Calcite	1.655	1.4864
Quartz	1.544	1.553
Tourmaline	1.64	1.62
Ice	1.306	1.307

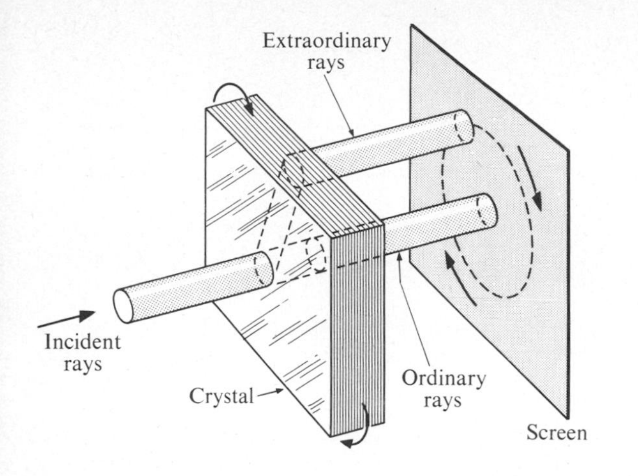

Fig. 43–7 *A narrow beam of natural light can be split into two beams by a doubly refracting crystal.*

optic axis, in which the velocity is a maximum or a minimum. Some values of n_O and n_E, the indices for the ordinary and extraordinary rays, are listed in Table 43–1.

Figure 43–6 is drawn for a crystal in which the velocity of the ellipsoidal waves is *greater* than that of the spherical waves, except in the direction of the optic axis. In some crystals the velocity of the ellipsoidal waves is *less* than that of the spherical waves except along the optic axis, where the two are equal. Both this type of crystal and that described above are called *uniaxial.* In some crystals there are two different directions in which the velocities are equal. These crystals are called *biaxial,* but since all of the doubly refracting crystals used in optical instruments (chiefly quartz and calcite) are uniaxial, we shall concentrate on this type.

43–4 POLARIZERS

Experiment shows that the ordinary and extraordinary waves in a doubly refracting crystal are linearly polarized in mutually perpendicular directions. Consequently, if some means can be found to separate one wave from the other, a doubly refracting crystal may be used to obtain linearly polarized light from natural light. There are several ways in which this separation may be accomplished:

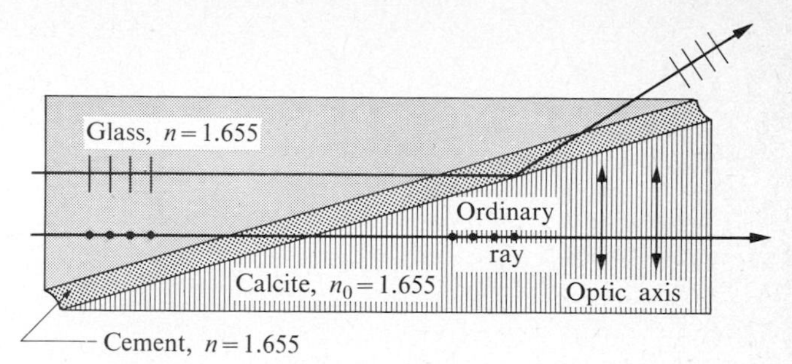

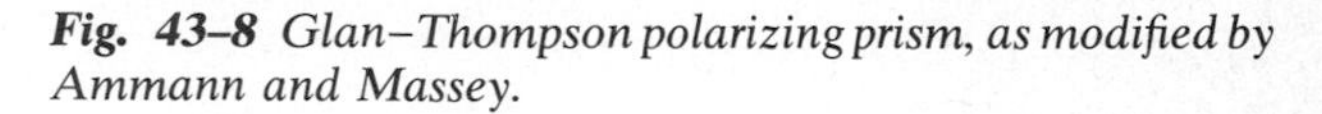

Fig. 43–8 *Glan–Thompson polarizing prism, as modified by Ammann and Massey.*

1. One of the rays may be made to undergo internal reflection and be deflected to one side, allowing the other to proceed undeflected.
2. Both rays may be separated slightly so that, at sufficient distance from the separating prism, only one ray is intercepted.
3. One of the rays may be absorbed while the other is unaffected.

Many ingenious composite prisms of doubly refracting crystals have been developed as polarizers in the last 150 years. They are known by the names of the discoverers: Nicol, Rochon, Wollaston, Glan, Foucault, Ahrens, and many others. The best known is the Nicol prism, invented in 1828 by the Scottish physicist William Nicol. It is complicated and difficult to construct, uses a large amount of optically clear calcite, produces a lateral displacement of the transmitted beam, gives rise to a distorted image, and produces less than 100% linearly polarized light.

A prism originally developed by Glan and Thompson and recently modified by Ammann and Massey (*Journal of the Optical Society of America,* November 1968) is constructed on a similar principle but avoids the defects of the Nicol prism. The modified Glan–Thompson prism consists of a glass prism with index of refraction 1.655 joined to a calcite prism with a cement that has the same index of refraction; see Fig. 43–8, where the thickness of the layer of cement is exaggerated. The natural unpolarized light incident at the left is equivalent to two equal linearly polarized beams. The direction of polarization of one (represented by dots) is perpen-

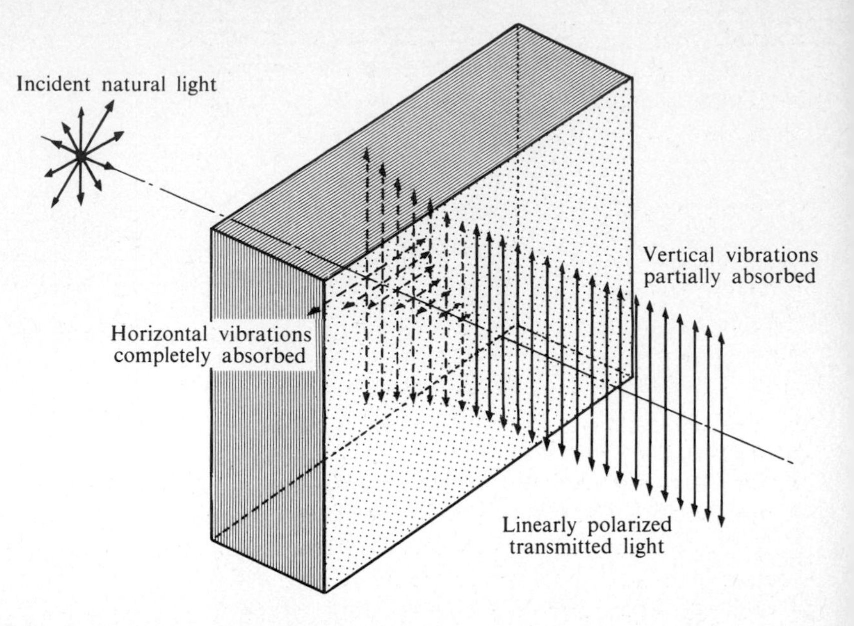

Fig. 43–9 *Linearly polarized light transmitted by a dichroic crystal.*

dicular to the page and that of the other (represented by short vertical lines) is parallel to the page. Both beams travel the same path with equal speed in the glass. In calcite, however, with the optic axis vertical as shown, perpendicular vibrations constitute the ordinary ray for which calcite has an index of refraction 1.6583. This value is so close to 1.655 that the perpendicular vibrations proceed without deflection from glass to cement to calcite and emerge into the air as 100% linearly polarized light.

The parallel vibrations, if they existed in the calcite crystal, would constitute the extraordinary ray with an index of refraction 1.4864. In traveling *in the cement* toward the calcite, this ray proceeds toward a medium in which its speed is greater, and therefore a critical angle exists. With the dimensions of the prism as shown, the angle of incidence of the ray with parallel vibrations exceeds the critical angle, and this ray is therefore internally reflected and deflected from its original direction. This polarizing prism functions properly for all incident rays which make an angle of not more than 10° above or below the horizontal.

Certain doubly refracting crystals exhibit *dichroism*; that is, one of the polarized components is *absorbed* much more strongly than the other. Hence, if the crystal is cut of the proper thickness, one of the components is practically extinguished by absorption, while the other is transmitted in appreciable amount, as indicated in Fig. 43–9. Tourmaline is one example of such a dichroic crystal.

An early form of Polaroid, invented by Edwin H. Land in 1928, consists of a thin layer of tiny needlelike dichroic crystals of herapathite (iodoquinine sulfate), in parallel orientation, embedded in a plastic matrix and enclosed for protection between two transparent plates. A more recent modification, developed by Land in 1938 and known as an H-sheet, is a molecular polarizer. It consists of long polymeric molecules of polyvinyl alcohol (PVA) that have been given a preferred direction by stretching, and have been stained with an ink containing iodine that causes the sheet to exhibit dichroism. The PVA sheet is laminated to a support sheet of cellulose acetate butyrate.

Polaroid disks do not polarize all wavelengths equally. When two such disks are crossed, small amounts of red and of violet (the two ends of the visible spectrum) are transmitted. When white light passes through one Polaroid sheet, the transmitted light is slightly colored. The large area of such plates, however, and their moderate cost more than compensate for these small deficiencies. The existence of

Polaroid sheets has stimulated the development and applications of polarized light to an extent that was out of the question when reflecting surfaces, Nicol prisms, and other costly devices had to be used.

43-5 PERCENTAGE POLARIZATION. MALUS' LAW

When light is incident on a polarizer, as in Fig. 43-10, only plane-polarized light is transmitted. The polarizer may be a pile of plates, a Nicol prism, or a sheet of Polaroid. It is represented as a Polaroid disk in Fig. 43-10. The dotted line across the polarizer indicates the direction of the electric vector of the transmitted light. This light falls on a photocell, and the current in a microammeter connected to the cell is proportional to the quantity of light incident on it.

If the incident light is unpolarized, then as the polarizer is rotated about the incident ray as an axis, the reading of the microammeter remains constant. The polarizer transmits the components of the incident waves in which the $\boldsymbol{E}$-vector is parallel to the transmission direction of the polarizer, and by symmetry the components are equal for all azimuths.

If there is any variation in the meter reading as the polarizer is rotated, the incident light is *not* natural light and is said to be *partially* polarized. Suppose the meter reading does vary. Let I_{max} and I_{min} represent the maximum and minimum values of the quantity of light incident on the photocell, or the maximum and minimum meter readings, since the two are proportional. The *percentage polarization* of

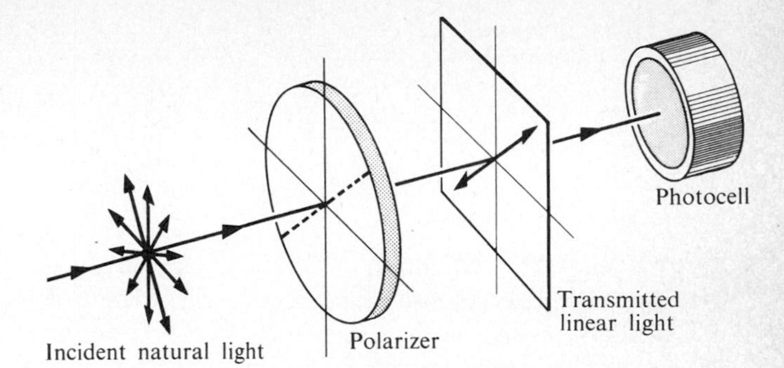

Fig. 43-10 *The intensity of the transmitted linear light is the same at all azimuths of the polarizer.*

the incident light is defined as

$$\text{Percent polarization} = \frac{I_{max} - I_{min}}{I_{max} + I_{min}} \times 100. \qquad (43\text{-}2)$$

Suppose now that a second Polaroid sheet is inserted in the light between polarizer and photocell, as in Fig. 43-11. Let the transmission direction of the second Polaroid sheet, or *analyzer*, be vertical, and let that of the polarizer make an angle θ with the vertical. The linear light transmitted by the polarizer may be resolved into two components as shown, one parallel and the other perpendicular to the transmission direction of the analyzer. Evidently only the parallel component, of amplitude $E \cos \theta$, will be transmitted by the analyzer. The transmitted light is maximum when $\theta = 0$, and is zero when $\theta = 90°$, or when polarizer and analyzer are *crossed*. At interme-

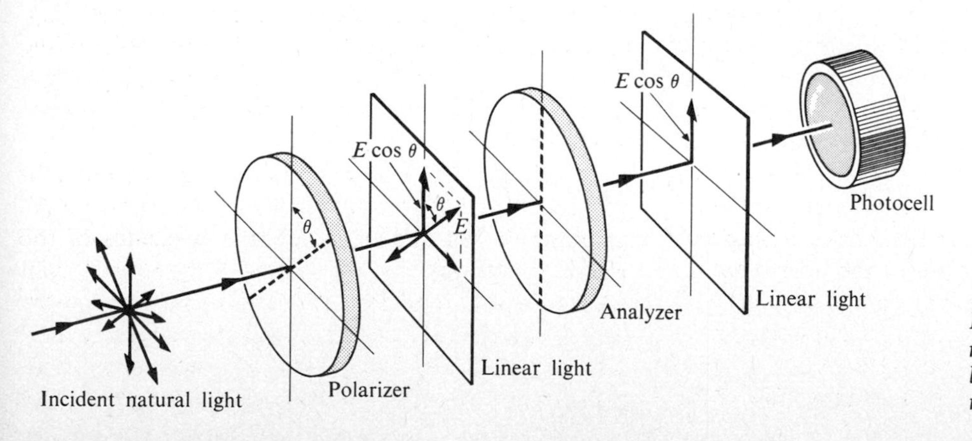

Fig. 43-11 *The analyzer transmits only the component of the linear light parallel to its transmission direction.*

diate angles, since the quantity of energy is proportional to the *square* of the amplitude, we have

$$I = I_{\max} \cos^2 \theta, \tag{43–3}$$

where $I_{\max}$ is the maximum amount of light transmitted and I is the amount transmitted at the angle θ. This relation, discovered experimentally by Etienne Louis Malus in 1809, is called *Malus' Law.*

The angle θ is the angle between the transmission directions of polarizer and analyzer. If either the analyzer or the polarizer is rotated, the amplitude of the transmitted beam varies with the angle between them according to Eq. (43–3).

Polaroid sheet is now widely used in sunglasses where, from the standpoint of its polarizing properties, it plays the role of the analyzer in Fig. 43–11. We have seen that when unpolarized light is reflected, there is a preferential reflection for light polarized perpendicular to the plane of incidence. When sunlight is reflected from a horizontal surface, the plane of incidence is vertical. Hence, in the reflected light there is a preponderance of light polarized in the horizontal direction, the proportion being greater the nearer the angle of incidence is to the polarizing angle. When such reflection occurs at smooth asphalt road surfaces, the surface of a lake, or a similar situation, it causes unwanted "glare," and vision is improved by eliminating it. The transmission direction of the Polaroid sheet in the sunglasses is vertical, so none of the horizontally-polarized light is transmitted to the eyes.

Apart from this polarizing feature, these glasses serve the same purpose as any dark glasses, absorbing 50% of the incident light, since even in an unpolarized beam, half the light can be considered as polarized horizontally and half vertically. Only the vertically polarized light is transmitted. The sensitivity of the eye is independent of the state of polarization of the light.

43–6 THE SCATTERING OF LIGHT

The sky is blue. Sunsets are red. Skylight is largely linearly polarized, as can readily be verified by looking at the sky directly overhead through a polarizing plate. It turns out that one and the same phenomenon is responsible for all three of the effects noted above.

In Fig. 43–12, sunlight (unpolarized) comes from the left along the z-axis and passes over an observer looking vertically upward along the y-axis. One of the molecules of the earth's atmosphere is

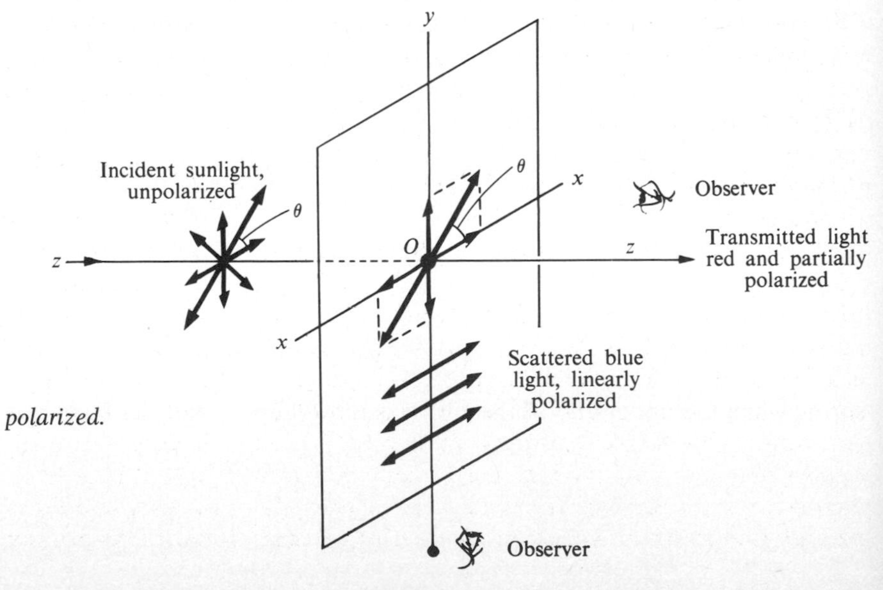

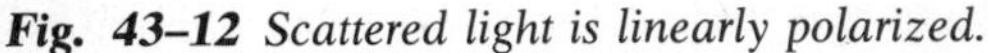
Fig. 43–12 *Scattered light is linearly polarized.*

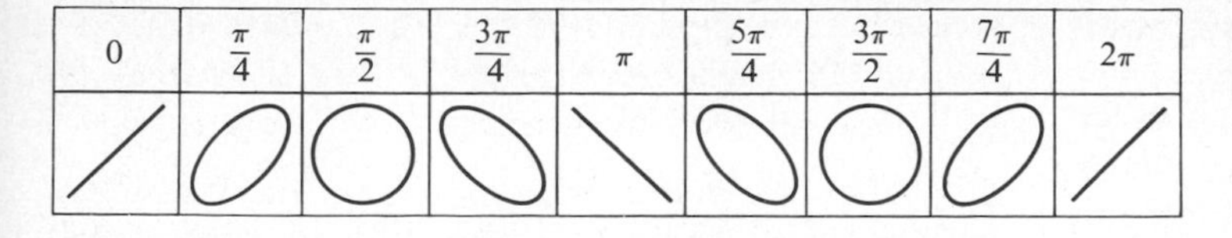

Fig. 43–13 Vibrations which result from the combination of a horizontal and a vertical simple harmonic motion of the same frequency and the same amplitude, for various values of the phase difference.

located at point O. The electric field in the beam of sunlight sets the electric charges in the molecule in vibration. Since light is a transverse wave, the direction of the electric field in any component of the sunlight lies in the xy-plane and the motion of the charges takes place in this plane. There is no field, and hence no vibration, in the direction of the z-axis.

An arbitrary component of the incident light, vibrating at an angle θ with the x-axis, sets the electric charges in the molecule vibrating in the same direction, as indicated by the heavy line through point O. In the usual way, we can resolve this vibration into two, one along the x-axis and the other along the y-axis. The result, then, is that each component in the incident light produces the equivalent of two molecular "antennas," oscillating with the frequency of the incident light, and lying along the x- and y-axes.

It has been explained in Section 37–6 that an antenna does not radiate in the direction of its own length. Hence, the antenna along the y-axis does not send any light to the observer directly below it. It does, of course, send out light in other directions. The only light reaching the observer comes from the component of vibration along the x-axis, and, as is the case with the waves from any antenna, this light is linearly polarized, with the electric field parallel to the antenna. The vectors on the y-axis below point O show the direction of vibration of the light reaching the observer.

The process described above is called *scattering*. The energy of the scattered light is abstracted from the original beam, which becomes weakened in the process. The vibration of the charges in the molecule is a *forced* vibration, like the vibration of a mass on a spring when the upper end of the spring is moved up and down with simple harmonic motion. It is well known that the amplitude of the forced vibrations increases as the driving frequency approaches the natural frequency of vibration of the spring–mass system. Now the natural frequencies of the electric charges in a molecule are in the same range as that of light in the ultraviolet region of the spectrum. The frequencies of the waves in visible light are *less* than the natural frequency, but the higher their frequency, or the shorter their wavelength, the closer is the driving frequency to the natural frequency, the greater the amplitude of vibration, and the greater the intensity of the scattered light. In other words, blue light is scattered more than red, with the result that the hue of the scattered light is blue.

Toward evening, when sunlight has to travel a large distance through the earth's atmosphere to reach a point over or nearly over an observer, a large proportion of the blue light in sunlight is removed from it by scattering. White light minus blue light is yellow or red in hue. Thus when sunlight, with the blue component removed, is incident on a cloud, the light reflected from the cloud to the observer has the yellow or red hue so commonly seen at sunset.

From the discussion above, it follows that if the earth had no atmosphere we would receive *no* skylight at the earth's surface, and the sky would appear as black in the daytime as it does at night. Thus, to an astronaut in a space ship or on the moon, the sky appears black, not blue.

43–7 CIRCULAR AND ELLIPTIC POLARIZATION

Linearly polarized light represents a special and relatively simple type of polarization. When the ordinary and extraordinary rays in a doubly refracting crystal are separated, each ray taken alone is linearly polarized, but with the directions of vibration at right angles. When, however, the crystal is cut with its faces parallel to the optic axis, so that light incident normally on one of its faces traverses the crystal in a direction perpendicular to the optic axis, as shown in Fig. 43–6(c), the ordinary and extraordinary rays are not separated. *They traverse the same path, but with different speeds.* Upon emerging from the second face of the crystal, the ordinary and extraordinary rays are out of phase with each other and give rise to either *elliptically* polarized, *circularly* polarized, or *linearly* polarized light, depending upon a number of factors which we shall proceed to discuss.

Since the direction of vibration in the ordinary ray is perpendicular to that in the extraordinary ray, we have to consider a fundamental problem which, for the sake of simplicity, may be discussed in mechanical terms: What sort of vibration results from the combination of two simple harmonic vibrations at right angles to each other and differing in phase? The solution may be reached in a variety of ways: (1) with the aid of mechanical equipment, (2) by using two vector diagrams for plotting two simple harmonic motions at right angles, (3) by establishing alternating potential differences on the horizontal and vertical plates of a cathode-ray oscilloscope, and (4) by mathematical calculation (see Problem 43–16).

In Fig. 43–13 are shown the results obtained by combining a horizontal and a vertical simple harmonic motion of the same frequency and the same amplitude, for nine different phase differences. It is at once evident that *two simple harmonic motions at right angles to each other never produce* ***destructive*** *interference, no matter what the phase difference.*

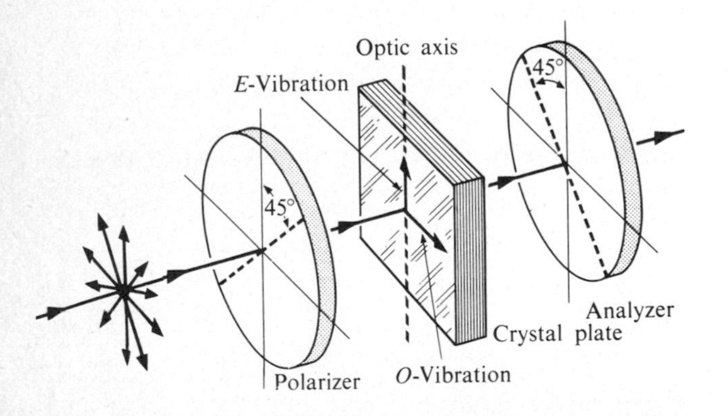

Fig. 43–14 *Crystal plate between crossed Polaroids. Since the optic axis makes an angle of 45° to the vibration direction of the linearly polarized light transmitted by the polarizer, the amplitudes of the O- and E-vibrations in the crystal are equal.*

1. When the phase difference is 0, 2π, or any even multiple of π, the result is a *linear* vibration at 45° to both original vibrations.

2. When the phase difference is π, 3π, or any *odd* multiple of π, the result is also a linear vibration, but at right angles to that corresponding to even multiples of π.

3. When the phase difference is $\pi/2$, $3\pi/2$, or any odd multiple of $\pi/2$, the resulting vibration is *circular.*

4. At all other phase differences, the resulting vibration is *elliptical.*

With these facts in mind, consider the optical apparatus shown in Fig. 43–14. After unpolarized light traverses the polarizer, it is linearly polarized, with the vibration direction along the dotted line drawn on the polarizer. This linear light then enters a crystal plate cut so that the light travels in a direction perpendicular to the optic axis. The crystal plate has been rotated about the light beam until the optic axis makes an angle of 45° with the direction of vibration of the linear light incident upon it. Since the E-vibration is, in this case, parallel to the optic axis, and the O-vibration is perpendicular to it, it follows that the amplitudes of the E- and O-beams are identical. The E- and O-beams travel through the crystal along the same path but with different speeds and, as they are about to emerge from the second crystal face, they combine to form one of the vibrations depicted in Fig. 42–13, depending on the phase difference.

The phase difference between the E- and O-vibrations at the second face of the crystal depends on the following: (1) the frequency of the light, (2) the indices of refraction of the crystal for E- and O-light, and (3) the thickness of the crystal.

If a given crystal has such a thickness as to give rise to a phase difference of $\pi/2$ for a given frequency, then, according to Fig. 43–13, a circular vibration results, and the light emerging from this crystal is said to be *circularly polarized light.* The crystal itself is called a *quarter-wave plate.* If the crystal plate shown in Fig. 43–14 is a quarter-wave plate, the intensity of light transmitted by the analyzer will remain constant as the analyzer is rotated. In other words, if an analyzer alone is used to analyze circularly polarized light, it will give the same result as when used to analyze unpolarized light.

A quarter-wave plate for, say, green light is not a quarter-wave plate for any other color. Other colors have different frequencies and different E- and O-indices. Hence the phase difference would not be $\pi/2$. (See Problem 43–13.)

If the crystal has such a thickness as to give rise to a phase difference of π for a given frequency, then, according to Fig. 43–13, a linear vibration *perpendicular* to the incident vibration direction results. The light that emerges from this crystal is linearly polarized, and by rotating the analyzer, we may find a position of the analyzer at which this light will be completely stopped. A crystal plate of this sort is called a *half-wave plate* for the given frequency of light. For any other frequency it would not be a half-wave plate.

If the phase difference produced by a crystal plate is such as to produce an elliptical vibration, the emerging light is said to be *elliptically polarized.*

43–8 PRODUCTION OF COLORS BY POLARIZED LIGHT

Consider a crystal plate which is a half-wave plate for red light. If red light, linearly polarized at 45° to the optic axis, is allowed to traverse the plate, it emerges from the plate as linearly polarized light with the vibration direction perpendicular to that of the incident light. An analyzer which is crossed with the polarizer will therefore transmit this red light.

Now suppose the incident radiation is *white* light. Only the red component of the white light will emerge from the half-wave plate as linearly polarized light. All the other wavelengths will emerge as either elliptically or circularly polarized light. When the analyzer is in a position to transmit the red, linearly polarized light completely, it will cut out part of every other wavelength. The light transmitted by the analyzer will therefore predominate in red, and will have a pinkish hue. When the analyzer is rotated through 90°, so as to cut the red light out completely, the other wavelengths will be transmitted to some extent and the resulting hue will be the complement of pink, that is, a blue-green.

Now suppose that we have a crystal plate of nonuniform thickness, such as a rough strip of selenite (gypsum). A small area of the strip may have the proper thickness to act as a half-wave plate for red light, another region may serve as a half-wave plate for yellow light, and so on. When a projecting lens is used to project an image of the selenite on a screen, the rest of the apparatus being the same as that in Fig. 43–14, the image will show patches of different colors, corresponding to regions of different thickness. These colors will change into the complementary values when the analyzer is rotated through 90°.

Ordinary cellophane, such as that used for wrapping cigarette packages, is doubly refracting. Striking color effects can be obtained by inserting various thicknesses of cellophane, or a crumpled ball of the material, between a polarizer and an analyzer.

43–9 OPTICAL STRESS ANALYSIS

When a polarizer and an analyzer are mounted in the "crossed" position, i.e., with their transmission directions at right angles to each other, no light is transmitted through the combination. But if a doubly refracting crystal is inserted between polarizer and analyzer, the light after passing through the crystal is, in general, elliptically polarized, and some light will be transmitted by the analyzer. Thus the field of view, dark in the absence of the crystal, becomes light when the crystal is inserted.

Some substances, such as glass, celluloid, and various plastics, while not normally doubly refracting, become so when subjected to mechanical stress. From a study of the specimen between crossed Polaroid disks, much information regarding the stresses can be obtained. Improperly annealed glass, for example, may be internally stressed to an extent which might cause it later to develop cracks. It is evidently important that optical glass should be free from such a condition before it is subjected to expensive grinding and polishing. Hence, such glass is always examined between crossed Polaroid sheets before grinding operations are begun.

The double refraction produced by stress is the basis of the science of *photoelasticity*. The stresses in opaque engineering materials, such as girders, boiler plates, gear teeth, etc., can be analyzed by constructing a transparent model of the object, usually of a plastic, and examining it between a polarizer and an analyzer in the crossed position. Very complicated stress distributions, such as those around a hole or a gear tooth, which it would be practically impossible to analyze mathematically, may thus be studied by optical methods. Figure 43–15 is a photograph of a photoelastic model under stress.

Fig. 43–15 *Photoelastic stress analysis. (Courtesy of Dr. W. M. Murray, Massachusetts Institute of Technology.)*

Liquids are not normally doubly refracting, but some become so when an electric field is established within them. This phenomenon is known as the *Kerr effect*. The existence of the Kerr effect makes it possible to construct an electrically controlled "light valve." A cell with transparent walls contains the liquid between a pair of parallel plates. The cell is inserted between crossed Polaroid disks. Light is transmitted when an electric field is set up between the plates and is cut off when the field is removed.

43–10 STUDY OF CRYSTALS

Polarized light is used extensively in the field of mineralogy. A transparent specimen of rock or crystal is cut into a thin plate and mounted on the stage of a *polarizing microscope*, which has a polarizer below the substage condenser and an analyzer above the objective lens. The substage condenser serves to converge the polarized light onto the specimen, from which it diverges to the objective lens. When monochromatic light is used, the image formed in the focal plane of the objective is found to have a pattern which indicates whether (a) the crystal is uniaxial or biaxial, (b) if uniaxial, where the optic axis is, and (c) if biaxial, the angle between the optic axes. Some of the polarization figures obtained with convergent polarized light are shown in Fig. 43–16. If white light is used, these patterns are brilliantly colored.

The figure obtained with a uniaxial crystal cut with its faces perpendicular to the optic axis consists of concentric colored rings; and when the polarizer and analyzer are crossed, a black cross is superimposed on the rings. The elements of the black cross are called brushes. When circularly polarized light is used, the cross may be eliminated.

Polarizing ring sights are constructed from a plate of crystal, two Polaroid sheets, two quarter-wave plates, and two protecting plates of glass. When mounted on a gun perpendicular to the barrel and viewed with the eye, colored rings appear in space and move with the gun. Aiming the gun is accomplished by sighting the target within the center of the ring system.

43–11 OPTICAL ACTIVITY

When a beam of linearly polarized light is sent through certain types of crystals and certain liquids, the direction of vibration of the emerging linearly polarized light is found to be different from the original direction. This phenomenon is called *rotation of the plane of polarization*, and substances which exhibit the effect are called *optically active*. Those which rotate the plane of polarization to the right, looking along the advancing beam, are called *dextrorotatory* or righthanded; those which rotate it to the left, *laevorotatory* or lefthanded.

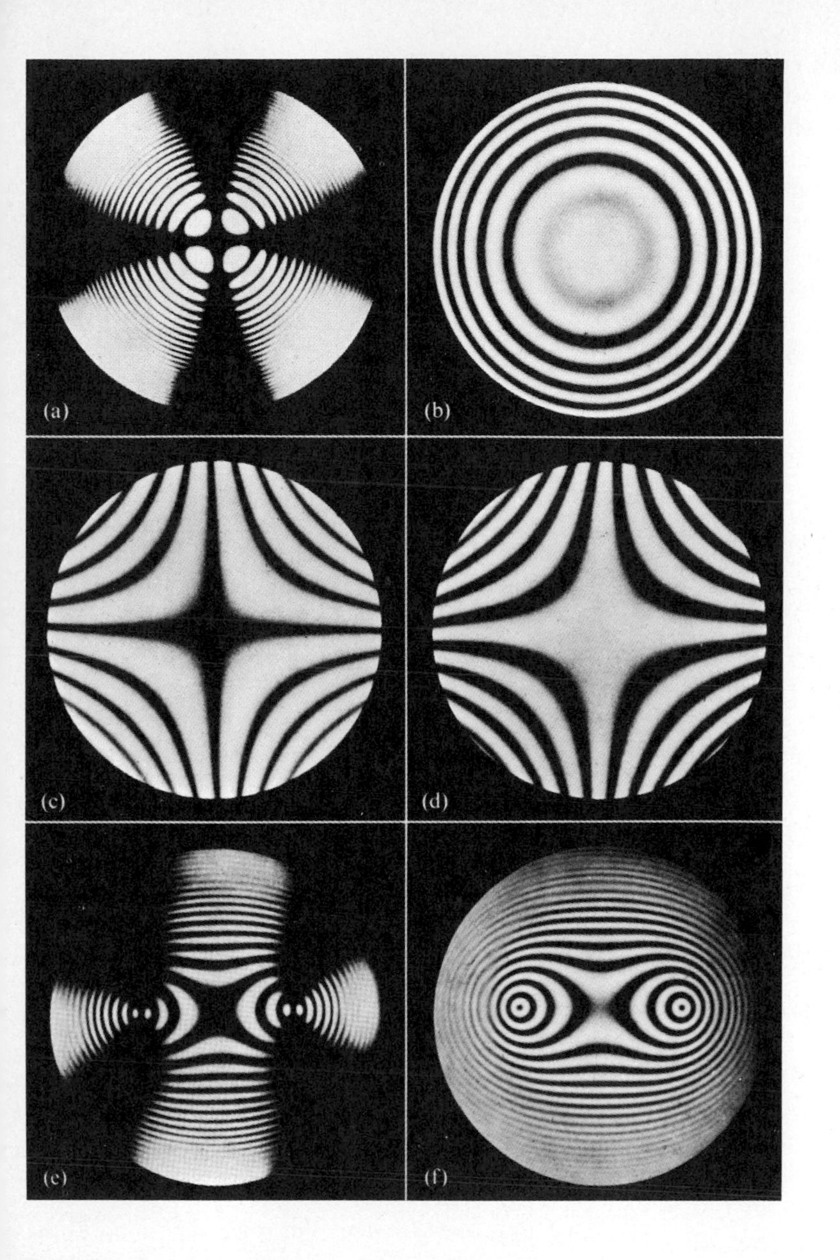

Fig. 43–16 *Figures obtained with convergent polarized light. (a) Uniaxial crystal cut perpendicular to the optic axis, crossed Polaroids. (b) Same crystal, parallel Polaroids, with quarter-wave plates to eliminate brushes. (c) Uniaxial crystal cut parallel to the optic axis, crossed Polaroids. (d) Same crystal, parallel Polaroids. (e) Biaxial crystal, crossed Polaroids. (f) Same crystal with quarter-wave plates to eliminate brushes. (Photographed by H. Hauswaldt, Magdeburg, 1902.)*

Optical activity may be due to an asymmetry of the molecules of a substance, or it may be a property of a crystal as a whole. For example, solutions of cane sugar are dextrorotatory, indicating that the optical activity is a property of the sugar molecule. The molecules of the sugars dextrose and levulose are mirror images, and their optical activities are opposite. The rotation of the plane of polarization by a sugar solution is used commercially as a method of determining the proportion of cane sugar in a given sample. Crystalline quartz is also optically active, some natural crystals being righthanded and others lefthanded. Here the optical activity is a consequence of the crystalline structure, since it disappears when the quartz is melted and allowed to resolidify into a glassy, noncrystalline state called fused quartz.

PROBLEMS

43–1 A beam of light is incident on a liquid of 1.40 refractive index. The reflected rays are completely polarized. What is the angle of refraction of the beam?

43–2 The critical angle of light in a certain substance is 45°. What is the polarizing angle?

43–3

a) At what angle above the horizontal must the sun be in order that sunlight reflected from the surface of a calm body of water shall be completely polarized?

b) What is the plane of the E-vector in the reflected light?

43–4 A parallel beam of "natural" light is incident at an angle of 58° on a plane glass surface. The reflected beam is completely linearly polarized.

a) What is the angle of refraction of the transmitted beam?

b) What is the refractive index of the glass?

43–5 The Glan–Thompson prism shown in Fig. 43–8 has a horizontal length of 10 cm.

a) What is the critical angle for a ray in the cement approaching the calcite?

b) If a horizontal ray approaching the calcite is to make an angle of incidence 10° larger than the critical angle, what should be the vertical height of the prism?

43–6 A parallel beam of linearly polarized light of wavelength 589 nm (in vacuum) is incident on a calcite crystal, as in Fig. 43–6(c). Find the wavelengths of the ordinary and extraordinary waves in the crystal.

43–7 A polarizer and an analyzer are oriented so that the maximum amount of light is transmitted. To what fraction of its maximum value is the intensity of the transmitted light reduced when the analyzer is rotated through (a) 30°, (b) 45°, (c) 60°?

43–8 A beam of linearly polarized light strikes a calcite crystal, the direction of the electric vector making an angle of 60° with the optic axis.

a) What is the ratio of the amplitude of the two refracted beams?

b) What is the ratio of their intensities?

43–9 Refer to Table 43–1 and draw Fig. 43–6(a), (c), and (d), for quartz.

43–10 Figure 43–17 represents a Wollaston prism made of two prisms of quartz cemented together. The optic axis of the righthand prism is perpendicular to the page, whereas that of the lefthand prism is parallel. The incident light is normal to the surface and gives rise to O- and E-beams which travel in the lefthand prism along the same path but with different speeds. Copy Fig. 43–17 and show on your diagram how the O- and E-beams are bent in going into the righthand prism and thence into the air.

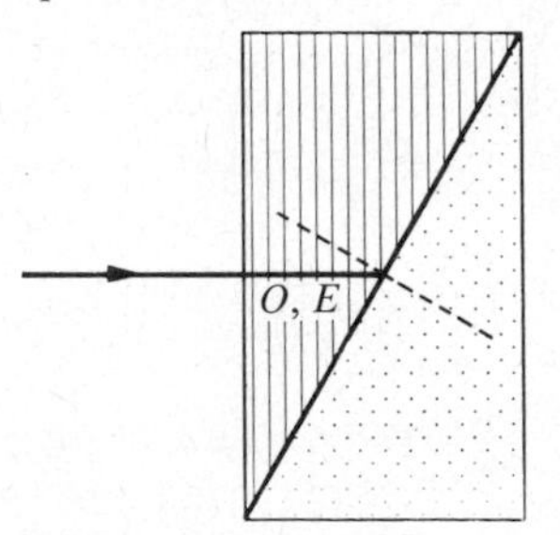

Fig. 43–17

43–11 A beam of light, after passing through the Polaroid disk P_1 in Fig. 43–18, traverses a cell containing a scattering medium. The cell is observed at right angles through another Polaroid disk P_2. Originally, the disks are oriented until the brightness of the field as seen by the observer is a maximum.

a) Disk P_2 is rotated through 90°. Is extinction produced?

b) Disk P_1 is now rotated through 90°. Is the field bright or dark?

c) Disk P_2 is then restored to its original position. Is the field bright or dark?

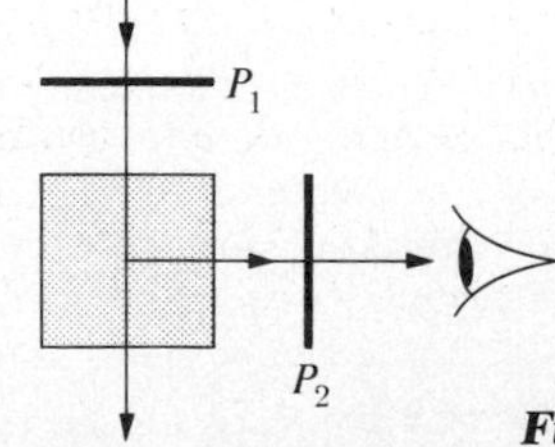

Fig. 43–18

43–12 In Fig. 43–19, A and C are Polaroid sheets whose transmission directions are as indicated. B is a sheet of doubly refractive material whose optic axis is vertical. All three sheets are parallel. Unpolarized light enters from the left. Discuss the state of polarization of the light at points 2, 3, and 4.

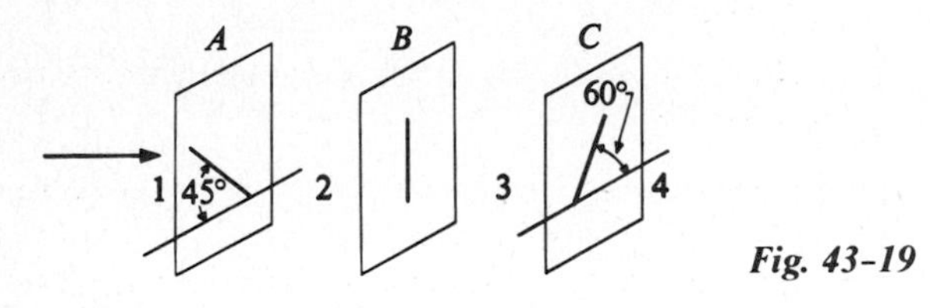

Fig. 43–19

43–13 The refractive index of a certain flint glass is 1.65. For what incident angle is light reflected from the surface of this glass completely polarized if the glass is immersed in (a) air? (b) water?

43–14 A certain birefringent material has a refractive index of 1.71 for the ordinary ray and 1.74 for the extraordinary ray, for 600-nm light. What thickness of material is needed for a quarter-wave plate?

43–15 It is desired to rotate the plane of polarization of plane-polarized light 90°, using two Polaroid filters. Explain how this can be done, and find the final intensity in terms of the incident intensity.

43–16 Three polarizing filters are stacked, with the polarizing axes of the second and third at 45° and 90°, respectively, with that of the first.

a) If unpolarized light of intensity I_0 is incident on the stack, find the intensity and state of polarization after each filter.

b) If the second filter is removed, how does the situation change?

43–17 Unpolarized light of intensity I_0 is incident on a polarizing filter, and the emerging light strikes a second polarizing filter with its axis at 45° to that of the first. Determine (a) the intensity of the emerging beam, and (b) its state of polarization.

43–18 A beam of right-circularly polarized light is reflected at normal incidence from a reflecting surface. Is the reflected beam right- or left-circularly polarized? Explain.

43–19 The phase difference δ between the E- and O-rays, after traversing a crystal plate such as that in Fig. 43–14, is given by

$$\delta = \frac{2\pi}{\lambda} t(n_O - n_E),$$

where λ is the wavelength in air and t is the thickness of the crystal.

a) Show that the minimum thickness of a quarter-wave plate is given by $t = \lambda/4(n_O - n_E)$.

b) What is this minimum thickness for a quarter-wave calcite plate and light of 400-nm wavelength?

43–20 What is the state of polarization of the light transmitted by a quarter-wave plate when the electric vector of the incident linearly polarized light makes an angle of 30° with the optic axis?

43–21 Assume the values of n_O and n_E for quartz to be independent of wavelength. A certain quartz crystal is a quarter-wave plate for light of wavelength 800 nm (in vacuum). What is the state of polarization of the transmitted light when linearly polarized light of wavelength 400 nm (in vacuum) is incident on the crystal, the direction of polarization making an angle of 45° with the optic axis?

43–22 Consider two vibrations, one along the y-axis,

$$y = a \sin(\omega t - \alpha),$$

and the other along the z-axis, of equal amplitude and frequency, but differing in phase,

$$z = a \sin(\omega t - \beta).$$

Let us write them as follows:

$$\frac{y}{a} = \sin \omega t \cos \alpha - \cos \omega t \sin \alpha, \qquad (1)$$

$$\frac{z}{a} = \sin \omega t \cos \beta - \cos \omega t \sin \beta. \qquad (2)$$

a) Multiply Eq. (1) by $\sin \beta$ and Eq. (2) by $\sin \alpha$ and then subtract the resulting equations.

b) Multiply Eq. (1) by $\cos \beta$ and Eq. (2) by $\cos \alpha$ and then subtract the resulting equations.

c) Square and add the results of (a) and (b).

d) Derive the equation $y^2 + z^2 - 2yz \cos \delta = a^2 \sin^2 \delta$, where $\delta = \alpha - \beta$.

e) Justify the diagrams in Fig. 43–13.

Chapter 44

Photons, Electrons, and Atoms

44–1 EMISSION AND ABSORPTION OF LIGHT

The past several chapters have been concerned with understanding various phenomena associated with the propagation of light, on the basis of an electromagnetic wave theory. The work of Hertz established the existence of electromagnetic waves and the fact that light is an electromagnetic wave. The phenomena of interference, diffraction, and polarization are easily understood on the basis of a wave model; and, when interference effects can be neglected, the further simplification of ray optics permits analysis of the behavior of lenses and mirrors. These phenomena are collectively referred to as *classical optics*, and insofar as the phenomena of classical optics are concerned, the electromagnetic wave theory is *complete.*

There are, however, many phenomena which are *not* so readily understood on this basis. An example is the emission of light from matter. The electromagnetic waves of Hertz, with frequencies of the order of 10^8 Hz, were produced by oscillations in a resonant L-C circuit similar to those studied in Chapter 34. Frequencies of visible light are much larger, of the order of 10^{15} Hz, far higher than the highest frequencies attainable with conventional electronic equipment.

In the mid-nineteenth century it was speculated that visible light might be produced by motion of electric charge within individual atoms rather than in macroscopic circuits. In fact, in 1862 Faraday placed a light source in a strong magnetic field in an attempt to determine whether the emitted radiation was changed by the field. He was not able to detect any change, but when his experiments were repeated thirty years later by Zeeman with greatly improved equipment, changes *were* observed.

Particularly puzzling was the existence of *line spectra.* Light emitted from atoms heated in a flame, or excited electrically in a glow tube such as the familiar neon sign or mercury-vapor light, does not contain a continuous spread of wavelengths, but only certain well-defined wavelengths. In a spectrometer using a narrow slit in conjunction with a prism or a diffraction grating, the spectrum pattern appears as a series of bright lines, and hence has come to be known as a *line spectrum.* It was learned early in the nineteenth century that each element emits a *characteristic spectrum,* suggesting that there is a direct relation between the characteristics and internal structure of an atom and its spectrum. Attempts to understand this relation on the basis of newtonian mechanics and classical electricity and magnetism were not successful, however.

There were other mysteries. The *photoelectric effect*, discovered by Hertz in 1887 during his investigations of electromagnetic wave propagation, is the liberation of electrons from the surface of a conductor when light strikes the surface. This phenomenon can be understood qualitatively on the basis that when light is absorbed by the surface it transfers energy to electrons near the surface, and that some of the electrons acquire enough energy to surmount the potential-energy barrier at the surface and escape from the material into space. More detailed investigation revealed some puzzling features which could *not* be understood on the basis of classical optics.

Still another area of unsolved problems centered around the production and scattering of *x-rays*, electromagnetic radiation with wavelengths shorter than those of visible light by a factor of the order of 10^4 and with correspondingly greater frequencies. These rays were produced in high-voltage glow discharge tubes, but the details of this process eluded understanding. Even worse, when these rays collided with matter, the scattered ray sometimes had a longer wavelength than the original ray. This is like directing a beam of blue light at a mirror and having it reflect as red!

All these phenomena, and several others, pointed forcefully to the conclusion that classical optics successful though it was in explaining ray optics, interference, and polarization, nevertheless had its limitations. Understanding the phenomena cited above would require at least some generalization of the classical theory. In fact, it has required something much more radical than that. All these phenomena are concerned with the *quantum* theory of radiation, which includes the assumption that despite the *wave* nature of electromagnetic radiation, it nevertheless has some properties akin to those of *particles.* In particular, the energy conveyed by an electromagnetic wave is always carried in units whose magnitude is proportional to the frequency of the wave. These units of energy are called *photons* or *quanta.*

Thus, electromagnetic radiation emerges as an entity with a dual nature, having both wave and particle aspects. The remainder of this chapter will be devoted to the applications of this duality to some of the phenomena mentioned above, and to study of this seemingly (but not actually) inconsistent nature of electromagnetic radiation.

44-2 THERMIONIC EMISSION

As a prelude to the discussion of the photoelectric effect, we consider briefly a related phenomenon, *thermionic emission*, discovered by Thomas Edison in 1883 during his experiments on electric light bulbs. A glassblower had sealed into the bulb of an ordinary filament lamp an extra metal electrode or plate, shown in Fig. 44-1. The glass bulb was then evacuated and the filament heated as usual. When the plate was connected through a galvanometer to the *positive* terminal of the 110-V dc source, a galvanometer deflection indicated the existence of a current, despite the fact that there was no conducting path from the plate to the other terminal of the battery. When the plate was connected to the *negative* terminal there was no current. Edison was very much interested in this phenomenon at the time but could not explain it.

The effect is caused by the escape of electrons from the hot filament. Ordinarily electrons in a conductor are prevented from escaping from the surface by a potential-energy barrier. When an electron starts to move away from the surface it induces a corresponding positive charge in the material, which tends to pull it back into the surface. To escape, the electron must somehow acquire enough energy to surmount this energy barrier; the minimum energy needed to escape is called the *work function*, and varies from one material to another. Typical

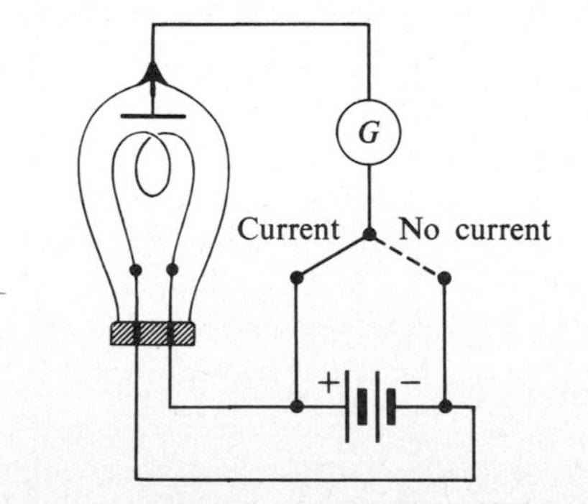

Fig. 44-1 *Edison's original thermionic-emission experiment.*

work functions are of the order of a few electron-volts. The electron-volt is defined in Section 26–8, and it would do no harm to review that section in preparation for this chapter.

At ordinary temperatures almost none of the electrons can acquire enough energy to escape, but when the filament is very hot the electron energies are greatly increased by thermal motion, and at sufficiently high temperatures considerable numbers are able to escape. Once out of the material, the electrons are attracted to the plate if it is positively charged, but repelled if it is negatively charged. The liberation of electrons from a hot wire is called *thermionic emission.*

The electrons which have escaped from the hot conductor form a cloud of negative charge near it, called a *space charge.* If a second conductor (the plate) is maintained by a battery at a higher potential than the first, the electrons in the cloud are attracted to it, and so long as the potential difference between the conductors is maintained, there will be a steady drift of electrons from the emitter or *cathode* to the other body, which is called the plate or *anode.*

In the common thermionic tube the cathode and anode (and often other electrodes as well) are enclosed within an evacuated glass or metal container, and leads to the various electrodes are brought out through the base or walls of the tube. The simplest thermionic tube, in which the only electrodes are the cathode and anode, is called a *diode.*

The diode is shown schematically in Fig. 44–2. The cathode and plate are represented by K and P. The cathode is often in the form of a hollow cylinder,

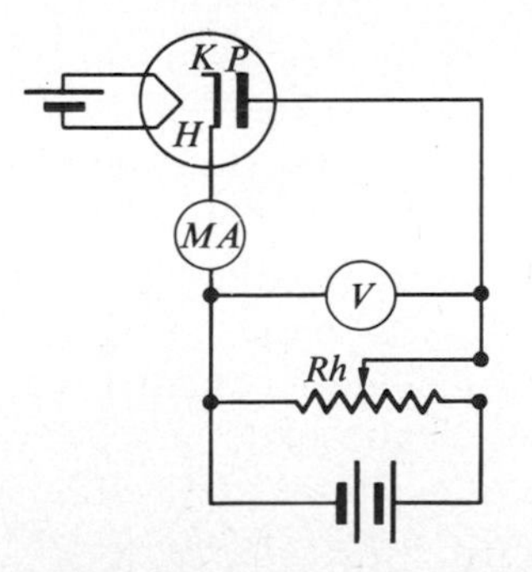

Fig. 44–2 *Circuit for measuring plate current and plate voltage in a vacuum diode.*

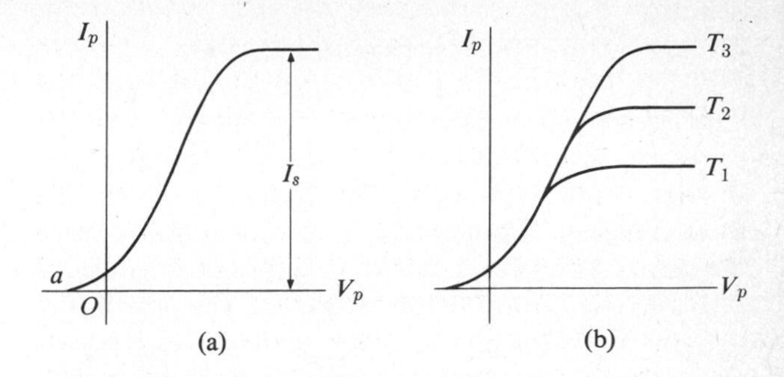

Fig. 44–3 *(a) Plate current-plate voltage characteristic of a diode. (b) Plate current curves at three different cathode temperatures.* $T_3 > T_2 > T_1$.

which is heated by a fine resistance wire H within it. Electrons emitted from the outer surface of the cathode are attracted to the plate, which is a larger cylinder surrounding the cathode and coaxial with it. The electron current to the plate is read on the milliammeter MA. The potential difference between plate and cathode can be controlled by the slide wire and read on voltmeter V.

If the potential difference between cathode and anode is small (a few volts), only a few of the emitted electrons reach the plate, the majority penetrating a short distance into the cloud of space charge and then returning to the cathode. As the plate potential is increased, more and more electrons are drawn to it, and with sufficiently high potentials (of the order of 100 V), *all* of the emitted electrons arrive at the plate. Further increase of plate potential does not increase the plate current, which is then said to become *saturated.*

A graph of plate current, I_p, versus plate potential, V_p, is shown in Fig. 44–3(a). Note that I_p is not zero even when V_p is zero. This is because the electrons leave the cathode with an initial velocity and the more rapidly moving ones may penetrate the cloud of space charge and reach the plate even with no accelerating field. In fact, a *retarding* field is necessary to prevent their reaching the plate, an effect which may be used to measure their velocities of emission.

The saturation current I_s, in Fig. 44–3(a) is equal to the current from the cathode, and for a given tube its magnitude increases markedly with cathode temperature. Figure 44–3(b) shows three plate current curves at three different temperatures, where $T_3 > T_2 > T_1$.

The work function ϕ of a surface may be considerably reduced by the presence of impurities. A small amount of thorium, for example, reduces the work function of pure tungsten by about 50%. Since the smaller the work function the larger the current density at a given temperature (or the lower the temperature at which a given emission can be attained), most vacuum tubes now use cathodes having composite surfaces.

Because thermionic emission occurs only at the heated cathode, not at the anode, electron flow occurs in only one direction, from cathode to anode. Most circuit applications of the diode make use of this one-directional characteristic; the simplest example is a *rectifier* for converting alternate current to direct current.

Lee de Forest discovered in 1907 that the electron flow can be modified by inserting a third electrode called a *grid* between cathode and anode. The grid is usually an open structure, either a screen or a coil of wire. When the grid is negative with respect to the cathode the resulting field enhances the space charge near the cathode and decreases the electron flow through the grid to the plate. Because of the grid's proximity to the cathode, the current is much more sensitive to changes in grid voltage than to changes in anode voltage, and thus the device can function as an *amplifier*. In the form just described it is called a *triode*. This basic device and various elaborations of it are very widely used in electronic equipment, although in the past two decades they have been supplanted in many applications by transistors, which will be discussed in Chapter 45.

44-3 THE PHOTOELECTRIC EFFECT

In the thermionic emission of electrons from metals, the energy needed by an electron to escape from the metal surface is furnished by the energy of thermal agitation. Electrons may also acquire enough energy to escape from a metal, even at low temperatures, if the metal is illuminated by light of sufficiently short wavelength. This phenomenon is called the *photoelectric effect*. It was first observed by Heinrich Hertz in 1887, who noticed that a spark would jump more readily between two spheres when their surfaces were illuminated by the light from another spark. The effect was investigated in detail in the following years by Hallwachs and Lenard.

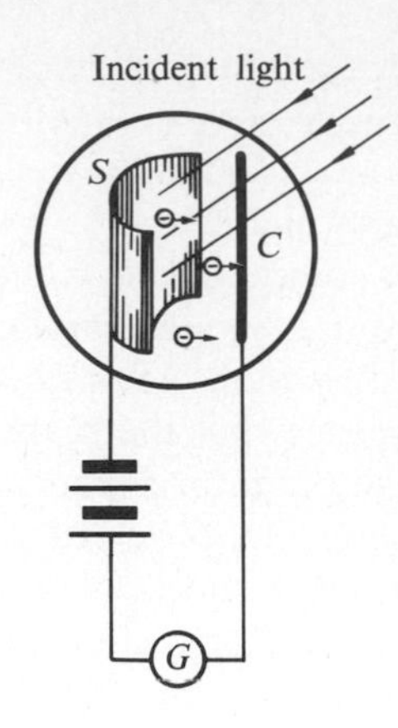

Fig. 44–4 *Schematic diagram of a photocell circuit.*

A modern phototube is shown schematically in Fig. 44–4. A beam of light, indicated by the arrows, falls on a photosensitive surface *S*. Electrons emitted by the surface are drawn to the collector *C*, normally maintained at a positive potential with respect to the emitter. Emitter and collector are enclosed in an evacuated container. The photoelectric current can be read on the galvanometer *G*.

It is found that with a given material as emitter, the wavelength of the light must be *shorter* than a critical value, different for different surfaces, in order for any photoelectrons at all to be emitted. The corresponding frequency *minimum* is called the *threshold frequency* of the particular surface. The threshold frequency for most metals is in the ultraviolet (critical wavelength 200 to 300 nm), but for potassium and cesium oxide it lies in the visible spectrum (400 to 700 nm).

Just as in the case of thermionic emission, the photoelectrons form a cloud of space charge around the emitter *S*. Some of the electrons are emitted with an initial velocity; this is shown by the fact that, even with no emf in the external circuit, a few electrons penetrate the cloud of space charge and reach the collector, causing a small current in the external circuit. The velocity of the most rapidly moving electrons can be deduced by measuring the reversed voltage (negative potential of collector) required to reduce the current to zero. This potential is known as the *stopping potential.*

A remarkable feature of photoelectric emission is the relation between the number and maximum velocity of escaping electrons, on the one hand, and the intensity and wavelength of the incident light, on the other. Surprisingly enough, it is found that the maximum velocity of emission is *independent of the intensity* of the light, but does depend on its wavelength. It is true that the photoelectric current increases as the light intensity is increased, but only because *more* electrons are emitted. With light of a given wavelength, no matter how feeble it may be, the maximum velocity of the photoelectrons from a given surface is always the same, provided, of course, that the frequency is above the threshold frequency.

The explanation of the photoelectric effect was given by Einstein in 1905, although his theory was so radical that it was not generally accepted until 1916, when it was confirmed by experiments performed by Millikan. Extending a proposal made two years earlier by Planck, Einstein postulated that a beam of light consisted of small bundles of energy which are now called *light quanta* or *photons.* The energy E of a photon is proportional to its frequency f, or is equal to its frequency multiplied by a constant. That is,

$$E = hf, \tag{44–1}$$

where h is a universal constant called Planck's constant whose value is 6.63×10^{-34} J s. When a photon collides with an electron at or just within the surface of a metal, it may transfer its energy to the electron. This transfer is an "all-or-none" process, the electron getting all the photon's energy or none at all. The photon then simply drops out of existence. The energy acquired by the electron may enable it to escape from the surface of the metal if it is moving in the right direction.

In leaving the surface of the metal, the electron loses energy in amount ϕ (the work function of the surface). Some electrons may lose more than this if they start at some distance below the metal surface, but the *maximum* energy with which an electron can emerge is the energy gained from a photon minus the work function. Hence the maximum kinetic energy of the photoelectrons ejected by light of frequency f is

$$\tfrac{1}{2}mv_{\max}^2 = hf - \phi. \tag{44–2}$$

This is Einstein's photoelectric equation, and it was in exact agreement with Millikan's experimental results.

44–4 LINE SPECTRA

We have seen how a prism or grating spectrograph functions to disperse a beam of light into a spectrum. If the light source is an incandescent solid or liquid, the spectrum is *continuous*; that is, light of all wavelengths is present. If, however, the source is a gas through which an electrical discharge is passing, or a flame into which a volatile salt has been introduced, the spectrum is of an entirely different character. Instead of a continuous band of color, only a few colors appear, in the form of isolated parallel lines. (Each "line" is an image of the spectrograph slit, deviated through an angle dependent on the frequency of the light forming the image.) A spectrum of this sort is termed a *line spectrum.* The wavelengths of the lines are characteristic of the element emitting the light. That is, hydrogen always gives a set of lines in the same position, sodium another set, iron still another, and so on. The line structure of the spectrum extends into both the ultraviolet and infrared regions, where photographic or other means are required for its detection.

It might be expected that the frequencies of the light emitted by a particular element would be arranged in some regular way. For instance, a radiating atom might be analogous to a vibrating string, emitting a fundamental frequency and its harmonics. At first sight there does not seem to be any semblance of order or regularity in the lines of a typical spectrum; for many years unsuccessful attempts were made to

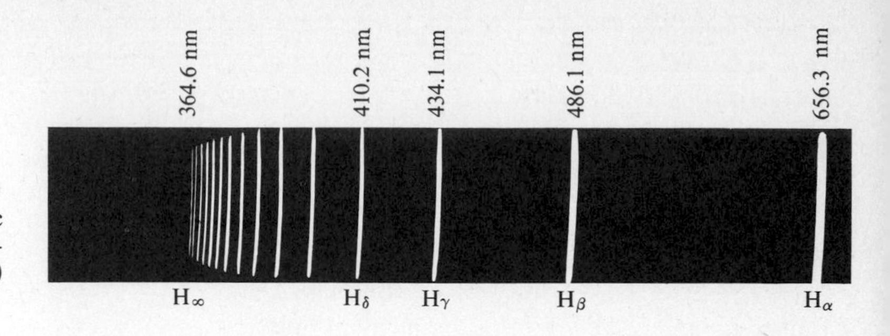

Fig. 44–5 The Balmer series of atomic hydrogen. (Reproduced by permission from Atomic Spectra and Atomic Structure *by Gerhard Herzberg. Copyright 1937 by Prentice-Hall, Inc.)*

correlate the observed frequencies with those of a fundamental and its overtones. Finally, in 1885, Johann Jakob Balmer (1825–1898) found a simple formula which gave the frequencies of a group of lines emitted by atomic hydrogen. Since the spectrum of this element is relatively simple, and fairly typical of a number of others, we shall consider it in more detail.

Under the proper conditions of excitation, atomic hydrogen may be made to emit the sequence of lines illustrated in Fig. 44–5. This sequence is called a *series*. There is evidently a certain order in this spectrum, the lines becoming crowded more and more closely together as the limit of the series is approached. The line of longest wavelength or lowest frequency, in the red, is known as H_α, the next, in the blue-green, as H_β, the third as H_γ, and so on. Balmer found that the wavelengths of these lines were given accurately by the simple formula

$$\frac{1}{\lambda} = R\left(\frac{1}{2^2} - \frac{1}{n^2}\right), \qquad (44\text{–}3)$$

where λ is the wavelength, R is a constant called the Rydberg constant, and n may have the integral values 3, 4, 5, etc. If λ is in meters,

$$R = 1.097 \times 10^7\ \text{m}^{-1}.$$

Letting $n = 3$ in Eq. (44–3), one obtains the wavelength of the H_α-line:

$$\frac{1}{\lambda} = 1.097 \times 10^7\ \text{m}^{-1}(1/4 - 1/9)$$

$$= 1.522 \times 10^6\ \text{m}^{-1},$$

whence

$$\lambda = 656.3\ \text{nm}.$$

For $n = 4$, one obtains the wavelength of the H_β-line, etc. For $n = \infty$, one obtains the limit of the series, at $\lambda = 364.6$ nm. This is the *shortest* wavelength in the series.

Other series spectra for hydrogen have since been discovered. These are known, after their discoverers, as the Lyman, Paschen, and Brackett series. The formulas for these are

Lyman series:

$$\frac{1}{\lambda} = R\left(\frac{1}{1^2} - \frac{1}{n^2}\right), \qquad n = 2, 3, \ldots,$$

Paschen series:

$$\frac{1}{\lambda} = R\left(\frac{1}{3^2} - \frac{1}{n^2}\right), \qquad n = 4, 5, \ldots,$$

Brackett series:

$$\frac{1}{\lambda} = R\left(\frac{1}{4^2} - \frac{1}{n^2}\right), \qquad n = 5, 6 \ldots.$$

The Lyman series is in the ultraviolet, and the Paschen and Brackett series are in the infrared. The Balmer series evidently fits into the scheme between the Lyman and the Paschen series.

The Balmer formula, Eq. (44–3), may also be written in terms of the frequency of the light, recalling that

$$c = f\lambda \qquad \text{or} \qquad \frac{1}{\lambda} = \frac{f}{c}.$$

Thus, Eq. (44–3) becomes

$$f = Rc\left(\frac{1}{2^2} - \frac{1}{n^2}\right) \quad (44\text{–}4)$$

or

$$f = \frac{Rc}{2^2} - \frac{Rc}{n^2}. \quad (44\text{–}5)$$

Each of the fractions on the right side of Eq. (44–5) is called a *term*, and the frequency of every line in the series is given by the difference between two terms.

There are only a few elements (hydrogen, singly ionized helium, doubly ionized lithium) whose spectra can be represented by a simple formula of the Balmer type. Nevertheless, it is possible to separate the more complicated spectra of other elements into series, and to express the frequency of each line in the series as the difference of two *terms*. The first term is constant for any one series, while the various values of the second term can be labeled by values of an integer index n analogous to the n appearing in Eq. (44–5). In a few simple cases the numerical values of the terms can be calculated from theoretical considerations, but for complex atoms they must be determined experimentally by analysis of spectra.

44–5 THE BOHR ATOM

Einstein invoked the concept of light quanta, or photons, to account for the experimental facts of photoelectric emission. We shall now see how the Danish physicist, Niels Bohr, in 1913, first applied the same ideas to the emission of light by atoms.

Experiments on the scattering of alpha particles by thin metallic foils were performed by Rutherford and his co-workers about 1906. These experiments, which will be discussed in Chapter 46, showed that an atom consists of a very small, massive, positively charged nucleus, surrounded by a swarm of electrons. To account for the fact that the electrons in an atom remain at relatively large distances from the nucleus, in spite of the electrostatic force of attraction of the nucleus for them, Rutherford postulated that the electrons *revolve* about the nucleus, the force of attraction providing the requisite centripetal force to retain them in their orbits.

This assumption, however, has an unfortunate consequence. A body moving in a circle is continuously accelerated toward the center of the circle and, according to classical electromagnetic theory, an accelerated electron radiates energy. The total energy of the electrons would therefore gradually decrease, their orbits would become smaller and smaller, and eventually they would spiral into the nucleus and come to rest. Furthermore, according to classical theory, the *frequency* of the electromagnetic waves emitted by a revolving electron is equal to the frequency of revolution. As the electrons radiated energy, their angular velocities would change continuously and they would emit a *continuous* spectrum (a mixture of all frequencies), in contradiction to the *line* spectrum actually observed.

Faced with the dilemma that electromagnetic theory predicted an unstable atom emitting radiant energy of all frequencies, while observation showed stable atoms emitting only a few frequencies, Bohr concluded that, in spite of the success of electromagnetic theory in explaining large-scale phenomena, it could not be applied to processes on an atomic scale. It became clear that a fairly radical departure from the established principles of classical mechanics and electromagnetism would be needed to understand the structure of atoms and the relation of atomic structure to atomic spectra.

In Bohr's new theory this departure took the form of two postulates. Bohr's first postulate was that *an electron in an atom can revolve in certain stable orbits without the emission of radiant energy*, contrary to the predictions of classical electromagnetic theory. According to this postulate, each atom has certain definite stable states in which it can exist, and each possible state has a definite total energy.

A *completely* stable atom, however, is as unsatisfactory as an unstable one, since atoms *do* emit radiant energy. Bohr's second postulate incorporated into atomic theory the quantum concepts that had been developed by Planck and applied by Einstein to the photoelectric effect. The second postulate was that *an electron may make a transition from one of its specified nonradiating orbits to another of lower energy. When it does so, a single photon is emitted having energy equal to the energy difference between the initial and final states, and with frequency f given by the relation*

$$hf = E_i - E_f, \qquad (44\text{–}6)$$

where h is Planck's constant and E_i and E_f are the energies of the initial and final states. For example, a photon of orange light of wavelength 600 nm has a frequency f given by

$$f = \frac{c}{\lambda} = \frac{3.00 \times 10^8 \text{ m s}^{-1}}{600 \times 10^{-9} \text{ m}} = 5.00 \times 10^{14} \text{ Hz}$$
$$= 5.00 \times 10^{14} \text{ s}^{-1}.$$

The corresponding photon energy is

$$E = hf = (6.62 \times 10^{-34} \text{ J s})(5.00 \times 10^{14} \text{ s}^{-1})$$
$$= 3.31 \times 10^{-19} \text{ J}$$
$$= 2.07 \text{ eV}.$$

Thus, this photon must be emitted in a transition between two states of the atom differing in energy by 2.07 eV.

This hypothesis, if correct, would shed new light on the analysis of spectra on the basis of *terms*, as described in Section 44–4. For example, Eq. (44–5) gives the frequencies of the Balmer series in the hydrogen spectrum. Multiplied by Planck's constant h, this becomes

$$hf = \frac{Rch}{2^2} - \frac{Rch}{n^2}. \qquad (44\text{–}7)$$

If we now identify $-Rch/n^2$ with the initial energy of the atom E_i and $-Rch/2^2$ with its final energy, E_f, before and after a transition in which a photon of energy $hf = E_i - E_f$ is emitted, then Eq. (44–7) takes on the same form as Eq. (44–6). More generally, if we assume that the possible energy levels for the hydrogen atom are given by

$$E_n = -\frac{Rch}{n^2}, \qquad n = 1, 2, 3, \ldots, \qquad (44\text{–}8)$$

then *all* the series spectra of hydrogen can be understood on the basis of transitions from one energy level to another. For the Lyman series the final state is always $n = 1$, for the Paschen series it is $n = 3$, and so on. Similarly, complex spectra of other elements, represented by *terms*, are understood on the basis that each term corresponds to an energy level; and a frequency, represented as a difference of two terms, corresponds to a transition between the two corresponding energy levels.

Although this scheme permits partial understanding of line spectra on the basis of energy levels of atoms, it is not yet complete because it provides no basis for *predicting* what the energy levels for any particular kind of atom should be. As we have already seen, classical mechanics and electromagnetic theory are inadequate for this task. Bohr devised a model which predicted correctly the energy levels of hydrogen, the simplest atom. He assumed that the electron in the hydrogen atom travels in a circular orbit, but that only certain orbit radii are permitted. Corresponding to each allowed radius is an allowed energy, and the permitted radii are determined by a new *quantum condition* which is outside the realm of classical mechanics. Bohr found that the allowed energy levels of atomic hydrogen, as computed from Eq. (44–8) were in agreement with observation, provided the electron was permitted to rotate about the nucleus *only in those orbits for which the angular momentum is some integral multiple of* $h/2\pi$. It will be recalled that the angular momentum of a particle of mass m, moving with tangential velocity v in a circle of radius r, is mvr. Hence the quantum condition above may be stated

$$\boxed{mvr = n\frac{h}{2\pi},} \qquad (44\text{–}9)$$

where $n = 1, 2, 3, \ldots$

We now incorporate this condition into the analysis of the hydrogen atom. This atom consists of a single electron of charge $-e$, rotating about a single proton of charge $+e$. The proton, being nearly 2000 times as massive as the electron, will be assumed stationary in this discussion. The electrostatic force of attraction between the charges,

$$F = \frac{1}{4\pi\varepsilon_0}\frac{e^2}{r^2},$$

provides the necessary centripetal force and, from Newton's second law,

$$\frac{1}{4\pi\varepsilon_0}\frac{e^2}{r^2} = \frac{mv^2}{r}. \qquad (44\text{–}10)$$

When Eqs. (44–9) and (44–10) are solved simultaneously for r and v, we obtain

$$r = \varepsilon_0 \frac{n^2h^2}{\pi me^2}, \qquad (44\text{–}11)$$

$$v = \frac{1}{\varepsilon_0}\frac{e^2}{2nh}.$$

Let

$$\varepsilon_0 \frac{h^2}{\pi me^2} = r_0. \qquad (44\text{–}12)$$

Then Eq. (44–11) becomes

$$r = n^2 r_0,$$

and the permitted, nonradiating orbits are of radii r_0, $4r_0$, $9r_0$, etc. The appropriate value of n is called the *quantum number* of the orbit.

The numerical values of the quantities on the left side of Eq. (44–12) are

$$\varepsilon_0 = 8.85 \times 10^{-12}\,\text{C}^2\,\text{N}^{-1}\,\text{m}^{-2},$$
$$h = 6.62 \times 10^{-34}\,\text{J s},$$
$$m = 9.11 \times 10^{-31}\,\text{kg},$$
$$e = 1.60 \times 10^{-19}\,\text{C}.$$

Hence r_0, the radius of the first Bohr orbit, is

$$r_0 = \frac{(8.85 \times 10^{-12})(6.62 \times 10^{-34})^2}{(3.14)(9.11 \times 10^{-31})(1.60 \times 10^{-19})^2}$$
$$= 5.3 \times 10^{-11}\,\text{m} = 0.53 \times 10^{-8}\,\text{cm}.$$

This is in good agreement with atomic diameters as estimated by other methods, namely, about 10^{-8} cm.

The kinetic energy of the electron in any orbit is

$$E_k = \tfrac{1}{2}mv^2 = \frac{1}{\varepsilon_0^2}\frac{me^4}{8n^2h^2},$$

and the potential energy is

$$E_p = -\frac{1}{4\pi\varepsilon_0}\frac{e^2}{r} = \frac{1}{\varepsilon_0^2}\frac{me^4}{4n^2h^2}.$$

The total energy, E, is therefore

$$\boxed{E = E_k + E_p = -\frac{1}{\varepsilon_0^2}\frac{me^4}{8n^2h^2}.} \qquad (44\text{–}13)$$

When numerical values of the constants are inserted, one obtains

$$E = -\frac{2.18 \times 10^{-18}\,\text{J}}{n^2} = -\frac{13.6\,\text{eV}}{n^2}.$$

The total energy has a negative sign because the reference level of potential energy is taken with the electron at an infinite distance from the nucleus. Since we are interested only in energy *differences*, this is not of importance.

The energy of the atom is *least* when its electron is revolving in the orbit for which $n = 1$, for then E has its largest negative value. For $n = 2, 3, \ldots$, the absolute value of E is smaller, hence the energy is progressively larger in the outer orbits. The *normal* state of the atom, called the *ground state*, is that of lowest energy, with the electron revolving in the orbit of smallest radius, r_0. The energy of this state is -13.6 eV. As a result of collisions with rapidly moving electrons in an electrical discharge, or for other causes, the atom may temporarily acquire sufficient energy to raise the electron to some larger orbit. The atom is then said to be in an *excited* state. From this state the electron can then fall back to a state of lower energy, emitting a photon in the process.

Let n be the quantum number of some excited state, and ℓ the quantum number of the lower state to which the electron returns after the emission process. Then E_i, the initial energy, is

$$E_i = -\frac{1}{\varepsilon_0^2}\frac{me^4}{8n^2h^2},$$

and E_f, the final energy, is

$$E_f = -\frac{1}{\varepsilon_0^2}\frac{me^4}{8\ell^2h^2}.$$

The decrease in energy, $E_i - E_f$, which we place equal to the energy hf of the emitted photon, is

$$E_i - E_f = hf = -\frac{1}{\varepsilon_0^2}\frac{me^4}{8n^2h^2} + \frac{1}{\varepsilon_0^2}\frac{me^4}{8\ell^2h^2},$$

or

$$f = \frac{1}{\varepsilon_0^2}\frac{me^4}{8h^3}\left(\frac{1}{\ell^2} - \frac{1}{n^2}\right). \qquad (44\text{–}14)$$

This equation is of precisely the same form as the Balmer formula (Eq. 44–4) for the frequencies in the hydrogen spectrum if we place

$$\frac{1}{\varepsilon_0^2}\frac{me^4}{8h^3} = Rc, \qquad (44\text{–}15)$$

and let $\ell = 1$ for the Lyman series, $\ell = 2$ for the Balmer series, etc. The Lyman series is therefore the group of lines emitted by electrons returning from some excited state to the ground state. The Balmer series is the group emitted by electrons returning from some higher state, but stopping in the *second orbit* instead of falling at once to that of lowest energy. That is, an electron returning from the third orbit ($n = 3$) to the second orbit ($\ell = 2$) emits the H_α-line. One returning from the fourth orbit ($n = 4$) to the second ($\ell = 2$) emits the H_β-line, etc. These transitions are shown in Fig. 44–6.

Every quantity in Eq. (44–15) may be determined quite independently of the Bohr theory, and apart from this theory we have no reason to expect these quantities to be related in this particular way. The quantities m and e, for instance, are found from experiments on free electrons, h may be found from the photoelectric effect, and R by measurements of wavelengths, while c is the velocity of light. However, if we substitute in Eq. (44–15) the values of these quantities, obtained by such diverse means, we find that it *does* hold exactly, within the limits of experimental error, providing direct confirmation of Bohr's theory.

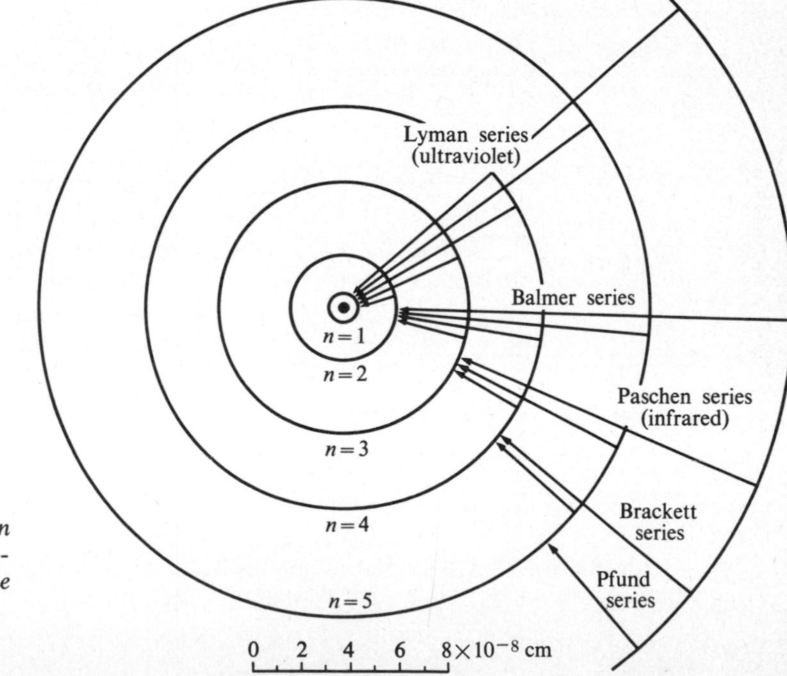

Fig. 44–6 *"Permitted" orbits of an electron in the Bohr model of a hydrogen atom. The transitions responsible for some of the lines of the various series indicated by arrows.*

44–6 ATOMIC SPECTRA

As we have seen in the past few sections, the key to the understanding of atomic spectra is the concept of atomic energy levels. Every spectrum line corresponds to a specific transition between two energy levels of an atom, and the corresponding frequency is given in each case by Eq. (44–6).

Thus the fundamental problem of the spectroscopist is to determine the energy levels of an atom from the measured values of the wavelengths of the spectral lines emitted when the atom proceeds from one set of energy levels to another. In the case of complicated spectra emitted by the heavier atoms, this is a task requiring tremendous ingenuity. Nevertheless, almost all atomic spectra have been analyzed, and the resulting energy levels have been tabulated or plotted with the aid of diagrams similar to the one shown for sodium in Fig. 44–7.

The lowest energy level of the atom is called the *ground state*, and all higher levels are called *excited states*. As we have seen, a spectral line is emitted when an atom proceeds from an excited state to a lower state. The only means discussed so far for raising the atom from the normal state to an excited state has been with the aid of an electric discharge. Let us consider now another method, involving absorption of radiant energy.

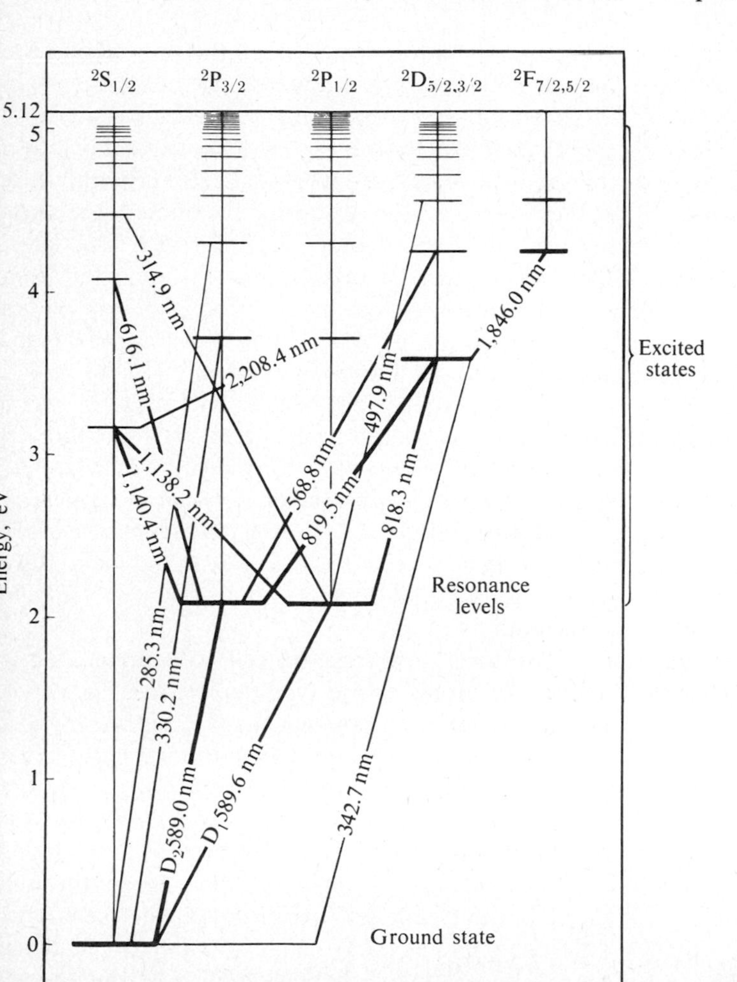

Fig. 44–7 *Energy levels of the sodium atom. Numbers on the lines between levels are wavelengths.*

From Fig. 44–7 it may be seen that a sodium atom emits the characteristic yellow light of wavelengths 5890 and 5896 Å (the D_1- and D_2- lines) when it undergoes the transitions from the two levels marked *resonance levels* to the ground state. Suppose a sodium atom in the ground state were to *absorb* a quantum of radiant energy of wavelength 589.0 or 589.6 nm. It would then undergo a transition in the opposite direction and be raised to one of the resonance levels. After a short time, the average value of which is called the *lifetime* of the excited state, the atom returns to the ground state and emits this quantum. For the resonance levels of the sodium atom, the lifetime is about 1.6×10^{-8} s.

This emission process is called *resonance radiation* and may be easily demonstrated as follows. A strong beam of the yellow light from a sodium arc is concentrated on a glass bulb which has been highly evacuated and into which a small amount of pure metallic sodium has been distilled. If the bulb is warmed to increase the sodium vapor pressure, resonance radiation will take place throughout the whole bulb, which glows with the yellow light characteristic of sodium.

A sodium atom in the ground state may absorb radiant energy of wavelengths other than the yellow resonance lines. All wavelengths corresponding to spectral lines *emitted* when the sodium atom returns to its normal state may also be *absorbed*. Thus, from Fig. 44–7, wavelengths 330.2 nm, 285.3 nm, etc., may be absorbed by a normal sodium atom. If, therefore, the continuous-spectrum light from a carbon arc is sent through an absorption tube containing sodium vapor, and then examined with a spectroscope, there will be a series of dark lines corresponding to the wavelengths absorbed, as shown in Fig. 44–8. This is known as an *absorption spectrum*.

The sun's spectrum is an absorption spectrum. The main body of the sun emits a continuous spectrum, whereas the cooler vapors in the sun's atmosphere emit line spectra corresponding to all the elements present. When the intense light from the main body of the sun passes through the cooler vapors, the lines of these elements are *absorbed*. The light *emitted* by the cooler vapors is so small compared with the unabsorbed continuous spectrum that the continuous spectrum appears to be crossed by many faint *dark* lines. These were first observed by Fraunhofer and are therefore called *Fraunhofer lines*. They may be observed with any student spectroscope pointed toward any part of the sky.

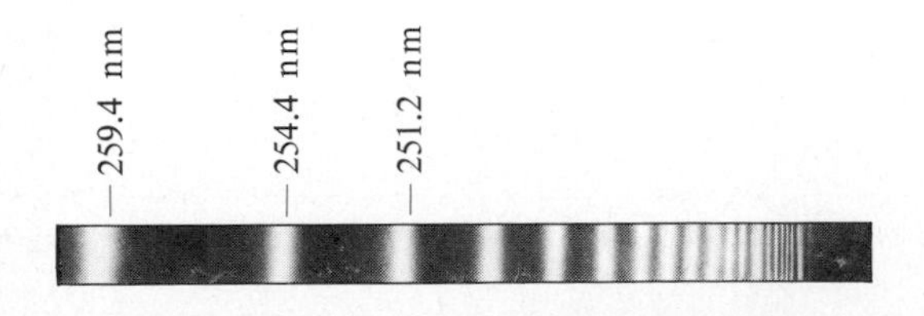

Fig. 44–8 *Absorption spectrum of sodium.*

44–7 THE LASER

If the energy difference between the normal and the first excited state of an atom is E, the atom is capable of absorbing a photon whose frequency f is given by the Planck equation $E = hf$. The *absorption* of a photon by a normal atom A is depicted schematically in Fig. 44–9(a). After absorbing the photon, the atom becomes an excited atom A^*. A short time later, *spontaneous emission* takes place and the excited atom becomes normal again by emitting a photon of the same frequency as that which was originally absorbed but in a random direction and with a random phase, as shown in Fig. 44–9(b). There is also a third process, first proposed by Einstein, called *stimulated emission*, shown schematically in Fig. 44–9(c). *Stimulated emission takes place when a photon encounters an excited atom and forces it to emit another photon of the same frequency, in the same direction, and in the same phase.* The two photons go off together as *coherent* radiation.

Consider an absorption cell containing a large number of atoms of the type depicted in Fig. 44–9. In the absence of an external beam of radiation, most of the atoms are in the ground state; there are only a few excited atoms present in the cell. The ratio of the number n_E of excited atoms to the number n_0 of normal atoms is extremely small.

Now suppose a beam of radiation is sent through the cell, with frequency f corresponding to the energy difference E. The ratio of the numbers n_E and n_0, that is, the ratio of the *populations* of the energy levels, is increased. Since the population of the normal state was so much larger than that of the excited

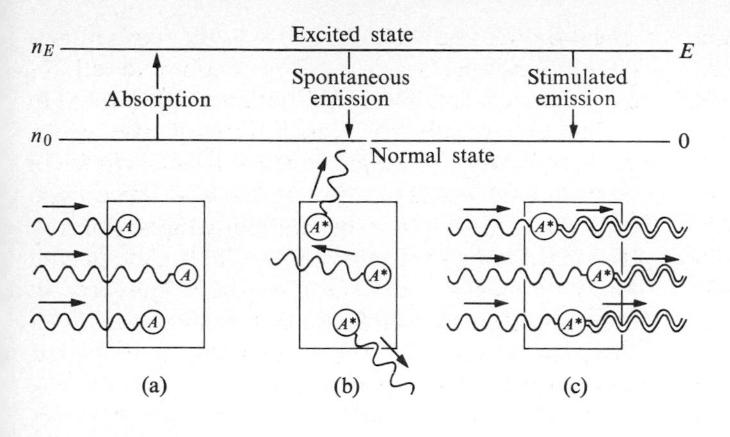

Fig. 44–9 *Three interaction processes between an atom and radiation.*

state, an enormously intense beam of light would be required to increase the population of the excited state to a value comparable to or greater than that of the normal state. Therefore, the rate at which energy is extracted from the beam by absorption of normal atoms far outweighs the rate at which energy is added to the beam by stimulated emission of excited atoms.

If a condition can be created in which n_E is substantially increased compared to the normal equilibrium value, creating a condition known as *population inversion*, the rate of energy radiation by stimulated emission may *exceed* the rate of absorption. The system then acts as a *source* of radiation with photon energy E. Furthermore, since the photons are the result of stimulated emission, they all have the same frequency, phase, polarization, and direction. The resulting radiation is therefore very much more *coherent* than light from ordinary sources, in which the emissions of individual atoms are *not* coordinated.

The necessary population inversion can be achieved in a variety of ways. As an example we consider the helium–neon laser, simple, inexpensive, and available in many undergraduate laboratories. A mixture of helium and neon, each typically at a pressure of the order of 10^2 Pa (or 10^{-3} atm), is sealed in a glass enclosure provided with two electrodes. When a sufficiently high voltage is applied, a glow discharge occurs. Collisions between ionized atoms and electrons carrying the discharge current excite atoms to various energy states.

Figure 44–10 shows an energy-level diagram for the system. Helium atoms excited to the $1s2s$ state cannot return to the ground state by emitting a 20.61-eV photon, as might be expected, because both the states have zero total angular momentum, while a photon must carry away at least one unit ($h/2\pi$) of angular momentum. Such a state in which radiative decay is impossible is called a *metastable state.*

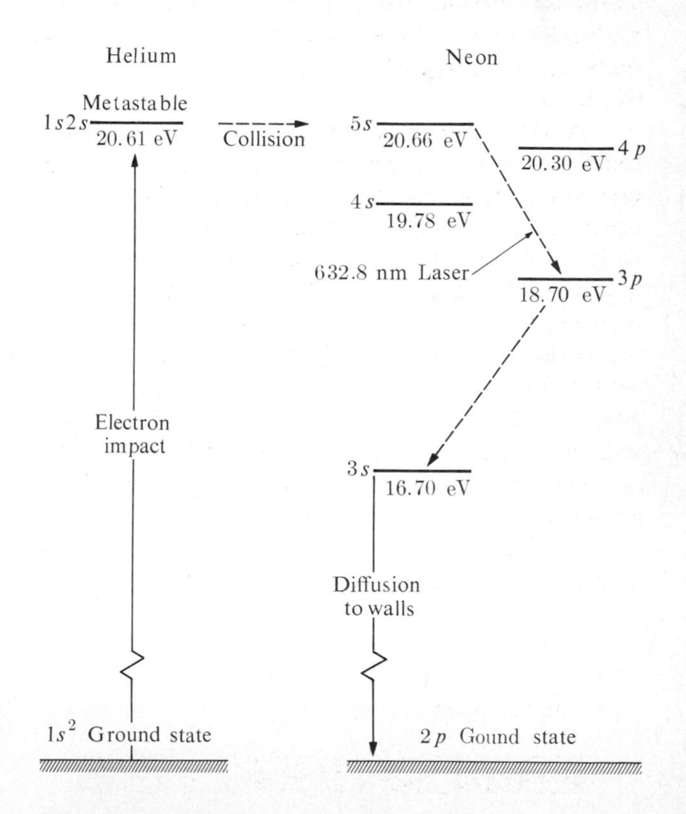

Fig. 44–10 *Energy-level diagram for helium-neon laser.*

The helium atoms *can*, however, lose energy by energy-exchange collisions with neon atoms initially in the ground state. A $1s2s$ helium atom, with its internal energy of 20.61 eV and a little additional kinetic energy, can collide with a neon atom in the ground state, exciting it to the $5s$ excited state at 20.66 eV and leaving the helium atom in the $1s^2$ ground state. Thus, we have the necessary mechanism for a population inversion in neon, with the population in the $5s$ state substantially enhanced. Stimulated emission from this state then results in the emission of highly coherent light at 632.8 nm, as shown on the diagram. In practice the beam is sent back and forth through the gas many times by a pair of parallel mirrors, so as to stimulate emission from as many excited atoms as possible. One of the mirrors is partially transparent, so a portion of the beam emerges as an external beam.

The net effect of all the processes taking place in a laser tube is a beam of radiation that is (1) very intense, (2) almost perfectly parallel, (3) almost monochromatic, and (4) spatially *coherent* at all points within a given cross section. To understand this fourth characteristic, we recall the simple double-slit interference experiment. A mercury arc placed directly behind the double slit would not give rise to interference fringes because the light issuing from the two slits would come from different points of the arc and would not retain a constant phase relationship. In the use of the usual laboratory arc lamp sources, it is necessary to use the light from a very small portion of the source to illuminate the double slit. The slightly diverging beam from a laser, however, may be allowed to fall directly on a double slit (or other interferometer) because the light rays from any two points of a cross section are in phase, and are said to exhibit "spatial coherence."

In recent years lasers have found a wide variety of practical applications. The high intensity of a laser beam makes it a convenient drill. A very small hole can be drilled in a diamond for use as a die in drawing very small-diameter wire. The ability of a laser beam to travel long distances without appreciable spreading makes it a very useful tool for surveyors, especially in situations where great precision is required over long distances, as in the case of a long tunnel drilled from both ends.

Lasers are finding increasing application in medical science. A laser can produce a very *narrow* beam with extremely *high intensity*, high enough to vaporize anything in its path. This property is used in the treatment of a detached retina; a short burst of radiation damages a small area of the retina, and the resulting scar tissue "welds" the retina back to the choroid from which it has become detached. Laser beams are also used in surgery; blood vessels cut by the beam tend to seal themselves off, making it easier to control bleeding. The use of laser radiation in the treatment of skin cancer is an active area of research.

44–8 X-RAY PRODUCTION AND SCATTERING

X-rays are produced when rapidly moving electrons which have been accelerated through potential differences of the order of 10^3 to 10^6 V, are allowed to strike a metal target. They were first observed by Wilhelm K. Roentgen (1845–1923) in 1895, and were originally called *Roentgen rays.*

X-rays are of the same nature as light or any other electromagnetic wave and, like light waves, they are governed by quantum relations in their interaction with matter. Hence, one may speak of x-ray photons or quanta, the energy of such a photon being given by the relation $E = hf$. Wavelengths of x-rays range from approximately 0.001 to 1 nm (10^{-12} to 10^{-9} m).

A common x-ray tube is the Coolidge type, invented by W. D. Coolidge of the General Electric laboratories in 1913. A Coolidge tube is shown in Fig. 44–11. A thermionic cathode and an anode are enclosed in a glass tube which has been pumped down to an extremely low pressure, so that electrons emitted from the cathode can travel directly to the anode with only a small probability of a collision on the way, reaching the anode with a speed corresponding to the full potential difference across the tube. X-radiation is emitted from the anode surface as a consequence of its bombardment by the electron stream.

Fig. 44–11 *Coolidge-type x-ray tube.*

Two distinct processes are involved in x-ray emission. Some of the electrons are stopped by the target and their kinetic energy is converted directly to x-radiation. Others transfer their energy in whole or in part to the atoms of the target, which retain it temporarily as "energy of excitation" but very shortly emit it as x-radiation. The latter is characteristic of the material of the target, while the former is not.

The atomic energy levels associated with x-ray excitation are rather different in character from those associated with visible spectra. To understand them we need some understanding of the arrangement of electrons in complex atoms, a topic to be discussed in greater detail in Chapter 45. For the present we state simply that in a many-electron atom the electrons are always arranged in concentric *shells* at increasing distances from the nucleus. These shells are labeled *K, L, M, N*, etc., the *K* shell being closest to the nucleus, the *L* shell next, and so on. For any given atom in the ground state there is a definite number of electrons in each shell.

For reasons to be discussed later, each shell has a maximum number of electrons it can accommodate, and we may speak of *filled shells* and *partially filled shells.* The *K* shell can contain at most two electrons. The next, the *L* shell, can contain eight. The third, the *M* shell, has a capacity for 18 electrons, while the *N* shell may hold 32. The sodium atom, for example, which contains 11 electrons, has two in the *K* shell, eight in the *L* shell, and a single electron in the *M* shell. Molybdenum, with 42 electrons, has two in the *K* shell, eight in the *L* shell, 18 in the *M* shell, and 14 in the *N* shell.

The *outer* electrons of an atom are the ones responsible for the optical spectra of the elements. Relatively small amounts of energy suffice to remove these to excited states, and on their return to their normal states, wavelengths in or near the visible region are emitted. The inner electrons, being closer to the nucleus, are more tightly bound and require much more energy to displace them from their normal levels. As a result, we would expect a photon of much larger energy, and hence much higher frequency, to be emitted when the atom returns to its normal state after the displacement of an inner electron. This is, in fact, the case, and it is the displacement of the inner electrons which gives rise to the emission of x-rays.

On colliding with the atoms of the anode, some of the electrons accelerated in an x-ray tube, provided they have acquired sufficient energy, will dislodge one of the inner electrons of a target atom, say one of the *K* electrons. This leaves a vacant space in the *K* shell, which is immediately filled by an electron from either the *L, M,* or *N* shell. The readjustment of the electrons is accompanied by a decrease in the energy of the atom, and an x-ray photon is emitted with energy just equal to this decrease. Since the energy change is perfectly definite for atoms of a given element, the emitted x-rays should have definite frequencies. In other words, the x-ray spectrum should be a *line spectrum.* We can predict further that there should be just three lines in the series, corresponding to the three possibilities that the vacant space may have been filled by an *L, M,* or *N* electron.

This is precisely what is observed. Figure 44–12 illustrates the so-called *K* series of the elements

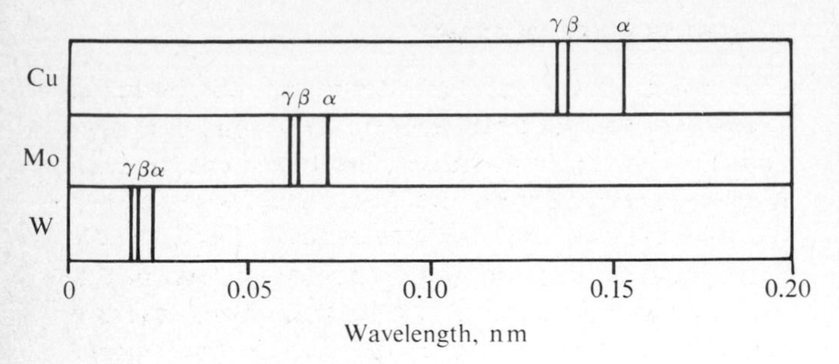

Fig. 44–12 *Wavelengths of the* K_α, K_β, *and* K_γ *lines of copper, molybdenum, and tungsten.*

tungsten, molybdenum, and copper. Each series consists of three lines, known as the K_α-, K_β-, and K_γ-lines. The K_α-line is produced by the transition of an L electron to the vacated space in the K shell, the K_β-line by an M electron, and the K_γ-line by an N electron.

In addition to the K series, there are other series known as the L, M, and N series, produced by the ejection of electrons from the L, M, and N shells rather than the K shell. As would be expected, the electrons in these outer shells, being farther away from the nucleus, are not held as firmly as those in the K shell. Consequently, the other series may be excited by more slowly moving electrons, and the photons emitted are of lower energy and longer wavelength.

In addition to the x-ray *line* spectrum there is a background of *continuous* x-radiation from the target of an x-ray tube. This is due to the sudden deceleration of those "cathode rays" (bombarding electrons) which do not happen to eject an atomic electron. The remarkable feature of the continuous spectrum is that while it extends indefinitely toward the *long* wavelength end, it is cut off very sharply at the *short* wavelength end. The quantum theory furnishes a simple explanation of the short-wave limit of the continuous x-ray spectrum.

A bombarding electron may be brought to rest in a single process if the electron happens to collide head on with an atom of the target; or it may make a number of collisions before coming to rest, giving up part of its energy each time. If we assume that the energy lost at each collision is radiated as an x-ray photon, these photons may have any energy up to a certain maximum, namely that of an electron which gives up *all* of its energy in a single collision. Hence there is a short-wave limit to the spectrum. The frequency of this limit is found by setting the energy of the electron equal to the energy of the x-ray photon:

$$\boxed{hf = \tfrac{1}{2}mv^2.} \qquad (44\text{–}16)$$

This is precisely the same equation as that for the photoelectric effect except for the work function term, which is negligible here since the energies of the x-ray photons are so large. In fact, the emission of x-rays may be described as an *inverse photoelectric effect.* In photoelectric emission the energy of a photon is transformed into kinetic energy of an electron; here, the kinetic energy of an electron is transformed into that of a photon.

Example Compute the potential difference through which an electron must be accelerated in order that the short-wave limit of the continuous x-ray spectrum shall be exactly 0.1 nm.

The frequency corresponding to 0.1 nm (10^{-10} m) is given by

$$f = \frac{c}{\lambda} = \frac{3 \times 10^8 \text{ m s}^{-1}}{10^{-10} \text{ m}} = 3 \times 10^{18} \text{ s}^{-1}$$
$$= 3 \times 10^{18} \text{ Hz}.$$

The energy of the photon is

$$hf = (6.62 \times 10^{-34} \text{ J s})(3 \times 10^{18} \text{ s}^{-1})$$
$$= 19.9 \times 10^{-16} \text{ J}.$$

This must equal the kinetic energy of the electron, $\frac{1}{2}mv^2$, which is also equal to the product of the electronic charge and the accelerating voltage, V:

$$\tfrac{1}{2}mv^2 = eV = 19.9 \times 10^{-16} \text{ J}.$$

Since

$$e = 1.60 \times 10^{-19}\ \text{C},$$

$$V = \frac{19.9 \times 10^{-16}\ \text{J}}{1.60 \times 10^{-19}\ \text{C}} = 12{,}400\ \text{V}.$$

A phenomenon called *Compton scattering*, first observed in 1924 by A. H. Compton, provides additional direct confirmation of the quantum nature of electromagnetic radiation. When x-rays impinge on matter, some of the radiation is *scattered*, just as visible light falling on a rough surface undergoes diffuse reflection. Observation shows that some of the scattered radiation has smaller frequency and longer wavelength than the incident radiation, and that the change in wavelength depends on the angle through which the radiation is scattered. Specifically, if the scattered radiation emerges at an angle ϕ with respect to the incident direction, and if λ and λ' are the wavelengths of the incident and scattered radiation, respectively, it is found that

$$\lambda' - \lambda = \frac{h}{mc}(1 - \cos\phi) \qquad (44\text{–}17)$$

where m is the electron mass.

Compton scattering cannot be understood on the basis of classical electromagnetic theory. On the basis of classical principles, the scattering mechanism is induced motion of electrons in the material, caused by the incident radiation. This motion must have the same frequency as that of the incident wave, and so the scattered wave radiated by the oscillating charges should have the same frequency. There is no way the frequency can be *shifted* by this mechanism.

The quantum theory, by contrast, provides a beautifully simple explanation. We imagine the scattering process as a collision of two particles, the incident photon and an electron initially at rest. The photon gives up some of its energy and momentum to the electron, which recoils as a result of this impact, and the final photon has less energy, smaller frequency, and longer wavelength than the initial one. Equation (44–17) can be derived from an analysis of this process.

44–9 WAVE MECHANICS

The Bohr model of the atom was successful in explaining the observed spectra of atomic hydrogen and of a few other elements, but for atoms having a large number of orbital electrons, and for molecules, the theory was not as satisfactory. Furthermore, there seemed to be no good justification, except that it led to the right answer, for the hypothesis that only those orbits are permitted for which the angular momentum is equal to some integral multiple of $h/2\pi$.

The next advance in atom building came in 1923, about 10 years after the Bohr theory. This was a suggestion by de Broglie that, since light is dualistic in nature, behaving in some aspects like waves and in others like particles, the same might be true of matter. That is, electrons and protons, which until that time had been thought to be purely corpuscular, might in some circumstances behave like *waves.* Specifically, de Broglie postulated that a free electron of mass m, moving with speed v, should have a wavelength λ given by

$$\boxed{\lambda = \frac{h}{mv},} \qquad (44\text{–}18)$$

where h is the same Planck's constant that appears in the frequency–energy relation for photons.

This wave hypothesis, unorthodox though it seemed at the time, almost immediately received very direct experimental confirmation. We have described in Chapter 42 how the layers of atoms in a crystal serve as a diffraction grating for x-rays. An x-ray beam is strongly reflected when it strikes a crystal at such an angle that the waves scattered from the atomic layers combine to reinforce one another. The point of importance here is that the existence of these strong reflections is evidence of the *wave* nature of x-rays.

In 1927, Davisson and Germer, working in the Bell Telephone Laboratories, were studying the nature of the surface of a crystal of nickel by directing a beam of *electrons* at the surface and observing the electrons reflected at various angles. It might be expected that even the smoothest surface attainable would still look rough to an electron, and that the electron beam would therefore be diffusely reflected. But Davisson and Germer found the electrons were reflected in almost the same way that x-rays would be reflected from the same crystal. The wavelengths of the electrons in the beam were computed from their known velocity, with the help of Eq. (44–18), and the angles at which strong reflection took place were found to be the same as those at which x-rays of the same wavelength would be reflected.

This wave hypothesis clearly requires sweeping revisions of our fundamental concepts regarding the description of matter. What we are accustomed to calling a *particle* actually looks like a particle only if we do not look too closely. In general, a particle has to be regarded as a spread-out entity which is not entirely localized in space, and at least in some cases this spreading out appears as a periodic pattern suggesting wavelike properties. The wave and particle aspects are not inconsistent, but the particle model is an *approximation* of a more general wave picture. We are reminded of the ray picture of geometrical optics, a special case of the more general wave picture of physical optics; indeed, there is a very close analogy between optics and the description of particles.

In a few years after 1923, the wave hypothesis of de Broglie was developed by Heisenberg, Schrödinger, and many others, into a complete theory called *wave mechanics* or *quantum mechanics.* A single section on "wave mechanics" cannot, of course, give the reader an adequate comprehension of this complex and highly mathematical subject, any more than the whole field of newtonian mechanics could be covered in the same amount of space. We can only point out the main lines of thought in a nonmathematical way, describe some of the experimental evidence for the wave nature of material particles, and show how the quantum numbers that were introduced in such an artificial way by Bohr now enter naturally into the problem of atomic structure.

One of the essential features of quantum mechanics is that material particles are no longer regarded as geometrical points, localized in space, but are intrinsically spread-out entities. The spatial distribution of a *free* electron may have a recurring pattern characteristic of a wave that propagates through space. Electrons *in atoms* are visualized as diffuse clouds surrounding the nucleus. The idea that the electrons in an atom move in definite orbits such as those in Fig. 44–6 has been abandoned. The orbits themselves, however, were never an essential part of Bohr's theory, since the quantities that determine the frequencies of the emitted photons are the *energies* corresponding to the orbits. The new theory still assigns definite energy states to an atom. In the hydrogen atom the energies are the same as those given by Bohr's theory; in more complicated atoms where the Bohr theory did not work, the quantum mechanical picture is in excellent agreement with observation.

We shall illustrate how quantization arises in atomic structure by an analogy with the classical mechanical problem of a vibrating string fixed at its ends. When the string vibrates, the ends must be nodes, but nodes may occur at other points also, and the general requirement is that the length of the string shall equal some *integer* number of half-wavelengths. The point of interest is that the solution of the problem of the vibrating string leads to the appearance of *integral numbers.*

In a similar way, the principles of quantum mechanics lead to a *wave equation* (Schrödinger's equation) that must be satisfied by an electron in an atom, subject also to certain boundary conditions. Let us think of an electron as a wave extending in a circle around the nucleus. In order that the wave may "come out even," the circumference of this circle must include some *integer number* of wavelengths. The wavelength of a particle of mass m, moving with a velocity v, is given, according to wave mechanics, by Eq. (44–18), $\lambda = h/mv$. Then if r is the radius and $2\pi r$ the circumference of the circle occupied by the wave, we must have $2\pi r = n\lambda$, where $n = 1, 2, 3,$ etc. Since $\lambda = h/mv$, this equation becomes

$$2\pi r = n\frac{h}{mv}, \qquad mvr = n\frac{h}{2\pi}.$$

But mvr is the angular momentum of the electron, so we see that the wave-mechanical picture leads naturally to Bohr's postulate that the angular momentum equals some integral multiple of $h/2\pi$.

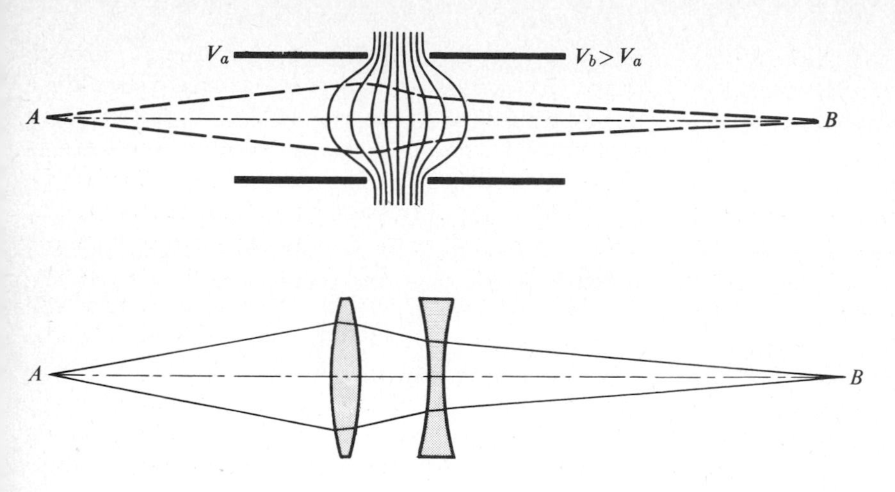

Fig. 44–13 *An electrostatic electron lens. The cylinders are at different potentials V_a and V_b. A beam of electrons diverging from point A is focused at point B.*

Fig. 44–14 *Optical analog of the electron lens in Fig. 44–13.*

44–10 THE ELECTRON MICROSCOPE

The shorter the wavelength, the smaller the limit of resolution of a microscope. The wavelengths of electron waves can easily be made very much *shorter* than the wavelengths of visible light. Hence, the limit of resolution of a microscope may be extended to a value several hundred times smaller than that obtainable with an optical instrument, by using electrons, rather than light waves, to form an image of the object being examined.

A beam of electrons can be focused by either a magnetic or an electric field of the proper configuration, and both types are used in electron microscopes. Figure 44–13 illustrates an electrostatic lens. Two hollow cylinders are maintained at different potentials. A few of the equipotentials are indicated, and the trajectories of a beam of electrons traveling from left to right are shown by the dashed lines. The optical analog of this electrostatic lens is shown in Fig. 44–14. It will be evident without going into further details that by the proper design of such lenses the elements of an optical microscope, such as its condensing lens, objective, and ocular, can all be duplicated electronically.

The source of electrons in an electron microscope is a heated filament. Electrons emitted by the filament are accelerated by an electron gun and strike the object to be examined. This must necessarily be a thin section so that some of the electrons can pass through it. The thicker portions of the section absorb more of the electron stream than do the thinner portions, just as would a slide in a slide projector. The entire apparatus must, of course, be evacuated.

The final image may be formed on a photographic plate, or on a fluorescent screen which can be examined visually or photographed with a still further gain in magnification. Commercial electron microscopes give satisfactory definition at an overall magnification (electronic followed by photographic) as great as 50,000 times. Figure 44–15 is an electron

Fig. 44–15 *Electron micrograph of aluminum oxide, magnified 53,500 times. (Courtesy of Radio Corporation of America.)*

micrograph of aluminum oxide, magnified 53,500 times.

It should be pointed out that the ability of the electron microscope to form an image does *not* depend on the wave properties of the electrons; their trajectories can be computed by treating them as charged particles, deflected by the electric or magnetic fields through which they move. It is only when considerations of *resolving power* arise that the electron wavelengths come into the picture. The situation is analogous to that in the optical microscope. The paths of light rays through an optical microscope can be computed by the principles of geometrical optics, but the resolving power of the microscope is determined by the wavelength of the light used.

PROBLEMS

$$e = 1.60 \times 10^{-19}\ \text{C}$$

$$m = 9.11 \times 10^{-31}\ \text{kg}$$

$$h = 6.63 \times 10^{-34}\ \text{J s}$$

$$N_0 = 6.02 \times 10^{26}\ \text{atoms kmol}^{-1}$$

Energy equivalent of 1 u = 931 MeV

$$\frac{e}{m} = 1.76 \times 10^{11}\ \text{C kg}^{-1}$$

$$k = 1.38 \times 10^{-23}\ \text{J K}^{-1}$$

$$1\ \text{eV} = 1.60 \times 10^{-19}\ \text{J}$$

$$1\ \text{u} = 1.66 \times 10^{-27}\ \text{kg}$$

$$\varepsilon_0 = 8.85 \times 10^{-12}\ \text{C}^2\ \text{N}^{-1}\ \text{m}^{-2}$$

44–1 In the photoelectric effect, what is the relation between the threshold frequency f_0 and the work function ϕ?

44–2 The photoelectric threshold wavelength of tungsten is 2.73×10^{-5} cm. Calculate the maximum kinetic energy of the electrons ejected from a tungsten surface by ultraviolet radiation of wavelength 1.80×10^{-5} cm. (Express the answer in electron volts.)

44–3 A photoelectric surface has a work function of 4.00 eV. What is the maximum velocity of the photoelectrons emitted by light of frequency 3×10^{15} Hz?

44–4 When ultraviolet light of wavelength 2.54×10^{-5} cm from a mercury arc falls upon a clean copper surface, the retarding potential necessary to stop the emission of photoelectrons is 0.59 V. What is the photoelectric threshold wavelength for copper?

44–5 When a certain photoelectric surface is illuminated with light of different wavelengths, the stopping potentials in the table below are observed:

Wavelength, nm	Stopping potential, V
366	1.48
405	1.15
436	0.93
492	0.62
546	0.36
579	0.24

Plot the stopping potential as ordinate against the frequency of the light as abscissa. Determine (a) the threshold frequency, (b) the threshold wavelength, (c) the photoelectric work function of the material, and (d) the value of Planck's constant h (the value of e being known).

44–6 The photoelectric work function of potassium is 2.0 eV. If light having a wavelength of 360 nm falls on potassium, find (a) the stopping potential, (b) the kinetic energy in electron volts of the most energetic electrons ejected, and (c) the velocities of these electrons.

44–7 What will be the change in the stopping potential for photoelectrons emitted from a surface if the wavelength of the incident light is reduced from 400 nm to 360 nm?

44–8 The photoelectric work functions for particular samples of certain metals are as follows: cesium, 2.00 eV; copper, 4.00 eV; potassium, 2.25 eV; and zinc, 3.60 eV.

a) What is the threshold wavelength for each metal?
b) Which of these metals could not emit photoelectrons when irradiated with visible light?

44–9 The light-sensitive compound on most photographic films is silver bromide, AgBr. A film is "exposed" when the light energy absorbed dissociates this molecule into its atoms. (The actual process is more complex, but the quantitative result does not differ greatly.) The energy of or heat of dissociation of AgBr is 23.9 kcal mol^{-1}. Find (a) the energy in electron volts, (b) the wavelength, and (c) the frequency of the photon which is just able to dissociate a molecule of silver bromide.

d) What is the energy in electron volts of a quantum of radiation having a frequency of 100 MHz?

e) Explain the fact that light from a firefly can expose a photographic film, whereas the radiation from a TV station transmitting 50,000 W at 100 MHz cannot.

f) Will photographic films stored in a light-tight container be ruined (exposed) by the radio waves constantly passing through them? Explain.

44–10

a) Show that the energy E (in electron volts) of a photon of wavelength λ (in nanometers) is given by E (eV) $= (1{,}240/\lambda)$(nm).

b) What is the energy in electron volts of a photon having a wavelength of 91.2 nm?

44–11 If 5% of the energy supplied to an incandescent light bulb is radiated as visible light, how many visible quanta are emitted per second by a 100-W bulb? Assume the wavelength of all the visible light to be 560 nm.

44–12 The directions of emission of photons from a source of radiation are random. According to the wave theory, the intensity of radiation from a point source varies inversely as the square of the distance from the source. Show that the number of photons from a point source passing out through a unit area is also given by an inverse-square law.

44–13 Show that the angular speed of the electron of a hydrogen atom in its orbit is $\omega = \pi m e^4/2\varepsilon_0^2 n^3 h^3$.

44–14 According to Bohr, the Rydberg constant R is equal to $me^4/8\varepsilon_0^2 h^3 c$. Calculate R in m^{-1} and compare with the experimental value.

44–15 Calculate (a) the frequency, and (b) the wavelength of the H_β-line of the Balmer series for hydrogen. This line is emitted in the transition from $n = 4$ to $\ell = 2$. Assume that the nucleus has infinite mass.

44–16

a) What is the least amount of energy in electron volts that must be given to a hydrogen atom so that it can emit the H_β-line (see Problem 44–15 and Fig. 44–6) in the Balmer series?

b) How many different possibilities of spectral line emission are there for this atom when the electron goes from $n = 4$ to the ground state?

44–17

a) Show that the frequency of revolution of an electron in its circular orbit in the Bohr model of the hydrogen atom is $f = me^4\varepsilon_0^2 n^3 h^3$.

b) Show that when n is very large, the frequency of revolution equals the radiated frequency calculated from Eq. (44–14) for a transition from

$$n = n' + 1$$

to

$$\ell = n'.$$

(This problem illustrates Bohr's *correspondence principle*, which is often used as a check on quantum calculations. When n is small, quantum physics gives results which are very different from those of classical physics. When n is large, the differences are not significant and the two methods then "correspond.")

44–18 A 10-kg satellite circles the earth once every 2 hr in an orbit having a radius of 8000 km.

a) Assuming that Bohr's angular momentum postulate applies to satellites just as it does to an electron in the hydrogen atom, find the quantum number of the orbit of the satellite.

b) Show from Bohr's first postulate and Newton's law of gravitation that the radius of an earth-satellite orbit is directly proportional to the square of the quantum number, $r = kn^2$, where k is the constant of proportionality.

c) Using the result from part (b), find the distance between the orbit of the satellite in this problem and its next "allowed" orbit.

d) Comment on the possibility of observing the separation of the two adjacent orbits.

e) Do quantized and classical orbits correspond for this satellite? Which is the "correct" method for calculating the orbits?

44–19 In Fig. 44–10, compute the energy difference for the $5s$–$3p$ transition; express your result in electron volts and in joules. Compute the wavelength of a photon having this energy, and compare your result with the observed wavelength of the laser light.

44–20 In the helium–neon laser, what wavelength corresponds to the $3p$–$3s$ transition in neon? Why is this not observed in the beam with the same intensity as the 632.8-nm laser line?

44–21 For crystal diffraction experiments, wavelengths of the order of 0.1 nm are often appropriate. Find the energy, in electron volts, for a particle with this wavelength if the particle is (a) a photon; (b) an electron.

44–22 Approximately what range of photon energies (in electron volts) corresponds to the visible spectrum? Approximately what range of wavelengths would electrons in this energy range have?

44–23

a) An electron moves with a velocity of 3×10^8 cm s^{-1}. What is its de Broglie wavelength?

b) A proton moves with the same velocity. Determine its de Broglie wavelength.

44–24 The average kinetic energy of a neutron is kT. What is the de Broglie wavelength associated with the neutrons in thermal equilibrium with matter at 300 K? (The mass of a neutron is approximately 1 u.)

44–25 What is the de Broglie wavelength of an electron which has been accelerated through a potential difference of 200 V? Would this electron exhibit particlelike or wavelike characteristics on meeting an obstacle or opening 1 mm in diameter?

44–26

a) What is the minimum potential difference between the filament and the target of an x-ray tube if the tube is to produce x-rays of wavelength 0.05 nm?

b) What is the shortest wavelength produced in an x-ray tube operated at 2×10^6 V?

44–27 An electron in a certain x-ray tube is accelerated from rest through a potential difference of 180,000 V in going from the cathode to the anode. When it arrives at the anode, what is (a) its kinetic energy in electron volts, (b) its relativistic velocity?

c) What is the velocity of the electron, calculated classically?

44–28 An x-ray tube is operating at 150,000 V and 10 mA.

a) If only 1% of the electric power supplied is converted into x-rays, at what rate is the target being heated in calories per second?

b) If the target has a mass of 300 g and a specific heat of 0.035 cal g^{-1}(°C)$^{-1}$, at what average rate would its temperature rise if there were no thermal losses?

c) What must be the physical properties of a practical target material? What would be some suitable target elements?

44–29 If electrons in a metal had the same energy distribution as molecules in a gas at the same temperature (which is not actually the case), at what temperature would the average electron kinetic energy equal 1 eV, typical of work functions of metals?

44–30 If hydrogen were monatomic, at what temperature would the average translational kinetic energy be equal to the energy required to raise a hydrogen atom from the ground state to the $n = 2$ excited state?

44–31 What is the de Broglie wavelength of an electron accelerated through 20,000 V, a typical voltage for color-television picture tubes?

44–32 The negative μ-meson (or muon) has a charge equal to that of an electron but a mass about 206 times as great. Consider a hydrogenlike atom consisting of a proton and a muon.

a) What is the ground-state energy?

b) What is the radius of the $n = 1$ Bohr orbit?

c) What is the wavelength of the radiation emitted in the transition from the $n = 2$ state to the $n = 1$ state?

44–33 For a hydrogen atom in the ground state, determine, in electron volts:

a) the kinetic energy;

b) the potential energy;

c) the total energy;

d) the energy required to remove the electron completely.

44–34 Referring to Problem 44–33, what are the energies if the atom is a singly-ionized helium atom (i.e., a helium atom with one electron removed)?

44–35 A singly-ionized helium ion (a helium atom with one electron removed) behaves very much like a hydrogen atom, except that the nuclear charge is twice as great.

a) How do the energy levels differ in magnitude from those of the hydrogen atom?

b) Which spectral series for He^+ have lines in the visible spectrum?

44–36 Refer to the examples at the end of Section 31–3. Suppose a hydrogen atom makes a transition from the $n = 3$ state to the $n = 2$ state (the Balmer Hα line at 656.3 nm) while in a magnetic field of magnitude 2 T. If the magnetic moment of the atom is parallel to the field in both the initial and final states,

a) by how much is each energy level shifted from the zero-field value;

b) by how much is the wavelength of the spectrum line shifted?

Chapter 45

Atoms, Molecules, and Solids

45–1 THE EXCLUSION PRINCIPLE

The hydrogen atom, discussed in Chapter 44, is the simplest of all atoms, containing one electron and one proton. Analysis of atoms with more than one electron increases in complexity very rapidly; each electron interacts not only with the positively charged nucleus but also with all the other electrons. In principle the motion of the electrons is governed by the Schrödinger equation, mentioned in Section 44–9, but the mathematical problem of finding appropriate solutions of this equation is so complex that it has not been accomplished exactly even for the helium atom, with two electrons.

Various approximation schemes can be used; the simplest (and most drastic) is to ignore completely the interactions between electrons, and regard each electron as influenced only by the electric field of the nucleus, considered to be a point charge. A less drastic and more useful approximation is to think of all the electrons together as making up a charge cloud which is, on the average, *spherically symmetric*, and to think of each individual electron as moving in the total electric field due to the nucleus and this averaged-out electron cloud. This is called the *central-field* approximation; it provides a useful starting point for the understanding of atomic structure.

An additional principle is also needed, the *exclusion principle.* To understand the need for this principle, we consider the lowest-energy state or *ground state* of a many-electron atom. The central-field model suggests that each electron has a lowest energy state (roughly corresponding to the $n = 1$ state for the hydrogen atom). It might be expected that, in the ground state of a complex atom, all the electrons should be in this lowest state. If this is the case, then when we examine the behavior of atoms with increasing numbers of electrons, we should find gradual changes in physical and chemical properties of elements as the number of electrons in the atoms increases.

A variety of evidence shows conclusively that this is *not* what happens at all. For example, the elements fluorine, neon, and sodium have, respectively, 9, 10, and 11 electrons per atom. Fluorine is a halogen and tends strongly to form compounds in which each atom acquires an extra electron. Sodium, an alkali metal, forms compounds in which it loses an electron, and neon is an inert gas, forming no compounds at all. This and many other observations show that in the ground state of a complex atom the electrons *cannot* all be in the lowest energy states.

The key to this puzzle, discovered by the Swiss physicist Wolfgang Pauli in 1925, is called the *Pauli exclusion principle.* Briefly, it states that *no two elec-*

trons can occupy the same quantum-mechanical state. Since different states correspond to different average distances from the nucleus, this means that, in a complex atom, there is not enough room for all the electrons to occupy states near the nucleus; some are forced into states farther away, having higher energies.

To understand the application of the Pauli principle to atomic structure, we need to use some results from quantum mechanics, the derivation of which is beyond our scope. First, the quantum-mechanical *state* of an electron in the central-field model is identified not by a single quantum number n, as in the Bohr model of the hydrogen atom, but by a set of *three* quantum numbers, all integers, usually called n, ℓ, and m. The first, n, is called the *principal quantum* number, corresponding to n for the hydrogen atom. It can be any positive integer (1,2,3,...). The energy of the state and its distance from the nucleus increase with n. The quantum number ℓ designates the magnitude of the angular momentum, L, according to the equation

$$L^2 = \ell(\ell + 1)\left(\frac{h^2}{4\pi^2}\right). \tag{45–1}$$

The value of ℓ can be zero or any positive integer up to and including $(n - 1)$. Finally, m designates the *component* of the angular momentum in a particular axis direction, usually taken to be the z-axis. Specifically,

$$L_z = \frac{mh}{2\pi}. \tag{45–2}$$

The value of m for any electron can be zero or any positive or negative integer up to and including $\pm\ell$.

This array of rules may seem bewildering, but some examples will help clarify matters. First, we observe that these results, obtained from the solutions of the Schrödinger equation, give *quantized* values for energy and angular momentum, just as the Bohr model did for the hydrogen atom. Second, we can make a list of the possible sets of quantum numbers and thus of the possible states of electrons in an atom. Such a list is given in Table 45–1, which also indicates two alternate notations. It is customary

Table 45–1

n	ℓ	m	Spectroscopic notation	Maximum number of electrons		Shell
1	0	0	1s	2		K
2	0	0	2s	2	8	L
2	1	−1				
2	1	0	2p	6		
2	1	1				
3	0	0	3s	2	18	M
3	1	−1				
3	1	0	3p	6		
3	1	1				
3	2	−2				
3	2	−1				
3	2	0	3d	10		
3	2	1				
3	2	2				
4	0	0	4s	2	32	N
4	1	−1				
4	1	0	4p	6		
4	1	1				
etc.						

to designate the value of ℓ by a letter, according to this scheme:

$\ell = 0$: s state

$\ell = 1$: p state

$\ell = 2$: d state

$\ell = 3$: f state

$\ell = 4$: g state

The origins of these letters are rooted in the early days of spectroscopy, and need not concern us. A state for which $n = 2$ and $\ell = 1$ is called a 2s state, and so on, as shown in Table 45–1. This table also shows the relation between values of n and the x-ray levels $(K, L, M, \ldots)$ described in Section 44–8. The $n = 1$ levels are designated as K, $n = 2$ as L, and so on. Because the average electron distance from the nucleus increases with n, each value of n corresponds roughly to a region of space around the nucleus in the form of a spherical shell. Hence, one speaks of the *L shell* as that region occupied by the electrons in the $n = 2$ states, and so on. States with the same n but different ℓ are said to form *subshells*, such as the $3p$ subshell.

We are now ready for a more precise statement of the exclusion principle: *In any atom, not more than two electrons can occupy any given quantum state.* That is, no more than two electrons can have the same set of three quantum numbers (n, ℓ, m). Since each quantum state corresponds to a certain distribution of the electron "cloud" in space, the principle says in effect that not more than two electrons can occupy the same region of space. This statement must be interpreted rather broadly, since the clouds describing electron distribution do not have definite, sharp boundaries, but the exclusion principle limits the degree of overlap of electron clouds that is permitted. The maximum numbers of electrons in each shell and subshell are shown in Table 45–1.

It may not be clear why *two* electrons should be permitted in each quantum state rather than just one. The reason is that we have not yet described the states completely. The missing component is *electron spin.* In the Bohr model, electron spin is added by visualizing the electron not as a point but as a small ball of charge spinning on its axis. Associated with this spin is angular momentum, and experiment shows that the component of this extra angular momentum in the direction of a specified axis (usually taken as the z-axis) is always one of the two values $h/4\pi$ or $-h/4\pi$. Thus, for a given set of values of (n, ℓ, m), there are two choices for the orientation of the spin angular momentum, corresponding to values ±1 of a *fourth* quantum number s, the *spin quantum number.* When two electrons occupy the same state they *must* have opposite values of s (opposite spin orientation).

Thus when we include electron spin, the Pauli exclusion principle can be restated: *In an atom, no two electrons can have all four quantum numbers the same.* When spin is included in the description of the quantum state, only one electron is permitted in each state.

In the modern quantum-mechanical picture, where we abandon the notion of a localized particle and speak instead of a diffuse charge cloud, there is no simple way to visualize electron spin. Nevertheless, the spin quantum number is an essential part of the description of a quantum-mechanical state. The exclusion principle, including the concept of electron spin, plays an essential role in the understanding of atomic structure. In the next section we shall see how the periodic table of the elements can be understood on the basis of this principle.

45–2 ATOMIC STRUCTURE

The number of electrons in an atom in its normal state is called the *atomic number*, denoted by Z. The nucleus contains Z protons and some number of neutrons. The proton and electron charges have the same magnitude but opposite sign, so in the normal atom the net electric charge is zero. Because the electrons are attracted to the nucleus, we expect the quantum states corresponding to regions near the nucleus to have lowest energy. We may imagine starting with a bare nucleus with Z protons, and adding electrons one by one until the normal complement of Z electrons for a neutral atom is reached. We expect the lowest-energy states, ordinarily those with the smallest values of n and ℓ, to fill first, and we use successively higher states until all electrons are accommodated.

The chemical properties of an atom are determined principally by interactions involving the outermost electrons, so it is of particular interest to find out how these are arranged. For example, when an atom has one electron considerably farther from the nucleus (on the average) than the others, this electron will be rather loosely bound; the atom will tend to lose this electron and form what chemists call an *electrovalent* or *ionic* bond, with valence $+1$. This behavior is characteristic of the alkali metals lithium, sodium, potassium, and so on.

We now proceed to describe ground-state electron configurations for the first few atoms (in order of increasing Z). For hydrogen the ground state is $1s$; the single electron is in the state $n = 1$, $\ell = 0$, $m = 0$, and $s = \pm1$. In the helium atom ($Z = 2$) *both* electrons are in $1s$ states, with opposite spins;

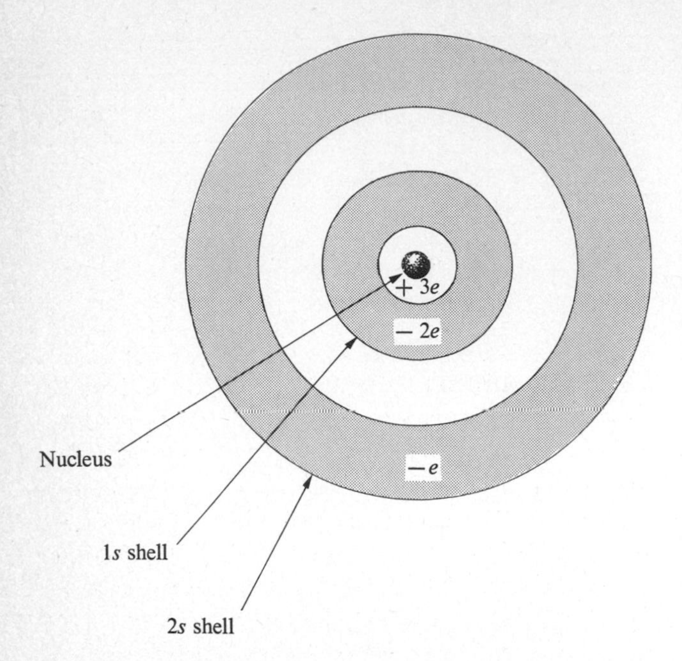

Fig. 45–1 *Schematic representation of charge distribution in lithium atom. The nucleus has a charge 3e; the two 1s electrons are closer to the nucleus than the 2s electron, which moves in a field approximately equal to that of a point charge 3e − 2e, or simply e.*

this state is denoted as $1s^2$. For helium, the K shell is completely filled and all others are empty.

Lithium ($Z = 3$) has three electrons; in the ground state two are in the $1s$ state, and one in a $2s$ state. We denote this state as $1s^22s$. On the average, the $2s$ electron is considerably farther from the nucleus than the $1s$ electrons, as shown schematically in Fig. 45–1. Thus, according to Gauss' law, the *net* charge influencing the $2s$ electron is $+e$, rather than $+3e$ as it would be without the $1s$ electrons present. Thus the $2s$ electron is loosely bound, as the chemical behavior of lithium suggests. An alkali metal, it forms ionic compounds with a valence of $+1$, in which each atom loses an electron.

Next is beryllium ($Z = 4$); its ground-state configuration is $1s^22s^2$, with two electrons in the L shell. Beryllium is the first of the *alkaline-earth* elements, forming ionic compounds with a valence of $+2$.

Table 45–2 shows the ground-state electron configurations of the first twenty elements. The L shell

Table 45–2 GROUND-STATE ELECTRON CONFIGURATIONS

Element	Symbol	Atomic number (Z)	Electron configuration
Hydrogen	H	1	$1s$
Helium	He	2	$1s^2$
Lithium	Li	3	$1s^22s$
Beryllium	Be	4	$1s^22s^2$
Boron	B	5	$1s^22s^22p$
Carbon	C	6	$1s^22s^22p^2$
Nitrogen	N	7	$1s^22s^22p^3$
Oxygen	O	8	$1s^22s^22p^4$
Fluorine	F	9	$1s^22s^22p^5$
Neon	Ne	10	$1s^22s^22p^6$
Sodium	Na	11	$1s^22s^22p^63s$
Magnesium	Mg	12	$1s^22s^22p^63s^2$
Aluminum	Al	13	$1s^22s^22p^63s^23p$
Silicon	Si	14	$1s^22s^22p^63s^23p^2$
Phosphorus	P	15	$1s^22s^22p^63s^23p^3$
Sulfur	S	16	$1s^22s^22p^63s^23p^4$
Chlorine	Cl	17	$1s^22s^22p^63s^23p^5$
Argon	Ar	18	$1s^22s^22p^63s^23p^6$
Potassium	K	19	$1s^22s^22p^63s^23p^64s$
Calcium	Ca	20	$1s^22s^22p^63s^23p^64s^2$

can hold a total of eight electrons, as the reader should verify from the rules in Section 45–1; at $Z = 10$ the K and L shells are filled and there are no electrons in the M shell. This is expected to be a particularly stable configuration, with little tendency to gain or lose electrons, and in fact this element is neon, an inert gas with no known compounds. The next element after neon is sodium ($Z = 11$), with filled K and L shells and one electron in the M shell. Thus its "filled-shell-plus-one-electron" structure resembles that of lithium; both are alkali metals. The element *before* neon is fluorine, with $Z = 9$. It has a vacancy in the L shell and might be expected to have an affinity for an electron, forming ionic compounds with a valence of -1. This behavior is characteristic of the *halogens* (fluorine, chlorine, bromine, iodine, astatine), all of which have "filled-shell-minus-one" configurations.

By similar analysis, one can understand all the regularities in chemical behavior exhibited by the periodic table of the elements, on the basis of electron configurations. With the M and N shells there is a slight complication because the $3d$ and $4s$ subshells ($n = 3$, $\ell = 2$, and $n = 4$, $\ell = 0$, respectively) overlap in energy. Thus argon ($Z = 18$) has all the $1s$, $2s$, $2p$, $3s$, and $3p$ states filled, but in potassium ($Z = 19$) the additional electron goes into a $4s$ rather than $3d$ level. The next several elements have one or two electrons in the $4s$ states and increasing numbers in the $3d$ states. These elements are all metals with rather similar properties, and form the first *transition series*, starting with scandium ($Z = 21$) and ending with zinc ($Z = 30$), for which the $3d$ levels are filled.

Thus, the similarity of elements in each *group* of the periodic table reflects corresponding similarity in electron configuration. All the inert gases (helium, neon, argon, krypton, xenon, and radon) have filled-shell configurations. All the alkali metals (lithium, sodium, potassium, rubidium, cesium, and francium) have "filled-shell-plus-one" configurations. All the alkaline-earth metals (beryllium, magnesium, calcium, strontium, barium, and radium) have "filled-shell-plus-two" configurations, and all the halogens (fluorine, chlorine, bromine, iodine, and astatine) have "filled-shell-minus-one" structures. And so it goes.

This whole theory can, of course, be refined to account for differences between elements in a group and to account for various aspects of chemical behavior. Indeed, some physicists like to claim that all of chemistry is contained in the Schrödinger equation! This statement may be a bit extreme, but we can see that even this qualitative discussion of atomic structure takes us a considerable distance toward understanding the atomic basis of many chemical phenomena. Next we shall examine in somewhat more detail the nature of the chemical bond.

45–3 DIATOMIC MOLECULES

As indicated in Section 45–2, the study of electron configurations in atoms provides valuable insight into the nature of the chemical bond, that is, the interaction that holds atoms together to form stable structures such as molecules and crystalline solids. There are a number of types of chemical bonds; the simplest to understand is the *ionic* bond, also called the *electrovalent* or *heteropolar* bond. The most familiar example is sodium chloride (NaCl), in which the sodium atom gives its $3s$ electron to the chlorine atom, filling the vacancy in the $3p$ subshell of chlorine. Energy is required to make this transfer if the atoms are far apart, but the result is two ions, one positively charged and one negatively, which attract each other. As they come together, their potential energy decreases, so that the final bound state of Na^+Cl^- has lower total energy than the state in which the two atoms are separated and neutral.

Removing the $3s$ electron from the sodium requires 5.1 eV of energy; this is called the *ionization energy* or *ionization potential.* Chlorine actually has an electron *affinity* of 3.8 eV. That is, the neutral chlorine atom can attract an extra electron which, once it takes its place in the $3p$ level, requires 3.8 eV for its removal. Thus, creating the separate Na^+ and Cl^- ions requires a net expenditure of only $5.1 - 3.8 = 1.3$ eV. The potential energy associated with the mutual attraction of the ions is about -5 eV, more than enough to repay the initial investment of creating the ions. This potential energy is determined by the closeness with which the ions can approach each other, and this in turn is determined by the exclusion principle, which forbids extensive overlap of the electron clouds of the two atoms.

Ionic bonds can involve more than one electron per atom. The alkaline-earth elements form ionic compounds in which each atom loses *two* electrons; an example is $Mg^{++}Cl_2^-$. Loss of more than two electrons is relatively rare, and instead a *different kind of bond* comes into operation.

The *covalent* or *homopolar* bond is characterized by a more nearly symmetric participation of the two atoms, as contrasted with the complete asymmetry involved in the electron-transfer process of the ionic bond. The simplest example of the covalent bond is the hydrogen molecule, a structure containing two protons and two electrons. As a preliminary to understanding this bond we consider first the interaction of two electric dipoles, as in Fig. 45–2. In (a) the dipoles are far apart; in (b) the like charges are farther apart than the unlike charges, and there is a net *attractive* force. In (c) and (d) the interaction is repulsive.

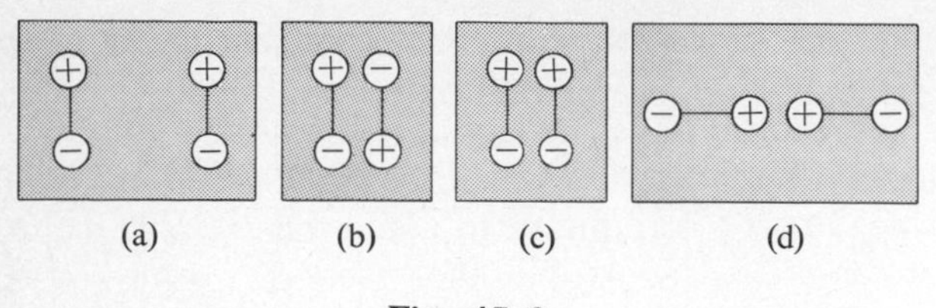

Fig. 45–2

In a molecule the charges are, of course, not at rest, but Fig. 45–2(b) at least makes it seem plausible that if the electrons are in the region between the protons, the attractive force the electrons exert on each proton may more than counteract the repulsive interactions of the protons on each other. That is, in the covalent bond in the hydrogen molecules the attractive interaction is supplied by a pair of electrons, one contributed by each atom, whose charge clouds are concentrated primarily in the region between the two atoms. Thus this bond may also be thought of as a shared-electron or electron-pair bond.

According to the exclusion principle, two electrons can occupy the same region of space only when they have opposite spin orientations. When the spins are parallel the state which would be most favorable from energy considerations is forbidden by the exclusion principle, and the lowest-energy state permitted is one in which the electron clouds are concentrated *outside* the central region between atoms. The nuclei then repel each other, and the interaction is repulsive rather than attractive. Thus, opposite spins are an essential requirement for an electron-pair bond, and no more than two electrons can participate in such a bond.

This is not to say, however, that an atom cannot have several electron-pair bonds. On the contrary, an atom having several electrons in its outermost shell can form covalent bonds with several other atoms. The bonding of carbon and hydrogen atoms, of central importance in organic chemistry, is an example. In the *methane* molecule (CH_4) the carbon atom is at the center of a regular tetrahedron, with a hydrogen atom at each corner. The carbon atom has four electrons in its L shell, and one of these electrons forms a covalent bond with each of the four hydrogen atoms, as shown in Fig. 45–3. Similar patterns occur in more complex organic molecules.

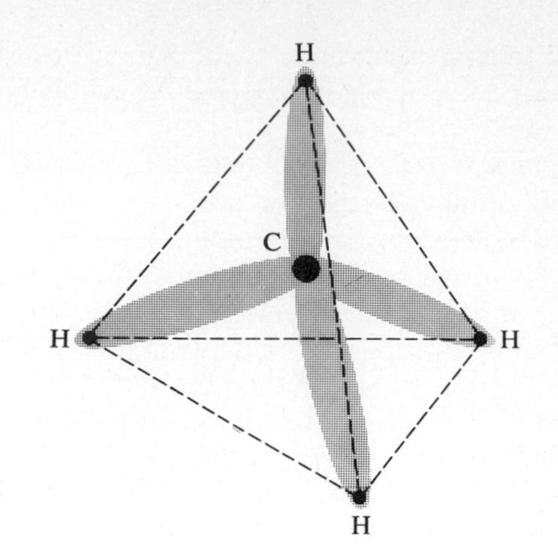

Fig. 45–3 *Methane* (CH_4) *molecule, showing four covalent bonds. The electron cloud between the central carbon atom and each of the four hydrogen nuclei represents the two electrons of a covalent bond. The hydrogen nuclei are at the corners of a regular tetrahedron.*

Both the ionic and covalent bonds are important in the structure of solids. In many respects a crystalline solid is really a very large molecule held together by chemical bonds. A third type of bonding, the *metallic* bond, is also important in the structure of some solids; we return to this subject in Section 45–5.

45–4 MOLECULAR SPECTRA

The energy levels of atoms are associated with the kinetic and potential energies of electrons with respect to the nucleus. Energy levels of molecules have additional features resulting from relative motion of the nuclei of the atoms, and there are characteristic spectra associated with these levels.

For the sake of simplicity we consider only diatomic molecules. Viewed as a rigid dumbbell, a diatomic molecule can *rotate* about an axis through its center of mass. Analysis of this rotation using the Schrödinger equation shows that the energy and angular momentum of this motion are *quantized.* Specifically, the energy of the rotational motion is found to be given by

$$E = \ell(\ell + 1)\left(\frac{h^2}{8\pi^2 I}\right) \qquad (\ell = 0, 1, 2, 3, \ldots), \tag{45–3}$$

where I is the moment of inertia of the molecule about an axis through the center of mass and perpendicular to the line joining the two nuclei. To indicate the magnitudes involved, we note that for an oxygen molecule $I = 5 \times 10^{-46}$ kg-m². Thus, the constant $h^2/8\pi^2 I$ in Eq. (45–3) is approximately equal to

$$\frac{h^2}{8\pi^2 I} = \frac{(6.6 \times 10^{-34}\ \text{J-s})^2}{8(3.14)^2(5 \times 10^{-46}\ \text{kg-m}^2)} = 10^{-23}\ \text{J}$$
$$= 0.6 \times 10^{-4}\ \text{eV}.$$

We note that this energy is much *smaller* than typical atomic energy levels associated with optical spectra. The energies of photons emitted and absorbed in transitions among rotational levels are correspondingly small, and they fall in the far infrared region of the spectrum.

The rigid-dumbbell model of a diatomic model suggests that the distance between the two nuclei is fixed; in fact, a more realistic model would represent the connection as a spring rather than a rigid rod. The atoms can undergo *vibrational* motion relative to the center of mass, and there is additional kinetic and potential energy associated with this motion. Application of the Schrödinger equation shows that the corresponding energy levels are given by

$$E = (n + 1/2)hf \qquad (n = 0, 1, 2, 3, \ldots), \tag{45–4}$$

where f is the frequency of the vibration. For typical diatomic molecules this turns out to be of the order of 10^{13} Hz; thus the constant hf in Eq. (45–4) is of the order of

$$hf = (6.6 \times 10^{-34}\ \text{J-s})(10^{13}\ \text{s}^{-1}) = 6.6 \times 10^{-21}\ \text{J}$$
$$= 0.04\ \text{eV}.$$

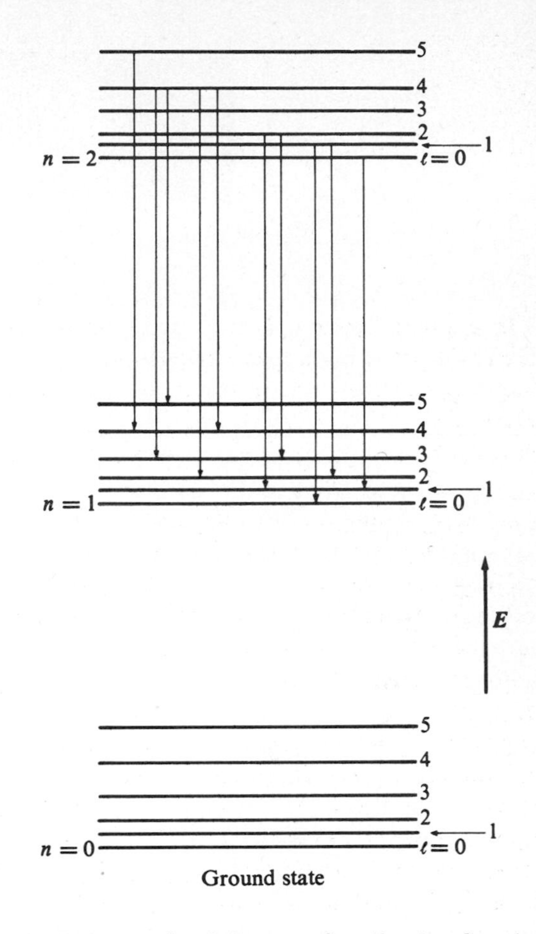

Fig. 45–4 *Energy-level diagram for vibrational and rotational energy levels of a diatomic molecule. For each vibrational level (n) there is a series of more closely spaced rotational levels (ℓ). Several transitions corresponding to a single band in a band spectrum are shown.*

Thus the vibrational energies, while still much smaller than those of the optical spectra, are typically considerably *larger* than the rotational energies.

An energy-level diagram for a diatomic molecule has the general appearance of Fig. 45–4. For each pair of values of n, there are many combinations of values of ℓ, leading to a series of closely spaced levels. Different pairs of n values give different series, and the resulting spectrum has the appearance of a series of *bands*; each band corresponds to a particular vibrational transition, and the individual

Fig. 45–5 Typical band spectrum. (Courtesy of R. C. Herman.)

lines in a band, to a particular rotational transition. A typical band spectrum is shown in Fig. 45–5.

The same considerations can be applied to more complex molecules. A molecule with three or more atoms has several different kinds or *modes* of vibratory motion, each with its own set of energy levels related to the frequency by Eq. (45–4). The resulting energy-level scheme and associated spectra can be quite complex, but the general considerations discussed above still apply. In nearly all cases the associated radiation lies in the infrared region of the electromagnetic spectrum. Analysis of molecular spectra has proved to be an extremely valuable analytical tool, providing a great deal of information about the strength and rigidity of molecular bonds and the structure of complex molecules.

45–5 STRUCTURE OF SOLIDS

At ordinary temperatures and pressures most materials are in the solid state, a condensed state of matter characterized by interatomic interactions of sufficient strength to give the material a definite volume and shape which change relatively little with applied stress. The separation of adjacent atoms in a solid is of the same order of magnitude as the diameter of the electron cloud around each atom.

A solid may be *amorphous* or *crystalline.* Crystalline solids have been discussed briefly in Section 20–7, and the reader would do well to review that discussion. A crystalline solid is characterized by *long-range order,* a recurring pattern in the arrangement of atoms; this pattern is called the *crystal structure* or the *lattice structure.* Amorphous solids have no long-range order but only short-range order, a state of affairs in which each atom has other atoms arranged around it in a more or less regular pattern, but without the recurring pattern characteristic of crystals. Amorphous solids have more in common with liquids than with crystalline solids; liquids also have short-range order but not long-range order.

The forces responsible for the regular arrangement of atoms in a crystal are, in some cases, the same as those involved in molecular binding. Corresponding to the classes of chemical bonds, there are ionic and covalent crystals. The alkali halides, of which ordinary salt (NaCl) is the most common variety, are the most familiar ionic crystals. The positive sodium ions and the negative chlorine ions occupy alternate positions in a cubic crystal lattice, as in Fig. 45–6. The forces are the familiar Coulomb's-law forces between charged particles; these forces are not directional, and the particular arrangement in which the material crystallizes is determined by the relative size of the two ions.

The simplest example of a *covalent* crystal is the *diamond structure,* a structure found in the diamond form of carbon and also in silicon, germanium, and tin, all elements in Group IV of the periodic table, with four electrons in the outermost shell. Each atom in this structure is situated at the center of a regular tetrahedron, with four nearest-neighbor atoms at the

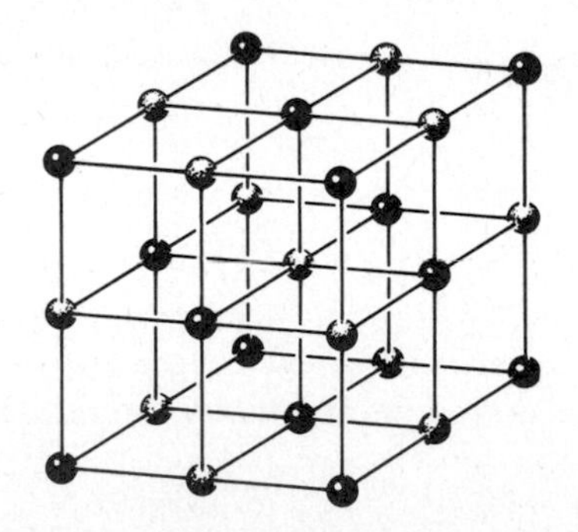

Fig. 45–6 Symbolic representation of a sodium chloride crystal, with exaggerated distances between ions.

(a)

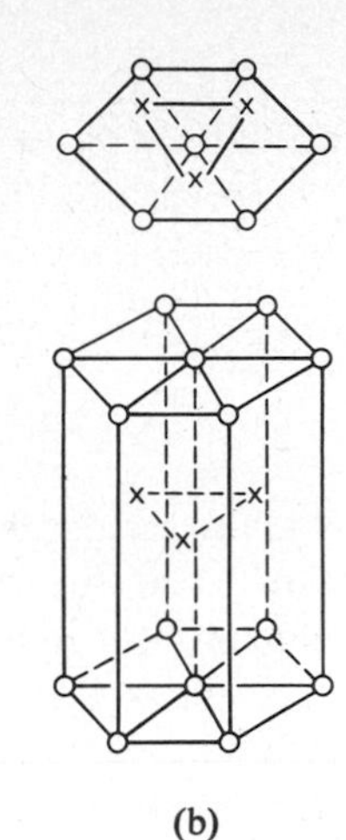

(b)

Fig. 45–7

corners, and it forms a covalent bond with each of these atoms. These bonds are strongly directional because of the asymmetrical electron distribution, leading to the tetrahedral structure.

A third crystal type which is less directly related to the chemical bond than ionic or covalent crystals is the *metallic crystal.* In this structure the outermost electrons are not localized at individual lattice sites but are detached from their parent atoms and free to move through the crystal. The corresponding charge clouds extend over many atoms. Thus a metallic crystal can be visualized roughly as an array of positive ions from which one or more electrons have been removed, immersed in a sea of electrons whose attraction for the positive ions holds the crystal together. This sea has many of the properties of a gas, and indeed one speaks of the *electron-gas model* of metallic solids.

In a metallic crystal the situation is as though the atoms would like to form shared-electron bonds but there are not enough valence electrons. Instead, electrons are shared among *many* atoms. This bonding is not strongly directional in nature, and the shape of the crystal lattice is determined primarily by considerations of *close packing,* i.e., the maximum number of atoms that can fit into a given volume.

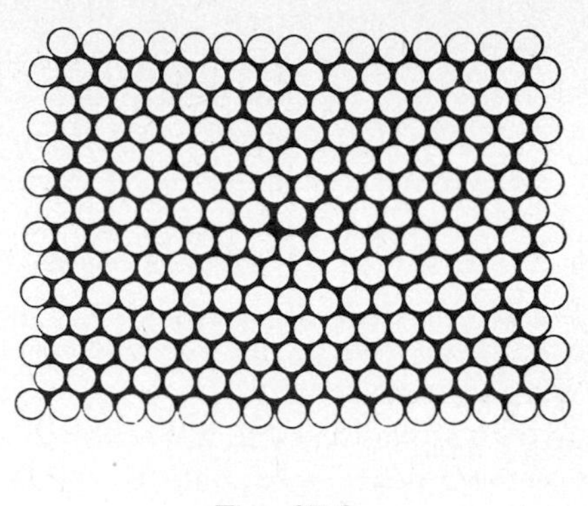

Fig. 45–8

The two most common metallic crystal lattices, the face-centered cubic and the hexagonal close-packed, are shown in Fig. 45–7. In each of these, each atom has 12 nearest neighbors.

The discussion in this section has centered around *perfect crystals,* crystals in which the crystal lattice extends uninterrupted through the entire material. Real crystals exhibit a variety of departures from this idealized structure. Materials are often *polycrystalline,* composed of many small single crystals bonded together at *grain boundaries.* Within a single crystal there may be *interstitial* atoms in places where they do not belong, and *vacancies,* lattice sites which should be occupied by an atom but are not. An imperfection of particular interest in semiconductors, to be discussed in Section 45–7, is the *impurity atom,* a foreign atom (e.g., arsenic in a silicon crystal) occupying a regular lattice site.

A more complex kind of imperfection is the *dislocation,* illustrated schematically in Fig. 45–8, in which one plane of atoms slips relative to another. The mechanical properties of metallic crystals are influenced strongly by the presence of dislocations; the ductility and malleability of some metals depends on the presence of dislocations which move through the lattice during plastic deformations.

45–6 PROPERTIES OF SOLIDS

Many *macroscopic* properties of solids, including mechanical, thermal, electrical, magnetic, and optical properties, can be understood by considering their relation to the *microscopic* structure of the material, and various aspects of the relation of structure to properties are part of the vigorous program of research in the physics of solids being carried out throughout the world. Although these topics cannot be discussed in detail in a book such as this one, a few examples will indicate the kinds of insights to be gained through study of the microscopic structure of solids.

We have already discussed, in Section 20–8, the subject of specific heats of crystals, using the same principle of equipartition of energy as in the kinetic theory of gases. A simple analysis permits understanding of the empirical rule of Dulong and Petit on the basis of a microscopic model. This analysis has its limitations, to be sure; it does not include the energy of electron motion, which in metals makes a small additional contribution to specific heat, nor does it predict the temperature dependence of specific heats resulting from the quantization of the lattice-vibration energy discussed in Section 20–8. But these additional refinements *can* be included in the model to permit more detailed comparison of observed macroscopic properties with theoretical predictions.

Electrical conductivity is understood on the basis of the mobility or lack of mobility of electrons in the material. In a metallic crystal the valence electrons are not bound to individual lattice sites but are free to move through the crystal. Thus we expect metals to be good conductors, and they usually are. In a covalent crystal the valence electrons are involved in the bonds responsible for the crystal structure and are therefore *not* free to move. Thus there are no mobile charges available for conduction, and such materials are expected to be insulators. Similarly, an ionic crystal such as NaCl has no charges that are free to move, and solid NaCl is an insulator. However, when salt is melted the ions are no longer locked to their individual lattice sites but are free to move, and *molten* NaCl is a good conductor.

There are, of course, no perfect conductors or insulators, although the resistivity of good insulators is greater than that of good conductors by an enormous factor, the order of at least 10^{15}. This great difference is one of the factors which makes extremely precise electrical measurements possible. In addition, the resistivity of all materials depends on temperature; in general, the small conductivity of insulators *increases* with temperature, but that of good conductors usually *decreases* at increased temperatures.

Two competing effects are responsible for this difference. In metals the *number* of electrons available for conduction is nearly independent of temperature, and the resistivity is determined by the frequency of collisions of electrons with the lattice. Roughly speaking, lattice vibrations increase with increased temperature, and the ion cores present a larger target area for collisions with electrons. In insulators, what little conduction does take place is due to electrons that have gained enough energy from thermal motion of the lattice to break away from their "home" atoms and wander through the lattice. The number of electrons able to acquire the needed energy is very strongly temperature-dependent; a twofold increase for a 10° temperature rise is typical. There is also increased scattering at higher temperatures, as with metals, but the increased number of carriers is a far larger effect; thus insulators invariably become better conductors at higher temperatures.

A similar analysis can be made for *thermal* conductivity, which involves transport of microscopic mechanical energy rather than electric charge. The wave motion associated with lattice vibrations is one mechanism for energy transfer, and in metals the mobile electrons also carry kinetic energy from one region to another. This effect turns out to be much larger than that of the lattice vibrations, with the result that metals are usually much better thermal conductors than insulators, which have at most very few free electrons available to transport energy.

Optical properties are also related directly to microscopic structure. Good electrical conductors cannot be transparent to electromagnetic waves, for the electric fields of the waves induce currents in the material, and these dissipate the wave energy into heat as the electrons collide with the atoms in the lattice. All transparent materials are very good insulators. Metals are, however, good *reflectors* of radiation, and again the reason is the presence of free electrons at the surface of the material, which can move in response to the incident wave and generate a reflected wave. Reflection from the polished surface of an insulator is a somewhat more subtle phenomenon, dependent on polarization of the material, but again it is possible to relate macroscopic properties to microscopic structure.

45–7 SEMICONDUCTORS

As the name implies, the electrical resistivity of a semiconductor is intermediate between that of good conductors and of good insulators. This is, however, only one aspect of the behavior of this important class of materials, so vital to present-day electronics. This discussion will be confined for simplicity to the elements germanium and silicon, the simplest semiconductors, but these examples serve to illustrate the principal concepts.

Silicon and germanium, having four electrons in the outermost subshell, crystallize in the diamond structure described in Section 45–5; each atom lies at the center of a regular tetrahedron, with four nearest neighbors at the corners and a covalent bond with each. Thus all the electrons are involved in the bonding and the materials should be insulators. However, an unusually small amount of energy is needed to break one of the bonds and set an electron free to roam around the lattice, 1.1 eV for silicon and only 0.7 eV for germanium. Thus even at room temperature a substantial number of electrons are dissociated from their parent atoms, and this number increases rapidly with temperature.

Furthermore, when an electron is removed from a covalent bond, it leaves a positively charged vacancy where there would ordinarily be an electron. An electron in a neighboring atom can drop into this vacancy, leaving the neighbor with the vacancy. In this way the vacancy, usually called a *hole*, can travel through the lattice and serve as an additional current-carrier. In a pure semiconductor, holes and electrons are always present in equal numbers; the resulting conductivity is called *intrinsic conductivity* to distinguish it from conductivity due to impurities, to be discussed later.

An analogy is often used to clarify the mechanism of conduction in a semiconductor. A crystal with no bonds broken is like a completely filled floor of a parking garage. No cars (electrons) can move because there is nowhere for them to go. But if one car is removed to the empty floor above, it can move around freely, and the vacancy it leaves also permits cars to move on the nearly filled floor; this motion is most easily described in terms of *motion of the vacant space* from which the car has been removed. The analogy can be drawn even more closely by considering the quantum states available to electrons in the solid, using the concept of energy bands, groups of closely spaced energy levels separated by "forbidden bands."

Now suppose we mix into melted germanium a small amount of arsenic, a Group V element having five electrons in its outermost subshell. When one of these electrons is removed, the remaining electron structure is essentially that of germanium; the only difference is that it is scaled down in size by the insignificant factor 32/33 because the arsenic nucleus contains a charge $+33e$ rather than $+32e$. Thus an arsenic atom can take the place of a germanium atom in the lattice. Four of its five valence electrons form the necessary four covalent bonds with the nearest neighbors; the fifth is very loosely bound (energy only about 0.01 eV) and even at ordinary temperature can very easily escape and wander about the lattice. The corresponding positive charge is associated with the nuclear charge and is *not* free to move, in contrast to the situation with electrons and holes in pure germanium.

Because at ordinary temperatures only a very small fraction of the valence electrons are able to escape their sites and participate in the conduction process, a concentration of arsenic atoms as small as one part in 10^{10} can increase the conductivity so drastically that conduction due to impurities becomes by far the dominant mechanism. In such a case the conductivity is due almost entirely to *negative* charge motion; the material is called an *n-type* semiconductor or is said to have *n*-type impurities.

Adding atoms of an element in Group III, with only three electrons in its outermost subshell, has an analogous effect. An example is gallium; placed in the lattice, the gallium atom would like to form four covalent bonds, but it has only three outer electrons.

It can, however, steal an electron from a neighboring germanium atom to complete the bonding. This leaves the neighbor with a *hole* or missing electron, and this hole can then move through the lattice just as with intrinsic conductivity. In this case the corresponding negative charge is associated with the deficiency of positive charge of the gallium nucleus ($+31e$ instead of $+32e$), so it is not free to move. This state of affairs is characteristic of *p-type* semiconductors, materials with p-type impurities. The two types of impurities, n and p, are also called *donors* and *acceptors*, respectively.

The assertion that in n and p semiconductors the current *is* actually carried by electrons and holes, respectively, can be verified using the Hall effect, discussed in Section 31–2. The direction of the Hall emf is opposite in the two cases, and measurements of the Hall effect in various semiconductor materials confirm the above analysis of the conduction mechanisms.

45–8 SEMICONDUCTOR DEVICES

The tremendous importance of semiconductor devices in practical electronic devices results directly from the fact that the conductivity can be controlled within wide limits and changed from one region of a device to another, by varying impurity concentrations. The simplest example is the *p-n junction*, a crystal of germanium or silicon with p-type impurities in one region and n-type in the other, with the two separated by a boundary region called the *junction*. The details of how such a crystal is produced need not concern us; one way, not necessarily the most economical, is to grow a crystal by pulling a seed crystal very slowly away from the surface of a melted semiconductor. By varying the concentration of impurities in the melt while the crystal is grown, one can make a crystal with two or more regions of varying conductivity.

When a p-n junction is connected in an external circuit as shown in Figure 45–9(a) and the potential V across the device is varied, the behavior of the current is as shown in Fig. 45–9(b). The device conducts much more readily in the direction $p \rightarrow n$ than in the reverse, in striking contrast to the behavior of

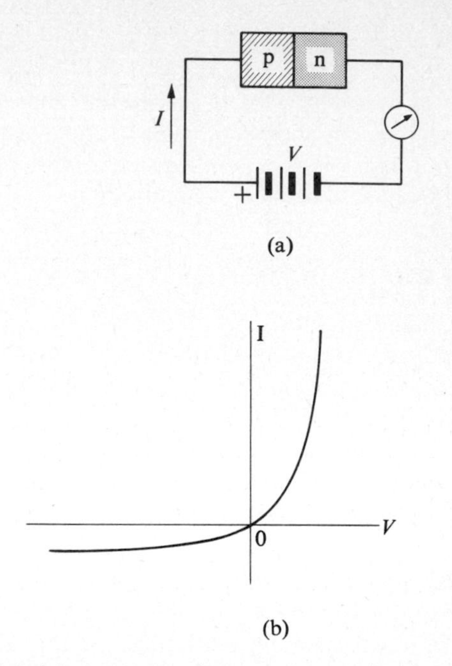

Fig. 45–9 *(a) A semiconductor p–n junction in a circuit; (b) graph showing the asymmetric voltage–current relationship, given by Eq. (45–5).*

materials that obey Ohm's law. In the language of electronics, such a one-way device is called a *diode*; a different kind of diode was discussed in Section 44–1 in connection with thermionic emission.

The behavior of a p-n junction diode can be understood at least roughly on the basis of the conductivity mechanisms in the two regions. When the p region is at higher potential than the n, holes in the p region flow into the n region and electrons in the n region into the p region, so both contribute substantially to current. When the polarity is reversed, the resulting electric fields tend to push electrons from p to n and holes from n to p. But there are very few electrons in the p region, only those associated with intrinsic conductivity and some that diffuse over from the n region. A similar condition prevails in the n region, and the current is much smaller than with the opposite polarity.

A more detailed analysis of this process, taking into account the effects of drift under the applied field and the diffusion that would take place even in

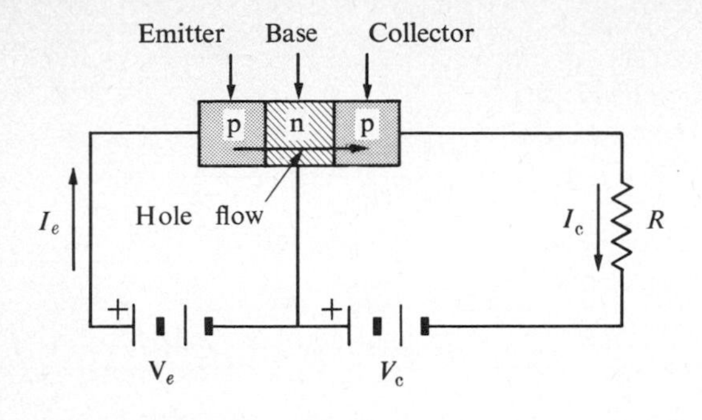

Fig. 45–10 Schematic diagram of a p-n-p transistor and circuit. When $V_e = 0$ the current in the collector circuit is very small. When a potential V_e is applied between emitter and base, holes travel from emitter to base, as shown; when V_c is sufficiently large, most of them continue into the collector. The collector current I_c is controlled by the emitter current I_e.

the absence of a field, shows that the voltage–current relationship is given by

$$I = I_0(e^{eV/kT} - 1), \tag{45–5}$$

where I_0 is a constant characteristic of the device, e is the electron charge, k is Boltzmann's constant, and T is absolute temperature.

The *transistor* includes two *p-n* junctions in a "sandwich" configuration which may be either *p-n-p* or *n-p-n*. One of the former type is shown in Fig. 45–10. The three regions are usually called the emitter, base, and collector, as shown. In the absence of current in the left loop of the circuit, there is only a very small current through the resistor R because the voltage across the base-collector junction is in the "reverse" direction, i.e., the direction of small current flow, as in a simple *p-n* junction. But when a voltage is applied between emitter and base, as shown, the holes traveling from emitter to base can travel *through* the base to the second junction where they come under the influence of the collector-to-base potential difference and thus flow on around through the resistor.

Thus the current in the collector circuit is *controlled* by the current in the emitter circuit. Furthermore, since V_c may be considerably larger than V_e, the power dissipated in R may be much larger than that supplied to the emitter circuit by the battery V_e. Thus the device functions as a power amplifier. If the

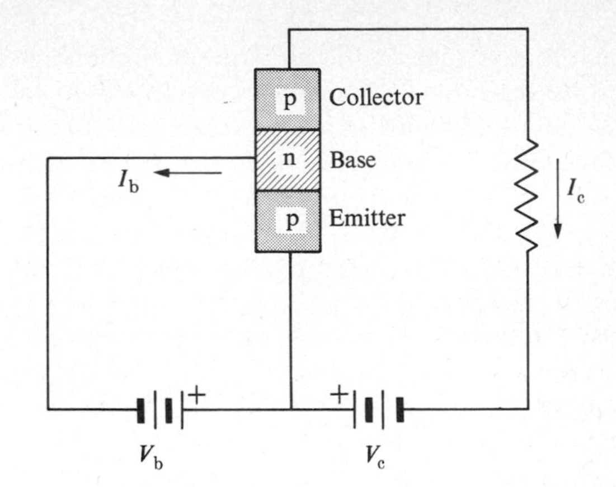

Fig. 45–11 A common-emitter circuit. When $V_b = 0$, I_c is very small and most of the voltage V_c appears across the base-collector junction. As V_b increases, the base-collector potential decreases and more holes can diffuse into the collector; thus I_c increases. Ordinarily I_c is much larger than I_b.

potential drop across R is greater than V_e, it may also be a voltage amplifier.

In this configuration the *base* is the common element between the "input" and "output" sides of the circuit. Another widely-used arrangement is the *common-emitter* circuit, shown in Fig. 45–11. In this circuit the current in the collector side of the circuit is much larger than in the base side, and the result is current amplification.

Transistors can perform many of the functions for which vacuum tubes have been used ever since the invention of the triode by de Forest in 1907; and where applicable they offer many advantages over vacuum tubes. These include mechanical ruggedness (since no vacuum container or fragile electrodes are involved), small size, long life (because they operate at relatively low temperatures and deteriorate very little with age), and efficiency (because no power is wasted heating a cathode for thermionic emission). Since their invention in 1948, semiconductor devices have completely revolutionized the electronics industry, including applications to communications, computer systems, control systems, and many other areas.

A further refinement in semiconductor technology is the *integrated circuit.* By successive depositing of layers of material and etching patterns to define current paths, one can combine the functions of several transistors, capacitors, and resistors on a single square of semiconductor material that may be no larger than 2 mm on a side. The impact of this concept on computer technology requiring enormously complex but still very compact circuitry is obvious, and nearly every day brings still more sophisticated devices.

PROBLEMS

45–1 Make a list of the four quantum numbers for each of the six electrons in the ground state of the carbon atom, and for each of the 14 electrons in the silicon atom.

45–2 Cobalt ($Z = 27$) has two electrons in $4s$ states. How many does it have in $3d$ states?

45–3 For germanium ($Z = 32$) make a list of the number of electrons in each state ($1s$, $2s$, $2p$, etc.)

45–4 Calculate the magnitude of the potential energy ($e^2/4\pi\epsilon_0 r$) of interaction of a charge +e with a charge −e (where e is the electron charge magnitude) at a distance of 2×10^{-10} m, roughly the magnitude of the lattice spacing in the NaCl crystal.

45–5 For the sodium chloride molecule discussed at the beginning of Section 45–3, what is the maximum separation of the ions for stability if they may be regarded as point charges? (The potential energy of interaction must be at least −1.3 eV.)

45–6 For magnesium, the first ionization potential is 7.6 eV; the second (additional energy required to remove a second electron) is almost exactly twice this, 15 eV, and the third ionization potential is much larger, about 80 eV. How can these numbers be understood?

45–7 The ionization potentials of the alkali metals are:

Li	5.4 eV	Rb	4.2 eV
Na	5.1	Cs	3.9
K	4.3		

Which of these would you expect to be most active chemically?

45–8 The rotation spectrum of HCl contains the following wavelengths:

60.4 μm
69.0
80.4
96.4
120.4

Find the moment of inertia of the HCl molecule.

45–9 Show that the frequencies in a pure rotation spectrum (disregarding vibrational levels) are all integer multiples of the quantity $h/2\pi I$.

45–10 If the distance between atoms in a diatomic oxygen molecule is 2×10^{-10} m, calculate the moment of inertia about an axis through the center of mass perpendicular to the line joining the atoms.

45–11 The dissociation energy of the hydrogen molecule (i.e., the energy required to separate the two atoms) is 4.72 eV. At what temperature is the average kinetic energy of a molecule equal to this energy?

45–12 The vibrational frequency of the hydrogen molecule is 1.29×10^{14} Hz.

a) What is the spacing of adjacent vibrational energy levels, in electron volts?
b) What is the wavelength of radiation emitted in the transition from the $n = 2$ to $n = 1$ vibrational state of the hydrogen molecule?
c) From what initial values of n do transitions to the ground state of vibrational motion lead to radiation in the visible spectrum?

45–13 The distance between atoms in a hydrogen molecule is 0.074 nm.

a) Calculate the moment of inertia of a hydrogen molecule about an axis perpendicular to the line joining the nuclei, at its center.
b) Find the energies of the $\ell = 0$, $\ell = 1$, and $\ell = 2$ rotational states.
c) Find the wavelength and frequency of the photon emitted in the transition from $\ell = 2$ to $\ell = 0$.

45–14 The spacing of adjacent atoms in a sodium-chloride crystal is 0.282 nm. Calculate the density of sodium chloride.

45–15 In the hexagonal close-packed crystal structure, regarding the atoms as rigid spheres, show that an atom can accommodate twelve nearest neighbors (all touching it) if six are placed around it in a plane hexagonal array, with three additional atoms above and three below the plane.

45–16 Suppose a piece of very pure germanium is to be used as a light detector by observing the increase in conductivity resulting from generation of electron-hole pairs by absorption of photons. If each pair requires 0.7 eV of energy, what is the maximum wavelength that can be detected? In what portion of the spectrum does it lie?

Chapter 46

Nuclear Physics

46–1 THE NUCLEAR ATOM

In previous chapters we have frequently made use of the fact that every atom contains a massive positively charged nucleus, much smaller than the overall dimensions of the atom but nevertheless containing most of the total mass of the atom. It is instructive to review the earliest experimental evidence for the existence of the nucleus, the *Rutherford scattering experiments.* The experiments were carried out in 1910–1911 by Sir Ernest Rutherford and two of his students, Hans Geiger and Ernest Marsden, at Cambridge, England.

The electron had been "discovered" in 1897 by Sir J. J. Thomson, and by 1910 its mass and charge were quite accurately known. It had also been well established that, with the sole exception of hydrogen, all atoms contain more than one electron. Thomson had proposed a model of the atom consisting of a relatively large sphere of positive charge (about 2 or 3×10^{-8} cm in diameter) within which were embedded, like plums in a pudding, the electrons.

What Rutherford and his coworkers did was to project other particles at the atoms under investigation, and from observations of the way in which the projected particles were deflected or *scattered,* they drew conclusions about the distribution of charge within the "target" atoms.

At this time the high-energy particle accelerators now in common use for nuclear physics research had not yet been developed, and Rutherford had to use as projectiles the particles produced in natural radioactivity, to be discussed later in this chapter. Some radioactive disintegrations result in the emission of *alpha particles*; these particles are now known to be identical with the nuclei of helium atoms, each consisting of two protons and two neutrons bound together, but without the two electrons normally present in a neutral helium atom. Alpha particles are ejected from unstable nuclei with speeds of the order of 10^7 m s^{-1}, and they can travel several centimeters through air, or the order of 0.1 mm through solid matter, before they are brought to rest by collisions.

The experimental setup is shown schematically in Fig. 46–1. A radioactive source at the left emits alpha particles. Thick lead screens stop all particles except those in a narrow *beam* defined by small holes. The beam then passes through a thin metal foil (gold, silver, and copper were used) and strikes a plate coated with zinc sulfide. A momentary flash or scintillation can be observed visually on the screen whenever it is struck by an alpha particle, and the number of particles that have been deflected through any angle from their original direction can therefore be determined.

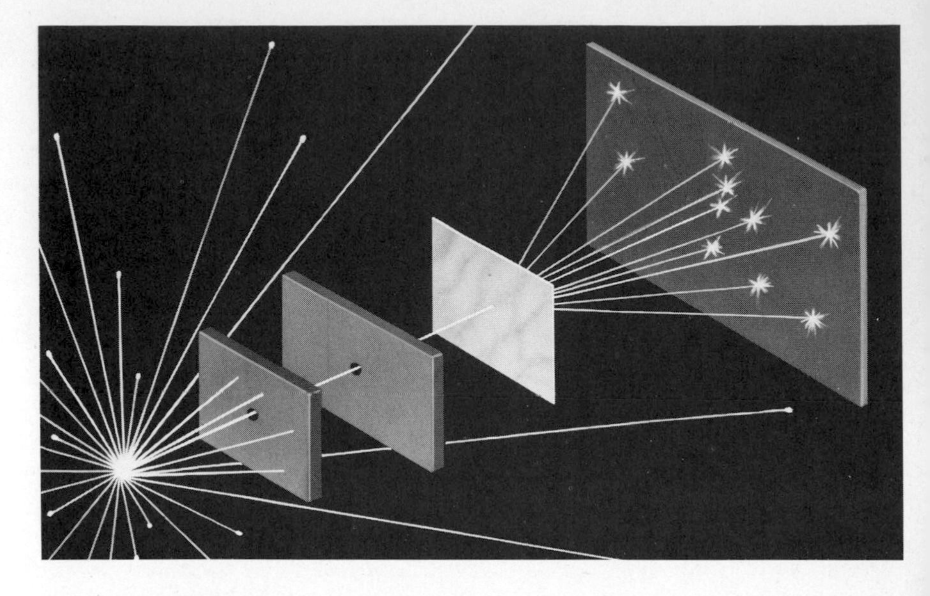

Fig. 46–1 *The scattering of alpha particles by a thin metal foil.*

According to the Thomson model, the atoms of a solid are packed together like marbles in a box. The experimental fact that an alpha particle can pass right through a sheet of metal foil forces one to conclude, if this model is correct, that the alpha particle is capable of actually penetrating the spheres of positive charge. Granted that this is possible, we can compute the deflection it would undergo. The Thomson atom is electrically neutral, so outside the atom no force would be exerted on the alpha particle. Within the atom, the electrical force would be due in part to the electrons and in part to the sphere of positive charge. However, the mass of an alpha particle is about 7400 times that of an electron, and from momentum considerations it follows that the alpha particle can suffer only a negligible scattering as a consequence of forces between it and the much less massive electrons. It is only interactions with the *positive* charge, which makes up most of the atomic mass, that can deviate the alpha particle.

The electrical force on an alpha particle within a sphere of positive charge is like the gravitational force on a mass point within a sphere, except that gravitational forces are attractive, while the force between two positive charges is a repulsion. The gravitational intensity within a sphere is zero at the center and increases linearly with distance from the center, because that part of the mass of the sphere lying *outside* any radius exerts no force at interior points. The alpha particle is therefore *repelled* from the center of the sphere with a force proportional to its distance from the center, and its trajectory can be computed for any initial direction of approach such as that in Fig. 46–2(a). On the basis of such calculations, Rutherford predicted the number of alpha particles that should be scattered at any angle with respect to the original direction.

The experimental results did not agree with the calculations based on the Thomson atom. In particular, many more particles were scattered through

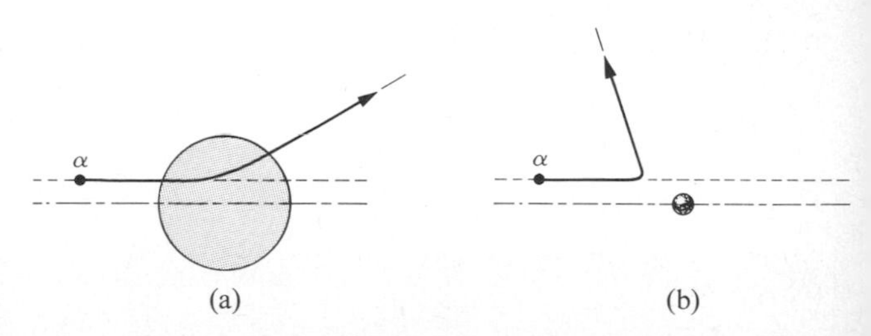

Fig. 46–2 *(a) Alpha particles scattered through a small angle by the Thomson atom. (b) Alpha particle scattered through a large angle by the Rutherford nuclear atom.*

large angles than were predicted. To account for the observed large-angle scattering, Rutherford concluded that the positive charge, instead of being spread through a sphere of atomic dimensions (2 or 3×10^{-8} cm) was concentrated in a much *smaller* volume, which he called a *nucleus.* When an alpha particle approaches the nucleus, the entire nuclear charge exerts a repelling effect on it down to extremely small separations, with the consequence that much larger deviations can be produced. Figure 46–2(b) shows the trajectory of an alpha particle deflected by a Rutherford nuclear atom, for the same original path as that in part (a) of the figure.

Rutherford again computed the expected number of particles scattered through any angle, assuming an inverse-square law of force between the alpha particle and the nucleus of the scattering atom. Within the limits of experimental accuracy, the computed and observed results were in agreement down to distances of approach of about 10^{-12} cm. These experiments thus indicate that the size of the nucleus is no larger than the order of 10^{-12} cm.

46–2 PROPERTIES OF NUCLEI

As indicated in Section 46–1, the most obvious feature of the atomic nucleus is its size, 100,000 times smaller than the atom itself. Since Rutherford's initial experiments, many additional scattering experiments have been performed, using high-energy protons, electrons, and neutrons as well as alpha particles. Although the "surface" of a nucleus is not a sharp boundary, these experiments determine an approximate radius for each nucleus. The radius is found to depend on the mass, which in turn depends on the total number A of neutrons and protons, usually called the *mass number.* The radii of most nuclei are represented fairly well by the empirical equation

$$r = r_0 A^{1/3} \qquad (46\text{–}1)$$

where r_0 is an empirical constant equal to 1.2×10^{-15} m, the same for all nuclei.

Since the volume of a sphere is proportional to r^3, Eq. (46–1) shows that the *volume* of a nucleus is proportional to A (i.e., to the total mass) and therefore that the mass per unit volume (proportional to A/r^3) is the same for all nuclei. That is, *all nuclei have approximately the same density.* This fact is of crucial importance in understanding nuclear structure.

Two additional important properties of nuclei are angular momentum and magnetic moment. The particles in the nucleus are in motion, just as the electrons in an atom are in motion; there is angular momentum associated with this motion, and, because circulating charge constitutes a current, there is also magnetic moment. Experimental evidence for the existence of nuclear angular momentum (often called *nuclear spin*) and magnetic moment came originally from spectroscopy. Some spectrum lines are found to be split into series of very closely spaced lines, called *hyperfine structure,* which can be understood on the basis of interactions between the electrons in an atom and the magnetic field produced by a nuclear magnetic moment. Detailed analysis of this phenomenon indicates that the nuclear angular momentum is *quantized,* just as it is for electrons and molecular rotation; the component of angular momentum in a specified axis direction is a multiple of $h/2\pi$, but some nuclei seem to require *integer* multiples (as with orbital angular momentum of electrons) and some *half-integer* multiples, as with electron spin. As we shall see later, the nuclear spin can be understood on the basis of the particles making up the nucleus.

Although it was once believed that nuclei were made of protons and electrons, the discovery of the neutron (discussed in Section 46–9) and many other experiments have established that the basic building blocks of the nucleus are the proton and the neutron. The total number of protons, equal in a neutral atom to the number of electrons, is the *atomic number* Z. The total number of *nucleons* (protons and neutrons) is called the *mass number* A. The number of neutrons, denoted by N, is called the *neutron number.* For any nucleus these are related by

$$A = Z + N \qquad (46\text{–}2)$$

Table 46–1 lists values of A, Z, and N for several nuclei. As the table shows, some nuclei have the same Z but different N. Since the electron structure, which determines the chemical properties, is determined by the charge of the nucleus, these are nuclei

Table 46–1 NUCLEAR PARTICLES

Nucleus	Mass number (total number of nuclear particles), A	Atomic number (number of protons), Z	Neutron number, $N=A-Z$
${}_{1}H^{1}$	1	1	0
${}_{1}D^{2}$	4	2	2
${}_{3}Li^{6}$	6	3	3
${}_{3}Li^{7}$	7	3	4
${}_{4}Be^{9}$	9	4	5
${}_{5}B^{10}$	10	5	5
${}_{5}B^{11}$	11	5	6
${}_{6}C^{12}$	12	6	6
${}_{6}C^{13}$	13	6	7
${}_{7}N^{14}$	14	7	7
${}_{8}O^{16}$	16	8	8
${}_{11}Na^{23}$	23	11	12
${}_{29}Cu^{65}$	65	29	36
${}_{80}Hg^{200}$	200	80	120
${}_{92}U^{235}$	235	92	143
${}_{92}U^{238}$	238	92	146

Table 46–2 ATOMIC DATA*

Element		Atomic number, Z	Neutron number, N	Atomic mass, u	Mass number, A
Hydrogen	H	1	0	1.00783	1
Deuterium	H	1	1	2.01410	2
Helium	He	2	1	3.01603	3
Helium	He	2	2	4.00260	4
Lithium	Li	3	3	6.01513	6
Lithium	Li	3	4	7.01601	7
Beryllium	Be	4	4	8.00508	8
Beryllium	Be	4	5	9.01219	9
Boron	B	5	5	10.01294	10
Boron	B	5	6	11.00931	11
Carbon	C	6	6	12.00000	12
Carbon	C	6	7	13.00335	13
Nitrogen	N	7	7	14.00307	14
Nitrogen	N	7	8	15.00011	15
Oxygen	O	8	8	15.99491	16
Oxygen	O	8	9	16.99913	17
Oxygen	O	8	10	17.99916	18

*American Institute of Physics Handbook, 1963

of the same element, but they have different masses and can be distinguished in precise experiments. The experimental investigation of isotopes through mass spectroscopy is discussed in Sections 30–6 and 30–7, and the reader would do well to review these sections. Nuclei of a given element having different mass numbers are called *isotopes* of the element, and a single nuclear species (single values of both Z and N) is called a *nuclide.*

Table 46–1 also shows the notation usually used to denote individual nuclides; the symbol of the element is used, with a pre-subscript equal to the atomic number Z and a post-superscript equal to the mass number A. This is of course redundant, since the element is determined by the atomic number, but the notation is a useful aid to memory.

The total mass of a nucleus is always *less* than the total mass of its constituent parts because of the mass-equivalent of the negative potential energy associated with the attractive forces that hold the nucleus together. In fact, the best way to determine the total potential energy, or *binding energy,* is to compare the mass of a nucleus with the masses of its constituents. The proton and neutron masses are

$$m_p = 1.673 \times 10^{-27}\ \text{kg}$$
$$m_n = 1.675 \times 10^{-27}\ \text{kg}.$$

Since these are nearly equal, it is not surprising that many nuclear masses are approximately integer multiples of the proton or neutron mass.

This observation suggests defining a new mass unit equal to the proton or neutron mass. Instead, it has been found more convenient to define the new unit, called the *atomic mass unit* (u), as $\frac{1}{12}$ the mass of the neutral carbon atom having mass number $A = 12$. It is found that

$$1\ \text{u} = 1.660566 \times 10^{-27}\ \text{kg}.$$

In atomic units, the masses of the proton, neutron,

and electron are found to be:

$$m_p = 1.007276\text{ u},$$
$$m_n = 1.008665\text{ u},$$
$$m_e = 0.000549\text{ u}.$$

The masses of some common atoms, including their electrons, are shown in Table 46–2. The masses of the bare nuclei are obtained by subtracting Z times the electron mass.

The forces that hold protons and neutrons together in the nucleus despite the electrical repulsion of the protons are not of any familiar sort but are unique to the nucleus. Some aspects of the nuclear force are still incompletely understood, but several qualitative features can be described. First, it does not depend on charge, since neutrons as well as protons must be bound. Second, it must be of short range, otherwise the nucleus would pull in additional protons and neutrons, but within the range it must be stronger than electrical forces, otherwise the nucleus could never be stable. Third, the nearly constant density of nuclear matter indicates that a given nucleon cannot interact simultaneously with all the other nucleons but only with those few in its immediate vicinity. This is again in contrast to the behavior of electrical forces, in which *every* proton in the nucleus repels every other one. This limitation on the maximum number of nucleons with which a nucleon can interact is called *saturation*; it is analogous in some respects to covalent bonding in solids. Finally, the nuclear forces appear to favor binding of pairs of particles (e.g., two protons with opposite spin, or two neutrons with opposite spin) and of *pairs of pairs*, a pair of protons and a pair of neutrons, with total spin zero. Thus the alpha particle is an exceptionally stable nuclear structure.

These qualitative features of the nuclear force are helpful in understanding the various kinds of nuclear instability, to be discussed in the following sections.

46–3 NATURAL RADIOACTIVITY

In studying the fluorescence and phosphorescence of compounds irradiated with visible light, Becquerel, in 1896, performed a crucial experiment which led to a deeper understanding of the properties of the nucleus of an atom. After illuminating some pieces of uranium–potassium sulfate with visible light, Becquerel wrapped them in black paper and separated the package from a photographic plate by a piece of silver. After several hours' exposure the photographic plate was developed and showed a blackening due to something that must have been emitted from the compound and was able to penetrate both the black paper and the silver.

Rutherford showed later that the emanations given off by uranium sulfate were capable of ionizing the air in the space between two oppositely charged metallic plates (an ionization chamber). The current registered by a galvanometer in series with the circuit was taken to be a measure of the "activity" of the compound.

A systematic study of the activity of various elements and compounds led Mme. Curie to the conclusion that this activity was an atomic phenomenon; and by the methods of chemical analysis, she and her husband, Pierre Curie, found that "ionizing ability" or "activity" was associated not only with uranium but with two other elements that they discovered, radium and polonium. The activity of radium was found to be more than a million times that of uranium. Since the pioneer days of the Curies, many more radioactive substances have been discovered.

The activity of radioactive material may be easily shown to be the result of three different kinds of emanations. An early experiment is shown in Fig. 46–3. A small piece of radioactive material is placed at the bottom of a long groove in a lead block. Some distance above the lead block a photographic plate is placed, and the whole apparatus is highly evacuated. A strong magnetic field is applied at right angles to the plane of the diagram. When the plate is developed, three distinct spots are found, one in the direct line of the groove in the lead block, one deflected to one side, and one to the other side. From a knowledge of the direction of the magnetic field, it is concluded that one of the emanations is positively charged (alpha particles), one is negatively charged (beta particles) and one is neutral (gamma rays).

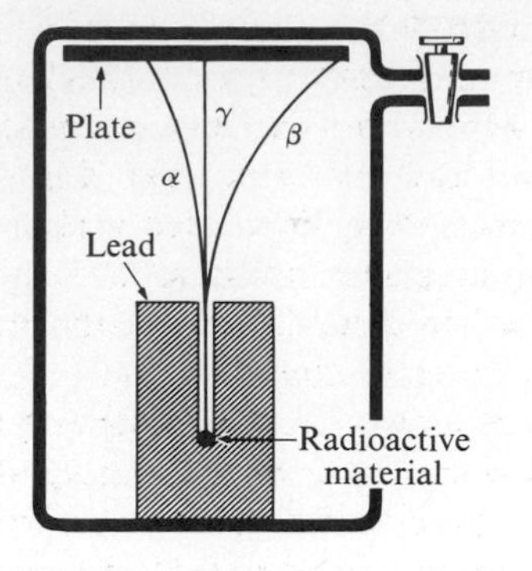

Fig. 46–3 The three emanations from a radioactive material and their paths in a magnetic field perpendicular to the plane of the diagram.

Further investigation showed that not all three emanations are emitted simultaneously by all radioactive substances. Some elements emit alpha particles, others emit beta particles, while gamma rays sometimes accompany one and sometimes the other. Furthermore, no simple macroscopic physical or chemical process, such as raising or lowering the temperature, chemical combination with other non-radioactive substances, etc., could change or affect in any way the activity of a given sample. As a result, it was suspected from the beginning that radioactivity is a *nuclear* process and that the emission of a charged particle from the nucleus of an atom results in leaving behind a different atom, occupying a different place in the periodic table. In other words, radioactivity involves the *transmutation of elements.*

The first measurement of the charge of the alpha particle used a device called a *Geiger counter,* still an important instrument of modern physics.

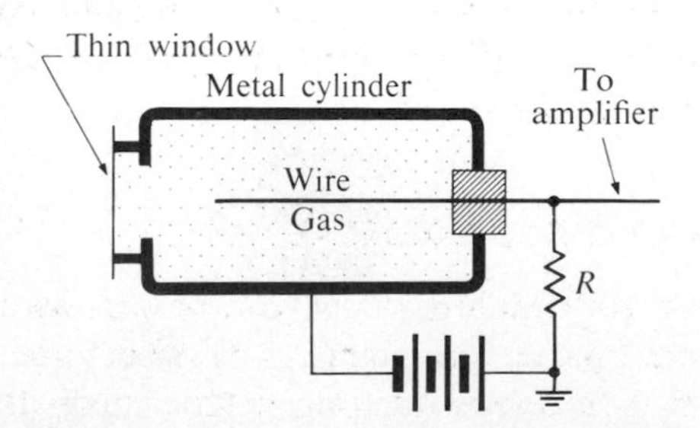

Fig. 46–4 Schematic diagram of a Geiger counter.

As shown in Fig. 46–4, a Geiger counter consists of a metal cylinder and a wire along the axis. The cylinder contains a gas such as air or argon, at a pressure of from 50 to 100 mm of mercury. A difference of potential slightly less than that necessary to produce a discharge is maintained between the wire and the cylinder wall. Alpha particles (or, for that matter, any particles to be studied) can enter through a thin glass or mica window. The particle entering the counter produces ionization of the gas molecules. These ions are accelerated by the electric field and produce more ions by collisions, causing the ionization current to build up rapidly. The current, however, decays rapidly since the circuit has a small time constant. There is therefore a momentary surge of current or a momentary potential surge across R which may be amplified and made to advance an electronic counter, or to activate a counting-rate meter.

Placing a known mass of radium a known distance from the window of a Geiger counter, Rutherford and Geiger counted the number of alpha particles emitted in a known time interval. They found that 3.57×10^{10} alpha particles were emitted per second per gram of radium. They then allowed the alpha particles from the same source to fall upon a plate and measured its rate of increase of charge. Dividing the rate of increase of charge by the number emitted per second, Rutherford and Geiger determined the charge on an alpha particle to be 3.19×10^{-19} C, or practically twice the charge on an electron, but opposite in sign.

The next problem was to determine the mass of an alpha particle. This was accomplished by measuring the ratio of charge to mass by the electric and magnetic deflection method described in Chapter 30. The ratio was found by Rutherford and Robinson to be 4.82×10^{7} C kg^{-1}. Combining this result with the charge of an alpha particle, the mass was found to be 6.62×10^{-27} kg, almost exactly four times the mass of a hydrogen atom.

Since a helium atom has a mass four times that of a hydrogen atom and, stripped of its two electrons (as a bare nucleus), has a charge equal in magnitude and opposite in sign to two electrons, it seemed certain that alpha particles were helium nuclei. To make the identification certain, however, Rutherford and Royds collected the alpha particles in a glass discharge tube over a period of about six days and then established an electric discharge in the tube. Examining the spectrum of the emitted light, they identified the characteristic helium spectrum and established without doubt that alpha particles *are* helium nuclei.

The speed of an alpha particle emitted from a given radioactive source such as radium may be measured by observing the radius of the circle traversed by the particle in a magnetic field perpendicular to the motion. Such experiments show that alpha particles are emitted with very high speeds. For example, the alpha particles emitted by radium, ${}_{88}Ra^{226}$, have speeds of about 1.5×10^7 m s^{-1}. The corresponding kinetic energy is

$$\begin{aligned} E_k &= \tfrac{1}{2}(6.62 \times 10^{-27}\ \text{kg})(1.5 \times 10^7\ \text{m s}^{-1})^2 \\ &= 7.5 \times 10^{-13}\ \text{J} \\ &= 4.7 \times 10^6\ \text{eV} \\ &= 4.7\ \text{MeV}. \end{aligned}$$

We note that the speed, although large, is only five percent of the speed of light, so the nonrelativistic kinetic-energy expression may be used. We also note that the energy is larger than typical energies of atomic electrons by a factor of the order of a million. Because of these large energies, alpha particles are capable of traveling several centimeters in air, or a few tenths or hundredths of a millimeter through solids, before they are brought to rest by collisions.

Beta particles are *negatively* charged and are therefore deflected in an electric or magnetic field. Deflection experiments similar to those described in Chapter 30 prove conclusively that beta particles have the same charge and mass as electrons. They are emitted with tremendous speeds, some reaching a value of 0.9995 that of light. Thus relativistic relations must be used in analyzing their motion. Unlike alpha particles, which are emitted from a given nucleus with one or a few definite velocities, beta particles are emitted with a *continuous* range of velocities, from zero up to a maximum which depends on the nature of the emitting nucleus. If the principles of conservation of energy and of momentum are to hold in nuclear processes, it is necessary to assume that the emission of a beta particle is accompanied by the emission of another particle of negligible rest mass and with no charge. This particle, called the *neutrino*, has zero rest mass and zero charge and therefore, even in traversing the densest matter, produces very little measurable effect. In spite of this, Reines and Cowan, in 1953 and again in 1956, succeeded in detecting its existence in a series of extraordinary experiments.

Since gamma rays are not deflected by a magnetic field, they cannot consist of charged particles. They are, however, diffracted at the surface of a crystal in a manner similar to that of x-rays but with extremely small angles of diffraction. Experiments of this sort lead to the conclusion that gamma rays are actually electromagnetic waves of extremely short wavelength, about 1/100 that of x-rays.

The gamma-ray spectrum of any one element is a line spectrum, suggesting that a gamma-ray photon is emitted when a nucleus proceeds from a state of higher to a state of lower energy. This view is substantiated in the case of radium by the following facts. When alpha particles are emitted from radium they are found to consist of two groups, those with a kinetic energy of 4.879 MeV and those with an energy of 4.695 MeV. When a radium atom emits an alpha particle of the smaller energy, the resulting nucleus (which corresponds to the element *radon*) has a *greater* amount of energy than if the higher-speed alpha particle had been emitted. This represents an excited state of the radon nucleus. If now the radon nucleus undergoes a transition from this excited state to the lower energy state, a gamma-ray photon of energy $(48.79 - 46.95) \times 10^5 = 1.84 \times 10^5$ eV should be emitted. The *measured* energy of the gamma-ray photon emitted by radium is 1.89×10^5 eV, in excellent agreement.

Thus, by correlating alpha-particle energies and gamma-ray energies, it is possible in some cases to construct *nuclear energy-level diagrams* similar to x-ray energy-level diagrams.

46–4 NUCLEAR STABILITY

Of about 1000 different nuclides now known, only about one-quarter are stable. The others are radioactive, with lifetimes ranging from a small fraction of a second to many years. The stable nuclei are

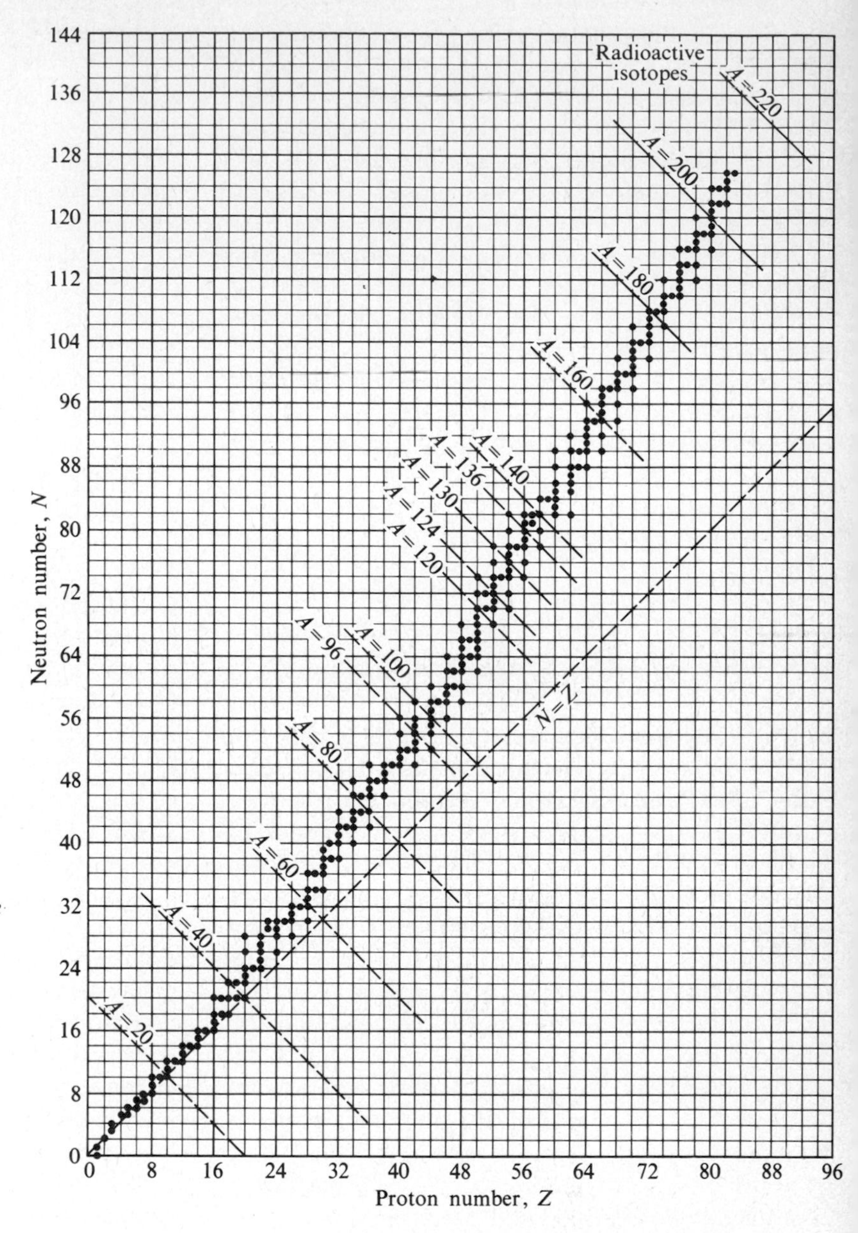

Fig. 46-5 *Neutron-proton diagram of stable nuclei.*

indicated on the graph in Fig. 46–5, where the neutron number N is plotted against the proton number Z. Since the mass number A is the sum $N + Z$, a curve of constant A is a straight line perpendicular to the line $N = Z$. In general, lines of constant A pass through only one or two stable nuclei (see $A = 20$, $A = 40$, $A = 60$, etc.), but there are four cases when such lines pass through three stable isotopes, namely at $A = 96, 124, 130,$ and 136. It is also an interesting fact that only four stable nuclei have both odd Z and odd N: ${}_1H^2$, ${}_3Li^6$, ${}_5B^{10}$, ${}_7N^{14}$. These are called *odd-odd nuclei.*

It is seen that the points representing stable isotopes define a rather narrow stability region. For low mass numbers, $N/Z = 1$. This ratio increases and becomes about 1.6 at large mass numbers. Points to the right of the stability region represent nuclei that have an excess of protons or a deficiency of neutrons; to the left of the stability region are the points representing nuclei with an excess of neutrons or a deficiency of protons. The graph also shows that there is a maximum value of A; no nucleus with A greater than 209 is stable.

The stability of nuclei can be understood qualitatively on the basis of the nature of the nuclear force and the competition between the attractive nuclear force and the repulsive electrical force. As pointed out in Section 46–2, the nuclear force favors pairs of nucleons, and pairs of pairs. In the absence of electrical interactions, the most stable nuclei would be those having equal numbers of neutrons and protons, $N = Z$. The electrical repulsion shifts the balance to favor greater numbers of neutrons, but a nucleus with *too many* neutrons is unstable because not enough of them are paired with protons. A nucleus with too many *protons* has too much repulsive electrical interaction, compared with the attractive nuclear interaction, to be stable.

Furthermore, as the number of nucleons increases, the total energy of electrical interaction increases faster than that of the nuclear interaction. To understand this, we recall the discussion of electrostatic energy in Section 27–4. The energy of a capacitor with a charge Q is proportional to Q^2. It can be shown that to place a total charge Q on the surface of a sphere of radius a requires a total energy $Q^2/8\pi\epsilon_0 a$.

Thus the (positive) electric potential energy in the nucleus increases approximately as Z^2, while the (negative) nuclear potential energy increases approximately as A, with corrections for pairing effects. Thus the competition of electric and nuclear forces accounts for the fact that the neutron–proton ratio in stable nuclei increases with Z, and also for the fact that there is a maximum A (and a maximum Z) for stability. At large A the electric energy *per nucleon* grows faster than the nuclear energy per nucleon, until the point is reached where stability is impossible. Unstable nuclei respond to these conditions in various ways; the next several sections discuss various types of decay of unstable nuclei.

46–5 RADIOACTIVE TRANSFORMATIONS

As noted in Section 46–2, every atom can be specified with the aid of three numbers:

1. The *atomic number* Z, or the positive charge of the nucleus, expressed as a multiple of the electronic charge.
2. The *atomic mass* or the mass of the atom expressed in atomic mass units (u).
3. The *mass number* A, the total number of nucleons.

These numbers are given in Table 46–2 for a few of the light elements.

When a radioactive atom emits an alpha particle, the atomic number is reduced by 2 and the mass number reduced by 4, since an alpha particle is a helium nucleus with a charge of 2 units and a mass number 4. On the other hand, when a beta particle is emitted, the atomic number is *increased* by 1 but the mass number remains the same, since a beta particle is an electron with a charge of $-e$ and a negligible mass. The emission of gamma rays leaves both the atomic number and the mass number unaltered. In natural radioactivity either an alpha particle or a beta particle is emitted, and gamma rays may accompany either process.

The number of radioactive nuclei in any sample of radioactive material decreases continuously as some of the nuclei disintegrate. The *rate* at which the number decreases, however, varies widely for different nuclei. Let N represent the number of radioactive nuclei in a sample at time t, and dN the number that undergo transformations in a short time interval dt. Since every transformation results in a *decrease* in the number N, the corresponding change in N is $-dN$ and the rate of change of N is $-dN/dt$. The larger the number of nuclei in the sample, the larger the number that will undergo transformations, so that the rate of change of N is proportional to N, or is equal to a constant λ multiplied by N. Therefore

$$\frac{dN}{dt} = -\lambda N.$$

It follows that the number of radioactive nuclei decreases *exponentially* with time. If N_0 is the number at some arbitrary time $t = 0$, the number N remaining at a later time t is

$$N = N_0 e^{-\lambda t}. \qquad (46\text{–}3)$$

The proportionality constant λ is called the *decay constant.*

The *half-life* t_h of a radioactive sample is defined as the time at which the number of radioactive nuclei has decreased to one-half the number at $t = 0$. At this time,

$$e^{-\lambda t} = \tfrac{1}{2}.$$

Taking natural logarithms of both sides and solving for t, we find

$$\lambda t_h = \ln 2,$$

$$t_h = \frac{\ln 2}{\lambda}.$$

Half of the original nuclei in a radioactive sample decay in a time interval t_h, half of those remaining at this time decay in a second interval t_h, and so on, so the number remaining after successive intervals of t_h is $\frac{1}{2}$, $\frac{1}{4}$, $\frac{1}{8}$, and so on. We also note that when $t = 1/\lambda$, N has decreased to $1/e$ of its initial value.

The *activity* of a sample is defined to be the number of disintegrations per unit time. A commonly used unit is the *curie*, defined to be 3.70×10^{10} decays per second. This is approximately equal to the activity of one gram of radium. Since the number of disintegrations is proportional to the number of radioactive nuclei in the sample, the activity decreases exponentially with time in the same way as the number N. Figure 46–6 is a graph of the activity of polonium, which has a half-life of 140 days.

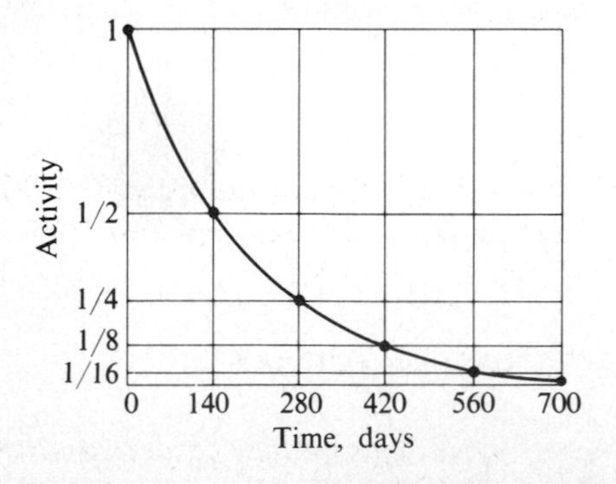

Fig. 46–6 *Decay curve for the radioactive element polonium. Polonium has a half-life of 140 days.*

It is not necessary to wait until half of a sample has decayed in order to measure the half-life. If the logarithm of the activity is plotted as a function of time, the resulting graph is a straight line having a (negative) slope equal to the decay constant λ.

In studying radioactivity the following questions are relevant:

1. What is the parent nucleus?
2. What particle is emitted from this nucleus?
3. What is the half-life of the parent nucleus?
4. What is the resulting nucleus (often called the *daughter nucleus*)?
5. Is the daughter nucleus radioactive and if so, what are the answers to questions 2, 3 and 4 for this nucleus, and so on?

Exhaustive investigations have been carried on in the last 60 years, and these questions have been answered for many nuclei. The results are most conveniently expressed on a diagram such as that shown in Fig. 46–7. The mass number A is plotted along the y-axis, and the atomic number Z along the x-axis. Unit increase of atomic number without change of mass number indicates emission of a beta particle; a decrease of two in atomic number accompanied by a decrease of four in mass number indicates emission of an alpha particle. The half-lives are given either in years (y), days (d), hours (h), minutes (m), or seconds (s).

Figure 46–7 shows the uranium series, starting with the abundant isotope U^{238}. Three other decay series are known; two of these occur in nature, one starting with the uncommon isotope U^{235}, the other with thorium (Th^{232}). The last series starts with neptunium (Np^{237}), an element not found in nature but produced in nuclear reactors. In each case the series continues until a stable nucleus is reached; for these series the final members are Pb^{206}, Pb^{207}, Pb^{208}, and Bi^{209}, respectively.

An interesting application of radioactivity is the dating of archeological and geological specimens by measurement of the concentration of radioactive isotopes. The most familiar example is carbon dating; C^{14}, an unstable isotope, is produced by nuclear reactions in the atmosphere caused by cosmic-ray bombardment, and there is a small proportion of C^{14}

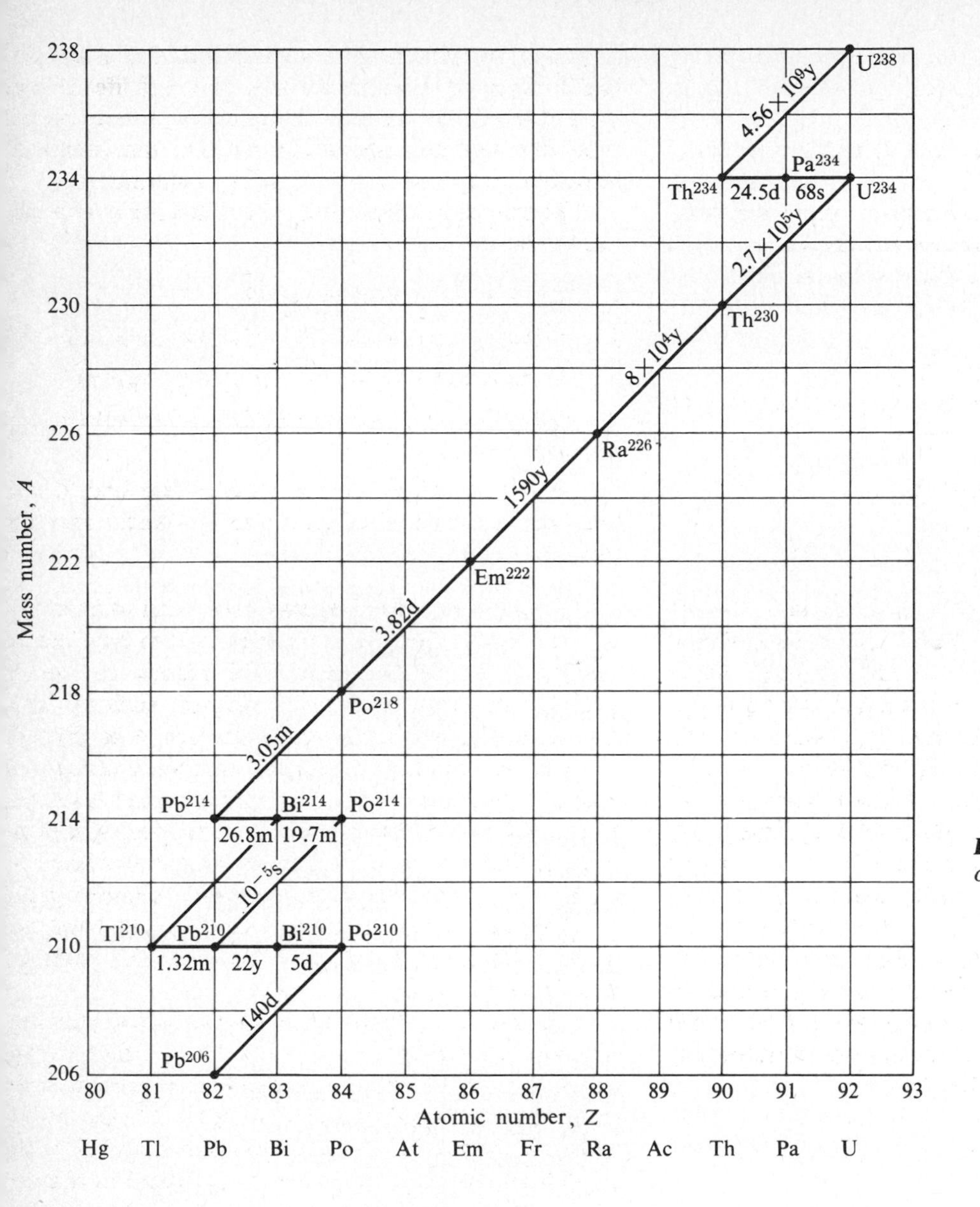

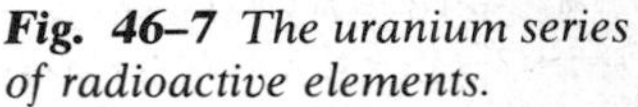

***Fig. 46–7** The uranium series of radioactive elements.*

in the CO_2 in the atmosphere. Plants which obtain their carbon from this source contain the same proportion of C^{14} as the atmosphere, but when a plant dies it stops taking in carbon, and the C^{14} it has already taken in decays, with a half-life of 5,568 years. Thus by measuring the proportion of C^{14} in the remains, one can determine how long ago the organism died. Similar techniques are used with other isotopes for dating of geologic specimens. A difficulty with carbon dating is that the C^{14} concentration in the atmosphere changes with time, over long intervals.

46–6 NUCLEAR REACTIONS

The nuclear disintegrations described up to this point have consisted exclusively of a natural, uncontrolled emission of either an alpha or beta particle. Nothing was done to initiate the emission, and nothing could be done to stop it. It occurred to Rutherford in 1919 that it ought to be possible to penetrate a nucleus with a massive high-speed particle such as an alpha particle and thereby either produce a nucleus with greater atomic number and mass number or induce an artificial nuclear disintegration. Rutherford was successful in bombarding nitrogen with alpha particles and obtaining as a result an oxygen nucleus and a proton, according to the reaction

$$_2He^4 + {}_7N^{14} \rightarrow {}_8O^{17} + {}_1H^1. \qquad (46\text{–}4)$$

In this symbolism, the number at the lower left corner represents the atomic number, and the upper right number represents the mass number. Thus, $_2He^4$ is an alpha particle and $_1H^1$ is a proton. Note that the sum of the initial atomic numbers is equal to the sum of the final atomic numbers, a condition imposed by conservation of charge. The sum of the initial mass numbers is also equal to the sum of the final mass numbers, but the initial rest mass is *not* equal to the final rest mass. The difference between the rest masses is equal to the *nuclear reaction energy*, according to Einstein's equation expressing the equivalence of mass and energy,

$$E = mc^2.$$

If the sum of the final rest masses *exceeds* the sum of the initial rest masses, energy is *absorbed* in the reaction. Conversely, if the final sum is less than the initial sum, energy is released in the form of kinetic energy of the final particles. (1 u = 931 MeV.)

For example, in the nuclear reaction represented by Eq. (46–4), the rest masses of the various particles, in u, are found from Table 46–2 to be

Initial		Final	
$_2He^4$ =	4.00260	$_8O^{17}$ =	16.99913
$_7N^{14}$ =	14.00307	$_1H^1$ =	1.00783
	18.00567		18.00696

The total rest mass of the final products exceeds that of the initial particles by 0.00129 u, which is equivalent to 1.20 MeV. This amount of energy is absorbed in the reaction. If the initial particles did not have this much kinetic energy, the reaction would not have taken place.

On the other hand, in the proton bombardment of lithium and consequent formation of two alpha particles,

$$_1H^1 + {}_3Li^7 \rightarrow {}_2He^4 + {}_2He^4, \qquad (46\text{–}5)$$

the sum of the final masses is smaller than the sum of the initial values, as shown by the following data:

Initial		Final	
$_1H^1$ =	1.00783	$_2He^4$ =	4.00260
$_3Li^7$ =	7.01601	$_2Hc^4$ =	4.00260
	8.02384		8.00520

Since the decrease in mass is 0.01864 u, energy of amount 17.3 MeV is liberated and appears as kinetic energy of the two separating alpha particles. This computation may be verified by observing the distance the alpha particles travel in air at atmospheric pressure before being brought to rest by collisions with molecules. The distance is found to be 8.31 cm. A series of independent experiments is then performed in which the range of alpha particles of known energy is measured. These experiments show that in order to travel 8.31 cm, an alpha particle must have an initial kinetic energy of 8.64 MeV. The energy of the two alphas together is therefore 2 $\times$ 8.64 = 17.28 MeV, in excellent agreement with the value 17.3 MeV obtained from the mass decrease.

Alpha particles and protons are not the only particles used to instigate artificial nuclear disintegration. The nucleus of a deuterium atom, known as the *deuteron* and represented by the symbol $_1H^2$, may be accelerated to high energy in one of a variety of particle accelerators. In order that positively charged particles such as the alpha particle, the proton, and the deuteron can be used to penetrate the nuclei of other atoms, they must travel with very high speeds to avoid being repelled or deflected by the positive charge of the nucleus they are approaching.

46–7 NUCLEAR FISSION

Up to this point, all nuclear reactions considered have involved the ejection of relatively light particles, such as alpha particles, beta particles, protons, or neutrons. That this is not always the case was discovered by Hahn and Strassman in Germany in 1939. These scientists bombarded uranium ($Z = 92$) with neutrons, and after a careful chemical analysis discovered barium ($Z = 56$) and krypton ($Z = 36$) among the products. Cloud-chamber photographs (see Fig. 8–11) showed the two heavy particles traveling in opposite directions with tremendous speed. The uranium nucleus is said to undergo *fission*. Measurement showed that an enormous amount of energy, 200 MeV, is released when uranium splits up in this way. Since the rest mass of a uranium atom exceeds the sum of the rest masses of the fission products, it follows, from the Einstein mass–energy relation, that the extra energy released during fission is transformed into kinetic energy of the fission fragments. Uranium fission may be accomplished by either fast or slow neutrons. Of the two most abundant isotopes of uranium, ${}_{92}U^{238}$ and ${}_{92}U^{235}$, both may be split by a fast neutron, whereas only ${}_{92}U^{235}$ is split by a slow neutron.

When uranium undergoes fission, barium and krypton are not the only products. Over 100 different isotopes of more than 20 different elements have been detected among fission products. All of these atoms are, however, in the middle of the periodic table, with atomic numbers ranging from 34 to 58. Because the neutron–proton ratio needed for stability in this range is much smaller than that of the original uranium nucleus, the *fission fragments*, as the residual nuclei are called, always have too many neutrons for stability. A few free neutrons are liberated during fission, and the fission fragments undergo a series of beta-decays (each of which increases Z by one and decreases N by one) until a stable nucleus is reached.

Discovery of the facts that 200 MeV of energy were released when uranium underwent fission, and that other neutrons were liberated from the uranium nucleus during fission, suggested the possibility of a *chain reaction*, that is, a self-sustaining series of events which, once started, will continue until all the uranium in a given sample is used up (provided the sample stays together). In the case of a uranium chain reaction, a neutron causes one uranium atom to undergo fission, during which a large amount of energy and several neutrons are emitted. These neutrons then cause fission in neighboring uranium nuclei, which also give out energy and more neutrons. The chain reaction may be made to proceed slowly and in a controlled manner, and the device for accomplishing this is called a *nuclear reactor*. If the chain reaction is fast and uncontrolled, the device is a bomb (called an *atomic bomb*), whose destructive ability is many thousands of times that of previously existing bombs. Nuclear reactors have many practical applications:

1. To produce a neutron beam of high intensity for nuclear bombardment studies.
2. To produce artificially radioactive isotopes for medical use and for biological research.
3. To provide a substitute for fossil fuel in a power plant.
4. To produce the fissionable element plutonium from ${}_{92}U^{238}$ for additional fuel or explosive purposes.

46-8 NUCLEAR FUSION

There are two types of nuclear reactions in which large amounts of energy may be liberated. In both types, the rest mass of the products is less than the original rest mass. The fission of uranium, already described, is an example of one type. The other involves the combination of two light nuclei to form a nucleus which is more complex but whose rest mass is less than the sum of the rest masses of the original nuclei. Examples of such energy-liberating reactions are as follows:

$$
\begin{aligned}
{}_1H^1 + {}_1H^1 &\rightarrow {}_1H^2 + {}_1e^0, \\
{}_1H^2 + {}_1H^1 &\rightarrow {}_2He^3 + \gamma\text{-radiation}, \\
{}_2He^3 + {}_2He^3 &\rightarrow {}_2He^4 + {}_1H^1 + {}_1H^1.
\end{aligned}
$$

In the first, two protons combine to form a deuteron and a positron. In the second, a proton and a deuteron unite to form the light isotope of helium. For the third reaction to occur, the first two reactions must occur twice, in which case two nuclei of light helium unite to form ordinary helium. These reactions, known as the *proton–proton chain*, are believed to take place in the interior of the sun and also in many other stars which are known to be composed mainly of hydrogen.

The positrons produced during the first step of the proton–proton chain collide with electrons; annihilation takes place, and their energy is converted into gamma radiation. The net effect of the chain, therefore, is the combination of four hydrogen nuclei into a helium nucleus and gamma radiation. The net amount of energy released may be calculated from the mass balance as follows:

$$\begin{aligned} \text{Rest mass of 4 hydrogen atoms} &= 4.03132\ \text{u} \\ \text{Rest mass of 1 helium atom} &= \underline{4.00260\ \text{u}} \\ \text{Difference in mass} &= 0.02872\ \text{u} \\ &= 26.7\ \text{MeV} \end{aligned}$$

In the case of the sun, 1 g of its mass contains about 2×10^{23} protons. Hence, if all of these protons were consumed, the energy released would be about 55,000 kWh. If the sun were to continue to radiate at its present rate, it would take about 30 billion years to exhaust its supply of protons.

Temperatures of millions of degrees are necessary to initiate the proton–proton chain. A star may achieve such a high temperature by contracting and consequently liberating a large amount of gravitational potential energy. When the temperature gets high enough, the reactions occur, more energy is liberated, and the pressure of the resulting radiation prevents further contraction. Only after most of the hydrogen has been converted into helium will further contraction and an accompanying increase of temperature result. Conditions are then suitable for the formation of heavier elements.

Temperatures and pressures similar to those in the interior of stars may be achieved on earth at the moment of explosion of a uranium or plutonium fission bomb. If the fission bomb is surrounded by proper proportions of the hydrogen isotopes, these may be caused to combine into helium and liberate still more energy. This combination of uranium and hydrogen is called a "hydrogen bomb."

Attempts are being made at this time all over the world to control the fusion of hydrogen isotopes and to utilize the resulting energy for peaceful purposes. The reactions whose control is being studied include the following:

$$_1H^2 + {}_1H^2 \rightarrow {}_1H^3 + {}_1H^1 + 4\ \text{MeV}, \tag{1}$$

$$_1H^3 + {}_1H^2 \rightarrow {}_2He^4 + {}_0n^1 + 17.6\ \text{MeV}, \tag{2}$$

$$_1H^2 + {}_1H^2 \rightarrow {}_2He^3 + {}_0n^1 + 3.3\ \text{MeV}, \tag{3}$$

$$_2He^3 + {}_1H^2 \rightarrow {}_2He^4 + {}_1H^1 + 18.3\ \text{MeV}. \tag{4}$$

In the first, two deuterons combine to form tritium and a proton. In the second, the tritium nucleus combines with another deuteron to form helium and a neutron. The result of both of these reactions is the liberation of 21.6 MeV of energy. Reactions (3) and (4) represent another pair that is about equal in probability to reactions (1) and (2), and which would be attended by the liberation of the same amount of energy. No one has as yet succeeded in producing these reactions under controlled laboratory conditions in such a way as to yield a net surplus of usable energy.

46–9 FUNDAMENTAL PARTICLES

The physics of fundamental particles has been a recognized field of research only in the past 40 years. The electron and the proton were known by the turn of the century, but the neutron was not established definitely until 1930; its discovery is an interesting story and a useful illustration of nuclear reactions.

In 1930, Bothe and Becker in Germany observed that when beryllium, boron, or lithium was bombarded by fast alpha particles, the bombarded material emitted something, either particles or electromagnetic waves, of much greater penetrating power than the original alpha particles. Further experiments in 1932 by Curie and Joliot in Paris confirmed these results, but all attempts to explain them in terms of gamma rays were unsuccessful. Chadwick in England repeated the experiments and found that they could be satisfactorily interpreted on the assumption that *uncharged* particles of mass approximately equal to that of the proton were emitted from the nuclei of the bombarded material. He called the particles *neutrons.* The emission of a neutron from a beryllium nucleus takes place according to the reaction

$$ {}_2He^4 + {}_4Be^9 \rightarrow {}_6C^{12} + {}_0n^1, \qquad (46\text{-}6) $$

where ${}_0n^1$ is the symbol for a neutron.

Since neutrons have no charge, they produce no ionization in their passage through gases. They are not deflected by the electric field around a nucleus and can be stopped only by colliding with a nucleus in a direct hit, in which case they may either undergo an elastic impact or penetrate the nucleus. It was shown in Chapter 8 that if an elastic body strikes a motionless elastic body of the same mass, the first is stopped and the second moves off with the same speed as the first. Since the proton and neutron masses are almost the same, fast neutrons are slowed down most effectively by collisions with the hydrogen atoms in hydrogenous materials such as water or paraffin. The usual laboratory method of obtaining slow neutrons is to surround the fast neutron source with water or blocks of paraffin.

Once the neutrons are moving slowly, they may be detected by means of the alpha particles they eject from the nucleus of a boron atom, according to the reaction

$$ {}_0n^1 + {}_5B^{10} \rightarrow {}_3Li^7 + {}_2He^4. \qquad (46\text{-}7) $$

The ejected alpha particle then produces ionization which may be detected in a Geiger counter or an ionization chamber.

The discovery of the neutron gave the first real clue to the structure of the nucleus. Before 1930 it has been thought that the total mass of a nucleus was due to protons only. We now know that a nucleus consists of both protons and neutrons (except hydrogen, whose nucleus consists only of one proton) and that (1) the mass number A equals the total number of nuclear particles and (2) the atomic number Z equals the number of protons.

The study of *cosmic rays* has been a very fertile field of particle physics. It has been known since the early years of this century that air and other gases are slightly ionized, and therefore slightly conductive, at all times, even in the absence of obvious causes of ionization such as x-rays, ultraviolet light, or radioactivity. For example, if a charged electroscope is left standing, it will eventually lose its charge no matter how well it is insulated. Ionization of air inside a vessel is decreased slightly if the vessel is lowered into a lake, but increases considerably if the vessel is transported in a balloon high into the stratosphere. Hess suggested that the ionization is due to some kind of penetrating waves or particles from outer space, and called them *cosmic rays.*

It is fairly certain that cosmic rays consist largely of high-speed protons with energies of the order of billions of electron volts. A collision between such a proton and the nucleus of a nitrogen or oxygen atom in the upper atmosphere gives rise to so many interesting secondary phenomena that the study of cosmic rays has become one of the richest sources of knowledge of the behavior and properties of fundamental particles.

An instrument devised by C. T. R. Wilson, called a *cloud chamber*, which is used extensively not only to study cosmic rays, but also to render visible the paths of the particles engaging in artificially produced nuclear reactions, was described in Section 18-8. Cloud chambers are sometimes made several feet in diameter, and when illuminated on the side, are photographed from above, at times with the aid of two cameras in order to obtain stereoscopic pictures. Very often the cloud chamber is placed in a magnetic field so that the ionizing particle travels perpendicular to the field. If the ionizing particle is charged, it will be deflected, and by measuring the radius of curvature of the path we may determine its momentum, if the charge is known. In recent years, the *bubble chamber*, also described in Section 18-8, has proved even more useful than the cloud chamber for studying ionizing particles.

The positive electron or *positron* was first observed during the course of an investigation of cosmic rays by Dr. Carl D. Anderson in 1932, in the cloud chamber photograph reproduced in Fig. 46-8. The photograph was made with the cloud chamber in a magnetic field perpendicular to the plane of the paper. A lead plate crosses the chamber and evidently the particle has passed through it. Since the curvature of the track is greater above the plate than below it, the velocity is less above than below; the inference is that the particle was moving upward,

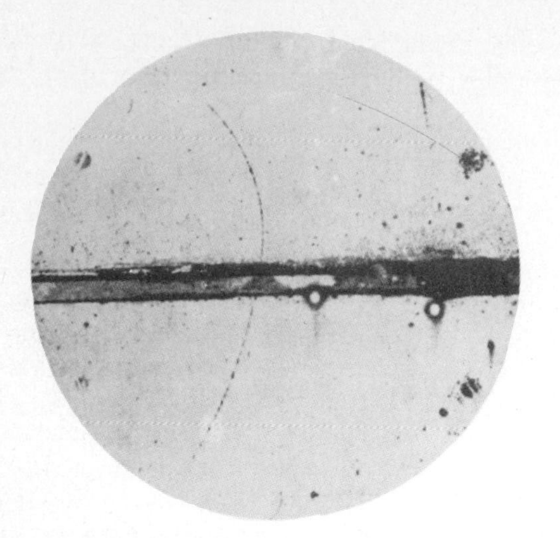

Fig. 46–8 Track of a positive electron traversing a lead plate 6 mm thick. (Photograph by C. D. Anderson.)

since it could not have *gained* energy going through the lead.

The *density* of droplets along the path is the same as would be expected if the particle were an electron. But the direction of the magnetic field and the direction of motion are consistent only with a particle of *positive* charge. Hence Anderson concluded that the track had been made by a positive electron or *positron.* Since the time of this discovery, many thousands of such tracks have been photographed and the positron's existence is now definitely established. Its mass equals that of a negative electron and its charge is equal but of opposite sign.

Positive electrons have only a transitory existence and do not form a part of ordinary matter. There are two known processes which result in positive electrons. They are ejected from the nuclei of certain artificially radioactive materials, and they spring into existence (along with a negative electron) in a process known as "pair production" in which a gamma ray is simultaneously annihilated. Charge is conserved in the process, since the particles have charges of opposite sign. The inverse process, *positron annihilation,* occurs when a positron collides with an ordinary electron; both particles disappear and two or three gamma-ray photons appear, with total energy $2mc^2$, where m is the electron (or positron) mass.

In 1935, the Japanese physicist Hideki Yukawa inferred, from theoretical considerations, the existence of a particle of mass intermediate between that of the electron and the proton. A particle of intermediate mass, but *not* identical with that predicted by Yukawa, was discovered one year later by Anderson and Neddermeyer as a component of cosmic radiation. This particle is now known as a *μ meson* (or *muon*). There are two muons, one positively charged and one negatively, with magnitude of charge equal to that of the electron. The two particles have equal mass, about 207 times the electron mass. The muons are unstable; each decays into an electron of the same sign, plus two neutrinos, with a half-life of about 2.3×10^{-6} s.

Yukawa first proposed the mesons as a basis for nuclear forces; he suggested that nucleons could interact by emitting and absorbing unstable particles, just as two basketball players interact by tossing the ball back and forth, or by snatching it away from each other. It was established soon after the discovery of the muons that they could not be Yukawa's particles because their interactions with nuclei were far too weak. But in 1947 another family of mesons was discovered; called *π mesons* or *pions,* these can be positive, negative, or neutral. The charged pions have masses of about 273 times the electron mass and decay into muons with the same sign, plus a neutrino, with a half-life of about 2.6×10^{-8} s. The neutral pion has a smaller mass, about 264 electron masses, and decays, with a very short half-life of about 2×10^{-16} s, into two gamma-ray photons.

In the years since 1947, *high-energy physics* has emerged as a distinct branch of physics. These years have witnessed the attainment of higher and higher energies in particle accelerators, the discovery of a whole array of new particles, and intensive efforts to understand the properties of these new particles and their interactions.

46–10 HIGH-ENERGY PHYSICS

It was recognized even in the early years of high-energy physics that the fundamental particles are not *permanent* entities but can be created and destroyed in interactions with other particles. The earliest such interaction to be observed was that of creation and destruction of electron–positron pairs. Such pairs are *created* in collisions of high-energy cosmic-ray particles with stationary targets; when an electron and a positron collide, both *disappear* and two or three gamma-ray photons are created to carry away the energy. This transitory nature of the fundamental particles may seem disturbing, but in one sense it is a welcome development. We have seen that photons and electrons (and indeed all particles) share the dual wave–particle nature discussed in Section 44–9, and photons are known to be created and destroyed (or emitted and absorbed) in atomic transitions. Thus it seems natural that other particles can also be created and destroyed.

As an example, it was speculated as early as 1932 that there might be an *antiproton*, bearing the same relation to the ordinary proton as the positron does to the electron, that is, a particle with the same mass as the proton but negatively charged. Finally in 1955 proton–antiproton pairs were created by impact on a stationary target of a beam of protons with kinetic energy 6 GeV (6×10^9 eV) from the Bevatron at the University of California at Berkeley.

Larger and higher-energy particle accelerators have played an essential role in the search for new particles and the exploration of their properties. The largest accelerator in use at present is the 500-GeV accelerator of the Enrico Fermi National Laboratory near Batavia, Illinois, which was built at a cost of approximately $500 million and went into operation in 1972. With this and similar accelerators, many new particles have been discovered and their properties investigated. Some of these are shown in the incomplete table in Fig. 46–9, which shows the symmetry of the array between particles and antiparticles.

The search for new particles has been accompanied by a corresponding effort to construct a theoretical framework for understanding their properties and interactions. Among the central elements of this framework are the classification of interactions into four main categories and the existence of several new conservation laws.

The four classes of interactions, in order of decreasing strength, are:

1. Strong interactions;
2. Electromagnetic interactions;
3. Weak interactions;
4. Gravitational interactions.

The particles having *strong* interactions are collectively called *hadrons*; these include the π and K mesons (pions and kaons) and some heavier mesons not shown in Fig. 46–9, and all the *nucleons* (protons and neutrons) and *hyperons* (the Λ, Σ, Ξ and Ω particles). Nucleons and hyperons are collectively known as *baryons*. The strong interactions are responsible for the emission and absorption of pions and heavy mesons by nucleons, and hence for the nuclear force. They are also responsible for the creation of pions, heavy mesons, and hyperons in high-energy collisions. Electrons, muons, and neutrinos, collectively called *leptons*, have no strong interactions.

The *electromagnetic* interactions are those associated directly with electric charge; as noted previously, the electromagnetic interaction between two protons is weaker at distances of the order of nuclear dimensions than the strong interaction, but it has longer range. Neutral particles have no electromagnetic interactions, with the exception of effects due to the magnetic moments of neutral baryons; these magnetic moments are believed to be associated with the emission and absorption of charged pions and heavy mesons.

The *weak* interaction is responsible for beta decay such as the conversion of a neutron into a proton, an electron, and a neutrino. It is also responsible for the *decay* of many unstable particles (pions into muons, muons into electrons, Λ particles into protons, and so on). The gravitational interaction, although of central importance for the large-scale structure of celestial bodies, is not believed to be of significance in the analysis of fundamental-particle interactions. For example, the gravitational attraction of two electrons is smaller than their electrical repulsion by a factor of about 2.4×10^{-43}.

Several conservation laws are believed to be obeyed by *all* the above interactions. These include the laws growing out of classical physics: energy, momentum, angular momentum, and electric charge. In addition, several new quantities having no classical analog have been introduced to help characterize the properties of particles. These include *baryon number* (the number of baryons minus the number of antibaryons), *isotopic spin* (a quantity used to describe the charge-independence of nuclear forces), *parity* (the comparative behavior of two systems that are mirror images of each other), and *strangeness* (a quantum number used to classify particle production and decay reactions). Baryon number is conserved in *all* interactions; isotopic spin is conserved in strong but not in electromagnetic or weak interactions. Parity and strangeness are conserved in strong and electromagnetic but not in weak interactions. Thus, the new conservation laws are not absolute but, instead, serve as a means for *classifying* interactions.

In recent years, various attempts have been made to understand the various particles on the basis of a comprehensive theory that would permit theoretical predictions of properties such as mass, charge, lifetime, and so on. Although this problem is by no means near solution, several interesting attempts have been made. One theory uses basic building blocks, three in number, called *quarks*, having fractional-unit charge and other unusual properties. The area of fundamental particles and their interactions is presently the subject of intense activity in both theoretical and experimental investigations, and it is certainly among the most vital, interesting, and fundamental areas of current investigation in physics.

46-11 RADIATION AND THE LIFE SCIENCES

The interaction of radiation with living organisms is a topic that grows daily in interest and usefulness. In this context we construe *radiation* to include radiation emitted as a result of nuclear instability (alpha, beta, gamma, and neutrons) as well as electromagnetic radiation such as microwaves and x-rays. Space permits only brief mention of a few examples; the two general classes of phenomena to be considered here are: (1) the use of radioactive isotopes as an analytical tool, and (2) the beneficial and harmful effects of radiation in changing living tissue.

Radioactive isotopes of an element have the same electron configuration as the stable isotopes and therefore exhibit the same chemical behavior. The location and concentration of radioactive isotopes can, however, be detected easily, even at a distance, by measuring the radiation emitted. For example, an unstable isotope of iodine, I^{131}, can be used to study thyroid function. It is known that nearly all the iodine in food which is not eliminated eventually reaches the thyroid; by feeding the patient minute, measured quantities of I^{131} and subsequently measuring the radiation from the thyroid, one can measure the activity of this organ.

More subtle applications occur in complex chemical reactions. By use of radioactive tracers one can "tag" specific parts of molecules and "follow" the radioactive atoms through complex reactions.

A different analytical technique, neutron activation analysis, makes use of the fact that when stable nuclei are bombarded with neutrons, some nuclei absorb neutrons, becoming radioactive beta-emitters. Each element has characteristic energies for the gamma radiation that usually follows beta emission, and measurement of these energies permits determination of the original elements present, even if the quantities are extremely minute.

The interactions of radiation with living tissue are extremely complex. It has been known for many years that excessive exposure to radiation, including sunlight, x-rays, and all the nuclear radiations, can cause destruction of tissues. In mild cases this destruction is manifested as a burn, as with common sunburn; greater exposure can cause very severe illness or death by a variety of mechanisms, one of

Fig. 46-9 *Some of the fundamental particles, showing classification scheme and the particle–antiparticle symmetry. Many of these particles have been discovered in the years since 1955. All the hyperons, the mesons, and the muons are unstable, and an antiparticle can be annihilated by a particle of the same class. The neutron is stable when bound in a stable nucleus, but the free neutron decays into a proton, an electron, and an antineutrino with a halflife of about 13 min. (Adapted from M. R. Wehr and J. A. Richards, Jr.,* Physics of the Atom. *Reading, Mass.: Addison-Wesley, 1967, Used by permission.)*

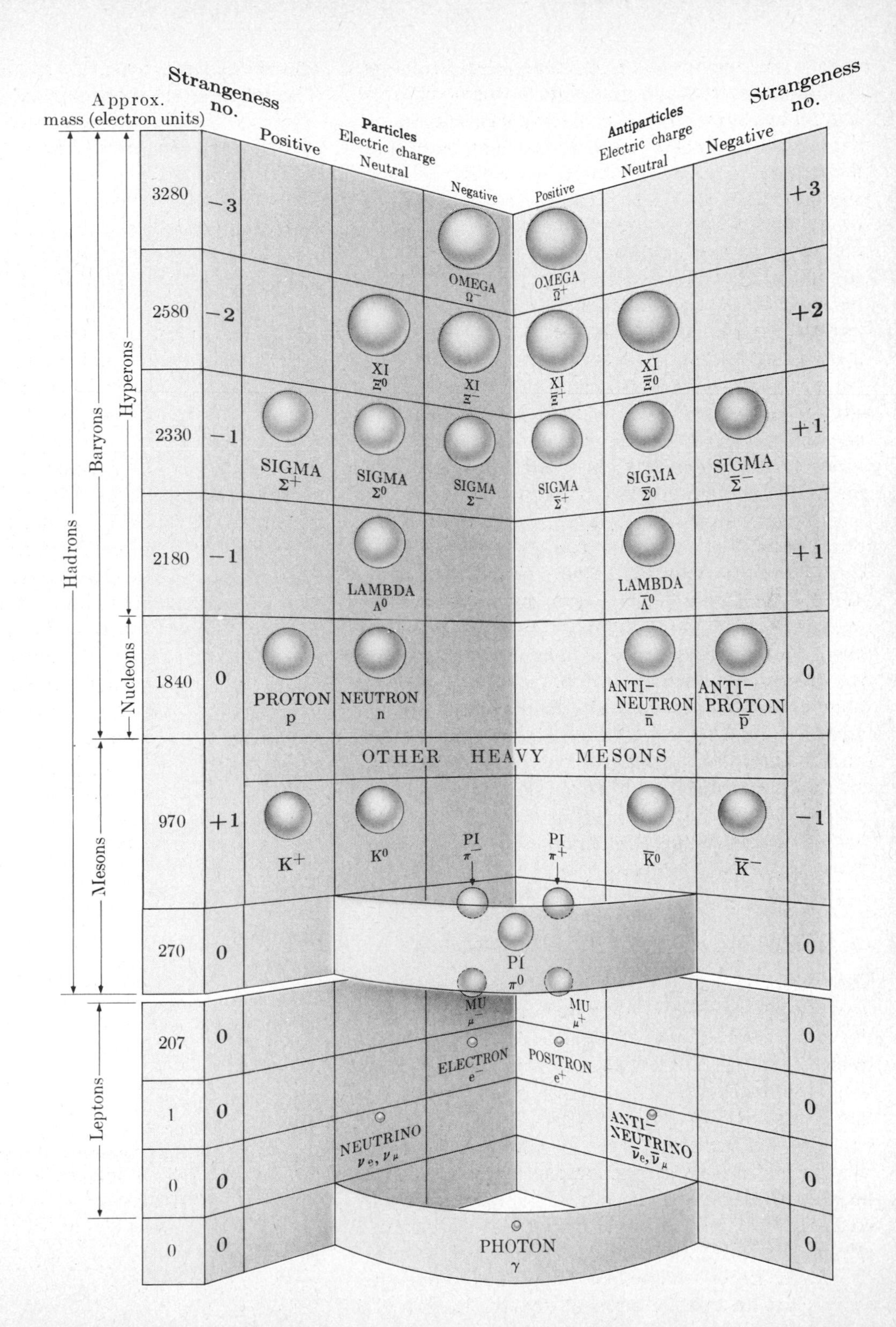
Strangeness no.
Approx. mass (electron units)
Particles
Electric charge
Positive
Neutral
Negative
Antiparticles
Electric charge
Positive
Neutral
Negative
Strangeness no.
3280
−3
OMEGA
Ω^-
OMEGA
$\overline{\Omega}^+$
+3
2580
−2
XI
Ξ^0
XI
Ξ^-
XI
$\overline{\Xi}^+$
XI
$\overline{\Xi}^0$
+2
2330
−1
SIGMA
Σ^+
SIGMA
Σ^0
SIGMA
Σ^-
SIGMA
$\overline{\Sigma}^+$
SIGMA
$\overline{\Sigma}^0$
SIGMA
$\overline{\Sigma}^-$
+1
2180
−1
LAMBDA
Λ^0
LAMBDA
$\overline{\Lambda}^0$
+1
1840
0
PROTON
p
NEUTRON
n
ANTI-NEUTRON
$\overline{n}$
ANTI-PROTON
$\overline{p}$
0
OTHER HEAVY MESONS
970
+1
K^+
K^0
PI
π^-
PI
π^+
$\overline{K}^0$
$\overline{K}^-$
−1
270
0
PI
π^0
0
207
0
MU
μ^-
MU
μ^+
0
ELECTRON
e^-
POSITRON
e^+
1
0
0
NEUTRINO
ν_e, ν_μ
ANTI-NEUTRINO
$\overline{\nu}_e, \overline{\nu}_\mu$
0
0
0
0
PHOTON
γ
0
0
0
Hadrons
Baryons
Hyperons
Nucleons
Mesons
Leptons

which is destruction of the components in bone marrow that produce red blood cells

On the other hand, radiation is present everywhere from sunlight, cosmic rays, and natural radioactivity, and some exposure to radiation is unavoidable. Exactly what constitutes a *safe* level of radiation exposure is open to considerable question, but the available evidence has been interpreted to show that exposure to the extent of 10 to 100 times that from natural sources is very rarely harmful.

There has been a great deal of hysteria concerning alleged radiation hazards from nuclear power plants. It is certainly true that the radiation level from these plants is *not* zero. But to make a meaningful evaluation of hazards one must compare these levels with the alternatives, such as coal-powered plants. The health hazards of coal smoke are serious and well documented, and even the radioactivity in the smoke from a coal-fired power plant is believed to be greater than that from a properly-operating nuclear plant of similar power capacity. It is clearly impossible to eliminate *all* hazards to health, and the next best alternative is an intelligent approach to the problem of *minimizing* hazards.

Radiation also has many *beneficial* effects; the great usefulness of x-rays in medical diagnosis is well known, and there is no doubt that, when x-rays are used properly, the benefits from this diagnostic usefulness almost always outweigh the small radiation hazard. Higher-energy radiation is used for intentional selective destruction of tissue, such as cancerous tumors. The hazards are considerable, but if the disease would be fatal without treatment, considerable hazard may be tolerable.

For sources of radiation for the treatment of cancers and related diseases, artificially produced isotopes are often used. One of the most commonly used is an isotope of cobalt, Co^{60}. This is prepared by bombarding the stable isotope Co^{59} with neutrons in a nuclear reactor. Neutron absorption leads to the unstable Co^{60}; with $Z = 27$ and $N = 33$, this is an "odd–odd" nucleus which decays to Ni^{60} by beta and gamma emission, with a half-life of about 5 years. Such artificial sources have several advantages over naturally radioactive sources. Having shorter half-lives, they are more *intense* sources. They do not emit alpha particles, which are usually not wanted, and the electrons emitted are easily stopped by thin metal sheets without appreciably attenuating the intensity of the desired gamma radiation.

PROBLEMS

46–1 Find the potential energy of an alpha particle 10^{-14} m away from a gold nucleus. Express your results in joules and in electron volts.

46–2 A 4.7-MeV alpha particle from a radium Ra^{226} decay makes a head-on collision with a gold nucleus. What is the distance of closest approach of the alpha particle to the center of the nucleus?

46–3 The radius of the uranium U^{238} nucleus is about 7.4×10^{-15} m. What energy is required to push an alpha particle to the "surface" of this nucleus?

46–4 Compute the approximate density of nuclear matter, and compare your result with typical densities of ordinary matter.

46–5 How much energy would be required to add a proton to a nucleus with $Z = 91$ and $A = 234$? Express your results in joules, and in MeV.

46–6 A carbon specimen found in a cave believed to have been inhabited by cave men contained 1/8 as much C^{14} as an equal amount of carbon in living matter. Find the approximate age of the specimen.

46–7 Radium (Ra^{226}) undergoes alpha emission, leading to radon (Rn^{222}). The masses, including all electrons in each atom, are 226.0254 u and 222.0163 u, respectively. Find the maximum kinetic energy that the emitted alpha particle can have.

46–8 The common isotope of uranium, U^{238}, has a half-life of 4.50×10^9 years, decaying by alpha emission.

a) What is the decay constant?

b) What mass of uranium would be required for an activity of one curie?

c) How many alpha particles are emitted per second by 1 g of uranium?

46–9 The radioactive isotope of cobalt Co^{60}, widely used in medicine, has a half-life of about 5.3 years. What mass of this isotope is required for an activity of 1 millicurie?

46–10 In the fission of U^{238}, 200 MeV of energy is released. Express this energy in joules per mole, and compare with typical heats of combustion.

46–11 Consider the fusion reaction.

$$H^2 + H^2 \rightarrow He^4 + \text{energy}.$$

Compute the energy liberated in this reaction, in MeV and in joules. Compute the energy *per mole* of deuterium, remembering that the gas is diatomic, and compare with the heat of combustion of hydrogen, about 2.9×10^5 joule mol^{-1}.

46–12 Why do elements with mass numbers of 210 and above decay by emission of alpha particles rather than single protons or neutrons?

46–13 If two gamma-ray photons are produced in positron annihilation, find the energy, frequency, and wavelength of each photon.

46–14 A neutral pion at rest decays into two gamma-ray photons. Find the energy, frequency, and wavelength of each photon.

NATURAL TRIGONOMETRIC FUNCTIONS

Angle					Angle				
Degree	Radian	Sine	Cosine	Tangent	Degree	Radian	Sine	Cosine	Tangent
0°	0.000	0.000	1.000	0.000					
1°	0.017	0.017	1.000	0.017	46°	0.803	0.719	0.695	1.036
2°	0.035	0.035	0.999	0.035	47°	0.820	0.731	0.682	1.072
3°	0.052	0.052	0.999	0.052	48°	0.838	0.743	0.669	1.111
4°	0.070	0.070	0.998	0.070	49°	0.855	0.755	0.656	1.150
5°	0.087	0.087	0.996	0.087	50°	0.873	0.766	0.643	1.192
6°	0.105	0.105	0.995	0.105	51°	0.890	0.777	0.629	1.235
7°	0.122	0.122	0.993	0.123	52°	0.908	0.788	0.616	1.280
8°	0.140	0.139	0.990	0.141	53°	0.925	0.799	0.602	1.327
9°	0.157	0.156	0.988	0.158	54°	0.942	0.809	0.588	1.376
10°	0.175	0.174	0.985	0.176	55°	0.960	0.819	0.574	1.428
11°	0.192	0.191	0.982	0.194	56°	0.977	0.829	0.559	1.483
12°	0.209	0.208	0.978	0.213	57°	0.995	0.839	0.545	1.540
13°	0.227	0.225	0.974	0.231	58°	1.012	0.848	0.530	1.600
14°	0.244	0.242	0.970	0.249	59°	1.030	0.857	0.515	1.664
15°	0.262	0.259	0.966	0.268	60°	1.047	0.866	0.500	1.732
16°	0.279	0.276	0.961	0.287	61°	1.065	0.875	0.485	1.804
17°	0.297	0.292	0.956	0.306	62°	1.082	0.883	0.469	1.881
18°	0.314	0.309	0.951	0.325	63°	1.100	0.891	0.454	1.963
19°	0.332	0.326	0.946	0.344	64°	1.117	0.899	0.438	2.050
20°	0.349	0.342	0.940	0.364	65°	1.134	0.906	0.423	2.145
21°	0.367	0.358	0.934	0.384	66°	1.152	0.914	0.407	2.246
22°	0.384	0.375	0.927	0.404	67°	1.169	0.921	0.391	2.356
23°	0.401	0.391	0.921	0.424	68°	1.187	0.927	0.375	2.475
24°	0.419	0.407	0.914	0.445	69°	1.204	0.934	0.358	2.605
25°	0.436	0.423	0.906	0.466	70°	1.222	0.940	0.342	2.748
26°	0.454	0.438	0.899	0.488	71°	1.239	0.946	0.326	2.904
27°	0.471	0.454	0.891	0.510	72°	1.257	0.951	0.309	3.078
28°	0.489	0.469	0.883	0.532	73°	1.274	0.956	0.292	3.271
29°	0.506	0.485	0.875	0.554	74°	1.292	0.961	0.276	3.487
30°	0.524	0.500	0.866	0.577	75°	1.309	0.966	0.259	3.732
31°	0.541	0.515	0.857	0.601	76°	1.326	0.970	0.242	4.011
32°	0.559	0.530	0.848	0.625	77°	1.344	0.974	0.225	4.332
33°	0.576	0.545	0.839	0.649	78°	1.361	0.978	0.208	4.705
34°	0.593	0.559	0.829	0.675	79°	1.379	0.982	0.191	5.145
35°	0.611	0.574	0.819	0.700	80°	1.396	0.985	0.174	5.671
36°	0.628	0.588	0.809	0.727	81°	1.414	0.988	0.156	6.314
37°	0.646	0.602	0.799	0.754	82°	1.431	0.990	0.139	7.115
38°	0.663	0.616	0.788	0.781	83°	1.449	0.993	0.122	8.144
39°	0.681	0.629	0.777	0.810	84°	1.466	0.995	0.105	9.514
40°	0.698	0.643	0.766	0.839	85°	1.484	0.996	0.087	11.43
41°	0.716	0.656	0.755	0.869	86°	1.501	0.998	0.070	14.30
42°	0.733	0.669	0.743	0.900	87°	1.518	0.999	0.052	19.08
43°	0.750	0.682	0.731	0.933	88°	1.536	0.999	0.035	28.64
44°	0.768	0.695	0.719	0.966	89°	1.553	1.000	0.017	57.29
45°	0.785	0.707	0.707	1.000	90°	1.571	1.000	0.000	

COMMON LOGARITHMS

N	0	1	2	3	4	5	6	7	8	9
10	0000	0043	0086	0128	0170	0212	0253	0294	0334	0374
11	0414	0453	0492	0531	0569	0607	0645	0682	0719	0755
12	0792	0828	0864	0899	0934	0969	1004	1038	1072	1106
13	1139	1173	1206	1239	1271	1303	1335	1367	1399	1430
14	1461	1492	1523	1553	1584	1614	1644	1673	1703	1732
15	1761	1790	1818	1847	1875	1903	1931	1959	1987	2014
16	2041	2068	2095	2122	2148	2175	2201	2227	2253	2279
17	2304	2330	2355	2380	2405	2430	2455	2480	2504	2529
18	2553	2577	2601	2625	2648	2672	2695	2718	2742	2765
19	2788	2810	2833	2856	2878	2900	2923	2945	2967	2989
20	3010	3032	3054	3075	3096	3118	3139	3160	3181	3201
21	3222	3243	3263	3284	3304	3324	3345	3365	3385	3404
22	3424	3444	3464	3483	3502	3522	3541	3560	3579	3598
23	3617	3636	3655	3674	3692	3711	3729	3747	3766	3784
24	3802	3820	3838	3856	3874	3892	3909	3927	3945	3962
25	3979	3997	4014	4031	4048	4065	4082	4099	4116	4133
26	4150	4166	4183	4200	4216	4232	4249	4265	4281	4298
27	4314	4330	4346	4362	4378	4393	4409	4425	4440	4456
28	4472	4487	4502	4518	4533	4548	4564	4579	4594	4609
29	4624	4639	4654	4669	4683	4698	4713	4728	4742	4757
30	4771	4786	4800	4814	4829	4843	4857	4871	4886	4900
31	4914	4928	4942	4955	4969	4983	4997	5011	5024	5038
32	5051	5065	5079	5092	5105	5119	5132	5145	5159	5172
33	5185	5198	5211	5224	5237	5250	5263	5276	5289	5302
34	5315	5328	5340	5353	5366	5378	5391	5403	5416	5428
35	5441	5453	5465	5478	5490	5502	5514	5527	5539	5551
36	5563	5575	5587	5599	5611	5623	5635	5647	5658	5670
37	5682	5694	5705	5717	5729	5740	5752	5763	5775	5786
38	5798	5809	5821	5832	5843	5855	5866	5877	5888	5899
39	5911	5922	5933	5944	5955	5966	5977	5988	5999	6010
40	6021	6031	6042	6053	6064	6075	6085	6096	6107	6117
41	6128	6138	6149	6160	6170	6180	6191	6201	6212	6222
42	6232	6243	6253	6263	6274	6284	6294	6304	6314	6325
43	6335	6345	6355	6365	6375	6385	6395	6405	6415	6425
44	6435	6444	6454	6464	6474	6484	6493	6503	6513	6522
45	6532	6542	6551	6561	6571	6580	6590	6599	6609	6618
46	6628	6637	6646	6656	6665	6675	6684	6693	6702	6712
47	6721	6730	6739	6749	6758	6767	6776	6785	6794	6803
48	6812	6821	6830	6839	6848	6857	6866	6875	6884	6893
49	6902	6911	6920	6928	6937	6946	6955	6964	6972	6981
50	6990	6998	7007	7016	7024	7033	7042	7050	7059	7067
51	7076	7084	7093	7101	7110	7118	7126	7135	7143	7152
52	7160	7168	7177	7185	7193	7202	7210	7218	7226	7235
53	7243	7251	7259	7267	7275	7284	7292	7300	7308	7316
54	7324	7332	7340	7348	7356	7364	7372	7380	7388	7396

N	0	1	2	3	4	5	6	7	8	9
55	7404	7412	7419	7427	7435	7443	7451	7459	7466	7474
56	7482	7490	7497	7505	7513	7520	7528	7536	7543	7551
57	7559	7566	7574	7582	7589	7597	7604	7612	7619	7627
58	7634	7642	7649	7657	7664	7672	7679	7686	7694	7701
59	7709	7716	7723	7731	7738	7745	7752	7760	7767	7774
60	7782	7789	7796	7803	7810	7818	7825	7832	7839	7846
61	7853	7860	7868	7875	7882	7889	7896	7903	7910	7917
62	7924	7931	7938	7945	7952	7959	7966	7973	7980	7987
63	7993	8000	8007	8014	8021	8028	8035	8041	8048	8055
64	8062	8069	8075	8082	8089	8096	8102	8109	8116	8122
65	8129	8136	8142	8149	8156	8162	8169	8176	8182	8189
66	8195	8202	8209	8215	8222	8228	8235	8241	8248	8254
67	8261	8267	8274	8280	8287	8293	8299	8306	8312	8319
68	8325	8331	8338	8344	8351	8357	8363	8370	8376	8382
69	8388	8395	8401	8407	8414	8420	8426	8432	8439	8445
70	8451	8457	8463	8470	8476	8482	8488	8494	8500	8506
71	8513	8519	8525	8531	8537	8543	8549	8555	8561	8567
72	8573	8579	8585	8591	8597	8603	8609	8615	8621	8627
73	8633	8639	8645	8651	8657	8663	8669	8675	8681	8686
74	8692	8698	8704	8710	8716	8722	8727	8733	8739	8745
75	8751	8756	8762	8768	8774	8779	8785	8791	8797	8802
76	8808	8814	8820	8825	8831	8837	8842	8848	8854	8859
77	8865	8871	8876	8882	8887	8893	8899	8904	8910	8915
78	8921	8927	8932	8938	8943	8949	8954	8960	8965	8971
79	8976	8982	8987	8993	8998	9004	9009	9015	9020	9025
80	9031	9036	9042	9047	9053	9058	9063	9069	9074	9079
81	9085	9090	9096	9101	9106	9112	9117	9122	9128	9133
82	9138	9143	9149	9154	9159	9165	9170	9175	9180	9186
83	9191	9196	9201	9206	9212	9217	9222	9227	9232	9238
84	9243	9248	9253	9258	9263	9269	9274	9279	9284	9289
85	9294	9299	9304	9309	9315	9320	9325	9330	9335	9340
86	9345	9350	9355	9360	9365	9370	9375	9380	9385	9390
87	9395	9400	9405	9410	9415	9420	9425	9430	9435	9440
88	9445	9450	9455	9460	9465	9469	9474	9479	9484	9489
89	9494	9499	9504	9509	9513	9518	9523	9528	9533	9538
90	9542	9547	9552	9557	9652	9566	9571	9576	9581	9586
91	9590	9595	9600	9605	9609	9614	9619	9624	9628	9633
92	9638	9643	9647	9652	9657	9661	9666	9671	9675	9680
93	9685	9689	9694	9699	9703	9708	9713	9717	9722	9727
94	9731	9736	9741	9745	9750	9754	9759	9763	9768	9773
95	9777	9782	9786	9791	9795	9800	9805	9809	9814	9818
96	9823	9827	9832	9836	9841	9845	9850	9854	9859	9863
97	9868	9872	9877	9881	9886	9890	9894	9899	9903	9908
98	9912	9917	9921	9926	9930	9934	9939	9943	9948	9952
99	9956	9961	9965	9969	9974	9978	9983	9987	9991	9996

PERIODIC TABLE OF THE ELEMENTS

Period	IA	IIA	IIIB	IVB	VB	VIB	VIIB	VIIIB		VIIIB	IB	IIB	IIIA	IVA	VA	VIA	VIIA	Noble gases
1	1 H 1.008																	2 He 4.003
2	3 Li 6.939	4 Be 9.012											5 B 10.811	6 C 12.011	7 N 14.007	8 O 15.999	9 F 18.998	10 Ne 20.183
3	11 Na 22.990	12 Mg 24.312											13 Al 26.982	14 Si 28.086	15 P 30.974	16 S 32.064	17 Cl 35.453	18 Ar 39.948
4	19 K 39.102	20 Ca 40.08	21 Sc 44.956	22 Ti 47.90	23 V 50.942	24 Cr 51.996	25 Mn 54.938	26 Fe 55.847	27 Co 58.933	28 Ni 58.71	29 Cu 63.54	30 Zn 65.37	31 Ga 69.72	32 Ge 72.59	33 As 74.922	34 Se 78.96	35 Br 79.909	36 Kr 83.80
5	37 Rb 85.47	38 Sr 87.62	39 Y 88.905	40 Zr 91.22	41 Nb 92.906	42 Mo 95.94	43 Tc (99)	44 Ru 101.07	45 Rh 102.91	46 Pd 106.4	47 Ag 107.87	48 Cd 112.40	49 In 114.82	50 Sn 118.69	51 Sb 121.75	52 Te 127.60	53 I 126.90	54 Xe 131.30
6	55 Cs 132.91	56 Ba 137.34	57 La 138.91	72 Hf 178.49	73 Ta 180.95	74 W 183.85	75 Re 186.2	76 Os 190.2	77 Ir 192.2	78 Pt 195.09	79 Au 196.97	80 Hg 200.59	81 Tl 204.37	82 Pb 207.19	83 Bi 208.98	84 Po (210)	85 At (210)	86 Rn (222)
7	87 Fr (223)	88 Ra (226)	89 Ac (227)	104 Rf (?) (259)	105 Ha (?) (260)													

58 Ce 140.12	59 Pr 140.91	60 Nd 144.24	61 Pm (145)	62 Sm 150.35	63 Eu 151.96	64 Gd 157.25	65 Tb 158.92	66 Dy 162.50	67 Ho 164.93	68 Er 167.26	69 Tm 168.93	70 Yb 173.04	71 Lu 174.97
90 Th 232.04	91 Pa (231)	92 U 238.03	93 Np (237)	94 Pu (242)	95 Am (243)	96 Cm (247)	97 Bk (249)	98 Cf (251)	99 Es (254)	100 Fm (253)	101 Md (256)	102 No (253)	103 Lr (257)

THE INTERNATIONAL SYSTEM OF UNITS

The Système International d'Unites, abbreviated SI, is the system developed by the General Conference on Weights and Measures and adopted by nearly all the industrial nations of the world. It is based on the mksa (meter-kilogram-second-ampere) system. The following material is reproduced from Publication SP-7012 of the Scientific and Technical Information Office of the National Aeronautics and Space Administration.

Names and Symbols of SI Units

Quantity	*Name of Unit*	*Symbol*	
	SI Base Units		
length	meter	m	
mass	kilogram	kg	
time	second	s	
electric current	ampere	A	
thermodynamic temperature	kelvin	K	
luminous intensity	candela	cd	
amount of substance	mole	mol	
	SI Derived Units		
area	square meter	m^2	
volume	cubic meter	m^3	
frequency	hertz	Hz	s^{-1}
mass density (density)	kilogram per cubic meter	kg/m^3	
speed, velocity	meter per second	m/s	
angular velocity	radian per second	rad/s	
acceleration	meter per second squared	m/s^2	
angular acceleration	radian per second squared	rad/s^2	
force	newton	N	$kg \cdot m/s^2$
pressure (mechanical stress)	pascal	Pa	N/m^2
kinematic viscosity	square meter per second	m^2/s	
dynamic viscosity	newton-second per square meter	$N \cdot s/m^2$	
work, energy, quantity of heat	joule	J	$N \cdot m$
power	watt	W	J/s
quantity of electricity	coulomb	C	$A \cdot s$
potential difference, electromotive force	volt	V	W/A
electric field strength	volt per meter	V/m	
electric resistance	ohm	Ω	V/A
capacitance	farad	F	$A \cdot s/V$
magnetic flux	weber	Wb	$V \cdot s$
inductance	henry	H	$V \cdot s/A$
magnetic flux density	tesla	T	Wb/m^2
magnetic field strength	ampere per meter	A/m	
magnetomotive force	ampere	A	
luminous flux	lumen	lm	$cd \cdot sr$
luminance	candela per square meter	cd/m^2	

Quantity	*Name of Unit*	*Symbol*	
illuminance	lux	lx	lm/m^2
wave number	1 per meter	m^{-1}	
entropy	joule per kelvin	J/K	
specific heat capacity	joule per kilogram kelvin	J/(kg · K)	
thermal conductivity	watt per meter kelvin	W/(m · K)	
radiant intensity	watt per steradian	W/sr	
activity (of a radioactive source)	1 per second	s^{-1}	
	SI Supplementary Units		
plane angle	radian	rad	
solid angle	steradian	sr	

Definitions of SI Units

meter (m)

The *meter* is the length equal to 1,650,763.73 wavelengths in vacuum of the radiation corresponding to the transition between the levels 2 p_{10} and 5 d_5 of the krypton-86 atom.

kilogram (kg)

The *kilogram* is the unit of mass; it is equal to the mass of the international prototype of the kilogram. (The international prototype of the kilogram is a particular cylinder of platinum-iridium alloy which is preserved in a vault at Sèvres, France, by the International Bureau of Weights and Measures.)

second (s)

The *second* is the duration of 9,192,631,770 periods of the radiation corresponding to the transition between the two hyperfine levels of the ground state of the cesium-133 atom.

ampere (A)

The *ampere* is that constant current which, if maintained in two straight parallel conductors of infinite length, of negligible circular cross section, and placed 1 meter apart in vacuum, would produce between these conductors a force equal to 2×10^{-7} newton per meter of length.

kelvin (K)

The *kelvin*, unit of thermodynamic temperature, is the fraction 1/273.16 of the thermodynamic temperature of the triple point of water.

candela (cd)

The *candela* is the luminous intensity, in the perpendicular direction, of a surface of 1/600,000 square meter of a blackbody at the temperature of freezing platinum under a pressure of 101,325 newtons per square meter.

mole (mol)

The *mole* is the amount of substance of a system which contains as many elementary entities as there are carbon atoms in 0.012 kg of carbon 12. The elementary entities must be specified and may be atoms, molecules, ions, electrons, other particles, or specified groups of such particles.

newton (N)

The *newton* is that force which gives to a mass of 1 kilogram an acceleration of 1 meter per second per second.

joule (J)

The *joule* is the work done when the point of application of 1 newton is displaced a distance of 1 meter in the direction of the force.

watt (W)

The *watt* is the power which gives rise to the production of energy at the rate of 1 joule per second.

volt (V)

The *volt* is the difference of electric potential between two points of a conducting wire carrying a constant current of 1 ampere, when the power dissipated between these points is equal to 1 watt.

ohm (Ω)

The *ohm* is the electric resistance between two points of a conductor when a constant difference of potential of 1 volt, applied between these two points, produces in this conductor a current of 1 ampere, this conductor not being the source of any electromotive force.

coulomb (C)

The *coulomb* is the quantity of electricity transported in 1 second by a current of 1 ampere.

farad (F)

The *farad* is the capacitance of a capacitor between the plates of which there appears a difference of potential of 1 volt when it is charged by a quantity of electricity equal to 1 coulomb.

henry (H)

The *henry* is the inductance of a closed circuit in which an electromotive force of 1 volt is produced when the electric current in the circuit varies uniformly at a rate of 1 ampere per second.

weber (Wb)

The *weber* is the magnetic flux which, linking a circuit of one turn, produces in it an electromotive force of 1 volt as it is reduced to zero at a uniform rate in 1 second.

lumen (lm)

The *lumen* is the luminous flux emitted in a solid angle of 1 steradian by a uniform point source having an intensity of 1 candela.

radian (rad)

The *radian* is the plane angle between two radii of a circle which cut off on the circumference an arc equal in length to the radius.

steradian (sr)

The *steradian* is the solid angle which, having its vertex in the center of a sphere, cuts off an area of the surface of the sphere equal to that of a square with sides of length equal to the radius of the sphere.

SI Prefixes

The names of multiples and submultiples of SI units may be formed by application of the prefixes:

Factor by which unit is multiplied	Prefix	Symbol
10^{12}	tera	T
10^{9}	giga	G
10^{6}	mega	M
10^{3}	kilo	k
10^{2}	hecto	h
10	deka	da
10^{-1}	deci	d
10^{-2}	centi	c
10^{-3}	milli	m
10^{-6}	micro	μ
10^{-9}	nano	n
10^{-12}	pico	p
10^{-15}	femto	f
10^{-18}	atto	a

PHYSICAL CONSTANTS

Based on values in the *American Institute of Physics Handbook* (1957), except where superseded by the determination of Taylor, Parker, and Langenberg, *Revs. Mod. Phys.* **41,** 375 (1969). The probable error for each value, properly considered part of the datum, has been omitted.

Name of Quantity	Symbol	Value
Speed of light in vacuum	c	2.9979×10^8 m s^{-1}
Charge of electron	q_e	-1.602×10^{-19} C
Rest mass of electron	m_e	9.10×10^{-31} kg
Ratio of charge to mass of electron	q_e/m_e	1.759×10^{11} C kg^{-1}
Planck's constant	h	6.626×10^{-34} J s
Boltzmann's constant	k	1.381×10^{-23} J K^{-1}
Avogadro's number (chemical scale)	N_0	6.023×10^{23} molecules mole^{-1}
Universal gas constant (chemical scale)	R	8.314 J mole^{-1} K^{-1}
Mechanical equivalent of heat	J	4.185×10^3 J kcal^{-1}
Standard atmospheric pressure	1 atm	1.013×10^5 N m^{-2}
Volume of ideal gas at 0° C and 1 atm (chemical scale)		22.415 liters mole^{-1}
Absolute zero of temperature	0 K	−273.15° C
Acceleration due to gravity (sea level, at equator)		9.78049 m s^{-2}
Universal gravitational constant	G	6.673×10^{-11} N · m^2 kg^{-2}
Mass of earth	m_E	5.975×10^{24} kg
Mean radius of earth		6.371×10^6 m = 3959 mi
Equatorial radius of earth		6.378×10^6 m = 3963 mi
Mean distance from earth to sun	1 AU	1.49×10^{11} m = 9.29×10^7 mi
Eccentricity of earth's orbit		0.0167
Mean distance from earth to moon		3.84×10^8 m = 60 earth radii
Diameter of sun		1.39×10^9 m = 8.64×10^5 mi
Mass of sun	m_S	1.99×10^{30} kg = 333,000 × mass of earth
Coulomb's law constant	$k = 1/4\pi\epsilon_0$	8.9874×10^9 N · m^2 C^{-2}
Faraday's constant (1 faraday)	F	96,487 C mole^{-1}
Mass of neutral hydrogen atom	m_{H^1}	1.007825 u
Mass of proton	m_p	1.007277 u
Mass of neutron	m_n	1.008665 u
Mass of electron	m_e	5.486×10^{-4} u
Ratio of mass of proton to mass of electron	m_p/m_e	1836.11
Rydberg constant for nucleus of infinite mass	R_∞	109,737 cm^{-1}
Rydberg constant for hydrogen	R_H	109,678 cm^{-1}
Wien displacement law constant		0.2898 cm K^{-1}

Numerical constants: $\pi = 3.142$; $e = 2.718$; $\sqrt{2} = 1.414$; $\sqrt{3} = 1.732$.

UNIT CONVERSION FACTORS

Length

1 m = 100 cm = 1000 mm = 10^6 μm = 10^9 nm

1 km = 1000 m = 0.6214 mi

1 m = 3.281 ft = 39.37 in.; 1 cm = 0.3937 in.

1 ft = 30.48 cm; 1 in. = 2.540 cm

1 mi = 5280 ft = 1.609 km

1 Å = 10^{-10} m = 10^{-8} cm = 10^{-1} nm

Area

1 cm^2 = 0.155 in^2; 1 m^2 = 10^4 cm^2 = 10.76 ft^2

1 in^2 = 6.452 cm^2; 1 ft^2 = 144 in^2 = 0.0929 m^2

Volume

1 liter = 1000 cm^3 = 10^{-3} m^3 = 0.0351 ft^3 = 61.02 $in.^3$

1 ft^3 = 0.02832 m^3 = 28.32 liters = 7.477 gallons

Time

1 min = 60 s; 1 hr = 3600 s

1 da = 86,400 s; 1 yr = 3.156×10^7 s

Velocity

1 cm s^{-1} = 0.03281 ft s^{-1}; 1 ft s^{-1} = 30.48 cm s^{-1}

1 mi min^{-1} = 60 mi hr^{-1} = 88 ft s^{-1}

1 km hr^{-1} = 0.2778 m s^{-1}; 1 mi hr^{-1} = 0.4470 m s^{-1}

Acceleration

1 m s^{-2} = 100 cm s^{-2} = 3.281 ft s^{-2}

1 cm s^{-2} = 0.01 m s^{-2} = 0.03281 ft s^{-2}

1 ft s^{-2} = 0.3048 m s^{-2} = 30.48 cm s^{-2}

1 mi hr^{-1} s^{-1} = 1.467 ft s^{-2}

Mass

1 kg = 10^3 g = 0.0685 slug

1 g = 6.85×10^{-5} slug

1 slug = 14.59 kg

1 u = 1.661×10^{-27} kg

Force

1 N = 10^5 dyn = 0.2247 lb

1 lb = 4.45 N = 4.45×10^5 dyn

Pressure

1 Pa = 1 N m^{-2} = 1.451×10^{-4} lb in^{-2} = 0.209 lb ft^{-2}

1 lb in^{-2} = 6891 Pa; 1 lb ft^{-2} = 47.85 Pa

1 atm = 1.013×10^5 Pa = 14.7 lb in^{-2} = 2117 lb ft^{-2}

Energy

1 J = 10^7 ergs = 0.239 cal; 1 cal = 4.186 J
(based on 15° calorie)

1 ft lb = 1.356 J; 1 Btu = 1055 J = 252 cal

1 eV = 1.602×10^{-19} J

Mass-Energy Equivalence

1 kg $\leftrightarrow$ 8.988×10^{16} J

1 u $\leftrightarrow$ 931.5 MeV

1 eV $\leftrightarrow$ 1.073×10^{-9} u

Power

1 W = 1 J s^{-1}

1 hp = 746 W = 550 ft lb s^{-1}

1 Btu hr^{-1} = 0.293 W

Answers to Odd-Numbered Problems

PART I

Chapter 1

1-1 230 N, 55° above negative x-axis
1-3 a) 19.3 N in a direction midway between the 10-N forces
b) 8.46 N in a direction midway between the 10-N forces
1-5 15 N, 53° above the x-axis
26 N, 28° below the x-axis
1-7 25.7 N, 30.6 N
1-9 a) 18.5 N b) 9.2 N
1-11 308 N, 25° above the x-axis
1-13 a) 7 N, 2.9 N b) 7.6 N c) 11 N
1-15 $F_x = 380$ N, west; $F_y = 960$ N, north
1-17 5 mi, 53.1° south of east
1-19 a) $N_x = -1$ cm, $N_y = 1$ cm
b) 1.41 cm, 135° counterclockwise from $+x$-axis

Chapter 2

2-1 a) The earth b) 4 N, the book c) No
d) 4 N, the earth, the book, upward
e) 4 N, the hand, the book, downward
f) First g) Third h) No i) No j) Yes
k) Yes l) One m) No n) Nothing
2-3 a) 10 N b) 20 N
2-5 a) 150 N in A, 180 N in B, 200 N in C
b) 200 N in A, 280 N in B, 200 N in C
c) 550 N in A, 670 N in B, 200 N in C
d) 167 N in A, 58 N in B, 125 N in C
2-7 a) Parts (b) and (c) can be solved.
b) In part (a), another side or angle is needed.
2-9 3150 N
2-11 a) 20 N b) 30 N
2-13 a) $w/(2 \sin \theta)$ b) $w/(2 \tan \theta)$
2-17 22 N
2-19 a) 76 N b) 24 N c) From 15.4 to 84. 6 N
2-21 $P = mg(\sin \phi + \mu \cos \phi)/(\cos \phi - \mu \sin \phi)$,
$\phi = \text{arc ctn } \mu$
2-23 1077 N
2-25 a) Held back b) 480 N
2-27 a) 3 N b) 4 N c) 5 N
2-29 b) 10 N c) 30 N

Chapter 3

3-3 b) 2.5 m c) 2.87 m
3-5 125 N, 100 N to the right, 25 N up
3-7 a) −30 N, 50 N b) 5/3 c) 60 N
d) 1 m from right end
3-9 a) 270 N b) 120 N
3-11 a) 325 N, 375 N b) 225 N c) 235 N d) 864 N
3-13 a) arctan 0.3 or 16.7° b) 24 cm c) 0.53

3-17 a) 215 N b) 185 N c) 295 N
3-19 a) 10 N b) 5 cm to right of c.g. c) 31 cm
3-23 $3\frac{1}{3}$ ft from end
3-25 a) Top: 100 N up, 66.7 N away from door; Bottom: 100 N up, 66.7 N toward door b) 120 N, 56° up from horizontal
3-27 4.6×10^6 m from center of earth

Chapter 4

4-1 a) 15 mi hr^{-1} b) 22 ft s^{-1} c) 671 cm s^{-1}
4-3 61 cm s^{-1}, 60.1 cm s^{-1}, 60.01 cm s^{-1}; 60 cm s^{-1}
4-5 a) 0, 6.3 m s^{-2}, −11.2 m s^{-2} b) 100 m, 230 m, 320 m
4-9 a) 3.95 ft s^{-2} b) 620 ft
4-11 a) 2.67 km hr^{-1} s^{-1} b) 7.5 s c) 83.4 m, 104.3 m
4-13 a) 12.5 cm s^{-2} b) 7840 cm
4-15 a) 100 m b) 20 m s^{-1}
4-17 a) 24 ft s^{-1}, 29 ft s^{-1}, 34 ft s^{-1} b) 2.5 ft s^{-1} c) 21.5 ft s^{-1} d) 8.6 s e) 23 ft f) 1 s g) 24 ft s^{-1}
4-19 a) 32 ft s^{-1} b) 2.83 s c) 16 ft d) 22.6 ft s^{-1} e) 16 ft s^{-1}
4-21 a) 8.7 s b) 37.5 m c) $v_A = 26$ m s^{-1}, $v_T = 17.5$ m s^{-1}
4-23 a) 94 ft s^{-1} b) 124 ft c) 53 ft s^{-1} d) 150 ft s^{-2} e) 1.96 s f) 93 ft s^{-1}
4-25 10.74 m
4-27 a) 48 ft s^{-1} b) 36 ft c) zero d) 32 ft s^{-2}, downward e) 80 ft s^{-1}
4-29 a) 32 ft b) 13 ft s^{-1}, −32 ft s^{-2} c) −51 ft s^{-1}, −32 ft s^{-2} d) 37 ft s^{-1}
4-31 a) 11.3 m b) 0.51 s c) 10 m s^{-1}, −9.8 m s^{-2}
4-33 a) 4 m s^{-2} b) 6 m s^{-1} c) 4.5 m
4-35 a) 64.8 s b) 101 mi
4-37 d) $k = 3.33 \times 10^{-3}$ ft^{-1} e) −1.33 ft s^{-2}
4-39 $v^2 = v_0^2 + K(y_0^2 - y^2)$
4-41 a) car A b) 2 s (−1 s) c) 1.22 s (0.549 s) d) 1.22 s (−0.549 s)
4-43 a) 50 s b) 150 s
4-45 40 min, 30 min
4-47 N 19.5° E, 170 mi hr^{-1}
4-49 a) Zero, 10 m s^{-1} westward b) 17.3 m s^{-1}, 20 m s^{-1}
4-51 a) E 41.7° S b) 2.24 m s^{-1} c) 446 s

Chapter 5

5-1 a) 0.102 kg b) 0.00102 g c) 0.0311 slug
5-3 The two forces in an action-reaction pair never act on the same body.
5-5 a) 0.5 slug b) 500 ft
5-7 a) 2 m s^{-2} b) 100 m c) 20 m s^{-1}
5-9 a) 19.6 N b) 2.55 s
5-11 a) 1.62×10^{-15} N b) 3.33×10^{-9} s c) 1.8×10^{15} m s^{-2}
5-13 6.2×10^{24} kg
5-15 1.9 m s^{-2}
5-17 21,600 N
5-19 a) 59.0 N b) Constant velocity c) 6.25 m s^{-1}
5-21 a) 40 lb b) 4 ft s^{-2} downward c) Zero
5-23 240 ft
5-25 $W = 2wa/(g + a)$
5-27 a) 37° b) 6.4 ft s^{-2} c) 2.5 s
5-29 3.13 s, 9.8 m
5-31 a) 2000 N, 1000 N b) 5000 N
5-33 a) 36.8 N b) 2.45 m s^{-2}
5-35 a) 2 lb b) 1.9 lb
5-37 a) 1.96 m s^{-2} b) 3.14 N
5-39 a) To the left b) 0.65 m s^{-2} c) 424 N
5-41 $2\,m_2g/(4m_1 + m_2)$, $m_2g/(4m_1 + m_2)$
5-43 a) 2.7 m s^{-2} b) 112.5 N c) 87.5 N
5-45 2.43 m
5-47 g/μ
5-49 a) Left b) Right c) Right d) Left e) Down slope f) Always perpendicular to plane
5-51 a) 133 lb b) 33 lb c) 33 lb
5-53 c) $v_T = (mg/k)^{1/2}$ d) $t_R = (m/kg)^{1/2}$ e) $v = v_T(e^{2t/t_R} - 1)/(e^{2t/t_R} + 1)$
5-55 a) Up b) Constant c) Constant d) Stop

Chapter 6

6-1 a) 0.5 s b) 12 ft s^{-1} c) 20 ft s^{-1}, 53° below horizontal
6-3 a) 20 s b) 6000 ft c) $v_x = 300$ ft s^{-1}. $v_y = 640$ ft s^{-1}
6-5 a) 30.6 m b) 62.5 m c) 25 m s^{-1}, 24.5 m s^{-1}; 35 m s^{-1}, 44.4° below horizontal
6-7 a) 0.022 m b) 0.087 m c) 0.196 m d) 4.9 m
6-9 a) 121 ft s^{-1} b) 57.2 ft c) 3.8 s
6-11 a) 1.01 m b) −3.91 m c) Yes
6-13 17 ft s^{-1}
6-15 a) 21.8 m s^{-1} b) 13.5 m s^{-1}, 36.2° below horizontal
6-17 2.00 s or 1.57 s
6-19 $R = 2v_0^2 \sin(\theta_0 - \alpha) \cos\theta_0 / g \cos^2\alpha$
6-21 Straight line, 27° to vertical
6-23 a) 25° above horizontal b) 650 mi c) 10,000 mi hr^{-1} d) 875 mi e) 5 min
6-25 a) $v_x = 2 + 3t^2$, $v_y = -2t$ b) 14.6 m s^{-1}, −15.9° c) 24 N, −4N d) 24 N, −9.46°

6-27 a) 6.5 cm s^{-1}, 25° E of N b) 3.25 cm s^{-2}, 25° E of N
6-29 a) 66,700 mi hr^{-1} b) 0.0193 ft s^{-2}
6-31 5.97°
6-33 22 m s^{-1}
6-35 a) 0.127 b) 2.25 in.
6-37 a) 6.73 s b) No
6-39 a) 52° b) No c) The bead remains at the bottom
6-41 a) 28° b) 1910 m c) 0.533 W
6-43 a) 1530 ft b) 1440 lb
6-45 a) 0.40 W b) 4.49 s c) 2 W d) He would strike the ground 9.9 m from the center of the wheel.
6-49 36,000 km above earth
6-51 a) 98 min b) 26.5 ft s^{-2}
6-53 9.8 oz

Chapter 7

7-1 6.0×10^7 J
7-3 9850 ft lb
7-5 a) 6 lb b) 48 lb c) 22.5 ft lb
7-7 a) 55,000 ft lb b) 4 times
7-9 4.5×10^{-17} J
7-11 9.8 J
7-13 a) 10 N, 20 N, 40 N b) 0.5 J, 2 J, 8 J
7-15 a) 33 J b) 13 J
7-17 a) 160 ft lb b) 160 ft lb
7-19 a) 2400 J b) 518 J c) 1412 J d) 470 J e) $b + c + d = a$
7-21 a) 240 N b) 235 J
7-23 a) 11.2 m s^{-1} b) 6.71 m s^{-1}
7-25 0.1 m
7-27 a) 0.27 b) 3.6 J
7-31 1.81 m s^{-1}
7-33 8.02 m s^{-1}
7-35 a) 5.51 m s^{-1} b) 0.0091 m
7-37 a) 4.42 m s^{-1} b) 14.7 N
7-39 a) 24.6 ft s^{-1} b) 27.2 ft s^{-1} c) 22.5 lb d) 10 ft s^{-1} e) 29° with vertical f) 11.8 ft s^{-1} g) Parabolic trajectory h) 12.64 ft s^{-1}
7-43 470 W, 0.47 kW
7-45 154 hp
7-47 $1.06
7-49 a) 1070 N b) 5 hp, 80 hp
7-51 a) 8200 J b) 8200 J c) 0.55 hp
7-53 a) 27,600 N b) 10.9 m s^{-1} c) 144,000 J d) 353,000 J e) 295 hp
7-55 a) 1070 N b) 41.9 hp c) 15.6 hp d) 9.1%
7-57 227 km^2
7-59 a) 1.8×10^{14} J b) 1.8×10^{20} W c) 1.83×10^{10} kg
7-61 a) 4.55×10^{-17} J b) 1.82×10^{-14} J, 2.79×10^{-14} J

Chapter 8

8-1 a) 2×10^5 kg m s^{-1} b) 40 m s^{-1} c) 28.2 m s^{-1}
8-3 a) 8×10^5 m s^{-2} b) 4×10^4 N c) 5×10^{-4} s d) 20 N s
8-5 No; No; transferred to earth
8-7 a) 0.67 m s^{-1} b) 13,333 J c) 1 m s^{-1}
8-9 780 ft s^{-1}
8-11 27.3 cm
8-13 a) 0.39 ft b) 1950 ft lb c) 3.9 ft lb
8-15 3.3 cm
8-17 17 mi hr^{-1}, 53° E of S
8-19 a) 10 cm s^{-1} b) 0.14 J c) −70 cm s^{-1}, 80 cm s^{-1}
8-21 5 cm s^{-1}, −25 cm s^{-1}
8-23 a) 50 cm s^{-1}, 53° below x-axis b) 0.038 J
8-25 d) $\rho A(v_J - v_B)$ e) $\rho A(v_J - v_B)^2$ f) $\rho A v_B(v_J - v_B)^2$
8-27 a) $v_A = 22.0$ m s^{-1}, $v_B = 15.6$ m s^{-1} b) No; 0.19
8-29 a) 21,400 m
8-31 a) 7.20 ft s^{-1} b) 375
8-33 a) 8.85 kg m s^{-1} b) 4430 N
8-35 $v_A = 26.0$ m s^{-1}, $v_B = 15$ m s^{-1}; 60° from initial direction of A.
8-37 0.91 m s^{-1}
8-39 a) 149° from direction of electron b) 1.06×10^{-20} kg m s^{-1} c) 1.45×10^{-16} J
8-41 16 N
8-43 25×10^6 dyn, 250 N b) Yes
8-45 15.9
8-47 4.23×10^7 m s^{-1}; greater
8-49 c) Measure mc^2 along the hypotenuse; excess length is E_k.
8-51 7.13×10^3 m s^{-1}; no

Chapter 9

9-1 a) 1.5 rad b) 1.57 rad, 90° c) 120 cm, 120 ft
9-3 471, 942, 1414 cm s^{-1}; 600, 1800, 5400 rev min^{-1}
9-5 −13 rad s^{-2}, 57 rev, 3.3 s
9-7 7.5 s
9-9 a) 20 rev s^{-1} b) 66 rev c) 63 m s^{-1} d) 7386 m s^{-2}
9-11 a) 15 cm s^{-2} b) 65 cm s^{-2} c) 126 cm s^{-2}

9-13 164,000 rev min^{-1}
9-15 b) $1/(12)^{1/2}$ rad
9-17 a) 2.67 kg m^2 b) 10.7 kg m^2
c) 1.6×10^{-3} kg m^2
9-19 $11m\ell^2/16$, $0.478\,\ell$
9-21 a) $mb^2/12$ b) $mb^2/3$
9-23 a) 50.5 N b) 40 N c) 9.42 s
9-25 0.59
9-27 a) 6.25 rad s^{-2} b) 250 J c) 4.74 rad s^{-2}; tension is less than weight
9-29 a) 21.3 lb b) 52.3 ft s^{-1} c) 2.45 s
9-31 a) 3.27 m s^{-2}, 0 rad s^{-2}, 26.1 N, 26.1 N
b) 0.754 m s^{-2}, 7.54 rad s^{-2}, 21.1 N, 36.2 N
9-33 a) 2.4 m s^{-1}, 3.2 m s^{-1} b) 0.05 kg m^2
c) 16 m s^{-2} d) 7.8 m s^{-2} e) 0.51 N, tension
9-35 a) 2 m s^{-2} b) 9.8 N
9-37 a) 1066 J b) 36.3 m
9-39 1.01 slug ft^2
9-41 a) 2×10^7 J b) 18 min
9-43 a) 2.96 rad s^{-2} b) 12.2 rad s^{-2} c) 50,200 J
9-45 1.45 J
9-47 a) 12 rad s^{-1} b) 0.027 J
9-49 $(mg \pm \mu Mg)/M\omega^2$
9-51 0.079 rev s^{-1}
9-53 a) -0.05 rad s^{-1} b) 32.7° c) 36.2°
9-55 a) 2 slug ft^2 b) 2620 ft lb
9-57 a) 1.8 N b) 4300 rev min^{-1}
9-59 a) 26,180 rad s^{-2} b) 1.00 N-m c) Clockwise, looking into page d) 3.5×10^{-3} N m

Chapter 10

10-1 4.8×10^{11} N m^{-2}
10-3 b) 14×10^6 lb in^{-2} c) 0.016×10^6 lb in^{-2}
10-5 4.1×10^8 N m^{-2}, 0.054 mm
10-7 62.8 N b) 5.7 mm c) 1.82×10^{-5} cm
10-9 1.66 mm
10-11 a) 1.82 m b) 1.5×10^8 N m^{-2}, 3.0×10^8 N m^{-2}
c) 1.36×10^{-3}, 1.49×10^{-3}
10-13 $\frac{1}{2} QP = P^2Y/2 = Q^2/2Y$
10-15 10^8 N m^{-2}, 5.0×10^{-4}, 2.5 mm, 0.95×10^{-4}
10-17 4.0×10^9 N m^{-2}, 2.5×10^{-10} m^2 N^{-1}
10-19 4.7×10^4 N
10-21 Steel: 0.64×10^{-6} atm^{-1}; water: 50×10^{-6} atm^{-1}; water is 78 times more compressible.
10-23 5.13 in.

Chapter 11

11-1 $B = A \cos \theta_0$, $C = A \sin \theta_0$
11-3 a) 9470 cm s^{-2}, 377 cm s^{-1}
b) 5680 cm s^{-2}, 301 cm s^{-1} c) 0.0368 s
11-5 a) 2.37×10^4 ft s^{-2} b) 740 lb c) 43 mi hr^{-1}
11-7 22.4 kg
11-9 a) 7.8 Hz b) 9.8×10^{10} N m^{-2}
11-11 a) 3.12 kg b) 4.31 cm below equilibrium, moving upward c) 36 N
11-13 a) 31 cm s^{-1} b) 49 cm s^{-2} c) 0.33 s
d) 100 cm
11-15 0.785 s
11-17 a) 5.6 lb b) 13.6 lb, 8 lb, 2.4 lb
c) 0.62 ft lb, 0.077 ft lb
11-19 a) $k = k_1 + k_2$ b) $k = k_1 + k_2$
c) $1/k = (1/k_1) + (1/k_2)$ d) $2^{1/2}$
11-21 a) 1.4 s, 3.5 cm b) Yes; friction c) No
11-23 0.248 m
11-25 4.8 s
11-27 a) 284 dyn cm rad^{-1} b) 5.8 C°
11-31 a) 0.21 m b) 0.70 rad s^{-1}
11-33 a) 81 ft
11-35 67 cm

Chapter 12

12-1 2.08×10^4 N m^{-2}, .206 atm
12-5 a) 1.077×10^5 N m^{-2} b) 1.037×10^5 N m^{-2}
c) 1.037×10^5 N m^{-2} d) 5 cm Hg
e) 68 cm water
12-7 a) 889 lb in^{-2} b) 25,132 lb
12-9 a) $\frac{1}{4}$ b) $\frac{3}{4}$
12-11 a) 98 N b) 1.95 N
12-13 3.62 m^2
12-15 8330 m^3, 9250 kg
12-17 a) 5 cm b) 490 N m^{-2}
12-19 a) 100 lb ft^{-3} b) *E*: 5 lb; *D*: 15 lb
12-23 100.87 g
12-25 2.12 *W* lb in
12-29 6.25×10^4 lb ft
12-31 a) $x = 30$ ft b) Yes
12-33 a) Yes b) For rubber, not for soap c) No
12-35 1372 N m^{-2}
12-37 a) 70.7 cm Hg b) 71.2 cm Hg c) 11 cm
12-39 4.3 cm
12-43 3.6×10^{-4} mm

Chapter 13

13-1 a) 14 m s^{-1} b) 4.4×10^{-3} m^3 s^{-1}
13-3 10 cm
13-5 39.6 ft s^{-1}
13-7 a) 1.21×10^{-3} m^3 s^{-1} b) 0.90 m
13-9 a) 16 ft s^{-1} b) 0.79 ft^3 s^{-1}
13-11 5.0×10^5 Pa

13–13 a) 6.4×10^5 Pa b) 0.265 $m^3\ s^{-1}$
13–15 0.0268 $m^3\ min^{-1}$
13–17 12 $ft^3\ min^{-1}$
13–19 .816 $m^3\ s^{-1}$
13–21 38.8 $m\ s^{-1}$
13–23 4.0×10^4 N
13–25 8 Pa
13–29 a) 0.77 $cm\ s^{-1}$ b) 1.89 $cm\ s^{-1}$
13–31 325 $ft\ s^{-1}$
13–33 a) Turbulent b) 5.3 $liter\ s^{-1}$
13–35 a) Yes b) No

Chapter 14

14–1 a) 4.3×10^{-8} s b) 10.4 m
14–3 The bolt at *A* appears to come first.
14–7 $\Delta s' = [(\Delta s)^2 - c^2\ (\Delta t)^2]^{1/2}$
14–9 5.77×10^{-9} s
14–11 $v/c = E/mc^2$
14–13 $2.60 \times 10^8\ m\ s^{-1}$
14–15 a) 0.5×10^{-4} b) 15% c) Factor of 2.30
14–17 $1.18 \times 10^{-18}\ m\ s^{-2}$
14–19 $2.988 \times 10^8\ m\ s^{-1}$
14–21 1.64×10^{-13} J
14–23 a) 4.55×10^{-15} J, 4.92×10^{-15} J
b) 18.2×10^{-15} J, 27.9×10^{-15} J
14–25 a) 1.8×10^{14} J b) 1.8×10^{20} W
c) 1.84×10^{10} kg

Chapter 15

15–1 600.45 K
15–3 a) 20°C b) 35°C c) −15°C
15–5 6.83 cm Hg
15–9 about 7%
15–11 235 cm
15–13 20 cm, 10 cm
15–15 250 ft
15–17 a) About 1 C° b) Lose
15–19 2.61×10^5 lb
15–21 $9.35 \times 10^7\ N\ m^{-2}$
15–23 a) 3×10^{-4} b) 9000 $lb\ in^{-2}$ c) 4.5×10^5 lb
d) 21,000 $lb\ in^{-2}$ e) 105° F
15–25 a) 893 atm b) 36.2° C
15–27 494 atm

Chapter 16

16–1 a) Yes b) No
16–3 4470 cal
16–5 370 years
16–7 80.7
16–9 a) 1:0.093:0.031 b) 1:0.83:0.35
16–11 0.1 Btu $lb^{-1}(F°)^{-1}$
16–13 0.092 cal $g^{-1}(C°)^{-1}$
16–15 a) 72°C b) Temperature of apparatus does not change.
16–17 a) 0.806 b) 84,280 cal
16–19 a) 273 J b) 3.42 J $mol^{-1}\ K^{-1}$
c) 10.9 J $mol^{-1}\ K^{-1}$
16–21 b) 3.01 J $g^{-1}\ (C°)^{-1}$
16–23 106 g
16–25 0° C with 0.2 g of ice left
16–27 539 cal g^{-1}
16–29 24° C
16–31 40°C
16–33 1.84 kg
16–35 a) 1602 ft^3 b) 310 ft^3
16–37 a) 12,000 Btu hr^{-1} b) 3515 W
16–39 a) 8.05×10^6 J b) 805 N

Chapter 17

17–1 4.32×10^5 J
17–3 a) 1.8 cal s^{-1} b) 20 cm
17–5 0.20 Btu in $hr^{-1}\ ft^{-2}\ (F°)^{-1}$
17–7 110° C
17–9 a) 40° C b) 2.4 cal s^{-1}
17–11 $(R_1R_2)^{1/2}$
17–13 a) 90.0° C b) 21.7 J s^{-1}
17–15 88.5 C°
17–17 a) 7.32×10^5 cal b) 8.05×10^5 cal
17–19 a) 460 $W\ m^{-2}$ b) $4.6 \times 10^6\ W\ m^{-2}$
17–21 0.0119 W
17–23 20.03 K
17–25 6.9 W
17–27 d) 31.4 C° cm^{-1} e) 29 cal s^{-1} f) Zero
g) $1.11 \times 10^{-4}\ m^2\ s^{-1}$ h) 10.9 C° s^{-1}
i) 3.33 s j) Decrease k) 5.4 C° s^{-1}

Chapter 18

18–1 0.3×10^5 Pa
18–3 a) 0.75 atm b) 2 g
18–5 $\rho = pM/RT$
18–7 3.25
18–9 a) 82 cm^3 b) 0.33 g
18–11 When the piston has moved 13.1 inches.
18–13 0.0023 g
18–15 b) 600 K c) 4 atm
18–17 0.286×10^5 Pa
18–19 Dry ice is 3.3 times as dense.
18–21 a) Falls b) Rises c) Approximately stationary

18-23 a) 37.2% b) 865 Pa or .0086 atm
c) 6.4 g m^{-3}
18-25 a) 11.6 g b) 29 g m^{-3} c) 110 g
18-27 1.3 lb hr^{-1}
18-29 0.929 kg

Chapter 19

19-1 a) Yes b) No c) Negative
19-3 a) No b) Yes c) Yes d) Work done on the resistor equals heat transferred to water.
19-5 171 m
19-7 10.8 ft^3
19-9 $U_1 = U_2$
19-11 a) 0.167×10^6 J b) 2.03×10^6 J
19-13 a) 1.1×10^{-2} ft^3 b) 650 ft lb c) 1800 Btu
d) 1.4×10^6 ft lb
19-15 $K \ln[(V_2 - b)/(V_1 - b)] - [(a/V_1) - (a/V_2)]$
19-17 b) 830 J c) On the atmosphere d) 2110 J
e) 2940 J f) 830 J (Same as (b))
19-21 a) 3.15 atm, 189 K b) 3.78 atm, 227 K
19-25 267° C
19-27 a) When piston is 3.13 in. from bottom b) 477 K
19-29 a) $p_1 = 1$ atm, $V_1 = 2.46\ l$;
$p_2 = 2$ atm, $V_2 = 2.46\ l$;
$p_3 = 1$ atm, $V_3 = 3.74\ l$
b) 180 J
19-31 18%
19-33 a) 320 K b) 20%
19-35 a) 900 cal b) 1600 cal c) 400 cal
19-37 a) 4.19×10^5 J b) 1306 J K^{-1} c) 50° C
d) 1306 J K^{-1} before, 1408 J K^{-1} after
19-39 a) 613 J K^{-1} b) −571 J K^{-1} c) 42 J K^{-1}
19-41 11.5 J K^{-1}
19-43 0.0503

Chapter 20

20-1 a) 3×10^{-7} cm b) About 10 times as great
20-3 1.95×10^{-4} cm
20-5 1000 atm
20-7 508 m s^{-1}; smaller by factor 1.45
20-9 1.00636
20-11 20.8 J mol^{-1} K^{-1} or 4.97 cal mol^{-1} K^{-1}
20-13 a) All equal b) 1:3.15:7.03
20-15 a) 1.1 cm s^{-1} b) No
20-17 Bi: 9.9 mm; Bi_2: 14 mm
20-19 2.82×10^{-10} m

Chapter 21

21-1 a) 2 cm b) 30 cm c) 100 Hz d) 3000 cm s^{-1}
21-5 320 m s^{-1}
21-7 a) 200 m s^{-1}, 20 m b) Increase by $2^{1/2}$
21-9 6320 cm s^{-1}
21-11 a) 0.5 Hz b) π rad s^{-1} c) $\pi/100$ cm^{-1}
d) $y = 10 \sin[(\pi x/100) - \pi t)]$ e) $y = -10 \sin \pi t$
f) $y = 10 \sin[(3\pi/2) - \pi t)]$ g) 10π cm s^{-1}
h) $y = 5\sqrt{2}$ cm, $v = -5\pi\sqrt{2}$ cm s^{-1}
21-13 48×10^{-6} atm^{-1}
21-15 a) 1.31 m, 4.80 m^{-1} b) 1644 s^{-1} c) 5.54 m
21-17 b) 290 m s^{-1}
21-19 a) 323 m s^{-1}, 1316 m s^{-1} b) 347 m s^{-1}
21-21 a) 135 Hz b) 1350 Hz c) 13,500 Hz
21-25 a) $F = Mv^2/L$ b) Stationary with respect to table or speed $2v$ c) Mv^2 d) $\frac{1}{2}Mv^2$
e) Sudden accelerations of successive elements of rope are analogous to a series of inelastic collisions; energy is not conserved.

Chapter 22

22-1 a) 200 Hz b) 49 (50th harmonic)
22-3 a) 36 m s^{-1} b) 65 N
22-5 1.28
22-7 a) 5000 m s^{-1} b) 340 m s^{-1}
22-9 Diatomic
22-11 a) 1140, 2280, 3420, 4560, 5700 Hz
b) 570, 1710, 2850, 3990, 5130 Hz
c) 16, 17
22-13 a) 420 Hz b) 415 Hz
22-15 2.9%
22-17 a) 0.655 m b) 56° C

Chapter 23

23-1 a) 9 times b) 4 times
23-3 a) 60 db b) 77 db
23-5 b) No; $\beta = 120 + 10 \log I$
23-7 10^{-6} W
23-9 a) 0.256 μm b) 0.293 N m^{-2} c) 50 m
23-11 a) π b) $I_A = 4 \times 10^{-6}$ W m^{-2},
$I_B = 12 \times 10^{-6}$ W m^{-2}
c) 2.1×10^{-6} W m^{-2}, 63.2 db
23-13 a) 0.029 m b) 21 cm c) 455 Hz
23-15 1.5000, 1.4983; just interval is 0.11% larger.
23-17 24 mi hr^{-1} or 10.7 m s^{-1}
23-19 a) 2.06 ft b) 2.46 ft c) 548 Hz d) 460 Hz
e) 572 Hz f) 439 Hz g) 2.12 ft, 2.40 ft, 547, 458, 572, and 437 Hz
23-21 a) 454 Hz b) 462 Hz c) 8 beats s^{-1}
23-23 a) 11 ft b) 1 ft c) 15 waves
d) 1100 ft s^{-1} e) 0.49 ft

PART II

Chapter 24

24-1 625
24-3 11.9 cm
24-5 a) 8.36×10^{19} N b) 8.36×10^{5} N
24-7 a) 2.75×10^{26} b) 6.58×10^{15} c) 2.39×10^{-11}
24-9 2.16×10^{-7} N
24-11 a) $2q^2/4\pi\epsilon_0 a^2$, vertically upward
b) $2q^2a/4\pi\epsilon_0(a^2 + x^2)^{3/2}$
24-13 a) 4×10^{6} b) 4.9×10^{-17}
24-15 a) 5.12×10^{5} N b) 1.45×10^{6} N
24-19 b) $0.41a$ to the right of q_2, $0.77a$ above q_2

Chapter 25

25-1 a) 4 N C^{-1} upward b) 6.4×10^{-19} N downward
25-3 a) 1010 N C^{-1} b) 2660 km s^{-1}
25-5 a) 1.42 cm b) 9.8 cm
25-7 -25×10^{-9} C
25-9 a) 1.8×10^{4} N C^{-1}, negative x-direction
b) 8×10^{3} N C^{-1}, positive x-direction
c) 3.3×10^{3} N C^{-1}, 70°, second quadrant
d) 6.4×10^{3} N C^{-1}, negative x-direction
25-11 41°
25-13 5.58×10^{-11} N C^{-1}
25-15 $\sqrt{2}k\lambda/r$
25-17 8.85×10^{-10} C
25-19 a) 2400 N m^2 C^{-1} b) Zero c) 100 N C^{-1}, negative x-direction
25-21 b) kq/r^2
25-25 b) $q/4\pi\epsilon_0 r^2$ or $\rho R^3/3\epsilon_0 r^2$

Chapter 26

26-1 a) -15×10^{-6} J b) 10^{5} N C^{-1}
26-3 b) 0.667 mm
26-5 a) 1800 V b) Zero c) 4.5×10^{-5} J
26-7 a) 3 m b) 0.2×10^{-6} C
26-9 b) $2kq/a$ e) $\pm a\sqrt{3}$
26-11 a) It remains at rest. b) It oscillates about the origin, along the y-axis. c) It accelerates away from the origin along the x-axis.
26-13 b) Zero c) $q/4\pi\epsilon_0[(x^2 + 2ax + 2a^2)^{-1/2} - (x^2 - 2ax + 2a^2)^{-1/2}]$
26-15 c) $4\,kqx(a^2 + x^2)^{-3/2}$
26-17 1.03×10^{7} m s^{-1}
26-19 c) E graph is negative slope of V graph.
26-21 c) $2k\lambda \ln (r_b/r_a)$
26-23 a) Outside $-(\lambda/2\pi\epsilon_0) \ln (r/R)$; inside $(\rho/4\epsilon_0)(R^2 - r^2)$
26-25 a) 4.34×10^{-4} m s^{-1} b) 2.74×10^{-4} m s^{-1} upward
26-27 a) $v = (2qV/m)^{1/2}$ b) 5.93×10^{5}
c) 2.24×10^{7} m s^{-1}
26-29 a) 0.9999999969 b) 4.8×10^{10} m s^{-1}
26-31 a) 0.047 MeV b) 0.294 MeV c) 7.2, 4.0
26-33 938.6 MeV
26-35 75.6 MeV
26-37 a) 0.704 cm b) 19.4° c) 4.92 cm
26-39 10.8 hp

Chapter 27

27-1 a) 1000 V b) 2000 V c) 0.5×10^{-3} J
27-3 a) 0.38×10^{-6} C b) 7600 V c) 1.43×10^{-3} J
d) 1.37×10^{-3} J
27-5 a) 1 μF b) 900 μC c) 100 V
27-7 a) 800 μC, 800 V; 800 μC, 400 V
b) 533 μC, 533 V; 1070 μC, 533 V
27-9 1.30×10^{-3} J
27-11 0.86 cal
27-13 a) $dC = -(\epsilon_0 A/x^2)\,dx$ b) $dW = -\frac{1}{2}(Q^2/C^2)\,dC$
c) $F = \frac{1}{2}QE$
27-15 a) 0.226 m^2 b) 1250 V
27-17 1770 pF
27-19 a) 26.6 μC m^{-2} b) 17.7 μC m^{-2}
c) 17.7 μC m^{-2} d) 26.6 μC m^{-2}
27-21 a) 3.43 b) 0.708×10^{-5} C m^{-2}
27-23 a) $V = 2k\lambda \ln (r_b/r_a)$
27-25 a) 17.7 pF b) 8.85×10^{-10} C
c) 2500 V m^{-1} d) 2.21×10^{-8} J

Chapter 28

28-1 a) 20 mA b) 2.75×10^{-6} m s^{-1}
28-3 1.3×10^{5} electrons mm^{-3}
28-5 a) 10^{-4} C s^{-1} b) 4×10^{-6} C m^{-2}
28-7 a) 604 C b) 121 A
28-9 a) 1.53 Ω b) 0.0958 Ω
28-11 a) 1.315×10^{-4} Ω b) 0.234 m
28-15 b) $V_{ab}r_a r_b/\rho r^2(r_b - r_a)$
28-17 a) 99.52 Ω b) 0.0148 Ω
28-19 278° C
28-21 60.9° C
28-23 2.47 mm
28-25 a) 0.5 Ω b) 10 V

28-27 Open: a) 12 V b) 0 c) 12 V
Closed: a) 11.1 V b) 11.1 V c) 0
28-29 a) 2.22 A b) 1.44 A c) 6.42 V
28-31 90 Ω
28-33 a) VI, I^2R, V^2/R b) 90 Ω
28-35 a) 24 W b) 4 W c) 20 W
28-37 a) −1.42 mV b) −12.1 μV $(C°)^{-1}$
c) −4.51 mV d) 0.0162 J

Chapter 29

29-3 a) 8 Ω b) 12 V
29-5 27 W
29-7 90 W
29-9 a) 36 Ω b) 3.33 A c) 2 A, 1.33 A
29-11 a) $\frac{1}{3}$ A b) 64 W, 16 W c) 25-W bulb
29-13 a) 8 Ω b) 72 V
29-15 a) −12 V b) 3 A c) −12 V d) (12/7) A from b to a e) 4.5 Ω f) 4.2 Ω
29-17 a) 18 V b) Point a c) 6 V d) 36 μC
29-19 $\mathscr{E}_1 = 18$ V, $\mathscr{E}_2 = 7$ V, $V_{ab} = 13$ V
29-21 a) 1200 Ω b) 30 V
29-23 10.9 V, 109.1 V
29-25 a) 99.0 V b) 0.0527
29-27 4.97 Ω
29-29 2985 Ω, 12,000 Ω, 135,000 Ω; 3000 Ω, 15,000 Ω, 150,000 Ω
29-33 a) 1450 Ω b) 4500, 1500, 500 Ω c) No
29-35 55 Ω
29-37 a) 0, 400, 630, 870, 1000 μC
b) 100, 60, 37, 14, 0 μA
c) 10 s d) 6.9 s
29-39 a) $\mathscr{E}Q$ b) $\frac{1}{2}$
29-41 $P = (Q_0{}^2/RC^2)\, e^{-2t/RC}$
29-45 a) 5×10^6 A m^{-2}
b) 3.34×10^{-10} A m^{-2}
29-47 a) $i/\pi R^2$ b) $q/\pi R^2\epsilon_0$ c) $(dq/dt)/\pi R^2$

Chapter 30

30-1 a) qvB along −z-axis b) qvB along +y-axis
c) Zero d) $qvB/\sqrt{2}$ parallel to −y-axis e) qvB in y-z plane at 45° to −y- and −z-axes
f) gvB along −y axis
30-3 3.27 T perpendicular to direction of $\mathbf{v}$
30-5 a) 1.14×10^{-3} T, away from reader
b) 1.57×10^{-8} s
30-7 Negative
30-9 8 mm
30-11 2.14×10^{-7} Wb
30-13 Higher velocity; collisions of gas molecules
30-15 a) Inner segment b) Outer segment
30-17 2.13×10^{-2} m
30-19 21
30-21 b) Yes
30-23 a) 1.32 T b) 4.22 MeV, 2.01×10^7 m s^{-1}

Chapter 31

31-1 $F_{ab} = 1.20$ N, −z-direction, $F_{bc} = 1.20$ N, −y-direction,
$F_{cd} = 1.70$ N, 45° up, parallel to y-z plane,
$F_{de} = 1.20$ N, −y-direction, $F_{ef} = 0$
31-3 6.9×10^{28} electrons m^{-3}
31-5 2.4 N
31-7 24 A
31-9 a) 0.16 N, 0.0083 N m
b) 0.16 N, 0.0048 N m c) Same torque
31-11 3.6×10^{-6} N m
31-13 a) 0.181 N m b) When the normal to the coil is at 30° to the field.
31-15 a) $VT/\pi R$ b) $VTNAB/\pi R(k'I')^{1/2}$
31-17 a) 0.5 A b) 4A c) 108 V d) 60 W
e) 48 W f) 540 W g) 71%

Chapter 32

32-1 a) 20 μT b) 7.1 μT c) 20 μT d) Zero
e) 5.4 μT
32-3 b) $B = 4k'Ia/(a^2 + x^2)$ d) Zero
32-5 a) $4k'I/a$ b) $4k'I/3a$ c) Zero d) $8k'I/3a$
32-7 a) 0.192×10^{-4} N m^{-1}, down
b) 0.192×10^{-4} N m^{-1}, up
32-9 1.02 cm
32-11 7.2×10^{-4} N, to left
32-15 1.91R
32-17 16 turns
32-19 6.91×10^{-3} T
32-21 a) $2k'I/r$ b) Zero
32-23 6.0×10^{-4} T
32-25 a) 7958 turns m^{-1} b) 1250 m

Chapter 33

33-1 a) 2 V m^{-1} upward b) 2 V m^{-1} downward
c) 3 V d) 3 V; upper end
33-3 a) 1 V from B to A b) 1.25 N c) 5 W
33-5 0.006 V
33-7 a) $FR/B^2\ell^2$ b) $v = v_T[1 - e^{-(B^2\ell^2/Rm)t}]$
33-9 a) 5.56 rad s^{-1} b) Zero
33-13 a) 0.2 V b) Zero c) 0.2 V
33-15 a) $\pi r_1{}^2\dot{B}$ b) $r_1\dot{B}/2$ c) $R^2\dot{B}/2r_2$
e) $\pi R^2\dot{B}/4$ f) $\pi R^2\dot{B}$ g) $\pi R^2\dot{B}$
33-17 a) Circles, clockwise b) 0.005 V m^{-1}, 3.14 mV
c) 1.57 mA d) Zero f) 3.14 mV

33–19 b) Zero c) 4 mV d) 2 mA e) 1 mV
33–21 a) Down b) Up c) Zero
33–23 a) $2k'i/r$ b) $2k'i\ell\, dr/r$ c) $2k'i\ell \ln (r_2/r_1)$ d) 5.55×10^{-8} V
33–25 a) a to b b) b to a c) b to a
33–27 a) 10^{-5} C
33–29 a) $B = mv/qR$ b) $\mathscr{E} = A\, dB/dt$
33–31 a) 0.737 rev s^{-1} b) 4320 N m
33–33 a) $0.02513 + 1.508 \times 10^{-3}\, t^2$ (volts) b) 0.352 mA

Chapter 34

34–1 0.790 mH
34–3 1.0 mH
34–5 a) $12\pi \cos 120\pi t$ (volts) b) $12\pi \cos 120\pi t$
34–7 $1508 \cos (120\pi\ s^{-1})t$ V; leads by 90°
34–9 a) 0.1 J b) 2 W
34–13 a) 0.05 A b) 1 A s^{-1} c) 0.5 A s^{-1} d) 0.23 s e) 0.0197, 0.0317, 0.0389, 0.0433 A
34–15 0.326 A to right
34–19 a) 2.39 mH b) 4.41 pF
34–23 a) 4 L b) L c) $\omega/2$

Chapter 35

35–1 0.194 K
35–3 a) 2000 A m^{-1} b) 7.97×10^5 A m^{-1} c) 399 d) 3.18×10^5 A e) 400
35–5 c) Diamagnetic d) 10 A, opposite e) It is repelled.
35–9 8 A
35–11 13.6 A m^{-1}
35–13 a) 70 A b) 175 A m^{-1} c) 6.2×10^{-3} Wb A^{-1} m^{-1} d) 1.29×10^5 A Wb^{-1} e) 5.42×10^{-4} Wb
35–15 a) 0.045 A b) 0.012 A c) 0.15 A, 0.23 A

Chapter 36

36–1 a) 127 Hz b) 7.97
36–3 a) 377 Ω b) 2.65 mH c) 2650 Ω d) 2650 μF
36–5 a) 5 mA b) 0.05 A c) 0.5 A
36–7 a) 0.05 A b) 5 mA c) 0.5 mA
36–11 a) 583 Ω b) 0.0858 A c) 25.8 V, 43.0 V d) 59°, lead
36–13 a) 626 Ω b) First larger, then smaller c) 61.4°, current leads
36–15 $V_1 = 23.9$ V, $V_2 = 35.9$ V, $V_3 = 79.8$ V, $V_4 = 43.9$ V, $V_5 = 50$ V
36–17 b) 745 rad s^{-1} or 118 Hz c) 1 d) $V_1 = 50/\sqrt{2}$ V, $V_2 = 112/\sqrt{2}$ V, $V_3 = 112/\sqrt{2}$ V, $V_4 = 0$, $V_5 = 50/\sqrt{2}$ V
36–19 a) 1590 Hz b) 12.4 V
36–21 0.0184 H
36–23 a) 25 W b) 25 W c) Zero d) Zero e) 0.127
36–25 a) 150 μF b) 917 W
36–27 a) 0.0316 (or 31.6) b) 3.16 V
36–29 a) 1591 Hz, 10,000 s^{-1} b) 1.0 A c) 1000 V d) 0.05 J
36–31 a) 150 b) 180 W c) 1.5 A

Chapter 37

37–1 a) 3×10^6 m s^{-1} b) 3 cm
37–3 a) 725 V m^{-1}, 1.92 A m^{-1}, 2.42×10^{-6} T b) 3.9×10^{26} W
37–5 About 4×10^4 V m^{-1}, 1.3×10^{-4} T
37–7 a) $\rho I/\pi a^2$, parallel to wire b) $Ir/2\pi a^2$, perpendicular to wire
37–9 12.2 V m^{-1}, 0.0326 A m^{-1}
37–11 a) $E = (\mu_0 a^2 n/2r)\,(dI/dt)$, tangent b) $S = (\mu_0 a^2 n^2 I/2r)\,(dI/dt)$, (radially inward)
37–13 300 m
37–15 a) 10^{10} Hz b) 1.0×10^{-7} T c) 1.19 W m^{-2} d) 1.98×10^{-9} N m^{-2}

Chapter 38

38–1 a) 1.5×10^{-9} m, 1.5×10^{-3} μm, 1.5 nm, 15 Å b) 5.37×10^{-7} m, 0.537 μm, 537 nm, 5370 Å
38–3 16.6 min
38–5 a) 400 nm b) 5.0×10^{14} Hz
38–13 1.732
38–15 a) 32° b) Independent of refractive index
38–17 4.54×10^{-5}
38–19 7.07 m
38–21 a) 44 lamps b) 44 kW c) 354 hp

Chapter 39

39–1 a) 2×10^8 m s^{-1} b) 333 nm
39–3 1.87
39–7 No
39–9 b) 90°
39–11 1.89
39–15 1.02 cm
39–17 30°
39–19 12.2°
39–21 a) 26.4° b) 0.352 cm
39–23 1.83°

Chapter 40

40–1 Half the observer's height

40-5 a) 9.88 cm b) 5 cm
40-7 a) 3 b) 30 cm behind mirror
40-9 1.71 cm, 0.143
40-11 b) 10 cm behind mirror, 4 cm, erect, virtual
40-13 a) $|s| < 5$ cm b) Erect
40-15 a) 45° b) 1.41 cm
40-17 4.47 cm
40-19 1.35
40-21 $s' = 30$ cm, $m = -1$
40-23 0.667 cm

Chapter 41

41-1 4R from center of sphere
41-3 a) The first image b) 30 cm c) Real d) At infinity
41-5 50 cm
41-7 1.5 cm
41-11

R_1	R_2	f
10 cm	20 cm	40 cm
10 cm	− 20 cm	13.3 cm
− 10 cm	20 cm	− 13.3 cm
− 10 cm	− 20 cm	− 40 cm

41-13 a) 6 cm or 12 cm from object b) −2, −0.5
41-15 60 cm to right of third lens
41-17 −7.2 cm
41-19 a) 20.0 cm b) 18.7 cm
41-21 b) 12 cm
41-23 0.375 in.
41-25 a) 35 mm b) 200 mm
41-27 2 cm
41-29 2 cm to left of first lens, 2 cm to right of second lens
41-31 a) + 3.2 diopters b) −2 diopters
41-33 a) 50 cm b) 2 m
41-35 a) 7.14 cm b) 3.5 mm
41-37 1/400 s
41-39 a) 120 mm b) 121.4 mm from lens
41-41 a) 842X b) 50X
41-43 380X
41-45 20X
41-49 25X, 35 cm from mirror

Chapter 42

42-1 1 mm
42-5 480 nm
42-7 0.045 mm
42-9 2.945×10^{-4} rad
42-11 2.79 mm
42-13 0.106 μm
42-15 545 nm
42-17 2.36 mm
42-19 0.5 mm
42-23 12.5°
42-25 500 nm
42-27 2000 ft
42-29 Smaller by a factor of $\frac{5}{6}$, different color
42-31 No difference

Chapter 43

43-1 35.5°
43-3 a) 37° b) Horizontal
43-5 a) 64° b) 2.87 cm
43-7 a) 0.75 b) 0.50 c) 0 25
43-11 a) Yes b) Dark c) Dark
43-13 a) 58.8° b) 51.1°
43-15 First filter at 45° to polarization of beam, second at 45° to first; $I_0/4$
43-17 a) $I_0/4$ b) Parallel to second filter
43-19 b) 0.582 μ
43-21 Linear polarization, rotated 90°

Chapter 44

44-1 $hf_0 = \phi$
44-3 1.72×10^8 cm s^{-1}
44-5 a) 4.58×10^{14}Hz b) 654 nm c) 1.90 eV d) 6.62×10^{-34} J s
44-7 0.34 V greater
44-9 a) 1.04 eV b) 1190 nm c) 2.52×10^{14} Hz d) 4.14×10^{-7} eV f) No
44-11 1.41×10^{19} s^{-1}
44-15 a) 6.16×10^{14} Hz b) 486 nm
44-19 1.96 eV, 3.14×10^{-19} J, 633 nm
44-21 a) 12,400 eV b) 150 eV
44-23 a) 0.242 nm b) 1.32×10^{-13} m
44-25 0.179 nm; particle
44-27 a) 1.80×10^5 eV b) 2.02×10^8 m s^{-1} c) 2.52×10^8 m s^{-1}
44-29 7729 K
44-31 8.67×10^{-12} m
44-33 a) 13.6 eV b) −26.2 eV c) −13.6 eV d) 13.6 eV
44-35 a) Larger by factor of 4 b) Series ending at $n = 4$, $n = 5$, and one line of series ending at $n = 3$.

Chapter 45

45-1 $(1s)^2(2s)^2(2p)^2$, $(1s)^2(2s)^2(2p)^6(3s)^2(3p)^6$
45-3 $(1s)^2(2s)^2(2p)^6(3s)^2(3p)^2(3d)^{10}(4s)^2(4p)^2$

45-5 1.11 nm
45-7 Cesium
45-11 36,480 K
45-13 a) 4.58×10^{-48} kg m^2 b) 0, 2.3×10^{-21} J, 7.28×10^{-21} J c) 27.3 μm, 1.10×10^{13} Hz

Chapter 46

46-1 22.9 MeV, 3.66×10^{-12} J
46-3 35.8 MeV, 5.74×10^{-12} J
46-5 17.6 MeV, 2.81×10^{-12} J
46-7 12.7 MeV
46-9 $.88 \times 10^{-3}$
46-11 23.8 MeV, 3.82×10^{-12} J, 2.3×10^{12} J mol^{-1}; larger by 10^7
46-13 8.19×10^{-14} J or 0.511 MeV, 1.24×10^{20} Hz, 2.43×10^{-12} m

Index